Springer Collected Works in Mathematics

For further volumes:
http://www.springer.com/series/11104

Spring 1975
at the University of California, Berkeley

Armand Borel

Oeuvres - Collected Papers II

1959 – 1968

Reprint of the 1983 Edition

 Springer

Armand Borel (1923 – 2003)
The Institute for Advanced Study
Princeton, NJ
USA

ISSN 2194-9875
ISBN 978-3-662-44310-1   (Softcover)
          978-3-540-12126-8   (Hardcover)
DOI 10.1007/978-3-662-44311-8
Springer Heidelberg New York Dordrecht London

Library of Congress Control Number: 2012954381

Printed on acid-free paper

Springer is part of Springer Science+Business Media (www.springer.com)

Préface

Ces trois volumes contiennent tous les articles que j'ai publiés, seul ou en collaboration, jusqu'en 1982, à l'exception de quelques exposés de Séminaire. N'y figurent pas non plus, bien entendu, les livres ou Notes de cours (à part une introduction). J'ai par contre inclus deux manuscrits restés inédits, ne serait-ce que pour fournir une référence originale à des travaux qui avaient été indirectement publiés, au moins en partie, à l'époque.

Les articles sont pour la plupart reproduits par photocopie. Quelques-uns d'entre eux cependant ont fait l'objet d'une composition typographique. Des corrections mineures ont été faites directement sur le texte. D'autres, plus longues, figurent dans les «Commentaires et corrections» à la fin de chaque volume. Ces derniers fournissent aussi des références à des résultats ultérieurs qui complètent ou généralisent des propositions du texte, ou encore répondent à des questions qui y sont posées. Quelques problèmes encore ouverts à ma connaissance sont aussi signalés. Ni ces renseignements ni les corrections ne prétendent à être exhaustifs.

On remarquera la forte proportion d'articles écrits en collaboration. Ils furent une source de discussions et échanges dont j'ai tiré grand profit, aussi suis-je heureux de saisir cette occasion pour exprimer ma gratitude aux coauteurs de ces travaux, non seulement pour m'avoir autorisé à les reproduire ici, mais avant tout pour la collaboration elle-même.

Enfin, je remercie vivement Springer-Verlag pour la proposition, si flatteuse, de faire figurer mes travaux dans sa série de «Collected Papers» et pour avoir mené à bien cette publication avec son habileté coutumière, en accédant volontiers à toutes mes demandes.

Princeton, Décembre 1982   A. Borel

Curriculum vitae

Né à la Chaux-de-Fonds, Neuchâtel, Suisse le 21 mai 1923

Études à la section de Mathématiques et Physique de l'École Polytechnique Fédérale, Zürich, 1942–1947. Diplôme en mathématiques au printemps 1947
Assistant à l'E.P.F., 1947–1949
Boursier du C.N.R.S., Paris, 1949–50
Remplaçant du professeur d'algèbre à l'Université de Genève, 1950–1952
Doctorat ès Sciences, Université de Paris, 1952
Membre de l'Institute for Advanced Study, Princeton, 1952–54
Visiting lecturer, University of Chicago, 1954–55
Professeur à l'E.P.F., Zürich, 1955–57
Professeur à l'Institute for Advanced Study, depuis 1957

Invitations de un mois ou plus:
MIT, Cambridge, Mass., printemps 1958, automne 1969
Tata Institute of Fundamental Research, Bombay, janvier-mars 1961, janvier 1968
Université de Paris, janvier-juin 1964
Université de Genève, printemps 1966
Cours à Columbia University, New York, printemps 1968
University of Utrecht, The Netherlands, printemps 1971
University of Buenos Aires, Argentina, août 1973
Institut des Hautes Etudes Scientifiques, Bures-sur-Yvette, automne 1973, juin 1978
Cours à Princeton University, octobre 1974–janvier 1975
University of California at Berkeley, printemps 1975
University of Chicago, printemps 1976
University of Washington, Seattle (Walker Ames Professorship), été 1976
Collège de France, Paris, mai 1977
Yale University, automne 1978
University of Amsterdam, mai 1978
University of Mexico, été 1953, été 1979
Mathematics Institute, Academia Sinica, Beijing, Chine, mai-juin 1981

Membre du Comité de rédaction: Annals of Mathematics 1962–1979; Inventiones Math., depuis 1979

Table des matières

Volume II

45.

(with F. Hirzebruch)

Characteristic classes and homogeneous spaces, II[*]

Amer. J. Math. **81** (1959) 315–382

Chapter VI. Applications to Todd Genera.[1]

20. Integration over the fibre in $(B_T, B_G, G/T, \rho(T, G))$.

Throughout § 20, the coefficients for cohomology are the real numbers and will not be mentioned explicitly.

20.1.   Let G be a compact connected Lie group, T a maximal torus, $2m$ the dimension of G/T, $\mathscr{C}$ an invariant almost complex structure on G/T, and $a_1, \cdots, a_m$ the roots of $\mathscr{C}$ (see 12.3, 13.4). $\mathscr{C}$ defines an *orientation* of G/T, and hence also an identification of $H^{2m}(G/T)$ with $\mathbf{R}$, which will always be used in this §. In the fibering $\xi = (B_T, B_G, G/T, \rho(T, G))$ (see [2, § 20] for its definition), the integration over the fibre is a linear map of $H^*(B_T)$ in $H^*(B_G)$ or of $H^{**}(B_T)$ into $H^{**}(B_G)$ which lowers degrees by $2m$, (see § 8).

The order q of $W(G)$ is equal to the Euler number $E(G/T)$ of G/T and the latter is equal to the value of the m-th Chern class on the fundamental cycle. Therefore, considering the a_i's as elements of $H^2(B_T)$, we have by 10.8

$$(1) \qquad (a_1 \cdots a_m)[G/T] = q = \text{order } W(G),$$

where the left side denotes the value of $a_1 \cdots a_m$ on an oriented fibre.

G/T is totally non-homologous to zero in ξ for real coefficients and q is also the dimension of $H^*(G/T)$, [2, § 26]. Therefore $\rho^*(T, G)$, which will be abbreviated by π^*, is injective, and we can choose homogeneous elements $h_1, \cdots, h_q \in H^*(B_T)$ with $h_q = a_1 \cdots a_m$, whose restrictions to a fibre form a basis of $H^*(G/T)$; an element $x \in H^*(B_T)$ can then be written in one and only one way in the form

$$(2) \qquad x = \pi^*(b_1)h_1 + \cdots + \pi^*(b_q)h_q \qquad (b_i \in H^*(B_G))$$

---

[*] Received June 19, 1958.

[1] Part I of this paper appeared earlier in this Journal, Vol. 80 (1958), pp. 458-538. We refer to it for the notation and a general introduction.

1

and we have by 8.4(1) and (1) above that

$$(3) \qquad\qquad x^{\natural} = q \cdot b_q.$$

For any $w \in W(G)$, the elements $w(h_1), \cdots, w(h_{q-1})$, $\operatorname{sgn} w \cdot h_q$ (see 2.6 for $\operatorname{sgn} w$), when restricted to a fibre, also form a base of $H^*(G/T)$. Therefore, if we apply w to (2) and use 8.4(1) again, we see that

$$(4) \qquad\qquad (w(x))^{\natural} = \operatorname{sgn} w \cdot x^{\natural}.$$

20.2. LEMMA. *Let* $x \in H^*(B_T)$ *be such that* $w(x) = \operatorname{sgn} w \cdot x$ *for all* $w \in W(G)$. *Then*

$$q \cdot x = \pi^*(x^{\natural}) \cdot a_1 \cdot \cdots \cdot a_m.$$

We may consider $H^*(B_T)$ as the ring of polynomials with real coefficients on the universal covering V_T of T. Let S_i be the symmetry to the hyperplane $a_i = 0$. Then, $S_i(x) = -x$ implies that x is zero on $a_i = 0$, and hence that x is divisible by a_i. It follows that $x = y \cdot a_1 \cdots a_m$ with $y \in H^*(B_T)$ and, in view of our assumption, invariant under $W(G)$. Therefore [2, §26], $y = \pi^*(b)$, $b \in H^*(B_G)$, and the lemma follows from (3).

20.3. THEOREM. *Let* $a_1, \cdots, a_m$ *be the roots of an invariant almost complex structure* $\mathcal{C}$ *on* G/T, *and* $\natural$ *be the integration over the fibre in* $(B_T, B_G, G/T, \pi)$ *with respect to the orientation defined by* $\mathcal{C}$. *Then for* $x \in H^*(B_T)$, *we have*

$$(5) \qquad\qquad \sum_{w \in W(G)} \operatorname{sgn} w \cdot w(x) = \pi^*(x^{\natural}) \cdot a_1 \cdot \cdots \cdot a_m.$$

Let y be the left-hand side of (5). Then $y^{\natural} = q \cdot x^{\natural}$ by (4). On the other hand, we have $w(y) = \operatorname{sgn} w \cdot y$ for any $w \in W(G)$; therefore 20.2 shows that

$$q \cdot y = \pi^*(y^{\natural}) \cdot a_1 \cdot \cdots \cdot a_m$$

which proves the theorem.

It follows from 20.3 that, if $x \in H^{**}(B_T)$, then

$$(6) \qquad\qquad \sum_{w \in W(G)} \operatorname{sgn} w \cdot w(x) = \pi^{**}(x^{\natural}) \cdot a_1 \cdots a_m.$$

21. Multiplicative sequences.

21.1. In this paragraph, ξ is a bundle in which F_ξ is a compact connected n-dimensional oriented manifold, G_ξ is a group of diffeomorphisms of F_ξ, and $\hat{\xi}$ is the bundle along the fibres (7.4). Γ is a commutative ring with unit and the cohomology groups of the fibres of ξ with respect to Γ are assumed to form a constant sheaf on B_ξ. Then $\natural = \natural_\xi$ is a map of $H^{**}(E_\xi, \Gamma)$ into $H^{**}(B_\xi, \Gamma)$ which lowers degrees by n. In particular, we have

$$(1) \qquad a^\natural = a[F_\xi] \cdot 1 \qquad\qquad (a \in H^n(E_\xi, \Gamma)),$$

where 1 is the unit of $H^{**}(B_\xi, \Gamma)$.

21.2. Let $\{K_j(p_1, \cdots, p_j)\}$ be a multiplicative sequence of polynomials in indeterminates p_i, with coefficients in Γ [19, §1]. If η is a real vector bundle, we put

$$(2) \qquad \mathscr{K}_\eta = \sum_{j \geqq 0} K_j(p_1(\eta), \cdots, p_j(\eta)).$$

We have $K_j(p_1(\eta), \cdots, p_j(\eta)) \in H^{4j}(B_\eta, \Gamma)$ and $\mathscr{K}_\eta \in H^{**}(B_\eta, \Gamma)$. If η is the tangent bundle to a compact oriented differentiable manifold X, then the genus $K(X)$ of X with respect to the sequence $\{K_j\}$ is defined by $K(X) = \mathscr{K}_\eta[X]$, i.e.

$$K(X) = K_r(p_1(\eta), \cdots, p_r(\eta))[X] \in \Gamma$$

if $4r = \dim X$, and $K(X) = 0$ if $\dim X \not\equiv 0 \bmod 4$.

21.3. DEFINITION. *Let $\hat{\xi}$ be the bundle along the fibres of ξ. The multiplicative sequence $\{K_j\}$ is said to be strictly multiplicative in ξ if and only if*

$$(i) \qquad\qquad (\mathscr{K}_{\hat{\xi}})^\natural \in H^0(B_\xi, \Gamma).$$

Let $\hat{p}_i$ be the Pontrjagin classes of $\hat{\xi}$. The condition (i) is equivalent to

$$(ii) \qquad\qquad K_j(\hat{p}_1, \cdots, \hat{p}_j)^\natural = 0 \qquad\qquad (4j > n).$$

The restriction of $\hat{\xi}$ to a fibre of ξ is the tangent bundle to the fibre. Therefore we have by (1) for any multiplicative sequence

$$(3) \qquad (K_j(\hat{p}_1, \cdots, \hat{p}_j))^\natural = K(F_\xi) \cdot 1 \qquad\qquad (4j = \dim F_\xi),$$

and therefore $\{K_j\}$ is strictly multiplicative in ξ if and only if

$$(4) \qquad\qquad (\mathscr{K}_{\hat{\xi}})^\natural = K(F_\xi) \cdot 1.$$

A multiplicative sequence is always strictly multiplicative in the product bundle, because in this case, $\hat{\xi}$ may be identified with the bundle induced from the tangent bundle to F_ξ by the projection of $E_\xi = B_\xi \times F_\xi$ onto F_ξ, and then (ii) is obviously true.

21.4. In addition to 21.1, we assume that ξ is a differentiable bundle (7.4), and that B_ξ, E_ξ are compact connected oriented manifolds, the

3

orientation of E_ξ being induced by those of B_ξ, F_ξ taken in this order. It follows then from the definition of the integration over the fibre (8.1) or from its equivalence with the Gysin homomorphism (8.3, remark) that

$$a[E_\xi] = a^\natural[B_\xi] \qquad\qquad (a \in H^*(E_\xi, \Gamma)).$$

Let η and η' be the tangent bundles to E_ξ and B_ξ respectively. We have an exact sequence (7.6):

$$(5) \qquad\qquad 0 \to \hat{\xi} \to \eta \to \pi^*\eta' \to 0, \qquad\qquad (\pi = \pi_\xi),$$

and the multiplication theorem (9.7) implies

$$p(\eta) = p(\hat{\xi}) \cdot \pi^* p(\eta') \qquad\qquad \operatorname{mod} \operatorname{Tors} H^*(E_\xi, \mathbf{Z}),$$

where $\operatorname{Tors} H^*(E_\xi, \mathbf{Z})$ is the torsion subgroup of $H^*(E_\xi, \mathbf{Z})$. By the fundamental property of multiplicative sequences [19, § 1.2], this yields

$$(6) \qquad\qquad \mathcal{K}_\eta = \mathcal{K}_{\hat{\xi}} \cdot \mathcal{K}_{\pi^*\eta'} = \mathcal{K}_{\hat{\xi}} \cdot \pi^*(\mathcal{K}_{\eta'}),$$

modulo the image of $\operatorname{Tors} H^*(E_\xi, \mathbf{Z}) \otimes \Gamma$ in $H^*(E_\xi, \Gamma)$.

If $E_\xi = B_\xi \times F_\xi$, then [19, § 5.2]

$$(7) \qquad\qquad K(E_\xi) = K(B_\xi) \cdot K(F_\xi).$$

More generally, we have

21.5. PROPOSITION. *Let ξ be a differentiable bundle satisfying the assumption 21.4 and let $\{K_j\}$ be a multiplicative sequence of polynomials with coefficients in Γ. If $\{K_j\}$ is strictly multiplicative in ξ, then $K(E_\xi) = K(B_\xi) \cdot K(F_\xi)$.*

Since $H^j(E_\xi, \mathbf{Z})$ has no torsion for $j = \dim E_\xi$, we get from 21.4 and 8.2

$$K(E_\xi) = \mathcal{K}_\eta[E_\xi] = (\mathcal{K}_{\hat{\xi}} \cdot \pi^* \mathcal{K}_{\eta'})^\natural[B_\xi] = \mathcal{K}_{\eta'}(\mathcal{K}_{\hat{\xi}})^\natural[B_\xi],$$

and 21.5 follows from 21.2, 21.3(4).

21.6. We repeat briefly this discussion for the case of Chern classes. Let $\{K_j(c_1, \cdots, c_j)\}$ be a multiplicative sequence of polynomials with coefficients in Γ, in indeterminates c_i. Given a complex vector bundle η, we introduce the elements $K_j(c_1(\eta), \cdots, c_j(\eta)) \in H^{2j}(X, \Gamma)$ and put

$$\mathcal{K}_\eta = \sum_{j \geq 0} K_j(c_1(\eta), \cdots, c_j(\eta)).$$

It is an element of $H^{**}(X, \Gamma)$. If η is the complex tangent bundle to a compact connected almost complex manifold (7.3), canonically oriented, the genus of X with respect to $\{K_j\}$ or its "K-genus" is $K(X) = \mathscr{K}_\eta[X]$. It is equal to $K_m(c_1(\eta), \cdots, c_m(\eta))[X]$ if m is the complex dimension of X.

21.7. Let ξ be as in 21.1. Assume moreover that $\hat{\xi}$ has been endowed with a complex structure $\hat{\xi}_C$ of the type considered in 7.4, that is, defined by means of an almost complex structure of F_ξ, invariant under G_ξ, and let $\hat{c}_i$ be its Chern classes. The multiplicative sequence $\{K_j\}$ is then said to be strictly multiplicative in ξ with respect to $\hat{\xi}_C$ if one of the three following equivalent conditions is fulfilled

(i) $$(\mathscr{K}_{\hat{\xi}_C})^\natural \in H^0(B_\xi, \Gamma)$$

(ii) $$(K_j(\hat{c}_1, \cdots, \hat{c}_j))^\natural = 0, \qquad\qquad (2j > \dim F_\xi)$$

(iii) $$(\mathscr{K}_{\hat{\xi}_C})^\natural = K(F_\xi) \cdot 1.$$

21.8. Let ξ and $\hat{\xi}_C$ be as before. Assume in addition that ξ is differentiable and that B_ξ carries an almost complex structure η'_C. Then an almost complex structure η_C of E_ξ is said to be *compatible with* η'_C and $\hat{\xi}_C$ if there is an exact sequence

$$(8) \qquad\qquad 0 \to \hat{\xi}_C \to \eta_C \to \pi^*(\eta'_C) \to 0, \qquad\qquad (\pi = \pi_\xi).$$

Since exact sequences of vector bundles of the type (5), (8) always split, η_C always exists and is determined up to isomorphism by η'_C and $\hat{\xi}_C$. The proof of the following propostition is exactly the same as in the case of Pontrjagin classes:

21.9. PROPOSITION. *We keep the assumptions of 21.8 and assume moreover that the multiplicative sequence $\{K_j\}$ is strictly multiplicative in ξ with respect to $\hat{\xi}_C$. Then $K(E_\xi) = K(B_\xi) \cdot K(F_\xi)$.*

22. The Todd genus of certain almost complex homogeneous spaces. Throughout this paragraph, all cohomology groups will be taken with real coefficients, and all characteristic classes which occur will be regarded as real classes unless otherwise mentioned. The symbol $\boldsymbol{R}$ will be omitted in real cohomology groups.

22.1. *Notation.* Let ξ and η be complex vector bundles with the same base space: $B = B_\xi = B_\eta$. We recall that the Todd multiplicative sequence $\{T_j(c_1, \cdots, c_j)\}$ has $x(1 - e^{-x})^{-1}$ as its characteristic power series [19, §1] and define the cohomology class $\mathscr{T}(\xi, \eta) \in H^{**}(B)$ by the equation

$$\mathscr{T}(\xi, \eta) = \mathrm{ch}(\eta) \sum_{j=0}^{\infty} T_j(c_1(\xi), \cdots, c_j(\xi)),$$

where $c_i(\xi) \in H^{2i}(B)$ is the i-th Chern class of ξ and $\mathrm{ch}(\eta)$ is the Chern character of η as defined in 9.1. For $d \in H^2(B)$, set

$$\mathcal{J}(\xi, d) = e^d \cdot \sum_{j=0}^{\infty} T_j(c_1(\xi), \cdots, c_j(\xi))$$

and for $d = 0$,

$$\mathcal{J}(\xi, 0) = \mathcal{J}(\xi).$$

It is clear that $\mathcal{J}(\xi, d) = \mathcal{J}(\xi, \eta)$ if η is the complex line bundle with d as its first Chern class.

More generally, as in [19, § 1], let $\{T_j(y; c_1, \cdots, c_j)\}$ be the generalized Todd sequence. This multiplicative sequence has

$$x(1 + y) / (1 - e^{-x(1+y)}) - yx$$

as its characteristic power series. We define $\mathcal{J}_y(\xi) \in H^{**}(B) \otimes \mathbf{R}[y]$ by the equation

$$\mathcal{J}_y(\xi) = \sum_{j=0}^{\infty} T_j(y; c_1(\xi), \cdots, c_j(\xi)).$$

Obviously, $\mathcal{J}_0(\xi) = \mathcal{J}(\xi)$.

If B is a compact almost complex manifold and if now ξ stands for the tangential complex vector bundle of B, then we set

$$\mathcal{J}(B, \eta) = \mathcal{J}(\xi, \eta), \qquad \mathcal{J}(B, d) = \mathcal{J}(\xi, d), \qquad \mathcal{J}_y(B) = \mathcal{J}_y(\xi).$$

Moreover, in agreement with the notations of [19, §§ 10, 11, 12], the following real numbers (respectively polynomials with real coefficients) are defined

$$T(B, \eta) = \mathcal{J}(B, \eta)[B],$$

$$T(B, d) = \mathcal{J}(B, d)[B],$$

$$T_y(B) = \mathcal{J}_y(B)[B] = \sum_{p=0}^{n} T^p(B) y^p, \quad \text{where} \ n = \dim_C B.$$

$T(B)$ denotes the Todd genus $(T(B) = T_0(B) = T(B, 0))$.

Finally, let us recall the following formal fact: If c_i is the i-th elementary symmetric function in $\gamma_1, \cdots, \gamma_n$ $(c_i = 0$ for $i > n)$, then

$$\sum_{j=0}^{\infty} T_j(c_1, \cdots, c_j) = e^{c_1/2} \prod_{i=1}^{n} \gamma_i / (2 \sinh(\gamma_i/2)).$$

22.2. Let G be a compact connected Lie group and T a maximal torus of G. Let V be the universal covering of T and $a_1, \cdots, a_m$ the *positive* roots of G with respect to an ordering $\mathcal{S}$ on V^* (2.4). Let ξ be a principal G-bundle and ξ_C the complex vector bundle along the fibres (7.4) of

$(E_\xi/T, B_\xi, G/T, \pi)$ which belongs to the invariant almost complex structure on G/T having $\epsilon_1 a_1, \epsilon_2 a_2, \cdots, \epsilon_m a_m$, $(\epsilon_i = \pm 1)$, as its roots (12.3 and 13.4). We orient G/T according to this almost complex structure and denote by $\natural$ the integration over the fibre with respect to this orientation (8.1). The total Chern class of $\hat{\xi}_C$ is given by the formula (10.8):

$$(1) \qquad c(\hat{\xi}_C) = (1 + \epsilon_1 a_1)(1 + \epsilon_2 a_2) \cdots (1 + \epsilon_m a_m).$$

Now let d be an arbitrary element of $H^1(T) = V^*$. We may regard d as an element of $H^2(E_\xi/T)$ under the negative transgression. Then, using (1) for the cohomology class $\mathcal{J}(\hat{\xi}_C, d)$ introduced in 22.1, we have

$$\mathcal{J}(\hat{\xi}_C, d) = e^d \cdot \prod \epsilon_j a_j / (1 - \exp(-\epsilon_j a_j)) = e^{(c_1/2) + d} \prod_{j=1}^m a_j / (2 \sinh(a_j/2)),$$

where $c_1 = c_1(\hat{\xi}_C) = \sum_{i=1}^m \epsilon_i a_i$. We are going to calculate $\mathcal{J}(\hat{\xi}_C, d)^\natural$ by 20.3. In view of 8.3, this is possible since ξ is induced from the universal bundle. First observe that $\prod a_j / (2 \sinh(a_j/2))$ is invariant under the operations of the Weyl group $W(G)$ since the roots are permuted up to sign and since $x/\sinh x$ is an even function in x. Thus we obtain, after setting $b = d + \frac{1}{2} c_1(\hat{\xi}_C)$ and $a = \sum_{j=1}^m a_j$,

$$\epsilon_1 \epsilon_2 \cdots \epsilon_m \pi^{**}(\mathcal{J}(\hat{\xi}_C, d)^\natural) = \sum_{w \in W(G)} \mathrm{Sgn}(w) e^{w(b)} / \prod_{i=1}^m 2 \sinh a_i/2$$

or, in the notations of 3.2:

$$(2) \qquad \epsilon_1 \epsilon_2 \cdots \epsilon_m \pi^{**}(\mathcal{J}(\hat{\xi}_C, d)^\natural) = E(b/2\pi(-1)^{\frac{1}{2}}) / E(a/4\pi(-1)^{\frac{1}{2}}).$$

It follows that $\mathcal{J}(\hat{\xi}_C, d)^\natural$ vanishes if b is singular. The right side of the preceding equation is a formal power series in $d, a_1, \cdots, a_m$ (regarded as elements of $H^2(E_\xi/T)$) and, as such, is an element of $H^{**}(E_\xi/T)$. On the other hand, $d, a_1, \cdots, a_m$ are originally elements of V^*, i.e. functions on V. If b is a non-singular weight, then $E(b)/E(a/2)$ is also a function on V, namely, up to a sign, the character of a certain irreducible representation of $\bar{G}$ (for $\bar{G}$, see 3.3). In fact, if b is a non-singular weight, then there is a unique element $w' \in W(G)$ such that $w'(b)$ is in the positive Weyl chamber (2.7) with respect to the ordering $\mathcal{S}$; i.e., $(w'(b), a_j) > 0$ for $1 \leq j \leq m$, and hence $w'(b) - a/2$ is the highest weight of an irreducible representation λ of $\bar{G}$ which is uniquely determined up to equivalence. According to 3.4, the function $E(b)/E(a/2)$ on V equals the character of λ as a function on V multiplied by $\mathrm{Sgn}(w')$. Thus we have seen that $\mathcal{J}(\xi_C, d)^\natural$ is essentially given by a character.

It is clear that $w'(b) - a/2$ is integral on the unit lattice of G if and only if d has this property. Assume now that d has the property just mentioned (in other words, that $d \in H^1(T, \mathbf{Z}) \subset H^1(T) = V^*$) and that $b = d + \frac{1}{2}c_1(\hat{\xi}_c)$ is non-singular. Then λ also defines a representation of G. The λ extension (6.5) $\lambda(\xi)$ of the principal G-bundle ξ is then defined. It is a $\mathbf{U}(n)$-bundle and we have

$$(3) \qquad \mathcal{I}(\hat{\xi}_c, d)^\natural = \mathrm{Sgn}(w')_{\epsilon_1 \epsilon_2} \cdots _{\epsilon_m} \cdot \mathrm{ch}(\lambda(\xi)).$$

Here ch is the Chern character as defined in 9.1. See also 10.2, 10.3.

22.3. In Sections 22.3 and 22.4, we shall apply the results of 22.2 to the very special case where B_ξ is a point; then $E_\xi = G$ and $E_\xi/T = G/T$. Every element of V^* may be regarded as an element of $H^2(G/T)$. Otherwise, we keep the notations of 22.2.

The homogeneous space G/T has 2^m invariant almost complex structures belonging to the 2^m possible choices of the signs $\epsilon_1, \epsilon_2, \cdots, \epsilon_m$. Now endow G/T with the invariant almost complex structure having the roots $\epsilon_1 a_1, \epsilon_2 a_2, \cdots, \epsilon_m a_m$. Then the first Chern class of G/T is

$$c_1 = c_1(G/T) = \epsilon_1 a_1 + \epsilon_2 a_2 + \cdots + \epsilon_m a_m.$$

Since $a/2$ is a weight, $c_1/2$ is also a weight, which implies by (10.1) that the integral first Chern class $c_1(G/T) \in H^2(G/T, \mathbf{Z})$ equals 0 when reduced to coefficients mod 2. *Thus the second Stiefel-Whitney class $w_2(G/T) \in H^2(G/T, \mathbf{Z}_2)$ vanishes.*

The Pontrjagin class p_i of G/T is the i-th elementary symmetric function in the a_j^2, and vanishes for $i > 0$ (10.9). Thus for $d \in H^2(G/T)$ (see end of 22.1),

$$\mathcal{I}(G/T, d) = \exp(d + \tfrac{1}{2}c_1) \in H^*(G/T),$$
$$(4) \qquad m! \cdot T(G/T, d) = ((c_1/2) + d)^m[G/T],$$
$$2^m m! \cdot T(G/T) = c_1{}^m[G/T].$$

On the other hand, in our special case, $T(G/T, d) \cdot 1 = \mathcal{I}(G/T, d)^\natural$ and we have, by 22.2, for an element $d \in V^*$, that

$$(5) \qquad T(G/T, d) = 0 \text{ if } d + (c_1/2) \text{ is singular,}$$

$$(6) \qquad T(G/T, d) = \pm \deg(\lambda),$$

if d is a non-singular weight, and if λ is a suitable representation.

By Theorem 4.3, the sum $c_1 = \epsilon_1 a_1 + \cdots + \epsilon_m a_m$ is non-singular if and only if $\epsilon_1 a_1, \cdots, \epsilon_m a_m$ is a positive system of roots of G. By 4.9 and 12.4, we get that $\epsilon_1 a_1 + \cdots + \epsilon_m a_m$ is non-singular if and only if the invariant almost complex structure on G/T with roots $\epsilon_1 a_1, \cdots, \epsilon_m a_m$ is integrable.

Putting the value 0 for d in (5), we see that the Todd genus of G/T endowed with a non-integrable invariant almost complex structure vanishes. If, however, the structure is integrable, i.e., c_1 is non-singular, then (for $d = 0$) we have in the notation of 22.2 that $w'(\tfrac{1}{2}c_1) = w'(b) = a/2$ and $\mathrm{Sgn}(w') = \epsilon_1 \epsilon_2 \cdots \epsilon_m$. Thus λ is the trivial representation of degree 1 and the Todd genus of G/T equals $\deg(\lambda) = 1$.

22.4. Let $a_1, a_2, \cdots, a_m$ be as before the positive system of roots of G with respect to some ordering $\mathcal{O}$ and let $a = \sum_{j=1}^{m} a_j$. Choose the integrable invariant almost complex (i.e. complex) structure on G/T which has $a_1, a_2, \cdots, a_m$ as its roots ($\epsilon_i = 1$) and let G/T be oriented accordingly. An arbitrary element $b \in V^*$ can be regarded as element of $H^2(G/T)$ and then the number $\delta(b) = b^m[G/T]/m!$ is defined. δ defines a homogeneous polynomial of degree m on V^*, which vanishes if $(b, a_j) = 0$, see (4) and (5). Since a_i and a_j are not proportional for $i \neq j$ and since $\delta(a/2) = 1$ by (4), we get

$$(7) \qquad b^m[G/T] = m! \prod_{j=1}^{m} (b, a_j)/(a/2, a_j), \qquad (b \in V^*).$$

Formula (7) shows that b *is singular if and only if* $b^m[G/T] = 0$.

Theorem 20.3 implies immediately that

$$(8) \qquad b^m[G/T] = (a_1 a_2 \cdots a_m)^{-1} \sum_{w \in W(G)} \mathrm{Sgn}(w) w(b)^m,$$

where the right side of this equation has to be regarded as a quotient of two homogeneous polynomials of degree m on V. Assume now that b is a weight. Then b is in the positive Weyl chamber ($(b, a_j) > 0$ for $1 \leq j \leq m$) if and only if $b - a/2$ is in the closure of the positive Weyl chamber ($(b, a_j) \geq 0$ for $1 \leq j \leq m$). By (3) and (6), we get:

If b is a weight contained in the positive Weyl chamber, then

$$(9) \qquad b^m[G/T]/m! = T(G/T, b - a/2) = \deg(\lambda),$$

where λ is the irreducible representation of $\bar{G}$ with main weight $b - a/2$, and a is the sum of the roots a_j of the invariant complex structure on G/T; i.e., a is the first Chern class of G/T endowed with this complex structure.

By (7) and (8), we get well known formulas for the degree of λ, see 3.4.

22.5. In this and the following Section, we shall use 22.2 to prove the strictly multiplicative behavior of the Todd sequence in certain fibre bundles. For this purpose, we take the value 0 for d in 22.2. Then $2b$ equals the first Chern class of the complex vector bundle $\hat{\xi}_c$ along the fibres of $(E_\xi/T, B_\xi, G/T, \pi)$. By 22.2 and 22.3, we see that the integration over the fibre (with respect to the orientation of G/T induced by $\hat{\xi}_c$) gives, when applied to $\mathcal{J}(\hat{\xi}_c)$, either 0 or $\mathrm{ch}(\lambda(\xi))$, where λ is the trivial 1-dimensional representation, and thus $\mathrm{ch}(\lambda(\xi)) = 1$. In either case, the integration over the fibre gives only a zero-dimensional term. According to the definition in 21.7, we get:

THEOREM. *Let ξ be a principle G-bundle. Choose an invariant almost complex structure on G/T and let $\hat{\xi}_c$ be the corresponding complex vector bundle along the fibres of $(E_\xi/T, B_\xi, G/T)$. Then the Todd sequence $\{T_j\}$ is strictly multiplicative in $(E_\xi/T, B_\xi, G/T)$ with respect to $\hat{\xi}_c$.*

For later use, we reformulate our result as follows: Let Ψ be a set of roots of G which contains for each root α exactly one of the roots α, $-\alpha$. Then Ψ is the set of roots of an invariant almost complex structure on G/T. Orient G/T by this structure and let $\natural(\Psi)$ be the integration over the fibre in $(E_\xi/T, B_\xi, G/T)$ with respect to that orientation. Then we have

$$(10) \qquad \left(\prod_{\alpha \in \Psi} \alpha/(1 - e^{-\alpha}) \right)^{\natural(\Psi)} = \begin{cases} 1, & \text{if } \Psi \text{ is a positive system} \\ 0, & \text{otherwise.} \end{cases}$$

Let ϵ be a map of Ψ into $\{1, -1\}$ and $s(\epsilon)$ the number of elements in Ψ which are mapped by ϵ on -1, and let $\mathrm{sgn}(\epsilon) = (-1)^{s(\epsilon)}$. We have

$$\prod_{\alpha \in \Psi} \epsilon(\alpha) \cdot \alpha/(1 - e^{-\epsilon(\alpha) \cdot \alpha}) = \exp\left(\tfrac{1}{2} \sum_{\alpha \in \Psi} \epsilon(\alpha) \cdot \alpha\right) \cdot \prod_{\alpha \in \Psi} \alpha/(2 \sinh(\alpha/2)).$$

Thus, by (10),

$$(10^*) \qquad \left(\exp\left(\tfrac{1}{2} \sum_{\alpha \in \Psi} \epsilon(\alpha) \cdot \alpha\right) \cdot \prod_{\alpha \in \Psi} \alpha/(2 \sinh(\alpha/2)) \right)^{\natural(\Psi)}$$

is equal to $\mathrm{sgn}(\epsilon) \cdot 1$, if $\{\epsilon(\alpha) \cdot \alpha \mid \alpha \in \Psi\}$ is a positive system, and to 0, otherwise.

We give two applications of formula (10): Let ξ be a principal G-bundle for which B_ξ is a compact *oriented* manifold. Let Ψ be a *positive* system of roots of G. Orient G/T accordingly and choose for E_ξ/T that orientation which is induced by those of B_ξ and G/T. Then, for $y \in H^*(E_\xi/T)$, we have (see 21.4):

$$y[E_\xi/T] = y^{\natural(\Psi)}[B_\xi].$$

This fact, together with (10) and 8.2, yields for every element x of $H^*(B_\xi)$ that

$$(11) \qquad x[B_\xi] = (\pi^*(x) \cdot \prod_{\alpha \in \Psi} \alpha/(1 - \exp(-\alpha)))[E_\xi/T],$$

where π is the projection of E_ξ/T on B_ξ.

22.6. Now let ξ be again an arbitrary principal G-bundle. For a closed connected subgroup U of G which contains a maximal torus T of G, consider the diagram

$$(12) \qquad E_\xi/T \xrightarrow{\ \pi_\nu\ } E_\xi/U \xrightarrow{\ \pi_\mu\ } B_\xi.$$

Orient G/U. Let Θ be a system of positive roots of U. Orient U/T by the invariant complex structure having Θ as its set of roots and give G/T the orientation induced by those of G/U and U/T.

According to the above diagram, we have the fibre bundles

$$\mu = (E_\xi/U, B_\xi, G/U, \pi_\mu), \qquad \nu = (E_\xi/T, E_\xi/U, U/T, \pi_\nu),$$

$$\bar{\xi} = (E_\xi/T, B_\xi, G/T, \pi_{\bar{\xi}}), \qquad \eta = (G/T, G/U, U/T, \pi_\eta),$$

which satisfy the assumptions of 8.4. (The ξ of 8.4 corresponds to the $\bar{\xi}$ here and the θ of 8.4 to η.) If we define the integrations over the fibres with respect to the orientations already chosen, then formula (2) of 8.4 holds. Thus we can infer from (10) and 8.2 the

PROPOSITION. *Let x be an arbitrary element of $H^{**}(E_\xi/U)$. Then in the foregoing notation*

$$x^{\natural\mu} = (\pi_\nu^{**}(x) \cdot \prod_{\alpha \in \Theta} \alpha/(1 - e^{-\alpha}))^{\natural\bar{\xi}}.$$

22.7. Let U be a closed connected subgroup of G containing a maximal torus T of G and assume that G/U has been endowed with an invariant almost complex structure $\mathcal{B}$. Thus G/U is oriented. Let Ψ be the set of the roots of $\mathcal{B}$ and Θ a system of positive roots of U according to which we orient U/T. Then G/T is oriented. Let ξ be again a principal G-bundle for which we consider the diagram (12) and the four fibre bundles μ, ν, $\bar{\xi}$, η. We wish to prove that the generalized Todd sequence (22.1) is strictly multiplicative (21.7) in μ with respect to the complex vector bundle $\hat{\mu}_C$ along the fibres arising from the given invariant almost complex structure on G/U. Let n

be the complex dimension of G/U, i. e., the number of roots in Ψ. First we observe that $\pi_\nu^{**}(\mathcal{J}_y(\hat{\mu}_C))$ is, in virtue of 10.8, equal to the element

$$(13) \qquad \prod_{\alpha \epsilon \Psi} ((1+y)\alpha/(1-e^{-(1+y)\alpha}) - y\alpha) \in H^{**}(E_\xi/T) \otimes \boldsymbol{R}[y]$$

which goes over into

$$(13^*) \qquad \prod_{\alpha \epsilon \Psi} (1+ye^{-\alpha})\alpha/(1-\exp(-\alpha)) \in H^{**}(E_\xi/T) \otimes \boldsymbol{R}[y]$$

if one multiplies the component of complex dimension i in (13) by $(1+y)^{n-i}$. We have to prove that $\mathcal{J}_y(\hat{\mu}_C)^{\natural\mu}$ is a zero-dimensional element of $H^{**}(B_\xi)$. In view of the proposition in 22.6 and the passage from (13) to (13^*), we must prove that the element

$$(\prod_{\alpha \epsilon \Psi} (1+ye^{-\alpha}) \cdot \prod_{\beta \epsilon \Theta \cup \Psi} \beta/(1-\exp(-\beta)))^{\natural\bar{\xi}}$$

which is equal to

$$(14) \qquad (\exp(\sum_{\beta \epsilon \Theta \cup \Psi} \beta/2) \cdot \prod_{\alpha \epsilon \Psi} (1+ye^{-\alpha}) \cdot \prod_{\beta \epsilon \Theta \cup \Psi} \beta/(2\sinh(\beta/2)))^{\natural\bar{\xi}}$$

is zero-dimensional. But this is a consequence of (10^*). The set $\Theta \cup \Psi$ plays here the role of Ψ in (10^*). Thus the element given in (14) equals the unit of $H^{**}(B_\xi)$ multiplied by $T_y(G/U)$, (G/U has the given almost complex structure). Here we have to use that in passing from (13) to (13^*), the component of complex dimension n is not changed. Using (10^*), we can obtain the value of $T_y(G/U)$. In order to formulate the final result more easily, we introduce the following definition.

DEFINITION. *Let U be a closed subgroup of G containing a maximal torus T of G. Let Ψ be a set of complementary roots of G with respect to U which contains for each complementary root α exactly one of the roots $\alpha, -\alpha$. Let Θ be a system of positive roots of U. Then $k^p(G/U, \Psi, \Theta)$ is defined as the number of those positive systems of roots of G which contain Θ and exactly $n-p$ roots of Ψ and thus p roots of $-\Psi$ $(0 \leq 2p \leq 2n = \dim_{\boldsymbol{R}} G/U)$.*

Using this definition, we have, in virtue of (10^*) and the fact that the element given in (14) equals $T_y(G/U) \cdot 1$, that

$$(15) \qquad T_y(G/U) = \sum_{p=0}^{n} T^p(G/U)y^p = \sum_{p=0}^{n} (-y)^p k^p(G/U, \Psi, \Theta).$$

We notice that $k^p(G/U, \Psi, \Theta)$ depends only on Ψ and not on the choice of the positive system Θ of U if Ψ is the set of roots of an invariant almost complex structure on G/U. The number $k^p(G/U, \Psi, \Theta)$ is also well defined if G/U does not admit an invariant almost complex structure.

Formula (15) states in particular that the Todd genus $T(G/U)$ equals $k^0(G/U, \Psi, \Theta)$. Thus $T(G/U)$ equals 1 if $\Psi \cup \Theta$ is a positive system of roots of G and is 0 otherwise. If the invariant almost complex structure on G/U with Ψ as the set of its roots is integrable, then $\Psi \cup \Theta$ is a positive system (13.7) and thus $T(G/U) = 1$. If $T(G/U) = 1$, then $\Psi \cup \Theta$ is a positive system, but $\Psi \cup -\Theta$ is also positive, since $-\Theta$ is a system of positive roots of U and $k^0(G/U, \Psi, -\Theta) = T(G/U) = 1$. Therefore, $T(G/U) = 1$ implies that $\Psi \cup \Theta$, $\Psi \cup -\Theta$ are positive and thus closed systems. $\Theta \cup -\Theta$ is closed, since it is the set of roots of a subgroup. Thus $T(G/U) = 1$ implies that $\Psi \cup \Theta \cup -\Theta$ is closed and that the given invariant almost complex structure is integrable (12.4).

We express the results of this section in the following theorem.

22.8. **Theorem.** *Let G be a compact Lie group and U a closed connected subgroup of maximal rank of G. The Todd genus of an invariant almost complex structure on G/U equals 1 (respectively 0) if the structure is integrable (respectively not integrable). With respect to a maximal torus T $(T \subset U \subset G)$, let Θ be a system of positive roots of U. Assume that G/U has been given an invariant almost complex structure $\mathcal{C}$ and that Ψ is the set of roots of $\mathcal{C}$. Then, letting n be the complex dimension of G/U and using the definition in 22.7, we have*

$$T_y(G/U) = \sum_{p=0}^{n} T^p(G/U) y^p = \sum_{p=0}^{n} (-y)^p k^p(G/U, \Psi, \Theta).$$

Let ξ be a principal G-bundle. The generalized Todd sequence

$$\{T_j(y; c_1, \cdots, c_j)\}$$

is strictly multiplicative in $(E_\xi/U, B_\xi, G/U)$ with respect to the complex vector bundle along the fibres $\hat{\xi}_C$ arising from $\mathcal{C}$. In particular, if B_ξ is a compact almost complex differentiable manifold, if ξ is differentiable and if the differentiable manifold E_ξ/U has been endowed with an almost complex structure compatible (21.8) with the almost complex structure of B_ξ and $\hat{\xi}_C$, then

$$T_y(E_\xi/U) = T_y(B_\xi) \cdot T_y(G/U).$$

22.9. For $y = 1$, the preceding theorem gives results on the index $\tau(G/U)$, see [19, §§ 8 and 10]. These results remain correct for an arbitrary G/U not necessarily almost complex:

Let G be a compact connected Lie group and U a closed connected subgroup of G of maximal rank. Let T be a maximal torus of U. Then, with respect to T, let Ψ be a set of complementary roots containing for each complementary root α exactly one of the roots $\alpha, -\alpha$. Let ξ be a principal G-bundle. The element $\prod_{\alpha \in \Psi} \alpha/\operatorname{tgh} \alpha \in H^{**}(E_\xi/T)$ is symmetric in the α^2 $(\alpha \in \Psi)$, and thus belongs to $\pi_\nu^{**}(H^{**}(E_\xi/U))$. According to 10.7,

$$\prod_{\alpha \in \Psi} \alpha/\operatorname{tgh} \alpha = \pi_\nu^{**} \left(\sum_{j=0}^{\infty} L_j(p'_1, p'_2, \cdots, p'_j) \right),$$

where the p'_i are the Pontrjagin classes of the real vector bundle $\hat{\mu}$ along the fibres of $(E_\xi/U, B_\xi, G/U)$. (For the L_j, see [19, §1].) We orient the bundle along the fibres and thus also G/U by requiring that

$$\prod_{\alpha \in \Psi} \alpha = \pi_\nu^{**}(W_{2n}),$$

where W_{2n} denotes the Euler class of $\hat{\mu}$ $(2n = \dim_R G/U)$. Then the same calculations as in 22.7 show that $\{L_j\}$ is strictly multiplicative in $(E_\xi/U, B_\xi, G/U)$ and that

$$\tau(G/U) = \sum_{p=0}^{n} (-1)^p k^p(G/U, \Psi, \Theta),$$

Θ being an arbitrary system of positive roots of U. As a consequence of the strictly multiplicative behavior, we have

$$\tau(E_\xi/U) = \tau(B_\xi) \cdot \tau(G/U),$$

in case B_ξ is a compact oriented differentiable manifold and ξ a differentiable bundle and after introducing convenient orientations.

In a similar way, under the assumptions of this section, we get by setting $y = -1$ for the Euler number $E(G/U)$ that

$$E(G/U) = \sum_{p=0}^{n} k^p(G/U, \Psi, \Theta).$$

22.10. The strictly multiplicative behavior (22.8) of the Todd sequence has certain formal consequences. We follow the notations of 22.7. Let ξ be a differentiable principal G-bundle over the compact almost complex differentiable manifold B_ξ and η a complex vector bundle over B_ξ. Consider the bundle $\mu = (E_\xi/U, B_\xi, G/U, \pi_\mu)$ and the complex vector bundle $\hat{\mu}_C$ along the fibres arising from a given invariant almost complex structure on G/U. Then endow E_ξ/U with an almost complex structure compatible with that of B_ξ and with $\hat{\mu}_C$. We have

$$(16) \qquad \mathcal{J}(E_\xi/U, \pi_\mu^*\eta)^{\natural\mu} = T(G/U) \cdot \mathcal{J}(B_\xi, \eta).$$

Proof. Since the complex tangent bundle of E_ξ/U is the Whitney sum of $\hat{\mu}_C$ and the complex tangent bundle of B_ξ lifted under π_μ, we obtain from the Whitney multiplication theorem (9.7) that

$$\mathcal{J}(E_\xi/U) = \mathcal{J}(\hat{\mu}_C) \cdot \pi_\mu{}^*\mathcal{J}(B_\xi).$$

Thus

$$\mathcal{J}(E_\xi/U, \pi^*{}_\mu\eta)^{\natural\mu} = (\pi_\mu{}^*(\mathrm{ch}(\eta)) \cdot \pi_\mu{}^*(\mathcal{J}(B_\xi)) \cdot \mathcal{J}(\hat{\mu}_C))^{\natural\mu}$$

$$= \mathcal{J}(\hat{\mu}_C)^{\natural\mu} \cdot \mathcal{J}(B_\xi, \eta) = T(G/U) \cdot \mathcal{J}(B_\xi, \eta),$$

which completes the proof.

As a consequence of (16), we get (see 21.4):

$$(17) \qquad T(E_\xi/U, \pi_\mu{}^*\eta) = T(G/U) \cdot T(B_\xi, \eta).$$

There is a formula for the generalized Todd sequence which is analogous to (16) and which follows from the strictly multiplicative behavior of the generalized Todd sequence. In order to write it down, we introduce the element $\mathrm{ch}_y(\eta)$ as the element obtained from $\mathrm{ch}(\eta)$ by multiplying its component of complex dimension j with $(1+y)^j$. The element $\mathrm{ch}_y(\eta)$ was denoted by $t_y(\eta)$ in [19, §12.2]. We have

$$(18) \qquad (\mathrm{ch}_y(\pi_\mu{}^*\eta) \cdot \mathcal{J}_y(E_\xi/U))^{\natural\mu} = T_y(G/U) \cdot \mathrm{ch}_y(\eta)\mathcal{J}_y(B_\xi),$$

which implies

$$(19) \qquad T_y(E_\xi/U, \pi_\mu{}^*\eta) = T_y(G/U) \cdot T_y(B_\xi, \eta).$$

For the definition of $T_y(B_\xi, \eta)$ and $T_y(E_\xi/U, \pi_\mu{}^*\eta)$, see [19, §12]. Compare also [19, §14.4].

22.11. Let G be a compact connected Lie group, U a closed connected subgroup of maximal rank and T a maximal torus of U. Let d be an element of $H^1(T, \mathbf{Z})$ which is orthogonal to all roots of U; i.e., d is invariant under all operations of the Weyl group of U. By the canonical isomorphism of $H^1(T, \mathbf{Z})$ with $\mathrm{Hom}(T, \mathbf{U}(1))$, the element d gives rise to an homomorphism of T in $\mathbf{U}(1)$ which has a unique extension to an homomorphism of U in $\mathbf{U}(1)$, also denoted by d. Now let ξ be a principal G-bundle. We extend the principal U-bundle $(E_\xi, E_\xi/U, U)$ by the homomorphism d of U in $\mathbf{U}(1)$. We get a principal $\mathbf{U}(1)$-bundle over E_ξ/U and the associated line bundle whose first Chern class we also denote by d. Following the notations of 22.6, it is clear that $\pi_\nu{}^*(d)$ is that element of $H^2(E_\xi/T, \mathbf{Z})$ which is obtained from the original element $d \in H^1(T, \mathbf{Z})$ by the negative transgression in $(E_\xi, E_\xi/T, T)$. Therefore, we may also denote $\pi_\nu{}^*d$ by d.

Now assume moreover that G/U carries an invariant complex structure. Let Ψ be the set of roots of this structure and let Θ be a system of positive roots of U. Then, by (13.7), $\Theta \cup \Psi$ is a positive system of G, to which belongs a positive Weyl chamber. Assume furthermore that d belongs to the closure of this Weyl chamber; i.e., $(d, \alpha) \geqq 0$ for $\alpha \in \Psi$. Let λ be the representation of G with main weight d and let d_ξ be the complex vector bundle over B_ξ associated with the λ-extension of ξ. Now, as in 22.10, we make the hypothesis that B_ξ is a compact almost complex manifold and that E_ξ/U has been given an almost complex structure compatible with the almost complex structure of B_ξ and the complex vector bundle $\hat{\mu}_C$ along the fibres of E_ξ/U arising from the given invariant complex structures on G/U. We have then, using the notations of 22.6, that

$$(20) \qquad \mathcal{J}(E_\xi/U, d)^{\natural \mu} = \mathcal{J}(B_\xi, d_\xi).$$

Proof.

$$\mathcal{J}(E_\xi/U, d)^{\natural \mu} = (\pi_\nu^*(e^d \cdot \pi_\mu^*(\mathcal{J}(B_\xi)) \cdot \mathcal{J}(\hat{\mu}_C)) \cdot \prod_{\alpha \in \Theta} \alpha/(1 - e^{-\alpha}))^{\natural \bar{\xi}}$$

$$= \mathcal{J}(B_\xi)(e^d \prod_{\alpha \in \Theta \cup \Psi} \alpha/(1 - e^{-\alpha}))^{\natural \bar{\xi}}.$$

Formula (20) follows then by applying 22.2.

Remarks. (1) Formulas (16) and (20) are closely related to Grothendieck's generalized Riemann-Roch formula (not yet published) ; compare also with [7b, p. 241].

(2) The representation λ induces a holomorphic map β of G/U into a complex projective space $\boldsymbol{P}_q(\boldsymbol{C})$ such that $d \in H^2(E_\xi/U)$ restricted to G/U equals $\beta^*(e^*)$, where $e^* \in H^2(\boldsymbol{P}_q(\boldsymbol{C}), \boldsymbol{Z})$ is the cohomology class dual to a hyperplane (see 14.4).

23. The A-genus of certain homogeneous spaces. Throughout this paragraph, all cohomology groups are taken with real coefficients and all characteristic classes which occur are regarded as real classes.

23.1. Let $\{\hat{A}_j(p_1, \cdots, p_j)\}$ be the multiplicative sequence of polynomials [19, §1] with $\frac{1}{2}z^{\frac{1}{2}}/\sinh \frac{1}{2}z^{\frac{1}{2}}$ as characteristic power series. The polynomials $\hat{A}_j$ are related to the A_j introduced in [19, §1.6] by the equation

$$A_j = 2^{4j}\hat{A}_j.$$

For a real vector bundle ξ, we define the cohomology class $\hat{a}(\xi) \in H^{**}(B_\xi)$ as follows:

$$\hat{a}(\xi) = \sum_{j=0}^{\infty} \hat{A}_j(p_1(\xi), \cdot \cdot \cdot, p_j(\xi)).$$

If ξ is the tangent bundle of a differentiable manifold X, then we set $\hat{a}(\xi) = a(X)$. The genus $\hat{A}(X)$ of a compact oriented differentiable manifold X is given by

$$\hat{A}(X) = \hat{a}(X)[X].$$

$\hat{A}(X)$ vanishes if the dimension of X is not divisible by 4. For $\dim X = 4k$, we have

$$\hat{A}(X) = \hat{A}_k(p_1(X), \cdot \cdot \cdot, p_k(X))[X],$$

and, obviously, $A(X) = 2^{4k}\hat{A}(X)$, where $A(X)$ is the genus corresponding to the power series $2z^{\frac{1}{2}}/\sinh 2z^{\frac{1}{2}}$, and which is called the A-genus of X. If X is almost complex with vanishing first Chern class, then its Todd genus equals its $\hat{A}$-genus; see 22.1 and [19, p. 15].

23.2. Let G be a compact connected Lie group, T a maximal torus of G and U a closed connected subgroup of G containing T. Choose an ordering $\mathcal{S}$ (2.4), let Θ be the set of those roots which are positive with respect to $\mathcal{S}$ and belong to U, and let Ψ be the set of positive complementary roots. Orient U/T and G/T by the invariant complex structures with root systems Θ and $\Theta \cup \Psi$ respectively. We orient G/U by its Euler class using Ψ (see 22.9). Then, in the fibre bundle $(G/T, G/U, U/T)$, the orientations of G/U and and U/T induce that of G/T. Let ξ be a principal G-bundle. We adhere now strictly to the notations given in 22.6. Consider the fibre bundle

$$\mu = (E_\xi/U, B_\xi, G/U, \pi_\mu).$$

We wish to calculate the value of $\hat{A}(G/U)$ and to investigate under which conditions the sequence $\{\hat{A}_j\}$ behaves strictly multiplicatively (21.3) in μ. Let $\hat{\mu}$ be the real vector bundle along the fibres of μ. Then, by 10.7,

$$\pi_\nu^{**}\hat{a}(\hat{\mu}) = \prod_{\alpha \in \Psi} \alpha/(2\sinh(\alpha/2))$$

and we get, by the proposition in 22.6, that

$$\hat{a}(\hat{\mu})^{\natural\mu} = (\prod_{\beta \in \Theta} \beta/(1 - e^{-\beta}) \prod_{\alpha \in \Psi} \alpha/(2\sinh(\alpha/2)))^{\natural\bar{\xi}}.$$

If we write s for the sum of all $\alpha \in \Theta$, then

$$\hat{a}(\hat{\mu})^{\natural\mu} = (e^{\frac{1}{2}s} \prod_{\alpha \in \Theta \cup \Psi} \alpha/(2\sinh(\alpha/2)))^{\natural\bar{\xi}}.$$

Let a be the sum of all roots in $\Theta \cup \Psi$. Since $\Theta \cup \Psi$ is a positive system of roots for G and since G/T is oriented by the invariant complex structure having $\Theta \cup \Psi$ as its set of roots, we get, in virtue of 22.2, that

$$(1) \qquad \pi_\xi^{**}(\hat{a}(\hat{\mu})^{\natural\mu}) = E(s/4\pi(-1)^{\frac{1}{2}})/E(a/4\pi(-1)^{\frac{1}{2}}).$$

The genus $\hat{A}(G/U)$ is given by the constant term on the right side of the preceding equation which, by 22.3 (4), is also equal to $(s^m/2^m m!)[G/T]$, where $m = \dim_C(G/T)$. Thus, by 22.4 (7),

$$(2) \qquad \hat{A}(G/U) = \prod_{\beta \in \Theta \cup \Psi} (s,\beta)/(a,\beta).$$

Now assume that U is the centralizer of a toral subgroup of G and that Ψ is the set of roots of an invariant complex structure on G/U. Then $a-s$ represents the first Chern class of G/U. The element $a-s$ is orthogonal to all roots of Θ (see 14.2 and 14.8). Therefore, $(s,\beta) = (a,\beta)$ for all $\beta \in \Theta$ and thus, by (2),

$$(3) \qquad \hat{A}(G/U) = \prod_{\beta \in \Psi} (s,\beta)/(a,\beta).$$

23.3. THEOREM. *Let G be a compact connected Lie group, T a maximal torus of G and U a closed connected subgroup of G containing T. Choose an ordering $\mathscr{S}$ and let Θ be the set of those roots which are positive with respect to $\mathscr{S}$ and belong to U. Let s denote the sum of all $\alpha \in \Theta$. Then the following holds:*

   i) *The genus $\hat{A}(G/U)$ vanishes if and only if s is a singular element.*

   ii) *If ξ is a principal G-bundle and $\hat{A}(G/U) = 0$, then the sequence $\{\hat{A}_j\}$ is strictly multiplicative in $(E_\xi/U, B_\xi, G/U)$. In particular, if B_ξ is a compact orientable differentiable manifold and ξ a differentiable bundle, then $\hat{A}(G/U) = 0$ implies that $\hat{A}(E_\xi/U) = 0$ also.*

   iii) *If $\hat{A}(G/U)$ is not zero, then U is the centralizer of a toral subgroup of G; i.e., G/U is homogeneous algebraic (§14). In particular, $\hat{A}(G/U)$ vanishes if the second Betti number of G/U is zero.*

   iv) *If ξ is the universal principal G-bundle and $U \neq G$, then $\{\hat{A}_j\}$ is strictly multiplicative in $(E_\xi/U, B_\xi, G/U)$ if and only if $\hat{A}(G/U) = 0$.*

Proof of i) *and* ii). The statement i) follows from formula (2). If s is singular, then $E(s/4\pi(-1)^{\frac{1}{2}}) = 0$ (see 3.2). Thus, ii) follows from i) and formula (1).

The proof of iii) will be preceded by the following lemma.

23.4. LEMMA. *If G is compact, connected, and semi-simple, if T is a maximal torus of G, and if U $(U \neq G)$ is a closed connected semi-simple* *subgroup of G which contains T, then the sum s of all roots of U which are positive with respect to a given ordering $\mathcal{S}$ on $V_T{}^*$ is singular.*

Proof. It is enough to prove the lemma for the case that U is a maximal connected subgroup of G, i. e., U is not contained in a closed connected subgroup of G different from U and G. In this case, we have ([7, p. 205], compare also 10.1)

$$G/U \cong (G_1/U_1) \times (G_2/U_2) \times \cdots \times (G_k/U_k),$$

where G_i is simple, rank $U_i = \operatorname{rank} G_i$, and U_i is a maximal connected subgroup of G_i. The lemma holds for G/U if it is true for at least one of the factors. Thus it suffices to prove the lemma for the case G simple and U a maximal connected subgroup of G. These spaces G/U were listed in [7, p. 219], see also 13.3. Because of i), it suffices to prove the lemma for one ordering on $V_T{}^*$, for then it is proved for all orderings on $V_T{}^*$. The proof will proceed by checking the various cases with G simple and U maximal.

If $G = \boldsymbol{B}_l$ and $U = \boldsymbol{B}_i \times \boldsymbol{D}_{l-i}$ $(0 \leqq i \leqq l-2)$, the positive roots of U with respect to a suitable ordering and a suitable maximal torus are

$$\pm x_r + x_t \ (1 \leqq r < t \leqq i), \ \pm x_r + x_t \ (i+1 \leqq r < t \leqq l), \ x_r \ (1 \leqq r \leqq l).$$

The sum s of these roots is

$$\sum_{j=1}^{i} (2j-1)x_j + 2 \sum_{j=1}^{l-i} (j-1)x_{i+j}$$

which is orthogonal to the root x_{i+1}, and thus s is singular.

If $G = \boldsymbol{C}_l$ and $U = \boldsymbol{C}_i \times \boldsymbol{C}_{l-i}$, the positive roots of U with respect to a suitable ordering are (16.4)

$$\pm x_r + x_t \ (1 \leqq r < t \leqq i), \ \pm x_r + x_t \ (i+1 \leqq r < t \leqq l), \ 2x_r \ (1 \leqq r \leqq l).$$

The sum s of these roots is $2 \sum_{j=1}^{i} jx_j + 2 \sum_{j=1}^{l-i} jx_{i+j}$ which is orthogonal to the root $-x_1 + x_{i+1}$, and thus s is singular. Next we check $\boldsymbol{F}_4/\boldsymbol{B}_4$ and $\boldsymbol{G}_2/\boldsymbol{A}_1 \times \boldsymbol{A}_1$. In these cases, one can choose a set of complementary roots containing for each complementary root α exactly one of the roots α, $-\alpha$ and such that the sum of all roots of this set is 0 (see 18.3 and 19.2). But then it is an immediate consequence of 4.4 that s is singular. The space $\boldsymbol{G}_2/\boldsymbol{A}_2$ has dimension 6 (in fact, it is the 6-sphere). Thus $\hat{A}(\boldsymbol{G}_2/\boldsymbol{A}_2) = 0$, and the lemma is correct in this case by i).

If G is simple and U maximal, one can choose orderings $\mathcal{S}$ and $\mathcal{S}'$ on $V_T{}^*$ such that each root of U simple with respect to $\mathcal{S}$ is a root of G which is either simple with respect to δ' or equals the negative dominant root of G with respect to δ' (see [7]). If $G = \boldsymbol{D}_l, \boldsymbol{E}_6, \boldsymbol{E}_7, \boldsymbol{E}_8$, then all δ'-simple roots of G and the corresponding dominant root of G have equal lengths and thus all simple roots of U with respect to the ordering δ have all the same length ρ in the Killing metric of G. Then the sum s of all roots of U which are positive with respect to δ has a representative contravariant vector lying in the principal diagonal of the δ-positive Weyl chamber of U [25a, p. 221], since $(s, \beta) = (\beta, \beta)$ for each δ-simple root β of U, by 3.1. According to [25a, Théorème 7], the principal diagonal contains only singular vectors, and thus s is singular.

It remains to check our lemma for the spaces $\boldsymbol{F}_4/\boldsymbol{A}_1 \times \boldsymbol{C}_3$ and $\boldsymbol{F}_4/\boldsymbol{A}_2 \times \boldsymbol{A}_2$. The Schläfli diagram of $\boldsymbol{F}_4$ (including the dominant root) is

$$
\begin{array}{ccccc}
1 & 1 & 2 & 2 & 2 \\
\bullet\!\!-\!\!\!-\!\!\!-\!\!\bullet\!\!=\!\!=\!\!\bullet\!\!-\!\!\!-\!\!\!-\!\!\bullet\!\!-\!\!\!-\!\!\!-\!\!\bullet \\
\phi_1 & \phi_2 & \phi_3 & \phi_4 & -\phi
\end{array}
$$

where the ϕ_i are the δ'-simple roots of $\boldsymbol{F}_4$ and where $\phi = 2\phi_1 + 4\phi_2 + 3\phi_3 + 2\phi_4$ is the δ'-dominant root [25a]. The integers 1, 2 indicate the values of (ϕ_i, ϕ_i) and (ϕ, ϕ).

The subgroup $U = \boldsymbol{A}_1 \times \boldsymbol{C}_3$ is represented by the following diagram which indicates the δ-simple roots of U.

$$
\begin{array}{cccc}
1 & 1 & 2 & 2 \\
\bullet\!\!-\!\!\!-\!\!\!-\!\!\bullet\!\!=\!\!=\!\!\bullet & \bullet \\
\phi_1 & \phi_2 & \phi_3 & -\phi
\end{array}
$$

By an easy calculation, we get for the sum s of all δ-positive roots of U

$$
s = 4\phi_1 + 6\phi_2 + 3\phi_3 - 2\phi_4
$$

which is orthogonal to the root $\phi_1 + 2\phi_2 + 2\phi_3 + \phi_4$. Thus s is singular.

Finally, we consider $\boldsymbol{F}_4/\boldsymbol{A}_2 \times \boldsymbol{A}_2$. This space is almost complex (13.3) with vanishing Todd genus (22.8). Since the second Betti number of $\boldsymbol{F}_4/\boldsymbol{A}_2 \times \boldsymbol{A}_2$ is zero, also the $\hat{A}$-genus vanishes (23.1) and thus the lemma is true in this case by (i).

The proof of the lemma is completed.[2]

23.5. *Proof of* iii). The proof proceeds by induction on the dimension of G/U. Assume iii) is proved for U', G' with $\dim G'/U' < n$. We prove it now for U, G with $\dim G/U = n$. If Q is the center of G, then G/U

---

[2] (Added in proof). Another proof of this lemma will be given in a forthcoming paper of the authors.

$= (G/Q)/(U/Q)$ and if U/Q is the centralizer of a toral subgroup in G/Q, then the same holds for U in G, and conversely. Thus, in proving iii) for U, G with $\dim G/U = n$, we may assume that G is semi-simple. Suppose $\hat{A}(G/U) \neq 0$. Then by i) and 23.4, the subgroup U is not semi-simple. Let T' be the identity component of its center and V the centralizer of T'. Then V is connected. $V \neq G$, since G is semi-simple. Thus $U \subset V \subsetneq G$. We have the fibre bundle $(G/U, G/V, V/U)$ and $\hat{A}(V/U) \neq 0$ by ii) . By the induction hypothesis, U is the centralizer of T' in V; but then $U = V$.

23.6. *Proof of* iv). Let ξ be the universal principal G-bundle. It follows immediately from formula (1) in 23.2 that the sequence $\{\hat{A}_j\}$ is strictly multiplicative in $(E_\xi, B_\xi, G/U)$ if and only if the function $E(s) \cdot E(a)^{-1}$, which is defined on V_T, is a constant (see 3.2). This happens in the following two cases and only then.

a) $E(s)$ is identically 0.   b) $E(s) = \pm E(a)$.

We shall show that b) is impossible, if $U \neq G$. If b) holds, then s is not singular and U is the centralizer of a toral subgroup of G, according to i) and iii). From b), we infer more precisely that s is a transform of a under the Weyl group of G; i.e., $s = w(a)$ for some $w \in W(G)$. Now we can define a new ordering by letting the element $x \in V_T{}^*$ be positive if $(s, x) > 0$. Since $s = w(a)$, we conclude that s equals the sum of all roots of G which are positive in this ordering. Since $(s, \beta) > 0$ for all $\beta \in \Theta$ (notations of 23.3), all $\beta \in \Theta$ are positive in the new ordering. Since s is the sum of all roots in Θ, we infer that the sum of all complementary roots positive in the new ordering is zero which is impossible if $U \neq G$. Therefore $\{\hat{A}_j\}$ is strictly multiplicative in $(E_\xi/U, B_\xi, G/U)$ if and only if a) holds, but this is the case if and only if s is singular (3.2). In virtue of 23.3, i), the element s is singular if and only if $\hat{A}(G/U)$ vanishes. This completes the proof.

23.7. *Remarks.*

1) As a corollary of 23.3, i), we mention that s is singular if the real dimension of G/U is not divisible by 4. Thus s can only be non-singular if G/U is homogeneous algebraic of even complex dimension, see 23.3, iii).

2) In view of 23.3, ii), one might formulate the following *conjecture*: Let ξ be a bundle for which F_ξ is a compact oriented differentiable manifold and G_ξ is a group of differentiable homeomorphisms of F_ξ. If $\hat{A}(F_\xi) = 0$, then $\{\hat{A}_j\}$ is strictly multiplicative in ξ. (For the notations see 21.1-21.3.)

3) If the first Chern class of a compact almost complex manifold X vanishes, then $\hat{A}(X)$ is equal to the Todd genus of X. Taking this into account, 23.3, iii) is in agreement with 13.2 and 22.8.

4) In the proof of Lemma 23.4, we used the theorem of de Siebenthal on the principal diagonal. de Siebenthal proved his theorem by "checking all cases." There exists a general proof of it (A. Borel, unpublished).

24. Applications to simply connected algebraic homogeneous spaces.

24.1. Let X be a non-singular n-dimensional projective manifold, whose cohomology classes with respect to complex coefficients of type (p, q) vanish if $p \neq q$. Then the $h^{p,q}$ of X satisfy (see for instance [19]):

$$\chi^p(X) = \sum_{q=0}^{n} (-1)^q h^{p,q} = (-1)^p h^{p,p},$$

(1)

$$\chi^p(X) = (-1)^p \sum_{r+s=2p} h^{r,s} = (-1)^p b_{2p},$$

where b_i is the Betti number of X in the real dimension i; it vanishes if i is odd. From (1) and [19, § 21.3], we conclude

$$(2) \qquad T_y(X) = \sum_{p=0}^{n} (-y)^p b_{2p}.$$

For $y = 1$ (respectively $y = -1$), $T_y(X)$ is equal to the index $\tau(X)$ (respectively the Euler number $E(X)$) of X ([19], pp. 84, 122); hence

$$(3) \qquad \tau(X) = \sum_{p=0}^{n} (-1)^p b_{2p}, \qquad E(X) = \sum_{p=0}^{n} b_{2p}.$$

24.2. Let G be a compact connected Lie group, T a maximal torus, and U the centralizer of a toral subgroup of T. Let $\mathcal{C}$ be an invariant complex structure on G/U, Ψ its root system and Θ a system of positive roots of U. Then (13.7), $\Theta \cup \Psi$ is a positive system of roots of G. The complex manifold G/U (with the structure $\mathcal{C}$) is projective and satisfies the assumptions of 24.1 (see 14.4, 14.10). It follows then from (2) and 22.7, in the notations of 22.7, that

$$(4) \qquad b_{2p}(G/U) = k^p(G/U, \Psi, \Theta).$$

As was recalled in 2.7, the map $w \to w(\Theta \cup \Psi)$ is a 1-1 correspondence between the Weyl group $W(G)$ of G and the systems of positive roots. Let $W(G/U, \Psi, \Theta)$ be the set of those elements in $W(G)$ for which $\Theta \subset w(\Theta \cup \Psi)$. Each *right* coset $w(U) \cdot w$ of $W(G)$ modulo $W(U)$ contains at most one element of $W(G/U, \Psi, \Theta)$, since only the identity of $W(U)$ transforms Θ onto Θ. Moreover, given $w \in W(G)$, the system $w(\Theta \cup \Psi)$ contains a system Θ' of positive roots of U; hence, if u is the element of $W(U)$, carrying Θ' onto Θ, we have $u \cdot w \in W(G/U, \Psi, \Theta)$. Thus $W(G/U, \Psi, \Theta)$ is a system of representatives for the right cosets of $W(G)$ modulo $W(U)$.

Given $w \in W(G)$, let $\mu(w)$ be the number of elements in $w(\Theta \cup \Psi) \cap (-\Psi)$. Then, clearly, $k^p(G/U, \Psi, \Theta)$ is the number of elements in $W(G/U, \Psi, \Theta)$ for which $\mu(w) = p$. Since Ψ is invariant under $W(U)$, (13.4, remark), we have $\mu(w) = \mu(w')$ if w and w' belong to the same right coset of $W(G)$ modulo $W(U)$. By (4) and 13.7, we have the:

24.3. **Theorem.** *Let U be the centralizer of a torus in the compact connected Lie group G. Let $\mathscr{S}$ be an ordering of the roots of G for which the set Ψ of positive complementary roots is closed. For $w \in W(G)$, let $\mu(w)$ be the number of positive roots whose image under w is a negative complementary root. Then we have, with $2n = \dim G/U$:*

$$(5) \quad \sum_{p=0}^{n} b_{2p} t^{2p} = (\operatorname{ord} W(U))^{-1} \sum_{w \in W(G)} t^{2\mu(w)},$$

$$\tau(G/U) = \sum_{p=0}^{n} (-1)^n b_{2p} = (\operatorname{ord} W(U))^{-1} \sum_{w \in W(G)} (-1)^{\mu(w)},$$

where $\tau(G/U)$ is the index of G/U, and b_{2p} its $2p$-th Betti number.

24.4. It follows in particular that $b_{2p}(G/T)$ (T maximal torus of G) equals the number of elements of $W(G)$ for which $w(\Psi)$ contains exactly p negative roots. Therefore, in the notations of 2.6, we have

$$\sum_{p=0}^{n} b_{2p}(G/T) t^{2p} = \sum_{w \in W(G)} t^{2s(w)}.$$

Theorem 24.3 was proved independently by R. Bott (Bull. Soc. Math. France 84 (1956), 251-281) in a slightly different formulation. 24.4 was also proved by C. Chevalley by means of a cellular decomposition (Tohoku Math. Journal 7 (1955), 14-66). The general case could also be read off from the cellular decomposition mentioned in [5].

24.5. *Kodaira's vanishing theorem.* Let X be a compact connected Kählerian manifold, n its complex dimension, and F a holomorphic complex line bundle over X. The bundle F is said to be negative of order $\geq k$ if its first Chern class $c_1(F)$ can be represented by a closed real $(1,1)$-form ω of class C^∞ which, around every point $x \in X$, can be written in the form

$$\omega = i \sum g_{\alpha\bar\beta} dz_\alpha \wedge d\bar z_\beta$$

where $(g_{\alpha\bar{\beta}})$ is a hermitian matrix with at least k negative eigenvalues. In particular, F is negative of order $\geqq n$ if and only if it is negative in the sense of Kodaira, that is, if and only if F^{-1} is positive in the sense of Kodaira. In [22], Kodaira has shown that *if F is negative, then the cohomology groups $H^q(X, F)$ of X with respect to the sheaf of germs of holomorphic sections of F vanish for $q < \dim_C X$.* By Serre's duality theorem, this is equivalent to the following statement: *Let K be the canonical bundle of X. If $F \otimes K^{-1}$ is positive, then $H^q(X, F) = 0$ for $q > 0$.*

Bott [7b, p. 231] has given a generalization of the first theorem in the case $q = 0$: *if F is negative of order $\geqq 1$, then $H^0(X, F) = 0$, that is, F does not admit a not identically zero holomorphic cross section.*

Remark. Bott formulates his theorem in a slightly different fashion; but, if one takes into account the lemma in [22, p. 1271], one gets Bott's theorem in the above form.

24.6. We keep the notations of 24.1 and 24.2. Since $H^{0,1}(G/U) = H^{0,2}(G/U) = 0$, by 14.10, the map assigning to a holomorphic line bundle over G/U its first Chern class defines an isomorphism between the group of isomorphism classes of line bundles and $H^2(G/U, \mathbf{Z})$. The negative transgression defines a homomorphism of the group A of weights which are orthogonal to the roots of U onto $H^2(G/U, \mathbf{Z})$ by 14.2. For any weight, we define $H^i(G/U, d)$ as the i-th cohomology group of G/U with respect to the sheaf of germs of holomorphic sections of a complex bundle with first Chern class d. $\chi(G/U, d)$ will denote the alternating sum of the dimensions of the $H^i(G/U, d)$.

24.7. THEOREM. *Let U be the centralizer of a torus in G, Ψ be the set of roots of an invariant complex structure on G/U, and Θ a set of positive roots for U. Let d be a weight orthogonal to the roots of U. If $(d, b) \geqq 0$ for all $b \in \Psi$, then $H^i(G/U, d) = 0$ $(i > 0)$ and $\dim_C H^0(G/U, d)$ equals $T(G/U, d)$, which is the degree of the irreducible representation of $\bar{G}$ (see 3.3) with main weight d (in the ordering which has $\Theta \cup \Psi$ as positive roots). If $(d, b) < 0$ for at least one $b \in \Psi$, then $H^0(G/U, d) = 0$.*

The last assertion follows from Bott's theorem (24.5) and 14.6. Let c_1 be the first Chern class of G/U. Then $(c_1, b) > 0$ for $b \in \Psi$ by 14.8, and $(d, b) \geqq 0$ for $b \in \Psi$ implies $(d + c_1, b) > 0$. Since $c_1 = -c_1(K)$, the vanishing of $H^i(G/U, d)$, $(i > 0)$, follows then from 14.6 and 24.5.

Assume G/T and U/T to be endowed with the invariant complex structures having as root systems $\Theta \cup \Psi$ and Θ respectively. Then (14.3),

$(G/T, G/U, U/T, v)$ is a complex analytic fibering; we have by 22.8 and 22.10

$$(6) \qquad T(G/U, d) = T(G/T, d),$$

and, therefore, by Riemann-Roch

$$(7) \qquad \chi(G/U, d) = \chi(G/T, d).$$

Since $H^i(G/U, d) = H^i(G/T, d) = 0$ for $i > 0$, we get

$$\dim_C H^0(G/U, d) = T(G/T, d),$$

and the remaining assertion of the theorem follows from 22.4.

Remarks. (1) Assume that $U = T$. Let $d \in H^2(G/T, \mathbf{Z})$ be such that $d + (c_1/2)$ is in the closure of the positive Weyl chamber, but not inside; therefore it is singular, $(d, b) < 0$ for at least one positive root b, and $(d + c_1, b) > 0$ for all positive roots b. By 14.6 and 24.5, it follows that *all* cohomology groups $H^i(G/T, d)$ vanish, in agreement with the fact (22.3(5)) that $T(G/T, d) = \chi(G/T, d) = 0$ if $d + (c_1/2)$ is singular.

(2) If the weight d is orthogonal to the roots of U, the element $d \in H^2(G/T, \mathbf{Z})$ is the first Chern class of a line bundle which is the image under v^* of a line bundle on G/U with first Chern class $d \in H^2(G/U, \mathbf{Z})$. Since $H^{p,q}(U/T) = 0$ for $p \neq q$ (14.10), one can deduce by a spectral argument applied to the fibering $(G/T, G/U, U/T, v)$ that, more generally than in the proof of 24.7, v^* induces an isomorphism of $H^i(G/U, d)$ onto $H^i(G/T, d)$ for all i and all d.

24.8. We assume here that G is semi-simple. Then $H^2(G/T, \mathbf{Z})$ is isomorphic to the group of weights of G.

The projective space associated to the vector space $H^0(G/T, d)$ can be identified with the complete linear system of all positive divisors whose homology class is dual to d. Thus the preceding results on $\dim H^0(G/T, d)$ are also consequences of the results of [7a] quoted in 14.4.

Bott [7b] has proved the following theorem, which had been conjectured by the authors in view of 22.2 and 24.7:

THEOREM (Bott). *Let d be a weight. Then all groups $H^i(G/T, d)$ vanish if and only if $d + (c_1/2)$ is singular. If $d + (c_1/2)$ is regular and if w is the unique element of $W(G)$ which brings $d + (c_1/2)$ into the positive Weyl chamber, then $H^i(G/T, d)$ is zero if $i \neq s(w)$, and is equal to the degree of the irreducible representation $\bar{G}$ with main weight $w(d + (c_1/2)) - (c_1/2)$ if $i = s(w)$ (see 2.6 for $s(w)$).*

24.9. *Degrees of embeddings.* We follow the preceding notations. Let d be a weight orthogonal to all roots of U for which moreover $(d, b) > 0$ for all $b \in \Psi$. Let Γ be the representation with main weight d and $\check{\Gamma}$ the contragredient representation. G/U is *strictly* associated (14.4) to Γ, and $\check{\Gamma}$ induces an embedding j of G/U in the complex projective space $\boldsymbol{P}_q(\boldsymbol{C})$, where $q + 1$ is the degree of the representation Γ. If $e^* = H^2(\boldsymbol{P}_q(\boldsymbol{C}), \boldsymbol{Z})$ is dual to a hyperplane of $\boldsymbol{P}_q(\boldsymbol{C})$, then $j^*(e^*) = d$ (d regarded now as element of $H^2(G/U, \boldsymbol{Z})$) and the value of the cohomology class d^n ($n = \dim_C G/U$) on the fundamental cycle of G/U is the degree of the embedding in the sense of algebraic geometry. The following formula is clear for an arbitrary $d \in H^2(G/U, \boldsymbol{R})$

$$d^n[G/U] = n! \lim_{r \to \infty} r^{-n} T(G/U, rd).$$

Let a be the sum of all roots in $\Theta \cup \Psi$, then (6) and 22.3(4) and 22.4(7) give

$$T(G/U, rd) = \prod_{c \in \Theta} (rd + a/2, c)/(a/2, c) \cdot \prod_{b \in \Psi} (rd + a/2, b)/(a/2, b).$$

Since $(d, c) = 0$ for $c \in \Theta$, the first product equals 1. Passing to the limit yields

$$(8) \qquad d^n[G/U] = n! \prod_{b \in \Psi} (d, b)/(a/2, b) \qquad \text{for } d \in H^2(G/U, \boldsymbol{R}).$$

24.10. **Theorem.** *Let G be a compact connected Lie group, T a maximal torus of G and U the centralizer of a toral subgroup of T. Endow G/U with an invariant complex structure, and let Ψ be the set of its roots. Choose an ordering $\mathcal{S}$ on V_T for which Ψ is the set of all positive complementary roots. Let a be the sum of all positive roots. Let d be a weight orthogonal to the roots of U and for which $(d, b) > 0$ for all $b \in \Psi$. The contragredient representation of the irreducible representation of $\bar{G}$ (3.3) with main weight d induces an embedding of G/U in a complex projective space (14.4). The degree of this embedding in the sense of algebraic geometry is*

$$(9) \qquad d^n[G/U] = n! \prod_{b \in \Psi} (d, b)/(a/2, b), \qquad (n = \dim_C G/U).$$

24.11. As an example, we take $G = \boldsymbol{U}(4)$ and $U = \boldsymbol{U}(2) \times \boldsymbol{U}(1) \times \boldsymbol{U}(1)$. In 13.9, two invariant complex structures $\mathcal{C}_1$, $\mathcal{C}_2$ on G/U were defined. We shall calculate the number $c_1^5[G/U]$ with respect to these two structures. Let $a^{(1)}$ (respectively $a^{(2)}$) be the sum of the positive roots with respect to the ordering $\mathcal{S}_1$ (respectively $\mathcal{S}_2$) defined in 13.9. We have

$$a^{(1)} = 3x_4 + x_1 - x_2 - 3x_3, \qquad a^{(2)} = 3x_1 + x_2 - x_3 - 3x_4.$$

The first Chern classes and the roots of these two structures have been given in 13.9. With respect to the coordinates x_i, the metric in the universal covering V_T of the maximal torus of $U(4)$ is the usual euclidean metric. Thus, in the formulas (8), (9), the scalar product is the ordinary one, and, by a straightforward computation, the Chern number $c_1^5[G/U]$ of G/U with respect to $\mathscr{E}_1$ (respectively $\mathscr{E}_2$) is 4860 (respectively 4500). Therefore we get an example of two 5-dimensional algebraic varieties which are C^∞-differentiably homeomorphic, but have different Chern numbers.

Chapter VII. Genera Defined by Pontrjagin Classes.

In this chapter, a real number s is said to be an integer exc 2, or integral exc 2, if there exists an integer k such that $2^k \cdot s$ is an integer. Analogously, a real cohomology class x is integral exc 2 if x, multiplied by a suitable power of 2, is the image of an integral cohomology class under the coefficient homomorphism induced by $Z \to R$.

25. The integrality of the A-genus.

25.1. Let $\{L_j(p_1, \cdot \cdot \cdot, p_j)\}$ and $\{A_j(p_1, \cdot \cdot \cdot, p_j)\}$ be the multiplicative sequences [19, § 1] with $z^{\frac{1}{2}}/\mathrm{tgh}\, z^{\frac{1}{2}}$ and $2z^{\frac{1}{2}}/\sinh 2z^{\frac{1}{2}}$ respectively as characteristic power series. The polynomials A_k have rational coefficients which, when written as quotients of relatively prime integers, do not contain the factor 2 in their denominators. It suffices to prove this for the coefficients a_k of the power series $2z^{\frac{1}{2}}/\sinh 2z^{\frac{1}{2}}$. The coefficient of z^k ($k \geq 1$) in this series is

$$a_k = (-1)^k 2^{2k+1} (2^{2k-1} - 1) B_k / (2k)!$$

and, by elementary number theory, $(2k)!$ is not divisible by 2^{2k}, whereas by von Staudt's theorem, the Bernoulli number B_k contains 2 exactly to the first power in its denominator, which proves the desired result.

25.2. If X is a compact oriented differentiable manifold, then the genera $L(X)$, $A(X)$, $\hat{A}(X)$ are defined (21.2, 23.1). They are rational numbers which vanish if the dimension of X is not divisible by 4. We have

$$A(X) = 2^{4k}\hat{A}(X) \qquad \text{for } \dim X = 4k.$$

$A(X)$ may be written with an odd denominator; by [19, Haupsatz 8.2.2], the rational number $L(X)$ equals the index $\tau(X)$ and thus is an integer. In this paragraph, we wish to prove in particular that the A-genus $A(X)$ is also an integer or, equivalently, that $\hat{A}(X)$ is integral exc 2.

For a compact almost complex manifold X with Chern classes $c_i \in H^{2i}(X, \mathbf{Z})$ and for elements $d_1, \cdots, d_s \in H^2(X, \mathbf{R})$, we define the virtual Todd genus as in [19, §11] by the formula

$$(1) \quad T(d_1, \cdots, d_s)_X = ((1 - e^{-d_1}) \cdots (1 - e^{-d_s}) \sum_{j=0}^{\infty} T_j(c_1, \cdots, c_j))[X].$$

This virtual Todd genus is a real number.

For $d \in H^2(X, \mathbf{R})$ the number $T(X, d)$ is defined by the formula

$$(2) \quad T(X, d) = (e^d \sum_{j=0}^{\infty} T_j(c_1, \cdots, c_j))[X], \qquad \text{(see 22.1).}$$

By [19, Satz 14.3.2], the Todd genus $T(X) = T(X, 0)$ is an integer exc 2 and the virtual Todd genus $T(d_1, \cdots, d_s)_X$ is integral exc 2 if $d_1, \cdots, d_s$ are images of integral cohomology classes. We have

$$T(X, d) = T(X) - T(-d)_X$$

and thus $T(X, d)$ is also integral exc 2 if d is the image of an integral class. We give now a slight generalization of these results.

25.3. PROPOSITION. *If the elements $d, d_1, \cdots, d_s$ of $H^2(X, \mathbf{R})$, (X compact almost complex), are integral exc 2, then $T(X, d)$ and the virtual Todd genus $T(d_1, \cdots, d_s)_X$ are integral exc 2.*

By (1) and (2), it is sufficient to prove that $T(X, d)$ is integral exc 2 if $2^k d$ is the image of an integral class for some positive integer k. This statement will be proved by induction on k. It is proved already for $k = 0$, and we assume it to be true for $k - 1$. We have

$$e^d = (1 - (1 - e^{-2d}))^{-\frac{1}{2}},$$

and therefore

$$T(X, d) = \left(\sum_{r=0}^{\infty} (-1)^r \binom{-\frac{1}{2}}{r} (1 - e^{-2d})^r \cdot \sum_{j=0}^{\infty} T_j(c_1, \cdots, c_j) \right)[X].$$

The coefficients $(-1)^r \binom{-\frac{1}{2}}{r} = 2^{-2r} \binom{2r}{r}$ are integers exc 2. If $2^k d$ is the image of an integral class, we see that $T(X, d)$ is a finite linear combination, with integers exc 2 as coefficients, of numbers $T(X, f)$, where f runs through certain elements of $H^2(X, \mathbf{R})$ for which $2^{k-1} f$ is the image of an integral class. By the induction assumption, it is therefore an integer exc 2.

The following theorem will include the integrality of the A-genus (25.2).

25.4. THEOREM. *Let X be a compact oriented differentiable manifold with the Pontrjagin classes $p_j \in H^{4j}(X, \mathbf{Z})$. Let the element d of $H^2(X, \mathbf{R})$*

be integral exc 2. Then the number $\hat{A}(X, d)$ defined by

$$\hat{A}(X, d) = (e^d \sum_{j=0}^{\infty} \hat{A}_j(p_1, \cdots, p_j))[X]$$

is integral exc 2.

The theorem is trivial if the dimension of X is odd. Therefore we may put $\dim X$ equal to $2q$. Let $\xi = (E, X, \boldsymbol{SO}(2q))$ be the principal tangent bundle of X. Let T be a maximal torus of $\boldsymbol{SO}(2q)$ and $(x_1, \cdots, x_q)$ a base of $H^1(T, \boldsymbol{Z})$, see 10.1. We consider the fibre bundle

$$\zeta = (E/T, X, \boldsymbol{SO}(2q)/T, \pi).$$

Then $\pi^*(\xi)$ is the Whitney sum of q principal $\boldsymbol{U}(1)$-bundles $\xi_1, \cdots, \xi_q$, where ξ_j is the extension of $(E, E/T, T)$ with respect to $t \to \exp 2\pi i x_j(t)$. The first Chern class of ξ_i is x_i if we regard x_i under the negative transgression of $(E, E/T, T)$ as an element of $H^2(E/T, \boldsymbol{Z})$. Let $a_1, \cdots, a_m$ $(m = q(q-1))$, be the positive roots of $\boldsymbol{SO}(2q)$ with respect to T and an ordering. The a_i are the roots of an invariant integrable almost complex structure (§ 12) on $\boldsymbol{SO}(2q)/T$, to which belongs a complex structure of the vector bundle along the fibres of ζ. Thus the principal bundle η along the fibres of ζ is restricted to $\boldsymbol{U}(m)$ and the corresponding principal $\boldsymbol{U}(m)$-bundle η' is the Whitney sum of m principal $\boldsymbol{U}(1)$-bundles $\eta_1, \eta_2, \cdots, \eta_m$ whose first Chern classes are $a_1, \cdots, a_m$ regarded as elements of $H^2(E/T, \boldsymbol{Z})$.

The principal tangent bundle of E/T is the Whitney sum of $\pi^*\xi$ and η; thus E/T admits an almost complex structure whose principal tangent $\boldsymbol{U}(m+q)$-bundle is the Whitney sum of $\eta_1, \cdots, \eta_m, \xi_1, \cdots, \xi_q$. Hence E/T is an almost complex split manifold [19, § 13.5] with total Chern class

$$c(E/T) = (1 + a_1)(1 + a_2) \cdots (1 + a_m)(1 + x_1)(1 + x_2) \cdots (1 + x_q).$$

By 22.5(11), we get for an arbitrary element $d \in H^2(X, \boldsymbol{R})$,

$$\hat{A}(X, d) = (\pi^*(e^d) \prod_{j=1}^{m} a_j/(1 - e^{-a_j}) \cdot \pi^*(\sum_{k=0}^{\infty} \hat{A}_k(p_1, \cdots, p_k)))[E/T].$$

Observing that

$$\pi^*(p(\xi)) = p(\pi^*\xi) = (1 + x_1^2)(1 + x_2^2) \cdots (1 + x_q^2)$$

and using the identity $x/(1 - e^{-x}) = (\tfrac{1}{2}x/\sinh \tfrac{1}{2}x) \cdot \exp(x/2)$, we obtain

$$\pi^*(\sum_{k=0}^{\infty} \hat{A}_k(p_1, \cdots, p_k))$$

$$= \prod_{i=1}^{q} (x_i/2)/\sinh(x_i/2) = e^{-\frac{1}{2}(x_1 + \cdots + x_q)} \prod_{i=1}^{q} x_i/(1 - e^{-x_i}).$$

Thus we see that

$$\hat{A}(X, d) = T(E/T, \pi^*(d) - \tfrac{1}{2}(x_1 + \cdots + x_q)),$$

and this, together with 25.3, proves 25.4.

25.5. THEOREM. *Let $\eta = (E, X, \mathbf{U}(k))$ be a principal bundle over a compact oriented differentiable manifold X ($\dim X = 2q$) and let p_j denote the Pontrjagin classes of X. Let $d \in H^2(X, \mathbf{R})$ be integral exc 2. Then the number $\hat{A}(X, d, \eta)$ defined by*

$$\hat{A}(X, d, \eta) = (e^d \operatorname{ch}(\eta) \cdot \sum_{j=0}^{\infty} \hat{A}_j(p_1, \cdots, p_j))[X]$$

is integral exc 2. (As in 9.1, $\operatorname{ch}(\eta)$ denotes the Chern character of η).

We consider the associated bundle $\zeta = (E/T, X, \mathbf{U}(k)/T, \pi)$ where T is the standard maximal torus of $\mathbf{U}(k)$. Let $(x_1, \cdots, x_k)$ be the standard base of $H^1(T, \mathbf{Z})$. Then $\pi^*(\eta)$ is the Whitney sum of k principal $\mathbf{U}(1)$-bundles whose first Chern classes are $x_1, \cdots, x_k$ if we consider $x_1, \cdots, x_k$ via negative transgression as elements of $H^2(E/T, \mathbf{Z})$. We note that

$$(3) \qquad \pi^* \operatorname{ch}(\eta) = \operatorname{ch}(\pi^* \eta) = e^{x_1} + e^{x_2} + \cdots + e^{x_k}.$$

Let $a_1, \cdots, a_m$ ($m = k(k-1)/2$) be the positive roots of $\mathbf{U}(k)$ with respect to T and an ordering. By 10.7, the Pontrjagin class $\hat{p}$ of the bundle ζ along the fibres of ζ is given by

$$(4) \qquad \hat{p} = \prod_1^m (1 + a_i^2)$$

and therefore

$$(5) \quad \sum \hat{A}_j(\hat{p}_1, \cdots, \hat{p}_j) = \prod (a_i/2)/\sinh(a_i/2) = e^{-(a_1 + \cdots + a_m)/2} \cdot \prod a_i/(1 - e^{-a_i}).$$

The tangent bundle to E/T is the direct sum of $\hat{\zeta}$ and of $\pi^* \sigma$ where σ is the tangent bundle to X (7.6). Thus if p'_i denotes the i-th Pontrjagin class of E/T, we have in view of (5)

$$(6) \quad \pi^*(\sum \hat{A}_j(p_1, \cdots, p_j)) \cdot \prod a_i/(1 - e^{-a_i}) = e^{(a_1 + \cdots + a_m)/2} \sum \hat{A}_j(p'_1, \cdots, p'_j).$$

On the other hand, it follows from 22.5(11) that

$$\hat{A}(X, d, \eta) = (\pi^*(e^d \cdot \operatorname{ch}(\eta) \cdot \sum \hat{A}_j(p_1, \cdots, p_j)) \cdot \prod a_i/(1 - e^{-a_i}))[E/T].$$

Together with (3) and (6), this gives

$$\hat{A}(X, d, \eta) = \sum_i \hat{A}(E/T, \pi^*(d) + x_i + (a_1 + \cdots + a_m)/2),$$

and the right hand side is an integer exc 2 by Theorem 25.4.

25.6. *Remarks.* The preceding theorem is the most general integrality theorem we give in this paper. All the theorems of integrality for the Todd genus, etc., $[19, \S 14.4, 2)]$ are formal consequences of it: Let X be a compact almost complex manifold of complex dimension q and η a principal $\boldsymbol{U}(k)$-bundle over X. We have $T(X, \eta) = \hat{A}(X, c_1/2, \eta)$. Thus $T(X)$ and $T(X, \eta)$ are integers exc 2. As a consequence, $T_y(X)$ and $T_y(X, \eta)$ are polynomials in y with integers exc 2 as coefficients $[19, \text{p. } 93, (7), (8)]$. The virtual T_y-characteristic $T_y(v_1, \cdots, v_r \,|, \eta)_X$ as defined in $[19, \text{p. } 95]$, $(v_1, \cdots, v_r$ are elements of $H^2(X, \boldsymbol{Z}))$, is a polynomial of degree $q - r$ in y which can be written as a formal power series in y, the coefficients being finite linear combinations with integral coefficients of polynomials $T_y(X, \xi)$, where ξ runs through certain unitary bundles depending on η, $v_1, \cdots, v_r$. This is purely formal (see also the analogous statement for the χ_y-theory, $[19, \text{p. } 132]$). Thus, $T_y(v_1, \cdots, v_r \,|, \eta)_X$ is also a polynomial with integers exc 2 as coefficients.

The proofs of 25.4 and 25.5 depend mainly on the strictly multiplicative behaviour of $x(1 - e^{-x})^{-1}$, and on Proposition 25.3 which we actually would need only for almost complex split manifolds. The theory of Thom enters implicitly in the proof of 25.3 (integrality of virtual indices, see $[19, \S 9$ and end of $\S 13]$).

We do not know how far in 25.5 "integral exc 2" could be replaced by "integral." We can only dare the following *conjectures* which are motivated by the theorem of Riemann-Roch (see $[18]$). Let X be a compact oriented differentiable manifold and η a principal $\boldsymbol{U}(k)$-bundle over X.

1) *Let w_2 denote the second Stiefel-Whitney class of X, $(w_2 \in H^2(X, \boldsymbol{Z}_2))$. If $d \in H^2(X, \boldsymbol{Z})$ reduced mod 2 is w_2, then $\hat{A}(X, d/2, \eta)$ is an integer.*

2) *If $w_2 = 0$ and $\dim X \equiv 4 \pmod 8$, then $\hat{A}(X)$ is an even integer.*

2*) *If $w_2 = 0$, $\dim X \equiv 4 \pmod 8$ and if the structural group of η can be reduced to $\boldsymbol{SO}(k)$, then $\hat{A}(X, 0, \eta)$ is an even integer.*

These conjectures would be generalizations of Rohlin's theorem $[24]$ that the Pontrjagin number $p_1[X]$ is divisible by 48 if $\dim X = 4$ and $w_2 = 0$. Rohlin's theorem goes over into conjecture 2 fro $\dim X = 4$.[3]

25.7. *Examples.* Putting the value 0 for d in 25.4 yields that $\hat{A}(X)$

---

[3] (Added in proof). A proof of (1), using the integrality of the Todd genus recently proved by Milnor (yet unpublished) will be given in the paper mentioned in footnote 2). For a different approach which proves (1) and (2*), see F. Hirzebruch, Séminaire Bourbaki, Exposé 177, Febr. 1959.

is integral exc 2 or, equivalently, that the A-genus of X (see 25.2) is an integer. This is non-trivial only if the dimension of X is divisible by 4. For $\dim X = 8$, we get [19, p. 14]

$$(5) \qquad (-4p_2 + 7p_1{}^2)[X] \equiv 0 \pmod{45}.$$

The integrality of the L-genus (index) gives

$$(6) \qquad (7p_2 - p_1{}^2)[X] \equiv 0 \pmod{45}.$$

The two congruences (5) and (6) are not independent of each other; (5) results if one multiplies (6) by -7. For $\dim X = 12$, the integrality of $A(X)$ and $L(X)$ respectively means

$$(7) \qquad (16p_3 - 44p_2p_1 + 31p_1{}^3)[X] \equiv 0 \pmod{945},$$

$$(8) \qquad (62p_3 - 13p_2p_1 + 2p_1{}^3)[X] \equiv \pmod{945}.$$

In this case, neither of the two congruences is a formal consequence of the other, since one can derive from (7) and (8) that

$$(9) \qquad (p_1p_2)[X] \equiv 0 \pmod{3}.$$

(8) is a formal consequence of (7) and (9), and (7) of (8) and (9). The congruence (9) can also be obtained by the use of Steenrod's reduced powers. In fact, by [15, Theorem 2.1],

$$p_1{}^3 \equiv -p_1(7p_2 - p_1{}^2) \pmod{3}.$$

Assume now that X is a compact connected oriented differentiable manifold of dimension $2q$, whose real Pontrjagin classes p_j vanish for $j \neq 0$, $\dim X$. Taking into account that $\hat{A}(X)$ is integral exc 2, we get

$$(10) \qquad d^q[X]/q! \text{ is integral exc 2 for all } d \in H^2(X, \mathbf{Z})$$

and also, by 25.5,

$$(11) \qquad \mathrm{ch}(\eta)[X] \text{ is integral exc 2 for every } \mathbf{U}(k)\text{-bundle over } X.$$

Let c_j be the Chern classes of η. We infer from the definition of the Chern character (9.1) that for $q \neq 0$, the $2q$-dimensional component $\mathrm{ch}(\eta)_q$ of $\mathrm{ch}\,\eta$ is of the form

$$(12) \qquad q!\,\mathrm{ch}(\eta)_q = (-1)^{q+1}q \cdot c_q + P(c_1, \cdots, c_{q-1}),$$

where P is a polynomial in $q-1$ indeterminates with integral coefficients. Therefore (11) and (12) prove in particular the following:

25.8. THEOREM. *Let ξ be a $U(k)$-bundle over S_{2q}, and let c_q be its q-th Chern class. Then $c_q[S_{2q}]/(q-1)!$ is an integer exc 2.*

The theorem is non trivial only for $k \geqq q$. For $q = k$, it implies that the spheres S_{2q} are not almost complex for $q \geqq 4$. (See also [18, §2.1].) This was proved by Borel-Serre by showing that $c_q[S_{2q}]$ is divisible by every prime p less than q and not dividing q, see [6, Propositions 12.4 and 15.1].

25.9. COROLLARY. *Let η be a principal $O(k)$-bundle $(Sp(k)$-bundle$)$ over the sphere S_{4q}. Let p_q (respectively e_q) be the q-th Pontrjagin class (q-th symplectic Pontrjagin class) of η. Then*

$$p_q[S_{4q}]/(2q-1)! \ or \ e_q[S_{4q}]/(2q-1)! \ respectively$$

is an integer exc 2.

For the proof, it is enough to observe that p_q (respectively e_q) is by definition up to sign the Chern class c_{2q} of the complex extension η' of η, with respect to the inclusion $O(k) \subset U(k)$ (respectively $Sp(k) \subset U(2k)$), and then to apply Theorem 25.8 to η'.

26. Applications to homotopy groups of Lie groups. In this paragraph, C_2 will be the class of finite commutative 2-groups.

26.1. The boundary homomorphism in the homotopy sequence of a bundle ξ will be denoted by ∂_ξ. We recall that there is a commutative diagram

$$
(1) \qquad
\begin{array}{ccccc}
\pi_i(B_\xi) & \longleftrightarrow & \pi_i(E_\xi \bmod F) & \xrightarrow{\ \partial_\xi\ } & \pi_{i-1}(F) \\
\downarrow \alpha & & \downarrow \alpha & & \downarrow \alpha \\
H_i(B_\xi, Z) & \longleftarrow & H_i(E_\xi \bmod F, Z) & \xrightarrow{\ \partial_*\ } & H_{i-1}(F, Z),
\end{array}
$$

where F is some fibre, ∂_* the boundary homomorphism of the relative homology sequence and α the Hurewicz homomorphism. Using the bottom line of (1) to define transgression in homology and the corresponding maps

$$H^i(B_\xi, Z) \longrightarrow H^i(E_\xi \bmod F, Z) \xleftarrow{\ \partial^*\ } H^{i-1}(F, Z)$$

to define the transgression τ_ξ in cohomology, we obtain readily the:

PROPOSITION. *Let $x \in \pi_i(B_\xi)$ and let $y \in H^{i-1}(F_\xi, Z)$ be transgressive. Then for any image $\tau_\xi(y) \in H^i(B_\xi, Z)$ of y by transgression, we have*

$$KI(\tau_\xi(y), \alpha(x)) = KI(y, \alpha\partial_\xi x),$$

where KI (for Kronecker index) denotes the standard pairing of homology and cohomology.

26.2. ι_n will denote a generator of $\pi_n(S_n)$ or its image in $H_n(S_n, \mathbf{Z})$, and ι_n^* the dual generator of $H^n(S_n, \mathbf{Z})$. We recall [26, § 18.5] that if we associate to a principal G-bundle ξ over S_n the element $\partial_\xi(\iota_n) \in \pi_{n-1}(G)$, we define a 1-1 correspondence between the set of equivalence classes of principal G-bundles over S_n and $\pi_{n-1}(G)$. Also, since for any finite dimension, we may take a differentiable manifold as classifying space for G, each equivalence class may be represented by a differentiable bundle, and we shall assume our bundles to be differentiable whenever convenient. Clearly, if $\lambda: G \to G'$ is a homomorphism and if the G-bundle ξ is represented by α, then its λ-extension is represented by $\lambda_0(\alpha)$, where $\lambda_0: \pi_{n-1}(G) \to \pi_{n-1}(G')$ is induced by λ.

We shall be interested in the cases $n = 2q$, $G = \mathbf{U}(m)$ $(m \geq q)$, $n = 4q$, $G = \mathbf{Sp}(r)$, $(r \geq q)$, $n = 4q$, $G = \mathbf{SO}(s)$ $(s \geq 2q + 1)$, and shall denote by c^*_k or $c^*_k(\xi)$ (respectively e^*_k or $e^*_k(\xi)$, respectively p^*_k or $p^*_k(\xi)$) the value of the k-th Chern (respectively symplectic Pontrjagin, respectively Pontrjagin) class on ι_n, where $n = 2k$ (respectively $n = 4k$, respectively $n = 4k$). It follows directly from the definition of the characteristic classes by means of classifying spaces that $\xi \to c^*_k(\xi)$ (respectively $\xi \to e^*_k(\xi)$, respectively $\xi \to p^*_k(\xi)$) is a homomorphism of the $(n-1)$-th homotopy group $\pi_{n-1}(G)$ of the structural group into $\mathbf{Z}$; hence this homomorphism depends only on $\pi_{n-1}(G)$ modulo torsion. Finally, we recall that the maps

$$\pi_{2q-1}(\mathbf{U}(r)) \to \pi_{2q-1}(\mathbf{U}(s)), \pi_{4q-1}(\mathbf{Sp}(r)) \to \pi_{4q-1}(\mathbf{Sp}(s)) \quad (s \geq r \geq q)$$

$$\pi_{4q-1}(\mathbf{SO}(r)) \to \pi_{4q-1}(\mathbf{SO}(s)) \quad (s \geq r \geq 4q + 1)$$

induced by the standard inclusions are isomorphisms [26, § 22.8, 25.2, 25.5] and that

$$\pi_{4q-1}(\mathbf{SO}(2r + 1)) \to \pi_{4q-1}(\mathbf{SO}(2s + 1)), \qquad (s \geq r \geq q),$$

is an isomorphism mod C_2; this last fact follows from the homotopy sequence of the fibering $\mathbf{SO}(2r + 1)/\mathbf{SO}(2r - 1) = \mathbf{W}_{4r-1}$, where $\mathbf{W}_{4r-1}$ is the manifold of unit tangent vectors to S_{2r}, and from the existence of a map $S_{4r-1} \to \mathbf{W}_{4r-1}$ which induces a C_2-isomorphism of $\pi_i(S_{4r-1})$ onto $\pi_i(\mathbf{W}_{4r-1})$ for all $i \geq 0$ (see [25], Chapitre IV, Prop. 2).

26.3. LEMMA. (a) *Let ξ be a principal $\mathbf{U}(q)$-bundle over S_{2q}, and let η be the associated bundle with fibre $S_{2q-1} = \mathbf{U}(q)/\mathbf{U}(q-1)$. Then*

$$\partial_n(\iota_{2q}) = \pm c^*_q(\xi) \cdot \iota_{2q-1}$$

(b) *Let ξ be a principal $\boldsymbol{Sp}(q)$-bundle over $\boldsymbol{S}_{4q}$, and η be the associated bundle with fibre $\boldsymbol{S}_{4q-1} = \boldsymbol{Sp}(q)/\boldsymbol{Sp}(q-1)$. Then $\partial_\eta \iota_{4q} = \pm\, e^*{}_q(\xi)\iota_{4q-1}$.*

The assertion (a) follows from the fact that $c_q(\xi)$ is the image by transgression of $\pm\, \iota_{2q-1}$ in η (see § 29) and from 26.1.

By definition, $e_q(\xi) = (-1)^q c_{2q}(\xi')$, where ξ' is the λ-extension of ξ under the inclusion $\lambda: \boldsymbol{Sp}(q) \to \boldsymbol{U}(2q)$. It is immediately seen that the pair inclusion $(\boldsymbol{Sp}(q-1), \boldsymbol{Sp}(q)) \to (\boldsymbol{U}(2q-1), \boldsymbol{U}(2q))$ induces a homeomorphism of $\boldsymbol{Sp}(q)/\boldsymbol{Sp}(q-1)$ onto $\boldsymbol{U}(2q)/\boldsymbol{U}(2q-1)$. As a consequence, the bundle $(\xi', \boldsymbol{S}_{4q-1})$ associated to ξ' is the λ-extension of η, and then (b) is implied by (a).

26.4. The result quoted at the end of 26.2 implies in particular that $\pi_{4q-1}(\boldsymbol{W}_{4q-1})$ is the direct sum of $\boldsymbol{Z}$ and of a finite 2-group, and that there exists an integer $2^{a(q)}$ such that the image of the Hurewicz homomorphism

$$\alpha:\ \pi_{4q-1}(\boldsymbol{W}_{4q-1}) \to H_{4q-1}(\boldsymbol{W}_{4q-1}, \boldsymbol{Z})$$

is generated by $2^{a(q)} \cdot j_q$, where j_q is a generator of $H_{4q-1}(\boldsymbol{W}_{4q-1}, \boldsymbol{Z})$.

LEMMA. *Let ξ be a principal $\boldsymbol{SO}(2q+1)$-bundle over $\boldsymbol{S}_{4q}$, η be the associated bundle with fibre $\boldsymbol{W}_{4q-1} = \boldsymbol{SO}(2q+1)/\boldsymbol{SO}(2q-1)$, and let γ_q be a generator of $\pi_{4q-1}(\boldsymbol{W}_{4q-1})$ mod 2-torsion. Then we have in the previous notation*

$$\partial_\eta \iota_{4q} = \pm\, 2^{-a(q)-1} p^*{}_q(\xi)\gamma_q \quad \text{modulo 2-torsion.}$$

Modulo 2-torsion, we have $\partial_\eta \iota_{4q} = c \cdot \gamma_q$, for some integer c, and therefore

$$\alpha \partial_\eta \iota_{4q} = 2^{a(q)} \cdot c \cdot j_q.$$

By § 30, $p_q(\xi)$ is the image by transgression in η of $\pm\, 2j_q{}^*$, where $j_q{}^*$ is the generator of $H^{4q-1}(\boldsymbol{W}_{4q-1}, \boldsymbol{Z})$ dual to j_q. Hence we have by 26.1

$$\pm\, p^*{}_q(\xi) = KI(2j_q{}^*,\ \alpha \partial_\eta \iota_{4q}) = 2^{a(q)+1} \cdot c$$

which proves the lemma.

26.5. THEOREM. There exists:

(a) *over $\boldsymbol{S}_{2q}$ a $\boldsymbol{U}(m)$-bundle with $c^*{}_q = (q-1)!$ for $m \geqq q$.*

(a*) *over $\boldsymbol{S}_{4q}$ a $\boldsymbol{SO}(n)$-bundle with $p^*{}_q = (2q-1)! \cdot 2$ for $n \geqq 4q$ and a $\boldsymbol{Sp}(m)$-bundle with $e^*{}_q = (2q-1)! \cdot 2$ for $m \geqq q$.*

(b) *over $\boldsymbol{S}_{4q}$, q even, a $\boldsymbol{SO}(n)$-bundle with $p^*{}_q = (2q-1)!$ for $n \geqq 4q+1$ (for $n \geqq 8$ if $q = 2$).*

(c) *over S_{4q}, q odd, a $Sp(m)$-bundle with $e^*_q = (2q-1)!$ for $m \geq q$.*

(d) *over S_{4q} a $SO(n)$-bundle with p^*_q equal, up to a power of 2, to the greatest odd factor of $(2q-1)!$ for $n \geq 2q+1$.*

By 26.2 and the end remark in 9.7, it is enough to prove (a), (b), (c) for one particular value of m or n; (d) follows from (a*). The case $q = 2$ in (b) will be dealt with in 26.6.

Let η be the principal $SO(2q)$-bundle of the tangential bundle to S_{2q}, and let $\lambda: Spin(2q) \to SO(2q)$ be the covering map. Since $w_2(\eta) = 0$, the bundle η may be λ-restricted to a principal $Spin(2q)$-bundle. In fact, $\rho(\lambda): B_{Spin(n)} \to B_{SO(n)}$ is (for any $n \geq 2$) a fibre map with fibre B_{Z_2} (see [2] §22 or [6] §1), i.e. an Eilenberg-MacLane space $K(Z_2, 1)$; its spectral sequence shows readily that the obstruction to a cross section is the universal second Stiefel-Whitney class w_2; then, by a standard argument, every map $\sigma: B \to B_{SO(n)}$ with $\sigma^*(w_2) = 0$ can be factorized through $\rho(\lambda)$, and this shows in our case the existence of a λ-restriction η' of η.

Let x_i, $(1 \leq i \leq q)$, be the standard basis of the usual maximal torus T of $SO(2q)$, and let T' be the inverse image of T in $Spin(2q)$; it is connected (see 10.1) and is a maximal torus of $Spin(2q)$; we shall also denote by x_i the image of x_i in $H^1(T', Z)$ under the covering map; these generate a subgroup of index 2 of $H^1(T', Z)$. Let $\beta: Spin(2q) \to U(2^{q-1})$ be one of the half-spinor representations, say the one with the highest weight $\frac{1}{2}(x_1 + \cdots + x_q)$, and let θ be the β-extension of η'). We want to prove

$$(2) \qquad\qquad c_q(\theta) = (-1)^{q-1}(q-1)! \cdot \iota_{2q}$$

which,[4] in view of our initial remark, will prove (a). Let ω_j $(1 \leq j \leq 2^{q-1})$, be the weights of β. It is known that these are just the linear forms

$$\tfrac{1}{2}(\epsilon_1 x_1 + \cdots + \epsilon_q x_q), \quad (\epsilon_i = \pm 1, (i = 1, \cdots, q), \textstyle\prod \epsilon_i = 1)$$

(in fact, these are all transforms under the Weyl group $W(SO(2q))$ of the highest weight, hence they must be weights; moreover, since there are 2^{q-1} of them, they represent all weights). Let ρ be the projection of $E_{\eta'}/T'$ onto S_{2q}. By (9.5), we have $\rho^*(W_{2q}(\eta)) = x_1 \cdots x_q$, and hence

$$(3) \qquad\qquad x_1 \cdots x_q = 2 \cdot \rho^*(\iota_{2q}).$$

By (10.3), $\rho^*(c_q(\theta))$ is the q-th elementary symmetric function in the ω_j. Since the lower symmetric functions are zero here (because $H^i(S_{2q}, Z) = 0$

---

[4] The other half-spinor representation yields a bundle whose q-th Chern class is $-c_q(\theta)$.

for $0 < i < 2q$), we get

$$(4) \qquad (-1)^{q-1} q \cdot \rho^*(c_q(\theta)) = \omega_1{}^q + \cdots + \omega_s{}^q, \qquad (s = 2^{q-1}).$$

Let $\omega_j = \tfrac{1}{2}(\epsilon_1 x_1 + \cdots + \epsilon_q x_q)$ be one particular weight. We have

$$\omega_j{}^q = q! \cdot 2^{-q} \cdot x_1 \cdots x_q + b_j,$$

where b_j is a sum of monomials in the x_i's, none of which contains all variables x_i, and therefore

$$\sum \omega_j{}^q = q! \cdot 2^{-1} \cdot x_1 \cdots x_q + \sum b_j$$

or, taking (3) into account,

$$\sum \omega_j{}^q = q! \cdot \rho^*(\iota_{2q}) + \sum b_j$$

so that (2) will follow from (4) if we show that

$$(5) \qquad\qquad\qquad \sum b_j = 0.$$

$W(\boldsymbol{SO}(2q))$ is the group of permutations of the x_i combined with an even number of changes of signs. Thus, the ring I_W of invariants of $W(\boldsymbol{SO}(2q))$ is generated by $x_1 \cdots x_q$ and by the symmetric functions in the $x_i{}^2$. The Weyl group permutes the ω_j, and therefore $\sum b_j \in I_W$; since no monomial in this sum contains all variables x_i's, b_j must then be a symmetric function in the $x_i{}^2$; but, by (9.3), it is then the image under ρ^* of a polynomial in the Pontrjagin classes of η. Since the Pontrjagin classes of $\boldsymbol{S}_{2q}$ are all zero, this proves (5).

Let now $q = 2s$ be even; let ξ be the $\boldsymbol{U}(2s)$-bundle over $\boldsymbol{S}_{4s}$ with Chern class $(2s-1)!$, and ξ^* be its extension under the contragredient representation (10.6). We have $c_{2s}(\xi) = c_{2s}(\xi^*)$, hence

$$c_{2s}(\xi \oplus \xi^*) = (2s-1)! \cdot 2 \cdot \iota_{4s},$$

but in $\boldsymbol{U}(4s)$, the matrices of the form $A \dot{+} \bar{A}$ ($A \in \boldsymbol{U}(2s)$) form a subgroup conjugate to a subgroup of $\boldsymbol{Sp}(2s)$ or of $\boldsymbol{SO}(4s)$, and $\xi \oplus \xi^*$ can be considered as the complexification of a $\boldsymbol{Sp}(2s)$- or of a $\boldsymbol{SO}(4s)$-bundle. This proves (a*).

It is known that the image group of a half-spinor representation of $\boldsymbol{SO}(4q)$ is conjugate to a subgroup of $\boldsymbol{SO}(2^{2q-1})$ (respectively $\boldsymbol{Sp}(2^{2q-2})$), if q is even (respectively odd). (See E. Cartan, Jour. Math. Pur. Appl. 10 (1914), 149-186, § XV, p. 173, or A. I. Malcev, Isv. Ak. Nauk. SSSR Ser. Math. 8 (1944), 143-174, A. M. S. Translation 33, pp. 29-30.) This implies that θ is the complexification of a $\boldsymbol{SO}(2^{2q-1})$- (respectively $\boldsymbol{Sp}(2^{2q-2})$-) bundle, for q even (respectively odd), and (b) and (c) follow from (2).

26.6. *Remark.* The image group of $\mathbf{Spin}(8)$ under a half-spinor representation is conjugate to $\mathbf{SO}(8)$, as is well known and follows also from the result just quoted. The standard representation of $\mathbf{SO}(8)$ and the two half-spinor representations provide three homomorphisms of $\mathbf{Spin}(8)$ onto $\mathbf{SO}(8)$ which are, up to equivalence, all the representations of degree 8 of $\mathbf{Spin}(8)$. They may be distinguished by the element or order 2 of the center of $\mathbf{Spin}(8)$, (isomorphic to $\mathbf{Z}_2 + \mathbf{Z}_2$), which they map onto the identity. They may be obtained from one another by performing on $\mathbf{Spin}(8)$ the automorphisms of the triality principle, (which are transitive on the elements of the center of $\mathbf{Spin}(8)$ different from the identity).

The last step of the proof of 26.5 shows therefore that θ is a $\mathbf{SO}(8)$-bundle over $\mathbf{S}_8$ with $p^*{}_2 = -6$, which ends the proof of (b). On the other hand, the projective lines on the Cayley plane $\mathbf{W}$ are homeomorphic to $\mathbf{S}_8$ and a generator u of $H^8(\mathbf{W}, \mathbf{Z})$ restricts on them to a fundamental cocycle; therefore 9.7 and 19.4 show that the normal bundle θ' to a projective line in $\mathbf{W}$ has also $p^*{}_2 = -6$. In fact, it can be shown directly that θ and θ' are isomorphic; we sketch the proof, using 19.1 and some information on $\mathbf{W}$ to be found for instance in [1]: Let $\mathbf{U}$ be a subgroup of $\mathbf{F}_4$ isomorphic to $\mathbf{Spin}(9)$, $\mathbf{V}$ a subgroup of $\mathbf{U}$ isomorphic to $\mathbf{Spin}(8)$, and P the point of $\mathbf{W}$ fixed under $\mathbf{U}$. Then $\mathbf{V}$ leaves exactly two other points Q, R fixed, and the projective line M joining Q, R is operated upon transitively by $\mathbf{U}$. Thus the fibering $\xi = (\mathbf{U}, \mathbf{U}/\mathbf{V}, \mathbf{V})$ may be identified with $(\mathbf{Spin}(9), \mathbf{S}_8, \mathbf{Spin}(8))$. The natural representation of $\mathbf{V}$ into the tangent space $\mathbf{W}_Q$ of $\mathbf{W}$ at Q decomposes into the representations ρ_1, ρ_2 into M_Q and into the subspace N_Q of $\mathbf{W}_Q$ orthogonal to M_Q. Thus the tangent (respectively normal) bundle to M is the ρ_1- (respectively ρ_2-) extension of ξ. The representation of $\mathbf{V}$ in $\mathbf{W}_Q$ is faithful, because its kernel belongs to the center of $\mathbf{F}_4$, which is reduced to $\{e\}$. Hence (see beginning of this section) ρ_1 and ρ_2 are not equivalent, the normal bundle to M and the bundle θ of 26.5 arise from the tangent bundle to $\mathbf{S}_8$ by the same construction.

26.7. **THEOREM.** *The fibrations*

$$U(q)/U(q-1) = S_{2q-1}, \qquad Sp(q)/Sp(q-1) = S_{4q-1},$$

$$SO(2q+1)/SO(2q-1) = W_{4q-1}, \qquad G_2/Sp(1) = W_{11}$$

give rise to the following sequences, which are exact modulo the class C_2 of finite commutative 2-groups:

(a) $0 \to \mathbf{Z}_{(q-1)!} \to \pi_{2q-2}(U(q-1)) \to \pi_{2q-2}(U(q)) \to 0$ $\qquad\qquad (q \geqq 2)$

(b) $\quad 0 \to \mathbf{Z}_{(2q-1)!} \to \pi_{4q-2}(\mathbf{Sp}(q-1)) \to \pi_{4q-2}(\mathbf{Sp}(q)) \to 0 \qquad (q \geqq 2)$

(c) $\quad 0 \to \mathbf{Z}_{(2q-1)!} \to \pi_{4q-2}(\mathbf{SO}(2q-1)) \to \pi_{4q-2}(\mathbf{SO}(2q+1)) \to 0 \qquad (q \geqq 2)$

(d) $\quad 0 \to \mathbf{Z}_{k15} \to \pi_{10}(\mathbf{S}_3) \to \pi_{10}(\mathbf{G}_2) \to 0$, *for some* $k \geqq 1.$[5]

(a) Let $\alpha \in \pi_{2q-1}(\mathbf{U}(q))$. Let ξ be a principal $\mathbf{U}(q)$-bundle over $\mathbf{S}_{2q}$ representing α, and η be the associated bundle with fibre $\mathbf{S}_{2q-1}$. Then $E_\eta = E_\xi/\mathbf{U}(q-1)$ and the restriction of the natural map $E_\xi \to E_\eta$ to a fibre is the projection map in the fibering $(\mathbf{U}(q), \mathbf{S}_{2q-1}, \mathbf{U}(q-1))$ hence we get a commutative diagram

$$\begin{array}{ccc}
\pi_{2q}(\mathbf{S}_{2q}) & \xrightarrow{\ \partial_\eta\ } & \pi_{2q-1}(\mathbf{S}_{2q-1}) \\
\uparrow \downarrow & & \uparrow{\scriptstyle \psi} \\
\pi_{2q}(\mathbf{S}_{2q}) & \xrightarrow[\ \partial_\xi\ ]{} & \pi_{2q-1}(\mathbf{U}(q)),
\end{array}$$

where ψ is part of the homotopy sequence of $(\mathbf{U}(q), \mathbf{S}_{2q-1}, \mathbf{U}(q-1))$. By definition, $\alpha = \partial_\xi(\iota_{2q})$, hence we have by 26.3a.

$$\psi(\alpha) = \pm c^*{}_q(\xi)\iota_{2q-1}$$

which is divisible by the greatest odd factor of $(q-1)!$ in virtue of 25.8. Since α is arbitrary, this shows that $\psi(\pi_{2q-1}(\mathbf{U}(q))$ is contained in the subgroup generated by $b(q-1) \cdot \iota_{2q-1}$, where $b(q-1)$ is the greatest odd factor of $(q-1)!$; on the other hand, the same argument together with (26.5), shows that $\psi(\pi_{2q-1}(\mathbf{U}(q)))$ contains $(q-1)! \cdot \iota_{2q-1}$, and the mod C_2 exactness of (a) follows.

The proofs for (b) and (c) are quite analogous, the sole difference being that one has to invoke 26.3b and 26.4 instead of 26.3a.

For the fibration $\mathbf{G}_2/\mathbf{Sp}(1) = \mathbf{W}_{11}$, we refer to [4, § 17]. Let $\alpha \in \pi_{11}(\mathbf{G}_2)$, let ξ be a principal $\mathbf{G}_2$-bundle representing α, and let η be the associated bundle with fibre $\mathbf{W}_{11}$. We have the commutative diagram

$$(6) \qquad \begin{array}{ccc}
\pi_{12}(\mathbf{S}_{12}) & \xrightarrow{\ \partial_\eta\ } & \pi_{11}(\mathbf{W}_{11}) \\
\uparrow \downarrow & & \uparrow{\scriptstyle \psi} \\
\pi_{12}(\mathbf{S}_{12}) & \xrightarrow[\ \partial_\xi\ ]{} & \pi_{11}(\mathbf{G}_2).
\end{array}$$

Now $\mathbf{G}_2$ is embedded in $\mathbf{SO}(7)$ and its action on $\mathbf{W}_{11}$ extends to that of $\mathbf{SO}(7)$; in other words, $\mathbf{G}_2$, as a subgroup of $\mathbf{SO}(7)$, acts transitively on $\mathbf{W}_{11} = \mathbf{SO}(7)/\mathbf{SO}(5)$ and $\mathbf{G}_2 \cap \mathbf{SO}(5) = \mathbf{Sp}(1)$. This means that if we

---

[5] It will be shown later that $k = 1$.

extend the structural group of η to $\boldsymbol{SO}(7)$, we get the associated bundle to the extension ξ' of ξ. Therefore we have by 26.4:

$$\partial_\eta(\iota_{12}) = \pm\, 2^{-a(3)-1}p^*{}_3(\xi')\gamma_3 \bmod 2\text{-torsion}$$

and (6) gives then: $\psi(\alpha) = \pm\, p^*{}_3(\xi')\cdot\gamma_3$, up to a power of two. Since (25.9) the number $p^*{}_3(\xi')$ is divisible by 15, and since this argument is valid for any $\alpha \in \pi_{11}(\boldsymbol{G}_2)$, the mod C_2 exactness of (d) is established.

26.8. PROPOSITION. *If we have*

$$(7) \qquad \pi_9(\boldsymbol{SO}(7)) \equiv 0, \qquad \pi_{10}(\boldsymbol{U}(5)) \equiv \boldsymbol{Z}_{15} \bmod C_2,$$

then the following congruences $\bmod C_2$ *are valid:* $\pi_{10}(\boldsymbol{G}_2) \equiv 0$, $\pi_{10}(\boldsymbol{SO}(5))$ $\equiv \boldsymbol{Z}_{15}$, $\pi_{10}(\boldsymbol{SO}(n)) \equiv 0$ $(n \geq 7, n \neq 8)$, $\pi_{10}(\boldsymbol{Sp}(n)) \equiv 0$ $(n \geq 3)$, $\pi_{10}(\boldsymbol{U}(n))$ $\equiv 0$ $(n \geq 6)$.

In this proof, all congruences are $\bmod C_2$. We use the following results:

$$(8) \qquad \pi_{n+1}(\boldsymbol{S}_n) \equiv \pi_{n+2}(\boldsymbol{S}_n) \equiv 0 \quad (n \geq 3), \qquad \pi_{n+3}(\boldsymbol{S}_n) \equiv \boldsymbol{Z}_3 \quad (n \geq 5),$$

$$(9) \qquad \pi_{10}(\boldsymbol{S}_3) \equiv \boldsymbol{Z}_{15}, \qquad \pi_9(\boldsymbol{S}_3) \equiv \boldsymbol{Z}_3.$$

(For the last equality of (8), see J.-P. Serre, Comm. Math. Helv. 27 (1953), 198-232, for the other ones, see [25]; as to (7), see 26.9 and 26.10 below.)

The first equality in (9) shows that in 26.7d, we have $k = 1$ and $\pi_{10}(\boldsymbol{G}_2) \equiv 0$. The fibering $\boldsymbol{G}_2/\boldsymbol{Sp}(1) = \boldsymbol{W}_{11}$, discussed in [4, § 17], together with (9) and the congruence $\pi_i(\boldsymbol{W}_{11}) \equiv \pi_i(\boldsymbol{S}_{11})$ shows that $\pi_9(\boldsymbol{G}_2) \equiv \boldsymbol{Z}_3$. Applying this and (7) to the exact homotopy sequence of the fibering $\boldsymbol{Spin}(7)/\boldsymbol{G}_2 = \boldsymbol{S}_7$ (see [1]), we get the mod C_2 exact sequence

$$0 \to \pi_{10}(\boldsymbol{SO}(7)) \to \boldsymbol{Z}_3 \to \boldsymbol{Z}_3 \to 0;$$

hence $\pi_{10}(\boldsymbol{SO}(7)) \equiv 0$. The mod C_2 exact sequence 26.7c yields then, for $q = 3$, that $\pi_{10}(\boldsymbol{SO}(5)) \equiv \boldsymbol{Z}_{15}$. Since, as is well known, the universal covering $\boldsymbol{Spin}(5)$ of $\boldsymbol{SO}(5)$ is isomorphic to $\boldsymbol{Sp}(2)$, we deduce from 26.7b that $\pi_{10}(\boldsymbol{Sp}(3)) \equiv 0$, and hence also $\pi_{10}(\boldsymbol{Sp}(n)) \equiv 0$ for $n \geq 3$.

Since $\boldsymbol{SO}(9)/\boldsymbol{SO}(7) = \boldsymbol{W}_{15}$ has, mod C_2, the homotopy groups of $\boldsymbol{S}_{15}$, we have $\pi_{10}(\boldsymbol{SO}(9)) \equiv \pi_{10}(\boldsymbol{SO}(7)) \equiv 0$ and then $\pi_{10}(\boldsymbol{SO}(n)) \equiv 0$ $(n \geq 9)$ follows from (8), the finiteness of $\pi_{10}(\boldsymbol{SO}(11))$ (see [25]) and the homotopy sequence of $\boldsymbol{SO}(n)/\boldsymbol{SO}(n-1) = \boldsymbol{S}_{n-1}$.

Finally, (7) and 26.7a give $\pi_{10}(\boldsymbol{U}(6)) \equiv 0$, and therefore $\pi_{10}(\boldsymbol{U}(n)) \equiv 0$ for $n \geq 6$.

26.9. The preceding results (found in Spring 1957) contradict several of those of [30]. Since then, Toda has made new computations whose outcome (yet unpublished) agrees with the above. They have also been confirmed by Bott (Proc. Nat. Ac. Sci. USA 43 (1957), pp. 933-935) who in particular determines all stable homotopy groups of the classical groups.

26.10. Bott has also shown (to be published) that the sequence 26.7(a) is exact also for the 2-primary components (since $\pi_{2q-2}(U(q)) = 0$, by Bott, loc. cit., 26.9, this gives $\pi_{2q-2}(U(q-1)) = Z_{(q-1)!}$). This implies (see 26.3 and the proof of 26.7) the following generalization of 25.8, 25.9:

THEOREM (Bott). *Let ξ be a $U(k)$-bundle over S_{2q}. Then $c^*_q(\xi)$ is divisible by $(q-1)!$. Let η be a $SO(k)$- (respectively $Sp(k)$-), bundle over over S_{4q}. Then $p^*_q(\eta)$ (respectively $e^*_q(\eta)$), is divisible by $(2q-1)!$.*

Using this theorem, Kervaire has proved more generally that $p^*_q(\eta)$ (respectively $e^*_q(\eta)$) is divisible by $(2q-1)! \cdot 2$ if q is odd (respectively even) (Amer. Jour. Math., vol. 80 (1958), pp. 632-638).

The stable homotopy groups $\pi_{2q-1}(U(k))$, $(k \geqq q)$, and $\pi_{4q-1}(SO(k))$ $(k \geqq 4q+1)$ are infinite cyclic according to Bott (loc. cit. in 26.9). The generator of the first group has the Chern number $c^*_q = \pm (q-1)!$, the generator of the second group has Pontrjagin number p^*_q equal to $\pm (2q-1)!$ if q is even and $\pm (2q-1)! \cdot 2$ if q is odd; similarly for the symplectic groups (with odd and even interchanged). This follows from the preceding theorem, the result of Kervaire and 26.5. The "spinor-method" in 26.5 gives an explicit construction for these generators.

The above theorem would follow by the same argument as 25.8, 25.9 if one could prove that the virtual Todd genus with respect to an *integral* class is an integer. In this respect, compare the conjectures in 25.6. The last one would contain Kervaire's result for the orthogonal groups.

Finally, we remark that the proof of 25.8, 25.9 also applies if the sphere is replaced by a compact connected oriented manifold X whose real Pontrjagin classes p_j vanish for $4j \neq 0$, dim X, and if ξ (respectively η) is a $U(k)$-bundle (respectively $O(k)$- or $Sp(k)$-bundle) whose real Chern (respectively Pontrjagin or symplectic Pontrjagin) classes vanish in all positive dimensions less than dim X.

26.11. Milnor and Kervaire, independently, have deduced from the result of Bott quoted in 26.10 that S_n, endowed with its usual differentiable structure, is not parallelizable if $n \neq 1, 3, 7$. An easy argument similar to 26.3 shows that if S_{2n-1} is parallelizable, that is if the fibering $SO(2n)/SO(2n-1)$ $= S_{2n-1}$ has a cross section, then there exists a $SO(2n)$-bundle η over S_{2n}

whose Euler-Poincaré class W_{2n} is equal to $1 \cdot \iota_{2n}$. Therefore the theorem of Milnor-Kervaire is a consequence of the

THEOREM (Milnor). *Let ξ be a $SO(2q)$-bundle over S_{2q}, where $q \neq 1$, $2, 4$. Then $W_{2q}(\xi)$ is divisible by 2.*

We want to give here a proof for this, different from Milnor's but also using 26.10.

Let η be a $SO(2q)$-bundle whose second Stiefel-Whitney class w_2 vanishes. Then (see beginning of the proof of 26.5), η has a λ-restriction η', where λ is the projection of $Spin(2q)$ onto $SO(2q)$. We denote again by θ the extension of η' by means of the half-spinor representation β. It is a $U(2^{q-1})$-bundle.

LEMMA. *In the previous notations, we have*

(i) $c_q(\theta)/(q-1)! = W_{2q}(\eta)/2$ *if q is odd,*

(ii) $c_{2k}(\theta)/(2k-1)!$

$$= -\,(tg^{(2k-1)}(0)/((2k-1)!4))p_k(\eta) - W_{4k}(\eta)/2 + R_{2k}(\eta)$$

$$\text{if } q = 2k \geqq 2,$$

where $R_{2k}(\eta)$ is a polynomial with rational coefficients in $p_1(\eta), \cdots p_{k-1}(\eta)$, and $tg^{(2k-1)}(0)$ denotes the $(2k-1)$-th derivative of $tg\,x$ at $x = 0$.

We keep the notations of 26.5. Since ρ^* is injective, we allow ourselves to omit the symbol ρ^*. The computations of 26.5 show first that

$$(10) \qquad \sum \omega_j^q = q! \cdot x_1 \cdots x_q/2 + q!\, r'_q(x_1^2, \cdots, x_q^2),$$

where r'_q has rational coefficients; this can be written

$$(11) \qquad \sum \omega_j^q = q!\, W_{2q}(\eta)/2 + q! \cdot r_q,$$

r_q being a polynomial in the $p_i(\eta)$ with rational coefficients. On the other hand, $c_q(\theta)$ is the q-th elementary symmetric function in the ω_j's, hence

$$\sum \omega_j^q = (-1)^{q-1} q \cdot c_q(\theta) + q!\, s_q,$$

where s_q is a polynomial in the $c_i(\theta)$ $(i < q)$ with rational coefficients. Now, θ is extension of a $Spin(2q)$-bundle η', hence its characteristic ring is contained in the characteristic ring of η'. Since we consider real cohomology, $\rho^*(\lambda) : H^*(B_{SO(2q)}) \to H^*(B_{Spin(2q)})$ is an isomorphism, and therefore the $c_i(\theta)$ belong to the characteristic ring of η, which is generated by the $p_i(\eta)$ $(i < q)$ and by $W_{2q}(\eta)$. Thus we get

$$(-1)^{q-1} c_q(\theta)/(q-1)! = W_{2q}(\eta)/2 + t_q,$$

where $t_q = r_q - s_q$ is a polynomial in the $p_i(\eta)$ $(2i < q)$ with rational coefficients. It is necessarily zero if q is odd, and this proves (i). In order to prove (ii), we have to compute the coefficient d_k of $p_k(\eta)$ in t_{2k}. By the above, d_k is equal to the coefficient of $p_k(\eta)$ in r_{2k}. Let S be defined by $S(x_1) = -x_1$ and $S(x_i) = x_i$ $(i \geqq 2)$ and put $\sigma_j = S(\omega_j)$. The set $\{\sigma_j\}$ is then the set of forms

$$(\epsilon_1 x_1 + \cdots + \epsilon_{2k} x_{2k})/2, \quad \prod \epsilon_i = -1,$$

(which are the weights of the second half spinor representation), and (10) yields

$$(12) \qquad \sum \sigma_j^{2k} = -(2k)! \, x_1 \cdots x_{2k}/2 + (2k)! \, r'_{2k}(x_1^2, \cdots, x_{2k}^2),$$

hence

$$(13) \qquad \sum (\omega_j^{2k} + \sigma_j^{2k}) = (2k)! \, 2 \cdot r_{2k},$$

so that $2d_k$ is the coefficient of p_k in

$$((2k)!)^{-1} \sum (\omega_j^{2k} + \sigma_j^{2k}).$$

We have clearly

$$\sum_{\epsilon_i = \pm 1} \exp(\epsilon_1 x_1 + \cdots + \epsilon_{2k} x_{2k})/2 = 2^{2k} \prod_{j=1}^{j=2k} \cosh(x_j/2).$$

Let $\{D_j(p_1, \cdots, p_j)\}$ be the multiplicative sequence with $\cosh z^{\frac{1}{2}}/2$ as characteristic power series (this is well defined since $\cosh x$ is an even function in x). Then $2^{-2k+1} d_k$ is for $k \geqq 1$ the coefficient of p_k in D_k. The formula 1.4(10) of [19] yields therefore, (with $2d_0 = 1$),

$$2 \sum_{j \geqq 0} d_j(-z/4)^j = \cosh(\tfrac{1}{2} z^{\frac{1}{2}}) \cdot d(z/\cosh(\tfrac{1}{2} z^{\frac{1}{2}}))/dz,$$

$$2 \sum_{j \geqq 0} d_j(-z/4)^j = 1 - (z^{\frac{1}{2}}/4) \operatorname{tgh}(\tfrac{1}{2} z^{\frac{1}{2}}).$$

Putting $z = -4x^2$, we get then

$$\sum_{j \geqq 1} d_j x^{2j-1} = (1/4) \operatorname{tg} x,$$

and this ends the proof of (ii).

Proof of the theorem. Since the base space of ξ is S_{2q} $(q \neq 1)$, we have $w_2(\xi) = 0$, hence we may apply the lemma to ξ. For q odd, the theorem follows then from (i) and from the divisibility theorem of Bott (26.10). Let now $q = 2k$ be even. We have $tg'(0) = 1$, $tg^{(3)}(0) = 2$, and it is well known, and easily checked, that $tg^{(2k-1)}(0)$ is an integer divisible by 4 for

$k \geq 3$. Moreover, in (ii) of the lemma we have $R_{2k} = 0$ since $B_\xi = S_{4k}$. Thus, for $q = 2k$, the theorem follows from the lemma and 26.10.

27. Multiplicative properties of the index and consequences.

27.1. *Notation.* Throughout this and the following paragraph, all cohomology groups will be taken with real coefficients and all characteristic classes which occur will be regarded as real classes unless otherwise mentioned.

$\mathcal{D}$ denotes the class of differentiable bundles ξ, where E_ξ, B_ξ, F_ξ are compact connected oriented differentiable manifolds, the orientation of E_ξ being induced by those of B_ξ, F_ξ taken in that order, and where the fundamental group of B_ξ operates trivially on $H^*(F_\xi)$.

We recall (21.5) that if $\{K_j\}$ is a multiplicative sequence with real coefficients which is strictly multiplicative in ξ, then

$$(1) \qquad\qquad K(E_\xi) = K(B_\xi) \cdot K(F_\xi) \qquad\qquad (\xi \in \mathcal{D}).$$

We wish to prove a theorem which is a sort of converse to 21.5.

27.2. THEOREM. *Let F be a compact connected oriented differentiable manifold. If $\{K_j\}$ is a multiplicative sequence of polynomials with real coefficients, for which (1) holds in every bundle $\xi \in \mathcal{D}$ such that $F_\xi = F$, then $\{K_j\}$ is strictly multiplicative in each of these bundles.*

The proof uses essentially the theorem of Thom [29, Corollaire II 30] that every real cohomology class of B_ξ is a finite linear combination of real cohomology classes representable by submanifolds. By definition, a real cohomology class is representable by a submanifold if and only if it corresponds by Poincaré duality to a real homology class containing the fundamental cycle of a compact oriented differentiable manifold differentiably imbedded in B_ξ.

The strictly multiplicative behavior of $\{K_j\}$ is obviously true if $\dim B_\xi = 0$. Let us make the induction hypothesis that it is proved for $\dim B_\xi < n$. Then we will prove it, using (1), for an n-dimensional manifold B_ξ. Let $\hat{p}_i$ be the Pontrjagin classes of the bundle along the fibres of ξ. We have to show that

$$\left(\sum_{j=0}^{\infty} K_j(\hat{p}_1, \cdot \cdot \cdot, \hat{p}_j) \right)^{\natural} = K(F_\xi) \cdot 1,$$

where $\natural : H^*(E_\xi) \to H^*(B_\xi)$ is the integration over the fibre (8.1). We can restrict the bundle ξ to every submanifold Y of B_ξ. Since integration over the fibre and restriction commute (8.3), we obtain by our induction hypothesis that for $\dim Y < \dim B_\xi$ the restriction $(\sum_{j=0}^{\infty} K_j(\hat{p}_1, \cdots, \hat{p}_j))^\natural$ to Y equals $K(F_\xi) \cdot 1$, where 1 denotes now the unit of the cohomology ring of Y. This, together with the above mentioned theorem of Thom, implies

$$(\sum_{j=0}^{\infty} K_j(\hat{p}_1, \cdots, \hat{p}_j))^\natural = K(F_\xi) \cdot 1 + c,$$

where $c \in H^n(B_\xi)$.

We denote the Pontrjagin classes of B_ξ by p'_i and those of E_ξ by p_i. Then (compare the proof of 21.5)

$$\begin{aligned}
K(E_\xi) &= (\sum_{j=0}^{\infty} K_j(p_1, \cdots, p_j))[E_\xi] \\
&= ((K(F_\xi) \cdot 1 + c) \cdot \sum_{t=0}^{\infty} K_t(p'_1, \cdots, p'_t))[B_\xi] \\
&= K(F_\xi)(\sum_{t=0}^{\infty} K_t(p'_1, \cdots, p'_t))[B_\xi] + c[B_\xi] \\
&= K(F_\xi) \cdot K(B_\xi) + c[B_\xi].
\end{aligned}$$

According to (1) we have $K(E_\xi) = K(B_\xi)K(F_\xi)$ and thus obtain $c = 0$, which completes the proof.

It was proved recently [12] for the index τ and a bundle $\xi \in \mathcal{D}$ that

$$(2) \qquad\qquad \tau(E_\xi) = \tau(B_\xi) \cdot \tau(F_\xi).$$

Since the index τ equals the genus L defined by the multiplicative sequence $\{L_j\}$, see [19, §8], we get in virtue of the preceding theorem the following result:

27.3. **Theorem.** *The sequence $\{L_j\}$ is strictly multiplicative in every* $\xi \in \mathcal{D}$.

If $\xi \in \mathcal{D}$ and $\natural$ is the integration over the fibre in ξ, we have therefore

$$(L_j(\hat{p}_1, \cdots, \hat{p}_j))^\natural = 0 \qquad\qquad (4j \neq \dim F_\xi),$$
$$(L_j(\hat{p}_1, \cdots, \hat{p}_j))^\natural = \tau(F_\xi) \cdot 1 \qquad\qquad (4j = \dim F_\xi).$$

27.4. *Examples.*

1) $\dim F_\xi = 2$. In this case, $\tau(F_\xi)$ vanishes. $L_j(\hat{p}_1, 0, \cdots, 0)$ is a non-zero multiple of $\hat{p}_1^j$, see [19, §1]. Since $\hat{p}_1 = \hat{W}_2^2$, where $\hat{W}_2$ is the Euler class of the bundle along the fibres, we get (in real cohomology)

$$(\hat{W}_2^{2j})^\natural = 0, \qquad\qquad j = 1, 2, 3, \cdots.$$

If B_ξ is of dimension $4k-2$ and hence E_ξ of dimension $4k$, then $\hat{W}_2{}^{2k}=0$.

2) $\dim F_\xi = 3$. We get for the Pontrjagin class $\hat{p}_1$ of the bundle along the fibres that $(\hat{p}_1{}^j)^\natural$ $(j=1,2,3,\cdots)$, vanishes in real cohomology.

3) $\dim F_\xi = 4$. We have $45 \cdot L_2(\hat{p}_1, \hat{p}_2) = 7\hat{p}_2 - \hat{p}_1{}^2$. Thus

$$(7\hat{p}_2 - \hat{p}_1{}^2)^\natural = 0.$$

This equation is proved in real cohomology and not known in integral cohomology. Again $\hat{p}_2 = \hat{W}_4{}^2$, where $\hat{W}_4$ is the Euler class of the bundle along the fibres. If $\dim B_\xi = 4$ (and hence $\dim E_\xi = 8$), then

$$7 \cdot \hat{W}_4{}^2 = \hat{p}_1{}^2.$$

27.5. *Remark.* The strict multiplicativity of $\{L_j\}$ was proved in 22.9 for bundles $\xi \in \mathcal{D}$ with $F_\xi = G/U$ (rank G = rank U) and G as structural group. It was also shown (§ 23) in this special case that the sequence $\{A_j\}$ is strictly multiplicative provided $A(F_\xi) = 0$. It might be conjectured that in every fibre bundle $\xi \in \mathcal{D}$ the vanishing of $A(F_\xi)$ implies that of $A(E_\xi)$. As a consequence, we would have (27.2) that $\{A_j\}$ is strictly multiplicative in every fibre bundle $\xi \in \mathcal{D}$ with $A(F_\xi) = 0$.

28. A uniqueness theorem on the index.

28.1. In this paragraph, we shall prove that the sequence $\{L_j\}$ is essentially the only multiplicative sequence with coefficients in a field of characteristic 0 giving rise to a genus which is multiplicative in fibre bundles. Throughout this paragraph, we keep the notations of 27.1.

Let M be a $4k$-dimensional compact oriented differentiable manifold and p_i its Pontrjagin classes, which may be written formally as elementary symmetric functions:

$$1 + p_1 + \cdots + p_k = (1+\beta_1) \cdots (1+\beta_k).$$

Then the number $s(M)$ is defined by

$$s(M) = (\beta_1{}^k + \cdots + \beta_k{}^k)[M], \qquad \text{(see } [19, \S 6.3]).$$

28.2. Lemma. *Let $\xi \in \mathcal{D}$ be a principal $U(q)$-bundle over a 4-dimensional base space and c_i its Chern classes. If $2q + 2 = 4k \geqq 8$, then*

$$s(E_\xi/U(1) \times U(q-1)) = (-q(q+2)c_2 + (\binom{q+1}{2} - 1)c_1{}^2)[B_\xi].$$

For the proof, we use the notations of 15.1. We always take into account that B_ξ is 4-dimensional, and hence, for example, $c_i = 0$ for $i > 2$ and

$$(1) \qquad \pi^*(1 + c_1 + c_2) = (1 + x_1)(1 + x_2) \cdots (1 + x_q).$$

Furthermore

$$(2) \qquad x_1{}^q - x_1{}^{q-1}\pi^*(c_1) + x_1{}^{q-2}\pi^*(c_2) = 0.$$

Letting $a = \sum\limits_{j=1}^{q} (x_j - x_1)^{q+1}$, we get by (1) that

$$(3) \qquad (-1)^q a = - q \cdot x_1{}^{q+1} + (q+1) x_1{}^q \cdot \pi^*(c_1)$$
$$- \binom{q+1}{2} x_1{}^{q-1} \cdot \pi^*(c_1{}^2 - 2c_2).$$

We infer readily from (2) and (3) that

$$(-1)^q a = (q+2)q \cdot x_1{}^{q-1}\pi^*(c_2) + \left(1 - \binom{q+1}{2}\right) \cdot x_1{}^{q-1} \cdot \pi^*(c_1{}^2).$$

We may put $a = \rho^*(b)$ and $x_1 = \rho^*(\gamma_1)$. Then the preceding formula yields

$$(4) \qquad b = - q(q+2)(-\gamma_1)^{q-1}\sigma^*(c_2) + \left(\binom{q+1}{2} - 1\right) \cdot (-\gamma_1)^{q-1}\sigma^*(c_1{}^2)$$

M will denote the total space of the bundle $\theta = (E_\xi/U(1) \times U(q-1),$ $B_\xi, P_{q-1}(C))$. The tangent bundle of M is the Whitney sum of $\hat\theta$, the bundle along the fibres of θ, and of the tangent vector bundle of B_ξ lifted by σ. We have for the total Pontrjagin classes

$$(5) \qquad p(M) = p(\hat\theta) \cdot \sigma^* p(B_\xi) = p(\hat\theta) \cdot (1 + \sigma^* p_1(B_\xi)).$$

In 15.1, one finds a formula for $\rho^* c(\eta')$ which yields

$$(6) \qquad \rho^* p(\hat\theta) = \prod_{j=1}^{q} (1 + (x_j - x_1)^2).$$

By (5) and (6), we get for $2q + 2 = 4k$

$$s(M) = (b + (\sigma^*(p_1(B_\xi))^k))[M].$$

Since $(p_1(B_\xi))^k = 0$ for $k \geq 2$, we have $s(M) = b[M]$ which, by (4), completes the proof because the value of $(-\gamma_1)^{q-1}$ on the oriented fibres of θ equals 1.

28.3.   We are going to construct a special base sequence for the algebra $\Omega \otimes Q$ of Thom [29]. Consider over $X = P_2(C)$ the differentiable principal $U(q)$-bundle $\xi(q)$ which is the Whitney sum of $q - 2$ trivial $U(1)$-bundles and of the two principal $U(1)$-bundles with g and $-g$ respectively as first

Chern classes, where g is the cohomology class dual to a complex projective line imbedded in X. The Chern classes of $\xi(q)$ are

$$c_1 = 0, \qquad c_2 = -g^2.$$

For $2q + 2 = 4k \geqq 8$, let E^{4k} be the $4k$-dimensional manifold fibred with $P_{q-1}(C)$ as fibre and associated to $\xi(q)$; i.e.

$$E^{4k} = E_{\xi(q)}/U(1) \times U(q-1), \qquad\qquad (2q + 2 = 4k \geqq 8).$$

According to the preceding lemma, we have

$$(7) \qquad\qquad s(E^{4k}) = (2k + 1)(2k - 1) = 4k^2 - 1 \neq 0.$$

For $k = 1$, we put $E^4 = P_2(C)$ and then (7) holds also in this case.

By [19, § 6], the sequence $\{E^{4k}\}$, $(k = 1, 2, 3, \cdots)$, of $4k$-dimensional manifolds is a base sequence of the algebra $\Omega \otimes Q$ of Thom; and we have in terms of the usual base sequence $P_{2k}(C)$

$$(8) \qquad E^{4k} = (2k - 1)P_{2k}(C) + \text{composite terms in the } P_{2j}(C),\, j < k.$$

28.4. THEOREM. *Let $\{K_r(p_1, \cdots, p_r)\}$ be a multiplicative sequence of polynomials with coefficients in a field of characteristic 0 and suppose that the corresponding genus K satisfies the equation $K(E) = K(B) \cdot K(F)$ for all differentiable fibre bundles in $\mathfrak{D}$, or equivalently (27.2), that $\{K_r(p_1, \cdots, p_r)\}$ is strictly multiplicative in $\mathfrak{D}$. Put $a = K(P_2(C))$. Then*

$$(9) \qquad\qquad K(Y) = a^r \cdot L(Y), \qquad\qquad (4r = \dim Y)$$

for all $4r$-dimensional compact oriented differentiable manifolds Y, and moreover

$$(10) \qquad K_r(p_1, \cdots, p_r) = a^r L_r(p_1, \cdots, p_r), \qquad\qquad (r = 1, 2, 3 \cdots).$$

We prove (9) by induction over r. It is true for $r = 1$ since $P_2(C)$ generates Ω^4. Suppose it is proved for all Y with $\dim Y < 4r$. The vector space $\Omega^{4r} \otimes Q$ over the rationals is generated by E^{4r} and "composite" manifolds M^{4r} which are cartesian products of lower dimensional manifolds. Since K and L are both multiplicative in cartesian products, (9) is true on the composite manifolds M^{4r} by induction hypothesis, and since K and L are also both multiplicative in differentiable fibre bundles, (9) is true on E^{4r} too (again we have used the induction hypothesis). Thus (9) holds on $\Omega^{4r} \otimes Q$.

This proves (9) in full generality which implies (10), [19, Satz 6.5.1].

Appendix I.

29. The different definitions of the Chern classes.

29.1. *Orientation conventions.* In the n-dimensional complex vector space V with coordinates $z_j = x_j + iy_j$, we take as usual the orientation defined by the order $x_1, y_1, \cdots, x_n, y_n$. This determines also an orientation for complex analytic manifolds as well as for the $2n-1$ dimensional sphere S_{2n-1} of unit vectors with respect to some hermitian metric on V. The image of the fundamental cycle of S_{2n-1} thus defined in $\pi_{2n-1}(S_{2n-1})$, or $H_{2n-1}(S_{2n-1}, Z)$, or $H_{2n-1}(V-0, Z)$, (0 being the origin in V), and the element of $H^{2n-1}(S_{2n-1}, Z)$ or $H^{2n-1}(V-0, Z)$ taking the value 1 on it will be called the canonical generator of the corresponding group.

Let $W_{n,n-q+1}$ be the complex Stiefel manifold of ordered systems of $n-q+1$ orthonormal vectors in C^n $(1 \leq q \leq n)$. We know that its first non-vanishing homotopy group is in dimension $2q-1$ and is infinite cyclic. Now $W_{n,n-q+1}$ is fibered by $W_{q,1} = S_{2q-1}$, with base $W_{n,n-q}$; the projection assigning to each $(n-q+1)$-frame the $(n-q)$-frame formed by its first $n-q$ elements. The fibre may thus be identified with the set of unit vectors in C^q and its injection in $W_{n,n-q+1}$ induces isomorphisms for the $(2q-1)$-st homotopy or homology or cohomology groups. The element corresponding to the canonical generator previously defined will also be called the canonical generator.

Similarly, let $W^*_{n,n-q+1}$ be the manifold of ordered systems of $n-q+1$ linearly independent vectors in C^n; it has $W_{n,n-q+1}$ as a deformation retract; let $e_1, \cdots, e_{n-q}$ be independent vectors, and let V be a q-dimensional subspace supplementary to the space spanned by the e_i's. The subspace U of $W^*_{n,n-q+1}$ made of the systems (f_j), $(j = 1, \cdots, n-q+1)$, for which $f_j = e_j$ $(j \leq n-q)$ and f_{n-q+1} is in V may be identified with $V - 0$. Its injection in the Stiefel manifold is an isomorphism for homotopy or homology in dimension $2q-1$ and we define as before the canonical generator of $H_{2q-1}(W^*_{n,n-q+1}, Z)$ and $\pi_{2q-1}(W^*_{n,n-q+1})$.

29.2. *The Hopf fibering.* (x_i), $(1 \leq i \leq n+1)$, are the coordinates of C^{n+1} and the homogeneous coordinates in the complex projective space P_n. By the Hopf fibering over P_n, we mean here $C^{n+1} - 0$ endowed with the usual $C^* = GL(n,1)$ bundle structure. e will denote a hyperplane with the positive orientation or the corresponding homology class and $e^* \in H^2(P_n, Z)$ will be the dual cohomology class. Let U_i be the set of points in P_n for which $x_i \neq 0$, $(1 \leq i \leq n+1)$; using the usual conventions for the transition functions of a bundle [19, § 3. 2. a] and of a line bundle associated to a divisor D [19, § 15. 2] we see that in $U_i \cap U_j$ the Hopf fibering is given by $f_{ij} = x_i/x_j$, whereas the bundle $\{e\}$ associated to e is given by $g_{ij} = x_j/x_i$; thus the Hopf fibering is the inverse of $\{e\}$. We recall that the Hopf fibering over $\boldsymbol{P}_n$ is a $2n$-universal bundle for $\boldsymbol{C}^*$ or $\boldsymbol{U}(1)$.

29.3. *The definitions of Chern classes.* Including the definition (9.1), there are apparently seven definitions of Chern classes, which we proceed to list now; $^i c_j$ will be the j-th Chern class according to the i-th definition, and $^i c$ the sum of the $^i c_j$.

(1) *The definition* (9.1) *of this paper.* It may also be formulated in the following way: in the Hopf fibering, we put $^1 c_1 = -\tau(x)$, where x is the canonical generator of $H^1(\boldsymbol{C}^*, \boldsymbol{Z})$; for a general $\boldsymbol{C}^*$-bundle, we use the characteristic map; for a general $\boldsymbol{GL}(n, C)$ bundle, we go over to the bundle of flags and take the elementary symmetric functions in the Chern classes of the different line bundles in which the lifted bundle decomposes.

(2) *The definition of* [19]: it is quite similar to (1), except that we put $^2 c_1 = -e^*$ in the Hopf fibering.

(3) *The obstruction definition.* Given a complex vector bundle $(E, B, \boldsymbol{C}^n)$, we consider the associated bundle $(E', B, \boldsymbol{W}_{n,n-q+1})$. The first obstruction to the construction of a cross section (B is supposed to be a complex here) is an element of $H^{2q}(B, \pi_{2q-1}(\boldsymbol{W}_{n,n-q+1}))$. We identify the coefficient group with $\boldsymbol{Z}$ by sending the canonical generator onto 1, and thus get a class $^4 c_q \in H^{2q}(B, \boldsymbol{Z})$.

This convention was introduced in [20], and was recalled at the beginning of [11] but was not made in [9], where consequently the obstruction classes are defined up to sign only.

(4) *The definition of* [6]: in the Hopf fibering, considered as universal bundle, we put $^4 c_1 = \tau(x)$, and then proceed as in (1).

(5) *Schubert systems.* Let $\boldsymbol{H}(n, N)$ be the complex Grassmann manifold of n-dimensional subspaces of $\boldsymbol{C}^{n+N}$. It is the base space of the $2N$-universal bundle $(\boldsymbol{U}(n+N)/\boldsymbol{U}(N), \boldsymbol{H}(n, N), \boldsymbol{U}(n))$ for $\boldsymbol{U}(n)$, where $\boldsymbol{U}(N)$, (respectively $\boldsymbol{U}(n)$), is the subgroup of $\boldsymbol{U}(n+N)$, leaving the n first (respectively N last) coordinate vectors fixed. For $n = 1$, we have the Hopf fibering. We take as universal $^5 c_q$ the dual class to the Schubert cycle of dimension $2(nN-q)$ represented by the symbol $(N-1, \cdots, N-1, N, \cdots, N)$,

(q times $N-1$, $(n-q)$ times N). By the intersection properties of Schubert varieties, ^5c_q is also the cohomology class taking the value 1 on the Schubert cycle $(0, \cdots 0, 1, \cdots, 1)$ $((n-q)$ times 0, q times 1), and zero on all other Schubert cycles of Ehresmann's cell decomposition of $\boldsymbol{H}(n, N)$; (Schubert cycles are defined for instance in [9, 10, 11].) This defines ^5c in the universal bundle; for a general bundle, we take its image by the characteristic map. For $n = 1$, the Schubert symbol $(N-1)$ represents the hyperplane of $\boldsymbol{P}_N(\boldsymbol{C}) = \boldsymbol{H}(1, N)$. Thus, in the Hopf fibering, we have $^5c_1 = e^*$.

(6) *Definition by means of differential forms.* This leads to real cohomology classes, defined for differentiable bundles. Let ξ be a differentiable principal $\boldsymbol{U}(n)$-bundle and let Ω_{ij} $(1 \leq i, j \leq n)$ be the curvature forms of a connexion on E_ξ. We then consider the (mixed) differential form:

$$\Psi = \sum \Psi_q = \det \left| \mathrm{Id} + (2\pi i)^{-1} \Omega_{ij} \right|$$

(Ψ_q of degree q; the product in the determinant is of course the exterior product). It defines a form on B_ξ, which is closed. The image of Ψ_q in $H^{2q}(B_\xi, \boldsymbol{R})$ is by definition ^6c_q. This definition is introduced in [9] (our Ψ_q is the Ψ_{n-q+1} of Chern), although in an apparently slightly more restrictive way. Chern considers only bundles of (tangential) orthonormal frames to a hermitian manifold and a special connexion characterized by a property of its torsion tensor [9, p. 111]. However, by a theorem of Weil, whose proof is reproduced in [10, pp. 58-59], the cohomology class of Ψ_q is independent of the particular connexion chosen in ξ.

(7) *Definition by transgression.* ^7c_q is the image by transgression of the canonical generator of $H^{2q-1}(\boldsymbol{W}_{n,n-q+1}, \boldsymbol{Z})$ in the bundle with fiber $\boldsymbol{W}_{n,n-q+1}$ associated to a given complex vector bundle.

The purpose of this Appendix is to prove the

29.4.   Theorem. *Let ^ic_j be the j-th Chern class of a bundle $(E, B, \boldsymbol{U}(n))$ with respect to the i-th definition $(j = 1, \cdots, n; i = 1, \cdots, 7)$. Then*
$$^1c_j = {}^2c_j = {}^3c_j = (-1)^j \cdot {}^4c_j = (-1)^j \cdot {}^5c_j = (-1)^j \cdot {}^6c_j = -{}^7c_j.$$

29.5.   *Remarks.* (a) All these definitions have the naturality property: if $f: \xi \to \eta$ is a homomorphism, then $\bar{f}^*(^ic(\eta)) = {}^ic(\xi)$, where $\bar{f}: B_\xi \to B_\eta$ is induced by f. This is obvious for $i = 1, 2, 4, 5, 7$, and standard for $i = 3$; for $i = 6$, it follows by the theorem of Weil quoted above, because if Ω_{ij} are the curvature forms of a connexion $\mathscr{C}$ on E_η, then their images on E_ξ under f will be the curvature forms of the connexion induced from $\mathscr{C}$ by f.

(b) Let a, b $(1 \leq a, b \leq 7)$ be given, and assume that ${}^a c_1 = \epsilon \cdot {}^b c_1$ $(\epsilon = \pm 1)$ and that ${}^a c$ obeys duality.[6] Then ${}^b c$ obeys duality if and only if ${}^a c_j = \epsilon^j \cdot {}^b c_j$, $(1 \leq j \leq n)$. This is readily seen by using the bundle of flags. Thus ${}^i c$ has the duality property for $i \leq 6$, but not for $i = 7$. For $i = 1, 2, 4$, the duality property follows immediately from an identity between elementary symmetric functions (see e. g. [6]). In the course of the proof of the theorem we shall *use* the fact that ${}^5 c$ obeys duality, which is proved in [11], and therefore we do not provide a new proof for it. We note in passing that in the introduction of [11], Chern classes are defined as obstructions (with signs) but the proof for duality is carried out for Schubert cocycles. However, our 29.4 shows that the obstruction classes obey duality, a fact for which there is, to our knowledge, no direct proof in the literature.

29.6. LEMMA. *In the Hopf fibering, we have* ${}^7 c_1 = + e^* = - {}^3 c_1$.

In view of a general fact about transgression and obstructions, recalled below (29.7), it is in fact enough to prove one equality, but both may be easily checked directly: As to the first one, we put $a = \sum x_i \cdot \bar{x}_i$ and consider

$$\Omega = (i/2\pi)\,(a^{-1} \cdot \sum dx_i \wedge d\bar{x}_i - a^{-2} \cdot \sum x_i \cdot \bar{x}_j\, dx_j \wedge d\bar{x}_i)\,;$$

it is the imaginary part, multiplied by $1/\pi$, of the Fubini metric

$$(1) \qquad ds^2 = a^{-1} \cdot \sum dx_i \cdot d\bar{x}_i - a^{-2} \cdot \sum x_i \bar{x}_j\, dx_j \cdot d\bar{x}_i.$$

Ω is a closed real form of type $(1,1)$ on $\boldsymbol{P}_n$ and we integrate it on the projective line $\boldsymbol{P}_1 : x_3 = \cdots = x_{n+1} = 0$ with homogeneous coordinates (x_1, x_2); leaving out $(0,1)$, we replace $\boldsymbol{P}_1$ by the cross section $x_1 = 1$ and get for the integral

$$(i/2\pi) \int_{\boldsymbol{P}_1} (1 + x_2 \cdot \bar{x}_2)^{-2} \cdot dx_2 \wedge d\bar{x}_2 = 1.$$

Hence Ω represents e^*. On the other hand, we have

$$\Omega = (i/2\pi)\, d\bar{\partial} \log a = d(\,(i/2\pi) a^{-1} \cdot \sum x_i \cdot d\bar{x}_i),$$

and the restriction of $(i/2\pi)\bar{\partial} \log a$ to the fibre $x_2 = \cdots = x_{n+1} = 0$ is $i(2\pi\bar{x}_1)^{-1} d\bar{x}_1$, whose integral on the positively oriented unit circle is again 1; this proves the first equality.

${}^3 c_1$ is the first obstruction to the construction of a cross section in the Hopf fibering and we only need to know its value on $\boldsymbol{P}_1$. Over this line, we consider the cross section defined (except at $(0,1)$) by $(x_1, x_2) \to (1, x_2/x_1)$.

---

[6] By duality we mean the multiplication theorem 9.7(6).

Around $(0,1)$, we take as product representation of the bundle the one which identifies 1 on the typical fibre with $(x_1/x_2, 1)$; then we see easily that around $(0,1)$, the map $\boldsymbol{P}_1 \to \boldsymbol{C}^*$ which defines the value of the obstruction cochain on a 2 simplex having $(0,1)$ in its interior, is of the form $z \to 1/z$; this value will therefore be -1, and so will be that of the obstruction cochain on $\boldsymbol{P}_1$, hence the second equality.

29.7. *Proof of the theorem.* By the definitions, the lemma and naturality, we have

$$^1c_1 = {}^2c_2 = {}^3c_1 = -\,{}^4c_1 = -\,{}^5c_1 = -\,{}^7c_1$$

since all these classes are equal to $-e^*$ in the Hopf fibering. For $i = 1, 2, 4, 5$, ^ic satisfies duality, and we get therefore (see 29.5b)

$$(2) \qquad {}^1c_i = {}^2c_i = (-1)^i \cdot {}^4c_i = (-1)^i \cdot {}^5c_i \qquad\qquad (1 \leqq i \leqq n).$$

The equality

$$(3) \qquad\qquad {}^3c_j = -\,{}^7c_j, \qquad\qquad (1 \leqq j \leqq n),$$

follows from the more general fact that in a fibre bundle, "transgression is minus obstruction" for a proof of which we refer to [26, § 37.16].

The $2N$-universal bundle for $\boldsymbol{U}(n)$ over $\boldsymbol{H}(n,N)$ is $(\boldsymbol{W}_{n+N,n}, \boldsymbol{H}(n,N), \boldsymbol{U}(n))$ which may also be written as $(\boldsymbol{U}(n+N)/\boldsymbol{U}(N), \boldsymbol{H}(n,N), \boldsymbol{U}(n))$, where $\boldsymbol{U}(N)$, (respectively $\boldsymbol{U}(n)$), is the subgroup of $\boldsymbol{U}(n+N)$ leaving the n first (respectively N last) coordinate vectors fixed. Let (u_{ij}), $(1 \leqq i,j \leqq n+N)$, be the standard coordinates in the matrix space. Following Chern [9], we denote by θ_{ab} the left invariant Maurer-Cartan form on $\boldsymbol{U}(n+N)$ which is equal to du_{ba} (and not du_{ab}) at the neutral element. In these notations, we have

$$(4) \qquad\qquad d\theta_{ab} = \sum_{1}^{n+N} \theta_{ai} \wedge \theta_{ib}; \qquad \theta_{ba} = -\,\bar{\theta}_{ba}.$$

It is easily seen that for $1 \leqq a, b \leqq n$, the forms are right invariant under $\boldsymbol{U}(N)$ and satisfy

$$\theta_{ab} \cdot u = \sum_{ij} u_{ia}\theta_{ij}(u^{-1})_{bj} \qquad\qquad (u \in \boldsymbol{U}(n), u = (u_{ij})),$$

hence they induce forms on $\boldsymbol{U}(n+N)/\boldsymbol{U}(N)$ which define there a connexion for the $\boldsymbol{U}(n)$-bundle structure. (4) shows that its curvature forms are

$$\Omega_{ij} = \sum_{k=n+1}^{k=N} \theta_{ik} \wedge \theta_{kj}, \qquad\qquad (1 \leqq i,j \leqq n).$$

Thus, by definition

$$\Psi = \sum \Psi_q = \det | \, \mathrm{Id} \, + (2\pi i)^{-1}\Omega_{ij} \, |$$

represents ^6c in the universal bundle. By a computation which we shall not reproduce, Chern ([9], Chap. II, §2, [10], p. 77) has shown that Ψ_q is equal to ^5c_q; hence

$$(5) \qquad\qquad {}^5c_j = {}^6c_j, \qquad\qquad\qquad (j=1,\cdots,n),$$

first in the universal bundle, and then by naturality in all differentiable $U(n)$-bundles.

To conclude the proof, we have to compare the obstruction classes with one of the six other types, and it will be enough to show that

$$(6) \qquad\qquad {}^3c_j = (-1)^j \cdot {}^6c_j, \qquad\qquad\qquad (1 \leqq j \leqq n).$$

We have first

$$ {}^3c_j = \epsilon_j \cdot {}^6c_j \qquad\qquad\qquad (\epsilon_j = \pm 1, j = 1, \cdots, n),$$

with ϵ_j depending only on j; by naturality, we have only to prove this in the universal bundle; there it follows from (2), (3), (5), and from the fact that ^7c_j and ^4c_j both generate the kernel of $\rho^*(U(j-1), U(n))$ in dimension $2j$, which is infinite cyclic. The proof of this is identical with that of a similar statement on Stiefel-Whitney classes [3, Lemme 5.1] and will be left to the reader.

To determine ϵ_j, it is then enough to compute the 2 classes ^3c and ^6c for one bundle with Chern classes not zero or of order 2. To this end, we take the tangent bundle of P_n. By [9, p. 118], we have

$$\Psi_j = (2\pi i)^{-j} \binom{n+1}{j} \Lambda^j$$

for the torsionless connexion associated to the Fubini metric, where Λ is the exterior form obtained from the metric (1) by replacing the symmetric products by exterior products; in the notation of 29.6, we have, therefore, $\Omega = i(2\pi)^{-1} \cdot \Lambda$ and

$$\Psi_j = \binom{n+1}{j} (-1)^j \Omega^j.$$

We have seen in the proof of 29.6 that Ω represents e^* and, consequently,

$$(7) \qquad\qquad {}^6c_j = \binom{n+1}{j}(-1)^j (e^*)^j, \qquad\qquad (1 \leqq j \leqq n),$$

as also follows from 15.1 and (2), (5). On the other hand, a direct construction of vector fields in [9, Theorem 13] shows that

$$(8) \qquad {}^{3}c_j = \binom{n+1}{j}(e^*)^j, \qquad\qquad (j = 1, \cdots, n),$$

whence $\epsilon_j = (-1)^j$ and (6). Of course we must check that in the proof of (8), the indices of singularities are counted with the proper sign conventions; this presents no difficulty, but for the sake of completeness, we outline Chern's construction. Let

$$\begin{bmatrix} 1 & 1 & \cdots & 1 \\ a_{11} & a_{12} & & a_{1,n+1} \\ & & \cdots & \\ a_{n1} & a_{n2} \cdots & & a_{n,n+1} \end{bmatrix}$$

be a matrix with constant coefficients and all minors of degree j $(1 \leqq j \leqq n+1)$ different from zero. Let H_j be the j-dimensional projective subspace defined by

$$\sum_k a_{ik}x_k = 0 \qquad\qquad (1 \leqq i \leqq n-j).$$

We denote by $\mathfrak{v}_j$ the vector field on $\boldsymbol{C}^{n+1} - 0$ defined by

$$\mathfrak{v}_j(x) = \sum_k a_{jk}x_k e_k, \qquad\qquad ((e_k) \text{ canonical basis of } \boldsymbol{C}^{n+1}).$$

It is invariant under $x \to a \cdot x$ $(a \in \boldsymbol{C}^*)$ and defines a vector field $\mathfrak{w}_j$ on $\boldsymbol{P}_n$. The vectors $\mathfrak{w}_j$ $(j \leqq r)$ are dependent at a point P with homogeneous coordinates (x_i) if and only if the $r+1$ vectors $\mathfrak{x} = \sum x_i e_i$ and $\mathfrak{v}_1, \cdots, \mathfrak{v}_r$ are dependent at $x = (x_i)$, which is equivalent with the vanishing of $n+1-r$ homogeneous coordinates of P.

Let now j be fixed and put $r = n - j + 1$. On H_j, $\mathfrak{w}_1, \cdots, \mathfrak{w}_{r-1}$ are independent everywhere whereas $\mathfrak{w}_1, \cdots, \mathfrak{w}_r$ are dependent at $\binom{n+1}{j}$ points. We use these vector fields to compute the value of ${}^3c_j$ on H_j; since we already know that it is equal to $\pm\binom{n+1}{j}$, we have only to show that the singular points have non-negative indices. Let Q be a singular point, and W a neighborhood of Q on H_j. Using the fields $\mathfrak{w}_1, \cdots, \mathfrak{w}_{r-1}$ which are also independent at Q we see immediately that the map $W - Q \to \boldsymbol{W}^*_{n,r}$ leading to the index is homotopic to a complex analytic map of $W - Q$ into a subspace of $\boldsymbol{W}^*_{n,r}$ of the type of the space U introduced in 29.1, and that the resulting map of $W - Q$ in $\boldsymbol{C}^j - 0$ extends analytically to Q and maps it onto the origin; hence it has a positive degree, and the index is $\geqq 0$ according to the conventions of 29.1 and 29.3(3).

29.8. *Remarks.* (a) The obstruction classes ${}^3c_j$ are the obstructions to the construction of *contravariant* vector fields. Hodge [20] uses covariant vector fields and, by 29.4 and 10.6a, or directly, the resulting classes are our ${}^4c_j$.

(b) According to Hodge [20] and Nakano (Mem. Coll. Sci. Kyoto 29 (1955), 145-149), the canonical classes of Eger-Todd are to be identified with the Schubert classes ${}^5c_j$.

(c) The lack of sign conventions for obstruction classes in [9] leads to a slight inaccuracy for the Chern classes of $\boldsymbol{P}_n$; Theorem 13 gives $c_j = \binom{n+1}{j}(e^*)^j$ and Theorem 12 gives the class of Ψ_j; as we have seen, these differ by $(-1)^j$.

29.9. Finally, to be complete, we list some properties or alternate definitions of the first Chern class. For the notations, and concepts used here, see [19].

(a) Let $\boldsymbol{C}^*_c$ be the sheaf of germs of continuous $\boldsymbol{C}^*$-valued functions on the space B. A complex line bundle over B is represented by an element $\xi \in H^1(B, \boldsymbol{C}^*_c)$ and we have

$$ {}^1c_1 = \delta\xi, $$

where δ is the coboundary operator $H^1(B, \boldsymbol{C}^*_c) \to H^2(B, \boldsymbol{Z})$ associated to the exact sequence $0 \to \boldsymbol{Z} \to \boldsymbol{C}_c \xrightarrow{e} \boldsymbol{C}^*_c \to 0$, where e is the exponential map [19, § 4.3.1].

(b) Let V be an oriented m-dimensional manifold, B an oriented $(m-2)$-dimensional submanifold, η the normal bundle oriented in such a way that orientation of B plus orientation of η gives the orientation of V. It has then a complex structure compatible with its orientation, determined up to isomorphism; its class ${}^1c_1$ is dual to the homology class defined by B [19, § 4.8.1].

(c) Kodaira has introduced the following definition for the Chern class c_1 of a holomorphic principal $\boldsymbol{C}^*$-bundle $\xi = (E, B, \boldsymbol{C}^*, \pi)$. Let (U_j) be a covering by coordinate neighborhoods, z_j the coordinate of the standard fibre over U_j, (f_{jk}) the transition functions, (a_j) a cross section of the bundle with fibre $\boldsymbol{R}^+$ defined by the transition functions $f_{jk} \cdot \bar{f}_{jk}$. Then $c_1(\xi)$ is the class of the form $\gamma = (i/2\pi)\partial\bar\partial \log a_j$. *This class is equal to* ${}^1c_1$.

Proof. We have $\pi^*(\gamma) = d\psi$, where ψ is a 1-form over E with the local representation $(-i/2\pi)\bar\partial \log(\bar z_j/a_j)$ over U_j; the restriction of ψ to a fibre is $(-i/2\pi)\bar z_j^{-1}d\bar z_j$ and its integral over the positively oriented unit circle is -1. Thus the Kodaira class is equal to $-\tau(x)$, where x is the canonical generator of $H^1(\boldsymbol{C}^*,\boldsymbol{Z})$, and is equal to $^1c_1$ by 29.4.

(d) On a complex manifold B of complex dimension n, the *canonical bundle* K is the line bundle of exterior forms of type $(n,0)$, i.e. the bundle of n-forms on the tangent bundle. By (10.6a, b), its Chern class is $-{}^1c_1(\theta)$, where θ is the tangential bundle. In particular, the determinant g of a positive non-degenerate hermitian metric provides a section of the bundle with fibre $\boldsymbol{R}^+$ and transition functions $|f_{jk}|^2$, (f_{jk} being the transition functions of K). Thus the Ricci form

$$(-i/2\pi)\partial\bar\partial \log g,$$

where $R_{\alpha\beta} = -\partial^2 \log g/\partial z_\alpha \partial\bar z_\beta$ is the Ricci tensor, represents $^1c_1(\theta)$ (see Kodaira, Annals of Math. 60 (1954), 28-48).

Appendix II.

30. Pontrjagin classes.

30.1. *Notation.* Tors A is the torsion subgroup of the commutative group A, and Tors$_p A$ its p-primary component.

Let V be a vector space graded by finite dimensional subspaces V^i ($i \geqq 0$). By $P(V,t)$, we denote its Poincaré polynomial

$$P(V,t) = \sum \dim V^i \cdot t^i,$$

and for a topological space X, we write $P_p(X,t)$ for $P(H^*(X,\boldsymbol{Z}_p),t)$.

f^*_p and $f^*_{\boldsymbol{Z}}$ denote the homomorphism of cohomology rings over $\boldsymbol{Z}_p$ and $\boldsymbol{Z}$ induced by a continuous map f.

Let ξ be a bundle with connected fibres, and A a commutative group. Then $T^i(F_\xi, A)$ or simply T^i_ξ denotes the subgroup of transgressive elements in $H^i(F_\xi, A)$. We recall that the transgression τ_ξ is a homomorphism of T^i_ξ into the quotient of $H^{i+1}(B_\xi, A)$ by a subgroup which will be denoted by $L^{i+1}(B_\xi, A)$ or L^{i+1}_ξ; we have $L^2_\xi = 0$.

30.2. *Cohomology* mod p of $B_{\boldsymbol{O}(n)}$ and $B_{\boldsymbol{SO}(n)}$. Let G be a compact Lie group, T a maximal torus. The ring of invariants of $W(G)$ operating in the usual way in $H^*(B_T, \boldsymbol{Z})$ is denoted by I_G. If $G = \boldsymbol{SO}(2n+1), \boldsymbol{O}(2n+1), \boldsymbol{O}(2n)$, (respectively $G = \boldsymbol{SO}(2n)$), and if (y_j) is the base induced by transgression in (E_G, B_T, T) from the basis of $H^1(T, \mathbf{Z})$ discussed in §9, then $W(G)$ is the group of permutations of the y_j's combined with an arbitrary (respectively even) number of changes of signs, and consequently $I_G = S(y_1^2, \cdots, y_n^2)$, (respectively is generated by $S(y_1^2, \cdots, y_n^2)$ and by the product of the y_i's).

PROPOSITION. *Let $G = SO(n)$ or $O(n)$, let T be its standard maximal torus, and let Q be the subgroup of diagonal matrices. Then*

(a) *For $p \neq 2$, $\rho^*_p(T, G)$ maps $H^*(B_G, \mathbf{Z}_p)$ isomorphically onto $I_G \otimes \mathbf{Z}_p$.*

(b) *$\rho^*_2(Q, O(n))$ maps $H^*(B_{O(n)}, \mathbf{Z}_2)$ isomorphically onto $S(u_1, \cdots, u_n)$, where (u_i) is a suitable basis of $H^1(B_Q, \mathbf{Z}_2)$.*

For (b), see [3, Théorème 5], where a similar statement is also proved for $SO(n)$. For $G = SO(n)$, the assertion (a) is proved in [2, §29]. For $G = O(n)$, it follows from the more general

30.3. PROPOSITION. *Let G be a compact Lie group, G_0 its largest connected subgroup, T a maximal torus. If $H^*(G_0, \mathbf{Z})$ has no p-torsion and if the order of G/G_0 is not divisible by p, then $\rho^*_p(T, G)$ is an isomorphism of $H^*(B_G, \mathbf{Z}_p)$ onto $I_G \otimes \mathbf{Z}_p$.*

For $p = 0$, see [2, Prop. 27.1]. For p prime, the proof is practically identical, in view of the absence of torsion on G_0/T, (14.2), and is left to the reader.

30.4. *A remark on the Bockstein homomorphism.* Let X be a space with finitely generated integral cohomology groups. Let r_i be the number of cyclic direct summands of the p-primary component of $H^i(X, \mathbf{Z})$. Then by the universal coefficient formula

$$P_p(X, t) - P_0(X, t) = (1 + 1/t) \sum r_i \cdot t^i.$$

Let β_p be the Bockstein homomorphism attached to the exact sequence

$$0 \to \mathbf{Z} \xrightarrow{\alpha} \mathbf{Z} \to \mathbf{Z}_p \to 0 \qquad\qquad (\alpha(x) = px, x \in \mathbf{Z})$$

followed by reduction mod p. Clearly $\beta_p \circ \beta_p = 0$, and

$$s_i = \dim \beta_p(H^{i-1}(X, \mathbf{Z}_p))$$

is the number of torsion coefficients of $\mathrm{Tors}_p H^i(X, \mathbf{Z})$ which are equal to p. Thus we see:

LEMMA. $\mathrm{Tors}_p H^*(X, \mathbf{Z})$ *consists of elements of order p if and only if*

$$P_p(X, t) - P_0(X, t) = (1 + 1/t) P(\beta_p(H^*(X, \mathbf{Z}_p)), t).$$

When this is the case, the kernel of β_p is the reduction $\bmod p$ of $H^*(X, \mathbf{Z})$, and its image is the reduction $\bmod p$ of $\mathrm{Tors}_p H^*(X, \mathbf{Z})$.

We recall that β_p is an antiderivation and that $\beta_2 = Sq^1$.

30.5. *Integral cohomology of $B_{\mathbf{O}(n)}$ and $B_{\mathbf{SO}(n)}$.* In this section, we shall prove that the torsion elements of the cohomology of these classifying spaces are all of order 2; first we insert a remark to be used in the proof.

Let H be an anticommutative graded algebra with unit over a field K, with $H^0 = K$, and let D be an antiderivation on H, raising degrees by one, of square zero. Let A be a graded subspace stable under D. We denote by N_A, M_A, I_A, J_A, respectively, the kernel of D in A, a graded supplementary subspace to N_A in A, the image of A under D, a graded supplementary subspace to I_A in N_A. Since D increases degrees by one and is an isomorphism of M_A onto I_A, we have

$$(1) \qquad P(A, t) = (1 + 1/t) P(I_A, t) + P(J_A, t).$$

Let now B be a second graded subspace stable under D, linearly disjoint from A over K; i.e., such that the subspace $A \cdot B$ spanned by the products $a \cdot b$ ($a \in A$, $b \in B$) is isomorphic to $A \otimes B$ under the natural map $a \otimes b \to a \cdot b$. Using the previous notations, we have, as an elementary special case of the Künneth formula

$$(2) \qquad P(J_{A \cdot B}, t) = P(J_A, t) \cdot P(J_B, t).$$

PROPOSITION. *The torsion elements of $H^*(B_{\mathbf{O}(n)}, \mathbf{Z})$ and $H^*(B_{\mathbf{SO}(n)}, \mathbf{Z})$ are of order 2.*

It follows from 30.2 that these spaces have only 2-torsion. By 30.4, there remains to prove that for $G = \mathbf{SO}(n), \mathbf{O}(n)$, we have

$$(3) \qquad P_2(B_G, t) - P_0(B_G, t) = (1 + 1/t) P(Sq^1(H^*(B_G, \mathbf{Z}_2)), t).$$

We have

$$(4) \qquad \begin{aligned} &H^*(B_{\mathbf{SO}(n)}, \mathbf{Z}_2) = \mathbf{Z}_2[w_2, \cdots, w_n], \quad Sq^1 w_i = (i+1) w_{i+1} \\ &H^*(B_{\mathbf{O}(n)}, \mathbf{Z}_2) = \mathbf{Z}_2[w_1, \cdots, w_n], \quad Sq^1 w_i = w_1 w_i + (i+1) w_{i+1} \end{aligned}$$

($w_i = i$-th Stiefel-Whitney class, $i \leq n$, $w_{n+1} = 0$), (see e.g. [3]). We first consider $\mathbf{SO}(n)$. By the foregoing, we may write

$$H^*(B_{SO(2m+1)}, \mathbf{Z}_2) \cong A_1 \otimes \cdots \otimes A_m,$$

$$H^*(B_{SO(2m)}, \mathbf{Z}_2) \cong A_1 \otimes \cdots \otimes A_{m-1} \otimes A'_m,$$

where

$$A_i = \mathbf{Z}_2[w_{2i}, w_{2i+1}] \quad (1 \leq i \leq m), \quad A'_m = \mathbf{Z}_2[w_{2m}]$$

are stable under the cup product and Sq^1, and A'_m is annihilated by Sq^1. By (4), the kernel of Sq^1 in A_i is spanned by the elements

$$w_{2i}{}^s \cdot w_{2i+1}{}^t \qquad\qquad (s \geq 0, s \text{ even}, t \geq 0),$$

and its image by the elements

$$w_{2i}{}^s \cdot w_{2i+1}{}^t \qquad\qquad (s \text{ even}, t > 0).$$

Thus, in the notations above, we may take as base for M_{A_i} (respectively J_{A_i}), the monomials

$$w_{2i}{}^s \cdot w_{2i+1}{}^t, \quad (s \text{ odd}, t \geq 0), \quad (\text{respectively } w_{2i}{}^s, s = 0, 2, 4, 6, \cdots)$$

and we obtain

$$P(J_{A_i}, t) = (1 - t^{4i})^{-1}, \qquad\qquad (1 \leq i \leq m)$$

and, since $Sq^1(A'_m) = 0$

$$P(J_{A'_m}, t) = P(A'_m, t) = (1 - t^{2m})^{-1}.$$

By iterated application of (2) and by (1), we get

$$P_2(B_{SO(2m+1)}, t) = (1 + 1/t)P(Sq^1(H^*(B_{SO(2m+1)}, \mathbf{Z}_2)), t) + \prod_{i=1}^{m}(1 - t^{4i})^{-1},$$

$$P_2(B_{SO(2m)}, t) = (1 + 1/t)P(Sq^1(H^*(B_{SO(2m)}, \mathbf{Z}_2)), t) + (1 - t^{2m})^{-1}\prod_{i=1}^{m-1}(1 - t^{4i})^{-1},$$

which proves (3) for the unimodular case, since in both formulas, the last term on the right is the rational Poincaré series in view of 30.2.

We now pass to $\mathbf{O}(n)$. Choose a new basis of $H^*(B_{O(n)}, \mathbf{Z}_2)$ by

$$w^*_1 = w_1, w^*_{2i} = w_{2i}, \qquad\qquad (i \leq [n/2]),$$

$$w^*_{2i+1} = w_{2i+1} + w_i \cdot w_{2i}, \qquad\qquad (i < [n/2]);$$

we have

$$H^*(B_{O(n)}, Z_2) = \mathbf{Z}_2[w^*_1, \cdots, w^*_n],$$

$$Sq^1 w^*_1 = (w^*_1)^2; \; Sq^1 w^*_{2i} = w^*_{2i+1} \qquad\qquad (i < [n/2]),$$

$$Sq^1 w^*_{2m} = w^*_1 \cdot w^*_{2m}, \qquad\qquad (n = 2m),$$

$$Sq^1 w^*_{2i+1} = Sq^1 Sq^1 \cdot w^*_{2i} = 0, \qquad\qquad (i \geq 1).$$

We put

$$H^*(B_{O(2m+1)}, \mathbf{Z}_2) = A_0 \otimes A_1 \otimes \cdots \otimes A_m,$$

$$H^*(B_{O(2m)}, \mathbf{Z}_2) = A'_0 \otimes A_1 \otimes \cdots \otimes A_{m-1},$$

where

$$A_i = \mathbf{Z}_2[w^*_{2i}, w^*_{2i+1}] \qquad (1 \leqq i \leqq m),$$

$$A_0 = \mathbf{Z}_2[w^*_1] = \mathbf{Z}_2[w_1],$$

$$A'_0 = \mathbf{Z}_2[w^*_1, w^*_{2m}] = \mathbf{Z}_2[w_1, w_{2m}].$$

These are stable under cup-product and Sq^1, and we have as above

$$P(J_{A_i}, t) = (1 - t^{4i})^{-1}, \qquad (1 \leqq i \leqq m).$$

In A_0, the kernel of Sq^1 is spanned by w_1^{2s} $(s \geqq 0)$; and the image by w_1^{2s} $(s \geqq 1)$; hence $P(J_{A_0}, t) = 1$. Let us now prove that

$$(5) \qquad P(J_{A'_0}, t) = (1 - t^{2m})^{-1}.$$

We have

$$Sq^1(w_1^s \cdot w_{2m}^t) = (s + t) w_1^{s+1} w_{2m}^t$$

which shows that the monomials $w_1^s w_{2m}^t$ with $s + t$ even (respectively $s + t$ even and $s > 0$) span $N_{A'_0}$ (respectively $I_{A'_0}$). Thus we may take the elements w_{2m}^t (t even) as a base for $J_{A'_0}$, and this proves (5). The remainder of the proof of (3) for $\mathbf{O}(n)$ is then the same as for $\mathbf{SO}(n)$.

30.6. COROLLARY. *Let* $G = \mathbf{O}(n)$ *or* $\mathbf{SO}(n)$. *Then the kernel of* Sq^1 *on* $H^*(B_G, \mathbf{Z}_2)$ *is the reduction* $\mathrm{mod}\, 2$ *of* $H^*(B_G, \mathbf{Z})$ *and its image is the reduction* $\mathrm{mod}\, 2$ *of* $\mathrm{Tors}\, H^*(B_G, \mathbf{Z})$. *An element of* $H^*(B_G, \mathbf{Z})$ *is completely determined by its canonical images in* $H^*(B_G, \mathbf{R})$ *and* $H^*(B_G, \mathbf{Z}_2)$.

The first assertion follows from 30.4 and 30.5. The second one is an elementary fact about spaces with torsion elements of order 2 only.

In connection with the integral cohomology of $B_{O(n)}$ and $B_{SO(n)}$, let us also mention the following

30.7. PROPOSITION. *Let* $G = \mathbf{O}(n)$, $\mathbf{SO}(n)$ *and let* T *be a maximal torus of* G. *Then* $\rho^*_\mathbf{Z}(T, G)$ *maps* $H^*(B_G, \mathbf{Z})$ *onto* I_G; *its kernel is the torsion subgroup of* $H^*(B_G, \mathbf{Z})$.

We have seen in 9.3 that $S(y_1^2, \cdots, y_m^2)$, $(m = [n/2])$, is contained in $\rho^*_\mathbf{Z}(T, G)$, which proves our statement for $G = \mathbf{SO}(2m+1)$, $\mathbf{O}(2m+1)$, $\mathbf{O}(2m)$. For $G = \mathbf{SO}(2m)$, we have to know moreover that $\rho^*_\mathbf{Z}(T, \mathbf{SO}(2m))$ contains the product of the y_j's, but this follows from 9.5.

30.8. *Pontrjagin classes.* We follow the notations of 9.2, 9.3. By 30.6, the equalities

$$(6) \qquad \rho^*_0(T, G)(\tilde{p}) = \prod(1 + y_i^2), \qquad\qquad (G = \boldsymbol{O}(n), \boldsymbol{SO}(n)),$$

(7) p_i (respectively $p_{i+\frac{1}{2}}$) reduced mod 2, is equal to w_{2i}^2 (respectively w_{2i+1}^2),

completely characterize the integral Pontrjagin classes. (6), over the integers, follows from 9.1, 9.3, 9.4. It implies that $p_{i+\frac{1}{2}}$ is a torsion element. (7) needs only to be proved for $G = \boldsymbol{O}(n)$ and, in view of 30.2(b), will follow from

$$(8) \qquad \rho^*_2(\boldsymbol{Q}(n), \boldsymbol{O}(n))(\tilde{p}) = \prod(1 + u_j^2).$$

We have

$$\rho^*_2(\boldsymbol{Q}(n), \boldsymbol{O}(n))(\tilde{p}) = \rho^*_2(\boldsymbol{Q}(n), \boldsymbol{U}(n))(c) = \rho^*_2(\boldsymbol{Q}(n), \boldsymbol{T}')(\prod(1 + x_i))$$

where $\boldsymbol{T}'$ is the subgroup of diagonal matrices of $\boldsymbol{U}(n)$ and (x_i) is the standard basis of $H^2(B_{\boldsymbol{T}'}, \boldsymbol{Z})$, and therefore (8) follows from [4, §11]:

$$\rho^*_2(\boldsymbol{Q}(n), \boldsymbol{T}')(x_i) = u_i^2 \qquad\qquad (1 \leqq i \leqq n).$$

30.9. *Remark on integral Stiefel-Whitney classes.* It follows also from 9.5, 30.6, (6), (7) that for an $\boldsymbol{SO}(2m)$-bundle, we have

$$(9) \qquad\qquad\qquad p_m = W_{2m}^2,$$

both sides being considered as integral classes. The relations $w_{2i+1} = Sq^1 w_{2i}$ for $\boldsymbol{SO}(n)$-bundles and (30.6), show that the universal w_{2i+1} is the reduction mod 2 of a uniquely determined element

$$W_{2i+1} \in H^{2i+1}(B_{\boldsymbol{SO}(n)}, \boldsymbol{Z})$$

of order 2, the *integral $2i + 1$-th Stiefel-Whitney class*, and that we also have

$$p_{i+\frac{1}{2}} = (W_{2i+1})^2$$

over the integers.

30.10. *Pontrjagin classes and transgression.* As usual, $V_{n,k}$ denotes the Stiefel-manifold of orthonormal k-frames in euclidean n-space. We recall [2, §10] that for n odd, $H^j(V_{n,n-2i+1}, \boldsymbol{Z})$ is equal to $\boldsymbol{Z}$ for $j = 0, 4i - 1$, to $\boldsymbol{Z}_2$ for j even running from $2i$ to $4i - 2$, and is zero for the other values of j which are $\leqq 4i - 1$. We denote by v_i a generator of $H^{4i-1}(\boldsymbol{V}_{n,n-2i+1}, \boldsymbol{Z})$.

The first non-vanishing integral cohomology group of strictly positive dimension of the complex Stiefel manifold $\boldsymbol{W}_{n,n-2i+1}$ is of dimension $4i - 1$ and is infinite cyclic. We denote its canonical generator (see 29.1) by t_i.

The inclusion of $O(n)$ in $U(n)$ induces a natural injective map of $V_{n,k} = O(n)/O(n-k)$ into $W_{n,k} = U(n)/U(n-k)$.

LEMMA. *Let n be odd, and $\lambda_{n,i}$ be the natural inclusion of $V_{n,n-2i+1}$ in $W_{n,n-2i+1}$. Let v_i and t_i be defined as above. Then $\lambda^*_{n,i}(t_i) = \pm 2v_i$.*

This lemma will be proved in the next section. Let ξ be a principal $O(n)$-bundle, and ξ' be its complex extension. The given homomorphism of ξ into ξ' induces in a natural way a homomorphism $\alpha: \eta \to \eta'$ of the associated bundles with respective typical fibres $V_{n,n-2i+1}$ and $W_{n,n-2i+1}$. By the transgression definition $(29.3\,(7))$, we have $c_{2i} = -\tau_{\eta'}(t_i)$. Hence 29.4 and the lemma imply the

THEOREM. *Let n be odd and t_i be the canonical generator of $H^{4i-1}(W_{n,n-2i+1}, Z)$ and choose $v_i \in H^{4i-1}(V_{n,n-2i+1}, Z)$ such that $2v_i = \lambda^*_{n,i}(t_i)$. Let ξ be a principal $O(n)$-bundle, η the associated bundle with fibre $V_{n,n-2i+1}$. Then $p_i(\xi) = (-1)^{i+1}\tau_\eta(2v_i)$ modulo $L^{4i}\eta$.*

For the notation $L^{4i}\eta$, see 30.1. Since the cohomology groups of $V_{n,n-2i+1}$ in positive dimensions $< 4i-1$ are 2-groups, the spectral sequence definition of $L^{4i}\eta$ [2, § 5] shows that $L^{4i}\eta$ is a 2-group, so that the theorem characterizes p_i up to 2-torsion. We remark also that v_i itself is not universally transgressive (see following proof).

Let n be odd and β be the natural projection of E_ξ onto $E_\eta = E_\xi/O(2i-1)$. Then we have of course

$$p_i(\xi) = (-1)^{i+1}\tau_\xi(\beta^*(2v_i)) \quad \mathrm{mod}\ L^{4i}_\xi.$$

By [2, § 10], $\beta^*(v_i)$ generates a direct summand of $H^{4i-1}(O(n), Z)$ or of $H^{4i-1}(SO(n), Z)$.

$30.11.$ *Proof of the lemma.* We first consider the case where $n = 2i+1$ and denote by ξ the universal bundle for $O(2i+1)$, by ξ' its complex extension, by η and η' the associated bundles with fibres $V_{2i+1,2}$ and $W_{2i+1,2}$ and by α the natural map of η in η'. We have $H^*(V_{2i+1,2}, Z_2) = \wedge(h_{2i-1}, h_{2i})$, with $Sq^1 h_{2i-1} = h_{2i}$ and $\tau_\eta(h_{2i-1}) = w_{2i}$, $\tau_\eta(h_{2i}) = w_{2i+1}$ (see for instance [3]); this implies easily that $h_{2i-1} \cdot h_{2i}$ is not universally transgressive. Since $h_{2i-1} \cdot h_{2i}$ is the reduction mod 2 of v_i, the latter is not universally transgressive in integral cohomology. For $V_{2i+1,2}$, we have $H^0 = H^{4i-1} = Z$, $H^{2i} = Z_2$ (see e.g. [2], § 10), therefore, the non-zero terms in the spectral sequence of η over the integers have fibre degrees 0, $2i$, $4i-1$ and those with fibre degree $2i$ have order 2. Therefore $2t_i$ is universally transgressive and $L^{4i}\eta$ consists of elements of order 2.

We know now that $2t_i$ generates $T^{4i-1}(V_{2i+1,2}, \mathbf{Z})$; this group contains necessarily $\lambda^*_{2i+1,i}(v_i)$ since v_i is transgressive in η'. Thus we have $\lambda^*_{2i+1,i}(v_i) = 2k \cdot t_i$ for some integer k, hence also

$$p_i = \pm k\tau_\eta(2t_i) \quad \mathrm{mod}\, L^{4i}\eta.$$

Let $\mathbf{T}$ be the standard maximal torus of $\mathbf{O}(2i+1)$. Then, in the notation of 9.3, we get from (1) in 9.3 and from the fact that $L^{4i}\eta$ is a 2-group:

$$y_1^2 \cdots \cdots y_i^2 = \pm k\rho^*_{\mathbf{Z}}(\mathbf{T}, \mathbf{O}(2i+1))(\tau_\eta(2t_i)).$$

Since $H^*(B_\mathbf{T}, \mathbf{Z})$ is the ring of polynomials in the y_j's, this yields $k = \pm 1$, and the lemma for $n = 2i+1$.

In the general case, we consider the commutative diagram

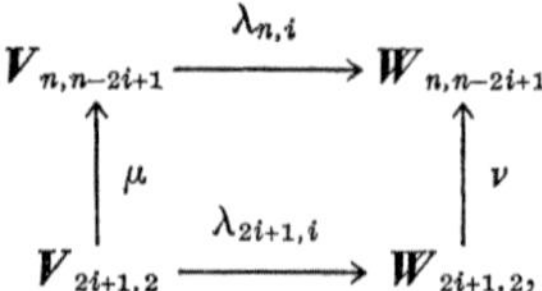

where μ, ν are the injection of a fibre in the standard fiberings. It follows from §§ 9, 10 in [2] that μ^*, ν^* are isomorphisms in dimension $4i-1$; therefore the general case of the lemma follows by commutativity of the above diagram from the particular case already proved.

30.12. *A property of $\rho^*_{\mathbf{Z}}(\mathbf{O}(r), \mathbf{O}(n))$.* We end this appendix by proving that $\rho^*_{\mathbf{Z}}(\mathbf{O}(r), \mathbf{O}(n))$ is surjective $(r \leqq n)$, a fact which is useful in the discussion of relative Pontrjagin classes (see M. Kervaire, Amer. J. Math. 79 (1957)).

LEMMA. *Let X, Y be two spaces with finitely generated integral cohomology groups and $f\colon X \to Y$ be a continuous map. We assume:*

(a) *The orders of the torsion elements of $H^*(X, \mathbf{Z})$ and $H^*(Y, \mathbf{Z})$ are square free.*

(b) *f^*_0 is surjective.*

(c) *For all primes p, f^*_p is a surjective map for the kernels of the Bockstein maps β_p (see 30.4).*

Then $f^*_\mathbf{Z}$ is surjective.

Let M_i be the image of $H^i(Y, \mathbf{Z})$ under f^*. By (b), it has a finite index, say g_i, in $H^i(X, \mathbf{Z})$. In view of (30.4), the assumptions (a), (c) imply that

$$f^*: H^i(Y, \mathbf{Z}) \otimes \mathbf{Z}_p \to H^i(X, \mathbf{Z}) \otimes \mathbf{Z}_p \qquad (i \geqq 0, p \text{ prime})$$

is surjective, or in other words, that

$$H^i(X, \mathbf{Z}) = p \cdot H^i(X, \mathbf{Z}) + M_i,$$

hence, by iteration,

$$H^i(X, \mathbf{Z}) = g_i \cdot H^i(X, \mathbf{Z}) + M_i = M_i.$$

PROPOSITION. *The homomorphism $\rho^*_{\mathbf{Z}}(\mathbf{O}(r), \mathbf{O}(n))$, $(r \leqq n)$, is surjective.*

By (30.5), $B_{\mathbf{O}(r)}$ and $B_{\mathbf{O}(n)}$ satisfy (a) of the lemma.

It follows from (9.3) and (30.2) that $H^*(B_{\mathbf{O}(n)}, \mathbf{Z}_p)$ is the ring of polynomials in the universal Pontrjagin classes $(i \geqq 1)$ for $p \neq 2$. Since these are preserved under the natural inclusion $\mathbf{O}(r) \subset \mathbf{O}(n)$ (see 9.7), it follows that $\rho^*_p(\mathbf{O}(r), \mathbf{O}(n))$ is surjective for $p \neq 2$. This shows that (b) is fulfilled and also, in view of (30.2), (30.5), that (c) is true for $p \neq 2$. Thus, in order to derive the proposition from the lemma, there remains to show that

(10) $\qquad \rho^*_2(\mathbf{O}(r), \mathbf{O}(n))$ *is surjective for the kernels of Sq^1.*

We write ρ^* for $\rho^*_2(\mathbf{O}(r), \mathbf{O}(n))$, denote by w_i (respectively $\bar{w}_i$) the universal Stiefel-Whitney classes for $\mathbf{O}(r)$- (respectively $\mathbf{O}(n)$-) bundles and define w^*_i, $\bar{w}^*_i$ as in the proof of (3).

The assertion (10) is clearly true on any subalgebra $A \subset H^*(B_{\mathbf{O}(n)}, \mathbf{Z}_2)$ which is stable under Sq^1 and mapped injectively by ρ^*; the latter is given by

$$\rho^*(\bar{w}_i) = w_i, \quad \rho^*(\bar{w}^*_i) = w^*_i, \quad (i \leqq r), \quad \rho^*(\bar{w}_j) = 0 \quad (j > r).$$

Therefore, by (4), this applies to

$$A = \mathbf{Z}_2[\bar{w}_1, \cdots, \bar{w}_i] \quad (i \leqq r, i \text{ odd}), \quad \mathbf{Z}_2[\bar{w}_1, \cdots, \bar{w}_{r-1}, \bar{w}_r^2] \quad (r \text{ even}),$$
$$\mathbf{Z}_2[\bar{w}_1^2, \bar{w}^*_2, \cdots, \bar{w}^*_{r-1}, \bar{w}_r^2] \quad (r \text{ even}).$$

This establishes (10) for r odd; for $r = 2m$ even, it reduces its proof to that of the following statement: given

$$x = w_{2m} \cdot P + Q, Sq^1 x = 0, (P, Q \in \mathbf{Z}_2[w_1, \cdots, w_{2m-1}, w_{2m}^2]),$$

there exists $y \in H^*(B_{\mathbf{O}(n)}, \mathbf{Z}_2)$ with the properties

            A. BOREL AND F. HIRZEBRUCH.

$$Sq^1 y = 0, \qquad \rho^*(y) = x.$$

$Sq^1 x = 0$ gives

$$w_{2m}(w_1 P + Sq^1 P) + Sq^1 Q = 0$$

and, since $\mathbf{Z}_2[w_1, \cdots, w_{2m-1}, w_{2m}{}^2]$ is stable under Sq^1,

$$w_1 P + Sq^1 P = Sq^1 Q = 0.$$

We may write

$$P = w_1 R + S, \qquad (R, S \in \mathbf{Z}_2[w_1{}^2, w^*_2, \cdots, w^*_{2m-1}, w_{2m}{}^2].$$

Hence

$$Sq^1 P = w_1{}^2 R + w_1 \cdot Sq^1 R + Sq^1 S,$$

$$0 = w_1 P + Sq^1 P = w_1(S + Sq^1 R) + Sq^1 S,$$

and, as before,

$$S + Sq^1 R = Sq^1 S = 0.$$

Now let $\bar{P}, \bar{Q}, \bar{R} \in H^*(B_{O(n)}, \mathbf{Z}_2)$ be the elements obtained from P, Q, R by barring the w_i's. Then a trivial computation shows that

$$y = \bar{w}_{2m} \cdot \bar{P} + \bar{Q} + \bar{w}_{2m+1}\, \bar{R}$$

has the desired properties.

The Institute for Advanced Study, Princeton
and
University of Bonn, Germany.

REFERENCES.

[1] A. Borel, " Le plan projectif des octaves et les sphères comme espaces homogènes," *Comptes Rendus des Séances de l'Académie des Sciences*, Paris, vol. 230 (1950), pp. 1378-1380.

[2] ———, " Sur la cohomologie des espaces fibrés principaux et des espaces homogènes de groupes de Lie compacts," *Annals of Mathematics*, vol. 57 (1953), pp. 115-207.

[3] ———, " La cohomologie mod 2 de certains espaces homogènes," *Commentarii Mathematici Helvetici*, vol. 27 (1953), pp. 165-191.

[4] ———, " Sur l'homologie et la cohomologie des groupes de Lie compacts connexes," *American Journal of Mathematics*, vol. 76 (1954), pp. 273-342.

[5] ———, " Kählerian coset spaces of semi-simple Lie groups," *Proceedings of the National Academy of Sciences of the United States of America*, vol. 40 (1954), pp. 1147-1151.

[6] —— and J.-P. Serre, "Groupes de Lie de puissances réduites de Steenrod," *American Journal of Mathematics*, vol. 75 (1953), pp. 409-448.

[7] —— and J. de Siebenthal, "Sur les sous-groupes fermés de rang maximum des groupes de Lie compacts connexes," *Commentarii Mathematici Helvetici*, vol. 23 (1949), pp. 200-221.

[7a] —— and A. Weil, "Représentations linéaires et espaces homogènes kähleriens des groupes de Lie compacts," *Séminaire Bourbaki*, May 1954 (Exposé by J.-P. Serre).

[7b] R. Bott, "Homogeneous vector bundles," *Annals of Mathematics*, vol. 66 (1957), pp. 203-248.

[8] E. Cartan, "La géométrie des groupes simples," *Annali di Matematica*, vol. 4 (1927), pp. 209-256, vol. 5 (1928), pp. 253-260.

[9] S. S. Chern, "Characteristic classes of hermitian manifolds," *Annals of Mathematics*, vol. 47 (1946), pp. 85-121.

[10] ——, "Topics in differential geometry," *Notes, Institute for Advanced Study*, Princeton, 1951.

[11] ——, "On the characteristic classes of complex sphere bundles and algebraic varieties," *American Journal of Mathematics*, vol. 75 (1953), pp. 565-597.

[12] —— F. Hirzebruch, J.-P. Serre, "On the index of a fibered manifold," *Proceedings of the American Mathematical Society*, vol. 8 (1957), pp. 587-596.

[13] E. B. Dynkin, "The structure of semi-simple Lie algebras," *Uspehi Matematiceskih Nauk (N. S.)*, vol. 2 (1947), pp. 59-127, translated in *American Mathematical Society Translations* 17, 1950.

[14] A. Frölicher, "Zur Differentialgeometrie der komplexen Strukturen," *Mathematische Annalen*, vol. 129 (1955), pp. 50-95.

[15] F. Hirzebruch, "On Steenrod's reduced powers, the index of inertia, and the Todd genus," *Proceedings of the National Academy of Sciences of the United States of America*, vol. 39 (1953), pp. 951-956.

[16] ——, "Todd arithmetic genus for almost complex manifolds," "On Steenrod's reduced powers in oriented manifolds," "The index of an oriented manifold and the Todd genus of an almost complex manifold," *Notes, Princeton University*, 1953.

[17] ——, "über die quaternionalen projektiven Raüme," *Bayerische Akademie der Wissenschaften* (1954), pp. 301-312.

[18] ——, "Some problems on differentiable and complex manifolds," *Annals of Mathematics*, vol. 60 (1954), pp. 213-236.

[19] ——, "Neue topologischen Methoden in der algebraischen Geometrie," *Ergebnisse der Mathematik und ihrer Grenzgebiete (N. F.)*, Heft 9, Berlin 1956.

[20] W. V. D. Hodge, "The characteristic classes on algebraic varieties," *Proceedings of the London Mathematical Society*, (3) vol. 1 (1951), pp. 138-151.

[21] H. Hopf und H. Samelson, "Ein Satz über die Wirkungsräume geschlossener Liescher Gruppen," *Commentarii Mathematici Helvetici*, vol. 13 (1940-41), pp. 240-251.

[22] K. Kodaira, "On a differential-geometric method in the theory of analytic stacks," *Proceedings of the National Academy of Sciences of the United States of America*, vol. 39 (1953), pp. 1268-1273.

[23] "Groupes et algèbres de Lie," *Séminaire S. Lie, Notes*, Paris, 1954.

[24] W. Rohlin, "New results in the theory of four-dimensional manifolds" (in Russian), *Doklady Akademii Nauk, S. S. S. R.*, vol. 84 (1952), pp. 221-224.

[25] J-P. Serre, "Groupes d'homotopie et classes de groupes abéliens," *Annals of Mathematics*, vol. 58 (1953), pp. 258-294.

[25a] J. de Siebenthal, "Sur les sous-groupes fermés connexes des groupes de Lie clos," *Commentarii Mathematici Helvetici*, vol. 25 (1951), pp. 210-256.

[26] N. Steenrod, *Topology of fibre bundles*, Princeton Mathematical Series, no. 14, 1951.

[27] E. Stiefel, "Ueber eine Beziehung zwischen geschlossenen Lie'schen Gruppen und . . .", *Commentarii Mathematici Helvetici*, vol. 14 (1941-42), pp. 350-380.

[28] ————, "Kristallographische Bestimmung der Charaktere der geschlossenen Lie'-schen Gruppen," *ibid.*, vol. 17 (1944-45), pp. 165-200.

[29] R. Thom, "Quelques propriétés globales des variétés différentiables," *ibid.*, vol. 28 (1954), pp. 17-86.

[30] H. Toda, "Quelques tables des groupes d'homotopie des groupes de Lie," *Comptes Rendus des Séances de l'Académie des Sciences*, Paris, vol. 241 (1955), pp. 922-923.

[31] H. C. Wang, "Closed manifolds with homogeneous complex structure," *American Journal of Mathematics*, vol. 76 (1954), pp. 1-32.

[32] H. Weyl, "Theorie der Darstellung kontinuierlicher halb-einfacher Gruppen durch lineare Transformationen I, II, III," *Mathematische Zeitschrift*, vol. 23 (1925), pp. 271-309, vol. 24 (1926), pp. 328-395.

46.

Fixed points of elementary commutative groups

Bull. Amer. Math. Soc. **65**(1959) 322–326

1   In this Note G is always a compact Lie group. A G-space is a topological space on which G acts continuously. We shall be mainly concerned with the case where G is an *elementary commutative p-group* (p prime or zero), that is, a direct product of a finite number of copies of the circle group T^1 if $p=0$ or of the cyclic group Z_p of order p if p is prime. In this study, a basic role is played by the space X_G, defined in §1. The proofs and additional results are given in the Notes of a seminar on transformation groups, to be published in the Annals of Mathematics Studies.

Although it is not always essential, we assume for simplicity that X is locally compact. $H_c^i(X;L)$ (resp. $H^i(X;L)$) is the ith cohomology group of X with compact (resp. closed) supports, coefficients in L. $\dim_L X$ (resp. $\dim_p X$) is the cohomological dimension [6] with respect to L (resp. a field K_p of characteristic p). The orbit space of X is denoted by X'; $\pi_X \colon X \to X'$ is the canonical projection, $G_x = \{g\in G,\, g\cdot x=x\}$ the stability group of X and $F(H; X)$ the fixed point set of a subgroup H of G.

1. The space X_G. Let X, Y be two G-spaces. The twisted product $X\times_G Y$ is the orbit space of $X\times Y$ under the "diagonal" action $g(x, y)=(g\cdot x, g\cdot y)$. The projections on the two factors induce maps π_1, π_2 of $X\times_G Y$ onto X' and Y' respectively. It is easily seen that given $x'\in X'$ and $x\in\pi_X^{-1}(x')$, the subspace $\pi_1^{-1}(x')$ may be identified with Y/G_x and similarly for π_2. Also, π_2 is a bundle map when Y is a principal G-bundle. We specialize Y to be a universal bundle E_G for G, hence $Y'=E_G/G$ is a classifying space B_G for G.

1.1. **LEMMA.** *Let $X_G=X\times_G E_G$. Then X_G has projections π_1, π_2 on X' and B_G respectively. π_2 is the projection in a fibre bundle with structural group G and typical fibre X. Let $x'\in X'$, then $\pi_1^{-1}(x')$ may be identified with B_{G_x} ($x\in\pi_X^{-1}(x')$). If G acts trivially on X, then $X_G=X\times B_G$. If $x\in F$, then $\pi_1^{-1}(x)=x\times B_G$ is a cross section for π_2.*

This space has occurred earlier in special cases (see [2; 7; 8] for instance). For discrete G, an algebraic analogue may be found in [11, Chapter V].

69

1.2. LEMMA. *Assume that* $\dim_L X$ *is finite and that* $H^*(B_{G_x}; L)$ *is trivial for* $x \notin F$. *Then the restriction map* $H_c^*(X_G; L) \to H_c^*(F \times B_G; L)$ *induced by the inclusion* $F_G \to X_G$ *is an isomorphism in degrees* $> \dim_L X$.

This applies in particular when $G = T^1$, $L = K_0$, or $G = L = Z_p$ (p prime) and yields easily the Smith theorem on homology spheres [13], its analogue for circle groups [8], and the dimensional parity theorems of Floyd [9] and Liao [12]. Also, dimensional parity holds when $G = Z_4$, $L = Z_2$ and G acts freely outside F.

2. **Fixed point theorems.** Elementary commutative p-group will be abbreviated by $[p]$-group. In this section, G is a $[p]$-group, X is a G-space which is compact, connected, of finite dimension over K_p, and the number of distinct stability groups of G on X is finite.

2.1. THEOREM. *Assume that* $\dim H^*(X; K_p)$ *is finite, and that* X *is totally nonhomologous to zero* $\bmod\ p$ *in* X_G. *Then* $\dim H^*(F; K_p) = \dim H^*(X; K_p)$. *In particular* $F \neq \phi$.

It is easy to see that if G is any p-group, we have $\dim H^*(X; K_p) \geq \dim H^*(F; K_p)$ provided G acts trivially on $H^*(X; K_p)$, so that the main point here is the reverse inequality. The latter is proved by a discussion of the Fary spectral sequence of π_1, for a suitable family of closed sets. A similar argument proves the:

2.2. THEOREM. *Assume that* X *is a cohomology* n-*sphere* $\bmod\ p$. *Let* G_i $(i \geq 1)$ *be the different subgroups of* G *which have index* p *if* $p \neq 0$, *or which are connected and of codimension 1 if* $p = 0$. *Let* n_i (*resp.* r) *be the integer such that the fixed point set of* G_i (*resp.* G) *is a cohomology* $n_i -$ (*resp.* $r -$) *sphere* $\bmod\ p$ *by the Smith theorem. Then* $n - r = \sum_i (n_i - r)$.

3. **Applications.** A compact connected manifold of dimension $2n$ is homologically Kählerian if there exists a class $Q \in H^2(X; K_0)$ such that the multiplication by Q^{n-s} is an isomorphism of $H^s(X; K_0)$ onto $H^{2n-s}(X; K_0)$ $(0 \leq s \leq n)$.

3.1. THEOREM. *Let* G *be a toral group acting on a compact connected homologically Kählerian manifold.* (a) *if* $F \neq \phi$, *then* $\dim H^*(F; K_0) = \dim H^*(X; K_0)$, (b) *if* $H'(X; K_0) = 0$, *then* $F \neq \phi$. (c) *If* X *has no torsion for some prime* p, *then* F *has no* p-*torsion.*

This follows from the above and Theorems II. 1.1, II. 1.2 of [1]. These results form a topological counterpart to the (more precise) ones obtained by Frankel [10] for toral groups of isometries on Kählerian manifolds.

3.2. Theorem. *Let K be a compact connected Lie group, U a closed subgroup, and p be a prime. Assume that $H^*(K/U; K_p)$ is equal to its characteristic algebra* [2, §18]. *Then every $[p]$-subgroup G of K is conjugate to a subgroup of U. In particular, if K or B_K has no p-torsion, then every $[p]$-subgroup of K lies in a torus of K.*

This is proved by showing that one can apply 2.1 to G operating by left translations on K/U or respectively on K/T, where T is a maximal torus of K. We mention that there is a converse to the second assertion of 3.2. For further results concerning p-torsion and $[p]$-subgroups of compact Lie groups, see a forthcoming Note of the author.

3.3. Theorem. *Let G be a $[p]$-group (p prime) acting on X. Assume that $H^*(X; K_p) = K_p[x]/(x^{s+1})$, (the degree k of x being even if $p \neq 2$). Then* $\dim H_c^*(F; K_p) = s+1$ *in each of the following cases*: (a) $(p, s+1)$ $=1$, (b) $k=4$, $p \neq s+1$, $p \neq 2$, (c) X *is the quaternionic projective space of real dimension $4s$ ($s \geq 1$) and $p \neq 2$.*

This follows from 2.1 and from the fact, that, in the cases (a), (b), (c), the space X is totally nonhomologous to zero mod p in any fibre bundle whose structural group acts trivially on $H^*(X; K_p)$. Examples of fixed point free $[p]$-groups on complex or quaternionic projective spaces or on the Cayley plane show that 3.3 probably cannot be improved significantly.

4. Cohomology manifolds. Given open subsets $U \subset V$ of X, we denote by j_{VU}^i the natural homomorphism $H_c^i(U; L) \to H_c^i(V; L)$ and by j_{VU}^* the sum of the j_{VU}^i. A space X is a *cohomology n-manifold* (an $n - cm$) over the principal ideal domain L of coefficients if it has finite dimension over L and if for each $x \in X$ there is an open neighborhood U_x of x and a free submodule of rank one A_x of $H_c^n(U_x; L)$ such that every $y \in U_x$ has a basic system of open neighborhoods V for which $A_x = \operatorname{Im} j_{U_x V}^*$. It is orientable if U_x may be taken as the connected component of x.

This definition corresponds to the locally orientable generalized manifold of Wilder [3; 4]. It is equivalent to that of Yang [14]. A $n - cm$ is cohomologically locally connected over L [3], and of dimension n over L. If it is orientable and connected, then $H_c^n(X; L) = L$. The Poincaré duality, proved in [3] when L is a field, is extended to the general case in [4]. In particular, if X is connected and orientable, there is an exact sequence

$$0 \to \operatorname{Ext}(H_c^{i+1}(X; L), L) \to H^{n-i}(X; L) \to \operatorname{Hom}(H_c^i(X; L), L) \to 0$$

where $H^i(X; L)$ denotes cohomology with closed supports.

4.1. Proposition. *Let X be a G-space which is a connected $n-cm$ over L. Assume that $H^*(B_{G_x}; L)$ is trivial for $x \notin F$. Let $U = U_0 \supset \cdots \supset U_n = V \supset \cdots \supset U_{2n}$ be a sequence of connected orientable invariant neighborhoods such that $j^i_{U_j U_j+1}$ is zero for $i \neq n$ and an isomorphism for $i = n$ ($0 \leq j \leq 2n-1$). Then $\operatorname{Im} j^i_{U \cap F, V \cap F} \cong H^{n-i}_c(B_G; H^n_c(U; L))$ and is a direct summand.*

Here is meant the cohomology of B_G with respect to the sheaf defined by the nth cohomology groups of the fibres in the fibering of U_G over B_G. This result is the main tool in proving the local theorems: F is a cohomology manifold over L in the cases $G = L = Z_p$ (Smith) or $G = T^1$, L a field or the integers (Conner-Floyd), and is orientable if X is. There are also dimensional parity theorems paralleling those mentioned at the end of §1, the first two of which were obtained first by Bredon [5] to which one may add the following: If $G = T^1$, $L = Z_2$, then the dimension over Z_2 of any component Y of the fixed point set X of the element of order 2 in G has the same parity as $\dim_2 X$ provided Y contains a point fixed under G. Finally, 2.2 has also a local analogue:

4.2. Theorem. *Let G be a $[p]$-group, X be a $n-cm$ over K_p and $x \in F$. The subgroups G_i being defined as in 2.2, let n_i (resp. r) be the dimension over K_p of the component of $F(G_i; X)$, (resp. $F(G; X)$) passing through x. Then $n - r = \sum_i (n_i - r)$.*

References

1. A. Blanchard, *Sur les variétés analytiques complexes*, Ann. Sci. Ecole Norm. Sup vol. 73, pp. 157–202.

2. A. Borel, *Sur la cohomologie des espaces fibrés principaux et des espaces homogènes de groupes de Lie compacts*, Ann. of Math. vol. 57 (1953) pp. 115–207.

3. ———, *The Poincaré duality in generalized manifolds*, Michigan Math. J. vol. 4 (1957) pp. 227–239.

4. A. Borel and J. C. Moore, *Homology in locally compact spaces and generalized manifolds* (to appear).

5. G. Bredon, *Orientation in generalized manifolds and applications to the theory of transformation groups*, to appear in Michigan Math. J.

6. H. Cohen, *A cohomological definition of dimension for locally compact Hausdorff spaces*, Duke Math. J. vol. 21 (1954) pp. 209–224.

7. P. E. Conner, *On the action of the circle group*, Michigan Math. J. vol. 4 (1957) pp. 241–247.

8. P. E. Conner and E. E. Floyd, *Orbit spaces of circle groups of transformations*, Ann. of Math. vol. 67 (1958) pp. 90–98.

9. E. E. Floyd, *On periodic maps and the Euler characteristic of associated spaces*, Trans. Amer. Math. Soc. vol. 72 (1952) pp. 138–147.

10. T. Frankel, *Fixed points and torsion on Kähler manifolds*, Ann. of Math. vol. 70 (1959) pp. 1–8.

11. A. Grothendieck, *Sur quelques points d'algèbre homologique*, Tôhoku Math. J. vol. 9 (1957) pp. 119–221.

12. S. D. Liao, *A theorem on periodic transformation of homology spheres*, Ann. of Math. vol. 39 (1938) pp. 68–83.

13. P. A. Smith, *Fixed points of periodic transformations*, Appendix B in Lefschetz, *Algebraic topology*, 1942.

14. C. T. Yang, *Transformation groups on a homological manifold*, Trans. Amer. Math. Soc. vol. 87 (1958) pp. 261–283.

INSTITUTE FOR ADVANCED STUDY

47.

(with F. Hirzebruch)

Characteristic classes and homogeneous spaces, III *

Amer. J. Math. **82** (1960) 491–504

This paper consists of three parts, related to each other only by the fact that they bring complements to [1].

In [1, §§ 25, 26], certain expressions ($\hat{A}$-genus, Chern characters of bundles over spheres, etc.) were proved to be integers "exc 2," that is, up to a power of two. This restriction came from the fact that the proofs relied heavily on the integrality "exc 2" of the Todd genus of an almost complex manifold proved in [5]. Since then Milnor [8, 12] has shown the Todd genus to be an integer. This fact will be used in § 3 to free our earlier results from the powers of two. For this, it will be necessary to generalize slightly the notion of almost complex manifold, and to introduce between vector bundles an equivalence relation (called here S-equivalence), in which the trivial bundles form one class. These preliminaries are dealt with in §§ 1, 2.

In [1, § 23.3], it was proved that the A-genus of a coset space G/U is zero when G and U are compact, connected, semi-simple, of the same rank. The proof made use of a lemma (23.4) stating that the sum of the positive roots of U is singular in G, which was proved essentially by case by case checking. § 4 brings an a priori proof of this lemma, in the framework of the theory of roots. When all roots of G have the same length, 23.4 is equivalent to a theorem of de Siebenthal [10] saying that the "main diagonal" of U is singular in G. We also give a general proof of this result, which is obtained in [10] by case by case checking.

Finally, § 5 gives two elementary sufficient conditions under which the Stiefel-Whitney class $w(M)$ or the Pontrjagin class $\bar{p}(M)$ (see [1, § 9.3]) of a compact manifold M reduces to 1, which are then applied to G/T.

The notation of [1] will be used freely.

1. S-classes of vector bundles.

1.1. *Notation.* L stands for the field either of real numbers $\boldsymbol{R}$, or of complex numbers $\mathfrak{C}$, or of quaternions $\boldsymbol{K}$. $\boldsymbol{GL}(n, L)$ (resp. $\boldsymbol{U}(n, L)$) is the

---

* Received June 16, 1959.

74

general linear group (resp. unitary group) in L^n. A bundle with typical fibre L^n and structural group $\boldsymbol{GL}(n, L)$ or $\boldsymbol{U}(n, L)$ is called an L-vector bundle.

1.2. DEFINITION. *Let X be a topological space. Two L-vector bundles ξ, η over X are said to be S-equivalent (suspension equivalent) if there exist trivial bundles α, β such that the Whitney sums $\xi \oplus \alpha$ and $\eta \oplus \beta$ are equivalent bundles in the usual sense. The S-equivalence class, or S-class, of ξ will be denoted by $[\xi]$, and $K'(X, L)$ will be the set of S-classes of L-vector bundles over X.*

Let ξ, η be two L-vector bundles. $[\xi] = [\eta]$ means that the associated principal bundles become equivalent after the standard extension of the structural group to $\boldsymbol{U}(N, L)$, for some N, or also that the associated unit sphere bundles become equivalent after iterated suspension of the fibres.

The Whitney sum is commutative, associative, and clearly compatible with S-equivalence. Therefore it defines in $K'(X, L)$ a commutative, associative, operation for which the S-class of the trivial bundle is a zero element.

Let $f \colon X \to Y$ be a continuous map. Then, if we associate to an L-vector bundle over Y the induced bundle on X, we define clearly a homomorphism of $K'(Y, L)$ into $K'(X, L)$.

1.3. PROPOSITION. *Let X be a locally compact, paracompact, finite dimensional space. Then $K'(X, L)$ is a commutative group (with respect to Whitney sum). The S-class of the trivial bundle is the zero element.*

There remains only to show the existence of the inverse. On the Grassmann manifold $\boldsymbol{U}(n + N, L)/\boldsymbol{U}(n, L) \times \boldsymbol{U}(N, L)$ there are two canonical L-vector bundles ξ, η with typical fibres L^n, L^N, whose sum is the trivial bundle. Hence $[\xi] + [\eta] = 0$. Since, by the classification theorem, any L-vector bundle with typical fibre L^n over X is induced from ξ by a map of X into the Grassmannian (for N suitably large), our assertion follows immediately.

1.4. The total Chern class of a complex vector bundle depends only on its S-class, as follows from the multiplication theorem [1, § 9.7]. It belongs to the set $\Gamma(X, \boldsymbol{Z})$ of elements of $H^*(X, \boldsymbol{Z})$ having a zero dimensional term equal to 1, and vanishing odd dimensional components. $\Gamma(X, \boldsymbol{Z})$ is a commutative group under the cup-product, and it is clear that assigning to each complex vector bundle its Chern class yields a group homomorphism of $K'(X, \boldsymbol{C})$ into $\Gamma(X, \boldsymbol{Z})$. An analogous remark can be made of course for the Pontrjagin, symplectic Pontrjagin and Stiefel-Whitney classes.

2. Weakly almost complex structures.

2.1. The standard inclusion of $GL(n, C)$ into $GL(2n, R)$ induces obviously a homomorphism of $K'(X, C)$ into $K'(X, R)$ to be denoted by λ. An S-class of real vector bundles is said to admit (to have) a complex structure if it belongs to the image of λ (and if an element of its inverse image has been chosen). A real vector bundle ξ admits (has) a *weak complex structure* if $[\xi]$ admits (has) a complex structure. Thus, a weak complex structure of a real vector bundle ξ is given by a trivial bundle α and a complex structure of $\xi \oplus \alpha$ in the usual sense $[1, \S 7.3]$. Finally, a manifold is *weakly almost complex* (admits a weak almost complex structure) if its tangent bundle has been endowed with (admits) a weak complex structure.

An orientation of a real vector bundle ξ with fibre R^q is a section of the associated bundle with $O(q)/SO(q)$ as fibre. Since

$$O(q)/SO(q) \to O(q+1)/SO(q+1)$$

is bijective, an orientation of ξ depends only on the S-class of ξ. Thus, a weak complex structure of ξ defines an orientation of ξ. In particular, *a weakly almost complex manifold is canonically oriented.*

2.2. *Chern classes.* The Chern class of a weakly almost complex manifold X is by definition the Chern class of the weak complex structure of its tangent bundle. If X is compact, of dimension $2n$, then $c_n[X]$ is not necessarily the Euler number. For instance, take for X the unit sphere in R^{2n+1}. The normal bundle and its Whitney sum with the tangent bundle are trivial. Therefore S_{2n} admits a weak almost complex structure defined by a trivial complex bundle. Then $c_n[S_{2n}] = 0$.

2.3. *Submanifolds of codimension 2.* Let X be a compact weakly almost complex manifold, ξ its real tangent bundle, and $[\xi']$ the complex structure of $[\xi]$. Let $d \in H^2(X, Z)$. According to Thom [11] there exists an oriented submanifold D of X, of codimension 2, whose normal bundle ν is a real vector bundle with structure group $SO(2)$ and characteristic class $i^*(d)$, where i is the embedding of D in X. Since $SO(2) = U(1)$, the bundle ν has a complex structure ν', whose total Chern class is $1 + i^*(d)$. Let δ be the real tangent bundle to D. Then we have in $K'(D, R)$ the equalities

$$[\delta] = [i^*\xi] - [\nu] = \lambda(i^*[\xi'] - [\nu']),$$

which show that $[\delta]$ admits a complex structure represented by a bundle δ' whose Chern class is

$$c(\delta') = i^*(c(\xi')) \cdot (1 + i^*(d))^{-1}.$$

This proves the following proposition.

2.4. PROPOSITION. *Let X be a compact weakly almost complex manifold and $c(X)$ be its total Chern class. Then every element $d \in H^2(X, \mathbf{Z})$ is representable by a submanifold D of codimension 2 which carries a weakly almost complex structure, whose total Chern class $c(D)$ is equal to $i^*(c(X) \cdot (1 + d)^{-1})$, where i is the embedding of D in X.*

2.5. Let X be a compact weakly almost complex manifold of even dimension, $d \in H^2(X, \mathbf{R})$, and η a complex vector bundle over X. The Todd genus $T(X)$, the virtual Todd genus $T(d)_X$ of d, and the number $T(X, \eta)$ are then defined in exactly the same way as for an almost complex manifold. If η is a complex line bundle, with first Chern class a, then $T(X, \eta) = T(X, a)$, where $T(X, a) = T(X) - T(-a)_X$. For all this, see [5, §§ 10-12]; the definitions given there were also recalled in [1, §§ 22.1, 25.1]. It follows then from 2.4 that for every element $d \in H^2(X, \mathbf{Z})$, *the virtual Todd genus $T(d)_X$ is equal to the Todd genus of some compact weakly almost complex manifold.*

2.6. Milnor [8] (see also [12]) has established a complex analogue of cobordism theory, and has proved that the *Todd genus of a weakly almost complex manifold is an integer.* This (and 2.5) yield the

PROPOSITION. *Let X be a compact weakly almost complex manifold. Then for every $d \in H^2(X, \mathbf{Z})$, the number $T(X, d)$ is an integer.*

3. Integrality theorems for differentiable manifolds.

For the definition of $\hat{A}(X, d)$ and $\hat{A}(X, d, \eta)$ we refer to [1, §§ 25.4, 25.5].

3.1. THEOREM. *Let X be a compact oriented differentiable manifold and d an element of $H^2(X, \mathbf{Z})$ whose restriction $\bmod 2$ is equal to $w_2(X)$. Then $\hat{A}(X, \tfrac{1}{2}d)$ is an integer.*

We use the notation of 25.4. Thus E/T is an almost complex manifold, $\pi : E/T \to X$ a fibre map, ξ the tangent bundle to X, and $x_1, \cdots, x_q \in H^2(E/T)$ are the roots of the Chern polynomial of a complex structure of $\pi^*(\xi)$. This implies that

$$\pi^*(w_2) = x_1 + \cdots + x_q \quad \bmod 2.$$

Furthermore, we have the equality

$$(1) \qquad \hat{A}(X, \tfrac{1}{2}d) = T(E/T, \tfrac{1}{2}(\pi^*(d) - (x_1 + \cdots + x_q))).$$

If now $d \equiv w_2 \bmod 2$, then $\pi^*(d) - (x_1 + \cdots + x_q) \equiv 0 \bmod 2$, therefore the real cohomology class $\frac{1}{2}(\pi^*(d) - (x_1 + \cdots + x_q))$ comes from an integral class under the coefficient homomorphism $\mathbf{Z} \to \mathbf{R}$. Theorem 3.1 follows then from (1) and 2.6.

3.2. COROLLARY. *Let X be a compact, oriented, differentiable manifold with vanishing second Stiefel-Whitney class. Then the genus $\hat{A}(X)$, belonging to the power series $\frac{1}{2}z^{\frac{1}{2}}/\sinh \frac{1}{2}z^{\frac{1}{2}}$, is an integer.*

In this case, we can replace d by 0 in 3.1. Since $\hat{A}(X, 0) = \hat{A}(X)$, the corollary follows.

3.3. *Examples.* The polynomial $\hat{A}_1$ is equal to $-p_1/24$. Thus, if $\dim X = 4$,

$$\hat{A}(X, d/2) = ((1 + d/2 + d^2/8)(1 - p_1/24))[X] = (d^2/8 - p_1/24)[X].$$

Since $p_1[X] = 3 \cdot \tau$, where τ is the index of X (see [5, § 0.7; 11, Cor. IV.13]), we get on a 4-dimensional oriented manifold the congruence

$$d^2[X] \equiv \tau \bmod 8 \, (d \in H^2(X, \mathbf{Z}), d \equiv w_2 \bmod 2).$$

This can also be formulated as a statement on the quadratic form of the manifold (F. Hirzebruch-H. Hopf, *Mathematische Annalen*, vol. 136 (1958), pp. 156-172).

Let now X be 6-dimensional. Then

$$\hat{A}(X, d/2) = ((1 + d/2 + d^2/8 + d^3/48)(1 - p_1/24))[X]$$

yields the congruence

$$d^3[X] \equiv (d \cdot p_1)[X] \bmod 48, \quad (d \in H^2(X, \mathbf{Z}), \ d \equiv w_2 \bmod 2).$$

3.4. The coefficient of p_k in $\hat{A}_k$ is $-B_k/((2k!) \cdot 2)$, where B_k is the k-th Bernouilli number [5, p. 13]. This can be easily deduced from [5, § 1]. In fact, by the usual expression of the Pontrjagin classes in terms of Chern classes [5, p. 12], $p_k = (-1)^k 2 \cdot c_{2k}$, modulo decomposable elements; since $A_k = 2^{4k} \hat{A}_k$, formula (12) of [5, p. 15] shows that the coefficient of p_k is $(-1)^k/2$ times the coefficient of c_{2k} in T_{2k}; but this coefficient is also the coefficient of $c_1{}^k$ in T_{2k} [5, Bemerkung 2, p. 15], and it follows readily from the first formula in [5, § 1.7, p. 15] that the latter is equal to $(-1)^{k-1}B_k/(2k)!$, whence our assertion. Together with 3.2, it implies:

3.5. THEOREM. *Let X be a compact oriented differentiable manifold* *of dimension $4k$ whose tangent bundle is trivial when restricted to the complement of some point in X. Then $B_k \cdot p_k[X]/((2k)! \cdot 2)$ is an integer.*

This theorem has found an interesting application to the stable homotopy groups of spheres [7]. (See also [6].)

3.6. **Theorem.** *Let X be a compact oriented differentiable manifold, η a complex vector bundle over X, and d an element of $H^2(X, \mathbf{Z})$ whose restriction* mod 2 *is equal to $w_2(X)$. Then $\hat{A}(X, \tfrac{1}{2}d, \eta)$ is an integer.*

We follow the notation of $[1, \S 25.5]$. Thus σ is the tangent bundle to X, E/T is the total space of a bundle ξ over X with fibre map π, and $a_j \in H^2(E/T, \mathbf{Z})$ $(1 \leq j \leq m)$ are cohomology classes whose sum a is the first Chern class of the complex vector bundle along the fibres. Therefore a, reduced mod 2, is the second Stiefel-Whitney class of the bundle along the fibres $\hat{\xi}$. Since the tangent bundle to E/T is the sum of $\pi^*(\sigma)$ and of $\hat{\xi}$, $[1, \S 7.6]$, we have

$$(2) \qquad\qquad \pi^*(d) + a \equiv w_2(E/T) \bmod 2.$$

In $[1, \S 25.5]$ it is proved that

$$\hat{A}(X, \tfrac{1}{2}d, \eta) = \sum_i \hat{A}(E/T, \tfrac{1}{2}(\pi^*(d) + 2x_i + a), \qquad (x_i \in H^2(X, \mathbf{Z})).$$

Therefore, 3.6 follows from 3.1 and (2).

3.7. *Applications.* Theorem 3.6 gives a positive answer to conjecture (1) in $[1, \S 25.6]$. Conjecture (2) ($\hat{A}(X)$ is *even* if $w_2(X) = 0$ and $\dim X \equiv 4 \bmod 8$) and also 3.6 have since been proved by a quite different method [6] which uses Bott's results [3] instead of Milnor's theorem. As a consequence, in 3.5, $B_k p_k[X]$ divided by $(2k)! \cdot 2$ is an even integer for odd k, which yields a slight sharpening of the Kervaire-Milnor theorem [7].

In $[1, \S 26.10]$ we mentioned the *theorem of Bott* [3] that the Chern class of a complex vector bundle over S_{2q} is divisible by $(q-1)!$, which had been proved in $[1, \S 25.8]$ only "exc 2." This and the corresponding divisibility property of Pontrjagin classes follow now from 3.6, in the same way as $[1, \S\S 25.8, 25.9]$ were derived from $[1, \S 25.5]$.

Let X be a compact almost complex manifold η a complex vector bundle over X. Then $T(X, \eta) = \hat{A}(X, \tfrac{1}{2}c_1, \eta)$. Therefore by 3.6, $T(X, \eta)$ *is an integer.* This gives an affirmative answer to the first question of Problem 22 in [4]. It follows also that all numbers introduced in connection with the Riemann-Roch theorem, which were proved to be integers "exc 2" in $[1, \S 25.6]$, are actually integers.

4. Some properties of roots of compact Lie groups. G will always be a compact connected semi-simple Lie group, T a maximal torus. No distinction is made between a point in the universal covering V of T and its image in T (or equivalently, between a point in the Lie algebra $\mathfrak{t}$ of T and its image in T under the exponential map) ; expressions like positive Weyl chamber, dominant root, simple roots are always understood with respect to some ordering. For the notation see [1, § 2].

4.1. If a, b are **roots, the** number $q(a,b) = 2\,(a,b)\,(b,b)^{-1}$ is an integer, and $0 \leqq q(a,b) \cdot q(b,a) \leqq 3$. **Consequences:** $q(a,b) = 0, 1, 2, 3$. If $(aa) < (bb)$ and $(ab) \neq 0$, then $q(ab) = \pm\,1$. If $q(a,b) = \pm\,1$, then $|\,q(a,b)| \leqq |\,q(b,a)|$, hence $(a,a) \leqq (b,b)$; if $(a,b) \neq 0$, and $(a,a) \geqq (b,b)$, then $(a,a)(b,b)^{-1} = 1$, $2, 3$. Let now G be simple. Then $W(G)$ is irreducible. Since the roots of a given length span a subspace invariant under $W(G)$, it follows that if G has a root of a certain length λ, then for any root a of G, there exists a root of length λ not orthogonal to a. Consequently, if the roots are normalized so that the minimal root length is one, then the other possible values are $2, 3$. More precisely, it is known that the values of (a,a) are $1, 3$ for $G = \boldsymbol{G}_2$, $1, 2$ for $G = \boldsymbol{B}_n, \boldsymbol{C}_n, \boldsymbol{F}_4$ and 1 in the other cases. We recall that if a, b are roots, then so is $c = a - q(a,b)\,b$, and clearly $(a,a) = (c,c)$.

4.2. *Let G be simple, a_i ($1 \leqq i \leqq l$) be a system of simple roots, and $d = d_1 a_1 + \cdots + d_l a_l$ be the highest root. Then d has maximal length.*

For completeness, we give a proof. There exists a positive root of maximal length $c = c_1 a_1 + \cdots + c_l a_l$ not orthogonal to d (4.1). If $c = d$, we are done, so assume $d \neq c$. Since d is dominant, $c + d$ is not a root, therefore $(c,d) > 0$, $q(c,d) = k \geqq 1$, and $c - kd$ is a root. The coefficient of a_i in $c - kd$ must then be smaller in absolute value than d_i, whence $k = 1$; but then $(c,c) \leqq (d,d)$ by 4.1 and d has maximal length.

4.3. Let G be simple, a_i ($1 \leqq i \leqq l$) be simple roots of G, $d = d_1 a_1 + \cdots + d_l a_l$ be the highest root, and U be a maximal connected semisimple subgroup containing T. Then there exists an index j such that d_j is prime, U is the centralizer of the point P_j defined by $d_j \cdot a_i(P_j) = \delta_{ij}$ ($1 \leqq i \leqq l$). The simple roots of U with respect to a suitable ordering are the a_i's ($i \neq j$) and $-d$. The roots of U are exactly the roots of G in which a_j has coefficient 0 or $\pm\, d_j$. They form a *closed* system (i. e. if a, b are 2 roots of U such that $a + b$ is a root of G, then $a + b$ is a root of U). If the center of G reduces to the identity, then P_j generates a cyclic group of order d_j which is the center of U. One obtains in this way all maximal connected semi-simple subgroups of maximal rank, up to inner automorphisms. For all this, see [2].

4.4. G being again semi-simple, let a_i $(1 \leqq i \leqq l)$ be its simple roots. Then the equations $a_1 = \cdots = a_l$ define a 1-dimensional subspace contained in the positive Weyl chamber, or also a 1-dimensional torus S in T, to be called the *main diagonal*. It belongs to a three dimensional simple group H, the *principal subgroup* of G in the sense of de Siebenthal, which is defined up to inner automorphisms by those conditions [10, § 13, Th. 2]. H is not contained in a proper subgroup of rank l [10, § 12, Th. 1].

4.5. Let $G = \boldsymbol{SU}(2)$, and Γ_n be the natural representation of degree n of G in the space of homogeneous forms of degree $n - 1$ in two variables. As is known, Γ_n is up to equivalence, the only irreducible representation of degree n of G. If n is odd, it is equivalent to a real representation and not faithful. For n even, Γ_n is faithful, equivalent to the complex conjugate representation but not to a real representation. This implies in particular: if Γ is a real representation of G whose restriction to a maximal torus does not contain the trivial representation, then Γ is faithful, breaks up in a sum of real irreducible representations each of which is complex equivalent to a sum $\Gamma_n + \Gamma_n$, n even, hence $\Gamma = \Delta + \Delta$, where Δ is a sum of representations Γ_n, n even.

4.6. THEOREM. *Let U be a proper connected semi-simple subgroup of G of maximal rank. Then the main diagonal of U is singular in G* [10, § 8, Théorème 7].

We may assume that the center of G is reduced to the identity. If G is a direct product $G_1 \times G_2$, then $U = U_1 \times U_2$, where $U_i = U \cap G$ is a subgroup of maximal rank of G_i (see e. g. [2]), and its main diagonal clearly projects onto the main diagonal of U_i $(i = 1, 2)$. Using this and induction, the proof of 4.6 is easily reduced to the case where G is simple, with center reduced to (e), and U is maximal connected. Assuming this from now on, we follow the notation of 4.3 admitting moreover the simple roots to be numbered in such a way that $j = 1$. Let c^* be the point of S defined by $a_i(c^*) = 1$ $(2 \leqq i \leqq l)$ and $d(c^*) = -1$. Then

$$(1) \qquad\qquad d_1 a_1(c^*) = -1 - d_2 - \cdots - d_l.$$

$a(c^*)$ is integral for all roots a of a system of simple roots of U, hence for all roots of U; therefore, c^* is an element of the center of U.

Assume now that, contrary to our assertion, S is regular. Then $\mathrm{Ad}_{\mathfrak{g}/\mathfrak{u}}S$ does not contain the trivial representation. Let H be the principal subgroup of U containing S. By 4.5, $H = \boldsymbol{SU}(2)$, $\mathrm{Ad}_{\mathfrak{g}/\mathfrak{u}}H$ is faithful, and is a sum of two equivalent representations. From this, and from standard facts about representations of the circle group, it follows that given a complementary root a, there exists a positive complementary root $b \neq \pm a$ such that $b(s) = \pm a(s)$ for all $s \in S$. In particular, taking $a = a_1$, there exists a positive root $b = b_1 a_1 + \cdots + b_l a_l$ not proportional to a_1, such that

$$(2) \qquad b_1 a_1(c^*) + b_2 + \cdots + b_l = \pm a_1(c^*)$$

$(b_i \geqq 0, \ (i \geqq 1), \ b_1 \geqq 1, \ (b_2, \cdots, b_l) \neq (0, \cdots, 0))$. Let z be the element $\neq e$ in the center of H. Its connected centralizer U_1 in G is $\neq G$, since G has no center, and contains H, T; the last assertion of 4.4 shows then that U_1 contains U, hence is equal to U, since the latter is maximal connected. Thus, by 4.3, $d_1 = 2$ and $b_1 = 1$. It follows that in the right hand side of (2) we must have the minus sign, and we get

$$(3) \qquad -2a_1(c^*) = b_2 + \cdots + b_l,$$

but this, together with $b_i \leqq d_i$, obviously contradicts (1). Therefore S is singular.

4.7. There exists therefore a positive complementary root $b = b_1 a_1 + \cdots + b_l a_l$ such that $b(s) = 0$ for all $s \in S$. In particular, $b_1 a_1(c^*) = -(b_2 + \cdots + b_l)$ is integral; since $0 < b_1 < d_1$ and d_1 is prime, this and (1) show that $a_1(c^*)$ is integral. We have proved:

CoROLLARY. *We keep the notation of 4.3 and assume d_j to be prime. Then $1 + d_1 + \cdots + d_l$ is divisible by d_j. If c is the linear form defined by $(a_i, c) = 1$ $(i \neq j)$, $(d, c) = -1$, then (a, c) is integral for all the roots a of G.*

Before stating our next theorem, we discuss some more properties of roots.

4.8. *Let G be simple, a_i $(1 \leqq i \leqq l)$ be the simple roots, and $c = c_1 a_1 + \cdots + c_l a_l$ be a root of G. Then $c_i(a_i, a_i) \cdot (c, c)^{-1}$ is an integer $(1 \leqq i \leqq l)$.*

It is known that if we perform an inversion with respect to a sphere of radius $2^{\frac{1}{2}}$ in V, then a system of roots is transformed into a system of roots (of a group G' which may or may not be isomorphic to G). Let $e \to \bar{e}$ be this transformation. Then $\bar{e} = 2e \cdot (e, e)^{-1}$, and in particular

$$\bar{c} = 2 \cdot (c, c)^{-1} \sum_i c_i (a_i, a_i) \cdot 2^{-1} \cdot \bar{a}_i,$$

$$\bar{c} = \sum_i c_i ((a_i, a_i) \cdot (c, c)^{-1}) \bar{a}_i.$$

This shows first that all roots in the new system are linear combinations with coefficients of the same sign of $\bar{a}_1, \cdots, \bar{a}_l$; hence $\bar{a}_i > 0$ defines a Weyl chamber for the new system, and the $\bar{a}_i$ are a simple system of roots. Therefore the coefficients of $\bar{c}$ are integers.

4.9. *Let G be simple, U be a maximal connected subgroup of maximal rank, and b be the sum of the positive roots of U. Then (b, a) is integral for all roots a of G, the minimal root length being assumed to be 1.*

Proof. Since $W(U) \subset W(G)$, it is enough to prove this for one particular ordering. Let us consider one, say $\mathscr{S}$, with respect to which the simple roots of U are, in the notation of 4.3, $-d$ and the a_i's $(i \neq j)$. We have then [1, § 3.1]

$$(4) \qquad\qquad (b, a_i) = (a_i, a_i) \ (i \neq j), \ (d, b) = -(d, d).$$

The minimal root length being assumed to be 1, these are all integers (4.1) and it is therefore sufficient to show that (b, a_j) is an integer. (4) yields

$$(5) \qquad\qquad d_j(b, a_j) = -(d, d) - \sum_{i \neq j} d_i(a_i, a_i),$$

hence $d_j(b, a_j)$ is an integer. By 4.1, (b, a_j) is at any rate a half integer, so that we are done if d_j is odd. If all scalar products (a, a) are equal to 1, our assertion follows from (4) and 4.7; there remains therefore the case where $d_j = 2$ and (4.1) there are two root lengths. By 4.8, $d_i(a_i, a_i)(d, d)^{-1}$ is integral and by 4.2, d has maximal length; thus if $(d, d) = 2$, each term on the right hand side of (5) is even, and (b, a_j) is integral. If now $(d, d) = 3$, then $G = \mathbf{G}_2$, $d = 3a_1 + 2a_2$, which implies $j = 2$, $(a_1, a_1) = 1$ and $d_2(b, a_2) = -6$.

4.10. *Let G be simple, and assume that there are two root lengths $s < t$. Then any root of length t is the sum of two roots of length s.*

Let a be a root of length t. Since $W(G)$ is irreducible, there exists at least one root b of length s, not orthogonal to a; then $c = b - q(b, a)a$ is a root of length s (4.1). Since $q(b, a) = \pm 1$ by 4.1, our assertion is proved.

4.11. **Theorem.** *Let U be a proper connected semi-simple subgroup of maximal rank of G and let b be the sum of the positive roots of U. Then b is singular in G.*

As in 4.6, it is first seen that it suffices to prove our assertion for some ordering, when G is simple and U is maximal connected.

Proof will be by contradiction. Assume that b is regular. Let then $\mathscr{d}$ be the ordering of the roots of G defined by $a > 0$ if and only if $(b, a) > 0$ [1, § 2.8]. On the roots of U, it coincides with the original ordering, with respect to which b had been defined, as follows from [1, § 3.1], hence b is also the sum of the roots of U which are positive for $\mathscr{d}$. Let us number the simple roots a_i $(1 \leq i \leq l)$ for $\mathscr{d}$ so that a_i is a root of U if and only if $i \leq j$. (A priori, it is conceivable that no a_i belongs to U, in which case we set $j = 0$.) The $l - j$ other simple roots of U will then be denoted by a_i' $(j + 1 \leq i \leq l)$. Of course $j \neq l$ since $U \neq G$. By [1, § 3.1]

$$(6) \qquad (b, a_i) = (a_i, a_i) \ \ (i \leq j), \qquad (b, a_i') = (a_i', a_i') \ \ (i \geq j + 1).$$

For a given a_i', there exist non negative integral c_j's such that $a_i' = c_1 a_1 + \cdots + c_l a_l$, therefore

$$(7) \qquad \begin{aligned} (a_i', a_i') = (b, a_i') &= c_1 (a_1, a_1) + \cdots + c_j (a_j, a_j) \\ &\quad + c_{j+1} (b, a_{j+1}) + \cdots + c_l (b, a_l). \end{aligned}$$

At least two c_i's are not zero; by the definition of $\mathscr{d}$ and 4.9 we have $(a_i', a_i') \geq 2$, hence a_i' has maximal length.

We want to prove now that $G \neq \boldsymbol{G}_2$. If G were equal to $\boldsymbol{G}_2$, then $a_i' = c_1 a_1 + c_2 a_2$ with $c_1 \cdot c_2 \neq 0$, $(a_i', a_i') = 3$, hence by 4.8 the coefficient of the root of length one would have to be a multiple of 3, but this contradicts (7) and the fact that all scalar products are integers ≥ 1.

Thus, $G \neq \boldsymbol{G}_2$, there are two root lengths, and $(a_i', a_i') = 2$. It also follows that two of the c_i's, say $c_{s(i)}, c_{t(i)}$ $(s(i) < t(i))$ are equal to 1, and the others to zero. Since a_i' is simple as a root of U, we must have $t(i) \geq j + 1$, and then also $s(i) \geq j + 1$ since the root system of U is closed (4.3); (4.8) implies then that $(a_k, a_k) = 2$ for $k = s(i), t(i)$. In particular, we see that all simple roots of G of length one belong to U.

Let now c be the first (with respect to $\mathscr{d}$) positive complementary root of length one. This exists in view of 4.10. In order to have a contradiction, it is enough to prove that $(b, c) \leq 0$ and this will follow if we show that

$$(8) \qquad (c, a_i) \leq 0 \qquad (i \leq j), \qquad (c, a_i') \leq 0 \qquad (i \geq j + 1).$$

By the above, c is not a simple root, therefore $c - q(c, a_i) a_i$, expressed as linear combination of the a_i $(1 \leq i \leq l)$, has some positive coefficient. If $(c, a_i') \neq 0$, then $q(c, a_i') = \pm 1$ by 4.1 and because of $(c, c) = 1$, $(a_i', a_i') = 2$; since

a_i' is the sum of two simple roots, it follows again that $c - q(c, a_i')a_i'$ has some positive coefficient. Therefore, the roots $c - q(c, a_i)a_i$ $(i \leqq j)$, and $c - q(c, a_i')a_i'$ $(i \geqq j + 1)$ are positive, and moreover complementary of length one since c is. By the choice of c, they must then be greater than c, in the sense of $\mathcal{S}$, and this implies (8).

5. The Stiefel-Whitney class of G/T.

5.1.   Let ξ be a differentiable bundle with connected fibres. Let $b \in B_\xi$ and $F = \pi_\xi^{-1}(b)$. The normal bundle to F in E_ξ is of course trivial, since it is induced by π_ξ from the tangent space of B_ξ at b. Therefore F is orientable if E_ξ is. Furthermore, the multiplication theorem $[1, \S 9.7]$ shows that $w(F)$ (resp. $\bar{p}(F)$) is the restriction to F of $w(E_\xi)$, (resp. $\bar{p}(E_\xi)$). In particular, it reduces to 1 if E_ξ is parallelizable. More precisely, if the S-class ($\S 1$) of the tangent bundle to E_ξ is zero, then the S-class of the tangent bundle to F is also zero. A similar observation is valid for Chern classes and S-equivalence class in a complex analytic (or almost complex) bundle.

5.2.   Assume now that ξ is a principal differentiable bundle. The bundle along the fibres $\hat{\xi}$ $[1, \S 7.4]$ is then parallelizable. Since the tangent bundle to E_ξ is the sum of $\hat{\xi}$ and of the bundle induced by π_ξ from the tangent bundle to B_ξ $[1, \S 7.6]$, its S-class will be zero if the S-class of the tangent bundle to B_ξ is zero. Furthermore, the multiplication theorem gives

$$\pi_\xi^*(w(B_\xi)) = w(E_\xi), \qquad \pi_\xi^*(\bar{p}(B_\xi)) = \bar{p}(E_\xi).$$

Hence if $w(B_\xi) = 1$ (resp. $\bar{p}(B_\xi) = 1$), then $w(E_\xi) = 1$ (resp. $\bar{p}(E_\xi) = 1$). If $w(E_\xi) = 1$ (resp. $\bar{p}(E_\xi) = 1$), and π_ξ^* is injective, then $w(B_\xi) = 1$ (resp. $\bar{p}(B_\xi) = 1$). (Coefficients in a field of characteristic two for the Stiefel-Whitney classes, arbitrary coefficients for the Pontrjagin classes.) Again a similar assertion is valid for Chern classes in the almost complex case.

5.3.   PROPOSITION. *Let G be a compact connected Lie group, S a toral subgroup. Then $w(G/S) = 1$ and $\bar{p}(G/S) = 1$.*

Let T be a maximal torus containing S. Then we have the principal fibering $(G/S, G/T, T/S)$, therefore (5.2) it is enough to prove 5.3 for $S = T$. For the Pontrjagin class, see $[1, \S 10.9]$. There remains to prove that $w(G/T) = 1$. Without loss of generality it may be assumed that G is semi-simple and simply connected. Let Q be the subgroup of elements of order two in T. Then $(G/Q, G/T, T/Q, \pi)$ is a principal fibering. The total space, being the quotient of a group by a finite subgroup, is parallelizable, therefore,

(5.2) it will be enough to show that π^* is injective in cohomology mod 2. Since G and G/T are simply connected, $\pi_1(G/Q) = Q$, and the map $\pi_1(T/Q) \to \pi_1(G/Q)$ defined by the inclusion i is surjective. It follows easily that $i^*: H^*(G/Q, \mathbf{Z}_2) \to H^*(T/Q, \mathbf{Z}_2)$ is an isomorphism in dimension 1. But T/Q is a torus, hence $H^*(T/Q, \mathbf{Z}_2)$ is generated by its element of degree $\leqq 1$; therefore i^* is surjective, the fibre is totally non homologous to zero in cohomology mod 2; as is well known, this implies that π^* is injective.

5.4. It can also be derived directly from 5.1, 5.2 that *the S-class* (§1) *of the tangent bundle to G/S (S toral subgroup of G) is zero.*

In view of 5.2 and of the existence of the principal fibering $(G/S, G/T, T/S)$ it is enough to prove 5.4 when $S = T$ is a maximal torus. Let $\mathfrak{g}$ be the Lie algebra of G, $\mathfrak{R}$ the set of regular elements, and let G operate on $\mathfrak{g}$ by the adjoint representation. Since the centralizer of a regular element $x \in \mathfrak{g}$ is the maximal torus containing the one-parameter subgroup spanned by x, the orbits of G in $\mathfrak{R}$ are homeomorphic to G/T. Moreover, it is classical, and easily checked, that these orbits are the fibres in a differentiable fibering of $\mathfrak{R}$. Since $\mathfrak{R}$ is parallelizable, as an open subset of $\mathfrak{g}$, our assertion follows from 5.1.

The nullity of $w_2(G/T)$ was noticed in [1, § 22.3] and, as remarked above, $\bar{p}(G/T) = 1$ was also proved in [1, § 10.9].

5.5. Without entering into details, let us mention a case containing the preceding one, in which 5.1 applies. Let G operate differentiably on a connected manifold M. Among the different stability groups $G_x = \{g \in G, g \cdot x = x\}$, let H be one of smallest dimension, which has the minimal number of connected components among stability groups of that dimension. Then the set of points whose stability group is conjugate to H is an open set in M, which is differentiably fibered by the orbits [9, pp. 221-222]. Those are homeomorphic to G/H, and are called the *main orbits*. 5.1 yields then the

PROPOSITION. *Let G be a compact Lie group acting on a connected manifold M, and let F be a main orbit. Then F is orientable if M is. $w(F)$ and $\bar{p}(F)$ are the restrictions to F of $w(M)$ and $\bar{p}(M)$. If the S-class of the tangent bundle to M is zero, then the S-class of the tangent bundle to F is zero. In particular, if G/H is homeomorphic to the main orbit of a linear representation then it is orientable, the S-class of its tangent bundle is zero, and $w(G/H) = 1$, $\bar{p}(G/H) = 1$.*

The proof given in 5.4 is the particular case of 5.5 corresponding to the adjoint representation, where the main orbits are homeomorphic to G/T.

INSTITUTE FOR ADVANCED STUDY,
UNIVERSITY OF BONN, GERMANY.

REFERENCES.

[1] A. Borel and F. Hirzebruch, "Characteristic classes and homogeneous spaces," Part I, *American Journal of Mathematics*, vol. 80 (1958), pp. 459-538, Part II, *ibid.*, vol. 81 (1959), pp. 315-382.

[2] —— and J. de Siebenthal, "Sur les sous-groupes fermés connexes des groupes de Lie clos," *Commentarii Mathematici Helvetici*, vol. 23 (1949), pp. 200-221.

[3] R. Bott, "The space of loops on a Lie group," *The Michigan Mathematical Journal*, vol. 5 (1958), pp. 35-61.

[4] F. Hirzebruch, "Some problems on differentiable and complex manifolds," *Annals of Mathematics*, vol. 60 (1954), pp. 213-236.

[5] ——, "Neue topologische Methoden in der algebraischen Geometrie," *Ergebnisse der Mathematik und ihrer Grenzgebiete (N. F.)* 9, Berlin, 1956.

[6] ——, "A Riemann-Roch theorem for differentiable manifolds," *Séminaire Bourbaki*, Exp. 177, Février 1959.

[7] M. Kervaire and J. Milnor, "Bernoulli numbers, homotopy groups and a theorem theorem of Rohlin," *Proceedings of the International Congres of Mathematicians*, Edinburgh, 1958 (to appear).

[8] J. Milnor, "On the cobordism ring, and a complex analogue," *American Journal of Mathematics*.

[9] D. Montgomery-L. Zippin, *Topological transformation groups*, New York, 1955.

[10] J. de Siebenthal, "Sur les sous-groupes fermés connexes des groupes de Lie clos," *Commentarii Mathematici Helvetici*, vol. 25 (1951), pp. 210-256.

[11] R. Thom, "Quelques propriétés globales des variétés différentiables," *ibid.*, vol. 28 (1954), pp. 17-85.

[12] ——, "Travaux de Milnor sur le cobordisme," *Séminaire Bourbaki*, Exp. 180, Février 1959.

48.

On the curvature tensor
of the hermitian symmetric manifolds

Ann. Math., (2) **71** (1960) 508–521

This note is a complement to a paper of E. Calabi and E. Vesentini [2]. In that work the authors attach to an irreducible bounded symmetric domain D a number $\gamma(D)$, depending on the curvature of D with respect to an invariant kählerian metric; this number is related to the vanishing of certain cohomology groups on the compact quotients of D by discrete groups of automorphisms. By definition $\gamma(D) = R/n\lambda_1$ where $n = \dim_{\mathbf{C}} D$, R is the (constant) scalar Riemannian curvature of D, and λ_1 the smallest eigenvalue of a certain linear operator Q, self-adjoint with respect to the metric, defined in terms of the curvature tensor. The trace of Q is equal to $R/2$. This number is computed in [2] for the four main classes of bounded symmetric domains, with the help of an explicit representation of the domain. Our purpose here is to discuss it from the point of view of Lie algebras. The computations to which we are thus led can be performed in all cases, but we shall confine ourselves to the two exceptional domains omitted in [2].

Certain facts will be proved *a priori*, one will be checked using the classification. The following statement will be true in all cases (2.6, 2.7):

The operator Q has at most two different eigenvalues. Let X be the compact hermitian symmetric manifold corresponding to D, and g be a suitable generator of the infinite cyclic group $H^2(X; \mathbf{Z})$. Then $\gamma(D) \cdot g$ is the first Chern class of X.

The first Chern class of X was discussed in [1a, §16]. $\gamma(D)$ has the values 12 and 18 in the two exceptional cases $\mathbf{E}_6/\mathbf{Spin}\,(10) \times \mathbf{T}^1$ and $\mathbf{E}_7/\mathbf{E}_6 \times \mathbf{T}^1$ [1]. It can also be seen that *the eigenspaces of Q are the only subspaces invariant under the isotropy group*; however we shall check this here only for the two exceptional cases (**3.3**), and use this to determine the eigenvalues of Q (**3.4**).

Q operates on the symmetric tensors of type (2,0). The curvature tensor also defines an operator S on the space of tensors of type (1,1). Its eigenvalues are determined in §4, where it is also shown (**4.3**) that the underlying space contains a subspace which, as a module over the Lie algebra

---

[1] We allow ourselves to use this notation, although in fact the isotropy group is in the first (second) case a quotient of the direct product written above by a cyclic group of order 4 (resp. 3) which is not contained in either factor.

$\mathfrak{k}$ of the isotropy group, is isomorphic to $\mathfrak{k}$.

1. Preliminaries

In this section we recall some known facts and fix the notation. We follow in general the notation of [1a]. For the proofs of the results mentioned without comment, see the references given in [1a, 2].

1.1. G will be a simple compact connected Lie group, with center reduced to $\{e\}$, and K the identity component of the fixed point set of an involutive automorphism of G. Let $\mathfrak{g}$, $\mathfrak{k}$ be the Lie algebras of G, K, $\mathfrak{g}_c$, $\mathfrak{k}_c$ their complexifications, and G_c be the complex Lie group with Lie algebra $\mathfrak{g}_c$ and center reduced to $\{e\}$. Thus G is a maximal compact subgroup of G_c and $\mathfrak{g}$ a compact real form of $\mathfrak{g}_c$.

Let $\mathfrak{p}$ be the orthogonal complement of $\mathfrak{k}$ in $\mathfrak{g}$ with respect to the Killing form and let $\mathfrak{g}_0 = \mathfrak{k} + \sqrt{-1}\,\mathfrak{p}$. It is a subalgebra over the real numbers $\mathbf{R}$ of $\mathfrak{g}_c$ and also a real form of $\mathfrak{g}_c$. The corresponding analytic subgroup G_0 of G_c (the latter being now considered as a Lie group over $\mathbf{R}$) is a simple non-compact Lie group, with center reduced to $\{e\}$, of which K is a maximal compact subgroup. The coset spaces G/K and G_0/K are Riemannian symmetric.

1.2. We shall always assume that *the center of K is not finite*. It is then one-dimensional. The quotient G/K (resp. G_0/K) is a *simply connected irreducible compact hermitian symmetric manifold* (resp. an *irreducible bounded symmetric domain* in the space of several complex variables [5,6]), and all such spaces are obtained in this way. Their classification is recalled in [2]; the spaces G/K (or G_0/K) are divided into four infinite classes and two exceptional spaces $\mathbf{E}_6/\mathbf{Spin}\,(10) \times \mathbf{T}^1$, $\mathbf{E}_7/\mathbf{E}_6 \times \mathbf{T}^1$; the latter ones, or their non-compact counterparts, are denoted E III and E VII in E. Cartan's enumeration of symmetric spaces [4]. The curvature tensors of G/K and G_0/K are opposite in sign so that, for our purpose, it will be enough to consider one of them. In order to fix the ideas and to be able to refer more readily to [1a] we shall deal mostly with G/K.

1.3. Let $\mathfrak{t}$ be a Cartan subalgebra of $\mathfrak{k}$ (i.e., the Lie algebra of a maximal torus of K). Under the assumption 1.2, it is also a Cartan subalgebra of $\mathfrak{g}$ or $\mathfrak{g}_0$, hence its complexification $\mathfrak{t}_c$ is a Cartan subalgebra of $\mathfrak{g}_c$. The scalar product of real valued linear forms a, b on $\mathfrak{t}$ with respect to the negative Killing form, is denoted (a, b), and $h_a \in \mathfrak{t}$ will be the contravariant representative of a. Let Σ be the system of roots of $\mathfrak{g}$ with respect to $\mathfrak{t}$, [1a, §1], and $\mathfrak{v}_a$ ($a \in \Sigma$) be the one-dimensional eigenspace of $\mathfrak{t}_c$ in $\mathfrak{g}_c$ corresponding to a. As is recalled in [1a] we have

$$[\mathfrak{v}_a, \mathfrak{v}_b] = \mathfrak{v}_{a+b} \qquad \text{if } a + b \text{ is a root} \qquad (a + b \neq 0)$$

(1)
$$[\mathfrak{v}_a, \mathfrak{v}_b] = 0 \qquad \text{if } a + b \text{ is not a root } (a + b \neq 0)$$

$$h_a \in [\mathfrak{v}_a, \mathfrak{v}_{-a}] \ .$$

Moreover, there are elements $e_a \in \mathfrak{v}_a \ (a \in \Sigma)$ such that:

(2)
$$[h, e_a] = 2\pi i \, a(h) e_a \qquad [e_a, e_{-a}] = (2\pi i)^{-1} h_a \ , \qquad (h \in \mathfrak{t}),$$

$\mathfrak{g}$ is spanned by $\mathfrak{t}$ and the vectors $e_a + e_{-a}$, $i(e_a - e_{-a})$, and

(3)
$$N_{a,b} = N_{b,c} = N_{c,a} \qquad (a + b + c = 0, \ a + b, \ b + c, \ c + a \neq 0)$$

$$N_{a,b} = -N_{-a,-b}$$

where $a, b, c \in \Sigma$ and where $N_{a,b} \ (a, b \in \Sigma, \ a + b \neq 0)$ is defined by

$$[e_a, e_b] = N_{a,b} e_{a+b} \ , \qquad (a, b \in \Sigma, \ a + b \neq 0).$$

Under the assumption **1.2**, we may choose an ordering of the roots and a numbering of the simple roots $a_1, \cdots, a_l \ (l = \dim_\mathbf{R} \mathfrak{t} = \operatorname{rank} G)$ such that the center $\mathfrak{c}$ of $\mathfrak{k}$ is given by $a_2 = \cdots = a_l = 0$. Let $\mu_i(a)$ denote the coefficient of $a_i \ (1 \leq i \leq l)$ in a linear form a on $\mathfrak{t}$, expressed as linear combination of the a_i's and $d(a) = \sum \mu_i(a)$. The roots of $\mathfrak{k}$ are then the roots of $\mathfrak{g}$ for which $\mu_1(a) = 0$; moreover we have $\mu_1(a) = \pm 1$ on the complementary roots (the roots of $\mathfrak{g}$ which are not roots of $\mathfrak{k}$). The set of positive complementary roots will be denoted by Ψ, and we put

$$\mathfrak{n}^+ = \sum_{a \in \Psi} \mathfrak{v}_a \qquad \mathfrak{n}^- = \sum_{-a \in \Psi} \mathfrak{v}_a \ .$$

The tangent space $(G/K)_0$ to G/K at K will be identified with $\mathfrak{p}$ in the usual manner. Then $(G/K)_0 \otimes \mathbf{C}$ is identified with $\mathfrak{n}^+ \oplus \mathfrak{n}^-$. Moreover $e_a \to e_{-a}$ is the complex conjugation of $\mathfrak{p} \otimes \mathbf{C}$ with respect to $\mathfrak{p}$ and $e_a \to e_a + e_{-a}$ defines an isomorphism of complex vector spaces of $\mathfrak{n}^+$ onto $\mathfrak{p}$ (the latter space being endowed with the complex structure stemming from the invariant complex structure of G/K). Accordingly we also write $e_{\bar{a}}$ for $e_{-a} \ (a \in \Psi)$, (see [1a, Chapter IV]), and take Ψ as set of coordinate indices.

1.4. LEMMA. *Let $a, b \in \Psi$. Then $a + b$ is not a root, hence $(a, b) \geq 0$, and the subspaces $\mathfrak{n}^+ = \sum_{c \in \Psi} \mathfrak{v}_c$ and $\mathfrak{n}^- = \sum_{-c \in \Psi} \mathfrak{v}_c$ are commutative subalgebras. The root a_1 has maximal length, hence the Cartan integer $q(a, a_1) = 2(a, a_1)(a_1, a_1)^{-1}$ is equal to 0 or 1 for $a \in \Psi$.*

The first part follows from the facts recalled above about $\mu_1(a)$ and a standard property of the roots [1a, §2].

The highest root d has maximal length [1c, §4.2] and $\mu_1(d)(a_1, a_1)(d, d)^{-1}$ is an integer [1c, §4.8]. Since $\mu_1(d) = 1$, the root a_1 has maximal length too. The last assertion follows then from the above and some known facts about the Cartan integers proved for instance, in [8, Exp. 13] and recalled in [1c, §4.1].

1.5. *The invariant metric.* Let ω_a $(a \in \Sigma)$ be the left invariant Maurer-Cartan form on G_c whose value at e is defined by

$$\omega_a(e_a) = 1 , \qquad \omega_a(e_b) = \omega_a(t) = 0 \qquad (a \neq b; t \in \mathfrak{t}_c)$$

and η_ν (ν a linear form on $\mathfrak{t}_c$), be the left-invariant Maurer-Cartan form which vanishes on the spaces $\mathfrak{v}_a$ $(a \in \Sigma)$ and induces ν on $\mathfrak{t}_c$. Let ϖ_1 be the first fundamental highest weight [1a, §3.4]. It is characterised by

$$(4) \qquad\qquad 2(\varpi_1, a_i) = \delta_{1i}(a_1, a_1) \qquad\qquad (i = 1, \cdots, l).$$

This and **1.4** imply that

$$(5) \qquad\qquad 2(\varpi_1, a) = (a_1, a_1) \qquad\qquad (a \in \Psi).$$

Let now $\sigma = -(2\pi(a_1, a_1))^{-1}\eta_{\varpi_1}$. It follows from (4), (5) and [1a, §14.6] that $ds^2 = \sum_{a \in \Psi} \omega_a \cdot \varpi_a$ (symmetric product) represents an invariant kählerian metric on G/K, whose imaginary part is $d\sigma$. In the sequel, we use this metric.

2. The curvature tensor and the operator Q

2.1. Let $x, y, z \in \mathfrak{p}$. It follows from the theory of symmetric spaces that the linear transformation $\Omega(x, y)$ of $\mathfrak{p}$ into itself assigned to x, y by the curvature tensor is given by

$$\Omega(x, y)(z) = \mathrm{ad}\,[y, x](z) = [[y, x], z]$$

(this result is due to E. Cartan; for a more modern treatment, see for instance [7]). In the kählerian case, using the conventions made in **1.3**, this gives for the components of the curvature tensor with respect to the base $(e_a)(a \in \Psi)$

$$R^b_{a\bar{c}d} = \Omega^b_a(e_{-c}, e_d) = [[e_d, e_{-c}], e_a]^b$$

where the superscript b means the coefficient of e_b in $[[e_d, e_{-c}], e_a]$. Since the e_a's form an orthonormal basis (**1.5**) we have

$$R^{bc}_{ad} = R^b_{a\bar{c}d} = [[e_d, e_{-c}], e_a]^b .$$

LEMMA. $[[e_d, e_{-c}], e_a]^b$ *is equal to* 0 *if* $a + d \neq b + c$, *to* (a, c) *if* $a = b$, $c = d$, *to* $N_{d,-c}N_{a,-b}$ *if* $a + d = b + c$, $a \neq b$.

If $c = d$, then

$$[[e_a, e_{-c}], e_a] = (2\pi i)^{-1}[h_c, e_a] = (a, c)e_a$$

which proves the first equality, in the case $c = d$, and the second one. Let now $d \neq c$. Then

$$[[e_a, e_{-c}], e_a] = N_{a,-c}[e_{a-c}, e_a] = N_{a-c}N_{a-c,a}e_{a-c,a} ;$$

this yields the first equality when $d \neq c$; if now $d - c + a = b$, then $N_{d-c,a} = N_{a,-b}$ by **1.3**(3), which completes the proof.

2.2. LEMMA. $R_{ad}^{bc} = R_{da}^{bc} = R_{ad}^{cb}$.

This fact is well-known in kählerian geometry but can also be checked directly here. If $a + d \neq b + c$, then all terms are zero, so assume $a + d = b + c$. By the Jacobi identity

$$[[e_{-c}, e_d], e_a] + [[e_a, e_{-c}], e_d] + [[e_d, e_a], e_{-c}] = 0 \qquad (a, b, c \in \Psi).$$

But $[e_d, e_a] = 0$ by **1.4** hence

$$[[e_a, e_{-c}], e_d] = [[e_a, e_{-c}], e_a]$$

which shows the symmetry in a, d. The second equality follows from **1.3** if $a = b$, from **2.1** if $a \neq b$.

2.3. Let S be the symmetrized tensor product of $(G/K)_0$ with itself. It may be identified with a quotient of $\mathfrak{n}^+ \otimes \mathfrak{n}^+$, and we write $x \cdot y$ for the image of $x \otimes y$ $(x, y \in \mathfrak{n}^+)$ in S. Thus $x \cdot y = y \cdot x$ and the (symmetric) products $e_a \cdot e_b$ form a basis of S when $\{a, b\}$ runs through the unordered pairs of elements of Ψ. In view of **2.2**,

$$(6) \qquad Q(e_b \cdot e_c) = \sum_{a, d \in \Psi} R_{bc}^{ad} e_a \cdot e_d$$

defines a linear transformation of S into itself; **2.1** and **1.3** (3) imply

$$(7) \qquad Q(e_b \cdot e_c) = \sum_{a+d=b+c}[[e_a, e_{-c}], e_a]^b e_a \cdot e_d \qquad (a, b, c, d \in \Psi)$$

as well as the following

LEMMA. *The coefficient of $e_b \cdot e_c$ in $Q(e_b \cdot e_c)$ is equal to $2(b, c)$ if $b \neq c$, to (c, c) if $b = c$. The coefficient of $e_a \cdot e_d$ in $Q(e_b \cdot e_c)$ is equal to zero if $a + d \neq b + c$, and to $2 \cdot N_{d,-c} \cdot N_{a,-b}$ if $a + d = b + c, a \neq b$. The matrix of Q with respect to the basis $(e_b \cdot e_c)$ is real symmetric.*

The operator Q of [2] is also defined by (6). Since the curvature tensors of G/K and of the corresponding bounded symmetric domain G_0/K (notation of **1.2**) are opposite in sign, we define, in analogy with [2] $\gamma(G/K)$ as the quotient of $2 \operatorname{Tr} Q/n$ by the maximal eigenvalue of Q. We have then $\gamma(G/K) = \gamma(G_0/K)$, where the right-hand side has the meaning of [2], recalled in the introduction.

2.4. PROPOSITION. *We keep the previous notation.*

(a) *Let $c \in \Psi$ be of maximal length. Then $e_c \cdot e_c$ is an eigenvector of Q, with eigenvalue (c, c).*

(b) *We have $\operatorname{Tr} Q = (s, s) = n(s, a_1) = (n/2)(m(G/K) + 2)(a_1, a_1)$ where*

$s = \sum_{a \in \Psi} a$, $n = \dim_{\mathbf{C}} G/K$, and $m(G/K)$ is the number of roots $a \in \Psi$, $a \neq a_1$, for which $a - a_1$ is a root of G.

As is known [8, Exp. 10], given two roots b, c, the integers r, such that $b + rc$ is a root, form an interval $[p, q]$, $(p \leq 0 \leq q]$ and $p + q = -q(b, c)$. Assume now b, $c \in \Psi$ and c of maximal length. Then $q = 0$ and $q(b, c) = 0,1$ by **1.4**. Thus either $(b, c) = 0$, $p = 0$ or $(b, c) \neq 0$ and $p = -1$. Now, if $e_c \cdot e_c$ were not an eigenvector, there should exist a, $d \in \Psi$ such that $a + d = 2c$ by (7), and then $a - 2c$ would be a root, contradicting what has just been said.

The equality $\operatorname{Tr} Q = (s, s)$ follows from **2.3**. The sum s is orthogonal to the simple roots $a_2, \cdots, a_l$ of $\mathfrak{k}$ (see the beginning of the proof of 14.8 in [1a]) and $\mu_1(a) = 1$ for $a \in \Psi$ **(1.2)**, therefore $(s, s) = n(s, a_1)$. Finally, by definition,

$$2(s, a_1) = 2(a_1, a_1) + \sum_{a \in \Psi, a \neq a_1} q(a, a_1) \cdot (a_1, a_1)$$

and the last equality follows from **1.4**.

2.5. PROPOSITION. *In the two exceptional cases* $\mathbf{E}_6/\mathbf{Spin}\,(10) \times \mathbf{T}^1$ *and* $\mathbf{E}_7/\mathbf{E}_6 \times \mathbf{T}^1$ *the number* $m(G/K)$ *defined in* **2.4**(b) *has the values* 10 *and* 16 *respectively.*

1st *case*: The positive roots[2] of $\mathbf{E}_6$ are

$$x_i - x_j \quad (i < j), \qquad x_i + x_j + x_k \quad (i < j < k), \qquad \sum_1^6 x_i ,$$

where $1 \leq i, j, k \leq 6$, and the simple roots are

$$a_i = x_i - x_{i+1} \quad (i = 1, \cdots, 5), \qquad a_6 = x_4 + x_5 + x_6 .$$

The complementary roots are

$$x_1 - x_i \quad (i = 2, \cdots, 6), \qquad x_1 + x_i + x_j \quad (2 \leq i < j \leq 6), \qquad \sum_1^6 x_i ,$$

and those whose difference with a_1 is a root are $x_1 - x_i$ $(i \geq 3)$ and $x_1 + x_i + x_j$ $(3 \leq i < j \leq 6)$.

2nd *case*: The positive roots[2] of $\mathbf{E}_7$ are

$$
\begin{aligned}
x_i - x_j \qquad & (1 \leq i < j \leq 7) \\
x_i + x_j + x_k \qquad & (1 \leq i < j < k \leq 7) \\
d - x_i \qquad & (d = \textstyle\sum_1^7 x_j \,;\; i = 1, \cdots, 7)\,;
\end{aligned}
$$

the simple roots are

$$a_i = x_i - x_{i+1} \quad (1 \leq i \leq 6), \qquad a_7 = x_5 + x_6 + x_7 .$$

The positive complementary roots are

---

[2] As given in E. Cartan's Thesis (Oeuvres Complètes, Part I, Vol. 1, pp. 137-286, see p. 217). The numbering of the simple roots agrees with the conventions made in 1.3.

$$x_1 - x_i \qquad (i \geq 2)$$
$$x_1 + x_i + x_j \qquad (2 \leq i < j \leq 7)$$
$$d - x_i \qquad (2 \leq i \leq 7) \ ;$$

those whose difference with a_1 is a root are $x_1 - x_i$ $(i \geq 3)$, $x_1 + x_i + x_j$ $(3 \leq i < j \leq 7)$ and $d - x_2$.

2.6. Lemma. *The operator Q has at most two eigenvalues.*

For the four main classes, this was shown in [2]. For the two exceptional cases, it will be checked in the next section (see **3.2**, **3.3**).

2.7. Theorem. *The maximal eigenvalue of Q is equal to (a_1, a_1). Consequently $\gamma(G/K)$ is equal to $2(s, a_1)(a_1, a_1)$, or also to $m(G/K) + 2$ in the notation of **2.4**. In particular $\gamma(\mathbf{E}_6/\mathbf{Spin}\,(10) \times \mathbf{T}^1) = 12$ and $\gamma(\mathbf{E}_7/\mathbf{E}_6 \times \mathbf{T}^1) = 18$.*

By **1.4** and **2.4**(a), the number (a_1, a_1) is an eigenvalue of Q. On the other hand, since $m(G/K) + 2 \leq n + 1$ by definition, **2.4**(b) shows that $\mathrm{Tr}\,Q \leq n(n + 1)(a_1, a_1)/2$. Since $n(n + 1)/2$ is the dimension of the space operated upon by Q, the other eigenvalue of Q is $\leq (a_1, a_1)$. The other assertions are consequences of **2.4** and **2.5**. Also, the relation between $\gamma(G/K)$ and the first Chern class of G/K mentioned in the introduction follows from the theorem and [1a, §16].

3. A representation of the isotropy group

3.1. Let K' be the semi-simple part of K. The Lie algebra $\mathfrak{k}$ is direct sum of its center $\mathfrak{c}$ and of the Lie algebra $\mathfrak{k}'$ of $\mathfrak{k}$, and $\mathfrak{t}$ is the direct sum of $\mathfrak{c}$ and of a Cartan subalgebra $\mathfrak{t}'$ of $\mathfrak{t}$. The restriction to $\mathfrak{t}'$ of a linear form a on $\mathfrak{t}$ will be denoted a'. Thus $(a_2', \cdots, a_l')$ is a system of simple roots for $\mathfrak{k}'$. Let Φ be the corresponding system of positive roots of $\mathfrak{k}'$ and $a_0' = \frac{1}{2}\sum_{a' \in \Phi} a'$. By H. Weyl's formula, the degree $d(\Gamma)$ of an irreducible representation of $\mathfrak{k}'$, with highest weight b', is given by

$$(8) \qquad d(\Gamma) = \prod_{a' \in \Phi}(b' + a_0', a')(a_0', a')^{-1} \ .$$

It is clearly enough to extend the product to the roots which are not orthogonal to b'. Moreover (see e.g., [1a, 3.1]), we have $2(a_0', a_i') = (a_i', a_i')$ $(i = 2, \cdots, l)$. Therefore, *if all the roots of $\mathfrak{k}'$ have the same length*, we have

$$2(a_0', a') = (a_2', a_2')d(a') \qquad (a' \in \Phi)$$

(notation of **1.3**), and (8) may be written

$$(9) \qquad d(\Gamma) = \prod_{a'}(2(b', a')(a_2', a_2')^{-1} + d(a'))\cdot d(a')^{-1} \ ,$$

where a' runs through the elements of Φ which are not orthogonal to b'.

3.2. Let Γ_1 be the restriction to K' of the isotropy representation. Its weights are the restriction to t' of the elements of Ψ. Let π_1 be the highest one. Γ_1 defines in a natural way a representation Γ_2 of K' in the symmetric square S of $\mathfrak{n}^+$, whose weights are the sums $a' + b'$ ($a, b \in \Psi$), the highest one being $2\pi_1$; the vector $e_a \cdot e_b$ belongs to $a' + b'$. Since the curvature tensor is invariant under Γ_1, it is clear that Γ_2 commutes with the operator Q introduced in §2. Therefore the eigenspaces of Q are stable under Γ_2 so that, in the two exceptional cases, **2.6** is a consequences of the following:

3.3. LEMMA. *Let G/K be one of the two exceptional compact hermitian symmetric manifolds, and π_1 be the highest weight of the restriction to the semi-simple part K' of K of the isotropy representation. Then the representation Γ_2 of K' induced by Γ_1 in the symmetric square of the tangent space at K to G/K is the sum of the irreducible representation Δ with highest weight $2\pi_1$ and of one other irreducible representation. Moreover the space of Δ belongs to the eigenvalue (a_1, a_1) of Q.*

(In fact, this lemma can be checked in all cases, but we do not know an *a priori* proof for it.)

In both cases all roots of K' or even of G have the same length, so that we may apply (9); moreover, $e_{\pi_1} \cdot e_{\pi_1}$ which belongs to the highest weight $2\pi_1$ of Γ_2, is in the space of Δ and, since all roots of G have the same length, is also by **2.4** an eigenvector of Q, with eigenvalue (a_1, a_1). This proves the last assertion of the lemma. The remaining part will be checked for the two cases separately.

First case. $G = \mathbf{E}_6$, $K' = \mathbf{Spin}\,(10)$, $\dim_{\mathbf{C}} G/K = 16$, (type E III of [4]). Using a standard notation, we write the positive roots of K' as $x_i - x_j$ ($5 \geq i > j \geq 1$) and take as scalar product the unit form. The simple roots are

$$a'_{i-1} = x_i - x_{i-1} \quad (2 \leq i \leq 5), \qquad a'_5 = x_2 + x_1 \,.$$

The linear isotropy representation is the half-spinor representation with weights

$$\frac{1}{2}(\varepsilon_1 x_1 + \cdots + \varepsilon_5 x_5) \,, \qquad \prod \varepsilon_i = -1$$

[4, p. 131], [3, p. 38], hence

$$\pi_1 = \frac{1}{2}(x_5 + x_4 + x_3 + x_2 - x_1)$$

and the weights of Γ_2 are

$$\varepsilon_1 x_1 + \varepsilon_2 x_2 + \cdots + \varepsilon_5 x_5 \quad (\textstyle\prod \varepsilon_i = -1) \qquad \text{with multiplicity 1}$$
$$(10) \quad \varepsilon_1 x_i + \varepsilon_2 x_j + \varepsilon_3 x_k \quad (i < j < k;\ \varepsilon_s = \pm 1,\ s = 1, 2, 3) \quad \text{with multiplicity 1}$$
$$\pm x_i \qquad (1 \leq i \leq 5) \qquad \text{with multiplicity 4.}$$

π_1 is "dual" to the simple root a'_5 and the latter occurs with the coefficient $+1$ in every positive root which is not orthogonal to π_1. Therefore

$$(2\pi_1, a') = (a'_5, a'_5)$$

for every positive root not orthogonal to π_1. Those are the roots

$$x_i + x_j \quad (5 \geq i > j \geq 2) , \qquad x_i - x_1 \quad (5 \geq i \geq 2) .$$

Expressing them in terms of the simple roots, we get

$$d(x_i + x_j) = i + j - 2 , \qquad d(x_i - x_1) = i - 1 .$$

The formula (9) implies then, by a straightforward computation, that $d(\Delta) = 126$. The list of weights (10) shows then that Γ_2 is the sum of Δ and of the representation as orthogonal group in 10 real variables.

Second case. $G = \mathbf{E}_7$, $K' = \mathbf{E}_6$, $\dim_{\mathbf{C}} G/K = 27$ (type E VII of [4]). The positive roots of $\mathbf{E}_6$ are

$$x_i - x_j \quad (i < j) , \qquad x_i + x_j + x_k \quad (i < j < k) , \qquad \textstyle\sum_1^6 x_i ,$$

where $1 \leq i, j, k \leq 6$, and the simple roots are

$$a'_i = x_i - x_{i+1} \quad (i = 1, \cdots, 5) , \qquad a'_6 = x_4 + x_5 + x_6 .$$

The isotropy representation Γ_1 is the fundamental representation g_1 of $\mathbf{E}_6$ [4, p. 131], whose highest weight is defined by

$$2(\pi_1, a'_i) = \delta_{1i}(a'_1, a'_1) \qquad (1 \leq i \leq 6) .$$

The root a'_1 occurs with the coefficient 1 in the highest root

$$d = x_1 + \cdots + x_6 = a'_1 + 2a'_2 + 3a'_3 + 2a'_4 + a'_5 + 2a'_6$$

therefore it has also the coefficient $+1$ in every positive root which is not orthogonal to π_1; hence

$$2(\pi_1, a') = (a'_1, a'_1) , \qquad (a' > 0 ,\ (a', \pi_1) \neq 0) .$$

It follows then by (9) that

$$d(\Delta) = \textstyle\prod_{a'}(2 + d(a')) \cdot d(a')^{-1}$$

where a' runs through the roots having a_1 with coefficient $+1$. These are readily seen to be

$$x_1 - x_i , \quad x_1 + x_i + x_j \quad (1 \leq i < j \leq 6) , \qquad x_1 + \cdots + x_6 .$$

A simple computation yields then $d(\Delta) = 351$. The degree of Γ_2 is

$27 \cdot 28/2 = 378$. Therefore Γ_2 is the sum of Δ and of a 27-dimensional representation. Now the weights of g_1 are

$$-x_i + (2/3)d , \quad -x_i - (d/3) , \quad x_i + x_j - (d/3) \qquad (1 \leqq i < j \leqq 6)$$

with $d = x_1 + \cdots + x_6$ [3, p. 39]. The weights of Γ_2 are the sums of those of Γ_1. Since 0 does not occur, and 27 is the smallest degree of a non-trivial irreducible representation of $\mathbf{E}_6$, Γ_2 must be the sum of Δ and one other irreducible representation. In fact, the weights show that the second component is the contragredient representation to g_1, (g_3 in the notation of [3, p. 40]).

3.4. PROPOSITION. *In the notation of §2, the eigenvalues of Q are (a_1, a_1) with multiplicity* 126, $-3(a_1, a_1)$ *with multiplicity* 10 *in the case $G/K = \mathbf{E}_6/\mathbf{Spin}\,(10) \times \mathbf{T}^1$, and (a_1, a_1) with multiplicity* 351, $-4(a_1, a_1)$ *with multiplicity* 27 *in the case $G/K = \mathbf{E}_7/\mathbf{E}_6 \times \mathbf{T}^1$.*

In the first case, (a_1, a_1) has at least multiplicity 126 by **3.3**. Since $\mathrm{Tr}\, Q = 96(a_1, a_1)$ by **2.4**, **2.5**, the second eigenvalue λ_2 is determined by

$$126(a_1, a_1) + 10\lambda_2 = 96(a_1, a_1)$$

whence the result. Similar proof in the second case.

As was noticed in **2.3**, the eigenvalues of Q for the bounded symmetric domain G_0/K corresponding to G/K are opposite in sign to those of Q for G/K. If moreover we norm the scalar product by requiring that the maximal root length (a_1, a_1) be equal to one, then the eigenvalues of Q for the first (resp. second) exceptional bounded symmetric domain are -1 with multiplicity 126, 3 with multiplicity 10 (resp. -1 with multiplicity 351, 4 with multiplicity 27) as was stated in [2].

4. The tensors of type $(1, 1)$

In [2] mention is made of the operator defined by the curvature on the forms of type $(1, 1)$. Although its eigenvalues are not needed in [2], we shall also discuss them here, since this can be done very conveniently in the framework of the previous sections.

4.1. The space of contravariant tensors of type $(1, 1)$ at the origin of G/K will be identified with $\mathfrak{n}^+ \otimes \mathfrak{n}^-$ and the base element $e_a \otimes e_{-b}\ (a, b \in \Psi)$ will be written $e_{a\bar{b}}$. This space is considered as a $\mathfrak{k}$-module *via* the adjoint action, which is characterised by

(11)
$$\mathrm{ad}\, t(e_{a\bar{b}}) = 2\pi i\,(a(t) - b(t))e_{a\bar{b}} \qquad (t \in \mathfrak{t})$$
$$\mathrm{ad}\, e_\mu(e_{a\bar{b}}) = N_{\mu,a}e_{a+\mu,\bar{b}} + N_{\mu,-b}e_{a,\overline{b-\mu}} \qquad (\mu \in \Theta\, ;\, a, b \in \Psi) .$$

Here and in the sequel Θ is the set of roots of $\mathfrak{k}$ with respect to $\mathfrak{t}$, and μ, ν

are elements of Θ. The curvature tensor defines a linear transformation S on $\mathfrak{n}^+ \otimes \mathfrak{n}^-$ by

$$S(e_{a\bar{b}}) = \sum_{c,d} R^{c\bar{d}}_{a\bar{b}} \, e_{c\bar{d}} \,.$$

Since in our coordinate system the metric is the unit form **(1.5)** we have

$$R^{c\bar{d}}_{a\bar{b}} = R^b_{acd} = [[e_a, e_{-c}], e_a]^b$$

and **(2.1)** implies

$$
\begin{aligned}
S(e_{a\bar{a}}) &= \sum_{c \in \Psi} (c, a) e_{c\bar{c}} \\
S(e_{a\bar{b}}) &= \sum_{c-d=a-b} N_{d,-c} N_{a,-b} e_{c\bar{d}} \,. \\
S(e_{a\bar{b}}) &= 0 \qquad\qquad (a - b \notin \Sigma \,, \, a \neq b) \,.
\end{aligned}
$$
(12)

We define

(13)
$$
\begin{aligned}
h^* &= 2\pi i \sum_{c \in \Psi} c(h) e_{c\bar{c}} &\qquad (t \in \mathfrak{t}) \\
e^*_\mu &= \sum_{a-b=\mu} N_{a,-b} e_{a\bar{b}} &\qquad (a, b \in \Psi \,, \, \mu \in \Theta) \,.
\end{aligned}
$$

4.2. LEMMA. *The vector $z^* = \sum_{c \in \Psi} e_{c\bar{c}}$ is an eigenvector of S, with eigenvalue $(s, a_1) = (m(G/K) + 2)/2$ in the notation of* **2.4.** *The transformation S maps the vector $e_{a\bar{b}}$ $(a - b = \mu \in \Theta)$ onto a scalar multiple of e^*_μ, and the eigenvalue of e^*_μ is $\lambda_\mu = \sum_{a-b=\mu} N^2_{a,-b}$, $(a, b \in \Psi)$.*

This is a direct consequence of (12).

4.3. THEOREM. *The linear transformation $f : \mathfrak{t} \to \mathfrak{n}^+ \otimes \mathfrak{n}^-$ defined by $f(h) = h^*$ $(h \in \mathfrak{t})$, $f(e_\mu) = e^*_\mu$ $(\mu \in \Theta)$ is a $\mathfrak{t}$-module isomorphism of $\mathfrak{t}$ onto $f(\mathfrak{t})$.*

By assumption, K has no subgroup $\neq \{e\}$ invariant in G, hence $\mathfrak{t}$ acts faithfully on the tangent space; this means, by (11), that given $h \in \mathfrak{t}$, $h \neq 0$, there exists $a \in \Psi$ such that $a(h) \neq 0$ and that for every $\mu \in \Theta$ there exists $a \in \Psi$ such that $\mu + a \in \Psi$. From this it follows that f is injective, and that $\lambda_\mu \neq 0$.

By (11), **1.3**, the center of $\mathfrak{t}$ annihilates $f(\mathfrak{t})$ and $\mathfrak{t}$ annihilates the vectors $e_{a\bar{a}}$. Therefore, in order to prove that f is a $\mathfrak{t}$-module homomorphism, it will be enough to check the following equalities.

(i) $\operatorname{ad} e_\mu(e^*_\nu) = N_{\mu,\nu} e^*_{\mu+\nu}$ $\qquad (\mu, \nu \in \Theta \,, \, \mu + \nu \neq 0)$

(ii) $\operatorname{ad} e_{-\nu}(e^*_\nu) = -(2\pi i)^{-1} h^*_\nu$

(iii) $\operatorname{ad} h(e^*_\nu) = 2\pi i \, \nu(h) e^*_\nu$

(iv) $\operatorname{ad} e_\mu(h^*) = -2\pi i \, \mu(h) e^*_\mu$ $\qquad (\mu, \nu \in \Theta) \,.$

We have by (11)

$$\operatorname{ad} e_\mu(e^*_\nu) = \sum_{a-b=\nu} N_{a,-b} (N_{\mu,a} e_{\mu+a,\bar{b}} + N_{\mu,-b} e_{a,\overline{b-\mu}}) \,.$$

This vector belongs to the subspace spanned by the $e_{c\bar{d}}$ with $c - d = \mu + \nu$, and the coefficient of $e_{c\bar{d}}$ is

$$N_{c-\mu,-a}N_{\mu,c-\mu} + N_{c,-a-\mu}N_{\mu,-a-\mu} \cdot$$

That it is equal to $N_{\mu,}\, N_{c,-a}$ follows then from **1.3**(3) and from the known relation [8, Exp. 11, Lemma 2]

$$N_{ab}N_{cd} + N_{ca}N_{bd} + N_{bc}N_{ad} = 0$$

($a, b, c, d \in \Sigma$, not opposite in sign, $a + b + c + d = 0$).

Using (11), **1.3**(3), it is easily seen that

$$u_\nu = \operatorname{ad} e_{-\nu}(e_\nu^*) = \sum_{a-b=\nu} N_{a,-b}^2 (e_{b\bar b} - e_{a\bar a}) \, ,$$

hence, by (12),

$$S(u_\nu) = \sum_{a-b=\nu} N_{a,-b}^2 \left(\sum_{c\in\Psi}(b, c)e_{c\bar c} - \sum_{c\in\Psi}(a, c)e_{c\bar c} \right) \, ,$$

$$S(u_\nu) = -\lambda_\nu \sum_{c\in\Psi}(\nu, c)e_{c,\bar c} \, , \qquad (\lambda_\nu = \sum_{a-b=\nu} N_{a,-b}^2) \, .$$

However, the isotropy group commutes with S, and e_ν^* is an eigenvector of S, with eigenvalue λ_ν by **4.2**. Therefore so is u_ν and the previous equality yields (ii).

(iii) is a direct consequence of (11). Finally, using (11), we see that $\operatorname{ad} e_\mu(h^*)$ is a linear combination of the vectors $e_{a\bar b}$ ($a - b = \mu$) and that the coefficient of $e_{a\bar b}$ is

$$2\pi i \, b(h)N_{\mu,b} + 2\pi i \, a(h)N_{\mu,-a} \, .$$

Since we have $N_{\mu,b} = -N_{\mu,-a} = N_{a,-b}$ in view of **1.3**(3) and $a - b - \mu = 0$, the equality (iv) follows.

4.4. COROLLARY. *If the roots μ, ν of $\mathfrak{k}$ belong to the same simple component $\mathfrak{m}$ of $\mathfrak{k}$, then*

$$\sum_{a-b=\mu} N_{a,-b}^2 = \sum_{a-b=\nu} N_{a,-b}^2 \qquad\qquad (a, b \in \Psi) \, .$$

In fact $f(\mathfrak{m})$ is a minimal invariant subspace of $\mathfrak{k}$ in $\mathfrak{n}^+ \otimes \mathfrak{n}^-$ by **4.3**; therefore it belongs to an eigenspace of S. In particular e_μ^* and e_ν^* are eigenvectors of S for the same eigenvalue, and **4.4** follows from **4.2**.

In the same connection, we remark that **4.2**, **4.3** also imply $S(h_\mu^*) = \lambda_\mu h_\mu^*$. Together with (11), this yields the following relation

$$(14) \qquad\qquad \lambda_\mu(\mu, c) = \sum_{a\in\Psi}(\mu, a)(a, c) \qquad\qquad (\mu \in \Theta \, ; c \in \Psi) \, .$$

4.5. THEOREM. *The operator S has the following eigenvalues: 0 with multiplicity $n^2 - \dim_C \mathfrak{k}$, $(n = \dim_C G/K)$; $(s, a_1) = (1/2)(a_1, a_1)(m(G/K) + 2)$ (notation of **2.4**) with multiplicity 1, eigenvector $\sum_{a\in\Psi}e_{a\bar a}$; for each simple component $\mathfrak{m}$ of $\mathfrak{k}$ the eigenvalue $\lambda_\mu = \sum_{a-b=\mu} N_{a,-b}^2$ with eigenspace $f(\mathfrak{m})$, where μ is any root of $\mathfrak{m}$. Moreover $\operatorname{Tr} S = (s, s) = \operatorname{Tr} Q$.*

We have already seen in the proof of **4.4** that $f(\mathfrak{m})$ is an eigenspace of

S, with eigenvalue λ_μ, where μ is any root of $\mathfrak{m}$. By (12), S maps the vectors $e_{a\bar{a}}$ $(a \in \Psi)$ into $f(\mathfrak{t})$. This and **4.2** imply the other assertions about the eigenvalues.

By **4.1**, $R_{a\bar{b}}^{a\bar{b}}$ is equal to (a, a) if $a = b$, to $N_{b,-a}N_{a,-b}$ if $a \neq b$. But

$$N_{b,-a}N_{a,-b} = (a, a)\, q(p - 1)/2$$

where $[p, q]$ is the interval of integers k such that $-b + ka$ is a root (see e.g., [8, Exp. 11, Lemma 3]). Since $-b - a$ is not a root (**1.4**), we have $p = 0$ hence [8, Exp. X, Prop. 1], $q = 2(a, b)(a, a)^{-1}$. Consequently, using also **1.3**(3),

$$(15) \qquad R_{a\bar{b}}^{a\bar{b}} = N_{a,-b}^2 = (a, b) = q(a, a)/2 \, ,$$

and the last assertion of **4.5** follows.

4.6. COROLLARY. *Assume that the derived algebra $\mathfrak{k}'$ of $\mathfrak{k}$ is simple. Then S has three eigenvalues $(0, (s, a_1), \lambda)$ with multiplicities $n^2 - \dim_C \mathfrak{k}$, 1, $\dim_C \mathfrak{k} - 1$ and we have, in the notation of* **2.4**

$$(16) \qquad \lambda \cdot (\dim_C \mathfrak{k} - 1) = (n - 1)(m(G/K) + 2)(a_1, a_1)/2 \, .$$

4.7. From the classification, it is known that, except in the case $U(m + m')/U(m) \times U(m')$ $(m, m' \geq 2)$, where $\mathfrak{k}'$ has two components, $\mathfrak{k}'$ is always simple. Using the known values of $m(G/K) + 2$ (see **2.7**, [1a, 2]), one finds for $\lambda/(a_1, a_1)$ the values $(n - 2)/2$, $(n + 2)/4$, 1, 2, 3 in the cases

$$SO(2n)/U(n), \ Sp(n)/U(n), \ SO(n+2)/SO(n) \times T^1, \ E_6/Spin\,(10) \times T^1, \ E_7/E_6 \times T^1.$$

The numbers λ_μ can of course also be obtained from (15). In the case $U(m + m')/U(m) \times U(m')$ a straightforward computation based on (15), yields for $\lambda_\mu/(a_1, a_1)$ the values $m'/2$ and $m/2$, with respective multiplicities $m^2 - 1$, $(m')^2 - 1$. It can furthermore be checked that the nullspace of S is the irreducible component corresponding to the highest weight of the representation of $\mathfrak{k}$ in $\mathfrak{n}^+ \otimes \mathfrak{n}^-$.

THE INSTITUTE FOR ADVANCED STUDY

REFERENCES

1. A. BOREL and F. HIRZEBRUCH, *Characteristic classes and homogeneous spaces*, (a) Part I, Amer. Jour. Math., 80 (1958), 459–538; (b) Part II, *Ibid.*, 81 (1959), 315–381; (c) Part III, to appear.
2. E. CALABI and E. VESENTINI, *On compact, locally symmetric Kähler manifolds*, Annals of Math., 71 (1960), 472–507.
3. E. CARTAN, *Les groupes projectifs qui ne laissent invariante aucune multiplicité plane*, Bull. Soc. Math. France, 41 (1913), 53–96; Oeuvres Complètes, Part I, Vol. 1, 355–398.

4. ———, *Sur une classe remarquable d'espaces de Riemann*, Bull. Soc. Math. France, 55 (1927), 114–134; Oeuvres Complètes, Part I, Vol. 2, 639–659.

5. ———, *Sur les domaines bornés homogènes de l'espace de n variables complexes*, Abh. Math. Sem. Hamburg, 11 (1935), 116–162; Oeuvres Complètes, Part I, Vol. 2, 1259–1305.

6. H. CHANDRA, *Representations of semi-simple Lie groups*, VI, Amer. J. Math., 78 (1956), 564–628, §7.

7. K. NOMIZU, *Invariant affine connections on homogeneous spaces*, Amer. J. Math., 76 (1954), 33–65.

8. SÉMINAIRE S. Lie, mimeographed notes, Paris, 1955.

49.

(with J. C. Moore)

Homology theory for locally compact spaces

Mich. Math. J. 7 (1960) 137–159

1   In this paper, we develop a homology theory for locally compact spaces. On compact metric spaces, our theory is equivalent to the Steenrod homology theory [8]. The main purpose of introducing it is to obtain a Poincaré duality theorem for co-homology manifolds (see Section 7). The subject matter of the present paper is essentially the same as that of [3, Chapter II]. However, more emphasis has been put on the homology theory, which is treated from a slightly different point of view. Cohomology manifolds, which were the main concern of [3, Chap. I, II], will here be discussed more briefly.

The notation will in general be that of [7]. We assume familiarity with sheaf theory [4], [7]. In particular, the following concepts and notation will be used. A *family of supports* Φ in the space X is a collection of closed subsets such that

i) if A, B ϵ Φ, then A$\cup$ B ϵ Φ, and

ii) if B ϵ Φ and A is a closed subset of B, then A ϵ Φ.

Given a sheaf $\mathscr{S}$ on X, the symbols $\Gamma(\mathscr{S})$, $\Gamma_\Phi(\mathscr{S})$, $\mathscr{S}(A)$ will denote respectively the sections of $\mathscr{S}$, the sections of $\mathscr{S}$ with support in Φ, and the sections of $\mathscr{S}$ over the subspace A of X. Note that by *section* we shall always mean continuous section.

If A is a subspace of X and $\mathscr{S}$ is a sheaf on X, we denote by $\mathscr{S}|A$ the restriction of $\mathscr{S}$ to A. Further, if A is locally closed, we denote by $\mathscr{S}_A$ the sheaf on X which induces $\mathscr{S}|A$ on A and zero on X - A. Recall that if A is open in X, the sequence

$$0 \to \mathscr{S}_A \to \mathscr{S} \to \mathscr{S}_{X\text{-}A} \to 0$$

is an exact sequence of sheaves on X.

A sheaf $\mathscr{S}$ on X is *flabby* is the restriction map $\Gamma(\mathscr{S}) \to \mathscr{S}(A)$ is surjective for every open subspace A of X; it is called *soft* if $\Gamma(\mathscr{S}) \to \mathscr{S}(A)$ is surjective for every closed subspace A of X. If Φ is a family of supports in X, then $\mathscr{S}$ is *soft relative to* Φ if $\mathscr{S}|A$ is soft for every A ϵ Φ. When the family Φ is paracompactifying [7], this is equivalent to saying that $\Gamma_\Phi(\mathscr{S}) \to \Gamma_\Phi(\mathscr{S}|A) = \mathscr{S}(A)$ is surjective for every A ϵ Φ. Thus the word "flabby" is the translation of *flasque*, and "soft" the translation of *mou*.

Throughout this paper, we shall assume that K is a Dedekind ring given once and for all. All modules are assumed to be K-modules, and $\otimes$, Tor, Hom, and Ext are taken over K. All sheaves in addition to their stated properties will be assumed to be sheaves of K-modules. Finally, all topological spaces will be assumed to be Hausdorff and, from Section 2 on, locally compact.

Received November 12, 1959.

The second author was partially supported by the Air Force Office of Scientific Research under Contract AF 18(600), during the period in which this work was done.

102

1. THE CANONICAL RESOLUTION OF A SHEAF

If M is a module, we denote by F(M) the free module generated by the nonzero elements of M. There is a canonical surjection $F(M) \to M$, and we denote by R(M) the kernel of this map. Now, since K is an integral domain, there exists a quotient field K* of K, and R(M) is canonically imbedded in $K^* \otimes F(M)$. Let $\overline{M}$ be the quotient of $K^* \otimes F(M)$ by R(M); then there exists a natural map $M \to \overline{M}$ which is injective. Moreover, $\overline{M}$ is an injective module. Thus we have chosen functorially an imbedding of every module M in an injective module $\overline{M}$.

1.1. DEFINITION. *If $\mathscr{S}$ is a sheaf on* X, *let* $I(\mathscr{S})$ *be the sheaf on* X *such that* $I(\mathscr{S})(A) = \Pi_{x \in A} \overline{\mathscr{S}_x}$ *for every open subset* A *of* X, *where* $\mathscr{S}_x$ *is the stalk of* $\mathscr{S}$ *over* x.

There is a canonical injection of $\mathscr{S}$ into $I(\mathscr{S})$, and the sheaf $I(\mathscr{S})$ is *injective*. This means that if $\mathscr{L}^!$ is a subsheaf of $\mathscr{L}$ and we are given a homomorphism of $\mathscr{L}^!$ into $I(\mathscr{S})$, it can be extended to a map of $\mathscr{L}$ into $I(\mathscr{S})$ [7, Chap. II, Section 7.1]. The condition that a sheaf be injective is stronger than that it be flabby [7, *loc. cit.*]. Consequently $I(\mathscr{S})$ is flabby.

1.2. LEMMA. *If $\mathscr{S}$ is an injective sheaf on* X, *and* Φ *a family of supports,* $\Gamma_{\Phi}(\mathscr{S})$ *is an injective module.*

Since $\mathscr{S}$ is injective, it is a direct summand of $I(\mathscr{S})$. Therefore it suffices to prove the lemma for $I(\mathscr{S})$. Now if $f \in \Gamma_{\Phi}(I(\mathscr{S}))$ and k is a nonzero element of K, there exists $g \in \Gamma_{\Phi}(I(\mathscr{S}))$ with exactly the same support as f such that $k \cdot g = f$. Thus $\Gamma_{\Phi}(I(\mathscr{S}))$ is divisible, hence injective [5, p. 134].

1.3. DEFINITION. *If $\mathscr{S}$ is a sheaf on* X, *define*

$$\mathscr{C}^0(X; \mathscr{S}) = I(\mathscr{S}), \qquad \mathscr{Z}^1(X; \mathscr{S}) = \mathscr{C}^0(X; \mathscr{S})/\mathscr{S},$$

$$\mathscr{C}^q(X: \mathscr{S}) = I(\mathscr{Z}^q(X; \mathscr{S})), \qquad \mathscr{Z}^{q+1}(X; \mathscr{S}) = \mathscr{C}^q(X; \mathscr{S})/\mathscr{Z}^q(X; \mathscr{S}) \qquad (q \geq 1).$$

Note that the sequence of sheaves

$$0 \to \mathscr{S} \to \mathscr{C}^0(X; \mathscr{S}) \to \mathscr{C}^1(X; \mathscr{S}) \to \cdots$$

is exact; it will be called the *canonical injective resolution* of $\mathscr{S}$ and will be denoted by $\mathscr{C}^*(X, \mathscr{S})$. If Φ is a family of supports on X, we put

$$C_{\Phi}^*(X; \mathscr{S}) = \Gamma_{\Phi}(\mathscr{C}^*(X; \mathscr{S}));$$

the elements of $C_{\Phi}^*(X; \mathscr{S})$ are the *standard cochains of* X, *with coefficients in* $\mathscr{S}$ *and supports in* Φ.

1.4. LEMMA. *If* f: $\mathscr{S}^! \to \mathscr{S}$ *is a homomorphism of sheaves on* X, *there exists a canonical map* f*: $\mathscr{C}^*(X; \mathscr{S}^!) \to \mathscr{C}^*(X; \mathscr{S})$ *such that the diagram*

$$0 \to \mathscr{S}^! \to \mathscr{C}^0(X; \mathscr{S}^!) \to \mathscr{C}^1(X; \mathscr{S}^!) \to \cdots$$

$$\downarrow f \qquad\qquad \downarrow \qquad\qquad\qquad \downarrow$$

$$0 \to \mathscr{S} \to \mathscr{C}^0(X; \mathscr{S}) \to \mathscr{C}^1(X; \mathscr{S}) \to$$

is commutative.

This follows from the fact that the assignment $\mathscr{S} \to I(\mathscr{S})$ is functorial.

1.5. LEMMA. *If the sheaf* $\mathscr{S}$ *is locally concentrated on* A, *and* A *is a locally closed subspace of* X, *then the natural map*

$$\mathscr{C}^*(X; \mathscr{S})\big|A \to \mathscr{C}^*(A; \mathscr{S}),$$

where $\mathscr{C}^*(A; \mathscr{S}) = \mathscr{C}^*(A; \mathscr{S}\big|A)$, *is bijective.*

Proof. For $x \in A$,

$$\mathscr{C}^0(X; \mathscr{S})(x) = \lim_{U \to x} \prod_{y \in U} \overline{\mathscr{S}}_y,$$

$$\mathscr{C}^0(A; \mathscr{S})(x) = \lim_{U \to x} \prod_{y \in A \cap U} \overline{\mathscr{S}}_y.$$

However, if $\underline{U}$ is a small enough neighborhood of x, and $y \in U - U \cap A$, we have $\mathscr{S}_y = 0$ and $\overline{\mathscr{S}}_y = 0$; this gives the desired result.

The preceding proof is an exact duplicate of a proof in [7, p. 187].

1.6. LEMMA. *If* A *is closed in* X *and* S *is concentrated on* A, *then*

$$\Gamma_\Phi(\mathscr{C}^*(X; \mathscr{S})) = \Gamma_{\Phi \cap A}(\mathscr{C}^*(A; \mathscr{S})).$$

This lemma is just a translation of a known theorem [7, p. 188], and it is equivalent to the assertion that $C^*_\Phi(X; \mathscr{S}) = C^*_{\Phi \cap A}(A; \mathscr{S})$.

1.7. DEFINITION. *Suppose* Φ *is a family of supports on* X, *and* Ψ *a family of supports on* Y; *then a map* f: $X \to Y$ *is proper relative to* Φ *and* Ψ *if* $f^{-1}(A) \in \Phi$ *for every* $A \in \Psi$. *It is proper if it is proper relative to the families of compact subsets.*

1.8. PROPOSITION. *If* Φ, Ψ *are families of supports on* X *and* Y, *and* f: $X \to Y$ *is a proper map relative to* Φ, Ψ, *then for any sheaf* $\mathscr{S}$ *on* Y, *there is a map* f*: $C^*_\Psi(Y; \mathscr{S}) \to C^*_\Phi(X; f^*(\mathscr{S}))$ *compatible with the natural map* $\Gamma_\Psi(\mathscr{S}) \to \Gamma_\Phi(f^*(\mathscr{S}))$ *given by* $t \to t \circ f$. *Moreover, the map* f* *is unique up to chain homotopy.*

Since $f^*(\mathscr{C}^*(Y; \mathscr{S}))$ is a resolution of $f^*(\mathscr{S})$, the inverse image of $\mathscr{S}$ [7, p. 145], there exists a map which is unique up to homotopy of the resolution $f^*(\mathscr{C}^*(Y; \mathscr{S}))$ into the resolution $\mathscr{C}^*(X; f^*(\mathscr{S}))$, and the result follows.

We end this section with a refinement of Theorem 3.6.1 of [7, Chap. II], which will be needed in Section 2. Recall that given a sheaf $\mathscr{S}$ on the space X, together with a section s of $\mathscr{S}$ and an open covering $(U_i)_{i \in I}$ of X, a *partition* of s *subordinated* to $(U_i)_{i \in I}$ is a family $(s_i)_{i \in I}$ of sections of $\mathscr{S}$ such that the support $|s_i|$ of s_i is contained in U_i, the subspaces s_i form a locally finite family, and $\Sigma_{i \in I} s_i = s$ ([7], p. 155).

1.9. PROPOSITION. *Let* Φ *be a paracompactifying family on* X, *let* $\mathscr{S}$ *be a* Φ-*soft sheaf on* X, *and let* $(U_i)_{i \in I}$ *be an open covering of* X. *Let* $s \in \Gamma_\Phi(\mathscr{S})$, *and let* Q *be an element of* Φ *not meeting the support* $|s|$ *of* s. *Then there exists a partition* $(s_i)_{i \in I}$ *of* s, *subordinated to* $(U_i)_{i \in I}$, *whose elements belong to* $\Gamma_\Phi(\mathscr{S})$ *and are zero on* Q. *Further, if* B *is a neighborhood of* $|s| \cup Q$ *belonging to* Φ, *which is contained in the union of the* U_i *where* i *runs through a subset* I' *of* I, *it may be assumed that* $s_i = 0$ *for* $i \notin I'$.

We assume first that X belongs to Φ, in other words, that Φ consists of all closed subsets of X, and prove the first assertion in that case. The proof follows Godement's, and we describe it briefly. We may assume (U_i) to be locally finite. Choose then a closed covering $(F_i)_{i \in I}$ of X with $F_i \subset U_i$. Consider the set E of families $(s_j)_{j \in J}$ $(J \subset I)$, where s_j is a section of $\mathscr{S}$ over X, with support in U_i, equal to zero on Q, and $\Sigma_{j \in J} \, s_j = s$ on $F_J = \bigcup_{j \in J} F_j$. The set E is nonempty, and, ordered by inclusion, is inductive. By Zorn's lemma, it is then enough to show that if $J \neq I$, $(s_j)_{j \in J}$ is not maximal in E. Let then $i \notin J$. There exists a section s_i' of $\mathscr{S}$ on $Q \cup (X - U_i) \cup F_J \cup F_i$ which is zero on $Q \cup (X - U_i)$ and is equal to $s - \Sigma_{j \in J} s_j$ on $F_J \cup F_i$. Since $\mathscr{S}$ is soft, it can be extended to a section s_i over X and then $(s_j)_{j \in J} \cup s_i$ is an element of E.

We now prove 1.9 in the general case. By the above, there exists on B a partition $(s_i')_{i \in I'}$ of $s|B$, subordinated to the covering $(U_i \cap B)_{i \in I'}$, whose elements are zero on Q and on (B - Int B). Being zero on the boundary of B, the element s_i', extended by zero outside of B, is a section of $\mathscr{S}$, whose support is in $U_i \cap B$, hence also in Φ. We put $s_i = 0$ if $i \notin I'$, and it is clear that all our conditions are fulfilled.

2. THE DUAL OF A DIFFERENTIAL GRADED SHEAF

In the preceding section, the assumption that all spaces are locally compact was never used. However, in this section, local compactness will play a fundamental role.

2.1. *Notation.* The family of compact subsets of any space will be denoted by c. Let $\mathscr{S}$ be a sheaf on X, and let $\mathscr{T}$ be the presheaf on X such that

$$\mathscr{T}(U) = \mathrm{Hom}\,(\Gamma_c(\mathscr{S}\,|\,U), A)\,,$$

where A is a fixed K-module, and such that the map $\mathscr{T}(U) \to \mathscr{T}(V)$ is induced by the inclusion map $\Gamma_c(\mathscr{S}\,|\,V) \to \Gamma_c(\mathscr{S}\,|\,U)$ for $V \subset U$ open in X. Denote by $\mathscr{J}$ the sheaf determined by the presheaf $\mathscr{T}$.

2.2. LEMMA. *If the sheaf $\mathscr{S}$ is c-soft and $\{U_i\}_{i \in I}$ is an open covering of X, then the sequence*

$$\sum_{i, j \in I} \Gamma_c(\mathscr{S}\,|\,U_i \cap U_j) \overset{g}{\to} \sum_{i \in I} \Gamma_c(\mathscr{S}\,|\,U_i) \overset{h}{\to} \Gamma_c(\mathscr{S}) \to 0$$

is exact, where

    1) $g(x_{i,j}) = y_j - y_i$, y_i *being the image of* $x_{i,j} \in \Gamma_c(\mathscr{S}\,|\,U_i \cap U_j)$ *in* $\Gamma_c(\mathscr{S}\,|\,U_i)$ *and*

    2) $h(y_i)$ *is the image of* $y_i \in \Gamma_c(\mathscr{S}\,|\,U_i)$ *in* $\Gamma_c(\mathscr{S})$.

Proof. It follows immediately from 1.9 that h is surjective.

The fact that $h \circ g$ is zero is immediate. Therefore it remains to show that $\mathrm{Ker}\, h \subset \mathrm{Im}\, g$. Let $y \in \mathrm{Ker}\, h$. We may write $y = y_i + z$, where $y_i \in \Gamma_c(\mathscr{S}\,|\,U_i)$ and $z \in \Sigma_{j \in J} \Gamma_c(\mathscr{S}\,|\,U_j)$, and where in turn J is some finite subset of I. Since $h(y) = 0$, the support of $h(z)$ is covered by the open sets $U_i \cap U_j$ $(j \in J)$. Using a partition of unity subordinated to this covering, we can find elements $z_{ij} \in \Gamma_c(\mathscr{S}\,|\,U_i \cap U_j)$ $(j \in J)$ which, viewed as elements of $\Gamma_c(\mathscr{S})$, have a sum equal to $h(z)$. Let us write $z_{ij}^{(k)}$ $(k = i, j)$ for the image of z_{ij} in $\Gamma_c(\mathscr{S}\,|\,U_k)$. Then

$$g\left(\sum_{j \in J} z_{ij}\right) = - \sum_{j \in J} z_{ij}^{(i)} + \sum_{j \in J} z_{ij}^{(j)}, \qquad h(z) = \sum_{j \in J} h\left(z_{ij}^{(i)}\right) = \sum_{j \in J} h\left(z_{ij}^{(j)}\right).$$

Since $h(y_i + z) = 0$, it follows immediately that $y_i + \sum_{j \in J} z_{ij}^{(i)} = 0$, hence

$$y_i + z - \sum_{j \in J} g(z_{ij}) = z - \sum_{j \in J} z_{ij}^{(j)},$$

so that it will be enough to prove that the right-hand side is contained in Im g. Since $z \in \Sigma_{j \in J} \Gamma_c(\mathscr{S}\,|\,U_j)$, our assertion follows then by a simple induction on the number of nonzero components of y.

2.3. PROPOSITION. *If* $\mathscr{S}$ *is c-soft, the natural map* $\mathscr{T}(U) \to \mathscr{J}(U)$ *is bijective. If moreover* A *is injective, the sheaf* $\mathscr{J}$ *is flabby.*

Proof. Let $\{U_i\}_{i \in I}$ be a family of open subsets of X, and let U be their union. Suppose that t, t $' \in \mathscr{T}(U)$ and that they have the same image in each $\mathscr{T}(U_i)$; then t = t'; for we see by 2.2 that $\Sigma \Gamma_c(\mathscr{S}\,|\,U_i) \to \Gamma_c(\mathscr{S}\,|\,U) \to 0$ is exact, and therefore $0 \to \mathscr{T}(U) \to \Pi_{i \in I} \mathscr{T}(U_i)$ is exact.

Now suppose $t_i \in \mathscr{T}(U_i)$ for $i \in I$, and t_i, t_j have the same image in $\mathscr{T}(U_i \cap U_j)$. From the exactness of

$$0 \to \mathscr{T}(U) \xrightarrow{h^*} \prod_{i \in I} \mathscr{T}(U_i) \xrightarrow{g^*} \prod_{i,j} \mathscr{T}(U_i \cap U_j)$$

it follows that there is a unique $t \in \mathscr{T}(U)$, since that $h^*(t) = \{t_i\}$. This proves the first assertion [7, p. 109].

If now A is injective, the map $\mathscr{T}(X) \to \mathscr{T}(U)$ (U open in X) induced by the injection $\Gamma_c(\mathscr{S}\,|\,U) \to \Gamma_c(\mathscr{S})$ is surjective; hence $\mathscr{J}$ is flabby.

2.4. PROPOSITION. *Let* F *be a closed subset of* X. *If* $\mathscr{S}$ *is c-soft and* A *is injective, the restriction map* $\mu\colon \Gamma_c(\mathscr{S}) \to \Gamma_c(\mathscr{S}\,|\,F)$ *is surjective and induces a monomorphism* $\nu\colon \mathrm{Hom}\,(\Gamma_c(\mathscr{S}\,|\,F),\,A) \to \mathscr{T}(X)$ *whose image consists of the elements of* $\mathscr{T}(X)$ *which have their support in* F. *In particular, an element* $t \in \mathscr{T}(X)$ *with support in* F *has the value zero on the elements of* $\Gamma_c(\mathscr{S})$ *which have their supports in* X - F.

Since $\mathscr{S}$ is c-soft, μ is surjective (see Note below); therefore we have an exact sequence

$$0 \to \Gamma_c(\mathscr{S})_{X-F} \to \Gamma_c(\mathscr{S}) \xrightarrow{\mu} \Gamma_c(\mathscr{S}\,|\,F) \to 0,$$

where $\Gamma_c(\mathscr{S})_{X-F}$ means the elements of $\Gamma_c(\mathscr{S})$ with support in X - F. The module A being injective, this yields an exact sequence

$$0 \to \mathrm{Hom}\,(\Gamma_c(\mathscr{S}\,|\,F),\,A) \xrightarrow{\nu} \mathscr{T}(X) \xrightarrow{\sigma} \mathrm{Hom}\,(\Gamma_c(\mathscr{S})_{X-F},\,A) \to 0.$$

The elements of Im ν clearly have their supports in F. Now, if $t \in \mathscr{T}(X)$ has its support in F, each $x \in X - F$ has a neighborhood V_x such that t is zero on the elements of $\Gamma_c(\mathscr{S})$ with supports in V_x. With the aid of partitions of unity, it follows that t is zero on all elements of $\Gamma_c(\mathscr{S})$ with supports in X - F; therefore $t \in \operatorname{Ker} \sigma$ and $t \in \operatorname{Im} \nu$.

Note. In the beginning of the above proof, we have used a special case of the following elementary fact: Let F be a closed subset of X, let Φ be a paracompactifying family on X, and let $\mathscr{S}$ be a Φ-soft sheaf. Then the restriction map

$$\mu : \Gamma_\Phi(\mathscr{S}) \to \Gamma_{\Phi|F}(\mathscr{S}|F)$$

is surjective. To see this, given an element t of $\Gamma_{\Phi|F}(\mathscr{S}|F)$ with support K, first extend its restriction to $F \cap V$ to an element of $\Gamma_{\Phi|V}(\mathscr{S}|V)$, where V is a neighborhood of K belonging to Φ, which is zero on V-Int V. Then this element, extended by zero outside V, belongs to $\mu^{-1}(t)$.

2.5. *Notation and conventions.* A *differential graded module* M is a sequence of modules M^q indexed on the integers, together with maps d: $M^q \to M^{q+1}$ such that $d^2 = 0$. The differential graded module M is *bounded below* if $M^q = 0$ for q less than some fixed integer n; it is *bounded above* if $M^q = 0$ for $q > n$. We also denote M^q by M_{-q}, and we let d: $M_q \to M_{q-1}$ stand for d: $M^{-q} \to M^{-q+1}$. If M, N are differential graded modules, Hom (M, N)$_r$ is the module of maps f: $M \to N$ such that $f(M^q) \subset N^{q-r}$. Moreover, Hom (M, N) is the differential graded module such that Hom (M, N)^r = Hom (M, N)$_{-r}$, and such that if $f \in$ Hom (M, N)$_r$, then

$$(df)(x) = d(f(x)) + (-1)^{r+1} f(dx) .$$

Let K* be the quotient field of K, and let R(K) be the differential graded module such that $R(K)^0 = K*$, $R(K)^1 = K*/K$, $R(K)^q = 0$ for $q \neq 0, 1$, and such that d: $R(K)^0 \to R(K)^1$ is the natural map $K* \to K*/K$. For any differential graded module M, let D(M) = Hom (M, R(K)). The differential graded module D(M) is called the *dual of* M. By definition,

$$D(M)_r = \operatorname{Hom}(M^r, K*) \oplus \operatorname{Hom}(M^{r+1}, K*/K) .$$

Moreover, since R(K) is an injective resolution of K, there exists a split exact sequence

$$0 \to \operatorname{Ext}(H^{q+1}(M), K) \to H_q(D(M)) \to \operatorname{Hom}(H^q(M), K) \to 0 .$$

Further, since R(K) is an injective K-module, we see that if $0 \to M' \xrightarrow{f} M \xrightarrow{g} M'' \to 0$ is an exact sequence of differential graded K-modules, then

$$0 \to D(M'') \to D(M) \to D(M') \to 0$$

is an exact sequence of differential graded K-modules. Notice that here we assume df = fd, dg = gd, and that f, g are of degree zero, in other words, that

$$f \in Z^0 \operatorname{Hom}(M', M)$$

and $g \in Z^0$ Hom (M, M'').

A differential graded sheaf $\mathscr{A}$ is a sequence of sheaves $\mathscr{A}^q$ together with maps d: $\mathscr{A}^q \to \mathscr{A}^{q+1}$ such that $d^2 = 0$. For any open set U in X and any family Φ of subsets of X, $\Gamma_\Phi(\mathscr{A}|U)$ is a differential graded module with the component $\Gamma_\Phi(\mathscr{A}^q|U)$ of degree q.

2.6. DEFINITION. *If $\mathscr{A}$ is any differential graded sheaf, we denote by* $D(\mathscr{A})$ *the differential graded sheaf determined by the presheaf* $U \to D(\Gamma_c(\mathscr{A}\,|\,U))$.

2.7. PROPOSITION. *If $\mathscr{A}$ is a differential graded sheaf on* X *which is soft relative to the family of compact subsets, then*

1) *for* U *open in* X, $D(\mathscr{A})(U) = D(\Gamma_c(\mathscr{A}\,|\,U))$,

2) $D(\mathscr{A})^q$ *is flabby for every integer* q, *and*

3) *for every integer* q, *there exists a split exact sequence*

$$0 \to \mathrm{Ext}\,(H^{q+1}(\Gamma_c(\mathscr{A})),\,K) \to H_q(\Gamma(D(\mathscr{A}))) \to \mathrm{Hom}\,(H^q(\Gamma_c(\mathscr{A})),\,K) \to 0\,.$$

4) *If $\mathscr{A}$ is injective,* $D(\mathscr{A})(U)$ *is torsion-free, for* U *open in* X.

Proof. Parts 1 and 2 of the proposition follow at once from 2.3. Part 3 is an immediate corollary of the equality $\Gamma(D(\mathscr{A})) = D(\Gamma_c(\mathscr{A}))$ and of the corresponding fact about differential modules. If $\mathscr{A}$ is injective, then $\Gamma_c(\mathscr{A}\,|\,U)$ is injective by 1.2, and (4) follows from (1). (Recall that if A is divisible, then $\mathrm{Hom}\,(A, B)$ is torsion-free for any module B.)

2.8. *Notation.* We have already adopted the convention that if A is a module, then we also denote by A the constant sheaf with stalk A. We shall denote by $\mathscr{K}$ the sheaf such that $\mathscr{K}(U) = \Pi_{x \in U}\, K_x$ for U open in X, where K_x is the stalk of K over $x \in X$.

Recall that $\mathscr{K}(U)$ is a sheaf of rings, and that it is flabby, hence Φ-fine for any paracompactifying family Φ [7, Chap. II, 3.7]. As a consequence, every sheaf which is a $\mathscr{K}$-Module is Φ-fine, and *a fortiori* Φ-soft. Together with 2.3, this implies the following

2.9. PROPOSITION. *Let* A *be a module, and let $\mathscr{G}$ be a $\mathscr{K}$-module. Then the presheaf* $U \to \mathrm{Hom}\,(\Gamma_c(\mathscr{G}\,|\,U),\, A)$ *is a sheaf which is Φ-fine for any paracompactifying family Φ.*

3. HOMOLOGY THEORY

3.1. For any space X, we define $\mathscr{C}_H(X;\,K)$ to be the differential graded sheaf $D(\mathscr{C}^*(X;\,K))$. If $\mathscr{G}$ is any sheaf on X, we denote by $\mathscr{C}_H(X;\,\mathscr{G})$ the sheaf $\mathscr{G} \otimes \mathscr{C}_H(X;\,K)$. The sheaf $\mathscr{C}_H(X;\,K)$ is called the *standard sheaf for homology on* X, and $\mathscr{C}_H(X;\,\mathscr{G})$ is called the *standard sheaf for homology with coefficients in* $\mathscr{G}$. If Φ is any family of supports on X, we let $C_H^\Phi(X;\,\mathscr{G}) = \Gamma_\Phi(\mathscr{C}_H(X;\,\mathscr{G}))$ and $H_n^\Phi(X;\,\mathscr{G}) = H_n(C_H^\Phi(X;\,\mathscr{G}))$. The module $H_n^\Phi(X;\,\mathscr{G})$ is called the n-*dimensional homology group of* X *with coefficients in $\mathscr{G}$ and supports in Φ*. If Φ is the family of all closed subsets of X, we write $H_n(X;\,\mathscr{G})$ for $H_n^\Phi(X;\,\mathscr{G})$.

The *stalk* at $x \in X$ of the derived sheaf $\mathscr{H}(\mathscr{C}_H(X;\,K))$ is the local homology group at x, and it will be denoted $H_*^x(X;\,K)$. The sheaf $\mathscr{H}(\mathscr{C}_H(X;\,K))$ is the sheaf of local homology groups, and it will be denoted $\mathscr{H}_*(X;\,K)$.

For any space, the sheaf $\mathscr{C}_H(X;\,K)_n$ is flabby, and $\mathscr{C}_H(X;\,K)(U)$ is a torsion-free differential graded module for U open in X, by 2.7. Clearly, $\mathscr{C}_H(X;\,\mathscr{G})$ is a $\mathscr{K}$-module; therefore it is Φ-fine for any paracompactifying family Φ on X. For U open in X, the restriction of cross sections induces a natural homomorphism

$$j_{*_{XU}} : H_*^\Phi(X;\,K) \to H_*^{\Phi \cap U}(U;\,K)\,,$$

A. BOREL and J. C. MOORE

whose restriction to $H_n^\Phi(X; K)$ will be denoted by j_{nXU}.

3.2. THEOREM. *Let* f: $X \to Y$ *be a proper map. Then for every integer* n *there exists a natural map* f_*: $H_n(X; K) \to H_n(Y; K)$.

Proof. Observe that $f^*(K) = K$, where K stands for the constant sheaf with stalk K. Now by 1.8 there exists a map f^*: $C_c^*(Y; K) \to C_c^*(X; K)$ which is unique up to homotopy. This induces a map

$$Df^*: D(C_c^*(X; K)) \to D(C_c^*(Y; K)),$$

and since $D(C_c^*(X; K)) = \Gamma(\mathscr{C}_H(X; K))$, $D(C_c^*(Y; K)) = \Gamma(\mathscr{C}_H(Y; K))$, the first assertion follows.

3.3. THEOREM. (a) *For each space* X *and each integer* i, *there exists a split exact sequence*

$$0 \to \mathrm{Ext}\,(H_c^{i+1}(X; K), K) \to H_i(X; K) \to \mathrm{Hom}\,(H_c^i(X; K), K) \to 0$$

which is compatible with restrictions to open subsets. Hence there exists for each $x \in X$ *an exact sequence*

$$0 \to \mathrm{dir\,lim}\,\mathrm{Ext}\,(H_c^{i+1}(U; K), K) \to H_i^x(X; K) \to \mathrm{dir\,lim}\,\mathrm{Hom}\,(H_c^i(U; K), K) \to 0,$$

where U *runs through the open neighborhoods of* x.

(b) *Let* A *be a* c-*soft resolution of* K. *Then* $H_q(\Gamma(D(\mathscr{A}))) = H_q(X; K)$ *and* $\mathscr{H}(D(\mathscr{A})) = \mathscr{H}_*(X; K)$ *(*$q \in Z$*)*.

The assertion (a) follows from 2.7, the definitions of $H_i(X; K)$, $H_i^x(X; K)$, and the fact that a direct limit of exact sequences is exact.

Since $\mathscr{A}$ is a resolution of K, there exists a homomorphism f: $\mathscr{A} \to \mathscr{C}^*(X; K)$ which extends the identity map on K. Since each $\mathscr{A}^q$ is c-soft,

$$H^q(\Gamma_c(\mathscr{A})) \to H^q(\Gamma_c(\mathscr{C}^*(X; K)))$$

is bijective, as follows from Theorems 4.7.1 and 4.7.2 of [7, Chap. II]. By 2.7, $H_q(\Gamma(D(\mathscr{C}^*(X; K)))) \to H_q(\Gamma(D(\mathscr{A})))$ is bijective for each q, which implies the first equality of (b). This isomorphism is obviously compatible with the restriction to an open set, and the second equality of (b) follows.

Remark. Let F be a closed subspace of X, and let $\mathscr{A}$ be a c-soft resolution of K on X. The homomorphism derived from the natural map $D(\Gamma_c(\mathscr{A}\,|F)) \to D(\Gamma_c(\mathscr{A}))$, transpose of the restriction, can be identified with f_*: $H_*(F; K) \to H_*(X; K)$, f being the inclusion of F in X. This is easily seen from the above and the commutative diagram

$$
\begin{array}{ccc}
\Gamma_c(\mathscr{A}) & \longrightarrow & \Gamma_c(\mathscr{A}\,|F) \\
\downarrow & & \downarrow \\
\Gamma_c(\mathscr{C}^*(X; K)) & \to & \Gamma_c(\mathscr{C}^*(X; K)\,|F).
\end{array}
$$

3.4. THEOREM. *Let* Φ *be a family of supports of* X. *Then*

$$H_q^\Phi(X; K) = \lim_{F \in \Phi} H_q(F; K),$$

the limit being taken with respect to the homomorphisms defined by inclusion. If $\mathcal{A}$ is a c-soft resolution of K on X, then $H_q^\Phi(X; K) = H_q(\Gamma_\Phi(D(\mathcal{A})))$ $(q \in Z)$.

By 2.4, $M(F) = \mathrm{Hom}\,(\Gamma_c(\mathscr{C}^*(X; K)\,|\,F), R(K))$ may be identified with the set of elements of $C_H(X; K)$ having their supports in F. Since the restriction of a c-soft sheaf to a closed subset is c-soft $[7,\ \mathrm{Chap.\ II},\ 3.3.1]$, it follows from 3.3 that $H(M(F)) = H_*(F; K)$. The theorem follows from this, 2.4, and the remark to 3.3.

3.5. THEOREM. (a) *Let* $f: X \to Y$ *be a map, and let* $\Phi,\ \Psi$ *be families of supports on* X *and* Y *respectively such that* $f(\Phi) \subset \Psi$. *Then, if the restriction of* f *to any* $F \in \Phi$ *is proper, there exists a natural homomorphism* $H_q^\Phi(X; K) \to H_q^\Psi(Y; K)$ $(q \in Z)$. *In particular, there exists a natural homomorphism* $f_*: H_q^c(X: K) \to H_q^c(Y; K)$.

(b) *Let* F *be a closed subset of* X, *and* Ψ *a family of supports on* F. *Then the natural homomorphism* $H_q^\Psi(F; K) \to H_q^\Psi(X; K)$ *is an isomorphism* $(q \in Z)$.

The first assertion follows from 3.2 and the first part of 3.4. By 2.4, $C_H^\Psi(X; K)$ may be identified with $M_\Psi(F) = \Gamma_\Psi(D(\mathscr{C}^*(X; K)\,|\,F))$; by 3.4, $H(M_\Psi(F)) = H_*^\Psi(F; K)$, and (b) follows.

Remark. Let f be the inclusion of a closed subspace F in X. Then

$$f_*: H_*^{\Phi|F}(F; K) \to H_*^\Phi(X; K)$$

will often be denoted by i_{*FX}, and its restriction to $H_q^{\Phi|F}(F; K)$ by i_{qFX}. By the above, it is induced by the inclusion of

$$M_{\Phi|F}\,(F) = \Gamma_{\Phi|F}\,(D(\mathscr{C}^*(X; K)\,|\,F))$$

into $C_H(X; K)$, and $M_{\Phi|F}(F)$ may be identified with the module of elements of $C_H^\Phi(X; K)$ having their supports in F. Applied to the inclusion $F \cap U \to U$, where U is open in X, and to the family of all closed subsets, this leads to an injective homomorphism

$$i_{FX}: (D(\mathscr{C}^*(X; K)\,|\,F))^X \to \mathscr{C}_H(X: K),$$

hence also to a homomorphism

$$i_{*FX}: \mathscr{H}_*(F; K)^X \to \mathscr{H}_*(X; K),$$

where $\mathcal{A}^X$ denotes the extension by zero of a sheaf $\mathcal{A}$ on F $[7,\ \mathrm{II},\ 2.9]$.

3.6. THEOREM. *If* Φ *is a paracompactifying family on* X, *and*

$$0 \to \mathscr{S}' \to \mathscr{S} \to \mathscr{S}'' \to 0$$

is an exact sequence of sheaves on X, *there exists an exact homology sequence*

$$\cdots \to H_n^\Phi(X; \mathscr{S}') \to H_n^\Phi(X; \mathscr{S}) \to H_n^\Phi(X; \mathscr{S}'') \to H_{n-1}^\Phi(X: \mathscr{S}') \to \cdots.$$

Proof. Since $\mathscr{C}_H(X; K)$ is torsion-free, there exists an exact sequence of sheaves

$$0 \to \mathscr{C}_H(X; \mathscr{S}') \to \mathscr{C}_H(X; \mathscr{S}) \to \mathscr{C}_H(X; \mathscr{S}'') \to 0.$$

Since $\mathscr{C}_H(X; \mathscr{S}')$ is Φ-fine (by 3.1), the sequence

$$0 \to C_H^\Phi(X; \mathcal{S}') \to C_H^\Phi(X; \mathcal{S}) \to C_H^\Phi(X; \mathcal{S}'') \to 0$$

is exact ([7], p. 154), and the theorem follows.

3.7. THEOREM. *If* B *is a module, there exists an exact sequence*

$$0 \to H_n^c(X; K) \otimes B \to H_n^c(X; B) \to \mathrm{Tor}\,(H_{n-1}^c(X; K), B) \to 0.$$

$\mathcal{C}_H(X; K)$ is c-fine and torsion-free by 3.1 and 2.6; therefore

$$\Gamma_c((\mathcal{C}_H(X; B)) = \Gamma_c(\mathcal{C}_H(X; K)) \otimes B$$

[1, Exp. V, lemma 3(e)], and the result follows.

3.8. THEOREM. *Let* F *be a closed subspace of* X, *and let* U = X - F. *Then there exists an exact sequence*

$$\cdots \to H_q(F; K) \xrightarrow{i_{qFX}} H_q(X; K) \xrightarrow{j_q^{XU}} H_q(U; K) \to H_{q-1}(F: K) \to \cdots.$$

Let $\mathcal{A}$ be a c-soft resolution of K on X. Then we have the exact sequence

$$0 \to \Gamma_c(\mathcal{A}\,|\,U) \to \Gamma_c(\mathcal{A}) \to \Gamma_c(\mathcal{A}\,|\,F) \to 0,$$

(from which the exact sequence for cohomology with compact supports is derived) as follows readily from the remark to 2.4. Since R(K) is injective, this yields an exact sequence

$$(1) \qquad\qquad 0 \to D(\Gamma_c(\mathcal{A}\,|\,F)) \to D(\Gamma_c(\mathcal{A})) \to D(\Gamma_c(\mathcal{A}\,|\,U)) \to 0.$$

On account of 3.3(b), the desired homology sequence is the homology sequence of (1).

3.9. THEOREM. *Let* X *be the union of two open subspaces*, X_1, X_2, *and let* $X_{12} = X_1 \cap X_2$. *Then there exists an exact sequence*

$$\cdots \to H_n^c(X_{12}; K) \xrightarrow{\alpha} H_n^c(X_1; K) + H_n^c(X_2; K) \xrightarrow{\beta} H_n^c(X; K) \xrightarrow{\partial} H_{n-1}^c(X_{12}; K) \to \cdots,$$

where α *is the difference and* β *the sum of the inclusion homomorphisms.*

The proof is straightforward; it is the same as that given in [2, Section 8] to show the existence of an exact sequence of the Mayer-Vietoris type in Φ-cohomology, and we leave it to the reader. We content ourselves to recall the definition of ∂. Let z be a cycle of $\Gamma_c(\mathcal{C}_H(X; K))$. Since $\mathcal{C}_H(X; K)$ is c-fine by 3.1, we may write $z = z_1 + z_2$, where the support of z_1 (respectively z_2) is contained in X_1 (respectively X_2). Then $dz_1 + dz_2 = 0$, hence the support of dz_1 is equal to the support of dz_2 and is contained in X_{12}. Then the image under ∂ of the class of z is by definition the class of dz_1.

3.10. THEOREM. *Let* X *be the union of two closed subspaces* X_1, X_2, *and let* $X_{12} = X_1 \cap X_2$. *Then there exists an exact sequence*

$$\cdots \to H_n(X_{12}; K) \xrightarrow{\alpha} H_n(X_1; K) + H_n(X_2; K) \xrightarrow{\beta} H_n(X: K) \xrightarrow{\partial} H_{n-1}(X_{12}; K) \to \cdots,$$

where $\alpha(x) = i_{*X_{12},X_1}(x) - i_{*X_{12},X_2}(x)$, $\beta(x_1 + x_2) = i_{*X_1,X}(x_1) + i_{*X_1,X}(x_2)$.

Let M_1, M_2, M_{12} be the submodules of elements of $C_H(X; K)$ having supports in X_1, X_2, and X_{12} respectively. Then, (see the remark to 3.4)

$$(2) \qquad H(M_i) = H_*(X_i; K) \qquad (i = 1, 2), \qquad H(M_{12}) = H_*(X_{12}; K).$$

Let $c \in C_H(X; K)$. Then c can be written in at least one way as a sum $c = c_1 + c_2$ with $c_i \in M_i$ ($i = 1, 2$). In fact, since $\mathscr{C}_H(X; K)$ is flabby, it has a section c_1 which is equal to c on $X_1 - X_{12}$ and to zero on $X_2 - X_{12}$. Then $c_1 \in M_1$ and $c_2 = c - c_1$ is in M_2. Let now c be a cycle. Then $dc_1 = -dc_2$ has support in X_{12}, hence is an element of M_{12}. By definition, the image of the homology class of c is the class of dc_1 in $H(M_{12}) = H(X_{12}; K)$. It is a routine matter to verify that this definition is legitimate and that the sequence of 3.10 is exact. We leave the details to the reader.

3.11. *Remark.* In view of 3.3(b), $H_*(X; K)$ may also be defined as the *hyper-homology invariant* of $\mathrm{Hom}(\Gamma_c(\mathscr{A}), K)$, where $\mathscr{A}$ is a c-soft resolution of K on X, in the sense of [5, XVII]. Here we have defined it as the homology of $\mathrm{Hom}(\Gamma_c(\mathscr{A}), L)$, where L is an injective resolution of K. It is also the homology of $\mathrm{Hom}(P, K)$, where P is a projective resolution of $\Gamma_c(\mathscr{A})$. As will be recalled in Section 6, this second procedure has been used by Steenrod [8] for compact metric spaces. For closed subsets of euclidean spaces, one can also resolve the first argument by taking a resolution $\mathscr{A}$ of K on X for which $\Gamma_c(\mathscr{A})$ is a *free grating* which is homotopically c-fine (see [3, II, Section 4]).

4. KÜNNETH THEOREMS

4.1. THEOREM. *If* X *and* Y *are spaces, then, for each integer* n, *there exists a split exact sequence*

$$0 \to \sum_{r+s=n+1} \mathrm{Ext}(H_c^r(X), H_s(Y)) \to H_n(X \times Y) \to \sum_{r+s=n} \mathrm{Hom}(H_c^r(X), H_s(Y)) \to 0.$$

Proof. Let $\mathscr{A}$ be a torsion-free c-soft resolution of K on X, $\mathscr{B}$ a torsion-free c-soft resolution of K on Y, and $\mathscr{C}$ an injective resolution of K on $X \times Y$. Now it is known that there exists a map $f: \Gamma_c(\mathscr{A}) \otimes \Gamma_c(\mathscr{B}) \to \Gamma_c(\mathscr{C})$ which induces a homology isomorphism. Thus, we may use $D(\Gamma_c(\mathscr{A}) \otimes \Gamma_c(\mathscr{B}))$ to compute the homology of $X \times Y$. However, $D(\Gamma_c(\mathscr{A}) \otimes \Gamma_c(\mathscr{B})) = \mathrm{Hom}(\Gamma_c(\mathscr{A}), D(\Gamma_c(\mathscr{B})))$. Since $\Gamma_c(\mathscr{B})$ is torsion-free, $\mathrm{Hom}(\Gamma_c(\mathscr{B}), R(K)) = D(\Gamma_c(\mathscr{B}))$ is injective [5, VII, 1.4], and the theorem follows.

4.2. COROLLARY. *If* X *is a compact contractible space, then* $H_n(X \times Y) = H_n(Y)$ *for any space* Y.

Proof. Since X is compact, we have $H_c^q(X) = H^q(X)$ for all q; and since X is contractible, $H^q(X) = 0$ for $q \neq 0$ and $H^0(X) = K$. Now the result follows at once from the preceding theorem.

4.3. COROLLARY. *If* I *is the unit interval,* $F: I \times X \to Y$ *is a proper map, and* $F_t: X \to Y$ *is defined by* $F_t(x) = f(t, x)$, *then, for every integer* n,

$$(F_0)_* = (F_1)_*: H_n(X) \to H_n(Y).$$

4.4. COROLLARY. *If* $f, g: X \to Y$ *are homotopic maps, then, for every integer* n, $f_* = g_*: H_n^c(X) \to H_n^c(Y)$.

Proof. This corollary follows at once from 4.3 and 3.4.

5. COMPARISON WITH OTHER HOMOLOGY THEORIES

Using Lemma 1.6, we see that if A is a closed subspace of X, then

$$\Gamma_c(\mathscr{C}^*(X; K_A)) = \Gamma_c(\mathscr{C}^*(A; K)).$$

Since $\Gamma_c(\mathscr{C}^*(X; K)) \to \Gamma_c(\mathscr{C}^*(X; K_A))$ is surjective, $C_H(A; K) \to C_H(X; K)$ is injective. This means that we may define $H_q(X, A)$ to be $H_q(\Gamma(\mathscr{C}_H(X; K))/\Gamma(\mathscr{C}_H(A; K))$; and we then obtain a homology theory defined on pairs, in the sense of Eilenberg and Steenrod. Moreover, the homology groups of a point are just the usual homology groups of a point with coefficients in K. Therefore, for a finite complex X and a K-module B, $H_q(X; B)$ is the ordinary simplicial homology of X with coefficients in B. More generally, if X is a finite complex, and A is a subcomplex, then $H_q(X, A; B)$ is the simplicial homology of X modulo A with coefficients in B.

Now Eilenberg and Steenrod [6, p. 258] have formulated a notion of a continuous homology theory, and they have proved that the Čech homology theory is continuous on the category of compact pairs, and that further any homology theory which is continuous on compact pairs is naturally isomorphic with the Čech theory. For our homology theory, we see that if L = K/M, where M is a maximal ideal in K, then $H_*(X; L) = \mathrm{Hom}\,(H_c^*(X; L), L)$; with the help of this fact, it is easy to see that our homology theory with coefficients in L is continuous on compact pairs. Therefore it coincides with the Čech theory with coefficients in L for compact spaces. This is however not in general true for compact spaces and arbitrary coefficients.

For compact spaces, the sheaf-theoretic cohomology is the same as the Čech cohomology. Further, it is possible to define the Cech cohomology by using Čech cochains [7, p. 223]. If we let $C^*(X; K)$ be the Čech cochains of X with coefficients in K, then the homology we have defined has the property that if X is compact, then $H_n(X; K) = H_n(D(\check{C}^*(X; K)))$. However, if P is a projective resolution of $\check{C}^*(X; K)$, then standard homological algebra shows that $H_n(\mathrm{Hom}\,(P, K)) = H_n(D(\check{C}^*(X; K)))$. For compact separable metric spaces, Steenrod [8] defined a homology theory by first constructing Čech cochains, then choosing a particular projective resolution P of his Čech cochains and calling $H_n(\mathrm{Hom}\,(P, K))$ the $(n + 1)$st homology of the space based on regular cycles. Thus for a compact separable metric space, the group $H_n(X; K)$ is simply Steenrod's $(n + 1)$st homology group.

6. LOCAL CONNECTEDNESS

6.1. DEFINITION. For any space X, the augmented q-dimensional cohomology group of X is $H^q(X; K)$ if $q > 0$, and is the cokernel of the natural map $K \to H^0(X)$ induced by the map of X into a point, for $q = 0$.

The augmented q-dimensional homology group with compact supports is $H_q^c(X; K)$ for $q > 0$, and it is the kernel of the natural map $H_0^c(X) \to K$ induced by the map of X into a point, for $q = 0$.

6.2. DEFINITION. If f: $X \to Y$ is a map, then f is cohomologically trivial in dimension q if the image of the augmented cohomology in dimension q of Y in the augmented cohomology of X is zero.

The map f is homologically trivial in dimension q if the image of the augmented q-dimensional compact homology of X in the augmented q-dimensional compact homology of Y is trivial.

6.3. DEFINITION. The space X is cohomologically locally connected (respectively, in dimension q), relative to K, if for every point $x \in X$ and every neighborhood U of x there is a neighborhood V of x in U such that the natural map $V \to U$ is cohomologically trivial (respectively, in dimension q), relative to K. We say then that X is clc_K (respectively, $q\text{-}clc_K$), and that X is clc_K^r if it is $q\text{-}clc_K$ for all $q \leq r$.

The space X is homologically locally connected (respectively, in dimension q) relative to K, if for every $x \in X$ and every neighborhood U of x there exists a neighborhood V of x in U such that $V \to U$ is homologically trivial (respectively, in dimension q), relative to K. We say then that X is hlc_K (respectively $q\text{-}hlc_K$), and that X is hlc_K^r if it is $q\text{-}hlc_K$ for all $q \leq r$.

We recall that X is locally connected if and only it is clc_K^0.

6.4. PROPOSITION. *Let* B *be a K-module, and let* N *be an injective resolution of* B *of finite degree. Let* $\mathscr{A}$ *be the differential graded sheaf defined on* X *by the presheaf* $U \to \mathrm{Hom}\,(\Gamma_c(\mathscr{C}_H^*(X; K)|U), N)$. *Then*

(a) $\mathscr{A}(U) = \mathrm{Hom}\,(\Gamma_c(\mathscr{C}_H^*(X; K)|U), N)$;

(b) *the sheaf* $\mathscr{A}$ *is bounded below;* $\mathscr{A}^q$ *is flabby for each* q;

(c) *if* X *is* hlc_K^r, *then* $\mathscr{H}^q(\mathscr{A}) = 0$ $(q \leq r;\ q \neq 0)$ *and* $\mathscr{H}^0(\mathscr{A}) = B$.

Proof. Parts (a) and (b) follow from 2.7. Further, by 2.7 (3), there exists, for each Y open in X, an exact sequence

$$(1) \qquad 0 \to \mathrm{Ext}\,(H_{q-1}^c(Y); B) \to H^q(\mathscr{A}(Y)) \to \mathrm{Hom}\,(H_q^c(Y); B) \to 0$$

which is natural with respect to inclusion maps. Let now X be hlc_K^r, let $x \in X$, and let $W \subset V \subset U$ be open neighborhoods of x such that $W \to V$ and $V \to U$ are homologically trivial in dimensions q - 1, q $(q \leq r)$. We then consider the commutative diagram whose rows are the exact sequences (1) for Y = U, V, W and whose vertical maps are defined by inclusions; simple diagram-chasing shows that

$$H^q(\mathscr{A}(U)) \to H^q(\mathscr{A}(W))$$

is trivial. This proves (c).

6.5. THEOREM. *If* X *is* hlc_K^r, *and* B *is a K-module, there exists for* $q \leq r$ *an exact sequence*

$$0 \to \mathrm{Ext}\,(H_{q-1}^c(X), B) \to H^q(X; B) \to \mathrm{Hom}\,(H_q^c(X), B) \to 0.$$

Proof. This follows at once from the preceding proposition and the standard spectral sequence of a differential graded sheaf [7, p. 176]. In particular, this last implies that $H^q(X; B) = H^q(\Gamma(\mathscr{A}))$, where $\mathscr{A}$ is the sheaf of the preceding proposition $(q \leq r)$.

6.6. THEOREM. *The space* X *is* clc_K *if and only if it is* hlc_K. *If* X *is* clc_K^r *(respectively* hlc_K^r), *then it is* hlc_K^{r-1} *(respectively* clc_K^r). *If* K *is a field, the properties* clc_K^r *and* hlc_K^r *are equivalent.*

Assume X to be clc_K (respectively clc_K^r). Let $x \in X$, and let U be a compact neighborhood of x. Choose compact neighborhoods $W \subset V \subset U$ of x such that $V \to U$ and $W \to V$ are cohomologically trivial in all dimensions (respectively, in dimensions $q + 1$, q, with $q \leq r - 1$). Then, by the argument used in proving 6.4(c), it follows from 3.3(a) that $W \to U$ is homologically trivial in all dimensions (respectively, in dimension q). The converse statement is proved similarly, with the use of open neighborhoods, and with reference 6.5 taking the place of 3.3(a).

6.7. LEMMA. *Let X be a connected and locally connected space, and let F be a compact subset of X. Then there exists a compact connected subset A of X such that* $F \subset A$.

This lemma is elementary and well known. It is a special case of Theorem 3.3 on p. 105 of [9].

6.8. PROPOSITION. *Let q be an integer. Assume that for each integer* $m \leq q$ *the space X has the following property.*

$(F_{m,K})$: *For each* $x \in X$ *and each neighborhood V of x, there exists a neighborhood U of x such that* Im $i_{m,UV} : H_m^c(U; K) \to H_m^c(V; K)$ *is a finitely generated module.*

Let $Q \subset P$ *be subspaces of X such that* $\overline{Q}$ *is compact and contained in the interior of* P. *Then* Im $i_{m,QP}$ *is a finitely generated module for* $m \leq q$.

The proof is essentially the same as the proof of Proposition 6.2 in [2], and we sketch it briefly. First, by use of suitable coverings of $\overline{Q}$ and of induction on the number of elements of such coverings, the proof is easily reduced to that of the following statement: let U_1, $U_2 \subset P$ be open and relatively compact in X, and such that the image of $H_m^c(U_i; K) \to H_m^c(P; K)$ is finitely generated for $m \leq q$, $i = 1, 2$. Let $V_i \subset U_i$ be open with $\overline{V}_i \subset U_i$ $(i = 1, 2)$, and let $V = V_1 \cup V_2$, $U = \overline{U}_1 \cup U_2$. Then Im $i_{m,V,P}$ is finitely generated for $m \leq q$.

Proceeding by induction on dimension, we may assume Proposition 6.8 to be true for $q - 1$. In particular, Im $i_{m,VP}$ $(m \leq q - 1)$ and the image of

$$H_{q-1}^c(V_1 \cap V_2; K) \to H_{q-1}^c(U_1 \cap U_2; K)$$

are finitely generated. That Im $i_{q,VP}$ is finitely generated follows then by inspection of the commutative diagram

$$
\begin{array}{ccccc}
H_q^c(V_1; K) + H_n^c(V_2; K) & \to & H_q^c(U_1; K) + H_q^c(U_2; K) & \searrow & \\
\downarrow & & \downarrow & \nearrow & H_q^c(P; K), \\
(1) \quad H_q^c(V; K) & \to & H_q^c(U; K) & & \\
\downarrow & & \downarrow & & \\
H_{q-1}^c(V_1 \cap V_2; K) & \to & H_{q-1}^c(U_1 \cap U_2; K) & &
\end{array}
$$

where the vertical maps are parts of the exact sequences introduced in 3.9 and where the other maps are defined by inclusions.

6.9. THEOREM. *Let X be compact and* hlc_K^r. *Then, for each integer* $q \leq r$, *the modules* $H_q(X; K)$ *and* $H^q(X; K)$ *are finitely generated,*

Ext $(H^{q+1}(X; K), K)$ *is the torsion submodule of* $H_q(X; K)$, *and*

Ext $(H_{q-1}(X; K), K)$ *is the torsion submodule of* $H^q(X; K)$.

By 6.8, $H_q(X; K)$ is finitely generated. The other assertions follow then from 3.3(a) and 6.5.

Remark. Since the property clc_K^{r+1} implies hlc_K^r by 6.6, we could also have derived 6.9 from 3.3(a), 6.5, and from the known fact that if X is compact and clc_K^r, then $H^q(X; K)$ is finitely generated for $q \leq r$. This last statement, or rather the more general analogue of 6.8 for cohomology with closed supports (see [2, Prop. 6.3] for references), can also be given a proof similar to that of 6.8, the exact sequences of the diagram (1) being replaced by standard Mayer-Vietoris sequences, and property $(F_{m,K})$ by

$(F_{m,K}^*)$: For each $x \in X$ and each neighborhood V of x, there exists a neighborhood U of x in V such that $H^m(V; K) \to H^m(U; K)$ has a finitely generated image.

It is well known that $(F_{m,K}^*)$ is equivalent to $m\text{-clc}_K$. If K is a field, then $H_m^c(X; K) = \mathrm{Hom}\,(H^m(X; K), K)$ for X compact; therefore $F_{m,K}$ and $F_{m,K}^*$ are equivalent; in view of 6.6, the condition $F_{m,K}$ for $m \leq r$ then implies the property hlc_K^r, hence is equivalent to it. We do not know whether $F_{m,K}$ for all $m \leq r$ is equivalent to hlc_K^r when K is not a field, or whether $F_{m,K}$ is equivalent to $m\text{-hlc}_K$. (According to the referee, $F_{m,K}$ is indeed equivalent to $m\text{-hlc}_K$.)

6.10. PROPOSITION. (1) *If X is locally connected, then*

$$H_{-1}(X; K) = H_{-1}^c(X; K) = 0 .$$

(2) *If X is* hlc_K^1, *then* $H_0^c(X; K)$ *is the free module generated by the components of* X.

If X is locally connected, then $H_c^0(X; K)$ is a free module (generated by the compact components of X); the equality $H_{-1}(X; K) = 0$ follows then from 3.4. If a space Y is compact and connected, then $H_c^0(Y; K) = K$, and therefore $H_{-1}^c(X; K) = 0$ follows from 3.4 and 6.7.

It is known that $H^1(X; K)$ is torsion-free for any compact space X. Let now X be hlc_K^1, hence also clc_K^1 (6.6). By a theorem recalled in the previous remark, if $Q \subset P \subset X$ and Q is compact and lies in the interior of P, then the image of the restriction map $H^1(P; K) \to H^1(Q; K)$ is finitely generated; being also torsion-free, it is then a projective module [5, VII, 4.1]. From this it follows immediately that

$$\mathrm{dir}\ \lim \mathrm{Ext}\,(H^1(P; K), K) = 0 ,$$

where P runs through the compact subsets of X. The assertion (2) is then a consequence of 3.3, 3.4, and 6.7.

Remark. If K is a field, it is clearly enough to assume in (2) that X is locally connected. We do not know whether this is true in the general case.

7. FINITE-DIMENSIONAL SPACES AND GENERALIZED MANIFOLDS

7.1. A topological space X is *finite-dimensional* over K if there exists an integer n such that for each sheaf $\mathscr{S}$ on X and each paracompactifying family Φ, we have $H_\Phi^q(X; \mathscr{S}) = 0$ for $q > n$. The least such integer is called the dimension of X over K and is denoted by $\dim_K X$.

The dimension of X over K is also the least integer n such that $H_c^{n+1}(U; K) = 0$ for each open subset of X. (See [3, Chap. I, Section 5]; the proof is given there for a principal ideal domain of coefficients, but it is also valid for a Dedekind ring.)

7.2. THEOREM. *Let* $\dim_K X \leq n$, *and let* Φ *be a family of supports on* X. *Then*

(a) $H_q^\Phi(X; K) = H_q^x(X; K) = 0 \quad (q \geq n + 1; x \in X)$.

(b) *If* $\mathscr{A}$ *is a c-soft resolution of* K, $\mathscr{S}$ *is a torsion-free sheaf on* X, *and* Φ *is paracompactifying, then*

$$H_*^\Phi(X; \mathscr{S}) = H_*(\Gamma_\Phi(D(\mathscr{A}) \otimes \mathscr{S})).$$

The assertion (a) follows from 3.3(a), applied to the elements of Φ, and from 3.4.

As was noticed in the proof of 3.3(b), there is a homomorphism

$$f: \mathscr{C}_H(X; K) \to D(\mathscr{A})$$

which induces an isomorphism of the derived sheaves. Since $\mathscr{S}$ is torsion-free, it follows from the Künneth rule that the homomorphism

$$f \otimes 1: \mathscr{C}_H(X; K) \otimes \mathscr{S} \to D(\mathscr{A}) \otimes \mathscr{S}$$

also induces an isomorphism of the derived sheaves. Replacing in both sheaves the degrees by the opposite ones, we deduce from Theorem 4.6.2 and Section 4.13 of [7, Chap. II] that $f \otimes 1$ yields an isomorphism of $H(C_H^\Phi(X; \mathscr{S}))$ onto $H(\Gamma_\Phi(D(\mathscr{A}) \otimes \mathscr{S}))$, which proves (b).

7.3. THEOREM. *Let* $\dim_K X \leq n$. *Let* $\mathscr{S}$ *be a sheaf on* X, *and let* Φ *be a family of supports on* X. *Then, if either* $\mathscr{S} = K$ *or* Φ *is paracompactifying, there exists a canonical isomorphism*

$$\triangle: H_n^\Phi(X; \mathscr{S}) \to H_\Phi^0(X; \mathscr{H}_n(X; K) \otimes \mathscr{S}) = \Gamma_\Phi(\mathscr{H}_n(X; K) \otimes \mathscr{S}).$$

Let us denote by $\mathscr{B}$ the sheaf $\mathscr{C}_H(X; \mathscr{S})$ with the modified grading

$$\mathscr{B}^q = \mathscr{C}_H(X; \mathscr{S})_{n-q}.$$

Then

(1) $$H^q(\Gamma_\Phi(\mathscr{B})) = H_{n-q}^\Phi(X; \mathscr{S})$$

by definition. Since $\mathscr{C}_H(X; K)$ is torsion-free (3.1), we have

$$\mathscr{H}^q(\mathscr{B}) = \mathscr{H}_{n-q}(X; K) \otimes \mathscr{S},$$

hence in particular, by 7.2(a),

(2) $$\mathscr{H}^0(\mathscr{B}) = \mathscr{H}_n(X; K) \otimes \mathscr{S}; \mathscr{H}^q(\mathscr{B}) = 0 \quad (q < 0).$$

The sheaf $\mathscr{B}$ is flabby if $\mathscr{S} = K$, and it is Φ-fine if Φ is paracompactifying (3.1). By [7, II, 4.4.3], we have therefore in both cases

(3) $$H_\Phi^p(X; \mathscr{A}) = 0 \quad (p \geq 1).$$

This last fact and [7, II, 4.6.1] imply the existence of a spectral sequence in which

$$E_2^{p,q} = H_\Phi^p(X; \mathscr{H}^q(\mathscr{B}))$$

and where E_∞ is the graded module associated with $H^*(\Gamma_\Phi(\mathscr{B}))$ suitably filtered. Since $\dim_K X$ is finite, the filtration of the underlying double complex may be assumed to be defined by a finite number of submodules [7, p. 195], hence we have regular convergence. By (2), we have

$$(4) \qquad E_2^{p,q} = 0 \quad (p, q \geq 0), \qquad E_2^{p,0} = H_\Phi^p(X;\, \mathscr{H}_n(X;\, K) \otimes \mathscr{S})\,;$$

therefore

$$E_2^{0,0} = E_\infty^{0,0} = H^0(\Gamma_\Phi(\mathscr{B}))\,,$$

and the theorem follows from (1).

7.4. *Remark.* If U is an open subspace of X, the restriction of cross sections to U defines a homomorphism of the above spectral sequences for X and U, and this implies the commutativity of the diagram

$$
\begin{array}{ccc}
H_n^\Phi(X;\, \mathscr{S}) & \overset{\triangle}{\to} & \Gamma_\Phi(\mathscr{H}_n(X;\, K) \otimes \mathscr{S}) \\
\downarrow & & \downarrow \\
H_n^{\Phi \cap U}(U;\, \mathscr{S}) & \overset{\triangle}{\to} & \Gamma_{\Phi \cap U}(\mathscr{H}_n(U;\, K) \otimes \mathscr{S}|U)
\end{array}
$$

when either $\mathscr{S} = K$ or Φ and $\Phi \cap U$ are paracompactifying in X and U respectively.

7.5. DEFINITION. *A topological space* X *is a homology* n-*manifold over* K (*briefly, an* n-hm_K) *if*

(1) X *is finite-dimensional over* K;

(2) *the sheaf* $\mathscr{H}_q(\mathscr{C}_H(X;\, K))$ *is zero for* $q \neq n$;

(3) *the sheaf* $\mathscr{H}_n(\mathscr{C}_H(X;\, K))$ *is locally isomorphic with the constant sheaf* K.

If X is an n-hm_K, the sheaf $\mathscr{H}_n(\mathscr{C}_H(X;\, K))$ will be denoted by $\mathscr{I}$ or $\mathscr{I}_X$, and it will be called the *orientation sheaf.* X is *orientable* if $\mathscr{I}_X$ is isomorphic to the constant sheaf. Such an isomorphism is called an *orientation* of X.

Note that in 7.5(3) it was assumed that $\mathscr{I}_X$ is locally isomorphic to K. Thus homology manifolds have been defined in a manner which makes them automatically locally orientable. If condition (3) is replaced by

(3)' for any point $x \in X$, the stalk of the sheaf $\mathscr{I}_X$ is isomorphic to K, we get the condition K - n of [3, II].

7.6. THEOREM. *Let* X *be a homology* n-*manifold over* K (*or a space satisfying condition* K - n *of* 7.5); *let* $\mathscr{I}$ *be its orientation sheaf,* Φ *a family of supports, and* $\mathscr{S}$ *a sheaf on* X. *If either* $\mathscr{S} = K$ *or* Φ *is paracompactifying, then, for all integers* q, *there exists a canonical isomorphism*

$$\triangle\colon H_{n-p}(X;\, \mathscr{S}) \to H^p(X;\, \mathscr{I} \otimes \mathscr{S})\,.$$

The proof is quite similar to that of 7.2. We have now, in the notation of that proof,

$$(6) \qquad \mathscr{H}^0(\mathscr{B}) = \mathscr{I} \otimes \mathscr{S}, \qquad \mathscr{H}^q(\mathscr{B}) = 0 \quad (q \neq 0)\,.$$

Therefore, in the spectral sequence considered there, we have

$$(7) \qquad E_2^{p,q} = 0 \quad (q \neq 0), \qquad E_2^{p,0} = H^p(X; \mathscr{T} \otimes \mathscr{S}),$$

whence

$$H_\Phi^p(X; \mathscr{T} \otimes \mathscr{S}) = E_2^{p,0} = E_\infty^{p,0} = H^p(\Gamma_\Phi(\mathscr{B})) = H_{n-p}^\Phi(X; \mathscr{S}).$$

7.7. PROPOSITION. *Let* X *be an* n-hm$_K$, *and* $\mathscr{T}$ *its orientation sheaf. Then there exists a sheaf* $\mathscr{L}$, *locally isomorphic to* K, *and an isomorphism*

$$\alpha(\mathscr{L}): \mathscr{L} \otimes \mathscr{T} \to K.$$

If $\mathscr{L}'$ *is another such sheaf and* $\alpha(\mathscr{L}'): \mathscr{L}' \otimes \mathscr{T} \to K$ *an isomorphism, there exists a unique isomorphism* $\beta: \mathscr{L} \to \mathscr{L}'$ *such that the diagram*

$$
\begin{array}{ccc}
\mathscr{L} \otimes \mathscr{T} & \xrightarrow{\ \beta \otimes 1\ } & \mathscr{L}' \otimes \mathscr{T} \\
& \alpha(\mathscr{L}') \searrow \quad \swarrow \alpha(\mathscr{L}') & \\
& K &
\end{array}
$$

is commutative.

A sheaf locally isomorphic to K may be thought of as a fibre bundle whose structural group is the group of units in K, so that our proposition is a standard fact about line bundles. We sketch the proof. Let $\{U_i\}_{i \in I}$ be an open covering such that $\mathscr{T}|U_i$ is isomorphic with K on U . Let $f_i: \mathscr{T}|U_i \to K \times U_i$ be such an isomorphism. The transition function $\theta_{ij}: U_i \cap U_j \to G$ (where G is the group of units of K) is then defined by $f_j f_i^{-1}(k, x) = (\theta_{ij}(x) \cdot k, x)$. The bundle $\mathscr{L}$ is then defined by means of the transition functions f_{ij}^{-1}, and $\alpha(\mathscr{L})$ is the obvious map. Details are left to the reader.

7.8. DEFINITION. *Let* X *be an* n-hm$_K$, *and* $\mathscr{T}$ *its orientation sheaf. The inverse of* $\mathscr{T}$ *is a sheaf* $\mathscr{L}$ *on* X, *locally isomorphic to* K, *together with an isomorphism* $\alpha(L): \mathscr{L} \otimes \mathscr{T} \to K.$

Note that if X is orientable, then both $\mathscr{L}$ and $\mathscr{T}$ are isomorphic to K.

7.9. THEOREM. *Let* X *be an* n-hm$_K$; *let* $\mathscr{S}$ *be a sheaf on* X, Φ *a paracompactifying family of supports,* $\mathscr{T}$ *the orientation sheaf, and* $\mathscr{L}$ *the inverse of* $\mathscr{T}$. *Then, for each integer* q, *there exists an isomorphism*

$$\alpha: H_\Phi^q(X; \mathscr{S}) \to H_{n-q}^\Phi(X; \mathscr{L} \otimes \mathscr{S}).$$

Using 7.6, and the associativity and commutativity of tensor products, we have

$$H_\Phi^q(X; \mathscr{S}) = H_\Phi^q(X; \mathscr{L} \otimes \mathscr{T} \otimes \mathscr{S}) = H_{n-q}^\Phi(X; \mathscr{L} \otimes \mathscr{S}).$$

7.10. DEFINITION. *A topological space* X *is a cohomology* n-*manifold over* K (*an* n-cm$_K$) *if*

(1) *it is finite-dimensional over* K;

(2) *for each open set* U *in* X *and each integer* $q \neq n$, *each point* $x \in U$ *has an open neighborhood* V *in* U *such that image* $H_c^q(V; K) \to H_c^q(U; K)$ *is zero;*

(3) *for each* $x \in X$ *and each neighborhood* U *of* x, *there exists an open neighborhood* V *of* x *and a free submodule* A *of* $H_c^n(V; K)$ *with a single generator, such that*

every point $y \in V$ *has a fundamental system of open neighborhoods* W *for which image* $H_c^n(W; K) \to H_c^n(U; K)$ *is equal to* A.

7.11. We recall that an n-cm$_K$ is clc$_K$ [2], [9] and has dimension n over K [3, I, 3.1]. It is said to be *orientable* if each connected component has the property assigned to V in (3). If X is connected and orientable, then $H_c^n(X; K) = K$ and for each open connected subset U, the map $H_c^n(U; K) \to H_c^n(X; K)$ is an isomorphism [2, I, 4.3]. With the above definition, cohomology manifolds are automatically locally orientable. If in (3) the requirement is made only for $y = x$, then one gets a condition paralleling (3') of 7.2. If K is a field, then the spaces satisfying (1), (2), and this weaker version of (3) are the generalized manifolds of Wilder [9] (see also [2], [3]). It is not known whether these generalized manifolds are always locally orientable.

7.12. THEOREM. *Let* X *be a topological space. Then the following two conditions are equivalent*

(1) X *is an (orientable)* n-cm$_K$;

(2) X *is an (orientable)* n-hm$_K$ *and is* hlc$_K$.

Let X be an n-cm$_K$. Then it is clc$_K$ (7.11), hence locally connected, and hlc$_K$ (6.6). That it is an n-hm$_K$ (and orientable if it is an orientable n-cm$_K$) follows then easily from 3.3 and 7.10.

Let now X satisfy (2). It is then clc$_K$ by 6.6. For U open and orientable, $H_c^q(U; K) = H_{n-q}^c(U; K)$ by 7.6, hence, if we take into account part (2) of 6.10, the condition that X be hlc$_K$ becomes exactly the condition that X be an n-cm$_K$. Let now X be an orientable n-hm$_K$. Using 6.10, we see that for each component Y of X, $H_c^n(Y; K) = K$. This implies that X is also orientable as a cohomology manifold [2, I, 4.3].

Remark. As was recalled in 7.5, an n-cm$_K$ is always clc$_K$. This is in fact a special case of a theorem relating cohomological local connectedness and local Betti numbers (see [2], [3] for more details). We do not know whether a similar theorem holds in homology, and in particular, whether a homology manifold is always hlc$_K$. Also, it follows from 7.1 and 7.6 that if X is an n-hm$_K$, then dim$_K X \leq n + 1$. We do not know whether this bound can be brought down to n, as in the case of cohomology manifolds.

7.13. *Remarks on* [2]. The discussion centering around Theorem 7.2 of [2], where the author tries to put into relations the duality theorems of [2] and [9] is marred by a mistake, pointed out by the referee of this paper, and by a misprint. Further, as was mentioned by the referee, a result of the present paper answers a question raised in [2]. We take this opportunity to rectify and complete that part of [2]. As in [2] let $h_r^*(X, K)$ be the projective limit of the groups $H^r(F; K)$, where F runs through the compact subsets of X, with respect to the usual restriction maps. If K is a field, and X is clc^r, then

$$(1) \qquad\qquad h_r^* = H^r(X; K).$$

In fact, we can take a cofinal set of compact subsets F_α which are the closures of their interiors Int F_α. Then

$$h_r^* = \lim_{\longleftarrow} H_\alpha^r(F_\alpha; K) = \lim_{\longleftarrow} H^r(\text{Int } F_\alpha; K).$$

Our assertion follows then from 3.4, 6.5, 6.6 and from the elementary fact that if a vector space V is the inductive limit of vector spaces V_α, then its dual V^* is the projective limit of the spaces V_α^*.

On line 4 of [2, p. 236], one should read "inductive" instead of "projective." Thus $h^r(X; K)$ is the *inductive* limit of the Čech homology groups $\check{H}_r(F; K)$ of the compact subsets of X. Now, K being again a field, $H^r(F; K)$ is the dual space of $\check{H}_r(F; K)$, hence, by the previous remark,

$$(2) \qquad h_r^* = \mathrm{Hom}\,(h^r(X; K), K)\,.$$

Assume now that X is a connected, paracompact, orientable $n\text{-cm}_K$, where K is a field. Then, using property (P, Q), we see as in [2], that $h^r(X; K)$ and

$$h_r(X; K) = H_c^r(X; K)$$

have at most countable dimension. By (1), (2), 6.5, and [2, 3.2], $h_r(X; K)$ and $h^{n-r}(X; K)$ have the same dual space, namely $H^{n-r}(X; K)$, hence they are isomorphic. Conversely, if $h_r(X; K) = h^{n-r}(X; K)$, then by (1), (2), $H^{n-r}(X; K) = \mathrm{Hom}\,(H_c^r(X; K), K)$. Consequently, for connected orientable paracompact $n\text{-cm}_K$ over a field K [2, 3.2] and [2, 7.2], which is the author's interpretation of a result of Wilder, are equivalent.

8. APPENDIX

In this appendix, we introduce and discuss some properties of a certain connected sequence of bifunctors on the category of sheaves, which yields the homology groups when the first variable is put equal to K. To start with, we assume only that our ground-ring K is an integral domain. Its field of fractions is denoted by K^*.

8.1. For any K-module B, consider B as an abelian group, imbed B in $\overline{B}$ as described in Section 1. Let $\widetilde{B} = \mathrm{Hom}_Z\,(K, \overline{B})$, where Z is the ring of integers. We now have an exact sequence

$$0 \to \mathrm{Hom}_Z\,(K, B) \to \mathrm{Hom}_Z\,(K, \overline{B})\,,$$

and $B = \mathrm{Hom}_K(K, B) \subset \mathrm{Hom}_K\,(K, \overline{B})$. So we have functorially imbedded any K-module B in an injective K-module $\widetilde{B}$.

For any sheaf $\mathscr{A}$, let $I(\mathscr{A})$ be the sheaf $U \to \Pi_{x \in U}\, \widetilde{\mathscr{A}}_x$. The sheaf $I(\mathscr{A})$ is injective, and as in Section 1, we may define $\mathscr{C}^*(X; \mathscr{A})$ to be the canonical injective resolution of $\mathscr{A}$.

8.2. Let R(K) be an injective resolution of K, considered as a K-module, chosen once and for all. If $\mathscr{L}$ is a sheaf, then $D(\mathscr{L})$ denotes the sheaf defined by the presheaf $U \to \mathrm{Hom}\,(\Gamma_c(\mathscr{L}\,|\,U), R(K))$. If $\mathscr{A}$ and $\mathscr{B}$ are sheaves, let $\mathscr{C}_H(X; \mathscr{A}, \mathscr{B})$ be the differential graded sheaf given by the presheaf

$$U \to \mathrm{Hom}\,(\Gamma_c(\mathscr{C}^*(X; \mathscr{A})\,|\,U), R(K)) \otimes \mathscr{B}(U)\,.$$

Thus $\mathscr{C}_H(X; \mathscr{A}, \mathscr{B}) = \mathscr{C}_H(X; \mathscr{A}, K) \otimes \mathscr{B}$, and $\mathscr{C}_H(X; K, K)$ is essentially the sheaf defined in 3.1 and denoted by $\mathscr{C}_H(X; K)$.

If $\mathscr{L}$ is c-soft, then $D(\mathscr{L})$ is flabby, $D(\mathscr{L})(U)$ is $\mathrm{Hom}\,(\Gamma_c(\mathscr{L}\,|\,U), R(K))$; if $\mathscr{L}$ is injective, $D(\mathscr{L})$ is torsion-free. The sheaf $\mathscr{C}_H(X; \mathscr{A}, K)$ is a $\mathscr{K}$-module, flabby, torsion-free, and Φ-fine for any paracompactifying family Φ; and

$$\mathscr{C}_H(X; \mathscr{A}, K) = \operatorname{Hom}(\Gamma_c(\mathscr{C}^*(X; \mathscr{A})\,|\,U), R(K)).$$

This follows from 2.3, 2.7, and 2.9 if we take into account the fact that these propositions and their proofs are valid without any change if K is an arbitrary integral domain.

8.3. Define

$$T_n(\mathscr{A}, \mathscr{B}) = H_n(\Gamma(\mathscr{C}_H(X; \mathscr{A}; \mathscr{B})), \qquad T_n^{\Phi}(\mathscr{A}, \mathscr{B}) = H_n(\Gamma_{\Phi}(\mathscr{C}_H(X; \mathscr{A}; \mathscr{B})),$$

where Φ is a paracompactifying family on X. Then,

$$T_n(K; \mathscr{B}) = H_n(X; \mathscr{B}), \qquad T_n^{\Phi}(K; \mathscr{B}) = H_n^{\Phi}(X; \mathscr{B}).$$

Note also that if $\mathscr{L}$ is any injective resolution of $\mathscr{A}$, we have

$$T_n(\mathscr{A}, \mathscr{B}) = H_n(\Gamma(D(\mathscr{L}) \otimes \mathscr{B})), \qquad T_n^{\Phi}(\mathscr{A}, \mathscr{B}) = H_n(\Gamma_{\Phi}(D(\mathscr{L}) \otimes \mathscr{B})).$$

In fact, there then exist homomorphisms

$$\alpha: \mathscr{C}^*(X; \mathscr{A}) \twoheadrightarrow \mathscr{L}, \qquad \beta: \mathscr{L} \to \mathscr{C}^*(X; \mathscr{A})$$

such that $\alpha \circ \beta$ and $\beta \circ \alpha$ are homotopic to the identity. They induce homomorphisms

$$\alpha': D(\mathscr{L}) \otimes \mathscr{B} \to \mathscr{C}_H(X; \mathscr{A}, \mathscr{B}), \qquad \beta': \mathscr{C}_H(X; \mathscr{A}, \mathscr{B}) \to D(\mathscr{L}) \otimes \mathscr{B}$$

such that $\alpha' \circ \beta'$ and $\beta' \circ \alpha'$ are homotopic to the identity, and our assertion follows.

Let $0 \to \mathscr{B}' \to \mathscr{B} \to \mathscr{B}'' \to 0$ be an exact sequence of sheaves on X. Since $\mathscr{C}_H(X; \mathscr{A})$ is torsion-free (8.2), the sequence

$$0 \to \mathscr{C}_H(X; \mathscr{A}, \mathscr{B}') \to \mathscr{C}_H(X; \mathscr{A}, \mathscr{B}) \to \mathscr{C}_H(X; \mathscr{A}, \mathscr{B}'') \to 0$$

is exact. Since $\mathscr{C}_H(X; \mathscr{A})$ is Φ-fine (8.2), the sheaf $\mathscr{C}_H(X; \mathscr{A}, \mathscr{B}')$ is Φ-soft; therefore [7, Théorème 3.5.2, p. 153] the sequence

$$0 \to \Gamma_{\Phi}\mathscr{C}_H(X; \mathscr{A}, \mathscr{B}') \to \Gamma_{\Phi}\mathscr{C}_H(X; \mathscr{A}, \mathscr{B}) \to \Gamma_{\Phi}\mathscr{C}_H(X; \mathscr{A}, \mathscr{B}'') \to 0$$

is exact and yields an exact sequence

$$(1) \qquad \cdots \to T_n^{\Phi}(\mathscr{A}, \mathscr{B}') \to T_n^{\Phi}(\mathscr{A}, \mathscr{B}) \to T_n^{\Phi}(\mathscr{A}, \mathscr{B}'') \to T_{n-1}^{\Phi}(\mathscr{A}, \mathscr{B}') \to \cdots.$$

Let now $0 \to \mathscr{A}' \to \mathscr{A} \to \mathscr{A}'' \to 0$ be an exact sequence of sheaves. It can be extended to an exact sequence $0 \to \mathscr{L}' \to \mathscr{L} \to \mathscr{C}^*(X; A'') \to 0$ of injective resolutions [7, p. 261]. Then, since R(K) is injective, the sequence

$$0 \to \mathscr{C}_H(X; \mathscr{A}'') \to D(\mathscr{L}) \to D(\mathscr{L}') \to 0$$

is also exact. Its elements are torsion-free (8.2), and therefore

$$0 \to \mathscr{C}_H(X; \mathscr{A}'') \otimes \mathscr{B} \to D(\mathscr{L}) \otimes \mathscr{B} \to D(\mathscr{L}') \otimes \mathscr{B} \to 0$$

158   A. BOREL and J. C. MOORE

is exact. The sheaf $\mathscr{C}_H(X; \mathscr{A}'')$ is also Φ-fine (8.2); therefore its tensor product
with $\mathscr{B}$ is Φ-soft, and [7, 3.5.2, p. 153] the sequence

$$0 \to \Gamma_\Phi(\mathscr{C}_H(X; \mathscr{A}'', \mathscr{B})) \to \Gamma_\Phi(D(\mathscr{L}) \otimes \mathscr{B})) \to \Gamma_\Phi(D(\mathscr{L}') \otimes \mathscr{B}) \to 0$$

is exact. In view of the remark made after the definition of $T_n(\mathscr{A}, \mathscr{B})$, this yields an
exact sequence

$$(2) \qquad \cdots \to T_n^\Phi(\mathscr{A}'', \mathscr{B}) \to T_n^\Phi(\mathscr{A}, \mathscr{B}) \to T_n^\Phi(\mathscr{A}', \mathscr{B}) \to T_{n-1}^\Phi(\mathscr{A}'', \mathscr{B}) \to \cdots .$$

We have therefore shown that the sequence of bifunctors $T_n^\Phi(\mathscr{A}, \mathscr{B})$ on the category of sheaves on X, contravariant in the first variable, covariant in the second
variable, is connected.

8.4. If the sheaf $\mathscr{A}$ is injective, we may consider $\mathscr{A}$ to be its own injective
resolution. Then $T_n^\Phi(\mathscr{A}, \mathscr{B}) = 0$ for $n > 0$, if $\mathscr{A}$ is injective. If the global dimension of the ring K is N, we may assume that $R(K)^q = 0$ for $q > N$; thus
$T_n^\Phi(\mathscr{A}, \mathscr{B}) = 0$ for $n < -N$, and the functor $T_{-N}^\Phi(\mathscr{A}, \mathscr{B})$ is right-exact.

If K is a field, we then have $T_q^\Phi(\mathscr{A}, \mathscr{B}) = 0$ for $q < 0$, and $T_q^\Phi(\mathscr{A}, \mathscr{B}) = 0$ for
$q \neq 0$ and $\mathscr{A}$ injective. This means that in this case, for $n \geq 0$, we may consider
$T_n^\Phi(\mathscr{A}, \mathscr{B})$ as the n-th left-derived functor of the functor $T_0^\Phi(\mathscr{A}, \mathscr{B})$, which is right-
exact.

If K is of global dimension 1, that is, if K is a Dedekind ring which is not a field,
then $T_q^\Phi(\mathscr{A}, \mathscr{B}) = 0$ for $q < -1$; and if in addition $\mathscr{A}$ is injective, then $T_q^\Phi(\mathscr{A}, \mathscr{B}) = 0$
for $q > 0$. Then, if we show that $T_0^\Phi(\mathscr{A}, \mathscr{B}) = 0$ for $\mathscr{A}$ injective, we may consider
$T_n^\Phi(\mathscr{A}, \mathscr{B})$ as the n-th left-derived functor of the right-exact functor $T_{-1}^\Phi(\mathscr{A}, \mathscr{B})$.
Then it remains to show that in this case $T_0^\Phi(\mathscr{A}, \mathscr{B}) = 0$ for $\mathscr{A}$ injective, which we
do next.

8.5. LEMMA. *Let* K *be a Dedekind ring which is not a field; let* Φ *be a paracompactifying family, and* $\mathscr{A}$ *an injective sheaf on* X. *Then* $T_0^\Phi(\mathscr{A}, \mathscr{B}) = 0$.

We have an exact sequence of presheaves which, for each open U in X is the
exact sequence

$$0 \to \mathrm{Hom}\,(\Gamma_c(\mathscr{A}|U), K) \to \mathrm{Hom}\,(\Gamma_c(\mathscr{A}|U), K^*) \to \mathrm{Hom}\,(\Gamma_c(\mathscr{A}|U), K^*/K)$$

$$\to \mathrm{Ext}\,(\Gamma_c(\mathscr{A}|U), K) \to 0 .$$

However since $\Gamma_c(\mathscr{A}|U)$ is injective, $\mathrm{Hom}\,(\Gamma_c(\mathscr{A}|U), K) = 0$, and thus we have an
exact sequence

$$0 \to \mathrm{Hom}\,(\Gamma_c(\mathscr{A}|U), K^*) \to \mathrm{Hom}\,(\Gamma_c(\mathscr{A}|U), K^*/K) \to \mathrm{Ext}\,(\Gamma_c(\mathscr{A}|U), K) \to 0 .$$

Since K^* and K^*/K are injective, the two left-hand presheaves are sheaves, and
they are flabby (2.3). Thus (by [7, Théorème 3.1.2, p. 148]) the right-hand presheaf
$U \to \mathrm{Ext}\,(\Gamma_c(\mathscr{A}|U), K)$ is also a sheaf, and is flabby. Let this exact sequence of
sheaves be denoted by

$$0 \to \mathscr{C}_0 \to \mathscr{C}_{-1} \to \mathscr{E} \to 0 .$$

Now we may consider $\mathscr{C} = \mathscr{C}_0 + \mathscr{C}_{-1}$ as a differential graded sheaf whose differential
is the natural map $\mathscr{C}_0 \to \mathscr{C}_{-1}$; and we may use $\mathscr{C}$ instead of $\mathscr{C}_H(X; \mathscr{A}, K)$. For any
sheaf $\mathscr{B}$, we have $H_q(\Gamma_\Phi(\mathscr{C} \otimes \mathscr{B})) = T_q^\Phi(\mathscr{A}, \mathscr{B})$, for each q. Now the stalks of the

sheaf $\mathscr{C}_0$ are modules over K^*, and they are injective. Therefore, for each point $x \in X$,

$$0 \to (\mathscr{C}_0)_x \to (\mathscr{C}_{-1})_x \to \mathscr{E}_x \to 0$$

is split exact, and since $(\mathscr{C}_{-1})_x$ is torsion-free, so is $\mathscr{E}_x$. This means that for each sheaf $\mathscr{B}$, the sequence

$$0 \to \mathscr{C}_0 \otimes \mathscr{B} \to \mathscr{C}_{-1} \otimes \mathscr{B} \to \mathscr{E} \otimes \mathscr{B} \to 0$$

is exact, and since $\mathscr{C}_0$ is flabby, that is, $\mathscr{C}_0 \otimes \mathscr{B}$ is Φ-soft for any paracompactifying family Φ, the sequence

$$0 \to \Gamma_\Phi(\mathscr{C}_0 \otimes \mathscr{B}) \to \Gamma_\Phi(\mathscr{C}_{-1} \otimes \mathscr{B}) \to \Gamma_\Phi(\mathscr{E} \otimes \mathscr{B}) \to 0$$

is exact. Consequently, if $\mathscr{A}$ is injective, the kernel $T_0^\Phi(\mathscr{A}, \mathscr{B})$ of

$$\Gamma_\Phi(\mathscr{C}_0 \otimes \mathscr{B}) \to \Gamma_\Phi(\mathscr{C}_{-1} \otimes \mathscr{B})$$

is zero, and $T_{-1}^\Phi(\mathscr{A}, \mathscr{B}) = \Gamma_\Phi(\mathscr{E} \otimes \mathscr{B})$.

REFERENCES

1. A. Borel, *Cohomologie des espaces localement compact d'après J. Leray*, Notes, Federal School of Technology, Zürich, Switzerland, Second Edition, 1957.

2. ———, *The Poincaré duality in generalized manifolds*, Michigan Math. J. 4 (1957) 337-239.

3. ———, Seminar on transformation groups, Annals of Mathematics Studies 46, Princeton, 1960.

4. H. Cartan, Séminaire de l'École Normale Supérieure, a) 1948-49; b) 1950-51.

5. H. Cartan and S. Eilenberg, *Homological algebra*, Princeton Mathematical Series No. 19, 1956.

6. S. Eilenbert and N. E. Steenrod, *Foundations of algebraic topology*, Princeton Mathematical Series No. 15, 1952.

7. R. Godement, *Topologie algébrique et théorie des faisceaux*, Actualités Sci. Ind. 1252, Hermann, Paris, 1958.

8. N. E. Steenrod, *Regular cycles of compact metric spaces*, Ann. of Math. (2) 41 (1940), 833-851.

9. R. L. Wilder, *Topology of manifolds*, Amer. Math. Soc. Colloquium Publications 32 (1949).

The Institute for Advanced Study
and
Princeton University

50.

Density properties for certain subgroups
of semi-simple groups without compact components

Ann. Math., (2) **72** (1960) 179–188

Let G be a connected semi-simple Lie group, all of whose simple factors are non-compact. It is in particular unimodular,[1] hence if H is a closed unimodular subgroup, the quotient space G/H carries a positive measure μ invariant under G [4]. Assuming $\mu(G/H)$ to be finite, we shall prove the following two "density" properties of H in G:

(i) for every linear representation ρ,[2] every element of $\rho(G)$ is a finite linear combination with constant coefficients of elements of $\rho(H)$;

(ii) the subgroup H is not contained in any proper closed subgroup having a finite number of connected components, in particular, if G is algebraic, then H is dense in the Zariski topology of G, viewed as a real algebraic set.

This will also imply that the centralizer of H in G is the center of G,[3] and that the identity component of H is invariant in G. Loosely speaking, this last fact means that H is the product of some components of G by a discrete subgroup in the product of the other simple factors of G.

By Lemma 1.4, due to Selberg, H satisfies in G the condition called property (S) in definition 1.1. It turns out that the proof of the above results makes explicit use only of property (S); further the latter is convenient to handle here notably because it is inherited by bigger subgroups and quotient groups (see 1.2); therefore the results will be directly stated and proved in §4 under that assumption. The main step consists in showing that if G is simple, any irreducible G-module is also an irreducible H-module. This is done in §3. From this we deduce that, G being again semi-simple, every subalgebra of the Lie algebra of G invariant under H is an ideal, and then derive easily the results mentioned above. For our purposes, the main property derived from the condition (S) is contained in Lemma 1.3.

---

[1] A locally compact group is said to be unimodular if every left invariant Haar measure is also right invariant.

[2] Throughout this paper, linear representations are finite-dimensional.

[3] When G is a classical real simple group, this last fact was first checked by A. Selberg (unpublished), using Lemma 1.4. The present paper grew out of an attempt to give a general proof of this result.

1. The property (S)

1.1. DEFINITION. *Let G be a topological group. A subgroup H will be said to have property (S) in G if for every neighborhood U of the identity in G and every element $g \in G$ there exists an integer $n > 0$ such that $g^n \in U \cdot H \cdot U$.*

1.2. We first list a few easy facts, to be used without further comment. H is a subgroup of the topological group G, having property (S) in G.

Any subgroup of G containing H has property (S) in G.

If $f : G \to G'$ is a (continuous) homomorphism, then $f(H)$ has property (S) in $f(G)$, viewed as a topological subgroup of G'.

If $f : G' \to G$ is a surjective open homomorphism with kernel central in G', then $f^{-1}(H)$ has property (S) in G'.

If G is compact, every subgroup has property (S).

If H is compact and G is a connected Lie group, then G is compact. In fact, if G is non-compact, there exists an isomorphism $t \to g(t)$ of the real line onto a closed subgroup of G [2, Th. 6]; now if U is compact, then so is $U \cdot H \cdot U$, therefore there exists $a > 0$ such that $g(t) \notin U \cdot H \cdot U$ for $|t| > a$. Then $g(a)^n \notin U \cdot H \cdot U$ for all $n \geq 1$. The same argument applies to a connected locally compact group, taking into account Th. 13 of [2] and the fact that such a group is a projective limit of Lie groups.

1.3. LEMMA. *Let G be a topological group, H be a subgroup having property (S) in G, and ρ be a linear representation of G in a vector space V over $\mathbf{R}$ (resp. $\mathbf{C}$). Let X be a nilpotent endomorphism of V such that $g = \exp X$ belongs to $\rho(G)$, and let k be an integer such that $X^{k+1} = 0$. Then X^k is a finite linear combination with real coefficients of elements of $\rho(H)$.*

PROOF. Since H has property (S) in G, there exist: a strictly increasing sequence n_j $(j = 1, 2, \cdots)$ of positive integers, elements $u_j, v_j \in G$ tending to e as $j \to \infty$, and elements $h_j \in H$, such that

$$(1) \qquad\qquad g^{n_j} = \rho(u_j \cdot h_j \cdot v_j) \qquad\qquad (j = 1, 2, \cdots) .$$

Now we have

$$(2) \qquad g^n = 1 + \frac{n}{1} X + \frac{n^2}{2!} X^2 + \cdots + \frac{n^k}{k!} X^k \qquad\qquad (n \in Z) ,$$

therefore

$$\lim_{j \to \infty}(g^{n_j}/n_j^k) = X^k/k! ,$$

the limit being meant of course in the space $E(V)$ of endomorphisms of

V endowed with the usual topology. In particular $\lim g^{n_j}/n_j^k$ exists. Then (1) shows that $\lim \rho(h_j)/n_j^k$ also exists and is equal to $\lim g^{n_j}/n_j^k$, hence to $X^k/k!$. Therefore $X^k/k!$ belongs to the real vector space spanned in $E(V)$ by the elements of $\rho(H)$, as was to be proved.

1.4. LEMMA (Selberg). *Let G be a locally compact group, H be a closed subgroup such that G/H carries a positive invariant measure μ with respect to which $\mu(G/H)$ is finite. Then for every neighborhood U there exists an integer $m \geq 1$ with the following property: for every $g \in G$ there is an integer n such that $1 \leq n \leq m$ and $g^n \in U \cdot H \cdot U^{-1}$. In particular, H has property (S) in G.*

Let $\pi : G \to G/H$ be the natural projection, and let m be some integer strictly greater than $\mu(G/H)/\mu(\pi(U))$. Then the open subsets $g^i(\pi(U))$ $(i = 1, \cdots, m)$ cannot be pairwise disjoint, whence the existence of i, j with $1 \leq j < i \leq m$ such that $g^i(\pi(U)) \cap g^j(\pi(U)) \neq \varnothing$. Then $g^i \cdot U \cap g^j \cdot U \cdot H \neq \varnothing$, and $g^{i-j} \in U \cdot H \cdot U^{-1}$.

2. Semi-simple Lie algebras

We collect here some facts about semi-simple Lie algebras. We say that a real semi-simple Lie algebra $\mathfrak{g}$ is compact if its adjoint group Ad $\mathfrak{g}$ is compact. By a classical result of H. Weyl, every connected Lie group with Lie algebra $\mathfrak{g}$ is then compact.

2.1. Let $\mathfrak{g}$ be a real non-compact Lie algebra, and $\mathfrak{k}$ be the Lie algebra of a maximal compact subgroup of the adjoint group Ad $\mathfrak{g}$ of $\mathfrak{g}$. Let $\mathfrak{c}$ be a maximal commutative algebra orthogonal to $\mathfrak{k}$ with respect to the Killing form. Then it is known that $\mathfrak{g} = \mathfrak{k} + \mathfrak{c} + \mathfrak{n}$ (direct) where $\mathfrak{n}$ is a nilpotent algebra whose normalizer contains $\mathfrak{c}$, having no element $\neq 0$ commuting with $\mathfrak{c}$. Further ad x is nilpotent if $x \in \mathfrak{n}$, and semi-simple with real eigenvalues if $x \in \mathfrak{c}$. (This can be deduced for example from [3, Exp. 11, p. 12].) This implies clearly that we can always find $c \in \mathfrak{c}$, $n \in \mathfrak{n}$, both non-zero, such that $[c, n] = n$.

2.2. For any linear representation ρ of a semi-simple Lie algebra, $\rho(x)$ is a semi-simple (resp. nilpotent) endomorphism of the underlying vector space whenever $\mathrm{ad}_\mathfrak{g} x$ is a semi-simple (resp. nilpotent) endomorphism of $\mathfrak{g}$ [1, §6, No. 7]. Thus, if $\mathfrak{g}$ is a real semi-simple Lie algebra without compact components and ρ is non-trivial, we can always find elements $c, n \in \mathfrak{g}$ such that $\rho(n)$ is nilpotent, not zero, and $[\rho(c), \rho(n)] = \rho(c) \cdot \rho(n) - \rho(n) \cdot \rho(c) = \rho(n)$. It follows then by induction on s that we have

$$[\rho(c), (\rho(n))^s] = \rho(c)(\rho(n))^s - (\rho(n))^s\rho(c) = s(\rho(n))^s \qquad (s = 1, 2, \cdots)\,.$$

2.3.   Let $\mathfrak{a}$ be a complex Lie algebra, of dimension n over **C**. By restriction of the scalars, it defines a $2n$-dimensional Lie algebra over **R**, to be denoted by $\mathfrak{a}_r$. Let $\mathfrak{b}$ be a real Lie subalgebra of $\mathfrak{a}_r$. Then the vector space spanned over **C** by $\mathfrak{b}$ is a complex Lie subalgebra of $\mathfrak{a}$, to be denoted by $\mathfrak{b}^*$. There is a natural surjective homomorphism μ of the complexification $\mathfrak{b}_c = \mathfrak{b}\otimes_{\mathbf{R}}\mathbf{C}$ of $\mathfrak{b}$ onto $\mathfrak{b}^*$. In particular, $\mathfrak{b}^*$ is semi-simple (resp. solvable) if $\mathfrak{b}$ is so.

LEMMA.   *Let $\mathfrak{a}$ be a complex simple Lie algebra; let $\mathfrak{b}$ be a proper real subalgebra of $\mathfrak{a}_r$ such that $\mathfrak{b}^* = \mathfrak{a}$. Then $\mathfrak{a} = \mathfrak{b}_c$, that is, $\mathfrak{b}$ is a real form of $\mathfrak{a}$.*

Let $\mathfrak{r}$ be a solvable ideal of $\mathfrak{b}$. Then $[\mathfrak{b}, \mathfrak{r}] \subset \mathfrak{r}$ implies $[\mathfrak{b}^*, \mathfrak{r}^*] \subset \mathfrak{r}^*$, hence $\mathfrak{r}^*$ is a solvable ideal of $\mathfrak{a}$. Thus $\mathfrak{r}^* = 0$, $\mathfrak{r} = 0$ and $\mathfrak{b}$ is semi-simple. If $\mathfrak{b} = \mathfrak{b}_1 \times \mathfrak{b}_2$ then $\mathfrak{a} = \mathfrak{b}_1^* + \mathfrak{b}_2^*$ with $[\mathfrak{b}_1^*, \mathfrak{b}_2^*] = 0$. Since $\mathfrak{a}$ is simple, this implies that $\mathfrak{b}_1$ or $\mathfrak{b}_2$ is zero, hence $\mathfrak{b}$ is simple. If now $\mathfrak{b}_c$ is not simple, then we have $\mathfrak{b}_c = \mathfrak{m}_1 \times \mathfrak{m}_2$ with $\mathfrak{m}_{1r} \cong \mathfrak{m}_{2r} \cong \mathfrak{b}$ and further $\mathfrak{b}_c \to \mathfrak{b}^* = \mathfrak{a}$ is surjective, with kernel equal to one of the ideals $\mathfrak{m}_1$, $\mathfrak{m}_2$. This implies that $\mathfrak{a}_r$ and $\mathfrak{b}$ have the same dimension and contradicts the assumption $\mathfrak{b} \neq \mathfrak{a}_r$. Therefore, $\mathfrak{b}_c$ is simple, but then $\mu : \mathfrak{b}_c \to \mathfrak{b}^* = \mathfrak{a}$ is an isomorphism.

2.4. LEMMA.   *Let $\mathfrak{g}$ be a complex semi-simple Lie algebra, ρ_1, ρ_2 two representations of $\mathfrak{g}$. We assume that there is at least one simple component $\mathfrak{s}$ of $\mathfrak{g}$ for which neither $\rho_1(\mathfrak{s})$, nor $\rho_2(\mathfrak{s})$ is $\{0\}$. Then $\rho_1 \otimes \rho_2$ is not irreducible.*

We choose once and for all an ordering of the roots of $\mathfrak{g}$ and denote by b half the sum of the positive roots of $\mathfrak{g}$. By H. Weyl's formula [3, Exp. 19, §3, formule (8)], the degree $d(\rho)$ of an irreducible representation ρ with highest weight π is given by

$$(1) \qquad\qquad d(\rho) = \prod_{a>0}(a, \pi + b)(a, b)^{-1}$$

where $(\ ,\ )$ is the scalar product defined by the Killing form and a runs through the positive roots of $\mathfrak{g}$.

Assume ρ_1, ρ_2 to be irreducible, since otherwise the lemma is obvious. Let π_i be the highest weight of ρ_i $(i = 1, 2)$. The highest weight of $\rho_1 \otimes \rho_2$ is $\pi_1 + \pi_2$, therefore $\rho_1 \otimes \rho_2$ contains as a component the irreducible representation with highest weight $\pi_1 + \pi_2$. We have to show that the latter is not all of $\rho_1 \otimes \rho_2$; since the degree of $\rho_1 \otimes \rho_2$ is $d(\rho_1) \cdot d(\rho_2)$, this, in view of (1), is equivalent to

$$(2) \quad \prod_{a>0}\left(\frac{(a,\pi_1)}{(a,b)}+1\right)\left(\frac{(a,\pi_2)}{(a,b)}+1\right) > \prod_{a>0}\left(\frac{(a,\pi_1)}{(a,b)}+\frac{(a,\pi_2)}{(a,b)}+1\right).$$

It follows from our assumption that there is at least one root, for instance the highest root of $\mathfrak{s}$, whose scalar products with π_1 and π_2 are both non-zero. Since all scalar products occurring in (2) are $\geqq 0$, the validity of (2) follows immediately.

3. A special case

3.1. *Notation.* Let V be a finite dimensional space over C. We denote by $E(V)$ and $GL(V)$ the algebra of endomorphisms of V and the group of automorphisms of V. The Lie algebra $\mathfrak{gl}(V)$ of V will be identified with the Lie algebra associated in the usual manner to the associative algebra $E(V)$, that is $\mathfrak{gl}(V)$ is identified with $E(V)$ endowed with the bracket operation $[A, B] = A \cdot B - B \cdot A$. For any $g \in GL(V)$, the map $X \to g \cdot X \cdot g^{-1}$ is an automorphism of $E(V)$ or $\mathfrak{gl}(V)$, to be denoted by g_{11}. The map $g \to g_{11}$ is a homomorphism of $GL(V)$ into $GL(E(V))$. It induces a homomorphism $X \to X_{11}$ of $\mathfrak{gl}(V)$ into $\mathfrak{gl}(E(V))$ and we have $X_{11}(Y) = \mathrm{ad}\, X(Y) = X \cdot Y - Y \cdot X$.

The group $GL(V)$ is a complex Lie group. Viewed as a real Lie group, it will be denoted by $GL(V)_r$.

3.2. PROPOSITION. *Let V be a finite dimensional vector space over C, and G an analytic subgroup of $GL(V)_r$. Assume that G is a simple non-compact connected real Lie group, and that V is an irreducible G-module. Let H be a subgroup of G having property (S) in G. Then V is an irreducible H-module.*

The proof is divided into three lemmas. The proposition itself follows from 3.3 (b) and 3.5. As before, $\mathfrak{g}$ denotes the Lie algebra of G.

3.3. LEMMA. *Let $\mathfrak{q}$ be the vector subspace of $\mathfrak{gl}(V)$ spanned by the elements $X^{k(x)}$ where X runs through the nilpotent elements of $\mathfrak{g}$ and $k(X)$ is the greatest integer such that $X^{k(x)} \neq 0$. Let $\mathfrak{m}$ be the smallest complex Lie subalgebra of $\mathfrak{gl}(V)$ containing $\mathfrak{q}$. Then*

(a) $[\mathfrak{g}, \mathfrak{q}] \subset \mathfrak{q}$ *and* $[\mathfrak{g}, \mathfrak{m}] \subset \mathfrak{m}$;

(b) *every subspace of V stable under H is stable under $\mathfrak{m}$.*

(c) $\mathfrak{m}$ *is semi-simple.*

If X is nilpotent and belongs to $\mathfrak{g}$, then so does $g \cdot X \cdot g^{-1}$ for any $g \in G$, and further $k(X) = k(g \cdot X \cdot g^{-1})$. Therefore $g_{11}(\mathfrak{q}) = \mathfrak{q}$ for any $g \in G$; hence also $g_{11}(\mathfrak{m}) = \mathfrak{m}$, by the definition of $\mathfrak{m}$. This implies $X_{11}(\mathfrak{q}) \subset \mathfrak{q}$, $X_{11}(\mathfrak{m}) \subset \mathfrak{m}$ for any $X \in \mathfrak{g}$, that is, $[\mathfrak{g}, \mathfrak{q}] \subset \mathfrak{q}$, $[\mathfrak{g}, \mathfrak{m}] \subset \mathfrak{m}$.

By 1.3, every subspace W of V stable under H is invariant under $X^{k(X)}$, where X is nilpotent in $\mathfrak{g}$ and $k(X)$ is as in 3.3, and therefore also under all endomorphisms of $\mathfrak{q}$. The complex subalgebra of $\mathfrak{gl}(V)$ leaving W invariant then contains $\mathfrak{q}$, hence also $\mathfrak{m}$, and this proves (b).

Let M be the complex analytic subgroup of $GL(V)$ with Lie algebra $\mathfrak{m}$. By the above, its normalizer contains G; therefore if T is a minimal invariant subspace of M in V, then so is $g(T)$ for any $g \in G$. Since V is irreducible under G, this implies at once that V is the direct sum of a finite number of transforms of T under G. Therefore $\mathfrak{m}$ is completely reducible, and we have $\mathfrak{m} = \mathfrak{z} \times \mathfrak{m}'$, where $\mathfrak{m}'$ is semi-simple and $\mathfrak{z}$ is commutative, and made up of semi-simple endomorphisms of V [1, §6, No. 5]. We claim that every nilpotent element $X \in \mathfrak{m}$ belongs to $\mathfrak{m}'$. In fact, we have $X = Z + Y(Z \in \mathfrak{z},\ Y \in \mathfrak{m}';\ [Z,\ Y] = 0)$; $\mathfrak{m}'$, being semi-simple, contains the semi-simple part S and the nilpotent part N of Y [1, §6, No. 7]; since both S and N commute with Z, it follows that $S + Z$ and N are the semi-simple part and the nilpotent part of X respectively. Thus $X = N$, that is $X \in \mathfrak{m}'$. But then $\mathfrak{q} \subset \mathfrak{m}'$, and $\mathfrak{m}' = \mathfrak{m}$ by definition of $\mathfrak{m}$. This proves (c).

3.4. **Lemma.** *Let $\mathfrak{g}^*$ be the smallest complex subalgebra of $\mathfrak{gl}(V)$ containing $\mathfrak{g}$. Then no simple component of $\mathfrak{g}^*$ commutes with $\mathfrak{q}$.*

By 2.2, $\mathfrak{g}$ does not commute with $\mathfrak{q}$, therefore the lemma is true when $\mathfrak{g}^*$ is simple. From now on, $\mathfrak{g}^*$ is assumed not to be simple. Since it is a quotient of the complexification $\mathfrak{g}_c$ of $\mathfrak{g}$ (2.3), $\mathfrak{g}_c$ is also not simple; $\mathfrak{g}$ being simple, we have $\mathfrak{g}_c = \mathfrak{g}_1 \times \mathfrak{g}_2$ with $\mathfrak{g}_1 \cong \mathfrak{g}_2$ and the projection $\pi_i : \mathfrak{g} \to \mathfrak{g}_i$ is bijective. Further, $\mathfrak{g}^*$ having at least two components, the natural map $\mu : \mathfrak{g}_c \to \mathfrak{g}^*$ is an isomorphism; we identify $\mathfrak{g}^*$ with $\mathfrak{g}_c$.

By assumption V is an irreducible $\mathfrak{g}$-module, hence also an irreducible $\mathfrak{g}^*$-module. Therefore V, as a $\mathfrak{g}^*$-module, is the tensor product $V_1 \otimes V_2$, where V_i is an irreducible $\mathfrak{g}_i$-module $(i = 1, 2)$.[4] The corresponding representations $\rho_i : \mathfrak{g}_i \to \mathfrak{gl}(V_i)$ and $\rho_i \cdot \pi_i : \mathfrak{g} \to \mathfrak{gl}(V_i)_r$ are then faithful. If $x \in \mathfrak{g}$, then $x = y_1 \otimes 1 + 1 \otimes y_2$, where $y_i = \rho_i \pi_i(x)$ is not zero if $x \neq 0$. Let $x \in \mathfrak{g}$ be nilpotent, not zero, and k be such that $x^k \neq 0$, $x^{k+1} = 0$. By definition, $x^k \in \mathfrak{q}$, therefore it is enough to show that $\mathfrak{g}_i$ does not commute with x^k. We carry this out for $i = 1$. We have

$$x^k = \sum_{0 \leq i \leq k} \binom{k}{i} y_1^i \otimes y_2^{k-i} \qquad (y_j = \rho_j \pi_j(x)\,;\, j = 1, 2)\,.$$

There is a value of $i \geq 1$ for which $y_1^i \otimes y_2^{k-i} \neq 0$, because otherwise, y_1

---

[4] This follows for instance from Theorem 2 in No. 4 and the corollary in No. 7 of Bourbaki, Algèbre, Ch. VIII, §7, applied to the associative algebras generated in $\mathfrak{gl}(V)$ by 1 and the two factors.

being $\neq 0$, we should have $y_2^{k-1} = 0$, hence $y_2^k = 0$ and finally $x^k = 0$. Denoting by s the smallest such integer, we may therefore write

$$x^k = 1 \otimes y_2^k + \sum_{s \leq i \leq k} \binom{k}{i} y_1^i \otimes y_2^{k-i} \qquad (s \geq 1) .$$

By 2.1, there exists $c \in \mathfrak{g}$ such that $[c, x] = x$. We have $\pi_1(c) = c' \otimes 1$ with $c' = \rho_1 \cdot \pi_1(c)$. By 2.2, $[c', y_1^i] = i y_1^i$ for $i \geq 1$, whence immediately

$$(3) \qquad [\pi_1(c), x^k] = \sum_{s \leq i \leq k} \binom{k}{i} \cdot i \cdot y_1^i \otimes y_2^{k-i} , \qquad (s \geq 1).$$

It is then not difficult to check that $[\pi_1(c), x^k] \neq 0$. In fact, let (e_j) $(j = 1, \cdots, m)$ be a basis of V_1 such that either $y_1(e_j) = e_{j'}$ with $j' > j$ or $y_1(e_j) = 0$, which exists since y_1 is nilpotent; let a be the smallest index such that $y_1^s(e_a) \neq 0$ and b be such that $y_1^s(e_a) = e_b$. Then $y_1^i(e_a) = e_{b'}$ with $b' > b$ if $i > s$. Therefore if we write the matrices of the endomorphisms occurring in (3) in blocks, with respect to the decomposition of V into the subspaces $Ce_j \otimes V_2$, we have

$$[\pi_1(c), x^k]_{ba} = \binom{k}{s} \cdot s \cdot (y_1^s \otimes y_2^{k-s})_{ba} \neq 0$$

which proves our contention.

3.5. **Lemma.** *In the notation of* 3.3, *we have* $\mathfrak{g} \subset \mathfrak{m}$.

We prove the equivalent assertion $\mathfrak{g}^* \subset \mathfrak{m}$. Let $\mathfrak{z}(\mathfrak{m})$ be the centralizer of $\mathfrak{m}$ in $\mathfrak{gl}(V)$. Since $\mathfrak{m}$ is semi-simple (3.3 (c)), $\mathfrak{z}(\mathfrak{m}) \cap \mathfrak{m} = 0$ and the normalizer of $\mathfrak{m}$ in $\mathfrak{gl}(V)$ is the direct product $\mathfrak{z}(\mathfrak{m}) \times \mathfrak{m}$ [1, §6, No.2]. By 3.3 (a) it contains $\mathfrak{g}$, and therefore also $\mathfrak{g}^*$. Let $\mathfrak{p}$ be the projection of $\mathfrak{g}^*$ into $\mathfrak{z}(\mathfrak{m})$, that is, the smallest subalgebra of $\mathfrak{z}(\mathfrak{m})$ such that $\mathfrak{g}^* \subset \mathfrak{p} \times \mathfrak{m}$; it is also semi-simple. Since V is an irreducible $\mathfrak{g}^*$-module, it is also an irreducible $\mathfrak{p} \times \mathfrak{m}$ module, and may therefore[4] be identified with the tensor product $V_1 \otimes V_2$ of a $\mathfrak{p}$-module V_1 and an $\mathfrak{m}$-module V_2. Thus the given representation of $\mathfrak{g}^*$ in V may be identified with a tensor product $\rho_1 \otimes \rho_2$, where ρ_1 (resp. ρ_2) is the composition of the projection of $\mathfrak{g}^*$ in $\mathfrak{p}$ (resp. $\mathfrak{m}$) with a representation σ_1 (resp. σ_2) of $\mathfrak{p}$ (resp. $\mathfrak{m}$) in $\mathfrak{gl}(V_1)$ (resp. $\mathfrak{gl}(V_2)$). The representations σ_1 and σ_2 are of course faithful. By 3.4, ρ_2 is then faithful. Assume now that, contrary to our contention, $\mathfrak{g}^* \not\subset \mathfrak{m}$. Then ρ_1 is non-trivial on at least one simple factor of $\mathfrak{g}^*$, but this contradicts 2.4. Hence $\mathfrak{g}^* \subset \mathfrak{m}$.

4. Density properties of subgroups

4.0. *In this section, G is a connected semi-simple real Lie group, all of whose simple components are non-compact, and H is a subgroup of G, having property (S) in G. Since the quotient of a semi-simple group is locally isomorphic to a direct factor of the group, any quotient of G,* which is $\neq \{e\}$, is also semi-simple, without compact simple components.

4.1. THEOREM. *If a subalgebra of the Lie algebra* $\mathfrak{g}$ *of* G *is stable under* H, *it is an ideal of* $\mathfrak{g}$.

It is enough to prove this when the center of G is reduced to $\{e\}$ (see 1.2). We distinguish three cases.

(i) The complexification $\mathfrak{g}_c$ of $\mathfrak{g}$ is simple. This is equivalent to saying that $\mathfrak{g}_c$ is an irreducible $\mathfrak{g}$-module under the adjoint representation. Now if $\mathfrak{b} \neq 0$ is a subalgebra of $\mathfrak{g}$ invariant under H, then $\mathfrak{b}_c$ is also invariant under H, hence equal to $\mathfrak{g}_c$ by 3.2. Therefore $\mathfrak{b} = \mathfrak{g}$.

(ii) $\mathfrak{g}$ is simple, but $\mathfrak{g}_c$ is not. Then, as is known, $\mathfrak{g} = \mathfrak{a}_r$ (notation of 2.3) where $\mathfrak{a}$ is a complex simple Lie algebra, and the adjoint representation of G in $\mathfrak{a}$ is irreducible. Let $\mathfrak{b} \neq 0$ be a subalgebra of $\mathfrak{g}$ invariant under H. By 3.2, the smallest complex subalgebra $\mathfrak{b}^*$ of $\mathfrak{a}$ containing $\mathfrak{b}$ is equal to $\mathfrak{a}$, therefore, by 2.3, $\mathfrak{b}$ either is equal to $\mathfrak{g}$ or is a real form of $\mathfrak{a}$. It remains to show that the second alternative is not possible. So assume $\mathfrak{a} = \mathfrak{b}_c$. Let $B = \mathrm{Ad}\,\mathfrak{b}$, and $A = \mathrm{Ad}\,\mathfrak{a}$. The latter, viewed as a real Lie group is just $\mathrm{Ad}\,\mathfrak{g}$. Let ι be the complex conjugation of $\mathfrak{a}$ with respect to $\mathfrak{b}$. It defines an automorphism of $\mathfrak{g} = \mathfrak{a}_r$, whose fixed point set is $\mathfrak{b}$. This automorphism extends to an automorphism of $\mathrm{Ad}\,\mathfrak{g}$, also to be denoted by ι, whose fixed point set F is a closed subgroup having B as identity component. If we take in $\mathfrak{a}$ a basis formed by elements of $\mathfrak{b}$, then it is easily seen that $\iota(a)$, $(a \in \mathrm{Ad}\,\mathfrak{g})$ is the complex conjugate matrix of a (in fact, this is true for the Lie algebra, hence for exponentials of elements in the Lie algebra, hence for finite products of exponentials, and these fill up $\mathrm{Ad}\,\mathfrak{g} = G$). Thus F is the subgroup of matrices of G having real coefficients. It is clear that any element of G leaving $\mathfrak{b}$ invariant also has real coefficients. Thus $H \subset F$, and F should have property (S) in G. However, let $x \in \mathfrak{a}$ be such that $\mathrm{ad}\,x$ is nilpotent. Let k be the greatest integer for which $(\mathrm{ad}\,x)^k \neq 0$ and c be a complex number such that c^k is not real. Then at least one of the two matrices $(\mathrm{ad}\,x)^k$, $(\mathrm{ad}\,cx)^k$ has some complex, not real, coefficients. Since $\mathrm{Ad}_\mathfrak{g} H$ consists of real matrices, this contradicts 1.3.

(iii) $G = G_1 \times \cdots \times G_k$, where the G_i are simple, non-compact, with center reduced to $\{e\}$. Let π_i be the projection of G onto G_i, and $H_i = \pi_i(H)$. Then (1.2), H_i has property (S) in G_i. Let $\mathfrak{b}$ be a subalgebra of $\mathfrak{g}$, invariant under H, and $\mathfrak{b}_i = \pi_i(\mathfrak{b})$, $\mathfrak{c}_i = \mathfrak{b} \cap \mathfrak{g}_i$. Then $\mathfrak{b}_i$ and $\mathfrak{c}_i$ are invariant under H_i, therefore by (i), (ii), are equal to 0 or $\mathfrak{g}_i$. It is enough to show that for each i, either $\mathfrak{b}_i = 0$ or $\mathfrak{c}_i = \mathfrak{g}_i$. Assume it is not so. Projecting on a suitable direct factor of G, and using 1.2, we easily see that, in order to prove this, it suffices to eliminate the case where $c_i = 0$, $\mathfrak{b} \cong \mathfrak{g}_i (i = 1, \cdots, k)$, and $\mathfrak{b}$ is the diagonal $\{(b, \cdots, b)\}$ of $\mathfrak{g}$. In that case, $G_i \cong B$, where B is the analytic subgroup of G with Lie algebra $\mathfrak{b}$, and B is the diagonal of $G = B \times \cdots \times B$. Since G has no center, it is clear that B is its own normalizer in G. Then it contains H, hence B has property (S) in G. However this is impossible for $k \geq 2$: in fact, G_1 contains a closed one-parameter group $(g(t))$, isomorphic to $\mathbf{R}$. Let then U_1 be a symmetric neighborhood of e in G_1 and t_0 be such that $g(t) \notin U_1$ for $t \geq t_0$. Let further V_1 be a symmetric neighborhood such that $V_1^4 \subset U_1$, $V = V_1 \times V_1 \times \cdots \times V_1$, and $g = (g(t_0), e, \cdots, e)$. Then $g^n = u \cdot h \cdot v \, (u, v \in V, \, h = (h', \cdots, h') \in H)$ implies $h' \in V_1^2$, hence $g(t_0)^n \in U_1$, a contradiction.

4.2. COROLLARY. *Let $\overline{H}$ be the closure of H in G, and let H_0 be the component of the identity in $\overline{H}$. Then $G = G_1 \cdot H_0$ where G_1 is connected invariant in G, commutes with H_0, $G_1 \cap H_0$ is discrete, and $\overline{H} = H_1 \cdot H_0$ where H_1 is a discrete subgroup of G_1 having property (S) in G_1. In particular, if G is simple, and $\overline{H} \neq G$, then $\overline{H}$ is discrete.*

By 4.1, the Lie algebra $\mathfrak{h}$ of $\overline{H}$ is invariant in $\mathfrak{g}$, therefore $\mathfrak{g} = \mathfrak{g}_1 \times \mathfrak{h}$. It follows then, G_1 being the invariant subgroup of G with Lie algebra $\mathfrak{g}_1$, that $G = G_1 \cdot H_0$, with $G_1 \cap H_0$ discrete; there exists therefore a surjective homomorphism $\mu : G' = G_1 \times H_0 \to G$, with kernel in the center of G', which is injective on each factor. Then (1.2) $H' = \mu^{-1}(\overline{H})$ has property (S) in G'. Its identity component is H_0, and hence $H' = H_1 \cdot H_0$ with H_1 discrete in G_1. Finally, H_1, being the projection of H' in G_1 has property (S) in G_1 by 1.2.

4.3. COROLLARY. *Let L be a closed subgroup of G containing H. If L has a finite number of connected components, then $L = G$. In particular, if G is real algebraic, every real algebraic subgroup of G containing H is equal to G.*

Let L_0 be the identity component of L. It is of course invariant under H, hence is also invariant in G by 4.1. Let $\pi : G \to G/L_0$ be the natural projection. Then $\pi(L)$ has property (S) in G/L_0 by 1.2, and is finite. Therefore (1.2), G/L_0 is compact, and $G = L_0$ by 4.0. The second assertion follows from the first one and from the fact that an affine real algebraic variety has a finite number of connected components, with respect to the usual topology.

4.4. COROLLARY. *The centralizer of H is the center of G.*

Let $g \in G$ be in the centralizer of G, $\pi : G \to \mathrm{Ad}\, G$ be the natural homomorphism and $g' = \pi(g)$. Then $\pi(H)$ has property (S) in $\mathrm{Ad}\, G$ and belongs to the centralizer Z of g'. The group $\mathrm{Ad}\, G$ is the identity component of a real algebraic group, namely the group $\mathrm{Aut}\, \mathfrak{g}$ of automorphisms of $\mathfrak{g}$, and the centralizer of g' in $\mathrm{Aut}\, \mathfrak{g}$ is then of course an algebraic subgroup. Consequently, Z is a closed subgroup of $\mathrm{Ad}\, G$, with a finite number of connected components, containing $\pi(H)$. By 4.3, $Z = \mathrm{Ad}\, G$, that is $g' = e$, and g belongs to the kernel of π. Thus g is central in G.

4.5. COROLLARY. *Let ρ be a linear representation of G in a real (resp. complex) vector space V. Then every element of $\rho(G)$ is a finite linear combination with real (resp. complex) coefficients of elements of $\rho(H)$.*

This of course can also be expressed by saying that the associative algebras generated in $E(V)$ by $\rho(G)$ and $\rho(H)$ are the same.

Let σ be the representation of G in $E(V)$ or $\mathfrak{gl}(V)$ defined by $g \to (\rho(g))_{11}$ (notation of 2.1), that is $\sigma(g)(x) = \rho(g) \cdot x \cdot \rho(g^{-1})$, $(g \in G,\ x \in \mathfrak{gl}(V))$. The group $\sigma(G)$ is a closed analytic subgroup of $GL(E(V))$. This follows for instance from the fact that $\sigma(\mathfrak{g})$, being semi-simple, is an algebraic Lie algebra, hence $\sigma(G)$ is the identity component of a real algebraic group of $GL(E(V))$ (resp. $GL(E(V))_r$). Let W be the vector subspace of $E(V)$ spanned by $\rho(H)$. Clearly $W \cdot W \subset W$ and W is invariant under $\sigma(H)$. Let M be the subgroup of $\sigma(G)$ leaving W invariant. It contains $\sigma(H)$, hence has property (S), and is algebraic, or rather, as the group Z in the preceding proof is of finite index in an algebraic subgroup. It has therefore a finite number of connected components, hence is equal to $\sigma(G)$ by 4.3. Thus W is stable under $\sigma(G)$. Let now $L = \{g \in G,\ \rho(g),\ \rho(g^{-1}) \in W\}$. This is a symmetric closed subset of G. Since $W \cdot W \subset W$, L is also stable under products, therefore it is a closed subgroup of G; it is invariant in G, since W is stable under G, and $\rho(g \cdot u \cdot g^{-1}) = \sigma(g)\,(\rho(u))\,(g,\ u \in G)$. It also contains H by definition of W. Thus the image of H in G/L is reduced to the identity and has property (S) in G/L. Consequently, G/L is compact (1.2); but then $G = L$ by 4.0, as was to be proved.

INSTITUTE FOR ADVANCED STUDY

REFERENCES

1. BOURBAKI, Groupes et algèbres de Lie. Chapitre I, Algèbres de Lie, Hermann éd. Paris, 1960.
2. K. IWASAWA, *On some types of topological groups*, Ann. of Math. 50 (1949), 507–558.
3. SÉMINAIRE S. LIE, Mimeographed Notes, Paris, 1955.
4. A. WEIL, L'intégration dans les groupes topologiques et ses applications, Hermann éd., Paris, 1940.

51.

Commutative subgroups and torsion in compact Lie groups

Bull. Amer. Math. Soc. **66** (1960) 285–288

In this note, G is a compact connected Lie group. We are concerned with the torsion of the cohomology ring $H^*(G; Z)$ of G over the integers, certain commutative subgroups of G, and relations between these two questions.

NOTATION. $E(m_1, \cdots, m_r)$ or $E_A(m_1, \cdots, m_r)$ denotes the exterior algebra over the ring A of a free A-module with r generators of respective degrees $m_1, \cdots, m_r$; p is a prime number, Z_p the field of integers mod p, Q the field of rational numbers. Tors H is the torsion subgroup of a group H, ord M the order of a finite group M. The identity component of a closed subgroup H if G is denoted by H_0, if L is a subset of G, its centralizer in G is denoted by $Z(L)$.

1. H-spaces with finitely generated cohomology groups. In this section, X is a compact connected H-space, for which $H^*(X; Z)$ is finitely generated. As is known $H^*(X; Q) = E(m_1, \cdots, m_r)$ with m_i odd; we assume $m_i \leqq m_j$ if $i \leqq j$. With this notation we have

PROPOSITION 1.1. (a) $H^*(X; Z)/\text{Tors } H^*(X; Z) = E_Z(m_1, \cdots, m_r)$. (b) *Let p be odd and K be a field of characteristic p. Then $H^*(X; K)$ contains a subalgebra isomorphic to $E_K(m_1, \cdots, m_r)$. Each system of generators of type (M) of $H^*(X; K)$ (in the sense of $[2, \S 6]$) contains at least r elements of odd degrees.*

Let $H = H^*(X; Z)/\text{Tors } H^*(X; Z)$. The transpose of the product map induces a homomorphism $H \to H \otimes H$ satisfying the conditions imposed on a Hopf algebra over Z. Hence $H \otimes A$ is a Hopf algebra over A for any ring A. Then (a) follows from the structure theorem $[2, \text{Théorème } 6.1]$ applied to $H \otimes L$, L being a field, by an easy induction. (b) follows from (a), the structure theorem, and the fact that the product of the generators of $E_K(m_1, \cdots, m_r)$ is nonzero.

PROPOSITION 1.2. *Let p be odd. Assume that each element of $H^*(X; K_p)$ has height $\leqq p$ (in the sense of $[2, \S 6]$) and that X is simply connected. Let k be the first integer such that $H^k(X; Z)$ has p-torsion. Then $p \cdot k \leqq m_r + p - 1$.*

This may be proved using the spectral sequence connecting $H^*(X; Z_p)$ to $H^*(X; Z)/\text{Tors } H^*(X; Z) \otimes Z_p$, whose differentials are the successive Bockstein operators. This result is sufficient for the applications mentioned in §2; however, since then, a more complete one has been obtained by Araki [1].

2. **On the p-torsion of compact Lie groups.** By E_i $(i=6, 7, 8)$ we mean here the compact simply connected group whose Lie algebra is denoted by that symbol in the standard Killing-Cartan classification. It was shown in [7] that E_i has no p-torsion for $i=6$, $p\geq 7$, and $i=7$, 8, $p\geq 11$. By Theorem V of [11], E_i has the homology of the Eilenberg-MacLane space $K(Z, 3)$ up to dimension m_2-1 (in the notation of §1). Together with the results of H. Cartan [12] on the homology of $K(\pi, n)$ this shows that E_6, E_7, E_8 have 2- and 3-torsion, and that E_8 also has 5-torsion. Combined with 1.1, 1.2, these results also allow one to prove the

PROPOSITION 2.1. *The group E_i has no p-torsion for $i=6$, 7, $p\geq 5$ and for $i=8$, $p\geq 7$. Furthermore*

$$H^*(E_6; Z_3) = E(3, 7, 9, 11, 15, 17) \otimes P(8; 3),$$

$$H^*(E_8; Z_5) = E(3, 11, 15, 23, 27, 35, 39, 47) \otimes P(12; 5)$$

where $P(a; p)=Z_p[x]/(x^p)$, with x of degree a.

Together with results of [4; 7], this yields a *verification* of the second assertion conjectured in [6] of the following:

THEOREM 2.2. (a) *G has no p-torsion if and only if B_G has no p-torsion.* (b) *Let G be simple and simply connected. Then if p does not divide the coefficients of the highest root, expressed as linear combination of the simple roots, then G has no p-torsion.*

For the "only if" part of (a) see [6, Proposition 5.1]. The "if" part is checked using 2.1 and [7]. The reduction to simple groups uses the fact that G is *homeomorphic* to the product of $Z(G)_0$ by its greatest semi-simple subgroup. More generally, *if $G=S\cdot G_1$ where S is a torus and G_1 a closed connected subgroup, both invariant, then G is homeomorphic to the product of a torus by G_1.* Using induction on dim S, it is enough to prove this when S is a circle. Then, if $G\neq G_1$, the group G is a principal G_1-bundle over a circle, and this bundle is trivial since G_1 is connected.

3. **Commutative subgroups and p-torsion.** As in [8], a $[p]$-group is a commutative group of type $(p, \cdots, p)$; its rank is the number of factors isomorphic to Z_p.

THEOREM 3.1. (a) *Let H be a compact commutative subgroup of G.*

Assume that $\mathrm{ord}(H/H_0)$ *is prime to the order of* Tors $\pi_1(G)$, *and that either H contains a regular element of G or* H/H_0 *is generated by two elements. Then H is contained in a torus of G.* (b) $\pi_1(G)$ *has no p-torsion if and only if every* $[p]$-*group of rank* $\leqq 2$ *belongs to a toral subgroup.*

The proof of 3.1 uses notably the end remark of §2, the fact that $G/Z(S)$ is torsion free when S is a torus [5; 10], and the fact that *if G is simply connected, the centralizer $Z(g)$ of any element of g is connected*, which can be proved using Proposition 2, p. 56 of [13] and some properties of singular elements. The assertion (b), in the direction not covered by (a), is checked by explicit construction of examples. The main property to verify is the following: Let G be simple, simply connected, and z be an element of order p of the center of G. Then there exist $u, v \in G$ with $uvu^{-1}v^{-1} = z$ and $u^p = v^p = e$ if p is odd, $u^p = v^p = z$ if $p = 2$.[1]

THEOREM 3.2. (a) *Let p be a prime. Then G has no p-torsion if and only if every* $[p]$-*subgroup is contained in some torus.* (b) *Let H be a compact commutative subgroup of G. Assume that* ord (H/H_0) *is prime to the order of* Tors $H^*(G, Z)$. *Then H is contained in a torus of G.*

The "only if" part of (a) was proved in [3, Chapter XIII]. The "if" part is checked by explicit constructions. These show that if G is simple, simply connected and has p-torsion, then G contains a $[p]$-subgroup of rank 3 not belonging to a torus. The proof of (b) proceeds by induction on dim G. It uses the following fact, which, when U is not the centralizer of a torus, had to be checked using the classification of subgroups of maximal rank [9].

PROPOSITION 3.3. *Let U be a connected closed subgroup of maximal rank of G. If G has no p-torsion, then U and G/U have no p-torsion.*

4. **Remarks about maximal** [2]-**subgroups.** It is known [2; 3] that $H^*(G; Z_2)$ has a simple system of universally transgressive generators when $G = SO(n)$, $Sp(n)$, $SU(n)$, $(n \geqq 1) G_2$, F_4, Spin(m) $(m \leqq 9)$. In the first four cases, it was also shown [3] that maximal [2]-subgroups are conjugate by inner automorphisms and play in cohomology mod 2 a role analogous to that of maximal tori in rational cohomology. This also holds in the remaining cases; more precisely:

THEOREM 4.1. *The maximal* [2]-*subgroups of* G_2, F_4, Spin (7), Spin (8), Spin (9) *are conjugate by inner automorphisms, their ranks*

---

[1] Theorem 3.1(b) was suggested in discussions with J. Wolf, to whom I am also indebted for an example in the case of Spin $(4n)$, $p = 2$.

are equal to 3, 5, 4, 5, 5 *respectively. If U is one of them, then G/U is totally nonhomologous to zero* mod 2 *in the fibering*

$$(B_U,\ B_G,\ G/U,\ \rho(U,\ G)).$$

REFERENCES

1. S. Araki, *A theorem of differential Hopf algebras and the cohomology* mod 3 *of the compact exceptional groups* E_7, E_8, (to appear).

2. A. Borel, *Sur la cohomologie des espaces fibrés principaux et des espaces homogènes de groupes de Lie compacts*, Ann. of Math. vol. 57 (1953) pp. 115–207.

3. ———, *La cohomologie modulo* 2 *de certains espaces homogènes*, Comment. Math. Helv. vol. 27 (1953) pp. 165–197.

4. ———, *Sur l'homologie et la cohomologie des groupes de Lie compacts connexes*, Amer. J. Math. vol. 76 (1954) pp. 273–342.

5. ———, *Kählerian coset spaces of semi-simple Lie groups*, Proc. Nat. Acad. Sci. U.S.A. vol. 40 (1954) pp. 1147–1151.

6. ———, *Topology of Lie groups and characteristic classes*, Bull. Amer. Math. Soc. vol. 61 (1955) pp. 397–432.

7. ———, *Sur la torsion des groupes de Lie*, J. Math. Pures Appl. vol. 35 (1956) pp. 127–139.

8. ———, *Seminar on transformation groups*, Annals of Mathematics Studies, no. 46, Princeton, 1960.

9. A. Borel and J. de Siebenthal, *Les sous-groupes fermés connexes de rang maximum des groupes de Lie clos*, Comment. Math. Helv. vol. 23 (1949–50) pp. 200–223.

10. R. Bott, *An application of the Morse theory to the topology of Lie groups*, Bull. Soc. Math. France vol. 84 (1956) pp. 251–281.

11. R. Bott and H. Samelson, *Applications of the Morse theory to symmetric spaces*, Amer. J. Math. vol. 80 (1958) pp. 964–1029.

12. H. Cartan, Séminaire E.N.S., 1954–1955 (mimeographed).

13. J. de Siebenthal, *Sur les groupes de Lie compacts non connexes*, Comment. Math. Helv. vol. 31 (1956) pp. 41–89.

INSTITUTE FOR ADVANCED STUDY

<h1 style="text-align:center">53.</h1>

<h1 style="text-align:center">Sous groupes commutatifs et torsion des groupes de Lie
compacts connexes</h1>

Tôhoku Math. J., (2) **13** (1961) 216–240

Ce travail apporte quelques renseignements sur la torsion du groupe de cohomologie entière d'un groupe de Lie compact connexe G, que nous appellerons, suivant l'usage, la torsion de G, sur les sous-groupes commutatifs de G, et met ces deux questions en relation. Nous nous intéresserons en particulier aux rapports qui existent entre la p-torsion (p nombre premier) et les sous-groupes commutatifs de type $(p,\ldots\ldots,p)$, que nous appellerons ici les $[p]$-sous-groupes. Ce travail a été résumé dans une Note au Bull. Amer. Math. Soc. 66 (1960), pp. 285-288.

En nous appuyant sur quelques remarques concernant les H-espaces dont la cohomologie entière est de type fini, faites dans le §1, et sur le Théorème V de [14], on verra que le groupe simplement connexe exceptionnel $\mathbf{E}_i$ n'a pas de p-torsion lorsque $p = 5$, $i = 6, 7$ et $p = 7$, $i = 7, 8$, et l'on déterminera aussi $H^*(\mathbf{E}_6; \mathbf{Z}_3)$, $H^*(\mathbf{E}_8; \mathbf{Z}_5)$, (Théor. 2.2, 2.3). Compte tenu de résultats connus [5, 8, 14], on pourra alors indiquer les nombres premiers intervenant dans les coefficients de torsion de tous les groupes simples et simplement connexes (2.5) et l'on *vérifiera* (2.6) le théorème suivant, conjecturé dans [7]:

THÉORÈME A. *Supposons G, simple et simplement connexe, et soit p un nombre premier ne divisant pas les coefficients de la plus grande racine de G, exprimée comme somme de racines simples. Alors G n'a pas de p-torsion.*

Dans un groupe de Lie compact connexe G il existe des $[p]$-sous-groupes évidents, les éléments d'ordre p d'un tore maximal, mais il peut y en avoir d'autres, (pour $p = 2$, les matrices diagonales de $\mathbf{SO}\,(n)$, $(n \geqq 3)$ par exemple). Cependant, d'après [9, XII, 5,3, 5.4], si G n'a pas de p-torsion, tout $[p]$-sous-groupe H fait partie d'un tore, ce qui précise un résultat de [11] affirmant que, sous l'hypothèse faite, le rang de H est au plus égal à la dimension des tores maximaux de G. Cette dernière condition est évidemment nécessaire pour que H fasse partie d'un tore de G, mais elle n'est pas suffisante en général. En effet, on démontrera:

(i) *Pour que le groupe fondamental $\pi_1(G)$ de G soit sans p-torsion, il faut et il suffit que tout $[p]$-sous-groupe de rang deux soit contenu dans un*

139

tore de G[1].

(ii) *Pour que G soit sans p-torsion, il faut et il suffit que tout $[p]$-sous-groupe de rang $\leqq 3$ soit contenu dans un tore.*

(*Voir* 3.12, 4.5.) On verra par construction d'exemples que ces conditions sont suffisantes; la nécessité de la condition de (i) (en fait, un énoncé plus général, cf. 3.9) résultera principalement de ce que *si $\pi_1(G)$ est sans torsion, le centralisateur de tout élément de G est connexe* (3.5)[2]; celle de la condition de (ii) est conséquence du théor. 5.3 de [9, XII].

Finalement, on aura prouvé, en partie par des raisonnements a priori, en partie par des vérifications utilisant la classification, le

THÉORÈME B. *Soit p un nombre premier et soit G un groupe de Lie compact connexe. Alors les quatre conditions suivantes sont équivalentes:*
(1) *G n'a pas de p-torsion.*
(2) *L'espace classifiant B_G de G n'a pas de p-torsion.*
(3) *Tout $[p]$-sous-groupe est contenu dans un tore.*
(4) *Tout $[p]$-sous-groupe de rang $\leqq 3$ est contenu dans un tore.*

Dans le §5, le théorème 5.3 de [9, XIII] déjà cité sera généralisé par le:

THÉORÈME C. *Soient H un sous-groupe compact commutatif de G, et H_0 la composante connexe de l'élément neutre de H. Si G n'a pas de p-torsion pour tout diviseur premier de l'ordre de H/H_0, alors H est contenu dans un tore.*

Pour l'établir, on s'appuiera en particulier sur le fait que si G est sans p-torsion, alors un sous-groupe connexe de rang maximum U et l'espace homogène correspondant G/U sont sans p-torsion (5.1).

Dans les démonstrations des résultats précédents, on se ramènera fréquemment au cas des groupes semi-simples grâce au fait qu'un groupe de Lie compact connexe est homéomorphe au produit d'un tore par sa partie semi-simple (3.2).

Le Théor. 6.2 montre que, si $G = \mathbf{F}_4$, **Spin** (n) $(n = 7,8,9)$, les $[2]$-sous-groupes maximaux de G sont conjugués par automorphismes intérieurs et l'analogie de [4] entre $[2]$-sous-groupes maximaux en cohomologie mod 2 et tores maximaux en cohomologie réelle subsiste. Nous ne savons pas si l'on peut en général mettre directement en relations la cohomologie mod p de G ou de B_G

---

1) Ce résultat a été suggéré par des discussions avec J. Wolf, à qui je dois aussi l'énoncé et une partie de la vérification du lemme 3.11.

2) Pour démontrer 3.5, on se ramène tout de suite au cas où G est semi-simple et simplement connexe. On verra alors que plus généralement l'ensemble des points fixes d'un automorphisme σ de G est connexe (3.4), théorème d'abord prouvé par R. Bott (non publié), de manière entièrement différente. Lorsque σ est involutif, ce résultat, dû à E. Cartan, est aussi établi dans [14, IV, Prop. 3.1].

avec les $[p]$-sous-groupes maximaux. Des exemples (voir 6.3-6.5) montrent que ces derniers peuvent former plus d'une classe de conjugaison.

Enfin, le § 7 apporte des compléments au § 4 de [9, XII], concernant les [3]-groupes opérant sur le plan des octaves.

Les résultats établis dans les §§ 1,2 avaient été communiqués antérieurement à S. Araki, qui les a complétés en décrivant notamment la cohomologie mod 3 de $\mathbf{E}_7$, $\mathbf{E}_8$ (voir [1]). Dans cet ordre d'idées, il reste donc en premier lieu à déterminer $H^*(\mathbf{E}_i;\ \mathbf{Z}_2)$, $(i = 6, 7, 8)$.

Les résultats de cet article mettent ainsi en évidence quelques liens entre la structure de groupe d'un groupe de Lie et sa topologie. Cependant, l'auteur regrette vivement d'en avoir été réduit à établir plusieurs propositions d'énoncé général à l'aide de vérifications utilisant la classification.

> "You have a low shopkeeping mind. You think of things that would never come into a gentleman's head."

1   "That's the Swiss national character, dear lady."

(B. Shaw)

0. Rappel et notations.

Dans tout ce travail, G désigne un groupe de Lie compact connexe.

0.1. Tors. H est le sous-groupe de torsion d'un groupe commutatif H, ord H est l'ordre d'un groupe fini H; si A est un sous-ensemble du groupe H, on note $Z(A)$ le centralisateur de A dans H. Un $[p]$-groupe est un groupe commutatif de type $(p,\ldots\ldots,p)$; son rang est le nombre de facteurs cycliques.

$\mathbf{Q}$ est le corps des rationnels, $\mathbf{R}$ celui des réels, $\mathbf{Z}_p$ (p premier) le corps des entiers mod p, ou le groupe cyclique d'ordre p.

$E(m_1,\ldots\ldots,m_k)$ dénote l'algèbre extérieure d'un module libre gradué sur un anneau A, avec r générateurs de degrés respectifs $m_1,\ldots\ldots,m_k$ (impairs si la caractéristique de A est $\neq 2$), $P(a, b)$ désigne une algèbre de polynomes tronquée, $A[x]/(x^b)$ engendrée par un élément x de degré a, de hauteur b.

La série de Poincaré d'un module M libre sur A, gradué par des sous-modules $M^i (i \geqq 0)$ de type fini, est

$$P(M, t) = \sum_{i \geqq 0} (\text{rang } M^i) \cdot t^i,$$

et si X est un espace topologique, on pose

$$P_p(X, t) = P(H^*(X;\ \mathbf{Z}_p), t), \quad P_0(X, t) = P(H^*(X;\ \mathbf{Q}), t).$$

0.2. La composante connexe de l'élément neutre d'un groupe de Lie H, que l'on appellera la *composante neutre de H*, sera notée H_0. On rappelle que si H est compact connexe, on a $H = Z(H)_0 \cdot H_{ss}$ avec $Z(H)_0 \cap H_{ss}$ fini, H_{ss} désignant le plus grand groupe semi-simple de H.

Soient G' un revêtement fini de $G, f: G' \to G$ la projection canonique, et N le noyau de f; le sous-groupe N est central, donc contenu dans tous les tores maximaux de G'. Par conséquent, si U est un sous-groupe connexe de rang maximum de G, $f^{-1}(U)$ est connexe. En particulier si T (resp. T') est un tore maximal de G (resp. G'), $f^{-1}(T)$ (resp. $f(T')$) est un tore maximal de G' (resp. G). Si $G = G_1 \cdot G_2$ (G_1, G_2 invariants connexes fermés d'intersection finie) alors $U = (G_1 \cap U) \cdot (G_2 \cap U)$ (cf. [12]). On déduit de cela que G/U s'identifie au quotient du revêtement universel de G_{ss} par un sous-groupe connexe, donc que G/U est simplement connexe.

0.3. Tout élément $g \in G$ est sur un sous-groupe à un paramètre, donc $g \in Z(g)_0$. Les tores maximaux de G sont conjugués par automorphismes intérieurs, et leur intersection est le centre de G. Etant donné un tore $S \subset G$ et un élément $g \in Z(S)$, il existe un tore T contenant g et S, donc le centralisateur $Z(S)$ d'un tore S est connexe, et un tore maximal est son propre centralisateur (pour tout cela, *voir* par exemple Sém. S. Lie, Paris (1954-55), Exp. 23). L'espace homogène $G/Z(S)$, où S est un tore de G, est sans torsion et ses nombres de Betti en dimensions impaires sont nuls [6, 13].

0.4. On convient de noter ici $\mathbf{G_2}, \mathbf{F_4}, \mathbf{E_6}, \mathbf{E_7}, \mathbf{E_8}$ les représentants *simplement connexes* des structures simples exceptionnelles. Leurs centres sont donc d'ordres respectifs 1, 1, 3, 2, 1. Les représentants simplement connexes des structures simples classiques sont les groupes unitaires unimodulaires complexes et quaternioniens $\mathbf{SU}(n)$, $\mathbf{Sp}(n)$ et les groupes de spineurs $\mathbf{Spin}(n)$ $(n \neq 2, 4)$. Le centre de $\mathbf{SU}(n)$ (resp. $\mathbf{Sp}(n)$, resp. $\mathbf{Spin}(2m+1)$, resp. $\mathbf{Spin}(4m+2)$) est cyclique d'ordre n (resp, 2, 2, 4), celui de $\mathbf{Spin}(4m)$ est isomorphe à $(\mathbf{Z_2})^2$.

Tout groupe simple compact connexe et simplement connexe est isomorphe à l'un des groupes sus-mentionnés.

1. *H*-espaces dont la cohomologie à un nombre fini de générateurs.

1.1. Dans ce paragraphe, X désigne un espace compact connexe, dont le groupe de cohomologie entière admet un nombre fini de générateurs, et qui est muni d'un produit de type $(1, 1)$, par exemple avec élément neutre à gauche et à droite. D'après le théorème de Hopf:

$$(1) \qquad H^*(X; \mathbf{Q}) = E(m_1, \ldots, m_l), \qquad (m_i \text{ impair}; \ 1 \leq i \leq l).$$

1.2. PROPOSITION. *Soit H le quotient de $H^*(X; \mathbf{Z})$ par son groupe de torsion. Alors H est l'algèbre extérieure d'un module libre à l générateurs de degrés $m_1, \ldots, m_l$.*

Soit $h: H^*(X; \mathbf{Z}) \to H^*(X \times X; \mathbf{Z})$ l'homomorphisme défini par le produit dans X. En utilisant la règle de Künneth, on voit que h induit un homomorphisme $h' : H \to H \otimes H$ qui vérifie la condition imposée aux algèbres de Hopf, donc pour tout anneau $K, H \otimes K$ est une algèbre de Hopf sur K. Cela étant, on voit que la démonstration de la Prop. 7.3 de [3], (où l'on prouve 1.2 en supposant X sans torsion) est aussi valable pour H.

1.3. PROPOSITION. *Soit p un nombre premier. Alors $H^*(X; \mathbf{Z}_p)$ contient un sous-module gradué isomorphe à $E(m_1,\ldots\ldots,m_l)$, annulé par tous les opérateurs de Bockstein, et qui est une sous-algèbre isomorphe à $E(m_1,\ldots\ldots,m_l)$ si p est impair.*

Soit T le sous-groupe de torsion de $H^*(X; \mathbf{Z})$, et soient $y_i \in H^*(X; \mathbf{Z})$ des éléments de $H^*(X; \mathbf{Z})$ dont les images dans H forment un système minimal de générateurs. Vu 1.2, les monômes

$$y_{i_1}\ldots\ldots y_{i_t} \qquad (i_1 < \ldots\ldots < i_t; \ t = 1,\ldots\ldots, l)$$

forment une base d'un supplémentaire de T, et sont donc linéairement indépendants dans $H^*(X; \mathbf{Z}) \otimes K$ pour tout corps K. Si $K = \mathbf{Z}_p$, ils sont annulés par les opérateurs de Bockstein, puisqu'ils proviennent de classes entières. Si de plus p est impair, les y_i sont de carré nul dans $H^*(X; \mathbf{Z}) \otimes \mathbf{Z}_p$ et notre assertion résulte de ce que $H^*(X; \mathbf{Z}) \otimes \mathbf{Z}_p$ s'identifie à une sous-algèbre de $H^*(X; \mathbf{Z}_p)$.

1.4. PROPOSITION. *Soient p impair et (u_i) un système de type (M) [3,§6] de générateurs de $H^*(X; \mathbf{Z}_p)$. Alors (u_i) contient au moins l générateurs de degrés impairs.*

L'élément y_i considéré dans la démonstration de 1.3 s'écrit comme un polynome $P_i(u_1,\ldots\ldots,u_m)$. Comme y_i est de degré impair,chaque monôme de P_i doit contenir au moins un u_j de degré impair. Puisque $u_j \cdot u_j = 0$ si $d^0 u_j$ est impair, notre assertion résulte du fait que le produit $y_1,\ldots\ldots y_l$ est non nul d'après 1.3.

1.5. Une algèbre de Hopf H sur $\mathbf{Z}_p$ est dite être de type (p) si tout élément de H est de hauteur $\leq p$ [1, §1]. Soit (u_i) un système de générateurs de type (M) de H, au sens de [3, § 6]. La sous-algèbre engendrée par 1 et u_i est donc égale à $E(d^0 u_i)$ si $d^0 u_i$ est impair, à $P(d^0 u_i; p)$ si $d^0 u_i$ est pair, et H est le produit tensoriel de ces sous-algèbres [3,Théor. 6.1]. On a alors la proposition suivante, due à S.Araki [1,Théorème 7]:

PROPOSITION. *Soit T le sous-groupe de torsion de $H^*(X; \mathbf{Z})$ et soit p un nombre premier impair. Si $H^*(X; \mathbf{Z}_p)$ est de type (p), il existe une suite finie d'algèbres de Hopf de type (p) $A_0 = (H^*(X; \mathbf{Z})/T) \otimes \mathbf{Z}_p, A_1,\ldots\ldots,A_m = H^*(X; \mathbf{Z}_p)$, telle que A_i s'obtienne à partir de A_{i-1} en remplaçant un facteur algèbre extérieure $E(m)$ par un produit $E(a) \otimes P(a + 1; p)$ où $m = p \cdot a + p - 1$.*

Soit k le premier entier pour lequel la composante p-primaire $T_{k,p}$ de H^k $(X; \mathbf{Z})$ est non nulle. D'après la règle des coefficients universels, $H^k(X; \mathbf{Z}_p)$ et $H^{k-1}(X; \mathbf{Z}_p)$ contiennent des sous-espaces U_k, U_{k-1} isomorphes à $T_{k,p} \otimes \mathbf{Z}_p \cdot$ Si X est simplement connexe, il résulte alors du théorème de structure des algèbres de Hopf que $H^*(X; \mathbf{Z}_p)$ possède un système de générateurs de type (M) contenant des bases de U_k, U_{k-1} et n'ayant en degrés $< k - 1$ que des éléments de degré impair. On déduit alors immédiatement de la proposition précédente:

1.6. COROLLAIRE. *Supposons X simplement connexe et $H^*(X; \mathbf{Z}_p)$ de type (p), p étant impair. Soit k le premier entier pour lequel la composante p-primaire de $H^k(X; \mathbf{Z})$ est non nulle. Alors $p \cdot k + p - 1 \leqq m_l$.*

Originellement, l'auteur avait déduit (2.2), (2.3) du Théor. V de [14], de (1.3) et de l'inégalité

$$(1.7) \qquad\qquad p \cdot k \leqq m_l + p - 1,$$

qui est un peu plus faible que 1.6, mais dont il n'y a plus lieu de donner une démonstration directe, puisque le résultat de [1] est beaucoup plus précis. Dans la suite, nous n'utiliserons 1.5 explicitement qu'à propos de $H^*(\mathbf{E}_8; \mathbf{Z}_5)$, où cela permet d'abréger les calculs.

2. Sur la torsion des groupes simplement connexes.

2.1. Rappelons que l'on a

$$H^*(\mathbf{E}_i; \mathbf{Q}) = E(m_{i1}\ldots\ldots,m_{ii}) \qquad (i = 6, 7, 8)$$

et que les valeurs $(m_{i1},\ldots\ldots,m_{ii})$ sont respectivement

$$(3, 9, 11, 15, 17, 23), \quad (3, 11, 15, 19, 23, 27, 35), \quad (3, 15, 23, 27, 35, 39, 47, 59)$$

(voir par exemple [13]). D'après [14, Théor. V], on a

$$\pi_3(\mathbf{E}_i) \cong \pi_{m_{i2}}(\mathbf{E}_i) \cong \mathbf{Z}, \qquad \pi_j(\mathbf{E}_i) = 0 \ (j \leqq m_{i2}, j \neq 3, m_{i2} \ ; i = 6, 7, 8).$$

Une application $f_i : \mathbf{E}_i \to K(\mathbf{Z}, 3)$ qui est un isomorphisme pour π_3 induit alors, d'après le théorème de J. H. C. Whitehead, [17, Chap III], un isomorphisme en cohomologie jusqu'au degré $m_{i2} - 1$ inclus.

2.2. THÉORÈME. *Le groupe $\mathbf{E}_i$ n'a pas de p-torsion si $i = 6, 7$, $p = 5$ et $i = 7, 8$, $p = 7$.*

Prenons par exemple $i = 8$, $p = 7$. On a $H^j(\mathbf{E}_8; \mathbf{Z}_7) = H^j(K(\mathbf{Z}, 3); \mathbf{Z}_7)$ pour $j \leqq 14$. D'après H.Cartan [15, Exp.9], $H^*(K(\mathbf{Z}, 3); \mathbf{Z}_7) = E(3)$ jusqu'au degré 14, donc, si $\mathbf{E}_8$ a de la 7-torsion en dimension k, on doit avoir $k \geqq 16$; mais d'autre part, 1.7 et 2.1 donnent $7 \cdot k \leqq 65$, donc $\mathbf{E}_8$ n'a pas de 7-torsion.

On raisonne de même dans les trois autres cas.

2.3. THÉORÈME. *On a*

(1) $\qquad H^*(\mathbf{E}_6; \mathbf{Z}_3) = \wedge (x_3, x_7, x_9, x_{11}, x_{15}, x_{17}) \otimes \mathbf{Z}_3 [x_8] / (x_8^3)$

avec $d^0 x_i = i$, $\beta_3(x_7) = x_8$, $\beta_3(x_i) = 0$ $(i \neq 7)$

(2) $\qquad H^*(\mathbf{E}_8; \mathbf{Z}_5) = \wedge (x_3, x_{11}, x_{15}, x_{23}, x_{27}, x_{35}, x_{39}, x_{47}) \otimes \mathbf{Z}_5 [x_{12}] / (x_{12}^5)$

avec $d^0 x_i = i$, $\beta_5(x_{11}) = x_{12}$, $\beta_5(x_i) = 0$ $(i \neq 11)$.

Ici β_ν désigne le 1er opérateur de Bockstein, c'est à dire l'homomorphisme cobord associé à la suite de coefficients $0 \to \mathbf{Z} \to \mathbf{Z} \to \mathbf{Z}_p \to 0$, suivi de la réduction mod p.

D'après [15] et (2.1), $H^*(\mathbf{E}_6; \mathbf{Z}_3)$ est isomorphe jusqu'au degré 8 inclus à

$$\wedge (x_3, x_7) \otimes \mathbf{Z}_3[x_8] \qquad (\beta(x_3) = \beta_3(x_8) = 0; \ \beta_3(x_7) = x_8).$$

Le théorème 6.1 de [3] donne alors

$$H^*(\mathbf{E}_6; \mathbf{Z}_3) = \wedge (x_3, x_7) \otimes \mathbf{Z}_3[x_8] / (x_8^{3^c}) \otimes E(a_1, \ldots, a_r) \otimes P(b_1, 3^{c_1})$$
$$\otimes \ldots \otimes P(b_s, 3^{c_s}).$$

Vu 1.4 et ce qui vient d'être dit, on doit avoir

(3) $\qquad r \geq 4$, $a_i \geq 9$ $(i = 1, \ldots, r)$; $b_i \geq 10$ si $c_i \geq 1$.

Le produit des trois premiers facteurs du deuxième membre est donc non nul en un degré $t \geq 46 + (3^c - 1) \cdot 8$. et $P(b, 3^c)$ est non nul en dimension $(3^c - 1) \cdot b$. On a donc

$$46 + (3^c - 1) \cdot 8 + \sum_i (3^{c_i} - 1) \cdot b_i \leq \dim \mathbf{E}_6 = 78$$

d'où $c = 1$, $c_i = 0$, et le théorème résulte aussitôt de 1.3 et 2.1.

On voit de même que

$$H^*(\mathbf{E}_8; \mathbf{Z}_5) = \wedge (x_3, x_{11}) \otimes \mathbf{Z}_5[x_{12}] / (x_{12}^{5^c}) \otimes E(a_1, \ldots, a_r) \otimes P(b_1, 5^{c_1})$$
$$\otimes \ldots \otimes P(b_s, 5^{c_s})$$

avec

$$\beta_5(x_3) = \beta_5(x_{12}) = 0, \qquad \beta_5(x_{11}) = x_{12},$$
$$r \geq 6, \ a_i \geq 15 \ (i = 1, \ldots, r); \ b_i \geq 16 \text{ si } c_i \geq 1.$$

On en tire

$$104 + (5^c - 1) \cdot 12 + \sum_i (5^{c_i} - 1) \cdot b_i \leq \dim \mathbf{E}_8 = 248$$

d'où $c = 1$, $c_i \leq 1$; par suite $H^*(\mathbf{E}_8; \mathbf{Z}_5)$ est de type (5) et on peut appliquer (1.5). Vu (2.1), chacune des substitutions mentionnées dans la Prop. 1.5 fait apparaitre un générateur de degré $\leqq 11$. D'après qui a été dit plus haut, il ne peut y en avoir qu'une, celle qui remplace $E(59)$ par $E(11) \otimes P(12,5)$, et le théorème pour $\mathbf{E}_8$ est alors immédiat.

2.4. REMARQUE. Il est clair que l'algèbre d'homologie de $H^*(\mathbf{E}_6; \mathbf{Z}_3)$ par rapport à β_3 est isomorphe à $E(m_{61}, \ldots, m_{66})$, dans les notations de 2.1, c'est à dire au produit tensoriel par $\mathbf{Z}_3$ du quotient de $H^*(\mathbf{E}_6; \mathbf{Z})$ par son sous-groupe de torsion. On a une assertion analogue pour $\mathbf{E}_8$ et $p = 5$. Il s'ensuit donc, comme on sait, que les coefficients de torsion de $H^*(\mathbf{E}_6; \mathbf{Z})$ (resp. $H^*(\mathbf{E}_8; \mathbf{Z})$) ne sont pas divisibles par 9 (resp. 25). Un phénomène analogue a été observé par Araki [1] pour $\mathbf{E}_7, \mathbf{E}_8$, $p = 3$. Joint aux résultats de [3, 5], cela vérifie l'assertion suivante:

Supposons G simple et simplement connexe. Alors aucun coefficient de torsion de $H^*(G; \mathbf{Z})$ n'est divisible par le carré d'un nombre premier p, sauf peut-être si $G = \mathbf{E}_6, \mathbf{E}_7, \mathbf{E}_8, p = 2$.

Cela conduit évidemment à se demander s'il existe un H-espace à cohomologie entière de type fini, *simplement connexe*, ayant un coefficient de torsion divisible par un carré. L'auteur n'en connait pas d'exemple.

2.5. THÉORÈME *Supposons G simple et simplement connexe, et soit p un nombre premier. Alors G a de la p-torsion exactement dans les cas suivants:* $p = 2$, $G = \mathbf{Spin}(n)$, $(n \geqq 7)$, $\mathbf{G}_2, \mathbf{F}_4, \mathbf{E}_6, \mathbf{E}_7, \mathbf{E}_8$, $p = 3$, $G = \mathbf{F}_4, \mathbf{E}_6, \mathbf{E}_7, \mathbf{E}_8$; $p = 5$, $G = \mathbf{E}_8$.

Pour $\mathbf{SU}(n)$, $\mathbf{Sp}(n)$, cf.[3]; pour $\mathbf{Spin}(n)$, $\mathbf{G}_2, \mathbf{F}_4$, voir [5]. D'après (2.3) et [8], $\mathbf{E}_6, \mathbf{E}_7$ n'ont pas de p-torsion pour $p \geqq 5$, et $\mathbf{E}_8$ est sans torsion pour $p \geqq 7$. D'autre part il résulte de (2.1) et des résultats de H.Cartan sur $K(\mathbf{Z}, 3)$ que les groupes $\mathbf{E}_i$ ont en fait de la p-torsion pour les valeurs indiquées dans 2.5, d'où le théorème[3]

2.6. THÉORÈME. *Soient G simple et simplement connexe et p un nombre premier ne divisant pas les coefficients de la racine dominante, exprimée comme combinaison linéaire des racines simples. Alors G n'a pas de p-torsion.*

Les coefficients en question sont 1 pour $\mathbf{SU}(n)$, $(1, 2)$ pour $\mathbf{Sp}(n)$, $\mathbf{Spin}(n)$, $(2, 3)$ pour $\mathbf{G}_2, \mathbf{F}_4$, $(1, 2, 3)$ pour $\mathbf{E}_6$, $(1, 2, 3, 4)$ pour $\mathbf{E}_7$ et $(2, 3, 5, 6)$ pour $\mathbf{E}_8$ (voir par exemple J. de Siebenthal, Comm. Math. Helv. 25 (1951), pp. 210-256). 2.6 résulte donc de 2.5. On remarquera que la réciproque de 2.6 n'est en défaut que dans les cas $G = \mathbf{Sp}(n)$, $p = 2$, $G = \mathbf{G}_2$, $p = 3$.

---

3) L'existence de p-torsion dans les cas qu'énumère 2.5 résulte aussi de 4.4 et de [9, XII, 5.3]; pour les groupes exceptionnels, elle a également été vérifiée, de manière différente, par Oniščik, Mat. Sbornik 51 (1960), pp. 273-6

3. Sous-groupes commutatifs de G et torsion de $\pi_1(G)$.

3.1. PROPOSITION. *Soit $G = S \cdot H$ un groupe de Lie compact connexe engendré par un tore central S et par un sous-groupe invariant connexe fermé H. Alors G est homéomorphe au produit de H par un tore.*

En procédant par récurrence sur la dimension de S, on voit qu'il suffit de traiter le cas où S est un cercle. Il n'y a rien à démontrer si $S \subset H$. Sinon G est fibré principal de groupe structural H sur $G/H = S/S \cap H$ qui est aussi un cercle. Puisque le groupe structural est connexe, cette fibration est triviale, [19, Cor. 18.6] et G est le produit de H par $S/S \cap H$.

3.2. COROLLAIRE. *Le groupe G est homéomorphe au produit de sa partie semi-simple G_{ss} par la composante neutre de son centre.*

Il résulte en particulier du corollaire que $\pi_1(G)$ ou $H^*(G; \mathbf{Z})$ est sans p-torsion si et seulement $\pi_1(G_{ss})$ ou $H^*(G_{ss}; \mathbf{Z})$ est sans p-torsion.

3.3. LEMME. *Supposons G semi-simple. Soient G' un groupe de Lie compact connexe, $f: G' \to G$ un homomorphisme surjectif de noyau N fini, et p un nombre premier. Alors les deux conditions suivantes sont équivalentes:*
(a) $\pi_1(G)$ *(resp. $H^*(G; \mathbf{Z})$) est sans p-torsion.*
(b) $(\operatorname{ord} N, p) = 1$ *et $\pi_1(G')$ (resp. $H^*(G'; \mathbf{Z})$) est sans p-torsion.*

La suite d'homotopie de $G'/N = G$ donne la suite exacte de groupes commutatifs finis

$$0 \to \pi_1(G') \to \pi_1(G) \to \pi_0(N) \to 0,$$

d'où l'équivalence de (a) et (b) pour les groupes fondamentaux. Comme $\pi_1(G)$ est commutatif fini, on a

$$\operatorname{Tors} \pi_1(G) = \operatorname{Tors} H^2(G; \mathbf{Z})$$

et de même pour G'. D'autre part, si p ne divise pas l'ordre de N, G et G' sont simultanément avec ou sans p-torsion [5, Cor. 9.3]; l'équivalence de (a) et (b) pour la cohomologie résulte immédiatement de là.

3.4. THÉORÈME. *Supposons G simplement connexe et soit σ un automorphisme de G. Alors l'ensemble F des points fixes de σ est connexe.*[4]

Soient T un tore maximal de G, V son revêtement universel, $\pi: V \to T$ la projection canonique, $\Gamma = \pi^{-1}(e)$ et $D(G)$ le diagramme de G, c'est à dire

---

4   4) Par une démonstration assez semblable on peut montrer que l'ensemble des points fixes d'un automorphisme semi-simple d'un groupe de Lie semi-simple complexe, connexe et simplement connexe, est connexe. *Voir* une Note ultérieure de l'auteur.

l'image réciproque de l'ensemble des éléments de T qui sont singuliers dans G. Le diagramme est réunion d'un nombre fini de familles d'hyperplans parallèles. On sait que les adhérences des composantes connexes de $V - D(G)$ sont des polyèdres congruents, permutés, de manière simplement transitive, par le groupe $W^*(G)$ engendré par les symétries aux plans de $D(G)$. Le groupe $W^*(G)$ est engendré par le groupe de Weyl $W(G)$ de G et par les translations du réseau Γ, en particulier, chacun des polyèdres sus-mentionnés rencontre Γ en exactement un point. (Pour tout cela, *voir* par exemple [20].)

L'ensemble F est un sous-groupe fermé de G et nous devons montrer que $x \in F$ entraine $x \in F_0$. Supposons tout d'abord x *régulier* dans G, et soit T l'unique tore maximal le contenant. Il est clair, que, dans les notations précédentes, σ induit un automorphisme de V laissant $D(G)$ et Γ invariants, que nous noterons aussi σ. Soit Δ l'adhérence d'une composante connexe de $t - D(G)$ contenant l'origine et un point x^* de $\pi^{-1}(x)$; le point x étant régulier, x^* est intérieur à Δ; comme $\sigma(x) = x$, on a $\sigma(x^*) - x^* = \gamma \in \Gamma$; ce qui a été rappelé plus haut montre alors que $\Delta + \gamma = \sigma(\Delta)$; par conséquent $\sigma(\Delta)$ contient l'origine et γ, d'où $\gamma = 0$ et x^* est aussi fixe par σ. La droite joignant l'origine à x^* est alors laissée fixe point par point par σ, et son image par π est un groupe à un paramètre de F contenant x, donc $x \in F_0$.

Soit maintenant x singulier. La partie semi-simple de $Z(x)_0$ est alors $\neq (e)$, donc [18, Chap. II, §2], un tore maximal S de $Z(x) \cap F$ est de dimension ≥ 1. Soit U la composante neutre de $Z(S) \cap Z(x)$. Elle est invariante par σ. Comme x et S font partie d'un tore maximal (0.3), U est de rang maximum et contient x et S dans son centre. Nous voulons maintenant montrer que U est un tore maximal de G. Si ce n'est pas le cas, alors la partie semi-simple U_{ss} de U est $\neq (e)$; elle est évidemment invariante par σ, donc [18, Chap. II, 2], $U_{ss} \cap F$ contient un tore S' de dimension ≥ 1; comme S est dans le centre de $U, S \cap S'$ est fini et $S \cdot S'$ est un tore de $Z(x) \cap F$, de dimension strictement plus grande que dim S, ce qui contredit l'hypothèse faite sur S.

Ainsi U est un tore maximal de G. Montrons que $x \cdot S$ contient un élément régulier de G. Nous écrivons T pour U et reprenons les notations du début de la démonstration. Soit M la composante neutre de $\pi^{-1}(S)$, et soit $x^* \in \pi^{-1}(x)$. Alors $x^* + M$ ne fait partie d'aucun plan de $D(G)$, car sinon, d'après la théorie des éléments singuliers $Z(x) \cap Z(S)$ serait de dimension au moins égale à dim $T + 2$, et U ne serait pas un tore. Il existe donc $s \in M$ tel que $y = x^* + s$ soit régulier. Alors $\pi(y) = x \cdot \pi(s)$ est un élément de F qui est régulier dans G. D'après ce qui a déjà été démontré, on a donc $\pi(y) \in F_0$, d'où aussi $x \in F_0$.

3.5. COROLLAIRE. *Dans un groupe de Lie compact connexe dont le groupe fondamental est sans torsion, le centralisateur d'un élément quelconque* *est connexe.*

Un élément $x \in G$ peut s'écrire sous la forme $x = s \cdot h$ $(s \in Z(G)_0;\ h \in G_{ss})$ et il est clair que $Z(x) = Z(G)_0 \cdot Z'(h)$, où $Z'(h)$ désigne le centralisateur de h dans G_{ss}. Vu 3.2, G_{ss} est simplement connexe, donc, d'après 3.4 appliqué à l'automorphisme intérieur défini par h, $Z'(h)$ est connexe, d'où le corollaire.

3.6. LEMME. *Soient A, A' des groupes, $f : A' \to A$ un homomorphisme dont le noyau N est un sous-groupe fini du centre de A', H un sous-groupe commutatif fini de $f(A')$, d'ordre premier à ord N. Alors A' contient un sous-groupe H' que f applique isomorphiquement sur H.*

Soient p un diviseur premier de ord H, H_p la composante p-primaire de H et S_p un p-sous-groupe de Sylow de $f^{-1}(H_p)$. Comme ord N est premier à p, f applique S_p isomorphiquement sur H_p et $f^{-1}(H_p) = S_p \cdot N$. Il s'ensuit aussi que les p-sous-groupes de Sylow de $f^{-1}(H_p)$ sont conjugués par N. Mais ce dernier est central, donc S_p est invariant dans $f^{-1}(H_p)$, le groupe $f^{-1}(H_p)$ est commutatif et S_p en est la composante p-primaire. Comme H_p est central dans H, il en résulte aussi que S_p est central dans $f^{-1}(H)$. On prend alors pour H' le produit des groupes S_p, où p parcourt les diviseurs premiers de ord H.

3.7. LEMME. *Soient K un groupe de Lie compact, $\pi : K \to K/K_0$ la projection canonique, et a un élément d'ordre une puissance d'un nombre premier p de K/K_0. Alors $\pi^{-1}(a)$ contient un élément dont l'ordre est une puissance (finie) de p.*

Soient $u \in \pi^{-1}(a)$ et U le plus petit sous-groupe fermé de K contenant u. Evidemment, $\pi(U) = U/(U \cap K_0)$ est engendré par a. Le groupe U est compact commutatif, donc est le produit direct d'un groupe fini M par U_0, et il est clair que $\pi^{-1}(a) \cap M$ contient un élément d'ordre une puissance de p.

3.8. THÉORÈME. *Soit p un nombre premier ne divisant pas l'ordre de Tors $\pi_1(G)$. Alors*

(a) *Pour tout $g \in G$, l'ordre de $Z(g)/Z(g)_0$ est premier à p.*

(b) *Si G est semi-simple et si F est le sous-groupe des points fixes d'un automorphisme de G, l'ordre de F/F_0 est premier à p.*

(a) On a $g = s \cdot h$, $(s \in S = Z(G)_0,\ h \in G_{ss})$, donc $Z(g) = S \cdot Z'(h)$, où $Z'(h)$ désigne le centralisateur de h dans G_{ss}. Evidemment, $Z(g)_0 = S \cdot Z'(h)_0$; comme $S \cap G_{ss}$ fait partie de $Z'(h)_0$, vu 0.3, il s'ensuit que $Z(g)/Z(g)_0 \cong Z'(h)/Z'(h)_0$. Compte tenu de 3.2, on voit que l'on est ramené à (b).

(b) Il suffit, vu 3.7, de faire voir que si x est un élément de F dont l'ordre m est une puissance de p, alors $x \in F_0$.

Soient G' le revêtement universel de G, $\pi : G' \to G$ la projection canonique et N le noyau de π. L'automorphisme σ induit un automorphisme σ' de G', dont l'ensemble des points fixes sera noté F', et l'on a $\pi \circ \sigma' = \sigma \circ \pi$. Comme p est premier à l'ordre de $\pi_1(G)$, il ne divise pas l'ordre de N (3.3) donc (3.6), $\pi^{-1}(x)$ contient un élément x' d'ordre m. On a

$$\sigma'(x') = x' \cdot y \qquad (y \in N)$$

d'où, si q est l'ordre de y, $\sigma(x'^q) = x'^q$. Par conséquent, x'^q est dans F'; mais F' est connexe d'après 3.4, donc $x^q \in F_0$; comme q est premier à m, il s'ensuit que $x \in F_0$.

3.9. COROLLAIRE. *Soit H un sous-groupe compact commutatif de G, tel que l'ordre de H/H_0 soit premier à l'ordre de* Tors $\pi_1(G)$. *Alors H fait partie d'un tore de G dans chacun des deux cas suivants:* (a) *H contient un élément régulier de G,* (b) *H/H_0 est engendré par deux éléments.*

Soient dans le cas (a), x un élément de H régulier dans G et, dans le cas (b), x, y des éléments de H dont les images dans H/H_0 engendrent ce groupe. $Z(x)$ contient H, et $H/(Z(x)_0 \cap H)$ est un quotient de H/H_0, donc est d'ordre premier à l'ordre de Tors $\pi_1(G)$. Mais un nombre premier p ne peut diviser l'ordre de $Z(x)/Z(x)_0$ que si $\pi_1(G)$ a de la p-torsion d'après 3.8, donc $H \subset Z(x)_0$. Dans le cas (a), $Z(x)_0$ est un tore; dans le cas (b), on a $y, H_0 \in Z(x)_0$, et y centralise H_0. D'après 0.3, $Z(x)_0$ possède un tore maximal T qui contient y, H_0 et x. Alors $T \supset H$.

3.10. LEMME. (a) *Soient G' un groupe de Lie compact connexe, $\pi : G' \twoheadrightarrow G$ un homomorphisme surjectif de noyau N fini, et H (resp. H') un sous-groupe de G (resp. G'). Si H (resp. H) n'est pas contenu dans un tore de G (resp. G'), alors $\pi^{-1}(H)$ (resp. $\pi(H')$) n'est pas contenu dans un tore de G'(resp. G).*

(b) *Soient G_1, G_2 des sous-groupes invariants connexes fermés de G, d'intersection finie, tels que $G = G_1 \cdot G_2$, et soit H_1 un sous-groupe de G_1. Si H_1 ne fait pas partie d'un tore de G_1, il n'est contenu dans aucun tore de G.*

L'assertion (a) est conséquence de (0.2).

Dans le cas (b), il suffit de remarquer que si T est un tore maximal de G, alors $T = (T \cap G_1) \cdot (T \cap G_2)$ et $T \cap G_i$ est un tore maximal de G_i $(i = 1,2)$ [12].

3.11. LEMME. *Supposons G semi-simple, simplement connexe. Soient p un nombre premier, z un élément d'ordre p du centre de G. Alors il existe des éléments $u, v \in G$ tels que $u \cdot v \cdot u^{-1} \cdot v^{-1} = z$ et qui sont d'ordre p si p est impair, de carré égal à z si $p = 2$.*

Soient $G_1, \ldots, G_s$ les facteurs simples de G. Si le lemme est vrai pour les

G_i il l'est pour G. En effet, si l'on pose $z = (z_1,\ldots,z_s)$, alors z_i est un élément d'ordre p du centre de G_i ou l'identité. On prend $u_i,v_i \in G_i$ vérifiant le lemme dans le premier cas, égaux à l'identité dans le second cas, et alors $u = (u_1,\ldots,u_s)$, $v = (v_1,\ldots,v_s)$ vérifient le lemme pour G et z. Nous pouvons donc supposer G simple. D'après la classification (0.4) on a les possibilités suivantes

    (a) p impair : $G = \mathrm{SU}(p.q)$, et en outre, pour $p = 3$, $G = \mathbf{E}_6$

    (b) $p = 2 : G = \mathrm{SU}(2 \cdot q)$, $\mathbf{Spin}(n)$, $\mathbf{Sp}(n)$, $\mathbf{E}_7$.

    Soit p impair, $G = \mathrm{SU}(p \cdot q)$. Alors $z = \sigma \cdot \mathrm{Id}$ où σ est une racine p-ième de l'unité et il suffit évidemment de considérer le cas où $q = 1$. Si $e_1,\ldots,e_p$ est la base canonique de $\mathbf{C}^p$, on prend pour u et v les transformations linéaires définies par

$$(1) \qquad\qquad u : e_i \to \sigma^i e_i, \qquad v : e_1 \to e_2 \to \ldots \to e_p \to e_1.$$

    Soit $p = 3$, $G = \mathbf{E}_6$. Le centre de G est d'ordre 3. Il résulte de [12, p. 219 et Remarque II, p. 220] que G contient un sous-groupe $H = H^*/N$ avec $H^* = \mathrm{SU}(3) \times \mathrm{SU}(3) \times \mathrm{SU}(3)$, $N = \mathbf{Z}_3$. Le centre de H^* est isomorphe à $\mathbf{Z}_3^3$, et son image dans $\mathbf{E}_6$ contient le centre de $\mathbf{E}_6$. Le lemme dans ce cas résulte donc de ce qui a été démontré plus haut.

    Dorénavant $p = 2$. Si $G = \mathrm{SU}(2n)$, alors $z = -\mathrm{Id}$, on peut se borner au cas $n = 1$, dans lequel on prend

$$u = \begin{pmatrix} i & 0 \\ 0 & -i \end{pmatrix}, \qquad v = \begin{pmatrix} 0 & 1 \\ -1 & 0 \end{pmatrix}.$$

Si $G = \mathbf{Sp}(n)$, on a encore $z = -\mathrm{Id}$, et l'on se ramène au cas précédent en considérant l'inclusion évidente de $\mathbf{Sp}(1) \times \ldots \times \mathbf{Sp}(1)$ (n facteurs) dans $\mathbf{Sp}(n)$.

    Soit $G = \mathbf{E}_7$. Alors G contient un sous-groupe $H = H'/N$, avec $N = \mathbf{Z}_3$, $H' = \mathrm{SU}(3) \times \mathrm{SU}(6)$, [12, p. 219, et Rem. II, p. 220]. Le centre de H' est isomorphe à $\mathbf{Z}_3 + \mathbf{Z}_6$ et son élément d'ordre deux s'envoie sur l'élément d'ordre deux du centre de G. On est donc ramené aux cas déjà traités.

    Il reste à considérer $G = \mathbf{Spin}(n)$, ($n \geqq 3$). Soit C_n l'algèbre de Clifford de $\mathbf{R}^n$, muni de la forme quadratique unité, et soit $e_1,\ldots,e_n$ la base canonique de $\mathbf{R}^n$. On a donc

$$e_i \cdot e_j + e_j \cdot e_i = 2 \cdot \delta_{ij},$$

le produit étant celui de C_n. Le groupe $\mathbf{Spin}(n)$ s'identifie au groupe des éléments pairs inversibles $x \in C_n$ pour lesquels $x \cdot \mathbf{R}^n \cdot x^{-1} = \mathbf{R}^n$. Il contient en particulier ± 1, et les produits $e_i \cdot e_j$ ($1 \leqq i,j \leqq n$) (cf. par exemple [16, Chap. II]).

    Soit tout d'abord $n = 2m + 1$ (resp. $n = 4m + 2$). Le centre de G est alors isomorphe à $\mathbf{Z}_2$ (resp. $\mathbf{Z}_4$) et l'on a nécessairement $z = -1$. On pose alors

$$u = e_1 \cdot e_2, \qquad v = e_2 \cdot e_3.$$

Soit $n = 4m$. Le centre de G est isomorphe à $(\mathbf{Z}_2)^2$ et est formé des éléments $\pm 1, \pm(e_1 \cdot e_2 \cdot \dots \cdot e_n)$. Si $m = 1$, alors $G \cong \mathrm{SU}(2) \times \mathrm{SU}(2)$, cas déjà traité. Le cas $m \geqq 1$ s'y ramène si l'on remarque que la répartition des vecteurs e_i en blocs de quatre $(e_{4k+1}, e_{4k+2}, e_{4k+3}, e_{4k+4})$ $(k = 0, \dots, m-1)$ donne lieu à un homomorphisme de $\mathbf{Spin}(4) \times \dots \times \mathbf{Spin}(4)$ (m facteurs) dans $\mathbf{Spin}(4m)$, appliquant le centre sur un sous-groupe contenant le centre.

3.12. THÉORÈME. *Soit p un nombre premier. Alors les deux conditions suivantes sont équivalentes*:

(a) *$\pi_1(G)$ n'a pas de p-torsion,*

(b) *tout sous-groupe H isomorphe à $\mathbf{Z}_p + \mathbf{Z}_p$ de G est contenu dans un tore.*

L'implication (a) $\to$ (b) est un cas particulier de 3.9. Il reste donc à montrer que si $\pi_1(G)$ a de la p-torsion, alors G possède un sous-groupe $H \cong \mathbf{Z}_p + \mathbf{Z}_p$ qui ne fait partie d'aucun tore de G: vu (3.2) et (3.10(b)), il suffit de le faire lorsque G est semi-simple. Soient G' le revêtement universel de G et N le noyau de la projection canonique $\pi: G' \to G$. Le groupe N est isomorphe à $\pi_1(G)$, donc contient un sous-groupe $M \cong \mathbf{Z}_p$. Vu 3.10 (a), il suffit de considérer le cas $G = G'/M$. Supposons donc que $N \cong \mathbf{Z}_p$. Soit z un générateur de N. Alors (3.11) il existe $u, v \in G'$ tels que $u \cdot v \cdot u^{-1} \cdot v^{-1} = z$, et que $u^p = v^p = e$ si p est impair, $u^p = v^p = z$ si $p = 2$. Les éléments u, v engendrent un sous-groupe H' d'ordre p^3 de G' contenant N, et dont l'image $H = H'/N$ dans G est isomorphe à $\mathbf{Z}_p + \mathbf{Z}_p$. Le sous-groupe H' n'est pas commutatif, donc ne peut faire partie d'un tore de G'. Par conséquent, (3.10 (a)) H n'est pas contenu dans un tore de G.

4. $[p]$-sous-groupes et p-torsion de G.

4.1. Soient U un sous-groupe fermé de G, A un anneau de coefficients. Par définition, $H^*(G/U; A)$ est égale à sa sous-algèbre caractéristique si G/U est totalement non homologue à zéro, relativement à A, dans la fibration universelle $(B_U, B_G, G/U)$. Dans ce cas, G/U est totalement non homologue à zéro, relativement à A, dans toute fibration $(E/U, B, G/U, \pi)$ ou (E, B, G, π) est une fibration principale, de groupe structural G [3, Cor. à la Prop. 18.3].

4.2. PROPOSITION. *Soient T un tore maximal de G et p un nombre premier. Alors les deux conditions suivantes sont équivalentes*:

(a) *B_G est sans p-torsion,*

(b) *$H^*(G/T; \mathbf{Z}_p)$ est égale à sa sous-algèbre caractéristique.*

G/T est sans torsion et ses nombres de Betti en dimensions impaires sont nuls (0.3). Si B_G est sans p-torsion, alors $H^i(B_G; \mathbf{Z}_p)$ est aussi nul pour i impair

[3, Théor. 19. 1], donc la suite spectrale de $(B_T, B_G, G/T)$ a son terme E_2 nul en degrés impairs, et est triviale, ce qui montre que (a) → (b). Réciproquement, si (b) est vraie, cette suite spectrale est triviale, et $\rho(T, G)^* : H^*(B_G; \mathbf{Z}_p) \to H^*(B_T; \mathbf{Z}_p)$ est injectif, donc $H^*(B_G; \mathbf{Z}_p)$ est nul en dimensions impaires et B_G n'a pas de p-torsion.

REMARQUE. Si G est simplement connexe, ces deux conditions sont aussi équivalentes à la suivante [8, Prop. 5. 5] :

(c) $H^*(G/T; \mathbf{Z}_p)$ est engendrée, comme algèbre, par 1 et $H^2(G/T; \mathbf{Z}_p)$.

4. 3. LEMME. *Soient U un sous-groupe fermé connexe de G qui soit la composante neutre du centralisateur d'un élément $a \in G$, et H un sous-groupe de U qui ne fasse partie d'aucun tore de U. Alors le sous-groupe engendré par a et H n'est pas contenu dans un tore de G.*

En effet, un tore contenant a et H ferait partie de la composante neutre du centralisateur de a.

4. 4. PROPOSITION. *Supposons G simple et simplement connexe et soit p un nombre premier. Alors G possède un sous-groupe $H \cong (\mathbf{Z}_p)^3$ ne faisant partie d'aucun tore de G dans chacun des cas suivants : $p = 2$, $G = \mathbf{Spin}(n)$ $(n \geq 7)$, $\mathbf{G}_2, \mathbf{F}_4, \mathbf{E}_6, \mathbf{E}_7, \mathbf{E}_8$; $p = 3$, $G = \mathbf{F}_4, \mathbf{E}_6, \mathbf{E}_7, \mathbf{E}_8$; $p = 5$, $G = \mathbf{E}_8$.*

Soit $G = \mathbf{Spin}(n)$, $(n \geq 7)$, $p = 2$. On reprend les notations de la démonstration de 3. 11 dans le cas de $\mathbf{Spin}(n)$. Soit H le sous-groupe de G engendré par les produits $(e_1 \cdot e_2 \cdot e_3 \cdot e_4)$, $(e_1 \cdot e_2 \cdot e_5 \cdot e_6)$ et $(e_1 \cdot e_3 \cdot e_5 \cdot e_7)$. Il est visiblement isomorphe à $(\mathbf{Z}_2)^3$.

L'image de e_i dans le groupe orthogonal *complet* $\mathbf{O}(n)$ (par la représentation χ de [16]) est la symétrie au plan $x_i = 0$. On en déduit immédiatement que le centralisateur connexe de l'image $\chi(H)$ de H dans $\mathbf{SO}(n)$ par la représentation canonique $\chi : \mathbf{Spin}(n) \to \mathbf{SO}(n)$ s'identifie à $\mathbf{SO}(n - 7)$. Comme le rang de $\mathbf{SO}(n - 7)$ est strictement plus petit que celui de $\mathbf{SO}(n)$, cela montre que $\chi(H)$ n'est pas contenu dans un tore maximal de $\mathbf{SO}(n)$, donc (3. 10 (a)), H ne fait pas partie d'un tore de G.

Soit $p = 2$, $G = \mathbf{G}_2$. D'après [12, p. 219, Rem. II. p. 220], $\mathbf{G}_2$ contient un sous-groupe $U \cong (\mathbf{SU}(2) \times \mathbf{SU}(2))/N$ qui est le centralisateur d'un élément a d'ordre deux. Il s'ensuit que N est d'ordre deux, donc (3. 11) que U contient un sous-groupe $H' = (\mathbf{Z}_2)^2$ ne faisant partie d'aucun tore de U. Vu (0. 3), $a \notin H'$, donc a et H' engendrent un sous-groupe $H \cong (\mathbf{Z}_2)^3$ qui (4.3) n'est pas contenu dans un tore de G; (pour une autre description d'un tel sous-groupe, voir [10, §17; 11]).

On voit de même, à l'aide des inclusions $(\mathbf{SU}(2) \times \mathbf{Sp}(3))/N \subset \mathbf{F}_4$, $(\mathbf{Spin}(16)/N) \subset \mathbf{E}_8$ et $(\mathbf{SU}(2) \times \mathbf{SU}(6))/N \subset \mathbf{E}_6$ que $\mathbf{F}_4$, $\mathbf{E}_8$ et le quotient Ad $\mathbf{E}_6$

de $\mathbf{E}_6$ par son centre contiennent un sous-groupe $H \cong (\mathbf{Z}_2)^3$ ne faisant partie d'aucun tore. Comme le centre de $\mathbf{E}_6$ est d'ordre trois, il en est alors de même pour $\mathbf{E}_6$ vu 3.10, 3.6.

$\mathbf{E}_7$ contient un sous-groupe $U \cong (\mathbf{E}_6 \times T^1)/N$, et U est le centralisateur dans $\mathbf{E}_7$ de la composante neutre S de son centre, qui est de dimension 1. Soit H un sous-groupe isomorphe à $(\mathbf{Z}_2)^3$ de sa partie semi-simple U_{ss}, non contenu dans un tore de cette dernière, dont l'existence vient d'être démontrée. H ne fait partie d'aucun tore de G. Sinon en effet, $Z(H)_0$ contiendrait H dans son centre, et S, donc (0.3), S et H feraient partie d'un même tore T; on aurait alors $T \subset Z(S) = U$, ce qui contredirait 3.10 (b).

Dans les cas $p = 3$, $G = \mathbf{F}_4, \mathbf{E}_7, \mathbf{E}_8$, $p = 5$, $G = \mathbf{E}_8$, on raisonne comme plus haut à partir des inclusions [12, p. 219]:

$$(\mathrm{SU}(3) \times \mathrm{SU}(3))/N \subset \mathbf{F}_4, \ (\mathrm{SU}(3) \times \mathrm{SU}(6))/N \subset \mathbf{E}_7, \ (\mathrm{SU}(3) \times \mathbf{E}_6)/N \subset \mathbf{E}_8,$$
$$(\mathrm{SU}(5) \times \mathrm{SU}(5))/N \subset \mathbf{E}_8.$$

Soit enfin $p=3$, $G=\mathbf{E}_6$. Alors G contient $U \cong (\mathrm{SU}(2) \times \mathrm{SU}(2) \times \mathrm{SU}(2))/N$ et $N = \mathbf{Z}_3$, $Z(U) = (\mathbf{Z}_3)^2$ [12, p. 219, Rem. II, p. 220]. D'après (3.12), U possède un sous-groupe $H' = (\mathbf{Z}_3)^2$ non contenu dans un tore de U, par suite, vu (0.3), $H' \cap Z(U) = (e)$. Soit $a \in Z(U)$, $a \notin Z(G)$, ce qui existe puisque le centre de G est d'ordre 3. Alors a et H' engendrent un sous-groupe $H = (\mathbf{Z}_3)^3$. D'autre part, comme U est un sous-groupe connexe maximal [12], $U = Z(a)_0$, donc, (4.3), H ne fait partie d'aucun tore de G.

4.5. THÉORÈME. *Soit p un nombre premier. Alors les quatre conditions suivantes sont équivalentes:*

    (1) *G est sans p-torsion;*
    (2) *B_G est sans p-torsion;*
    (3) *tout $[p]$-sous-groupe est contenu dans un tore;*
    (4) *tout $[p]$-sous-groupe de rang $\leqq 3$ est contenu dans un tore.*

$(1) \to (2)$ d'après [8, 5.1]. $(2) \to (3)$ d'après [9, XII, 5.4] et (4.2). (3) entraîne évidemment (4).

Pour terminer, il suffira de prouver que $(4) \to (1)$ et pour cela de faire voir: (*) Si G a de la p-torsion, alors G contient un $[p]$-groupe de rang $\leqq 3$ ne faisant partie d'aucun tore. Vu (3.2) et (3.10(b)), on peut supposer G semi-simple. Si $\pi_1(G)$ a de la p-torsion, (*) résulte de 3.12. Supposons donc $(\mathrm{ord}\,\pi_1(G), p) = 1$ et soit G' le revêtement universel de G. Alors, (3.3), G' a de la p-torsion, et vu 3.10, il suffit de montrer (*) pour G'. On peut donc se borner à G simplement connexe, donc aussi (3.10) à G simple et simplement connexe. Notre assertion résulte alors de (2.5) et (4.4).

4.6. REMARQUE. Les implications $(1) \to (2) \to (3)$ ont été établies ici par des raisonnements *a priori*. C'est la démonstration de $(4) \to (1)$ qui repose sur des vérifications. Nous ne savons pas non plus démontrer $(3) \to (1)$ ou $(2) \to (1)$ sans passer par ces vérifications. Nous avons aussi utilisé les $[p]$-sous-groupes pour voir que $H^*(B_G; \mathbf{Z})$ a de la p-torsion lorsque $p = 2, 3$, $G = \mathbf{E}_6, \mathbf{E}_7, \mathbf{E}_8$, $p = 5$, $G = \mathbf{E}_8$. Pour $p = 3$, cela résulte également de la détermination partielle de $H^*(B_G; \mathbf{Z}_3)$ $(G = \mathbf{F}_4, \mathbf{E}_6, \mathbf{E}_7, \mathbf{E}_8)$ faite par Araki, "On the non-commutativity of Pontrjagin rings mod 3 of some compact exceptional groups", (à paraître).

5. Sous-groupes commutatifs et p-torsion de G.

5.1. PROPOSITION. *Soit U un sous-groupe fermé connexe de rang maximum de G, et soit p un nombre premier. Si G est sans p-torsion, alors U et G/U sont sans p-torsion.*

Si U est sans p-torsion, il en est de même pour G/U ([3, Prop. 30.1], compte tenu du fait que G/T et U/T sont sans torsion [6, 13]). Supposons maintenant G et G/U sans p-torsion. Alors B_G est sans p-torsion [8, Prop. 5.1], donc la fibre et la base de la fibration universelle $(B_U, B_G, G/U)$ ont des nombres de Betti mod p nuls en dimensions impaires; il en est alors de même pour B_U, donc B_U est sans p-torsion et (4.5) U est sans p-torsion.

Il suffira donc de prouver que si G est sans p-torsion, alors U ou G/U est sans p-torsion. On procède par récurrence sur $\dim G$. Le sous-groupe U contient $Z(G)_0$, on a $G/U = G_{ss}/(U \cap G_{ss})$ et $U \cap G_{ss}$ a même rang que G_{ss} (0.3); compte tenu de (3.2), on peut donc supposer G semi-simple. Soient G' son revêtement universel et U' l'image réciproque de U dans G'. On a $G'/U' = G/U$, le sous-groupe U' est connexe (0.3) et G' est sans p-torsion (3.3). Comme U' est le produit de ses intersections avec les facteurs simples de G' [12] on est ramené au cas où G est simple, simplement connexe.

Supposons de plus U connexe maximal. S'il n'est pas semi-simple, il est nécessairement égal au centralisateur de la composante neutre de son centre, et G/U est sans torsion (0.3). Soit donc U semi-simple. Toutes les inclusions de ce type sont énumérées dans [12, p. 219]. De plus, (0.4) et la remarque II de [12, p. 220] permettent de calculer $\pi_1(U)$. On vérifie alors, à l'aide de (2.5) et (3.3) que U est sans p-torsion si G est sans p-torsion.

Supposons enfin que U ne soit pas connexe maximal dans G et soit V un sous-groupe propre connexe maximal le contenant. Alors V est sans p-torsion, d'après ce qui vient d'être dit, et U est sans p-torsion d'après l'hypothèse d'induction appliquée à l'inclusion $U \subset V$.

5   **5.2.** THÉORÈME. *Soit H un sous-groupe compact commutatif de G. Sup-*

posons que pour tout diviseur premier p de $\mathrm{ord}(H/H_0)$, le groupe G soit sans p-torsion. Alors H est contenu dans un tore de G.

Démonstration par récurrence sur $\dim G$. Si H est central, il fait partie de tout tore maximal de G (0.3). Supposons donc que $H \not\subset Z(G)$ et soit x un élément de H non contenu dans le centre de G. On a $H \subset Z(x)$ et $H/(Z(x)_0 \cap H)$ est un quotient de H/H_0. Vu l'hypothèse, l'ordre de $H/(Z(x)_0 \cap H)$ est donc premier à celui de $\mathrm{Tors}\ \pi_1(G)$, et 3.8 montre que $H \subset Z(x)_0$. Le sous-groupe $Z(x)_0$ est de rang maximum (0.3). Vu 5.1 et l'hypothèse d'induction, H fait partie d'un tore de $Z(x)$.

REMARQUE. La démonstration précédente n'utilise pas le Théor. 5.3 de [9,XII] affirmant que si G est sans p-torsion, tout $[p]$-sous-groupe est contenu dans un tore, qu'elle établit donc à nouveau. Cependant, elle a l'inconvénient de s'appuyer sur 5.1, qui a été vérifié à l'aide de la classification.

6. Les $[p]$-sous-groupes maximaux de quelques groupes de Lie.

Dans ce paragraphe, nous considérons exclusivement la cohomologie mod *2, et omettons les coefficients.* $\mathbf{P}_n$ *désigne l'espace projectif réel de dimension n, $\mathbf{S}_n$ la sphère de dimension n.*

6.1. Rappelons que si $G = \mathbf{G}_2, \mathbf{F}_4, \mathbf{Spin}(n)$, $(n = 7, 8, 9)$, alors $H^*(B_G)$ est une algèbre de polynomes à générateurs de degrés respectifs

$$(4, 6, 7),\ (4, 6, 7, 16, 24),\ (4, 6, 7, 8),\ (4, 6, 7, 8.\ 8),\ (4, 6, 7, 8, 16)$$

(voir [5]). Si $G = \mathbf{G}_2$, les [2]-sous-groupes maximaux sont de rang 3, conjugués par automorphismes intérieurs [10, §17], et si Q est l'un d'eux, $H^*(G/Q)$ est égale à sa sous-algèbre caractéristique [4, §13]. Le théorème 6.2 montrera qu'il en est de même dans les autres cas. Si l'on tient compte des résultats de [3,4], on voit que (6.2) vaut dans tous les cas connus où $H^*(B_G)$ est une algèbre de polynomes. Dans ces cas, on a donc la "formule de Hirsch mod 2" (6.2(1)) et l'analogie de [4] entre tores maximaux en cohomologie réelle et [2]-sous-groupes maximaux en cohomologie mod 2.

6.2. THÉORÈME. *Soit $G = \mathbf{Spin}(n)$ $(n = 7, 8, 9)$, $\mathbf{F}_4$. Alors les [2]-sous-groupes maximaux de G sont conjugués par automorphismes intérieurs et sont de rangs respectifs $4, 5, 5, 5$. Si Q est l'un deux, $H^*(G/Q)$ est égale à son algèbre caractéristique, par conséquent [2, §4] : $\rho(Q,G)^* : H^*(B_G) \to H^*(B_Q)$ est injectif, $H^*(G/Q)$ s'identifie au quotient de $H^*(B_Q)$ par l'idéal engendré par les éléments de degré > 0 de l'image de $\rho(Q, G)^*$ et*

(1) $\qquad P_2(G/Q,t) = (1-t)^{-r}/P_2(B_G,t) \qquad (r = \text{rang de } Q).$

On a la fibration $\mathbf{Spin}(7)/\mathbf{G}_2 = \mathbf{S}_7$ (voir [2]). Le centre M de $\mathbf{Spin}(7)$ est d'ordre deux, non contenu dans $\mathbf{G}_2$, donc $\mathbf{Spin}(n)$ contient un sous-groupe $V \cong \mathbf{Z}_2 \times \mathbf{G}_2$. La fibration $\mathbf{Spin}(7)/\mathbf{G}_2 = \mathbf{S}_7$ est définie par l'intermédiaire de la représentation des spineurs [2], donc l'élément $z \neq e$ de M agit par l'involution antipodique sur $\mathbf{S}_7$, et $\mathbf{Spin}(7)/V = \mathbf{S}_7/M = \mathbf{P}_7$ On peut évidemment identifier B_V au produit $B_M \times B_{\mathbf{G}_2}$, donc

$$P_2(B_V,t) = P_2(B_{\mathbf{G}_2},t)\cdot(1-t)^{-1}.$$

Comme $P_2(\mathbf{P}_7,t) = (1-t^8)\cdot(1-t)^{-1}$, on déduit de (6.1) que

$$P_2(B_V,t) = P_2(\mathbf{P}_7,t)\cdot P_2(B_{\mathbf{Spin}(7)}, t),$$

par conséquent [4, Prop. 2.1], $\mathbf{P}_7$ est totalement non homologue à zéro dans la fibration universelle $(B_V, B_{\mathbf{Spin}(7)}, \mathbf{P}_7)$. Le théor. 5.2 de [9, XII] montre alors que tout [2]-sous-groupe de $\mathbf{Spin}(7)$ est conjugué à un sous-groupe de V. Mais $V = \mathbf{Z}_2 \times \mathbf{G}_2$ donc, vu (6.1), ses [2]-sous-groupes maximaux sont de rang 4, et conjugués par automorphismes intérieurs. Il en est alors de même dans $\mathbf{Spin}(7)$.

Soit Q un [2]-sous-groupe maximal de $\mathbf{Spin}(7)$, contenu dans V. On a $Q = M \times Q'$, où Q' est un [2]-sous-groupe maximal de $\mathbf{G}_2$. Comme $\mathbf{G}_2/Q'$ est totalement non homologue à zéro dans la fibration $(B_{Q'}, B_{\mathbf{G}_2}, \mathbf{G}_2/Q')$, il est immédiat que $(\mathbf{G}_2 \times M)/Q = \mathbf{G}_2/Q'$ est aussi totalement non homologue à zéro dans la fibration $(B_Q, B_V, V/Q)$; par suite (4.1), V/Q est aussi totalement non homologue à zéro dans la fibration $(\mathbf{Spin}(7)/Q, \mathbf{P}_7, V/Q)$, d'où

$$P_2(\mathbf{Spin}(7)/Q, t) = P_2(\mathbf{P}_7, t)\cdot P_2(\mathbf{G}_2/Q',t);$$

mais on a [4, §13]:
$$P_2(\mathbf{G}_2/Q', t) = (1-t^4)\cdot(1-t^6)\cdot(1-t^7)\cdot(1-t)^{-3}$$

d'où, vu (6.1),

$$P_2(B_Q, t) = P_2(B_{\mathbf{Spin}(7)},t)\cdot P_2(\mathbf{Spin}(7)/Q, t)$$

ce qui, compte tenu de [4, Prop. 2.1], entraine finalement que $\mathbf{Spin}(7)/Q$ est totalement non homologue à zéro dans la fibration $(B_Q, B_{\mathbf{Spin}(7)}, \mathbf{Spin}(7)/Q)$, autrement dit que $H^*(\mathbf{Spin}(7)/Q)$ est égale à sa sous-algèbre caractéristique.

La fibration classique $\mathbf{SO}(8)/\mathbf{SO}(7) = \mathbf{S}_7$ donne lieu à la fibration $\mathbf{Spin}(8)/\mathbf{Spin}(7) = \mathbf{S}_7$. Le centre de $\mathbf{Spin}(8)$ est isomorphe à $(\mathbf{Z}_2)^2$; il contient un élément z induisant l'involution antipodique de $\mathbf{S}_7$, et z ne fait évidemment pas partie de $\mathbf{Spin}(7)$, qui a un point fixe sur $\mathbf{S}_7$. Le groupe $\mathbf{Spin}(8)$ contient donc un sous-groupe $V = \mathbf{Z}_2 \times \mathbf{Spin}(7)$ tel que $\mathbf{Spin}(8)/V = \mathbf{P}_7$.

D'après [2], on a une fibration $\mathbf{Spin}(9)/\mathbf{Spin}(7) = \mathbf{S}_{15}$, définie par la représentation des spineurs. Le centre de $\mathbf{Spin}(9)$ contient alors un élément z induisant l'involution antipodique, donc ne faisant pas partie de $\mathbf{Spin}(7)$, d'où une inclusion $V \subset \mathbf{Spin}(9)$, avec $V \cong \mathbf{Z}_2 \times \mathbf{Spin}(7)$ et $\mathbf{Spin}(9)/V = \mathbf{P}_{15}$.

Cela étant, la démonstration de (6.2) pour $G = \mathbf{Spin}(8)$, $\mathbf{Spin}(9)$ est tout à fait analogue à celle donnée pour $G = \mathbf{Spin}(7)$, et est laissée au lecteur.

Dans le cas où $G = \mathbf{F}_4$, nous partons de la fibration $\mathbf{F}_4/\mathbf{Spin}(9) = \mathbf{W}$, où $\mathbf{W}$ est le plan projectif des octaves [2]. D'après [9, XII, 4.2], tout [2]-groupe d'homéomorphismes de $\mathbf{W}$ admet un point fixe; cela entraine en particulier que tout [2]-sous-groupe de $\mathbf{F}_4$ est conjugué à un sous-groupe de $\mathbf{Spin}(9)$, d'où la première assertion de (6.2) pour $\mathbf{F}_4$.

Soit alors Q un [2]-sous-groupe maximal de $\mathbf{F}_4$ contenu dans $\mathbf{Spin}(9)$. D'après ce qui a été dit plus haut, $H^*(\mathbf{Spin}(9)/Q)$ est égale à sa sous-algèbre caractéristique donc, (4.1), $\mathbf{Spin}(9)/Q$ est totalement non homologue à zéro dans la fibration $(\mathbf{F}_4/Q, \mathbf{W}, \mathbf{Spin}(9)/Q)$ et l'on a

$$P_2(\mathbf{F}_4/Q, t) = P_2(\mathbf{W}, t) \cdot P_2(\mathbf{Spin}(9)/Q, t)$$

$$P_2(\mathbf{F}_4/Q, t) = (1 - t^{24})(1 - t^8)^{-1} \cdot (1-t)^{-5}/P_2(B_{\mathbf{Spin}(9)}, t),$$

ce qui, vu (6.1), donne

$$(1 - t)^{-5} = P_2(B_{\mathbf{F}_4}, t) \cdot P_2(\mathbf{F}_4/Q, t).$$

Par conséquent [4, Prop. 2.1], $\mathbf{F}_4/Q$ est totalement non homologue à zéro dans la fibration universelle $(B_Q, B_{\mathbf{F}_4}, \mathbf{F}_4/Q)$.

La suite de ce paragraphe est consacrée à quelques exemples de [p]-sous-groupes maximaux non conjugués.

6.3. *Les* [2]-*sous-groupes de* $\mathbf{Spin}(10)$. Soit $Q(n)$ le sous-groupe des matrices diagonales de $\mathbf{O}(n)$. On sait que les [2]-sous-groupes de $\mathbf{O}(n)$ et de $\mathbf{SO}(n)$ sont conjugués à des sous-groupes de $Q(n)$ et $SQ(n) = \mathbf{SO}(n) \cap \mathbf{Q}(n)$ respectivement. D'autre part le centralisateur d'un [2]-sous-groupe de $\mathbf{SO}(n)$ ou de $\mathbf{O}(n)$ est un produit de groupes orthogonaux et de groupes à deux éléments, donc ses [2]-sous-groupes maximaux sont aussi conjugués. Il s'ensuit aisément que si deux sous-groupes de $SQ(n)$ sont conjugués par un automorphisme intérieur de $\mathbf{SO}(n)$, ils sont aussi conjugués par un automorphisme intérieur laissant $SQ(n)$ invariant.

Soit $\pi : \mathbf{Spin}(n) \to \mathbf{SO}(n)$ la projection canonique et soit $Q' = \pi^{-1}(SQ(n))$. Comme tout [2]-sous-groupe maximal de $\mathbf{Spin}(n)$ contient le noyau de π, les faits rappelés dans l'alinéa précédent entrainent évidemment que tout [2]-sous-groupe de $\mathbf{Spin}(n)$ est conjugué à un [2]-sous-groupe de Q', que les [2]-sous-groupes maximaux de Q' sont des [2]-sous-groupes maximaux de $\mathbf{Spin}(n)$ et enfin que deux [2]-sous-groupes maximaux de Q' conjugués dans $\mathbf{Spin}(n)$ sont aussi conjugués par un élément du normalisateur de Q'.

Dans les notations de la démonstration de 3.11, Q' est engendré par les produits $e_i \cdot e_j$; ses éléments d'ordre deux sont les produits de $4k\,(k = 1, 2, \ldots\ldots)$ facteurs e_j, deux à deux distincts avec ± 1.

Cela étant, il n'y a aucune difficulté à vérifier que les [2]-sous-groupes maximaux de **Spin**(10) sont de rang 5, et forment deux classes de conjugaison représentées par les sous-groupes H_1, H_2 de Q' engendrés respectivement par

$$- 1, \quad e_1 \cdot e_2 \cdot e_3 \cdot e_4, \quad e_3 \cdot e_4 \cdot e_5 \cdot e_6, \quad e_5 \cdot e_6 \cdot e_7 \cdot e_8, \quad e_7 \cdot e_8 \cdot e_9 \cdot e_{10}$$

et par

$$- 1, \quad e_1 \cdot e_2 \cdot e_3 \cdot e_4, \quad e_3 \cdot e_4 \cdot e_5 \cdot e_6, \quad e_5 \cdot e_6 \cdot e_7 \cdot e_8, \quad e_1 \cdot e_3 \cdot e_5 \cdot e_7$$

H_1 est l'ensemble des éléments d'ordre deux d'un tore maximal, H_2 contient le sous-groupe H de 4.4, donc ne fait pas partie d'un tore.

6.4. *Les $[p]$-sous-groupes de* **PSU**(p), (p premier impair). Soit $G = $ **PSU**(p) le groupe projectif unitaire. On a donc $G = $ **SU**$(p)/\mathbf{Z}_p$. Nous voulons montrer que G possède deux classes de conjugaison de $[p]$-sous-groupes, représentées l'une par l'ensemble des éléments d'ordre p d'un tore maximal, l'autre par le sous-groupe de rang deux construit à l'aide de 3.11. Le rang d'un $[p]$-sous-groupe maximal est donc égal à $(p - 1)$ ou à deux.

Soient T' le tore maximal formé des matrices diagonales de **SU**(p) et T son image dans G. Soient v l'élément du normalisateur de T' défini par la permutation cyclique des coordonnées (cf. 3.11(1)) et v^* son image dans G. Un calcul immédiat montre que $Z(v^*) \cap T$ est un groupe cyclique d'ordre p, engendré par l'image u^* de l'élément u de 3.11(1). En particulier, l'automorphisme intérieur Int $v^*: x \to v^* \cdot x \cdot v^{*-1}$ n'a qu'un nombre fini de points fixes dans T donc $t \to v^* \cdot t \cdot v^{*-1} \cdot t^{-1}$ est un endomorphisme surjectif de T, ce qui entraine immédiatement que étant donnés $x, y \in T$, il existe $t \in T$ tel que Int $t(v^* \cdot y) = v^* \cdot x$.

Soit alors H un $[p]$-sous-groupe maximal de G. D'après [11, Théor. 1], on peut supposer $H \subset N$, où N désigne le normalisateur de T. Le seul cas à considérer est évidemment celui où $H \not\subset T$. Alors $H/(H \cap T)$ est un $[p]$-sous-groupe du groupe de Weyl $W(G) = N/T$. Ce dernier est le groupe des permutations de p objets : ses p-groupes de Sylow sont cycliques d'ordre p, l'un d'eux est engendré par l'image de v^*. Par conséquent, après conjugaison par un élément convenable de N, on peut supposer que $H \cap (v^* \cdot T) \neq \phi$. Mais on a remarqué plus haut que deux éléments de $v^* \cdot T$ sont conjugués par un élément de T; on peut donc supposer que $v^* \in H$. Alors H est engendré par v^* et par les éléments d'ordre p de $Z(v^*) \cap T$, donc, vu ce qui a été dit plus haut, par les images des éléments u, v de 3.11 (1).

6.5. On tire facilement de 6.4 que les $[p]$-sous-groupes maximaux de

$(\mathbf{SU}(p) \times \mathbf{SU}(p))/N$, $(N = \mathbf{Z}_p)$ sont de rang $\leqq 2(p-1)$ et forment au moins deux classes de conjugaison. Nous voulons en déduire que les [3]-sous-groupes maximaux de $\mathbf{F}_4$ sont de rangs $\leqq 4$ et se répartissent en au moins deux classes de conjugaison. Soit H un [p]-sous-groupe maximal. D'après [11, Théor. 1], il fait partie du normalisateur d'un tore maximal T, et $H/(H \cap T)$ est un p-groupe du groupe de Weyl $W(\mathbf{F}_4)$. Mais [12], $\mathbf{F}_4$ contient un sous-groupe $U = (\mathbf{SU}(3) \times \mathbf{SU}(3))/N$ $(N = \mathbf{Z}_3)$, et on peut supposer $U \supset T$. Les groupes de Weyl de $\mathbf{F}_4$ et U sont d'ordres respectifs $2^6 \cdot 3^2$ et 36, donc $W(U)$ contient un 3-groupe de Sylow de $W(\mathbf{F}_4)$; cela entraine qu'après automorphisme intérieur par un élément convenable du normalisateur de T, on peut supposer que $H \subset U$. D'autre part, tout [3]-sous-groupe maximal de U contient le centre de U, qui est cyclique d'ordre 3, donc (4.3), si un tel sous-groupe ne fait pas partie d'un tore de U, il n'est pas non plus contenu dans un tore de $\mathbf{F}_4$; d'où notre assertion.

On voit que même, à l'aide de l'inclusion $(\mathbf{SU}(5) \times \mathbf{SU}(5))/N \subset \mathbf{E}_8$ que les [5]-sous-groupes maximaux de $\mathbf{E}_8$ sont de rang $\leqq 8$ et forment au moins deux classes de conjugaison.

7. Les [3]-groupes opérant sur le plan des octaves.

7.1. Soit $\mathbf{W}$ le plan projectif des octaves. On a donc $H^*(\mathbf{W}; \mathbf{Z}) = Z[x]/(x^3)$, avec x de degré 8. Le Théor. 4.2 de [9,XII] montre que, si $p \neq 3$, tout [p]-groupe opérant sur $\mathbf{W}$, ou sur un espace compact connexe ayant même cohomologie mod p que $\mathbf{W}$, a un point fixe au moins, (en fait que dim $H^*(F; \mathbf{Z}_p) = 3$, F étant l'ensemble des points fixes). Nous voulons ici justifier et compléter la remarque du §4 de [9, XII] concernant le cas où $p = 3$.

7.2. PROPOSITION. (a) *Le plan projectif des octaves* $\mathbf{W}$ *possède un groupe de collinéations* $H \cong (\mathbf{Z}_3)^3$ *sans point fixe.*

(b) *Soit X un espace compact connexe tel que* $H^*(X; \mathbf{Z}_3) = \mathbf{Z}_3[x]/(x^3)$ $(d^0 x = 8)$. *Soient H un groupe isomorphe à* $(\mathbf{Z}_3)^2$ *opérant sur X, et F l'ensemble de ses points fixes. Alors* dim $H^*(F; \mathbf{Z}_3) = 3$. *En particulier, F n'est pas vide.*

Pour vérifier (a), on considère de nouveau la fibration $\mathbf{F}_4/\mathbf{Spin}(9) = \mathbf{W}$. Comme $\mathbf{Spin}(9)$ n'a pas de 3-torsion, tout [3]-sous-groupe de $\mathbf{Spin}(9)$ est contenu dans un tore, par conséquent, un sous-groupe $H \cong (\mathbf{Z}_3)^3$ de $\mathbf{F}_4$, ne faisant pas partie d'un tore, ce qui existe d'après 4.4, n'a pas de point fixe sur $\mathbf{W}$. Par ailleurs, on sait que $\mathbf{F}_4$ s'identifie à un sous-groupe (compact maximal) du groupe des collinéations de $\mathbf{W}$, d'où (a).

(b) Soit (E_r) la suite spectrale mod 3 de la fibration (X_H, B_H, X, π_2), [9, IV, 3]. Comme un 3-groupe agit trivialement sur $\mathbf{Z}_3$, H agit trivialement sur $H^*(X; \mathbf{Z}_3)$; pour établir (b), il suffit donc, vu [9, XII, 3.5], de faire voir que

$E_2 = E_\infty$, donc que $d_9(x) = 0$. Or on a

$$H^*(B_H;\ \mathbf{Z}_3) = \wedge\, (a_1, a_2) \otimes \mathbf{Z}_3[b_1, b_2]\ (d^0 a_i = 1,\ b_i = \beta_3(a_i),\ i = 1, 2)$$

(où β_3 est, comme dans 2.3, le premier opérateur de Bockstein); comme évidemment $E_2 = E_9$, on peut écrire

$$y = d_9(x) = a_1 \cdot Q_1(b_1, b_2) + a_2 \cdot Q_2(b_1, b_2)$$

où $Q_i(b_1, b_2)$ est un polynôme homogène de degré quatre en b_1, b_2. L'espace $H^*(X;\ \mathbf{Z}_3)$ étant nul en dimensions 9 et 12, on a

$$(1) \qquad\qquad \beta_3(x) = \mathcal{P}_3^1(x) = 0,$$

$\mathcal{P}_3^1$ désignant naturellement la première puissance réduite de Steenrod; β_3 et $\mathcal{P}_3^1$ commutent à la transgression, par conséquent

$$(2) \qquad\qquad \kappa_{10}^9(\beta_3 y) = \kappa_{13}^9(\mathcal{P}_3^1 y) = 0,$$

κ_j^9 étant, comme d'habitude, l'homomorphisme $H^i(B_H;\ \mathbf{Z}_3) = E_9^{i,0} \to E_j^{i,0}$, $(i = 1, 2, 3, \ldots\ldots)$. Ici le noyau de κ_j^9 $(10 \leqq j \leqq 17)$ est visiblement l'intersection de $H^i(B_H;\ \mathbf{Z}_3)$ avec l'idéal (y) engendré par y. Il nous suffira donc de montrer que $\beta_3(y),\ \mathcal{P}_3^1(y) \in (y)$ entrainent $y = 0$. Evidemment $(y) \cap H^{10}(B_H;\ \mathbf{Z}_3) \subset (a_1 \cdot a_2)$ et $\beta_3(y) \in \mathbf{Z}_3[b_1, b_2]$, donc $\beta_3(y) \in (y)$ équivaut à $\beta_3(y) = 0$. Nous devons donc prouver

$$(3) \qquad \beta_3(y) = 0,\ \text{et}\ \mathcal{P}_3^1(y) \in (y)\ \text{entrainent}\ y = 0.$$

Ecrivons y sous la forme

$$(4) \qquad y = a_1 \cdot \sum_{0 \leqq i \leqq 4} u_i \cdot b_1^i \cdot b_2^{4-i} + a_2 \cdot \sum_{0 \leqq i \leqq 4} v_i \cdot b_1^i \cdot b_2^{4-i}\ (u_i, v_i \in \mathbf{Z}_3).$$

On sait que $\mathcal{P}_3^1$ est une dérivation et que $\mathcal{P}_3^1(a_i) = 0$, $\mathcal{P}_3^1(b_i^s) = s \cdot b_i^{s+2}$; par conséquent $\mathcal{P}_3^1(y)$ est somme de

$$a_1 \cdot (u_4 \cdot b_1^6 + u_3 \cdot b_1^3 \cdot b_2^3 + u_2 \cdot 2 \cdot (b_1^4 \cdot b_2^2 + b_1^2 \cdot b_2^4) + u_1 \cdot b_1^3 \cdot b_2^3 + u_0 \cdot b_2^6)$$

et de

$$a_2 \cdot (v_4 \cdot b_1^6 + v_3 \cdot b_1^3 \cdot b_2^3 + v_2 \cdot 2 \cdot (b_1^2 \cdot b_2^4 + b_1^4 \cdot b_2^2) + v_1 \cdot b_1^3 \cdot b_2^3 + v_0 \cdot b_2^6),$$

mais $\mathcal{P}_3^1(y) \in (y)$ s'écrit:

$$\mathcal{P}_3^1(y) = (q \cdot b_1^2 + r \cdot b_1 \cdot b_2 + s \cdot b_2^2) \cdot y \quad (q, r, s \in \mathbf{Z}_3),$$

d'où, par comparaison de coefficients, le système d'équations

$$q \cdot u_4 = u_4;\ q \cdot u_3 + r \cdot u_4 = 0;\ q \cdot u_2 + r \cdot u_3 + s \cdot u_4 = 2 \cdot u_2;$$

$$(5u) \quad q \cdot u_1 + r \cdot u_2 + s \cdot u_3 = u_3 + u_1;\ q \cdot u_0 + r \cdot u_1 + s \cdot u_2 = 2 \cdot u_2;$$

$$r \cdot u_0 + s \cdot u_1 = 0;\ s \cdot u_0 = u_0,$$

et un système similaire, où v_i remplace u_i, que nous désignerons par (5v). β_3 est une antidérivation, annulant b_i et appliquant a_i sur b_i ($i = 1, 2$). La condition $\beta_3(y) = 0$ entraine alors

$$(6) \qquad u_4 = u_3 + v_4 = u_2 + v_3 = u_1 + v_2 = u_0 + v_1 = v_0 = 0.$$

Supposons $q \neq 0$. Comme $u_4 = 0$, on tire de (5u) que $u_3 = 0$, donc, vu (6) $v_4 = 0$, puis, vu (5v), $v_3 = 0$ et, vu (6), $u_2 = 0$. On obtient alors le système

$$u_1 + v_2 = 0; \quad q \cdot v_2 = 2 \cdot v_2; \quad q \cdot u_1 = u_1$$

qui entraine $u_1 = v_2 = 0$, d'où finalement, en utilisant à nouveau (5u), (6), $u_0 = v_1 = 0$. Par conséquent, $q \neq 0$ entraine $y = 0$. De même, $s \neq 0$ entraine $y = 0$. Enfin, on voit de la même manière, en utilisant alternativement (5u), (5v), et (6) que $q = s = 0$ entraine $y = 0$, ce qui démontre (3).

7.3. REMARQUE. Nous profitons de cette occasion pour signaler une erreur dans la démonstration du Théor. 3.6 de [9, XII]. L'égalité $\beta(c) = 0$ n'implique pas, a priori, que $\beta(d_3(c)) = 0$ mais seulement que $\beta(d_3(c))$ appartient à l'idéal M engendré par $d_3(H^2(X))$. Cependant, on a $\beta(d_3(c)) = 0$ *si tout* $c \in H^2(X)$ *est la restriction d'une classe entière transgressive*, ce qui a lieu en particulier quand la cohomologie entière de X est de type fini. La démonstration du texte vaut sans changement sous cette condition supplémentaire. Le théorème 3.6 lui-même vaut au moins si p est impair. En effet, le raisonnement du texte montre que $d_3(c)$ est somme de monômes de la forme $a_i \cdot a_j \cdot a_k$ ($i < j < k$); il est alors immédiat que $\beta(d_3(c)) \in M$ équivaut à $d_3(c) = 0$.

BIBLIOGRAPHIE

[1] S. ARAKI, A theorem of differential Hopf algebras and the cohomology mod 3 of the compact exceptional groups E_7 and E_8, to appear.

[2] A. BOREL, Le plan projectif des octaves et les sphères comme espaces homogènes, C. R. Acad. Sci., Paris 230 (1950), 1378-1380.

[3] __________, Sur la cohomologie des espaces fibrés principaux et des espaces homogènes de groupes de Lie compacts, Annals of Math. 57 (1953), 115-207.

[4] __________, La cohomologie mod 2 de certains espaces homogènes, Comm. Math. Helv. 27 (1953), 165-197.

[5] __________, Sur l'homologie et la cohomologie des groupes de Lie compacts connexes, Amer. Jour. Math. 76 (1954), 273-342.

[6] __________, Kählerian coset spaces of semi-simple Lie groups, Proc. Nat. Acad. Sci., U.S.A. 40 (1954), 1147-1151.

[7] __________, Topology of Lie groups and characteristic classes, Bull. Amer. Math. Soc. 61 (1955), 397-432.

[8] __________, Sur la torsion des groupes de Lie, Jour. Math. pur. appl. (9) 35(1956), 127-139.

[9] __________, Seminar on transformation groups, Annals of Math. Studies, No. 46, Princeton, 1960.

A. BOREL

[10] A. BOREL AND F. HIRZEBRUCH, Characteristic classes and homogeneous spaces I, Amer. J. Math. LXXX (1958), 458-538.

[11] A. BOREL AND J-P. SERRE, Sur certains sous-groupes des groupes de Lie compacts, Comm. Math. Helv. 27 (1953), 128-139.

[12] A. BOREL ET J. DE SIEBENTHAL, Les sous-groupes fermés connexes de rang maximum des groupes de Lie clos, ibid. 23 (1949-50), 200-221.

[13] R. BOTT, An application of the Morse theory to the topology of Lie groups, Bull. Soc. Math. France, 84 (1956), 251-282.

[14] R. BOTT AND H. SAMELSON, Application of the theory of Morse to symmetric spaces, Amer. J. Math. LXXX (1958), 964-1029.

[15] H. CARTAN, Seminaire E. N. S , Paris, 1954-55.

[16] C. CHEVALLEY, The algebraic theory of spinors, Columbia University Press, New York, 1954.

[17] J. P. Serre, Groupes d'homotopie et classes de groupes abéliens, Annals of Math. 58 (1953), 258-294.

[18] J. DE SIEBENTHAL, Sur les groupes de Lie compacts non connexes, Comm. Math. Helv. 31 (1956), 41-89.

[19] N. STEENROD, Topology of fibre bundles, Princeton, 1951.

[20] E. STIEFEL, Ueber eine Beziehung zwischen geschlossenen Lie'schen Gruppen und, ⋯ Comm. Math. Helv. 14 (1941-42), 350-380.

THE INSTITUTE FOR ADVANCED STUDY, PRINCETON, N. J., U. S. A.

54.

(with Harish-Chandra)

Arithmetic subgroups of algebraic groups

Bull. Amer. Math. Soc. **67** (1961) 579–583

A complex algebraic group G is in this note a subgroup of $GL(n, C)$, the elements of which are all invertible matrices whose coefficients annihilate some set of polynomials $\{P_\mu[X_{11}, \cdots, X_{nn}]\}$ in n^2 indeterminates. It is said to be defined over a field $K \subset C$ if the polynomials can be chosen so as to have coefficients in K. Given a subring B of C, we denote by G_B the subgroup of elements of G which have coefficients in B, and whose determinant is a unit of B. Assume in particular G to be defined over $\mathbf{Q}$. Then $G_{\mathbf{Z}}$ is an "arithmetically defined discrete subgroup" of $G_{\mathbf{R}}$, or, more briefly, an *arithmetic subgroup* of $G_{\mathbf{R}}$. A typical example is the group of units of a nondegenerate integral quadratic form, and as a matter of fact, the main results stated below generalize facts known in this case from reduction theory. The proofs will be published elsewhere.

1. **Reductive groups.** A complex algebraic group G is an *algebraic torus* (a torus in the terminology of [1]) if it is connected and can be diagonalized or, equivalently, if it is birationally isomorphic to a product of groups C^* [1, Chapter II]. The group G is *reductive* if its identity component G^0 may be written as $G^0 = T \cdot G'$, where T is a central algebraic torus, and G' is an invariant connected semi-simple group, or, equivalently, if all rational representations of G are fully reducible.

LEMMA 1. *Let $G_1 \supset \cdots \supset G_m$ be reductive algebraic subgroups of $GL(n, C)$, defined over $\mathbf{R}$. Then there exists $a \in SL(n, \mathbf{R})$ such that the groups $a \cdot G_{i\mathbf{R}} \cdot a^{-1}$ are stable under $x \to {}^t x$ $(i = 1, \cdots, m)$.*

This lemma, formulated in a somewhat different terminology, is due to G. D. Mostow [4]. Lemma 1, for $m = 1$, implies easily that the usual properties of maximal compact subgroups and of the Iwasawa decompositions (see [7] for instance) are valid for real algebraic reductive groups.

LEMMA 2. *Let G be a connected reductive complex algebraic group, H an algebraic subgroup. Then G/H is an affine variety if and only if H is reductive. If G and H are defined over $\mathbf{Q}$, and H is reductive, there exists a rational representation $\pi: G \to GL(m, C)$, defined over $\mathbf{Q}$, such*

164

that there is a point $v \in Z^m$ whose orbit under G is closed and whose isotropy group is H.

The fact that if G/H is an affine variety (or more generally a Stein manifold) then H is reductive, is due to Matsushima [3] (whose proof can be simplified using ordinary cohomology with complex coefficients). The converse is stated in [3] (and attributed to Iwahori and Sigiura); it can be proved by realizing G/H as a closed orbit in a suitable linear representation.

2. **Siegel domains.** Let $G \subset GL(n, C;)$ be an algebraic reductive group defined over R, $G_R = K \cdot A \cdot N$ an Iwasawa decomposition of G_R, and $\mathfrak{g}$, $\mathfrak{k}$, $\mathfrak{a}$, $\mathfrak{n}$ the Lie algebras of G_R, K, A, N respectively. K is a maximal compact subgroup, A is real diagonalizable, connected, N is unipotent, $\mathfrak{a}$ is orthogonal to $\mathfrak{k}$ with respect to the Killing form. Let further $\psi \subset \mathfrak{a}^*$ be the set of roots of $\mathfrak{g}$ with respect to $\mathfrak{a}$ (the restricted roots) and, for $\alpha \in \psi$, let $\mathfrak{g}_\alpha = \{x \in \mathfrak{g}, [h, x] = \alpha(h) \cdot x, h \in \mathfrak{a}\}$. Then, for a suitable ordering on $\mathfrak{a}^*$, we have $\mathfrak{n} = \sum_{\alpha > 0} \mathfrak{g}_\alpha$. Let $A_t = \{a \in A, \alpha(\log a) \leq t, \alpha \in \psi, \alpha > 0\}$, $(t > 0)$. A *Siegel domain* of G_R, with respect the given Iwasawa decomposition, is a subset $\mathfrak{S}_{t,\omega} = K \cdot A_t \cdot \omega$, where ω is a compact set in N. The $\mathfrak{S}_{t,\omega}$'s, ordered by inclusion, form a filtered set, and their union is G_R. One would obtain equivalent families by letting α run only through the *simple* restricted roots in the definition of A_t, or by replacing A_t by $a_0 \cdot A^-(a_0 \in A)$, A^- being the exponential of the negative Weyl chamber. It is easily seen that if G is semi-simple, a Siegel domain has finite Haar measure.

A *standard Siegel domain* $\mathfrak{S}$ in $GL(n, R)$ is a Siegel domain with respect to the usual Iwasawa decomposition (where $K = O(n)$, A is the group of diagonal matrices with strictly positive entries, N the group of upper triangular unipotent matrices) such that $GL(n, R) = \mathfrak{S} \cdot SL(n, Z)$. The existence of such domains is classical. By a well known theorem of Siegel [8], given $x \in GL(n, Q)$, the set of $y \in SL(n, Z)$ such that $\mathfrak{S} \cap \mathfrak{S} \cdot y \cdot x \neq \emptyset$ (resp. $\mathfrak{S} \cap \mathfrak{S} \cdot x \cdot y \neq \emptyset$) is finite. $\mathfrak{S} \cap SL(n, R)$ will be called a standard Siegel domain of $SL(n, R)$.

LEMMA 3. *Let $\pi: SL(n, C) \to GL(m, C)$ be a right rational representation, defined over Q, $v \in R^m$ be a point whose orbit under $SL(n, R)$ is closed and whose isotropy group in $SL(n, R)$ is stable under $x \to {}^t x$, and let $\mathfrak{S}$ be a standard Siegel domain of $SL(n, R)$. Then $v \cdot \pi(\mathfrak{S}) \cap Z^m$ is finite.*

This lemma can be proved more generally for a rational representation of a reductive group, a suitable Iwasawa decomposition and a Cartan involution compatible with it, but the above special case suffices for the applications. Applied to the natural representation of $SL(n, C)$ in the space of quadratic forms, it yields the finiteness of the number of reduced integral forms with a given nonzero determinant, stated first by Hermite [2].

3. Fundamental sets for arithmetic subgroups. As is well known, Hermite has used the result just mentioned to construct a fundamental set for the group of units of a nondegenerate quadratic form F in the space of majorizing forms of F [2]. The construction of the set U below is in a sense a generalization of his procedure.

THEOREM 1. *Let G be a connected complex algebraic group defined over Q. Then there exists an open set U in G_R with the following properties*: (i) $G_R = U \cdot G_Z$; (ii) $K \cdot U = U$ *for a suitable maximal compact subgroup of G_R*; (iii) *For any $x \in G_Q$, $U^{-1} \cdot U \cap (x \cdot G_Z \cup G_Z \cdot x)$ is finite*; (iv) *if G has no nontrivial rational character defined over Q, U has finite Haar measure.*

The group G is the semi-direct product of a reductive group and an invariant unipotent group N, both defined over Q. Since N_R/N_Z is compact, the proof of Theorem 1 is easily reduced to the case where G is reductive. Assuming moreover, as we may, $G \subset SL(n, C)$, we take a right rational representation $\pi: SL(n, C) \to GL(m, C)$ such that $\pi(SL(n, Z)) \subset SL(m, Z)$, and that there exists $v \in Z^m$ whose orbit is closed and whose isotropy group is G. The existence of π follows mainly from Lemma 2. Let $a \in SL(n, R)$ be such that $a \cdot G_R \cdot a^{-1}$ is stable under $x \to {}^t x$ (Lemma 1) and $\mathfrak{S}$ be an open neighborhood of a standard Siegel domain of $SL(n, R)$ contained in a standard Siegel domain. By Lemma 3, there exists a finite number of elements $b_1, \cdots, b_m \in SL(n, Z)$ such that

$$(1) \qquad v \cdot \pi(SL(n, Z)) \cap v \cdot \pi(a^{-1}\mathfrak{S}) = \{v \cdot \pi(b_1^{-1}), \cdots, v \cdot \pi(b_m^{-1})\}.$$

We have $G_R = a^{-1} \cdot H$, where H is the set of elements in $SL(n, R)$ which map $v \cdot \pi(a^{-1})$ onto v. From this, (1) and the equality $SL(n, R) = \mathfrak{S} \cdot SL(n, Z)$ it follows easily that $U = \bigcup_i (G_R \cap a^{-1} \cdot \mathfrak{S} \cdot b_i)$ satisfies (i), (ii). Property (iii) is then a consequence of the theorem of Siegel recalled in §2. When G is semi-simple, property (iv) follows from the following lemma:

LEMMA 4. *Let $G_1 \subset G$ be algebraic semi-simple groups. Assume that there are Iwasawa decompositions $G_{1R} = K_1 \cdot A_1 \cdot N_1$, $G_R = K \cdot A \cdot N$ of G_{1R} and G_R such that $K_1 \subset K$, $A_1 \subset A$, $N_1 \subset N$, and that a positive root of the Lie algebra of G_R with respect to the Lie algebra of A, for the ordering defined by N, restricts to a positive root of the Lie algebra of G_{1R}, for* *the ordering defined by N_1, Let $\mathfrak{S}$ be a Siegel domain of G_R with respect to the Iwasawa decomposition $K \cdot A \cdot N$, and $x \in G_R$. Then $\mathfrak{S} \cdot x \cap G_{1R}$ is contained in a finite number of translates of a Siegel domain of G_{1R}.*

By an elementary argument, property (iii) implies the

COROLLARY. *Let G be a complex algebraic group defined over $\mathbf{Q}$. Then G_Z is finitely generated.*

The result on reduced forms stated at the end of §2 implies the finiteness of the number of classes of integral forms with a given non-zero determinant. The latter has the following generalization.

THEOREM 2. *Let G be a connected reductive algebraic group defined over $\mathbf{Q}$, and $\pi: G \rightarrow \mathbf{GL}(m, \mathbf{C})$ a rational representation defined over $\mathbf{Q}$, and $H = G_Z \cap \pi^{-1}(\mathbf{GL}(m, \mathbf{Z}))$. Then for any closed orbit X of G in $\mathbf{C}^m$, the integral points of X form a finite number of orbits of H.*

COROLLARY. *Let G, G' be connected algebraic groups, defined over $\mathbf{Q}$, and $\mu: G \rightarrow G'$ a rational surjective homomorphism with finite kernel, defined over $\mathbf{Q}$ (an isogeny). Then $\mu(G_Z)$ and G'_Z are commensurable.*

4. **Arithmetic subgroups with compact fundamental sets.** The quotient G_R/G_Z is compact for instance when G is the orthogonal group of a form which does not represent zero. This fact has the following generalization, which had been conjectured by R. Godement:

THEOREM 3. *Let G be a complex algebraic group defined over Q. Then the following conditions are equivalent: (i) G_R/G_Z is compact: (ii) the identity component of G has no nontrivial rational character defined over $\mathbf{Q}$, and every unipotent element of G_Q (or, equivalently, of G_Z) belongs to the radical of G_Q.*[1]

REMARKS. (1) Theorem 1, its corollary and Theorem 3 were known essentially for the classical groups (see [6; 8; 9]). Theorem 3 for algebraic tori is proved in [5]. As is known, it is easy to derive from them similar results on groups of matrices with coefficients in the ring of integers of a number field K, which belong to an algebraic group defined over K.

(2) Theorem 3 and known properties of semi-simple Lie algebras imply easily that a connected semi-simple Lie group G always has discrete subgroups H such that G/H is compact.

---

[1] We understand that another proof of Theorem 3 has since been given by G. D. Mostow and T. Tamagawa.

Bibliography

1. A. Borel, *Groupes linéaires algébriques*, Ann. of Math. vol. 64(1956) pp. 20–80.

2. C. Hermite, *Oeuvres complètes*, Vol. 1, Paris, Gauthier-Villars, 1905.

3. Y. Matsushima, *Espaces homogènes de Stein des groupes de Lie complexes*, Nagoya Math. J. vol. 16 (1960) pp. 205–218.

4. G. D. Mostow, *Self-adjoint group*, Ann. of Math. vol. 62 (1955), pp. 44–55.

5. T. Ono, *Sur une propriété arithmétique des groupes algébriques commutatifs*, Bull. Soc. Math. France vol. 85 (1957) pp. 307–323.

6. K. G. Ramanathan, *Unit of fixed points in involutorial algebras*, Proceedings of the International Symposium on Algebraic Number Theory, Tokyo, 1955.

7. Séminaire S. Lie, *Théorie des algèbres de Lie, Topologie des groupes de Lie*, Paris, 1954–1955.

8. C. L. Siegel, *Einheiten quadratischer Formen*, Abh. Math. Sem. Univ. Hamburg vol. 13 (1939) pp. 209–239.

9. A. Weil, *Discontinuous subgroups of classical groups*, Notes, University of Chicago, 1958.

The Institute for Advanced Study and
Columbia University

55.

Some properties of adele groups attached to algebraic groups

Bull. Amer. Math. Soc. **67** (1961) 583–585

Communicated by Deane Montgomery, July 22, 1961

This note is a sequel to the previous one [1], and is devoted to some applications of the results of the latter to adele groups. The results are valid for linear algebraic groups defined over number fields, but this case is easily reduced to that of groups defined over Q [3, Chapter I], to which we shall limit ourselves for simplicity.

The notation of [1] is freely used. For the unexplained notions concerning adeles, see [2; 3].

1. **Adeles.** Let G be a connected algebraic linear group defined over Q. The adele group attached to G is denoted by G_A. The group G_Q is identified with the subgroup of principal adeles of G_A; it is discrete. We put

$$G_A^0 = G_R \times \prod_{p \text{ prime}} G_{Z_p} \qquad (Z_p: \text{ring of } p\text{-adic integers}).$$

By definition, G_A^0, endowed with the product topology, is an open subgroup of G_A. The group G is said to be of type (F) if G_A is the union of a *finite* number of double cosets $G_A^0 \cdot x \cdot G_Q (x \in G_A)$ [2].

169

Let $X_Q(G)$ be the group of rational characters of G, defined over Q. Each $\chi \in X_Q(G)$ induces a continuous homomorphism $\chi_A: G_A \to I(Q)$, where $I(Q)$ is the idele group of Q. Composed with the norm mapping $\eta: I(Q) \to R^+$, it yields a continuous homomorphism $\eta \circ \chi_A: G_A \to R^+$. Let $_N G_A$ be the intersection of the kernels of the homomorphisms $\eta \circ \chi_A (\chi \in X_Q(G))$. It is unimodular [2, Proposition 11] and contains G_Q. The group G is said to be of type (M) (resp. (C)), if $_N G_A/G_Q$ has finite invariant measure (resp. is compact), [2].

THEOREM 1. *Let G be a connected algebraic group, defined over Q. Then G is of type (F) and of type (M). It is of type (C) if and only if every unipotent element of G_Q belongs to the radical of G_Q. The quotient G_A/G_Q is compact if and only if G is of type (C) and $X_Q(G) = 1$.*

In view of the results of [2], it is enough to prove this for reductive groups. To this end, a fundamental set for G_Q in G_A is constructed by a method which is quite analogous to the one outlined in [1, §3]. This yields condition (F), and reduces the other assertions to Theorems 1, 3 of [1].

2. Let $GL(m, A)$ and $SL(m, A)$ be the adele-groups attached to $GL(m, Q)$ and $SL(m, Q)$ respectively. These are groups of automorphisms of A^m, the free module of rank m over the adele ring A of Q. A rational homomorphism $\pi: G \to GL(m, C)$, defined over Q, induces a continuous homomorphism $\pi_A: G_A \to GL(m, A)$, for which the following analogue to Theorem 2 in [1] holds:

THEOREM 2. *Let G be a connected reductive algebraic group, defined over Q, $\pi: G \to GL(m, C)$ a rational representation, defined over Q, and $v \in Q^m$ a point whose orbit is closed. Then $v \cdot \pi_A(G_A) \cap Q^m$ is the union of a finite number of orbits of G_Q.*

(In this statement Q^m is identified with the principal adeles of A^m.)

3. **Principal homogeneous spaces.** Let G be a connected algebraic group defined over Q. A principal homogeneous space V of G, over Q, is a variety defined over Q, on which G acts as an algebraic transformation group by means of a morphism $G \times V \to V$ defined over Q, and such that for any $v \in V$, the map $g \to g \cdot v$ is a birational biregular map of G onto V. (Since we are in characteristic zero, this last condition is equivalent to: G is simply transitive on V.) Two such spaces V, V' are isomorphic over Q if there exists a G-equivariant birational biregular map of V onto V', defined over Q.

THEOREM 3. *Let G be a connected reductive group defined over Q. Then the principal homogeneous spaces over Q which have rational points in all completions of Q form a finite number of isomorphism classes.*

Assume $G \subset SL(n, C)$. Let $\pi: SL(n, C) \to GL(m, C)$ be a rational representation defined over Q, for which there exists a point $v \in Q^m$ whose orbit V is closed and whose isotropy group is G [1, Lemma 2]. It is not difficult to show that the orbits of $SL(n, Q)$ in $v \cdot \pi_A(SL(n, A))$ are in 1-1 correspondence with the isomorphism classes of principal homogeneous spaces over Q which have rational points in all completions of Q, hence Theorem 3 follows from Theorem 2.

REMARKS. (1) Theorem 1 was known for solvable groups [2] and for many classical groups [3].

(2) Theorem 3 is well known for algebraic tori. In this case, the principal homogeneous spaces in question may form several classes. It is not known to the author whether there may be more than one class when G is semi-simple (and connected). For further remarks on this question, see S. Lang, Bull. Amer. Math. Soc. vol. 66 (1960) pp. 240–249.

BIBLIOGRAPHY

1. A. Borel and Harish-Chandra, *Arithmetic subgroups of algebraic groups*, Bull. Amer. Math. Soc. vol. 67 (1961) pp. 579–583.

2. T. Ono, *On some arithmetic properties of linear algebraic groups*, Ann. of Math. vol. 70 (1959) pp. 266–290.

3. A. Weil, *Adeles and algebraic groups*, Notes, The Institute for Advanced Study, Princeton, 1961.

THE INSTITUTE FOR ADVANCED STUDY

56.

(avec A. Haefliger)

La classe d'homologie fondamentale d'un espace analytique

Bull. Soc. Math. France **89** (1961) 461–513

On sait qu'une variété analytique complexe de dimension complexe n possède une classe d'homologie fondamentale de dimension $2n$, liée à son orientation naturelle, et que toute variété analytique réelle (ou même topologique) a une classe fondamentale mod 2. Notre but ici est de démontrer qu'il en est de même pour un espace analytique quelconque, complexe ou réel, de mettre en relations l'intersection de cycles analytiques d'une variété analytique avec l'intersection de leurs classes d'homologie ([1]), et d'établir quelques liens entre le réel et le complexe.

Pour éviter le recours à la triangulation ou à des propriétés de rétraction locale, nous utiliserons la théorie de l'homologie des espaces localement compacts de [2]. Toutes les propriétés des groupes d'homologie au sens de [2] dont nous avons besoin sont rappelées au paragraphe 1. Les plus utiles pour notre propos sont la nullité de ces groupes en dessus de la dimension topologique et l'existence d'une suite exacte liant l'homologie d'un espace, d'un sous-espace fermé et de l'ouvert complémentaire. Ces

---

([1]) Dans le cas analytique complexe, il s'agit là de résultats « bien connus », qui ont été souvent utilisés, et qu'on déduit classiquement du théorème de triangulation des espaces analytiques. L'existence d'une classe fondamentale mod 2 pour un ensemble algébrique réel a été conjecturée par Thom, qui en a donné aussi une esquisse de démonstration dans [22]. C'est du reste cette question qui est à l'origine du présent article.

172

groupes vérifient les axiomes d'Eilenberg-Steenrod et coïncident sur les variétés avec les groupes d'homologie singulière.

Bien que nous n'ayons à considérer que des espaces localement analytiques, il est commode d'introduire une notion générale de variété à singularités et de classe d'homologie fondamentale d'un tel espace; un espace X sera dit être de type VS_n s'il est de dimension n et s'il contient un fermé F de dimension $\leq n-1$ dont le complémentaire est une variété de dimension (pure) n. Une classe d'homologie fondamentale de X sur l'anneau de coefficients K, est alors un élément de $H_n(X; K)$ qui induit en chaque point de X-F un générateur du n-ième groupe d'homologie locale en ce point. Le paragraphe 2 est consacré à quelques remarques générales sur ces notions, qui montrent en particulier que le problème de l'existence d'une classe fondamentale est essentiellement de nature locale (pour autant que X-F soit orientable).

Le paragraphe 3 démontre l'existence d'une classe fondamentale entière (resp. mod 2) pour un espace analytique complexe (resp. localement analytique réel). Elle est immédiate si l'ensemble des points singuliers est de codimension ≥ 2, ce qui a toujours lieu dans le cas complexe. Lorsque X est un germe d'ensemble analytique réel irréductible, on se ramène à ce cas en appliquant au complexifié de X le théorème de normalisation d'Oka (ou simplement la normalisation affine si X est algébrique).

Dans le paragraphe 4, on attache à chaque composante propre de l'intersection de deux cycles analytiques X, Y d'une variété analytique complexe V un entier $i(X.Y, C)$, par considération de la classe d'homologie de $X \cap Y$, et l'on montre que $i(X.Y, C)$ vérifie les propriétés caractéristiques du symbole d'intersection en géométrie algébrique ([23], app. III). Il s'ensuit en particulier que si V est projective, l'application qui associe à chaque cycle X sa classe d'homologie $h(X)$ est un homomorphisme de l'anneau de Chow de $A(V)$ de V dans $H_*(V; Z)$.

Le paragraphe 5 considère une variété projective non singulière V définie sur les réels et l'application qui associe à un cycle X de dimension complexe s la classe d'homologie $\rho(X) \in H_s(V^0; Z_2)$ de la partie réelle de X, où V^0 est la partie réelle de V, et donne notamment une condition suffisante pour que ρ induise un isomorphisme de $H_*(V; Z_2)$ sur $H_*(V^0; Z_2)$ diminuant le degré de moitié. Le point essentiel est le fait que si X et Y sont des cycles de V se coupant proprement, alors $\rho(X.Y) = \rho(X).\rho(Y)$, ce qui est démontré plus généralement lorsque V est une variété analytique complexe dont V^0 est la partie réelle en un sens convenable (5.1-5.8).

Le paragraphe 6 est consacré à deux applications. L'une établit une assertion de Thom concernant le polynôme en classes de Stiefel-Whitney attaché à un type de singularité S et le polynôme en classes de Chern attaché au type de singularité complexifié de S (6.2). Dans 6.3 et 6.4, on montre que si X admet une décomposition cellulaire dont les cellules ont une classe fondamentale, alors ces classes fondamentales, forment une base de l'homologie de X, et dans 6.5 on décrit une classe d'espaces homogènes algébriques possédant des décompositions cellulaires et auxquels on peut appliquer les résultats du paragraphe 5.

Enfin, le paragraphe 7 apporte quelques compléments à [2], qui sont utilisés dans ce travail, et concernent surtout le cap-produit.

1. Homologie des espaces localement compacts.

Ce paragraphe énumère les propriétés des groupes d'homologie dont nous aurons besoin, et fixe les notations. Son contenu sera souvent utilisé sans référence.

1.1. — K désigne toujours un anneau principal (qui pourrait en fait être de Dedekind, mais seuls les cas où K est l'anneau $\mathbf{Z}$ des entiers ou un corps nous intéresseront); $\mathbf{Z}_p$ (p premier), $\mathbf{Q}$, $\mathbf{R}$, $\mathbf{C}$ sont comme d'habitude le corps des entiers mod p, des nombres rationnels, réels et complexes respectivement.

On rappelle qu'une famille de supports Φ sur un espace topologique X est un ensemble de fermés de X ayant les deux propriétés suivantes : si A, $B \in \Phi$, alors $A \cup B \in \Phi$; si $A \in \Phi$ et si B est fermé dans A, alors $B \in \Phi$. Elle est dite paracompactifiante si de plus tout élément de Φ est paracompact et possède un voisinage appartenant à Φ ([10], chap. I, § 2.5 et 3.2).

Dans la suite, les espaces topologiques sont toujours supposés localemen compacts.

1.2. — Soient X un espace, $\mathcal{S}$ un faisceau de K-modules sur X et Φ une famille de supports sur X. On note $H_i^\Phi(X; \mathcal{S})$ [resp. $H_\Phi^i(X; \mathcal{S})$] le i-ième groupe d'homologie (resp. de cohomologie) de X, à coefficients dans $\mathcal{S}$ et à supports dans Φ, au sens de [2] (resp. [10]), et $H_*^\Phi(X; \mathcal{S})$ [resp. $H_\Phi^*(X; \mathcal{S})$] est la somme directe des $H_i^\Phi(X; \mathcal{S})$ [resp. $H_\Phi^i(X; \mathcal{S})$]. Ce sont des K-modules. Comme d'habitude on écrit aussi K pour le faisceau constant $X \times K$. Dans ce travail, nous n'aurons essentiellement à considérer que l'homologie et la cohomologie à coefficients dans K. Dans la notation précédente on supprime Φ si Φ est l'ensemble des fermés de X, on remplace Φ par c ou $c(X)$ si Φ est la famille des compacts de X, et Φ par F si Φ est l'ensemble des fermés d'un sous-espace fermé F de X. En particulier, si F est réduit à un point $x \in X$, $H_i^x(X; K)$ désignera ici le i-ième groupe d'homologie de X à supports dans x. Ce n'est pas en général le i-ième groupe d'homologie locale en x (*voir* plus bas) qui sera noté $\mathcal{H}_i(X; K)_x$; sur ce point nous nous écartons donc de la notation de [1] et [2].

On a $H^i(X; \mathcal{S}) = 0$ si $i \leq -1$ et $H_i(X; \mathcal{S}) = 0$ si $i \leq -2$. Si X est un polyèdre fini, $H_i(X; K)$ est le i-ième groupe d'homologie simpliciale de X et si Y est un sous-complexe de X, $H_i(X - Y; K)$ s'identifie au i-ième groupe d'homologie simpliciale relative de X mod Y ([2], § 5).

1.3. — Pour tout entier i, on a la suite exacte

$$0 \to \mathrm{Ext}(H_c^{i+1}(X\,;\,K),\,K) \to H_i(X\,;\,K) \to \mathrm{Hom}(H_c^i(X\,;\,K),\,K) \to 0.$$

Elle montre en particulier que $H_{-1}(X\,;\,K)$ est nul si $H_c^0(X\,;\,K)$ est libre, ce qui a lieu notamment lorsque X est localement connexe (et aussi lorsque X possède une base dénombrable des ouverts, d'après un résultat non publié de F. RAYMOND). La forme bilinéaire sur $H_i(X\,;\,K) \times H_c^i(X\,;\,K)$ définie par l'homomorphisme de $H_i(X\,;\,K)$ dans $\mathrm{Hom}(H_c^i(X\,;\,K),\,K)$ sera notée $\langle a,\,b \rangle \,[a \in H_i(X\,;\,K),\ b \in H_c^i(X\,;\,K)]$. Si K est un corps, elle est non dégénérée.

Si X est somme de sous-espaces disjoints $X_\alpha\,(\alpha \in I)$, $H_c^*(X\,;\,K)$ est la somme directe des modules $H_c^*(X_\alpha\,;\,K)$, et la suite exacte ci-dessus montre donc que $H_*(X\,;\,K)$ s'identifie au *produit direct* des modules $H_*(X_\alpha\,;\,K)$.

1.4. — $H_i(X\,;\,\mathcal{S})$ est défini comme le groupe dérivé du module des sections d'un certain faisceau différential gradué $\mathcal{C}_H(X\,;\,\mathcal{S})$ ([2], § 3); si U est ouvert dans X, la restriction des sections définit un homomorphisme naturel

$$j_*^{XU}: \quad H_*^\Phi(X\,;\,\mathcal{S}) \to H_*^{\Phi \cap U}(U\,;\,\mathcal{S}).$$

Si Ψ est une famille de supports contenant Φ, l'inclusion de $\Gamma_\Phi(C_H(X\,;\,\mathcal{S}))$ dans $\Gamma_\Psi(\mathcal{C}_H(X\,;\,\mathcal{S}))$ définit un homomorphisme de $H_*^\Phi(X\,;\,\mathcal{S})$ dans $H_*^\Psi(X\,;\,\mathcal{S})$ qu'on appellera un homomorphisme d'agrandissement de la famille des supports.

Le faisceau dérivé de $\mathcal{C}_H(X\,;\,K)$ est le *faisceau d'homologie locale* $\mathcal{H}(X\,;\,K)$ *de* X et sa fibre $\mathcal{H}_*(X\,;\,K)_x$ en un point x est le *groupe d'homologie locale de* X *en* x. On a

$$\mathcal{H}_i(X\,;\,K)_x = \lim \mathrm{dir}\,(H_*(U\,;\,K),\,j_*^{UV}),$$

où U parcourt les voisinages ouverts de x.

L'homomorphisme j_*^{UX} est transposé de l'homomorphisme

$$j_{XU}^*: \quad H_c^*(U\,;\,K) \to H_c^*(X\,;\,K)$$

et **1.3** est compatible avec j_{XU}^*, j_*^{XU} en un sens évident. Comme les suites exactes commutent aux limites inductives, on en déduit en particulier qu'il y a une suite exacte

$$0 \to \lim \mathrm{dir}\,\mathrm{Ext}(H_c^{i+1}(U\,;\,K),\,K) \to \mathcal{H}_i(X\,;\,K)_x \to \lim \mathrm{dir}\,\mathrm{Hom}(H_c^i(U\,;\,K),\,K) \to 0.$$

Par définition, $H_*^\Phi(X\,;\,K) = H_*(\Gamma_\Phi(\mathcal{C}_H(X\,;\,K)))$; il existe donc un homomorphisme naturel

$$(\mathrm{I}) \qquad H_*^\Phi(X\,;\,K) \to H_\Phi^0(X\,;\,\mathcal{H}_*(X\,;\,K)) = \Gamma_\Phi(\mathcal{H}_*(X\,;\,K)).$$

En le composant avec l'homomorphisme $\Gamma_\Phi(H_*(X; K)) \to H_*(X; K)_x$ qui associe à une section sa valeur en $x \in X$, on obtient donc un homomorphisme

$$(2) \qquad \mu_x : H_*^\Phi(X; K) \to H_*(X; K)_x \qquad (x \in X).$$

1.5. — Soient X, Y des espaces, $f : X \to Y$ une application continue Φ, Ψ des familles de supports sur X et Y respectivement. Si $f(\Phi) \subset \Psi$ et si la restriction de f à tout $F \in \Phi$ est une application propre, alors f induit un homomorphisme $f_{*\Phi, \Psi}$ ou f_* de $H_*^\Phi(X; K)$ dans $H_*^\Psi(Y; K)$, préservant les degrés. Si $g : Y \to Z$ est une application continue de Y dans un espace Z, et si $g_{*\Psi, \Theta}$ est définie, alors $(g \circ f)_{*\Phi, \Theta}$ l'est, et est égale à $g_{*\Psi, \Theta} \circ f_{*\Phi, \Psi}$. Soient V un ouvert de Y et $U = f^{-1}(V)$. Si $f_{*\Phi, \Psi}$ est définie alors $f_{*\Phi \cap U, \Psi \cap V} : H_*^{\Phi \cap U}(U; K) \to H_*^{\Psi \cap V}(Y; K)$ l'est aussi, et l'on a

$$f_{*\Phi \cap U, \Psi \cap V} \circ j_*^{XU} = j_*^{YV} \circ f_{*\Phi, \Psi}.$$

Toute application continue $f : X \to Y$ induit un homomorphisme $f_* : H_*^c(X; K) \to H_*^c(Y; K)$. Si f est propre, elle définit un homomorphisme $f_* : H_*(X; K) \to H_*(Y; K)$. L'inclusion d'un sous-espace fermé F de X dans X induit un homomorphisme $i_{*FX} : H_*^{\Phi \cap F}(F; K) \to H_*^\Phi(X; K)$ pour toute famille de supports Φ sur X. Si Φ est l'ensemble des fermés de F, cet homomorphisme est un isomorphisme au moyen duquel on identifiera souvent $H_*(F; K)$ avec $H_*^F(X; K)$. La famille Φ étant de nouveau quelconque, le module $H_*^\Phi(X; K)$ s'identifie à la limite directe des modules $H_*(F; K)$, par rapport aux homomorphismes d'inclusion, F parcourant Φ ([2]; 3.3). On a l'axiome d'homotopie sous la forme suivante ([2], 4.3 et 4.4). Si $f, g : X \to Y$ sont homotopes, alors

$$f_* = g_* : \quad H_*^c(X; K) \to H_*^c(Y; K).$$

Soient I l'intervalle unité, $F : X \times I \to Y$ une application propre et f, g les restriction de F à $X \times (o)$ et $X \times (\mathrm{I})$ respectivement. Alors

$$f_* = g_* : \quad H_*(X; K) \to H_*(Y; K).$$

1.6. — Soit F un sous-espace fermé de X et soit $U = X - F$. Alors on a une suite exacte ([2], § 3.8)

$$\ldots \to H_i(F; K) \xrightarrow{i_{*FX}} H_i(X; K) \xrightarrow{j_*^{XU}} H_i(U; K) \to H_{i-1}(F; K) \dashrightarrow \ldots$$

1.7. — Supposons que X soit réunion d'une famille localement finie de sous-espaces fermés $X_\alpha (\alpha \in I)$, et soit Y l'espace somme des X_α. Alors les injections $i_\alpha : X_\alpha \to X$ définissent une application $\mu : Y \to X$ qui est évidemment continue et propre. Comme $H_*(Y; K)$ est le produit direct des modules $H_*(X_\alpha; K)$ (cf. 1.3) on en déduit un homomorphisme

$$\mu_* : \prod_{\alpha \in I} H_*(X_\alpha; K) \to H_*(X; K),$$

dont la restriction à un nombre fini de facteurs $X_\alpha (\alpha \in J)$ est la somme des applications $i_{\alpha\star}(\alpha \in J)$. L'image a d'un élément $(a_\alpha)_{\alpha \in I}$ s'appellera la somme (infinie, localement finie) des éléments a_α. Si U est un ouvert de X ne rencontrant qu'un nombre fini de X_α, on a

$$j_\star^{XU}(a) = \sum_{\alpha \in I} j_\star^{XU} \circ i_{\star X_\alpha, X}(a_\alpha),$$

somme qui a un sens puisqu'elle ne comprend qu'un nombre fini de termes non nuls, vu l'hypothèse faite sur U. Cette opération de somme infinie, localement finie, est visiblement compatible avec la restriction à un ouvert et avec les applications continues.

1.8. — Supposons que X soit réunion de deux sous-espaces fermés A, B et soit $C = A \cap B$. Alors on a une suite exacte (la suite de Mayer-Vietoris) :

$$\ldots \xrightarrow{\partial} H_n(C; K) \xrightarrow{\alpha} H_n(A; K) + H_n(B; K) \xrightarrow{\beta} H_n(X; K) \xrightarrow{\partial} H_{n-1}(C; K) \to \ldots,$$

où $\alpha = i_{\star C, A} - i_{\star C, B}$, $\beta = i_{\star A, X} + i_{\star B, X}$ ([**2**], **3.10**).

1.9. — Il existe un accouplement, le cap-produit

$$\cap : \quad H_s^\Phi(X; K) \times H_\Psi^t(X; K) \to H_{s-t}^{\Phi \cap \Psi}(X; K),$$

compatible avec la restriction à un ouvert et l'agrandissement des familles de supports, vérifiant

$$(a \cap b) \cap c = a \cap (b \cup c) \qquad [a \in H_s^\Phi(X; K), b \in H_\Psi^t(X; K), c \in H_\Theta^u(X; K)],$$

les deux membres de cette égalité étant envisagés comme des éléments de $H_{s-t-u}^{\Phi \cap \Psi \cap \Theta}(X; K)$ (*voir* appendice). En particulier, $H_\star^\Phi(X; K)$ est un module à droite sur l'anneau $H_\Phi^\star(X; K)$.

Soit $f : X \to Y$ une application continue. Soient Φ, Φ' (resp. Ψ, Ψ') des familles de supports sur X (resp. Y). Supposons que $f(\Phi) \subset \Psi$, $f(\Phi') \subset \Psi'$, $f^{-1}(\Psi') \subset \Phi'$ et que la restriction de f à tout $F \in \Phi$ soit une application propre. Alors

$$f_\star : \quad H_\star^\Phi(X; K) \to H_\star^\Psi(Y; K), \qquad H_\star^{\Phi \cap \Phi'}(X; K) \to H_\star^{\Psi \cap \Psi'}(Y; K),$$

$$f^\star : \quad H_{\Psi'}^\star(Y; K) \to H_{\Phi'}^\star(X; K)$$

sont définies (pour $f^\star$, *voir* [**10**], II, **4.16**), et l'on a

$$(\text{1}) \qquad f_\star(a \cap f^\star b) = f_\star a \cap b \qquad [a \in H_s^\Phi(X; K), b \in H_{\Psi'}^t(X; K)],$$

les deux membres de cette égalité faisant partie de $H_{s-t}^{\Psi \cap \Psi'}(Y; K)$, (*voir* § 7).

1.10. — On note $\dim_K X$ la dimension cohomologique de X, relativement à K. Les conditions suivantes sont équivalentes :

(i) $\dim_K X \leqq n$;

(ii) pour toute famille paracompactifiante Φ et tout faisceau $\mathcal{S}$ de K-modules, on a $H_\Phi^{n+1}(X; \mathcal{S}) = 0$;

(iii) pour tout ouvert $U \subset X$, on a $H_c^{n+1}(U; K) = 0$.

Elles entraînent la nullité de $H^i{}_\Phi(X; \mathcal{S})$ et $H_c^i(U; K)$ pour tout $i \geqq n + 1$. [$Voir$ [1], exp. I, ou Grothendieck, $Tohoku$ $math.$ $J.$, t. 9, 1957; dans [1] on ne considère que les familles contenant tous les points de X, mais l'extension mentionnée ici est immédiate, puisque $H^i{}_\Phi(X; \mathcal{S}) = H_\Phi^i(U; \mathcal{S})$, U étant l'ensemble des points contenus dans Φ.] Si X est séparable métrique, de dimension classique n, alors $\dim_K X \leqq n$ pour tout K. On a $\dim_K Y \leqq \dim_K X$ si Y est un sous-espace localement compact de X. $\mathrm{Dim}_K X$ est la borne supérieure des dimensions des sous-espaces compacts de X et a un caractère local : $\dim_K X \leqq n$ si et seulement si tout point est contenu dans un ouvert de dimension $\leqq n$.

Supposons $\dim_K X \leqq n$. Alors, vu 1.3 et 1.4, on a

$$(1) \qquad H_i(X; K) = \mathcal{H}_i(X; K)_x = 0 \qquad (i \geqq n + 1, \ x \in X).$$

De plus l'homomorphisme naturel

$$(2) \qquad \Delta : \ H_n(X; K) \to H^0(X; \mathcal{H}_n(X; K))$$

($cf.$ 1.4) est un isomorphisme, quelle que soit la famille de supports Φ ([2], 7.3).

1.11. — Supposons que X soit une variété de dimension n. (En fait, ce qui suit vaut lorsque X est une variété cohomologique ou homologique relativement à K, [2], §7). Alors $\dim_K X = n$, $\mathcal{H}_i(X; K) = 0$ pour $i \neq n$, et $\mathcal{H}_n(X; K)$ est localement isomorphe au faisceau constant $X \times K$; le faisceau $\mathcal{H}_n(X; K)$ est le $faisceau$ $d'orientation$ de X et sera noté $\mathcal{C}$. Il est constant si et seulement si X est orientable. Une $orientation$ de X est le choix d'un isomorphisme de $\mathcal{C}$ sur $X \times K$. On a pour tout entier i et toute famille Φ de supports un isomorphisme canonique (la dualité de Poincaré), compatible avec la restriction à un ouvert

$$\Delta : \ H_i{}^\Phi(X; K) \to H_\Phi^{n-i}(X; K).$$

Il existe un et, à un isomorphisme unique près, un seul faisceau $\mathcal{C}'$ tel que $\mathcal{C} \otimes \mathcal{C}'$ soit le faisceau constant $X \times K$ ([2], 7.7). Si $K = Z$, Z_2 ou si X est orientable, les seuls cas intéressants pour la suite, alors $\mathcal{C} = \mathcal{C}'$.

X est localement connexe, donc $H^0(X; K)$ s'identifie canoniquement à la somme directe des anneaux $H^0(X_\alpha; K)$, où X_α parcourt les composantes connexes de X, anneaux qui sont eux-mêmes canoniquement isomorphes à K. Soient ι_α l'élément neutre de $H^0(X_\alpha; K)$ et $\iota = \sum_\alpha \iota_\alpha$. Alors $\Delta^{-1}(\iota)$ est un élément bien déterminé de $H_n(X; \mathcal{C}')$ la *classe fondamentale de X*. Si X est orientable et orientée, alors $\Delta^{-1}(\iota)$ s'identifie à un élément de $H_n(X; K)$, la *classe fondamentale de la variété orientée X*, qui sera souvent noté $[X]$. Par l'isomorphisme Δ, la classe $[X]$ peut s'envisager comme une section de $\mathcal{C}$ qui induit en chaque point $x \in X$ un générateur (de K-module) de la fibre en x. Soit $x_\alpha \in X_\alpha$. Alors $[X]$ est déterminée par ses valeurs aux points x_α; réciproquement, si X est orientable, tout choix de générateurs dans les fibres $\mathcal{C}_{x_\alpha}$ correspond à une orientation. Lorsque $K = Z_2$, X est toujours orientable sur K et admet exactement une orientation.

Si X est une variété et Φ est paracompactifiante, $H_i^\Phi(X; K)$ s'identifie au $i^{\text{ème}}$ groupe d'homologie singulière, à supports dans Φ. Cela résulte de la dualité de Poincaré en homologie singulière ([4], exp. XX, §3), du théorème de dualité rappelé ci-dessus, et du fait que si X est une variété, ou plus généralement un espace HLC, $H_\Phi^i(X; K)$ est canoniquement isomorphe au $i^{\text{ème}}$ groupe de cohomologie singulière, à supports dans Φ([4], exp. XX, §1).

1.12. — Supposons X orientée. Alors l'isomorphisme Δ^{-1} est le cap-produit avec $[X]$. Soient $a \in H_i^\Phi(X; K)$, $b \in H_j^\Psi(X; K)$. Leur produit d'intersection sera noté $a.b$. Par définition

$$a.b = \Delta^{-1}(\Delta a \cup \Delta b);$$

c'est un élément de $H_{i+j-n}^{\Phi \cap \Psi}(X; K)$ $(n = \dim_K X)$. On a aussi $a.b = a \cap \Delta b$. Ce produit est associatif et anticommutatif (*cf.* §7), et compatible avec l'agrandissement des familles de supports.

Supposons que a et b contiennent des cycles a', b' dont les supports A et B ont une intersection de dimension sur K strictement plus petite que $i + j - n$. Alors $a.b = o$. En effet, on peut envisager a et b comme des éléments de $H_i^A(X; K)$ et $H_i^B(X; K)$, donc $a.b$ comme un élément de $H_{i+j-n}^{A \cap B}(X; K)$. Mais (*cf.* 1.5) ce dernier groupe s'identifie à $H_{i+j-n}(A \cap B; K)$, qui est nul si $\dim_K A \cap B < i + j - n$ (**1.10**).

2. Définition et propriétés générales de la classe fondamentale.

2.1. Définition d'un espace de type VS_n. — Un ouvert d'un espace topologique de dimension n sera dit *épais* si son complémentaire est de dimension $\leq n - 1$.

Un espace de type VS_n (variété de dimension n avec singularités) est un espace localement compact de dimension n possédant un ouvert épais homéomorphe à une variété de dimension n (c'est-à-dire dont toutes les composantes connexes sont de dimension n).

Soit X de type VS_n. Les points au voisinage desquels X est homéomorphe à $\mathbf{R}^n$ forment un ouvert épais X_R dont le complémentaire sera noté X_S. Les points de X_R (resp. X_S) sont les points réguliers (resp. singuliers) de X.

Cette notion est de caractère local. Tout ouvert de dimension n d'un espace de type VS_n est de type VS_n. Si un espace localement compact X possède un recouvrement par des ouverts U_i $(i \in I)$ de type VS_n, il est lui-même de type VS_n; en effet, $X_R = \bigcup_{i \in I} U_{iR}$ et $X_S \cap U_i = U_{iS}$ $(i \in I)$; l'ensemble X_S est bien de $\dim_Z \leq n-1$ puisque chacun de ses points possède un voisinage de $\dim_Z \leq n-1$ $(cf.\ 1.10)$.

Le produit d'un espace de type VS_n par un espace de type VS_m est un espace de type VS_{n+m}.

On définit de même la notion d'espace de type VS_n sur l'anneau principal K. C'est un espace localement compact X de $\dim_K$ égale à n et possédant un ouvert épais homéomorphe à une variété cohomologique sur K de dimension n $([2],\ \S\ 7)$. X_R est alors l'ensemble des points possédant un voisinage homéomorphe à une variété cohomologique de $\dim_K$ égale à n, et $X_S = X - X_R$. Les remarques faites ci-dessus sont encore valables, compte tenu du fait qu'un produit de variétés cohomologiques est une variété cohomologique $(cf.\ [1],\ I,\ 4.10\ a)$. Un espace de type VS_n est « de type VS_n sur K » quel que soit K.

2.2 Définition de la classe fondamentale. — Soit X un espace de type VS_n sur K. Une classe d'homologie fondamentale pour X, à coefficients dans K, est un élément de $H_n(X;K)$ dont l'image dans le groupe d'homologie locale $\mathcal{H}_n(X;K)_x$ par l'homomorphisme μ_x de 1.4 (2) est un générateur de ce groupe pour tout point régulier x.

On verra ci-dessous qu'un élément de $H_n(X;K)$ est une classe fondamentale s'il satisfait à la condition précédente en tout point d'un ensemble épais, ou même simplement d'un ensemble rencontrant chaque composante connexe de X_R. Par exemple, si X est un ensemble analytique ou algébrique, il suffira de vérifier cette condition aux points simples.

Il est clair qu'une classe fondamentale définit un isomorphisme du faisceau constant $X_R \times K$ sur $\mathcal{H}_n(X_R;K)$. Pour que X admette une classe fondamentale à coefficients dans K, il est donc nécessaire que X_R soit orientable relativement à K $(cf.\ 1.10)$; cette condition est évidemment suffisante lorsque X est une variété.

La restriction à un ouvert U de dimension n de X d'une classe fondamentale de X est évidemment une classe fondamentale de U.

2.3. PROPOSITION. — *Soient X un espace de type VS_n sur K et U un ouvert épais de X. Alors X possède au plus une classe fondamentale dont la restriction à U soit donnée. Si $K = \mathbf{Z}_2$, X possède au plus une classe fondamentale. Si U possède une classe fondamentale et si $\dim_K(X-U) \leq n-2$, alors X possède une classe fondamentale dont la restriction à U est la classe fondamentale donnée.*

En effet, dans la suite exacte (*cf.* **1.6**) :

$$H_n(X - U; K) \to H_n(X; K) \xrightarrow{j} H_n(U; K) \to H_{n-1}(X - U; K),$$

le premier terme est nul puisque $\dim_K(X - U) \leq n - 1$, donc j est injectif. De plus, si $K = \mathbf{Z}_2$, on sait que la variété $X_R \cap U$ ne possède qu'une classe fondamentale; comme $X_R \cap U$ est aussi un ensemble épais, cela établit la deuxième assertion. Enfin, si $\dim_K(X - U) \leq n - 2$, alors le dernier terme de la suite exacte est aussi nul, donc j est un isomorphisme.

Remarque. — Si U est un ouvert épais, connexe formé de points réguliers, et si X possède une classe fondamentale, alors $H_n(X; K) \simeq K$ et $i_n : H_n(X; K) \to H_n(U; K)$ est un isomorphisme. En effet, j_n est injectif comme on l'a vu. D'autre part, comme U est connexe, orientable, l'application $\mu_x : H_n(U; K) \to \mathcal{H}_n(U; K)_x$ [*cf.* **1.4** (2)] est un isomorphisme. Puisque $\mathcal{H}_n(U; K)$ s'identifie à $\mathcal{H}_n(X; K)_x$. et que $\mu_x \circ j_n$ est surjectif par définition de la classe fondamentale, il s'ensuit bien que j_n est surjective.

2.4. Corollaire. — *Soient X un espace de type VS_n sur K, A un sous-ensemble de X_R rencontrant chaque composante connexe de X_R, et $c \in H_n(X; K)$. Pour que c soit une classe fondamentale il faut et il suffit que sa valeur en tout point $x \in A$ soit un générateur de $\mathcal{H}_n(X; K)_x$, et c est complètement déterminée par ses valeurs aux points de A.*

Cela résulte de **2.3** et du fait que, $\mathcal{H}_n(X_R; K)$ étant localement isomorphe à $X_R \times K$, l'ensemble des points $x \in X_R$ où une section de ce faisceau est égale à un générateur de $\mathcal{H}_n(X; K)_x$ est ouvert et fermé dans X_R.

2.5. Proposition. — *Soient X et Y des espaces de type VS_n sur K et f une application propre de X dans Y. Supposons que Y_R contienne un ouvert U rencontrant chaque composante de Y_R et tel que f soit un homéomorphisme de $f^{-1}(U)$ sur U, et soit $c \in H_n(X, K)$ une classe fondamentale de X. Alors $f_* c$ est une classe fondamentale de Y.*

Puisque f est propre, $f_* c$ est défini (*cf.* **1.5**). La proposition résulte de **2.4** et du fait que f_* commute avec les restrictions à $f^{-1}(U)$ et U.

2.6. Corollaire. — *Soit X un espace de type VS_n sur K. Supposons que $X = \bigcup_{i \in I} X_i$ est une réunion localement finie de sous-espaces fermés X_i et que $X_i \cap X_j$ soit contenu dans X_S quels que soient $i, j \in I$, $i \neq j$, ce qui entraîne que X_i est de type VS_n sur K ou que $\dim_K X_i < n$. Soit $c_i \in H_n(X_i; K)$ une classe fondamentale de X_i si $\dim_K X_i = n$, l'élément nul sinon. Alors la somme des c_i (cf. **1.7**) est une classe fondamentale de X.*

Soit Y l'espace somme des X_i. Le module $H_*(X; K)$ s'identifie au produit direct des modules $H_*(X_i; K)$ d'après **1.4**. Il est clair que Y est de type VS_n sur K et que $(c_i)_{i\in I}$ en est une classe fondamentale. Le corollaire résulte alors de 1.7 et de 2.5.

2.7. Proposition. — *Soit $(U_i)_{i\in I}$ un recouvrement ouvert d'un espace X de type VS_n sur K et soit $J \subset I$ l'ensemble des i pour lesquels $\dim_K U_i = n$:*

(a) Si pour tout $i \in J$, U_i possède une classe fondamentale c_i sur K et si c_i et c_j ont même restriction à $U_i \cap U_j$ $(i, j \in J)$, alors X possède une classe fondamentale dont la restriction à U_i est c_i $(i \in J)$.

(b) Si $K = \mathbf{Z}_2$ et si U_i $(i \in J)$ possède une classe fondamentale sur K, alors X possède une classe fondamentale sur K.

Posons $c_i = 0$ si $i \notin J$, $i \in I$. Vu 1.10 (2), 2.3 et l'hypothèse, les sections de $\mathcal{H}_n(X; K)$ définies dans $U_i \cap U_j$ par c_i et c_j $(i, j \in I)$ coïncident. Elles définissent alors une section de $\mathcal{H}_n(X; K)$ sur X, donc un élément de $H_n(X; K)$ (*cf.* 1.10), qui a évidemment les propriétés requises.

2.8 Les n$^\text{os}$ 2.8 à 2.10 ont pour but de mettre en relation la notion algébrique d'orientation d'un espace vectoriel sur R avec la notion topologique d'orientation de V considéré comme une variété, et de fixer les conventions d'orientation. Ils ne prétendent pas apporter du nouveau au lecteur.

Convention d'orientation. — Soit V un espace vectoriel sur $\mathbf{R}$ de dimension n. Une orientation de V est déterminée par une base de V, deux bases définissant la même orientation si le déterminant de la matrice de passage d'une base à l'autre est positif.

La *somme* ordonnée $A + B$ de deux sous-espaces vectoriels orientés A et B de V, tels que $A \cap B = 0$, est munie de l'orientation déterminée par une base dont les premiers vecteurs forment une base de A et les derniers une base de B.

Un sous-espace vectoriel B *complémentaire* d'un sous-espace vectoriel orienté A de l'espace vectoriel orienté V est orienté de sorte que l'orientation donnée de V soit la somme de celle de A et de celle de B.

Enfin si A et B sont deux sous-espaces vectoriels orientés de V orienté tels que $A + B = V$, *l'intersection* $C = A \cap B$ est orientée de la manière suivante : soient $(a_1, \ldots, a_r)$ et $(b_1, \ldots, b_s)$ des bases définissant l'orientation de sous-espaces complémentaires de A et B respectivement, alors la base $(a_1, \ldots, a_r, b_1, \ldots, b_s)$ définit l'orientation d'un sous-espace complémentaire de C.

Soit V un espace vectoriel complexe de dimension n. Considéré comme un espace vectoriel réel (de dimension $2n$), il est muni d'une orientation naturelle : si $e_1 \ldots, e_n$ est une base de V sur $\mathbf{C}$, la base $e_1, ie_1, \ldots, e_n, ie_n$ définira l'orientation naturelle de V. Les conventions précédentes appliquées à des sous-espaces vectoriels complexes de V munis de l'orientation naturelle redonnent toujours l'orientation naturelle.

2.9. Classe fondamentale associée à une orientation d'un espace vectoriel. — Soit $c \in H_1(\mathbf{R}, \mathbf{Z})$ la classe fondamentale de la droite $\mathbf{R}$ telle que, dans la suite exacte associée au segment $\mathrm{I} = [0, 1]$ et à sa frontière (*cf.* 1.6), le bord de la restriction de c à $]0, 1[$ soit $[1] - [0]$, où $[1]$ et $[0]$ désignent les générateurs canoniques de l'homologie des points 1 et 0. Soit c^* le générateur de $H_0^1(\mathbf{R}, \mathbf{Z})$ tel que $c \cap c^*$ soit le générateur canonique de $H_0^0(\mathbf{R}, \mathbf{Z})$.

Soit p_k la projection de $\mathbf{R}^n$ sur $\mathbf{R}$ associant à un point sa $k^{\text{ième}}$ coordonnée. D'après la formule de Künneth, $p_1^*(c^*) \cup p_2^*(c^*) \cup \ldots \cup p_n^*(c^*)$ est un générateur c_n^* de $H_0^n(\mathbf{R}^n, \mathbf{Z})$. L'élément c_n de $H_n(\mathbf{R}^n, \mathbf{Z})$ défini par $c_n \cap c_n^* = 1$ [générateur canonique de $H_0^0(\mathbf{R}^n, \mathbf{Z})$] sera par définition la classe fondamentale de $\mathbf{R}^n$ associée à l'orientation naturelle de $\mathbf{R}^n$ (définie par sa base canonique).

Soient h et h' des isomorphismes de $\mathbf{R}^n$ sur un espace vectoriel V; les éléments $h_*(c_n)$ et $h'_*(c_n)$ sont égaux ou opposés suivant que les orientations de V images par h et h' de l'orientation naturelle de $\mathbf{R}^n$ coïncident ou non. La classe fondamentale de V orienté par h sera par définition $h_*(c_n)$.

2.10. Étant donné un sous-espace vectoriel W orienté, de dimension r, d'un espace vectoriel V on notera $w \in H_r(W; \mathbf{Z})$ sa classe fondamentale, w^* l'élément de $H_0^r(W; \mathbf{Z})$ tel que $w \cap w^*$ soit le générateur canonique de $H_0^0(W; \mathbf{Z})$, i_{*W} l'isomorphisme de $H_*(W; \mathbf{Z})$ sur $H_*^W(V, \mathbf{Z})$ induit par l'injection i_W de W dans V. Soit W' un supplémentaire et soit p_W (resp. $p_{W'}$) la projection de V sur W (resp. W') parallèle à W' (resp. W). Si l'on oriente W' et l'on munit V de l'orientation somme des orientations de W et W', alors il résulte de **2.9** et de la règle de Künneth que

$$(1) \qquad v^* = p_W^*(w^*) \cup p_{W'}^*(w'^*).$$

Proposition. — *Soient A et B deux sous-espaces vectoriels orientés de l'espace vectoriel réel orienté V, de dimensions r et s, qui engendrent V. Alors :*

(a) Si $A \cap B = 0$ et si l'orientation de V est la somme des orientations de A et B, on a

$$i_A^* \Delta i_{*B} b = a^*, \qquad i_B^* \Delta i_{*A} a = (-1)^{r \cdot s} b^*.$$

(b) Si $C = A \cap B$ est orienté suivant la convention de 2.8, on a

$$i_{*A}(a) \cdot i_{*B}(b) = i_{*C}(c).$$

On désignera par 1 le générateur canonique de $H_0^0(A; \mathbf{Z})$ ou de $H_0^0(V; \mathbf{Z})$ et par $[V]$ la classe fondamentale de V. On a

$$1 = a \cap a^* = a \cap i_A^* \cdot p_A^*(a^*),$$

puisque $p_A \circ i_A$ est l'identité, d'où, vu **1.9**,

$$1 = i_{*A} 1 = i_{*A}(a \cap i_A^{*} p_A^{*}(a^{\star})) = i_{*A} a \cap p_A^{\star}(a^{\star}),$$
$$1 = ([V] \cap \Delta i_{*A}(a)) \cap p_A(a^{\star}) = [V] \cap (\Delta i_{*A}(a) \cup p_A^{\star}(a^{\star})).$$

D'autre part, (1) donne

$$[V] \cap (p_A^{\star}(a^{\star}) \cup p_B^{\star}(b^{\star})) = 1,$$

d'où

(2)
$$\Delta i_{*A} a = (-1)^{rs} p_B^{\star} b^{\star},$$
$$i_B^{*} \Delta i_{*A} a = (-1)^{rs} i_B^{*} p_B^{\star} b^{\star} = (-1)^{rs} b^{\star}.$$

On démontre de même

(3)
$$\Delta i_{*B} b = p_A^{\star} a^{\star}$$

et la première égalité de (a).

Pour établir (b) supposons V orienté, et soient A', B' des complémentaires de A et B respectivement, orientés suivant (2.8), et $C' = A' + B'$. Comme $p_{A'} \circ i_{C'}$ et $p_{B'} \circ i_{C'}$ sont les projections de C' sur A' et B', parallèles à B' et A', on a d'après (1)

$$c'^{\star} = i_{C'}^{*}(p_{A'}^{\star} a'^{\star} \cup p_{B'}^{\star} b'^{\star}),$$

d'où, vu (a) et (2)

$$(-1)^{t(n-t)} \Delta i_{*C} c = (-1)^{r(n-r)+s(n-s)}(\Delta i_{*C} a \cup \Delta i_{*B} b)$$

$(t = \dim C)$, ou encore,

$$\Delta i_{*C} c = \Delta i_{*A} a \cup \Delta i_{*B} b,$$

ce qui équivaut à (b).

2.11. Proposition. — *Soient A et B des espaces de type VS_n et VS_m sur K possédant des classes fondamentales a et b respectivement sur K. Le produit $A \times B$ est alors de type VS_{n+m} sur K et possède une classe fondamentale sur K.*

Soient $\alpha \in \mathrm{Hom}(H_c^n(A; K), K)$ et $\beta \in \mathrm{Hom}(H_c^m(B; K); K)$ les éléments déterminés par a et b respectivement ($cf.$ **1.3**). Si l'on identifie par la formule de Künneth $H_c^n(A; K) \otimes H_c^m(B; K)$ à $H_c^{m+n}(A \times B; K)$, l'élément $\gamma = \alpha \otimes \beta$ appartient à $\mathrm{Hom}(H_c^{n+m}(A \times B; K), K)$ et détermine par **1.3** un élément unique c de $H_{m+n}(A \times B; K)$, car $H_c^{m+n+1}(A \times B; K) = 0$. Cet élément c est la classe fondamentale cherchée. Pour le voir, il suffit de le vérifier pour la restriction de c à tout ouvert de la forme $U \times V$, où U et V sont des ouverts connexes contenus dans A_R et B_R respectivement. Soient a_0, b_0 et c_0 les restrictions de a, b et c à U, V et $U \times V$ respectivement, et soient α_0, β_0 et γ_0 les classes associées comme plus haut par **1.3** à a_0, b_0 et c_0 respectivement. Par hypothèse, a_0 et b_0 sont des générateurs des groupes auxquels ils appartiennent; il en est donc de même de α_0 et β_0 et de γ_0 qui s'identifie par la règle de Künneth à $\alpha_0 \otimes \beta_0$; donc c_0 est aussi un générateur de $H_{n+m}(U \times V; K)$.

Si A_R et B_R sont des variétés topologiques, U et V homéomorphes à des espaces numériques et si $K = \mathbf{Z}$, il est aisé de vérifier à l'aide de **2.10** que l'orientation de $U \times V$ déterminée par c est la somme de l'orientation de U et de celle de V déterminée par a et b respectivement.

2.12. Corollaire. — *Soit E un espace fibré localement trivial dont la base A est de type VS_n sur K et la fibre B de type VS_m sur K. Supposons que A possède une classe fondamentale a sur K et B une classe fondamentmentale b sur K invariante par le groupe de structure. Alors E possède une classe fondamentale c.*

Soit p la projection de E sur A et (U_i) un recouvrement ouvert de A tel que $p^{-1}(U_i) = U_i \times B$. Par la proposition précédente, on construit une classe fondamentale c_i dans $p^{-1}(U_i)$; l'hypothèse que le groupe structural laisse a invariante entraîne que c_i et c_j coïncident aux points réguliers de $p^{-1}(U_i \cap U_j)$; Il suffit alors d'appliquer **2.6**.

2.13. Caractérisation de la classe fondamentale d'une variété plongée. — Soit V une variété de classe C^k (k entier ≥ 0, ou $k = \infty$, ω) de dimension q et soit X un sous-espace fermé de V de type VS_n. On dira que X est *bien plongé* en un point $x \in X$ s'il existe un voisinage U de x dans V et des coordonnées locales d'origine x valables dans U, telles que $X \cap U$ soit le plan $x_i = 0$ ($n + 1 \leq i \leq q$). Le plan P défini par $x_j = 0$ ($1 \leq i \leq n$) sera appelé *plan tranverse* en x. Supposons V orientée, et soit Δ l'isomorphisme de dualité dans V (*cf.* **1.11**); soit r^* l'homomorphisme de $H_X^*(V; K)$ dans $H_x^*(P; K)$ induit par l'injection de P dans V.

Proposition. — *On conserve les notations précédentes. On suppose que X possède une classe fondamentale $c \in H_n(X; K)$ qu'on identifie à un élément de $H_n^X(V; K)$ (cf. **1.5**). Alors :*

(a) $r^* \Delta c$ est un générateur de $H_x^{l-n}(P; K)$.

(b) Si $K = \mathbf{Z}$, l'orientation donnée de U est la somme (prise dans cet ordre) de l'orientation de $X \cap U$ associée à c et de l'orientation de P associée à $r^* \Delta c$.

Réciproquement, supposons que X soit bien plongé en tous les points d'un ensemble A rencontrant chaque composante connexe de X_R, et soit c un élément de $H_n^X(V; K)$ tel que $r^* \Delta c$ soit un générateur de $H_x^{-u}(P; K)$ pour tout point $x \in A$, P étant un plan transverse en x. Alors c, vu comme un élément de $H_n(X; K)$, est une classe fondamentale de X.

Cette proposition résulte du fait que Δ commute avec la restriction à U et qu'on peut alors appliquer **2.10**; on tient compte de plus de **2.4** pour la réciproque.

2.14. Application transverse. — Soient V, W des variétés de classe C^k et $f : V \to W$ une application de classe C^k. Soit Y un sous-espace fermé de W, de type VS_n. On dira que f est *transverse à* Y *en* $v \in f^{-1}(Y_R)$ si l'espace Y est bien plongé dans W en $f(v)$ (*cf.* 2.13) et si l'espace tangent $W_{f(v)}$ à W en $f(v)$ est engendré par $f(V_v)$ et par l'espace tangent à Y en $f(v)$. On sait que dans ce cas l'image réciproque X par f d'un voisinage assez petit de $f(v)$ dans Y est une sous-variété dont la codimension est égale à la codimension de Y dans W. Plus précisément, on peut trouver des coordonnées locales (x_i) $(1 \leq i \leq s)$ et $(y_j)(1 \leq j \leq t)$ dans V et W, d'origines respectives v et $f(v)$, ayant les propriétés suivantes : X est représenté par les équations $x_i = 0$ $(i \geq s - t + n + 1)$, Y par les équation $y_j = 0$ $(j \geq n + 1)$, et f est un homéomorphisme du plan tranverse $x_i = 0$ $(i \leq s - t + n)$ sur le plan transverse $y_j = 0$ $(j \leq n)$.

2.15. Proposition. — *Soient V (resp. W) une variété de classe C^k de dimension s (resp. t) et f une application de classe C^k de V dans W. Soit Y un sous-espace fermé de W qui soit de type VS_n et qui soit bien plongé (2.13) en tout point d'un ensemble épais U dans Y. Supposons que f soit transverse à Y en tout point de $f^{-1}(U)$ et que $\dim_Z f^{-1}(Y - U) < s - t + n$. Alors $X = f^{-1}(Y)$ est un espace de type VS_m, avec $m = s - t + n$. Si V et W sont orientées relativement à un anneau de coefficients K et si Y possède une classe fondamentale $[Y]$, alors X possède une classe fondamentale $[X]$ telle que, i et j désignant les injections de X et Y dans V et W respectivement, on ait*

$$(4) \qquad \Delta i_*[X] = f^* \Delta j_*[Y].$$

[Dans cette proposition, les familles Φ et Ψ de supports dans V et W sont telles que $f^{-1}(\Psi) \subset \Phi$, $X \in \Phi$ et $Y \in \Psi$. Si Φ et Ψ sont les familles des fermés contenus dans X et Y, $[X]$ est déterminé d'une manière unique par (4).]

Les faits rappelés en 2.14 montrent que $f^{-1}(U)$ est une variété de dimension m, donc X est de type VS_m.

Soit $f^* : H_Y^*(W; K) \to H_X^*(V; K)$. La classe unique $[X]$ telle que $i_*[X] = \Delta^{-1} f^* \Delta j_*[Y]$ est la classe fondamentale cherchée en vertu de 2.4 et de 2.13.

Remarque. — Soit $K = \mathbf{Z}$. En utilisant 2.10, on voit que les orientations définies par $[X]$ et $[Y]$ sont telles que les orientations des plans transverses (*cf.* 2.8) en v et $f(v)$ se correspondent par f.

3. Existence de la classe fondamentale d'un espace analytique.

3.1. — On notera $d(A_a)$ la dimension complexe d'un germe d'ensemble analytique complexe A en un point a de $\mathbf{C}^m$ (*cf.* [5], exp. VI; [19], déf. 3.2 et 3.3). Par $\dim_{\mathbf{Z}} A_a$ on entendra la dimension cohomologique sur $\mathbf{Z}$

au voisinage de a d'un ensemble analytique de $\mathbf{C}^{m}$ induisant le germe A_a en a. On sait que $\dim_{\mathbf{Z}} A_a = 2.d(A_a)$. Rappelons-en brièvement une démonstration : on procède par récurrence sur $d(A)_a$. Si a est un *point simple*, alors A_a est par définition le germe en a d'une variété analytique complexe de dimension complexe $d(A_a)$. Sinon, un ensemble analytique A induisant A_a en a est la réunion d'un ensemble analytique S tel que $d(S_a) < d(A_a)$ ([19], lemma 3.11), et d'une variété complexe de dimension complexe $d(A_a)$. Comme cette dernière est réunion dénombrable de fermés, on peut appliquer le théorème somme de la théorie de la dimension.

Il s'ensuit que, si X est un espace analytique complexe (au sens de [20]) de dimension complexe $d(X) = n$, $d(X)$ étant la borne supérieure des dimensions des germes de X en ses différents points, alors $\dim_{\mathbf{Z}} X = 2n$, et réciproquement. Les points simples de X forment un ouvert épais dont le complémentaire est de dimension sur $\mathbf{Z}$ au plus égale à $2n - 2$. Donc X est de type VS_{2n} et, en vertu de 2.3, on a le théorème suivant :

3.2. Théorème. — *Un espace analytique complexe X de dimension complexe n possède une et une seule classe fondamentale entière induisant l'orientation naturelle en chaque point simple.*

3.3. — Nous utiliserons à plusieurs reprises le fait que *l'ensemble des points simples d'un espace analytique complexe X irréductible est connexe.* Rappelons que ceci résulte de ce que la normalisation $\tilde{X}$ de X (définie comme l'espace des germes irréductibles de X, *cf.* [5] et [19], 6.11) est connexe si et seulement si X est irréductible et que le sous-ensemble de $\tilde{X}$ qui se projette sur l'ensemble de points singuliers de X ne sépare pas X (*cf.* [19], th. 6.9). Il s'ensuit plus généralement que le complémentaire $X - F$ d'un sous-ensemble analytique F est connexe. En effet, l'ensemble U des points simples de X est dense dans X, donc $U - (F \cap U)$ est dense dans $X - F$. Mais U est une variété analytique connexe et $F \cap U$ est un sous-ensemble analytique de U; donc $U - (F \cap U)$ est connexe (*cf.* [19], th. 2.2) et son adhérence l'est aussi.

Proposition. — *Soit X un espace analytique irréductible de dimension complexe n. Alors $H_{2n}(X;\mathbf{Z})$ est isomorphe à $\mathbf{Z}$ et est engendré par la classe fondamentale. Si X est non compact, $H^{c}_{2n}(X;\mathbf{Z}) = 0$ ([2]).*

Soient S l'ensemble des points singuliers de X et $U = X - S$. On a vu que $\dim_{\mathbf{Z}} S \leqq 2n - 2$. La première assertion résulte alors du fait que U est connexe et de l'isomorphisme $H_{2n}(X;\mathbf{Z}) = H_{2n}(U;\mathbf{Z})$ (*cf.* 2.3). Vu 1.10, cet isomorphisme signifie que l'homomorphisme de restriction $\Gamma(\mathcal{H}_{2n}(X;\mathbf{Z})) \to \Gamma(\mathcal{H}_{2n}(U;\mathbf{Z}))$ est un isomorphisme. Comme U est connexe, orientable, $\mathcal{H}_{2n}(U;\mathbf{Z})$ est le faisceau constant $U \times Z$; par conséquent, une

---

([2]) La deuxième assertion répond à une question de K. Stein.

section de $\mathcal{H}_{2n}(X; \mathbf{Z})$ qui est nulle en un point de U est identiquement nulle. Si maintenant X n'est pas compact, alors U n'est pas relativement compact et la remarque précédente montre que $\Gamma_c(\mathcal{H}_{2n}(X; \mathbf{Z})) = 0$. D'après 1.10, cela équivaut à la deuxième assertion.

3.4. Relation avec le courant associé à un sous-ensemble analytique. — Soit X un sous-ensemble analytique M de dimension complexe m, les composantes irréductibles de X ayant la même dimension complexe q.

D'après P. Lelong [15] (*cf.* aussi G. de Rham [18]), on peut associer à X un courant fermé T de M dont la valeur sur toute forme différentielle φ à support compact ne rencontrant pas les points singuliers de X est l'intégrale de φ sur la sous-variété des points simples de X.

La classe d'homologie de M représentée par X est par définition l'image de la classe fondamentale de X par l'homomorphisme $i_* : H_*(X) \to H_*(M)$ induit par l'injection de X dans M.

PROPOSITION. — *La classe de cohomologie de M représentée par le courant T associé à X est la classe duale à la classe d'homologie de M représentée par X et réduite en coefficients réels.*

Si X est non singulier, la proposition est essentiellement démontrée dans G. de Rham [17]. Le cas général se ramène à celui-ci. Soit S l'ensemble des points non simples de X; le courant T_0 associé à $X_0 = X - S$, dans $M_0 = M - S$ est la restriction de T à M_0. Ainsi la classe de cohomologie de M_0 représentée par T_0 est la restriction à M_0 de la classe de cohomologie de M représentée par T. De même la classe duale à X_0 dans M_0 est la restriction à M_0 de la classe duale à X dans M (*cf.* 2.13); donc sa réduction en coefficients réels est la classe de T_0. Comme $\dim_{\mathbf{Z}} S \leq \dim_{\mathbf{Z}} X - 2$, l'application de restriction $H_{2n-2q}(M) \to H_{2n-2q}(M_0)$ est un isomorphisme, ce qui établit la proposition.

Le cas analytique réel. — Pour établir l'analogue de 3.2 dans le cas réel, nous aurons besoin de quelques propriétés des germes d'ensembles analytiques réels que nous rassemblons dans les deux numéros qui suivent.

3.5. — Soit A_a le germe d'un ensemble analytique réel A de $\mathbf{R}^7$ en un point a de $\mathbf{R}^7$ et soit A_a^c son complexifié (*cf.* [7], p. 91). Par définition $d(A_a) = d(A_a^c)$. Comme en 3.1, on définit $\dim_{\mathbf{Z}} A_a$ comme étant la dimension sur $\mathbf{Z}$ de A au voisinage de a. On sait ([7], prop. 9) que les composantes irréductibles de A_a^c sont les complexifiées des composantes irréductibles de A_a. Si A_a est irréductible, il existe un ensemble analytique réel $A' \subset A$, avec $d(A_a') < d(A_a)$, tel que dans tout voisinage U suffisamment petit de a, $A - A'$ soit une sous-variété analytique réelle non vide de dimension égale à $d(A_a)$ (*cf.* [7], prop. 10); l'ensemble $U - U \cap A'$ est formé des points simples de U. Le germe A_a étant de nouveau non nécessairement irréductible, cela entraîne comme en 3.1 que $\dim_{\mathbf{Z}} A_a = d(A_a)$ et que tout voisinage suffisamment petit de a dans A est un espace de type VS_n, avec $n = d(A_a)$.

Soit B_a un germe d'ensemble analytique complexe de $\mathbf{C}^q$ en a. Sa « partie réelle » $B_a \cap \mathbf{R}^q$ sera notée B_a^0; c'est un germe analytique réel. Évidemment $(B_a^0)^c \subset B_a$, et si $B_a = A_a^c$, alors $B_a^0 = A_a$. Des résultats précités on déduit immédiatement que

$$\dim_{\mathbf{Z}}(B_a^0) \leqq d(B_a),$$

et que si B_a est irréductible, il y a égalité si et seulement si B_a est le complexifié de B_a^0.

3.6. Normalisation. — On note H_q l'anneau des séries entières convergentes au voisinage de l'origine a de $\mathbf{R}^q$, à coefficients réels.

Soit A_a un germe analytique réel irréductible au point a. L'idéal I des éléments de H_q qui s'annulent sur A_a est premier et l'anneau quotient $E = H_q/I$ est d'intégrité. Soient K le corps des fractions de E et F la clôture intégrale de E dans K.

Considérons $\mathbf{R}^q$ plongé canoniquement dans $\mathbf{C}^q$. Répétons les mêmes constructions en remplaçant A_a par son complexifié A_a^c, le corps des réels par celui des complexes et utilisons les mêmes notations en ajoutant l'indice supérieur c. Les isomorphismes suivants sont canoniques : $H_q^c = H_q \otimes \mathbf{C}$, $E^c = E \otimes \mathbf{C}$, $F^c = F \otimes \mathbf{C}$ et H_q, E et F sont identifiés à des sous-anneaux de H_q^c, E^c et F^c respectivement.

L'anneau $E^c = H_q^c/I^c$ est un anneau analytique ([5], exp. VIII *bis*), engendré analytiquement par les images u_i ($1 \leqq i \leqq q$) des fonctions coordonnées $z_i \in H_q^c$ de $\mathbf{C}^q$. L'anneau F^c est aussi un anneau analytique et un E^c-module de type fini (*cf.* [5], exp. X, th. 1 et exp. VII, th. 1; [19], th. 6.10). Il admet donc un système fini de générateurs $v_1, v_2, \ldots, v_s$, qu'on peut choisir dans F et tels que $u_i = v_i$ ($1 \leqq i \leqq q$). On a donc $F = H_s/J$ et $F^c = H_s^c/J^c$, où J et J^c sont des idéaux premiers de H_s et H_s^c respectivement.

Soit B (resp. B^c) un ensemble analytique dans un ouvert de $\mathbf{R}^s$ (resp. $\mathbf{C}^s$) dont le germe B_b (resp. B_b^c) à l'origine b de $\mathbf{R}^s$ (resp. $\mathbf{C}^s$) est défini par l'annulation de J (resp. J^c). Soit A (resp. A^c) un ensemble analytique dans $\mathbf{R}^q$ (resp. $\mathbf{C}^q$) induisant le germe A_a (resp. A_a^c) en a. On peut supposer que A et B sont les parties réelles de A^c et B^c respectivement. Le transposé de l'inclusion φ^c de E^c dans F^c est le germe d'une application analytique ψ^c de B^c dans A^c définie au voisinage de b et qui envoie le point de coordonnées $(z_i)_{1 \leqq i \leqq s}$ sur le point de coordonnées $(z_j)_{1 \leqq j \leqq q}$. L'application ψ^c est donc définie par des équations réelles et sa restriction ψ à B est une application d'un voisinage de b dans A.

Le lemme suivant s'appuie essentiellement sur le théorème d'Oka affirmant l'existence de la normalisation.

Lemme. — *Avec les notations précédentes, il existe des voisinages X et Y de a et b dans A et B respectivement tels que :*

1° *X et Y sont des espaces de type VS_n avec $n = d(A_a)$ et $\dim_\mathbf{Z} Y_S \leqq n - 2$.*

2° *Ψ définit une application propre de Y dans X; l'espace X possède un ouvert épais U, non vide, contenu dans X_R, tel que Ψ soit un homéomorphisme de $\Psi^{-1}(U)$ sur U.*

Soit $\tilde{A}^c$ l'espace des germes irréductibles de A^c en ses différents points, muni de sa topologie et de sa projection naturelle μ sur A^c (*cf.* [5], exp. VII); l'espace $\tilde{A}^c$ est localement compact et la projection μ est propre (*cf.* [19], th. 6.9). D'après le théorème d'Oka ([19], th. 7.2), l'espace $\tilde{A}^c$ peut être muni d'une structure d'espace analytique normal de sorte que la projection μ soit holomorphe; l'anneau local en un point $\tilde{a}$ de $\tilde{A}^c$ est la clôture intégrale de l'anneau local du germe irréductible $\tilde{a}$ de A^c au point $\mu(\tilde{a})$. L'espace $\tilde{A}^c$ muni de la projection μ est appelé la normalisation de A.

Ainsi l'anneau local de $\tilde{A}^c$ au point $\tilde{a}$ correspondant au germe de A^c en a est isomorphe à l'anneau local F^c de B^c en b; les germes de ces espaces en ces points sont donc isomorphes. Il en résulte qu'en prenant des voisinages convenables X^c et Y^c assez petits de a et b dans A^c et B^c, on peut considérer que (Y^c, Ψ^c) est la normalisation de X^c. Si X^c est choisi assez petit, Y^c et X^c sont des espaces analytiques de dimension $n = d(A_a)$. Les points non simples de Y^c forment un sous-ensemble analytique S de dimension complexe $\leqq n - 2$ (*cf.* [5], exp. XI, th. 2; [19], th. 7.4).

Soient Y et X les parties réelles de Y^c et X^c; ce sont des espaces de type VS_n (*cf.* 3.5). De plus S contient la complexification du sous-ensemble analytique des points non simples de Y; il en résulte que $\dim_\mathbf{Z} Y_S \leqq n - 2$, ce qui est notre première assertion.

Il est clair que la restriction Ψ à Y de l'application propre Ψ^c est aussi propre. Soit x un point simple de X; c'est aussi un point simple de X^c (*cf.* [7], p. 92); son image réciproque par Ψ^c est un point simple unique. Comme Ψ^c est définie par des équations réelles, ce point doit aussi appartenir à Y, et il est aussi simple sur Y. Ainsi, U désignant l'ouvert épais de points simples de X, l'application Ψ est un homéomorphisme de $\Psi^{-1}(U)$ sur U, ce qui prouve 2°.

3.7. Théorème. — *Soit X un espace de type VS_n. On suppose que tout point de X possède un voisinage homéomorphe à un sous-ensemble analytique réel de dimension $\leqq n$ d'un ouvert d'un espace numérique. Alors X possède une et une seule classe fondamentale mod 2.*

Vu **2.7**, il suffit de voir que, dans un ensemble analytique réel, tout point a un voisinage qui possède une classe fondamentale mod 2. Puisqu'un germe d'ensemble analytique réel A_a est réunion finie de germes irréductibles, on peut se borner, d'après **2.6**, au cas d'un germe irréductible. Il suffit donc de montrer que l'espace X du lemme précédent possède une classe fondamentale, donc, vu 2.5, que Y a une classe fondamentale; ceci résulte de l'inégalité $\dim_{\mathbf{Z}} Y_S \leq n - 2$ et de 2.3.

3.8. — Disons qu'un espace X de type VS_n est localement algébrique réel si tout point $x \in X$ possède un voisinage homéomorphe à un sous-ensemble algébrique réel de dimension $\leq n$ d'un ouvert d'un espace numérique. X vérifie en particulier l'hypothèse de 3.7 et possède donc une et une seule classe fondamentale mod 2. Cependant, on peut donner de ce fait une démonstration directe analogue à la précédente, mais dans laquelle le théorème de normalisation d'Oka est remplacé par la normalisation d'un ensemble algébrique affine irréductible, qui est de nature élémentaire. Nous voulons la décrire ici brièvement, car ce cas est en fait le seul que nous rencontrerons dans les applications.

On utilisera les analogues suivants des résultats de 3.5 qui sont démontrés dans [24] ou bien résultent directement des résultats de [24].

(*) Soit $X \subset \mathbf{R}^m$ un ensemble algébrique réel et soit X^c son complexifié. Alors les composantes irréductibles de X sont les parties réelles des composantes irréductibles de X^c, donc X est réunion d'un nombre fini de composantes irréductibles, et X est irréductible si et seulement si X^c l'est. Soit Y un sous-ensemble algébrique de $\mathbf{C}^m$, défini sur $\mathbf{R}$, i.e. par des équations polynomiales à coefficients réels, et soit $Y^0 = Y \cap \mathbf{R}^m$ sa partie réelle. Alors

$$\dim_{\mathbf{Z}} Y^0 \leq d(Y) = 2^{-1} . \dim_{\mathbf{Z}} Y.$$

Supposons Y irréductible. Alors il y a égalité si et seulement si $(Y^0)^c = Y$, et Y^0 est alors irréductible.

Soit X un espace de type VS_n qui soit localement algébrique réel. Nous voulons montrer qu'il possède une et une seule classe fondamentale mod 2. D'après 2.7, la question est locale, et il suffit de considérer le cas où X est un ouvert d'un ensemble algébrique d'un espace numérique, donc aussi où X est un ensemble algébrique affine de dimension n de $\mathbf{R}^m$. 2.6 et (*) montrent alors qu'on peut supposer X, donc X^c, irréductible. Soit E l'anneau des coordonnées de X, c'est-à-dire le quotient II/I de l'anneau des polynômes à coefficients réels en les coordonnées de $\mathbf{R}^m$ par l'idéal I de E. Alors $E^c = (II/I) \otimes C$ s'identifie à l'anneau des coordonnées de X^c. Soit F la clôture intégrale de E dans son corps des fractions. Alors $F^c = F \otimes C$ est la clôture intégrale de E^c dans son corps des fractions. On sait (*cf.* par exemple [14], V, § 1) que E^c est l'anneau de coordonnées d'une variété algébrique affine irréductible normale $Y \subset \mathbf{C}^q$, définie sur $\mathbf{R}$, et qu'il existe une application birationnelle partout définie $\mu : Y \to X^c$, définie sur $\mathbf{R}$, ayant les propriétés suivantes : μ est surjective, propre, l'image réciproque de tout point $x \in X$ est formée d'un nombre fini de points, μ est un homéomorphisme birégulier de $\mu^{-1}(U)$ sur U, où U est l'ensemble des points simples de X^c. Compte tenu de (*), on voit que μ est une application propre de $Y^0 = (Y \cap \mathbf{R}^m)$ dans X, et qu'il existe un ouvert épais U^0 de X tel que μ soit un homéomorphisme de $\mu^{-1}(U^0)$ sur U^0. De plus Y_S^0 est contenu dans la partie réelle de l'ensemble Z des points singuliers de Y. Mais, Y étant normale, on a $d(Z) \leqq n - 2$ ([14], p. 122) donc, vu (*), $\dim_{\mathbf{Z}} Y_S^0 \leqq n - 2$. L'existence de la classe fondamentale résulte alors de 2.3 et 2.5.

4. Intersection de sous-ensembles analytiques complexes.

4.1. Rappel. — Soit V un espace analytique irréductible, de dimension complexe n. Tout sous-ensemble analytique X de V admet une représentation minimale comme réunion de sous-ensembles analytiques irréductibles X, ses composantes irréductibles, qui forment un recouvrement localement fini de X. Inversement toute réunion localement finie de sous-ensembles analytiques irréductibles de V est un sous-ensemble analytique de V. La dimension $d(X)$ d'un sous-ensemble analytique X est le maximum des dimensions de ses composantes irréductibles. Sa codimension est $n - d(X)$. On dira que X est de dimension pure p si chaque composante irréductible de X est de dimension p.

L'intersection de deux sous-ensembles analytiques X et Y de V est un sous-ensemble analytique. Si X et Y sont irréductibles de codimensions respectives p et q, alors toute composante irréductible de $X \cap Y$ contenant un point simple de V est de codimension au plus $p + q$.

4.2. Le groupe des cycles analytiques de V. — Soit V un espace analytique et soit $(X_i)_{i \in I}$ l'ensemble des sous-ensembles analytiques irréductibles de V. Le groupe $\mathfrak{Z}(V)$ des cycles analytiques de V est le groupe commutatif dont les éléments sont les combinaisons linéaires formelles $Z = \sum_{i \in I} n_i X_i$ à coefficients entiers, dans lesquelles les X_i pour lesquels $n_i \neq 0$ forment une famille localement finie dans V. Ces éléments X_i sont les composantes de Z. Le support $|Z|$ de Z est le sous-ensemble analytique réunion des composantes de Z. Un cycle Z est homogène de dimension p si toutes ses composantes sont de dimension p.

Tout sous-ensemble analytique irréductible X de V, de dimension p, représente une classe d'homologie $h(X) \in H_{2p}^{X}(V; Z)$. C'est l'image, par l'homomorphisme induit par l'inclusion de X dans V, de la classe fondamentale de X (cf. 3.2). Si $Z = \sum n_i X_i$ est un cycle analytique de V, alors $h(Z) \in H_*^{|Z|}(V; Z)$ désignera la somme $\sum n_i h(X_i)$ au sens de 1.7.

Si Φ est une famille de supports sur V contenant $|Z|$, on notera $h^\Phi(Z)$ l'image de $h(Z)$ par l'homomorphisme d'agrandissement des supports. En particulier, $h^V(Z)$ est l'image de $h(Z)$ dans $H_*(V; Z)$. Il est clair que si $(Z_j)_{j \in J}$ est une famille de cycles analytiques dont les supports forment une famille localement finie, on a, en posant $|Z| = \bigcup_{j \in J} |Z_j|$,

$$h^{|Z|}\left(\sum_{j \in J} Z_j\right) = \sum_{j \in J} h(Z_j);$$

par conséquent, l'application $Z \to h^V(Z)$ est un hòmomorphisme de $\mathfrak{Z}(V)$ dans $H_*(V; \mathbf{Z})$.

4.3. Lemme. — *Soit X un espace analytique de dimension pure p et soient X_λ ses composantes irréductibles. L'application qui fait correspondre à une famille d'éléments $a_\lambda \in H_{2p}(X_\lambda; \mathbf{Z})$ leur somme $\sum a_\lambda$ (cf. 1.7) est un isomorphisme j de $\prod_\lambda H_{2p}(X_\lambda, \mathbf{Z})$ sur $H_{2p}(X; \mathbf{Z})$.*

Pour tout λ, soit X'_λ le complémentaire de la réunion des composantes irréductibles de X distinctes de X_λ; soit X' la réunion des X'_λ. On a le diagramme commutatif suivant :

$$
\begin{array}{ccc}
H_{2p}(X\;; \mathbf{Z}) & \xrightarrow{\;\varphi\;} & H_{2p}(X'\;; \mathbf{Z}) \\
\big\uparrow j & & \big\uparrow j' \\
\prod_\lambda H_{2p}(X_\lambda; \mathbf{Z}) & \xrightarrow{\;\Pi\varphi_\lambda\;} & \prod_\lambda H_{2p}(X'_\lambda; \mathbf{Z}),
\end{array}
$$

où j' est défini comme j, et où φ et φ_λ sont les homomorphismes de restriction. Il est clair que j' est bijectif, car les X'_λ sont ouverts et fermés disjoints dans X'; de même φ et φ_λ sont bijectifs car $\dim_{\mathbf{Z}}(X_\lambda - X'_\lambda) \leq \dim_{\mathbf{Z}}(X - X') \leq 2p - 2$. Donc j est un isomorphisme.

4.4. Définition de la multiplicité d'intersection. — Soient X et Y des sous-ensembles analytiques de dimension pure de l'espace analytique irréductible V. On dira qu'une composante irréductible C de $X \cap Y$ est propre si elle contient un point simple de V et si $\operatorname{codim} C = \operatorname{codim} X + \operatorname{codim} Y$.

Soient $C_\lambda (\lambda \in I)$ les composantes propres de $X \cap Y$, M leur réunion, N la réunion des composantes non propres de $X \cap Y$, et $p = d(C_\lambda)$. Comme $\dim_{\mathbf{Z}}(M \cap N) \leq 2p - 2$, la suite exacte de Mayer-Vietoris (1.8), (1.5) et (1.10) montrent que $H_{2p}^{|X| \cap |Y|}(V; \mathbf{Z})$ est isomorphe à la somme directe $H_{2p}^{M}(V; \mathbf{Z}) \oplus H_{2p}^{N}(V; \mathbf{Z})$. Vu 4.3, on a donc un isomorphisme canonique

$$\prod_{\lambda \in I} H_{2p}^{C_\lambda}(V; \mathbf{Z}) \oplus H_{2p}^{N}(V; \mathbf{Z}) = H_{2p}^{|X| \cap |Y|}(V; \mathbf{Z}).$$

Tout élément $c \in H_{2p}^{|X| \cap |Y|}(V; \mathbf{Z})$ s'écrit d'une et d'une seule manière comme somme (au sens de 1.7)

$$c = \sum_{\lambda \in I} c_\lambda + n \qquad [c_\lambda \in H_{2p}^{C_\lambda}(V; \mathbf{Z}), \ n \in H_{2p}^N(N; \mathbf{Z})].$$

Supposons d'abord que V soit une variété analytique complexe connexe. L'intersection $h(X).h(Y)$ des classes d'homologie représentées par X et Y est un élément de $H_{2p}^{|X| \cap |Y|}(V; \mathbf{Z})$; vu 3.3, sa composante dans $H_{2p}^{C_\lambda}(V; \mathbf{Z})$ est un multiple entier de $h(C_\lambda)$. Cet entier est par définition la *multiplicité de la composante* (propre) C_λ *dans l'intersection de* X *et* Y. Il sera noté $i(X.Y, C_\lambda)$. Lorsque V est un espace analytique irréductible, on définit $i(X.Y, C_\lambda)$ comme étant la multiplicité d'intersection $i(X'.Y', C_\lambda')$, où X', Y' et C_λ' sont les intersections de X, Y et C_λ avec l'ouvert des points simples de V. Comme $h(X).h(Y) = h(Y).h(X)$, on a

$$i(X.Y, C_\lambda) = i(Y.X, C_\lambda) \qquad (\lambda \in I).$$

Il résulte immédiatement de la définition que si X_μ (resp. Y_ν) sont les composantes irréductibles de X (resp. Y) qui contiennent une composante propre C_λ de $X \cap Y$, on a

$$i(X.Y, C_\lambda) = \sum_{\mu, \nu} i(X_\mu.Y_\nu, C_\lambda).$$

4.5. Proposition. — *Soit U un voisinage ouvert d'un point x, simple sur V, d'une composante irréductible propre C de l'intersection de deux sous-ensembles X et Y purement dimensionnels. Soit $C_U(x)$ une composante irréductible propre de $C \cap U$ contenant x. Alors*

$$i(X.Y, C) = i((X \cap U).(Y \cap U), C_U(X)).$$

Ceci résulte du fait que l'application h commute avec la restriction à U, et montre que la notion de multiplicité d'intersection est *locale*.

Cette proposition pourrait tout aussi bien se formuler en termes de germes irréductibles de X, Y, et C au point x et des classes d'homologie locale qu'ils représentent.

4.6. Intersection des cycles. — Soient X et Y des sous-ensembles analytiques irréductibles de V. Si toutes les composantes irréductibles C_λ de $X \cap Y$ sont propres, l'intersection des cycles X et Y sera par définition le cycle $X.Y = \sum_\lambda i(X.Y, C_\lambda) C_\lambda$. L'intersection de deux cycles analytiques est définie par linéarité si l'intersection de chaque composante du premier avec chaque composante du second est définie.

Lorsque V est sans singularités, et que Z_1 et Z_2 sont deux cycles de V dont l'intersection $Z_1 \cdot Z_2$ est définie, on a

$$h^{|Z_1| \cap |Z_2|}(Z_1 \cdot Z_2) = h(Z_1) \cdot h(Z_2).$$

C'est vrai lorsque Z_1 et Z_2 sont irréductibles d'après la définition de la multiplicité d'intersection (4.4); c'est donc aussi vrai par linéarité dans le cas général.

Propriétés de l'intersection.

4.7. Associativité. — Soit X, Y, Z trois sous-ensembles analytiques de dimension pure d'un espace analytique irréductible V et soit C une composante propre de $X \cap Y \cap Z$, c'est-à-dire que C contient un point simple de V et que

$$\operatorname{codim} C = \operatorname{codim} X + \operatorname{codim} Y + \operatorname{codim} Z;$$

soient C_λ les composantes de $X \cap Y$ contenant C, et C_μ celles de $Y \cap Z$ contenant C. Pour tout λ, C_λ est une composante propre de $X \cap Y$ et C une composante propre de $C_\lambda \cap Z$; pour tout μ, C_μ est une composante propre de $Y \cap Z$ et C une composante propre de $X \cap C_\mu$. En effet

$$\operatorname{codim} C \leq \operatorname{codim} C_\lambda + \operatorname{codim} Z \leq \operatorname{codim} X + \operatorname{codim} Y + \operatorname{codim} Z;$$

on a donc égalité; de même pour μ. On a la *formule d'associativité* :

$$\sum_\lambda i(X \cdot Y, C_\lambda)\, i(C_\lambda \cdot Z, C) = \sum_\mu i(X \cdot C_\mu, C)\, i(Y \cdot Z, C_\mu),$$

et la valeur commune des deux membres sera notée $i(X \cdot Y \cdot Z, C)$.

En effet, d'après 4.5, il suffit de vérifier la formule dans l'ouvert complémentaire de la réunion des points singuliers de V et de toutes les composantes de $X \cap Y$ et $Y \cap Z$ ne contenant pas C; elle découle alors immédiatement de l'associativité du produit d'intersection des classes d'homologie.

4.8. Critère de multiplicité 1. — Supposons qu'une composante propre C de l'intersection de deux sous-ensembles analytiques de dimension pure X et Y de V contienne un point qui soit simple à la fois sur V, X et Y et où les espaces tangents à X et Y sont transversaux (c'est-à-dire engendrent l'espace tangent à V). Alors $i(X \cdot Y, C) = 1$.

On peut en effet trouver un voisinage de ce point et des coordonnées locales dans U telles que $X \cap U$ et $Y \cap U$ soient représentés par des sous-espaces linéaires transversaux; le critère est alors une conséquence de **2.10** et **4.5**.

4.9. Formule de projection. — Soient V et W des espaces analytiques irréductibles, X un sous-ensemble analytique de dimension pure de V, et Y

un sous-ensemble analytique de dimension pure de $V \times W$ dont l'image $p(Y)$ par la projection $p : V \times W \to V$ soit un sous-ensemble analytique de V. On suppose que $p(Y)$ possède un ouvert épais U dont le complémentaire dans $p(Y)$ soit analytique, qui rencontre X, et tel que p soit un isomorphisme de $p^{-1}(U) \cap Y$ sur U. Soit C un sous-ensemble analytique irréductible de Y contenant un point simple de $V \times W$, rencontrant $p^{-1}(U)$ et tel que $p(C)$ soit analytique. Alors C est une composante propre de $(X \times W) \cap Y$ si et seulement si $p(C)$ est une composante propre de $X \cap p(Y)$ et l'on a dans ce cas la *formule de projection* :

$$(1) \qquad i(X \cdot p(Y), p(C)) = i((X \times W) \cdot Y, C).$$

Vu **4**.5, l'égalité à démontrer est de caractère local. On peut donc supposer V et W sans singularité et, après avoir remplacé V par le complémentaire de $p(Y) - U$, on peut admettre que p est un isomorphisme de Y sur $p(Y)$. La classe duale à $h(X \times W)$ dans $V \times W$ est l'image par $p^\star$ de la classe duale ξ à $h(X)$ dans V (*cf.* **2**.15). Or, d'après **1**.9, on a

$$(p_\star h(Y)) \cap \xi = p_\star(h(Y) \cap p^{\star}\xi),$$

ce qui, vu **1**.12, peut s'écrire

$$(p_\star h(Y)) \cdot h(X) = p_\star(h(Y)) \cdot h(X \times W).$$

Comme p applique Y homéomorphiquement sur $p(Y)$, on a

$$p_\star(h(Y)) = h(p(Y)) \qquad \text{et} \qquad p_\star(h(C)) = h(p(C)),$$

et **(1)** résulte alors de la définition du symbole i.

4.10. **Point d'intersection isolé.** — Soient X un sous-ensemble analytique dans un ouvert de $\mathbf{C}^n$, de dimension pure p, L un $(n - p)$-plan et P un point isolé de $X \cap L$. On peut alors trouver un voisinage ouvert connexe U de P tel que $U \cap X \cap L = P$, et une famille de $(n - p)$-plans L_t, parallèles à L, dépendant continûment d'un paramètre réel $t (0 \leq t \leq 1)$, tels que :

(1) $L_0 = L$;

(2) L_1 coupe transversalement X dans U ;

(3) l'intersection de $X \cap U$ avec la réunion F des L_t est un compact ;
(cela résulte aisément des théorèmes 3 et 4 de [11]). Nous voulons montrer *que $i(X \cdot L, P)$ est égal au nombre r des points d'intersections de X avec L_1 contenus dans U.*

Soit Φ la famille des fermés de $F \cap U$. Écrivons λ_0, λ_1, ξ pour $h(L_0 \cap U)$, $h(L_1 \cap U)$ et $h(X \cap U)$. Ce sont donc des classes d'homologie de U, à supports dans la famille des fermés de $L_0 \cap U$, $L_1 \cap U$ et $X \cap U$ respectivement. Soient encore $(\lambda_i \cdot \xi)^c$, l'image de $\lambda_i \cdot \xi (i = 0, 1)$ dans le $0^{\text{ième}}$ groupe d'homologie à supports compacts $H_0^c(U)$ de U. Évidemment $(\lambda_i \cdot \xi)^c = (\lambda_i^\Phi \cdot \xi)^c (i = 0, 1)$.

Mais, en vertu de l'axiome d'homotopie (*cf.* 1.5), on a $\lambda_0^\Phi = \lambda_1^\Phi$, donc $(\lambda_0.\xi)^c = (\lambda_1.\xi)^c$. Le groupe $H_0^c(U; Z)$ est cyclique infini, et engendré par un élément η qui est la classe d'homologie de tout point de U. Par suite $(\lambda_t.\xi)^c = i(X.L, P).\eta$ et $(\lambda_1.\xi)^c = r.\eta$, d'où l'égalité annoncée.

4.11. Image directe d'un cycle. — Toute application analytique f propre d'un espace analytique V dans un espace analytique W induit un homomorphisme $f_{\mathcal{Z}}$ de $\mathcal{Z}(V)$ dans $\mathcal{Z}(W)$ défini de la manière suivante. L'image de tout sous-ensemble analytique X de V est un sous-ensemble analytique fX de W et $d(fX) \leq d(X)$ (*cf.* [16], th. 20). Supposons que X soit irréductible de dimension p; alors $f_* h(X) \in H_{2p}^{fX}(W; Z)$ est zéro si $d(fX) < p$ vu 1.10(1) et un multiple δ positif de $h(fX)$ si $d(fX) = p$ vu 3.3. Par définition $f_{\mathcal{Z}}(X)$ sera le cycle zéro dans le premier cas et le cycle $\delta(fX)$ dans le second; $f_{\mathcal{Z}}$ se définit par linéarité pour tout cycle de V.

Il résulte de la définition que, pour tout cycle Z de V, on a

$$f_* h(Z) = h^{r|Z|}(f_{\mathcal{Z}}(Z)).$$

Remarquons que, pour définir $f_{\mathcal{Z}}(Z)$, il suffit de supposer que la restriction de f au support de Z est propre.

4.12. Image inverse d'un cycle. — Soit f une application analytique d'un espace analytique V de dimension pure n dans un espace analytique W de dimension pure m. Soit F le graphe de f dans $V \times W$ et soit $\mathcal{Z}_f(W)$ le sous-groupe de $\mathcal{Z}(W)$ engendré par les sous-ensembles irréductibles X de W tels que $(V \times X).F$ soit défini; soit alors $f_{\mathcal{Z}}^{-1} X$ le cycle de V image du cycle $(V \times X).F$ par l'isomorphisme canonique de F sur V; c'est un cycle homogène positif de codimension égale à celle de X et de support $f^{-1}(X)$. Par linéarité on définit alors un homomorphisme $f_{\mathcal{Z}}^{-1}$ de $\mathcal{Z}_f(W)$ dans $\mathcal{Z}(V)$.

Lorsque V et W sont des variétés analytiques, on a pour **tout** cycle $Z \in \mathcal{Z}_f(W)$,

$$(1) \qquad \Delta h(f_{\mathcal{Z}}^{-1} Z) = f^* \Delta h(Z).$$

Soient en effet p et q les projections de $V \times W$ sur V et W respectivement et f^0 l'application $v \to (v, f(v))$ de V sur F. D'après 2.15, on a $\Delta h(q_{\mathcal{Z}}^{-1} Z) = q^* \Delta h(Z)$; comme $h(F) = f_*^0([V])$, la définition de $f_{\mathcal{Z}}^{-1} Z$ et 1.9 donnent alors

$$p_* f_*^0 h(f_{\mathcal{Z}}^{-1} Z) = p_*(f_*^0([V]) \cap q^* \Delta h(Z)) = p_* f_*^0([V]) \cap f^{0*} q^* \Delta h(Z)),$$

d'où, puisque $p \circ f^0$ est l'identité et que $f = q \circ f^0$.

$$h(f_{\mathcal{Z}}^{-1} Z) = [V] \cap f^* \Delta h(Z),$$

ce qui équivaut à (1).

4.13. Classe d'homologie d'un diviseur. — Soit D un diviseur positif sur une variété complexe V; D peut être défini par un ensemble $(f_i)_{i \in I}$ de fonctions holomorphes f_i définies sur les ouverts U_i d'un recouvrement de V, telles que sur $U_i \cap U_j$, $i, j \in I$, $f_{ij} = f_i/f_j$ soit une fonction holomorphe ne prenant pas la valeur zéro. Les f_{ij} sont les fonctions de transition d'un fibré vectoriel $[D]$ de rang 1 sur V : $[D]$ est le quotient de l'espace somme $\sum_i U_i \times \mathbf{C}$ par la relation d'équivalence qui identifie $(x_i, y_i) \in U_i \times \mathbf{C}$ avec $(x_j, y_j) \in U_j \times \mathbf{C}$ si $x_i = x_j$ et $y_i = f_{ij} y_j$. Les graphes $(x, f_i(x))$ des fonctions f_i se recollent dans $[D]$ et définissent une section holomorphe f de $[D]$. D'autre part les équations locales $f_i = 0$ définissent un sous-ensemble analytique $|D|$ de codimension 1 ; $|D|$ est le support d'un cycle analytique Z_D : si x est un point simple de $|D|$, la multiplicité de la composante irréductible de $|D|$ contenant x est l'ordre du zéro d'une fonction $f_i \in D$ en x.

Proposition. — *La classe d'homologie de V représentée par le cycle Z_D d'un diviseur D est duale à la classe de Chern du fibré vectoriel $[D]$ associé à D.*

Soit en effet s la section nulle de $[D]$ et posons $V_0 = s(V)$. La classe de Chern c de $[D]$, définie comme la première obstruction à construire une section de $[D]$ partout différente de zéro est aussi l'image par s^* de la classe duale à V_0 dans $[D]$: $c = s^* \Delta h(V_0)$ (*cf.* [21]). Comme f est homotope à s, on a aussi $c = f^* \Delta h(V_0)$. D'après 4.12, il suffit de vérifier que $Z_D = f_{\mathscr{Z}}^{-1}(V_0)$, ce qui résulte de la définition de Z_D.

4.14. Équivalence analytique. — Soient Z_1 et Z_2 deux cycles analytiques sur une variété analytique V. On dira que Z_1 et Z_2 sont analytiquement équivalents dans V s'il existe une variété analytique connexe W, un cycle analytique Z dans $V \times W$ et deux points w_1, $w_2 \in W$ tels que les cycles $(V \times w_1).Z$ et $(V \times w_2).Z$ soient définis et aient Z_1 et Z_2 pour images par la projection naturelle de $V \times W$ sur V.

Proposition. — *Soient Z_1, Z_2 deux cycles analytiquement équivalents de la variété analytique V. Alors $h^V(Z_1) = h^V(Z_2)$.*

Soient p et q les projections de $V \times W$ sur V et W respectivement et soit Φ la famille des fermés de $V \times W$ auxquels la restriction de p est propre. La projection q induit un homomorphisme q^* de la cohomologie à supports compacts de W dans la Φ-cohomologie de $V \times W$. Vu 2.15, les classes de cohomologie duales de $h^\Phi(V \times w_1)$ et $h^\Phi(V \times w_2)$ sont les images par q^* des classes duales à w_1 et w_2 dans W. Elles sont donc égales, ce qui entraîne

$$p_*(h^\Phi(V \times w_1).h^\Phi(Z)) = p_*(h^\Phi(V \times w_2).h^\Phi(Z))$$

et la proposition résulte de 4.6 et 4.11.

4.15. Cas de la géométrie algébrique. — Soit V une variété algébrique complexe. Si X et Y sont des sous-ensembles algébriques, $i(X.Y, C)$ a la même valeur que le coefficient d'intersection de la théorie des intersections en géométrie algébrique car ce dernier est complètemeut caractérisé par les propriétés 4.7 à 4.10 ([23], appendix III)·

Supposons V non singulière. Soit $A(V)$ le groupe des classes d'équivalence rationnelle de cycles algébriques de V. Comme deux cycles rationnellement équivalents sont aussi analytiquement équivalents, 4.2 et 4.14 montrent que $Z \to h^V(Z)$ induit un homomorphisme $h_*: A(V) \to H_*(V; \mathbf{Z})$. Un morphisme propre de V dans une variété algébrique non singuliére W induit un homomorphisme $f_*: A(V) \to A(W)$ et l'on a, $h_* f_* = f_* h_*$ vu 4.11.

Si V est non singulière et quasi projective, $A(V)$ est un anneau (l'anneau de Chow de V) et, vu 4.6, h_* est un homomorphisme d'anneaux. Un morphisme f de V dans une variété quasi projective non singulière W induit un homomorphisme $f_{\overline{\partial}}^{-1}: A(W) \to A(V)$ et 4.12 implique qu'on a $f^0 h_* = h_* f_{\overline{\partial}}^{-1}$, où $f^0 = \Delta^{-1} f^* \Delta$ est l'homomorphisme inverse (Umkehrungshomomorphismus) de Hopf.

4.16. Cas de la géométrie analytique. — On peut aussi, suivant le modèle de la géométrie algébrique, faire une théorie de l'intersection des cycles analytiques d'un espace analytique, et montrer que le symbole d'intersection $i(X.Y, C)$ est complètement caractérisé par les propriétés 4.7 à 4.10. Comme nous n'en ferons pas usage, nous nous contentons d'énoncer ce fait « bien connu », pour lequel nous ne pouvons malheureusement pas fournir de référence. Remarquons simplement que la démonstration consiste essentiellement à faire voir que tout symbole $i(X.Y, C)$ est égal à une multiplicité d'intersection ponctuelle du type considéré en 4.10, ce qui montre en particulier que $i(X.Y, C)$ *est toujours un entier* > 0; et que les principaux pas de cette réduction seront répétés dans la démonstration de 5.3.

5. Intersection de sous-ensembles analytiques réels.

5.1. — Soit V une *complexification* d'une variété analytique réelle V^0 de dimension n. On entend par là que V est une variété analytique complexe de dimension complexe n, dont V^0 est une sous-variété analytique réelle, ayant les propriétés suivantes :

(i) Pour tout point $x^0 \in V^0$, on peut trouver des coordonnées locales $z^1, \ldots, z^n$ dans un voisinage U de x^0 telles que $V^0 \cap U$ soit défini par l'annulation des parties imaginaires des z^l; un tel système de coordonnées sera appelé réel.

(ii) V^0 est l'ensemble des points fixes d'une involution J de V qui est antiholomorphe, c'est-à-dire telle que si f est une fonction holomorphe dans un ouvert W, alors la fonction conjuguée de $f(Jw)(w \in W)$ est holomorphe dans JW.

Soit X un sous-ensemble analytique de V. Alors $\bar{X} = JX$ est aussi un sous-ensemble analytique, irréductible si X l'est. La partie réelle X^0 de X est le sous-ensemble $X \cap V^0$ de V^0, et est égale à la partie réelle de $\bar{X}$. D'après **3.5**, $\dim_{\mathbf{Z}} X^0 \leqq d(X)$.

Supposons X *irréductible*. Si $X \neq \bar{X}$; alors $d(X \cap \bar{X}) < d(X)$, donc

$$\dim_{\mathbf{Z}} X^0 = \dim_{\mathbf{Z}} (X \cap \bar{X})^0 < d(X).$$

Si X est de dimension pure, est égal à $\bar{X}$, et $\dim_{\mathbf{Z}}(X^0) = d(X)$, alors l'ensemble des points de X^0 qui sont simples sur X forment un ensemble épais dans X^0, et sont simples sur X^0. En effet, soit S le sous-ensemble analytique des points singuliers de X. Alors $d(S) < d(X)$, donc $\dim_{\mathbf{Z}} S^0 < \dim_{\mathbf{Z}} X^0$, ce qui montre que X^0 contient au moins un point simple de X. Notre assertion résulte donc du fait que si $x \in X^0$ est simple sur X, alors X^0 est au voisinage de x une variété analytique réelle de dimension égale à $d(X)$ ([**7**]).

5.2. — Soit ρ l'application qui associe à un ensemble analytique irréductible X de V l'image dans $H_p^{X^0}(V; \mathbf{Z}_2)$ de la classe fondamentale de X^0 si $p = \dim_{\mathbf{Z}} X^0 = d(X)$, l'élément nul de $H_*^{X^0}(V; \mathbf{Z}_2)$ si $\dim_{\mathbf{Z}} X^0 < d(X)$. L'application ρ s'étend par linéarité au groupe $\mathcal{Z}(V)$ des cycles de V : au cycle $Z = \sum_i n_i X_i$ de composantes X_i et de support $|X| = \bigcup_i X_i$, on associe la somme localement finie $\sum n_i \rho(X_i)$ (**1.7**), qui est un élément de $H_*^{X^0}(V; \mathbf{Z}_2)$. Vu **5.1**, cette application est nulle sur les ensembles irréductibles non invariants par J. Nous en considérerons sa restriction au groupe $\mathcal{Z}_{\mathbf{R}}(V)$ des cycles analytiques de V qui sont invariants par J. De même qu'en **4.2**, on voit que, Φ étant une famille de supports sur V, ρ est un homomorphisme (de groupes commutatifs) dans $H_*^{\Phi \cap V^0}(V^0; \mathbf{Z}_2)$ du sous-groupe des éléments de $\mathcal{Z}_{\mathbf{R}}(V)$ ayant leur support dans Φ.

5.3. Théorème. — *Soient X, Y deux cycles analytiques de V invariants par J dont l'intersection est définie. Alors*

$$(\mathrm{I}) \qquad \rho^{|X \cap Y|^0}(X.Y) = \rho(X).\rho(Y).$$

Montrons tout d'abord que les deux membres de (I) sont nuls si $X = Z + \bar{Z}$ avec Z irréductible, $Z \neq \bar{Z}$. On a $\rho(X) = 0$ d'après **5.1** et la définition de ρ, donc

$$\rho(X).\rho(Y) = 0.$$

D'autre part

$$\rho(X.Y) = \rho(Z + \bar{Z}.Y) = \rho(Z.Y + \overline{Z.Y}) = 2\rho(Z.Y) = 0.$$

Comme $\mathcal{Z}_{\mathbf{R}}(V)$ est engendré par les sous-ensembles analytiques irréductibles invariants par J et par les cycles $Z + \overline{Z}$, où Z est irréductible, on pourra se borner au cas où X et Y sont invariants par J et irreductibles. $X \cap Y$ est donc de dimension pure $s = a + b - n$ $[a = d(X), b = d(Y), n = d(V)]$. Supposons tout d'abord que $\dim_{\mathbf{Z}}(X \cap Y)^0 < s$. Alors on a $\dim_{\mathbf{Z}} C_i^0 < s$ pour chaque composante irréductible C_i de $X \cdot Y$, donc $\rho(X \cdot Y) = 0$ par définition. Comme $\rho(X) \in H_a(V^0; \mathbf{Z}_2)$, $\rho(Y) \in H_b(V^0; \mathbf{Z}_2)$, le deuxième membre est aussi nul d'après **1.12**. Compte tenu de **5.1**, nous sommes donc ramenés au cas où X et Y sont irréductibles, invariants par J, où $\dim_{\mathbf{Z}}(X \cap Y)^0 = d(X \cap Y) = s$, et donc où les points réels de $X \cap Y$ qui sont simples sur $X \cap Y$ forment un ensemble épais de $(X \cap Y)^0$.

5.4. — Montrons qu'il suffit de vérifier (1) au voisinage de chaque point réel simple de $X \cap Y$. Supposons donc que pour tout point réel P de $X \cap Y$ il existe un voisinage U tel que

$$\rho((X \cap U) \cdot (Y \cap U)) = \rho(X \cap U) \cdot \rho(Y \cap U).$$

Soit J_U l'homomorphisme de restriction $H_*^{\Phi}(V^0; \mathbf{Z}_2) \to H_*^{\Phi \cap U^0}(U^0; \mathbf{Z}_2)$. Pour tout cycle Z, on a $\rho(Z \cap U) = j_U \rho(Z)$ (pour Φ convenable), d'où

$$j_U \rho(X \cdot Y) = \rho((X \cap U) \cdot (Y \cap U)) = \rho(X \cap U) \cdot \rho(Y \cap U)$$
$$= j_U \rho(X) \cdot j_U \rho(Y) = j_U(\rho(X) \cdot \rho(Y)).$$

Ainsi, $\rho(X \cdot Y)$ et $\rho(X) \cdot \rho(Y)$, vus comme éléments de $H_s((X \cap Y)^0; \mathbf{Z}_2)$ par l'isomorphisme de **1.5**, induisent la même classe d'homologie locale en tout point d'un ouvert épais de $(X \cap Y)^0$. Vu **1.6**, **1.10**(1) et **1.10**(2), ils coïncident.

Nous allons tout d'abord vérifier (1) dans trois cas particuliers au voisinage d'un point réel simple $P \in X \cap Y$.

5.5. — Supposons que X et Y se coupent transversalement en P. En se restreignant à un voisinage U assez petit de P invariant par J et en introduisant des coordonnées locales réelles convenables dans U, on est ramené au cas où X et Y sont les intersections d'un ouvert U' de $\mathbf{C}^n$ avec des sous-espaces affines, définis par des équations réelles, se coupant transversalement au point réel P, donc aussi en tout autre point de $C = X \cap Y$. On a alors $X \cdot Y = C$ d'après **4.8**. D'autre part, X^0 et Y^0 sont des sous-espaces affines transversaux de $\mathbf{R}^n$ car ils sont définis par les mêmes équations que X et Y, d'où $\rho(X) \cdot \rho(Y) = \rho(C)$, vu **2.10**, ce qui établit (1) au voisinage de P.

5.6. — Supposons que P soit un point isolé de $X \cap Y$, simple sur $X \cap Y$ et sur Y. A l'aide d'une carte locale réelle, on peut se ramener au cas où X est un ensemble analytique de dimension p dans un ouvert connexe U de $\mathbf{C}^n$, $U \cap \mathbf{R}^n = U^0$ étant aussi connexe, et où Y est l'insersection de U avec un sous-espace linéaire L de dimension n-p coupant X au seul point réel P, X et L

étant définis par des équations réelles. On peut trouver une famille de $(n-p)$-plans L_l définis par des équations réelles jouissant des mêmes propriétés que dans 4.10.

Comme dans 4.10, toutes les classes d'homologie de dimension o étant considérées à support les compacts de U^0, on montre que

$$\rho(X).\rho(X) = \rho(X).\rho(L_1).$$

Comme L_1 coupe transversalement X, 5.3 et 5.4 entraînent

$$\rho(X.L_1) = \rho(X).\rho(L_1).$$

ll reste donc à vérifier que $\rho(X.L) = \rho(X.L_1)$. Comme $X.L = i(X.L, P) P$, $\rho(X.L)$ est égal à $i(X.L, P)$ fois le générateur de $H_0^c(U^0; \mathbf{Z}_2)$; d'autre part (4.8), $\rho(X.L_1)$ est égal à r^0 fois le générateur de $H_0^c(U^0; \mathbf{Z}_2)$, où r^0 est le nombre des points réels de $X \cap L_1$; or r^0 est congru mod 2 au nombre r des points de $X \cap L_1$, car les points non réels sont conjugués deux à deux. D'après 4.10, $i(X.L, P)$ est égal à r, donc congru mod 2 à r^0.

5.7. — Supposons que P soit un point réel simple de Y et $X \cap Y$. A l'aide de coordonnées locales convenables au voisinage de P, on se ramène au cas où Y et $X \cap Y = C$ sont des sous-espaces linéaires dans un ouvert de $\mathbf{C}^n$ (X et Y étant définis par des équations réelles et $P \in \mathbf{R}^n$). Soit alors L un plan de $\mathbf{C}^n$ défini par des équations réelles, de codimension égale à la dimension s de C et coupant transversalement C en P. L'élément $\rho(X).\rho(Y)$ de $H_s^{C^0}(\mathbf{R}^n; \mathbf{Z}_2)$ est un multiple k du générateur $\rho(C)$. De $\rho(X).\rho(Y) = k\rho(C)$, on déduit

$$\rho(X).\rho(Y).\rho(L) = k\rho(C).\rho(L)$$

et d'après 5.5,

$$\rho(X).\rho(Y.L) = k\rho(C.L).$$

En vertu de 5.6, on a

$$\rho(X.Y.L) = k\rho(P) \qquad \text{ou} \qquad i(X.Y.L, P)\rho(P) = k\rho(P);$$

or

$$i(X.Y.L, P) = i(X.Y, C)i(C.L, P) = i(X.Y, C)$$

en vertu de 4.7, 4.8, d'où $i(X.Y, C) = k \bmod 2$. Donc

$$\rho(X.Y) = \rho(X).\rho(Y).$$

En vertu de 5.4, le théorème est démontré si Y est non singulier.

5.8. — Nous passons maintenant au cas mentionné à la fin de 5.3.

Le produit $V \times V$ et une complexification de $V^0 \times V^0$ et la diagonale D de $V \times V$ coupe $V^0 \times V^0$ suivant la diagonale. Soient p_1 et p_2 les projections de $V^0 \times V^0$ sur le premier et le deuxième facteur. Nous allons montrer d'abord que

$$p_{1*}[\rho(X \times Y).\rho(D)] = \rho(X).\rho(Y).$$

Les deux cycles $X \times V$ et $V \times Y$ se coupent transversalement suivant $X \times Y$; d'après 5.5, $\rho(X \times Y) = \rho(X \times V).\rho(V \times Y)$. D'autre part, Δ désignant l'isomorphisme de dualité, on a (*cf.* **2.15**)

$$\Delta\rho(X \times V) = p_1^* \Delta\rho(X) \qquad \text{et} \qquad \Delta\rho(V \times X) = p_2^* \Delta\rho(Y).$$

Donc

$$p_{1*}[\rho(X \times Y).\rho(D)] = p_{1*}[\rho(X \times V).\rho(V \times Y).\rho(D)]$$
$$= p_{1*}[\rho(D) \cap (p_1^* \Delta\rho(X) \cup p_2^* \Delta\rho(Y))].$$

Si d est l'application diagonale $V^0 \to V^0 \times V^0$, on sait que

$$d^*[p_1^* \Delta\rho(X) \cup p_2^* \Delta\rho(Y)] = \Delta\rho(X) \cup \Delta\rho(Y);$$

de plus $\rho(D)$ est l'image par d_* de la classe fondamentale $[V^0]$ de V^0; en appliquant la formule (1) de **1.9**, on a donc

$$p_{1*}[\rho(X \times Y).\rho(D)] = p_{1*} d_*[[V^0] \cap (\Delta\rho(X) \cup \Delta\rho(Y))],$$

et comme $p_1 \circ d$ est l'identité, ceci donne $\rho(X).\rho(Y)$.

D'autre part (**5.7**) : $\rho(X \times Y).\rho(D) = \rho[(X \times Y).D]$. Or on a

$$(1) \qquad\qquad (X \times Y).D = \sum_\lambda i(X.Y, C_\lambda) C_\lambda^D,$$

où les C_λ sont les composantes irréductibles de $X \cap Y$ et les C_λ^D leurs images par l'application diagonale; de plus $p_{1*}\rho(C_\lambda^D) = \rho(C_\lambda)$. Donc

$$p_{1*}[\rho(X \times Y).\rho(D)] = \sum_\lambda i(X.Y, C_\lambda)\rho(C_\lambda) = \rho(X.Y),$$

C. Q. F. D.

Rappelons comment se démontre la formule (1). Tout d'abord par **4.8**,

$$X \times Y = (X \times V).(V \times Y).$$

Ainsi, vu **4.7** et **4.8**,

$$i[(X \times V).D, C_\lambda^D] = i[(X \times V).(V \times Y).D, C_\lambda^D]$$
$$= i[(X \times V).Y^D, C_\lambda^D] \, i[(V \times Y).D, Y^D];$$

or $i[(V \times Y).D.Y^D] = 1$ par **4.8** et $i[(X \times V).Y^D, C_\lambda^D] = i(X.Y, C_\lambda)$ par **4.9**, ce qui démontre (1).

5.9. Remarque. — Il résulte de ce théorème que si X et $Y \in \mathcal{J}_{\mathbf{R}}(V)$ sont irréductibles et si $\dim_{\mathbf{Z}}(X^0) < d(X)$, alors toute composante propre C de $X \cap Y$ telle que $\dim_{\mathbf{Z}}(C^0) = d(C)$ a une multiplicité paire. En effet, on peut supposer que $X.Y$ est défini (en enlevant au besoin les composantes non propres); d'après le théorème; $\rho(X).\rho(Y) = \rho(X.Y)$, et comme $\rho(X) = 0$, on a

$$\rho(X.Y) = 0,$$

d'où notre assertion.

5.10. — Soit $\mathbf{P}_n(\mathbf{C})$ l'espace projectif complexe de dimension n considéré comme une complexification de l'espace projectif réel $\mathbf{P}_n(\mathbf{R})$. Soit V une sous-variété algébrique localement fermée non singulière de $\mathbf{P}_n(\mathbf{C})$ qui soit une complexification de sa partie réelle $V^0 = V \cap \mathbf{P}_n(\mathbf{R})$; cela signifie que l'adhérence $\bar{V}$ de V coïncide avec l'adhérence $\bar{V}^0$ de V^0 au sens de la topologie de Zariski de $\mathbf{P}_n(\mathbf{C})$ et que $\bar{V} - V$ est un ensemble algébrique défini sur $\mathbf{R}$ [en d'autres termes, tout polynôme qui s'annule sur V^0 s'annule aussi sur V, $\bar{V} - V$ est invariant par l'involution $z \to \bar{z}$ de $\mathbf{P}_n(\mathbf{C})$].

Une *sous-variété* X de V *définie* sur $\mathbf{R}$ est un sous-ensemble algébrique de V défini par des équations réelles et qui est irréductible sur $\mathbf{R}$.

Les *cycles algébriques de V définis sur* $\mathbf{R}$ entiers (ou mod 2) sont les combinaisons linéaires formelles à coefficients entiers (ou mod 2) d'un nombre fini de sous-variétés de V définies sur $\mathbf{R}$. Ils forment un groupe abélien noté $\mathcal{J}_{\mathbf{R}}(V)$ [ou $\mathcal{J}_{\mathbf{R}}(V; \mathbf{Z}_2)$].

Une *sous-variété réelle* X de V^0 est un sous-ensemble de X de V^0 tel que son adhérence $\bar{X}$ dans V (au sens de la topologie de Zariski) est irréductible et que $\bar{X} \cap V^0 = X$; on dira que $\bar{X}$ est la *complexification* de X. On remarquera que X est un espace de type VS_p, où $p = d(X)$. Les *cycles algébriques réels de* V^0 sont les combinaisons linéaires formelles à coefficients les entiers mod 2 d'un nombre fini de sous-variétés réelles de V^0. Ils forment un groupe abélien noté $\mathcal{J}(V^0; \mathbf{Z}_2)$.

L'application faisant correspondre à toute sous-variété réelle de V^0 sa complexification induit un homomorphisme γ de $\mathcal{J}(V^0; \mathbf{Z}_2)$ dans $\mathcal{J}_{\mathbf{R}}(V; \mathbf{Z}_2)$. Inversement l'application faisant correspondre à toute sous-variété X de V définie sur $\mathbf{R}$ le cycle réel $X^0 = X \cap V^0$ si X est la complexification de X^0 [ou ce qui revient au même si $\dim_{\mathbf{Z}} X^0 = d(X)$], et zéro dans le cas contraire, induit un homomorphisme α de $\mathcal{J}_{\mathbf{R}}(V; \mathbf{Z}_2)$ sur $\mathcal{J}(V^0; \mathbf{Z}_2)$. Il est clair que $\alpha \circ \gamma$ est l'identité.

5.11. Intersection des cycles algébriques réels. — On dira qu'une composante irréductible C de l'intersection de deux sous-variétés réelles X de Y de V^0 est propre si toutes les composantes irréductibles de $\gamma(X) \cap \gamma(Y)$ contenant C sont propres. Ceci implique en particulier que

$$\dim_{\mathbf{Z}} C \leq \dim_{\mathbf{Z}} X + \dim_{\mathbf{Z}} Y - n.$$

La *multiplicité* $i(X.Y, C)$ *de la composante propre* C de $X \cap Y$ sera par définition l'entier mod 2 congru à $i(\gamma(X).\gamma(Y), \gamma(C))$ si $\gamma(C)$ est une composante propre de $\gamma(X) \cap \gamma(Y)$, et zéro dans le cas contraire. Si toutes les composantes C_λ de $X \cap Y$ sont propres, l'intersection des cycles X et Y sera par définition le cycle réel $X.Y = \sum_\lambda i(X.Y, C_\lambda)\, C_\lambda$; l'intersection de deux cycles réels, lorsqu'elle a un sens, se définit par linéarité.

PROPOSITION. — *L'homomorphisme* $\alpha : \mathfrak{Z}_{\mathbf{R}}(V; \mathbf{Z}_2) \to \mathfrak{Z}(V^0; \mathbf{Z}_2)$ *est multiplicatif*: *si* X *et* $Y \in \mathfrak{Z}_{\mathbf{R}}(V; \mathbf{Z}_2)$ *sont tels que* $X.Y$ *soit défini, alors* $\alpha(X).\alpha(Y)$ *est défini et* $\alpha(X.Y) = \alpha(X).\alpha)Y)$.

La proposition découle immédiatement de la définition de $X.Y$ si X et Y sont les complexifications de leur partie réelle. Lorsque X et Y sont irréductibles et que X n'est pas la complexification de sa partie réelle X^0, alors $\dim_{\mathbf{Z}} X^0 < d(X)$ et $\alpha(X) = 0$; d'autre part toute composante C de $X \cap Y$ qui est la complexification de sa partie réelle C^0 [donc telle que $\dim_{\mathbf{Z}} C^0 = d(C)$] a une multiplicité paire d'après 5.9 (on pourrait aussi le vérifier directement en se ramenant à des intersections isolées); donc $0 = \alpha(X.Y) = \alpha(X).\alpha(Y)$.

La théorie de l'intersection des cycles réels peut se développer comme celle des cycles algébriques complexes; on y démontre la formule d'associativité (conséquence du fait que α est multiplicatif), le critère de multiplicité 1, etc.

REMARQUE : — On peut construire sur ce même modèle une théorie de l'intersection mod 2 des ensembles C-analytiques d'une variété analytique réelle V^0 (au sens de [3]); les germes d'ensembles analytiques complexes au voisinage de V^0 dans une complexification de V joueront le même rôle que les cycles algébriques complexes.

5.12. Classes d'homologie représentées par des cycles algébriques. — Soit r l'homomorphisme $\mathfrak{Z}(V^0; \mathbf{Z}_2)$ dans $H_*(V^0; \mathbf{Z}_2)$ faisant correspondre à toute sous-variété réelle X de V^0 de dimension p la classe d'homologie mod 2 de V^0 qu'elle représente : $r(X)$ est l'image par l'homomorphisme induit par l'inclusion de X dans V^0 de la classe fondamentale mod 2 de X.

Désignons encore par h (resp. ρ) les homomorphismes de $\mathfrak{Z}_{\mathbf{R}}(V; \mathbf{Z}_2)$ dans $H_*(V; \mathbf{Z}_2)$ [resp. $H_*(V^0; \mathbf{Z}_2)$] faisant correspondre à toute sous-variété X de V définie sur $\mathbf{R}$ la classe d'homologie $h(X)$ (*cf.* 4.2), réduite mod 2, et à support tous les fermés de V [resp. la classe $\rho(X)$ (*cf.* 5.2) à support les fermés de V^0].

On a le diagramme commutatif suivant :

$$
\begin{array}{ccc}
\mathfrak{Z}_{\mathbf{R}}(V\ ; \mathbf{Z}_2) & \overset{h}{\to} & H_*(V\ ; \mathbf{Z}_2) \\
\alpha \downarrow \quad \uparrow \gamma & \searrow \rho & \\
\mathfrak{Z}(V^0; \mathbf{Z}_2) & \overset{r}{\to} & H_*(V^0; \mathbf{Z}_2)
\end{array}
$$

A part γ, tous les homomorphismes considérés sont multiplicatifs (*cf.* 4.6, 5.3 et 5.11).

Soit $A_{\mathbf{R}}(V)$ le groupe des classes d'équivalence rationnelle sur $\mathbf{R}$ de cycles de V définis sur $\mathbf{R}$. Rappelons que deux cycles Z_1 et $Z_2 \in \mathcal{Z}_{\mathbf{R}}(V)$ sont dits rationnellement équivalents sur $\mathbf{R}$ s'il existe un cycle Z défini sur $\mathbf{R}$ dans $V \times \mathbf{C}$ tels que $(V \times \mathrm{1}).Z$ et $(V \times \mathrm{o}).Z$ soient définis et que leurs images par la projection naturelle sur V soient égales respectivement à $Z_1 - Z_2$ et à o.

L'homomorphisme h est compatible avec la relation d'équivalence rationnelle (*cf.* 4.15); il en est de même pour ρ (et donc pour r) :

5.13. **Proposition** : — *Soient Z_1 et Z_2 deux cycles de V rationnellement équivalents sur* $\mathbf{R}$. *Alors les classes d'homologie* mod 2, $\rho(Z_1)$ *et* $\rho(Z_2)$ *de* V^0 (*à support les fermés de* V^0) *sont égales.*

La démonstration est analogue à celle de 4.14. Soient p et q les projections de $V^0 \times \mathbf{R}$ sur V^0 et $\mathbf{R}$ respectivement ($\mathbf{R}$ étant canoniquement plongé dans $\mathbf{C}$) et soit Φ la famille des fermés de $V^0 \times \mathbf{R}$ tels que la restriction de p à chacun d'eux soit propre. Les classes $\rho^{\Phi}(V \times \mathrm{1})$ et $\rho^{\Phi}(V \times \mathrm{o})$ sont égales car elles sont duales aux images par q^* des classes duales aux points $\mathrm{1}$ et o dans $\mathbf{R}(2.15)$. Comme

$$\rho^{\Phi}((V \times i).Z) = \rho^{\Phi}((V \times i)).\rho(Z) \qquad (i = \mathrm{1}, \mathrm{o})$$

d'après 5.3, il résulte que

$$p_* \rho^{\Phi}((Z_1 - Z_2) \times \mathrm{1}) = \rho(Z_1 - Z_2) = \mathrm{o}.$$

Chow a montré [9] qu'étant donnés deux cycles Z_1 et $Z_2 \in \mathcal{Z}_{\mathbf{R}}(V; \mathbf{Z}_2)$, il est toujours possible de trouver un cycle Z_2' rationnellement équivalent à Z_2 sur $\mathbf{R}$ tel que $Z_1.Z_2'$ soit défini. Il en résulte en particulier que les images de h, ρ ou r sont des sous-anneaux de $H_*(V; \mathbf{Z}_2)$ et $H_*(V^0; \mathbf{Z}_2)$ respectivement.

5.14. **Proposition.**

(1) *Supposons que toute classe d'homologie* mod 2 *de dimension paire de V, à support compact, puisse être représentée par un cycle algébrique défini sur* $\mathbf{R}$ (*à support compact*). *Alors h s'annule sur le noyau de* ρ.

(2) *V^0 étant connexe, supposons que toute classe d'homologie* mod 2 *de V^0, à support compact, puisse être représentée par un cycle algébrique réel* (*à support compact*). *Alors ρ s'annule sur le noyau de* h.

Démonstration.

(1) Soit $Z \in \mathcal{Z}_{\mathbf{R}}(V; \mathbf{Z}_2)$ un cycle homogène de dimension p tel que $\rho(Z) = \mathrm{o}$; pour montrer que $h(Z) = \mathrm{o}$, il suffit de vérifier que l'intersection de $h(Z)$ avec toute classe d'homologie à supports compacts, de dimension

$2n - 2p$, de V est nulle. Soient $C \in \mathfrak{Z}_{\mathbf{R}}(V; \mathbf{Z}_2)$ un cycle à support compact homogène de dimension $2n - 2p$ et $h(C)$ la classe d'homologie à supports compacts qu'il représente. D'après Chow [9], il existe un cycle Z' rationnellement équivalent à Z sur $\mathbf{R}$ tel que $Z'.C$ soit défini. On a $h(Z) = h(Z')$ (*cf.* 4.15), il suffit donc de prouver que $h(Z').h(C) = 0$. Or

$$\rho(Z').\rho(C) = \rho(Z'.C),$$

les deux membres étant des classes à supports compacts (*cf.* 5.3); comme $0 = \rho(Z) = \rho(Z')$ (*cf.* 5.13), on a $\rho(Z'.C) = 0$. Cela signifie que le nombre des points réels de $Z'.C$, comptés avec leur multiplicité, est pair; puisque $Z'.C$ est invariant par l'involution, il s'ensuit qu'il est formé d'un nombre pair de points; comme V est connexe, cela entraîne que $h(Z'.C) = 0$, d'où $h(Z').h(C) = 0$.

(2) Soit $Z \in \mathfrak{Z}_{\mathbf{R}}(V; \mathbf{Z}_2)$ un cycle homogène de dimension p tel que $h(Z) = 0$; soit C^0 un cycle réel compact de dimension $n - p$ et soit C sa complexification. Il faut prouver que $\rho(Z).\rho(C) = 0$. Il est possible de trouver un cycle Z' rationnellement équivalent à Z sur $\mathbf{R}$ tel que $\mathbf{Z}'.C$ soit défini. Comme $h(\mathbf{Z}') = h(\mathbf{Z}) = 0$, on a

$$h(Z'.C) = h(Z').h(C) = 0,$$

les deux membres étant à support compact. Il en résulte comme dans (1) que $\rho(Z'.C) = 0$ (on a supposé V^0 connexe); donc

$$0 = \rho(Z'.C) = \rho(Z').\rho(C) = \rho(Z).\rho(C).$$

5.15. Conséquences. — La conclusion de (1) signifie qu'en faisant correspondre à la classe d'homologie représentée par un cycle réel de V^0 la classe d'homologie mod 2 représentée par sa complexification, on définit un homomorphisme multiplicatif de $r[\mathfrak{Z}_*(V^0; \mathbf{Z}_2)]$ dans $H_*(V; \mathbf{Z}_2)$ doublant les dimensions.

La conclusion de (2) signifie qu'en faisant correspondre à la classe d'homologie mod 2 représentée par un cycle Z de V défini sur $\mathbf{R}$ la classe d'homologie représentée par sa partie réelle $\alpha(Z)$, on définit un homomorphisme multiplicatif de $h_*(\mathfrak{Z}_{\mathbf{R}}(V; \mathbf{Z}_2))$ dans $H_*(V^0; \mathbf{Z}_2)$ divisant les dimensions par deux.

En particulier, on a la

PROPOSITION. — *Supposons que toute classe d'homologie mod 2 de V (resp. V^0), à support compact ou non, peut être représentée par un cycle algébrique de V défini sur $\mathbf{R}$ (resp. un cycle réel de V^0). Alors l'application faisant correspondre à un cycle réel de V^0 sa complexification induit un isomorphisme d'anneaux de $H_*(V^0, \mathbf{Z}_2)$ sur $H_*(V, \mathbf{Z}_2)$ doublant les dimensions.*

5.16. Exemple des variétés grassmaniennes. — Soit $\widetilde{G}_{n,N}$ la grassmanienne complexe des n-plans passant par l'origine de $\mathbf{C}^N$; c'est une variété algébrique définie sur $\mathbf{R}$, sa partie réelle s'identifiant à la grassmanienne $G_{n,N}$ des n-plans de $\mathbf{R}^N$. L'application faisant correspondre à un n-plan ses coordonnées plückériennes est un plongement défini sur $\mathbf{R}$ de $\widetilde{G}_{n,N}$ dans un espace projectif.

Il est bien connu (*cf.* par exemple [8]) que toute classe d'homologie mod 2 de $\widetilde{G}_{n,N}$ ou de $G_{n,N}$ peut être représentée par une combinaison linéaire de variétés de Schubert définies sur $\mathbf{R}$. Les hypothèses de la proposition précédente sont donc vérifiées. De plus la classe de Stiefel-Whitney universelle $w_i \in H^i(G_{n,N}; \mathbf{Z}_2)$ et la classe de Chern universelle $c_i \in H^{2i}(\widetilde{G}_{n,N}; \mathbf{Z})$ sont duales à des variétés de Schubert représentées par le même symbole, ce qui prouve que la seconde est la complexification de la première.

On a donc un isomorphisme d'anneaux de $H_*(G_{n,N}; Z_2)$ sur $H_*(\widetilde{G}_{n,N}; Z_2)$ qui envoie la classe duale à W_i sur la classe duale à C_i réduite mod 2.

5.17. Supposons les hypothèses de 5.14 (2) vérifiées et soit $B(V) \subset H^*(V; \mathbf{Z}_2)$ l'image de $h_*(\mathfrak{Z}_\mathbf{R}(V; \mathbf{Z}_2))$ par la dualité de Poincaré. C'est une sous-algèbre, et, vu 5.14, $\sigma^* = \Delta \circ \rho \circ h^{-1} \circ \Delta^{-1}$ est un homomorphisme multiplicatif de $B(V)$ dans $H^*(V^0; \mathbf{Z}_2)$, qui divise les dimensions par 2. Puisqu'il est multiplicatif, on a $\sigma^*(\mathrm{Sq}^{2i} x) = \mathrm{Sq}^i \sigma^*(x)$ pour $x \in B(V)$ de degré $2i$. On peut se demander si l'on n'a pas plus généralement

$$(1) \qquad \sigma^*(\mathrm{Sq}^{2j} x) = \mathrm{Sq}^j \sigma^*(x) \qquad [x \in B(V); j \geqq 0].$$

Il résultera de 5.19 que cela est vrai si $x = \Delta h(X)$, où X peut être représentée par une combinaison linéaire de sous-variétés non singulières, définies sur $\mathbf{R}$. Même si cette hypothèse faite sur X est restrictive (ce que nous ignorons), la validité de (1) en général semble plausible, mais nous ne savons pas l'établir. En vertu des résultats connus, (1) est vraie dans les grassmanniennes, ou aussi dans les variétés de drapeaux considérées à la fin de 6.5.

5.18. Soient X une variété projective non singulière définie sur $\mathbf{R}$, E un fibré vectoriel de rang p défini sur $\mathbf{R}$, de base X. Nous suivons les constructions et notations de [12]. Soient donc $P(E)$ le fibré en espaces projectifs associé à E, L_E le fibré linéaire de rang 1 sur $P(E)$, et $\xi_E \in A(P(E))$ la classe fondamentale de L_E ([12], p. 138 et 140). Les classes de Chern $c_i(E) \in A^{2i}(X)$ de E sont caractérisées par l'égalité

$$(1) \qquad \sum_{i=0}^{i=p} c_i(E)\, \xi_E^{p-i} = 0, \qquad c_0(E) = [X].$$

Nous *admettrons* ici que

$$(2) \qquad c_i(E) \in A_{\mathbf{R}}^{i2}(X) \quad (i = 0, \ldots, p), \; \xi_E \in A_{\mathbf{R}}(P(E)) \quad (^3).$$

Soient maintenant E^c, $P(E_0)$ et L_E^0 les parties réelles de E, $P(E)$ et L_E respectivement. $P(E^0)$ est donc le fibré en espaces projectifs réels associé au fibré vectoriel réel E, et L_E^0 le fibré de groupe structural $\mathbf{R}^*$ de base $P(E^0)$. Un raisonnement analogue à 4.13 montre que $\rho_*(\xi_E)$ est duale à la classe caractéristique $\mu \in H^1(P(E^0); \mathbf{Z}_2)$ de· L_E^0. D'après la formule de Hirsch-Wu $(^4)$, les classes de Stiefel-Whitney, $w_i(E) \in H^i(X^0; \mathbf{Z}_2)$ de E^0 sont caractérisées par

$$(3) \qquad \sum_{i=0}^{i=p} w_i(E^0) \cdot \mu^{p-i} = 0, \qquad w_0(E^0) = 1.$$

D'après 5.3 et 5.13, l'égalité (1) entraîne

$$(4) \qquad \sum_{i=0}^{i=p} \rho_*(c_i(E)) \, \rho_*(\xi_E)^{p-i} = 0$$

ce qui, vu (3), montre par dualité que $\rho_*(c_i(E)) = \Delta^{-i} w_i(E_0)$, donc prouve la

PROPOSITION. — *Soit E un fibré vectoriel algébrique de rang p défini sur $\mathbf{R}$ dont la base est une variété quasi projective non singulière X, et soient $c_i(E)$ les classes de Chern de E, vues comme classes d'équivalence rationnelle sur $\mathbf{R}$. Alors $\Delta\rho_*(c_i(E))$ est la $i^{\text{ième}}$ classe de Stiefel-Whitney du fibré partie réelle de E.*

5.19. Soient X une sous-variété non singulière définie sur $\mathbf{R}$, d'une variété quasi projective non singulière V définie sur $\mathbf{R}$, et E le fibré normal à X dans V. La partie réelle E^0 de E est le fibré normal à la partie réelle X^0 de X dans la partie réelle V^0 de V.

PROPOSITION. — *Soit c_i la $i^{\text{ième}}$ classe de Chern du fibré normal à X dans V, réduite mod 2, et soit j_* ; $A_{\mathbf{R}}(X) \to A_{\mathbf{R}}(V)$ l'homomorphisme induit par l'injection j de X dans V. Alors on a*

$$\mathrm{Sq}^{2i} \Delta h(X) = \Delta h_*(j_*(c_i)), \qquad \mathrm{Sq}^i \Delta\rho(X) = \Delta\rho_*(j_*(c_i)).$$

$(^3)$ Cette assertion sera certainement justifiée lorsque quelqu'un aura pris la peine de faire la théorie des classes de Chern en géométrie algébrique « avec corps de base » (ou avec « schéma de base »). Comme il s'agit ici de remarques incidentes, dont le reste de ce travail ne dépend pas, nous ne chercherons pas à justifier (2).

$(^4)$ Cette formule peut se démontrer à partir du théorème de multiplication de Whitney exactement comme CHERN l'a fait pour l'analogue cohomologique de (1) [On the characteristic classes of complex sphere bundles and algebraic varieties, *Amer. J. of Math.*, t. 75, 1953, p. 565-597].

Désignons en effet par ν et par ν^0 les classes mod 2 duales à $h(X)$ et $\rho(X)$, par $w_{2i} \in H^{2i}(X; \mathbf{Z}_2)$ et $w_i^0 \in H^i(X^0; \mathbf{Z}_2)$ les classes de Stiefel-Whitney de E et E_0 ($i = 0, 1, \ldots$), et par j_* et j_*^0 les homomorphismes de Gysin associés aux injections $j : X \to V$ et $j^0 : X^0 \to Y^0$. D'après Thom [21], on a

$$\mathrm{Sq}^{2i} \nu = j_*(w_{2i}), \qquad \mathrm{Sq}^i \nu^0 = j_*^0(w_i^0).$$

La proposition résulte donc du fait que w_{2i} est la réduction mod 2 de la $i^{\text{ième}}$ classe de Chern et de la proposition précédente.

6. Applications.

Singularités complexes et réelles.

6.1. Nous reprenons les définitions et notations des exposés 7 et 8 de [6]. Ainsi $L_{n,p}^r$ (resp. $\tilde{L}_{n,p}^r$) est l'espace des jets d'ordre r à l'origine de $\mathbf{R}^n$ (resp. $\mathbf{C}^n$) des applications différentiables (resp. holomorphes) de $\mathbf{R}^n$ dans $\mathbf{R}^p$ (resp. $\mathbf{C}^n$ dans $\mathbf{C}^p$) appliquant origine sur origine; L_m^r (resp. $\tilde{L}_m^r$) est le groupe des jets d'ordre r à l'origine O de $\mathbf{R}^m$ (resp. $\mathbf{C}^m$) des isomorphismes différentiables (resp. analytiques complexes) de voisinages de l'origine sur voisinages de l'origine laissant fixe O. Remarquons que $\tilde{L}_{n,p}^r$ est l'espace affine des suites de p polynômes à coefficients complexes à n variables, de degré $\leq r$, et sans terme constant; sa partie réelle est $L_{n,p}^r$; le groupe $\tilde{L}_m^r$ est formé des suites de m polynômes complexes à m variables de degré $\leq r$, sans terme constant, la matrice des termes de degré 1 étant non singulière; sa partie réelle est L_m^r. Enfin $\tilde{L}_n^r \times \tilde{L}_p^r$ agit algébriquement sur $\tilde{L}_{n,p}^r$ (par la loi de composition des jets, c'est-à-dire par changement de variables); cette action est algébrique, définie sur $\mathbf{R}$; sa restriction à la partie réelle est l'action de $L_n^r \times L_p^r$ sur $L_{n,p}^r$.

Un *type de singularité* réelle ou complexe est défini par une sous-variété algébrique réelle S ou complexe $\tilde{S}$ de $L_{n,p}^r$ ou $\tilde{L}_{n,p}^r$, invariante par $L_n^r \times L_p^r$ ou $\tilde{L}_n^r \times \tilde{L}_p^r$.

Soit G_m la variété grassmanienne des m-plans passant par l'origine d'un espace numérique réel de grande dimension N, et soit V_m la variété de Stiefel des m-repères de cet espace numérique. Le fibré $V_n \times V_p$ de base $G_n \times G_p$ est un espace universel pour le groupe $L_n^1 \times L_p^1$ (pour une certaine dimension).

Identifions L_m^1 au sous-groupe de L_m^r des jets d'ordre r des automorphismes linéaires de $\mathbf{R}^m$; ainsi $L_n^1 \times L_p^1$ agit algébriquement sur $L_{n,p}^r$. Soit alors $E_{n,p}^r$ le fibré associé à $V_n \times V_p$ de fibre $L_{n,p}^r$ et soit S_E le sous-fibré de fibre S.

En remplaçant le corps $\mathbf{R}$ par $\mathbf{C}$, la même construction donne un espace fibré algébrique $\tilde{E}_{n,p}^r$ défini sur $\mathbf{R}$ de base $\tilde{G}_n \times \tilde{G}_m$ (produit des grassmaniennes complexes) dont la partie réelle est $E^r_{n,p}$. De même $\tilde{S}$ définit un sous-fibré $\tilde{S}_{\tilde{E}}$ de fibre $\tilde{S}$, défini sur $\mathbf{R}$ si et seulement si S l'est.

La fibre $L^r_{n,p}$ est un espace numérique; la projection de $E^r_{n,p}$ sur $G_n \times G_p$ induit donc un isomorphisme pour la cohomologie, par lequel nous identifions $H^*(E^r_{n,p})$ et $H^*(G_n \times G_p)$. Le sous-fibré E_S est une sous-variété algébrique réelle qui représente donc une classe d'homologie mod 2. Sa classe duale est donc exprimée (N étant supposé assez grand) comme un polynôme $P_S(W_i, W'_j)$ dans les classes de Stiefel-Whitney, W_i de G_n et W'_j de G_p, appelé le *polynôme universel* associé au type de singularité S.

Le même argument dans le cas complexe associe à $\tilde{S}$ un polynôme universel $P_{\tilde{S}}(C_i, C'_j)$ à coefficients entiers dans les classes de Chern C_i de $\tilde{G}_n$ et C'_j de $\tilde{G}_p$.

6.2. Théorème. — *Soit $P_S(W_i, W'_j)$ le polynôme universel associé au type de singularité réelle S et soit $P_{\tilde{S}}(C_i, C'_j)$ le polynôme universel associé au type de singularité complexe $\tilde{S}$, où $\tilde{S}$ est le complexifié de S. Alors $P_S(W_i, W'_j)$ s'obtient à partir de $P_{\tilde{S}}(C_i, C'_j)$ par réduction mod 2 des coefficients et remplacement de C_i par W_i et C'_j par W'_j.*

Comme $\tilde{S}_{\tilde{E}}$ est la complexification de S_E, le théorème est une conséquence de 5.15 et 5.16, si l'on peut vérifier qu'il existe un plongement défini sur $\mathbf{R}$ de $\tilde{E}^r_{n,p}$ dans un espace projectif. Or le fibré vectoriel $E^r_{n,p}$ est naturellement plongé dans l'espace fibré en espaces projectifs obtenu en complétant chaque fibre par son hyperplan à l'infini. Il suffit donc de montrer que si V est un fibré algébrique défini sur $\mathbf{R}$, de fibre $\mathbf{P}_n(\mathbf{C})$, groupe structural $\mathbf{GL}(n, \mathbf{C})$, base une variété projective non singulière B, alors X admet un plongement projectif défini sur $\mathbf{R}$. Or X admet un plongement projectif sur $\mathbf{C}$ d'après [13], th. 8, donc aussi un plongement projectif défini sur $\mathbf{R}$ ([14], chap. VI, cor. 3).

Décompositions cellulaires.

6.3. — Dans la suite, nous appellerons *décomposition cellulaire* d'un espace localement compact X une partition de X en sous-espaces X_i ($i \in I$), ayant les propriétés suivantes :

(i) $(X_i)_{i \in I}$ est un recouvrement localement fini de X, par des sous-espaces disjoints, homéomorphes à des espaces numériques.

(ii) X_i est ouvert dans son adhérence $\overline{X}_i$ et $\overline{X}_i - X_i$ est réunion d'un nombre fini de X_j de dimensions $< \dim X_i$.

Les sous-espaces $\overline{X}_i$ sont les *cellules* de la décomposition envisagée, les X_i en sont les « cellules ouvertes ». Il est clair que $\overline{X}_i$ est de type VS_{n_i} ($n_i = \dim X_i$); si X est de dimension finie, alors X est de type VS_n

($n = \max \dim X_i$) et la réunion des X_i de dimension n est un ensemble épais de points réguliers.

6.4. Proposition. — *Soit X un espace de type VS_n admettant une décomposition cellulaire $(X_i)_{i \in I}$. Supposons que $\overline{X}_i$ possède une classe fondamentale c_i sur l'anneau K pour tout $i \in I$. Alors :*

(a) $H_*(X; K)$ s'identifie au produit direct des K-modules monogènes libres engendrés par les classes c_i.

(b) Si $\overline{X}_i$ est compact ($i \in I$), $H_*^c(X; K)$ s'identifie au K-module libre ayant comme base les classes c_i.

Dans la démonstration, l'homologie est à coefficients dans K.

Soit $J \subset I$ l'ensemble des indices j pour lesquels $\dim X_j = n$, et soit Y la réunion des $X_j (j \in J)$; ces derniers sont disjoints, homéomorphes à R^n, donc (**1.3**)

$$(\mathrm{1}) \qquad H_n(Y) = \prod_{j \in J} H_n(X_j), \qquad H_q(Y) = \mathrm{o} \quad (q \neq n).$$

Comme $\overline{X}_j$ possède une classe fondamentale, la restriction

$$H_n(\overline{X}_j) \to H_n(X_j)$$

est un isomorphisme (**2.3**, remarque). Vu **1.7**, il s'ensuit que l'application $H_n(X) \to H_n(Y)$ est surjective. L'espace $X - Y$ étant de dimension $\leq n - \mathrm{I}$, la suite exacte d'homologie de $X \bmod X - Y$ montre alors

$$(\mathrm{2}) \qquad H_n(X) = H_n(Y), \qquad H_q(X) = H_q(X - Y) \quad (q \neq n).$$

Comme les $X_i (i \in I, i \notin J)$ forment une décomposition cellulaire de $X - Y$, on en déduit (a) par récurrence sur n.

Supposons maintenant $\overline{X}_i$ compact pour tout $i \in I$. Soit $\Phi = c(X) \cap Y$ la famille de supports sur Y formée des intersections de Y avec les compacts de X. Évidemment, Φ est l'ensemble des fermés de Y qui ne rencontrent qu'un nombre fini de $X_j (j \in J)$, d'où

$$H_n^{\Phi}(Y) = \sum_{j \in J} H_n(X_j), \qquad H_q^{\Phi}(Y) = \mathrm{o} \quad (q \neq n).$$

L'assertion (b) se démontre alors comme (a), à l'aide de la suite exacte [**7.10**, (1)]

$$\ldots \to H_m^c(X - Y) \to H_m^c(X) \to H_m^{\Phi}(Y) \to H_{m-1}^c(X - Y) \to \ldots$$

Cas particuliers. — Si X est un espace analytique complexe, et les $\overline{X}_i$ sont les sous-ensembles analytiques, alors la proposition s'applique pour tout K, vu **3.2**. Soit $K = \mathbf{Z}_2$. D'après **3.7**, la proposition s'applique si les

$\overline{X}_i$ sont localement analytiques réels, ou encore (2.5) s'il existe pour chaque $i \in I$ une application propre $f_i : M_i \to \overline{X}_i$ d'un espace localement analytique réel M_i dans $\overline{X}_i$ qui soit un homéomorphisme de $f_i^{-1} X_i$ sur X_i. Un exemple de ce cas est fourni par les décompositions cellulaires de J. H. C. Whitehead des variétés de Stiefel.

6.5. Espaces homogènes algébriques.

— A titre d'illustration, nous décrivons ici sans donner de démonstrations des espaces homogènes algébriques admettant une décomposition cellulaire. Nous supposons le lecteur familier avec la théorie des groupes semi-simples ([5]).

Soit G un groupement semi-simple réel, linéaire connexe, non compact et soit $G = K.A.N$ une décomposition d'Iwasawa de G, où K est compact maximal, A commutatif formé de matrices diagonalisables et N nilpotent simplement connexe formé de matrices à valeurs propres toutes égales à 1. Soient encore M et M^* le centralisateur et le normalisateur de A dans K. Alors $W = M^*/M$ est un groupe fini. Notons g_w un représentant de $w \in W$ dans M^*. D'après le lemme de Bruhat, G est réunion des doubles classes $N g_w AN$, qui sont deux à deux disjointes; de plus, il existe un sous-groupe connexe N_w de N tel que $N g_w AN = N_w g_w . AN$ et que tout élément de $N g_w AN$ s'écrive d'une seule manière comme produit $x g_w ay$ $(x \in N_w, y \in N, a \in A)$. Soient alors $X = G/U$, où U est le produit semi-direct $M.A.N$ et $\pi : G \to X$ la projection canonique. On a aussi $X = K/M$, donc X est compact. Il résulte alors de théorèmes généraux sur les groupes algébriques que X est une variété projective réelle, sur laquelle G opère par des transformations projectives et que les orbites $X_w = N.\pi(g_w) = N_w.\pi(g_w)$ des points $\pi(g_w)$ sous l'action de N forment une décomposition cellulaire de X, dont les adhérences topologiques $\overline{X}_w$ sont des ensembles algébriques réels.

On peut généraliser : soient θ le système des racines simples de l'algèbre de Lie g de G, par rapport à l'algèbre de Lie a de A, pour l'ordre défini par N. θ contient donc r éléments, $r = \dim A$. Pour tout sous-ensemble I de θ, désignons par a_I la sous-algèbre de a sur laquelle les racines $\alpha \in I$ s'annulent, par A_I le sous-groupe correspondant de A, M et M_I^* le centralisateur et le normalisateur de A_I dans K et $U_I = M_I.A.N$. Alors

$$X_I = G/U_I = K/M_I.$$

On montre que X_I admet un plongement projectif réel, sur lequel G opère par des transformations projectives, que les orbites de N forment une décomposition cellulaire, et que leurs adhérences sont des ensembles algébriques réels. Ces cellules sont en correspondance biunivoque avec les

---

([5]) *Voir*, par exemple, *Séminaire Sophus Lie*, t. 1, 1954-1955. Pour le lemme de Bruhat, *cf.* HARISH-CHANDRA, On a lemma of F. Bruhat, *J. Math. pures et appl.*, Série 9, t. 35, 1956, p. 203-210.

classes de restes de W mod le sous-groupe W_I engendré par les symétries
aux plans $\alpha = 0$. Vu **6.4** ces cellules forment une base de l'homologie
mod 2 de X_I. Si I est vide, on retrouve l'espace $X = G/U$.

Ces décompositions sont les analogues réels des décompositions cellulaires
analytiques complexes bien connues des espaces homogènes algébriques projectifs complexes des groupes de Lie semi-simples complexes ([6]).

Supposons maintenant que dim A soit égale au rang de G. Cela équivaut à
dire que G est une « forme normale réelle » de sa complexification G^c au
sens de E. CARTAN. Alors A^c est un sous-groupe de Cartan de G^c, M^{*c}/M^c
est le groupe de Weyl de G^c et $B^c = A^c.N^c$ est un sous-groupe connexe
résoluble maximal de G^c. On a $G^c/B^c = L/T$ où L est compact maximal
dans G^c et T est un tore maximal de L. On déduit alors du lemme de Bruhat
dans le cas analytique complexe que les orbites de N^c dans G^c/B^c forment
une décomposition cellulaire; de plus G^c/B^c est une complexification de X,
les cellules ont comme parties réelles les cellules X_{w} considérées plus haut.
Alors **5.15** et **6.3** montrent que ρ est un isomorphisme de $H_*(X^c; \mathbf{Z}_2)$ sur
$H_*(X; \mathbf{Z}_2)$ diminuant les degrés de moitié.

Toujours sous l'hypothèse dim A = rang G, on peut également complexi-
fier X_I et sa décomposition cellulaire. M_I^c est alors le centralisateur de A_I
dans G^c, $M_I^c \cap L$ s'identifie au centralisateur Z_I d'un tore de L, et

$$G^c/U_I^c = L/Z_I$$

est une complexification de X_I, admettant une décomposition cellulaire, formée des orbites de N^c, qui est la complexification de la décomposition cellulaire de X_I mentionnée plus haut, et **5.15** s'applique. L'exemple classique
est celui où $G = \mathbf{SL}(n, \mathbf{R})$, $K = \mathbf{SO}(n)$, N est le sous-groupe des matrices
triangulaires supérieures à valeurs propres égales à 1, et A le sous-groupe
des matrices diagonales de valeurs propres positives, de déterminant 1. Alors
$G^c = \mathbf{SL}(n, \mathbf{C})$, $L = \mathbf{U}(n)$, B^c est le groupe des matrices triangulaires, et
$r = n - 1$. L'espace X_I (resp. X_I^c) est la variété des drapeaux réels non
orientés

$$F(d_1, \ldots, d_k, n) \, [\text{resp. complexes } F^c(d_1, \ldots, d_k, n)]$$
$$(1 \leq d_1 \leq \ldots \leq d_k \leq n - 1; \; k = 1, \ldots, n - 1).$$

$F(d_1, \ldots, d_k, n)$ [resp. $F^c(d_1, \ldots, d_k, n)$] est donc l'ensemble des systèmes
de k sous-espaces emboîtés de $\mathbf{R}^n$ (resp. $\mathbf{C}^n$) de dimensions respectives
$d_1, \ldots, d_k$. Si I est vide, on retrouve les variétés de drapeaux usuelles
(type $1, 2, \ldots, n - 1$). Si I comprend $n - 2$ éléments, on retrouve les grassmanniennes, etc.

([6]) *Voir*, par exemple, BOREL (Armand), Kählerian coset spaces of semisimple Lie
groups, *Proc. Nat. Acad. Sc. U. S, A.*, t. 40, 1954, p. 1147-1151. Ces espaces sont
également décrits dans : BOREL-HIRZEBRUCH, Characteristic classes and homogeneous
spaces I, *Amer. J. of Math.*, t. 80, 1958, p. 458-538 (en particulier § 14) ou dans :
TITS, Sur certaines classes d'espaces homogènes de groupes de Lie, *Acad. royale Belg.
Mém.*, Cl. Sc., t. 29, 1955, n° 3, 268 pages (en particulier chapitre III).

7. **Appendice : le cap-produit.**

Ce paragraphe apporte quelques compléments à [2]. Nous supposons ce dernier connu du lecteur et en utilisons librement les notations.

7.1 Le cap-produit. — Soit $\mathcal{F}^*$ une résolution de K sur X, qui soit un faisceau c-mou d'algèbres différentielles graduées, unitaires, sans torsion. $\mathcal{F}^*$ est donc aussi c-fin ([10], II, 3.7). On peut prendre pour $\mathcal{F}^*$ par exemple le faisceau des germes de cochaînes d'Alexander-Spanier ou la résolution canonique simpliciale $\mathcal{F}^*(X; K)$ de [10], II, 6.4. Le produit dans $\mathcal{F}^*$ sera noté $\cup$. Pour toute famille Φ de supports, $\Gamma_\Phi(\mathcal{F}^*)$ est une algèbre différentielle graduée sans torsion. Si $\mathcal{F}^*$ est Φ-mou, Φ étant paracompactifiante, ou si $\mathcal{F}^*$ est flasque, alors $H^*(\Gamma_\Phi(\mathcal{F}^*)) = H_\Phi^*(X; K)$, et le produit induit par $\cup$ dans $H^*(X; K)$ est le cup-produit.

Soit U un ouvert de X. On définit une application bilinéaire.

$$\cap : \quad D(\mathcal{F}^*)_s(U) \times \Gamma_c(\mathcal{F}^l \,|\, U) \to D(\mathcal{F}^*)_{s-l}(U)$$

par la formule

$$(a \cap b)(z) = a(b \cup z) \qquad [a \in D(\mathcal{F}^*)_s(U),\ b \in \Gamma_c(\mathcal{F}^* \,|\, U),\ z \in \Gamma_c(\mathcal{F}^* \,|\, U)].$$

Cet accouplement est compatible avec la restriction à un ouvert, donc définit aussi un accouplement de $D(\mathcal{F}^*)_s$ et $\mathcal{F}^l$ à $D(\mathcal{F}^*)_{s-l}$ et par linéarité, un accouplement de $D(\mathcal{F}^*)$, $\mathcal{F}^*$ à $D(\mathcal{F}^*)$. Il a les propriétés suivantes :

$$(1) \qquad a \cap (b \cup c) = (a \cap b) \cap c \qquad [a \in D(\mathcal{F}^*),\ b \in \mathcal{F}^*,\ c \in \mathcal{F}^*],$$

$$(2) \qquad d(a \cap b) = da \cap b + (-1)^s\, a \cap db \qquad [a \in D(\mathcal{F}^*)_s,\ b \in \mathcal{F}^*].$$

En effet,

$$(a \cap (b \cup c))(z) = a((b \cup c) \cup z) = a(b \cup (c \cup z))$$
$$= (a \cap b)(c \cup z) = ((a \cap b) \cap c)(z),$$

quel que soit $z \in \mathcal{F}^*$, d'où (1). D'autre part, on a par définition ([2], **2.4**) :

$$(3) \qquad (df)(z) = d(f(z)) + (-1)^{r+1} f(dz) \qquad [f \in D(\mathcal{F}^*)_r,\ z \in \mathcal{F}^*],$$

donc en particulier

$$(4) \quad (d(a \cap b))(z) = d((a \cap b)(z)) + (-1)^{s-l+1}(a \cap b)(dz),$$
$$d((a \cap b)(z)) = d(a(b \cup z)) = (da)(b \cup z) + (-1)^s a(d(b \cup z)),$$
$$d((a \cap b)(z)) = (da \cap b)(z) + (-1)^s a(db \cup z) + (-1)^{s+l} a(b \cup dz),$$

$$(5) \quad d((a \cap b)(z)) = (da \cap b)(z) + (-1)^s (a \cap db)(z) + (-1)^{s+l}(a \cap b)(dz),$$

et (2) résulte de (4) et (5). L'accouplement $\cap$ définit aussi une application bilinéaire de $\Gamma_\Phi D(\mathcal{F}^*)) \times \Gamma_\Psi(\mathcal{F}^*)$ dans $\Gamma_{\Phi \cap \Psi}(D(\mathcal{F}^*))$ quelles que soient les familles de supports, qui, vu (2), passe à l'homologie, et donne lieu à une application bilinéaire

$$H_s(\Gamma_\Phi(D(\mathcal{F}^\star))) \times H^t(\Gamma_\Psi(\mathcal{F}^\star)) \to H_{s-t}(\Gamma_{\Phi\cap\Psi}(D(\mathcal{F}^\star)))$$

vérifiant

$$(6) \quad \begin{cases} a \cap (b \cup c) = (a \cap b) \cap c \\ [a \in H_s(\Gamma_\Phi(D(\mathcal{F}^\star))), \ b \in H^t(\Gamma_\Psi(\mathcal{F}^\star)), \ c \in H^u(\Gamma_\Theta(\mathcal{F}^\star))]. \end{cases}$$

7.2. Théorème. — *Soit $\mathcal{F}^\star(X; K)$ la résolution canonique simpliciale de K sur X de* [10], *chap.* II, 6.4. *La construction précédente appliquée à $\mathcal{F}^\star(X; K)$ définit un accouplement, le cap-produit,*

$$\cap : \quad H_s^\Phi(X; K) \times H_\Psi^t(X; K) \to H_{s-t}^{\Phi\cap\Psi}(X; K),$$

quels que soient s, $t \in \mathbf{Z}$ et les familles de supports, Φ, Ψ. On a

$$a \cap (b \cup c) = (a \cap b) \cap c \quad [a \in H_t^\Phi(X; K), \ b \in H_\Psi^s(X; K), \ c \in H_\Theta^u(X; K)].$$

Le cap-produit est compatible avec la restriction à un ouvert et avec l'agrandissement des familles de supports.

Dans cet énoncé on a admis implicitement que $\mathcal{F}^\star(X; K)$ est un faisceau d'algèbres différentielles graduées unitaires, ce qui résulte de [10], II, 6.4 (les formules de multiplication et de différentiation étant les mêmes que pour les cochaînes d'Alexander-Spanier). Comme $\mathcal{F}^\star(X; K)$ est flasque (*loc. cit.*) on a $H^*(\Gamma_\Psi(\mathcal{F}^\star(X; K))) = H_\Psi^*(X; K)$ d'après [10], II, th. 4.7.1 et $H_*^\Phi(X; K) = H_*(\Gamma_\Phi(D(\mathcal{F}^\star(X; K))))$ d'après [2], 3.5, donc **7.2** résulte de **7.1**.

7.3. Remarque. — Soit de nouveau $\mathcal{F}^\star$ la résolution c-molle de K considérée en **7.1**. On a encore $H_*^\Phi(X; K) = H(\Gamma_\Phi(D(\mathcal{F}^\star)))$ par [2], 3.5. Si Ψ est *paracompactifiante*, on a aussi

$$(7) \qquad\qquad H_\Psi^*(X; K) = \mathrm{H}(\Gamma_\Psi(\mathcal{F}^\star)),$$

d'après [10], II, 4.7.1, donc **7.1** donne lieu à un cap-produit entre $H_*^\Phi(X; K)$ et $H_\Psi^*(X; K)$. Ce cap-produit est le même que celui de **7.2**; autrement dit, le cap-produit ne change pas si l'on remplace $\mathcal{F}^\star$ par une autre résolution $\mathcal{G}^\star$ de K qui soit aussi un faisceau c-mou d'algèbres différentielles graduées unitaires. En effet, on a des homomorphismes évidents $\mathcal{F}^\star \to \mathcal{F}^\star \otimes \mathcal{G}^\star$ et $\mathcal{G}^\star \to \mathcal{F}^\star \otimes \mathcal{G}^\star$; d'autre part $\mathcal{F}^\star \otimes \mathcal{G}^\star$ est une résolution de K, qui est aussi un faisceau d'algèbres différentielles graduées unitaires c-fin, puisque les facteurs le sont. Il suffit donc d'établir l'indépendance annoncée lorsqu'il existe un homomorphisme (de faisceaux différentiels gradués d'algèbres) $\mu : \mathcal{F}^\star \to \mathcal{G}^\star$, ce qui est immédiat. En effet, la construction de **7.1**, faite à partir de $\mathcal{F}^\star$ et de $\mathcal{G}^\star$ est évidemment compatible avec μ; d'autre part l'homomorphisme $H^*(\Gamma_\Psi(\mathcal{F}^*)) \to H^*(\Gamma_\Psi(\mathcal{G}^*))$ est un isomorphisme d'après [10], II, th. 4.6.2, et la démonstration de 3.5 (b) dans [2] montre qu'il en est de même pour $H(\Gamma_\Phi(D(\mathcal{G}^*))) \to H(\Gamma_\Phi(D(\mathcal{F}^*)))$.

Il est plausible que dans le cas général où Ψ n'est pas supposée paracompactifiante, le cap-produit ne change pas si l'on remplace $\mathcal{F}^*(X;K)$ par une résolution flasque d'algèbres différentielles unitaires, mais nous ne savons pas le démontrer.

7.4. Soient M un sous-espace fermé de l'espace Z, $\mathcal{S}$ une résolution par des faisceaux c-mous de K sur Z. L'homomorphisme de restriction $\Gamma_c(\mathcal{S}) \to \Gamma_c(\mathcal{S}\,|\,M)$ sera noté $i^0_{Z,M}$. Il est surjectif, son transposé

$$i_0^{M,Z}: \quad D(\Gamma_c(\mathcal{S}\,|\,M)) \to D(\Gamma_c(\mathcal{S}))$$

est injectif, et l'image de $i_0^{M,Z}$ est l'ensemble des éléments de $D(\Gamma_c(\mathcal{S}))$ dont le support est dans M ([2], § 2.4). On a

$$H_*(M;K) = H(D(\Gamma_c(\mathcal{S}\,|\,M))), \qquad H_*(Z;K) = H(D(\Gamma_c(\mathcal{S}))),$$

et $i_0^{M,Z}$ induit l'homomorphisme $i_*^{M,Z}$ associé à l'inclusion $M \to Z$ ([2], 3.5, remarque).

Étant donné un faisceau $\mathcal{A}$ sur Z, on note $\mathcal{C}^*(Z;\mathcal{A})$ le faisceau des germes de sections (continues ou non) de $\mathcal{A}$ ([10], II, 4.3).

Soit $f: X \to Y$ une application continue, et soient $\mathcal{B}$ un faisceau sur Y, $f^*\mathcal{B}$ son image réciproque sur X ([10], II, 1.12). Il résulte immédiatement des définitions qu'on a une inclusion $f^*(\mathcal{C}^*(Y;\mathcal{B})) \to \mathcal{C}^*(X;f^*\mathcal{B})$. Comme la résolution canonique simpliciale de [10] est obtenue en itérant le foncteur $\mathcal{A} \to \mathcal{C}^*(Z;\mathcal{A})$, il s'ensuit que f induit un homomorphisme

$$f^*: \quad \mathcal{F}^*(Y;K) \to \mathcal{F}^*(X;K),$$

donc aussi un homomorphisme

$$f^0: \quad \Gamma_c(\mathcal{F}^*(Y;K)) \to \Gamma(\mathcal{F}^*(X;K))$$

et plus généralement, si Φ, Ψ sont des familles de supports sur X et Y respectivement telles que $f^{-1}(\Psi) \subset \Phi$, un homomorphisme

$$f^0: \quad \Gamma_\Psi(\mathcal{F}^*(Y;K)) \to \Gamma_\Phi(\mathcal{F}^*(X;K)).$$

Si les éléments de $\Gamma_\Psi(\mathcal{F}^q(Y;K))$ et $\Gamma_\Phi(\mathcal{F}^q(X;K))$ sont interprétés comme des fonctions sur Y^{q+1} et X^{q+1} respectivement ([10], II, 6.4 (b)), on a

$$(f^0\varphi)(x_0,\ldots,x_q) = \varphi(f(x_0),\ldots,f(x_q)) \qquad [\varphi \in \Gamma_\Psi(\mathcal{F}^q(Y;K))],$$

comme dans le cas des cochaînes d'Alexander-Spanier. En particulier, f_0 est multiplicatif.

Il résulte aisément des théorèmes d'unicité que f^0 induit, par passage aux groupes dérivés, l'homomorphisme $f^* : H^*_{\Psi}(Y; K) \to H^*_{\Phi}(X; K)$ associé à f. Plus généralement, soient $\mathscr{X}$, $\mathscr{Y}$ des résolutions flasques de K sur X et Y (ou des résolutions respectivement Φ-molles et Ψ-molles si Φ, Ψ sont paracompactifiantes) pour lesquelles il existe un homomorphisme $\alpha : f^* \mathscr{Y} \to \mathscr{X}$, alors l'homomorphisme induit de

$$H(\Gamma_{\Psi}(\mathscr{Y})) = H^*_{\Psi}(Y; K) \qquad \text{dans} \quad H(\Gamma_{\Phi}(\mathscr{X})) = H^*_{\Phi}(X; K)$$

est f^*.

7.5 Théorème. — *Soient* $f : X \to Y$ *une application continue,* Φ, Φ' (*resp.* Ψ, Ψ') *des familles de supports sur* X (*resp.* Y), *telles que la restriction de* f *à tout élément de* Φ *soit une application propre et que* $f(\Phi) \subset \Psi$, $f(\Phi \cap \Phi') \subset \Psi \cap \Psi''$ *et* $f^{-1}(\Psi') \subset \Phi'$. *Alors on a*

$$(8) \qquad f_*(a \cap f^* b) = (f_* a) \cap b \qquad [\, a \in H^{\Phi}_*(X, K),\ b \in H^*_{\Psi'}(Y; K)].$$

Dans cet énoncé, on a utilisé implicitement le fait que, sous les conditions indiquées, f induit des homomorphismes

$$H^{\Phi}_*(X; K) \to H^{\Psi}_*(Y; K) \qquad H^{\Phi \cap \Phi'}_*(X; K) \to H^{\Psi \cap \Psi'}_*(Y; K)$$

(*voir* [2], 3.5), ainsi qu'un homomorphisme

$$f^* : \quad H^*_{\Psi'}(Y; K) \to H^*_{\Phi'}(X; K),$$

comme on l'a rappelé ci-dessus. Soit $M \in \Phi$ et soit $M' = f(M)$. On a un homomorphisme naturel

$$(f \,|\, M)^0 : \quad \Gamma_c(\mathscr{F}^*(Y; K) \,|\, M') \to \Gamma_c(\mathscr{F}^*(X; K) \,|\, M),$$

tel que

$$(9) \qquad\qquad (f \,|\, M)^0 \circ i^0_{Y, M'} = i^0_{X, M} \circ f^0,$$

(notations de 7.4), et dont le transposé est un homomorphisme

$$(f \,|\, M)_0 : \quad D(\Gamma_c(\mathscr{F}^*(X; K) \,|\, M)) \to D(\Gamma_c(\mathscr{F}^*(Y; K) \,|\, M')).$$

Si $N \in \Phi$ contient M, il est clair que $(f \,|\, M)_0$ et $(f \,|\, N)_0$ sont compatibles avec l'inclusion; comme d'autre part

$$\Gamma_{\Phi}(D(\mathscr{F}^*(X; K))) = \bigcup_{F \in \Phi} D(\Gamma_c(\mathscr{F}^*(X; K) \,|\, F)),$$

$$\Gamma_{\Psi}(D(\mathscr{F}^*(Y; K))) = \bigcup_{G \in \Psi} D(\Gamma_c(\mathscr{F}^*(Y; K) \,|\, G))$$

([2], 2.4), on en déduit un homomorphisme

$$f_0 : \quad \Gamma_{\Phi}(D(\mathscr{F}^*(X; K))) \to \Gamma_{\Psi}(D(\mathscr{F}^*(Y; K)))$$

qui, vu (9), vérifie

$$(10) \qquad f_0 \circ i_0^{M,X} = i_0^{M',Y} \circ (f|M)_0 \qquad [M \in \Phi,\ M' = f(M)].$$

L'homomorphisme dérivé de f_0 s'identifie à l'application

$$f_* : \quad H_*^\Phi(X;K) \to H_*^\Psi(Y;K)$$

introduite dans [2], 3.5. On peut évidemment faire des remarques analogues pour $\Phi \cap \Phi'$ et $\Psi \cap \Psi'$. Pour établir (8), il suffit de démontrer

$$(11) \quad f_0(a \cap f^0 b) = (f_0 a) \cap b \qquad [a \in \Gamma_\Phi(D(\mathcal{F}^*(X;K))),\ b \in \Gamma_{\Psi'}(\mathcal{F}^*(Y;K))].$$

(i) Supposons tout d'abord que f soit propre et que Φ soit l'ensemble des fermés de X. Alors f_0 est l'homomorphisme transposé de

$$f^0 : \quad \Gamma_c(\mathcal{F}^*(Y;K)) \to \Gamma_c(\mathcal{F}^*(X;K)).$$

Par conséquent, pour tout $z \in \Gamma_c(\mathcal{F}^*(Y;K))$, on a

$$(f_0(a \cap f^0 b))(z) = (a \cap f^0 b)(f^0 z),$$

d'où, par définition du cap-produit,

$$(f_0(a \cap f^0 b))(z) = a(f^0(b \cup z)) = (f_0 a)(b \cup z) = (f_0 a \cap b)(z),$$

ce qui établit (8) dans le cas particulier envisagé.

Cette démonstration vaut naturellement aussi lorsque $\mathcal{F}^*(Y;K)$ et $\mathcal{F}^*(X;K)$ sont remplacées par des résolutions $\mathcal{Y}$, $\mathcal{X}$ c-molles de K pour lesquelles il existe un homomorphisme $f^* \mathcal{Y} \to \mathcal{X}$.

(ii) Prouvons maintenant (8) dans le cas général. Soit M le support de a, et soit $M' = f(M)$. On pourra appliquer (i) en particulier aux inclusions $M \to X$, $M' \to Y$ et à $(f|M) : M \to M'$. On peut écrire

$$a = i_0^{M,X} a' \qquad [a' \in D(\Gamma_c(\mathcal{F}^*(X;K)|M))],$$

donc

$$f_0(a \cap f^0 b) = f_0(i_0^{M,X} a' \cap f^0 b).$$

Vu (i) et (9), (10), on a

$$i_0^{M,X} a' \cap f^0 b = i_0^{M,X}(a' \cap i_{X,M}^0 . f^0 b) = i_0^{M,X}(a' \cap (f|M)^0 i_{Y,M'}^0(b)),$$
$$f_0(i_0^{M,X} a' \cap f^0 b) = f_0 . i_0^{M,X}(a' \cap (f|M)^0 . i_{Y,M'}^0(b)),$$
$$f_0(i_0^{M,X} a' \cap f^0 b) = i_0^{Y,M'} . (f|M)_0(a' \cap (f|M)^0 . i_{Y,M'}^0(b)).$$

Appliquant (i) à $(f|M)_0$, puis à l'inclusion $M' \to Y$, on en tire

$$f_0(a \cap f^0 b) = i_0^{M',Y}((f|M)_0 a' \cap i_{Y,M'}^0(b)),$$
$$f_0(a \cap f^0 b) = i_0^{M',Y} . (f|M)_0(a') \cap b,$$

d'où finalement, vu (10),

$$f_0(a \cap f^0 b) = (f_0 . i_0^{M,X}(a') \cap b) = (f_0 a) \cap b.$$

7.6. Proposition. — *Soient $\mathscr{F}^*(X; K)$ la résolution simpliciale canonique de K sur X, $\mathcal{Q}$ un faisceau sur X et Φ une famille paracompactifiante de supports sur X. Alors :*

(i) $H_\Phi^*(X; \mathcal{Q}) = H(\Gamma_\Phi(\mathscr{F}^*(X; K) \otimes \mathcal{Q}))$.

(ii) *Si $\mathcal{Q}$ est sans torsion et si $\dim_K X < \infty$, on a*

$$H_*^\Phi(X; \mathcal{Q}) = H(\Gamma_\Phi(D(\mathscr{F}^*(X; K)) \otimes \mathcal{Q})).$$

Il résulte visiblement de la définition de $\mathscr{F}^*(X; K)$ que ce dernier est sans torsion, par conséquent $\mathscr{F}^*(X; K) \otimes \mathcal{Q}$ est une résolution de $\mathcal{Q}$. Soit $\mathscr{K} = \mathcal{C}(X; K)$ le faisceau des germes de sections (continues ou non) du faisceau constant $X \times K$. Il est évidemment flasque, donc ([10], II, 3.5) Φ-mou pour toute famille paracompactifiante Φ. Il est immédiat que $\mathscr{F}^*(X; K)$, et $\mathscr{F}^*(X; K) \otimes \mathcal{Q}$ sont des $\mathscr{K}$-modules, donc sont Φ-mous ([10], II, th. 3.7.1). L'assertion (i) est alors conséquence du théorème 4.7.1 de [10], chap. II.

Il existe un homomorphisme de $\mathscr{F}^*(X; K)$ dans la résolution canonique injective $\mathcal{C}^*(X; K)$ de K introduite dans [2], d'où des homomorphismes

$$\alpha : \quad \mathcal{C}_H(X; K) = D(\mathcal{C}^*(X; K)) \to D(\mathscr{F}^*(X; K)),$$
$$\alpha' : \quad \mathcal{C}_H(X; K) \otimes \mathcal{Q} \to D(\mathscr{F}^*(X; K)) \otimes \mathcal{Q}.$$

α induit un isomorphisme de faisceaux dérivés ([2], 3.3(b)). Si $\mathcal{Q}$ est sans torsion, alors on a, par la règle de Künneth :

$$\mathscr{H}(\mathcal{C}_H(X; K) \otimes \mathcal{Q}) = \mathscr{H}(\mathcal{C}_H(X; K)) \otimes \mathcal{Q},$$
$$\mathscr{H}(D(\mathscr{F}^*(X; K)) \otimes \mathcal{Q}) = \mathscr{H}(D(\mathscr{F}^*(X; K))) \otimes \mathcal{Q},$$

donc α' induit aussi un isomorphisme des faisceaux dérivés. D'autre part, $D(\mathscr{F}^*(X; K))$ est aussi un K-module, donc $D(\mathscr{F}^*(X; K)) \otimes \mathcal{Q}$ est Φ-fin. De même $\mathcal{C}_H(X; K) \otimes \mathcal{Q}$ est Φ-fin ([2], 3.1). (ii) est donc un cas particulier du théoréme 4.6.2 de [10], chap. II.

7.7. — Nous voulons maintenant étendre la définition du cap-produit dans certains cas où les coefficients forment un faisceau. Il est à espérer que cette extension est possible dans le cas général et que les restrictions faites ci-dessous ne sont dues qu'à une insuffisance technique. Elle n'est pas utilisée dans le présent travail

Soient $\mathcal{L}$, $\mathfrak{M}$ des faisceaux sur X. Il est clair que 7.1 donne aussi lieu à une application

$$(D(\mathscr{F}^*) \otimes \mathcal{L}) \times (\mathscr{F}^* \otimes \mathfrak{M}) \to D(\mathscr{F}^*) \otimes \mathcal{L} \otimes \mathfrak{M},$$

vérifiant (1), (2), d'où une application bilinéaire

$$H_s(\Gamma_\Phi(D(\mathscr{F}^\star)\otimes\mathscr{L}))\times H^t(\Gamma_\Psi(\mathscr{F}^\star\otimes\mathfrak{M}))\to H_{s-t}(\Gamma_{\Phi\cap\Psi}(D(\mathscr{F}^\star)\otimes\mathscr{L}\otimes\mathfrak{M}))$$

qu'on appellera aussi le cap-produit. A l'aide de **7.6**, on en tire tout de suite le

THÉORÈME. — *Soient $\mathscr{L}$, $\mathfrak{M}$, $\mathfrak{N}$ des faisceaux sans torsion sur X, Φ, Ψ, Θ des familles de supports sur X. On suppose :*

(i) *$\mathscr{L}=X\times K$, ou bien Φ paracompactifiante et $\dim_K X<\infty$;*
(ii) *$\mathfrak{M}=X\times K$, ou bien Ψ paracompactifiante ;*
(iii) *$\mathscr{L}\otimes\mathfrak{M}=X\times K$ ou bien $\Phi\cap\Psi$ paracompactifiante et $\dim_K X<\infty$.*
Alors il existe un accouplement, le cap-produit

$$\cap\ :\quad H^\Phi_s(X\,;\,\mathscr{L})\times H^t_\Psi(X\,;\,\mathfrak{M})\to H^{\Phi\cap\Psi}_{t-s}(X\,;\,\mathscr{L}\otimes\mathfrak{M})\qquad (s,\,t\in\mathbf{Z})$$

qui est compatible avec l'agrandissement des familles de supports et avec les homomorphismes de faisceaux de coefficients. Si de plus :

(iv) *$\mathfrak{N}=X\times K$ ou bien Θ est paracompactifiante ;*
(v) *$\mathfrak{M}\otimes\mathfrak{N}=X\times K$ ou bien $\Psi\cap\Theta$ est paracompactifiante ;*
(vi) *$\mathscr{L}\otimes\mathfrak{M}\otimes\mathfrak{N}=X\times K$ ou bien $\Phi\cap\Psi\cap\Theta$ est paracompactifiante et $\dim_K X<\infty$, alors on a*

$$(a\cap b)\cap c=a\cap(b\cup c)\qquad [a\in H^\Phi_s(X\,;\,\mathscr{L}),\ b\in H^t_\Psi(X\,;\,\mathfrak{M}),\ c\in H^u_\Theta(X\,;\,\mathfrak{N})].$$

7.8. — Dans ce numéro et le suivant, X désigne une variété homologique ou cohomologique [**2**], § **7** sur K, de dimension (cohomologique sur K) égale à n, $\mathscr{C}=\mathscr{H}_n(X\,;\,K)$ le faisceau d'orientation de X, et $\mathscr{C}'$ un inverse de $\mathscr{C}$. Le faisceau $\mathscr{C}'$ est donc localement isomorphe à $X\times K$ et l'on a un isomorphisme de $\mathscr{C}\otimes\mathscr{C}'$ sur $X\times K([\mathbf{2}],\ \mathbf{7.7})$. Pour tout entier q, tout faisceau $\mathscr{L}$ sur X et toute famille de supports Φ, paracompactifiante si $\mathscr{L}\neq X\times K$, la dualité de Poincaré définit un isomorphisme

$$\Delta\ :\quad H^\Phi_q(X\,;\,\mathscr{L})\to H^{n-q}_\Phi(X\,;\,\mathscr{C}\otimes\mathscr{L}),$$

donc aussi un isomorphisme

$$\Delta\ :\quad H^\Phi_q(X\,;\,\mathscr{C}'\otimes\mathscr{L})\to H^{n-q}_\Phi(X\,;\,\mathscr{L}).$$

Si X est orientable, ou bien paracompacte, on a en particulier un isomorphisme

$$\Delta\ :\quad H_n(X\,;\,\mathscr{C}')\to H^0(X\,;\,K)$$

et $\Delta^{-1}(1)$, [1 étant élément neutre de $H^\star(X\,;\,K)$] est la classe fondamentale de X.

7.9. THÉORÈME. — *On conserve les notations de 7.8. On suppose X orientable ou bien paracompacte et l'on pose $\xi=\Delta^{-1}(1)$. Soient $\mathscr{L}$ un faisceau sans torsion sur X, et Φ une famille de supports, qui est para-*

compactifiante si $\mathcal{L}$, $\mathcal{C} \neq X \times K$. *Alors pour tout entier* q, $a \to \xi \cap a$ *est un isomorphisme de* $H_{\Phi}^{q}(X; \mathcal{L})$ *sur* $H_{n-q}^{\Phi}(X; \mathcal{C}' \otimes \mathcal{L})$; *son inverse est* Δ.

Soit ξ' un cycle de $\Gamma(D(\mathcal{F}^{*}(X; K)) \otimes \mathcal{C}')$ représentant ξ; l'application $a \to \xi' \cap a$ est un homomorphisme

$$\alpha : \quad \mathcal{F}^{q}(X; K) \otimes \mathcal{L} \to D(\mathcal{F}^{*}(X; K))_{n-q} \otimes \mathcal{C}' \otimes \mathcal{L}.$$

On pose $\mathcal{B}^{q} = D(\mathcal{F}^{*}(X; K))_{n-q}$ et l'on munit $\mathcal{B} = \sum \mathcal{B}^{q}$ de la différentielle $(-1)^{n}d$, où d est la différentielle de $D(\mathcal{F}^{*}(X; K))$; l'application α peut alors s'envisager comme une application de $\mathcal{F}^{*}(X; K) \otimes \mathcal{L}$ dans $\mathcal{B}^{*} \otimes \mathcal{C}' \otimes \mathcal{L}$ qui conserve le degré et qui commute aux différentielles vu **7.1** (2) et le fait que ξ' est un cycle. On a, puisque $\mathcal{L}$ et $\mathcal{C}'$ sont sans torsion :

$$\mathcal{H}(\mathcal{F}^{*}(X; K) \otimes \mathcal{L}) \simeq \mathcal{L},$$
$$\mathcal{H}(\mathcal{B}^{*} \otimes \mathcal{C}' \otimes \mathcal{L}) \simeq \mathcal{H}(\mathcal{B}^{*}) \otimes \mathcal{C}' \otimes \mathcal{L} \simeq \mathcal{C} \otimes \mathcal{C}' \otimes \mathcal{L} \simeq \mathcal{L}.$$

Il s'ensuit immédiatement que α induit un isomorphisme des faisceaux dérivés, donc ([**10**], II, t. 4.6.2), α induit un isomorphisme

$$\alpha^{*} : \quad H^{*}(\Gamma_{\Phi}(\mathcal{F}^{*}(X; K) \otimes \mathcal{L})) \to H^{*}(\Gamma_{\Phi}(\mathcal{B}^{*} \otimes \mathcal{C}' \otimes \mathcal{L})).$$

d'où la première assertion.

Soit $\mathcal{S}^{*}$ une résolution flasque de $\mathcal{B}^{*} \otimes \mathcal{C}' \otimes \mathcal{L}$, et soit $\beta : \mathcal{B}^{*} \otimes \mathcal{C}' \otimes \mathcal{L} \to \mathcal{S}^{*}$ l'augmentation. Il résulte de la démonstration du théorème de dualité ([**2**], **7.3**) et des remarques faites dans [**10**], II, § 4.5 que Δ est l'homomorphisme composé

$$H_{n-q}^{\Phi}(X; \mathcal{C}' \otimes \mathcal{L}) = H^{q}(\Gamma_{\Phi}(\mathcal{B}^{*} \otimes \mathcal{C}' \otimes \mathcal{L})) \overset{\mu}{\to} H^{q}(\Gamma_{\Phi}(\mathcal{S}^{*})) \overset{\nu}{\to} H_{\Phi}^{q}(X; \mathcal{L}),$$

où μ est défini par β et où ν est l'identification canonique. Il s'ensuit que le composé

$$H^{q}(\Gamma_{\Phi}(\mathcal{F}^{*}(X; K) \otimes \mathcal{L})) \to H_{n-q}^{\Phi}(X; \mathcal{C}' \otimes \mathcal{L}) \to H_{\Phi}^{q}(X; \mathcal{L})$$

est l'identification canonique, d'où la deuxième assertion.

7.10. Remarques sur [2].

(a) Dans l'énoncé de **7.3**, il faut ajouter « and if Tor $(\mathcal{H}_{*}(X; K)_{x}, \mathcal{S}_{x}) = 0$ $(x \in X)$ » après « paracompactifying ». Dans (3), remplacer $\mathcal{A}$ par $\mathcal{B}$, et dans (4), remplacer $(p, q \geq 0)$ par $(p < 0$ ou $q < 0)$.

(b) La suite exacte d'homologie **3.8** vaut aussi avec une famille de supports. Plus précisément on a la

PROPOSITION. — *Soient* F *un sous-espace fermé de* X, $U = X - F$, *et* Φ *une famille de supports sur* X. *Alors on a une suite exacte*

$$(1) \quad \ldots \to H_{q}^{\Phi | F}(F; K) \to H_{q}^{\Phi}(X; K) \to H_{q}^{\Phi | U}(U; K) \to H_{q-1}^{\Phi | F}(F; K) \to \ldots.$$

En effet soit $\mathcal{C} = \mathcal{C}_H(X; K)$ le faisceau fondamental pour l'homologie de X ([2], §3). Alors $\mathcal{C} \mid U$ est le faisceau fondamental pour U. La restriction définit un homomorphisme $\rho : \Gamma_\Phi(\mathcal{C}) \to \Gamma_{\Phi \cap U}(\mathcal{C} \mid U)$. Montrons qu'il est surjectif. Soit donc c une section de $\mathcal{C}$ sur U, à support dans $L \cap U$ où $L \in \Phi$. Comme c est nulle dans $U - (L \cap U)$ il est clair qu'il existe une section c' de $\mathcal{C}$ sur $U \cup (X - L)$ qui est égale à c sur U et est nulle sur $X - L$. Comme $\mathcal{C}$ est flasque, cette section se prolonge en une section c_1 sur X, dont le support est contenu dans L, donc dans Φ, d'où notre assertion. Si N est le noyau de r on a donc une suite exacte

$$(2) \qquad 0 \to N \to \Gamma_\Phi(\mathcal{C}) \to \Gamma_{\Phi \cap U}(\mathcal{C} \mid U) \to 0.$$

Mais $H_*(N) = H_*^{\Phi \mid F}(F; K)$ d'après ([2], 3.5, Remark); pour obtenir (1), il suffit donc de prendre la suite exacte d'homologie associée à (2).

BIBLIOGRAPHIE

[1] BOREL (Armand). — *Seminar on transformation groups.* — Princeton, Princeton University Press, 1960 (*Annals of Mathematics Studies*, 46).

[2] BOREL (F.) and MOORE (J. C.). — Homology theory for locally compact spaces, *Mich. math. J.*, t. 7, 1960, p. 137-159.

[3] BRUHAT (F.) et WHITNEY (H.). — Quelques propriétés fondamentales des ensembles analytiques réels, *Comment. Math. Helvet.*, t. 33, 1959, p. 132-160.

[4] CARTAN (Henri). — Cohomologie des groupes, suite spectrale, faisceaux, 2ᵉ édition, *Séminaire Cartan*, t. 3, 1950-1951. — Paris, Secrétariat mathématique, 1955 (multigraphié).

[5] CARTAN (Henri). — Variétés analytiques complexes et fonctions automorphes, *Séminaire Cartan*, t. 6, 1953-1954.

[6] CARTAN (Henri). — Quelques questions de topologie, *Séminaire Cartan*, t. 9, 1956-1957. — Paris, Secrétariat mathématique, 1958 (multigraphié).

[7] CARTAN (Henri). — Variétés analytiques réelles et variétés analytiques complexes, *Bull. Soc. math. France*, t. 85, 1957, p. 77-100.

[8] CHERN (S. S.). — *Topics in differential geometry.* — Princeton, Institute for advanced Study, 1951 (multigraphié).

[9] CHOW (W. L.). — On equivalence classes of cycles in an algebraic variety, *Annals of Math.*, t. 64, 1956, p. 450-472.

[10] GODEMENT (Roger). — *Topologie algébrique et théorie des faisceaux.* — Paris, Hermann, 1958 (*Act scient. et ind.*, 1252; *Publ. Inst. Math. Univ. Strasbourg*, 13).

[11] GRAUERT (H.) und REMMERT (R.). — Komplexe Räume, *Math. Annalen*, t. 136, 58, p. 245-318.

[12] GROTHENDIECK (Alexander). — La théorie des classes de Chern, *Bull. Soc. math. France*, t. 86, 1958, p. 137-154.

[13] KODAIRA (K.). — On Kähler varieties of restricted type (an intrinsic characterization of algebraic varieties), *Annals of Math.*, t. 60, 1954, p. 28-48.

[14] LANG (Serge). — *Introduction to algebraic geometry.* — New-York, Interscience Publishers, 1958 (*Interscience Tracts in pure and applied Mathematics*, 5).

[15] LELONG (Pierre). — Intégration sur un ensemble analytique complexe, *Bull. Soc. math. France*, t. 85, 1957, p. 239-262.

[16] REMMERT (R.). — Projektionen analytischer Mengen, *Math. Annalen*, t. 130, 1956, p. 410-441.

223

[17] DE RHAM (Georges). — *Variétés différentiables, formes, courants, formes harmoniques*. — Paris, Hermann, 1955 (*Act. scient. et ind.*, 1222; *Publ. Inst. math. Univ. Nancago*, 3).

[18] DE RHAM (Georges). — On currents in an analytic complex manifold, *Seminars on analytic functions*, t. 1, p. 54-64. — Princeton, Institute for advanced Study, 1957.

[19] ROSSI (H.). — *Analytic spaces*, I and II. — Princeton University, 1960 (multigraphié).

[20] SERRE (Jean-Pierre). — Géométrie algébrique et géométrie analytique, *Ann. Inst. Fourier*, Grenoble, t. 6, 1955-1956, p. 1-42.

[21] THOM (René). — Espaces fibrés en sphères et carrés de Steenrod, *Ann. scient. Ec. Norm. Sup.*, t. 69, 1952, p. 109-181.

[22] THOM (René). — Un lemme sur les applications différentiables, *Bol. Soc. Mat. Mex.*, Série, 2, t. 1, 1956, p. 59-71.

[23] WEIL (André). — *Foundations of algebraic geometry*. — New-York, American mathematical Society, 1946 (*Amer. math. Soc. Coll. Publ.*, 29).

[24] WHITNEY (H.). — Elementary structure of real algebraic varieties, *Annals of Math.*, t. 66, 1957, p. 545-556.

(Manuscrit reçu le 26 mai 1961.)

Armand BOREL,
Institute for advanced Study,
Princeton, N. J. (États-Unis).

André HAEFLIGER,
Faculté des Sciences,
Université de Genève,
Genève (Suisse).

<h1 style="text-align:center">57.</h1>

(mit R. Remmert)

Über kompakte homogene Kählersche Mannigfaltigkeiten

Math. Ann. 145 (1962) 429–439

1. Mit V wird stets eine zusammenhängende kompakte komplexe Mannigfaltigkeit bezeichnet. V heißt homogen, wenn es zu je 2 Punkten $v_1, v_2 \in V$ eine biholomorphe Abbildung g von V auf sich mit $g(v_1) = v_2$ gibt. Eine kompakte Kählersche Mannigfaltigkeit heiße homogen, wenn sie — aufgefaßt als komplexe Mannigfaltigkeit — homogen ist. Das Ziel dieser Note ist der Beweis von

Satz I: *Jede zusammenhängende kompakte homogene Kählersche Mannigfaltigkeit ist das direkte Produkt aus einem komplexen Torus und einer projektiv-rationalen Mannigfaltigkeit*[1]).

Spezialfälle dieser Aussage sind bekannt. Für projektiv-algebraische Mannigfaltigkeiten wurde sie bereits in [4], § 16, angekündigt; der Beweis, welcher für Grundkörper beliebiger Charakteristik gilt, soll an anderer Stelle dargestellt werden. Unter der zusätzlichen Voraussetzung, daß eine kompakte Gruppe von Holomorphismen transitiv auf der gegebenen Kählerschen Mannigfaltigkeit wirkt, bewies Y. MATSUSHIMA den Satz (vgl. [10]); im Falle, daß eine mit V gleichdimensionale komplexe Liesche Gruppe von Holomorphismen transitiv wirkt, gab H. C. WANG (vgl. [15]) einen Beweis (siehe hierzu Folgerung zum Satz 1).

Der Beweis von Satz I wird in den Abschnitten 2—5 vorbereitet. Wir zeigen, daß jede kompakte homogene komplexe Mannigfaltigkeit V sich stets auf zwei Weisen fasern läßt: einmal (Satz 1) ist *V ein holomorphes Faserbündel über dem Albanesetorus von V*; zum anderen (Satz 7) kann V in Verallgemeinerung eines Satzes von GOTO in kanonischer Weise als *ein holomorphes Faserbündel über einer projektiv-rationalen homogenen Mannigfaltigkeit mit einer komplex-parallelisierbaren Faser* dargestellt werden. Im Kählerschen Fall lassen sich über diese beiden Faserbündel unter Heranziehung der Theorie der komplexen Lieschen Gruppen weitere Aussagen machen, die dann im Abschnitt 6 einen einfachen Beweis von Satz I ermöglichen.

*) Unterstützt durch Directorate of Mathematical Sciences, AFOSR, European Office of Aerospace Research, US Air Force, Grant No. AF-EOAR-61-50.

[1]) Eine kompakte komplexe Mannigfaltigkeit heißt *projektiv-rational*, wenn sie projektiv-algebraisch ist und einen (über C) rationalen Körper meromorpher Funktionen besitzt. Es sei daran erinnert, daß nach einem Satz von CHOW eine analytische Menge in einer projektiv-algebraischen Mannigfaltigkeit selbst stets projektiv-algebraisch ist.

Unter einem komplexen Torus verstehen wir stets einen komplexen Periodentorus, d. h. eine Quotientenmannigfaltigkeit eines Zahlenraumes C^n nach einem reell $2n$-dimensionalen Gitter.

Liesche Gruppen werden vorwiegend mit $G, H, M, N, \ldots$ bezeichnet. Für ihre zusammenhängenden Komponenten der 1 schreiben wir entsprechend $G^0, H^0, M^0, N^0, \ldots$; $G(V)$ bezeichnet die Gruppe aller Holomorphismen von V.

2. Das Albanesebündel einer homogenen kompakten komplexen Mannigfaltigkeit. — Ausgangspunkt ist die folgende wohlbekannte Tatsache (vgl. [1], p. 163):

(a) *Es gibt zu jeder zusammenhängenden kompakten komplexen Mannigfaltigkeit V einen komplexen Torus $A(V)$ und eine holomorphe Abbildung $\alpha: V \to A(V)$, so daß jede holomorphe Abbildung $\beta: V \to B$ von V in irgendeinen komplexen Torus B sich in der Form $\beta = \tau \circ \alpha$ darstellen läßt, wobei $\tau: A(V) \to B$ eine durch β eindeutig bestimmte holomorphe affine Abbildung von $A(V)$ in B ist. Es gilt stets:* $\dim_R A(V) \leq b_1(V)$; *ist V kählersch, so gilt:* $\dim_R A(V) = b_1(V)$.

Der Torus $A(V)$ und die Abbildung α sind durch V eindeutig bestimmt, d. h. sind A' ein weiterer Torus und $\alpha': V \to A'$ eine weitere holomorphe Abbildung mit der obigen Universalitätseigenschaft, so gibt es eine biholomorphe affine Abbildung ι von A' auf $A(V)$, so daß $\alpha = \iota \circ \alpha'$.

Man nennt $A(V)$ den *Albanesetorus* zu V und entsprechend $\alpha: V \to A(V)$ die *Albaneseabbildung*. Die Zahl $a(V) := \dim_C A(V)$ nennen wir die *Albanesezahl* von V; ist V projektiv-algebraisch, so ist $a(V)$ die Irregularität von V.

Die Gruppe $G(V)$ aller Holomorphismen von V kann nach [2] in kanonischer Weise als eine komplexe Transformationsgruppe von V aufgefaßt werden. Falls $G(V)$ transitiv auf V wirkt, so auch $G(V)^0$. Mit $T(V)$ wollen wir die komplexe Liesche Gruppe der Translationen von $A(V)$ bezeichnen; es gilt $T(V) = G(A(V))^0$. Die Torusgruppe $T(V)$ ist — aufgefaßt als komplexe Mannigfaltigkeit — zu $A(V)$ biholomorph äquivalent. Wir werden wesentlich benutzen:

(a') *Es gibt einen holomorphen Gruppenhomomorphismus $\gamma: G(V) \to G(A(V))$, so daß für alle $g \in G(V)$ gilt:* $\alpha \circ g = \gamma(g) \circ \alpha$. *Es gilt* $\gamma(G(V)^0) \subset T(V)$. *Kern γ ist eine abgeschlossene komplexe Liesche Untergruppe von $G(V)$, die auf jeder α-Faser holomorph wirkt. Ein Holomorphismus $g \in G(V)^0$ gehört genau dann zu Kern γ, wenn es einen Punkt $v_0 \in V$ gibt, so daß v_0 und $g(v_0)$ in derselben α-Faser liegen.*

Ist insbesondere H eine Untergruppe von $G(V)^0$, die transitiv auf V wirkt, so wirkt $H \cap (\text{Kern } \gamma)$ transitiv auf jeder α-Faser.

Die Existenz von γ wurde von A. Blanchard bewiesen, vgl. [1], Proposition I. 2.1. Die Holomorphie von γ wird in [1] nicht explizit hervorgehoben, folgt jedoch aus der Definition von γ (vgl. [1], p. 165). Die übrigen Aussagen von (a') sind unmittelbar zu verifizieren.

In diesem Abschnitt soll nun bewiesen werden:

Satz 1: *Ist V eine zusammenhängende kompakte homogene komplexe Mannigfaltigkeit, so gilt:*

a) *Die Albaneseabbildung $\alpha: V \to A(V)$ und der Homomorphismus $\gamma: G(V)^0 \to T(V)$ sind surjektiv.*

b) *V ist bzgl. α ein holomorphes Faserbündel über $A(V)$. Die typische Faser F ist eine zusammenhängende bzgl. Kern γ homogene kompakte komplexe Mannigfaltigkeit; man hat eine exakte Sequenz*

$$0 \to \pi_1(F) \to \pi_1(V) \to \pi_1(A(V)) \to 0 .$$

Beweis: ad a). Wir setzen abkürzend $G^0 := G(V)^0$. Die Gruppe G^0 wirkt transitiv auf V. Wird $a_0 \in \alpha(V)$ fest gewählt, so gilt also wegen (a'): $\alpha(V) = \{t(a_0), t \in \gamma(G^0)\}$. Faßt man $A(V)$ als eine komplexe Torusgruppe mit a_0 als Nullpunkt auf, so ist mithin $\alpha(V)$ eine Untergruppe von $A(V)$. Da $\alpha(V)$ abgeschlossen in $A(V)$ liegt und die Bahn eines Punktes unter einer komplexen Transformationsgruppe von $A(V)$ ist, so ist $\alpha(V)$ eine analytische Menge in $A(V)$ (auch direkt nach [13], Satz 24). Mithin ist $\alpha(V)$ sogar ein komplexer Untertorus von $A(V)$. Nun besitzt dieser Torus $\alpha(V)$ zusammen mit der durch α induzierten holomorphen Abbildung $\alpha' : V \to \alpha(V)$ ersichtlich auch die Universalitätseigenschaft des Albanesetorus. Daher muß gelten $\alpha(V) = A(V)$. Dies impliziert weiter $\gamma(G^0) = T(V)$.

ad b). Wir bezeichnen den Kern von $\gamma : G^0 \to T(V)$ mit M. Ist dann H die Isotropiegruppe eines Punktes $v_0 \in V$ bzgl. G^0, so gilt $H \subset M$ nach (a'). Identifizieren wir V mit G^0/H und $A(V)$ mit G^0/M (letzteres ist nach a) möglich), so erscheint V bzgl. α als ein holomorphes Faserbündel über $A(V)$ mit der homogenen kompakten komplexen Mannigfaltigkeit M/H als typischer Faser F. Um zu zeigen, daß M/H zusammenhängend ist, betrachten wir die kleinste offene Untergruppe M' von M, die H enthält. M' besteht offensichtlich gerade aus denjenigen zusammenhängenden Komponenten von M, deren Durchschnitt mit H nicht leer ist. Daher ist M'/H zusammenhängend.

Die holomorphe Abbildung $\alpha : V \to A(V)$ ist nun das Produkt der holomorphen Abbildungen $\alpha_1 : V = G^0/H \to G^0/M'$ und $\alpha_2 : G^0/M' \to G^0/M = A(V)$. Die kompakte komplexe Mannigfaltigkeit G^0/M' ist bzgl. α_2 eine unverzweigte kompakte Überlagerung von $A(V)$ und also selbst ein komplexer Torus. Aus der Universalitätseigenschaft von $A(V)$ folgt, daß α_2 biholomorph ist. Dies impliziert $M = M'$, d. h. die typische Faser $F = M/H$ von α ist zusammenhängend.

Man hat jetzt die folgende exakte Homotopiesequenz

$$\cdots \to \pi_2(A(V)) \to \pi_1(F) \to \pi_1(V) \to \pi_1(A(V)) \to 0 .$$

Da $A(V)$ ein Torus ist, so gilt $\pi_2(A(V)) = 0$. Da weiter M nach (a') transitiv auf jeder α-Faser wirkt, ist Satz 1 vollständig bewiesen.

Aus Satz 1 ergibt sich sofort die folgende von H. C. WANG (vgl. [15], Theorem 1 und Corollary 2) herrührende Aussage:

Folgerung: *Auf einer kompakten Kählerschen Mannigfaltigkeit V kann genau dann eine mit V gleichdimensionale komplexe Liesche Gruppe G holomorph und transitiv wirken, wenn V ein komplexer Torus ist.*

Es ist klar, daß ein komplexer Torus diese Eigenschaften hat. Sei umgekehrt V eine kompakte Kählersche Mannigfaltigkeit und G eine mit V gleichdimensionale komplexe Liesche Gruppe, die holomorph und transitiv auf V wirkt. Ohne Einschränkung der Allgemeinheit sei G zusammenhängend. Dann ist V zu einer komplexen Quotientenmannigfaltigkeit G/H äquivalent, wo H eine diskrete Untergruppe von G ist. Ist nun $n := \dim_C G$, so gibt es n linear unabhängige (bzgl. der Rechtstranslationen von G) invariante holomorphe Differentialformen 1. Ordnung auf G (Maurer-Cartansche Formen). Diese Formen induzieren, da die natürliche Projektion $G \to G/H$ eine lokal-biholomorphe Abbildung ist, n linear unabhängige holomorphe Differentialformen 1. Ordnung auf V. Da V kählersch ist, so gilt also $b_1(V) \geq 2n$, so daß $A(V)$ mindestens mit V gleichdimensional ist. Nach Satz 1 ist dann aber $\alpha : V \to A(V)$ biholomorph.

3. Homogene projektiv-rationale Mannigfaltigkeiten. — Das Ziel dieses Abschnittes ist der Beweis von Satz 4. Wir zeigen vorbereitend

Satz 2: *Jede holomorphe Abbildung einer homogenen projektiv-rationalen Mannigfaltigkeit Q in sich hat einen Fixpunkt.*

Beweis: Nach [3] gestattet Q eine „analytische Zellenzerlegung": die $2s$-dimensionalen abgeschlossenen Zellen sind sämtlich s-dimensionale irreduzible algebraische Mengen in Q und bilden eine Homologiebasis $\{\gamma_1^{2s}, \ldots, \gamma_{j_s}^{2s}\}$ der $2s$-ten Homologiegruppe von Q, $s = 1, \ldots, \dim_C Q$. Aus Resultaten von W. L. Chow (vgl. [7]) folgt überdies, daß jeder irreduzible s-dimensionale analytische Zyklus in Q zu einem Zyklus $\sum\limits_{i=1}^{j_s} n_i \cdot \gamma_i^{2s}$ homolog ist, wobei stets $n_i \geq 0$. Da jede holomorphe Abbildung τ von Q in sich irreduzible analytische Mengen auf ebensolche abbildet (vgl. [13], Satz 24), so sind bzgl. der Basen $\{\gamma_i^{2s}\}$ die Spuren aller Homomorphismen, die von τ in den Homologiegruppen gerader Dimension induziert werden, nicht negativ. Da die Spur in der Dimension 0 jedenfalls 1 ist und alle Homologiegruppen ungerader Dimension verschwinden, so ist mithin die Wechselsumme über alle Spuren positiv. Dann hat die holomorphe Abbildung τ aber nach der Lefschetz-Hopfschen Formel einen Fixpunkt in Q, q. e. d.

Jede komplexe Liesche Gruppe G der komplexen Dimension n läßt sich in natürlicher Weise als eine $2n$-dimensionale reelle Liesche Gruppe auffassen. Wir bezeichnen diese durch G bestimmte reelle Liesche Gruppe durchweg mit G_r. Dann gilt:

Hilfssatz 3: *Es sei X eine projektiv-algebraische Mannigfaltigkeit mit $b_1(X) = 0$, M sei eine analytische Menge in X. Weiter sei H eine auflösbare zusammenhängende (reelle) Liesche Untergruppe von $G(X)_r$, so daß $h(M) = M$ für alle $h \in H$. Dann gibt es wenigstens einen Punkt $x_0 \in M$, so daß $h(x_0) = x_0$ für alle $h \in H$.*

Beweis: Da $b_1(X) = 0$, so gibt es nach dem théorème principal I aus [1] eine projektive Einbettung von X in einen P_s, so daß $G(X)^0$ mit der größten zusammenhängenden Untergruppe von $G(P_s)$, die X invariant läßt, identifiziert werden kann. Die Gruppe H ist dann eine auflösbare zusammenhängende (reelle) Liesche Untergruppe von $G(P_s)_r$. Es sei H_c diejenige zusammenhängende komplexe Liesche Untergruppe von $G(P_s)$, deren Liesche Algebra die Komplexifizierung der (reellen) Lieschen Algebra von H ist. H_c umfaßt H

und ist auflösbar. Nach einem klassischen Satz von LIE[2]) gibt es daher eine „Fahne" $T_0 \subset T_1 \subset \cdots \subset T_{s-1}$, wobei T_i eine i-dimensionale analytische Ebene im P_s ist, so daß $h(T_i) = T_i$ für alle $h \in H$ und $i = 0, 1, \ldots, s - 1$. Da M analytisch und also auch algebraisch im P_s ist, können wir den Index j, $0 \leq j \leq s - 1$, so wählen, daß $M \cap T_j$ wenigstens einen isolierten Punkt x_0 enthält. Da $H \subset H_c$ zusammenhängend ist und $M \cap T_j$ invariant läßt, gilt also $h(x_0) = x_0$ für alle $h \in H$, w. z. b. w.

Bemerkung: Anstelle von H_c könnte man im vorstehenden Beweis auch die kleinste algebraische Untergruppe H' von $G(P_s)$ betrachten, die H enthält. Sie ist auflösbar, zusammenhängend und läßt M invariant und besitzt daher nach [5], p. 64, einen Fixpunkt auf M.

Es folgt nun schnell

Satz 4: *Ist Q eine homogene projektiv-rationale Mannigfaltigkeit, so ist jede zusammenhängende reelle Liesche Untergruppe U von $G(Q)_r$, die transitiv auf Q wirkt, halb-einfach. Überdies besteht der Zentralisator von U in $G(Q)$ nur aus der Identität.*

Beweis: Es bezeichne $R(U)$ das Radikal[3]) von U. Da $b_1(Q) = 0$[4]), so hat die Gruppe $R(U)$ nach Hilfssatz 3 einen Fixpunkt auf Q. Da $R(U)$ ein Normalteiler von U ist und U transitiv auf Q wirkt, ist dann jeder Punkt von Q ein Fixpunkt von $R(U)$. Da U effektiv wirkt, folgt hieraus $R(U) = 1$. Die Gruppe U ist also halb-einfach.

Es bleibt zu zeigen, daß die Identität der einzige Holomorphismus von Q ist, der mit allen Elementen $u \in U$ vertauschbar ist. Sei $g_0 \in G(Q)$ so beschaffen, daß $g_0 \cdot u = u \cdot g_0$ für alle $u \in U$. Dann läßt U die Fixpunktmenge F_0 von g_0 invariant. Da F_0 nach Satz 2 nicht leer ist und U transitiv auf Q wirkt, folgt $F_0 = Q$. Da $G(Q)$ effektiv wirkt, ist g_0 also die Identität. — Satz 4 ist bewiesen.

4. Beweis der Trivialität gewisser Faserbündel. — In diesem Abschnitt wird gezeigt:

Satz 5: *Es sei Q eine zusammenhängende, projektiv-rationale Mannigfaltigkeit und E ein holomorphes Faserbündel über einem komplexen Torus A mit Q als typischer Faser. E ist genau dann eine homogene komplexe Mannigfaltigkeit, wenn Q homogen und E als Bündel über A biholomorph äquivalent zum direkten Produkt $A \times Q$ ist.*

Wir stützen uns auf das folgende

---

[2]) Der Satz von LIE lautet: *Es sei H eine auflösbare zusammenhängende komplexe Liesche Gruppe und $\varrho : H \to GL(n, C)$ ein holomorpher Homomorphismus. Dann gibt es ein Element $a \in GL(n, C)$, so daß alle Matrizen $a \cdot \varrho(H) \cdot a^{-1}$ Dreiecksmatrizen sind (alle Koeffizienten unter der Diagonalen sind 0).* — In unserem Falle wird dieser Satz auf die komplexe Untergruppe von $GL(s + 1, C)$ angewendet, die die Komponente der 1 des Urbildes von H_c unter dem Standardhomomorphismus $GL(s + 1, C) \to G(P_s)$ ist.

[3]) Das Radikal $R(G)$ einer Lieschen Gruppe G ist laut Definition die größte zusammenhängende auflösbare invariante Liesche Untergruppe von G. $R(G)$ ist stets abgeschlossen in G.

[4]) Es ist wohlbekannt, daß jede projektiv-rationale Mannigfaltigkeit einfach-zusammenhängend ist. Siehe etwa [8] und [14].

Lemma 6: *Es sei G eine zusammenhängende reelle oder komplexe Liesche Gruppe und S eine zusammenhängende abgeschlossene invariante reelle oder komplexe Liesche Untergruppe von G, die halb-einfach und ohne Zentrum ist. Die Faktorgruppe G/S sei auflösbar. Dann ist G ein direktes Produkt aus S und dem Radikal $R(G)$. Ist G/S insbesondere abelsch, so stimmt $R(G)$ mit dem Zentrum $Z(G)$ von G überein.*

Beweis: Das Radikal $R(G)$ wird (nach [6], p. 76, Corollaire 3, sowie [12], p. 278, Theorem 84), vermöge der natürlichen Projektion $G \to G/S$ auf das Radikal von G/S und daher auf G/S selbst abgebildet. Es gilt mithin $G = R(G) \cdot S$. Der Durchschnitt $R(G) \cap S$ ist eine abgeschlossene, invariante auflösbare Untergruppe von S und also, da S halb-einfach ist, diskret. Als diskreter Normalteiler liegt $R(G) \cap S$ sogar im Zentrum von S. Daraus ergibt sich $R(G) \cap S = 1$ und mithin: $G = R(G) \times S$. Ist nun G/S noch abelsch, so folgt weiter, da jedes Element von $R(G)$ mit jedem Element von S vertauschbar und $R(G)$ zu G/S isomorph ist, $R(G) = Z(G)$, q. e. d.

Wir beweisen nun Satz 5. Ist Q homogen und gilt $E = A \times Q$, so ist auch E homogen. Sei umgekehrt E homogen. Wir schreiben E in der Form G^0/H, wo $G^0 := G(E)^0$ und H die Isotropiegruppe eines Punktes von E in G^0 ist. Wir bezeichnen mit $\pi : E \to A$ die holomorphe Bündelprojektion und mit $\alpha : E \to A(E)$ die Albaneseabbildung. Da jede π-Faser F projektiv-rational ist und also einen einpunktigen Albanesetorus hat, so ist $\alpha \mid F$ immer konstant. Es gibt mithin eine holomorphe Abbildung $\iota : A \to A(E)$, so daß $\alpha = \iota \circ \pi$. Wegen der Universalität von $A(E)$ ist ι notwendig biholomorph und affin, so daß wir A mit $A(E)$ und π mit α identifizieren dürfen. Auf Grund von Satz 1, b) dürfen wir noch setzen $Q = M/H$, wenn M den Kern von $\gamma : G^0 \to T(E)$ bezeichnet. Die Gruppen M und M^0 sind abgeschlossene invariante komplexe Liesche Untergruppen von G, beide wirken transitiv auf jeder α-Faser. Für jeden Punkt $a \in A$ sei $M_a := \{g \in M, g \mid \overset{-1}{\pi}(a) = \text{Identität}\}$ und $M'_a := M_a \cap M^0$. Die Gruppen M_a bzw. M'_a sind jeweils Normalteiler von M bzw. M^0 und es gilt: $\underset{a \in A}{\cap} M_a = 1$, da M effektiv auf E wirkt. Die zusammenhängende komplexe Liesche Gruppe M^0/M'_a wirkt nun holomorph, transitiv und effektiv auf der Faser $\overset{-1}{\pi}(a)$. Nach Satz 4 ist daher M^0/M'_a stets halb-einfach und ohne Zentrum. Dann ist aber M^0 selbst halb-einfach und ohne Zentrum: das Radikal bzw. Zentrum von M^0 wird nämlich vermöge eines jeden Homomorphismus $M^0 \to M^0/M'_a$, $a \in A(E)$, in das Radikal bzw. Zentrum von M^0/M'_a, d. h. auf die 1 abgebildet. Radikal und Zentrum von M^0 sind daher in jeder Gruppe M'_a enthalten und bestehen folglich nur aus dem neutralen Element.

Da G^0/M^0 als zusammenhängende Überlagerungsgruppe der abelschen Gruppe $T(E)$ selbst abelsch ist, so folgt aus Lemma 6, daß $G^0 = Z \times M^0$, wobei Z das Zentrum von G^0 bezeichnet. Speziell gilt also $M = (M \cap Z) \times M^0$. Da $M \cap Z$ eine zentrale komplexe Liesche Untergruppe von M ist und M/M_a als eine transitiv und effektiv auf $\overset{-1}{\pi}(a)$ wirkende komplexe Liesche Untergruppe von $G(\overset{-1}{\pi}(a))$ aufgefaßt werden kann, so wird $M \cap Z$ vermöge $M \to M/M_a$

in den Zentralisator von M/M_a in $G(\overset{-1}{\pi}(a))$ und daher wegen Satz 4 auf die 1 abgebildet. Es gilt folglich $M \cap Z \subset M_a$ für jedes $a \in A$ und daher $M \cap Z = 1$. Dies bedeutet $M = M^0$, d. h. $G^0 = Z \times M$. Da H als Isotropiegruppe in M enthalten ist und $Z = G/M$ mit A identifiziert werden darf, so folgt hieraus $E = G^0/H = Z \times (M/H) = A \times Q$, w. z. b. w.

5. Verallgemeinerung eines Satzes von Goto. — In [9] wird u. a. bewiesen (vgl. proposition 3, p. 818), daß jede kompakte homogene komplexe Mannigfaltigkeit V mit endlicher Fundamentalgruppe ein holomorphes Faserbündel über einer projektiv-rationalen Mannigfaltigkeit mit einem komplexen Torus als Faser ist. Wir zeigen hier:

Satz 7: *Jede kompakte homogene komplexe Mannigfaltigkeit V ist in natürlicher Weise ein holomorphes Faserbündel über einer projektiv-rationalen (einfach-zusammenhängenden) homogenen Mannigfaltigkeit B mit einer zusammenhängenden komplex-parallelisierbaren Faser P.*

Dabei möge allgemein eine r-dimensionale komplexe Mannigfaltigkeit X *komplex-parallelisierbar* genannt werden, wenn es r holomorphe Vektorfelder auf X gibt, die in jedem Punkt von X komplex linear-unabhängig sind. Es ist klar, daß X stets dann komplex-parallelisierbar ist, wenn es eine r-dimensionale komplexe Liesche Gruppe gibt, die holomorph und transitiv auf X wirkt. Nach H. C. Wang [15] gilt die Umkehrung, wenn X zusammenhängend und kompakt ist.

Im Beweise von Satz 7 sowie später werden wir wesentlich benutzen:

Satz *: *Jede zusammenhängende kompakte homogene Kählersche Mannigfaltigkeit mit endlicher Fundamentalgruppe ist projektiv-rational und insbesondere einfach-zusammenhängend.*

Vergleiche hierzu [3] und [9]. Der Beweis des erstgenannten Verfassers für den einfachen Zusammenhang ist im Vortragsexposé von J. P. Serre (Séminaire Bourbaki, Mai 1954) dargestellt.

Um nun Satz 7 zu beweisen, bezeichnen wir mit H die Isotropiegruppe eines Punktes von V in $G(V)^0$ und mit N den Normalisator von H^0 in $G(V)^0$. N ist eine abgeschlossene komplexe Liesche Untergruppe von G, die H umfaßt. Daher gibt es eine natürliche holomorphe Faserbündelabbildung von $V = G/H$ auf die homogene kompakte komplexe Mannigfaltigkeit $B := G/N$ mit $P := N/H$ als typischer Faser. Nach Definition ist H^0 invariant in N, so daß auch gilt: $P = (N/H^0)/(H/H^0)$. Da H/H^0 diskret in N/H^0 liegt, ist P komplex-parallelisierbar. Um zu zeigen, daß dieses Faserbündel auch die weiteren im Satz 7 behaupteten Eigenschaften hat, genügt es also zu beweisen, daß N zusammenhängend und B projektiv-rational ist. Dies folgt aber unmittelbar aus

Satz 7': *Es sei G eine zusammenhängende komplexe Liesche Gruppe und H eine abgeschlossene komplexe Liesche Untergruppe von G, so daß G/H kompakt ist. Dann gilt:*

1. *Der Normalisator N von H^0 in G enthält das Radikal $R(G)$ von G[5].*
2. *N ist zusammenhängend und G/N ist projektiv-rational.*

---

[5]) Diese erste Aussage wurde uns samt Beweis von J. Tits mündlich mitgeteilt.

Beweis: Es genügt offenbar, den Satz für den Fall, daß G einfach-zusammenhängend ist, zu beweisen. Sei $n := \dim_C G$, $k := \dim_C H$. Wir fassen die Liesche Algebra $\mathfrak{h}$ von H als einen Punkt der Grassmannschen Mannigfaltigkeit $M_{n,k}$ der k-dimensionalen Untervektorräume der Lieschen Algebra $\mathfrak{g}$ von G auf. Vermöge der adjungierten Darstellung wirkt G in natürlicher Weise holomorph auf $M_{n,k}$, und die Isotropiegruppe von $\mathfrak{h}$ in G ist genau N. Da $H \subset N$, so gibt es eine holomorphe Abbildung von G/H auf die Bahn $B \approx G/N$ von $\mathfrak{h} \in M_{n,k}$ unter G. Da G/H nach Voraussetzung kompakt ist, so ist B eine analytische Menge in $M_{n,k}$, auf welcher G transitiv wirkt. Da $M_{n,k}$ eine projektiv-algebraische Mannigfaltigkeit mit verschwindender erster Bettischer Zahl ist, so hat $R(G)$ nach Hilfssatz 3 einen Fixpunkt auf B. Da $R(G)$ ein Normalteiler von G ist und G transitiv auf B wirkt, ist dann jeder Punkt von B, insbesondere auch $\mathfrak{h}$, ein Fixpunkt von $R(G)$. Hieraus folgt aber $R(G) \subset N$, womit die Aussage 1. bereits bewiesen ist.

Nach dem Satz von Levi-Malcev (vgl. [6], théorème 5, p. 89; sowie [12], theorem 84, p. 278) ist G das halb-direkte Produkt $R(G) \cdot S$ von $R(G)$ mit einer abgeschlossenen halb-einfachen (i. a. nicht invarianten) komplexen Lieschen Untergruppe S von G. Mit G sind auch $R(G)$ und S einfach-zusammenhängend. Da $R(G) \subset N^0$, so gilt

$$N = R(G) \cdot L, \ N^0 = R(G) \cdot L^0, \text{ wo } L := N \cap S \ .$$

Hieraus ergibt sich für $B = G/N$ die neue Quotientendarstellung $B = S/L$.

Die Gruppe S ist als halb-einfache komplexe Liesche Gruppe algebraisch, und die durch die adjungierte Darstellung von G induzierte Darstellung $x \to A d_{\mathfrak{g}} x$ von S ist rational (siehe Bemerkung am Ende des Beweises). Da $L = \{x \in S, A d_{\mathfrak{g}} x(\mathfrak{h}) = \mathfrak{h}\}$ nach Definition von L, so ist L also eine algebraische Untergruppe von S. Als solche zerfällt L in nur endlich viele zusammenhängende Komponenten, d. h. L/L^0 ist eine endliche Gruppe. Da $B = S/L$ und S zusammenhängend und einfach-zusammenhängend ist, so hat man $\pi_1(B) \cong$ $\cong L/L^0$ nach der Homotopiesequenz. Daher ist die Fundamentalgruppe der homogenen komplexen Mannigfaltigkeit B endlich. Da nach dem oben Bewiesenen B als analytische Menge in der Grassmannschen Mannigfaltigkeit $M_{n,k}$ liegt, ist B auch projektiv-algebraisch. Es folgt dann aus Satz *, daß B sogar projektiv-rational ist. Insbesondere gilt also $L/L^0 = \pi_1(B) = 1$, d. h. $L = L^0$. Da dies $N = N^0$ impliziert, so ist Satz 7' vollständig bewiesen.

Bemerkung: Die oben benutzte Eigenschaft von S ist wohlbekannt; sie kann etwa wie folgt bewiesen werden. Jede zusammenhängende halb-einfache komplexe Liesche Gruppe G besitzt eine treue holomorphe Darstellung (siehe Séminaire Sophus Lie, Paris 1954; Exp. 22). Da eine zusammenhängende lineare komplexe Liesche Gruppe, deren Liesche Algebra mit ihrer Kommutatoralgebra übereinstimmt, stets algebraisch ist (vgl. C. Chevalley: Théorie des groupes de Lie, Bd. 2, Groupes algébriques, Paris 1951; p. 177), so darf man also G mit einer algebraischen Untergruppe von $GL(m, C)$ identifizieren, wenn m genügend groß gewählt wird. Sei nun $\varrho : G \to GL(n, C)$ irgendeine holomorphe Darstellung von G. Der Graph $\Gamma_\varrho := \{(g, \varrho(g))\}$ von ϱ ist eine halb-einfache komplexe Liesche Untergruppe von $GL(m, C) \times GL(n, C)$ und mithin nach dem oben Bemerkten algebraisch. Die Projektion von $GL(m, C) \times GL(n, C)$ auf den ersten Faktor induziert eine bijektive rationale Abbildung von Γ_ϱ auf G. Da G und Γ_ϱ singularitätenfrei sind, ist diese Abbildung $\Gamma_\varrho \to G$ sogar biholomorph und birational (etwa nach einem elementaren Spezialfall von Zariskis "Main theorem", vgl. z. B. S. Lang: Introduction to algebraic geometry, Interscience Publishers, New York, 1958, Chap. V). Dies impliziert, daß auch ϱ selbst eine rationale Abbildung ist.

Corollar 1 (Goto): *Ist $\pi_1(V)$ endlich, so ist P ein komplexer Torus.*

Beweis: Es sei G eine universelle Überlagerungsgruppe von $G(V)^0$. Wir behalten die Bezeichnungen des Beweises von Satz 7' bei. Dann ist die Fundamentalgruppe von V zur Gruppe H/H^0 isomorph. Auf Grund der gemachten Voraussetzung ist also H/H^0 endlich. Aus $P = (N/H^0)/(H/H^0)$ folgt daher, da P kompakt ist, daß N/H^0 eine kompakte komplexe Liesche Gruppe und also eine komplexe Torusgruppe ist. Dann ist aber P selbst notwendig ein komplexer Torus, w. z. b. w.

Corollar 2: *Jede zusammenhängende kompakte komplexe Mannigfaltigkeit V, auf der eine auflösbare komplexe Liesche Gruppe G holomorph und transitiv wirkt, ist komplex-parallelisierbar.*

In der Tat! Ohne Einschränkung der Allgemeinheit dürfen wir G als eine zusammenhängende Gruppe voraussetzen, die effektiv auf V wirkt. Setzen wir wieder $V = G/H$, so folgt nach Satz 7', 1), da G auflösbar ist, daß H^0 ein Normalteiler von G ist. Da G effektiv auf V wirkt, folgt hieraus $H^0 = 1$. V ist mithin Quotient von G nach der diskreten Gruppe H und folglich komplex-parallelisierbar.

6. Beweis von Satz I. Folgerungen. — Nach den in den Abschnitten 2—5 getroffenen Vorbereitungen ist nun der Beweis von Satz I einfach zu führen. Wir betrachten für die gegebene kompakte homogene Kählersche Mannigfaltigkeit V zunächst die im Satz 7 beschriebene natürliche holomorphe Faserung. Da die Faser P ebenfalls eine Kählersche Mannigfaltigkeit ist, so ist P nach der Folgerung aus Satz 1 ein komplexer Torus. Insbesondere ist $\pi_1(P)$ abelsch. Aus der exakten Homotopiesequenz

$$\cdots \to \pi_1(P) \to \pi_1(V) \to \pi_1(B) = 0$$

ergibt sich dann, daß auch $\pi_1(V)$ abelsch ist. Nach Satz 1 ist V weiter bzgl. der Albaneseabbildung α ein holomorphes Faserbündel über $A(V)$ mit einer zusammenhängenden typischen Faser F. Aus der exakten Sequenz $0 \to \pi_1(F) \to \ \to \pi_1(V) \to \pi_1(A(V)) \to 0$ folgt, daß F ebenfalls eine abelsche Fundamentalgruppe hat und weiter, daß für die 1. Bettischen Zahlen von V, $A(V)$, F gilt: $b_1(V) = b_1(A(V)) + b_1(F)$. Da V eine Kählersche Mannigfaltigkeit ist, so hat man $b_1(V) = \dim_R A(V) = b_1(A(V))$. Es gilt mithin $b_1(F) = 0$, wodurch sich die Fundamentalgruppe von F als endlich herausgestellt hat. Da auch F eine homogene kompakte Kählersche Mannigfaltigkeit ist, so ist F nach Satz * projektiv-rational. Aus Satz 5 folgt dann $V = A(V) \times F$, womit Satz I bewiesen ist.

Wir notieren abschließend noch drei Folgerungen aus Satz I.

Folgerung 1: *Ist V eine zusammenhängende kompakte homogene Kählersche Mannigfaltigkeit, so ist $G(V)^0$ reduktiv und es wirkt bereits jede maximale kompakte Untergruppe von $G(V)^0$ transitiv auf V. Insbesondere kann man V stets so mit einer Kählerschen Metrik versehen, daß die Gruppe aller holomorphen Isometrien transitiv auf V wirkt.*

Der Beweis ergibt sich unmittelbar aus Satz I und der Tatsache, daß auf einer zusammenhängenden und einfach-zusammenhängenden kompakten Mannigfaltigkeit die maximalen kompakten Untergruppen einer transitiv wirkenden Lieschen Gruppe von Homöomorphismen selbst transitiv wirken (vgl. [11]).

Folgerung 2: *Es sei V eine zusammenhängende kompakte Kählersche Mannigfaltigkeit und $H \subset G(V)_r$ eine reelle Liesche Gruppe, die transitiv auf V wirkt. Dann gilt:*

   a) *Ist H auflösbar, so ist V ein komplexer Torus.*

   b) *Ist H halb-einfach, so ist V projektiv-rational.*

   c) *Gilt $\dim_R H = \dim_R V$, so ist V ein komplexer Torus.*

Beweis: Wir schreiben nach Satz I: $V = A \times F$, wo A der Albanesetorus von V und F projektiv-rational ist. Dann gilt: $G(V)^0 = G(A)^0 \times G(F)^0$ (vgl. [1], p. 161, Corollaire); dabei ist $G(A)^0$ eine komplexe (abelsche) Torusgruppe und $G(F)^0$ nach Satz 4 eine halb-einfache komplexe Liesche Gruppe. Sind $\gamma: G(V)^0 \to G(A)^0$ und $\delta: G(V)^0 \to G(F)^0$ die kanonischen holomorphen Epimorphismen, so betrachten wir die Gruppen $\gamma(H^0)$ und $\delta(H^0)$. Da H^0 ebenfalls transitiv auf V wirkt, so wirken $\gamma(H^0)$ bzw. $\delta(H^0)$ transitiv auf A bzw. F; insbesondere gilt also: $\gamma(H^0) = G(A)^0$.

Ist nun H auflösbar, so ist $\delta(H^0)$ eine zusammenhängende auflösbare reelle Liesche Untergruppe von $G(F)_r$. Nach Satz 4 ist $\delta(H^0)$ aber auch halb-einfach. Daraus folgt $\delta(H^0) = 1$; d. h. F besteht aus genau einem Punkt.

Ist H halb-einfach, so ist $\gamma(H^0) = G(A)^0$ eine abelsche, halb-einfache reelle Liesche Gruppe. Das bedeutet $G(A)^0 = 1$, d. h. A ist ein einziger Punkt.

Es sei jetzt $\dim_R H = \dim_R V$. Die reelle Liesche Gruppe $N := G(F)^0 \cap H^0$ ist der Kern von $\gamma \mid H^0 : H^0 \to G(A)^0$ und wirkt deshalb transitiv auf F. Weiter gilt:

$$\dim_R N + \dim_R G(A)^0 = \dim_R H^0 \,,$$

woraus mit

$$\dim_R H^0 = \dim_R V = \dim_R A + \dim_R F, \ \dim_R A = \dim_R G(A)^0$$

folgt, daß N mit F gleichdimensional ist. F ist mithin der Quotient von N^0 nach einer diskreten Untergruppe. Da F einfach-zusammenhängend und N^0 zusammenhängend ist, so sind F und N^0 topologisch äquivalent; insbesondere ist N^0 eine kompakte reelle Liesche Gruppe. Nun gilt, da F homogen und projektiv-rational ist, $\chi(F) > 0$, wenn $\chi(F)$ die Euler-Poincarésche Charakteristik von F bezeichnet (vgl. [3], [9]). Andererseits verschwindet χ für jede zusammenhängende, nicht einpunktige kompakte reelle Liesche Gruppe, da es fixpunktfreie Homöomorphismen gibt, die der Identität homotop sind. Aus $\chi(F) = \chi(N^0)$ folgt daher, daß F einpunktig ist, q. e. d.

Folgerung 3: *Jede homogene kompakte Kählersche Mannigfaltigkeit, deren universelle Überlagerung eine Zelle ist, ist ein komplexer Torus.*

Denn die universelle Überlagerung von V ist nach Satz I das Produkt aus der universellen Überlagerung von $A(V)$ mit F.

Literatur

[1] BLANCHARD, A.: Sur les variétés analytiques complexes. Ann. sci. école norm. super. **73**, 157—202 (1956).

[2] BOCHNER, S., and D. MONTGOMERY: Groups on analytic manifolds. Ann. Math. **48**, 659—669 (1947).

[3] BOREL, A.: Kählerian coset spaces of semisimple Lie groups. Proc. Nat. Acad. Sci. U. S. **40**, 1147—1151 (1954).

[4] BOREL, A.: Topology of Lie groups and characteristic classes. Bull. Am. Math. Soc. **61**, 397—432 (1955).

[5] BOREL, A.: Groupes linéaires algébriques. Ann. Math. **64**, 20—82 (1956).

[6] BOURBAKI, N.: Groupes et Algèbres de Lie, Chap. I. Paris: Hermann 1960.

[7] CHOW, W. L.: Algebraic varieties with rational dissections. Proc. Nat. Acad. Sci. U. S. **42**, 116—119 (1956).

[8] EHRESMANN, CH.: Sur la topologie de certains espaces homogènes. Ann. Math. **35**, 396—443 (1934).

[9] GOTO, M.: On algebraic homogeneous spaces. Am. J. Math. **76**, 811—818 (1954).

[10] MATSUSHIMA, Y.: Sur les espaces homogènes kählériens d'un groupe de Lie reductif. Nagoya Math. J. **11**, 53—60 (1957).

[11] MONTGOMERY, D.: Simply connected homogeneous spaces. Proc. Am. Math. Soc. **1**, 467—469 (1950).

[12] PONTRJAGIN, L.: Topological groups. Princeton University Press 1946.

[13] REMMERT, R.: Holomorphe und meromorphe Abbildungen komplexer Räume. Math. Ann. **133**, 328—370 (1957).

[14] SERRE, J. P.: On the fundamental group of a unirational variety. J. London Math. Soc. **34**, 481—484 (1959).

[15] WANG, H. C.: Complex parallisable manifolds. Proc. Am. Math. Soc. **5**, 771—776 (1954).

(Eingegangen am 11. September 1961)

58.

(with Harish-Chandra)

Arithmetic subgroups of algebraic groups

Ann. Math., (2) **75** (1962) 485–535

TABLE OF CONTENTS

Introduction

A complex matric group $G \subset \mathrm{GL}(n, \mathbf{C})$ is algebraic, defined over $\mathbf{Q}$, if it consists of all invertible matrices whose coefficients annihilate some set of polynomials on $\mathbf{M}(n, \mathbf{C})$ with rational coefficients. In this case, let $G_{\mathbf{Z}}$ be the subgroup of elements of G which have integral coefficients, determinant ± 1, and $G_{\mathbf{R}} = G \cap \mathrm{GL}(n, \mathbf{R})$. Then $G_{\mathbf{Z}}$ is an arithmetically defined subgroup of $G_{\mathbf{R}}$, or more briefly, an *arithmetic subgroup* of $G_{\mathbf{R}}$. Typical examples are $\mathrm{SL}(n, \mathbf{Z}) \subset \mathrm{SL}(n, \mathbf{R})$, Siegel's modular group, or the group of units of a non-degenerate rational quadratic form; and the main purpose of this paper is to generalize facts known in these and other cases involving classical groups, chiefly from reduction theory. In particular, we shall prove that $G_{\mathbf{Z}}$ is finitely generated, and give necessary and sufficient conditions under which $G_{\mathbf{R}}/G_{\mathbf{Z}}$ is compact, or of finite invariant measure. In analogy with the terminology used in the classical cases, we shall also call $G_{\mathbf{Z}}$ the *group of units* of $G_{\mathbf{R}}$.

In view of known facts about algebraic groups and algebraic tori, the main case to investigate is that of semi-simple groups, and this paper is mainly concerned with the latter. However, it turns out that the reductive groups (that is fully reducible groups, or groups whose identity component is isogenous to the product of an algebraic torus by a semi-simple group) form the natural domain of validity for some results, and part of the discussion will be carried out directly for reductive groups. The two following properties will be particularly useful for our purposes:

(A) Let $G_1 \supset \cdots \supset G_m$ be reductive real algebraic subgroups of $\mathbf{GL}(n, \mathbf{R})$. Then there exists $a \in \mathbf{SL}(n, \mathbf{R})$ such that the groups $a \cdot G_i \cdot a^{-1}$ are all stable under $x \to {}^t x$ (1.9).

(B) Let G be a connected complex algebraic reductive group defined over $\mathbf{Q}$, H a closed subgroup defined over $\mathbf{Q}$. Then H is reductive if and only if G/H can be realized as the closed orbit of a rational point in a rational representation defined over $\mathbf{Q}$ (3.8).

The statement (A) is known [21, § 7], and (B) is a slight sharpening of known facts. Proofs have been included for the sake of completeness. However, the techniques used in them will seldom occur elsewhere in the paper, so that the reader who wishes to proceed as directly as possible to the main part of the paper may skip 1.5 to 1.9, the proofs of 1.10, 1.11, and 3.4 to 3.8 without serious inconvenience. Section 2 is also preliminary, and collects some basic facts and notions about algebraic groups.

Section 4 introduces Siegel domains. Let G be a real algebraic semi-simple, or reductive group, and $G = K \cdot A \cdot N$ an Iwasawa decomposition of G (see 1.11). A Siegel domain $\mathfrak{S}_{t,\omega}(t > 0; \omega$ a compact set in $N)$ is the set of elements $K \cdot A_t \cdot \omega$, where A_t is the set of exponentials of elements on which the simple restricted roots have values smaller than t. Their chief properties are:

(i) the set of elements $a \cdot n \cdot a^{-1}(a \in A_t, n \in \omega)$ is relatively compact (4.2);

(ii) if G is semi-simple, $\mathfrak{S}_{t,\omega}$ has finite Haar measure (4.3).

When $G = \mathbf{GL}(n, \mathbf{R})$ and $K \cdot A \cdot N$ is the standard Iwasawa decomposition, it is classical that $\mathfrak{S}_{t,\omega}$ meets only a finite number of its right translates under $x \cdot G_\mathbf{Z} \cup G_\mathbf{Z} \cdot x$ $(x \in G_\mathbf{Q})$, and that $G = \mathfrak{S}_{t,\omega} \cdot \mathbf{SL}(n, \mathbf{Z})$ for t, ω big enough (see 2.5 for references). These facts will be used in the present paper.

Section 5 is devoted to a finiteness lemma (5.3, 5.4), which generalizes the finiteness of the number of integral reduced forms with a given non-zero determinant. This lemma, together with (A), (B), is used in 6.5 to show that if G is reductive, defined over $\mathbf{Q}$, there exist open sets U in $G_\mathbf{R}$ such that $G_\mathbf{R} = U \cdot G_\mathbf{Z}$, $K \cdot U = U$ for a suitable maximal compact subgroup K; and $U^{-1} \cdot U \cap (x \cdot G_\mathbf{Z} \cdot y)$ is finite if $x, y \in G_\mathbf{Q}$. The construction of U is analogous to Hermite's procedure to obtain a fundamental domain for the group of units of an indefinite rational form in the space of majorizing forms; however, we shall operate directly in the group, instead of using the symmetric space G/K. The finite generation of $G_\mathbf{Z}$ follows immediately. It is also shown (6.9) that in a rational representation of G defined over $\mathbf{Q}$, the integral points contained in a closed orbit form a finite number of orbits of $G_\mathbf{Z}$. Applied to the natural representation of $\mathbf{SL}(n, \mathbf{Z})$

in the space of quadratic forms, this yields the finiteness of the number of classes of integral forms with a given non-zero determinant. Theorem 6.5 is extended to general algebraic groups in 6.12. In 6.11 it is proved that if $\mu: G \to G'$ is an isogeny, defined over $\mathbf{Q}$, then $\mu(G_\mathbf{Z})$ and $G'_\mathbf{Z}$ are commensurable.

The finiteness of the measure of $G_\mathbf{R}/G_\mathbf{Z}$ is proved in § 7 when G is semi-simple, in § 9 when the identity component G^0 of G has no non-trivial rational character defined over $\mathbf{Q}$ (see 7.8, 9.4). In § 7, the basic lemma is 7.5, which says roughly that if $\mathfrak{S}$ is a Siegel domain of a real algebraic semi-simple Lie group, and G_1 is a suitably embedded semi-simple subgroup of G, then $\mathfrak{S} \cdot x \cap G_1$ is contained in the union of a finite number of translates of a Siegel domain of G_1. Theorem 9.4 follows easily from 7.8, a result of Ono [22] on algebraic tori, and some remarks on rational characters made in § 8.

Section 10 is a preliminary to § 11, and discusses closed conjugacy classes. It is shown that if G is an algebraic group, the conjugacy class of an element x in G, (or in the Lie algebra of G), is closed if x is semi-simple, and not closed if G is reductive and x not semi-simple (10.1).

Section 11 gives a necessary and sufficient condition for $G_\mathbf{R}/G_\mathbf{Z}$ to be compact. When G is semi-simple, the condition is that $G_\mathbf{Q}$ consists of semi-simple elements. It generalizes the compactness of $G_\mathbf{R}/G_\mathbf{Z}$ when $G_\mathbf{Q}$ is the multiplicative group of elements of norm 1 in a division algebra over $\mathbf{Q}$, or when G is the orthogonal group of a rational form which does not represent zero. This condition had been conjectured by R. Godement. Its necessity follows easily from 10.1. Conversely, if $G_\mathbf{Q}$ consists of semi-simple elements, then 10.1 implies the existence of a locally faithful rational representation defined over $\mathbf{Q}$ (the adjoint representation) in which all rational points have closed orbits. The main part of the proof starts from that fact, and is a suitable adaptation of a known argument used in the classical case.

The definition of groups of units given above can be generalized by replacing $\mathbf{Q}$ and $\mathbf{Z}$ by a number field K and the ring of algebraic integers of K respectively. However this case is reduced to the previous one by the well-known operation of "restriction of the ground field", and the main results of the paper extend automatically to groups over number fields, as will be seen in § 12.

In § 13, we have relegated some remarks on algebraic groups, not used in the present paper, but which may be viewed as natural complements to some auxiliary results proved in §§ 1, 8.

The main results of this paper have been summarized in [3]. The applications to adele groups announced in [2] will be published elsewhere.

Notation. Linear representations will always be right representations. $H\backslash G$ (resp. G/H) is the space of cosets Hg (resp. gH) of a group G modulo a subgroup H. The identity component of a topological group G is denoted G^0. If a group G operates on a space M, the subgroup of G leaving a given point $m \in M$ fixed, (the *isotropy group* of m), is denoted by G_m. diag $(a_1, \cdots, a_n)$ is a diagonal $n \times n$ matrix with diagonal coefficients $a_1, \cdots, a_n$. The Lie algebra of a Lie group $G, H, M, \cdots$ is denoted by the corresponding German letter.

1. Reductive real algebraic groups

1.1. Let $\mathfrak{g}$ be a real semi-simple Lie algebra, $\mathfrak{g} = \mathfrak{k} + \mathfrak{p}$ a Cartan decomposition of $\mathfrak{g}$ and $\theta\colon k + p \to k - p$ the corresponding Cartan involution. It is convenient to allow $\mathfrak{g}$ to be compact, then $\mathfrak{p} = 0$ and θ is the identity. The Cartan involutions and Cartan decompositions are conjugate under Ad$\mathfrak{g}$ (see [20], for instance). The Cartan involutions of $\mathfrak{sl}(n, \mathbf{R})$ are the transformations $x \to -x^*$, where x^* is the adjoint of x with respect to some positive non-degenerate quadratic form on $\mathbf{R}^n$. They are also involutive automorphisms of $\mathfrak{gl}(n, \mathbf{R})$, to be called the Cartan involutions of $\mathfrak{gl}(n, \mathbf{R})$. Similarly, the automorphisms $x \to (x^{-1})^*$ of $\mathbf{SL}(n, \mathbf{R})$ or $\mathbf{GL}(n, \mathbf{R})$ will be called the Cartan involutions of these groups. The restriction of a Cartan involution of $\mathfrak{gl}(n, \mathbf{R})$ to a semi-simple subalgebra $\mathfrak{g}$, which is stable under it, is a Cartan involution of $\mathfrak{g}$.

Given a subspace $\mathfrak{m}$ of $\mathfrak{g}$ or of $\mathfrak{gl}(n, \mathbf{R})$, and a Cartan involution θ of $\mathfrak{g}$ or $\mathfrak{gl}(n, \mathbf{R})$, we put

$$\mathfrak{m}_{\mathfrak{k}} = \mathfrak{m} \cap \mathfrak{k}, \qquad \mathfrak{m}_{\mathfrak{p}} = \mathfrak{m} \cap \mathfrak{p}.$$

If $\mathfrak{m} = \theta(\mathfrak{m})$, then $\mathfrak{m} = \mathfrak{m}_{\mathfrak{k}} + \mathfrak{m}_{\mathfrak{p}}$ (and conversely), and $\mathfrak{m}$ is spanned by semi-simple elements.

It is standard that, given a real or complex representation of $\mathfrak{g}$ in a finite dimensional vector space V, there exists a Hilbert space structure on V with respect to which the elements of $\rho(\mathfrak{k})$ (resp. $\rho(\mathfrak{p})$) are skew-hermitian (resp. hermitian). The algebra $\rho(\mathfrak{g})$ and the analytic group it generates are then self-adjoint; this also implies that $\rho(x)$ is a semi-simple endomorphism of V with purely imaginary (resp. real) eigenvalues whenever $x \in \mathfrak{k}$ (resp. $x \in \mathfrak{p}$). If ρ is the adjoint representation, such a scalar product is given by $B^*(x, y) = -B(x, \theta(y))(x, y \in \mathfrak{g})$, where B is the Killing form of $\mathfrak{g}$.

If $\mathfrak{g} \subset \mathfrak{gl}(n, \mathbf{R})$, and ρ is the identical representation, the preceding remarks show that $\mathfrak{g}$ is stable under some Cartan involution of $\mathfrak{gl}(n, \mathbf{R})$. By the conjugacy of Cartan involutions, it follows that the Cartan involutions of $\mathfrak{g}$ are the restrictions to $\mathfrak{g}$ of the Cartan involutions of $\mathfrak{gl}(n, \mathbf{R})$ leaving $\mathfrak{g}$ invariant.

1.2. A subalgebra $\mathfrak{g}$ of a Lie algebra $\mathfrak{m}$ is *reductive in* $\mathfrak{m}$ if $\mathrm{ad}_{\mathfrak{m}}\mathfrak{g}$, (the image of $\mathfrak{g}$ in the adjoint representation of $\mathfrak{m}$), is completely reducible. A Lie algebra $\mathfrak{g}$ is *reductive* if it is so in itself. This is the case if and only if $\mathfrak{g}$ is the direct product of its center by a semi-simple ideal, which is then necessarily equal to the derived algebra $\mathscr{D}\mathfrak{g}$ of $\mathfrak{g}$ [5, § 6, No. 4].

LEMMA. *Let G be a closed subgroup of* $\mathbf{GL}(n, \mathbf{R})$ *with a finite number of connected components. Then the following conditions are equivalent:*

(i) *G is completely reducible;*

(ii) *$\mathfrak{g}$ is completely reducible;*

(iii) *$\mathfrak{g}$ is reductive in $\mathfrak{gl}(n, R)$;*

(iv) *$\mathfrak{g}$ is reductive, and its center consists of semi-simple endomorphisms.*

The identity component G^0 of G is invariant in G, of finite index. By an elementary fact (see e.g. G. D. Mostow, Amer. J. Math. 78 (1956), 200–221, Lemma 3.1), G is completely reducible if and only if G^0 is, hence (i) is equivalent to (ii). For the other equivalences, see [5, § 6, Nos. 3–6].

1.3. *A real algebraic group* is a subgroup of $\mathbf{GL}(n, \mathbf{R})$ which consists of all invertible matrices whose coefficients annihilate some set of polynomials with real coefficients, in n^2 indeterminates. A subalgebra of $\mathfrak{gl}(n, \mathbf{R})$ is algebraic if it is the Lie algebra of a real algebraic group. A real algebraic group is *reductive* if it is a completely reducible linear group.

1.4. LEMMA. *Let $\mathfrak{m}$ be an algebraic commutative fully reducible subalgebra of $\mathfrak{gl}(n, \mathbf{R})$. Then $\mathfrak{m} = \mathfrak{m}' + \mathfrak{m}''$, where $\mathfrak{m}'$ (resp. $\mathfrak{m}''$) consists of all elements of $\mathfrak{m}$ with purely imaginary (resp. real) eigenvalues, and is an algebraic subalgebra. $\mathfrak{m}$ is invariant under a Cartan involution of $\mathfrak{gl}(n, \mathbf{R})$. If θ is a Cartan involution of $\mathfrak{gl}(n, \mathbf{R})$ leaving $\mathfrak{m}$ invriant, then $\mathfrak{m}_{\mathfrak{k}} = \mathfrak{m}'$, $\mathfrak{m}_{\mathfrak{p}} = \mathfrak{m}''$, and θ leaves invariant every algebraic subalgebra of $\mathfrak{m}$.*

The algebra $\mathfrak{m}$ consists of semi-simple elements, and is therefore contained in a Cartan subalgebra $\mathfrak{c}$ of $\mathfrak{gl}(n, \mathbf{R})$ [4, 2.7]. It is known (see e.g. [11, p. 107]) that a Cartan subalgebra of a semi-simple Lie algebra $\mathfrak{g}$ is invariant under a Cartan involution of $\mathfrak{g}$. If we apply this to $\mathfrak{sl}(n, \mathbf{R})$, and recall that $\mathfrak{c}$ is the product of the center of $\mathfrak{gl}(n, \mathbf{R})$ by a Cartan subalgebra of $\mathfrak{sl}(n, \mathbf{R})$, we see that $\mathfrak{c}$ is invariant under a Cartan involution θ' of $\mathfrak{gl}(n, \mathbf{R})$. Using 1.1, we have then $\mathfrak{c} = \mathfrak{c}_{\mathfrak{k}} + \mathfrak{c}_{\mathfrak{p}}$ where $\mathfrak{c}_{\mathfrak{k}}$ (resp. $\mathfrak{c}_{\mathfrak{p}}$) is the set of elements of $\mathfrak{c}$ with purely imaginary (resp. real) eigenvalues.

Let $x \in \mathfrak{m}$. We have $x = k + p\,(k \in \mathfrak{c}_\mathfrak{k},\ p \in \mathfrak{c}_\mathfrak{p})$, and in order to prove our first contention, it is enough to show that k, $p \in \mathfrak{m}$ or also, since k and p are real matrices, that k, $p \in \mathfrak{m}_c$. After a suitable complex change of coordinates, we may assume $\mathfrak{c}$ to be diagonal, and write

$$k = \operatorname{diag}\,(i\mu_1, \cdots, i\mu_n) \qquad p = \operatorname{diag}\,(\lambda_1, \cdots, \lambda_n)$$

$$(\lambda_i,\ \mu_i \in \mathbf{R},\ i = 1, \cdots, n)\ .$$

But [7a, p. 160] the smallest algebraic algebra in $\mathfrak{gl}(n, \mathbf{C})$ containing x is the set of matrices $\operatorname{diag}\,(a_1, \cdots, a_n)$, where $(a_1, \cdots, a_n)$ annihilates all linear forms with integral coefficients which are zero on $(\lambda_1 + i\mu_1, \cdots, \lambda_n + i\mu_n)$. It contains therefore k and p; a fortiori k, $p \in \mathfrak{m}_c$.

If now θ is a Cartan involution of $\mathfrak{gl}(n, \mathbf{R})$ leaving $\mathfrak{m}$ invariant, and $\mathfrak{a}$ is an algebraic subalgebra of $\mathfrak{m}$, then we have $\mathfrak{m}' = \mathfrak{m}_\mathfrak{k}$, $\mathfrak{m}'' = \mathfrak{m}_\mathfrak{p}$ by 1.1, hence also $\mathfrak{a}' \subset \mathfrak{m}_\mathfrak{k}$, $\mathfrak{a}'' \subset \mathfrak{m}_\mathfrak{p}$, which shows that $\theta(\mathfrak{a}) = \mathfrak{a}$.

1.5. LEMMA. *Let $\mathfrak{g}$ be either semi-simple or equal to $\mathfrak{gl}(n, \mathbf{R})$ and $\mathfrak{m}$ a subalgebra of $\mathfrak{g}$. Then if $\mathfrak{m}$ is stable under a Cartan involution θ of $\mathfrak{g}$, it is reductive in $\mathfrak{g}$, and the restriction of θ to $\mathscr{D}\mathfrak{m}$ is a Cartan involution of $\mathscr{D}\mathfrak{m}$. Conversely, if $\mathfrak{m}$ is reductive and algebraic, it is stable under some Cartan involution of $\mathfrak{g}$.*

Using the last assertion of 1.1, and identifying $\mathfrak{g}$ with an algebraic subalgebra of $\mathfrak{gl}(n, \mathbf{R})$ $(n = \dim \mathfrak{g})$ by means of the adjoint representation, we see first that it is enough to prove the first part when $\mathfrak{g} = \mathfrak{gl}(n, \mathbf{R})$.

Let $\mathfrak{m}$ be stable under a Cartan involution θ of $\mathfrak{g}$. The ideal $\mathfrak{n}$ formed by the nilpotent matrices of the radical of $\mathfrak{m}$ is then also stable under θ, hence (1.1) spanned by semi-simple matrices; thus $\mathfrak{n} = 0$ and $\mathfrak{m}$ is fully reducible [5; § 6, Théorème 4].

Let now $\mathfrak{m}$ be fully reducible, and algebraic. The existence of a Cartan involution θ of $\mathfrak{g}$ leaving $\mathfrak{m}$ invariant is known if $\mathfrak{m}$ is semi-simple [20], or if $\mathfrak{m}$ is commutative (1.4). Let now $\mathfrak{m}$ be neither semi-simple nor commutative. Its center $\mathfrak{c}$ is fully reducible, algebraic, therefore invariant under a Cartan involution θ of $\mathfrak{g}$. The centralizer $\mathfrak{z}(\mathfrak{c})$ of $\mathfrak{c}$ in $\mathfrak{g}$ is reductive in $\mathfrak{g}$ [4, §§ 3, 4], stable under θ, and θ induces a Cartan involution of $\mathscr{D}\mathfrak{z}(\mathfrak{c})$. By [20] there exists a Cartan involution θ' of $\mathfrak{g}$ which leaves $\mathscr{D}\mathfrak{z}(\mathfrak{c})$ and $\mathscr{D}\mathfrak{m}$ invariant. The restriction of θ and θ' to $\mathscr{D}\mathfrak{z}(\mathfrak{c})$ are Cartan involutions of $\mathscr{D}\mathfrak{z}(\mathfrak{c})$, and the analytic group generated by $\mathscr{D}\mathfrak{z}(\mathfrak{c})$ contains an element g such that $\theta'' = \operatorname{Ad}g \circ \theta \circ \operatorname{Ad}g^{-1}$ has the same restriction to $\mathscr{D}\mathfrak{z}(\mathfrak{c})$ as θ' (1.1). The Cartan involution θ'' leaves then invariant $\mathfrak{c}$, $\mathscr{D}\mathfrak{m}$, hence also $\mathfrak{m}$.

1.6. LEMMA. *Let $\mathfrak{g} \supset \mathfrak{g}'$ be algebraic subalgebras reductive in $\mathfrak{gl}(n, \mathbf{R})$. Then there exists a Cartan involution θ of $\mathfrak{gl}(n, \mathbf{R})$ leaving $\mathfrak{g}$ invariant. Any such Cartan involution is conjugate under an element of the analytic* *group G generated by $\mathfrak{g}$ to a Cartan involution leaving also $\mathfrak{g}'$ invariant.*

Let $\mathfrak{c}$ and $\mathfrak{c}'$ be the centers of $\mathfrak{g}$ and $\mathfrak{g}'$, and θ a Cartan involution of $\mathfrak{gl}(n, \mathbf{R})$ leaving $\mathfrak{g}$ invariant. The algebra $\mathfrak{c} + \mathfrak{c}'$ is commutative, algebraic, reductive in $\mathfrak{gl}(n, \mathbf{R})$, hence so is $\mathfrak{c}'' = \mathscr{D}\mathfrak{g} \cap (\mathfrak{c} + \mathfrak{c}')$, and $\mathfrak{c}'' + \mathscr{D}\mathfrak{g}'$ is algebraic, reductive in $\mathscr{D}\mathfrak{g}$. By 1.5, $\mathfrak{c}'' + \mathscr{D}\mathfrak{g}'$ is stable under a Cartan involution of $\mathscr{D}\mathfrak{g}$. Using 1.1, this shows the existence of $g \in G$ such that $\theta' = \mathrm{Ad}g \circ \theta \circ \mathrm{Ad}g^{-1}$ leaves $\mathfrak{c}'' + \mathscr{D}\mathfrak{g}'$ invariant; since $\mathfrak{g}$ centralizes $\mathfrak{c}$, we also have $\theta'(\mathfrak{c}) = \mathfrak{c}$, hence $\theta'(\mathfrak{c} + \mathfrak{c}'') = \mathfrak{c} + \mathfrak{c}''$, and, by 1.4, $\theta'(\mathfrak{c}') = \mathfrak{c}'$; therefore $\theta'(\mathfrak{g}') = \mathfrak{g}'$.

This proves the second assertion of 1.6. The first one is the special case where $\mathfrak{g} = \mathfrak{gl}(n, \mathbf{R})$ and $\mathfrak{g}' = \mathfrak{g}$.

1.7. **Lemma.** *Let M be reductive algebraic subgroup of $\mathbf{GL}(n, \mathbf{R})$, θ a Cartan involution of $\mathbf{GL}(n, \mathbf{R})$ leaving the Lie algebra $\mathfrak{m}$ of M invariant, $\mathfrak{gl}(n, \mathbf{R}) = \mathfrak{k} + \mathfrak{p}$ the corresponding Cartan decomposition. Then $M = L \cdot \exp(\mathfrak{m}_\mathfrak{p})$ where L is a compact subgroup with Lie algebra $\mathfrak{m}_\mathfrak{k}$, which is connected if M is so.*

Let $\mathfrak{c}$ be the center of $\mathfrak{m}$, and Q the normalizer of $(\mathscr{D}\mathfrak{m})_\mathfrak{k}$ in M. The group Q normalizes $\mathfrak{c}$, hence also $\mathfrak{c}_\mathfrak{k}$ and $\mathfrak{c}_\mathfrak{p}$ (1.4), and $(\mathscr{D}\mathfrak{m})_\mathfrak{p}$, which is the orthogonal complement of $(\mathscr{D}\mathfrak{m})_\mathfrak{k}$ in $\mathscr{D}\mathfrak{m}$ with respect to the Killing form. By a standard result, $(\mathscr{D}\mathfrak{m})_\mathfrak{k}$ is equal to its normalizer in $\mathscr{D}\mathfrak{m}$; therefore the Lie algebra $\mathfrak{q}$ of Q is equal to $\mathfrak{c} + (\mathscr{D}\mathfrak{m})_\mathfrak{k}$. The algebras $\mathfrak{c}_\mathfrak{k} = \mathfrak{c} \cap \mathfrak{k}$ and $(\mathscr{D}\mathfrak{m})_\mathfrak{k} = \mathfrak{k} \cap \mathscr{D}\mathfrak{m}$, being algebraic, generate closed, hence compact, subgroups A, B of Q^0, with A central. $\mathfrak{c}_\mathfrak{p}$ is the Lie algebra of a vector subgroup V, which is invariant in Q. It follows immediately that $Q^0 = A \cdot B \cdot V$, and hence Q^0/V is compact. Since Q^0 has finite index in Q, the quotient Q/V is also compact. By Iwasawa's theorem (see e.g., [27, Exp. 22, Théorème 1]), there exists a compact subgroup L of Q, containing $A \cdot B$, such that Q is the semi-direct product of L and V. The Lie algebra of L is therefore equal to $\mathfrak{m}_\mathfrak{k}$. By the conjugacy of Cartan decompositions of $\mathscr{D}\mathfrak{m}$ under $\mathrm{Ad}\mathscr{D}\mathfrak{m}$, the group Q meets every connected component of M, therefore $M = Q \cdot M^0$. By a classical result of E. Cartan (see [20], [12, Lemma 31] for instance), the analytic subgroup generated by $\mathscr{D}\mathfrak{m}$ is equal to $B \cdot \exp((\mathscr{D}\mathfrak{m})_\mathfrak{p})$, therefore $M^0 = A \cdot B \cdot V \cdot \exp((\mathscr{D}\mathfrak{m})_\mathfrak{p}) = A \cdot B \cdot \exp \mathfrak{m}_\mathfrak{p}$, and $M = L \cdot V \cdot A \cdot B \cdot \exp(\mathfrak{m}_\mathfrak{p}) = L \cdot \exp(\mathfrak{m}_\mathfrak{p})$. Moreover $L = A \cdot B$ if $M = M^0$.

1.8. **Lemma.** *Let $G \supset G'$ be reductive algebraic subgroups of $\mathbf{GL}(n, \mathbf{R})$. Then there exists a Cartan involution of $\mathbf{GL}(n, \mathbf{R})$ leaving G invariant. Every such Cartan involution is conjugate by an element $\mathrm{Ad}g(g \in G^0)$ to a Cartan involution leaving also G' invariant.*

As in 1.6, the first assertion is a special case of the second one. By 1.6, a Cartan involution leaving G invariant is conjugate by an automorphism

$\mathrm{Ad}x(x \in G^0)$ to a Cartan involution θ leaving also the Lie algebra $\mathfrak{g}'$ of G' invariant. Let $\mathfrak{gl}(n, \mathbf{R}) = \mathfrak{k} + \mathfrak{p}$ and $\mathbf{GL}(n, \mathbf{R}) = K \cdot P$ be the corresponding decompositions of $\mathfrak{gl}(n, \mathbf{R})$ and $\mathbf{GL}(n, \mathbf{R})$. By 1.7, $G' = L \cdot \exp(\mathfrak{g}'_\mathfrak{p})$, where L is a compact subgroup with Lie algebra $\mathfrak{g}'_\mathfrak{k}$. The involution θ clearly leaves invariant G^0, L^0, but not necessarily L. Assume that we have found $y \in G^0$ such that

$$(1) \qquad K' = y \cdot K \cdot y^{-1} \supset L \qquad y \cdot \exp(\mathfrak{g}'_\mathfrak{k}) \cdot y^{-1} = \exp(\mathfrak{g}'_\mathfrak{p}) \ .$$

Then $\theta' = \mathrm{Ad}y \circ \theta \circ \mathrm{Ad}y^{-1}$ will do. In fact it is conjugate under $y \cdot x$ to the initial Cartan involution, acts trivially on L, and acts by $p \to p^{-1}$ on $\exp(\mathfrak{g}'_\mathfrak{p})$, hence it leaves G' invariant.

There remains to find $y \in G^0$ satisfying (1). Let M be the normalizer of $\mathfrak{g}'_\mathfrak{p}$ in G. It contains L, and is stable under θ. We view P as usual as a Riemannian symmetric space of $\mathbf{GL}(n, \mathbf{R})$ under the operations $p \to x \cdot p \cdot x^*$, where $x^* = \theta(x^{-1})$. Then $M_\mathfrak{p} = \exp \mathfrak{m}_\mathfrak{p}$ is a totally geodesic submanifold, and, by E. Cartan's fixed point argument, every compact subgroup of $\mathbf{GL}(n, \mathbf{R})$ leaving $M_\mathfrak{p}$ invariant has a fixed point on $M_\mathfrak{p}$ (for all this, see for instance [20]). In particular L has a fixed point a on $M_\mathfrak{p}$. Let y be the square root of a contained in $M_\mathfrak{p}$. Then, for any $x \in L$,

$$(y^{-1} \cdot x \cdot y)(y^{-1} \cdot x \cdot y)^* = y^{-1} \cdot x \cdot y \cdot y \cdot x^* \cdot y^{-1}$$
$$= y^{-1}x \cdot a \cdot x^* \cdot y^{-1} = y^{-1} \cdot y^2 \cdot y^{-1} = e \ ,$$

and therefore $y^{-1} \cdot L \cdot y \subset K$, which is the first part of (1). Since M normalizes $\mathfrak{g}'_\mathfrak{p}$, we have $[\mathfrak{m}_\mathfrak{p}, \mathfrak{g}'_\mathfrak{p}] \subset \mathfrak{g}'_\mathfrak{p}$; but $[\mathfrak{m}_\mathfrak{p}, \mathfrak{g}'_\mathfrak{p}] \subset [\mathfrak{p}, \mathfrak{p}] \subset \mathfrak{k}$, hence $[\mathfrak{m}_\mathfrak{p}, \mathfrak{g}'_\mathfrak{p}] = 0$, and y commutes elementwise with $\exp \mathfrak{g}'_\mathfrak{p}$. Thus y also verifies the second equality of (1).

1.9. THEOREM (Mostow [21]). *Let $G_1 \supset \cdots \supset G_m$ be reductive real algebraic subgroups of $\mathbf{GL}(n, \mathbf{R})$. Then there exists $a \in \mathbf{SL}(n, \mathbf{R})$ such that the groups $a \cdot G_i \cdot a^{-1} (i = 1, \cdots, m)$ are self-adjoint.*

By 1.1, this theorem is equivalent to the existence of a Cartan involution of $\mathbf{GL}(n, \mathbf{R})$ leaving the G_i's invariant. By 1.8, there exists a Cartan involution leaving G_1 invariant. Assume $\theta(G_i) = G_i$ for $1 \leq i \leq k < m$. By 1.8, θ is conjugate under an element $\mathrm{Ad}x(x \in G^0_k)$ to a Cartan involution θ' leaving G_{k+1} stable. Since the G_i's $(i \leq k)$ are still invariant under θ', the theorem follows by induction.

The following proposition strengthens 1.7, and extends to reductive real algebraic groups well-known properties of semi-simple Lie groups.

1.10. PROPOSITION. *Let G be a reductive algebraic subgroup of $\mathbf{GL}(n, \mathbf{R})$, θ a Cartan involution of $\mathbf{GL}(n, \mathbf{R})$ leaving G invariant (1.6), $\mathfrak{gl}(n, \mathbf{R}) = \mathfrak{k} + \mathfrak{p}$, and $\mathbf{GL}(n, \mathbf{R}) = K \cdot P$ the corresponding decompositions* *of* $\mathfrak{gl}(n, \mathbf{R})$ *and* $\mathbf{GL}(n, \mathbf{R})$. *Then the set L of fixed points of θ in G is a maximal compact subgroup with Lie algebra $\mathfrak{g}_{\mathfrak{t}}$. Every compact subgroup of G is conjugate by an element* $\mathrm{Ad}x(x \in \exp(\mathfrak{g}_{\mathfrak{p}}))$ *to a subgroup of L. The map $(x, y) \to x \cdot \exp(y)$ of $L \times \mathfrak{g}_{\mathfrak{p}}$ into G is an analytic homeomorphism.*

Let $g = k \cdot p(k \in K, p \in P)$ be an element of G. Since $\theta(g^{-1}) \cdot g = p^2$, we have $p^2 \in G$, hence also $p^{2m} \in G$, $(m \in \mathbf{Z})$. In a suitable coordinate system, $\log p = \mathrm{diag}(\lambda_1, \cdots, \lambda_n)$, $(\lambda_i$ real$)$, and $p^{2m} = \mathrm{diag}(\exp 2m\lambda_1, \cdots, \exp 2m\lambda_n)$. It is then an elementary fact that every polynomial with real coefficients over the space of $n \times n$ matrices which is annihilated by the elements $p^{2m}(m \in \mathbf{Z})$ is also annihilated by the diagonal matrices $\mathrm{diag}(\exp t\lambda_1, \cdots, \exp t\lambda_n)$ $(t$ real$)$. Since G is algebraic, this shows that G contains the 1-parameter group generated by $\log p$. Therefore $\log p \in \mathfrak{g}_{\mathfrak{p}}$, $p \in \exp \mathfrak{g}_{\mathfrak{p}}$, and $k \in G \cap K = L$. Thus $G = L \cdot \exp \mathfrak{g}_{\mathfrak{p}}$. That every compact subgroup of G is conjugate under $\exp \mathfrak{g}_{\mathfrak{p}}$ to a subgroup of L is proved by the fixed point argument of E. Cartan used in 1.8. The proof of the last statement for semi-simple Lie groups given in [12, Lemma 31] is valid without change here.

1.11. We now extend Iwasawa's decomposition to reductive algebraic groups. Let $G \subset \mathbf{GL}(n, \mathbf{R})$ be reductive, algebraic, θ a Cartan involution of $\mathbf{GL}(n, \mathbf{R})$ leaving G invariant, K its fixed point set, $\mathfrak{a}$ a maximal subalgebra of $\mathfrak{g}_{\mathfrak{p}}$. For $\lambda \in \mathfrak{a}^*$, let $\mathfrak{g}_\lambda = \{x \in \mathfrak{g}, [a, x] = \lambda(a)x, a \in \mathfrak{a}\}$. Choose some order on $\mathfrak{a}^*$ and let $\mathfrak{n} = \sum_{\lambda > 0} \mathfrak{g}_\lambda$.

PROPOSITION. *We keep the previous notation. Then $\mathfrak{n}$ generates a closed unipotent subgroup N invariant under $A = \exp \mathfrak{a}$, and $G = K \cdot A \cdot N$. The map $(k, a, n) \to k \cdot a \cdot n$ of $K \times A \times N$ onto G is an analytic homeomorphism.*

Let $\mathfrak{c}$ be the center of $\mathfrak{g}$. Then $\mathfrak{g} = \mathfrak{c} \times \mathscr{D}\mathfrak{g}$, and $\mathfrak{a} = \mathfrak{c}_{\mathfrak{p}} \times \mathfrak{a}_1$, where $\mathfrak{a}_1$ is a maximal subalgebra of $(\mathscr{D}\mathfrak{g})_{\mathfrak{p}}$. Clearly, for $\lambda \neq 0, \mathfrak{g}_\lambda = \{x \in \mathscr{D}\mathfrak{g}, [a_1, x] = \lambda(a_1)x, a_1 \in \mathfrak{a}_1\}$. By the usual Iwasawa decomposition, the analytic group M generated by $\mathscr{D}\mathfrak{g}$ is of the form $M = K_1 \cdot A_1 \cdot N$ where $K_1 = K \cap M$, $A_1 = \exp \mathfrak{a}_1$, and N is closed, unipotent, with Lie algebra $\mathfrak{n}$. As was remarked in 1.7, $\mathfrak{c}_{\mathfrak{t}}$ and $\mathfrak{c}_{\mathfrak{p}}$ generate respectively a torus T and a vector subgroup V, therefore

$$G^0 = T \cdot K_1 \cdot V \cdot A_1 \cdot N = T \cdot K_1 \cdot A \cdot N.$$

But $G = K \cdot G^0$, with $K \supset T \cdot K_1$, hence $G = K \cdot A \cdot N$. Since V is central, and N is stable under conjugation by elements of A_1, the group N is also invariant under A. The second assertion is proved by standard arguments, as in the usual case [27; Exp. 11].

An Iwasawa decomposition $G = K \cdot A \cdot N$ and a Cartan involution θ

whose fixed point set is K, and which induces $x \to x^{-1}$ on A, will be called *compatible*.

Let G' be an open subgroup of G. Since $G = K \cdot G^\circ$ and θ leaves K pointwise fixed, we have $\theta(G') = G'$. The automorphism of G' induced by θ and the decomposition $G' = (K \cap G') \cdot A \cdot N$ will also be called a Cartan involution and an Iwasawa decomposition of G' respectively. Clearly, 1.8 to 1.11 are also valid for open subgroups of reductive real algebraic groups.

2. Algebraic groups

In this paragraph and the next we assume some familiarity with the elementary theory of affine algebraic sets and algebraic groups, and shall recall only some of the relevant definitions and facts. For more details, see e.g. [1, 8, 17].

2.1. A *complex algebraic group*, or simply, an *algebraic group*, is a subgroup G of $\mathbf{GL}(n, \mathbf{C})$, which consists of all invertible matrices $g = (g_{ij})$ whose coefficients annihilate some set of polynomials $\{P_\alpha[X_{11}, \cdots, X_{nn}]\}$ with complex coefficients. In other words, G is the intersection of $\mathbf{GL}(n, \mathbf{C})$ with an affine algebraic set in the space $\mathbf{M}(n, \mathbf{C})$ of $n \times n$ complex matrices. The group is said to be defined over a subfield k of $\mathbf{C}$ if the P_α may be chosen so as to have coefficients in k.[1] The intersection of all the fields of definition is also a field of definition. An algebraic group is also a complex Lie group; it is connected as a manifold if and only if it is irreducible as an algebraic set, or if and only if it is connected in the Zariski topology.

Let G be an algebraic group, and A a subring of $\mathbf{C}$. Then G_A will denote the subgroup of elements of G whose coefficients are in A and whose determinant is a unit of A. If A is a field of definition, then G_A is an algebraic A-group in the terminology of [2], an algebraic group over A in [7], and if G is connected, G_A is Zariski dense in G [26, p. 44]. Conversely, if H is an algebraic group over A, in the sense of [7], then $H = G_A$, where G is the smallest complex algebraic group containing H.

An algebraic group G is not necessarily a closed subset of $\mathbf{M}(n, \mathbf{C})$; however, if we add one coordinate and put $g_{n+1,n+1} = (\det g)^{-1}$, $g_{n+1,i} = g_{i,n+1} = 0$ $(i = 1, \cdots, n)$, then G becomes an algebraic subgroup of $\mathbf{SL}(n + 1, \mathbf{C})$, which is closed in $\mathbf{M}(n + 1, \mathbf{C})$. This operation clearly does not change G_A.

---

[1] This is the definition of [1], with the universal field specialized to $\mathbf{C}$. Being in characteristic zero, we need not distinguish between Zariski k-closed and defined over k (or between defined and quasi-defined over k [1]).

2.2. A *rational representation* $\pi\colon G \to \mathrm{GL}(V)$ is a homomorphism of G into $\mathrm{GL}(V)$ (V finite dimensional complex vector space) whose restriction to each connected component of G is a rational map of G into the space of endomorphisms $E(V)$ of V. The coefficients of $\pi(g)$ with respect to a basis of V are regular functions on each connected component of G. If G is a closed set in $\mathbf{M}(n, \mathbf{C})$, they are therefore polynomials in the coefficients of g, whose coefficients are constant on each connected component of G. From this and the end remark in 2.1, it follows that in general the coefficients of $\pi(g)$ are polynomials in those of g and in $(\det g)^{-1}$.

The rational representation π will be said to be defined over k if each component of G is, and if there exists a basis (v_i) of v such that the coefficients of $\pi(g)$ with respect to that basis are regular functions defined over k on each connected component of G. In that case, we shall denote by V_A the set of linear combinations of the v_i's with coefficients in $A \supset k$.

Given a rational representation $\pi\colon G \to \mathrm{GL}(V)$, we shall most often write $v \cdot g$ for $v \cdot \pi(g)$ ($v \in V,\ g \in G$). An orbit $v \cdot G$ is always open and everywhere dense in its Zariski closure (smallest affine algebraic set containing it), therefore the latter coincides with the closure in the ordinary topology. If G is connected, and π is defined over k, then $v \cdot G$, its closure and the isotropy group G_v of v are defined over $k(v)$.

Let G be a connected algebraic group, H an algebraic subgroup, both defined over k. Then $H \backslash G$ is in a canonical way an algebraic variety defined over k. Given a rational representation $\pi\colon G \to \mathrm{GL}(V)$ and a point $v \in V$ for which $G_v = H$, the map $g \to v \cdot \pi(g)$ induces a regular bijective map of $H \backslash G$ onto $v \cdot G$, defined over k if π is. Since we are in characteristic zero and these varieties are non-singular, this map is in fact birational and biregular.

2.3. PROPOSITION. *Let G be a connected algebraic group defined over* $\mathbf{R}$, $\pi\colon G \to \mathrm{GL}(V)$ *a rational representation of G defined over* $\mathbf{R}$, *and X an orbit of G in V. Then $X_{\mathbf{R}} = X \cap V_{\mathbf{R}}$ is the union of a finite number of orbits of $(G_{\mathbf{R}})^0$, which are closed if X is so.*

$\dim_{\mathbf{R}}$ and $\dim_{\mathbf{C}}$ will denote the topological and the complex dimension, respectively. Let $s = \dim_{\mathbf{C}} G$, $t = \dim_{\mathbf{C}} X$. The $s - t = \dim_{\mathbf{C}} G_x (x \in X)$. We assume $X_{\mathbf{R}}$ to be non-empty, hence X and its closure $\bar{X}$ are defined over $\mathbf{R}$. The latter is an irreducible algebraic set of complex dimension t. The group $G_{\mathbf{R}}$ being Zariski-dense in G, the set $\bar{X}$ is the smallest algebraic set containing $(\bar{X})_{\mathbf{R}}$, hence $(\bar{X})_{\mathbf{R}}$ is a real algebraic irreducible set of dimension t, and $(\bar{X})_{\mathbf{R}} = A \cup B$, where A is a manifold of dimension t, the set of real simple points of $\bar{X}$, and B a real algebraic set of dimension $< t$ [31, §§ 10, 11]. In view of their characterizations, A and B are both invariant under G_R. Let now $x \in X_\mathrm{R}$. It is a simple point of $\bar{X}$, hence $x \in A$, $x \cdot G_\mathrm{R}$ is an open submanifold of A, and $x \cdot G_\mathrm{R}^0$ is a connected component of A. But A, being the complement of a real algebraic set in a real algebraic set, has only a finite number of components [31, Theorem 4], which proves the first part of our assertion. If moreover $X = \bar{X}$, then $(\bar{X})_\mathrm{R} = X_\mathrm{R} \subset A$, hence B is empty, A is a closed submanifold with a finite number of components, which must then also be closed.

2.4. PROPOSITION. *Let G be a connected algebraic group, H an algebraic subgroup, k a field of definition for G and H, and assume $H\backslash G$ to be an affine algebraic set. Then there exists a rational representation $\pi: G \to \mathbf{GL}(V)$ defined over k and a point $v \in V_k$ such that $G_v = H$, $v \cdot G$ is closed, and $g \to v \cdot g$ induces a biregular birational map of $H\backslash G$ onto $v \cdot G$.*[2]

Let A and B be rings of regular (i.e. rational, everywhere defined) functions, defined over k, of G and $H\backslash G$ respectively. The variety $H\backslash G$ is defined over k, non-singular and affinely imbeddable. It is therefore also biregularly homeomorphic over k to an affine algebraic set defined over k (see A. Weil, Amer. J. Math. 78 (1950), p. 509–524, Theorem 7). Therefore

(i) B is a finitely generated k-algebra and separates the points of $H\backslash G$.[3]

We let G operate on $A \otimes \mathbf{C}$ on the right by $(f \cdot g)(x) = f(g \cdot x)(f \in A \otimes \mathbf{C}$; $g, x \in G)$. The transpose of the canonical projection $G \to H\backslash G$ identifies then $B \otimes \mathbf{C}$ with the ring of invariants of H.

Let $b_1, \cdots, b_s$ be a finite system of generators of B, P_i a finite dimensional vector subspace of A such that $P_i \otimes \mathbf{C}$ is invariant under G and $b_i \in P_i$ [25, Theorem 12], P the direct sum of the P_i, $v = (b_1, \cdots, b_s)$, and π the natural representation of G in $V = P \otimes \mathbf{C}$. From (i) it is clear that $G_v = H$. Let $X = v \cdot G$, and μ be the map of $H\backslash G$ into the Zariski closure $\bar{X}$ of X defined by $g \to v \cdot g$. It is rational, defined over k, everywhere regular. Let B' be the ring of regular functions on $\bar{X}$. Since X is Zariski dense in $\bar{X}$, the elements of B' are determined by their restriction to X, and μ induces an injective homomorphism ${}^t\mu$ of B' into $B \otimes \mathbf{C}$. Let us prove now that ${}^t\mu$ is surjective. For this, it is enough to show that $b_i \in {}^t\mu(B')$ $(i = 1, \cdots, s)$. In P_i, we may always find a basis $f_1, \cdots, f_t$ such that $f_j(e) = 0(j \geqq 2)$. We have $b_i \cdot g = \sum_j a_j(g) \cdot f_j$, where the a_j's are regular functions on G, therefore

---

[2] Propositions 2.4, 2.5 and their proofs are also valid if $\mathbf{C}$ is replaced by a universal field of arbitrary characteristic.

[3] In our case, this also follows directly from the fact that G_k is Zariski dense in G [26].

$$b_i(g) = (b_i \cdot g)(e) = \sum_j a_j(g)f_j(e) = a_1(g) \cdot f_1(e) \; .$$

This implies that $b_i = {}^t\mu(x_1 \cdot f_1(e))$, where x_1 is the first coordinate function with respect to the basis (f_j). Thus $b_i \in {}^t\mu(B')$.

Thus, v is a regular map of $H\backslash G$ into $\bar{X}$, defined over k, whose transpose is an isomorphism of the rings of regular functions. Since $H\backslash G$ and $\bar{X}$ are affine, this implies that v is a biregular, bijective, hence also that $X = \bar{X}$.   q.e.d.

The following proposition is a slight strengthening of Proposition 5 in [8, Exp. 10] and is proved in the same way. It will be used only in 3.7.

2.5. PROPOSITION. *Let G be a connected algebraic group, H an algebraic subgroup, and k a field of definition for G and H. Then there exists a rational representation $\pi\colon G \to \mathrm{GL}(V)$, defined over k, a point $v \in V_k$, $(v \neq 0)$, such that H is the set of elements of G which leave the 1-dimensional subspace $[v]$ spanned by v invariant.*[2]

Again let A be as in 2.4, and I the ideal of H in $A \otimes C$. It is invariant under H, and has a finite system of generators belonging to A. By [25, Theorem 12] there exists a finite dimensional subspace P of A such that $P \otimes C$ is invariant under G, and that $P \cap I$ contains a system of generators of I. Let $d = \dim P \cap I$ and π the natural representation of G in $V = \bigwedge^d (P \otimes C)$. Then $v = P \cap I$ fulfills our condition.

3. Reductive algebraic groups. Affine homogeneous spaces

3.1. An *algebraic torus* (a torus, in the terminology of [1]) is an algebraic group which is birationally isomorphic to a direct product of groups C^* (multiplicative group of non-zero complex numbers), or equivalently [1], a connected group which is diagonalizable. A connected subgroup of $\mathrm{GL}(n, C)$ is an algebraic torus if and only if its Lie algebra is algebraic, commutative, and consists of semi-simple elements (or is reductive in $\mathfrak{gl}(n, C)$). An algebraic group G will be called *reductive* if $G^0 = T \cdot G'$ where T is an algebraic torus, central in G^0, and G' is semi-simple. For an algebraic group $G \subset \mathrm{GL}(n, C)$ the following conditions are equivalent:

(i) G is reductive;

(ii) G is fully reducible;

(iii) $\mathfrak{g}$ is reductive in $\mathfrak{gl}(n, C)$.

In fact, G is reductive, or fully reducible, if and only if G^0 is, and this equivalence follows from the previous remarks and the characterization of fully reducible subalgebras of $\mathfrak{gl}(n, C)$ [5, § 6, No. 5]. Let π be a rational representation of G. If G is a torus, then so is $\pi(G)$, [1; 9.4], therefore, $\pi(G)$ is reductive if G is. The reductive algebraic groups are the algebraic groups, all of whose rational representations are completely reducible.

3.2. Proposition. *Let G be an algebraic group defined over* **R.** *Then the following conditions are equivalent:*

(i) *G is reductive;*

(ii) *$G_{\mathbf{R}}$ is reductive;*

(iii) *$\mathfrak{g}$ is the complexification of the (real) Lie algebra of a maximal compact subgroup of G^0.*

Let $\mathfrak{g}$ and $\mathfrak{g}_{\mathbf{R}}$ be the Lie algebras of G and $G_{\mathbf{R}}$, over **C** and **R** respectively. Then $\mathfrak{g} = \mathfrak{g}_{\mathbf{R}} \otimes \mathbf{C}$. Thus $\mathfrak{g}_{\mathbf{R}}$ is fully reducible if and only if $\mathfrak{g}$ is, and the equivalence of (i) and (ii) follows from 3.1 and 1.2. Let now $G^0 = T \cdot G'$ be reductive, A and B be maximal compact subgroups of T and G' respectively. Then $A \times B$ is a maximal compact subgroup of $T \times G'$, which moreover contains the (finite) kernel of the natural homomorphism of $T \times G'$ onto G^0, therefore $A \cdot B$ is a maximal compact subgroup of G^0. Since the Lie algebras $\mathfrak{a}$ and $\mathfrak{b}$ of A and B are real forms of the Lie algebras of T and G', this shows that (i) $\Rightarrow$ (iii). If now $\mathfrak{g}$ is the complexification of the Lie algebra of a compact group, it is certainly fully reducible, hence (iii) $\Rightarrow$ (i).

3.3. Proposition. *Let G be a reductive connected algebraic group, $\pi: G \to \mathbf{GL}(V)$ a rational representation. Then the invariant polynomials separate the closed orbits of G.*

Let X, Y be two distinct closed orbits. These are two affine subvarieties with empty intersection. There exists therefore a polynomial P vanishing on X, and a polynomial Q vanishing on Y such that $P + Q = 1$. We have then $P \cdot g + Q \cdot g = 1$ for any $g \in G$. Let K be a maximal compact subgroup of G, dk the Haar measure on K with total measure 1, and

$$P^* = \int_K P \cdot k \, dk \qquad Q^* = \int_K Q \cdot k \, dk \ .$$

We still have $P^* + Q^* = 1$, and $P^*(x) = Q^*(y) = 0 (x \in X, y \in Y)$, hence $P^* = 1$ on Y. Moreover P^* and Q^* are invariant under K. Since the natural representation of G on the space of polynomials on V of a given degree is rational, and G is the smallest algebraic subgroup of G containing K (3.2), P^* and Q^* are invariant under G, whence our contention.

3.4. Remark. The property (iii) also characterizes reductive groups among connected complex Lie groups. In fact, let G be complex, connected, K a maximal compact subgroup, and assume that $\mathfrak{g} = \mathfrak{k} \otimes \mathbf{C}$. Of course $\mathfrak{k} = \mathfrak{c} \times \mathscr{D}\mathfrak{k}$ with $\mathfrak{c}$ the Lie algebra of a torus (usual sense) and $\mathscr{D}\mathfrak{k}$ a compact semi-simple Lie algebra, therefore $\mathfrak{g} = (\mathfrak{c} \otimes \mathbf{C}) \times (\mathscr{D}\mathfrak{k} \otimes \mathbf{C})$, $G = T \cdot G'$, with G' complex semi-simple, T isomorphic to a product of $\mathbf{C}^*$'s, and $T \cap G'$ central, hence finite. $T \cap G'$ is also the intersection of any two maximal compact subgroups of T and G', therefore is uniquely characterized by the maximal compact subgroup K. This implies that if G and G^* are complex, connected, reductive, and have isomorphic maximal compact subgroups, then they are isomorphic (as complex Lie groups at first). Moreover, it is known that a reductive group has only one structure of algebraic group compatible with the given complex analytic structure. (For a more general theorem, see G. Hochschild and G. D. Mostow, Amer. J. Math. 88 (1961), p.p. 111–136, §8.) Therefore a complex analytic isomorphism between two reductive groups is necessarily birational, biregular.

3.5. **Theorem.** *Let G be a connected reductive algebraic group, and H an algebraic subgroup. Then G/H is an affine algebraic variety if and only if H is reductive.*

This result is not new. Since a non-singular affine variety is a Stein manifold, it is a consequence of the two following assertions, where G and H are as in the theorem:

(a) If G/H is a Stein manifold, then H is reductive.

(b) If H is reductive, G/H is affine.

The assertion (a) is due to Matsushima [18]. Assertion (b) is stated in [18] without proof, and attributed to Iwahori-Sugiura. For the sake of completeness, we insert here a proof of (b), and also one for (a), which is shorter than that of [18], although based on a similar idea. A consequence of (a) and (b) is that a quotient G/H, with G connected, reductive, is Stein if and only if it is an affine variety. We also remark that the statement 3.5 makes sense in arbitrary characteristic, but we do not know whether it holds true.

3.6. **Proof of (a).** G/H^0 is a finite Galois covering of G/H, hence is a Stein manifold [6], and we may assume H to be connected. We denote by $H_i(X)$ the ith singular homology group of the space X, with complex coefficients.

Let $n = \dim_{\mathbf{C}} G$, $m = \dim_{\mathbf{C}} H$. We claim first that H reductive is equivalent to $H_m(H) \neq 0$. As a manifold H is the topological product of a euclidean space by a maximal compact subgroup, hence H reductive implies $H_m(H) \neq 0$. A complex analytic group being the quotient by a finite group of the semi-direct product of a semi-simple Lie by a solvable group, it is enough to prove the converse for H solvable. A maximal compact subgroup K of H is then commutative, of real dimension $\geq m$. Since H is a complex linear group, $\mathfrak{k}$ and $i \cdot \mathfrak{k}$ are linearly independent over $\mathbf{R}$, hence $\mathfrak{h} = \mathfrak{k} + i\mathfrak{k}$ (direct), and $\mathfrak{h} \cong \mathfrak{k} \otimes \mathbf{C}$.

G is a locally trivial fibering, with typical fibre H, base G/H, and structural group H, therefore the Betti numbers of G are majorized by those of $(G/H) \times H$ (as follows from the existence of a spectral sequence). We have $H_n(G) \neq 0$; since G/H is Stein, $H_i(G/H) = 0$ ($i > n - m = \dim_{\mathbf{C}} G/H$) [6], hence by the Künneth rule, $H_i(H) \neq 0$ for some $i \geq m$. But G is Stein (it is algebraic), hence so is H (as a closed submanifold of G), and $H_i(H) = 0$ for $i > m$; we must then have $H_m(H) \neq 0$. As was already proved, this implies that H is reductive.

3.7. PROOF OF (b). Let $K \supset L$ be maximal compact subgroups of G and H respectively and ρ a faithful representation of K in $\mathbf{GL}(n, \mathbf{R})$. It follows from 3.4 that ρ extends to a biregular birational isomorphism of G onto the smallest complex algebraic group containing $\rho(K)$. We identify G with the latter. Thus G is defined over $\mathbf{R}$, and H, the smallest complex algebraic group containing L, is also defined over $\mathbf{R}$. The group $G_{\mathbf{R}}$, being real algebraic, with identity component K, is compact and consequently $G_{\mathbf{R}} = K$. By 2.5, there exists a rational representation $\pi \colon G \to \mathbf{GL}(V)$ defined over $\mathbf{R}$, and a non-zero element $v \in V_{\mathbf{R}}$ such that H is the subgroup of G leaving $\mathbf{C} \cdot v$ invariant. This yields a real 1-dimensional representation of L. Since L is compact, its image has at most 2 elements. By 1,10, applied to H, viewed as real reductive algebraic group, we have $H/H^0 \cong L/L^0$, the subgroup H' of H leaving v fixed has index ≤ 2, and G/H' is either equal to G/H or a two-fold covering of G/H. But the quotient of an affine variety by a finite group is an affine variety (see Serre, Symposium de Topologia Algebrica, Mexico, 1958, 24–53, § 13), therefore it is enough to prove that G/H' is affine. Changing slightly the notation, we may assume H to be the isotropy group of v. Let now $X = v \cdot G$, and $\bar{X}$ its closure. We want to prove that $X = \bar{X}$. The set $Y = \bar{X} - X$ is algebraic, defined over $\mathbf{R}$. Let $S_1, \cdots, S_q$ be polynomials with real coefficients which generate the ideal of Y and $S = S_1^2 + \cdots + S_q^2$. On $V_{\mathbf{R}}$, this polynomial vanishes on $Y_{\mathbf{R}}$ only. Therefore, for $x \in X_{\mathbf{R}}$, S has a constant sign on $x \cdot K$. We let G operate in the usual fashion on the ring A of polynomials over V, and consider as in 3.3 the average S^* of S over K:

$$S^* = \int_K S \cdot k \, dk \,,$$

where dk is a Haar measure on K. It is invariant under K, hence also under G; consequently, S^* is constant on X, and therefore on $\bar{X}$. Since S does not vanish on the connected set $x \cdot K (x \in X_{\mathbf{R}})$, S^* is not zero on $\bar{X}$; on the other hand, Y is invariant under K, and S vanishes on Y, therefore S^* vanishes on Y. This is a contradiction, unless $Y = \varnothing$, $X = \bar{X}$. Thus X is an affine algebraic set. The map $g \to v \cdot g$ induces an isomorphism of $H\backslash G$ onto X and allows one to identify $H\backslash G$ with X (2.2).

For future reference, we state here a consequence of 2.4, 2.5, 3.4 to

3.7, the only one to be used in the sequel.

3.8. THEOREM. *Let G be a connected reductive algebraic group, defined over $\mathbf{Q}$, and H an algebraic subgroup. Then the two following conditions are equivalent:*

(i) *H is reductive, defined over $\mathbf{Q}$;*

(ii) *there exists a rational representation $\pi\colon G \to \mathbf{GL}(V)$ of G, defined over $\mathbf{Q}$, and an element $v \in V_{\mathbf{Q}}$, such that $v \cdot G$ is closed and $G_v = H$.*

In fact if (i) is true, then $H\backslash G$ is affine by 3.5, and (ii) follows from 2.4. If (ii) is true, then $H\backslash G$ may be identified with $v \cdot G$ (see end of the preceding proof) hence is affine, and H is reductive by 3.6. Finally, since $H = G_v$, $(v \in V_{\mathbf{Q}})$, H is defined over $\mathbf{Q}$.

4. Siegel domains

4.1. Let G be an open subgroup of a real algebraic, reductive group, $G = K \cdot A \cdot N$ an Iwasawa decomposition of G (1.11), and Σ the set of simple restricted roots, in the ordering for which the roots of $\mathrm{ad}_{\mathfrak{g}}\mathfrak{a}$ in $\mathfrak{n}$ are positive. Let $A_t = \{a \in A, \lambda(\log a) \leq t; \lambda \in \Sigma\}$. A *Siegel domain* in G is a subset of the form $K \cdot A_t \cdot \omega$ where ω is compact in N. It will usually be denoted by $\mathfrak{S}_{t,\omega}$, or simply by $\mathfrak{S}$ if t, ω need not be specified. Clearly, $\mathfrak{S}_{t,\omega} \subset \mathfrak{S}_{t',\omega'}$ if $t \leq t'$, $\omega \subset \omega'$.

It is understood once and for all that the choice of a Siegel domain pre-supposes an Iwasawa decomposition which, unless otherwise stated, will be written $K \cdot A \cdot N$ or equivalently, pre-supposes a Cartan involution θ, a maximal subalgebra on which $\theta = -\mathrm{Id}$, and an ordering of the restricted roots.

The union of the Siegel domains, with respect to a fixed Iwasawa decomposition, is G, and any finite union of such Siegel domains is contained in a Siegel domain.

Let us say that two families A, B of subsets of G are equivalent if every element of A (resp. B) is contained in an element of B (resp. A). It is clear that we get a family equivalent to the set of $\mathfrak{S}_{t,\omega}$ if we replace above A_t by $a \cdot A_0 (a \in A)$, or if we allow λ to run through all the positive roots in the definition of A_t.

4.2. PROPOSITION. *Let G be an open subgroup of a real algebraic, reductive group, $G = K \cdot A \cdot N$ an Iwasawa decomposition of G, and ω a compact set in N. Then the set of elements $a \cdot n \cdot a^{-1}(a \in A_t, n \in \omega)$ is relatively compact in N.*

Let $(x_i)(1 \leq i \leq m = \dim \mathfrak{n})$ be a basis of $\mathfrak{n}$, such that $[h, x_i] = \lambda_i(h)x_i$ $(h \in \mathfrak{a})$, where λ_i is a linear form on $\mathfrak{a}$ which is >0 for the given ordering $(1 \leq i \leq m)$, and that $\lambda_i \leq \lambda_j$ if $i \leq j$. For each j, the elements $x_i(i \geq j)$ span then an ideal of $\mathfrak{n}$ and it follows from standard properties of simply connected nilpotent groups that $(t_1, \cdots, t_m) \to$ $\exp(t_1 x_1) \cdots \exp(t_m x_m)$ is an analytic homeomorphism of $\mathbf{R}^m$ onto N. Further, if $n = \prod_i \exp(t_i x_i)$ and $a \in A$, then

$$a \cdot n \cdot a^{-1} = \prod_i \exp t_i \lambda_i (\log a) x_i .$$

The λ_i's are positive roots, therefore linear combinations with positive coefficients of the simple roots. Hence if $a \in A_t$ there exists a constant t' such that $\lambda_i(\log a) \leq t'$ for all i. If further $n \in \omega$, then the t_i's are also bounded, and the proposition is proved.

4.3. Proposition. *Let G be an open subgroup of a real algebraic, semisimple Lie group. Then any Siegel domain has finite Haar measure.*

Let $G = K \cdot A \cdot N$ be the underlying Iwasawa decomposition, and dk, da, dn Haar measures on K, A and N. Then, in the notation of 4.2,

$$dg = \exp[\sigma(\log a)] \, dk \cdot da \cdot dn \qquad (\sigma = \lambda_1 + \cdots + \lambda_m)$$

is a Haar measure on G [9, Lemma 35], hence

$$\int_{\mathfrak{S}_{t,\omega}} dg = c \cdot \int_{\lambda(\log a) \leq t, \lambda \in \Sigma} \exp[\sigma(\log a)] da .$$

Let $\lambda_1, \cdots, \lambda_r$ be the simple roots. Then $\sigma = m_1 \lambda_1 + \cdots + m_r \lambda_r$ with $m_i > 0 (1 \leq i \leq r)$. Since G is semi-simple, $r = \dim \mathfrak{a}$, the λ_i form a coordinate system on $\mathfrak{a}$, and the exponential $\mathfrak{a} \to A$ carries the euclidean measure on the Haar measure, we have, up to a constant factor

$$\int_{\lambda(\log a) \leq t, \lambda \in \Sigma} \exp[\sigma(\log a)] da = \prod_i \int_{-\infty}^t \exp(m_i \lambda_i) d\lambda_i < \infty .$$

4.4. *Siegel domains in* $\mathbf{GL}(n, \mathbf{R})$. In $\mathbf{GL}(n, \mathbf{R})$ we consider the usual Iwasawa decomposition where $K = \mathbf{O}(n)$, A is the subgroup of diagonal matrices with positive coefficients, and N the group of upper triangular matrices with diagonal coefficients equal to 1. If we denote the diagonal coefficients $a_1 \cdots a_n$ of an element $a \in A$ by $\exp \mu_1, \cdots, \exp \mu_n$, the simple positive roots are the linear forms $\mu_i - \mu_{i+1}(i = 1, \cdots, n-1)$. Therefore the subset $\mathfrak{S}_{t,u} = \{k \cdot a \cdot \nu\}$ where $k \in \mathbf{O}(n)$, $a = \mathrm{diag}(a_1, \cdots, a_n)$, $a_i \leq t \cdot a_{i+1}(i = 1, \cdots, n-1)$, $\nu = (n_{ij}) \in N$, $|n_{ij}| \leq u(i < j)$ is a Siegel domain in the sense of 4.1. Clearly $\mathfrak{S}_{t,u} \subset \mathfrak{S}_{t',u'}$ if $t \leq t'$, $u \leq u'$ and $\mathbf{GL}(n, \mathbf{R})$ is the union of the $\mathfrak{S}_{t,u}(t, u > 0)$. If t and u are big enough (in fact $t > 2 \cdot 3^{-1/2}$, $u > 1/2$ will do), $\mathfrak{S}_{t,u}$, or its intersection with $\mathbf{SL}(n, \mathbf{R})$, or with the group of matrices of determinant ± 1, will be called a *standard Siegel domain* of the group in question. These domains will be important for us, because of the following result; for which references are given below.

4.5. Proposition. *Let $\mathfrak{S}$ be a standard Siegel domain of* $\mathbf{GL}(n, \mathbf{R})$. *Then*

(a) $\mathbf{GL}(n, \mathbf{R}) = \mathfrak{S} \cdot \mathbf{SL}(n, \mathbf{Z})$.

(b) *For $x, y \in \mathbf{GL}(n, \mathbf{Q})$, the intersection $\mathfrak{S}^{-1} \cdot \mathfrak{S} \cap x \cdot \mathbf{SL}(n, \mathbf{Z}) \cdot y$ is finite.*

Let P_n be the space of real symmetric positive non-degenerate $n \times n$ matrices, and $\pi \colon X \to {}^t X \cdot X$ the usual projection of $\mathbf{GL}(n, \mathbf{R})$ onto P_n. We let as usual $\mathbf{GL}(n, \mathbf{R})$ act on the right on P_n by $F \to F[X] = {}^t X \cdot F \cdot X$ ($F \in P_n$, $X \in \mathbf{GL}(n, \mathbf{R})$). For $t, u > 0$, the *Siegel domain* $\mathfrak{S}'_{t,u}$ in P_n is the set of matrices $D[T]$ where

$$D = \operatorname{diag}(d_1, \cdots, d_n) \qquad (d_i \leqq t \cdot d_{i+1}; i = 1, \cdots, n-1)$$
$$T = (t_{ij}) \in N \qquad (|t_{ij}| \leqq u, i < j).$$

Clearly, $\mathfrak{S}_{t,u} = \pi^{-1}(\mathfrak{S}'_{t^2,u})$; the Siegel domains of $\mathbf{GL}(n, \mathbf{R})$ defined above are the inverse images of the Siegel domains in P_n. From the definitions, it is clear that $q \cdot \mathfrak{S}_{t,u} = \mathfrak{S}_{t,u}$, and $\mathfrak{S}'_{t,u}[q \cdot I] = \mathfrak{S}'_{t,u}$ for $q > 0$. Since $g \in \mathbf{GL}(n, \mathbf{Q})$ may be written as $g = q \cdot g'$, where g' has integral coefficients, and $\pi(X \cdot Y) = \pi(X)[Y]$, Proposition 4.5 follows from

(a') *For $t \geqq 4/3$, $u \geqq 1/2$, $P_n = \bigcup_{S \in \mathbf{SL}(n,\mathbf{Z})} \mathfrak{S}'_{t,u}[S]$.*

(b') *For each positive number q, the set of integral matrices S with determinant smaller than q in absolute value for which $\mathfrak{S}'_{t,u}[S] \cap \mathfrak{S}'_{t,u} \neq \varnothing$, is finite.*

The statement (a'), which is more or less implicit in Hermite's work [14], is proved in [16]; in fact S is allowed in [16] to have determinant -1, but it is obvious from the proof that it may be taken in $\mathbf{SL}(n, \mathbf{Z})$. The second assertion is a well known theorem of Siegel [28].

Clearly, the interior of a standard Siegel domain contains a standard Siegel domain, a finite union of standard Siegel domains is contained in a standard Siegel domain, and every element of $\mathbf{GL}(n, \mathbf{R})$ belongs to a standard Siegel domain.

5. A finiteness lemma

5.1. Let G be a locally compact group, acting continuously on the right on a locally compact space M, and G_x the isotropy group of a point $x \in M$. Then $\mu \colon g \to x \cdot g$ induces a continuous bijective map of the space $G_x \backslash G$ of cosets $G_x \cdot g$ onto the orbit $x \cdot G$ of x. It follows from a theorem of Arens [19, p. 65] that if G is countable at infinity, and $x \cdot G$ is *closed*, then μ is a homeomorphism of $G_x \backslash G$, endowed with the quotient topology, onto $x \cdot G$, endowed with the induced topology. Since a compact set of $G_x \backslash G$ is always the image of a compact set of G, this can also be expressed by saying that, under the assumptions made, the set of $g \in G$ for which $x \cdot g$ belongs to a fixed compact set of M is of the form $G_x \cdot \Omega$, with Ω compact in G. Of this we shall need the following special case.

5.2. PROPOSITION. *Let G be a Lie group with finitely many connected components, π a continuous complex or real linear representation of G in a finite dimensional vector space V, $v \in V$ a point whose orbit $v \cdot \pi(G)$ is closed, and Q a compact subset of V. Then $\{g \in G, v \cdot \pi(g) \in Q\} = G_x \cdot \Omega$, with Ω compact.*

By a *lattice* Γ in a finite dimensional vector space over $\mathbf{Q}$ or $\mathbf{R}$ we mean as usual a discrete additive subgroup which is generated by a vector space basis of V. A subspace W of V is said to be rational with respect to Γ if $W \cap \Gamma$ is a lattice in W, or, equivalently, if W is defined by linear equations with integral coefficients in a coordinate system in which Γ is the lattice of integral points.

5.3. LEMMA. *Let G be a subgroup of finite index in a semi-simple real algebraic group, θ a Cartan involution of G, $G = K \cdot A \cdot N$ an Iwasawa decomposition of G compatible with θ, and $\mathfrak{S}$ a Siegel domain of G (with respect to the decomposition $K \cdot A \cdot N$). Let π be a real representation of G in a finite dimensional vector space V, and Γ a lattice in V with respect to which the maximal eigenspaces V_i of $\pi(A)$ are rational. Let $v \in V$ be a point whose orbit is closed and whose isotropy group G_v is stable under θ. Then $v \cdot \pi(\mathfrak{S}) \cap \Gamma$ is finite.*

We denote by $|v|$ the norm of $v \in V$ for a euclidean norm with respect to which the elements of $\pi(A)$ are self-adjoint (1.1). The space V is the direct sum of the V_i's, which are mutually orthogonal. We denote by E_i the orthogonal projection of V onto V_i, and by μ_i the weight of $\mathfrak{a}$ in V_i. Since V_i is rational with respect to Γ, the subgroup of Γ spanned by the intersections $\Gamma \cap V_i$ has finite index, and $E_i(\Gamma)$ is also a *lattice* in V_i. There exists therefore a constant $c > 0$ such that $w \in \Gamma$, $E_i(w) \neq 0$ implies $|E_i(w)| \geqq c$.

Let now $x = k_x \cdot a_x \cdot n_x \in \mathfrak{S}$, and $y_x = x \cdot a_x^{-1}$, $z_x = y_x \cdot a_x^{-1}$. In the sequel, we shall drop π, and write $w \cdot g$ instead of $w \cdot \pi(g)(w \in V, g \in G)$.

Let $w = v \cdot x$, and $w_i = E_i(w)$. Then

(1) $\quad E_i(v \cdot y_x) = \exp\left[-\mu_i(\log a_x)\right] \cdot w_i, \quad E_i(v \cdot z_x) = \exp\left[-2\mu_i(\log a_x)\right]w_i.$

It follows from 4.2 that

$$\{y_x\}_{x \in \mathfrak{S}} = K \cdot \{a \cdot n \cdot a^{-1}\}_{a \cdot n \in \mathfrak{S}}$$

is relatively compact. Consequently, $\{v \cdot y_x\}_{x \in \mathfrak{S}}$ is relatively compact and there exists a constant c' such that $|w \cdot a_x^{-1}| = |v \cdot y_x| \leqq c'$. Hence

(2) $\qquad\qquad |E_i(v \cdot y_x)| = |w_i| \exp\left[-\mu_i(\log a_x)\right] \leqq c'.$

Assume now $w \in \Gamma$, and fix an i. There are two possibilities:

(i) $w_i = 0$; then $E_i(v \cdot y_x) = E_i(v \cdot z_x) = 0$;

(ii) $w_i \neq 0$; then, by the above, $|w_i| \geq c$, hence, using (1),

$$(3) \qquad |E_i(v \cdot z_x)| = |E_i(v \cdot y_x)|^2/|w_i| \leq c'^2/c .$$

In both cases, $|E_i(v \cdot z_x)|$ has a bound which is independent of i and of $x \in \mathfrak{S}$. This proves therefore that if $v \cdot x \in \Gamma$, $x \in \mathfrak{S}$, then $v \cdot z_x$ belongs to some compact set of V. Since $v \cdot G$ is closed by assumption, 5.2 shows the existence of a compact subset Q of G such that

$$(4) \qquad v \cdot x \in \Gamma, \, x \in \mathfrak{S} \Rightarrow z_x \in G_v \cdot Q .$$

We now fix such an x, and drop the index x. We have

$$z = k \cdot a \cdot n \cdot a^{-2} = k \cdot a^{-1} \cdot a^2 \cdot n \cdot a^{-2} \in G_v \cdot Q .$$

But $\theta(k) = k$, $\theta(a) = a^{-1}$, and, by assumption, $\theta(G_v) = G_v$, hence

$$(5) \qquad \begin{aligned} \theta(z) &= k \cdot a \cdot \theta(a^2 \cdot n \cdot a^{-2}) = x \cdot n^{-1} \cdot \theta(a^2 \cdot n \cdot a^{-2}) \in G_v \cdot \theta(Q) \\ x &\in G_v \cdot \theta(Q) \cdot \theta(a^2 \cdot n \cdot a^{-2})^{-1} \cdot n . \end{aligned}$$

By definition of $\mathfrak{S}$, the element n lies in a fixed compact set. Further if we have $\mathfrak{S} = \mathfrak{S}_{t,\omega}$ in the notation of 4.1, then $a^2 \in A_{2t}$ and therefore, by 4.2, $a^2 \cdot n \cdot a^{-2}$ lies in a fixed compact set. Then so does $\theta(a^2 \cdot n \cdot a^{-2})^{-1}$, and (5) shows the existence of a compact set $Q' \subset G$ such that

$$v \cdot x \in \Gamma, \, x \in \mathfrak{S} \Rightarrow x \in G_v \cdot Q' .$$

But then $v \cdot x$ belongs to $\Gamma \cap v \cdot Q'$, which is finite.

REMARK. The lemma and its proof are valid if G is reductive, provided $\pi(A)$ is real diagonalizable, as will always happen if π is a rational representation. In fact we shall use the following form of the lemma (only for $G = \mathbf{GL}(n, \mathbf{C})$, and we could also limit ourselves to the case $G = \mathbf{SL}(n, \mathbf{C})$.

5.4. LEMMA. *Let G be a reductive complex algebraic group defined over $\mathbf{Q}$, θ a Cartan involution of $G_{\mathbf{R}}$, $G_{\mathbf{R}} = K \cdot A \cdot N$ an Iwasawa decomposition of $G_{\mathbf{R}}$ compatible with θ, where A is contained in an algebraic torus T of G which is defined over $\mathbf{Q}$ and is isomorphic over $\mathbf{Q}$ to a product of groups $\mathbf{C}^*$, and $\mathfrak{S}$ a Siegel domain. Let $\pi \colon G \to \mathbf{GL}(V)$ be a rational representation of G which is defined over $\mathbf{Q}$, and Γ a lattice of $V_{\mathbf{Q}}$. Let v be a point of $V_{\mathbf{R}}$ whose orbit under G is closed and whose isotropy group in $G_{\mathbf{R}}$ is stable under θ. Then $v \cdot \pi(\mathfrak{S}) \cap \Gamma$ is finite.*

By 2.3, $v \cdot \pi(G_{\mathbf{R}})$ is closed. Moreover $\pi(T)$ is diagonalizable *over* $\mathbf{Q}$ [26, Proposition 5], which means that a maximal eigenspace of $\pi(A)$ has a basis in $V_{\mathbf{Q}}$, hence that it is "rational with respect to Γ". Thus, taking the above remark into account, all conditions under which the proof of 5.3 is valid, are fulfilled.

5.5. EXAMPLE. Let $G = \mathbf{SL}(n, \mathbf{R})$, θ be the involution $g \to {}^t g^{-1}$, and $\mathfrak{S}$ a standard Siegel domain (4.4). Let V be the space of symmetric real $n \times n$ matrices, and Γ the lattice of integral symmetric matrices. For π, we take the usual representation $F \to {}^t X \cdot F \cdot X = F[X]$. Here A is the group of positive diagonal matrices with det 1, and $\pi(A)$ is diagonal with respect to the standard basis of Γ; the maximal eigenspaces of $\pi(A)$ (which are in fact 1-dimensional) are then rational with respect to Γ. Let c be a strictly positive number, p, q two positive integers whose sum is n, and v the diagonal matrix with p entries equal to c, and q entries equal to $-c$. Then G_v is the proper orthogonal group of the quadratic form v, and is clearly stable under θ. Moreover, the orbit $v \cdot G$ of v consists of all quadratic forms with determinant $(-1)^q \cdot c^n$ and signature (p, q), hence is closed. The lemma applies, and shows that $v[\mathfrak{S}] \cap \Gamma$ is finite. But the matrices $v[g](g \in \mathfrak{S})$ are those of the forms of determinant $(-1)^q \cdot c^n$ and signature (p, q), which are reduced in the sense of Hermite (if we agree to call reduced those positive quadratic forms whose matrix is of the type ${}^t X \cdot X$ (with $X \in \mathfrak{S}$). This is a somewhat bigger set than the one considered by Hermite). The lemma in this case means therefore that the integral reduced forms with given non-zero determinant and signature are finite in number, a well known statement of Hermite [14, p. 127]. Since $\mathbf{SL}(n, \mathbf{R}) = \mathfrak{S} \cdot \mathbf{SL}(n, \mathbf{Z})$, it implies that the number of proper classes of integral quadratic forms with a given non-zero determinant is finite.

If we consider, instead of π, the natural representation of $\mathbf{SL}(n, \mathbf{R})$ in the space of homogeneous forms of degree $m \geq 2$, then the lemma also applies, and shows that the number of classes of integral forms belonging to some closed orbit is finite. Such generalizations of Hermite's result have been given by C. Jordan (Journal Ec. Polytechnique, 48 (1880), pp. 151–168), and H. Poincaré (*ibidem*, 51 (1882), pp. 45–91).

6. A fundamental set for arithmetic subgroups

6.1. Let G be an algebraic group defined over $\mathbf{Q}$. Then $G_\mathbf{Z}$ is a discrete subgroup of $G_\mathbf{R}$. It is known [29], and will follow from 6.3, that if we change the imbedding, that is, if we replace G by its image G' under a rational injective homomorphism ρ, defined over $\mathbf{Q}$, then $G'_\mathbf{Z}$ is commensurable with $\rho(G_\mathbf{Z})$. (We recall that two subgroups of a group are commensurable if their intersection has finite index in both.) Therefore the commensurable class of $G_\mathbf{Z}$ in $G_\mathbf{R}$ has an intrinsic meaning.

6.2. PROPOSITION. *Let G be a connected algebraic group defined over*

Q, $\pi\colon G \to \mathbf{GL}(V)$ *a rational representation defined over* **Q**. *Then the subgroup of* $G_\mathbf{Z}$ *leaving a lattice* Γ *in* $V_\mathbf{Q}$ *invariant has finite index in* $G_\mathbf{Z}$.

Let H be the identity component of $G \cap \mathbf{SL}(n, \mathbf{C})$. Then $H_\mathbf{Z}$ has finite index in $G_\mathbf{Z}$, we may therefore assume $G \subset \mathbf{SL}(n, \mathbf{C})$. In this case, the coefficients $\pi(g)_{\mu\nu}$ of $\pi(g)$, with respect to a basis of Γ, are polynomials in those of g, with rational coefficients (2.2). Let $g'_{ij} = g_{ij} - \delta_{ij}$. Then

$$y_{\mu\nu}(g) = \pi(g)_{\mu\nu} - \delta_{\mu\nu} = P_{\mu\nu}(g'_{11}, \cdots, g'_{nn}) , \qquad (1 \leqq \mu, \nu \leqq \dim V) ,$$

where the $P_{\mu\nu}$ are polynomials with rational coefficients and no constant term. Let m be a common multiple of the denominators of the $P_{\mu\nu}$, and M be the congruence subgroup of the elements in $G_\mathbf{Z}$ which are $\equiv \mathrm{Id} \bmod m$. Then M has finite index in $G_\mathbf{Z}$, and $\pi(g)_{\mu\nu} \in \mathbf{Z}$ for $g \in M$.

6.3. COROLLARY. *There exists a lattice in* $V_\mathbf{Q}$ *containing* Γ *which is invariant under* $G_\mathbf{Z}$. *If* π *is faithful,* $G_\mathbf{Z}$ *is commensurable with the subgroup of* G *leaving* Γ *invariant.*

We keep the notation of the previous proof. The subgroup M is invariant in $G_\mathbf{Z}$, therefore, for any $g \in G_\mathbf{Z}$, the lattice $\Gamma \cdot \pi(g)$ is invariant under M. Then the sum of the lattices $\Gamma \cdot \pi(g_i)$, where g_i runs through a system of representatives of $G_\mathbf{Z}/M$, is a lattice invariant under $G_\mathbf{Z}$. The second assertion follows from 6.2 applied to π and to π^{-1}.

6.4. COROLLARY. *Let* $G = H \cdot N$ *be the semi-direct product of a subgroup* H *and of an invariant subgroup* N, *both defined over* **Q**. *Then* $H_\mathbf{Z} \cdot N_\mathbf{Z}$ *has finite index in* $G_\mathbf{Z}$.

We may assume G to be connected. The map $g = h \cdot n \to h(h \in H, n \in N)$ is a rational homomorphism defined over **Q**, hence (6.2) $G_\mathbf{Z}$ has a subgroup M of finite index whose image is in $H_\mathbf{Z}$. Then $M \subset H_\mathbf{Z} \cdot N_\mathbf{Z}$.

6.5. THEOREM. *Let* $G \subset \mathbf{GL}(n, \mathbf{C})$ *be a reductive algebraic group defined over* **Q**, $a \in \mathbf{SL}(n, \mathbf{R})$ *such that* $a \cdot G_\mathbf{R} \cdot a^{-1}$ *is self-adjoint (see 1.9), and* $\mathfrak{S}$ *a standard Siegel domain of* $\mathbf{GL}(n, \mathbf{R})$ *(see 4.4). Then there exist finitely many elements* $b_1, \cdots, b_m \in \mathbf{SL}(n, \mathbf{Z})$ *such that the interior* U *of* $\bar{U} = \bigcup_{i=1}^{i=m} (a^{-1} \cdot \mathfrak{S} \cdot b_i) \cap G_\mathbf{R}$ *has the following properties:*

(i) $G_\mathbf{R} = U \cdot G_\mathbf{Z}$;

(ii) $K \cdot U = U$ *for a suitable maximal compact subgroup* K *of* $G_\mathbf{R}$;

(iii) $U^{-1} \cdot U \cap x \cdot G_\mathbf{Z} \cdot y$ *is finite for any* $x, y \in G_\mathbf{Q}$.

The group $G_\mathbf{Z}$ *is finitely generated.*

By 3.8, we may find a rational representation $\pi\colon \mathbf{GL}(n, \mathbf{C}) \to \mathbf{GL}(V)$, defined over **Q**, for which there exists $v \in V_\mathbf{Q}$ whose isotropy group is G and whose orbit is closed. Using 6.3, we take in $V_\mathbf{Q}$ a lattice Γ invariant under $\mathbf{GL}(n, \mathbf{Z})$; replacing v by a multiple if necessary, we many assume $v \in \Gamma$. The group $G_\mathbf{R}$ is real algebraic, reductive (3.2); by 1.9, there exists $a \in \mathbf{SL}(n, \mathbf{R})$ such that $a \cdot G_\mathbf{R} \cdot a^{-1} = G'_\mathbf{R}$ is self-adjoint. Let $v' = v \cdot \pi(a^{-1})$. Then $v' \cdot \pi(\mathbf{GL}(n, \mathbf{C})) = v \cdot \pi(\mathbf{GL}(n, \mathbf{C}))$ is closed, hence (2.3) so is $v' \cdot \pi(\mathbf{GL}(n, \mathbf{R}))$. The isotropy group of v' in $\mathbf{GL}(n, \mathbf{R})$ is $G'_\mathbf{R}$, and is by construction invariant under the Cartan involution $\theta: g \to {}^t g^{-1}$, which underlies the definition of the standard Siegel domain $\mathfrak{S}$. Moreover, in this case, A is the subgroup of diagonal matrices with positive real eigenvalues, and it belongs to the algebraic torus $D(n)$ of all diagonal matrices of $\mathbf{GL}(n, \mathbf{C})$, which is defined over $\mathbf{Q}$. Consequently, all conditions of 5.4 are fulfilled, and we may assert that $v' \cdot \pi(\mathfrak{S}) \cap \Gamma$ is finite. *A fortiori,* $v' \cdot \pi(\mathfrak{S}) \cap v \cdot \pi(\mathbf{SL}(n, \mathbf{Z}))$ is finite. Let then $b_1, \cdots, b_m \in \mathbf{SL}(n, \mathbf{Z})$ be such that

$$(1) \qquad v \cdot \pi(\mathfrak{S}) \cap v \cdot \pi(\mathbf{SL}(n, \mathbf{Z})) \subset \{v \cdot \pi(b_1^{-1}), \cdots, v \cdot \pi(b_m^{-1})\} \ .$$

Let now

$$(2) \qquad H = \{g \in \mathbf{GL}(n, \mathbf{R}) \mid v' \cdot \pi(g) = v\} \ .$$

Then

$$(3) \qquad H = a \cdot G_\mathbf{R} = G'_\mathbf{R} \cdot a \ .$$

Let $h \in H$. By the classical reduction theory (4.5),

$$h = s \cdot b \qquad\qquad \left(s \in \mathfrak{S},\, b \in \mathbf{SL}(n, \mathbf{Z})\right) \ .$$

The equality $v' \cdot \pi(h) = v$ yields $v' \cdot \pi(s) = v \cdot \pi(b^{-1})$, hence, by (1), there exists an index i, $(1 \leqq i \leqq m)$ such that

$$v' \cdot \pi(s) = v \cdot \pi(b^{-1}) = v \cdot \pi(b_i^{-1}) \ .$$

We have then $b_i^{-1} \cdot b \in G \cap \mathbf{SL}(n, \mathbf{Z}) \subset G_\mathbf{Z}$, hence $b \in b_i G_\mathbf{Z}$, and

$$(4) \qquad \begin{aligned} &H \subset \bigcup_i \mathfrak{S} \cdot b_i \cdot G_\mathbf{Z} \ , \\ &G_\mathbf{R} = a^{-1} \cdot H \subset \bigcup_i a^{-1} \cdot \mathfrak{S} \cdot b_i \cdot G_\mathbf{Z} \ , \\ &G_\mathbf{R} = \bar{U} \cdot G_\mathbf{Z}, \qquad\qquad \left(\bar{U} = \bigcup_i (a^{-1} \cdot \mathfrak{S} \cdot b_i) \cap G_\mathbf{R}\right) \ . \end{aligned}$$

The equality (4) is *a fortiori* true if we replace $\mathfrak{S}$ by a standard Siegel domain containing $\mathfrak{S}$ in its interior, hence (i) is proved.

The group $K' = G'_\mathbf{R} \cap O(n)$ is maximal compact in $G'_\mathbf{R}$ (1.10), and $K = a^{-1} \cdot K' \cdot a$ is maximal compact in $G_\mathbf{R}$. Clearly, $K' \cdot (\mathfrak{S} b_i \cap G'_\mathbf{R}) \subset \mathfrak{S} b_i \cap G'_\mathbf{R}$, and therefore $K \cdot \bar{U} = \bar{U}$, $K \cdot U = U$.

Let now $x, y \in G_\mathbf{Q}$, and $u \in U^{-1} \cdot U \cap x \cdot G_\mathbf{Z} \cdot y$. There exist, then, two indices $i, j (1 \leqq i, j \leqq m)$ such that

$$u \in (a^{-1} \cdot \mathfrak{S} \cdot b_i)^{-1} \cdot (a^{-1} \cdot \mathfrak{S} \cdot b_j) = b_i^{-1} \cdot \mathfrak{S}^{-1} \cdot \mathfrak{S} \cdot b_j \ ,$$

whence

$$b_i \cdot u \cdot b_j^{-1} \in \mathfrak{S}^{-1} \cdot \mathfrak{S} \, ,$$

and the finiteness of the number of the possible u's follows from Siegel's theorem (4.5). For $x = y = e$, this implies that $U^{-1} \cdot U \cap G_{\mathbf{Z}}$ is finite, in other words that U meets only a finite number of its right translates under $G_{\mathbf{Z}}$. Since $G_{\mathbf{Z}} \cap (G_{\mathbf{R}})^0$ is of finite index in $G_{\mathbf{Z}}$, the finite generation of $G_{\mathbf{Z}}$ follows from the following well known elementary lemma:

6.6. Lemma. *Let H be a group which operates on a connected topological space M, and U an open set such that $U \cdot H = M$. Then $J = \{h \in H \mid U \cdot h \cap U \neq \varnothing\}$ is a set of generators for H.*

In fact let H' be the subgroup generated by J. Then $U \cdot H'$ is open. If $U \cdot h \cap U \cdot h' \neq \varnothing (h \in H, h' \in H')$, then $U \cdot h \cdot h'^{-1} \cap U \neq \varnothing$, hence $h \in J \cdot h' \subset H'$, from which it follows first that $U \cdot H' = M$, and then that $H' = H$.

6.7. Remarks.

(1) Let us call fundamental set an open subset of $G_{\mathbf{R}}$ satisfying properties (i), (ii), (iii). In view of the inclusion properties of standard Siegel domains listed at the end of 4.5, the class C of fundamental sets constructed in 6.5 has the following properties: it covers $G_{\mathbf{R}}$, any finite union of such sets is contained in a set of C, any such set contains the closure of an element of C.

The main part of 6.5 will be extended to algebraic groups in 6.12. It will be shown later that, at any rate for a suitable a, the set U has finite Haar measure when G is semi-simple.

(2) Let K be a maximal compact subgroup such that $K \cdot U = U$, and π the natural projection of $G_{\mathbf{R}}$ onto $P = K \backslash G_{\mathbf{R}}$. Clearly, $U' = \pi(U)$ has the following properties: (i') $P = U' \cdot G_{\mathbf{Z}}$; (iii') for x, $y \in G_{\mathbf{Q}}$, the set of $g \in G_{\mathbf{Z}}$ for which $U' \cap U' \cdot x \cdot g \cdot y \neq \varnothing$ is finite. Conversely, the inverse image of a set U' in P having the properties (i'), (iii') has the properties (i), (ii), (iii), so that the construction of fundamental sets in $G_{\mathbf{R}}$ or in $K \backslash G_{\mathbf{R}}$ are essentially equivalent questions. In the special case where G is the orthogonal group of an integral non-degenerate indefinite quadratic form F, $K \backslash G_{\mathbf{R}}$ is the space of majorizing positive forms of F in the sense of Hermite. If we take, furthermore, the natural representation of $\mathbf{GL}(n, \mathbf{C})$ in the space of symmetric matrices, then our construction reduces to that of Hermite (always with the minor difference that instead of using an arbitrary Siegel domain $\mathfrak{S}$ of $\mathbf{GL}(n, \mathbf{R})$, Hermite uses the inverse image of the space of reduced positive forms in his sense, that is, the domain given by 4.5a'). In this case, the two properties of U' above, and the resulting finite generation of $G_{\mathbf{Z}}$, are also proved by Hermite [14, pp. 201–

233, § VIII].

6.8. LEMMA. *Let $M \subset G$ be reductive subgroups of $\mathbf{GL}(n, \mathbf{C}) = G'$, both defined over $\mathbf{Q}$. Let $\pi: G' \to \mathbf{GL}(W)$ be a rational representation, defined over $\mathbf{Q}$, Γ a lattice in $W_{\mathbf{Q}}$ invariant under $G'_{\mathbf{Z}}$, $w' \in \Gamma$ a point whose orbit under G' is closed and whose isotropy group in G' is M. Then $w' \cdot G_{\mathbf{R}} \cap \Gamma$ consists of a finite number of orbits of $G_{\mathbf{Z}}$.*

By 1.9, there exists $a \in \mathbf{SL}(n, \mathbf{R})$ such that $a \cdot G_{\mathbf{R}} \cdot a^{-1}$ and $a \cdot M_{\mathbf{R}} \cdot a^{-1}$ are self-adjoint. By 6.5, there exists a finite number of elements $b_i \in G'_{\mathbf{Z}}$ such that $G_{\mathbf{R}} = \bigcup_i (G_{\mathbf{R}} \cap a^{-1} \cdot \mathfrak{S} \cdot b_i) \cdot G_{\mathbf{Z}}$ where $\mathfrak{S}$ is a standard Siegel domain of $\mathbf{GL}(n, \mathbf{R})$. It suffices therefore to show that

$$w \cdot (G_{\mathbf{R}} \cap a^{-1} \cdot \mathfrak{S} \cdot b_i) \cap \Gamma$$

is finite, hence, *a fortiori*, that $w \cdot a^{-1} \cdot \mathfrak{S} \cap \Gamma$ is finite. Let $w' = w \cdot a^{-1}$. Then $G'_{w'} = a \cdot M \cdot a^{-1}$, hence $(G'_{w'})_{\mathbf{R}}$ is self-adjoint. Further, the group A of diagonal matrices which underlies the definition of $\mathfrak{S}$ belongs to an algebraic torus which is diagonal over $\mathbf{Q}$ hence 5.4 applies and yields the finiteness of $w \cdot a^{-1} \cdot \mathfrak{S} \cap \Gamma = w' \cdot \mathfrak{S} \cap \Gamma$.

6.9. THEOREM. *Let G be a reductive algebraic group defined over $\mathbf{Q}$, $\pi: G \to \mathbf{GL}(V)$ a rational representation defined over $\mathbf{Q}$, Γ a lattice in $V_{\mathbf{Q}}$ invariant under $G_{\mathbf{Z}}$, and X a closed orbit of G. Then $X \cap \Gamma$ consists of a finite number of orbits of $G_{\mathbf{Z}}$.*

Since G^0 has finite index in G, we may assume G to be connected. We assume $X \cap \Gamma \neq \varnothing$ (otherwise there is nothing to prove), take $v \in \Gamma \cap X$, and put $H = G_v$. The group H and the orbit $X = v \cdot G$ are defined over $\mathbf{Q}$. Since X is closed, H is reductive (3.8). We prove first:

(*) There exists a rational representation $\pi': G \to \mathbf{GL}(W)$ defined over $\mathbf{Q}$, a point $w \in W_{\mathbf{Q}}$ such that $G_w = H$, $w \cdot \pi'(G) = X'$ is closed, and that $X' \cap \Gamma'$ consists of a finite number of orbits of $G_{\mathbf{Z}}$ for any lattice $\Gamma' \subset W_{\mathbf{Q}}$ invariant under $G_{\mathbf{Z}}$.

Let $G' = \mathbf{GL}(n, \mathbf{C})$. The group H being reductive, defined over $\mathbf{Q}$, there exists by 3.8 a rational representation $\rho: G' \to \mathbf{GL}(W)$ defined over $\mathbf{Q}$, and a point $w \in W_{\mathbf{Q}}$ such that $w \cdot \rho(G')$ is closed and $H = G'_w$. We claim that the restriction π' of ρ to G fulfills our conditions. The orbit $X' = w \cdot \pi'(G)$ is closed because if we identify $w \cdot \rho(G')$ with $H \backslash G'$, the orbit X' is the inverse image of a point in the projection $H \backslash G' \to G \backslash G'$. Let now $\Gamma' \subset W_{\mathbf{Q}}$ be a lattice invariant under $G_{\mathbf{Z}}$. It is contained in a lattice invariant under $G'_{\mathbf{Z}}$ (6.3), therefore it is enough to consider $X' \cap \Gamma'$ for lattices invariant under $G'_{\mathbf{Z}}$. Of course, $X' \cap \Gamma' \subset X' \cap W_{\mathbf{R}}$, which consists of a finite number of closed orbits of $G_{\mathbf{R}}$ (2.3); therefore, it suffices to show that for $w' \in X' \cap \Gamma'$, the intersection $w' \cdot \pi'(G_{\mathbf{R}}) \cap \Gamma'$ consists of a finite number of orbits of $G_{\mathbf{Z}}$. There exists $g \in G$ such that $w' = w \cdot \pi'(g)$, hence $G'_{w'} = g^{-1} \cdot H \cdot g \subset G$. The group $G'_{w'}$ is reductive, since H is, and defined over $\mathbf{Q}$, since $w' \in W_{\mathbf{Q}}$. Moreover $w' \cdot \rho(G') = w \cdot \rho(G')$ is closed. The fact that $w' \cdot \pi'(G_{\mathbf{R}}) \cap \Gamma'$ is the union of a finite number of orbits of $G_{\mathbf{Z}}$ is now a consequence of 6.8 (with $M = G'_{w'}$).

The maps $g \to v \cdot \pi(g)$ and $g \to w \cdot \pi'(g)$ induce isomorphisms of $H\backslash G$ with X and X', defined over $\mathbf{Q}$, whence an equivariant isomorphism $\varphi \colon X \to X'$, defined over $\mathbf{Q}$. Let $x_1, \cdots, x_r$ and $y_1, \cdots, y_s$ be coordinates in V and W with respect to bases of Γ and Γ' respectively. The function $y_i(\varphi(x))(x \in X)$ is a regular function, defined over $\mathbf{Q}$. Since X is an affine algebraic set, $y_i(\varphi(x))$ may be written as a polynomial $P_i(x_1, \cdots, x_r)$ in the x_i's, with rational coefficients $(1 \leq i \leq s)$. If q is a common multiple of the denominators of those coefficients, then $q \cdot y_i(\varphi(x))$ is integral whenever $x_1, \cdots, x_r$ are, therefore

$$\varphi(X \cap \Gamma) \subset X' \cap \frac{1}{q}\Gamma' \, .$$

The lattice $(1/q)\Gamma'$ is of course also invariant under $G_{\mathbf{Z}}$, therefore $(1/q)\Gamma' \cap X'$ consists of a finite number of orbits of $G_{\mathbf{Z}}$ by (*). Since φ is G-equivariant, its restriction to $X \cap \Gamma$ is $G_{\mathbf{Z}}$-equivariant, hence $X \cap \Gamma$ is also a finite number of orbits of $G_{\mathbf{Z}}$.

6.10. In order to go from the reductive to the general case in the two following sections, we shall use the following facts: a connected algebraic group G defined over a field k (of characteristic 0) is the semi-direct product of a reductive group H and of an invariant unipotent group N, both defined over k.[4] If G is unipotent, defined over $\mathbf{Q}$, then $G_{\mathbf{R}}/G_{\mathbf{Z}}$ is compact. The latter fact is completely elementary; in fact, G being nilpotent, defined over $\mathbf{Q}$, its Lie algebra has a basis (x_i) $(i = 1, \cdots, \dim G)$ consisting of matrices with integral coefficients, such that for each $j \geq 1$, the elements $x_i(i \geq j)$ span an ideal. Let, further, m be the degree of the ambient linear group. Then $x_i^m = 0$ $(i = 1, \cdots, \dim G)$, and $g_i = \exp(m! \cdot x_i) \in G_{\mathbf{Z}}$. Using induction on $\dim G$, it is immediately seen that the quotient of $G_{\mathbf{R}}$ by the subgroup generated by the g_i's is compact.

6.11. COROLLARY. *Let G, G' be algebraic groups defined over $\mathbf{Q}$ and $\mu \colon G \to G'$ a surjective rational homomorphism defined over $\mathbf{Q}$, with finite kernel (an isogeny). Then $\mu(G_{\mathbf{Z}})$ and $G'_{\mathbf{Z}}$ are commensurable.*

Here again, it is enough to prove this when G and G' are connected. It

---

[4] The corresponding statement for Lie algebras is proved in [7b, Chap. V, § 4, Prop. 5]. For the global version, see G. D. Mostow, Amer. J. Math. 78 (1956), 200-221, Theorem 6.1.

follows from 6.3 that we may assume $\mu(G_Z) \subset G'_Z$. By adding one coordinate (see 2.1), we may have G' closed in $\mathbf{M}(n', \mathbf{C})$.

Let now G be reductive. We let G act on $V = \mathbf{M}(n', \mathbf{C})$ by right translations $x \to x \cdot \mu(g)$, and get in this way a rational representation $G \to GL(V)$ defined over $\mathbf{Q}$. The subgroup G'_Z is the intersection of the closed orbit G' of G with the lattice $\mathbf{M}(n', \mathbf{Z})$ which is invariant under G_Z, and belongs to $V_\mathbf{Q}$. By 6.9, G'_Z consists of a finite number of cosets of $\mu(G_Z)$, hence $[G'_Z : \mu(G_Z)] < \infty$.

If G is unipotent, $G_\mathbf{R}/G_Z$ is compact; then so is $G'_\mathbf{R}/\mu(G_Z)$; this space is a covering of $G'_\mathbf{R}/G'_Z$ with a discrete fibre which has $[G'_Z : \mu(G_Z)]$ elements. This number must be finite.

In the general case, $G = H \cdot N$, $G' = H' \cdot N'$ with H, $H' = \mu(H)$ reductive, defined over $\mathbf{Q}$, N, $N' = \mu(N)$ unipotent, invariant, defined over $\mathbf{Q}$ (6.10). By the above $\mu(H_Z \cdot N_Z)$ has finite index in $H'_Z \cdot N'_Z$, and 6.11 follows from 6.4.

6.12. Theorem. *Let G be an algebraic group defined over $\mathbf{Q}$. There exists an open set U in $G_\mathbf{R}$ such that $U \cdot G_Z = G_\mathbf{R}$, and that for any $x, y \in G_\mathbf{Q}$, the intersection $U^{-1} \cdot U \cap x \cdot G_Z \cdot y$ is finite. The group G_Z is finitely generated.*

It is enough to prove this for G connected. If G is reductive, see 6.5. If G is unipotent, then (6.10) there exists a relatively compact open subset U such that $G_\mathbf{R} = U \cdot G_Z$. Then U clearly satisfies our second condition.

In the general case, we use the decomposition $G = H \cdot N$ of 6.10. Let A and B be open subsets in H and N satisfying our two conditions in H and N, with B, moreover, relatively compact. We assert that $U = A \cdot B$ verifies our contention. In fact

$$G = H \cdot N = A \cdot H_Z \cdot N \subset A \cdot N \cdot H_Z \subset A \cdot B \cdot N_Z \cdot H_Z \subset A \cdot B \cdot G_Z,$$

which proves the first condition. The group $H_Z \cdot N_Z$ has finite index in G_Z (6.4). Therefore, our second assertion is equivalent to the finiteness of $U^{-1} \cdot U \cap (x \cdot H_Z \cdot N_Z \cdot y)$ for arbitrary $x, y \in G_\mathbf{Q}$. The group G being isomorphic to $H \cdot N$ over $\mathbf{Q}$, we also have $G_\mathbf{Q} = H_\mathbf{Q} \cdot N_\mathbf{Q}$, and may write $x = a \cdot b$, $y = c \cdot d$ ($a, c \in H_\mathbf{Q}$, $b, d \in N_\mathbf{Q}$). Let $h \in H_Z$, $n \in N_Z$. We have

$$A \cdot B \cdot x \cdot h \cdot n \cdot y = A \cdot a \cdot h \cdot c \cdot B' \cdot n'$$

$$(B' = (a \cdot h \cdot c)^{-1} \cdot B \cdot a \cdot b \cdot h \cdot c \; ; \quad n' = c^{-1} \cdot n \cdot c \cdot d),$$

hence $A \cdot B \cap A \cdot B \cdot x \cdot h \cdot n \cdot y \neq \varnothing$ is equivalent to

$$(1) \qquad A \cdot a \cdot h \cdot c \cap A \neq \varnothing, \qquad B' \cdot n' \cap B \neq \varnothing.$$

By our assumption on A, the possible h's are finite in number. Since B is relatively compact, $B'^{-1} \cdot B$ is also relatively compact, hence its intersection with the discrete set $c^{-1} \cdot N_Z \cdot c \cdot d$ is finite. Thus there is a finite number of possibilities for h, and for each of them, a finite number of possible n's.

The above implies in particular that $U^{-1} \cdot U \cap G_Z$ is finite. The finite generation of G_Z then follows from 6.6, as in 6.5.

6.13. REMARK. As was mentioned in the introduction, the construction of U in 6.5 is a generalization of Hermite's procedure in the case of indefinite quadratic forms [14]. As is well known, the latter had been adapted, notably by Siegel, to many other cases, the most inclusive one being that of the automorphism group of a rational involutorial semi-simple algebra [24, 29]. This case represents essentially all classical groups with center reduced to the identity. However, U is constructed there in the symmetric space $K \backslash G_R$, rather than in G_R, but this is a minor difference (see 6.7). This implies, of course, the finite generation of G_Z. The finiteness of the volume of G_R/G_Z, or, equivalently, of $(K \backslash G_R)/G_Z$, in that case is also proved in [24], generalizing earlier results of Siegel.

In order to construct U, we take a rational representation of the ambient linear group $\mathbf{GL}(n, \mathbf{C})$, such that the representation space has a rational point with closed orbit and isotropy group G, whose existence follows from 2.5 and 3.5b. In this respect, we point out that 3.5a is used only in proving 6.9, and there only to ascertain that H is reductive. This last fact is obvious in 6.11, where $H = (e)$ and follows from [4] in 11.6, where H runs through the centralizers of semi-simple elements; as these are the only two applications of 6.9 made in this work, we see that, except for 6.9 in full generality, this paper can be made independent of 3.5a.

7. The finiteness of the volume for semi-simple groups

In this paragraph, we often write b^a for $a \cdot b \cdot a^{-1}$, where a, b are elements of a group.

For the sake of reference, we first sketch the proof of an elementary lemma on nilpotent Lie groups (see also [13, Lemma 1]):

7.0. LEMMA. *Let N be a connected, simply connected, real or complex nilpotent Lie group, $\mathfrak{n}^{(i)}$ a strictly decreasing sequence of ideals of $\mathfrak{n}$ such that $[\mathfrak{n}, \mathfrak{n}^{(i)}] \subset \mathfrak{n}^{(i+1)}$, $\mathfrak{n}_1$ and $\mathfrak{n}_2$ two mutually complementary subspaces such that $\mathfrak{n}^{(i)} = \mathfrak{n}_1 \cap \mathfrak{n}^{(i)} + \mathfrak{n}_2 \cap \mathfrak{n}^{(i)}$ $(i = 0, \cdots; \mathfrak{n}^{(0)} = \mathfrak{n})$. Then $(x, y) \to \exp x \cdot \exp y (x \in \mathfrak{n}_2, y \in \mathfrak{n}_1)$ is an analytic homeomorphism of $\mathfrak{n}_2 \times \mathfrak{n}_1$ onto N.*

Let s be the biggest index such that $\mathfrak{n}^{(s)} \neq 0$. Then $\mathfrak{n}^{(s)}$ is central, generates a closed simply connected central subgroup $N^{(s)}$ of N, and

$N/N^{(s)}$ is simply connected. Let $\mathfrak{n}_1'$ and $\mathfrak{n}_2'$ be supplementary subspaces to $\mathfrak{n}^{(s)} \cap \mathfrak{n}_1$ and $\mathfrak{n}^{(s)} \cap \mathfrak{n}_2$ in $\mathfrak{n}_1$ and $\mathfrak{n}_2$ respectively. Proceeding by induction on s, we may assume the lemma to be true for $\mathfrak{n}_1/(\mathfrak{n}_1 \cap \mathfrak{n}^{(s)})$ and $\mathfrak{n}_2/(\mathfrak{n}_2 \cap \mathfrak{n}^{(s)})$ in $\mathfrak{n}/\mathfrak{n}^{(s)}$. This implies immediately that $(x, y) \to \exp x \cdot \exp y$ is an analytic homeomorphism of $\mathfrak{n}_2' \times \mathfrak{n}_1'$ onto $M = \exp \mathfrak{n}_2' \cdot \exp \mathfrak{n}_1'$, and that $(m, z) \to m \cdot z$ is a homeomorphism of $M \times N^{(s)}$ onto N. The lemma then follows readily from the following facts; the exponential is an analytic homeomorphism of $\mathfrak{n}$ onto N, $\exp(a + b) = \exp a \cdot \exp b$ if $[a, b] = 0$ $(a, b \in \mathfrak{n})$, and $\mathfrak{n}^{(s)}$ is central.

7.1. Let G be a real algebraic semi-simple Lie group, θ a Cartan involution of G, and $G = K \cdot A \cdot N$ an Iwasawa decomposition of G compatible with θ (1.11). We use the notation of 4.1, except for the fact that Σ will denote the set of all positive roots for the ordering defined by N. Moreover, given $x = k \cdot a \cdot n (k \in K, a \in A, n \in N)$, we put $H(x) = \log a$ and $\nu(x) = n$.

Let M be the centralizer and M^* the normalizer of A in K. Then $M^*/M = W$ is a finite group, the "restricted" Weyl group of G. It acts by inner automorphisms on A or $\mathfrak{a}$, and this representation is faithful.

7.2. LEMMA. *We keep the above notation. Then $M^*/(M^* \cap G^0) = G/G^0$ and each coset of M^* modulo $M^* \cap G^0$ contains an element which normalizes N. The group G is the disjoint union of the subsets $N m_w M_w A N$, where m_w runs through a system of representatives of the elements w of W.*

By Bruhat's lemma, [10], G^0 is the disjoint union of the subsets $N m_w (M \cap G^0) A \cdot N$, where w runs through $M(M^* \cap G^0)/M$. Our second assertion follows from this and the first assertion. By a theorem of E. Cartan (see e.g. [12, Lemma 33]), given $k \in K$, there exists $k' \in K^0$ such that $k A k^{-1} = k' A k'^{-1}$. Therefore M^* meets each connected component of G. Moreover, $M(M^* \cap G^0)/M$ is transitive on the Weyl chambers of $\mathfrak{a}$, therefore each coset of M^* modulo $M^* \cap G^0$ contains an element which leaves the positive Weyl chamber invariant, hence normalizes N. This proves the first assertion.

REMARK. The above is also valid in a semi-simple Lie group with finitely many connected components, whose identity component has a finite center.

7.3. Let $w \in W$, $\Sigma(w)$ be the set of positive roots whose transforms under w are negative, and $\Phi(w)$ the set of roots which are linear combinations of elements of $\Sigma(w) \cup (-\Sigma(w))$. Then

$$\mathfrak{g}_w = \sum_{\alpha \in \Phi(w)} \left(\mathfrak{g}_\alpha + [\mathfrak{g}_\alpha, \mathfrak{g}_{-\alpha}] \right)$$

is clearly a subalgebra. Let $\mathfrak{n}_w = \mathfrak{g}_w \cap \mathfrak{n}$, $\mathfrak{a}_w = \mathfrak{g}_w \cap \mathfrak{a}$, $\mathfrak{m}_w = \mathfrak{g}_w \cap \mathfrak{m}$, $\mathfrak{k}_w = \mathfrak{g}_w \cap \mathfrak{k}$. As usual, for $\alpha \in \mathfrak{a}^*$, h_α is the element of $\mathfrak{a}$ defined by $(h_\alpha, h) = \alpha(h)$, where $(\ ,\ )$ is the scalar product defined by the Killing form. Then $\mathfrak{g}_w$ is semi-simple,

$$(1) \qquad \mathfrak{g}_w = \mathfrak{k}_w + \mathfrak{a}_w + \mathfrak{n}_w$$

is an Iwasawa decomposition of $\mathfrak{g}_w$, and

$$(2) \qquad \mathfrak{a}_w = \sum_{\alpha \in \Sigma(w)} \mathbf{R} \cdot h_\alpha = \sum_{\alpha \in \Phi(w)} \mathbf{R} \cdot h_\alpha \ .$$

In fact, it is elementary and known that if $x \in \mathfrak{g}_\alpha$, $y \in \mathfrak{g}_{-\alpha}$, then

$$(3) \qquad \begin{aligned} & B\big(x, \theta(x)\big) \neq 0 && (x \neq 0) \ , \\ & [x, \theta(x)] = h_\alpha B\big(x, \theta(x)\big), \quad [x, y] - h_\alpha B(x, y) \in \mathfrak{m} \ , \end{aligned}$$

where B is the Killing form, from which we deduce (2) and

$$(4) \qquad \mathfrak{m}_w + \mathfrak{a}_w = \sum_{\alpha \in \Phi(w)} [\mathfrak{g}_\alpha, \mathfrak{g}_{-\alpha}], \quad \mathfrak{g}_w = \mathfrak{m}_w + \mathfrak{a}_w + \sum_{\alpha \in \Phi(w)} \mathfrak{g}_\alpha \ ,$$

so that (1) follows from

$$(5) \qquad \mathfrak{g}_\alpha + \mathfrak{g}_{-\alpha} = \mathfrak{g}_\alpha + \theta(\mathfrak{g}_\alpha) = \mathfrak{g}_\alpha + (\mathfrak{g}_\alpha + \mathfrak{g}_{-\alpha}) \cap \mathfrak{k} \ .$$

The algebra $\mathfrak{g}_w$ is stable under θ, hence reductive (1.4). By (4), $\mathfrak{m}_w + \mathfrak{a}_w$ belongs to the derived algebra $\mathscr{D}\mathfrak{g}_w$ of $\mathfrak{g}_w$. By (2), for each $\alpha \in \Phi(w)$, there exists $h \in \mathfrak{a}_w$ which does not annihilate α, hence $\mathfrak{g}_\alpha = [h, \mathfrak{g}_\alpha] \subset \mathscr{D}\mathfrak{g}$. Thus $\mathfrak{g}_w = \mathscr{D}\mathfrak{g}_w$, and $\mathfrak{g}_w$ is semi-simple.

Up to 7.7, G is a real algebraic semi-simple Lie group, θ a Cartan involution of G, $G = K \cdot A \cdot N$ an Iwasawa decomposition compatible with θ, Σ the set of the positive roots in the ordering defined by N. The Siegel domains of G are always defined with respect to the given Iwasawa decomposition.

7.4. Lemma. *Let $\mathfrak{S}$ be a Siegel domain of G, $x \in G$ and t a positive real number. Then $\mathfrak{S}x \cap K \cdot A_t \cdot N$ is contained in a Siegel domain of G.*

For any Siegel domain $\mathfrak{S}'$ and elements $a \in A$, $n \in N$, the sets $\mathfrak{S}' \cdot a$ and $\mathfrak{S}' \cdot n$ belong to Siegel domains. In view of Bruhat's lemma (7.2) we may assume $x = m^{-1}$, $m \in M^*$. Let w be the element of W defined by m. We use the notation of 7.3. Let further $\mathfrak{n}'_w$ be the sum of the $\mathfrak{g}_\alpha$ where $\alpha \in \Sigma$, $\alpha \notin \Phi(w)$, $\mathfrak{a}'_w$ be the subspace of $\mathfrak{a}$ on which the elements of $\Phi(w)$ are all zero. Let A_w, A'_w, N_w, N'_w be the exponentials of $\mathfrak{a}_w, \mathfrak{a}'_w, \mathfrak{n}_w, \mathfrak{n}'_w$. Then $A = A_w \cdot A'_w$, and the group A'_w centralizes the analytic group $G(w)$ with Lie algebra $\mathfrak{g}_w$. Let further $\alpha_1, \cdots, \alpha_s$ be the positive roots arranged in increasing order. Then

$$\mathfrak{n}^{(i)} = \sum_{j \geq i} \mathfrak{g}_{\alpha_j} \ , \qquad \mathfrak{n}_1 = \mathfrak{n}_w, \mathfrak{n}_2 = \mathfrak{n}'_w \ ,$$

satisfy the assumptions of 7.0, hence $(n, n') \to \exp n \cdot \exp n'$ is an analytic homeomorphism of $\mathfrak{n}_w \times \mathfrak{n}'_w$ onto $N = N_w \cdot N'_w$. Let $y \in \mathfrak{S}$ and $z = y \cdot m^{-1}$. We may write

$$
\begin{aligned}
&y = k \cdot a \cdot n &&(k \in K, a \in A, n \in N), \\
(1)\qquad &a = a_1 \cdot a_2, \quad n = n_1 \cdot n_2 &&(a_1 \in A_w, a_2 \in A'_w, n_1 \in N_w, n_2 \in N'_w).
\end{aligned}
$$

In the sequel, we say that an element, which is a function of y, *is bounded* if it stays within some compact set when y varies subject to our conditions. We prove first that a_1 is bounded. We have

$$
\begin{aligned}
z &= k \cdot a \cdot n \cdot m^{-1} = k \cdot n^a \cdot a \cdot m^{-1} = k \cdot n^a \cdot m^{-1} \cdot m \cdot a \cdot m^{-1}, \\
z &= k \cdot m^{-1} \cdot n^{ma} \cdot w(a),
\end{aligned}
$$

hence

$$
(2)\qquad\qquad H(z) = H(n^{ma}) + \log w(a).
$$

By 4.2, n^a is bounded, hence so is n^{ma}. There exists therefore $t_1 \geq t$ such that

$$
(3)\qquad\qquad \alpha(\log w(a)) \leq t_1 \qquad\qquad (\alpha \in \Sigma).
$$

Let now $\alpha \in \Sigma(w)$. Then ${}^t w(\alpha) < 0$, and, by (3)

$$
(4)\qquad\qquad -\alpha(\log a) = -{}^t w(\alpha)(\log w(a)) \leq t_1.
$$

By definition of a Siegel domain, there exists t_0 such that $\alpha(\log a) \leq t_0$, for $\alpha > 0$. For t'' big enough, we have therefore

$$
(5)\qquad\qquad |\alpha(\log a)| \leq t'' \qquad\qquad (\alpha \in \Sigma(w)).
$$

But $\alpha(\log a) = \alpha(\log a_1)$ for $\alpha \in \Sigma(w)$, and the roots $\alpha \in \Sigma(w)$ span the dual of $\mathfrak{a}_w$, therefore (5) implies that $\log a_1$ varies in a compact set. Thus a_1 is bounded, as was contended. But then, $w(a_1)$ is also bounded, and (3) shows the existence of t_3 such that

$$
(6)\qquad\qquad \alpha(\log w(a_2)) \leq t_3 \qquad\qquad (\alpha \in \Sigma).
$$

Since a_2 commutes with n_1, we may write

$$
(7)\qquad z = k \cdot m^{-1} \cdot (m \cdot a_1 \cdot n_1 \cdot m^{-1}) \cdot w(a_2) \cdot n'_2 \qquad (n'_2 = m \cdot n_2 \cdot m^{-1}).
$$

The element n is bounded (by definition of a Siegel domain), hence so are n_1, n_2 and n'_2. The product $a_1 \cdot n_1$ belongs to $G(w)$ and ${}^t w(\Sigma(w)) = -\Sigma(w^{-1})$; hence

$$
m \cdot a_1 \cdot n_1 \cdot m^{-1} \in m \cdot G(w) \cdot m^{-1} = G(w^{-1}),
$$

$$
m \cdot a_1 \cdot n_1 \cdot m^{-1} = k_3 \cdot a_3 \cdot n_3,
$$

$$
(8)\qquad (k_3 \in K \cap G(w^{-1}), a_3 \in A \cap G(w^{-1}), n_3 \in N \cap G(w^{-1})),
$$

where a_3 and n_3 are bounded, since the left hand side is. The element a_2

commutes with $G(w)$, hence $w(a_2)$ commutes with $G(w^{-1})$, and we have by (7)

$$z = k \cdot m^{-1} \cdot k_3 \cdot a_3 \cdot w(a_2) \cdot n_3 \cdot n_2' \; ;$$

therefore,

$$H(z) = \log a_3 + \log w(a_2) \; .$$

We have already seen that a_3, n_3 and n_2' are bounded. Taking (6), into account, this shows the existence of $s > 0$ and of a compact set ω' in N such that

$$z \in K \cdot A_s \cdot \omega' \; ,$$

which proves the lemma.

7.5. **Lemma.** *Let G_1 be an algebraic semi-simple subgroup of G, and assume it satisfies the following condition:* (a) $\theta(G_1) = G_1$, $G_1 = K_1 \cdot A_1 \cdot N_1(K_1 = K \cap G_1, A_1 = A \cap G_1, N_1 = N \cap G_1)$ *is an Iwasawa decomposition of G_1, and the restrictions to $\mathfrak{a}_1$ of the positive roots on $\mathfrak{a}$ are $\geqq 0$ for the ordering associated to $\mathfrak{n}_1$. Let $\mathfrak{S}$ be a Siegel domain of G, and $x \in G$. Then there exists a Siegel domain $\mathfrak{S}_1$ of G_1 (for the Iwasawa decomposition $K_1 \cdot A_1 \cdot N_1$) and a finite number of elements $x_1, \cdots, x_q \in G_1$ such that $\mathfrak{S}x \cap G_1 \subset \bigcup_i \mathfrak{S}_1 \cdot x_i$.*

In this proof, the Siegel domains of G_1 are defined with respect to the decomposition $K_1 \cdot A_1 \cdot N_1$. Let

$$\mathfrak{n}_0 = \sum\nolimits_{\alpha \in \Sigma; \alpha(\mathfrak{a}_1)=0} \mathfrak{g}_\alpha \; , \qquad \mathfrak{n}' = \sum\nolimits_{\alpha \in \Sigma; \alpha(\mathfrak{a}_1)\neq 0} \mathfrak{g}_\alpha \; .$$

Then, clearly, $\mathfrak{n}_0$ is a subalgebra and $\mathfrak{n}'$ an ideal of $\mathfrak{n}$. By 7.0, $(n, n') \to \exp n \cdot \exp n'$ is an analytic homeomorphism of $\mathfrak{n}_0 \times \mathfrak{n}'$ onto N. For any $\alpha \in \Sigma$ whose restriction to $\mathfrak{a}_1$ is not zero, let $\mathfrak{g}_\alpha'$ be a supplementary subspace to $\mathfrak{g}_1 \cap \mathfrak{g}_\alpha$ in $\mathfrak{g}_\alpha$, and let $\mathfrak{n}_2$ be the sum of all the $\mathfrak{g}_\alpha'$. Then $\mathfrak{n}' = \mathfrak{n}_1 + \mathfrak{n}_2$, where $\mathfrak{n}_1$ is a subalgebra, and it follows from 7.0 that $(n_2, n_1) \to \exp n_2 \cdot \exp n_1$ is an analytic homeomorphism of $\mathfrak{n}_2 + \mathfrak{n}_1$ onto N'. Thus $(n_0, n_2, n_1) \to \exp n_0 \cdot \exp n_2 \cdot \exp n_1$ is an analytic homeomorphism of $\mathfrak{n}_0 + \mathfrak{n}_2 + \mathfrak{n}_1$ onto $N = N_0 \cdot N_2 \cdot N_1$.

By 7.2, we have $x = a \cdot u \cdot m^{-1} \cdot v(a \in A; u, v \in N; m \in M^*)$. Since $\mathfrak{S} \cdot a \cdot u$ is contained in a Siegel domain, we may assume $x = m^{-1} \cdot v$. Let us write $v = v' \cdot v''(v' \in N_0 \cdot N_2; v'' \in N_1)$. Then $\mathfrak{S}x \cap G_1 = (\mathfrak{S} \cdot m^{-1} \cdot v' \cap G_1) \cdot v''$. We may therefore assume $v = v' \in N_0 \cdot N_2$. In this case, 7.5 will follow from the more precise lemma:

7.6. **Lemma.** *We keep the above notation and assume*

$$x = m^{-1} \cdot v \qquad\qquad (m \in M^*, v \in N_0 \cdot N_2) \; .$$

Let M_1^* be the normalizer and M_1 the centralizer of A_1 in K_1, and

$x_i(1 \leq i \leq \operatorname{ord} M_1^*/M_1)$ a set of representatives of the cosets of M_1^* modulo M_1. Then there exists a Siegel domain $\mathfrak{S}_1$ of G_1 such that $\mathfrak{S} \cdot x \cap G_1 \subset \bigcup_i \mathfrak{S}_1 \cdot x_i$.

Let $y = k \cdot a \cdot n (k \in K, a \in A, n \in N)$ be an element of $\mathfrak{S}$ such that $z = y \cdot x \in G_1$. Then

$$z = k \cdot a \cdot n \cdot m^{-1} \cdot v = k_1 \cdot a_1 \cdot n_1 \qquad (k_1 \in K_1,\ a_1 \in A_1,\ n_1 \in N_1) \ .$$

We first show that $a_1 \cdot n_1 \cdot a_1^{-1}$ is bounded. We have

$$z = k \cdot n^a \cdot a \cdot m^{-1} \cdot v = k \cdot m^{-1} \cdot m \cdot n^a \cdot m^{-1} \cdot m \cdot a \cdot m^{-1} v$$
$$= k \cdot m^{-1} \cdot n^{ma} \cdot a^m \cdot v \ ,$$
$$z \in K \cdot n^{ma} \cdot a^m \cdot v = K \cdot e^{H(u)} \cdot \nu(u) \cdot a^m \cdot v \qquad\qquad (u = n^{ma}) \ ,$$

where $\nu(u)$ is the component in N of u (see 7.1), whence

$$(1) \qquad\qquad a_1 = e^{H(u)} \cdot a^m \ , \qquad n_1 \cdot v^{-1} = (a^m)^{-1} \cdot \nu(u) \cdot a^m \ .$$

By 4.2, n^a is bounded, hence so are $u = n^{ma}$, $\nu(u)$, and therefore by (1), also $a_1 \cdot (a^m)^{-1}$, and

$$(2) \qquad\qquad a_1 \cdot n_1 \cdot v^{-1} \cdot a_1^{-1} = (a_1 \cdot n_1 \cdot a_1^{-1}) \cdot (a_1 \cdot v^{-1} \cdot a_1^{-1}) \ .$$

It is clear from the definitions that a_1 normalizes N_1 and $N_0 \cdot N_2$, therefore the two factors on the right hand side belong to N_1 and $N_0 \cdot N_2$ respectively. Since the map $(n', n'') \to n' \cdot n''$ is a homeomorphism of $N_1 \times N_0 \cdot N_2$ onto N, we see that both factors on the right hand side of (2) are bounded.

Let, for $c > 0$,

$$V_{i,c} = \{z \in \mathfrak{S}x \cap G_1 \,|\, z \cdot x_i^{-1} \in K \cdot A_c \cdot N\} \ .$$

We want to prove the existence of $t > 0$ such that

$$(3) \qquad\qquad \mathfrak{S}x \cap G_1 \subset \bigcup_i V_{i,t} \ .$$

The restricted Weyl group W_1 of G_1 is transitive on the Weyl chambers, therefore, given z, there exists an i such that $x_i \cdot a_1 \cdot x_i^{-1} \in A_1^-$, where $A_1^- = A_{1,0}$ denotes the exponential of the negative Weyl chamber. We have

$$z \cdot x_i^{-1} = k_1 \cdot a_1 \cdot n_1 \cdot x_i^{-1} = k_1 \cdot n_1^{a_1} \cdot a_1 \cdot x_i^{-1} \in K_1 \cdot n_1^{x_i \cdot a_1} \cdot a_1^{x_i} \ ,$$
$$z \cdot x_i^{-1} \in K_1 \cdot e^{H(u)} \cdot a_1^{x_i} \cdot \left(a_1^{x_i^{-1}} \cdot \nu(u) \cdot a_1^{x_i}\right) \ , \qquad\qquad (u = n_1^{x_i \cdot a_1}) \ ,$$

which implies

$$H(z \cdot x_i^{-1}) = H(u) + H(x_i \cdot a_1 \cdot x_i^{-1}) \ .$$

But $x_i \cdot a_i \cdot x_i^{-1} \in A_1^-$ by assumption, hence, taking 7.5(a) into account, $\alpha(x_i \cdot a_1 \cdot x_i^{-1}) \leq 0$ for $\alpha \in \Sigma$. Moreover, $n_1^{a_1}$ is bounded, as was proved above, hence u is bounded. There exists, therefore, $t_i > 0$ such that $H(z \cdot x_i^{-1}) \subset \log A_{1, t_i}$ for all $z \in \mathfrak{S}x \cap G_1$ for which

$$x_i \cdot a \, x_i^{-1} = x_i \cdot e^{H(z)} \cdot x_i^{-1} \in A_1^- \ .$$

This proves (3). In view of the condition (a) of 7.5, $A_{1,t} \subset A_t$, and it follows from 7.4 that we may find a Siegel domain $\mathfrak{S}_i'$ of G such that $z \cdot x_i^{-1} \in \mathfrak{S}_i'$ when $z \in V_{i,t}$; we have then $V_{i,t} \subset \mathfrak{S}_i' \cdot x_i$ or equivalently

$$(4) \qquad\qquad V_{i,t} \subset (\mathfrak{S}_i' \cap G_1) \cdot x_i \ .$$

But in view of 7.5(a), $\mathfrak{S}_i' \cap G_1$ is a Siegel domain of G_1. Since a finite union of Siegel domains (with respect to a fixed Iwasawa decomposition) is contained in a Siegel domain, the lemma follows from (3), (4).

7.7. LEMMA. *Let G' be a real algebraic semi-simple subgroup of G. Then there exists $a \in G$ such that $G_1 = a \cdot G' \cdot a^{-1}$ verifies condition* (a) *of 7.5.*

Let $\mathfrak{g} = \mathfrak{k} + \mathfrak{p}$ be the Cartan decomposition associated to θ. By 1.1 and 1.8, there exists $b \in G$ such that $G_2 = b \cdot G' \cdot b^{-1}$ is stable under θ. Then $\mathfrak{g}_2 = (\mathfrak{k} \cap \mathfrak{g}_2) + (\mathfrak{p} \cap \mathfrak{g}_2)$ is a Cartan decomposition of the Lie algebra $\mathfrak{g}_2$ of G_2. Let $\mathfrak{a}_2$ be a maximal subalgebra of $\mathfrak{p} \cap \mathfrak{g}_2$ and $\mathfrak{a}'$ a maximal subalgebra of $\mathfrak{p}$ containing $\mathfrak{a}_2$. By E. Cartan's conjugacy theorem (see e.g. [12, Lemma 33]), there exists $k \in K$ such that $k \cdot \mathfrak{a}' \cdot k^{-1} = \mathfrak{a}$. The group $G_3 = k \cdot G_2 \cdot k^{-1}$ is then still invariant under θ, and $\mathfrak{g}_3 \cap \mathfrak{a} = \mathfrak{a}_3$ is a maximal subalgebra of $\mathfrak{p}_3 = \mathfrak{g}_3 \cap \mathfrak{p}$. Let us now choose orderings on $\mathfrak{a}^*$ and $\mathfrak{a}_3^*$ such that the restriction to $\mathfrak{a}_3$ of a positive element of $\mathfrak{a}^*$ is $\geqq 0$, for instance take the lexicographic orderings with respect to a basis whose first elements span $\mathfrak{a}_3$. Let then $m \in M^*$ be such that $\mathrm{Ad}\, m$ transforms the positive Weyl chamber of $\mathfrak{a}$ for this ordering into the positive Weyl chamber for the ordering defined by N. Then $a = m \cdot k \cdot b$ fulfills our conditions.

7.8. THEOREM. *Let G be a semi-simple algebraic Lie group, defined over* **Q**. *Then $G_{\mathbf{R}}/G_{\mathbf{Z}}$ has finite Haar measure.*

We may assume G to be contained in $\mathbf{SL}(n, \mathbf{C})$ (see 2.1), and also to be connected, since $(G^0)_{\mathbf{R}}$ has finite index in $G_{\mathbf{R}}$. We now apply 7.7, with G' and G replaced respectively by $G_{\mathbf{R}}$ and $\mathbf{SL}(n, \mathbf{R})$, θ and $K \cdot A \cdot N$ by the standard Cartan involution and Iwasawa decomposition of $\mathbf{SL}(n, \mathbf{R})$, and take $a \in \mathbf{SL}(n, \mathbf{R})$ such that $G' = a \cdot G \cdot a^{-1}$ is self-adjoint and satisfies condition (a) of 7.5. Let $\mathfrak{S}$ be a standard Siegel domain of $\mathbf{SL}(n, \mathbf{R})$. By 6.5, there exist finitely many elements $b_i \in \mathbf{SL}(n, \mathbf{Z})$ such that $G_{\mathbf{R}} = \bar{U} \cdot G_{\mathbf{Z}}$ with

$$\bar{U} = \bigcup_i (a^{-1} \cdot \mathfrak{S} \cdot b_i) \cap G_{\mathbf{R}} \ .$$

We want to prove that $\bar{U}$ has finite Haar measure on $G_{\mathbf{R}}$. It is enough to show that $(a^{-1} \cdot \mathfrak{S} \cdot b_i) \cap G_{\mathbf{R}}$ has finite measure, hence also that $\mathfrak{S} \cdot b_i \cdot a^{-1} \cap G_{\mathbf{R}}'$ has finite Haar measure on $G_{\mathbf{R}}'$. By 7.5 (with G and G_1 replaced by

$\mathrm{SL}(n, \mathbf{R})$ and $G'_{\mathbf{R}}$), $\mathfrak{S} \cdot b_i \cdot a \cap G'_{\mathbf{R}}$ is contained in a finite union of translates of a Siegel domain of $G'_{\mathbf{R}}$. Since a Siegel domain of a semi-simple Lie group has finite volume (4.3), our contention is proved.

8. Remarks on characters of algebraic groups

8.1. Let G be an algebraic group. The group of *rational characters* of G, that is of rational homomorphisms of G into $\mathbf{C}^*$, is denoted by $X(G)$. It is a finitely generated commutative group, free when G is connected [23; 26]. If K is a field of definition for G, then $X_K(G)$ will be the group of rational characters defined over K.

A rational homomorphism $f: G \to G'$ obviously induces a homomorphism $f^0: X(G') \to X(G)$, which maps $X_K(G')$ into $X_K(G)$ if K is a field of definition for G, G' and f.

Let G be reductive, connected, defined over K, $S = Z(G)^0$ the identity component of its center, and G' the derived group of G. Then the restriction map $X(G) \to X(S)$ is injective, and identifies $X(G)$ and $X_K(G)$ with subgroups of finite index of $X(S)$ and $X_K(S)$ respectively. In fact, we have $X(G') = 1$, since G' is semi-simple, hence the restriction is injective. Conversely, given χ in $X(S)$ or in $X_K(S)$, χ^m ($m = $ order of $G' \cap S$) is trivial on $S \cap G'$, and therefore extends to a character of G.

8.2. Let now $G = T$ be an algebraic torus, K a field of definition for T. Let $\Gamma(T)$ be the group of rational homomorphisms of $\mathbf{C}^*$ into T, and $\Gamma_K(T)$ the group of rational homomorphisms of $\mathbf{C}^*$ into T which are defined over K. Both $X(T)$ and $\Gamma(T)$ are free, of rank $n = \dim T$. Given $a \in \Gamma(T)$, $b \in X(T)$, the homomorphism $b \circ a: \mathbf{C}^* \to \mathbf{C}^*$ has the form $x \to x^m (m \in \mathbf{Z})$. The map $(a, b) \to \langle a, b \rangle = m$ is an integral valued non-degenerate bilinear form on $\Gamma(T) \times X(T)$, which puts these two groups in duality [8; Exp. 9, No. 5]. A rational homomorphism $f: T \to T'$ induces a homomorphism $f_0: \Gamma(T) \to \Gamma(T')$, and we have $\langle f_0 a, b \rangle = \langle a, f^0 b \rangle$ $(a \in \Gamma(T), b \in X(T'))$.

8.3. An algebraic torus T is said to *split over* K if it is defined over K and isomorphic over K to a product of groups $\mathbf{C}^*$. If $T \subset \mathbf{GL}(n, \mathbf{C})$, this is equivalent to the existence of $x \in \mathbf{GL}(n, K)$ such that $x \cdot T \cdot x^{-1}$ is diagonal [26, Prop. 5]. If T splits over K, then $X_K(T) = X(T)$, $\Gamma_K(T) = \Gamma(T)$, the subtori and the homomorphic images (over K) of T split over K. Therefore, if $a \in \Gamma_K(T)$, $a \neq 0$, then $a(\mathbf{C}^*)$ is a one-dimensional subtorus of T which splits over K, and if $\Gamma_K(T) = \Gamma(T)$, then T splits over K. That the two latter conditions are also equivalent to $X_K(T) = X(T)$ follows from the following lemma:

8.4. LEMMA. *Let T be an algebraic torus, K a field of definition for T. Then*

(a) *Given any subtorus S of T, defined over K, there exists a subtorus S', defined over K, such that $S \cdot S' = T$, and $S \cap S'$ is finite.*

(b) *$X_K(T)$ and $\Gamma_K(T)$ have equal ranks. In particular, $X_K(T) \neq \{0\}$ if and only if T contains a subtorus $S \neq (e)$ which splits over K.*

The torus T certainly splits over $\bar{K}$ [1, Chap. II], therefore there exists a finite Galois extension K' of K over which T splits. Let A be the Galois group of K' over K. It operates in a natural fashion on $\Gamma(T)$, $X(T)$, and we have

$$\langle \sigma(a), \sigma(b) \rangle = \langle a, b \rangle \qquad (a \in \Gamma(T), b \in X(T), \sigma \in A) .$$

The corresponding linear representations ρ, ρ' of A in $\Gamma(T) \otimes \mathbf{Q}$ and $X(T) \otimes \mathbf{Q}$ are therefore contragredient to each other. The fixed points of A in $\Gamma(T)$ and $X(T)$ are $\Gamma_K(T)$ and $X_K(T)$ respectively. Consequently, the ranks of $\Gamma_K(T)$ and $X_K(T)$ are equal to the dimension of the fixed point sets of ρ, ρ'. Since ρ and ρ' are contragredient to each other, these dimensions are equal, whence the first part of (b). The second one follows then from 8.3. Let now S be a subtorus of T, defined over K. Then $\Gamma(S)$ may be identified with a submodule of $\Gamma(T)$, which is invariant under A. Since A is finite, there is a subspace V of $\Gamma(T) \otimes \mathbf{Q}$, supplementary to $\Gamma(S) \otimes \mathbf{Q}$, and invariant under A. The images $\gamma(\mathbf{C}^*)$, $(\gamma \in \Gamma(T) \cap V)$, are subtori defined over K' which are permuted by A. They generate a subtorus S' of T, invariant under A, hence defined over K, and such that $\Gamma(S') = V \cap \Gamma(T)$. We have then $\Gamma(S) \cap \Gamma(S') = (0)$, and $\Gamma(S) + \Gamma(S')$ has finite index in $\Gamma(T)$, whence (a).

9. The finiteness of the volume

9.1. Let G be a Lie group, H a discrete subgroup. Then a Haar measure on G induces on G/H a measure μ, such that $g(\mu) = \chi(g) \cdot \mu$, where $\chi(g) = |\det \mathrm{Ad}\, g|$. If $\mu(G/H) < \infty$, then $\chi(g) = 1 (g \in G)$, μ is invariant and G is unimodular (any left invariant Haar measure is right invariant).

9.2. LEMMA. *Let G^*, G be connected algebraic groups defined over $\mathbf{Q}$, $\pi: G^* \to G$ an isogeny over $\mathbf{Q}$. Then $G_\mathbf{R}^*$ is unimodular and $G_\mathbf{R}^*/G_\mathbf{Z}^*$ has finite invariant measure, if and only if $G_\mathbf{R}$ is unimodular, and $G_\mathbf{R}/G_\mathbf{Z}$ has finite invariant measure.*

It is clear that $G_\mathbf{R}$ is unimodular if and only if $G_\mathbf{R}^*$ is so. Let N be the kernel of π. By 6.11, $G_\mathbf{Z}^*$ has a subgroup M of finite index, whose image $\pi(M)$ is a subgroup of finite index of $G_\mathbf{Z}$. We have $G_\mathbf{R}^*/M \cdot N \cong G_\mathbf{R}/\pi(M)$, and

$G_\mathbf{R}^*/G_\mathbf{Z}^*$ (resp. $G_\mathbf{R}/G_\mathbf{Z}$) has finite measure if and only if $G_\mathbf{R}^*/M \cdot N$ (resp. $G_\mathbf{R}/\pi(M)$) has, whence the lemma.

9.3. Let T be an algebraic torus, defined over $\mathbf{Q}$. We shall use here and in § 10 the fact that if $X_\mathbf{Q}(T) = 1$, then $T_\mathbf{R}/T_\mathbf{Z}$ is compact, proved by Ono [22]. (It is not stated explicitly there, but is an immediate consequence of the compactness of the quotient $J(T)/T_\mathbf{Q}$ of the idele group of T by the principal ideles.)

On the other hand, if T is diagonal, then $T_\mathbf{Z}$ is obviously finite. By 8.3, it follows that $T_\mathbf{Z}$ is finite whenever T splits over $\mathbf{Q}$. In that case, $\mu(T_\mathbf{R}/T_\mathbf{Z})$ is of course infinite ($T \neq (e)$).

9.4. **THEOREM.** *Let G be an algebraic group, defined over $\mathbf{Q}$. Then $G_\mathbf{R}$ is unimodular, and $G_\mathbf{R}/G_\mathbf{Z}$ of finite invariant measure if and only if* $X_\mathbf{Q}(G^0) = 1$.

It is clearly enough to prove this when G is connected. Assume first that $G = T$ is a torus. If $X_\mathbf{Q}(G) = 1$, then $T_\mathbf{R}/T_\mathbf{Z}$ is compact by Ono's result (9.3). If $X_\mathbf{Q}(G) \neq 1$, then, by 8.4, we have $T = S \cdot S'$, where S, S' are subtori of strictly positive dimension, and S splits over $\mathbf{Q}$. We have a natural isogeny $S \times S' \to T$, and it follows from 9.2, 9.3 that $\mu(T_\mathbf{R}/T_\mathbf{Z})$ is infinite.

Let now G be reductive, $G = S \cdot G'$ its standard decomposition, where $S = Z(G)^0$, and G' is the derived group of G. Therefore G is the quotient of $S \times G'$ by a finite group. By 7.8, $G_\mathbf{R}'/G_\mathbf{Z}'$ has finite invariant measure. By 8.1, $X_\mathbf{Q}(G)$ and $X_\mathbf{Q}(S)$ have equal ranks. Our assertion in this case follows therefore from the above and 9.2.

In the general case, $G = H \cdot N$ is the semi-direct product of a reductive group H and of an invariant unipotent group N, both defined over $\mathbf{Q}$, and $N_\mathbf{R}/N_\mathbf{Z}$ is compact (6.10). Of course, $X(N) = 1$, hence $X(G) = X(H)$, $X_\mathbf{Q}(G) = X_\mathbf{Q}(H)$.

Let $X_\mathbf{Q}(G) = 1$. Then $X_\mathbf{Q}(H) = 1$, and $H_\mathbf{R}/H_\mathbf{Z}$ has finite invariant measure by the above. Moreover, the determinant of $\mathrm{Ad}\, h \mid \mathfrak{n}(h \in H)$ is one, since $h \to \det(\mathrm{Ad}\, h \mid \mathfrak{n})$ is an element of $X_\mathbf{Q}(H)$. Therefore, $G_\mathbf{R}$ is unimodular, and a Haar measure on $G_\mathbf{R}$ is the product of Haar measures on $H_\mathbf{R}$ and $N_\mathbf{R}$. We have $H_\mathbf{R} = A \cdot H_\mathbf{Z}$, $N_\mathbf{R} = B \cdot N_\mathbf{Z}$ with A, B open, of finite Haar measure, whence $G_\mathbf{R} = A \cdot N \cdot H_\mathbf{Z} = A \cdot B \cdot G_\mathbf{Z}$ with $A \cdot B$ of finite measure.

Assume now $G_\mathbf{R}$ to be unimodular, and $G_\mathbf{R}/G_\mathbf{Z}$ to have finite invariant measure. Since, $G_\mathbf{R}, H_\mathbf{R}, N_\mathbf{R}$ are unimodular, we must have $\det(\mathrm{Ad}\, h \mid \mathfrak{n}) = 1(h \in H)$, and the Haar measure on $G_\mathbf{R}$ is the product of Haar measures on $H_\mathbf{R}$ and $N_\mathbf{R}$. The subgroup $H_\mathbf{Z} \cdot N_\mathbf{Z}$ has finite index in $G_\mathbf{Z}$ (6.4), and $G_\mathbf{R}/H_\mathbf{Z} \cdot N_\mathbf{Z}$ has finite measure. The projection $G_\mathbf{R}/N_\mathbf{Z} \to G_\mathbf{R}/N_\mathbf{R} \cong H_\mathbf{R}$ is a fiber map with compact fibre $N_\mathbf{R}/N_\mathbf{Z}$, which commutes with the action of

H_Z defined by right translations. Therefore H_R/H_Z must also have finite invariant measure; by the above, this implies $X_Q(H) = 1$, hence $X_Q(G) = 1$.

19. Closed conjugacy classes

10.1. PROPOSITION. *Let G be a real or complex algebraic group, $x \in G$, $y \in \mathfrak{g}$, and $C(x) = \{g \cdot x \cdot g^{-1}, g \in G\}$, $C(y) = \operatorname{Ad} G(y)$ the conjugacy classes of x and y. Then*

(a) *If x (resp. $\operatorname{ad} y$) is semi-simple, $C(x)$ (resp. $C(y)$) is closed.*

(b) *If G is reductive, and x (resp. $\operatorname{ad} y$) is not semi-simple, $G(x)$ (resp. $C(y)$) is not closed.*

PROOF OF (a) Going over to the complexification, if necessary, it is enough, by 2.3, to consider the case where G is complex algebraic. For an endomorphism A of a vector space, we denote by $C(A, \lambda)$ its characteristic polynomial $\det(A - \lambda \cdot \operatorname{Id.})$ and by $M_A(\lambda)$ its minimal polynomial, where λ is an indeterminate. Let

$$P_y = \{z \in \mathfrak{g}, C(\operatorname{ad} z, \lambda) = C(\operatorname{ad} y, \lambda), M_{\operatorname{ad} y}(\operatorname{ad} z) = 0\}.$$

This is an algebraic subset of $\mathfrak{g}$, clearly invariant under G. The minimal polynomial of $\operatorname{ad} z (z \in P_y)$ divides the minimal polynomial of $\operatorname{ad} y$, hence has only simple factors, and $\operatorname{ad} z$ is also semi-simple. In particular, the dimension of the centralizer $Z(z)$ of z in G is equal to the multiplicity of the eigen-value zero of $\operatorname{ad} z$. But $\operatorname{ad} z$ has the same eigenvalues as $\operatorname{ad} y$, hence $\dim Z(z) = \dim Z(y)$. The orbit $\operatorname{Ad} G(z)$ of z, whose dimension is equal to $\dim G - \dim Z(z)$, has therefore the same dimension as $\operatorname{Ad} G(y)$, and P_y is a disjoint union of orbits of the same dimension. Since the boundary of an orbit is a union of orbits of strictly smaller dimension [1, § 15], it follows that $\operatorname{Ad} G(z)$ is closed for every $z \in P_y$.

The proof for x is entirely analogous. We introduce the set

$$P_x = \{z \in G, C(\operatorname{Ad} z, \lambda) = C(\operatorname{Ad} x, \lambda), M_{\operatorname{Ad} x}(\operatorname{Ad} z) = 0\}.$$

This is an algebraic set, invariant under G. The dimension of $Z(z)$ will be equal to the multiplicity of the eigenvalue one of $\operatorname{Ad} z$, hence equal to $\dim Z(x)$, and P_x consists again of orbits of the same dimension.

PROOF OF (b) Let first $G = \mathbf{SL}(2, \mathbf{R})$ or $\mathbf{SL}(2, \mathbf{C})$ and $x \neq 1$ be unipotent (resp. $y \neq 0$ be nilpotent). After a suitable inner automorphism, we may assume

$$x = \begin{pmatrix} 1 & 1 \\ 0 & 1 \end{pmatrix} \quad \left(\text{resp. } y = \begin{pmatrix} 0 & 1 \\ 0 & 0 \end{pmatrix}\right).$$

Let $g_t = \operatorname{diag}(t, t^{-1})$. Then

$$g_t \cdot x \cdot g_t^{-1} = \begin{pmatrix} 1 & t^2 \\ 0 & 1 \end{pmatrix} \quad \left(\text{resp. Ad } g_t(y) = \begin{pmatrix} 0 & t^2 \\ 0 & 0 \end{pmatrix} \right),$$

hence the identity (resp. the origin) belongs to $\bar{C}(x)$ (resp. $\bar{C}(y)$).

Let now G be reductive, and ad y be not semi-simple. We may write $y = z + y'$, with z central, $y' \in \mathscr{D}\mathfrak{g}$, hence ad $y' =$ ad y not semi-simple. It is clearly enough to show that the conjugacy class of y' in the semi-simple part of $\mathfrak{g}$ is not closed; we may therefore assume G to be semi-simple. We have then $y = s + n$, with $[s, n] = 0$, ad s semi-simple, ad n nilpotent and not zero [5, § 6, No. 3]. The centralizer $\mathfrak{z}(s)$ of s in $\mathfrak{g}$ is reductive, its center consists of semi-simple elements [4, Prop. 4.1], $\mathscr{D}\mathfrak{z}(s)$ is semi-simple, and contains n. By the Jacobson-Morosow theorem [15], there exists a three-dimensional subalgebra $\mathfrak{m}$, isomorphic to $\mathfrak{sl}(2, \mathbf{R})$, or $\mathfrak{sl}(2, \mathbf{C})$, containing n. The analytic subgroup M generated by $\mathfrak{m}$ in Ad $\mathfrak{g}$ is a homomorphic, locally isomorphic, image of $\mathbf{SL}(2, \mathbf{R})$ or $\mathbf{SL}(2, \mathbf{C})$. By the above, the conjugacy class of n in $\mathfrak{m}$ has zero in its closure; since M centralizes s, it follows that the semi-simple element s belongs to the closure of $\bar{C}(y)$.

For x, the proof is similar. We write $x = z \cdot x'$, with z central, semi-simple, and x' in the semi-simple part of G, and not semi-simple. It is enough to show that $C(x')$ is not closed, hence we may assume G to be semi-simple. We have $x = x_s \cdot x_u$ with $x_s \cdot x_u = x_u \cdot x_s$, x_s semi-simple, $x_u \neq e$ unipotent [1, § 8]. Applying the Jacobson-Morosow theorem to $\log x_u$, in the derived algebra of $\mathfrak{z}(x_s)$, we see that x_u belongs to a group M, which is a homomorphic, locally isomorphic image of $\mathbf{SL}(2, \mathbf{R})$ or $\mathbf{SL}(2, \mathbf{C})$, and centralizes x_s. By the above, the conjugacy class of x_u in M contains e, hence $\bar{C}(x)$ contains x_s.

10.2. REMARK. The proof of (b) shows more precisely that the semi-simple part of x (resp. y) belongs to $\bar{C}(x)$ (resp. $\bar{C}(y)$), and, in the real case, that it belongs to the closure of the conjugacy class of x (resp. y) with respect to the identity component G^0 of G, hence with respect to any open subgroup. Now, if G is a real (resp. complex) semi-simple Lie group, then Ad $\mathfrak{g}$ is of finite index in a real algebraic group (resp. is an algebraic group). If G is moreover linear, it is of finite index in a real (resp. complex) algebraic group. Consequently, 10.1 yields the following:

10.3. COROLLARY. *Let G be a real or complex semi-simple Lie group, and $y \in \mathfrak{g}$. Then* Ad $G(y)$ *is closed if and only if* ad y *is semi-simple. If G is linear, the conjugacy class in G of an element $x \in G$ is closed if and only if x is semi-simple.*

11. Groups of units with compact fundamental sets

11.1. Lemma. *Let G be a locally compact separable group, M a locally compact separable space on which G operates continuously on the right, $m \in M$, and H a closed subgroup. If $m \cdot H$ is closed, and $H\backslash G$ is compact, then $m \cdot G$ is closed.*

It follows from our assumption that $G = H \cdot K$ with K compact. If $m \cdot h_j \cdot k_j \to p(h_j \in H, k_j \in K)$, then, assuming, as we may, $k_j \to k \in K$, we have $m \cdot h_j \to p \cdot k^{-1}$. Since $m \cdot H$ is closed, there exists $h \in H$ such that $p \cdot k^{-1} = m \cdot h$, whence $p = m \cdot h \cdot k \in m \cdot G$.

11.2. Let G act on itself by inner automorphisms. If H is discrete, the conjugacy class in H of $h \in H$ is of course a closed subset of G; therefore, if $H\backslash G$ is compact, then the conjugacy class in G of any element of H is closed. Lemma 11.1 was suggested by this remark, which is due to Selberg. Together with 10.3, it shows that if G is a linear semi-simple Lie group, H a discrete subgroup such that $H\backslash G$ is compact, then any element of H is semi-simple.

11.3. Proposition. *Let G be a connected algebraic group defined over $\mathbf{Q}$, and $\pi: G \to \mathbf{GL}(V)$ a rational representation of G defined over $\mathbf{Q}$. If $G_{\mathbf{R}}/G_{\mathbf{Z}}$ is compact, then the orbit under $G_{\mathbf{R}}$ of an element $v \in V_{\mathbf{Q}}$ is closed.*

By 6.3, there exists a lattice $\Gamma \subset V_{\mathbf{Q}}$ which contains v and is invariant under $G_{\mathbf{Z}}$. Therefore, $v \cdot \pi(G_{\mathbf{Z}})$ is a discrete set, and 11.3 follows from 11.1.

11.4. Lemma. *Let G be a connected reductive algebraic group, k a field of definition for G. Then the following conditions are equivalent:*

(a) $X_k(G) = 1$, and $\mathfrak{g}_k$ consists of semi-simple elements;

(b) $X_k(G) = 1$, and G_k consists of semi-simple elements:

(c) $X_k(S) = 1$ *for every algebraic subtorus S of G which is defined over k.*

If k is a number field, and J its ring of integers, these conditions are equivalent to:

(d) $X_k(G) = 1$, and G_J consists of semi-simple elements.

(a) $\Rightarrow$ (b). The group G, being algebraic, contains the unipotent and semi-simple parts of its elements [1, Chap. II], therefore, if G_k contains a non-semi-simple element, it also contains a unipotent element $g \neq e$. But then $\log g$ is a non-zero nilpotent element of $\mathfrak{g}_k$.

(b) $\Rightarrow$ (c). Assume (c) to be false. By 8.4, there exists then a one-dimensional subtorus S of G which splits over k. It is not central, since otherwise 8.4 and 8.1 would imply $X_k(G) \neq 1$. The image of S in $\mathrm{Aut}\,\mathfrak{g}$

then also splits, which means that it can be diagonalized over k. The weights of the Lie algebra $\hat{s}_k$ of S_k in $\mathfrak{g}_k$ are then elements of $(\hat{s}_k)^*$, and $\mathfrak{g}_k$ is a direct sum of subspaces $\mathfrak{g}_{k,\alpha}(\alpha \in (\hat{s}_k{}^*))$, where as usual, $\mathfrak{g}_{k,\alpha} = \{x \in \mathfrak{g}_k \mid [s, x] = \alpha(s)x,\ s \in \hat{s}_k\}$. Since S is not central, $\mathfrak{g}_{k,\alpha} \neq 0$ for at least one $\alpha \neq 0$. But then it is well known that an element $x \in \mathfrak{g}_{k,\alpha}$ is nilpotent. If $x \neq 0$, then e^x is a unipotent element in G_k, different from the identity, which contradicts (b).

(c) $\Rightarrow$ (a). By 8.1, (c) implies that $X_k(G) = 1$. Assume now that $\mathfrak{g}_k$ does not consist of semi-simple elements. Being algebraic, it contains then a nilpotent element $x \neq 0$ [7a, p. 165], which necessarily belongs to $\mathscr{D}\mathfrak{g}$. Applying the Jacobson-Morosow theorem to $\mathscr{D}\mathfrak{g}_k$, we get a three dimensional subalgebra $\mathfrak{m} \subset \mathscr{D}\mathfrak{g}_k$, containing x, isomorphic over k to the Lie algebra of $\mathbf{SL}(2, k)$. There exists therefore in G an algebraic subgroup M, with Lie algebra $\mathfrak{m} \otimes \mathbf{C}$, defined over k, and which is the image of $\mathbf{SL}(2, \mathbf{C})$ under a rational homomorphism with finite kernel, defined over k. The image of the group of diagonal matrices in $\mathbf{SL}(2, \mathbf{C})$ is then a one dimensional subtorus of G, which splits over k. Let now k be a number field, J the ring of integers of k. Clearly, $(a) \Rightarrow (d)$. Assume now (a) to be false. As remarked at the beginning of the proof, there exists then $x \in \mathfrak{g}_k$ which is nilpotent and not zero. Let s be a positive rational integer such that $x^s = 0$, t an element of J such that $t \cdot x$ has integral coefficients, and $m = s! \cdot t$. Then e^{mx} is a unipotent element, different from the identity contained in G_J, in contradiction with (d).

11.5. COROLLARY. *Let G be a connected reductive algebraic group. If G satisfies the conditions of 11.4, then every connected subgroup of G which is defined over k is reductive, and satisfies those conditions.*

Let M be a connected algebraic subgroup of G, defined over k. As recalled in 6.10, $M = H \cdot N$ is the semi-direct product of a reductive group H and of a unipotent invariant subgroup N, both defined over k. In our case $N_k = (e)$ by the condition (b), hence $N = (e)$, and $M = H$ is reductive. Then it clearly satisfies (c).

11.6. THEOREM. *Let G be a reductive algebraic group, defined over $\mathbf{Q}$. Then $G_{\mathbf{R}}/G_{\mathbf{Z}}$ is compact if and only if $X_{\mathbf{Q}}(G^0) = 1$ and $G_{\mathbf{Q}}$ consists of semi-simple elements.*

The group G^0 has finite index in G, therefore $G_{\mathbf{R}}/G_{\mathbf{Z}}$ is compact if and only if $(G^0)_{\mathbf{R}}/(G^0)_{\mathbf{Z}}$ is. In the sequel, we assume G to be connected. Our assertion is then that the conditions (a) to (d) of 11.4 are equivalent to the compactness of $G_{\mathbf{R}}/G_{\mathbf{Z}}$.

Let first $G_{\mathbf{R}}/G_{\mathbf{Z}}$ be compact. If $X_{\mathbf{Q}}(G) \neq 1$, there exists a non-trivial one dimensional rational representation $\pi : G \to \mathbf{C}^*$, defined over $\mathbf{Q}$. But then $\pi(G) = \mathbf{C}^*$, and $\pi(G_\mathbf{R})$ contains the multiplicative group $\mathbf{R}^+$ of strictly positive real numbers. The orbit under $G_\mathbf{R}$ of any element in $\mathbf{Q}^*$ (in fact in $\mathbf{C}^*$) has the origin in its closure, and is not closed, in contradiction with 11.3. Thus $X_\mathbf{Q}(G) = 1$. The condition on $G_\mathbf{Q}$ follows from 10.1 and 11.3.

From now on, G is assumed to verify the conditions of 11.4, and we prove the compactness of $G_\mathbf{R}/G_\mathbf{Z}$ by induction on $\dim G$. There is nothing to prove in dimension zero, therefore we may assume 11.6 to be true for all groups (connected or not) of dimension strictly smaller than $\dim G$. In particular, in view of 11.5, $H_\mathbf{R}/H_\mathbf{Z}$ is compact for every proper algebraic subgroup H of G which is defined over $\mathbf{Q}$.

We have $G = S \cdot G'$ with $S = Z(G)^0$, G' semi-simple invariant, S and G' defined over $\mathbf{Q}$. If $0 < \dim S < \dim G$, then, by the above $S_\mathbf{R} = A \cdot S_\mathbf{Z}$, $G'_\mathbf{R} = B \cdot G'_\mathbf{Z}$ with A and B compact, hence $S_\mathbf{R} \cdot G'_\mathbf{R} = A \cdot B \cdot S_\mathbf{Z} \cdot G'_\mathbf{Z}$. Since $S_\mathbf{R} \cdot G'_\mathbf{R}$ has finite index in $G_\mathbf{R}$, and $S_\mathbf{Z} \cdot G'_\mathbf{Z} \subset G_\mathbf{Z}$, it follows that $G_\mathbf{R}/G_\mathbf{Z}$ is compact. If $G = S$, see 9.3.

Let now $G = G'$ be semi-simple. Every element of $\mathfrak{g}_\mathbf{Q}$ is semi-simple (11.4), and therefore its conjugacy class under G is closed (10.1). The adjoint representation is consequently a locally faithful rational representation of G, defined over $\mathbf{Q}$, in which all rational points have closed orbits. Since, as pointed out above, $H_\mathbf{R}/H_\mathbf{Z}$ is compact for every proper algebraic subgroup of G which is defined over $\mathbf{Q}$, it will be enough, in order to conclude the proof of 11.6, to prove the following lemma.

11.7. LEMMA. *Let G be a connected semi-simple algebraic group defined over $\mathbf{Q}$. Assume that $H_\mathbf{R}/H_\mathbf{Z}$ is compact for every proper algebraic subgroup of G defined over $\mathbf{Q}$, and that there exists a locally faithful rational representation $\pi\colon G \to \mathrm{GL}(V)$ defined over $\mathbf{Q}$ in which all points of $V_\mathbf{Q}$ have closed orbits. Then $G_\mathbf{R}/G_\mathbf{Z}$ is compact.*

The representation π being fully reducible, we may take out the trivial representations, and assume

$$(1) \qquad\qquad G_x \neq G \qquad\qquad (x \in V, x \neq 0) .$$

We fix a lattice Γ in $V_\mathbf{Q}$ which is invariant under $G_\mathbf{Z}$ (see 6.3), and take coordinates in V with respect to a basis of Γ. Let P be the set of polynomials on V which are invariant under G. Since π is defined over $\mathbf{Q}$, we have $P = P' \otimes \mathbf{C}$, where P' is the set of invariant polynomials with rational coefficients. By the theorem of invariants, applied to $\pi\colon \mathfrak{g}_\mathbf{Q} \to \mathfrak{gl}(V_\mathbf{Q})$,, [5, § 6, No. 9], P' is a finitely generated algebra over $\mathbf{Q}$. It is therefore generated by 1, and by finitely many homogeneous polynomials $P_1, \cdots, P_s$, which we may assume to have integral coefficients and degrees ≥ 1.

Let $\sigma: V \to \mathbf{C}^s$ be the map $v \to (P_1(v), \cdots, P_s(v))$. It is continuous, maps Γ, $V_\mathbf{Q}$, $V_\mathbf{R}$ in $\mathbf{Z}^s$, $\mathbf{Q}^s$, $\mathbf{R}^s$ respectively. By assumption, any $v \in V_\mathbf{Q}$ has a closed orbit. Since the P_i's, together with 1, generate P over $\mathbf{C}$, and the invariant polynomials separate the closed orbits (3.3), we see that

$$(2) \qquad v, v' \in V_\mathbf{Q}, \sigma(v) = \sigma(v') \Rightarrow v' \in v \cdot G .$$

The origin is a closed orbit, hence is the only closed orbit on which all P_i's are zero, and

$$(3) \qquad \sigma(v) \neq 0 \qquad\qquad (v \in V_\mathbf{Q} - 0) .$$

The group G being semi-simple, connected, $\pi(G)$ consists of transformations of determinant one, and $\pi(G_\mathbf{R})$ leaves invariant the euclidean measure on V, identified with $\mathbf{R}^n$ by means of the basis chosen above. By Minkowski's classical idea, there exists a compact set C in $V_\mathbf{R}$ containing zero such that $C \cdot \pi(g) \cap \Gamma \neq \{0\}$ for any $g \in G_\mathbf{R}$. (For the sake of completeness, we recall the proof: let C_0 be a compact set in $V_\mathbf{R}$ with measure strictly greater than 1. Then the projection $V_\mathbf{R} \to V_\mathbf{R}/\Gamma$ is not injective on $C_0 \cdot \pi(g)$ whence the existence of $z \in \Gamma - 0$ such that $C_0 \cdot \pi(g) \cap (C_0 \cdot \pi(g) + z)$ $\neq \varnothing$. Therefore $C = \{x - y; x, y \in C_0\}$ fulfills our condition.) The image $\sigma(C)$ of C is compact, $\sigma(C) \cap \mathbf{Z}^s$ is finite, and we may find finitely many elements $w_j \in \Gamma - 0 (1 \leq j \leq t)$ such that

$$\sigma(C) \cap \sigma(\Gamma - 0) = \{\sigma(w_1), \cdots, \sigma(w_t)\} .$$

By (2), the intersection $\sigma^{-1}(\sigma(v)) \cap \Gamma (v \in V_\mathbf{Q})$, belongs to the orbit of v, which is closed by assumption, hence (6.9) consists of a *finite* number of orbits of $G_\mathbf{Z}$. There exists therefore a finite number of elements $v_1, \cdots, v_m \in \Gamma - 0$ such that

$$(4) \qquad \bigcup_i v_i \cdot \pi(G_\mathbf{Z}) = \sigma^{-1}(\sigma(C) \cap \sigma(\Gamma - 0)) \cap \Gamma .$$

Let now $g \in G_\mathbf{R}$. There exist $c \in C$ and $v \in \Gamma - 0$ such that $c \cdot \pi(g^{-1}) = v$. The polynomials P_i being invariant under G, we have then

$$\sigma(v) = \sigma(c \cdot \pi(g^{-1})) = \sigma(c) \in \sigma(C) \cap \sigma(\Gamma - 0) ,$$

hence $v \in \sigma^{-1}(\sigma(C) \cap \sigma(\Gamma - 0))$; by (4), there exists an index $i(1 \leq i \leq m)$ such that

$$c \cdot \pi(g^{-1}) = v \in v_i \cdot \pi(G_\mathbf{Z}) .$$

Given $g \in G_\mathbf{R}$, we have thus found an index i, and $b \in G_\mathbf{Z}$, such that $v_i \cdot \pi(b \cdot g) \in C$. This shows that

$$G_\mathbf{R} = \bigcup_i G_\mathbf{Z} \cdot X_i \qquad (X_i = \{g \in G_\mathbf{R} \mid v_i \cdot \pi(g) \in C\}) .$$

Let G_i be the isotropy group of $v_i (1 \leq i \leq m)$. The orbit $v_i \cdot \pi(G)$ is closed, hence so is $v_i \cdot \pi(G_\mathbf{R})$ (see 2.3); there exists a compact set $B_i \subset G_\mathbf{R}$ such that

$X_i = G_{i\mathbf{R}} \cdot B_i$ (5.2). We have then

$$G_{\mathbf{R}} = \bigcup_i G_{\mathbf{Z}} \cdot G_{i\mathbf{R}} \cdot B_i \qquad (B_i \text{ compact}) .$$

But G_i is defined over $\mathbf{Q}$, since $v_i \in V_{\mathbf{Q}}$, and is a proper subgroup since $v_i \neq 0$ (see (1)). Therefore $G_{i\mathbf{R}} = G_{i\mathbf{Z}} \cdot C_i$, with C_i compact, and finally $G_{\mathbf{R}} = G_{\mathbf{Z}} \cdot A$ with $A = \bigcup_i C_i \cdot B_i$ compact.

REMARK. The preceding proof is an adaptation of A. Weil's argument, pertaining to groups of automorphisms of algebras with involution "which do not represent zero" [30, Theorem 4.1.1], and was in part suggested by it.

11.8. THEOREM. *Let G be an algebraic group defined over* $\mathbf{Q}$. *Then* $G_{\mathbf{R}}/G_{\mathbf{Q}}$ *is compact if and only if* $X_{\mathbf{Q}}(G^0) = 1$, *and every unipotent element of* $G_{\mathbf{Z}}$, *or, equivalently, of* $G_{\mathbf{Q}}$, *belongs to the radical of* G.[5]

As in 11.6, we may restrict ourselves to the case of a connected group G. Then $G = H \cdot N$ is the semi-direct product of a reductive group H and of an invariant unipotent group N, both defined over $\mathbf{Q}$, and $N_{\mathbf{R}}/N_{\mathbf{Z}}$ is compact (see 6.10 for references). In view of 11.6, it will be enough to show:

(i) $G_{\mathbf{R}}/G_{\mathbf{Z}}$ is compact if and only if $H_{\mathbf{R}}/H_{\mathbf{Z}}$ is compact.

(ii) H verifies the conditions of 11.4 if and only if $X_{\mathbf{Q}}(G) = 1$, and every unipotent element of $G_{\mathbf{Q}}$ (resp. of $G_{\mathbf{Z}}$) belongs to N.

PROOF OF (i). Let $G_{\mathbf{R}}/G_{\mathbf{Z}}$ be compact. Since $H_{\mathbf{Z}} \cdot N_{\mathbf{Z}}$ is of finite index in $G_{\mathbf{Z}}$ (6.4), we have $G_{\mathbf{R}} = K \cdot H_{\mathbf{Z}} \cdot N_{\mathbf{Z}}$, with K compact, whence $H_{\mathbf{R}} = \pi(K) \cdot H_{\mathbf{Z}}$, where π is the natural projection of G onto H. Let now $H_{\mathbf{R}} = A \cdot H_{\mathbf{Z}}$ with A compact. Then

$$G_{\mathbf{R}} = A \cdot H_{\mathbf{Z}} \cdot N_{\mathbf{R}} = A \cdot N_{\mathbf{R}} \cdot H_{\mathbf{Z}} = A \cdot B \cdot N_{\mathbf{Z}} \cdot H_{\mathbf{Z}} ,$$

with B compact, since $N_{\mathbf{R}}/N_{\mathbf{Z}}$ is compact, whence $G_{\mathbf{R}} = K \cdot G_{\mathbf{Z}}$, with $K = A \cdot B$ compact.

PROOF OF (ii). A unipotent group has only the trivial rational character, hence $X(G)$ and $X_{\mathbf{Q}}(G)$ are naturally isomorphic to $X(H)$ and $X_{\mathbf{Q}}(H)$. Assume H to verify the conditions of 11.4. Then $X_{\mathbf{Q}}(G) = 1$. Moreover, since a rational representation maps unipotent elements into unipotent elements [1, Chap. II], the unipotent elements of $G_{\mathbf{Q}}$ belong to the kernel of π, that is to N.

Assume now that $X_{\mathbf{Q}}(G) = 1$, and that every unipotent element of $G_{\mathbf{Z}}$ belongs to N. We have then $X_{\mathbf{Q}}(H) = 1$, by the initial remark of the proof of (ii). If $x \in G_{\mathbf{Q}}$ is unipotent, then a suitable power x^m of x is a

---

[5] Another proof of Theorem 11.8 has been given by G. D. Mostow and T. Tamagawa, to appear in Ann. of Math.

unipotent element of G_Z (see the proof of (d) $\Rightarrow$ (a) in 11.4). It belongs to N by assumption, hence so does the one parameter group generated by $\log x^m = m \cdot \log x$, and therefore $x \in N$. Thus every unipotent element of G_Q is in N. Since H_Q contains the semi-simple and unipotent parts of its elements [1, Chap. II], we see that H_Q has only semi-simple elements.

12. Groups over number fields

12.1. *Notation.* K is a number field, J the ring of algebraic integers of K, Φ the set of distinct isomorphisms of K into $\mathbf{C}$, and $\bar{\sigma}$ the composition of $\sigma \in \Phi$ with the complex conjugation. Φ' will be a subset of Φ which contains exactly one representative of each pair $(\sigma, \bar{\sigma})$. As usual, x^σ is the image of $x \in K$ under σ, and K^σ the image of K. The completion of K^σ with respect to the absolute value in $\mathbf{C}$ is denoted by L_σ. Thus $L_\sigma = \mathbf{C}$ if $\sigma \neq \bar{\sigma}$, and $L_\sigma = \mathbf{R}$ otherwise. This notation is used throughout this paragraph.

12.2. Let $G \subset \mathbf{GL}(n, \mathbf{C})$ be a connected algebraic group defined over K, and I the ideal of polynomials on $\mathbf{M}(n, \mathbf{C})$, with coefficients in K, which vanish on G. The algebraic group defined by the ideal I^σ is denoted by G^σ. It has K^σ as a field of definition. For a subset ψ of Φ, we put

$$(1) \qquad G_\psi = \prod_{\sigma \in \psi} G^\sigma , \qquad G_{\psi,r} = \prod_{\sigma \in \psi} G^\sigma_{L_\sigma} .$$

In G_ψ, or in $G_{\psi,r}$ we identify G_J (resp. G_K) with the set of elements $(x^\sigma)_{\sigma \in \psi}(x \in G_J,$ resp. $x \in G_K)$.

There exists an algebraic group $G' = R_{K/Q}G$, the *group obtained from G by restriction of the groundfield from K to* $\mathbf{Q}$, which is defined over $\mathbf{Q}$, and is isomorphic over $\bar{K}$ to G_Φ. It is essentially unique up to isomorphism over $\mathbf{Q}$ [30, Chap. I], and is isomorphic to G_Φ by an isomorphism μ' of the form $(\mu^\sigma)_{\sigma \in \Phi}$, where $\mu : G' \to G$ is a rational homomorphism defined over K, which verifies

$$(2) \qquad \mu'(G'_Z) = G_J , \qquad \mu'(G'_Q) = G_K \qquad (G' = R_{K/Q}G) .$$

The homomorphism μ also induces in a natural way an isomorphism of $X_K(G)$ onto $X_Q(G')$. Let $\sigma \neq \bar{\sigma}$. The standard embedding of $\mathbf{GL}(n, \mathbf{C})$ into $\mathbf{GL}(2n, \mathbf{R})$ induces an isomorphism of $G^\sigma_{L_\sigma}$ onto the set of real points of an algebraic group $R_{C/R}G^\sigma \subset \mathbf{GL}(2n, \mathbf{C})$; from this, one deduces the existence of an isomorphism of real algebraic groups $\beta : G'_R \to G_{\Phi',r}$, which also maps G'_Z and G'_Q onto G_J and G_K respectively.

In general, G_J is not discrete in $G^\sigma_{L_\sigma}$, however, it is clear from the above, that G_J is discrete in G_Φ or in $G_{\Phi'}$. *More generally, if ψ contains all $\sigma \in \Phi'$ for which $G^\sigma_{L_\sigma}$ is not compact, then G_B is discrete in $G_{\psi,r}$.*

To see this, it is enough to show that given $\delta > 0$, there are only finitely many $b \in J$ which occur as coefficients of matrices in G_J, and which verify $|b^\sigma| < \delta (b \in \psi)$. If $\bar\sigma \in \psi$, then $|b^\sigma| < \delta$. If $\sigma, \bar\sigma \notin \psi$, then $G_{L_\sigma}^\sigma$ is compact, hence $|b^\sigma| < \varepsilon$, where ε depends only on $G_{L_\sigma}^\sigma$. Our assertion follows then from the familiar fact that J has only finitely many elements b, all of whose conjugates $b^\sigma (b \in \Phi)$ are in absolute value under a given bound.

When this condition is fulfilled, G_J is called an arithmetically defined subgroup, or a group of units of $G_{\psi,r}$. In view of the possibility of restricting the groundfield, it is clear that there is no essential loss in generality in limiting oneself to the case $K = \mathbf{Q}$, $J = \mathbf{Z}$, and that the main results of the preceding paragraphs extend automatically to the groups of units considered here. We state this formally for some of them for the convenience of reference, and leave the reformulation of the others to the reader.

12.3. Theorem. *We keep the notation of 11.1. Let G be a connected algebraic group defined over K, and ψ a subset of Φ' containing all σ for which $G_{L_\sigma}^\sigma$ is not compact. Then*

    (a) *G_J is finitely generated.*

    (b) *$G_{\psi,r}$ is the union of open subsets U having the following properties:*

      (i) *$G_{\psi,r} = U \cdot G_J$;*

      (ii) *$K \cdot U = U$ for a suitable maximal compact subgroup of $G_{\psi,r}$;*

      (iii) *$U^{-1} \cdot U \cap x \cdot G_J \cdot y$ is finite for $x, y \in G_K$.*

    (c) *$G_{\psi,r}/G_J$ has finite Haar measure if and only if $X_K(G) = 1$; it is compact if and only if $X_K(G) = 1$, and every unipotent element of G_K or, equivalently, of G_J, belongs to the radical of G.*

The first assertion follows from 12.2 and 6.5. When $\psi = \Phi'$, (b) and (c) follow from 12.2, 6.5, 6.7, 9.4 and 11.8. Let now $\psi \neq \Phi'$ and θ be the complement of ψ in Φ'. Then $G_{\theta,r}$ is compact, and $G_{\Phi',r} = G_{\psi,r} \times G_{\theta,r}$. Let U be an open set verifying the properties (i) to (iii) for $\psi = \Phi'$. Since $G_{\theta,r}$ is compact, and invariant, it belongs to all maximal compact subgroups of $G_{\Phi',r}$, and the maximal compact subgroups of $G_{\Phi',r}$ are the products of $G_{\theta,r}$ with the maximal compact subgroups of $G_{\psi,r}$. By (ii) we have $G_{\theta,r} \cdot U = U$, hence $U = G_{\theta,r} \times U'$ with U' open in $G_{\psi,r}$; it is then obvious that U' has the properties (i) to (iii).

Let us write A and B for the images of G_J in $G_{\psi,r}$ and $G_{\Phi',r}$ under the canonical imbeddings. We have then clearly $A \cdot G_{\theta,r} = B \cdot G_{\theta,r}$, therefore $G_{\psi,r}/A = G_{\Phi',r}/A \cdot G_{\theta,r} = G_{\Phi',r}/B \cdot G_{\theta,r}$ is the base space of a fibration of $G_{\Phi',r}/B$ with fibre $G_{\theta,r}$. Since $G_{\theta,r}$ is compact, $G_{\psi,r}/A$ is compact (or of finite measure) if and only if $G_{\Phi',r}/B$ is, and (c) follows from the above.

12.4. Corollary. *Let G be a connected algebraic group defined over*

K. We keep the assumptions of 11.3, and assume moreover that there is at least one $\sigma \in \Phi$ for which $G^{\tau}_{L\sigma}$ is compact. Then $G_{\psi,r}/G_J$ is compact if and only if $X_K(G) = 1$.

If $G^{\tau}_{L\sigma}$ is compact, then it is reductive, and all its elements are semi-simple. Of course, $x \in G_K$ is semi-simple if and only if x^{σ} is semi-simple ($\sigma \in \Phi$), in our case, G_K consists therefore of semi-simple elements, and 12.4 follows from 12.3(c).

13. Appendix: Remarks on algebraic groups

This appendix contains some remarks about algebraic groups which are actually not needed in the paper, but bring natural complements to some auxiliary results proved in §§ 1, 8. In A, unlike in § 2, the universal field underlying the definition of an algebraic group may have arbitrary characteristic.

A. Algebraic tori

13.1. Let T be an algebraic torus, K a field of definition for T. Then there always exists a separable finite Galois extension K' of K over which T splits (see Ono, Ann. of Math. 74 (1961), 61–139, Prop. 1.2.1). This being taken into account, it is clear that 8.2, 8.3, and 8.4 go over without change to the general case. It follows from 8.4 that if S is a subtorus of T, defined over K, and if $\chi \in X_K(S)$, then there exists an integer m such that χ^m extends to a rational character of T. In fact we take S' as in 8.4a, and m such that χ^m is trivial on the finite group $S \cap S'$. Since the character groups are finitely generated, this can also be expressed by saying that the injection $i : S \to T$ induces a homomorphism i^0 of $X(T)$ (resp. $X_K(T)$) onto a subgroup of finite index of $X(S)$ (resp. $X_K(S)$).

13.2. PROPOSITION. *Let T be an algebraic torus defined over a field K. Then T contains two subtori T_c, T_a defined over K such that $X_K(T_c) = 1$, T_a splits over K, $T_c \cap T_a$ is finite and $T = T_c \cdot T_a$. If S is an algebraic torus and $f : S \to T$ a rational homomorphism, both defined over K, then $f(S_c) \subset T_c$ and $f(S_a) \subset T_a$.*

Let K' be a Galois extension of K over which T splits, and A the Galois group of K' over K. Let T_c be the identity component of the intersection of the kernels of the characters defined over K, and T_a be the subtorus generated by the images of the elements $\gamma \in \Gamma_K(T)$. They are defined over K', and invariant under A. Since K' is separable over K, T_c and T_a are defined over K. By 8.2, T_a splits over K. By 13.1, $X_K(T_c) = 1$. Thus T_c has no non-trivial subtorus which splits over K (8.4), and $T_c \cap T_a$ is finite. Lemma 8.4 also implies that $\dim T_c + \dim T_a = \dim T$, whence

$T = T_a \cdot T_c$. Let now $f\colon S \to T$ be a rational homomorphism defined over K. We have $f_0(\Gamma_K(S)) \subset \Gamma_K(T)$, hence $f(S_a) \subset T_a$. If $\chi \in X_K(T)$, then $\chi \circ f \in X_K(S)$, hence $S_c \in \mathrm{Ker}\,(\chi \circ f)$, and $f(S_c) \in \mathrm{Ker}\,\chi$, which implies $f(S_a) \subset T_a$.

B. Real algebraic reductive groups

13.3. PROPOSITION. *Let* $T \subset \mathrm{GL}(n, \mathbf{C})$ *be an algebraic torus defined over* $\mathbf{R}$, *and* $\mathfrak{t}_\mathbf{R}$ *the Lie algebra of* $T_\mathbf{R}$. *Then* $T_{c,\mathbf{R}}$ *is a torus,* $T_{a,\mathbf{R}}$ *is real diagonalizable, the Lie algebra of* $T_{c,\mathbf{R}}$ *(resp.* $T_{a,\mathbf{R}}$*) is the set of elements of* $\mathfrak{t}_\mathbf{R}$ *with purely imaginary (resp. real) eigenvalues.*

Let $\mathfrak{t}'$ (resp. $\mathfrak{t}''$) be the set of elements of $\mathfrak{t}_\mathbf{R}$ with real (resp. purely imaginary) eigenvalues. Then $\mathfrak{t}_\mathbf{R} = \mathfrak{t}' + \mathfrak{t}''$, and $\mathfrak{t}', \mathfrak{t}''$ are algebraic (see 1.4). The irreducible real algebraic group T' with Lie algebra $\mathfrak{t}'$ is then a real algebraic torus which splits over $\mathbf{R}$, and is contained in $T_{a,\mathbf{R}}$. The analytic subgroup of $\mathrm{GL}(n, \mathbf{R})$ generated by $\mathfrak{t}''$ is closed, and belongs to a compact group hence is a torus in the usual sense (compact connected commutative Lie group). Since a compact linear group is algebraic [7b, p. 230], T'' is also the irreducible real algebraic subgroup of $T_\mathbf{R}$ with Lie algebra $\mathfrak{t}''$. Clearly, every element of $X_\mathbf{R}(T)$ is trivial on T'', therefore $T' = T_{a,\mathbf{R}}$, $T'' = T_{c,\mathbf{R}}$.

13.4. The preceding proposition shows that the decomposition $\mathfrak{m} = \mathfrak{m}_t + \mathfrak{m}_p$ of a fully reducible commutative algebraic Lie algebra corresponds to the global decomposition of 13.3. In particular (13.3) it is compatible with rational representations, and has therefore an intrinsic meaning, independent of the imbedding. From this and 1.10, 1.11, it follows that the notion of Cartan involution of a real reductive algebraic group is independent from the imbedding in $\mathrm{GL}(n, \mathbf{R})$, up to birational isomorphism. Since a real representation of a semi-simple Lie algebra is always rational, the following proposition generalizes a fact mentioned in 1.1:

13.5. PROPOSITION. *Let* G *be a real algebraic, reductive group,* θ *a Cartan involution of* G, *and* $\rho\colon G \to \mathrm{GL}(m, \mathbf{R})$ *a rational representation. Then there exists a Cartan involution* θ' *of* $\mathrm{GL}(m, \mathbf{R})$ *such that* $\rho(\theta(g)) = \theta'(\rho(g))$.

Let $\mathfrak{c}$ be the center of $\mathfrak{g}$. The image $\mathfrak{g}'$ of $\mathfrak{g}$ is the direct product of $\rho(\mathfrak{c})$ and of $\rho(\mathscr{D}\mathfrak{g}) = \mathscr{D}\mathfrak{g}'$. By 13.4, $\rho(\mathfrak{c})$ is completely reducible, and therefore $\mathfrak{g}'$ is reductive in $\mathfrak{gl}(m, \mathbf{R})$. By a general theorem [7a, p. 140], $\mathfrak{g}'$ is algebraic. Let G' be the real algebraic subgroup of $\mathrm{GL}(m, \mathbf{R})$ with Lie algebra $\mathfrak{g}'$, and M be the centralizer of $\rho(\mathfrak{c})$. The group M is algebraic, and its Lie algebra $\mathfrak{m}$ is reductive [4]. Thus (1.2) G' and M are real algebraic, reductive. As was recalled in 1.1, the image of a Cartan decomposition of $\mathscr{D}\mathfrak{g}$

is a Cartan decomposition of $\mathscr{D}\mathfrak{g}'$. It follows therefore from 1.6 and the conjugacy of Cartan decompositions of $\mathscr{D}\mathfrak{g}'$ that we may find a Cartan involution θ'' of $\mathbf{GL}(m, \mathbf{R})$ which leaves M, G' invariant and induces the given Cartan involution of $\mathscr{D}\mathfrak{g}'$. By 13.4 and 1.4, the $\mathfrak{k}$- and $\mathfrak{p}$-parts of $\rho(\mathfrak{c})$ with respect to θ'' are necessarily the images of the $\mathfrak{k}$- and $\mathfrak{p}$-parts of θ. The Cartan involution θ'' verifies therefore $\rho \circ \theta = \theta'' \circ \rho$ on $\mathfrak{g}$, hence also on G^0, and leaves G' invariant. It is defined by a positive non-degenerate quadratic form F which is invariant under the identity component of $\rho(K)$, where K is the fixed point set of θ in G. Let F' be the average of F, over $\rho(K)$. The corresponding Cartan involution θ' will then fulfill our conditions.

INSTITUTE FOR ADVANCED STUDY
COLUMBIA UNIVERSITY

BIBLIOGRAPHY

1. A. BOREL, *Groupes linéaires algébriques*, Ann. of Math., 64 (1956), 20–80.
2. ———, *Some properties of the adele groups attached to algebraic groups*, Bull. Amer. Math. Soc., 67 (1961), 583–585.
3. A. BOREL and HARISH-CHANDRA, *Arithmetic subgroups of algebraic groups*, Bull. Amer. Math. Soc., 67 (1961), 579–583.
4. A. BOREL and G. D. MOSTOW, *On semi-simple automorphisms of Lie algebras*, Ann. of Math., 61 (1955), 389–405.
5. N. BOURBAKI, Groupes et algèbres de Lie, Chap. I: Algèbres de Lie, Actualités Sci. Ind. 1285, Paris, 1960.
6. H. CARTAN, Séminaire E.N.S., 1951–52, Exp. XX (by J-P. Serre).
7. C. CHEVALLEY, Théorie des groupes de Lie, (a) T. II, Groupes algébriques, Actualités Sci. Ind. 1152, Paris, 1951; (b) T. III, Théorèmes généraux sur les algèbres de Lie, Actualités Sci. Ind. 1155, Paris, 1955.
8. ———, Séminaire sur la classification des groupes de Lie algébriques, Paris, 1958.
9. HARISH-CHANDRA, *Representations of a semi-simple Lie group on a Banach space* I, Trans. Amer. Math. Soc., 75 (1953), 185–243.
10. ———, *On a lemma of Bruhat*, J. math. pures appl. Série g, T. 35 (1956), 203–210.
11. ———, *The characters of semi-simple Lie groups*, Trans. Amer. Math. Soc., 83 (1956), 98–163.
12. ———, *Representations of semi-simple Lie groups* VI, Amer. J. Math., 78 (1956), 564–628.
13. ———, *A formula for semi-simple Lie groups*, Amer. J. Math., 79 (1957), 733–760.
14. C. HERMITE, Oeuvres complètes, Volume 1, Gauthier-Villars, Paris, 1905.
15. N. JACOBSON, *Completely reducible Lie algebras of linear transformations*, Proc. Amer. Math. Soc., 2 (1951), 105–113.
16. A. KORKINE and G. ZOLOTAREFF, *Sur les formes quadratiques*, Math. Ann., 6 (1873), 366–389.
17. S. LANG, Introduction to algebraic geometry, Interscience Tracts in Pure and Applied Math. 5, New York, 1958.
18. Y. MATSUSHIMA, *Espacse homogènes de Stein des groupes de Lie complexes*, Nagoya Math. J., 16 (1960), 205–218.

19. D. MONTGOMERY and L. ZIPPIN, Topological transformation groups, Interscience Tracts in Pure and Applied Math. 1, New York, 1955.

20. G. D. MOSTOW, *Some new decomposition theorems for semi-simple groups*, Mem. Amer. Math. Soc., 14 (1955), 31-54.

21. ———, *Self-adjoint group*, Ann. of Math., 62 (1955), 44-55.

22. T. ONO, *Sur une propriété arithmétique des groupes commutatifs*, Bull. Soc. Math. France 85 (1957), 307-323.

23. ———, *On some arithmetic properties of linear groups*, Ann. of Math., 70 (1959), 266-290.

24. K. G. RAMANATHAN, *Quadratic forms over involutorial division algebras* II, Math. Ann., 143 (1961), 293-332.

25. M. ROSENLICHT, *Some basic theorems on algebraic groups*, Amer. J. Math., 78 (1956), 401-443.

26. ———, *Some rationality questions on algebraic groups*, Annali di Matematica Ser. IV, T. 43, 1957, 25-50.

27. SÉMINAIRE S. LIE, Groupes et algèbres de Lie, Paris, 1955.

28. C. L. SIEGEL, *Einheiten quadratischer Formen*, Abh. Math. Sem. Hamburg, 13 (1939), 209-239.

29. A. WEIL, Discontinuous subgroups of classical groups, Notes, Chicago University, 1958.

30. ———, Adeles and algebraic groups, Notes, The Institute for Advanced Study, Princeton, 1961.

31. H. WHITNEY, *Elementary structure of real algebraic varieties*, Ann. of Math., 66 (1957), 545-556.

59.

Ensembles fondamentaux pour les groupes arithmétiques

Colloque sur la Théorie des Groupes Algébriques, Bruxelles 1962, 23–40

1. Introduction

1   Soit G un groupe algébrique linéaire semi-simple, défini sur le corps Q des nombres rationnels. G étant identifié à un groupe matriciel, et B étant un sous-anneau de C, on note, suivant l'usage, par G_B le sous-groupe des éléments de G dont les coefficients sont dans B et dont le déterminant est une unité de B. En particulier G_R est un groupe de Lie semi-simple réel, ayant un nombre fini de composantes connexes, et G_Z est un sous-groupe discret, le *groupe des unités de G*. Deux réalisations matricielles de G, définies sur Q, donnent lieu à des groupes d'unités commensurables ([1]). Les résultats de ce travail concernent essentiellement les groupes d'unités; cependant, il est commode de considérer aussi les sous-groupes d'indice fini des groupes d'unités, ou, ce qui revient au même, les sous-groupes de G_Q commensurables aux groupes d'unités, que nous appellerons dans ce travail *groupes arithmétiques* ([2]).

Soient K un sous-groupe compact maximal de G_R, $D = K \backslash G_R$ l'espace riemannien symétrique à courbure négative de G_R, et Γ un sous-groupe arithmétique de G. On appellera *ensemble fondamental* pour Γ dans D un sous-ensemble Ω de D ayant les deux propriétés suivantes :

([1]) Rappelons que deux sous-groupes A et B d'un groupe G sont dits être commensurables si $A \cap B$ est d'indice fini dans chacun d'eux.

([2]) Pour abréger, et bien qu'il existe des sous-groupes d'indice fini de $SL(2, Z)$ qui ne contiennent aucun sous-groupe de congruence de $SL(2, Z)$, donc qui ne sont pas de nature arithmétique au sens classique.

287

(i) $\Omega \cdot \Gamma = D$.

(ii) («propriété de Siegel») Quel que soit $a \in G_Q$, l'ensemble des éléments $x \in \Gamma$ pour lesquels $\Omega \cdot a \cap \Omega x \neq \varnothing$ est fini.

L'existence d'ensemble fondamentaux ouverts ou fermés, de *mesure invariante finie*, classique dans de nombreux cas particuliers, est établie de façon générale dans [3]. Nous nous proposons ici de décrire des ensembles fondamentaux, en général différents de ceux de [3] qui, lorsque D/Γ n'est pas compact, peuvent s'envisager comme réunion d'un compact et de «pointes» (cf. §2) et d'étudier, du point de vue topologique, la compactification de D/Γ (cf §4). On verra que les pointes jouent un rôle assez analogue à celui des pointes paraboliques des domaines fondamentaux de groupes fuchsiens. Nous aurons besoin pour cela, d'une part de certaines propriétés des groupes algébriques définis sur un corps parfait résumées dans le §1, d'autre part de résultats de Satake sur la compactification de D associée à une représentation linéaire de G rappelés au §3. Dans le §4, on montre aussi que les pointes d'un ensemble fondamental représentent les doubles classes de G_Q modulo Γ et H_Q, où H est un sous-groupe parabolique minimal défini sur Q, et l'on donne d'autres interprétations des pointes dans certains cas particuliers.

Dans cette Note, les démonstrations sont omises, ou au plus esquissées. Elles seront publiées ultérieurement.

Remarque. La condition (ii) n'est applicable telle quelle qu'aux groupes arithmétiques. Pour lui donner une portée plus générale, on peut soit l'affaiblir en demandant simplement que l'ensemble des $x \in \Gamma$ pour lesquels $\Omega \cap \Omega \cdot x \neq \varnothing$ soit fini, soit la renforcer en exigeant (ii) pour tout élément a du «*groupe des transformations*» $\widetilde{\Gamma}$ de G_R/Γ [9]. Rappelons que $\widetilde{\Gamma}$ est par définition l'ensemble des $a \in G_R$ tels que $a \cdot \Gamma \cdot a^{-1}$ soit commensurable à Γ; il contient G_Q mais peut en être distinct. Cependant on peut montrer que *si N est le plus grand sous-groupe normal défini sur Q de G tel que N_R soit compact, alors $\widetilde{\Gamma}$ est l'image réciproque de $(G/N)_Q$ par la projection canonique.* On en déduit facilement que les ensembles fondamentaux de [3], ou de 2.4 vérifient encore (ii) ainsi renforcée, (donc que les groupes arithmétiques sont «minkowskiens» dans dans la terminologie de [9]), et que les résultats de ce travail valent aussi pour les sous-groupes de G_R (et non pas seulement de G_Q) commensurables aux groupes d'unités.

§ 1. Groupes algébriques sur un corps parfait [3]

1.1. Soient T un tore algébrique, k un corps de définition de T. On note $X(T)$ le groupe des caractères rationnels de T (homomorphismes rationnels de T dans le groupe multiplicatif du corps universel). On sait que $X(T)$ est un groupe commutatif libre, de rang égal à la dimension de T, et que les éléments de $X(T)$ sont tous définis sur k et seulement si T est *décomposé sur k*, c'est-à-dire est isomorphe sur k à un produit de groupes multiplicatifs du corps universel.

1.2. Dorénavant, k est un corps parfait. Les tores d'un groupe algébrique G défini sur k qui sont maximaux parmi les tores décomposés sur k sont conjugués par des éléments de G_k. Leur dimension commune sera appelée le k–rang $r_k(G)$ de G. Soient $_kT$ un tore décomposé sur k maximal de G, $\mathscr{N}(_kT)$ et $\mathscr{Z}(_kT)$ son normalisateur et son centralisateur dans G. Le groupe $_kW(G) = \mathscr{N}(_kT)/\mathscr{Z}(_kT)$ est le *groupe de Weyl de G relatif à k*. Il est fini. Les caractères non triviaux de $_kT$ dans la représentation adjointe sont les *racines de G relatives à k*, ou les *k–racines de G*. L'ensemble des racines relatives à k sera noté $_k\Sigma(G)$ ou $_k\Sigma$. Le groupe $_kW(G)$ opère de façon naturelle sur $X(_kT)$ et laisse $_k\Sigma$ invariant. Supposons G connexe, semi-simple. Alors $X(_kT) \otimes \boldsymbol{Q}$ est engendré, en tant qu'espace vectoriel sur $\boldsymbol{Q}$, par $_k\Sigma$. Pour chaque $\alpha \in {}_k\Sigma$, il existe un élément $s_\alpha \in {}_kW$, nécessairement unique, qui transforme α en $-\alpha$ et induit l'identité sur $X(_kT)/\boldsymbol{Z} \cdot \alpha$, et $_kW$ est engendré par les s_α. Un ordre étant choisi sur $X(_kT)$, toute k-racine est combinaison linéaire à coefficients entiers de même signe de $r_k(G)$ racines positives, appelées les k-racines *simples*, et dont l'ensemble sera noté $_k\Delta$. En particulier, le groupe de Weyl relatif à k est toujours cristallographique. Il existe un sous-groupe unipotent $_kN$ défini sur k, dont le normalisateur contient $_kT$ et dont l'algèbre de Lie est la somme des espaces propres de $_kT$ correspondant aux

[3] Beaucoup des résultats de ce paragraphe ont été aussi obtenus, indépendamment, parfois sous des hypothèses plus générales quant aux corps de définition, par J. Tits [10]. Les démonstrations de quelques-uns d'entre-eux, communiqués antérieurement à R. Godement, ont été incorporées dans son exposé [6]. On suppose ici le lecteur familier avec la théorie des groupes linéaires algébriques sur un corps algébriquement fermé.

k-racines positives, et l'on a le «lemme de Bruhat» pour G_k :

$$(1) \qquad G_k = {}_kN_k \cdot \mathcal{N}({}_kT)_k \cdot {}_kN_k, \quad \mathcal{N}({}_kT) = \mathcal{N}({}_kT)_k \cdot \mathcal{Z}({}_kT).$$

1.3. Un sous-groupe fermé H du groupe algébrique connexe G est *parabolique* si G/H est une variété complète. Si H est aussi défini sur k, la fibration de G par H possède des sections locales rationnelles définies sur k, et la projection $G_k \to (G/H)_k$ est surjective. Deux sous-groupes paraboliques définis sur k et conjugués dans G sont aussi conjugués par un élément de G_k. Les sous-groupes paraboliques définis sur k minimaux sont conjugués. Un sous-groupe parabolique est égal à son normalisateur.

Supposons G semi-simple. Les sous-groupes paraboliques définis sur k minimaux sont conjugués au sous-groupe $H = \mathcal{Z}({}_kT) \cdot {}_kN$ (cf. 1.2), produit semi-direct de $\mathcal{Z}({}_kT)$ et du sous-groupe invariant ${}_kN$. Soient θ une partie, éventuellement vide, de ${}_k\Delta$, et T_θ la composante neutre de l'intersection des noyaux des caractères $\alpha \in \theta$ de ${}_kT$. Alors ${}_kN$ et le centralisateur $\mathcal{Z}(T_\theta)$ de T_θ engendrent un sous-groupe parabolique H_θ défini sur k, et tout sous-groupe parabolique défini sur k est conjugué à un H_θ, et à un seul.

1.4. Jusqu'à la fin du § 1, k est de caractéristique zéro, et G est un groupe algébrique semi-simple, connexe, défini sur k. Alors le sous-groupe ${}_kN$ de 1.2 est maximal parmi les sous-groupes unipotents définis sur k, et tout sous-groupe unipotent défini sur k est conjugué, par un élément de G_k, à un sous-groupe de ${}_kN$. Les sous-groupes paraboliques définis sur k minimaux sont les normalisateurs des sous-groupes unipotents définis sur k maximaux. Le groupe $\mathcal{Z}({}_kT)$ contient un unique sous-groupe algébrique connexe ${}_kM$, défini sur k, tel que $\mathcal{Z}({}_kT) = {}_kM \cdot {}_kT$ et que ${}_kM \cap {}_kT$ soit fini. Le groupe ${}_kM$ est *anisotrope*, ce qui signifie que ${}_kM_k$ est formé d'éléments semi-simples et que ${}_kM$ ne possède aucun caractère rationnel non trivial défini sur k. Le k-rang de G est nul si et seulement si G est anisotrope. Les k-racines vérifient la condition d'entiers : si $(\ ,\ )$ désigne un produit scalaire positif non dégénéré sur $X({}_kT) \otimes \mathbf{Q}$, invariant par ${}_kW$, le nombre $2(\alpha, \beta)/(\alpha, \alpha) = q(\beta, \alpha)$ est entier quels que soient $\alpha, \beta \in {}_k\Sigma$.

1.5. ([4]) Soient $_kT$ un tore décomposé sur k maximal de G, T un tore maximal de G défini sur k et contenant $_kT$, et k' une extension galoisienne finie de k sur laquelle T se décompose. Alors le groupe de Galois Gal (k'/k) de k' sur k opère sur $X(T)$ et sur l'ensemble Σ des racines (absolues) de G par rapport à T. On suppose des ordres sur $X(T)$ et $X(_kT)$ choisis de manière à ce que

(1) $\alpha \geqq 0 \Rightarrow \alpha \mid _kT \geqq 0$ $(\alpha \in X(T))$,

(2) $\alpha > 0,\ \alpha \mid _kT \neq 0 \Rightarrow s(\alpha) > 0$ $(\alpha \in X(T),\ s \in \mathrm{Gal}(k'/k))$.

Si ϱ est une représentation linéaire *à droite* de G dans un espace vectoriel de dimension finie V, on pose, pour tout $\lambda \in X(_kT)$:

$$V_\lambda = \{\, v \in V \mid v \cdot \varrho(t) = \lambda(t) \cdot v, \qquad (t \in {}_kT)\}.$$

Le caractère λ est un *k-poids*, ou un *poids relatif à k, de* ϱ si $V_\lambda \neq 0$. Supposons ϱ irréductible, soient μ_ϱ son poids minimal, par rapport à T et à l'ordre choisi, et λ_ϱ la restriction de μ_ϱ à $_kT$. Alors tout k-poids de ϱ est de la forme

(3) $\lambda = \lambda_\varrho \cdot a_1^{m_1} \cdot \ldots \cdot a_r^{m_r} (m_i \in \boldsymbol{Z};\ m_i \geqq 0 \quad (1 \geqq i \geqq r);$
$$\{\, a_1, \ldots, a_r \} = {}_k\Delta).$$

Soit $c(\lambda)$ l'ensemble des $a_i \in {}_k\Delta$ pour lesquels $m_i \neq 0$. Alors une partie θ de $_k\Delta$ est de la forme $c(\lambda)$ si et seulement si $\theta \cup \lambda_\varrho$ est *connexe*, c'est-à-dire n'est pas réunion de deux sous-ensembles non vides, disjoints, mutuellement orthogonaux. Pour un tel sous-ensemble θ, soit $V_\theta = \Sigma_{c(\lambda) \subset \theta}\, V_\lambda$, et soit θ' l'ensemble des éléments de $_k\Delta$ orthogonaux à $\theta \cup \lambda_\varrho$. Alors

$$\mathcal{N}(V_\theta) = \{\, g \in G \mid V_\theta \cdot \varrho(g) = V_\theta\} = H_{\theta \cup \theta'}\,,$$

(notation de 1.3). Soit $\mathfrak{g}_a$ le sous-espace propre de l'algèbre de Lie $\mathfrak{g}$ de G correspondant à $\alpha \in {}_k\Sigma$. La somme des sous-espaces $\mathfrak{g}_a + [\mathfrak{g}_a, \mathfrak{g}_{-a}]$, où α parcourt les k-racines qui sont combinaisons linéaires d'éléments de θ, est l'algèbre de Lie d'un sous-groupe algébrique semi-simple connexe L_θ, défini sur k, de k-rang égal

---

([4]) Si k est algébriquement fermé, les k-poids ne sont autres que les poids au sens usuel, et les résultats de n°s 1.5 et 1.6 sont classiques. Ils sont cependant formulés en général de manière différente, car ici les racines sont des caractères de $_\varrho T$, et non des formes linéaires sur son algèbre de Lie, et nous considérons des représentations *à droite*. Dans le cas où $k = \boldsymbol{R}$, les résultats de 1.5 sont dûs à Satake [7].

au nombre d'éléments de θ, et l'on a

$$(4) \qquad \mathcal{N}(V_\theta) = L_\theta \cdot \mathcal{Z}(V_\theta) \cdot \mathcal{Z}({}_kT),$$

où $\mathcal{Z}(V_\theta)$ est l'ensemble des éléments de $\mathcal{N}(V_\theta)$ dont la restriction à V_θ est un multiple scalaire de l'identité. $\mathcal{Z}(V_\theta)$ est un sous-groupe invariant de $\mathcal{N}(V_\theta)$ dont l'intersection avec L_θ est finie. Le groupe $\mathcal{Z}({}_kT)$ est dans le normalisateur de L_θ; le quotient $\mathcal{N}(V_\theta)/\mathcal{Z}(V_\theta)$ est isogène au produit de L_θ par un sous-groupe invariant de ${}_kM$; son k-rang est donc égal à celui de L_θ.

1.6. Il existe $r = r_k(G)$ représentations rationnelles irréductibles $\varpi_1, \ldots, \varpi_r$ de G, définies sur k, telles que les représentations rationnelles irréductibles de G, définies sur k, et de poids minimal (pour l'ordre choisi plus haut) défini sur k, soient celles dont le poids minimal est combinaison linéaire à coefficients entiers $\geqq 0$ des poids minimaux μ_i des représentations ϖ_i. On a

$$2(a_i, \lambda_j) = e_j \cdot (a_i, a_i) \cdot \delta_{ij} \quad (1 \leq i, j \leq r; \ e_j \in \mathbf{Z}, \ e_j > 0).$$

1.7. Soit $k = \mathbf{R}$. Alors ${}_kT$ est maximal parmi les tores de G qui sont définis sur $\mathbf{R}$ et dont tous les éléments réels ont toutes leurs valeurs propres réelles, et $N = {}_{\mathbf{R}}N_{\mathbf{R}}$ est un sous-groupe unipotent maximal de $G_{\mathbf{R}}$. Soient A la composante connexe (en topologie ordinaire) de l'élément neutre dans ${}_{\mathbf{R}}T_{\mathbf{R}}$ et soit K un sous-groupe compact maximal de $G_{\mathbf{R}}$ dont l'algèbre de Lie $\mathfrak{k}$ est orthogonale à celle $\mathfrak{a}$ de A, par rapport à la forme de Killing. Alors $G_{\mathbf{R}} = K.A.N$, et plus précisément l'application $(k, a, n) \to k.a.n$ ($k \in K$, $a \in A$, $n \in N$) est un homéomorphisme analytique de $K \times A \times N$ sur $G_{\mathbf{R}}$ (*décomposition d'Iwasawa* de $G_{\mathbf{R}}$). Rappelons encore que si $\mathfrak{p}$ est le complément orthogonal de $\mathfrak{k}$ dans $\mathfrak{g}_{\mathbf{R}}$, par rapport à la forme de Killing, alors $(k. p) \to k. \exp. p.$ ($k \in K$, $p \in \mathfrak{p}$) est un homéomorphisme analytique de $K \times \mathfrak{p}$ sur $G_{\mathbf{R}}$ (*décomposition de Cartan* de $G_{\mathbf{R}}$), et que $\mathfrak{a}$ est une sous-algèbre maximale de $\mathfrak{p}$. Il existe un et un seul automorphisme involutif s de $G_{\mathbf{R}}$, dont K est l'ensemble des points fixes; il applique tout élément x de $P = \exp \mathfrak{p}$ sur son inverse; c'est *l'involution de Cartan de $G_{\mathbf{R}}$* associée à $\mathfrak{k}$.

2.1. Soient G un groupe semi-simple algébrique défini sur Q et Γ un sous-groupe arithmétique de G. Un sous-ensemble $\Omega \subset G_R$ sera dit *fondamental pour* Γ s'il vérifie les conditions suivantes :

(i′) $\Omega \cdot \Gamma = G$.

(ii′) («propriété de Siegel») Quel que soit $a \in G_R$, l'ensemble des $x \in \Gamma$ tels que $\Omega \cdot a \cap \Omega \cdot x \neq \varnothing$ est fini.

(iii′) $K \cdot \Omega = \Omega$ pour un sous-groupe compact maximal K de G_R.

Soit π la projection naturelle de G_R sur $D = K \backslash G_R$. Il est clair que $\pi(\Omega)$ est un ensemble fondamental pour Ω dans D, au sens de l'introduction, et que l'image réciproque d'un ensemble fondamental dans D est un ensemble fondamental dans G_R. Par ailleurs G_R est fibré sur D, avec des fibres compactes; par conséquent Ω est compact, ou fermé, ou de mesure invariante finie, si et seulement si $\pi(\Omega)$ l'est. Il revient donc au même d'étudier les ensembles fondamentaux dans G_R ou dans D. Dans ce paragraphe, nous considérerons G_R.

2.2. Soient $_Q T$ un tore décomposé sur Q maximal, H un sous-groupe parabolique défini sur Q minimal contenant $_Q T$, et K un sous-groupe compact maximal de G_R dont l'algèbre de Lie $\mathfrak{k}$ est orthogonale, par rapport à la forme de Killing, à celle de $_Q T_R$ (on dira parfois que K et $_Q T$ sont *adaptés* l'un à l'autre). On reprend les notations de 1.2, 1.3, et l'on écrit H sous la forme $H = {}_Q M \cdot {}_Q T \cdot {}_Q N$, où $_Q N$, le radical unipotent de H, est un sous-groupe unipotent défini sur Q maximal de G, et où $_Q M$ est le plus grand sous-groupe anisotrope connexe de $\mathscr{Z}({}_Q T)$. On note $_Q \varDelta$ l'ensemble des Q-racines simples relativement à l'ordre associé à $_Q N$, i.e. pour lequel les poids de $_Q T$ dans l'algèbre de Lie de $_Q N$ sont positifs.

Soit $_Q A$ la composante connexe, en topologie ordinaire, de l'élément neutre dans $_Q T_R$. Pour tout $t \in R$, on pose

(1) $$_Q A_t = \{\, a \in {}_Q A \mid \ \alpha(a) \leqq e^t, \ \ \alpha \in {}_Q \varDelta \}.$$

On appellera *domaine de Siegel généralisé* de G_R (resp. de $K \backslash G_R$) tout sous-ensemble

(2) $$\mathfrak{S}_{t,\eta,\omega} = K \cdot \eta \cdot {}_R A_t \cdot \omega \, (t \in R; \quad \eta \subset {}_Q M_R; \quad \omega \subset {}_Q N_R;$$
$$\eta, \omega \ \text{compacts})$$

(resp. l'image de $\mathfrak{S}_{t,\eta,\omega}$ dans $K\backslash G_{\boldsymbol{R}}$ par la projection canonique).

Il est immédiat que si $h \in H_{\boldsymbol{R}}$, l'ensemble $\mathfrak{S}_{t,\eta,\omega} \cdot h$ est contenu dans un domaine de Siegel généralisé.

En fait, on devrait parler de domaine de Siegel généralisé relativement à $_{\boldsymbol{Q}}T$, H, K, puisque la définition précédente présuppose le choix de ces trois groupes, le dernier étant adapté au premier; cependant nous omettrons cette précision lorsque cela n'entraînera pas de confusion.

2.3. Si l'on prend tout « relatif à $\boldsymbol{R}$ » dans la définition ci-dessus, on retrouve les domaines de Siegel au sens usuel [3,8] : soit $G_{\boldsymbol{R}} = K \cdot A \cdot N$ une décomposition d'Iwasawa de $G_{\boldsymbol{R}}$ (cf. 1.7) telle que $A \supset A_{\boldsymbol{Q}}$ et $N \supset {}_{\boldsymbol{Q}}N_{\boldsymbol{R}}$, ce qui existe toujours; un domaine de Siegel est un ensemble

$$\mathfrak{S}_{t,\omega} = K \cdot A_t \cdot \omega \qquad (\omega \text{ compact dans } N),$$

où

$$A_t = \{\, a \in A \mid \alpha(a) \leqq e^t, \quad \alpha \in {}_{\boldsymbol{R}}\varDelta \,\},$$

(cf. [3, § 4]). Si $A = {}_{\boldsymbol{Q}}A$, ce qui équivaut à dire que $_{\boldsymbol{Q}}T$ est aussi un tore décomposé sur $\boldsymbol{R}$ maximal, alors $_{\boldsymbol{Q}}M_{\boldsymbol{R}} \subset K$, et les domaines de Siegel généralisés ne sont autres que les domaines de Siegel au sens usuel. Dans le cas général, on voit facilement que l'algèbre de Lie $_{\boldsymbol{Q}}\mathfrak{m}_{\boldsymbol{R}}$ de $_{\boldsymbol{Q}}M_{\boldsymbol{R}}$ est orthogonale à celle de $A_{\boldsymbol{Q}}$. Il en résulte que $Z_{\boldsymbol{R}}$ est invariant par l'involution de Cartan s de $G_{\boldsymbol{R}}$ dont K est l'ensemble des points fixes, et que

$${}_{\boldsymbol{Q}}M_{\boldsymbol{R}} = (K \cap {}_{\boldsymbol{Q}}M) \cdot (A \cap {}_{\boldsymbol{R}}M) \cdot (N \cap {}_{\boldsymbol{Q}}M)$$

est une décomposition d'Iwasawa de $_{\boldsymbol{Q}}M_{\boldsymbol{R}}$ (cf. [3, § 1]). Comme η est compact, et que $_{\boldsymbol{Q}}M$ centralise $_{\boldsymbol{R}}T$, on en déduit aussitôt que $\mathfrak{S}_{t,\eta,\omega}$ est contenu dans un domaine de Siegel de $G_{\boldsymbol{R}}$ relatif à la décomposition $K \cdot A \cdot N$. L'ensemble $\mathfrak{S}_{t,\eta,\omega}$ est donc de *mesure de Haar finie* [3, §4.3].

Remarquons que, comme pour les domaines de Siegel au sens usuel [3, § 4.1], on pourrait dans 2.2. (1) faire parcourir à α toutes les racines positives, ou encore remplacer $_{\boldsymbol{Q}}A_t$ par $a \cdot {}_{\boldsymbol{Q}}A_0$ $(a \in {}_{\boldsymbol{Q}}A)$ sans rien changer d'essentiel.

2.4. Théorème. *Soient G un groupe algébrique semi-simple défini sur $\boldsymbol{Q}$ et Γ un sous-groupe arithmétique de G. Alors il existe* *un domaine de Siegel généralisé* $\mathfrak{S}_{t,\eta,\omega}$ *de* $G_{\boldsymbol{R}}$ *et un nombre fini d'éléments* $x_i \in G_{\boldsymbol{Q}}$ $(1 \leqq i \leqq m)$ *tels que*

(A) $$\Omega = \bigcup_{i=1}^{i=m} \mathfrak{S}_{t,\eta,\omega} \cdot x_i \qquad (x_i \in G \; ; \; i = 1, ..., m),$$

soit un ensemble fondamental pour Γ *dans* $G_{\boldsymbol{R}}$. *Tout sous-ensemble de la forme* (A) *est de mesure de Haar finie et possède la propriété de Siegel.*

La démonstration prend comme point de départ l'existence d'ensembles fondamentaux du type indiqué dans [3, Theorem 6.5] et la validité de la conjecture de Godement [3, Theorem 11.6]. Elle utilise quelques-uns des résultats mentionnés dans le § 1 et des raisonnements en partie analogues à ceux du § 7 de [3].

Notons $_{\boldsymbol{Q}}A_{s,t}$ $(s < t)$ l'ensemble des éléments de $_{\boldsymbol{Q}}A$ sur lesquels les $\boldsymbol{Q}$-racines simples prennent des valeurs comprises entre e^s et e^t. Alors Ω est réunion du compact $\cup\, K \cdot \eta \cdot {}_{\boldsymbol{Q}}A_{s,t} \cdot \omega \cdot x_i$ et des ensembles $\mathfrak{S}_{s,\eta,\omega} \cdot x_i$. Ce sont ces derniers que nous appellerons les *pointes de* Ω, étant entendu que s prend une valeur négative, suffisamment grande en valeur absolue.

2.5. *Exemple.* Soit G le groupe orthogonal d'une forme quadratique rationnelle, non dégénérée, isotrope, dont on suppose la matrice F mise sous la forme

(1) $$F = \begin{pmatrix} 0 & 0 & S \\ 0 & F_0 & 0 \\ S & 0 & 0 \end{pmatrix} \qquad S = \begin{pmatrix} 0 & & 1 \\ & \cdot & \\ & \cdot & \\ 1 & & 0 \end{pmatrix},$$

et où F_0 ne représente pas zéro rationnellement. Le degré d de S est donc la dimension des sous-espaces isotropes maximaux de F qui sont définis sur $\boldsymbol{Q}$. On peut alors prendre pour H l'ensemble des matrices de G qui ont la forme

(2) $$X = \begin{pmatrix} A_1 & A_2 & A_3 \\ 0 & B_2 & B_3 \\ 0 & 0 & C_3 \end{pmatrix},$$

avec A_1 triangulaire supérieure. Il est élémentaire, et bien connu, que A_1 et A_2 sont arbitraires, que C_3 est triangulaire supérieure, et déterminée par A_1, que B_2 parcourt le groupe orthogonal $0(F_0)$ de F_0, et que A_1, A_2, B_2 déterminent aussi A_3 et B_3. Le groupe H

n'est autre que le groupe de stabilité du drapeau isotrope maximal $E_1 \subset \cdots \subset E_d$ où E_i est l'espace isotrope sous-tendu par les i premiers vecteurs base. Les groupes paraboliques minimaux définis sur Q sont les groupes de stabilité des drapeaux isotropes maximaux définis sur Q. On peut prendre pour $_QT$ l'ensemble des matrices diagonales de H pour lesquelles

$$(3) \quad B_2 = \mathrm{Id}. \quad A_1 = \begin{pmatrix} x_1 & & & 0 \\ & x_2 & \cdot & \\ & & \cdot & \\ & & & \cdot \\ 0 & & & x_d \end{pmatrix} \quad C_3 = \begin{pmatrix} x_d^{-1} & & & 0 \\ & x_{d-1}^{-1} & \cdot & \\ & & \cdot & \\ & & & \cdot \\ 0 & & & x_1^{-1} \end{pmatrix}.$$

On a alors

$$(4) \qquad \mathscr{L}(_QT) = {_QT} \times {_QM} \quad , \quad {_QM} = 0(F_0).$$

Les Q-racines sont les caractères $x_i \cdot x_j^{\pm 1}$ et leurs inverses $(1 \le i < j \le d)$, auxquels il faut encore ajouter $x_i^{\pm 1}$ $(1 \le i \le d)$ lorsque $F_0 \ne 0$. Pour les Q-racines simples, relatives à l'ordre défini par H, et pour la définition de $_QA_t$, on a deux cas à distinguer :

(i) $F_0 = 0$. Les Q-racines simples sont

$$x_i \cdot x_{i+1}^{-1} \quad (1 \le i \le d - 1), \qquad x_{d-1} \cdot x_d,$$

et l'on a

$${_QA_t} = \{\, a \in {_QA} \mid x_i \le e^t \cdot x_{i+1}\, (1 \le i \le d - 1);\ x_{d-1} \cdot x_d \le e^t \}.$$

(ii) $F_0 \ne 0$. Les Q-racines simples sont :

$$x_i \cdot x_{i+1}^{-1} \quad (1 \le i \le d - 1), \qquad x_d,$$

et l'on a

$${_QA_t} = \{\, a \in {_QA} \mid x_i \le e^t \cdot x_{i+1}\, (1 \le i \le d - 1);\ x_d \le e^t \}.$$

Pour K, il faut prendre l'intersection de G avec le groupe orthogonal d'une majorante de F, (au sens de Hermite), dont l'algèbre de Lie soit orthogonale à celle de $_QA$. Ecrivons l'espace V_0 de F_0 comme somme directe de deux sous-espaces orthogonaux V_1, V_2 tels que la restriction F_i de F_0 à V_i soit positive pour $i = 1$, négative pour $i = 2$. Alors on peut prendre pour K le produit du groupe orthogonal de $-F_2$ dans V_2 par le groupe orthogonal de la somme de F_1 dans V_1 avec la forme unité dans l'espace des d premières et d dernières coordonnées. Si F_0 est positive ou négative, alors $_QM_R$ est contenu dans K, $_QT$ est aussi un tore décomposé sur R maximal, et les domaines de Siegel généralisés coïncident avec les domaines de Siegel au sens usuel (cf. 2.3).

§ 3. Composantes de la frontière d'un espace symétrique

Dans tout ce paragraphe, G est un groupe algébrique semi-simple défini sur R, $G_R = K \cdot P$ une décomposition de Cartan de G_R et s l'involution de Cartan correspondante (cf. 1.7). K est donc un sous-groupe compact maximal, $P = \exp \mathfrak{p}$, où $\mathfrak{p}$ est le complément orthogonal de l'algèbre de Lie $\mathfrak{k}$ de K dans $\mathfrak{g}_R$, relativement à la forme de Killing, et $s(x) = x$ si $x \in K$, $s(x) = x^{-1}$ si $x \in P$. On note A la composante neutre, pour la topologie ordinaire, de l'ensemble des points réels d'un tore décomposé sur R maximal et l'on suppose, ce qui est loisible, que $A \subseteq P$. On choisit un ordre sur les R-racines et on note N la partie réelle du sous-groupe unipotent correspondant aux racines positives.

3.1. Soit ϱ une représentation irréductible à droite, localement fidèle de G_R dans un espace vectoriel (réel ou complexe) V de dimension finie. On suppose muni d'une structure hilbertienne telle que $\varrho(x)$ soit unitaire (resp. self-adjoint positif) si $x \in K$ (resp. $x \in P$), ce qui est toujours possible, et on note a^* l'adjoint d'un endomorphisme a de V. Soient $\mathscr{ES}(V)$ l'espace vectoriel des endomorphismes self-adjoints de V, et $P\mathscr{ES}(V)$ l'espace projectif correspondant. Le groupe G_R opère sur $\mathscr{ES}(V)$ et sur $P\mathscr{ES}(V)$ par les applications : $a \to a \cdot \varrho_0(g) = \varrho(g)^* \cdot a \cdot \varrho(g)$. On écrira aussi parfois $a \cdot g$ au lieu de $a \cdot \varrho_0(g)$. Comme $\varrho(P)$ est formée d'endomorphismes self-adjoints positifs de déterminant un, l'application $g \to \varrho(g)^* \cdot \varrho(g)$ définit un plongement équivariant de $D = K \backslash G_R$ dans $P\mathscr{ES}(V)$, que nous noterons aussi ϱ_0. L'adhérence $\varrho_0(D)$ de D, qui sera aussi notée D_ϱ, est la *compactification de D associée à ϱ*. C'est donc un espace de transformations pour G_R. Satake [7] a montré que $D_\varrho - D$ est réunion d'un nombre fini d'orbites de G_R; de plus, chacune de ces orbites est fibrée en espaces symétriques, plongés dans des sous-variétés linéaires de $P\mathscr{ES}(V)$ comme D l'est dans $P\mathscr{ES}(V)$ [5], appelés les *composantes de la*

---

[5] A cela près cependant que ces plongements peuvent être définis à partir de représentations non irréductibles, mais en tout cas somme de représentations irréductibles équivalentes.

frontière de D_ϱ. Il sera aussi commode de considérer D comme une composante (impropre) de D_ϱ. Pour être complets, nous allons rappeler la description des composantes de la frontière.

3.2. On suppose $\varrho(A)$ mis sous forme diagonale. Les termes diagonaux de $\varrho(a)$ ($a \in A$) sont donc de la forme 1.5 (3) (avec $k = \boldsymbol{R}$). Tout point de la frontière est équivalent modulo K à un point de $\overline{\varrho_0(A)}$ ou même, puisque K contient des représentants du groupe de Weyl relatif $_{\boldsymbol{R}}W$, à un point de $\overline{\varrho_0(A^-)}$, où A^- désigne la chambre de Weyl négative (ensemble des points de A sur lesquels les racines positives ont des valeurs $\leqq 1$). Soit (a_n) une suite divergente de points de A^-. Comme on ne s'intéresse qu'à la limite de $\varrho(a_n)^*$. $\varrho(a_n)$ dans $P\mathscr{E}\mathscr{S}(V)$, on peut dans 1.5 (3) faire abstraction du facteur commun λ_ϱ; la limite de $\varrho(a_n)^*$. $\varrho(a_n)$ aura ses coordonnées dans V_λ nulles si et seulement si il existe $\alpha \in (\lambda)$ telle que $\lim \alpha(a_n) = 0$. De cela, et des résultats de 1.5, on déduit facilement ce qui suit :

Soit Ψ_ϱ l'ensemble des parties de $_{\boldsymbol{R}}\varDelta$ qui forment avec λ_ϱ un ensemble connexe. Alors $\overline{\varrho_0(A^-)}$ est contenue dans la réunion des espaces $P\mathscr{E}\mathscr{S}(V_\psi)$ ($\psi \in \Psi_\varrho$). L'image de $\mathscr{N}(V_\psi)_{\boldsymbol{R}}/\mathscr{Z}(V_\psi)_{\boldsymbol{R}}$ dans $P\mathscr{E}\mathscr{S}(V_\psi)$ par l'application $x \rightarrow \varrho(x)^*$. $\varrho(x)$ est l'espace symétrique F_ψ à courbure négative de $L_{\psi,\boldsymbol{R}}$. Les composantes de D_ϱ sont les transformés par $G_{\boldsymbol{R}}$ des espaces $F_\psi (\psi \in \Psi_\varrho)$. Si ψ est le plus grand élément de Ψ_ϱ pour la relation d'inclusion, alors $V_\psi = V$, et $F_\psi = D$. Si ψ est vide, alors $V_\psi = V_{\lambda_\varrho}$ et L_ψ est réduit à (e); la composante F_ψ est un point. Si $\psi \subset \psi'$ alors F_ψ fait partie de l'adhérence de $F_{\psi'}$.

3.3. Soit F une composante de D. On pose

$$\mathscr{N}(F) = \{\, g \in G_{\boldsymbol{R}} \mid \quad F \cdot \varrho_0(g) = F \},$$
$$\mathscr{Z}(F) = \{\, g \in \mathscr{N}(F) \mid \quad f \cdot \varrho_0(g) = f \quad (f \in F)\},$$

et $G(F) = \mathscr{N}(F)/\mathscr{Z}(F)$. Les groupes $\mathscr{N}(F)$ et $\mathscr{Z}(F)$ sont donc conjugués à des groupes $\mathscr{N}(V_\psi)_{\boldsymbol{R}}$ et $\mathscr{Z}(V_\psi)_{\boldsymbol{R}}$ respectivement ($\psi \in \Psi_\varrho$). Le groupe $G(F)$ est isogène à L_ψ. On a de plus

$$(2) \qquad \mathscr{N}(F) = \{\, g \in G_{\boldsymbol{R}} \mid \quad F \cdot \varrho_0(g) \cap F \neq \varnothing \,\}.$$

Deux composantes de D_ϱ sont dites être de même type si elles font partie de la même orbite de $G_{\boldsymbol{R}}$. Le groupe K opère transitivement sur l'ensemble des composantes d'un type donné F, lesquelles correspondent biunivoquement aux points de $G_{\boldsymbol{R}}/\mathscr{N}(F)$.

§ 4. Compactifications de D/Γ

Dans tout ce paragraphe, G est un groupe algébrique connexe semi-simple défini sur $\boldsymbol{Q}$, Γ un sous-groupe arithmétique de G, K un sous-groupe compact maximal de $G_{\boldsymbol{R}}$, et $D = K \backslash G_{\boldsymbol{R}}$.

4.1. Définition. Soit ϱ une représentation irréductible de $G_{\boldsymbol{R}}$. Une composante F de D_ϱ (cf. 3.2) est *rationnelle* si son normalisateur $\mathscr{N}(F)$ et son centralisateur $\mathscr{Z}(F)$ (cf. 3.3) sont des sous-groupes algébriques réels de $G_{\boldsymbol{R}}$ définis sur $\boldsymbol{Q}$.

4.2. Lemme. *Soient F une composante rationnelle de D_ϱ. Alors l'image $\Gamma(F)$ de $\mathscr{N}(F) \cap \Gamma$ dans $G(F)$ par la projection canonique est un groupe arithmétique.*

Le quotient $\mathscr{N}(F)/\mathscr{Z}(F)$ étant semi-simple, le groupe $\mathscr{N}(F)$ contient un sous-groupe semi-simple L défini sur $\boldsymbol{Q}$ tel que $L \cap \mathscr{Z}(F)$ soit fini et que $\mathscr{N}(F) = L \cdot \mathscr{Z}(F)$. La restriction de π : $\mathscr{N}(F) \to G(F)$ à L est donc une isogénie, définie sur $\boldsymbol{Q}$. D'autre part, il résulte de $[^3; 6.3, 6.11]$ que $\pi(L_{\boldsymbol{Z}})$ est un sous-groupe arithmétique de $G(F)$, et que $\mathscr{N}(F)_{\boldsymbol{Z}}$ est commensurable à $L_{\boldsymbol{Z}} \cdot \mathscr{Z}(F)_{\boldsymbol{Z}}$, d'où le lemme.

4.3. Théorème. *Soient ϱ une représentation rationnelle irréductible localement fidèle, définie sur $\boldsymbol{Q}$, de G et D_ϱ la compactification de D associée à ϱ (3.1). Soit Ω un ensemble fondamental pour Γ dans D, convenablement choisi. Alors $\overline{\varrho_0(\Omega)} = \Omega_\varrho$ est contenu dans la réunion d'un nombre fini de composantes rationnelles de D_ϱ. Le quotient $S_\varrho = \Omega_\varrho \cdot \Gamma/\Gamma$ est réunion de D/Γ et d'un nombre fini de quotients $F_i/\Gamma(F_i)$, où les F_i sont des composantes rationnelles de D_ϱ; il admet une topologie d'espace compact séparé induisant sur D/Γ et sur $F_i/\Gamma(F_i)$ la topologie naturelle, telle que D/Γ soit ouvert dans S_ϱ et que la projection $\Omega_\varrho \cdot \Gamma \to S_\varrho$ induise une application continue de Ω_ϱ sur S_ϱ.*

Esquisse de la démonstration $(^6)$: on reprend les notations

---

$(^6)$ Fortement inspirée de $[^8]$.

de 1.5, 2.2. λ_ϱ est donc le $\boldsymbol{Q}$-poids minimal de ϱ, et Θ_ϱ est l'ensemble des parties de $_{\boldsymbol{Q}}\varDelta$ qui forment avec λ_ϱ un ensemble connexe. Pour étudier les composantes de D_ϱ, on peut utiliser une décomposition d'Iwasawa $G_{\boldsymbol{R}} = K \cdot A \cdot N$ de $G_{\boldsymbol{R}}$ telle que $A \supset {}_{\boldsymbol{Q}}A$ et $N \supset {}_{\boldsymbol{Q}}N_{\boldsymbol{R}}$. Alors la restriction à $_{\boldsymbol{R}}A$ d'une $\boldsymbol{R}$-racine positive (pour l'ordre défini par N) est $\geqq 0$, et $_{\boldsymbol{Q}}A^- \subset A^-$. On en tire aisément que l'espace V_θ (1.5) peut aussi s'écrire V_ψ où $\psi \subset {}_{\boldsymbol{R}}\varDelta$ forme avec le $\boldsymbol{R}$-poids minimal de ϱ un ensemble connexe. Cela signifie, vu 3.2, que l'image F_θ de $\mathscr{N}(V_\theta)_{\boldsymbol{R}}/\mathscr{Z}(V_\theta)_{\boldsymbol{R}}$ dans $P\mathscr{E}\mathscr{S}(V_\theta)$ par ϱ_0 est une composante de D_ϱ. Comme on a supposé ϱ définie sur $\boldsymbol{Q}$, l'espace V_θ et les groupes $N(V_\theta)$, $Z(V_\theta)$ le sont aussi, F_θ est une composante rationnelle de D_ϱ, et $\Gamma(F_\theta)$ un sous-groupe arithmétique de $G(F_\theta)$.

Soit $\mathfrak{S}$ un domaine de Siegel généralisé de D, défini relativement à K, $_{\boldsymbol{Q}}T$, H (cf. 2.2). En supposant $\varrho(_{\boldsymbol{Q}}T)$ mis sous forme diagonale, et en utilisant 1.5, on voit sans peine,

$$(1) \qquad\qquad \varrho_0(\mathfrak{S}) \subset \bigcup\nolimits_{\theta \in \Theta_\varrho} F_\theta,$$

que $\varrho_0(\mathfrak{S}) \cap F_\theta$ ($\theta \in \Theta_\varrho$) est un domaine de Siegel généralisé de F_θ, défini relativement aux images de $K \cap \mathscr{N}(V_\theta)$, $_{\boldsymbol{Q}}T$ et H par la projection naturelle, et que tout sous-ensemble de F_θ de cette nature est dans l'adhérence de l'image par ϱ_0 d'un domaine de Siegel généralisé de D. Soit

$$\Omega = \bigcup\nolimits_{i=1}^{i=m} \mathfrak{S} \cdot x_i \ (x_i \in G_{\boldsymbol{Q}}),$$

un ensemble fondamental pour Γ dans D (cf. 2.4), $\mathfrak{S}$ étant donc supposé suffisamment grand. L'inclusion (1) entraîne :

$$(2) \qquad \Omega_\varrho = \overline{\varrho_0(\Omega)} \subset \bigcup\nolimits_{1 \leqslant i \leqslant m, \theta \in \Theta_\varrho} F_\theta \cdot \varrho_0(x_i),$$

ce qui établit la première assertion.

En appliquant 2.4 aux groupes d'automorphismes de $F_\theta \cdot \varrho_0(x_i)$, on montre ensuite, que si Ω est suffisamment grand, on a

$$(3) \qquad\qquad F_\theta \cdot \Gamma = (F_\theta \cap \overline{\varrho_0(\Omega)}) \cdot \Gamma.$$

Comme $F_\theta \cap \mathfrak{S}$ a la propriété de Siegel, on tire de cela l'existence d'un nombre fini d'éléments $\gamma_j (j \in J)$ de Γ ayant la propriété suivante : si $\gamma \in \Gamma$ est tel que $\Omega_\varrho \cdot \varrho_0(\gamma) \cap \Omega_\varrho \neq \varnothing$, alors il existe $j \in J$ tel que $x \cdot \varrho_0(\gamma) = x \cdot \varrho_0(\gamma_j)$ pour tout $x \in \Omega_\varrho \cap \Omega_\varrho \cdot \varrho_0(\gamma^{-1})$. On voit alors que l'on se trouve dans les conditions d'application du Théorème 1' de [8], ce qui entraine la deuxième assertion.

Il est clair que $\cup_i F_\theta \cdot \varrho_0(x_i \cdot \Gamma)$ est formée de composantes rationnelles de type F_θ. D'autre part, si $x \in G_Q$, alors, d'après la propriété de Siegel, $\Omega \cdot x$ est contenu dans la réunion d'un nombre *fini* de translatés de Ω par des éléments de Γ. Par conséquent, $\Omega_\varrho \cdot \varrho_0(x) \subset \Omega_\varrho \cdot \varrho_0(\Gamma)$, d'où l'on déduit que $\cup_i F_\theta \cdot \varrho_0(x_i \cdot \Gamma)$ est la réunion de *toutes* les composantes rationnelles de type F_θ, et notre troisième assertion.

4.4. THÉORÈME. *Soient $_QT$ un tore décomposé sur Q maximal de G, H un sous-groupe parabolique défini sur Q minimal de G contenant $_QT$, et supposons K adapté à T. Soit $\mathfrak{S}_{t,\eta,\omega}$ un domaine de Siegel généralisé de G_R. Une réunion finie $\Omega = \cup_{1 \leq i \leq m} \mathfrak{S}_{t,\eta,\omega} \cdot x_i$ de translatés de $\mathfrak{S}_{t,\eta,\omega}$ par des éléments $x_i \in G_Q$ est un ensemble fondamental pour Γ, pour t, η, ω assez grands, si et seulement si l'ensemble des x_i contient un représentant au moins des doubles classes $H_Q \backslash G_Q / \Gamma$.*

Cet énoncé implique en particulier que le nombre de doubles classes $H_Q \backslash G_Q / \Gamma$ est fini, et est égal à la borne inférieure du nombre de pointes d'un ensemble fondamental, du type 2.4, pour Γ dans G_R.

Pour la démonstration, on choisit tout d'abord une représentation irréductible ϱ, définie sur Q, localement fidèles de G, dont le Q-poids minimal λ_ϱ ne soit orthogonal à aucune Q-racine simple. L'existence de ϱ résulte de 1.6. Vu 1.5, si θ est l'ensemble vide, V_θ est l'espace propre correspondant à $\lambda_\varrho, \mathcal{N}(V_\theta) = H$, et $\mathcal{N}(V_\theta) \subset_Q N$. Il existe donc une composante rationnelle F de D_ϱ dont le normalisateur soit H_R. On voit alors, comme à la fin de la démonstration de 4.4, que si Ω est un ensemble fondamental suffisamment grand, $\cup_i F \cdot \varrho_0(x_i \cdot \Gamma)$ est la réunion de toutes les composantes rationnelles de type F, d'où immédiatement

$$(1) \qquad G_Q = \bigcup_{1 \leq i \leq m} H_Q \cdot x_i \cdot \Gamma,$$

ce qui montre que $\{x_i\}$ contient un représentant au moins de chaque double classe de G_Q modulo H_Q et Γ.

Soit réciproquement $\{y_j\}_{1 \leq j \leq q}$ un système de représentants de ces doubles classes. On peut donc écrire

$$x_i = h_i \cdot y_{j(i)} \cdot \gamma_i \ (h_i \in H_Q, \gamma_i \in \Gamma; i = 1, ..., m)$$

Comme $\mathfrak{S}_{t,\eta,\omega} \cdot h$ fait partie d'un domaine de Siegel généralisé pour tout $h \in H_R$, on voit que

$$(2) \qquad \bigcup_{1 \leqslant i \leqslant m} \mathfrak{S}_{t,\eta,\omega} \cdot x_i \subset \bigcup_{1 \leqslant j \leqslant q} \mathfrak{S}_{t',\eta',\omega'} \cdot y_j \cdot \Gamma,$$

pour t', η', ω' suffisamment grands, d'où le théorème.

4.5. COROLLAIRE. *Soit P un sous-groupe parabolique défini sur Q de G. Alors les points rationnels de G/P forment un nombre fini d'orbites de Γ.*

Soit H un sous-groupe parabolique défini sur Q minimal et contenu dans P. L'application canonique $(G/H)_Q \rightarrow (G/P)_Q$ est surjective, puisque $G_Q \rightarrow (G/P)_Q$ l'est (cf. 1.3). On est donc ramené au cas où $H = P$, qui a été traité dans 4.4.

4.6. *Remarques.* (1) En utilisant 1.6, on voit facilement que le corollaire 4.5 équivaut à l'assertion suivante : Soit $\varrho : G \rightarrow \boldsymbol{GL}(V)$ une représentation rationnelle définie sur Q de G. Soit L un réseau de V_Q invariant par Γ et X un cône de V tel que $X - 0$ soit une orbite de G. Alors les vecteurs primitifs de $L \cap X$ forment un nombre fini d'orbites de Γ.

Prenons en particulier pour G le groupe orthogonal d'une forme quadratique entière F non dégénérée, pour Γ le groupe des unités de F, et pour X le cône isotrope. $X - 0$ est une orbite de G d'après le théorème de Witt. On retrouve donc le fait, classique, que les vecteurs primitifs isotropes forment un nombre fini de classes d'équivalence modulo le groupe des unités de F.

(2) R. Godement m'a indiqué une démonstration plus directe de 4.5, basée sur le fait que le groupe adélique G_A attaché à G vérifie la condition (F) (cf. [2]).

4.7. En utilisant quelques propriétés des réseaux [5; §§ 9, 12], ou des groupes adéliques attachés à G et H, on peut dans certains cas donner d'autres expressions du nombre d'éléments de $H_Q \backslash G_Q / \Gamma$, que nous noterons ici $\nu(G, \Gamma)$, et qui est aussi d'après 4.5 le nombre minimal de pointes d'un ensemble fondamental. Nous nous bornerons aux exemples suivants, sans démonstration :

(a) Soit $G = 0(F)$, où F est la forme quadratique 2.5 (1), mais on l'on suppose de plus que les points entiers forment un réseau maximal pour F. Alors $\nu(G, G_Z)$ est égal au nombre de classes d'équivalence de plans isotropes maximaux définis sur Q modulo G_Z, ou au nombre de classes dans le genre de F_0.

(b) Soient k un corps de nombres, h le nombre de classes d'idéaux de k. Supposons que $G = R_{k/Q}G'$ s'obtienne par restriction des scalaires, de k à Q au sens de [11, Chap. I] à partir de $G'=SL_{n+1}, Sp_{2n}$, vu comme groupe défini sur k. Alors $v(G, G_Z)=h^n$. Si $k = Q$, on a une seule pointe, résultat bien connu, compte tenu du fait que les domaines de Siegel généralisés sont les domaines de Siegel au sens habituel. Si k est totalement réel, et si $G' = SL_2$, le groupe G_Z est un groupe de Hilbert-Blumenthal, et l'on retrouve l'analogue d'un résultat classique de H. Maass (Sitzungsberichte Heidelberg. Akad. Wiss., Math-naturwiss. Klasse 1940, 2. Abhandlung).

(c) Plus généralement que dans (b), supposons que $G=R_{k/Q}G'$, où G' est un groupe de type normal (ou «de Chevalley») sur k, c'est-à-dire possède un tore maximal décomposé sur k. Alors il est pratiquement certain que l'on a $v(G, G_Z) \leq h^r$, et $v(G,G_Z)=h^r$ si G' est simplement connexe, pour une réalisation matricielle convenable de G' [7], r étant le rang de G'. Si $k = Q$, cela signifie qu'un domaine de Siegel convenable est un ensemble fondamental, un résultat démontré antérieurement de manière très différente par Harish-Chandra, dans le cadre de sa Note sur les formes automorphes (Proc. Nat. Ac. Sci. U.S.A. 45 (1959), 570-573).

The Institute for advanced study, Princeton N.J. U.S.A.

[7] Il suffit probablement de partir d'une représentation matricielle de G' dans laquelle le réseau des points à coordonnées dans les entiers de k est admissible au sens de Chevalley (Sém. Bourbaki, Exp. 219, 1961). En fait, il suffirait ici de savoir que les représentations de SL_2 dans G' définies par les groupes radiciels sont données par des polynomes à coefficients entiers en les coefficients des éléments de SL_2.

BIBLIOGRAPHIE

[1] A. BOREL, «Groupes linéaires algébriques», *Ann. of Math.* **64** (1956), 20-80.

[2] ——, «Some proporties of the adele groups attached to algebraic groups», *Bull. Amer. Math. Soc.* **67** (1961), 583-585.

[3] A. BOREL and HARISH-CHANDRA, «Arithmetic subgroups of algebraic groups», *Ann. of Math.* **75** (1962), 485-535.

[4] C. CHEVALLEY, Séminaire sur la classification des groupes de Lie algébriques», Paris, 1958.

[5] M. EICHLER, Quadratische Formen und orthogonale Gruppen, Grundlehren d.m. Wiss. LXIII, Springer, 1952.

[6] R. GODEMENT, «Groupes linéaire algébriques sur un corps parfait», Sém. Bourbaki, Exp. 206, Décembre 1960.

[7] I. SATAKE, «On representations and compactifications of symmetric Riemannian spaces» *Ann. of Math.* **71** (1960), 77–110.

[8] ——, «On compactifications of the quotient spaces for arithmetically defined discontinuous groups», *ibid.* **72** (1960), 555-580.

[9] Séminaire H. Cartan 1957-58. Exposé II par A. Weil.

[10] J. TITS, Groupes semi-simples isotropes, ce Recueil, p. 137.

[11] A. WEIL, Adeles and algebraic groups. Notes, The Institute for advanced Study, Princeton 1961.

60.

Some finiteness properties of adele groups over number fields

Publ. Math., Inst. Hautes Etud. Sci. **16** (1963) 5–30

The formalism of adeles and ideles, introduced in algebraic number theory by Chevalley, has been recently applied to more general situations. In particular, it allows one to associate to an algebraic matric group G defined over a finite number field k a locally compact group G_A, the adele group of G, in which the group G_k of rational points over k of G is naturally imbedded as a discrete subgroup, in the same way as the multiplicative group k^* of non-zero elements of k is imbedded in the idele group I_k of k.

When $k = \mathbf{Q}$, the pair G_A, G_k may be viewed roughly as the global counterpart of the pair $G_\mathbf{R}$, $G_\mathbf{Z}$, where, as usual, $G_\mathbf{R}$ is the group of real matrices of G, and $G_\mathbf{Z}$ the group of integral matrices with determinant ± 1 contained in G (the units of G); and, in fact, the chief purpose of this work is to prove for G_A, G_k the analogues of the main results of [4] on $G_\mathbf{R}$ and $G_\mathbf{Z}$.

This paper is divided into eight paragraphs. The first one fixes the notation and collects some definitions and elementary facts pertaining to the adele groups. The second one contains some remarks on the double cosets modulo G_A^∞ and G_k, where G_A^∞ is the stability group of the lattice of integral points in the underlying space k^n, chiefly: relation with the ideal classes of k if $G = \mathbf{GL}_n$, with the classes in the genus of a rational quadratic form F, if G is the orthogonal group of F, with strong approximation, and behaviour in semi-direct products. § 3 reviews, reformulates in part, and strengthens slightly some results of [4].

In § 4 we construct, in much the same way as in [4], *fundamental sets* for G_k in G_A (i.e. subsets of G_A which intersect only finitely many of their right translates under G_k and which meet every left coset $x.G_k$), when G is connected. The case of non-connected groups is dealt with in 5.2.

§ 5 proves first that the number of distinct double cosets $G_A^\infty.x.G_k$ is finite. This makes it clear that G_A/G_k is of finite invariant measure, or is compact, if and only if G_∞/G_0 is so; here G_∞ is the product of the groups G_{k_v}, where k_v runs through the completions of k with respect to the archimedean absolute values, and G_0 is the group of units of G, (if $k = \mathbf{Q}$, then $G_\infty = G_\mathbf{R}$, $G_0 = G_\mathbf{Z}$). Theorem 12.3 of [4] then shows that G_A/G_k is of finite invariant volume if and only if the identity component G^0 of G has no non-trivial rational character over k, and is compact if and only if, moreover, every unipotent element of G_k belongs to the radical of G_k. In analogy with Theorem 6.9

305

of [4], it is also proved in § 5 that if G is reductive, H is a reductive subgroup of G, and $\sigma : G \rightarrow G/H$ the natural projection, then $\sigma(G_A) \cap (G/H)_k$ consists of finitely many orbits of G_k.

As a consequence of this last result and of some elementary facts about Galois cohomology, we show in § 6 that if G is reductive, connected, the principal homogeneous spaces over k of G which have rational points over all completions of k form finitely many isomorphism classes.

§ 7 is concerned with the double cosets of G_k modulo G_0 and H_k, where H is a parabolic subgroup (i.e. is such that G/H is a projective variety) over k. Their number is finite and, under suitable assumptions, equal to the number of double cosets $H_A^\infty . x . H_k$ in H_A.

Finally, § 8 extends to S-units and certain subgroups of G_A the results of [4] and of §§ 4, 5 of the present paper pertaining to fundamental sets and closed orbits in rational representations.

The main results of this paper have been announced in [1, 3]. Some of them have already been established for solvable groups in [11] and for certain classical groups in [14]. The compactness criterion for G_A/G_k mentioned above is the adele version of Godement's conjecture and has also been given another proof by G. D. Mostow and T. Tamagawa (*Annals of Math.*, 76 (1962), p. 446-463).

§ 1. Preliminaries.

In this paragraph, we fix the notation and recall some notions and facts about adele groups attached to linear algebraic groups over number fields. This is not meant as a self-contained introduction to the subject; for more details, see [14, Chap. I]. The notation of the sections 1.1 to 1.4 will be used throughout.

1.1. k is an algebraic number field of finite degree, $\mathfrak{o}$ the ring of integers of k, V the set of primes (or of equivalence classes of absolute values) of k, $P \subset V$ the set of finite primes of k, $x \rightarrow ||x||_v$ the normalized absolute value associated to $v \in V$, k_v the completion of k with respect to $||x||_v$, $\mathfrak{o}_\mathfrak{p}$ $(\mathfrak{p} \in P)$ the ring of p-adic integers of $k_\mathfrak{p}$, and A_k or A the ring of adeles of k. If $v \in V - P$ is an infinite (or archimedean) prime, we put $\mathfrak{o}_v = k_v$. If S is a subset of V, then $\mathfrak{o}(S) = \{x \in k \,|\, x \in \mathfrak{o}_v \,(v \in V - S)\}$. In particular, $\mathfrak{o}(S) = \mathfrak{o}$ if S is contained in $V - P$.

1.2. The letter G will always stand for an algebraic matric group over k, n for its degree, and G^0 for its identity component (see 1.11). Given a subring B of an overfield of k, G_B denotes the group of elements of G with coefficients in B and determinant invertible in B. We shall write G_v for G_{k_v}.

G_{A_k} or G_A is the adele group of G, and

$$G_U = \prod_{\mathfrak{p} \in P} G_{\mathfrak{o}_\mathfrak{p}},$$

$$G_{A(S)} = \prod_{v \in S} G_v \times \prod_{v \in V - S} G_{\mathfrak{o}_v}, \qquad\qquad (S \subset V; \; S \text{ finite}).$$

We recall that G_A is by definition a subgroup of the direct product of the G_v's. An element $x = (x_v)_{v \in V}$ of that product is an adele of G if and only if $x_v \in G_{o_v}$ for almost all (i.e. for all but a finite number of) v's. The group G_A contains therefore $G_{A(S)}$. The group G_v $(v \in V)$ is locally compact with respect to the v-adic topology, and G_{o_p} $(p \in P)$ is open and compact in G_p; $G_{A(S)}$ is locally compact and G_U is compact for the product topology. G_A itself is a locally compact group, once endowed with the inductive limit topology of the $G_{A(S)}$, where S runs through the finite subsets of V ordered by inclusion. $G_{A(S)}$ is open in G_A; every compact subset of G_A is contained in the union of finitely many translates of G_A^∞, hence belongs to some group $G_{A(S)}$.

For any $S \subset V$ we put $G_{A(S)} = \{g = (g_v) \in G_A \,|\, g_v \in G_{o_v} \ (v \in V - S)\}$. If S is finite, this definition of $G_{A(S)}$ coincides with the previous one. π_S will denote the projection of G_A in $\prod_{v \in S} G_v$, G_S the image of G_A, and j_S the natural inclusion of G_S in G_A as the subgroup of adeles whose components outside S are equal to e; often, we identify G_S with $j_S(G_S)$ and G_U with $j_P(G_U)$. We have $G_A \cong G_S \times G_{V-S}$. If S is finite, then $G_S = \prod_{v \in S} G_v$.

If $S = V - P$, we write $G_0, G_\infty, G_A^\infty, \pi_\infty, j_\infty$ for $G_{o(S)}, G_S, G_{A(S)}, \pi_S, j_S$.

To any element $a \in G_k$ corresponds an adele of G, all of whose components are equal to a, called a *principal adele* of G. When viewed as subgroups of G_A, the groups G_k, $G_{o(S)}$, G_0 will, unless otherwise said, be identified with groups of principal adeles. They are discrete and $G_{o(S)} = G_{A(S)} \cap G_k$, in particular $G_0 = G_A^\infty \cap G_k$. If $S = V$, $G_{A(S)} = G_A$, $G_{o(S)} = G_k$. If S is empty or contained in $V - P$, then $G_{A(S)} = G_A^\infty$, $G_{o(S)} = G_0$. The group G_0 is the group of units, and $G_{o(S)}$ the group of S-units of G.

The (possibly infinite) number of double cosets $G_A^\infty . x . G_k$ $(x \in G_A)$ in G_A will be denoted by $c(G)$.

1.3. A rational homomorphism $f : G \to G'$ over k of G into another algebraic matric group over k induces a continuous homomorphism $f_v : G_v \to G'_v$ mapping G_{o_p} into G'_{o_p} for almost all p's, and consequently also induces continuous homomorphisms $f_A : G_A \to G'_A$, $f_S : G_S \to G'_S$, which are isomorphisms if f is.

The rational homomorphisms over k of G into $\mathbf{GL}_1$, usually called the rational characters defined over k of G, form a commutative group, to be denoted $X_k(G)$. Each element $\chi \in X_k(G)$ defines a homomorphism of G_A into $\mathbf{GL}_{1,A}$. The latter group is just the idele group I_k of k.

1.4. When $k = \mathbf{Q}$, the primes of k are of course the rational prime numbers and an infinite prime ∞. As is well-known, and as will be seen repeatedly in this paper, the general case of a number field can in many instances be reduced to the case where $k = \mathbf{Q}$ by the use of the so-called restriction of the groundfield [14, Chap. I]. We recall that, once a vector-space basis (α_i) $(1 \leqslant i \leqslant d = [k : \mathbf{Q}])$ of K over $\mathbf{Q}$ is chosen, there is associated to G an algebraic matric group $R_{k/\mathbf{Q}}G = G'$ over $\mathbf{Q}$, of degree $n.d$, and a rational homomorphism $\mu : G' \to G$ over k such that

$$(1) \qquad \mu^0 = (\mu^{\sigma_1}, \ldots, \mu^{\sigma_d}) : G' \to G^{\sigma_1} \times \ldots \times G^{\sigma_d}$$

is an isomorphism over $\bar{k}$, where $\sigma_1, \ldots, \sigma_d$ are the distinct isomorphisms of k into $\mathbf{C}$. The pair (G', μ) is, up to isomorphism, independent of the choice of the basis (α_i), and can also, up to isomorphism, be characterized by a universal property *(loc. cit.)*. The map μ also induces isomorphisms

$$(2) \qquad G'_{\mathbf{Q}} \cong G_k; \qquad G'_{\mathbf{R}} \cong G_\infty; \qquad G'_{A_{\mathbf{Q}}} \cong G_{A_k},$$

and, if (α_i) is a module basis of $\mathfrak{o}$ over $\mathbf{Z}$, as we shall always assume, then

$$(3) \qquad G'_{\mathbf{Z}} \cong G_{\mathfrak{o}}, \qquad G'^{\infty}_{A_{\mathbf{Q}}} \cong G^{\infty}_{A_k}, \qquad G'_{U} \cong G_{U}.$$

1.5. There exists a rational homomorphism $\nu : G \to G'$ over k such that

$$\nu^0 = (\nu^{\sigma_1}, \ldots, \nu^{\sigma_d}) : G^{\sigma_1} \times \ldots \times G^{\sigma_d} \to G',$$

is the inverse of μ^0. In fact, if say σ_1 is the identity, the intersection G_1 of the kernels of the homomorphisms μ^{σ_i} $(i \neq 1)$ is stable under every isomorphism of $\mathbf{C}$ over k, hence is defined over k, and it is mapped isomorphically onto G by μ. The map ν is then just the inverse of the restriction of μ to G_1. It clearly has the following properties

$$(4) \qquad \mu \circ \nu = id. \quad \mu^{\sigma_i} \circ \nu(G) = (e) \qquad\qquad (i \neq 1),$$

$$(5) \qquad \prod_{1 \leq i \leq d} (\nu^{\sigma_i}(\mu^{\sigma_i}(x'))) = x' \qquad\qquad (x' \in G').$$

Proposition. — *The map* $\alpha : X_{\mathbf{Q}}(G') \to X_k(G)$ *defined by* $\chi \to \nu \circ \chi$ $(\chi \in X_{\mathbf{Q}}(G'))$ *is an isomorphism.*

Let $\chi \in X_k(G)$. Then $\mu \circ \chi \in X_k(G')$, therefore the product in $X_k(G')$ of the characters $(\mu \circ \chi)^{\sigma_i}$ is defined over $\mathbf{Q}$, whence a map $\beta : X_k(G) \to X_{\mathbf{Q}}(G')$. A routine verification based on (4), (5), which we omit, shows then that $\alpha \circ \beta$ and $\beta \circ \alpha$ are both the identity map.

1.6. *Proposition.* — *Let* $G = H.N$ *be the semi-direct product of a subgroup* H *and a normal subgroup* N, *both algebraic, over* k. *Then* N_A *is normal in* G_A, *and* G_A *is the semi-direct product of* H_A *and* N_A.

Let $g = (g_v)_{v \in V}$ be an element of G_A. Since G_v is the semi-direct product of H_v and N_v, we may write uniquely $g_v = h_v . n_v$ $(h_v \in H_v, n_v \in N_v)$. We have $h_v = f_v(g_v)$ where $f : G \to G/N = H$ is the natural projection, hence $(h_v)_{v \in V} \in H_A$, which implies $(n_v)_{v \in V} \in N_A$ and $G_A \subset H_A . N_A$. The proposition then follows readily.

1.7. *Proposition.* — *Let* G' *be an algebraic matric group over* k, *and* $\Phi : G_A \to G'_A$ *an isomorphism which maps* G_v *onto* G'_v *for every* $v \in V$. *Then* $\Phi(G_{A(S)})$ *is commensurable with* $G'_{A(S}$ *for every subset* S *of* V.

(We recall that two subgroups of a group are commensurable with each other if their intersection has finite index in both of them.)

The group $\Phi(G_{\mathfrak{o}_p})$ $(p \in P)$ is open and compact in G'_p, hence is commensurable with $G'_{\mathfrak{o}_p}$. The groups $\Phi(G_U)$ and $\Phi^{-1}(G'_U)$, being compact, are contained respectively in $G'_{A(T)}$ and $G_{A(T)}$ for a suitable finite subset T of V; hence $\Phi(G_{\mathfrak{o}_p}) = G'_{\mathfrak{o}_p}$ for $p \in V - T$, and the proposition.

Remark. — This proposition applies notably when $\Phi = f_A$, where f is an isomorphism over k of G onto G′, or when $G = G'$ is normal in an algebraic matric group H over k, and Φ is the restriction to G_A of an inner automorphism of H_A. In the former case, it shows that $f(G_{o(S)})$ is commensurable with $G'_{o(S)}$ for any $S \subset V$.

1.8. *Proposition.* — *Let* S *be a subset of* V *containing the infinite primes. Then* $\pi_S(G_{o(S)})$ *is a discrete subgroup of* G_S *and* $G_{A(S)}/G_{o(S)}$ *is a principal fibration over* $G_S/\pi_S(G_{o(S)})$, *with compact structural group* $\prod_{p \in V-S} G_{o_p}$.

Let $M = \prod_{p \in V-S} G_{o_p}$. The projection $\pi_S : G_A \to G_S$ induces a homomorphism of $G_{A(S)}$ onto G_S, with kernel M. Since M is compact, $\pi_S(G_{o(S)})$ is discrete. Moreover, we have obviously

$$(6) \qquad j_S \cdot \pi_S(G_{o(S)}) \cdot M = \pi_S^{-1}(\pi_S(G_{o(S)})) \cap G_{A(S)} = G_{o(S)} \cdot M.$$

The group $N = G_{o(S)} \cdot M$ is closed and, since M is normal in $G_{A(S)}$, the space $G_{A(S)}/G_{o(S)}$ is a principal fibration over

$$G_{A(S)}/N = (G_S \times M)/N \cong G_S/\pi_S(G_{o(S)}),$$

with structural group

$$N/G_{o(S)} \cong M/(M \cap G_{o(S)}) \cong M.$$

1.9. *Proposition.* — *The group* $G_A/(G^0)_A$ *is compact.*

We view G/G^0 as a o-dimensional cycle over k in $\mathbf{GL}_n/G^0$, and denote by σ the restriction to G of the natural projection of $\mathbf{GL}_n$ onto $\mathbf{GL}_n/G^0$. Let $x \in G/G^0$ and $X = \sigma^{-1}(x)$. We show first:

(*) There exists a finite subset S_x of V, such that if $v \notin S_x$, and X_v is not empty, then $X_{o_v} = G_{o_v} \cap X_v$ is not empty.

The field $k' = k(x)$ is algebraic, of finite degree over k. Let V′ the set of primes of k', S′ the set of finite primes of k', and o′ the ring of integers of k.

Since X is defined over k', and is irreducible, non-singular, there exists a finite subset S'_x of V′ such that X contains an integral $\mathfrak{P}$-adic point for every $\mathfrak{P} \notin S'_x$ [1]. We take then for S_x the set of primes of k which are divided by some element of S'_x. It is finite. Let now $p \in P$, $p \notin S_x$. If X_p is not empty, and $y \in X_p$, then $k(x) = k(\sigma(y)) \subset k_p$, hence $k' \subset k_p$. This implies the existence of $\mathfrak{P} \in P' \cap V'$ such that $\mathfrak{P}|p$ and $k'_{\mathfrak{P}} = k_p$, $o'_{\mathfrak{P}} = o_p$. By definition of S_x, we have $\mathfrak{P} \notin S'_x$, therefore X contains a point with coordinates in $o'_{\mathfrak{P}}$, hence in o_p.

For $p \in P$, let us choose for each $x \in G/G^0$ an element of $(\sigma^{-1}(x))_v$ if $(\sigma^{-1}(x))_v \neq \emptyset$, belonging to $(\sigma^{-1}(x))_{o_v}$ if the latter is not empty, and let $C_{(v)}$ the set of these elements. It is finite, and belongs to G_{o_v} if v is outside the union of the sets S_x $(x \in G/G^0)$, therefore

$$C = \{ g = (g_v) \in G_A \mid g_v \in C_{(v)} \ (v \in V) \}$$

is compact. Since clearly $G_A = C \cdot (G^0)_A$, the proposition is proved.

---

[1] See S. Lang, *Bull. Amer. Math. Soc.*, 66 (1960), p. 240-249, end of introduction.

1.10. Let W be a finite dimensional vector space over k. Then $W_A = W \otimes A$ is the adele space attached to W. More generally, an adele space X_A may be attached to any algebraic variety X defined over k [14, Chap. I]. However, this will occur only incidentally in this paper, and we refer directly to [14] for the relevant facts.

1.11. *Remark.* — The notion of algebraic matric group over k, of degree n, which is meant here, differs slightly in presentation, but of course not in substance, from that of " algebraic matric group defined over k ", as used for instance in [4], or in the author's paper on linear algebraic groups (*Annals of Math.*, 64 (1956), p. 20-80), where the matrix coefficients and the fields of definition are assumed to belong to some universal field Ω, given once and for all. This convention would not be consistent with the consideration of all the G_v's. It will be more convenient to say that an algebraic matric group H over k, of degree n, is given by an ideal $\mathfrak{I} \subset k[X_{11}, X_{12}, \ldots, X_{nn}]$ such that for some (and hence for every) algebraic closure $\bar{k}$ of k, the set $H_{\bar{k}}$ of elements of $\mathbf{GL}(n, \bar{k})$ whose coefficients annihilate $\mathfrak{I}$ is a group. H is said to be connected if $H_{\bar{k}}$ is connected in the Zariski topology. If B is as in 1.2, then H_B is the group of elements of $\mathbf{GL}(n, B)$ whose coefficients annihilate $\mathfrak{I}$. In agreement with this point of view, we shall write $\mathbf{GL}_n$, $\mathbf{SL}_n$ rather than $\mathbf{GL}(n, \Omega)$, $\mathbf{SL}(n, \Omega)$ and then use indifferently $\mathbf{GL}_{n,B}$ and $\mathbf{GL}(n, B)$ or $\mathbf{SL}_{n,B}$ and $\mathbf{SL}(n, B)$ [1].

An isomorphism class over k of such groups corresponds to an affine algebraic group over k, of which these groups are matrix realizations. Properties of, or notions relative to, algebraic matric groups over k which are invariant under isomorphisms over k belong in fact to the underlying affine algebraic group. This is the case for G_A, $R_{k/\mathbf{Q}}G$, or also, by 1.7, for the finiteness of $c(G)$, but *not* the case for G_o, G_A^∞ or the actual value of $c(G)$.

We leave it to the reader who feels the need for it to make similar adjustments in the few occasions where we shall have to consider affine or projective varieties.

1.12. It will be convenient to be allowed to consider sometimes rational representations over k and homogeneous spaces of a non-necessarily connected matric algebraic group. In the cases of interest in this paper, this does not present any difficulty but, there being seemingly no handy reference for it, we feel compelled to devote a few sentences to that question.

Let L be an algebraic matric group, of degree m, over a field F. The ring F[L] of regular functions, defined over k, over L, the " coordinate ring " over F of L, is generated by 1, the matrix coefficients and $(\det x)^{-1}$ $(x \in L)$. A rational representation $\rho : L \to \mathbf{GL}_q$ is over F is the coefficients of $\rho(x)$ $(x \in L)$ belong to F[L].

Let M be an algebraic subgroup over F of L. Then $M \backslash L$ will be identified with its image in $M \backslash \mathbf{GL}_m$. The regular functions over F on $M \backslash L$ may be identified with the ring $F[L]^M$ of elements of F[L] which are invariant under left

---

[1] This way of introducing algebraic matric groups, which in a way brings us back to the pre-universal field days, was suggested to me by P. CARTIER; see his paper in the Proceedings of the *Colloque sur la Théorie des Groupes Algébriques*, Bruxelles, 1962 (to appear) for a more complete discussion, and a comparison with the point of view of schemes.

translations by elements of M. If $M\backslash L$ is an affine algebraic set, this determines completely its structure over F. In particular, if $\rho : L \to \mathbf{GL}_q$ is a right rational representation over F, and $w \in F^q$ a point whose orbit X is closed and whose isotropy group is M, then $g \to x . \rho(g)$ induces an isomorphism over F of $M\backslash L$ onto X.

1.13. Let F be a field of characteristic zero, L an algebraic matric group over F. For the sake of reference, we recall that if L_F is Zariski dense in L, in particular [12, Theorem 3] if L is connected, then $L = H . N$ is the semi-direct product of its unipotent radical N by a reductive algebraic subgroup H over F. This globalization of a known fact on algebraic Lie algebras is due to Mostow (*Amer. J. M.*, *78* (1956), 200-221, Theorem 6), see also G. Hochschild (*Illinois J. M.*, *5* (1961), 492-519, Section 3).

Remark (added in proof). — In fact, the preceding result is also valid if L is not connected (see [5]). This allows some simplifications in the sequel. In particular, in 3.6, the second part of the proof, from « In the general case » on, is superfluous ; in 4.6, the proof given is also valid in the non-connected case.

§ 2. The double cosets modulo G_A^∞ and G_k.

This paragraph is devoted to some simple examples and remarks concerning the double cosets $G_A^\infty . x . G_k$. In the sequel, 2.3 will not be used and 2.2 will be needed only when $k = \mathbf{Q}$.

2.1. In this section, we recall some facts on lattices, of which Proposition 2.2 will be an easy consequence.

Let X be a lattice in k^n. Then $\mathfrak{o}_\mathfrak{p} . X = X_\mathfrak{p}$ $(\mathfrak{p} \in P)$ is a lattice in $k_\mathfrak{p}^n$, and $X_\mathfrak{p} = \mathfrak{o}_\mathfrak{p}^n$ for almost all $\mathfrak{p}$'s. Conversely, given lattices $X'_\mathfrak{p}$ $(\mathfrak{p} \in P)$ such that $X'_\mathfrak{p} = \mathfrak{o}_\mathfrak{p}^n$ for almost all $\mathfrak{p}$, then $X' = \bigcap_{\mathfrak{p} \in P} X'_\mathfrak{p}$ is a lattice in k^n and we have $(X')_\mathfrak{p} = X'_\mathfrak{p}$ for all $\mathfrak{p} \in P$.

A lattice X in k^n is isomorphic to the direct sum of $\mathfrak{o}^{n-1}$ and a fractionary ideal of k, whose ideal class $\gamma(X)$ depends only on X. Two lattices X, X' are isomorphic if and only if $\gamma(X) = \gamma(X')$. (For all this, see e.g. [6, §§ 12, 13]).

To an ordered pair X, X' of lattices in k^n (resp. in $k_\mathfrak{p}^n$ $(\mathfrak{p} \in P)$), there is associated a fractionary ideal $\chi(X, X')$ of k (resp. $k_\mathfrak{p}$) [13, Chap. III, § 1]. If for example $k = \mathbf{Q}$ and $X' \subset X$, then $\chi(X, X')$ is the ideal generated by the index of X' in X. If $u \in \mathbf{GL}(n, k)$, then $\chi(X, u(X))$ is the ideal generated by $\det u$ (*loc. cit.*, Prop. 2, p. 58). Since $\chi(\mathfrak{o}, \mathfrak{a}) = \mathfrak{a}$ for any ideal $\mathfrak{a}$ of $\mathfrak{o}$ [13, p. 27], the above and the formal properties of $\chi(X, X')$ [13, Prop. 1, p. 58] imply that the ideal class of $\chi(\mathfrak{o}^n, X)$ is equal to $\gamma(X)$. We note finally that

$$\chi(X, X')_\mathfrak{p} = \chi(X_\mathfrak{p}, X'_\mathfrak{p}) \qquad\qquad (\mathfrak{p} \in P),$$

as follows directly from the definition of $\chi(X, X')$.

2.2. *Proposition.* — *The number $c(G)$ is equal to the class number of k if $G = \mathbf{GL}_n$, to one if $G = \mathbf{SL}_n$.*

Let $G = \mathbf{GL}_n$, and $g = (g_v) \in G_A$. By 2.1, given a lattice X in k^n, then $g(X) = \bigcap_{\mathfrak{p} \in P} g_\mathfrak{p}(X_\mathfrak{p})$ is a lattice, and G_A operates in this way transitively on the set of all lattices in k^n. The isotropy group of $\mathfrak{o}^n$ is just G_A^∞, hence the orbits of G_k in G_A/G_A^∞ are in $1-1$ correspondence with the isomorphism classes of lattices in k^n and therefore, by 2.1, with the ideal classes of k.

Let $G=\mathbf{SL}_n$, $g=(g_v)\in G_A$, and $X=g(\mathfrak{o}^n)$. Then

$$\chi(\mathfrak{o}_\mathfrak{p}^n, X_\mathfrak{p}) = (\det g_\mathfrak{p}) = \mathfrak{o}_\mathfrak{p}$$

for every $\mathfrak{p}\in P$, hence $\chi(\mathfrak{o}^n, X)=\mathfrak{o}$. Since $(\chi(\mathfrak{o}^n, X))=\gamma(X)$, there exists then $u\in\mathbf{GL}(n, k)$ such that $X=u(\mathfrak{o}^n)$, and since $(\det u)=\chi(\mathfrak{o}^n, X)$, the determinant of u is a unit of $\mathfrak{o}$. Multiplying u by an element of $\mathbf{GL}(n, \mathfrak{o})$ whose determinant is the inverse of $\det u$, we get an element $u'\in\mathbf{SL}(n, k)$ which brings $\mathfrak{o}^n$ onto X. Thus $\mathbf{SL}(n, k)(\mathfrak{o}^n)=\mathbf{SL}_{n, A}(\mathfrak{o}^n)$, which proves our assertion.

2.3. *Proposition. — Let $k=\mathbf{Q}$, F be a non-degenerate quadratic form on $\mathbf{Q}^n$, and $G=O(F)$ the orthogonal group of F. Then the elements of $G_A^\infty\backslash G_A/G_\mathbf{Q}$ are in $1-1$ correspondence with the classes in the genus of F.*

As is usual, we shall write $L[M]$ for $^tM.L.M$, where L, M are two $n\times n$ matrices.

Let S, T be two rational quadratic forms on $\mathbf{Q}^n$. We recall that S, T belong to the same class if there exist matrices $B, C\in\mathbf{GL}(n, \mathbf{Z})$ such that

$$(1)\qquad\qquad\qquad T=S[B], \quad S=T[C],$$

and that S, T belong to the same genus if for every prime v of $\mathbf{Q}$ there exist matrices $B_v, C_v\in\mathbf{GL}(n, \mathfrak{o}_v)$ such that

$$(2)\qquad\qquad\qquad T=S[B_v], \quad S=T[C_v].$$

In the latter case, the quotient $\det S/\det T$ is >0 and is a p-adic unit for every finite prime p, hence $\det S=\det T$, and $\det B_v=\pm\det C_v=\pm 1$ for every v. Using the fact that a p-adic quadratic form always has a unit of determinant -1, we see that if S, T are in the same genus, then (2) has solutions $B_v, C_v\in\mathbf{SL}(n, \mathbf{Q}_v)$.

Let now $U\in G_A$. By 2.2, we may write

$$(3)\qquad U=M.N^{-1} (M=(M_v)\in\mathbf{SL}_{n, A}^\infty), \quad N\in\mathbf{GL}(n, \mathbf{Q}), \qquad \det N=\pm 1.$$

We have then $F[N]=F[M_v]$, therefore $T=F[N]$ is a rational form in the genus of F. A straightforward verification shows that the class of T does not change if we use another decomposition of U similar to (3) or if we replace U by an element of $G_A^\infty.U.G_\mathbf{Q}$. Therefore $U\to T$ defines a map Φ of $G_A^\infty\backslash G_A/G_\mathbf{Q}$ into the set of classes in the genus of F. There remains to show that Φ is bijective.

Let T be as above, and $T'=F[M_v']=F[N']$ be obtained similarly from F. It T' belongs to the class of T, then there exists $L\in\mathbf{GL}(n, \mathbf{Z})$ such that $T=T'[L]$; we have

$$F[N']=F[N.L], \quad F[M_v']=F[M_v.L],$$

hence

$$M'\in G_A^\infty.M.L, \quad N'\in G_\mathbf{Q}.N.L, \quad M'.N'^{-1}\in G_A^\infty.M.N^{-1}.G_\mathbf{Q},$$

which shows that Φ is injective. Let now T be in the genus of F, and $M=(M_v)\in\mathbf{SL}(n, A)^\infty$ be such that $T=F[M_v]$ for every v. By Hasse's theorem, there exists $N\in\mathbf{GL}(n, \mathbf{Q})$

such that $T = F[N]$.   We have then $M.N^{-1} \in G_A$, $\det N = \pm 1$, and the class of T belongs to the image of Φ.

2.4.   *Proposition. — Assume that $G = H.N$ is the semi-direct product of a subgroup H and a normal subgroup N, both algebraic, over k, and that $c(N) = 1$.   Then $c(G)$ is finite if $c(H)$ so is, and is equal to one if $c(H)$ so is.*

Let $(x_i)_{i \in I}$ be a set of representatives of the double cosets $H_A^\infty.x.H_k$ $(x \in H_A)$.   We have $G_A = H_A.N_A$ by 1.5, and $N_A = N_A^\infty.N_k$ by assumption, whence

$$G = \bigcup_{i \in I} H_A^\infty.x_i.H_k.N_A = \bigcup_{i \in I} H_A^\infty.x_i.N_A^\infty.x_i^{-1}.x_i.G_k.$$

Since $x_i.N_A^\infty.x_i^{-1}$ is commensurable with N_A^∞ (1.7), there exist finitely many elements $y_{ij} \in N_A$ such that $x_i.N_A^\infty.x_i^{-1} \subset \bigcup_j N_A^\infty.y_{ij}$.   The set of products $(y_{ij}.x_i)$ contains then representatives of all double cosets $G_A^\infty.x.G_k$, which proves the first part of 2.4.   If moreover $c(H) = 1$, then I has one element, we may assume $x_i = e$, whence $y_{ij} = e$, and $c(G) = 1$.

2.5.   *Corollary. — If G is unipotent, then $c(G) = 1$.*

This follows from the approximation theorem if $\dim G = 1$, and from 2.4 by induction in the general case.

2.6.   The group G is said to have the *strong approximation property* if $G_{V-P}.G_k$ is dense in G_A [8, 9].   This property is clearly invariant under isomorphisms over k.

Let G have the strong approximation property.   If U is an open subgroup of G_A containing G_{V-P}, then $U.G_k = G_A$, for, given $g \in G_A$, we have $U.g \cap G_{V-P}.G_k \neq \emptyset$, whence $g \in U.G_{V-P}.G_k = U.G_k$.   In particular, $G_A = G_A^\infty.G_k$, and $c(G) = 1$.   This applies e.g. when $G = \mathbf{SL}_n$ or G is unipotent [8, 9].

2.7.   *Proposition. — Let $G = H.N$ be the semi-direct product of a subgroup H and a normal subgroup N, both algebraic over k.   Assume that N has the strong approximation property. Then every double coset $G_A^\infty.x.G_k\,(x \in G_A)$ intersects H_A, hence $c(G) \leqq c(H)$.   If moreover $G_A^\infty = H_A^\infty.N_A^\infty$, then $c(G) = c(H)$.*

For any $x \in G_A$, the group $x^{-1}.N_A^\infty.x$ is open in N_A and contains N_{V-P}, therefore, by 2.6, we have $N_A = N_A^\infty.N_k = x^{-1}.N_A^\infty.x.N_k$, hence also

$$(1) \qquad\qquad x.N_A^\infty.N_k = N_A^\infty.x.N_k \qquad\qquad (x \in G_A).$$

Let now $(x_i)_{i \in I}$ be a set of representatives of $H_A^\infty \backslash H_A / H_k$.   We have

$$G_A = \bigcup_i H_A^\infty.x_i.H_k.N_A = \bigcup_i H_A^\infty.x_i.N_A^\infty.N_k.H_k = \bigcup_i H_A^\infty.x_i.N_A^\infty.G_k,$$

and, using (1)

$$G_A = \bigcup_i H_A^\infty.N_A^\infty.x_i.G_k = \bigcup_i G_A^\infty.x_i.G_k,$$

which shows that $(x_i)_{i \in I}$ intersects each double coset $G_A^\infty.x.G_k$, and proves the first assertion.

Assume now that $G_A^\infty = H_A^\infty . N_A^\infty$, and let $x, y \in H_A$ be such that $x \in G_A^\infty . y . G_k$. Then we may write

$$x = a.b.y.u.v \qquad (a \in H_A^\infty,\ b \in N_A^\infty,\ u \in N_k,\ v \in H_k).$$

The element $x' = a^{-1}.x.v^{-1}$ of H_A belongs to the same double coset mod H_A^∞, H_k as x, and is also equal to $b.y.u$, hence, by (1)

$$x' = y.b'.u' \qquad (b' \in N_A^\infty,\ u' \in N_k).$$

Since G_A is the semi-direct product of H_A and N_A, this yields $x' = y$, and our second assertion.

§ 3. **Siegel domains in $\mathbf{GL}_n$ and fundamental sets in G_∞.**

In this paragraph G_0 is identified with $\pi_\infty(G_0)$.

3.1. A subset B of G_∞ is a fundamental set for G_0 if it satisfies the following conditions

$(F\ 0)_\infty$: $K.B = B$ for some maximal compact subgroup K of G.

$(F\ 1)_\infty$: $B.G_0 = G_\infty$.

$(F\ 2)_\infty$: For any $a, b \in G_k$, the set of $x \in G_0$ such that $B.a.x.b \cap B \neq \varnothing$ is finite.

The condition $(F\ 0)_\infty$ has been included chiefly to remain in agreement with [2, 4], but will play no role here. Since G_0 is discrete in G_∞, $(F\ 2)_\infty$ is certainly true if B is relatively compact. Therefore compact, or open and relatively compact, fundamental sets always exist when G_∞ / G_0 is compact.

It is clear that the condition $(F\ 2)_\infty$ is not changed if x is allowed to run through any finite union of right or left cosets of G_0 in G_k. For future use, we mention one such apparent strengthening of $(F\ 2)_\infty$:

Lemma. — The condition $(F\ 2)_\infty$ for a subset B of G_∞ is equivalent to:

$(F\ 2)'_\infty$: *For any $a, b \in G_k$, and any non-zero algebraic integer $r \in \mathfrak{o}$, the set of* $x \in G_r = \{ g \in G_k \mid r.g,\ r.g^{-1} \in \mathbf{M}(n, \mathfrak{o}) \}$ *such that $B.a.x.b \cap B \neq \varnothing$ is finite.*

Let $x \in G_r$. Then

$$(r).\mathfrak{o}^n \subset x(\mathfrak{o}^n) \subset (r^{-1}).\mathfrak{o}^n.$$

But there are only finitely many lattices between two lattices $\Gamma \subset \Gamma'$ of k^n (since Γ'/Γ is a finite group); therefore G_r is the union of finitely many left cosets modulo the isotropy group of $\mathfrak{o}^n$ in G_k, that is modulo G_0, and G_r is of course equal to G_0 if $r = 1$, whence the lemma.

3.2. The following property of a subset $B \subset G_\infty$, relatively to a given algebraic subgroup H over k of G, plays a considerable role in [4], although it is not stated explicitly there:

$(F\ 3)_\infty$: For any rational integer $m \geq 1$, any rational (right) representation $\rho : G \to \mathbf{GL}_m$ over k, any point $w \in k^m$ with a closed orbit and isotropy group H, and any lattice $\Gamma \subset k^m$, the intersection $w.\rho(B) \cap \Gamma$ is finite.

In view of the properties of $R_{k/\mathbf{Q}}$, it is clear that if $(F\,3)_\infty$ is true for $R_{k/\mathbf{Q}}G$, $R_{k/\mathbf{Q}}H$, then it is true for G, H.

3.3.   The standard Siegel domain $\mathfrak{S}_{t,u}$ $(t, u \in \mathbf{R}, t, u > 0)$ of $\mathbf{GL}(n, \mathbf{R})$ is by definition the set of products $k.a.n$ where $k \in \mathbf{O}(n)$, $a = \mathrm{diag}\,(a_1, \ldots, a_n)$ $(a_i \leqslant t.a_{i+1}; i = 1, \ldots, n-1)$ and $n = (n_{ij})$ is upper triangular, with ones in the diagonal, and $|n_{ij}| \leqslant u$ $(i < j)$, where the bounds are sufficiently large so that $\mathbf{GL}(n, \mathbf{R}) = \mathfrak{S}_{t,u}.\mathbf{SL}(n, \mathbf{Z})$, say $t > 4/3, u > \dfrac{1}{2}$ (see [4, § 4.5] for references). It has the property $(F\,2)_\infty$ by a well-known theorem of Siegel recalled in [4, § 4.5] and is therefore a fundamental set for $\mathbf{GL}\,(n, \mathbf{Z})$ in $\mathbf{GL}(n, \mathbf{R})$.

3.4.   *Lemma.* — *Let* $k = \mathbf{Q}$, $\mathfrak{S}$ *a standard Siegel domain of* $\mathbf{GL}(n, \mathbf{R})$, I *a finite subset of* $\mathbf{GL}(n, \mathbf{Q})$. *Let* G *be a reductive group,* $a \in \mathbf{SL}(n, \mathbf{R})$ *such that* $a.G_{\mathbf{R}}.a^{-1}$ *is self-adjoint* [1], *and* $B = \bigcup_{c \in I} (a^{-1}.\mathfrak{S}.c \cap G_{\mathbf{R}})$. *Then* B *has property* $(F\,2)_\infty$. *If* H *is an algebraic subgroup over* k *and* $a.H_{\mathbf{R}}.a^{-1}$ *is also self-adjoint, then* B *has property* $(F\,3)_\infty$ *with respect to* H.

Property $(F\,2)_\infty$ follows from the facts recalled in 3.3. As to $(F\,3)_\infty$, 3.4 is in this case chiefly a restatement of results proved in [4]. In fact, an easy commensurability argument, as given at the end of 6.9 in [4, p. 511], shows that it is enough to prove the existence of one rational representation $\rho' : G \to \mathbf{GL}_m$ over $\mathbf{Q}$ and of one point $w' \in \mathbf{Q}^m$ with closed orbit X' and isotropy group H such that $w'.\rho'(B) \cap \Gamma'$ is finite for any lattice $\Gamma' \subset \mathbf{Q}^m$. We start then with a representation $\rho' : \mathbf{GL}_n \to \mathbf{GL}_m$ over $\mathbf{Q}$ for which there exists $w' \in k^m$ whose orbit under $\mathbf{GL}_n$ is closed and whose isotropy group in $\mathbf{GL}_n$ is H. This exists by [4, 3.8], taking into account the fact that, $a.H_{\mathbf{R}}.a^{-1}$ being self-adjoint, H is reductive [4, § 1]. The orbit of w under G is then also closed, and it is *a fortiori* enough to show that $w'.\rho'(a^{-1}.\mathfrak{S}'.b) \cap \Gamma'$ is finite for any lattice $\Gamma' \subset \mathbf{Q}^m$ and any $b \in \mathbf{GL}(m, \mathbf{Q})$. This amounts to proving the finiteness of $w''.\rho'(\mathfrak{S}) \cap \Gamma'.\rho'(b^{-1})$, where $w'' = w'.\rho(a^{-1})$. Since $\Gamma'.\rho'(b^{-1})$ is also a lattice in $\mathbf{Q}^m$, and the isotropy group $a.H_{\mathbf{R}}.a^{-1}$ of w'' in $\mathbf{GL}(n, \mathbf{R})$ is self-adjoint, the finiteness in question is a consequence of [4, 5.4].

3.5.   It follows from 3.4 and [4] that if G is reductive, and H an algebraic subgroup over k of G, then G_∞ has open or closed fundamental sets which also verify $(F\,3)_\infty$.

In fact, using the restriction of the ground field, we may assume $k = \mathbf{Q}$. If $(F\,3)_\infty$ is not an empty condition relatively to H, then $H\backslash G$ admits a realization as a closed orbit, hence H is reductive [4, 3.8], and there exists $a \in \mathbf{SL}(n, \mathbf{R})$ such that $a.G_{\mathbf{R}}.a^{-1}$ and $a.H_{\mathbf{R}}.a^{-1}$ are self-adjoint (by a result of Mostow, also proved in [4, § 1]). Then, by [4, 6.5], $G_\infty = G_{\mathbf{R}} = B.G_{\mathbf{Z}}$, where B is as in 3.4, and satisfies therefore all our conditions.

3.6.   Similarly, we may strengthen slightly [4, 12.3] and assert that G_∞ contains closed or open fundamental sets.

---

(1) As in [4], a subgroup of $\mathbf{GL}(n, \mathbf{R})$ is said to be self-adjoint if it is invariant under the map $g \to {}^t\!g$, where ${}^t\!g$ is the transpose matrix of g.

If G is unipotent, then G_∞/G_0 is compact, and G_∞ has compact, or open and relatively compact, fundamental sets (see 3.1).

If G_k is Zariski-dense in G, we use the decomposition $G = H.N$ of 1.13. In view of the above, it suffices to show that if $B \subset H_\infty$ and $C \subset N_\infty$, with C relatively compact, are fundamental sets for H_0 and N_0 in H_∞ and N_∞ respectively, then B.C is a fundamental set for G_0 in G_∞. For $(F\ 0)_\infty$, this is clear because the maximal compact subgroups of H_∞ are also maximal compact in G_∞; as to $(F\ 1)_\infty$, this follows from

$$G_\infty = B.H_0.N_\infty = B.N_\infty H_0 = B.C.N_0.H_0.$$

The condition $(F\ 2)_\infty$ is invariant under a rational change of coordinates, since the latter replaces G_0 by a commensurable group (1.7, Remark). We may therefore assume that k^n is the direct sum of subspaces which are stable under H_k and acted upon trivially by N_k, and that N_k is upper triangular. Then $G_0 = H_0.N_0$, and the proof given in [4, 6.12] holds here too.

In the general case, let G_1 be the subgroup of G formed by all connected components of G which intersect G_k. By the theorem of Rosenlicht [12, Theorem 3], G_{1k} is Zariski dense in G_1. By the above, there exists an open or closed fundamental set Ω_1 for G_{10} in $G_{1\infty}$. We denote by K_1 a maximal compact subgroup of $G_{1\infty}$ such that $K_1.\Omega_1 = \Omega_1$.

Let (a_i) $(1 \leq i \leq s)$ be a set of representatives in G_∞ for the cosets $G_\infty/G_{1\infty}$. Then $\Omega = \bigcup_{1 \leq i \leq m} a_i.\Omega_1$ obviously verifies $(F\ 1)_\infty$. If now $\Omega.axb \cap \Omega \neq \varnothing$ $(a, b \in G_k, x \in G_0)$, then there exist i, j $(1 \leq i, j \leq m)$ such that

$$a_i\Omega_1.a.x.b. \cap a_j\Omega_1 \neq \varnothing.$$

By definition, $G_{1\infty}$ contains Ω_1, a, x, b, therefore a_i and a_j are in the same coset modulo $G_{1\infty}$, hence $i = j$, $\Omega_1 a.x.b \cap \Omega_1 \neq \varnothing$, and the possible x's are finite in number. Thus Ω verifies $(F\ 2)_\infty$.

In order to have Ω verify $(F\ 0)_\infty$ also, we take a maximal compact subgroup K of G_∞ containing K_1. Then K intersects all connected components (usual topology) of G_∞, hence we may take the $a_i's$ in K. We have then $K = \bigcup_i a_i K_1$,

$$\Omega = \bigcup_i a_i.\Omega_1 = \bigcup_i a_i.K_1.\Omega_1 = K.\Omega_1,$$

and therefore $K.\Omega = \Omega$.

Remark. — In the last part of the proof we have used the fact that the usual properties of maximal compact subgroups in a connected Lie group are also true in a Lie group L with finitely many connected components, in particular: every compact subgroup is contained in a maximal compact one, the maximal compact subgroups are conjugate by inner automorphisms, and the quotient of L by one of them is homeomorphic to a euclidean space (see G. D. Mostow, *Annals of Math.*, 62 (1955), p. 44-55).

§ 4. Fundamental sets for G_k in G_A.

4.1. *Definition.* — A subset $\Omega \subset G_A$ is a *fundamental set* for G_k if it verifies the two conditions:

$$(\mathrm{F}\ 1)_A : G_A = \Omega . G_k,$$

$$(\mathrm{F}\ 2)_A : \Omega^{-1} . \Omega \cap G_k \quad \text{is finite.}$$

Since G_k is discrete, $(\mathrm{F}\ 2)_A$ is true for any relatively compact set Ω; therefore, if G_A/G_k is compact, there always exist compact, or open and relatively compact, fundamental sets.

4.2. In analogy with 3.2, we also introduce the following condition for a subset Ω of G_A, relatively to a given algebraic subgroup H over k of G.

$(\mathrm{F}\ 3)_A$: For any rational integer $m \geq 1$, any rational representation $\rho : G \to \mathbf{GL}_m$ over k, and any $w \in k^m$ with closed orbit and isotropy group H, the set $w . \rho_A(\Omega) \cap k^m$ is finite.

4.3. *Lemma.* — *Let* B *be a subset of* G_∞ *and* C *a relatively compact subset of* G_P.

a) *If* B *verifies* $(\mathrm{F}\ 2)_\infty$, *then* B.C *verifies* $(\mathrm{F}\ 2)_A$.

b) *If* B *verifies* $(\mathrm{F}\ 3)_\infty$ *relatively to a subgroup* H, *then* B.C *verifies* $(\mathrm{F}\ 3)_A$ *relatively to* H.

Let $x \in G_k \cap (\mathrm{BC})^{-1} . \mathrm{BC}$. Then, for any $\mathfrak{p} \in P$, the elements x and x^{-1} belong to the relatively compact sets $\pi_\mathfrak{p}(\mathrm{C}^{-1} . \mathrm{C})$ and $\pi_\mathfrak{p}(\mathrm{C} . \mathrm{C}^{-1})$ respectively. The basic properties of the adele topology (see § 1) show then the existence of $r \in \mathfrak{o}, r \neq 0$, such that $rx, rx^{-1} \in \mathbf{M}(n, \mathfrak{o}_\mathfrak{p})$ for every $\mathfrak{p} \in P$, hence such that $rx, rx^{-1} \in \mathbf{M}(n, \mathfrak{o})$. For the components of infinity, we have $\mathrm{B} . x \cap \mathrm{B} \neq \emptyset$, whence our assertion (see 3.1).

The proof of *b)* is quite similar. Let ρ, w be as in $(\mathrm{F}\ 3)_A$ and $x \in w . \rho_A(\mathrm{B} . \mathrm{C}) \cap k^m$. For every $\mathfrak{p} \in P$, the element x belongs to the relatively compact set $\pi_\mathfrak{p}(w . \rho(\mathrm{C}))$; there exists then $r \in \mathfrak{o}, r \neq 0$, such that $r . x \in \mathfrak{o}_\mathfrak{p}^m$ for every $\mathfrak{p} \in P$, therefore such that $r . x \in \mathfrak{o}^m$. Thus

$$x \in w . \rho(\mathrm{B} . \mathrm{C}) \cap k^m \Rightarrow x \in \Gamma,$$

where $\Gamma = (r^{-1}) . \mathfrak{o}^m$ is a lattice in k^m. Projecting at infinity, we get $x \in w . \rho_\infty(\mathrm{B}) \cap \Gamma$, which is finite by assumption.

Remark. — The proof of $(\mathrm{F}\ 3)_\infty \to (\mathrm{F}\ 3)_A$ shows in fact that if B verifies $(\mathrm{F}\ 3)_\infty$ for one representation ρ, one point $w \in k^m$, and any lattice $\Gamma \subset k^m$, then B.C verifies $(\mathrm{F}\ 3)_A$ for ρ and w.

4.4. *Lemma.* — *Let* $k = \mathbf{Q}$, $G = \mathbf{GL}_n$, *and* $\mathfrak{S}$ *a standard Siegel domain of* $\mathbf{GL}(n, \mathbf{R})$. *Then* $G_A = \mathfrak{S} . G_U . G_\mathbf{Q}$.

This follows from $G_A = G_A^\infty . G_\mathbf{Q}$ (see 2.2) and 1.8 (6). In fact what is proved here is that if $c(G) = 1$, and $\mathrm{B} . \pi_\infty(G_\mathfrak{o}) = G_\infty$ ($\mathrm{B} \subset G_\infty$), then $G_A = \mathrm{B} . G_U . G_k$.

4.5. *Theorem.* — *Let* $k = \mathbf{Q}$, G *be reductive. Let* $\mathfrak{S}$ *be a standard Siegel domain of* $\mathbf{GL}(n, \mathbf{R})$ *and* $a \in \mathbf{SL}(n, \mathbf{R})$ *be such that* $a . G_\mathbf{R} . a^{-1}$ *is self-adjoint* [4, § 1]. *Then there*

exist finitely many elements $b_1, \ldots, b_s \in \mathbf{GL}(n, \mathbf{Q})$ *such that* $\Omega = \overset{s}{\underset{i=1}{\cup}} (a^{-1}.\mathfrak{S}.\mathbf{GL}_{n,\mathrm{U}}.b_i \cap \mathrm{G_A})$
is a fundamental set for $\mathrm{G_Q}$ *in* $\mathrm{G_A}$. *If* H *is an algebraic subgroup over* k *of* G *and* $a.\mathrm{H_R}.a^{-1}$
is self-adjoint, then Ω *verifies* (F 3)$_\mathrm{A}$ *relatively to* H.

Here, $a^{-1}.\mathfrak{S}.\mathbf{GL}_{n,\mathrm{U}}$ stands for

$$\{g = (g_v) \in \mathbf{GL}_{n,\mathrm{A}} \,|\, g_\infty \in a^{-1}.\mathfrak{S}, \, g_p \in \mathbf{GL}(n, \mathbf{Z}_p), \, (p \text{ prime})\}.$$

By [4, 3.8] there exists a rational representation $\rho : \mathbf{GL}_n \to \mathbf{GL}_m$ defined over $\mathbf{Q}$ and
a point $w \in \mathbf{Q}^m$ whose orbit under $\mathbf{GL}(n, \mathbf{C})$ is closed and whose isotropy group is G.

Let w' be the point of A^m defined by $w'_\infty = w.\rho(a^{-1})$, $w'_p = w_p$ (p prime). The
orbit of w'_∞ under $\mathbf{GL}(n, \mathbf{C})$ is the same as that of w, hence is closed, and the isotropy
group of w'_∞ in $\mathbf{GL}(n, \mathbf{R})$ is $a.\mathrm{G_R}.a^{-1}$, hence is self-adjoint. Lemma 3.4 (with G
and H replaced by $\mathbf{GL}_n$ and G respectively) and 4.3 show the existence of finitely many
elements $b_1, \ldots, b_s \in \mathbf{GL}(n, \mathbf{Q})$ such that

$$(1) \qquad w'.\rho_\mathrm{A}(\mathfrak{S}.\mathbf{GL}_{n,\mathrm{U}}) \cap w.\rho_\mathrm{A}(\mathbf{GL}(n, \mathbf{Q}) \subset \{w.\rho_\mathrm{A}(b_1^{-1}), \ldots, w.\rho_\mathrm{A}(b_s^{-1})\}.$$

Let us put

$$\mathrm{H} = \{x \in \mathbf{GL}_{n,\mathrm{A}} \,|\, w'.\rho_\mathrm{A}(x) = w\}.$$

Then, clearly,

$$(2) \qquad\qquad\qquad\qquad \mathrm{H} = j_\infty(a).\mathrm{G_A}.$$

Let now $x \in \mathrm{H}$. Using 4.4, we may write

$$x = s.b^{-1} \qquad (s \in \mathfrak{S}.\mathbf{GL}_{n,\mathrm{U}}, \, b \in \mathbf{GL}(n, \mathbf{Q}))$$

and the relation $w'.\rho_\mathrm{A}(x) = w$ gives

$$w'.\rho_\mathrm{A}(s) = w.\rho_\mathrm{A}(b);$$

which, by (1), shows that $w.\rho_\mathrm{A}(b) = w.\rho_\mathrm{A}(b_i^{-1})$ for some $i(1 \le i \le s)$. This yields $b.b_i \in \mathrm{G_Q}$,

$$\mathrm{H} \subset \overset{s}{\underset{i=1}{\cup}} (\mathfrak{S}.\mathbf{GL}_{n,\mathrm{U}}.b_i.\mathrm{G_Q})$$

which, together with (2), proves that $\mathrm{G_A} = \Omega.\mathrm{G_Q}$.

We may write $\Omega \subset \mathrm{B.C}$, where

$$(3) \qquad\qquad\qquad \mathrm{B} = \overset{s}{\underset{i=1}{\cup}} (a^{-1}.\mathfrak{S}.\pi_\infty(b_i) \cap \mathrm{G_\infty})$$

verifies (F 2)$_\infty$ by 3.4, and where

$$(4) \qquad\qquad\qquad \mathrm{C} = \overset{s}{\underset{i=1}{\cup}} (\pi_\mathrm{P}(b_i).\mathbf{GL}_{n,\mathrm{U}} \cap \mathrm{G_U})$$

is compact. The validity of (F 2)$_\mathrm{A}$, (F 3)$_\mathrm{A}$ then follows from 3.4 and 4.3.

4.6. *Theorem.* — $\mathrm{G_A}$ *contains closed or open fundamental sets for* $\mathrm{G_k}$ *if* G *is connected.*
If G *is reductive, and* H *is an algebraic subgroup over* k *of* G, *then* $\mathrm{G_A}$ *contains closed or open*
fundamental sets verifying (F 3)$_\mathrm{A}$ *relatively to* H.

By use of the restriction of the scalars, the proof is readily reduced to the case where $k = \mathbf{Q}$.

Let first G be reductive. Then 4.4 gives closed fundamental sets. But we can of course replace $\mathfrak{S}$ by the interior of a standard Siegel domain, and $\mathbf{GL}_{n,\mathrm{U}}$ by a bigger open, relatively compact subset of $\mathbf{GL}_{n,\mathrm{P}}$, and then we get open fundamental sets. If now $H\backslash G$ is realized as a closed orbit in a vector space, it is an affine variety, hence H is reductive [4, 3.8], and there exists $a \in \mathbf{SL}(n, \mathbf{R})$ such that both $a.\mathrm{G}_{\mathbf{R}}.a^{-1}$ and $a.\mathrm{H}_{\mathbf{R}}.a^{-1}$ are self-adjoint [4, § 1]. With this choice of a, the set Ω of 4.5 verifies $(\mathrm{F}\,3)_{\mathrm{A}}$ in virtue of 4.3 $b)$.

If now G is connected, we have $\mathrm{G} = \mathrm{H}.\mathrm{N}$, with H reductive, N unipotent and normal in G (1.13). By [11, Prop. 15] $\mathrm{N}_{\mathrm{A}}/\mathrm{N}_{\mathbf{Q}}$ is compact, hence 4.6 is true for $\mathrm{G} = \mathrm{N}$, and it is enough to show that if $\mathrm{B} \subset \mathrm{H}_{\mathrm{A}}$ and $\mathrm{C} \subset \mathrm{N}_{\mathrm{A}}$ are fundamental set for $\mathrm{H}_{\mathbf{Q}}$ and $\mathrm{N}_{\mathbf{Q}}$, with C relatively compact, then B.C is a fundamental set for $\mathrm{G}_{\mathbf{Q}}$ in G_{A}.

We have

$$\mathrm{G}_{\mathrm{A}} = \mathrm{H}_{\mathrm{A}}.\mathrm{N}_{\mathrm{A}} = \mathrm{B}.\mathrm{H}_{\mathbf{Q}}.\mathrm{N}_{\mathrm{A}} = \mathrm{B}.\mathrm{N}_{\mathrm{A}}.\mathrm{H}_{\mathbf{Q}} = \mathrm{B}.\mathrm{C}.\mathrm{G}_{\mathbf{Q}}.$$

Let $g = h.u$ $(h \in \mathrm{H}_{\mathbf{Q}}, u \in \mathrm{N}_{\mathbf{Q}})$ belong to $(\mathrm{BC})^{-1}.\mathrm{BC}$. This is the case if and only if

$$\mathrm{B}.h \cap \mathrm{B} \neq \varnothing, \quad h^{-1}.\mathrm{C}.h.u \cap \mathrm{C} \neq \varnothing.$$

Thus there are only finitely many possibilities for h, and, C being relatively compact, only finitely many possible u's for a given h, which ends the proof.

Remark. — The first part of 4.6 will be extended to non-connected groups in 5.2.

§ 5. Finiteness theorems for G_{A}, G_{k}.

5.1. *Theorem.* — *The number $c(\mathrm{G})$ of distinct double cosets $\mathrm{G}_{\mathrm{A}}^{\infty}.x.\mathrm{G}_{k}$ $(x \in \mathrm{G}_{\mathrm{A}})$ is finite.*

Since a compact subset C of G_{A} is contained in the union of finitely many right translates of $\mathrm{G}_{\mathrm{A}}^{\infty}$ (see 1.2), 5.1 is equivalent to the existence of a compact subset C of G_{A} such that

$$(1) \qquad\qquad \mathrm{G}_{\mathrm{A}} = \mathrm{G}_{\mathrm{A}}^{\infty}.\mathrm{C}.\mathrm{G}_{k}.$$

Let first G be connected. Using 1.13, 2.4, 2.5, we may assume G to be reductive. By restriction of the ground field (1.4), it is enough to consider the case where $k = \mathbf{Q}$. But then (1) is a consequence of 4.5 (3), (4).

In the general case, there exists a compact set D of G_{A} such that $\mathrm{G}_{\mathrm{A}} = \mathrm{D}.\mathrm{G}_{\mathrm{A}}^{0}$ (1.9). By 1.2 and the above, we may find finite subsets I, J of G_{A} and $\mathrm{G}_{\mathrm{A}}^{0}$ such that

$$\mathrm{G}_{\mathrm{A}} = \bigcup_{x \in \mathrm{I}, y \in \mathrm{J}} \mathrm{G}_{\mathrm{A}}^{\infty}.x.\mathrm{G}_{\mathrm{A}}^{0\,\infty}.y.\mathrm{G}_{k}.$$

The theorem follows now from the fact that $x.\mathrm{G}_{\mathrm{A}}^{\infty}.x^{-1}$ is commensurable with $\mathrm{G}_{\mathrm{A}}^{\infty}$ (1.7).

5.2. *Corollary.* — *Let B a fundamental set for G_{0} in G_{∞}. Then there exists a compact subset $\mathrm{C} \subset \mathrm{G}_{\mathrm{P}}$ such that $\mathrm{B}.\mathrm{C}'$ is a fundamental set for G_{k} in G_{A} for any relatively compact subset C' of G_{P} containing C.*

By 5.1, there exists a compact set C_0 in G_P such that $G_A = G_A^\infty . C_0 . G_k$.   Let

(2) $$C = G_U . C_0 . G_U.$$

We have

$$G_A^\infty . C = G_\infty . C = B . (j_\infty . \pi_\infty (G_0)) . C;$$

using (2) and 1.8, we get

$$G_A^\infty . C = B . C . (j_\infty . \pi_\infty (G_0)) = B . C . G_0,$$

hence

$$G_A = G_A^\infty . C . G_k = B . C . G_k.$$

The corollary then follows from 4.3.   Since B always exists (3.5), this produces fundamental sets also when G is not connected.

5.3.   Let G be reductive connected and H an algebraic subgroup over k.   When $k = \mathbf{Q}$, it is shown in [4, 3.8] that H is reductive if and only if there exists a rational representation $\rho : G \to GL(W)$ over k and a point $w \in W_k$ whose orbit X is closed and whose isotropy group is H.

This is also valid over a number field, with G not necessarily connected.   In fact let $G' = R_{k/\mathbf{Q}}G$, $H' = R_{k/\mathbf{Q}}H$, $X' = R_{k/\mathbf{Q}}X$.   We have then $X' = H' \backslash G'$.   If H is reductive, then so is H', hence X' is affine ([4, 3.8] and 1.12), X is affine, and the existence of ρ follows from [4, 2.4, footnote 2] ([1]).   If ρ exists, then X is affine, hence so is X', H' is reductive [4, 3.8], and therefore H is reductive.

5.4.   *Theorem.* — *Let G be reductive, H an algebraic subgroup over k of G, and σ the natural projection of G onto $X = H \backslash G$.   Assume that H is reductive or, equivalently* (5.3) *that X is an affine algebraic set.   Then* $\sigma_A(G_A) \cap X_k$ *is the union of finitely many orbits of* G_k ([2]).

Let $\rho : G \to \mathbf{GL}_m$ be a right rational representation over k and $w \in k^m$ a point whose orbit X' is closed and whose isotropy group is H (see 5.3).   There is an equivariant isomorphism $\Phi : X \to X'$ over k which maps $\sigma(H)$ onto w, whence an equivariant homeomorphism $\Phi_A : X_A \to X'_A$ which sends X_k onto X'_k and $\sigma_A(G_A)$ onto $w . \rho(G_A)$. Thus we are reduced to proving that $w . \rho(G_A) \cap k^m$ consists of finitely many orbits of G_k, or that $w . \rho(\Omega) \cap k^m$ is finite for a suitable fundamental set Ω for G_k in G_A; but this follows from 4.6.

5.5.   *Lemma.* — *The following conditions are equivalent*: a) G_A *is unimodular*, b) G_A^0 *is unimodular*, c) G_A^∞ *is unimodular*, d) G_∞ *is unimodular*, e) *every left invariant rational exterior differential form on* G^0, *of degree* $s = \dim G$, *defined over k, is right invariant.   If* $X_k(G^0) = 1$, *then* G_A *is unimodular*.

By 1.9, the quotient group G_A/G_A^0 is compact, therefore $(a) \Leftrightarrow (b)$.   The quotient

---

([1]) We do not need to know whether the result of Weil used in the proof of [4, 2.4] holds good for non-irreducible varieties, because $H' \backslash G'$ is embedded in $H' \backslash \mathbf{GL}_n$, which is irreducible.   An affine embedding over k of the latter then yields one for the former.

([2]) As in [4], we have been led by our conventions on Siegel domains to consider right representations in § 3.   This is why 5.8 is formulated for right coset spaces; but it is of course equivalent to the corresponding statement for G/H.

G_∞/G_∞^0 is finite, and 1.9 implies that $G_A^\infty/G_A^{0\infty}$ is compact. Therefore G_A^∞ (resp. G_∞) is unimodular if and only if $G_A^{0\infty}$ (resp. G_∞^0) is so. It suffices therefore to prove 5.5 when G is connected.

Let ω be a rational left-invariant differential form of degree s on G, defined over k. Then $\omega . g = \chi(g) . \omega$ where $\chi \in X_k(G)$, hence $X_k(G) = 1$ implies $e)$.

The form ω induces a left-invariant form ω_v on G_v for every $v \in V$. There exist convergence factors $(\lambda_v)_{v \in V}$ such that the product of the measures $\lambda_v \omega_v$ is defined on $G_{A(S)}$ ($S \subset V$, S finite) and is a Haar measure, and the inductive limit of these Haar measures is a Haar measure on G_A [14, Chap. II]. From this, the equivalence of $e)$ with any of $b)$, $c)$, $d)$ is clear.

5.6. *Theorem.* — (i) G_A/G_k *carries an invariant measure and has a finite volume for that measure if and only if* $X_k(G^0) = 1$.

(ii) G_A/G_k *is compact if and only if* $X_k(G^0) = 1$ *and every unipotent element of* G_k *belongs to the radical of* G_k.

By 5.1, there exists a finite subset I of G_A such that

$$G_A = \bigcup_{x \in I} G_A^\infty . x . G_k .$$

We have then

$$(3) \qquad G_A/G_k = \bigcup_{x \in I} x . \{(x^{-1}.G_A^\infty.x)/(x^{-1}.G_A^\infty.x \cap G_k)\}.$$

Since $x^{-1}.G_A^\infty.x$ is commensurable with G_A^∞ (1.7), the group $x^{-1}.G_A^\infty.x \cap G_k$ is commensurable with $G_A^\infty \cap G_k = G_0$. Taking 5.5 into account, we see that (i) and (ii) are respectively equivalent to the same assertions for G_A^∞/G_0. Since the latter space is fibered over G_∞/G_0 with compact fibers (1.8), (i) and (ii) are also equivalent to the statements obtained from them by writing G_∞/G_0 instead of G_A/G_k, but these follow from [4, 12.3] and from the fact that $(G^0)_\infty$ has a finite index in G_∞.

(Of course, we could also reduce the proof to the case where $k = \mathbf{Q}$ by use of 1.4, 1.5 and then refer to [4, 9.4, 11.8] rather than to [4, 12.3].)

5.7. Let $m_k : I_k \to \mathbf{R}^+$ be the continuous homomorphism of the idele group of k into $\mathbf{R}^+$ which associates to an idele $x = (x_v)_{v \in V}$ its idele module $\prod_{v \in V} \|x_v\|_v$. To each character $\chi \in X_k(G)$ corresponds a continuous homomorphism $m_k \circ \chi_A : G_A \to \mathbf{R}^+$; we put

$$(4) \qquad {}_m G_A = \bigcap_{\chi \in X_k(G)} \ker(m_k \circ \chi_A).$$

If s is a rational integer, then, clearly, $\ker m_k \circ \chi_A^s = \ker m_k \circ \chi_A$. Therefore

$$(5) \qquad {}_m G_A = \bigcap_{\chi \in M} \ker(m_k \circ \chi_A),$$

whenever M is a subgroup of finite index of $X_k(G)$.

Let

$$(6) \qquad L = \{g \in G \mid \chi(x) = \pm 1, (\chi \in X_k(G))\}.$$

This is an algebraic normal subgroup over k of G, whose unipotent radical coincides with that of G. Since G_U is compact, it belongs to $_m G_A$, whence

$$(7) \qquad\qquad {}_m G_A \cap G_A^\infty = L_\infty \times G_U,$$

which implies

$$(8) \qquad\qquad \pi_\infty(G_0) \subset L_\infty, \quad \pi_\infty(G_0) = \pi_\infty(L_0).$$

For connected groups, 5.6 admits the following generalization:

5.8. *Theorem. — Let* G *be connected. Then* $_m G_A = \bigcap\limits_{\chi \in X_k(G)} \ker(m_k \circ \chi)$ *is unimodular, contains* G_k, *the space* $_m G_A/G_k$ *has a finite invariant measure, and* $_m G_A/G_k$ *is compact if and only if every unipotent element of* G_k *belongs to the radical of* G_k.

A classical special case of 5.8 not included in 5.6 is the compactness of the quotient I_k^0/k^* of the group of ideles of k of idele-module one, by k^*.

By the product formula, $_m G_A$ contains G_k. Let J be a set of representatives for the distinct double cosets $G_A^\infty . x . G_k$ which meet $_m G_A$. It is finite by 5.1, and

$$_m G_A = \bigcup\limits_{x \in J} ({}_m G_A \cap G_A^\infty) . x . G_k.$$

The group $x^{-1}({}_m G_A \cap G_A^\infty) x = {}_m G_A \cap x^{-1} . G_A^\infty . x$ is commensurable with $_m G_A \cap G_A^\infty$ and therefore, as in 5.6, our theorem is equivalent to the corresponding assertion for $_m G_A \cap G_A^\infty$ and G_0. Using 5.7 (6), (7), (8) and 1.8, we see that $({}_m G_A \cap G_A^\infty)/G_0$ is fibered over $L_\infty/\pi_\infty(L_0)$ with compact typical fiber G_U. We are thus reduced to proving the statement 5.8 with $_m G_A$ and G_0 replaced by L_∞ and L_0, the latter group being now viewed as a subgroup of L_∞. This modified statement follows from 5.5 and [4, 9.4, 11.8] provided that we show that $X_k(L^0) = 1$. But this last fact is a consequence of the more general lemma:

5.9. *Lemma. — Let* F *be a field of characteristic zero,* M *a connected algebraic matric group over* F, *and* B *a connected normal algebraic subgroup over* k *of* M. *Then the cokernel of the restriction homomorphism* $X_F(M) \to X_F(B)$ *is finite.*

Let R(M) and R(B) be the radicals of M and B. The group R(B) is normal in R(M) and M (resp. B) is isogeneous to the semi-direct product of R(M) (resp. R(B)) with a semi-simple group. The restriction defines therefore an injective homomorphism $X_F(M) \to X_F(R(M))$ (resp. $X_F(B) \to X_F(R(B))$) with finite cokernel, and we may assume M and B to be solvable. The group M is then the semi-direct product, over k, of its unipotent radical N by a maximal algebraic torus T, whence a natural isomorphism $X_F(M) \cong X_F(T)$. Assuming, as we may, that $S = T \cap B$ is a maximal torus of B, we have similarly $B = (T \cap B) . (N \cap B)$, $X_F(B) = X_F(T \cap B)$, so that it is enough to prove 5.9 when M is an algebraic torus; in that case it follows for instance from [4, 8.4 $a)$].

5.10. *Remark. —* In Ono's terminology [11], G is of type (F) if $c(G)$ is finite, of type (C) (resp. (M)) if $_m G_A/G_k$ is compact (resp. of finite invariant measure), and G has no defect if the restriction homomorphism $X_k(G) \to X_k(G^0)$ has a finite cokernel. If G has no defect then the intersection L' of the kernels of the elements of $X_k(G^0)$ has finite index in the group L of 5.7 (6), hence L'_∞ has a finite index in L_∞ and 5.7 (5), (7) show that $_mG_{A}/_mG_A^0$ is compact. Thus $_mG_A$ is unimodular if and only if $_mG_A^0$ is so. Together with 5.1, 5.6, this proves: G is of type (F), a group with no defect is of type (M), a group with no defect is of type (C) if and only if every unipotent element of G_k belongs to the radical of G_k. For G solvable and G_k Zariski dense in G, this was already proved in [11].

§ 6. Application to principal homogeneous spaces.

Our aim in this paragraph is to give an application of 5.4 to Galois cohomology. As to the latter, we limit ourselves to a minimum of preliminaries, and refer to [5, 10, 13] for more details.

6.1. Let L be a group, and B a group on which L operates on the left. Being interested here only in Galois cohomology, we assume that all the orbits of L in B are finite and we define a 1-cocycle of L, with values in B, as a map $s \to b_s$ of L in B with a *finite image*, such that $b_{st} = s(b_t)$. Two cocycles (b_s), (b'_s) are cohomologous if there exists $c \in B$ such that $b'_s = c^{-1}.b_s.s(c)$. This is an equivalence relation. The set of such equivalence classes is the first Galois cohomology set of L with coefficients in B, denoted $H^1(L; B)$. If B is commutative, it is a commutative group; in general $H^1(L; B)$ is just a set with a distinguished zero element, the class of the coboundaries $s \to b^{-1}.s(b)$ $(b \in B)$. If L operates on another group C, again with finite orbits, then any equivariant map $f : B \to C$ induces a map $f^1 : H^1(L; B) \to H^1(L; C)$ sending the zero element onto the zero element; by definition, the kernel of f^1 is the inverse image of the zero element of $H^1(L; C)$.

The set B^L of fixed-points of L in B is, by definition, the o-cohomology group $H^0(L; B)$ of L in B, and the following lemma is in fact a part of the exactness of a cohomology sequence ([13, p. 133], [5]).

6.2. *Lemma.* — *We keep the previous notation, and assume that* B *is a subgroup of* C. *Then there exists a natural map* $\delta : (C/B)^L \to H^1(L; B)$ *which induces a bijection* δ_0 *of the set of orbits of* C^L *in* $(C/B)^L$ *onto the kernel* N *of the map* $i^1 : H^1(L; B) \to H^1(L; C)$ *induced by the injection of* B *in* C.

Let $\pi : C \to C/B$ be the canonical projection, and $o = \pi(B)$. Let $x \in (C/B)^L$, and choose $c \in \pi^{-1}(x)$. Then $x \in (C/B)^L$ implies $c^{-1}.s(c) \in B$, and $s \to c^{-1}.s(c)$ is a cocycle of L in B, whose class does not change when c varies in $\pi^{-1}(x)$, and is $\delta(x)$ by definition. Clearly $\delta(x) \in N$; it is easily checked that δ is constant on the C^L orbits.

If $\delta(x) = \delta(y)$ $(x, y \in (C/B)^L)$, then there exist $c, d \in C, z \in B$ such that

$$c.o = x, \quad d.o = y, \quad d^{-1}.s(d) = z^{-1}.c^{-1}.s(c).s(z) \quad (s \in L),$$

whence $c.z.d^{-1} \in C^L$. Since $c.z.d^{-1}.y = x$, this shows that δ_0 is injective. If now $(b_s) \in N$, then there exists $c \in C$ such that $b_s = c^{-1}.s(c)$ $(s \in L)$, whence $\pi(c) \in (C/B)^L$ and $(b_s) = \delta(\pi(c))$. Thus $\mathrm{Im}\,\delta \supset N$.

6.3. We shall be concerned here only with the case where L is the Galois group

Gal(F/k) over k of the field F of all algebraic numbers, and where $B = G_F$; as is usual, we write $H^1(k, G)$ for $H^1(\mathrm{Gal}\ (F/k); G_F)$.   This set may also be viewed as the inductive limit of the cohomology sets $H^1(\mathrm{Gal}\ (k'/k); G_{k'})$, where k' runs through the finite algebraic normal extensions of k.   Since $H^1(\mathrm{Gal}\ (k'/k), G_{k'}) = 0$ when $G = \mathbf{GL}_n$ (Theorem of Speiser, see e.g. [13, p. 159]), we also have

$$(1) \qquad\qquad\qquad H^1(k, \mathbf{GL}_n) = 0.$$

Together with 6.2, this implies the

Lemma. — *The map* $\delta : (\mathbf{GL}_n/G)_k \to H^1(k, G)$ *induces a bijection* δ_0 *of the set of orbits of* $\mathbf{GL}(n, k)$ *in* $(\mathbf{GL}_n/G)_k$ *onto* $H^1(k, G)$.

6.**4**.   A *principal homogeneous* space for G is an affine algebraic set X over which G operates, say on the right, as an algebraic transformation group, so that for each $x \in X$ the map $g \to x.g$ is a biregular map of G onto X.   Since we are in characteristic zero, it would be equivalent to require that G is simply transitive on X.   The principal homogeneous space is over k if X and the action of G on X are defined over k.

A principal homogeneous space over k is said to split over an extension k' of k if it has a point rational over k'.   If this is the case, and if $x \in X_{k'}$ then $g \to x.g$ identifies, over k', X with G operating on itself by right translations.

Two principal homogeneous space X, X' for G, over k, are isomorphic over k if there exists an equivariant birational biregular map over k of X onto X'.   It is well known that there is a natural $1 - 1$ correspondence Φ between these isomorphism classes and $H^1(k; G)$ (see [10] for instance).   We sketch the proof:

A principal homogeneous space X over k for G always contains an algebraic point x.   Given $s \in \mathrm{Gal}\ (F/k)$, there is a unique $g_s \in G_F$ such that $x.g_s = s(x)$.   It is readily checked that $s \to g_s$ is a 1-cocycle whose class depends only on the isomorphism class over k of X, and that the map Φ thus obtained is injective.   That Φ is surjective follows from " field-descent " but, in our case, may be deduced from 6.3: in fact, if $x \in (\mathbf{GL}_n/G)_k$, and if X_x is the inverse image of x in $\mathbf{GL}_n$, viewed in the obvious manner as a principal homogeneous space for G, then it follows immediately from the definitions that $\delta(x) = \Phi(X_x)$.

6.**5**.   A principal homogeneous space X over k is said to split locally everywhere if it splits over all completions of k.   Since an irreducible variety over k has integral p-adic points for almost all $p \in P$ (see 1.9 for a reference), so does X, if G is connected.

Let now G be connected, G' be an algebraic matric group over k containing G as an algebraic subgroup, and $\sigma : G' \to G'/G$ the natural projection. σ induces a continuous map $\sigma_A : G'_A \to (G'/G)_A$.   Let $x \in (G'/G)_k$ and $X = \sigma^{-1}(x)$ be the corresponding principal homogeneous space.   If $x \in \sigma_A(G'_A)$, then X obviously splits everywhere; the remark made earlier in this section shows that the converse is true if G is connected.   Together with 6.3, 6.4, this proves the following.

6.**7**.   Lemma. — *We keep the previous notation, and assume* G *to be connected.*   *Then the elements of* $\ker\ (H^1(k, G) \to H^1(k, G'))$ *which, viewed as principal homogeneous spaces, split* *locally everywhere, are in* $1 - 1$ *correspondance with the orbits of* G'_k *in* $\sigma_A(G'_A) \cap (G'/G)_k$.

If now G is reductive and $G' = \mathbf{GL}_n$, then $\sigma_A(G'_A) \cap (G'/G)_k$ is the union of finitely many orbits of G'_k by 5.4. Combined with 6.3, this proves the following:

6.8. *Theorem. — Let G be reductive, connected. Then the number of isomorphism classes over k of principal homogeneous spaces over k for G which split locally everywhere is finite.*

This theorem will be generalized in [5], where it will be shown that given a principal homogeneous space X over k for G (where G is subject only to our standing assumptions) and a finite subset $S \subset V$, the principal homogeneous spaces over k for G which are isomorphic to X over k_v for all $v \in V - S$ form a finite number of isomorphism classes over k.

§ 7. Application to parabolic subgroups.

7.1. Let G be connected. An algebraic subgroup H of G is *parabolic* if G/H is a complete variety. H is then connected, equal to its normalizer. If, moreover, H is defined over k, the fibering of G by H admits local rational cross sections defined over k (for all this, see [7]); consequently [14, p. 27]:

$$(1) \qquad (G/H)_A = G_A/H_A, \qquad (G/H)_k = G_k/H_k,$$

and, of course, $(G/H)_A$ is compact.

7.2. *Lemma. —* (Godement). *Let G be connected, and H a parabolic subgroup. Then there exists a finite subset* $I \subset G_A$ *such that*

$$G_A = \bigcup_{x \in I} G_A^\infty . x . H_k.$$

By (1) and the compactness of $(G/H)_A$, there exists a compact subset C of G_A such that $G_A = C . H_A$. The set C is covered by finitely many translates of G_A^∞ (1.2) and H_A by finitely many double cosets modulo H_A^∞ and H_k (5.1). Therefore G_A is a finite union of subsets $y . G_A^\infty . z . H_k$ ($y \in G_A$, $z \in H_A$). Since $y . G_A^\infty . y^{-1}$ is commensurable with G_A^∞ (1.7), it is contained in the union of finitely many right translates of G_A^∞, whence the lemma.

7.3. *Theorem. — Let G be connected and H be a parabolic subgroup, defined over k. Then* $(G/H)_k$ *is the union of finitely many orbits of* G_o.

Lemma 7.2 implies the existence of a finite subset I of G_k such that

$$G_k \subset \bigcup_{x \in I} G_A^\infty . x . H_k,$$

whence
$$G_k = \bigcup_{x \in I} (G_A^\infty \cap G_k) . x . H_k,$$

$$(2) \qquad G_k = \bigcup_{x \in I} G_o . x . H_k,$$

which is equivalent to our assertion.

7.4. *Remark. —* The above proof is due to Godement. Another one is given in [2]. In fact, the theorem is proved there only when $k = \mathbf{Q}$, and G is semi-simple. However the general case can easily be reduced to that one by use of the restriction of the scalars, of the surjectivity of $G_k \rightarrow (G/H)_k$, and of the fact that H contains the radical of G. We refer to [2, 4.6] for another formulation of this theorem.

We conclude this section with a proposition which is used in proving the assertions made in [2, 4.7], as will be shown elsewhere.

7.5. *Proposition.* — *Let* G *be connected, and* H *a parabolic subgroup. Assume that* $G_A = G_A^\infty . G_k$ *and that* $G_p = G_{o_p} . H_p$ *for every* $p \in P$. *Then the number* $\nu(G, H)$ *of double cosets of* G_k *modulo* G_o *and* H_k *is equal to* $c(H)$.

The assumption on the G_p ($p \in P$) implies

$$G_A = G_A^\infty . H_A,$$

whence

$$(3) \qquad G_A = \bigcup_{1 \le i \le m} G_A^\infty . h_i . H_k,$$

where (h_i) ($1 \le i \le m$) is a set of representatives in H_A for the distinct double cosets $H_A^\infty . h . H_k$. Since $G_A = G_A^\infty . G_k$ by assumption, there exists $x_i \in G_k$ such that $h_i \in G_A^\infty . x_i$, whence

$$G_k = \bigcup_{1 \le i \le m} (G_A^\infty \cap G_k) . x_i . H_k = \bigcup_{1 \le i \le m} G_o . x_i . H_k,$$

and $\nu(G, H) \le c(H)$.

If now $x_j \in G_o . x_i . G_k$, then $h_j \in G_A^\infty . h_i . H_k$, and therefore

$$h_j \in (G_A^\infty \cap H_A) . h_i . H_k = H_A^\infty . h_i . H_k,$$

which shows that $\nu(G, H) \ge c(H)$.

§ 8. Generalization to the groups of S-units.

The results of the previous paragraphs pertain to G_k, G_A, those of [4] to G_o, G_∞. Here we discuss some extensions to $G_{o(S)}$, $G_{A(S)}$, G_S ($S \subset V$).

As usual, S denotes a subset of V. *Unless otherwise stated, S is assumed to contain the infinite primes*, but it is not necessarily finite.

We shall often make no notational distinction between G_k and $\pi_S(G_k)$, or between $G_{o(S)}$ and $\pi_S(G_{o(S)})$. Otherwise said, when we view G_k, $G_{o(S)}$ as subgroups of G_S, then we mean $\pi_S(G_k)$, $\pi_S(G_{o(S)})$.

8.1. A subset $\Omega \subset G_{A(S)}$ is a *fundamental set* for $G_{o(S)}$ if it satisfies the two conditions:

$(F\ 1)_{A(S)} : \Omega . G_{o(S)} = G_{A(S)}$,

$(F\ 2)_{A(S)} : \Omega^{-1} . \Omega \cap G_{o(S)}$ is finite.

We could of course define a fundamental set Ω for $G_{o(S)}$ in G_S in the same way, but the case where $S = V - P$ suggests rather to require

$(F\ 1)_S : \Omega . G_{o(S)} = G_S$.

$(F\ 2)_S :$ For any $a, b \in G_k$, the set of elements $x \in G_{o(S)}$ such that $\Omega . a . x . b \cap \Omega \ne \emptyset$ is finite.

8.2. It will be convenient to have at one's disposal the analogue of $(F\ 2)'_\infty$ in 3.1. To this effect, we remark first that if Γ is a lattice in k^n, then

$$(1) \qquad o(S) . \Gamma = \bigcap_{p \notin S} (\Gamma_p \cap k^n), \qquad (\Gamma_p = o_p . \Gamma).$$

In fact, Γ being the direct sum of $\mathfrak{o}^{n-1}$ and of a fractional ideal of k [6, Satz 12.5], the proof of (1) reduces to the one-dimensional case, where it follows from ideal theory. Since $\Gamma_\mathfrak{p}=(\mathfrak{o}_\mathfrak{p})^n$ for almost all $\mathfrak{p}$'s, this implies that if Γ and Γ' are two lattices in k^n, then $\mathfrak{o}(S).\Gamma$ and $\mathfrak{o}(S).\Gamma'$ are commensurable, and therefore the set of groups $\mathfrak{o}(S)\Gamma''$, where Γ'' runs through the lattices in k^n such that

$$(2) \qquad\qquad \mathfrak{o}(S).\Gamma \subset \mathfrak{o}(S).\Gamma'' \subset \mathfrak{o}(S).\Gamma',$$

is finite.

Lemma. — The condition $(\mathrm{F}\ 2)_S$ *for a subset* Ω *of* G_S *is equivalent to* $(\mathrm{F}\ 2)'_S$: *For any* a, $b \in \mathrm{G}_k$ *and any non-zero algebraic integer* $r \in \mathfrak{o}$, *the set of elements* $x \in \mathrm{G}_k$ *such that* $r.x$, $r.x^{-1} \in \mathbf{M}(n, \mathfrak{o}(S))$ *and that* $\Omega.a.x.b \cap \Omega \neq \varnothing$ *is finite.*

Let $\mathrm{G}_{r,S}=\{g \in \mathrm{G}_k \mid r.g, r.g^{-1} \in \mathbf{M}(n, \mathfrak{o}(S))\}$ and $x \in \mathrm{G}_{r,S}$. Then

$$\mathfrak{o}(S).(r).\mathfrak{o}^n \subset \mathfrak{o}(S).x(\mathfrak{o}^n) \subset \mathfrak{o}(S).(r^{-1}).\mathfrak{o}^n.$$

By the above, $\mathrm{G}_{r,S}$ is then a finite union of left cosets modulo the isotropy group of $\mathfrak{o}(S).\mathfrak{o}^n$ in $\mathrm{G}_{r,S}$, which is nothing but $\mathrm{G}_{\mathfrak{o}(S)}$, whence the lemma.

8.3. The following condition for a subset $\Omega \subset \mathrm{G}_S$, relatively to an algebraic subgroup H over k, admits $(\mathrm{F}\ 3)_\infty$ and $(\mathrm{F}\ 3)_A$ as special cases.

$(\mathrm{F}\ 3)_S$: For any integer $m \geq 1$, any rational representation $\rho : \mathrm{G} \to \mathbf{GL}_m$, over k, any $w \in k^m$ with isotropy group H and closed orbit, and any lattice $\Gamma \subset k^m$, the set $w.\rho_A(\Omega) \cap \mathfrak{o}(S).\Gamma$ is finite.

8.4. *Lemma. — Let* $\mathrm{B} \subset \mathrm{G}_\infty$, C *a relatively compact subset of* $\mathrm{G}_{\mathrm{P} \cap S}$. *If* B *verifies* $(\mathrm{F}\ 2)_\infty$ *(resp.* $(\mathrm{F}\ 3)_\infty$ *relatively to a subgroup* H), *then* $\mathrm{B}.\mathrm{C}$ *verifies* $(\mathrm{F}\ 2)_S$ *(resp.* $(\mathrm{F}\ 3)_S$ *relatively to* H).

Let a, $b \in \mathrm{G}_k$ and x be as in $(\mathrm{F}\ 2)_S$. For $\mathfrak{p} \in \mathrm{P} \cap S$, the element x then belong to the relatively compact set $\pi_\mathfrak{p}(\mathrm{C}^{-1}.\mathrm{C})$, whence the existence of $s \in \mathfrak{o}$ such that $s.x$, $s.x^{-1} \in \mathbf{M}(n, \mathfrak{o}_\mathfrak{p})$ for every $\mathfrak{p} \in S$. Since x, $x^{-1} \in \mathbf{GL}(n, \mathfrak{o}_\mathfrak{p})$ for $\mathfrak{p} \notin S$ by assumption, we get sx, $sx^{-1} \in \mathbf{M}(n, \mathfrak{o})$. Since $\mathrm{B}x \cap \mathrm{B} \neq \varnothing$, the part of the lemma concerning F 2 follows from 3.1.

Let now $\rho : \mathrm{G} \to \mathbf{GL}_m$, $w \in k^m$ be as in $(\mathrm{F}\ 3)_S$, and $x \in w.\rho_S(\mathrm{B}.\mathrm{C}) \cap \mathfrak{o}(S).\Gamma$. In particular, $x \in \pi_\mathfrak{p}(w.\rho_S(\mathrm{C}))$ for $\mathfrak{p} \in \mathrm{P} \cap S$, whence again the existence of $r \in \mathfrak{o}$ such that $rx \in \mathfrak{o}_\mathfrak{p}^m$ for all $\mathfrak{p} \in \mathrm{P} \cap S$. Two lattices in k^n being commensurable, there exists $s \in \mathfrak{o}$ such that $(s)\Gamma \in \mathfrak{o}^m$; since $x \in \mathfrak{o}(S).\Gamma$, we have then $x \in \Gamma_\mathfrak{p}$ for $\mathfrak{p} \notin S$, hence $s.x \in \mathfrak{o}_\mathfrak{p}^m$ for $\mathfrak{p} \notin S$. Altogether, we get $r.s.x \in \mathfrak{o}^m$, which shows that x belongs to the lattice $\Gamma=(s^{-1}.r^{-1})\mathfrak{o}^m$, and, combined with $x=w.\rho_\infty(\mathrm{B})$, ends the proof.

Remark. — The remark to 4.3 *also applies to the proof of* $(\mathrm{F}\ 3)_\infty \Rightarrow (\mathrm{F}\ 3)_S$.

8.5. *Theorem. — The groups* $\mathrm{G}_{A(S)}$ *and* G_S *contain open or closed fundamental sets for* $\mathrm{G}_{\mathfrak{o}(S)}$, *which, in the case of* G_S, *verify* $(\mathrm{F}\ 3)_S$ *relatively to* H, *if* G *is reductive, and* H *an algebraic subgroup over* k *of* G.

In view of 8.4, 3.4, 3.5, it is enough to show that if $\mathrm{B} \subset \mathrm{G}_\infty$ satisfies $(\mathrm{F}\ 1)_\infty$, then there exists a compact set $\mathrm{C} \subset \mathrm{G}_{\mathrm{P} \cap S}$ such that $\mathrm{B}.\mathrm{C}$ and $\mathrm{B}.\mathrm{C}.\mathrm{M}$, where $\mathrm{M}=\prod_{\mathfrak{p} \notin S} \mathrm{G}_{\mathfrak{o}_\mathfrak{p}}$, verify $(\mathrm{F}\ 1)_S$ and $(\mathrm{F}\ 1)_{A(S)}$ respectively.

By 5.1, there exists a compact (in fact finite) subset $I \subset G_A$ such that

$$G_{A(S)} \subset \bigcup_{x \in I} G_A^\infty . x . G_k.$$

We may of course limit ourselves to the double cosets which intersect $G_{A(S)}$, and therefore may assume that $I \subset G_{A(S)}$. Then

$$(1) \qquad G_{A(S)} = \bigcup_{x \in I} G_A^\infty . x . (G_{A(S)} \cap G_k) = \bigcup_{x \in I} G_A^\infty . x . G_{o(S)},$$

which shows the existence of a compact set $C \subset G_{P \cap S}$ such that

$$(2) \qquad G_{A(S)} = G_A^\infty . C . M . G_{o(S)} \quad (M = \prod_{p \notin S} G_{o_p})(^1).$$

Since $L = \prod_{p \in S \cap P} G_{o_p}$ is compact, we may, and shall, assume

$$(3) \qquad\qquad\qquad L . C . L = C.$$

By assumption, $G_\infty = B . \pi_\infty(G_o)$, whence

$$G_{A(S)} = B . C . M . \pi_\infty(G_o) . G_{o(S)},$$

and by (3),

$$G_{A(S)} = B . C . G_U . \pi_\infty(G_o) G_{o(S)}.$$

Using 1.8 (6), we get

$$G_{A(S)} = B . C . G_U . G_o . G_{o(S)},$$
$$G_{A(S)} = B . C . G_U . G_{o(S)},$$

and, again by (3),

$$G_{A(S)} = B . C . M . G_{o(S)},$$

hence also

$$G_S = B . C . G_{o(S)},$$

which proves our assertion.

8.6. *Remarks.* — (1) In fact, this proves the existence of open or closed fundamental sets of the form $B . C . M$ or $B . C$, where B is a fundamental set for G_o in G_∞, C is relatively compact in $G_{P \cap S}$ and $M = \prod_{p \notin S} G_{o_p}$, which satisfy $(F\ 3)_S$ relatively to H if G is reductive, and H a given algebraic subgroup over k of G.

(2) If $k = \mathbf{Q}$, and G is reductive, then, in the notation of 4.5, there are fundamental sets for G_o in $G_{A(S)}$ and in G_S of the form

$$\Omega = \bigcup_{i=1}^{m} (a^{-1} \mathfrak{S} \mathbf{GL}_{n, U} . b_i \cap G_{A(S)}) \quad (b_i \in \mathbf{GL}(n, \mathbf{Z}(S))$$

and $\pi_S(\Omega)$ respectively. The proof is the same as that of 4.5, except for the fact that it uses the equality

$$(4) \qquad\qquad \mathbf{GL}_{n, A(S)} = \mathfrak{S} . \mathbf{GL}_{n, U} . G_{\mathbf{Z}(S)},$$

which is an obvious consequence of 4.4, rather than 4.4 itself.

8.7. Let Γ be a lattice in k^n, $G_{p, \Gamma}$ the stability group of $\Gamma_p = o_p . \Gamma$ in G_p $(p \in P)$,

---

and $G_{A, \Gamma} = G_A^\infty \cdot \prod_{\mathfrak{p} \in P} G_{\mathfrak{p}, \Gamma}$. The group $G_{\mathfrak{p}, \Gamma}$ is open and compact in $G_\mathfrak{p}$ and $G_{A, \Gamma}$ is of course the isotropy group of Γ in G_A, where G_A operates on the set of lattices in k^n by $\Gamma \to g(\Gamma) = \bigcap_{\mathfrak{p} \in P} g_\mathfrak{p}(\Gamma_\mathfrak{p})$ $(g = (g_v) \in G_A)$ (see 2.2). The facts on lattices recalled in 2.1 show that $G_{\mathfrak{p}, \Gamma} = G_{\mathfrak{o}_\mathfrak{p}}$ for almost all $\mathfrak{p} \in P$, hence that $G_{A, \Gamma}$ is commensurable with G_A^∞, or with $G_{A, \Gamma'}$ for any lattice $\Gamma' \subset k^n$.

Let now $\rho : G \to \mathbf{GL}_m$ be a rational representation over k. Then any lattice Γ in k^m is contained in a lattice $\Gamma' \subset k^m$ stable under $\rho(G_A^\infty)$. In fact, $\rho(G_U)$ is compact, and therefore $\rho(G_U) \cap \mathbf{GL}_{m, A, \Gamma}$ is commensurable with $\rho(G_U)$. Consequently, the set of transforms of Γ under $\rho(G_A^\infty)$ is finite, and the sum Γ' of those transforms is the desired lattice.

8.8. *Theorem.* — *Let* G *be reductive* $\rho : G \to \mathbf{GL}_m$ *a rational representation over* k, $w \in k^m$ *a point whose orbit* X *is closed, and* Γ *a lattice in* k^m. *Then* $w \cdot \rho_A(G_{A(S)}) \cap k^m$ (*resp.* $w \cdot \rho_S(G_S) \cap \mathfrak{o}(S) \cdot \Gamma$) *is contained in the union of finitely many orbits of* $G_{\mathfrak{o}(S)}$.

In view of 8.7, we may assume Γ to be invariant under G_A^∞, hence $\mathfrak{o}(S) \cdot \Gamma$ to be invariant under $G_{\mathfrak{o}(S)}$. The theorem follows then from the existence of fundamental sets in $G_{A(S)}$ (resp. G_S) which verify $(\mathrm{F}\,3)_{A(S)}$ (resp. $(\mathrm{F}\,3)_S$).

If $S = V$, then both parts of 8.8 reduce to 5.4.

8.9. In the next section, we shall need the following fact, whose proof uses a result of [5].

Lemma. — *Let* S *be any finite subset of* V. *Let* X *be a homogeneous space over* k *for* G. *Then* $X_S = \prod_{v \in S} X_v$ *is the union of finitely many orbits of* G_S.

It is enough to prove this when S consists of one element v. If $k_v = \mathbf{C}$, then G_v is transitive on X_v; if $k_v = \mathbf{R}$, this assertion is elementary (see [4, 2.3]). If v is finite, the only proof known to the author is a cohomological one, to be given in [5].

8.10. *Theorem.* — *Let* G *be reductive,* $\rho : G \to \mathbf{GL}_m$ *a rational representation over* k *of* G, X *a closed orbit and* Γ *a lattice in* k^m. *If* S *is finite, then* $X_S \cap \mathfrak{o}(S) \cdot \Gamma$ *is contained in finitely many orbits of* $G_{\mathfrak{o}(S)}$.

By 8.9, $X_S \cap \mathfrak{o}(S) \cdot \Gamma$ is contained in finitely many sets of the form $w \cdot \rho_S(G_S)$, with $w \in k^m$, to each of which we can apply 8.8.

Here, it would be in fact enough to prove this for G connected, since $(G_0)_S$ has finite index in G_S. We give two applications of 8.10.

8.11. *Corollary.* — *Let* S *be finite. Then the number of orbits of* $\mathbf{SL}(n, \mathfrak{o}(S))$ *in the set of all symmetric* $n \times n$ *matrices, with coefficients in* $\mathfrak{o}(S)$, *and a given non-zero determinant, is finite.*

This follows from 8.10 applied to the case where ρ is the natural representation of $\mathbf{SL}_n$ in the space of symmetric $n \times n$ matrices and X the set of symmetric matrices with the given determinant. If $S = V - P$, $k = \mathbf{Q}$, then $\mathfrak{o}(S) = \mathbf{Z}$, and we get of course the well-known finiteness of the number of classes of integral quadratic forms with a given non-zero determinant.

8.12. *Proposition.* — *Let* $\mu : G \to G'$ *be an isogeny over* k *of* G *onto an algebraic matric group* G' *over* k. *Let* S *be finite. Then* $\mu(G_{\mathfrak{o}(S)})$ *is commensurable with* $G'_{\mathfrak{o}(S)}$.

It is clearly enough to prove this when G is connected. Adding one variable if necessary, we may, by a familiar trick, (see [4, 2.1] for instance), assume that G′ consists of matrices of determinant one, hence is closed. Let n' be the degree of G′. Assume first that G is reductive. We let then operate G on the space of $n' \times n'$ matrices by $x \to x \cdot \mu(g)$ and define in this way a rational representation $\rho : G \to \mathbf{GL}_m$ $(m = n^2)$ over k. Our assertion follows then from 8.10 applied to ρ and to G′, viewed as the orbit of e.

If μ is an isomorphism, then 8.11 is elementary, and is a consequence of 1.7. This is necessarily the case if N is unipotent. In general, we use the semi-direct product decompositions $G = H.N$, $G' = H'.N'$ of 1.13. Since we are allowed to replace G (resp. G′) by a group isomorphic to G (resp. G′) over k, we may assume that $\mathbf{C}^n$ (resp. $\mathbf{C}^{n'}$) is the direct sum of subspaces spanned by canonical basis vectors, stable under H (resp. H′), on which N (resp. N′) acts trivially, and that N (resp. N′) is upper triangular. Then $G_{o(S)} = H_{o(S)} \cdot N_{o(S)}$ (resp. $G'_{o(S)} = H'_{o(S)} \cdot N'_{o(S)}$) and we are reduced to the two already discussed special cases.

Remark. — This extension to S-units of [4, 6.11] is not really new. K. Honda (*Jap. J. Math.*, *30* (1960), 84-101) has proved it when S is big enough (for group varieties, not necessarily linear ones, in fact), and Serre (unpublished) has removed the assumption on S by a suitable modification of Honda's method.

8.12. By the argument used in 5.6, it is easily derived from 8.5 (1) that $G_{A(S)}/G_{o(S)}$, G connected, is the union of finitely many open subsets, isomorphic to quotients G_A^∞/H_i, where H_i is commensurable with G_o. Thus the criteria of 5.6 for the finiteness of the volume or the compactness of G_A/G_k also apply to $G_{A(S)}/G_{o(S)}$ and $G_S/G_{o(S)}$.

BIBLIOGRAPHY

[1] A. Borel, Some properties of adele groups attached to algebraic groups, *Bull. A.M.S.*, *67*, (1961), p. 583-585.

[2] —, Ensembles fondamentaux pour les groupes arithmétiques, *Colloque sur la théorie des groupes algébriques*, Bruxelles, 1962.

[3] —, Arithmetic properties of algebraic groups, *Proc. Int. Congress of Math.*, Stockholm, 1962.

[4] — and Harish-Chandra, Arithmetic subgroups of algebraic groups, *Annals of Math.* (2), *75* (1962), p. 485-535.

[5] — and J.-P. Serre, *Théorèmes de finitude en cohomologie galoisienne* (en préparation).

[6] M. Eichler, Quadratische Formen und orthogonale Gruppen, *Grund. Math. Wiss.*, LXIII, Springer, Berlin, 1952.

[7] R. Godement, Groupes linéaires algébriques sur un corps parfait, *Sem. Bourbaki*, n° 206, Paris, 1960-61.

[8] M. Kneser, Approximation in algebraischen Gruppen, *Colloque sur la théorie des groupes algébriques*, Bruxelles, 1962.

[9] —, Einfach zusammenhängende algebraische Gruppen in der Arithmetik, *Proc. Int. Congress Math.*, Stockholm, 1962.

[10] S. Lang and J. Tate, Principal homogeneous spaces over abelian varieties, *Amer. Jour. Math.*, *80* (1958), p. 659-684.

[11] T. Ono, On some arithmetic properties of linear algebraic groups, *Annals of Math.* (2), *70* (1959), p. 266-290.

[12] M. Rosenlicht, Some rationality questions on algebraic groups, *Annali di Mat. pura ed applic.* (IV), *63* (1957), p. 25-50.

[13] J.-P. Serre, *Corps locaux*, Publ. Math. Inst. Nancago VIII, Hermann, Paris, 1962.

[14] A. Weil, *Adeles and algebraic groups* (Notes by M. Demazure and T. Ono), The Institute for Advanced Study, Princeton, N.J., 1961.

Reçu le 15 novembre 1962.

61.

Arithmetic properties of linear algebraic groups

Proc. Int. Congr. Mathematicians Stockholm 1962, Uppsala 1963, 10–22

The theory of Lie groups in the large has been developed chiefly from two points of view: the transcendental one, shaped by the work of H. Weyl and E. Cartan, which emphasizes maximal compact subgroups, Riemannian symmetric spaces, and is a meeting ground for Lie algebra methods, differential geometry, topology; and the more recent point of view of algebraic groups, based on the use of algebraic geometry, which has produced results in arbitrary characteristic but has also given new information and fruitful ideas in the classical case. All these, combined, have provided a suitable framework for a further development, linking arithmetic and algebraic groups, which is the subject matter of the present report.

The impetus for the recent work in that area came to a large extent from two sources. First, it was noticed some years ago that the formalism of adeles and ideles introduced in algebraic number theory by Chevalley could also be applied to linear algebraic groups. This allowed one to give new and simple formulations to a number of classical results, in particular on quadratic forms, which made sense for more general algebraic groups. They suggested therefore new problems as well as the possibility of a more unified treatment of the cases considered so far. The other incentive has been the wish to develop in a more general setting the theory of automorphic functions of several variables.

Important in both cases is the study of fundamental domains for certain discrete subgroups of real algebraic groups, called arithmetic groups or groups of units, to which I shall devote about the first half of this talk. The second part will be concerned with problems involving also p-adic algebraic groups or adele groups. Automorphic functions are related to these questions, but my time is limited and, more important, some of the main developments in that area will be discussed in other lectures, notably in those of Selberg [24] and Gel'fand [11]. I shall therefore leave them aside and concentrate on more algebraic questions, or on the more algebraic aspects of these questions.

1. Groups of units

I would like now first to define the groups of units in general, and then give a few examples and state some of their properties.

A subgroup G of $\mathbf{GL}(n, \mathbf{C})$, the group of $n \times n$ complex invertible matrices, is *algebraic* if it is the set of all invertible matrices whose coefficients annihilate some collection of polynomials in n^2 indeterminates, with complex coefficients; it is said to be defined over a subfield k of $\mathbf{C}$ if those polynomials may be chosen so as to have their coefficients in k. The intersection of the fields of definition is also a field of definition. If B is a subring of $\mathbf{C}$, then G_B denotes the group of elements of G which have their coefficients in B

331

and whose determinant is a unit of B, i.e. is invertible in B. In the sequel, unless otherwise said, G will stand for an algebraic subgroup of $\mathbf{GL}(n, \mathbf{C})$ which is defined over the field $\mathbf{Q}$ of rational numbers. Its *group of units* is then $G_{\mathbf{Z}}$, the group of integral matrices with determinant ± 1 contained in G. A group of units is thus attached to a matric group, its definition presupposes the choice of a basis. If we perform a rational change of coordinates or, more generally, if we replace G by its image under a faithful rational representation defined over $\mathbf{Q}$, then the new group of units is in general different from the first one, but is at any rate commensurable with it. Thus the commensurability class of groups of units has a more intrinsic meaning. I recall that two subgroups of a group are commensurable if their intersection has finite index in both.

The group $G_{\mathbf{R}}$ of real matrices of G is a real Lie group (i.e. a real analytic manifold endowed with a group structure for which the product and the inverse are real analytic functions of their arguments) with finitely many connected components. Therefore, its maximal compact subgroups are conjugate to each other by inner automorphisms and the quotient $D = K \backslash G_{\mathbf{R}}$ $= \{Kg, \, g \in G_{\mathbf{R}}\}$ of $G_{\mathbf{R}}$ by one of them is homeomorphic to a euclidean space. If G is semi-simple, i.e. if it has no infinite invariant commutative subgroup or, equivalently, if its Lie algebra is the direct product of its simple ideals, then D is a Riemannian symmetric space with negative curvature without flat component; and, conversely, every such space can be obtained in this way.

The group $G_{\mathbf{Z}}$ acts by right translations on $G_{\mathbf{R}}$, D, and is *properly discontinuous*: this means that every compact set meets only a finite number of its translates by $G_{\mathbf{Z}}$; in particular the set of transforms of a given point has no accumulation point. Our problem is then to find suitable sets of representatives for the orbits of $G_{\mathbf{Z}}$ in $G_{\mathbf{R}}$ or in D. These are essentially equivalent questions; for the proofs, it is often more convenient to work in $G_{\mathbf{R}}$, but the consideration of D brings us closer to the classical reduction theory. I shall therefore give the preference to D in this lecture.

2. Examples

(*a*) $G = \mathbf{GL}(n, \mathbf{C})$, $K = \mathbf{O}(n)$, the group of real $n \times n$ orthogonal matrices, and D is the space of positive non-degenerate real symmetric $n \times n$ matrices, on which $G_{\mathbf{R}} = \mathbf{GL}(n, \mathbf{R})$ acts in the standard fashion by:

$$A \to {}^{t}X \cdot A \cdot X \quad (A \in D, \; X \in \mathbf{GL}(n, \mathbf{R}), \; {}^{t}X \text{ transpose matrix of } X).$$

Here $G_{\mathbf{Z}}$ is the group of integral matrices with determinant ± 1, and its orbits are just the *classes of positive non-degenerate quadratic forms*. The construction of sets of representatives is the task of the reduction theory of postive quadratic forms which goes back to Gauss for $n = 2$, to Hermite [12] for $n \geqslant 3$.

(*b*) Let F be a rational non-degenerate quadratic form on $\mathbf{C}^{n}$ and $G = \mathbf{0}(F)$ the orthogonal group of F, i.e. the group of linear transformations leaving F invariant. $G_{\mathbf{Z}}$ is then classically called the group of units of F. In this case, D is the space of majorizing forms of F, in the sense of Hermite [12]: these are the minimal elements among the positive forms S which majorize

F in the obvious sense (i.e. verify $|F(x,x)| \leqslant S(x,x)$ for every $x \in \mathbf{R}^n$); they can also be defined as those positive quadratic forms whose matrices satisfy the relation $F \cdot S^{-1} \cdot F = S$. The maximal compact subgroups of $G_\mathbf{R}$ are the subgroups of $G_\mathbf{R}$ which leave invariant one of those majorizing forms.

(c) G is the symplectic group $\mathbf{Sp}(2n)$, the group of $2n \times 2n$ matrices leaving invariant the antisymmetric bilinear form

$$F(x,y) = \sum_{1 \leqslant i \leqslant n} (x_i \cdot y_{n+i} - y_i \cdot x_{n+i}).$$

The group $G_\mathbf{Z}$ is then Siegel's modular group and D is the generalized upper half-plane, the space of complex $n \times n$ symmetric matrices with positive non-degenerate imaginary part [28].

3. Fundamental sets for groups of units

A fundamental domain for a group acting on a set is, strictly speaking, a subset containing exactly one representative of each orbit. Here, we shall be concerned only with suitable approximations to fundamental domains in this strict sense, to be called fundamental sets.

A subset $\Omega \subset D$ is a *fundamental set* for $G_\mathbf{Z}$ if first it meets each orbit of $G_\mathbf{Z}$, i.e. if:

(i) $\Omega \cdot G_\mathbf{Z} = D$.

A second natural requirement is that Ω should intersect only a finite number of its translates $\Omega \cdot x$ by elements of $G_\mathbf{Z}$; however, when dealing with arithmetic groups, it is important to have at one's disposal the stronger condition:

(ii) For any $a \in G_\mathbf{Q}$, the set of $x \in G_\mathbf{Z}$ for which $\Omega \cdot a \cap \Omega \cdot x$ is not empty is finite,

and this will be our second condition for fundamental sets.

In our first example above, we may take as fundamental sets the so-called *Siegel domains*. The Siegel domain $\mathfrak{S}_{t,u}$ $(t, u > 0)$ in the space of positive $n \times n$ real symmetric matrices is the set of products ${}^t N \cdot A \cdot N$, where $N = (n_{ij})$ is upper triangular, with ones in the diagonal, and $|n_{ij}| \leqslant u$ $(i < j)$, and where A is diagonal, with positive diagonal entries $a_1, \ldots, a_n$ verifying $a_i \leqslant t \cdot a_{i+1}$ $(1 \leqslant i \leqslant n-1)$. It satisfies our second condition by a well known result of Siegel [26]; that (i) is fulfilled when $t \geqslant 4/3$, $u \geqslant 1/2$ is implicit in Hermite's work [12], and made explicit in [15].

The existence of closed or open fundamental sets in the general case, which have a finite volume with respect to an invariant measure when G is semi-simple, (or more generally when the identity component of G has no non-trivial rational character defined over $\mathbf{Q}$) is proved in a paper written jointly by Harish-Chandra and myself [7]. The crucial case is when G is semi-simple, for which the construction of fundamental sets generalizes in a way the procedure used by Hermite [12] in the case of indefinite rational quadratic forms (our second example): D being realized as a submanifold of the space D' of all positive non-degenerate real symmetric matrices of some degree n', these fundamental sets are the intersection of D with a finite number of translates, by elements of $\mathbf{GL}(n', \mathbf{Z})$, of a Siegel domain of D'.

The paper [7] also gives a necessary and sufficient condition under which $D/G_\mathbf{Z}$ is compact. When G is semi-simple, the condition, which had been conjectured by Godement, is that either $G_{\mathbf{Z}}$ or $G_{\mathbf{Q}}$ consists of semi-simple matrices (matrices which can be diagonalized over the complex numbers). In our second example above, this is equivalent to the non-existence of non-trivial rational zeros of F. From this it can be deduced that every Riemannian symmetric space with negative curvature has discontinuous groups of isometries with relatively compact fundamental domains which act freely (i.e. no element different from the identity has fixed points). In other words, every such space has compact Clifford-Klein forms [32]. Another proof of Godement's conjecture is given in [33].

4. Fundamental sets with cusps

The construction of fundamental sets in semi-simple groups defined over $\mathbf{Q}$ briefly mentioned previously depends chiefly on $G_{\mathbf{R}}$ and has comparatively little to do with $G_{\mathbf{Q}}$ or $G_{\mathbf{Z}}$. However, starting from this first approximation to fundamental domains, it is possible to construct a second, and usually better, one when $D/G_{\mathbf{Z}}$ is not compact, by taking also into account some properties of $G_{\mathbf{Q}}$. I have in mind here mainly the facts which center around the so-called Bruhat lemma, or rather around its generalization to rational points over non-algebraically closed fields, proved independently by Tits and myself ([5], [30]).

These new fundamental sets are better suited to the study of automorphic forms and of the compactification of $D/G_{\mathbf{Z}}$. They are the union of finitely many unbounded sets which behave more or less like the parabolic cusps of the fundamental domains for fuchsian groups in the upper half-plane, and which shall also be called cusps. These cusps are the translates by elements of $G_{\mathbf{Q}}$ of subsets which are defined roughly in the same way as the Siegel domains in $G_{\mathbf{R}}$ [7; § 4], but with the maximal subgroups which can be put in triangular or in diagonal form over $\mathbf{R}$ replaced by subgroups enjoying similar properties with respect to $\mathbf{Q}$ (see [5] for a precise description).(1)

The cusps correspond to the equivalence classes modulo $G_{\mathbf{Z}}$ of the maximal unipotent subgroups of $G_{\mathbf{Q}}$.(2) In some cases, they have other interpretations. For instance, if G is the orthogonal group of a rational quadratic form F, and if the integral points form a maximal lattice for F, they correspond to the equivalence classes modulo $G_{\mathbf{Z}}$ of maximal isotropic subspaces defined over $\mathbf{Q}$. If $G_{\mathbf{Z}}$ is the Hilbert-Blumenthal group associated to a number field k, they correspond to the ideal classes of k.

These fundamental sets can be used to show that the compactification of $D/G_{\mathbf{Z}}$ associated to a linear representation of G by Satake's procedure [23] consists of finitely many quotient spaces of the same type if the representation is defined over $\mathbf{Q}$. This result is valid for any semi-simple group defined over $\mathbf{Q}$ and yields only topological compactifications. It does not say whether some of those compactifications are normal complex analytic spaces when D is a bounded domain in the space of several complex variables. It does not seem too unlikely that this or related questions on automorphic functions

---

(1) It seems rather likely that this implies that the arithmetic groups are regular in the sense of Gel'fand [11].

(2) A matrix group is unipotent if it consists of matrices all of whose eigenvalues are equal to one; it can then be put in triangular form and is nilpotent.

can be handled with the help of the above fundamental sets and of certain results of Harish-Chandra (*Amer. J. Math.*, 78 (1956), 564–628), either by the Andreotti-Grauert method [2], or by showing the validity of certain conditions which are roughly analogous to the normality conditions discussed by Pyatetski-Šapiro in his book on classical domains and automorphic functions [22]. However, this has not yet been done, as far as I know.

5. Some finiteness theorems for groups of units

I would like now to state some properties of groups of units whose proofs make use of fundamental sets, although their statements do not.

(a) $G_{\mathbf{Z}}$ is finitely generated, finitely presented, and its finite subgroups form a finite number of conjugacy classes.

(Since only the finite generation of $G_{\mathbf{Z}}$ is proved in [7], I sketch the proof of the two other properties: the subgroup N of $G_{\mathbf{Z}}$ which acts trivially on D is finite, being contained in K; it is therefore sufficient to prove them for $G_{\mathbf{Z}}/N$ [3]; the finite presentation follows then from the existence of open fundamental sets and from the connectedness and simple connectedness of D [3; Satz 2]. Let Ω be a fundamental set and C the set of elements $c \in G_{\mathbf{Z}}$ for which $\Omega \cap \Omega \cdot c \neq \emptyset$. By our condition (ii), C is finite. Let now H be a finite subgroup of $G_{\mathbf{Z}}$. It belongs to a maximal compact subgroup of $G_{\mathbf{R}}$, hence has a fixed point $P \in D$; if $g \in G_{\mathbf{Z}}$ is such that $P \cdot g \in \Omega$, then $g^{-1} \cdot H \cdot g$ leaves a point of Ω fixed, hence belongs to C.)

(b) Let G be semi-simple, connected, and f a rational representation, defined over $\mathbf{Q}$, of G in a finite dimensional vector space V. Take coordinates in $V_{\mathbf{Q}}$ with respect to which $G_{\mathbf{Z}}$ is represented by integral matrices, which is always possible [7; 6.3], and let X be an orbit of G. Then if X is closed, the integral points of X form a finite number of orbits of $G_{\mathbf{Z}}$ [7; Theorem 6.9]; if the union of X and the origin is a cone, the primitive integral vectors contained in X form a finite number of orbits of $G_{\mathbf{Z}}$ [5].

Applications. (a) Let $G = \mathbf{SL}(n, \mathbf{C})$, the group of complex $n \times n$ matrices with determinant 1, f be the natural representation of G in the space of $n \times n$ symmetric matrices. The set X of symmetric matrices with a given non-zero determinant is then a closed orbit of G, and the first assertion shows that it contains only a finite number of classes of integral quadratic forms, a well known statement of Hermite [12].

(b) Let $G = 0(F)$ be the orthogonal group of the rational indefinite non-degenerate quadratic form F, and f be the identical representation of G. By Witt's theorem, the cone of isotropic vectors minus the origin is an orbit of G. The second assertion implies therefore that the primitive integral isotropic vectors form a finite number of orbits for the group of units of F, which is also well-known from reduction theory.

If G is an algebraic matric group defined over a number field k, then its group of units is by definition G_B, where B is the ring of algebraic integers of k. This case can be reduced to the one considered here ($k = \mathbf{Q}$, $B = \mathbf{Z}$) by the so-called "restriction of the scalars" [31; § 1], and *mutatis mutandis*, the previous results also apply to these apparently more general groups of units [7, § 12]. Clearly, they are also valid for subgroups of finite index of groups of units.

6. Adele groups

So far we have essentially considered $G_\mathbf{R}$, $G_\mathbf{Z}$ and $G_\mathbf{Q}$. Now $G_\mathbf{R}$ is attached to G and to the completion $\mathbf{R}$ of $\mathbf{Q}$ with respect to the usual absolute value. Similarly, we may associate a group G_p to G to the p-adic valuation of $\mathbf{Q}$, where p is any prime. G_p is then the group of $n \times n$ invertible matrices with coefficients in the field Q_p of p-adic numbers, which satisfy the same equations as the elements of G; this makes sense, since these are polynomial equations with coefficients in $\mathbf{Q}$, hence in $\mathbf{Q}_p$. Endowed with the p-adic topology, G_p is a locally compact, totally disconnected group. The subgroup $G_{\mathbf{Z}_p}$ of elements of G_p with coefficients in the ring $\mathbf{Z}_p$ of p-adic integers and whose determinant is a p-adic unit is then compact and open. The systematic consideration of the p-adic algebraic groups G_p besides the real algebraic groups $G_\mathbf{R}$ has considerably widened the scope of the theory of linear algebraic groups, and I shall try to illustrate this with some examples.

Following a procedure familiar in number theory, we may associate to G a certain subgroup of the direct product of $G_\mathbf{R}$ by the $G_p's$, the *adele group* attached to G, to be denoted by G_A. By definition, an element x of this direct product is an adele of G if and only if for all but a finite number of p's, the component of x in G_p belongs actually to $G_{\mathbf{Z}_p}$. Thus G_A contains in particular the product $G_A^\infty = G_\mathbf{R} \times \Pi_p\, G_{\mathbf{Z}_p}$ of the locally compact group $G_\mathbf{R}$ by the compact groups $G_{\mathbf{Z}_p}$; the group G_A^∞ is therefore locally compact for the product topology. G_A itself admits one structure of topological, locally compact group in which G_A^∞ is an open subgroup.

If we apply this construction to the cases where G is isomorphic over $\mathbf{Q}$ to the additive group of $\mathbf{C}$, or to the multiplicative group of non-zero complex numbers, we get respectively the adele and idele groups of $\mathbf{Q}$, in the sense of Chevalley.

The group $G_\mathbf{Q}$ is contained in $G_\mathbf{R}$ and in all G_p's, and any $a \in G_\mathbf{Q}$ belongs to $G_{\mathbf{Z}_p}$ for all but a finite number of p's, those which occur in the denominators of the coefficients or in the determinant of x; therefore, the element of the direct product $G_\mathbf{R} \times \Pi_p\, G_p$ all of whose components are equal to a is an adele of G, called a *principal adele*. Thus we may (and shall) identify $G_\mathbf{Q}$ with a subgroup of G_A, so to say diagonally embedded, which is easily seen to be discrete.

Adele groups were introduced by Ono [19, 20], Tamagawa, and also by M. Kneser in the case of orthogonal groups. They can also be associated to groups defined over finite number fields, or over function fields of one variable with finite constant field [31]. As above, the number field case can be reduced to the one considered here by restriction of the scalars [31].

7. Finiteness theorems for adele groups

The definition of G_A puts on almost equal footing the p-adic completions of $\mathbf{Q}$ and $\mathbf{R}$ and, roughly speaking, the pair $G_\mathbf{R}, G_\mathbf{Z}$ appears as the localization, at the infinite prime spot, of the pair $G_A, G_\mathbf{Q}$. However, this symmetry is not a true one, the infinite prime spot plays a special role among the primes, and the results discussed in the previous sections imply rather easily some global properties of adele groups [4, 6], proved for solvable groups first by Ono [20], in particular:

(*a*) If G is connected, the number of double cosets $G_A^\infty \cdot x \cdot G_Q$ of G_A is finite.

If G is the proper orthogonal group of a non-degenerate rational quadratic form F, these double cosets correspond to the proper classes in the genus of F. If G is obtained by restriction of the scalars from the multiplicative group of a number field k, they correspond to the ideal classes of k.

(*b*) There exist closed or open fundamental sets for G_Q in G_A, which have finite Haar measure when G is semi-simple or, more generally, when the identity component of G has no non-trivial rational character defined over Q.

A fundamental set is here, in analogy with the definition given in § 2, a subset of G_A which meets every coset $x \cdot G_Q$ $(x \in G_A)$ and intersects only a finite number of its right translates by elements of G_Q. Under the last condition of (*b*), every left invariant Haar measure on G_A is also right invariant (i.e. G_A is unimodular) and defines an invariant measure on G_A/G_Q; our assertion implies in this case that G_A/G_Q has finite volume for any such measure.

(*c*) G_A/G_Q is compact if G is semi-simple and G_Q consists of semi-simple elements.

The condition for compactness is the same as for G_R/G_Z. Another proof of (*c*) is given in [33].

8. Tamagawa numbers

Let G be connected, unimodular, and ω a left-invariant rational differential m-form ($m = \dim G$) defined over Q. Any other such form is a rational multiple of ω. The latter defines in a natural way Haar measures ω_∞ and ω_p on G_R and G_p (p any prime). Assume that the product of the volumes v_p of the groups G_{Z_p} for those measures converges absolutely. Then the product measure of ω_∞ and the ω_p's is defined on G_A^∞ and gives rise to a Haar measure ω_A on G_A, which does not change if ω is a replaced by a rational multiple of itself. Weil has proposed to call the volume $\tau(G)$ of G_A/G_Q with respect to this normalized Haar measure the *Tamagawa number of G*. The above convergence condition is fulfilled when G is semisimple ([3]); if it is not, we may still multiply the ω_p's by suitable convergence factors λ_p and define the Tamagawa number of G with respect to the λ_p's (for all this, see [31, Chapter 2]).

It was noticed by T. Tamagawa and M. Kneser that when G is the proper orthogonal group of a non-degenerate rational quadratic form F in at least three variables, the equality $\tau(G) = 2$ is essentially equivalent to some of the main results obtained by Siegel in his famous series of papers on the analytic theory of quadratic forms [26].([4]) Briefly, this comes about in the following way: it follows in a straightforward fashion from the definitions, for any semi-simple group, that $\tau(G) = v_\infty \cdot \Pi_p v_p$, where v_p (p prime) is as

---

([3]) This was noticed by Serre some years ago. It follows from the relation between v_p and the number of rational points of the reduction mod p of G when the reduction is "good" [31] and from the formulas giving the number of points of a simple algebraic group defined over a finite field.

([4]) It corresponds to the case $m = n$, $S = \mathcal{T}$ of Siegel (representations of a form into itself). As to the passage from there to the general case, see M. Kneser (*Math. Zeit.* 77 (1961), 188–194), A. Weil, (*Colloque sur la Théorie des Groups Algébriques*, Bruxelles, 1962).

above and where v_∞ is the sum of the volumes of $G_\mathbf{R}/G_\mathbf{Z}$ and of finitely many quotients $G_\mathbf{R}/\Gamma$, where Γ runs through groups of units for certain lattices which correspond to representatives of the double cosets $G_A^\infty \cdot x \cdot G_\mathbf{Q}$. If $\tau(G)$ is then equal to two, or to some simple more or less universal constant, which, so to say, comes from the outside, this means that $1/v_\infty$ may be expressed as an infinite product of factors v_p, where v_p depends only on G_p. In the orthogonal group case, v_∞ is Siegel's generalization of the measure of the genus of a positive form and v_p may be also interpreted as a p-adic density of numbers of units of F modulo high powers of p, whence the classical formulation of Siegel's theorem.

Similarly, if G is the symplectic group $\mathbf{Sp}(2n)$, then v_∞ is the volume of $G_\mathbf{R}/G_\mathbf{Z}$ and the equality $\tau(G)=1$ is equivalent to a formula of Siegel [28] expressing the volume of the fundamental domain for the modular group as an infinite product.

Thus the Tamagawa number has considerable interest. It has been computed in various cases by Tamagawa, Weil, Demazure [31], Ono [21, 34], but our knowledge of it is still very incomplete. In a number of cases, it is equal to the order of the fundamental group of G. However, Ono [34] has given an example of a connected semi-simple group whose Tamagawa number is not integral, although still rational. It is not known whether $\tau(G)$ is always a rational number. It is conjectured to be equal to one when G is connected and simply connected.

The known computations of Tamagawa numbers use induction, residues of certain analogues of zeta-functions, and the Poisson summation formula. This requires a mixture of additive and multiplicative features, and so far the method could be applied mainly to groups of invertible elements in certain algebras defined over $\mathbf{Q}$.

As to adele groups defined over function fields, Tamagawa numbers have been computed and analogues of (a), (b), (c) in § 7 have been proved only for certain classical groups [31]. For the finiteness theorems, the methods of [4, 6] are of no avail, in view of the lack of archimedean primes. A further difference is the need of a theory of algebraic groups defined over non-perfect fields. However the situation there might improve soon, since Grothendieck has recently obtained results of a "foundational" character on semi-simple groups defined over arbitrary fields, in fact on schemes of semi-simple groups, some of which are already used in [30].

9. Principal homogeneous spaces

Another problem involving p-adic algebraic groups as well as real algebraic groups and of considerable interest for arithmetic is the existence of rational points in principal homogeneous spaces.

A principal homogeneous space is so to say what remains of a group when you disregard the identity. A simple example is the affine space, as compared to the additive group of a vector space. More precisely, in the context of algebraic geometry, a principal homogeneous space is an algebraic variety V on which an algebraic group G acts, say on the right, by means of a map $G \times V \to V$ which is regular in the sense of algebraic geometry, so that for every $v \in V$, the map $g \to v \cdot g$ is a birational biregular map of G onto V. Here, contrary to our general convention, G need not be linear, defined over $\mathbf{Q}$, and may be any group variety. The principal homogeneous space is said to be defined over a field k if k is a field of definition for G, V and the action of G onto V. Of course, once we have picked up a point $v \in V$, we may identify V with G acting on itself by means of right translations, but an important question is precisely over which fields we can choose rational points.

For example, let $\mathfrak{g}, \mathfrak{g}'$ be two Lie algebras over k which become isomorphic after extension of the groundfield to an algebraic closure $\bar{k}$ of k. Then the set of isomorphisms of $\mathfrak{g} \otimes \bar{k}$ onto $\mathfrak{g}' \otimes \bar{k}$ is in an obvious way a principal homogeneous space for the group of automorphisms of $\mathfrak{g}' \otimes \bar{k}$. It is defined over k, and if it has a rational point over an extension k' of k, this means that $\mathfrak{g} \otimes k'$ is isomorphic to $\mathfrak{g}' \otimes k'$. Principal homogeneous spaces occur in this fashion when studying isomorphism classes of geometric objects whose automorphism group is algebraic.

There is an obvious notion of isomorphism over k for principal homogeneous spaces of a group G defined over k. These isomorphism classes are in 1–1 correspondence with the elements of a Galois cohomology set, usually denoted $H^1(k, G)$. This is the first cohomology set of the Galois group of the separable closure k_s of k, with coefficients in G_{k_s}. It G is commutative, $H^1(k, G)$ is a commutative group; otherwise it is just a set with a zero element, the isomorphism class of principal homogeneous spaces which "split over k", i.e. which have rational points over k and are therefore isomorphic over k to G acting on itself by right translations.

Coming now to the case where $k = \mathbf{Q}$, a particularly interesting question is whether two geometric objects defined over $\mathbf{Q}$ which are isomorphic over every completion of $\mathbf{Q}$ are already isomorphic over $\mathbf{Q}$. That for instance a statement of this kind is true for rational quadratic forms is a classical theorem of Hasse. In our present set up, the question is whether a principal homogeneous space defined over $\mathbf{Q}$ and having rational points for all completions of $\mathbf{Q}$ also has a point rational over $\mathbf{Q}$; if not, what can be said about the number, call it $\nu(G)$, of classes of such principal homogeneous spaces? Hasse's theorem implies that $\nu(G) = 1$ when G is an orthogonal group. It can be shown that $\nu(G)$ is *finite for any linear algebraic group G* (see [4, 6] when G is reductive, connected, [8] for the extension to the general case). It has been known for a long time that $\nu(G)$ may be $\neq 1$ when G is an algebraic torus, or is not connected; recently Serre has provided an example where G is semi-simple and connected (yet unpublished). It is conjectured that $\nu(G) = 1$ when G is connected and simply connected; but this would not contain Hasse's theorem, and the situation is not well understood at all. Another conjecture relating local and global properties is that, G being simple over $\mathbf{Q}$, if $G_\mathbf{R}$ and all G_p's contain unipotent elements $\neq e$, then so does $G_\mathbf{Q}$. This would generalize the fact, (due to Hasse, too), that a rational quadratic form which represents zero rationally over every completion of $\mathbf{Q}$ also does so over $\mathbf{Q}$.

The two quoted theorems of Hasse on quadratic forms play a role in the computation of Tamagawa numbers, and it is somewhat tempting, at least to me, to think that there is some connection between the failure of $\nu(G)$ to be one and the failure of $\tau(G)$ to be an integer; but there is so far little evidence to back this up.

In our previous example with Lie algebras, the finiteness of $\nu(G)$ implies that the isomorphism classes of Lie algebras over $\mathbf{Q}$ which are isomorphic to a given Lie algebra after extension of the groundfield to all completions of $\mathbf{Q}$ are finite in number. An example of Landherr [16] shows that there may be more than one class. That there is only one such class for Lie algebras of type G_2 or F_4 [1] is positive evidence for the above conjecture on simply connected groups.

The problem discussed so far is a global one, or rather relates local and global properties, and it is natural to complete it by local questions, over a p-adic field k (which is assumed here to be of characteristic zero). It is known that $H^1(k, G)$ is finite when G is linear, defined over k; moreover, the number of forms of G, i.e. the number of isomorphism classes over k of algebraic groups defined over k and isomorphic to G over an algebraic closure of k, is also finite [8]; the latter finiteness statement is not a special case of the former one, since the group of automorphisms of an algebraic group, unlike that of a Lie algebra, is not always an algebraic group. It is conjectured, and has been checked in many cases, that $H^1(k, G) = 0$ when G 6 is connected and simply connected.

There are a number of partial results and conjectures on principal homogeneous spaces for linear groups, for which we refer to [14, 25]. The problems on principal homogeneous spaces are often meaningful for general group varieties, not necessarily matric groups, and in fact have attracted much attention in the case of abelian varieties (see e.g. [17, 29]).

10. Groups of S-units

When G is the group obtained by restriction of the scalars from the multiplicative group of a finite number field k, $G_{\mathbf{Z}}$ is nothing but the group of units of k, and its finite generation is part of the Dirichlet unit theorem. It is well known that this was generalized by Hasse and Chevalley to the group of S-units of k, where S is a finite set of primes, i.e. to the group of elements of k^* which are p-adic units for all primes p not in S. If now G is as before an algebraic matric group defined over $\mathbf{Q}$, we may similarly introduce the group $G_{\mathbf{Z}_s}$ of S-units of G. These are the elements of $G_{\mathbf{Q}}$ whose determinant is, up to sign, a power product of primes of S, and whose coefficients have denominators which are products of elements of S. If S is the empty set, then $G_{\mathbf{Z}_s} = G_{\mathbf{Z}}$.

M. Kneser has shown that if G_p has a compact set of generators for every $p \in S$, then $G_{\mathbf{Z}_s}$ is finitely generated. The proof uses the finite generation of $G_{\mathbf{Z}}$ and (a) of § 7. The condition on the G_p's is fulfilled, for any p, when G is reductive (i.e. fully reducible or, equivalently, isogeneous to the product of a semi-simple group by an algebraic torus), as follows from the generalized Bruhat lemma of [5, 30]. Earlier, H. Behr [3] had shown for certain classical groups that $G_{\mathbf{Z}_s}$ is finitely generated and finitely presented. To my knowledge it is not proved, although of course very likely, that the latter property 7 is true for any reductive group G.

It is shown in [7] that if $f : G \to G'$ is an isogeny (a rational surjective homorphism with finite kernel) defined over $\mathbf{Q}$ of G onto an algebraic matric group G' also defined over $\mathbf{Q}$, then $f(G_{\mathbf{Z}})$ is commensurable with $G'_{\mathbf{Z}}$. Honda [13] has proved the analogous statement for $G_{\mathbf{Z}_s}$ when S is big enough (a restriction which Serre has recently removed (unpublished), see also [6]), and has also considered similar questions for abelian varieties.

Let us mention finally a problem which, as far as I know, is unsolved in the general case for groups of units (and *a fortiori* for groups of S-units): whether the commutator subgroup of $G_{\mathbf{Z}}$ has finite index in $G_{\mathbf{Z}}$ when G is semi-simple. It follows from results of Matsushima [18] that this is true when D (notation of § 1) is a product of irreducible bounded symmetric domains none of which is isomorphic to a unit ball and $D/G_{\mathbf{Z}}$ is compact. According to Steinberg, it is also true when G is the adjoint group of a semi-simple Lie algebra in Chevalley's normal form (unpublished).

11. Compact subgroups of p-adic groups

All this has aroused much interest in p-adic Lie groups and, to round off this report, we would like to mention some unsolved problems on them. An algebraic matric p-adic group is at the same time a p-adic Lie group (p-adic analytic manifold endowed with a group structure for which product and inverse are analytic functions). From the two approaches to Lie group theory mentioned at the beginning of the introduction, the second one has yielded results valid for p-adic as well as for real groups and has been so far more successful than the first one. In particular our knowledge of maximal compact subgroups is still scanty. Although they do not always exist (e.g. in the additive group of a p-adic field), it seems rather likely that a semi-simple algebraic matric group L always has some; this has been checked by Bruhat [9, 10], Satake for many classical groups and groups in Chevalley's normal form([5]). However, they may form several conjugacy classes [10].

Let B be a maximal irreducible subgroup of L which can be put in triangular form. An important question is to know whether there exists a compact subgroup K of L such that

$$L = K \cdot B, \tag{1}$$

$$L = K \cdot A \cdot K, \tag{2}$$

where A is a Cartan subgroup of B. In the real case, (1) is the Iwasawa decomposition, and (2) follows from a well-known theorem of E. Cartan. (1) and (2) have been checked for many classical groups by Bruhat [9], Satake, and play a basic role in the study of infinite dimensional representations and of zonal functions. In general, it is only known that L is the union of a finite number of double cosets $K \cdot x \cdot B$ for any compact and open subgroup K.

12. Concluding remarks

I hope that this report, although not exhaustive,([6]) has given an idea of the general trends and aims of these investigations, and of some of their connections with arithmetic. The general results obtained so far are chiefly finiteness theorems, that is of a qualitative nature. They have yet to be matched by quantitative results of similar scope, e.g. on Tamagawa numbers, or on the true extent of the generalization of Hasse's theorem. One

---

([5]) (Added in proof): A general proof has been recently indicated to me by R. P. Langlands.

([6]) In particular, we refer to [14] for approximation problems in adele groups.

possible obstacle may be the lack of a good device to make use of class field theory, or rather of some consequences of it which occur frequently when dealing with classical groups. But, more likely, what is required first of all are new developments of an analytic nature, notably on Eisenstein series, automorphic functions, representation theory, and some of this not only for real algebraic groups, but also for p-adic and adele groups. No doubt this is a vast program, whose completion does not seem in sight at present. However these questions have recently been and are being studied in several quarters from different angles. There are substantial results and promising lines of attack. It is therefore not too unlikely that further progress will be achieved in the not too distant future and that the arithmetic properties of algebraic groups surveyed here will become part of a general arithmetic and analytic theory of algebraic groups.

REFERENCES

[1]. ALBERT, A. A. & JACOBSON, N., On reduced exceptional simple Jordan algebras. *Ann. Math.*, 66 (1957), 400–417.

[2]. ANDREOTTI, A. & GRAUERT, H., Algebraische Körper von automorphen Funktionen. *Nachr. Akad. d. Wiss. Göttingen, math.-phys. Klasse 1961*, Heft 3, 40–48.

[3]. BEHR, H., Über die endliche Erzeugbarkeit verallgemeinerter Einheitsgruppen. (To appear.)

[4]. BOREL, A., Some properties of the adele groups attached to algebraic groups. *Bull. Amer. Math. Soc.*, 67 (1961), 583–585.

[5]. —— Ensembles fondamentaux pour les groupes arithmétiques. *Coll. Théorie des Groupes Algébriques*. Bruxelles, 1962.

[6]. —— Finiteness theorems for adele groups over number fields. (To appear.)

[7]. BOREL, A. & HARISH-CHANDRA, Arithmetic subgroups of algebraic groups. *Ann. Math.*, 75 (1962), 485–535.

[8]. BOREL, A. & SERRE, J.-P., Théorèmes de finitude en cohomologie galoisienne. (En préparation.)

[9]. BRUHAT, F., Sur les représentations des groupes classiques p-adiques I, II, *Amer. J. Math.* 83 (1961), 321–338, 343–368.

[10]. —— Sur les sous-groupes compact maximaux des groupes algébriques p-adiques. *Coll. Théorie des Groupes Algébriques*. Bruxelles, 1962.

[11]. GEL'FAND, I. M., Automorphic functions and theory of representations. *Proc. Internat. Congress Math. Stockholm*, 1962. Uppsala 1963.

[12]. HERMITE, C., *Œuvres Complètes*, Vol. 1. Gauthier-Villars, Paris, 1905.

[13]. HONDA, K., Isogenies, rational points and section points of group varieties. *Japan J. Math.*, 30 (1960), 84–101.

[14]. KNESER, M., Einfach zusammenhängende algebraische Gruppen in der Arithmetik. *Proc. Internat. Congress Math. Stockholm*, 1962. Uppsala 1963.

[15]. KORKINE, A. & ZOLOTAREFF, G., Sur les formes quadratiques. *Math. Ann.*, 6 (1873), 366–389.

[16]. LANDHERR, W., Liesche Ringe vom Typus A über einem algebraischen Zahlkörper (die lineare Gruppe) und hermitesche Formen über einem Schiefkörper. *Abh. Math. Sem. Hamburg*, 12 (1938), 200–241.

[17]. LANG, S., Some theorems and conjectures in diophantine equations. *Bull. Amer. Math. Soc.*, 66 (1960), 240–249.

[18]. Matsushima, Y., On the first Betti number of compact quotient spaces of higher dimensional symmetric spaces, *Ann. Math.*, 75 (1962), 312–330.

[19]. Ono, T., Sur une propriété arithmétique des groupes commutatifs. *Bull. Soc. Math. France*, 85 (1957), 307–323.

[20]. —— On some arithmetic properties of linear groups. *Ann. Math.*, 70 (1959), 266–290.

[21]. —— Arithmetic of algebraic tori, ibid. 74 (1961), 101–139.

[22]. Pyatetski-Šapiro, *Geometry of Classical Domains and Automorphic Functions*. Moscow, 1961.

[23]. Satake, I., On compactifications of the quotient spaces for arithmetically defined discontinuous groups. *Ann. Math.*, 72 (1960), 555–580.

[24]. Selberg, A., Discontinuous groups and harmonic analysis. *Proc. Internat. Congress Math. Stockholm*, 1962, 177–189. Uppsala 1963.

[25]. Serre, J-P., Cohomologie galoisienne des groupes algébriques linéaires. *Coll. Théorie des Groupes Algébriques*. Bruxelles, 1962.

[26]. Siegel, C. L., Über die analytische Theorie der quadratischen Formen. *Ann. Math.*, 36 (1935), 527–606; 37 (1936), 230–263; 38 (1937), 212–291.

[27]. —— Einheiten quadratischer Formen. *Abh. Math. Sem. Hamburg.* 13 (1939), 209–239.

[28]. —— Symplectic geometry. *Amer. J. Math.*, 65 (1943), 1–80.

[29]. Tate, J., *WC*-groups over p-adic fields. *Séminaire Bourbaki*, Exp. 157, Paris, Décembre 1957.

[30]. Tits, J., Groupes semi-simples isotropes, *Coll. Théorie des Groupes Algébriques*. Bruxelles, 1962.

[31]. Weil, A., *Adeles and Algebraic Groups*. Notes by M. Demazure & T. Ono. The Institute for Advanced Study, Princeton, 1961.

(Added in proof):

[32]. Borel, A., Compact Clifford-Klein forms of symmetric spaces. (To appear in *Topology*.)

[33]. Mostow, G. D. & Tamagawa, T., On the compactness of arithmetically defined homogeneous spaces, *Ann. Math.*, 76 (1962), 446–463.

[34]. Ono, T., On the Tamagawa number of algebraic tori. (To appear.)

62.

Compact Clifford-Klein forms of symmetric spaces

Topology **2** (1963) 111–122

§ 1. INTRODUCTION

1.1. A CLIFFORD–KLEIN form of a connected and simply connected Riemannian manifold M is a Riemannian manifold M' whose universal Riemannian covering (universal covering endowed with the metric lifted from the metric of M') is isomorphic to M. The main purpose of this Note is to prove the following:

THEOREM A. *A simply connected Riemannian symmetric space M always has a compact Clifford–Klein form. Any such form M' has a finite Galois covering which is proper, unless $M' \cong M$.*

We recall that a Riemannian manifold X is symmetric, in the sense of Cartan, if it is connected and if every point $x \in X$ is an isolated fixed point of an involutive isometry s_x of X. The map s_x is then the unique isometry of X leaving x fixed and whose differential at x is $-\mathrm{Id}$. The group $I(X)$ of isometries of X is transitive, and X may be identified with the quotient $I(X)^\circ/K$ of the identity component $I(X)^\circ$ of $I(X)$ by a compact subgroup.

Among the simply connected symmetric spaces with negative curvature are to be found the bounded symmetric domains in $\mathbf{C}^n$ [8]. A bounded domain $M \subset \mathbf{C}^n$ is symmetric if every point $x \in M$ is an isolated fixed point of an involutive complex analytic homeomorphism of M. It is then Riemannian symmetric with respect to the Riemannian metric defined by the Bergmann metric, and its group of complex analytic homeomorphisms contains $I(M)^\circ$. A Clifford–Klein form M' of M will be called complex analytic if it is a complex analytic manifold and if the natural projection of M onto M' is holomorphic. Since the group of complex analytic homeomorphisms of M has finite index in $I(M)$, Theorem A clearly implies the:

COROLLARY. *A bounded symmetric domain always has a compact complex analytic Clifford–Klein form. Any such form has a proper finite Galois covering.*

Combined with some results of Hirzebruch [12], this shows that certain bounded symmetric domains, in particular the open unit ball $B^n = \{z \in \mathbf{C}^n | z_1 \bar{z}_1 + \ldots + z_n \bar{z}_n < 1\}$

344

in $\mathbf{C}^n$ for n even, have compact complex analytic Clifford–Klein forms with arbitrary large index (see §5). Since all these manifolds are projective by a theorem of Kodaira [15], they yield, when $M = B^2$, counter-examples to a conjecture of Zappa [21]†.

1.2. The Clifford–Klein forms of the simply connected Riemannian manifold M are just, up to isomorphism, the quotients M/Γ, where Γ runs through the groups of isometries of M which act *freely* (i.e. only the identity has fixed points) and are *properly discontinuous*. We recall that a group of homeomorphisms of a locally compact space is properly discontinuous if every compact set meets only a finite number of its transforms; in particular, the stability group of a point is finite, and therefore a torsion-free properly discontinuous group of homeomorphisms always acts freely‡.

Similarly, the complex analytic Clifford–Klein forms of a bounded symmetric domain are the quotients of M by properly discontinuous torsion-free groups of complex analytic homeomorphisms.

In either case, the finite Galois coverings of M/Γ are, up to isomorphism, the quotients M/Γ', where Γ' runs through the normal subgroups of finite index of Γ. If $I(M)$ is transitive on M, then $M = I(M)/K$, where K is compact, therefore a subgroup $\Gamma \subset I(M)$ is properly discontinuous on M if and only if it is discrete in $I(M)$, as follows from Proposition 5 in [5, Chap. 3, §4, No. 2], and M/Γ is compact if and only if $I(M)/\Gamma$ is. Consequently, Theorem A is implied by

THEOREM B. *Let M be a simply connected Riemannian symmetric manifold, $I(M)$ its group of isometries. Then:*

(i) *$I(M)$ contains a discrete uniform subgroup‖.*

(ii) *If Γ is a discrete uniform subgroup of $I(M)$ or, more generally, if Γ is a finitely generated subgroup of $I(M)$, different from (e), then Γ has a proper normal torsion-free subgroup of finite index.*

The proofs of (i) and (ii) are quite different from each other: (ii) is a consequence of some rather general, essentially known, facts about linear groups (see §2), while (i) is easily derived from:

THEOREM C. *A connected semi-simple Lie group G always has a discrete uniform subgroup*

which is a straightforward application of the compactness criterion of [3, §11] (Godement's conjecture), once it is known that a real non-compact semi-simple Lie algebra $\mathfrak{g}$ has a form defined over a totally real number field $E \neq \mathbf{Q}$, all of whose conjugates are

---

† That suitable Clifford–Klein forms of B^2 would allow one to disprove this conjecture was pointed out several years ago by Hirzebruch, whose request for such manifolds is in fact at the origin of the present note.

‡ Conversely, if M has negative curvature (the only case of interest in this paper), a group of isometries acting freely is torsion free, since in this case every finite, (or compact), group of isometries has a fixed point [9, Note III, No. 19].

‖ A closed subgroup H of a topological group L is uniform if L/H is compact.

compact (3.8); that fact in turn will be deduced from the existence of a form of $\mathfrak{g}$ defined over $\mathbf{Q}$ which has a Cartan involution also defined over $\mathbf{Q}$ (3.7).

Theorem C was announced in [2]. The contents of this paper have been the subject matter of two lectures given at the University of Bonn in the Spring of 1961.

1.3. As is well-known, both the geometric and the arithmetic methods allow one to construct compact Clifford–Klein forms of the Poincaré plane, in other words, compact Riemann surfaces of genus ≥ 2, (see, e.g., [11]). It has been known for a long time that the latter one could also be used for many classical spaces of negative curvature, including the n-dimensional hyperbolic space ($n \geq 2$). In its usual form, it consists in constructing the fundamental group of the Clifford–Klein form by applying (2.2) to the groups of units of certain quadratic or hermitian forms, defined over suitable number fields, which do not represent zero rationally over their field of definition. We refer to [11, Abschnitt III, Kap. 3], [14] and [20] for various applications of this principle, and to (4.3) for a typical example. Löbell [16] and Seifert and Weber [17] have given some examples of geometrically defined compact Clifford–Klein forms of the hyperbolic 3-space. It is not known to the author whether the fundamental groups of these examples are commensurable with arithmetic groups or whether the geometric method can be used in other spaces.

§2. MATRIX GROUPS. PROOF OF THEOREM B(ii)

Propositions (2.2) and (2.3) are known, up to a slight strengthening, which does not require any substantial change in the proof, but is nevertheless needed here. For the sake of completeness, we include full proofs.

2.1. The characteristic polynomial of a $n \times n$ matrix x in the indeterminate λ, will be denoted $C(x, \lambda)$.

Let J be a ring, with identity. Then $\mathbf{GL}(n, J)$ denotes the group of $n \times n$ matrices with coefficients in J whose determinant is a unit of J.

Let in particular J be the ring of integers of a number field k of finite degree, and $\mathfrak{p}$ a prime ideal of J. The set of matrices of $\mathbf{GL}(n, J)$ which are $\equiv 1$ modulo $\mathfrak{p}$ is called a congruence subgroup of $\mathbf{GL}(n, J)$. It is a normal subgroup of finite index, since it is the kernel of the natural homomorphism of $\mathbf{GL}(n, J)$ into the finite group $\mathbf{GL}(n, J/\mathfrak{p})$.

2.2. PROPOSITION. *Let k be a finite number field, J the ring of integers of k, Γ a subgroup of $\mathbf{GL}(n, J)$ and $\gamma \neq e$ an element of Γ. Then there exists a torsion-free normal subgroup of finite index of Γ which does not contain γ.*

It is enough to show the existence of a torsion-free congruence subgroup of $\mathbf{GL}(n, J)$ which does not contain γ.

The eigenvalues of an element $x \in \mathbf{GL}(n, J)$ are algebraic of degree $\leq n$ over k, hence of degree $\leq n.[k : \mathbf{Q}]$ over $\mathbf{Q}$. If moreover x has finite order, its eigenvalues are roots of unity, hence there are only a finite number of possibilities for them. There exists therefore a finite number of unitary polynomials $P_1, \ldots, P_m \in J[\lambda]$, of degree n, all different from

$(1 - \lambda)^n$, such that if $x \in \mathbf{GL}(n, J)$ has finite order and is $\neq e$, then $C(x, \lambda) = P_i(\lambda)$ for some i. The congruence subgroup corresponding to a prime ideal $\mathfrak{p}$ of J such that

$$\gamma \not\equiv 1(\mathfrak{p}), \qquad P_j(\lambda) \not\equiv (1 - \lambda)^n \qquad (\mathfrak{p}) \; (1 \leqq j \leqq m), \tag{1}$$

will then not contain γ and be torsion-free.

2.3. PROPOSITION. *Let Γ be a finitely generated subgroup of* $\mathbf{GL}(n, \mathbf{C})$. *Then*

(1) *There exist a finite number field E, and a homomorphism $f: \Gamma \to \mathbf{GL}(n, E)$ whose kernel is torsion-free*.

(2) Γ *contains a normal torsion-free subgroup of finite index, which is proper if* $\Gamma \neq (e)$.†

Let $(x_i)_{1 \leqq i \leqq 2t}$ be a system of generators of Γ, where $x_{i+t} = x_i^{-1} (1 \leqq i \leqq t)$, and let us put

$$x_i = (a_{k,i}^j) \qquad (1 \leq i \leq 2t; 1 \leqq j, k \leqq n).$$

We view the point $A = (a_{k,i}^j)$ of $\mathbf{C}^m (m = 2t \cdot n^2)$ as a generic point over the field F of all algebraic numbers of an affine algebraic set V. Let $B = (b_{k,i}^j)$ be a point of V with co-ordinates in F, which exists by Hilbert's Nullstellensatz, and let $y_i = (b_{k,i}^j)$ be the corresponding specialisation of x_i. A relation between the x_i's is expressed by the vanishing at A of a polynomial in m indeterminates $X_{k,i}^j$ with coefficients in F (in fact, in $\mathbf{Z}$), and will therefore also be satisfied by the y_i's. Therefore the map $x_i \to y_i (1 \leqq i \leqq 2t)$ extends to a well-defined homomorphic mapping $f: \Gamma \to GL(n, E)$, where E is the field generated over $\mathbf{Q}$ by the co-ordinates of B.

Let $\gamma \in \Gamma$ be an element of finite order, different from e. Then

$$C(\gamma, \lambda) \in F[\lambda], \qquad C(\gamma, \lambda) \neq (1 - \lambda)^n.$$

The specialisation $A \to B$ being over F, we have $C(f(\gamma), \lambda) = C(\gamma, \lambda)$, hence $C(f(\gamma), \lambda) \neq (1 - \lambda)^n$, and $f(\gamma) \neq e$, which ends the proof of (1). Let $\Gamma \neq (e)$. We may then assume that $f(\Gamma) \neq (e)$: in fact, if the point A above is algebraic, then f is the identity map, if not, then $\dim V \geqq 1$, V contains at least two algebraic points, and for one of them at least, the corresponding specialisations of the x_i's are not all equal to the identity matrix. Since $\ker f$ is torsion-free, it is then enough to prove (2) for $f(\Gamma)$; thus we may assume $\Gamma \subset \mathbf{GL}(n, E)$, where E is a finite number field. Let then $\gamma \neq e$ be an element of Γ, $\mathfrak{p}$ a prime ideal of E which satisfies (1) in (2.2) and does not divide the denominators of the a_{ik}^j, and E' the finite residue field $J_\mathfrak{p}/\mathfrak{p} \cdot J_\mathfrak{p}$, where $J_\mathfrak{p}$ is the valuation ring of $\mathfrak{p}$. Then $\Gamma \subset \mathbf{GL}(n, J_\mathfrak{p})$, and the reduction mod $\mathfrak{p}$ of the coefficients yields a homomorphism $\Gamma \to \mathbf{GL}(n, E')$ whose kernel satisfies our conditions.

2.4. *Proof of Theorem B(ii)*. If M/Γ is compact, then Γ is finitely generated, since in this case there exists a compact set $C \subset M$ such that $M = \Gamma \cdot C$, and then the finitely many $\gamma \in \Gamma$ for which $\gamma \cdot C \cap C \neq \phi$ generate Γ by a well-known elementary lemma (see, e.g., [3; 6.6]). In this proof, we shall therefore only assume that Γ is finitely generated. Since

---

† Except for the last assertion: "which is proper if $\Gamma \neq (e)$", this proposition is exactly Lemma 8 of [18].

a subgroup Γ' of finite index of Γ is finitely generated (by a known result of Schreier and contains a normal subgroup of finite index (the intersection of all the conjugates of Γ' in Γ), we may assume $\Gamma \subset I(M)^\circ$.

Let $M = M_0 \times M_1 \times \ldots \times M_t$ be the de Rham decomposition of M as the Riemannian product of a euclidean space M_0 by irreducible non-flat Riemannian manifolds $M_1, \ldots, M_t$. The M_i's are all symmetric and, by the standard theory of the Riemannian symmetric spaces [1, 6, 7], there is for each of them two possibilities: either M_i has positive curvature, is compact, $G_i = I(M_i)^\circ$ is compact semi-simple, or M_i has negative curvature, is homeomorphic to a euclidean space, G_i is simple non-compact with center reduced to the identity, and $M_i = G_i/K_i$, where K_i is a maximal compact subgroup of G_i. In both cases, G_i has a faithful linear representation: in the former one, because it is compact, in the latter one, because it may be identified with the group of inner automorphisms of its Lie algebra. Since the group $G_0 = I(M_0)^\circ$ of proper motions of a euclidean space is also linear, we see that $I(M)^\circ = G_0 \times \ldots \times G_t$ admits a faithful linear representation. The theorem follows now from (2.3).

§3. CERTAIN FORMS OF REAL SEMI-SIMPLE LIE ALGEBRAS

3.1. In this paragraph, the reader is assumed to be familiar with the theory of real or complex semi-simple Lie algebras [13, 19]. The following notation will be used.

$\mathfrak{g}_c$, a complex semi-simple (finite dimensional) Lie algebra,

$\mathfrak{h}_c$, a Cartan subalgebra of $\mathfrak{g}_c$,

$B(x, y) = tr(\text{ad } x^\circ \text{ ad } y)$, $(x, y \in \mathfrak{g}_c)$, the Killing form of $\mathfrak{g}_c$,

(h, h'), the scalar product induced on $\mathfrak{h}_c$ or on the dual h_c^* of $\mathfrak{h}_c$ by the Killing form,

h_a $(a \in h_c^*)$, the element of $\mathfrak{h}_c$ such that $(h, h_a) = a(h)$ for any $h \in h_c$, $h_a^* = 2.h_a/(a, a)$,

Φ, the system of roots of $\mathfrak{g}_c$ with respect to $\mathfrak{h}_c$,

Δ, the set of simple roots with respect to some chosen ordering,

$\mathfrak{h}^\circ$, the real subspace of $\mathfrak{h}_c$ on which the roots take real values,

$\mathfrak{g}_u$, a compact form of $\mathfrak{g}_c$.

We note that $\mathfrak{g}_u \cap \mathfrak{h}_c$ is a Cartan subalgebra of $\mathfrak{g}_u$ if and only if

$$\mathfrak{g}_u \cap \mathfrak{h}_c = \sqrt{-1} \cdot \mathfrak{h}^\circ. \tag{1}$$

3.2. It is known that given $\mathfrak{h}_c$ and $\mathfrak{g}_u$ subject to (1), there exists a basis $(h_a^*, x_b)(a \in \Delta, b \in \Phi)$ of $\mathfrak{g}_c$ satisfying the following conditions:

$$[h, x_a] = a(h) \cdot x_a \qquad (h \in \mathfrak{h}_c \, ; \, a \in \Phi), \tag{2}$$

$$[x_a, x_{-a}] = -h_a^* \qquad (a \in \Phi), \tag{3}$$

$$[x_a, x_b] = 0 \qquad \text{if } a, b \in \Phi, a + b \notin \Phi, a + b \neq 0, \tag{4}$$

$$[x_a, x_b] = N_{a,b} x_{a+b} \qquad \text{if } a, b, a + b \in \Phi, \text{ where} \tag{5}$$

$$N_{a,b} = N_{-a,-b} = \pm (p + 1), \tag{6}$$

and $p \geq 0$ is the greatest integer such that $a - p.b \in \Phi$.

---

† *Abh. math. Sem. Hamburg Univ.* **5** (1927), 161–184 (§3).

$$\mathfrak{g}_u = \sqrt{(-1)} \cdot \mathfrak{h}^\circ + \sum_{a \in \Phi} (\bar{\mu}_a x_a + \mu_a \cdot x_{-a}) \qquad (\mu_a \in C). \tag{7}$$

The existence of a basis verifying (2) to (5), with h_a^* in (3) replaced by h_a, and the first equality of (6), is a classical result of Weyl. That it can be modified so as to satisfy (2) to (6) was shown by Chevalley [10], and we shall call Chevalley basis a basis fulfilling those conditions. For any such basis, the right hand side of (7) is a compact form of $\mathfrak{g}_c$. The fact that $\mathfrak{h}_c$ and $\mathfrak{g}_u$, subject to (1), may be prescribed in advance, follows from Chevalley's result and from the conjugacy of the compact forms of $\mathfrak{g}_c$ and of the Cartan subalgebras of $\mathfrak{g}_u$ by inner automorphisms of $\mathfrak{g}_c$ and $\mathfrak{g}_u$ respectively. A basis satisfying (2) to (7) for given $\mathfrak{h}_c$ and $\mathfrak{g}_u$ will be said to be adapted to $\mathfrak{h}_c$ and $\mathfrak{g}_u$. Since

$$a(h_b^*) = 2(a, b)/(b, b) \qquad (a, b \in \Phi) \tag{8}$$

is an integer, (2) to (6) imply that the structure constants of $\mathfrak{g}_c$ with respect to a Chevalley basis are integers [10].

3.3. Let $t \in \mathfrak{h}_c$ and $A = \exp \operatorname{ad} t$. Then $A(h) = h(h \in \mathfrak{h}_c)$ and the $h_a^*(a \in \Delta)$ together with the elements $y_b = A(x_b) = (\exp b(t)) \cdot x_b(b \in \Phi)$ again form a Chevalley basis. If moreover $t \in \sqrt{(-1)} \cdot \mathfrak{h}^\circ$, that is if the roots take purely imaginary values on t, then $A(\mathfrak{g}_u) = \mathfrak{g}_u$ and the new basis is still adapted to $\mathfrak{h}_c$ and $\mathfrak{g}_u$. This follows immediately from (3.2) and the definitions.

3.4. Let θ be an automorphism of $\mathfrak{g}_c$ leaving $\mathfrak{h}_c$ invariant. It induces an automorphism $'\theta$ of h_c^* leaving Φ invariant. Writing a' for $'\theta^{-1}(a)$, we have

$$(a + b)' = a' + b', \qquad (-a)' = -(a'), \qquad (a, b \in \Phi), \tag{9}$$

$$\theta(h_a^*) = h_{a'}^*, \qquad \theta(x_a) = c_a \cdot x_{a'} \qquad (a \in \Phi; c_a \in C). \tag{10}$$

Let now $a, b, a + b \in \Phi$. Then, applying θ to both sides of (5), we get

$$c_a \cdot c_b [x_{a'}, x_{b'}] = N_{a,b} \cdot c_{a+b} \cdot x_{a'+b'}$$

$$c_a \cdot c_b \cdot N_{a',b'} = N_{a,b} \cdot c_{a+b}.$$

Since $a \to a'$ is an automorphism of Φ, (6) shows that $N_{a',b'} = \pm N_{a,b} \neq 0$, whence

$$c_a \cdot c_b = \pm c_{a+b} \qquad (a, b, a + b \in \Phi). \tag{11}$$

We note also that we have, in view of (3), (10),

$$c_a \cdot c_{-a} = 1, \qquad (a \in \Phi). \tag{12}$$

If moreover $\theta(\mathfrak{g}_u) = \mathfrak{g}_u$, then the relation $\theta(x_a + x_{-a}) = c_a \cdot x_{a'} + c_{-a} \cdot x_{-a'}$ yields

$$c_a = \bar{c}_{-a}, \qquad (a \in \Phi), \tag{13}$$

hence, using (12),

$$c_a \cdot \bar{c}_a = 1, \qquad (a \in \Phi). \tag{14}$$

3.5. LEMMA. *Let θ be an involutive automorphism of $\mathfrak{g}_c$ which keeps $\mathfrak{h}_c$ and $\mathfrak{g}_u$ invariant and*

leaves fixed an element of $\mathfrak{h}^\circ$ *regular in* $\mathfrak{g}_c$. *Then there exists a Chevalley basis* $(h_a^*, y_b)(a \in \Delta,$
$b \in \Phi)$ *of* $\mathfrak{g}_c$ *adapted to* $\mathfrak{h}_c$ *and* $\mathfrak{g}_u$ *such that*

$$\theta(y_b) = \pm y_b \quad \text{if} \quad b = b', \qquad \theta(y_b) = y_{b'} \quad \text{if} \quad b \neq b'(b \in \Phi;\ b' = {}^t\theta(b)).\dagger$$

We keep the previous notation. From $\theta^2 = \text{Id.}$, we get

$$(a')' = a, \qquad c_a \cdot c_{a'} = 1 (a \in \Phi). \tag{15}$$

Since θ leaves a regular element contained in $\mathfrak{h}^\circ$ fixed, it also leaves a Weyl chamber invariant. Let then Φ^+ be the set of positive roots and Δ the set of simple roots for the ordering associated to this Weyl chamber; they are both invariant under ${}^t\theta^{-1}$. By (14), the c_a have module one. In view of (15), it is therefore possible to find $t \in \sqrt{(-1)}. \mathfrak{h}^\circ$ such that

$$a(t) = 0 \quad \text{if} \quad a = a', \qquad a(t) + a'(t) = 0 \quad \text{if} \quad a \neq a'(a \in \Delta) \tag{16}$$

$$e^{-2a(t)} = c_a \qquad (a \in \Delta). \tag{17}$$

Let then $A = \exp ad\ t$ and

$$z_b = A(x_b) = e^{b(t)} \cdot x_b \qquad (b \in \Phi).$$

The z_b's are still part of a Chevalley basis adapted to $\mathfrak{g}_u$ and $\mathfrak{h}_c$ (see 3.3). If $a \in \Delta, a = a'$, then $c_a = \pm 1$ by (15), $z_a = x_a$, and $\theta(z_a) = \pm z_a$. If $a \in \Delta, a \neq a'$, then we have, by (15), (16), (17):

$$\theta(z_a) = e^{a(t)} \cdot c_a \cdot x_{a'} = e^{-a(t)} \cdot x_{a'} = e^{a'(t)} \cdot x_{a'} = z_{a'}.$$

We have therefore $c_a = \pm 1$ if $a \in \Delta$, and, by (12), also if $-a \in \Delta$. Using induction on the ordering of the roots and (11), we then see that $\theta(z_b) = \pm z_{b'}$ for any $b \in \Phi$.

From each pair of positive roots permuted by θ, let us choose one. Put $y_b = z_b$, $y_{-b} = z_{-b}$ for that root, and $y_{b'} = \theta(z_b)$, $y_{-b'} = \theta(z_{-b})$. Let moreover $y_b = z_b$ if $b = b'$, $(b \in \Phi)$. It is then clear that the y_b's satisfy all our conditions.

3.6. Let $\mathfrak{g}$ be a real non-compact semi-simple Lie algebra, θ a Cartan involution of $\mathfrak{g}$. We have then the Cartan decomposition

$$\mathfrak{g} = \mathfrak{k} + \mathfrak{p}, \qquad \theta(k + p) = k - p \qquad (k \in \mathfrak{k}, p \in \mathfrak{p}), \tag{18}$$

whence

$$[\mathfrak{k}, \mathfrak{k}] \subset \mathfrak{k}, \qquad [\mathfrak{k}, \mathfrak{p}] \subset \mathfrak{p}, \qquad [\mathfrak{p}, \mathfrak{p}] \subset \mathfrak{k}. \tag{19}$$

$\mathfrak{k}$ is the Lie algebra of a maximal compact subgroup of Ad $\mathfrak{g}$, and

$$\mathfrak{g}_u = \mathfrak{k} + \sqrt{(-1)} \cdot \mathfrak{p} \tag{20}$$

is a compact form of the complexification $\mathfrak{g}_c = \mathfrak{g} \otimes C$ of $\mathfrak{g}$. Clearly, θ extends to an involution of $\mathfrak{g}_c$ leaving $\mathfrak{g}_u$ stable, and we have (18), (19) with $\mathfrak{p}$ replaced by $\sqrt{(-1)}.\mathfrak{p}$, and $\mathfrak{g}$ by $\mathfrak{g}_u$.

Let $(e\lambda)$ be a basis of $\mathfrak{g}$ consisting of elements belonging either to $\mathfrak{k}$ or to $\mathfrak{p}$. Following Cartan [7], we agree to let i, j, k and α, β, γ stand for indices of the subbases of $\mathfrak{p}$ and $\mathfrak{k}$

$\dagger$ More general lemmas of this type may be found in F. Gantmacher, (*Mat. Sbornik* **5** (1939), 101–144). They are formulated there in terms of a Weyl basis; the passage to a Chevalley basis is of course immediate, and it is again mainly for the sake of completeness that a full proof is given.

respectively, and λ, μ, ν for arbitrary indices. The structure constant $c^{\nu}_{\lambda,\mu}$ and $c'^{\nu}_{\lambda,\mu}$ of $\mathfrak{g}$ and $\mathfrak{g}_u$ with respect to the bases (e_α, e_i) and $(e_\alpha, \sqrt{(-1)}.e_i)$ are related by

$$c^{\gamma}_{\alpha,\beta} = c'^{\gamma}_{\alpha,\beta}, \quad c^{j}_{\alpha,i} = c'^{j}_{\alpha,i}, \quad c^{\alpha}_{i,j} = -c'^{\alpha}_{i,j}, \tag{21}$$

and it is clear by (19) that

$$c^{i}_{\alpha,\beta} = c'^{i}_{\alpha,\beta} = c^{\beta}_{\alpha,i} = c'^{\beta}_{\alpha,i} = c^{k}_{i,j} = c'^{k}_{i,j} = 0. \tag{22}$$

3.7. PROPOSITION. *Let $\mathfrak{g}$ be a real semi-simple Lie algebra. Then there exists a form of $\mathfrak{g}$ over $\mathbf{Q}$ which admits a Cartan involution defined over $\mathbf{Q}$.*

We have to show the existence of a basis of $\mathfrak{g}$, made up of eigenvectors of a Cartan involution of $\mathfrak{g}$, with respect to which the constants of structure are all rational numbers.

Let first $\mathfrak{g}$ be compact. Its Cartan involutions are then all equal to the identity. We identify $\mathfrak{g}$ with a compact form $\mathfrak{g}_u$ of $\mathfrak{g}_c = \mathfrak{g} \otimes \mathbf{C}$. Let $(h^*_a, y_b)(a \in \Delta, b \in \Phi)$ be a Chevalley basis of $\mathfrak{g}_c$ adapted to $\mathfrak{g}_u$. Then the elements

$$\sqrt{(-1)} \cdot h^*_a (a \in \Delta), \qquad u_b = y_b + y_{-b}, \qquad v_b = \sqrt{(-1)}(y_b - y_{-b})(b \in \Phi, b > 0), \tag{23}$$

form a basis of $\mathfrak{g}_u$, and it follows at once from (3.2) that the constants of structure for this basis are all integers.

Let now $\mathfrak{g}$ be non-compact, θ a Cartan involution of $\mathfrak{g}$. We use the notation of (3.6). By [4, Theorem 4.5], $\mathfrak{k}$ contains an element t regular in $\mathfrak{g}$, hence in $\mathfrak{g}_c$. Let $\mathfrak{h}$ be the unique Cartan subalgebra of $\mathfrak{g}$ containing t, and $\mathfrak{h}_c$ its complexification. $\mathfrak{h}$ is invariant under θ, hence $\mathfrak{h} = \mathfrak{h} \cap \mathfrak{k} + \mathfrak{h} \cap \mathfrak{p}$, which gives

$$\mathfrak{h}_c \cap \mathfrak{g}_u = \mathfrak{k} \cap \mathfrak{h} + \sqrt{(-1)}(\mathfrak{p} \cap \mathfrak{h}),$$

and shows that $\mathfrak{h}_c \cap \mathfrak{g}_u$ is a Cartan subalgebra of $\mathfrak{g}_u$. Since $\sqrt{(-1)}.t$ is a regular element of $\mathfrak{g}_c$ contained in $\mathfrak{h}^\circ$ and fixed under θ, we see that all assumptions of (3.5) are satisfied by θ, or rather by the extension of θ to $\mathfrak{g}_c$. Let then (h^*_a, y_b) be the Chevalley basis of (3.5). We divide the set Φ^+ of positive roots for the ordering considered in the proof of (3.5) into three disjoint subsets:

$$A = \{a \in \Phi^+ | a' = a, \theta(y_a) = y_a\},$$
$$B = \{b \in \Phi^+ | b' = b, \theta(y_b) = -y_b\},$$
$$C = \{c \in \Phi^+ | c' \neq c\}.$$

Then, in the notation of (23), $\mathfrak{k}$ is spanned by

$$\sqrt{-1} \cdot h^*_a \quad (a \in \Delta \cap (A \cup B)), \qquad \sqrt{-1} \cdot (h^*_c + h^*_{c'}) \quad (c \in \Delta \cap C),$$
$$u_a, v_a \quad (a \in A), \qquad u_c + u_{c'}, v_c + v_{c'} \quad (c \in C),$$

and $\mathfrak{p}$ by

$$h^*_c - h^*_{c'}(c \in \Delta \cap C),$$
$$\sqrt{-1} \cdot u_b, \sqrt{-1} \cdot v_b \quad (b \in B), \qquad \sqrt{-1}(u_c - u_{c'}), \qquad \sqrt{-1}(v_c - v_{c'}) \quad (c \in C).$$

Since the elements of (23) form a rational basis of $\mathfrak{g}_u$, it follows from (19) that the structure constants of $\mathfrak{g}$ for the basis just described are rational numbers. Since it consists of elements belonging either to $\mathfrak{k}$ or to $\mathfrak{p}$, (3.7) is proved.

3.8. PROPOSITION. *Let $\mathfrak{g}$ be a non-compact semi-simple real Lie algebra. Let $E \subset \mathbf{R}$ be a totally real number field of finite degree > 1, and S the set of distinct isomorphisms of E into $\mathbf{R}$ which are different from the identity. Then there exists a form of $\mathfrak{g}$ defined over E all of whose conjugates $\mathfrak{g}^s (s \in S)$ are isomorphic to a compact form $\mathfrak{g}_u$ of $\mathfrak{g}_c$.*

In other words, there exists a Lie algebra $\mathfrak{m}$ over E, such that

$$\mathfrak{m} \otimes R \cong \mathfrak{g}, \qquad \mathfrak{m}^s \otimes R \cong \mathfrak{g}_u \qquad (s \in S). \tag{24}$$

We may write $E = \mathbf{Q}(u)$, where u is a positive number all of whose conjugates are negative. Let v be the positive square root of u, and v_s the positive square root of $-u^s (s \in S)$. We take in $\mathfrak{g}$ a basis (e_λ) as in (3.6), but such that moreover the structure constants are rational numbers, which is possible by (3.7). Let then $\mathfrak{m}$ be the vector space over E spanned by the elements e_α and $v.e_i$, and $\mathfrak{m}^s$ the vector space over E^s spanned by the elements e_α and $\sqrt{-1}.v_s.e_i (s \in S)$. The relations

$$[e_\alpha, v \cdot e_i] = v \cdot [e_\alpha, e_i], \qquad [v \cdot e_i, v \cdot e_j] = u \cdot [e_i, e_j],$$

$$[e_\alpha, \sqrt{-1} \cdot v_s \cdot e_i] = \sqrt{-1} \cdot v_s \cdot [e_\alpha, e_i], \qquad [\sqrt{-1} \cdot v_s \cdot e_i, \sqrt{-1} \cdot v_s \cdot e_j] = u^s [e_i, e_j],$$

and (19), show that $\mathfrak{m}$ (resp. $\mathfrak{m}^s$) is a Lie algebra over E (resp. $E^s)(s \in S)$, and that the constants of structure of $\mathfrak{m}^s$ are the conjugates by s of the structure constants of $\mathfrak{m}$, for the bases considered here. Thus $\mathfrak{m}^s$ is indeed the conjugate by s of $\mathfrak{m}$, which justifies the notation. Since $m \otimes \mathbf{R} = \mathfrak{g}$ and $m^s \otimes \mathbf{R} = \mathfrak{g}_u (s \in S)$ by definition, (3.8) is proved.

§4. PROOFS OF THEOREM B(i) AND C

4.1. *Proof of Theorem C.* Let $\mathfrak{g}$ be the Lie algebra of G and π the natural projection of G onto Ad $\mathfrak{g}$. If Γ is a discrete subgroup of Ad g, then $\pi^{-1}(\Gamma)$ is discrete in G, and of course $G/\pi^{-1}(\Gamma)$ is homeomorphic to (Ad $\mathfrak{g})/\Gamma$. We may therefore assume that $G = $ Ad $\mathfrak{g}$, with $\mathfrak{g}$ non-compact. The group Ad $\mathfrak{g}$ being open, of finite index, in the group Aut $\mathfrak{g}$ of all automorphisms of $\mathfrak{g}$, we have to show the existence of a discrete uniform subgroup Γ in Aut $\mathfrak{g}$.

We now use (3.8), write $\mathfrak{g} = \mathfrak{m} \otimes \mathbf{R}$, where $\mathfrak{m}$ satisfies (24) of §3, and take a basis of $\mathfrak{g}$ contained in $\mathfrak{m}$. This allows us to identify $L = $ Aut $\mathfrak{g}_c$ with an algebraic subgroup of $\mathbf{GL}(n, \mathbf{C})$, $(n = \dim \mathfrak{g})$, defined over E, and Aut $\mathfrak{g}$ with the group $L_\mathbf{R}$ of the real matrices contained in L. The group $(L^s)_\mathbf{R}$ $(s \in S)$ is then the group of automorphisms of the compact semi-simple Lie algebra $m^s \otimes \mathbf{R} \cong \mathfrak{g}_u$, hence is compact. Let J be the ring of algebraic integers of E, and $\Gamma = L_J$ the group of units of L, i.e. the group of matrices in L with coefficients in J and determinant invertible in J. Since $(L^s)_\mathbf{R}$ is compact for any $s \in S$, and S is not empty, it follows from [3, 12.2] that Γ is discrete in Aut $\mathfrak{g}$ and from [3, 12.4] that (Aut $\mathfrak{g})/\Gamma$ is compact.

4.2. *Proof of Theorem B(i).* As was remarked in (2.4), the group $I(M)^\circ$ is the product of a semi-simple group by the group of proper motions of a euclidean space. The existence of discrete uniform subgroups has just been proved for the former one, and is obvious for the latter one. Thus $I(M)^\circ$ also has discrete uniform subgroups; since $I(M)^\circ$ is open, of finite index, in $I(M)$, the same is true for $I(M)$.

4.3. When G is isomorphic to the real orthogonal group of a non-degenerate indefinite quadratic form of signature (a, b) on $\mathbf{R}^n (n = a + b; a, b \geq 1)$ the procedure to construct discrete uniform subgroups alluded to in (1.3) amounts to identify G with the real orthogonal group of a quadratic form F over the totally real number field E, of signature (a, b), all of whose conjugates are positive, and to take for Γ the group G_J of units of F; in the notation of (3.8), we may for instance take the form F given by

$$F(x, x) = x_1^2 + \ldots + x_a^2 - u(x_{a+1}^2 + \ldots + x_{a+b}^2), \tag{1}$$

in which case, by the way, G_J is commensurable to the inverse image of the group Γ constructed in (4.1), for a suitable basis.

For $b = 1$, the identity component of G may be identified with the identity component of the group of isometries of the a-dimensional hyperbolic space.

We may proceed similarly when G is the unitary group of an indefinite hermitian form of signature (a, b). For instance, we identify G with the unitary group of the hermitian form H given by

$$H(z, z) = z_1 \cdot \bar{z}_1 + \ldots + z_a \cdot \bar{z}_a - u(z_{a+1} \cdot \bar{z}_{a+1} + \ldots + z_n \cdot \bar{z}_n) ; \tag{2}$$

by means of the map,

$$z = x + \sqrt{-1} \cdot y \rightarrow \begin{pmatrix} x & y \\ -y & x \end{pmatrix},$$

we realize G as the group of real points of an algebraic subgroup $G' \subset \mathbf{GL}(2n, \mathbf{C})$ defined over E. The group $(G'^s)_\mathbf{R}$ is then isomorphic to the unitary group of the positive non-degenerate hermitean form H^s, hence is compact. The group G_B, (B ring of integers of $E(\sqrt{-1})$) of units of H, which corresponds to G'_J under the isomorphism $G \rightarrow G'_\mathbf{R}$, is then discrete and uniform. When $b = 1$, G modulo its one-dimensional center is isomorphic to the group of automorphisms of the unit ball $B^b \subset \mathbf{C}^b$.

§5. REMARKS ON BOUNDED SYMMETRIC DOMAINS

5.1. Let $M = G/K$ be a bounded symmetric domain, where G is connected, semi-simple, with center reduced to (e), and K is a maximal compact subgroup of G. To M corresponds a compact hermitian symmetric space M_u, containing an open subset bi-holomorphically homeomorphic to M, and which is a projective rational variety. If $\mathfrak{g}$ is the Lie algebra of G, then $M_u = G_u/K$, with $G_u = \mathrm{Ad}\, \mathfrak{g}_u$ in the notation of §4. (See [12] for references).

Hirzebruch [12] has shown that if M' is a compact complex analytic Clifford–Klein form of M, then the Chern numbers of M' are proportional to those of M_u. The proportionality factor is the arithmetic genus of M', or also the quotient $c_1^n[M'] : c_1^n[M_u]$, ($n = \dim_C M$), and is >0 if n is even, <0 otherwise. If M'' is a finite Galois covering of M', with μ sheets, then the Chern numbers of M'' are equal to those of M' multiplied by μ. The Corollary to Theorem A implies therefore that M admits compact Clifford–Klein forms whose arithmetic genus is arbitrarily large in absolute value and whose Chern numbers are the product of $(-1)^n$ by arbitrarily large multiples of the Chern numbers of M_u.

5.2. It is also shown in [12; §3] that the index $\tau(M')$ of M' is zero, unless M is the product of irreducible bounded symmetric domains whose compact counterpart belongs to the following list

$$\mathbf{U}(p + 2r)/\mathbf{U}(p) \times \mathbf{U}(2r), \qquad \mathbf{SO}(4k + 2)/(\mathbf{SO}(4k) \times \mathbf{SO}(2)), \qquad \mathbf{E}_6/(\mathrm{Spin}(10) \times T^1), \qquad (1)$$

in which case $\tau(M') \geqq 1$. Since going over to a μ-sheeted Galois covering multiplies the index by μ, Corollary to Theorem A also shows that if M belongs to the latter category, then it admits compact complex analytic Clifford–Klein forms with arbitrarily large index.

For $p = 1$, $\mathbf{U}(p + 2r)/\mathbf{U}(p) \times \mathbf{U}(2r)$ is the complex projective space $\mathbf{P}_{2r}(\mathbf{C})$, and the corresponding bounded domain is the open unit ball $B^{2r} \subset \mathbf{C}^{2r}$.

5.3. It is conjectured in [21] that if X is an algebraic surface, then

$$\rho + \rho_o \geqq 4 \cdot p_g + 1, \tag{2}$$

where p_g is the geometric genus of X, and ρ (resp. ρ_0) the number of linearly independent algebraic (resp. transcendental) cycles of X. If $h^{p,q}$ denotes as usual the dimension of the space of harmonic forms of type (p, q), then $\rho + \rho_0 = h^{1,1} + h^{2,0} + h^{0,2}$, $p_g = h^{2,0}$, and the Hodge formula

$$\tau(X) = \sum_{p,q}(-1)^p h^{p,q}$$

shows that (2) is equivalent to

$$\tau(X) \leqq 1. \tag{3}$$

However, the compact Clifford–Klein forms of bounded symmetric domains being all projective algebraic varieties by Theorem 6 of [15], (3) contradicts 5.2 applied to the case where $M = B^2$, $M_u = \mathbf{P}_2(\mathbf{C})$.

REFERENCES

2   1. A. BOREL: Semi-simple Lie groups and symmetric spaces, in preparation.
2. A. BOREL and HARISH-CHANDRA: Arithmetic subgroups of algebraic groups, *Bull. Amer. Math. Soc.* **67** (1961), 579–583.
3. A. BOREL and HARISH-CHANDRA: Arithmetic subgroups of algebraic groups, *Ann. Math., Princeton* **75** (1962), 485–535.
4. A. BOREL and G. D. MOSTOW: On semi-simple automorphisms of Lie algebras, *Ann. Math., Princeton* **61** (1955), 389–405.
5. N. BOURBAKI: *Topologie Générale*, 2ème édition, Chaps. 3–4, *Act. Sci. Ind.* 1143, Hermann, Paris, 1960.
6. E. CARTAN: Sur certaines formes riemanniennes remarquables des géométries à groupe fondamental simple, *Annales E.N.S.* **44** (1927), 345–467.
7. E. CARTAN: La théorie des groupes finis et continus et l'Analysis Situs, *Mémor. Sci. Math.* No. 42, (1930).
8. E. CARTAN: Sur les domaines bornés homogènes de n variables complexes, *Abh. Math. Sem. Hamburg Univ.* **11** (1935), 116–162.

9. E. Cartan: *Leçons sur la géométrie des espaces de Riemann*, 2ème édition. Gauthier-Villars, Paris, 1946.

10. C. Chevalley: Sur certains groupes simples, *Tohoku Math. J.* (2) **7** (1955), 14–66.

11. R. Fricke und F. Klein: *Vorlesungen über die Theorie der automorphen Funktionen, Band I*, Teubner, Leipzig, 1897.

12. F. Hirzebruch: Automorphe Formen und der Satz von Riemann–Roch, *Symposium Internacional de Topologia Algebraica* (Mexico, 1958) pp. 129–144.

13. N. Jacobson: Lie algebras, *Interscience Tracts in Pure and Applied Mathematics, Vol. 10*, New York, 1961.

14. H. Klingen: Diskontinuerliche Gruppen in symmetrischen Räumen—I II; *Math. Ann.* **129** (1955), 345–369; **130** (1955), 137–146.

15. K. Kodaira: On Kähler varieties of restricted type, *Ann. Math., Princeton* **60** (1954), 28–48.

16. F. Löbell: Beispiele geschlossener dreidimensionaler Clifford–Kleinscher Formen Räume negativer Krümmung, *Ber. sächs. Ges. (Akad.) Wiss.* **83** (1931), 167–174.

17. H. Seifert und C. Weber: Die beiden Dode kaederräume, *Math. Z.* **37** (1933), 237–253.

18. A Selberg: On discontinuous groups in higher dimensional symmetric spaces, *Contributions to function theory*, Tata Institute of Fundamental Research, Bombay, 1960.

19. Séminaire S. Lie, Groupes et algèbres de Lie, Paris, 1955.

20. C. L. Siegel: Symplectic geometry, *Amer. J. Math.* **65** (1943), 1–80.

21. G. Zappa: Sopra una probabile diseguaglianza tra i caratteri invariantivi di una superficie algebrica, *Rend. mat. e appl.* (5) **14** (1955), 455–464.

The Institute for Advanced Study,

Princeton, N.J.,

U.S.A.

63.

(with W. L. Baily Jr.)

On the compactification of arithmetically defined quotients of bounded symmetric domains

Bull. Amer. Math. Soc. **70** (1964) 588–593

In previous papers [13], [2], [16], [3], [11], the theory of automorphic functions for some classical discontinuous groups Γ, such as the Siegel or Hilbert-Siegel modular groups, acting on certain bounded symmetric domains X, has been developed through the construction of a natural compactification of X/Γ, which is a normal analytic space, projectively embeddable by means of automorphic forms. The purpose of this note is to announce similar results for general arithmetic groups, which include the earlier ones as special cases.

1. For algebraic groups, we follow the notation and conventions of [4], [5]. $G \subset GL(m, \mathbf{C})$ will be a semisimple linear algebraic group defined over $\mathbf{Q}$, and, for every subring B of $\mathbf{C}$, we put as usual $G_B = G \cap GL(m, B)$. We let Γ be an *arithmetic subgroup* of G, i.e., a subgroup of $G_\mathbf{Q}$ commensurable with the group $G_\mathbf{Z}$ of units of G. The group acts on the right, in a properly discontinuous manner, on the symmetric space $X = K \backslash G_R$, where K is a maximal compact subgroup of G_R. Here we assume X to be *hermitian symmetric*, and therefore [7] equivalent to a bounded symmetric domain. The quotient $V = X/\Gamma$ then carries a natural ringed structure with which it becomes an irreducible normal analytic space.

There is no loss in generality in assuming G to be connected and centerless, and we shall do so. Although this is not essential, we shall here for simplicity assume that G is *simple over* $\mathbf{Q}$, in other words, that it has no proper invariant subgroup $N \neq (e)$ defined over $\mathbf{Q}$. We are, of course, interested here only in the case where V is not compact; this means that G_R has no compact factor $\neq (e)$ and that G contains a torus $S \neq (e)$ which splits over $\mathbf{Q}$ [5].

2. **Rational boundary components.** The space X has a "natural" compactification, obtained by taking the closure $\overline{X}$ of X in the Harish-Chandra realization of X as a bounded symmetric domain [7], [8]. The boundary is then the union of finitely many orbits of G_R, each of which has a fibration, whose fibers are locally closed

---

[1] Partial support by N.S.F. grant GP91 for the first-named author, by N.S.F. grant GP2403 for the second-named author.

356

analytic subvarieties of the ambient complex vector space, and are symmetric bounded domains for certain subgroups of G_R [9], [10], [11]. These fibers are the *boundary components* of X.

DEFINITION 1. A boundary component F of X is called *rational* if: (i) The complexification $\mathfrak{N}(F)_C$ of the normalizer $\mathfrak{N}(F) = \{g \in G_R \mid F \cdot g = F\}$ of F is an algebraic subgroup defined over Q of G; (ii) $\mathfrak{N}(F)_C$ contains a normal subgroup M' defined over Q such that $M'_R \subset Z(F) = \{g \in \mathfrak{N}(F) \mid x \cdot g = x \ (x \in F)\}$, and that $Z(F)/M'_R$ is compact.

These conditions imply readily that the image $\Gamma(F)$ of $\mathfrak{N}(F) \cap \Gamma$ in $\mathfrak{N}(F)/M'_R$ is an arithmetic group, acting properly discontinually on F. The previous definition can of course be given for any semisimple group L over Q, and a boundary component of the corresponding symmetric space Y with respect to any Satake compactification of Y [14]. It is then slightly weaker than the definition used in [4, §4]. Coming back to X, it is known that its natural compactification (with its boundary components) is equivalent to the Satake compactification associated to a certain class of representations [10]. The proofs of the results below make strong use of that fact.

THEOREM 1. *The map* $F \to \mathfrak{N}(F)_C$ *yields a* 1-1 *correspondence between rational boundary components and proper maximal parabolic subgroups defined over* Q *of* G.

We recall that an algebraic subgroup H of G is parabolic if G/H is a projective variety. The theorem implies in particular that, in the natural compactification, condition (ii) of Definition 1 is a consequence of (i). This theorem is proved by a rather detailed investigation of the R-roots and Q-roots of G, and on how the former restrict on the latter, which, at some points, makes use of the classification. In particular, the following, which generalizes results of [15, §5], is proved:

PROPOSITION 1. *Let* $_QT$ *be a maximal* Q*-trivial torus of* G. *Then the system* $_Q\Sigma$ *of* Q*-roots (cf.* [4]*) is either of type* C_s: $\pm(1/2)(\gamma_i \pm \gamma_j)$ $(1 \leq i, j \leq s = \dim {}_QT)$ *or of type* $BC_s = C_s \cup \{(1/2)\gamma_i\}_{i \leq s}$. *If we choose compatible orderings on* $_R\Sigma$ *and* $_Q\Sigma$, *then each simple* Q*-root is the restriction of exactly* q *simple* R*-roots, where* q *is the number of simple factors of* G_R.

3. **A topological compactification of** X/Γ. Let Ω be a sufficiently large fundamental set for Γ in X, as in [4, p. 31]. Using Theorem 1, it may be shown, as in [4], that the closure $\bar{\Omega}$ of Ω is contained in the union of Ω and of finitely many rational components, and meets every

equivalence class under Γ of rational boundary components. In what follows, we let X^* denote the union of X and of all rational boundary components. It is acted upon by G_Q. Moreover, we may apply some results of [15] and obtain the

THEOREM 2. *There exists a topology on X^*, which induces the natural topology on Ω, in which every element of G_Q operates continuously, and such that $V^* = X^*/\Gamma$, endowed with the quotient topology, is a compact Hausdorff space.*

The construction shows that $V^* = V \cup V_1 \cup \cdots \cup V_t$, where $V = X/\Gamma$ is open, everywhere dense, $V_i = F_i/\Gamma(F_i)$, and the F_i's are such that the groups $\mathfrak{N}(F_i)_C$ form a system of representatives for the equivalence classes, modulo inner automorphisms by elements of Γ, of proper maximal parabolic subgroups over Q of G. Moreover, V_i carries a natural structure of irreducible normal analytic space, and $\overline{V}_i$ is the union of V_i and of some V_j's of strictly smaller dimension $(1 \leqq i \leqq t)$. We may also prove the

LEMMA. *Each point $v \in V^*$ has a fundamental system of neighborhoods $\{U_\iota\}$ such that $U_\iota \cap V$ is connected for each ι.*

4. **The ringed structure on V^*.** A complex-valued function f defined on an open subset U of V^* will be called an $\mathfrak{X}$-function on U if it is continuous and if its restrictions to $V \cap U$ and to $V_i \cap U$ are analytic in the given analytic structures $(1 \leqq i \leqq t)$. The sheaf of germs of $\mathfrak{X}$-functions, to be denoted by $\mathfrak{X}$, provides V^* with a ringed structure. Our next goal is to prove that $(V^*, \mathfrak{X})$ is a normal analytic space, whose analytic structure is compatible with the given ones on V and the V_i's. It is easily deduced from the above lemma (as in [16, Exposé 11]) that the stalk $\mathfrak{X}_v$ of $\mathfrak{X}$ at $v \in V^*$ is integrally closed for every v. Using this fact, the properties of the V_i's stated above, the Remmert-Stein theorem on removable singularities of analytic sets [12], and arguments similar to those of [3, pp. 870–872] (see also [16, Exposé 11]), one shows that it is enough to check the two following properties of $\mathfrak{X}$: (a) For each i $(1 \leqq i \leqq t)$, the sheaf of germs of restrictions to V_i of $\mathfrak{X}$-functions is the structural sheaf of V_i; (b) each point $v \in V^*$ has a neighborhood U_v such that the H-functions on U_v separate the points of U_v. In order to do this, we study certain automorphic forms introduced below.

5. **Poincaré-Eisenstein series.** To construct these, we need some results of Pyateckiĭ-Shapiro [11] and of Korányi and Wolf [9] about unbounded realizations of X. Given a boundary component F of X,

there exists a "partial Cayley transform" ν_F of X onto a Siegel domain of the third kind (in the sense of [11]), given by

$$\operatorname{Im} z - \operatorname{Re} L_t(u, u) \in V \qquad (t \in F)$$

(for notation and definitions, see [11, pp. 26–35]) in such a way that the elements of $\mathfrak{N}(F)$ become quasi-linear transformations in the sense of [11], and those of $Z(F)$ take the form

$$z \to A \cdot z + a(u, t), \qquad u \to B \cdot u + b(t), \qquad t \to t.$$

Assume now that F is a rational boundary component. Let $J_F(w, g)$ be the functional determinant of g at w in the coordinates (z, u, t) of the unbounded realization associated to F. It can be shown that $J_F(w, g \cdot g') = J_F(w, g) \cdot \chi(g')$ if $g' \in Z(F)$, χ being a rational character on $Z(F)$ which takes the values ± 1 on $M' \cap \Gamma$. Then, given a polynomial P in the coordinates t on F, and a positive integer m, we may form the following series, to be called a Poincaré-Eisenstein series

$$E_P(w) = \sum_{\gamma \in \Gamma/(\Gamma \cap M')} P(t(w \cdot \gamma)) J_F(w, \gamma)^{2m},$$

where $w = (z, u, t)$, $t(w) = t$, and M' is the subgroup of $Z(F)_C$ occurring in (ii) of Definition 1. It can be shown that, for sufficiently large m, such a series converges uniformly on every compact set. By the use of some properties of J_F, and of standard facts on Poincaré series, this convergence proof is reduced to that of ordinary Eisenstein series (associated to parabolic subgroups over Q of G), where we apply an unpublished criterion of Godement. We then study the behaviour of such series as we approach a boundary component (the effect of the operator Φ of [16]). It can be shown that the "limit" of E_P on any rational boundary component F' may be viewed as an automorphic form for $\Gamma(F')$, which is a Poincaré series if F' is equivalent to F, and is zero if $\dim F' \leq \dim F$, and F' is not equivalent under Γ to F. In particular, every Poincaré series with respect to the bounded realization of X is a cusp form (i.e., has limit zero on all proper rational boundary components).

6. **The main results.** These facts, and well-known properties of Poincaré series [6], [16], show that the sheaf $\mathfrak{IC}$ verifies the conditions (a), (b) of §4. In fact, we see that, in (b), the points of U_ν may be separated by suitable ratios of Poincaré-Eisenstein series. From this follows, as in [2], [3], [16],

THEOREM 3. *The compactification V^* of $V = X/\Gamma$ carries a natural analytic structure, compatible with the analytic structure on any rational* *boundary component (including X), with which it becomes a normal analytic space. The normal analytic space V^* can be projectively embedded as a projectively normal algebraic variety by means of a set of automorphic forms for Γ, of some suitable high weight.*

It is easily seen that if dim $G > 3$, then $V^* - V$ has complex codimension ≥ 2. Making use of Levi's and H. Kneser's theorems on removable singularities of holomorphic or meromorphic functions, we derive from Theorem 3 the following corollaries:

COROLLARY 1. *Let* dim $G > 3$. *Then the field of meromorphic functions on $V = X/\Gamma$ is an algebraic function field of transcendence degree equal to* dim $_cX$. *Any such function is the quotient of automorphic forms of the same weight.*

We recall that we have assumed G to be simple over $\mathbf{Q}$. However, Theorem 3 is valid without any such restriction, and Corollaries 1 and 2 are valid if G has no invariant subgroup defined over $\mathbf{Q}$ of dimension three.

COROLLARY 2. *Let* dim $G > 3$. *If f is an automorphic form of even weight, its limit (in the sense mentioned above for Poincaré-Eisenstein series) on any rational boundary component exists. The space of automorphic forms of a given weight $m > 0$ is finite-dimensional.*

Corollary 1 and the finite-dimensionality of the space of automorphic forms may also be proved by showing that Γ is pseudo-concave in the sense of Andreotti-Grauert [1]; this approach will be discussed elsewhere by one of the authors of the present note. The first part of Corollary 2, which is an extension of Koecher's theorem, can also be proved, independently of Theorem 3, by the methods of [11, pp. 134–143], together with some of the results of [5] giving conditions for the finiteness of the volume of H_R/H_Z for an algebraic group H over $\mathbf{Q}$. Moreover, we note that Corollary 2 extends to vector-valued automorphic forms, with a suitable automorphy factor of the type considered in [16]. This follows from Theorem 3, and from an unpublished result of J.-P. Serre which, together with [6], implies that if A is the sheaf of germs of such automorphic forms, then the direct image i_*A of A, with respective to the inclusion $i: V \to V^*$, is an algebraic coherent sheaf on V^*. The finite-dimensionality is also of course a consequence of pseudo-concavity.

REFERENCES

1. A. Andreotti and H. Grauert, *Algebraische Körper von automorphen Funktionen*, Nachr. Akad. Wiss. Göttingen Math.-Phys. Kl. II (1961), 39–48.

2. W. L. Baily, Jr., *Satake's compactification of V_n*, Amer. J. Math. **80** (1958), 348–364.

3. ——, *On the Hilbert-Siegel modular space*, Amer. J. Math. **81** (1959), 846–874.

4. A. Borel, *Ensembles fondamentaux pour les groupes arithmétiques*, Colloque sur la théorie des groupes algébriques, CBRM, Brussels (1962), 23–40.

5. A. Borel and Harish-Chandra, *Arithmetic subgroups of algebraic groups*, Ann. of Math. (2) **75** (1962), 485–535.

6. H. Cartan, *Fonctions automorphes et séries de Poincaré*, J. Analyse Math. **6** (1958), 169–175.

7. Harish-Chandra, *Representations of semi-simple Lie groups.* VI, Amer. J. Math. **78** (1956), 564–628

8. S. Helgason, *Differential geometry and symmetric spaces*, Academic Press, New York, 1962.

9. A. Korányi and J. Wolf, *Generalized Cayley transforms of bounded symmetric domains*, Ann. of Math. [**81** (1965), 265–288].

10. C. C. Moore, *Compactifications of symmetric spaces.* II. *The Cartan domains*, Amer. J. Math. [**86** (1964), 358–378].

11. I. I. Pyateckiĭ-Shapiro, *The geometry of the classical domains and the theory of automorphic functions*, Gosizdat, Moscow, 1961. (Russian)

12. R. Remmert and K. Stein, *Ueber die wesentlichen Singularitäten analytischer Mengen*, Math. Ann. **126** (1953), 263–306.

13. I. Satake, *On the compactification of the Siegel space*, J. Indian Math. Soc. **20** (1956), 259–281.

14. ——, *On representations and compactifications of symmetric spaces*, Ann. of Math. (2) **71** (1960), 77–110.

15. ——, *On compactifications of the quotient spaces for arithmetically defined discontinuous groups*, Ann. of Math. (2) **72** (1960), 555–580.

16. Seminaire Henri Cartan, 10ème année, Fonctions Automorphes (multigraphé), Paris, 1958, 2 volumes.

UNIVERSITY OF CHICAGO,
THE INSTITUTE FOR ADVANCED STUDY AND
UNIVERSITY OF PARIS

64.

(avec J-P. Serre)

Théorèmes de finitude en cohomologie galoisienne

Comment. Math. Helv. **39** (1964) 111–164

Introduction

Les propriétés de finitude que nous avons en vue concernent un groupe algébrique G défini sur un corps parfait k. Il s'agit essentiellement de prouver:

(A) la finitude du nombre de classes d'espaces homogènes principaux sur k de G (resp. du nombre de k-formes de G) lorsque k est un corps p-adique et G est linéaire (resp. G est soit linéaire, soit une variété abélienne);

(B) la finitude du nombre de classes d'espaces homogènes principaux sur k de G (resp. du nombre de k-formes de G) qui sont isomorphes à un espace homogène principal donné (resp. isomorphes à G) sur toute complétion k_v de k, lorsque k est un corps de nombres algébriques, et G est linéaire (resp. G est soit linéaire, soit une variété abélienne).

On sait que les classes d'espaces homogènes principaux sur k de G correspondent aux éléments du premier ensemble de cohomologie $H^1(k, G)$. De même, les classes de k-formes de G correspondent aux éléments de $H^1(k, \operatorname{Aut} G)$, où $\operatorname{Aut} G$ désigne le foncteur des automorphismes de G. Les énoncés (A) et (B) sont donc équivalents à des propriétés de finitude de $H^1(k, G)$ (resp. de $H^1(k, \operatorname{Aut} G)$), et c'est sous cette forme qu'ils seront établis (cf. §§ 6, 7).

L'assertion (B) est un cas particulier de (A) lorsque $\operatorname{Aut} G$ est lui-même un groupe algébrique linéaire; mais ce n'est pas toujours le cas, comme le montre déjà l'exemple d'un tore G de dimension n, pour lequel $\operatorname{Aut} G$ peut être identifié au groupe discret $\mathbf{GL}(n, \mathbf{Z})$. Cela nous a amené à étudier la structure de $\operatorname{Aut} G$, lorsque G est linéaire, et à démontrer les propriétés de finitude de H^1 qui nous intéressent directement pour une classe de groupes, que nous appelons de type (ALA), classe qui comprend les groupes algébriques linéaires, leurs groupes d'automorphismes (du moins en caractéristique zéro), et les groupes d'automorphismes des variétés abéliennes. Le résultat en est un article dont la plus grande partie (§§ 1 à 5) est, à notre grand regret, consacrée à des préliminaires variés. Toutefois le lecteur qui ne s'intéresse qu'à la cohomologie des groupes linéaires peut omettre la lecture des §§ 3 à 5; de même, les §§ 4 et 5 sont inutiles pour la discussion des formes de variétés abéliennes.

362

Le contenu des différents paragraphes est le suivant:

Le § 1 donne les définitions et les principales propriétés des ensembles de cohomologie en dimension 0,1 d'un groupe topologique $\mathfrak{g}$, à valeurs dans un groupe discret A sur lequel $\mathfrak{g}$ opère continûment. Il s'agit surtout d'éta blir l'existence de suites exactes, et d'analyser les fibres de certaines applicat ions figurant dans ces suites exactes. Les résultats ne sont guère qu'une transcrip tion de faits connus dans le cas topologique (voir en particulier [8] dont nous n ous sommes largement inspirés). Dans la suite, on ne s'intéressera qu'au cas où $\mathfrak{g}$ est un groupe de Galois; ce cas est discuté dans le § 2, qui contient en outre di vers résultats techniques utilisés plus loin.

Les §§ 3 et 4 introduisent trois classes de groupes: les $\mathfrak{g}$-*groupes de t ype arithmétique*, groupes isomorphes à un sous-groupe d'indice fini du gro upe d'unités $G_{\mathbf{Z}}$ d'un groupe algébrique linéaire G défini sur $\mathbf{Q}$, sur lequel le gro upe profini $\mathfrak{g}$ opère au moyen d'automorphismes; les *groupes localement algébri ques sur k*, extensions de groupes discrets par des groupes algébriques; enfin , les *groupes de type* (ALA), extensions de $\mathfrak{g}$-groupes de type arithmétique par des groupes linéaires algébriques.

Le § 4 donne en outre un critère pour que le foncteur d'automorphismes d 'une variété algébrique (ou d'un groupe algébrique) soit localement algébrique . Ce critère est utilisé au § 5 pour montrer que, si G est un groupe algébrique liné aire défini sur un corps de caractéristique zéro, Aut G est un groupe de type (AL A). *Grosso modo*, on peut dire que la partie algébrique de Aut G est formée des automorphismes qui agissent trivialement sur le plus grand tore central S de la composante neutre de G, et que le quotient de type arithmétique correspon d à l'image de Aut G dans Aut S.

Le § 6 prouve la finitude de $H^1(k, A)$ lorsque A est de type (ALA) et qu e k est p-adique, ou plus généralement, k est un corps parfait n'ayant qu'un nom bre fini d'extensions de degré donné. Parmi les applications signalons, outre (A), la finitude du nombre de classes de conjugaison de sous-groupes de Cart an définis sur k dans un groupe linéaire G, ainsi que la finitude du nombre des orbite s de $G(k)$ opérant dans l'ensemble des points à valeurs dans k d'un espace homogène de G. Enfin, le § 7 traite le cas global. Dans les deux cas, on procède par «dévissage». Les démonstrations utilisent notamment deux résultats, qui sont conséquences de l'existence de domaines fondamentaux convenables pour les groupes d'unités: la finitude du nombre de classes de conjugaison de sous-groupes finis d'un groupe de type arithmétique, et le cas particulier de (B) où G est réductif connexe, et où les espaces homogènes principaux considérés ont des points à valeurs dans toutes les complétions de k.

Notations

Si k est un corps, on note $\overline{k}$ (resp. k_s) une clôture algébrique (resp. une clôture séparable) de k.

Si k' est une extension galoisienne d'un corps k, on note $g(k'/k)$ le groupe de Galois de cette extension; c'est un groupe profini (i. e. compact totalement discontinu).

§ 1. Cohomologie non commutative

Dans tout ce paragraphe, la lettre g désigne un *groupe topologique*. Le lecteur qui ne s'intéresse qu'aux applications à la cohomologie galoisienne pourra supposer que g est *profini*.

1.1. g-ensembles. Un g-ensemble est un espace topologique *discret* E sur lequel g opère à gauche de façon continue (i. e. de telle sorte que le groupe de stabilité de tout point de E soit ouvert dans g). Le transformé de $x \epsilon E$ par $s \epsilon g$ sera noté $s(x)$, $s \cdot x$ ou encore $^s x$ (mais jamais x^s pour éviter l'horrible formule $x^{(st)} = (x^t)^s$). Si E et E' sont deux g-ensembles, un *morphisme* de E dans E' est une application $f : E \to E'$ qui commute aux opérations de g. Les g-ensembles forment une catégorie.

Si E est un g-ensemble, on note $H^0(g, E)$, ou E^g, l'ensemble des $x \epsilon E$ tels que $^s x = x$ pour tout $s \epsilon g$. On dit que c'est le *0-ième ensemble de cohomologie de g à valeurs dans E*.

1.2. g-groupes. Un g-groupe A est un groupe dans la catégorie des g-ensembles. En d'autres termes, c'est un g-ensemble qui est muni d'une structure de groupe invariante par g; on a donc $^s(xy) = {}^s x \, {}^s y$ pour $x, y \epsilon A$ et $s \epsilon g$. Un g-groupe commutatif est parfois appelé un g-*module*.

Soit A un g-groupe. On appelle *cochaîne de g à valeurs dans A* toute application continue $s \mapsto a_s$ de g dans A. On appelle *cocycle de g à valeurs dans A* toute cochaîne $a = (a_s)$ qui vérifie l'identité:

$$a_{st} = a_s \cdot {}^s a_t \quad (s, t \epsilon g). \tag{1}$$

Si une application $a : s \mapsto a_s$ vérifie (1), on a:

$$a_1 = 1, \; a_s \cdot {}^s a_{s^{-1}} = 1 \quad (s \epsilon g). \tag{2}$$

L'ensemble h des éléments $s \epsilon g$ tels que $a_s = 1$ est donc un sous-groupe, et l'on a $a_{sh} = a_s$ pour $s \epsilon g$, $h \epsilon h$. Il s'ensuit que a est continue si et seulement si h est ouvert.

L'ensemble des cocycles de $\mathbf{g}$ à valeurs dans A est noté $Z^1(\mathbf{g}, A)$. Deux cocycles x, $y \epsilon Z^1(\mathbf{g}, A)$ sont dits *cohomologues* s'il existe $a \epsilon A$ tel que $x_s = a^{-1} y_s{}^s a$ pour tout $s \epsilon \mathbf{g}$. C'est là une relation d'équivalence dans $Z^1(\mathbf{g}, A)$; l'ensemble quotient est noté $H^1(\mathbf{g}, A)$; on l'appelle le *premier ensemble de cohomologie de $\mathbf{g}$ à valeurs dans A*. Si A est commutatif, $H^1(\mathbf{g}, A)$ est un groupe commutatif, le produit étant défini à partir du produit des valeurs de cocycles représentatifs. Dans le cas général, $H^1(\mathbf{g}, A)$ est seulement un *ensemble pointé*: il contient un élément distingué, la classe du cocycle unité; les cocycles appartenant à cette classe sont appelés *cobords*; ils sont de la forme $s \mapsto c^{-1} \cdot {}^s c$, avec $c \epsilon A$. L'élément distingué de $H^1(\mathbf{g}, A)$ sera noté indifféremment 0 ou 1.

Lorsque $\mathbf{g}$ est profini, $H^1(\mathbf{g}, A)$ s'identifie canoniquement à la limite inductive des $H^1(\mathbf{g}/\mathbf{h}, A^{\mathbf{h}})$, pour $\mathbf{h}$ parcourant l'ensemble des sous-groupes ouverts distingués de $\mathbf{g}$.

1.3. Fonctorialités. Soient $\mathbf{g}$ et $\mathbf{g}'$ deux groupes topologiques, E un $\mathbf{g}$-ensemble et E' un $\mathbf{g}'$-ensemble. Un homomorphisme continu $\lambda \colon \mathbf{g} \to \mathbf{g}'$ et une application $\mu \colon E' \to E$ seront dits *compatibles* si l'on a

$$\mu(\lambda(s) \cdot x') = s \cdot \mu(x') \quad \text{pour} \ \ s \epsilon \mathbf{g}, \ x' \epsilon E'.$$

Il est clair que μ applique alors $E'^{\mathbf{g}'}$ dans $E^{\mathbf{g}}$, ce qui définit une application

$$(\lambda, \mu)^0_* \colon H^0(\mathbf{g}', E') \to H^0(\mathbf{g}, E) \,.$$

Supposons de plus que E (resp. E') soit un $\mathbf{g}$-groupe (resp. un $\mathbf{g}'$-groupe) et que μ soit un homomorphisme. Soit $a' = (a'_{s'}) \epsilon Z^1(\mathbf{g}', E')$, et soit $a_s = \mu(a'_{\lambda(s)})$. On vérifie tout de suite que $a = (a_s)$ est un élément de $Z^1(\mathbf{g}, E)$. De plus, l'application $(\lambda, \mu)^1_* \colon Z^1(\mathbf{g}', E') \to Z^1(\mathbf{g}, E)$ ainsi définie est compatible avec les relations d'équivalence de ces deux ensembles, d'où, par passage aux quotients, une application de $H^1(\mathbf{g}', E')$ dans $H^1(\mathbf{g}, E)$ qui sera encore notée $(\lambda, \mu)^1_*$.

Lorsque $\mathbf{g} = \mathbf{g}'$ et que λ est l'identité, on écrit μ^i_* pour $(\lambda, \mu)^i_*$, $i = 0,1$.

Autre cas particulier: E a même ensemble sous-jacent que E', et $\mathbf{g}$ opère sur E par la formule $s(x) = \lambda(s) \cdot x$; on écrit alors $E = \lambda^* E'$, et l'on peut prendre pour μ l'application identique. Si en outre λ est l'inclusion d'un sous-groupe (resp. la projection de $\mathbf{g}$ sur un groupe quotient), $(\lambda, \mu)^1_*$ est appelée l'application de *restriction* (resp. d'*inflation*), et on la note Res (resp. Inf).

1.4. Torsion. Soient A un $\mathbf{g}$-groupe et E un $\mathbf{g}$-ensemble. On dit que A opère sur E à gauche, de façon compatible avec $\mathbf{g}$, si l'on s'est donné une application $(a, x) \mapsto a \cdot x$ de $A \times E$ dans E vérifiant les conditions suivantes:

$$^s(a \cdot x) = {^s}a \cdot {^s}x \quad (a \in A,\ s \in \mathfrak{g},\ x \in E) \tag{3}$$

$$a \cdot (b \cdot x) = (a \cdot b) \cdot x,\quad 1 \cdot x = x \quad (a,\ b \in A,\ x \in E)\,. \tag{4}$$

Soit alors $a = (a_s) \in Z^1(\mathfrak{g}, A)$. Pour tout $s \in \mathfrak{g}$ et tout $x \in E$, posons:

$$^{s'}x = a_s \cdot {^s}x\,.$$

On vérifie immédiatement que cette formule définit une nouvelle opération de $\mathfrak{g}$ sur E, et que cette opération est continue. Le $\mathfrak{g}$-ensemble ainsi obtenu est noté E_a. On dit que E_a *s'obtient en tordant* E *au moyen du cocycle* a. Il a même ensemble sous-jacent que E.

Si $b = (b_s) \in Z^1(\mathfrak{g}, A)$ est cohomologue à a, le $\mathfrak{g}$-ensemble E_b est isomorphe à E_a. En effet, soit $c \in A$ tel que $b_s = c^{-1}a_s{^s}c$. On a alors

$$c \cdot (b_s{^s}x) = a_s \cdot {^s}(c \cdot x) \qquad (s \in \mathfrak{g},\ x \in E)\,,$$

ce qui montre que l'application $x \mapsto c \cdot x$ est un $\mathfrak{g}$-isomorphisme de E_b sur E_a. [On notera cependant qu'il n'y a pas en général d'isomorphisme *canonique* entre E_a et E_b, et par suite il n'est pas possible *d'identifier* ces deux ensembles, ni de définir un «E_α» pour $\alpha \in H^1(\mathfrak{g}, A)$.]

L'opération de torsion jouit d'un certain nombre de propriétés élémentaires:

(a) E_a est fonctoriel en E (pour des A-morphismes $E \to E'$).

(b) On a $(E \times E')_a = E_a \times E'_a$.

(c) Si un $\mathfrak{g}$-groupe B opère à droite sur E (de façon à commuter à l'action de A), B opère aussi sur E_a.

(d) Si E est muni d'une structure de $\mathfrak{g}$-groupe, et si les éléments de A définissent des automorphismes de E, E_x est un $\mathfrak{g}$-groupe. Ceci s'applique notamment au cas $E = A$, le groupe A opérant sur lui-même par *automorphismes intérieurs*. On a alors:

$$^{s'}x = a_s{^s}xa_s^{-1} \quad (s \in \mathfrak{g},\ x \in A)$$

et $(A_a)^{\mathfrak{g}}$ est l'ensemble des $x \in A$ tels que $a_s{^s}x = xa_s$. La cohomologie du groupe tordu A_a est isomorphe à celle de A. Plus précisément:

1.5. Proposition. *Soit* $a = (a_s) \in Z^1(\mathfrak{g}, A)$ *et soit* A_a *le $\mathfrak{g}$-groupe obtenu en tordant* A *au moyen de* a (le groupe A opérant sur lui-même par automorphismes intérieurs). *Si* $b = (b_s) \in Z^1(\mathfrak{g}, A_a)$, *on a* $(b_sa_s) \in Z^1(\mathfrak{g}, A)$, *et l'on obtient ainsi une bijection*

$$t_a : Z^1(\mathfrak{g}, A_a) \to Z^1(\mathfrak{g}, A)\,.$$

Par passage au quotient, t_a *définit une bijection*

$$\tau_a : H^1(\mathfrak{g}, A_a) \to H^1(\mathfrak{g}, A)\,,$$

qui transforme l'élément neutre de $H^1(\mathbf{g}, A_a)$ en la classe α de a.

Puisque $(b_s) \in Z^1(\mathbf{g}, A_a)$, on a $b_{st} = b_s {}^{s'}b_t = b_s a_s{}^s b_t a_s^{-1}$, d'où

$$b_{st} a_{st} = b_s a_s{}^s b_t a_s^{-1} \cdot a_s{}^s a_t = b_s a_s \cdot {}^s(b_t a_t),$$

ce qui montre bien que $(b_s a_s)$ appartient à $Z^1(\mathbf{g}, A)$. Il est alors immédiat que t_a est une bijection. Le fait que t_a, ainsi que son inverse, soit compatible avec la relation « être cohomologues » se vérifie par un calcul analogue.

Exemple: $a^{-1} = (a_s^{-1})$ appartient à $Z^1(\mathbf{g}, A_a)$, et l'on a $t_a(a^{-1}) = 1$.

Remarque. Lorsque A est *abélien*, on a $A_a = A$, et τ_a est la *translation* par la classe α de a.

1.6. Transitivité de l'opération de torsion. Soient A un g-groupe et E un g-ensemble. Supposons que A opère à gauche sur E, de façon compatible avec $\mathbf{g}$ (cf. nº 1.4). Soit $a \in Z^1(\mathbf{g}, A)$. On peut tordre à la fois A (comme au nº précédent) et E au moyen du cocycle a. Le g-groupe A_a *opère sur* E_a de façon compatible avec $\mathbf{g}$:

$$ {}^{s'}(a \cdot x) = a_s{}^s(a \cdot x) = a_s{}^s a {}^s x = a_s{}^s a a_s^{-1} a_s x = {}^{s'}a \cdot {}^{s'}x .$$

Si $b \in Z^1(\mathbf{g}, A_a)$, on peut donc *tordre E_a au moyen de b*; il est immédiat que le résultat $(E_a)_b$ est simplement $E_{t_a(b)}$ (les notations étant celles de la prop. 1.5). En particulier $(E_a)_{a^{-1}} = E$.

1.7. Espaces homogènes principaux. Soit A un g-groupe. Un espace homogène principal (à droite) sur A est un g-ensemble non vide P, sur lequel A opère à droite (de façon compatible avec $\mathbf{g}$), et qui vérifie la condition suivante :

Pour tout couple $(x, y) \in P \times P$, il existe un élément $a \in A$ et un seul tel que $x = y \cdot a$. (On dit encore que A opère de façon *simplement transitive* sur P.)

La notion *d'isomorphisme* de deux espaces homogènes principaux se définit de manière évidente. Un espace homogène principal P est dit *trivial* s'il est isomorphe à A (opérant sur lui-même par translations à droite); il revient au même de dire que $P^{\mathbf{g}}$ est non vide.

Soit P un espace homogène principal, et soit $x \in X$. Pour tout $s \in \mathbf{g}$, il existe un élément unique $a_s \in A$ tel que ${}^s x = x \cdot a_s$. On a:

$$ {}^{st}x = {}^s(x \cdot a_t) = {}^s x \cdot a_s{}^s a_t , $$

d'où $a_{st} = a_s{}^s a_t$; de plus, (a_s) ne dépend que de la classe à gauche de s modulo le groupe de stabilité de x. Il s'ensuit que $a = (a_s)$ appartient à $Z^1(\mathbf{g}, A)$. Si x est remplacé par $x \cdot c$, avec $c \in A$, le cocycle a_s est remplacé par le cocycle cohomologue $c^{-1}a_s{}^sc$. On a donc associé à P un élément bien déterminé $\varepsilon(P)$ de $H^1(\mathbf{g}, A)$.

1.8. Proposition. *Soit $P(A)$ l'ensemble des classes d'isomorphisme d'espaces homogènes principaux sur A. L'application*

$$\varepsilon : P(A) \to H^1(\mathbf{g}, A)$$

définie ci-dessus est une bijection.

Montrons que ε est *injective*: soient $P, Q \epsilon P(A)$ tels que $\varepsilon(P) = \varepsilon(Q)$. D'après ce qui précède, on peut choisir des points $x \epsilon P$, $y \epsilon Q$ correspondant au même cocycle (a_s). On vérifie alors immédiatement que l'application $x \cdot a \mapsto y \cdot a$ est un isomorphisme de P sur Q.

Montrons que ε est *surjective*. Soit $a \in Z^1(\mathbf{g}, A)$, et soit X l'espace homogène principal trivial A. Le groupe A opère par translations *à gauche* sur X. Soit X_a l'espace obtenu en tordant X au moyen de a. On a ${}^{s'}x = a_s{}^sx$ pour tout $x \epsilon X_a$. Si l'on fait opérer A sur X_a au moyen des translations *à droite*, il est immédiat que l'on obtient un espace homogène principal. De plus, le cocycle associé à l'élément unité $1 \epsilon X_a$ n'est autre que a. Cela montre bien que ε est surjective.

Remarque. En fait, la démonstration précédente prouve que les éléments de $Z^1(\mathbf{g}, A)$ correspondent bijectivement aux *classes de couples* (P, x) où P est un espace homogène principal sur A, et x un point de P.

1.9. Relations entre torsion et espaces homogènes principaux.

Soit A un g-groupe, soit P un espace homogène principal sur A, et soit E un g-ensemble sur lequel A opère à gauche (de façon compatible avec g). Sur $P \times E$, considérons la relation d'équivalence qui identifie un couple (p, x) aux couples $(p \cdot a, a^{-1} \cdot x)$, $a \epsilon A$. Cette relation est compatible avec l'action de g, et le quotient est un g-ensemble, que nous noterons $P \times^A E$, ou encore E_P. Un élément de $P \times^A E$ s'écrit sous la forme $p \cdot x$, $p \epsilon P$, $x \epsilon E$ et l'on a $(pa) \cdot x = p(ax)$ pour tout $a \epsilon A$, ce qui justifie la notation. Pour tout $p \epsilon P$ l'application $x \mapsto p \cdot x$ est une bijection de E sur E_P; on dit que E_P est obtenu *en tordant E au moyen de P.*

La liaison avec le point de vue du n° 1.4 est facile à faire: si $p \epsilon P$, on a vu au n° 1.7 que p définit un cocycle (a_s) tel que ${}^sp = p a_s$. La bijection $x \mapsto p \cdot x$ introduite ci-dessus est alors un *isomorphisme du* g-*ensemble E_a sur le* g-*ensemble E_P.* On a en effet:

$$p \cdot {}^{s'}x = p \cdot a_s{}^sx = {}^sp \cdot {}^sx = {}^s(p \cdot x) \, .$$

1.10. Utilisation de la torsion. Soient A et B deux g-groupes, et soit $u : A \to B$ un g-homomorphisme (i. e. un homomorphisme compatible avec l'action de g sur A et B). On a vu au n° 1.3 que u définit une application d'ensembles pointés

$$v = u_*^1 : H^1(\mathbf{g}, A) \to H^1(\mathbf{g}, B).$$

Soit $\alpha \epsilon H^1(\mathbf{g}, A)$, et proposons-nous de décrire la *fibre* de v passant par α, i. e. l'ensemble $v^{-1}(v(\alpha))$. Soit $a \epsilon Z^1(\mathbf{g}, A)$ un représentant de α, et soit b son image dans B. L'homomorphisme u définit un g-homomorphisme

$$u_a : A_a \to B_b,$$

d'où une application $v_a = u_{a*}^1 : H^1(\mathbf{g}, A_a) \to H^1(\mathbf{g}, B_b)$.

Considérons le diagramme suivant (où les lettres τ_a et τ_b désignent les bijections définies dans la prop. 1.5):

$$H^1(\mathbf{g}, A_a) \overset{v_a}{\to} H^1(\mathbf{g}, B_b)$$
$$\tau_a \downarrow \qquad\qquad \tau_b \downarrow$$
$$H^1(\mathbf{g}, A) \overset{v}{\to} H^1(\mathbf{g}, B).$$

On vérifie immédiatement que ce diagramme est *commutatif*. Comme τ_b transforme l'élément neutre de $H^1(\mathbf{g}, B_b)$ en $v(\alpha)$, on en déduit que τ_a *induit une bijection du noyau de* v_a *sur la fibre* $v^{-1}(v(\alpha))$. En d'autres termes, la torsion permet de transformer toute fibre de v en un *noyau*; ce procédé de réduction sera utilisé plusieurs fois dans la suite.

1.11. Suite exacte associée à un sous-groupe. Soit B un g-groupe, et soit A un sous-g-groupe de B. Soit B/A l'espace homogène des classes à gauche de B suivant A; c'est un g-ensemble, et $H^0(\mathbf{g}, B/A)$ est défini. On notera $i : A \to B$ et $p : B \to B/A$ les applications canoniques évidentes. De plus, si $x \epsilon H^0(\mathbf{g}, B/A)$, l'image réciproque X de x dans B est un espace homogène principal sur A; sa classe dans $H^1(\mathbf{g}, A)$ sera noté $\delta(x)$. Si $b \epsilon B$ est tel que $p(b) = x$, on a $b^{-1} \cdot {}^s b = a_s \epsilon A$ pour tout $s \epsilon \mathbf{g}$, et il est clair que $a = (a_s)$ est un cocycle appartenant à la classe de $\delta(x)$. Avec ces notations, on a:

1.12. Proposition. *Soit A un sous-g-groupe de B. La suite d'applications d'ensembles pointés*:

$$0 \to H^0(\mathbf{g}, A) \overset{i_*^0}{\to} H^0(\mathbf{g}, B) \overset{p_*^0}{\to} H^0(\mathbf{g}, B/A) \overset{\delta}{\to} H^1(\mathbf{g}, A) \overset{i_*^1}{\to} H^1(\mathbf{g}, B)$$

est exacte. De plus, δ induit une bijection de l'ensemble des orbites de $B^{\mathbf{g}}$ dans $(B/A)^{\mathbf{g}}$ sur le noyau de i_*^1.

(On rappelle qu'une suite d'applications d'ensembles pointés:

$$\ldots \to X_{n-1} \to X_n \to X_{n+1} \to \ldots$$

est dite *exacte* si l'image réciproque de l'élément neutre de X_{n+1} est égale à l'image de X_{n-1} dans X_n.)

Il est clair que i_*^0 est injectif, et que le noyau de p_*^0 est égal à l'image de i_*^0. Par définition, le noyau de δ est formé des éléments $x \in (B/A)^{\mathfrak{g}}$ dont l'image réciproque X dans B contient un point invariant par $\mathfrak{g}$; on a donc bien $\mathrm{Ker}(\delta) = \mathrm{Im}(p_*^0)$. L'inclusion $\mathrm{Im}(\delta) \subset \mathrm{Ker}(i_*^1)$ résulte de la définition de δ au moyen de cocycles; pour prouver l'inclusion réciproque, soit $\alpha \in \mathrm{Ker}(i_*^1)$, et soit $a \in Z^1(\mathfrak{g}, A)$ un représentant de α. Puisque $\alpha \in \mathrm{Ker}(i_*^1)$, il existe $b \in B$ tel que $a_s = b^{-1} \cdot {}^s b$ pour tout $s \in \mathfrak{g}$; si $x = p(b) \in B/A$, il est clair que x est invariant par $\mathfrak{g}$, et que $\delta(x) = \alpha$. Cela achève de prouver l'exactitude de la suite considérée.

Soient maintenant $x, y \in (B/A)^{\mathfrak{g}}$; il nous faut prouver que $\delta(x) = \delta(y)$ si et seulement si x et y sont dans la même orbite de $B^{\mathfrak{g}}$. Supposons d'abord que $\delta(x) = \delta(y)$. On peut alors trouver des représentants b, c de x, y dans B tels que:

$$b^{-1} \cdot {}^s b = a_s = c^{-1} \cdot {}^s c \quad (s \in \mathfrak{g}, \, a_s \in A).$$

On en déduit $bc^{-1} = {}^s(bc^{-1})$, d'où $bc^{-1} \in B^{\mathfrak{g}}$; comme l'élément bc^{-1} transforme y en x, on voit bien que x et y appartiennent à la même orbite. La réciproque se démontre par un calcul analogue.

1.13. Corollaire. *Soit $a \in Z^1(\mathfrak{g}, A)$, et soit α la classe de a dans $H^1(\mathfrak{g}, A)$. Les éléments de $H^1(\mathfrak{g}, A)$ qui ont même image que α dans $H^1(\mathfrak{g}, B)$ correspondent bijectivement aux orbites du groupe $(B_a)^{\mathfrak{g}}$ opérant dans $(B_a/A_a)^{\mathfrak{g}}$.*

Cela résulte par *torsion* de la proposition précédente (cf. nº 1.10).

Remarque. On vérifie immédiatement que l'espace homogène B_a/A_a peut être identifié au $\mathfrak{g}$-ensemble $(B/A)_a$ obtenu en tordant l'espace homogène B/A grâce au cocycle a (le groupe A opérant sur B/A par translations à gauche).

1.14. Si $f : X \to Y$ est une application, nous dirons que f est *propre* si l'image réciproque de tout point de Y par f est finie. (Cela revient à dire que f est une application propre au sens topologique lorsqu'on munit X et Y de la topologie discrète.)

Corollaire. *Pour que $i_*^1 : H^1(\mathfrak{g}, A) \to H^1(\mathfrak{g}, B)$ soit propre (resp. injective), il faut et il suffit que, pour tout $a \in Z^1(\mathfrak{g}, A)$, les orbites de $(B_a)^{\mathfrak{g}}$ dans $(B_a/A_a)^{\mathfrak{g}}$ soient en nombre fini (resp. que $(B_a)^{\mathfrak{g}}$ opère transitivement sur $(B_a/A_a)^{\mathfrak{g}}$).*

Cela résulte immédiatement du corollaire 1.13.

L'image de i_*^1 est donnée par le résultat suivant:

1.15. Proposition. *Soit $b = (b_s) \in Z^1(\mathfrak{g}, B)$ et soit β la classe de b dans $H^1(\mathfrak{g}, B)$. Pour que β appartienne à l'image de*

$$i_*^1 : H^1(\mathbf{g}, A) \to H^1(\mathbf{g}, B),$$

il faut et il suffit que le $\mathbf{g}$-ensemble $(B/A)_b$ obtenu en tordant B/A au moyen de b ait un point invariant par $\mathbf{g}$.

(Bien entendu, on fait opérer B *à gauche* sur B/A; c'est ce qui permet de définir $(B/A)_b$.)

Pour que $\beta \in \mathrm{Im}\,(i_*^1)$, il faut et il suffit qu'il existe $b \in B$ tel que $b^{-1}b_s\,{}^s b$ appartienne à A pour tout $s \in \mathbf{g}$. Si l'on pose $p(b) = c$, cela revient à dire que $c = b_s\,{}^s c$ dans B/A, ou encore que $c \in H^0(\mathbf{g}, (B/A)_b)$, d'où la proposition.

Remarque. D'après le n° 1.6, le groupe tordu B_b opère à gauche sur $(B/A)_b$, et il est clair qu'il opère transitivement.

1.16. Suite exacte associée à une extension de g-groupes.

Soit B un $\mathbf{g}$-groupe, soit A un sous-$\mathbf{g}$-groupe distingué de B, et soit C le $\mathbf{g}$-groupe quotient. Comme au n° 1.11, nous noterons i l'injection canonique de A dans B, et p la projection $B \to C$. On a donc une *suite exacte de $\mathbf{g}$-groupes*

$$0 \to A \xrightarrow{i} B \xrightarrow{p} C \to 0.$$

Nous allons voir que le groupe $C^{\mathbf{g}}$ *opère canoniquement à droite sur* $H^1(\mathbf{g}, A)$. Soit donc $c \in C^{\mathbf{g}}$, et soit $b \in B$ tel que $p(b) = c$. On a vu au n° 1.11 qu'il existe un cocycle $a = (a_s)$ de A tel que ${}^s b = b \cdot a_s$ pour tout $s \in \mathbf{g}$. L'application $x \mapsto b^{-1}xb$ est un isomorphisme de A sur le groupe tordu A_a; cela résulte de la formule:

$$b^{-1} \cdot {}^s x \cdot b = a_s\,{}^s(b^{-1}xb) \cdot a_s^{-1}.$$

Par passage à la cohomologie, on obtient ainsi une bijection

$$H^1(\mathbf{g}, A) \to H^1(\mathbf{g}, A_a).$$

En la composant avec la bijection canonique $\tau_a : H^1(\mathbf{g}, A_a) \to H^1(\mathbf{g}, A)$, on obtient une bijection

$$\varrho_b : H^1(\mathbf{g}, A) \to H^1(\mathbf{g}, A).$$

Si ξ est la classe d'un cocycle $x = (x_s)$, $\varrho_b(\xi)$ est la classe du cocycle

$$(b^{-1}x_s b a_s) = (b^{-1}x_s\,{}^s b).$$

Il est clair que, si $b, b' \in p^{-1}(C^{\mathbf{g}})$, on a $\varrho_{b'b} = \varrho_b \cdot \varrho_{b'}$; d'autre part, si $b \in A$, ϱ_b est l'identité. Il en résulte que ϱ_b ne dépend que de $p(b) = c$, et on peut le noter ϱ_c; on a $\varrho_{c'c} = \varrho_c \cdot \varrho_{c'}$, ce qui signifie bien que $C^{\mathbf{g}}$ opère *à droite* sur $H^1(\mathbf{g}, A)$.

1.17. Proposition. *Soit* $0 \to A \xrightarrow{i} B \xrightarrow{p} C \to 0$ *une suite exacte de $\mathbf{g}$-homomorphismes de $\mathbf{g}$-groupes. Alors:*

(i) *La suite:*

$$0 \to H^0(\mathbf{g}, A) \overset{i^0_*}{\to} H^0(\mathbf{g}, B) \overset{p^0_*}{\to} H^0(\mathbf{g}, C) \overset{\delta}{\to} H^1(\mathbf{g}, A) \overset{i^1_*}{\to} H^1(\mathbf{g}, B) \overset{p^1_*}{\to} H^1(\mathbf{g}, C)$$

est exacte.

(ii) *Si* $c \in C^{\mathbf{g}}$, *on a* $\delta(c) = \varrho_c(1)$, *en notant* 1 *l'élément neutre de* $H^1(\mathbf{g}, A)$.

(iii) *L'application* i^1_* *induit une bijection de l'ensemble des orbites de* $C^{\mathbf{g}}$ *dans* $H^1(\mathbf{g}, A)$ *sur le noyau de* p^1_*.

(iv) *Soit* $x = (x_s) \in Z^1(\mathbf{g}, A)$, *et soit* ξ *sa classe dans* $H^1(\mathbf{g}, A)$. *Soit* $c \in C^{\mathbf{g}}$. *Pour que* $\varrho_c(\xi) = \xi$, *il faut et il suffit que* c *appartienne à l'image de l'homomorphisme* $(p_x)^0_* : (B_x)^{\mathbf{g}} \to C^{\mathbf{g}}$.

(On note B_x le groupe obtenu en tordant B au moyen du cocycle x, étant entendu que A opère sur B par automorphismes intérieurs.) Pour prouver (i), il suffit (vu la prop. 1.12) de montrer que $\mathrm{Ker}\,(p^1_*)$ est égal à $\mathrm{Im}\,(i^1_*)$. L'inclusion $\mathrm{Im}\,(i^1_*) \subset \mathrm{Ker}\,(p^1_*)$ résulte de ce que le composé $p \circ i$ est trivial. Inversement, soit $b = (b_s)$ un cocycle dont la classe β appartient à $\mathrm{Ker}\,(p^1_*)$. Il existe alors $u \in B$ tel que $b_s \equiv u.{}^s u^{-1} \bmod \cdot A$, d'où $u^{-1} b_s {}^s u = a_s \in A$; le cocycle (a_s) a pour classe un élément $\alpha \in H^1(\mathbf{g}, A)$, et il est clair que $i^1_*(\alpha) = \beta$, ce qui démontre (i). L'assertion (ii) est immédiate sur la définition de ϱ_c. D'autre part, soient $x = (x_s)$ et $x' = (x'_s)$ deux éléments de $Z^1(\mathbf{g}, A)$ qui sont cohomologues dans B; il existe $b \in B$ tel que $x'_s = b^{-1} x_s {}^s b$ pour tout $s \in \mathbf{g}$; si $c = p(b)$, on a ${}^s c = c$, d'où $c \in C^{\mathbf{g}}$. Il est clair que ϱ_c transforme la classe de x en celle de x'; la réciproque se démontre de même, ce qui établit (iii). Enfin, les notations étant celles de (iv), choisissons $b \in p^{-1}(c)$. L'équation $\varrho_c(\xi) = \xi$ équivaut alors à l'existence de $a \in A$ tel que $x_s = a^{-1} b^{-1} x_s {}^s b {}^s a$, ce qui s'écrit encore $ba = x_s^s (ba) x_s^{-1}$, ou encore $ba \in (B_x)^{\mathbf{g}}$; d'où (iv).

1.18. Corollaire. *Soit* $b \in Z^1(\mathbf{g}, B)$ *et soit* β *la classe de* b *dans* $H^1(\mathbf{g}, B)$. *Les éléments de* $H^1(\mathbf{g}, B)$ *qui ont même image que* β *dans* $H^1(\mathbf{g}, C)$ *correspondent bijectivement aux orbites du groupe* $(C_b)^{\mathbf{g}}$ *opérant dans* $H^1(\mathbf{g}, A_b)$.

(Le groupe A_b (resp. C_b) est défini par torsion au moyen de b, le groupe B opérant sur A (resp. C) par restriction (resp. passage au quotient) des automorphismes intérieurs.)

Cela résulte par torsion de la partie (iii) de la proposition précédente.

1.19. Corollaire. *Si* $C^{\mathbf{g}}$ *et* $H^1(\mathbf{g}, B)$ *sont finis* (resp. triviaux), *il en est de même de* $H^1(\mathbf{g}, A)$.

Cela résulte de la partie (iii) de la proposition précédente.

1.20. Corollaire. *Pour que* $p^1_* : H^1(\mathbf{g}, B) \to H^1(\mathbf{g}, C)$ *soit propre* (resp. *injective*), *il faut et il suffit que, pour tout* $b \in Z^1(\mathbf{g}, B)$, *les orbites de* $(C_b)^{\mathbf{g}}$ *dans* $H^1(\mathbf{g}, A_b)$ *soient en nombre fini* (resp. *que* $(C_b)^{\mathbf{g}}$ *opère transitivement sur* $H^1(\mathbf{g}, A_b)$).

Cela résulte du corollaire 1.18.

Remarque. La condition du corollaire 1.20 est notamment vérifiée lorsque tous les $H^1(\mathfrak{g}, A_b)$ sont finis (resp. nuls).

1.21. Cas d'un sous-groupe abélien distingué.

On garde les notations et hypothèses des n°s 1.16 à 1.20, et on suppose en outre que A est *abélien*. On note additivement le groupe $H^1(\mathfrak{g}, A)$. D'autre part, l'homomorphisme $C^{\mathfrak{g}} \to \text{Aut}\,(A)$ permet de faire opérer $C^{\mathfrak{g}}$ à gauche sur $H^1(\mathfrak{g}, A)$. Nous noterons $c \cdot \alpha$ cette opération $(c \in C^{\mathfrak{g}}, \alpha \in H^1(\mathfrak{g}, A))$.

1.22. Proposition. *Soit B un $\mathfrak{g}$-groupe, soit A un sous-$\mathfrak{g}$-groupe abélien distingué de B, et soit $C = B/A$. On a:*

$$\varrho_c(\alpha) = c^{-1} \cdot \alpha + \delta(c) \quad pour\ tout\ \alpha \in H^1(\mathfrak{g}, A)\ et\ tout\ c \in C^{\mathfrak{g}}.$$

Soit b un élément de B relevant c. On a $^sb = b \cdot x_s$, et $x = (x_s)$ est un cocycle de classe $\delta(c)$. D'autre part, si $a = (a_s)$ est un cocycle de classe α, on peut prendre pour représentant de $\varrho_c(\alpha)$ le cocycle $b^{-1}a_s{}^sb$, et pour représentant de $c^{-1} \cdot \alpha$ le cocycle $b^{-1}a_sb$. La proposition résulte alors de la formule:

$$b^{-1}a_s{}^sb = b^{-1}a_sb \cdot x_s\,.$$

1.23. Corollaire. (a) *On a $\delta(c'c) = \delta(c) + c^{-1} \cdot \delta(c')$ pour $c, c' \in C^{\mathfrak{g}}$.*

(b) *Si A est contenu dans le centre de B, l'application $\delta : C^{\mathfrak{g}} \to H^1(\mathfrak{g}, A)$ est un homomorphisme, et l'on a $\varrho_c(\alpha) = \alpha + \delta(c)$ pour $c \in C^{\mathfrak{g}}, \alpha \in H^1(\mathfrak{g}, A)$.*

Soit $\alpha \in H^1(\mathfrak{g}, A)$. On sait que $\varrho_{c'c}(\alpha) = \varrho_c(\varrho_{c'}(\alpha))$. La proposition 1.22 permet de calculer les deux membres de cette formule. Pour le premier, on trouve $c^{-1}c'^{-1}\alpha + \delta(c'c)$, et pour le second:

$$c^{-1}c'^{-1}\alpha + c^{-1}\delta(c') + \delta(c)\,.$$

En comparant, on obtient (a). L'assertion (b) résulte immédiatement de 1.22 et de (a).

Remarque. Lorsque $\mathfrak{g}$ est un groupe profini (ou un groupe discret), on peut aller plus loin: on définit pour tout $c \in Z^1(\mathfrak{g}, C)$ un élément $\Delta(c) \in H^2(\mathfrak{g}, A_c)$, et l'on montre que $\Delta(c) = 0$ si et seulement si la classe de c appartient à $\text{Im}\,(p_*^1)$. Comme nous n'aurons pas besoin de ce résultat dans la suite, nous nous bornerons à renvoyer le lecteur à GROTHENDIECK [8], ou à [19], Chap. I, n°s 5.6 et 5.7.

1.24. Action des automorphismes intérieurs. Soit A un $\mathfrak{g}$-groupe, et soit $t \in \mathfrak{g}$. Définissons des applications

$$\lambda_t : \mathfrak{g} \to \mathfrak{g}\ \text{ et }\ \mu_t : A \to A$$

par les formules:

$$\lambda_t(s) = t^{-1}st\,, \quad \mu_t(x) = {}^tx\,.$$

Ces applications sont *compatibles* entre elles, au sens du n° 1.3. Elles définissent donc des applications

$$(\lambda_t, \mu_t)^i_* : H^i(\mathbf{g}, A) \to H^i(\mathbf{g}, A), \quad i = 0,1.$$

1.25. Proposition. *Pour tout* $t \epsilon \mathbf{g}$, $(\lambda_t, \mu_t)^i_*$ *est l'application identique de* $H^i(\mathbf{g}, A)$ *sur lui-même.*

Pour $i = 0$, on a $H^0(\mathbf{g}, A) = A^{\mathbf{g}}$, et si $x \epsilon A^{\mathbf{g}}$, on a $\mu_t(x) = x$. D'où le fait que $(\lambda_t, \mu_t)^0_*$ est l'identité.

Soit maintenant $\alpha \epsilon H^1(\mathbf{g}, A)$, et soit $a = (a_s)$ un cocycle représentant α. La classe de cohomologie $(\lambda_t, \mu_t)^1_*(\alpha)$ est celle du cocycle $b = (b_s)$ défini par la formule :

$$b_s = {}^t(a_{t^{-1}st}).$$

Utilisant le fait que a est un cocycle, on obtient :

$$b_s = {}^t(a_{t^{-1}} \cdot {}^{t^{-1}}a_{st}) = {}^t a_{t^{-1}} \cdot a_{st} = a_t^{-1} \cdot a_{st} = a_t^{-1} \cdot a_s \cdot {}^s a_t,$$

ce qui montre que b_s est cohomologue à a_s, d'où la proposition.

1.26. Conservons les notations des deux n^{os} ci-dessus, et soit $\mathbf{h}$ un *sous-groupe distingué* de $\mathbf{g}$. Si $t \epsilon \mathbf{g}$, l'homomorphisme λ_t applique $\mathbf{h}$ dans lui-même. Les applications

$$\lambda_t : \mathbf{h} \to \mathbf{h}, \quad \mu_t : A \to A$$

sont encore compatibles. On en déduit comme ci-dessus des applications

$$(\lambda_t, \mu_t)^i_* : H^i(\mathbf{h}, A) \to H^i(\mathbf{h}, A)$$

que nous noterons σ_t pour abréger. Il est clair que $\sigma_{tt'} = \sigma_t \circ \sigma_{t'}$. D'autre part, la proposition 1.25, appliquée au groupe $\mathbf{h}$, montre que $\sigma_t = 1$ si $t \epsilon \mathbf{h}$. On peut donc définir σ_t pour $t \epsilon \mathbf{g}/\mathbf{h}$, et l'on voit que $\mathbf{g}/\mathbf{h}$ *opère à gauche sur* $H^i(\mathbf{h}, A)$. Pour $i = 0$, on retrouve les opérations évidentes de $\mathbf{g}/\mathbf{h}$ sur $A^{\mathbf{h}}$.

Remarque. Supposons que tout sous-groupe ouvert de $\mathbf{g}$ contienne un sous-groupe ouvert distingué (ce qui est notamment le cas lorsque $\mathbf{g}$ est *profini*). Alors $\mathbf{g}/\mathbf{h}$ opère *continûment* sur $H^1(\mathbf{h}, A)$. (Bien entendu, $\mathbf{g}/\mathbf{h}$ opère toujours continûment sur $H^0(\mathbf{h}, A)$.)

1.27. Proposition. *Soit* $\mathbf{h}$ *un sous-groupe distingué de* $\mathbf{g}$*, et soit* A *un* $\mathbf{g}$*-groupe. On a alors une suite exacte :*

$$0 \to H^1(\mathbf{g}/\mathbf{h}, A^{\mathbf{h}}) \xrightarrow{\alpha} H^1(\mathbf{g}, A) \xrightarrow{\beta} H^1(\mathbf{h}, A),$$

où α *est l'application associée à la projection* $\mathbf{g} \to \mathbf{g}/\mathbf{h}$ *et à l'inclusion* $A^{\mathbf{h}} \to A$*, et où* β *est la restriction. De plus* α *est injectif et l'image de* β *est contenue dans l'ensemble des éléments de* $H^1(\mathbf{h}, A)$ *invariants par* $\mathbf{g}/\mathbf{h}$ *(pour les opérations* σ_t *définies ci-dessus).*

Soient $a = (a_s)$ et $b = (b_s)$ deux cocycles de $\mathbf{g}/\mathbf{h}$ dans $A^{\mathbf{h}}$ qui sont cohomologues dans $Z^1(\mathbf{g}, A)$. Il existe donc $c \in A$ tel que

$$b_s = c^{-1} a_s{}^s c \,.$$

Si l'on prend s dans $\mathbf{h}$, on a $a_s = b_s = 1$, d'où ${}^s c = c$, ce qui montre que c appartient à $A^{\mathbf{h}}$; les cocycles a et b sont donc cohomologues dans $Z^1(\mathbf{g}/\mathbf{h}, A^{\mathbf{h}})$, ce qui prouve que α est injectif.

L'inclusion $\operatorname{Im}(\alpha) \subset \operatorname{Ker}(\beta)$ est évidente. Pour prouver l'inclusion opposée, soit $a = (a_s)$ un cocycle de $\mathbf{g}$ dont la restriction à $\mathbf{h}$ est un cobord. Il existe $c \in A$ tel que $a_s = c^{-1} \cdot {}^s c$ si $s \in \mathbf{h}$; quitte à remplacer (a_s) par le cocycle cohomologue $(c^{-1} a_s{}^s c)$, on peut donc supposer que $a_s = 1$ pour $s \in \mathbf{h}$. L'identité $a_{st} = a_s{}^s a_t$ montre alors que a_s ne dépend que de la classe de $s \bmod \cdot \mathbf{h}$, et appartient à $A^{\mathbf{h}}$. Le cocycle a provient donc d'un cocycle de $\mathbf{g}/\mathbf{h}$ à valeurs dans $A^{\mathbf{h}}$, ce qui achève de prouver que $\operatorname{Im}(\alpha) = \operatorname{Ker}(\beta)$.

Enfin, il est clair que $\beta : H^1(\mathbf{g}, A) \to H^1(\mathbf{h}, A)$ est compatible avec l'action de $\mathbf{g}$ sur ces deux ensembles. D'après la proposition 1.25, $\mathbf{g}$ opère trivialement sur $H^1(\mathbf{g}, A)$. Il opère donc aussi trivialement sur $\operatorname{Im}(\beta)$, d'où la proposition.

1.28. Induction. Soit $\mathbf{h}$ un sous-groupe de $\mathbf{g}$, et soit E un $\mathbf{h}$-ensemble. Soit E^* l'ensemble des applications $f : \mathbf{g} \to E$ qui vérifient les deux conditions suivantes:

(a) f est constante sur les classes à gauche modulo un sous-groupe ouvert de $\mathbf{g}$,

(b) $f(h \cdot x) = {}^h(f(x))$ pour $x \in \mathbf{g}$, $h \in \mathbf{h}$.

(Lorsque $\mathbf{g}$ est *profini*, ou *discret*, la condition (a) signifie simplement que f est continue.)

Si $f \in E^*$ et $g \in \mathbf{g}$, soit ${}^g f : \mathbf{g} \to E$ l'application définie par la formule:

$$({}^g f)(x) = f(xg) \,.$$

On constate sans difficultés que ${}^g f$ vérifie les conditions (a) et (b) et que ${}^g({}^{g'} f) = {}^{(gg')} f$. Le groupe $\mathbf{g}$ opère donc à gauche sur E^*, et la propriété (a) assure que cette opération est continue. Ainsi, E^* est muni d'une structure de $\mathbf{g}$-ensemble; on dit que c'est le $\mathbf{g}$-ensemble *induit* du $\mathbf{h}$-ensemble E. Lorsque E est un $\mathbf{h}$-groupe, E^* est un $\mathbf{g}$-groupe.

D'après (b), l'application $\mu : E^* \to E$ qui associe à tout $f \in E^*$ sa valeur à l'élément neutre est compatible (au sens du n° 1.3) avec l'homomorphisme d'inclusion $\lambda : \mathbf{h} \to \mathbf{g}$.

1.29. Proposition. *Soit $\mathbf{h}$ un sous-groupe de $\mathbf{g}$, soit A un $\mathbf{h}$-groupe, et soit A^* le $\mathbf{g}$-groupe induit correspondant. Supposons que les sous-groupes ouverts distin-*

gués de **g** *forment un système fondamental de voisinages de l'élément neutre. Alors les applications*

$$(\lambda, \mu)^i_* : H^i(\mathbf{g}, A^*) \to H^i(\mathbf{h}, A) \quad (i = 0,1)$$

sont bijectives.

Le cas $i = 0$ est trivial (et ne nécessite d'ailleurs aucune hypothèse sur **g**). Nous supposerons donc que $i = 1$; nous noterons φ l'application de $Z^1(\mathbf{g}, A^*)$ dans $Z^1(\mathbf{h}, A)$ définie par (λ, μ), et Φ l'application correspondante de $H^1(\mathbf{g}, A^*)$ dans $H^1(\mathbf{h}, A)$.

Soit $F = (F_s) \epsilon Z^1(\mathbf{g}, A^*)$. On a:

$$(1) \qquad\qquad F_s(h \cdot x) = {}^h(F_s(x)) \quad (h \epsilon \mathbf{h},\, x,\, s \epsilon \mathbf{g})\,,$$

$$(2) \qquad\qquad F_{st}(x) = F_s(x) \cdot F_t(xs) \quad (s,\, t,\, x \epsilon \mathbf{g})\,.$$

(3) Il existe un sous-groupe ouvert distingué **n** de **g** tel que $F_s(x) = 1$ si $s \epsilon \mathbf{n}$. Vu (2), cela équivaut à dire que $F_s(x)$ *ne dépend que des classes de* s *et de* x *modulo* **n**.

Ces trois propriétés *caractérisent* les éléments de $Z^1(\mathbf{g}, A^*)$.

Si $F \epsilon Z^1(\mathbf{g}, A^*)$ on notera F' l'application de **g** dans A définie par la formule:

$$F'(s) = F_s(1)\,.$$

Il résulte de (2) que l'on a:

$$(4) \qquad\qquad F_s(x) = F'(x)^{-1} F'(xs) \quad (x,\, s \epsilon \mathbf{g})\,,$$

ce qui montre que F' détermine F. De plus (1) et (2) impliquent:

$$(5) \qquad\qquad F'(h \cdot x) = F'(h)\, {}^h(F'(x)) \quad (h \epsilon \mathbf{h},\, x \epsilon \mathbf{g})\,,$$

et (3) implique:

(6) Il existe un sous-groupe ouvert distingué **n** de **g** tel que $F'(x)$ ne dépende que de la classe de x modulo **n**.

Inversement, si $F' : \mathbf{g} \to A$ vérifie (5) et (6), l'application F définie par la formule (4) vérifie les conditions (1), (2), (3): la vérification est immédiate. De plus $\varphi(F)$ est la restriction de F' à **h**.

Si $F, G \epsilon Z^1(\mathbf{g}, A^*)$ sont cohomologues, il existe $c \epsilon A^*$ tel que:

$$(7) \qquad\qquad G'(s) = c(1)^{-1} F'(s) c(s) \quad (s \epsilon \mathbf{g})\,,$$

et réciproquement, s'il existe $c \epsilon A^*$ vérifiant (7), la formule (4) montre que F et G sont cohomologues.

Montrons maintenant que Φ est *surjective*. Soit $z \epsilon Z^1(\mathbf{h}, A)$, et soit **n** un sous-groupe ouvert distingué de **g** tel que z soit constant sur les classes de **h** modulo $\mathbf{h} \cap \mathbf{n}$. Soit $(s_i)_{i \epsilon I}$ un système de représentants des classes à droite de **g** modulo $\mathbf{h} \cdot \mathbf{n}$; nous supposerons que ce système contient l'élément 1.

Si l'on a

$$h \cdot n \cdot s_i = h' \cdot n' \cdot s_j \quad (h, h' \epsilon \mathbf{h},\, n, n' \epsilon \mathbf{n},\, i, j \epsilon I)\,,$$

on a évidemment $i = j$ et $h \equiv h' \bmod \cdot \mathbf{h} \cap \mathbf{n}$, d'où $z_h = z_{h'}$. Il existe donc une application $F' : \mathbf{g} \to A$ telle que:

$$(8) \qquad\qquad F'(h \cdot n \cdot s_i) = z_h \quad \text{si} \quad h \epsilon \mathbf{h}, n \epsilon \mathbf{n}, i \epsilon I .$$

Il est clair que F' vérifie (5) et (6). De plus, la restriction de F' à $\mathbf{h}$ est égale à z. Compte tenu de ce qui a été dit plus haut, cela prouve que φ est surjective, donc aussi Φ.

Il reste à prouver que Φ est *injective*. Soient $F, G \epsilon Z^1(\mathbf{g}, A^*)$ tels que $\varphi(F)$ et $\varphi(G)$ soient cohomologues. Il existe alors $a \epsilon A$ tel que l'on ait:

$$(9) \qquad\qquad G'(h) = a^{-1} F'(h)^h a \quad (h \epsilon \mathbf{h}) .$$

Soit $\mathbf{n}$ un sous-groupe ouvert distingué de $\mathbf{g}$ tel que a soit invariant par $\mathbf{n}$, et que F' et G' soient constantes sur les classes de $\mathbf{g}$ modulo $\mathbf{n}$. Soit de nouveau $(s_i)_{i \epsilon I}$ un système de représentants des classes à droite de $\mathbf{g}$ modulo $\mathbf{h} \cdot \mathbf{n}$; on suppose encore que ce système contient l'élément 1. Si $i \epsilon I$, soit a_i l'élément de A défini par la formule:

$$(10) \qquad\qquad G'(s_i) = a^{-1} F'(s_i) a_i .$$

Si $t \epsilon \mathbf{h} \cap \mathbf{n}$, on a $^t(G'(x)) = G'(x)$ pour tout $x \epsilon \mathbf{g}$, et de même pour F'. En appliquant ceci à $x = s_i$, on voit que $^t a_i = a_i$; les a_i sont donc *invariants par* $\mathbf{h} \cap \mathbf{n}$. On en déduit comme ci-dessus l'existence d'une application $c : \mathbf{g} \to A$ telle que:

$$(11) \qquad\qquad c(h \cdot n \cdot s_i) = {}^h a_i \quad \text{si} \quad h \epsilon \mathbf{h}, n \epsilon \mathbf{n}, i \epsilon I .$$

Il est clair que c est constante sur les classes modulo $\mathbf{n}$, et vérifie l'identité $c(h \cdot x) = {}^h(c(x))$ pour $h \epsilon \mathbf{h}, x \epsilon \mathbf{g}$; on a donc $c \epsilon A^*$. Comme $F'(1) = G'(1) = 1$, on a en outre $c(1) = a$.

Soit maintenant $x = h \cdot n \cdot s_i$ $(h \epsilon \mathbf{h}, n \epsilon \mathbf{n}, i \epsilon I)$ un élément de $\mathbf{g}$. On a:

$$
\begin{aligned}
G'(x) &= G'(h) \cdot {}^h(G'(n \cdot s_i)) = G'(h) \cdot {}^h(G'(s_i)) \\
&= a^{-1} F'(h)^h a^h(G'(s_i)) \\
&= a^{-1} F(h)^h(F'(s_i)a_i) = a^{-1} F'(h \cdot s_i)^h a_i \\
&= c(1)^{-1} F'(x) c(x) .
\end{aligned}
$$

D'après (7), cela signifie que F et G sont cohomologues, ce qui achève la démonstration.

§ 2. Cohomologie galoisienne

2.1. Cohomologie d'un foncteur.

Soit k un corps et soit S la catégorie des extensions algébriques séparables de k (ou bien la catégorie de toutes les extensions de k). Soit A un foncteur de S

dans la catégorie *Ens* des ensembles (resp. dans la catégorie *Gr* des groupes). Nous supposerons que A vérifie les trois conditions suivantes:

(1) Si $K \to K'$ est un morphisme $(K, K' \epsilon S)$, le morphisme correspondant $A(K) \to A(K')$ est injectif.

(2) Si K' est une extension galoisienne de K $(K, K' \epsilon S)$, $A(K)$ s'identifie au sous-ensemble de $A(K')$ formé des éléments invariants par les opérations du groupe de GALOIS $g(K'/K)$.

(3) Pour tout $K \epsilon S$, on a $A(K) = \varinjlim \cdot A(K_\alpha)$, pour K_α parcourant l'ensemble des sous-extensions de K qui sont de type fini sur k.

Soit K'/K une extension galoisienne (avec $K, K' \epsilon S$). La condition (3) montre que $g(K'/K)$ opère continûment sur $A(K')$; ainsi $A(K')$ est muni d'une structure de $g(K'/K)$-*ensemble* (resp. de $g(K'/K)$-*groupe*, lorsque le foncteur A est à valeurs dans *Gr*). Le i-ième ensemble de cohomologie $H^i(g(K'/K), A(K'))$ est donc défini pour $i = 0$ (resp. pour $i = 0,1$), cf. § 1; on le notera $H^i(K'/K, A)$ et l'on écrira de même $Z^1(K'/K, A)$ à la place de $Z^1(g(K'/K), A(K'))$. D'après (2), on a:

$$H^0(K'/K, A) = A(K).$$

Remarque. Lorsque A prend ses valeurs dans la catégorie des groupes *commutatifs*, on peut définir pour tout $i \geqslant 0$ des groupes de cohomologie $H^i(K'/K, A)$, cf. [13] ainsi que [19], Chap. II.

2.2. Propriétés fonctorielles. Il est clair que les $H^i(K'/K, A)$ sont fonctoriels par rapport à A.

D'autre part, pour A fixé, considérons deux extensions galoisiennes K'_1/K_1 et K'_2/K_2, de groupes de GALOIS g_1 et g_2. Soit $f : K_1 \to K_2$ un morphisme dans la catégorie S. Supposons qu'il *existe* un morphisme $F : K'_1 \to K'_2$ prolongeant f. D'après la théorie de GALOIS, le corps $F(K'_1)$ ne dépend pas du chois de F, et l'extension $F(K'_1) \cdot K_2/K_2$ est galoisienne. Pour tout $s \epsilon g_2$, il existe un unique élément $\lambda_F(s) \epsilon g_1$ tel que $s \circ F = F \circ \lambda_F(s)$. L'application $\lambda_F : g_2 \to g_1$ ainsi définie est un homomorphisme, qui est compatible avec l'application

$$\mu_F : A(K'_1) \to A(K'_2)$$

définie par F. On en déduit (cf. n° 1.3) des applications

$$(\lambda_F, \mu_F)^i_* : H^i(K'_1/K_1, A) \to H^i(K'_2/K_2, A).$$

Lemme. *Les applications définies ci-dessus ne dépendent pas du choix de F.*

Soit F' un morphisme de K'_1 dans K'_2 prolongeant f. D'après la théorie de GALOIS, il existe $t \epsilon g_1$ tel que $F' = F \circ t$. On a $\lambda_{F'} = \lambda_t \circ \lambda_F$, où λ_t désigne l'automorphisme intérieur de $\mathfrak{g}_1$ défini par t^{-1}, et $\mu_{F'} = \mu_F \circ \mu_t$. Le lemme résulte alors du fait que $(\lambda_t, \mu_t)_*^i$ est l'identité (cf. n° 1.25).

Ainsi, les applications $H^i(K_1'/K_1, A) \to H^i(K_2'/K_2, A)$ ne dépendent que de f; pour $i = 0$, on obtient évidemment l'application de $A(K_1)$ dans $A(K_2)$ définie par f.

Prenons en particulier $K_1 = K_2 = k$ (f étant l'application identique) et pour K_1' et K_2' deux clôtures séparables de k. Le lemme ci-dessus donne une bijection *canonique* de $H^i(K_1'/k, A)$ sur $H^i(K_2'/k, A)$. On identifiera entre eux ces ensembles, et on les notera $H^i(k_s/k, A)$, ou simplement $H^i(k, A)$. Ce sont des foncteurs par rapport à k.

2.3. Exemples. (a) Soit X un *schéma* sur k. Pour toute extension K/k, nous noterons X_K, ou $X(K)$, l'ensemble des points de X à valeurs dans K (rappelons que c'est l'ensemble des k-morphismes de $\mathrm{Spec}\,(K)$ dans X). On définit ainsi un foncteur sur S à valeurs dans *Ens*, que l'on note encore X. Ce foncteur vérifie les conditions (1) et (2) du n° 2.1; il vérifie la condition (3) pourvu que l'on suppose que X est localement de type fini sur k (cf. [9], Chap. IV). Lorsque X est un *schéma en groupes* sur k, le foncteur correspondant prend ses valeurs dans la catégorie des groupes, et $H^i(K'/K, X)$ est défini pour $i = 0,1$.

(b) Soit X un schéma de type fini sur k. Pour toute extension K/k, soit $\mathrm{Aut}\,X(K)$ le groupe des K-automorphismes du schéma étendu $X \otimes_k K$. On obtient ici encore un foncteur $\mathrm{Aut}\,X : S \to Gr$, vérifiant les conditions (1), (2), (3) du n° 2.1. Il en est de même lorsque X est un schéma en groupes de type fini sur k, et que l'on se borne aux automorphismes respectant cette structure de groupe.

Nous verrons dans les §§ 4 et 5 que, dans certains cas, le foncteur $\mathrm{Aut}\,X$ provient d'un schéma en groupes sur k.

2.4. Conjugaison. Soit k'/k une extension galoisienne, de groupe de Galois $\mathfrak{g}$, et soit X' un k'-schéma. Si $s \in \mathfrak{g}$, on note $^sX'$ le schéma obtenu à partir de X' par le changement de base $s : k' \to k'$; si $x \in X'(k')$, on note ^sx le point correspondant de $(^sX)(k')$; si $f' : X' \to Y'$ est un morphisme de k'-schémas, on note $^sf'$ le morphisme de $^sX'$ dans $^sY'$ déduit de f' par changement de base. On a la formule:

$$(^sf')(^sx) = {}^s(f'(x)) , \quad \text{pour } s \in \mathfrak{g} , \ x \in X(k') .$$

Si $X' = X \otimes_k k'$ et $Y' = Y \otimes_k k'$, on peut identifier $^sX'$ à X' et $^sY'$ à Y'; on a alors $^sf' = f'$ pour tout $s \in \mathfrak{g}$ si et seulement si f' est «défini sur k», i. e. s'il existe un morphisme $f : X \to Y$ tel que $f' = f \otimes 1$; le morphisme f est alors unique.

2.5. Torsion. Soit k'/k une extension galoisienne, de groupe de GALOIS $\mathfrak{g}$, et soit X un schéma de type fini sur k (resp. un schéma en groupes de type fini sur k). Nous supposerons que X est *quasi-projectif*. Soit Aut X le foncteur d'automorphismes de X (cf. n° 2.3).

2.6. Proposition. *Soit* $a = (a_s) \in Z^1(k'/k,\ \text{Aut } X)$. *Il existe un k-schéma de type fini (resp. un k-schéma en groupes de type fini) Y et un k'-isomorphisme* $f' : X \otimes_k k' \to Y \otimes_k k'$ *tels que l'on ait:*

$$^s f' = f' \circ a_s \quad pour\ tout\quad s \in \mathfrak{g}\,.$$

Le couple (Y, f') est unique, à isomorphisme unique près.

La continuité du cocycle a permet de se ramener au cas où l'extension k'/k est *finie*. On applique alors à (a_s) le théorème de «descente du corps de base» de WEIL (cf. par exemple [17], Chap. V., § 4, n° 20). Cela revient simplement à ceci:

Si l'on pose $b_s = a_s \circ (1 \otimes s)$, on a $b_{st} = b_s \circ b_t$, et l'on peut faire opérer $\mathfrak{g}$ sur $X \otimes_k k'$ au moyen des b_s; ces opérations sont compatibles avec celles de $\mathfrak{g}$ sur k'. Il en résulte que $\mathfrak{g}$ opère librement, et l'on prend pour Y le schéma quotient $(X \otimes_k k')/\mathfrak{g}$ (ce quotient existe grâce au fait que X est supposée quasi-projective). On vérifie immédiatement que Y a les propriétés voulues.

Le schéma Y ainsi obtenu sera noté X_a; on dit qu'il s'obtient *en tordant* X *au moyen de* a. L'ensemble des isomorphismes de $X \otimes_k k'$ sur $Y \otimes_k k'$ est un *espace homogène principal* de Aut $X(k')$, et la classe de cohomologie correspondante est celle de a. On en déduit que X_a est isomorphe à X_b si et seulement si a et b sont cohomologues. *Les éléments de $H^1(k'/k,\ \text{Aut } X)$ correspondent donc bijectivement aux classes d'isomorphisme de k-schémas* (resp. de k-schémas en groupes) Y tels que $Y \otimes_k k'$ *soit isomorphe à* $X \otimes_k k'$. (Un tel k-schéma Y est appelé une *k'/k-forme* de X – lorsque $k' = k_s$, on dit simplement une *k-forme*.)

Exemple. Soit A un schéma en groupes de type fini sur k, et soit $a = (a_s) \in Z^1(k'/k, A)$. Si l'on fait opérer A à gauche sur lui-même (par translations), on peut tordre A au moyen de a. Le schéma P_a ainsi obtenu est un *espace homogène principal* sur A (cf. par exemple [17], Chap. V, n° 21 ou [13]) tel que $P_a(k') \neq \varnothing$. On retrouve ainsi la correspondance bien connue entre les *classes* de tels espaces et les *éléments* de $H^1(k'/k, A)$.

La notion de «torsion» définie plus haut est compatible avec celle du § 1. De façon précise:

2.7. Proposition. *Soient X, Y, a, f' vérifiant les hypothèses de 2.6. L'application $f' : X(k') \to Y(k')$ est un isomorphisme du $\mathbf{g}$-ensemble tordu $X(k')_a$ sur le $\mathbf{g}$-ensemble $Y(k')$.*

Cela résulte de la formule:

$$f'(a_s(^sx)) = (f' \circ a_s)(^sx) = (^sf')(^sx) = {}^s(f'(x)) \quad (x \in X(k'),\, s \in \mathbf{g}) .$$

2.8. Restriction des scalaires. Soient k_1 une extension finie séparable de k, et k' une extension galoisienne de k contenant k_1; notons $\mathbf{g}$ (resp. $\mathbf{h}$) le groupe de Galois de k'/k (resp. de k'/k_1); l'ensemble des k-morphismes de k_1 dans k' s'identifie canoniquement à l'espace homogène $\mathbf{g}/\mathbf{h}$. Soit X un schéma sur k_1; si $s \in \mathbf{g}$, le conjugué ^sX de X ne dépend que de la classe de s modulo $\mathbf{h}$. Soit d'autre part Y un schéma sur k, et soit $p : Y \otimes_k k_1 \to X$ un k_1-morphisme. La collection des conjugués $(^sp)_{s \in \mathbf{g}/\mathbf{h}}$ définit un morphisme

$$(^sp) : Y \otimes_k k' \to \overline{\prod_{s \in \mathbf{g}/\mathbf{h}}} {}^sX \otimes k' .$$

Lorsque (^sp) est un isomorphisme, on dit (cf. Weil [21], n° 1.3) que Y s'obtient à partir de X *par restriction des scalaires de k_1 à k*; le couple (Y, p) est alors unique, à un isomorphisme unique près. On écrit $Y = R_{k_1/k}(X)$ ou encore $Y = \overline{\prod_{k_1/k}}(X)$, cette dernière notation étant celle de Grothendieck. *L'existence de* $R_{k_1/k}(X)$ peut se démontrer, par exemple, lorsque X est quasi-projectif sur k.

Supposons que $Y = R_{k_1/k}(X)$ existe; soit E le $\mathbf{h}$-ensemble $X(k')$, et soit E^* le $\mathbf{g}$-ensemble induit (cf. n° 1.28). Nous allons voir que E^* *peut être identifié à* $Y(k')$. De façon plus précise, soit $y \in Y(k')$, et soit $\vartheta(y)$ l'application de $\mathbf{g}$ dans E définie par la formule:

$$\vartheta(y)(s) = {}^s((^{s^{-1}}p)(y)) ,$$

formule qui a un sens, puisque $(^{s^{-1}}p)(y)$ appartient à $^{s^{-1}}X(k')$.

2.9. Proposition. *Pour tout $y \in Y(k')$, on a $\vartheta(y) \in E^*$. L'application $\vartheta : Y(k') \to E^*$ ainsi définie est un isomorphisme de $\mathbf{g}$-ensembles.*

Il faut d'abord prouver que $\vartheta(y)$ vérifie les conditions (a) et (b) du n° 1.28. La condition (a) (continuité) est immédiate. La condition (b) s'écrit:

$$\vartheta(y)(h \cdot s) = {}^h(\vartheta(y)(s)) , \quad h \in \mathbf{h},\, s \in \mathbf{g} .$$

Elle résulte du fait que $^hp = p$. L'application ϑ est donc bien définie. On vérifie sans difficultés que c'est une bijection. Le fait que ϑ soit un morphisme de $\mathbf{g}$-ensembles résulte du calcul suivant:

$$\vartheta(y)(s \cdot g) = {}^{sg}((^{g^{-1}s^{-1}}p)(y))$$
$$= {}^{sg}(^{g^{-1}}((^{s^{-1}}p)(^{g}y))) = {}^{s}((^{s^{-1}}p)(^{g}y))$$
$$= \vartheta(^{g}y)(s) \ .$$

Lorsque X est un *schéma en groupes*, il en est de même de Y, et l'application ϑ est compatible avec les structures de groupes de $Y(k')$ et de E^*. Compte tenu de la prop. 1.29, cela donne:

2.10. Corollaire. *Si* $Y = R_{k_1/k}(X)$, *il existe une bijection canonique de* $H^i(k'/k, Y)$ *sur* $H^i(k'/k_1, X)$, $i = 0,1$.

En particulier, prenant $k' = k_s$, on voit qu'on peut *identifier* $H^i(k, Y)$ et $H^i(k_1, X)$.

Remarque. Pour $i = 0$, on retrouve la bijection canonique de $Y(k)$ sur $X(k_1)$, cf. WEIL, *loc. cit.* Pour $i = 1$, on peut interpréter la bijection $H^1(k'/k_1, X) \to H^1(k'/k, Y)$ de la manière suivante: à tout espace homogène principal P sur X, ayant un point dans k', on associe l'espace homogène principal $R_{k_1/k}(P)$ sur Y.

2.11. Variétés algébriques et groupes algébriques. Soit k un corps. Nous appellerons *variété algébrique définie sur* k (ou simplement *k-variété*) tout schéma X de type fini sur k qui est *absolument réduit*. Rappelons que cette dernière condition signifie que, pour toute extension k'/k, le schéma étendu $X \otimes_k k'$ est réduit (son faisceau structural n'a pas d'éléments nilpotents); en particulier X est réduit; la réciproque est vraie lorsque k est *parfait*, ce qui sera toujours le cas à partir du § 4. Lorsque k est *algébriquement clos*, la notion de « variété algébrique » définie ici est équivalente à celle de FAC.

De même, nous appellerons *groupe algébrique* défini sur k tout schéma en groupes G sur k qui est une variété algébrique (pour la structure de schéma sous-jacente). Il revient au même de dire que G est de type fini sur k, et *simple* sur k (au sens de GROTHENDIECK [10] – autrement dit «lisse» au sens de GROTHENDIECK [9], Chap. IV).

Lemme. (i) *Si* V *est une variété algébrique définie sur* k, $V(k_s)$ *est dense dans* V.

(ii) *Soient* V *et* W *deux k-variétés, et soient* f *et* f' *deux morphismes de* V *dans* W. *Si* f *et* f' *définissent la même application de* $V(k_s)$ *dans* $W(k_s)$, *on a* $f = f'$.

L'assertion (i) est bien connue, cf. par exemple [17], Chap. V, n° 23. L'assertion (ii) résulte de (i) et du fait que V est un schéma réduit.

Application. Soit P un espace homogène principal sur un groupe algébrique A ; il existe une extension k' de k telle que $P \otimes_k k'$ soit isomorphe à $A \otimes_k k'$ (par exemple $k' = \bar{k}$). Comme A est une variété algébrique, il en est de même de P, et le lemme ci-dessus montre que $P(k_s) \neq \varnothing$. On en conclut que P *est défini par un élément de* $H^1(k, A)$, cf. n° 2.6.

2.12. Un lemme de descente du corps de base. Soient k un corps, k_1 une extension finie de k contenue dans k_s, et $\mathbf{g}$ le groupe de Galois de k_s/k. Soit V une k_1-variété algébrique. On suppose $V(k_s)$ muni d'une structure de $\mathbf{g}$-ensemble, et l'on note $s(x)$ le transformé de $x \in V(k_s)$ par $s \in \mathbf{g}$. On se propose de donner des conditions pour qu'il existe une variété algébrique W définie sur k et un isomorphisme $\vartheta : V \otimes_{k_1} k_s \to W \otimes_k k_s$ tels que l'application correspondante $\vartheta : V(k_s) \to W(k_s)$ soit un isomorphisme de $\mathbf{g}$-ensembles.

Soit $s \in \mathbf{g}$. Soit ${}^s V$ le schéma sur ${}^s k_1$ obtenu à partir de V par le changement de base $s : k_1 \to {}^s k_1$. On a une bijection canonique $x \mapsto {}^s x$ de $V(k_s)$ sur $({}^s V)(k_s)$, d'où une application bien définie $f_s : ({}^s V)(k_s) \to V(k_s)$ vérifiant la formule :

$$f_s({}^s x) = s(x) \, .$$

Lemme. *Supposons vérifiées les trois conditions suivantes:*

(a) *Pour tout* $s \in \mathbf{g}$, *il existe un morphisme* $F_s : {}^s V \otimes k_s \to V \otimes k_s$ *dont l'action sur les points à valeurs dans* k_s *coïncide avec* f_s.

(b) *Il existe une extension finie séparable* k_2/k_1 *telle que* $f_s = 1$ *si* $s \in \mathbf{g}(k_s/k_2)$.

(c) *Il existe un recouvrement de* $V \otimes k_s$ *par des ouverts affines* U_α *tels que* $U_\alpha(k_s)$ *soit stable par* $\mathbf{g}$ *pour tout* α.

Il existe alors une k-variété W et un isomorphisme $\vartheta : V \otimes_{k_1} k_s \to W \otimes_k k_s$ *tels que* $\vartheta : V(k_s) \to W(k_s)$ *soit un isomorphisme de $\mathbf{g}$-ensembles. Le couple* (W, ϑ) *est unique, à un isomorphisme unique près.*

Il est clair que l'on peut supposer que $k_1 = k_2$, et que ce corps est une extension galoisienne de k. Le groupe $\mathbf{h} = \mathbf{g}(k_s/k_2)$ est alors un sous-groupe ouvert distingué de $\mathbf{g}$. Soient $s, t \in \mathbf{g}$ et soit $x \in V(k_s)$. On a :

$$F_{st}({}^{st}x) = st(x) = s(F_t({}^t x)) = F_s({}^s(F_t({}^t x))) = (F_s \circ {}^s F_t)({}^{st}x) \, .$$

On en conclut (cf. n° 2.11) que $F_{st} = F_s \circ {}^s F_t$. La condition (b) montre d'autre part que $F_s = 1$ si $s \in \mathbf{h}$. Ainsi F_s ne dépend que de la classe de s modulo $\mathbf{h}$, et est défini sur k_1. On peut alors appliquer le critère de descente du corps de base aux F_s (cf. par exemple [17], Chap. V, § 4, n° 20). On en déduit l'existence d'une k-variété algébrique W et d'un isomorphisme $\vartheta : V \otimes k_s \to W \otimes k_s$ tel que $F_s = \vartheta^{-1} \circ {}^s \vartheta$ $(s \in \mathbf{g})$. Si $x \in V(k_s)$, on a $F_s({}^s x) = s(x)$, d'où:

$$({}^s \vartheta)({}^s x) = \vartheta(s(x)), \text{ i. e. } {}^s(\vartheta(x)) = \vartheta(s(x)) \, ,$$

ce qui montre que le couple (W, ϑ) répond à la question posée. L'unicité est immédiate.

Remarque. Il est facile de vérifier que les conditions (a), (b), (c) du lemme sont non seulement suffisantes, mais aussi *nécessaires*.

Donnons pour terminer un résultat inédit de SPRINGER dont nous aurons besoin au § 6:

2.13. Proposition. *Soient k un corps parfait, A un k-groupe algébrique H un sous-groupe algébrique de A, et N le normalisateur de H dans A. Soit $a = (a_s) \epsilon Z_1(\overline{k}/k, A)$, et soit $\alpha \epsilon H^1(k, A)$ la classe de cohomologie correspondante. Soit A_a le groupe algébrique obtenu en tordant A au moyen de a (le groupe A opérant sur lui-même par automorphismes intérieurs). Les deux conditions suivantes sont équivalentes:*

(i) *α appartient à l'image de $H^1(k, N) \to H^1(k, A)$.*

(ii) *Il existe un sous-groupe algébrique H' de A_a tel que $H' \otimes \overline{k}$ et $H \otimes \overline{k}$ soit conjugués par un automorphisme intérieur défini par un élément de $A(\overline{k})$.*

(La condition (ii) a un sens, grâce au fait que $A_a \otimes \overline{k}$ peut être canoniquement identifié à $A \otimes \overline{k}$.)

Posons $\mathbf{g} = \mathbf{g}(\overline{k}/k)$, soit $X = A(\overline{k})/N(\overline{k})$, et soit X_a le g-ensemble obtenu en tordant X au moyen de a (le groupe $A(\overline{k})$ opérant *à gauche* sur X). A tout $x \epsilon X_a$, associons le sous-groupe $H_x = \text{Int}(x)(H \otimes \overline{k})$ de $A_a \otimes \overline{k}$. On vérifie sans difficultés que l'on obtient ainsi un isomorphisme de X_a sur le g-ensemble des sous-groupes algébriques H'_1 de $A_a \otimes \overline{k}$ qui sont conjugués de $H \otimes \overline{k}$ par un automorphisme intérieur de $A(\overline{k})$. La condition (ii) équivaut donc à dire que $H^0(\mathbf{g}, X_a)$ est *non vide*, et son équivalence avec la condition (i) résulte de la proposition 1.15.

2.14. Corollaire. *Supposons que A soit linéaire, et que H soit un sous-groupe de CARTAN de A. L'application canonique:*

$$H^1(k, N) \to H^1(k, A)$$

est alors surjective.

Soit $a \epsilon Z^1(\overline{k}/k, A)$. On sait (cf. [16]) que A_a possède un sous-groupe de CARTAN H' (défini sur k, bien entendu); de plus la conjugaison des sous-groupes de CARTAN (cf. par exemple [6], 7.01) montre que $H \otimes \overline{k}$ et $H' \otimes \overline{k}$ sont conjugués par un élément de $A(\overline{k})$. La proposition précédente s'applique, et l'on en déduit que la classe de cohomologie de a appartient à l'image de $H^1(k, N) \to H^1(k, A)$, d'où le corollaire.

§ 3. Groupes de type arithmétique

3.1. Groupes de type arithmétique. Soit G un groupe algébrique linéaire défini sur $\mathbf{Q}$, et soit $G_\mathbf{Q}$ le groupe de ses points rationnels. Si r est un plongement de G dans un groupe $\mathbf{GL}_n$, nous noterons $G_\mathbf{Z}(r)$ le sous-groupe de $G_\mathbf{Q}$ formé des éléments g tels que $r(g) \in \mathbf{GL}_n(\mathbf{Z})$. Un sous-groupe Γ de $G_\mathbf{Q}$ est dit *de type arithmétique dans* $G_\mathbf{Q}$ s'il existe un plongement r tel que $G_\mathbf{Z}(r)$ soit *commensurable* à Γ; rappelons que cela signifie que $\Gamma \cap G_\mathbf{Z}(r)$ est d'indice fini à la fois dans Γ et dans $G_\mathbf{Z}(r)$. On sait que, si cette condition est vérifiée pour un plongement r, elle l'est pour tout plongement. De plus, si Γ est de type arithmétique dans G, on peut choisir r de telle sorte que Γ soit *contenu dans* $G_\mathbf{Z}(r)$; en effet, les transformés par les éléments $\gamma \in \Gamma$ du réseau $\mathbf{Z}^n$ sont en nombre fini, et engendrent un réseau X stable par Γ; si $u \in \mathbf{GL}_n(\mathbf{Q})$ transforme $\mathbf{Z}^n$ en X, le plongement r' défini par $r'(g) = u^{-1}r(g)u$ est tel que $\Gamma \subset G_\mathbf{Z}(r')$.

Un groupe Γ sera dit *de type arithmétique* s'il est possible de le plonger comme sous-groupe de type arithmétique dans un groupe algébrique linéaire défini sur $\mathbf{Q}$.

3.2. Exemples. Le groupe $\mathbf{Z}$, le groupe $\mathbf{GL}_n(\mathbf{Z})$, un groupe fini, sont des groupes de type arithmétique.

Le produit d'un nombre fini de groupes de type arithmétique est de type arithmétique.

Soit R un anneau qui soit un $\mathbf{Z}$-module libre de type fini, et soit R^* le groupe des éléments inversibles de R. Le groupe R^* est *de type arithmétique*. En effet, si l'on identifie R à $\mathbf{Z}^n$, on voit facilement qu'il existe un sous-groupe algébrique G de $\mathbf{GL}_n$ dont les points rationnels correspondent bijectivement aux éléments inversibles de l'algèbre $R \otimes \mathbf{Q}$. Comme $R^* = G_\mathbf{Z}$, cela montre bien que R est de type arithmétique. [En fait, G peut même être défini comme *schéma en groupes sur* $\mathbf{Z}$ grâce au foncteur correspondant $G(A) = (R \otimes A)^*$, A parcourant la catégorie des anneaux commutatifs.]

3.3. Proposition. *Soit Γ un groupe de type arithmétique.*

(i) *Γ est de présentation finie* (autrement dit, Γ peut être défini par un nombre fini de générateurs et de relations).

(ii) *Les sous-groupes finis de Γ forment un nombre fini de classes modulo conjugaison.*

Pour la démonstration, voir [2] et [4].

Remarque. On peut se demander si la propriété (ii) est simplement une conséquence de la propriété (i). Lazard nous a fait observer qu'il n'en est rien: il existe en effet un groupe de présentation finie qui contient des sous-groupes finis de tout ordre (cf. Higman, Proc. Royal Soc. London, *262*, 1961).

3.4. g-groupes de type arithmétique. Soit **g** un groupe profini et soit Γ un g-groupe (cf. nº 1.2). Nous dirons que Γ est *de type arithmétique* si l'on peut trouver:

(a) un groupe algébrique linéaire G défini sur **Q**,

(b) un plongement $f: \Gamma \to G_{\mathbf{Q}}$,

(c) un homomorphisme continu de **g** dans le groupe $\operatorname{Aut} G(\mathbf{Q})$ des automorphismes de G, le groupe $\operatorname{Aut} G(\mathbf{Q})$ étant muni de la topologie discrète (l'image de **g** dans ce groupe est donc finie, ce qui permet de considérer $G_{\mathbf{Q}}$ comme un g-groupe),

ces données étant assujetties aux deux conditions suivantes:

(i) $f(\Gamma)$ est un sous-groupe de type arithmétique de $G_{\mathbf{Q}}$ (cf. nº 3.1),

(ii) le plongement $f: \Gamma \to G_{\mathbf{Q}}$ est compatible avec les structures de g-groupes de Γ et de $G_{\mathbf{Q}}$.

Puisque l'homomorphisme $\mathbf{g} \to \operatorname{Aut} G(\mathbf{Q})$ est continu, son noyau **h** est un sous-groupe ouvert distingué de **g**, et il est clair que le g/h-groupe Γ est de type arithmétique.

On notera également que, lorsque **g** est réduit à l'identité, on retrouve la notion définie au nº 3.1.

3.5. Exemples. Tout g-groupe fini, tout produit fini de g-groupes de type arithmétique est de type arithmétique.

Soit R un anneau qui soit un **Z**-module libre de type fini, et supposons que le groupe profini **g** opère continûment sur R. Le groupe R^* est alors un g-*groupe de type arithmétique.* Cela se voit en reprenant la construction du nº 3.2, et en remarquant que **g** opère sur le groupe algébrique G construit à cet endroit.

Soit C une variété abélienne définie sur un corps k, et soit **g** le groupe de Galois de k_s/k. Le groupe $\operatorname{Aut} C(k_s)$ des automorphismes de $C \otimes_k k_s$ est un g-*groupe de type arithmétique.* En effet, soit R l'anneau des endomorphismes de $C \otimes_k k_s$; on sait que R est un **Z**-module libre de type fini, et l'on a $R^* = \operatorname{Aut} C(k_s)$; on est alors ramené à l'exemple précédent. (Le même argument s'applique, plus généralement, à tout groupe algébrique commutatif qui est extension d'une variété abélienne par un tore.)

3.6. Proposition. *Soit Γ un g-groupe de type arithmétique, et soit **h** un sous-groupe de **g**. Le groupe $\Gamma^{\mathbf{h}}$ des éléments de Γ invariants par **h** est un groupe de type arithmétique. Si de plus **h** est un sous-groupe distingué fermé de **g**, le g/h-groupe $\Gamma^{\mathbf{h}}$ est de type arithmétique.*

Soit $f: \Gamma \to G_{\mathbf{Q}}$ un plongement vérifiant les conditions du nº 3.4. Soit H le sous-groupe de G formé des éléments invariants par **h**. C'est un groupe algébrique défini sur **Q**, et l'on a $f(\Gamma^{\mathbf{h}}) = f(\Gamma) \cap H_{\mathbf{Q}}$, ce qui montre que $\Gamma^{\mathbf{h}}$

est un sous-groupe de type arithmétique de $H_{\mathbf{Q}}$. Si de plus $\mathfrak{h}$ est un sous-groupe distingué fermé de $\mathfrak{g}$, le groupe $\mathfrak{g}/\mathfrak{h}$ opère sur H, et la restriction de f à $\Gamma^{\mathfrak{h}}$ est compatible avec les structures de $\mathfrak{g}/\mathfrak{h}$-groupes de $\Gamma^{\mathfrak{h}}$ et de $H_{\mathbf{Q}}$, ce qui achève de démontrer la proposition.

3.7. Proposition. *Soit Γ un $\mathfrak{g}$-groupe de type arithmétique, et soit $a = (a_s)$ un élément de $Z^1(\mathfrak{g}, \Gamma)$. Le $\mathfrak{g}$-groupe Γ_a obtenu en tordant Γ au moyen de a (cf. n° 1.4) est de type arithmétique.*

Soit $f : \Gamma \to G_{\mathbf{Q}}$ un plongement vérifiant les conditions du n° 3.4.

Pour tout $s \in \mathfrak{g}$, soit r_s l'automorphisme intérieur de G défini par $f(a_s)$. Faisons opérer $\mathfrak{g}$ sur G au moyen des $r_s \circ s$; le sous-groupe $f(\Gamma)$ de $G_{\mathbf{Q}}$ est stable par ces opérations, et l'action de $\mathfrak{g}$ sur Γ ainsi définie est celle du groupe tordu Γ_a. D'où la proposition.

3.8. Proposition. *Soit $\mathfrak{g}$ un groupe fini, et soit Γ un $\mathfrak{g}$-groupe de type arithmétique. Alors :*

(i) Le produit semi-direct Γ' de Γ par $\mathfrak{g}$ est un groupe de type arithmétique.

(ii) $H^1(\mathfrak{g}, \Gamma)$ est fini.

Soit $f : \Gamma \to G_{\mathbf{Q}}$ un plongement vérifiant les conditions du n° 3.4. Convenons de noter $\mathfrak{g}$ le groupe algébrique de dimension zéro défini canoniquement par $\mathfrak{g}$ (en tant que schéma, c'est simplement $\mathfrak{g}$ muni du faisceau d'anneaux constant $\mathbf{Q}$). Comme $\mathfrak{g}$ opère sur G, on peut former le produit semi-direct G' de G par $\mathfrak{g}$; c'est un groupe algébrique linéaire défini sur $\mathbf{Q}$. Le groupe $G'_{\mathbf{Q}}$ est produit semi-direct de $G_{\mathbf{Q}}$ par $\mathfrak{g}$; le groupe Γ' se plonge donc de façon naturelle dans $G'_{\mathbf{Q}}$. Comme de plus Γ est d'indice fini dans Γ', on voit que Γ' est un sous-groupe de type arithmétique de $G'_{\mathbf{Q}}$, d'où (i).

Soit maintenant $a = (a_s)$ un élément de $Z^1(\mathfrak{g}, \Gamma)$. Si nous convenons d'écrire les éléments du produit semi-direct Γ' sous la forme $\gamma \cdot s$, avec $\gamma \in \Gamma$ et $s \in \mathfrak{g}$, le cocycle a définit un *homomorphisme* $\vartheta_a : \mathfrak{g} \to \Gamma'$ par la formule $\vartheta_a(s) = a_s s$. L'image C_a de $\mathfrak{g}$ par ϑ_a est un sous-groupe fini de Γ', dont la connaissance équivaut à celle de a. Comme Γ' est de type arithmétique, la proposition 3.3 montre que les sous-groupes C_a se déduisent d'un nombre fini d'entre eux par conjugaison par les éléments de Γ', donc aussi par ceux de Γ, puisque Γ'/Γ est fini. Ainsi, il existe $a_1, \ldots, a_n \in Z^1(\mathfrak{g}, \Gamma)$ tels que, pour tout $a \in Z^1(\mathfrak{g}, \Gamma)$, il existe un indice i et un élément $\gamma \in \Gamma$ tels que $C_{a_i} = \gamma C_a \gamma^{-1}$; on tire de là :

$$\vartheta_{a_i}(s) = \gamma \vartheta_a(s) \gamma^{-1}, \quad \text{d'où} \quad a_s = \gamma^{-1}(a_i)_s {}^s\gamma,$$

ce qui montre que a et a_i sont cohomologues. D'où la finitude de $H^1(\mathfrak{g}, \Gamma)$.

§ 4. Groupes localement algébriques et groupes d'automorphismes

Dans tout ce paragraphe, la lettre k désigne un corps *parfait*. On note $\mathfrak{g}$ le groupe de GALOIS $\mathfrak{g}(\overline{k}/k)$.

4.1. Schémas localement algébriques. Soit X un schéma sur k. Nous dirons que X est *localement algébrique* s'il est réunion de sous-schémas ouverts et fermés qui sont des variétés algébriques au sens du n° 2.11. Un tel schéma est absolument réduit, et localement de type fini sur k; il résulte de 2.11 que $X(\overline{k})$ est dense dans X.

Dans ce qui suit, nous noterons C_k la catégorie des schémas localement algébriques sur k. Si k' est une extension de k, le foncteur $X \to X \otimes k'$ définit une équivalence de C_k sur une sous-catégorie de $C_{k'}$.

Les propriétés de la conjugaison données dans 2.4 s'étendent immédiatement aux schémas localement algébriques. En particulier, si X, $Y \in C_k$, et si $f \in \mathrm{Hom}\,(X \otimes \overline{k}, Y \otimes \overline{k})$, le morphisme f est « défini sur k » si et seulement si ${}^s f = f$ pour tout $s \in \mathfrak{g}$. D'après le lemme 2.11, il faut et il suffit pour cela que l'on ait:

$$f({}^s x) = {}^s(f(x)) \quad \text{pour} \quad s \in \mathfrak{g},\ x \in X(\overline{k}) \,.$$

4.2. Groupes localement algébriques. Soit G un schéma en groupes sur k. Nous dirons que G est un *groupe localement algébrique* si sa structure de schéma sous-jacente est localement algébrique (autrement dit si c'est un groupe dans la catégorie C_k). Un tel groupe est *simple* sur k (cf. [10]); sa composante neutre G^0 est un groupe *algébrique* connexe, qui est ouvert et fermé dans G.

Un groupe localement algébrique G tel que $G^0 = 1$ sera dit *discret*, ou *de dimension zéro*. Un tel groupe est déterminé à isomorphisme près par le $\mathfrak{g}$-groupe $G(\overline{k})$, et nous nous permettrons parfois l'abus de langage consistant à identifier G à $G(\overline{k})$.

Si G est un groupe localement algébrique quelconque, le quotient G/G^0 est défini: c'est le groupe de dimension zéro qui correspond à $G(\overline{k})/G^0(\overline{k})$. On a ainsi une suite exacte:

$$0 \to G^0 \to G \to G/G^0 \to 0 \,,$$

avec G^0 algébrique connexe et G/G^0 discret. Il est clair que tout k-sous-groupe de G/G^0 définit un sous-groupe ouvert de G; en particulier, les k-sous-groupes *finis* de G/G^0 correspondent aux sous-groupes *algébriques* ouverts de G.

4.3. Groupes de type (ALA). Un groupe localement algébrique A sur k sera dit *de type* (ALA) si sa composante neutre A^0 est un groupe algébrique *linéaire*, et si le $\mathfrak{g}$-groupe $A(\overline{k})/A^0(\overline{k})$ est extension d'un $\mathfrak{g}$-*groupe de type* *arithmétique* Γ (cf. n⁰ 3.4) par un g-*groupe fini* N. L'image réciproque A_1 de N dans A est un sous-groupe algébrique linéaire de A, qui est distingué dans A, et $A(\bar{k})/A_1(\bar{k}) = \Gamma$. On obtient ainsi une suite exacte

$$0 \to A_1 \to A \to \Gamma \to 0\,,$$

où A_1 est algébrique linéaire, et où Γ est un g-groupe de type arithmétique. Réciproquement, il est clair que tout groupe localement algébrique A qui est extension d'un g-groupe de type arithmétique par un groupe algébrique linéaire, est de type (ALA); le sigle choisi rappelle ce fait (Algébrique Linéaire — Arithmétique).

4.4. Le foncteur Aut X. Soit X une k-variété algébrique (resp. un k-groupe algébrique), et soit $Y \in C_k$. On dit que Y *agit sur* X si l'on s'est donné un morphisme $f: Y \times X \to X$ tel que $(pr_1, f): Y \times X \to Y \times X$ soit un isomorphisme (resp. un isomorphisme respectant la structure de groupe des fibres de pr_1). L'ensemble des actions de Y sur X sera noté Aut $X(Y)$; c'est également l'ensemble des automorphismes de $X \times Y$, considéré comme Y-schéma (resp. comme Y-schéma en groupes); en particulier, Aut $X(Y)$ a une structure naturelle de *groupe*. Si $Y, Y' \in C_k$, et si $q \in \mathrm{Hom}\,(Y, Y')$, on définit de façon évidente Aut $X(q): \mathrm{Aut}\,X(Y') \to \mathrm{Aut}\,X(Y)$. Ainsi Aut X est un *foncteur contravariant* de C_k dans la catégorie Gr des groupes.

Soit Y un groupe localement algébrique sur k. On dit que Y *opère sur* X s'il agit sur X et si cette action $f: Y \times X \to X$ est telle que le diagramme

$$Y \times Y \times X \overset{id \cdot \times f}{\to} Y \times X$$
$$q \times id \downarrow \qquad\qquad f \downarrow$$
$$Y \times X \overset{f}{\to} \quad X\,,$$

où q désigne la loi de composition de Y, est commutatif. Cela entraîne que l'élément neutre de Y agit trivialement sur X.

Nous dirons que *le foncteur* Aut X *est localement algébrique* (ou encore que Aut X *est un groupe localement algébrique*) si ce foncteur est *représentable* dans C_k. Cela signifie qu'il existe $A \in C_k$ et une action α de A sur X telle que, pour tout $Y \in C_k$, l'application de $\mathrm{Hom}\,(Y, A)$ dans Aut $X(Y)$ définie par α soit une bijection. On sait que le couple (A, α) est alors unique (à isomorphisme unique près), et que A est un groupe localement algébrique opérant sur X au moyen de α. On identifiera le plus souvent Aut X à A.

Soit Red_k la catégorie des k-schémas *réduits* (non nécessairement localement algébriques). On a:

4.5. Proposition. *Soit* X *une variété algébrique (resp. un groupe algébrique) sur* k, *et supposons que* Aut X *soit représentable dans* C_k *par un groupe localement algébrique* A. *Alors* A *représente également le foncteur* Aut X *dans la catégorie* Red_k.

Il faut montrer que, pour tout $S \in Red_k$, l'homomorphisme canonique

$$\alpha_S : \mathrm{Hom}\,(S,\,A) \to \mathrm{Aut}\,X\,(S)$$

est un isomorphisme. En localisant, on se ramène au cas où S est *affine*; soit Λ la k-algèbre correspondante. Soit (Λ_i) la famille des sous-algèbres de Λ qui sont de type fini sur k, et posons $S_i = \mathrm{Spec}\,(\Lambda_i)$. D'après GROTHENDIECK (cf. [9], Chap. IV, § 8), on a:

$$\mathrm{Hom}\,(S,\,A) = \varinjlim.\ \mathrm{Hom}\,(S_i,\,A) \ \text{et}\ \ \mathrm{Aut}\,X\,(S) = \varinjlim.\ \mathrm{Aut}\,X\,(S_i)\,.$$

Comme les S_i appartiennent à C_k, les homomorphismes

$$\alpha_{S_i} : \mathrm{Hom}\,(S_i,\,A) \to \mathrm{Aut}\,X\,(S_i)$$

sont des isomorphismes. Il en est donc de même de α_S.

4.6. Corollaire. *Soit* K *une extension de* k. *Alors*:

(a) *Le foncteur* Aut $(X \otimes K)$ *est représentable par* $A \otimes K$. *En particulier le groupe* Aut $X\,(K)$ *s'identifie à* $A\,(K)$.

(b) *Si* X *est quasi-projective, les* K-*formes de* $X \otimes K$ *correspondent bijectivement aux éléments de* $H^1(K,\,A)$.

L'assertion (a) est une conséquence immédiate de la proposition précédente (en considérant C_K comme sous-catégorie de Red_k). L'assertion (b) résulte de (a) et de 2.6.

4.7. Opérations effectives. Soit X une variété algébrique (resp. un groupe algébrique) sur k, soit A un groupe algébrique sur k, et supposons que A opère sur X. On dit que A opère *effectivement* s'il existe une famille finie $(x_i)_{i \in I}$ de points de $X\,(\bar{k})$ telle que les relations $a \cdot x_i = a' \cdot x_i$ pour tout $i \in I$, avec $a,\,a' \in A\,(k)$, entraînent $a = a'$. Si $Y \in C_k$, l'application canonique de $\mathrm{Hom}\,(Y,\,A)$ dans $\mathrm{Aut}\,X\,(Y)$ est alors *injective*.

Lemme. *Supposons* k *de caractéristique zéro. Soient* X *une variété algébrique (resp. un groupe algébrique) sur* k, A *un groupe algébrique opérant effectivement sur* X, *et* Y *un élément de* C_k *agissant sur* X. *On suppose que, pour tout* $y \in Y\,(\bar{k})$, *l'automorphisme correspondant de* $X \otimes \bar{k}$ *peut être défini par un élément* de $A(\overline{k})$. *L'action de Y sur X provient alors d'un morphisme de Y dans A, et ce morphisme est unique.*

Soit $(x_i)_{i\in I}$ une famille finie de points de $X(\overline{k})$ telle que les relations $a \cdot x_i = a' \cdot x_i$ pour tout $i \in I$ entraînent $a = a'$. Considérons l'application $m : A(\overline{k}) \to X(\overline{k})^I$ définie par la formule

$$m(a) = (a \cdot x_i)_{i\in I}\,.$$

Par hypothèse cette application est injective. Son image est l'orbite d'un groupe algébrique d'automorphismes de $X(\overline{k})^I$, donc est simple (nous commettons ici l'abus de langage consistant à identifier le schéma $X \otimes \overline{k}$ avec l'ensemble de ses points à valeurs dans $\overline{k}$, et de même pour A). Comme la caractéristique de k est nulle, on obtient ainsi un *isomorphisme* de $A \otimes \overline{k}$ sur une sous-variété algébrique de $(X \otimes \overline{k})^I$. Soit alors q l'application de $Y(\overline{k})$ dans $X(\overline{k})^I$ définie par la formule

$$q(y) = (y \cdot x_i)_{i\in I}\,.$$

Par hypothèse, l'image de q est contenue dans celle de m. On en déduit l'existence d'un morphisme $f : Y \otimes \overline{k} \to A \otimes \overline{k}$ tel que

$$f(y) \cdot x_i = y \cdot x_i \text{ pour tout } i \in I \;\; (y \in Y(\overline{k}))\,.$$

Si $y \in Y(\overline{k})$, il existe par hypothèse un élément $F(y) \in A(k)$ tel que $y \cdot x = F(y) \cdot x$ pour tout $x \in X(\overline{k})$. En appliquant ceci à $x = x_i$ $(i \in I)$, on voit que $F(y) = f(y)$. On a donc:

$$f(y) \cdot x = y \cdot x \text{ pour tout } y \in Y(\overline{k}) \text{ et tout } x \in X(\overline{k})\,.$$

Il reste à voir que f est défini sur k, autrement dit (cf. n° 4.1) que $f({}^s y) = {}^s(f(y))$ pour tout $s \in \mathfrak{g}$ et tout $y \in Y(\overline{k})$. Or, si $x \in X(\overline{k})$, on a:

$$f({}^s y) \cdot {}^s x = {}^s y \cdot {}^s x = {}^s(y \cdot x) = {}^s(f(y) \cdot x) = {}^s(f(y)) \cdot {}^s x\,,$$

ce qui montre que $f({}^s y)$ et ${}^s(f(y))$ opèrent de la même manière sur $X(\overline{k})$. Ils sont donc égaux.

4.8. Proposition. *Supposons k de caractéristique zéro. Soient X une variété algébrique (resp. un groupe algébrique) sur k, et A_0 un groupe algébrique connexe sur k opérant effectivement sur X, de manière à vérifier la propriété universelle suivante:*

(*) Si V est une variété algébrique connexe sur $\overline{k}$ agissant sur $X \otimes \overline{k}$, et si, pour un $v \in V(\overline{k})$, l'automorphisme correspondant de $X \otimes \overline{k}$ peut être défini par un élément de $A_0(\overline{k})$, il en est de même pour tous les éléments de $V(\overline{k})$.

Alors Aut X *est un groupe localement algébrique sur* k, *dont* A_0 *est la composante neutre.*

(De façon plus précise, le foncteur Aut X est représentable par un groupe localement algébrique A dont la composante neutre s'identifie canoniquement à A_0, en tant que groupe opérant sur X.)

Supposons tout d'abord k algébriquement clos. Le groupe $A_0(k)$ s'identifie à un sous-groupe de Aut $X(k)$. Soit $b \in$ Aut $X(k)$. On peut faire opérer A_0 sur X en faisant correspondre à tout $a \in A_0(k)$ l'automorphisme bab^{-1} de X; lorsque $a = 1$, on a $bab^{-1} = 1$. L'hypothèse (*) et le lemme 4.7 montrent alors qu'il existe un unique automorphisme $\sigma_b : A_0 \to A_0$ tel que $\sigma_b(a) = bab^{-1}$ pour tout $a \in A_0(k)$; en particulier, $A_0(k)$ est un sous-groupe *distingué* de Aut $X(k)$ et les automorphismes de A_0 induits par les automorphismes intérieurs de Aut $X(k)$ respectent la structure algébrique de $A_0(k)$. Chaque classe à gauche de Aut $X(k)$ modulo $A_0(k)$ possède donc une unique structure de k-variété telle que la translation $a \mapsto b \cdot a$ $(a \in A_0(k),\ b \in$ Aut $X(k))$ soit un isomorphisme de la variété $A_0(k)$ sur $b \cdot A_0(k)$; en faisant la somme disjointe de ces variétés, on obtient un groupe localement algébrique A, opérant sur X, de composante neutre A_0, et tel que $A(k) =$ Aut $X(k)$. Il reste à voir que A représente le foncteur Aut X. Pour cela, considérons d'abord une variété connexe V agissant sur X. Si $v \in V(k)$, notons v' l'élément de $A(k)$ défini par v. Soient u et v_0 deux éléments de $V(k)$. L'hypothèse (*) entraîne que $(v_0^{-1}u)' \in A_0(k)$; d'après 4.7 il existe donc un unique morphisme $f : V \to A_0$ tel que $(v_0^{-1}u)' = (v_0^{-1})'f(u)$ pour tout $u \in V(k)$, d'où un morphisme $g : V \to A$ tel que $u' = g(u)$ pour tout $u \in V(k)$. Soit Y un élément quelconque de C_k agissant sur X. Comme Y est réunion disjointe de sous-variétés ouvertes du type ci-dessus, ce qui précède montre qu'il existe un unique morphisme $g : Y \to A$ tel que Y agisse sur X par l'intermédiaire de g. On a donc bien démontré que A représente Aut X.

Revenons maintenant au cas général, où k n'est pas nécessairement algébriquement clos. On vient de prouver l'existence d'un groupe localement algébrique $\bar{A}$ sur $\bar{k}$, de composante neutre $A_0 \otimes \bar{k}$, qui représente Aut $(X \otimes \bar{k})$. Admettons pour l'instant que $\bar{A}$ peut s'écrire sous la forme $\bar{A} = A \otimes \bar{k}$, où A est un groupe localement algébrique sur k, l'identification $A(\bar{k}) =$ Aut $X(\bar{k})$ étant un isomorphisme de g-groupes. Vu 4.1, cette dernière condition implique que l'opération de $\bar{A}$ sur $X \otimes \bar{k}$ provient, par extension des scalaires, d'une opération de A sur k. Montrons que A représente Aut X. Soit $Y \in C_k$ agissant sur X. Il existe alors un unique morphisme $\bar{g} : Y \otimes \bar{k} \to \bar{A}$ tel que $Y \otimes \bar{k}$ agisse sur $X \otimes \bar{k}$ à travers $\bar{g}$. On voit alors exactement comme au n° 4.7 que $^s\bar{g} = \bar{g}$ pour tout $s \in$ g; donc $\bar{g}$ provient d'un morphisme $g : Y \to A$, ce qui montre bien que

A représente Aut X. De plus, il est clair que la composante neutre de A s'identifie à A_0. La proposition sera donc démontrée si nous prouvons l'existence de A.

Le groupe $A_0(\bar{k})$ opère par translations à gauche sur $\bar{A}(\bar{k})$; ses orbites sont des ensembles ouverts et fermés disjoints, et pour tout $b \in \bar{A}(\bar{k})$, le groupe de stabilité de b est réduit à l'élément neutre. Il est clair que la structure algébrique de $\bar{A}$ est complètement déterminée par celle de A_0 et par les conditions précédentes. L'existence de A résulte alors de la proposition suivante:

4.9. Proposition. *Soit B un k-groupe algébrique, et soit M un ensemble sur lequel $B(\bar{k})$ et $\mathbf{g}$ opèrent. Supposons vérifiées les deux conditions suivantes:*

(i) *Pour tout $m \in M$, le groupe de stabilité de m dans $B(\bar{k})$ est le groupe des $\bar{k}$-points d'un sous-groupe algébrique H_m de $B \otimes \bar{k}$.*

(ii) $\mathbf{g}$ *opère continûment sur M, et l'on a $^s(b \cdot m) = {}^sb \cdot {}^sm$ pour $s \in \mathbf{g}, b \in B(\bar{k})$, et $m \in M$.*

Il existe alors $Y \in C_k$ ayant les propriétés suivantes: B opère sur Y; l'ensemble $Y(\bar{k})$ s'identifie à M de façon compatible avec les opérations de $B(\bar{k})$ et de $\mathbf{g}$; pour tout $y \in Y(\bar{k})$, correspondant à $m \in M$, l'orbite $B(\bar{k}) \cdot y$ de y est ouverte et fermée dans $Y(\bar{k})$, et isomorphe à $(B/H_m)(\bar{k})$.

(On applique cette proposition en prenant $B = A_0$, $M = \bar{A}(\bar{k}) = \text{Aut } X(\bar{k})$.)

L'ensemble M est réunion disjointe d'ensembles de la forme $B(k) \cdot \mathbf{g}(m)$, $m \in M$, où $\mathbf{g}(m)$ désigne l'orbite de m par $\mathbf{g}$. Vu (ii), $\mathbf{g}(m)$ est fini. Il suffit donc de considérer le cas où le quotient X de M par $B(\bar{k})$ est *fini*. L'ensemble X est muni d'une structure de $\mathbf{g}$-ensemble, quotient de celle de M; si $x \in X$, on notera ^sx son transformé par $s \in \mathbf{g}$. Pour tout $x \in X$, choisissons un représentant $\alpha(x)$ de x dans M; il existe une extension galoisienne finie k_1/k telle que $^s\alpha(x) = \alpha(x)$ pour tout $s \in \mathbf{h}$, où $\mathbf{h} = \mathbf{g}(\bar{k}/k_1)$; le groupe $\mathbf{h}$ opère alors trivialement sur X. Il résulte des hypothèses (i) et (ii) que, si $x \in X$ et $s \in \mathbf{g}$, on a $^s(H_{\alpha(x)}) = H_{s\alpha(x)}$; en particulier (prenant $s \in \mathbf{h}$) on voit que les $H_{\alpha(x)}$ sont définis sur k_1. De plus, $^s\alpha(x)$ et $\alpha(^sx)$ font partie de la même orbite de $B(\bar{k})$; il existe donc $u_{s,x} \in B(\bar{k})$ tel que $^s\alpha(x) = u_{s,x}\alpha(^sx)$.

Soit maintenant V la k_1-variété somme des espaces homogènes $(B \otimes k_1)/H_{\alpha(x)}$, $x \in X$. On identifie $V(\bar{k})$ à M en faisant correspondre à $b' \in B(\bar{k})/H_{\alpha(x)}(\bar{k})$ le transformé $b \cdot \alpha(x)$ de $\alpha(x)$ par un représentant b de b'. Cette identification permet de transporter à $V(\bar{k})$ la structure de $\mathbf{g}$-ensemble de M; nous noterons $s(v)$ le transformé, pour cette structure, de l'élément $v \in V(\bar{k})$ par l'élément

$s \in \mathbf{g}$. Si $v \in B(\overline{k})/H_{\alpha(x)}(\overline{k})$, et si b est un représentant de v, on vérifie immédiatement la formule:

$$s(v) = {}^s b \cdot u_{s,x} \cdot \alpha({}^s x) \,. \tag{1}$$

Nous allons voir que les conditions (a), (b), (c) du lemme 2.12 sont vérifiées par la variété V et par la structure de $\mathbf{g}$-ensemble de $V(\overline{k})$. La variété ${}^s V\,(s \in \mathbf{g})$ est somme disjointe des espaces homogènes $(B \otimes k_1)/{}^s H_{\alpha(x)} = (B \otimes k_1)/H_{s\alpha(x)}$, et, si v, b sont comme ci-dessus, ${}^s v$ est l'image de ${}^s b$ par la projection naturelle de $B(\overline{k})$ sur $B(\overline{k})/{}^s H_{\alpha(x)}$. Soit $f_s : ({}^s V)(\overline{k}) \to V(\overline{k})$ l'application définie dans 2.12. Il résulte de (1) que la restriction de f_s à $B(\overline{k})/{}^s H_{\alpha(x)}(\overline{k})$ est l'application de cet espace sur $B(\overline{k})/H_{\alpha({}^s x)}(\overline{k})$ induite par la translation $b \mapsto b \cdot u_{s,x}$. Elle est donc bien définie à partir d'un morphisme $F_s : {}^s V \otimes \overline{k} \to V \otimes \overline{k}$, ce qui montre que (a) est vérifiée. Si $s \in \mathbf{h}$, on a ${}^s x = x$, d'où $u_{s,x} \in H_{\alpha({}^s x)}(\overline{k})$, et $F_s = 1$, ce qui établit (b). Enfin, la condition (c) provient de ce que tout sous-ensemble fini d'un espace homogène est contenu dans un ouvert affine (cf. [17], p. 111). On peut donc descendre le corps de base de V de k_1 à k, et la variété Y ainsi obtenue vérifie les conditions voulues.

Indiquons une application de la proposition 4.9:

4.10. Construction de variétés de sous-groupes. Soit B un k-groupe algébrique et soit M un ensemble non vide de sous-groupes algébriques de $B \otimes \overline{k}$. Faisons les deux hypothèses suivantes:

(a) Si $C \in M$, on a ${}^s C \in M$ pour tout $s \in \mathbf{g}$.

(b) Le groupe $B(\overline{k})$ opère transitivement (par conjugaison) sur M. (En d'autres termes, si $C_0 \in M$, les éléments de M sont les sous-groupes de la forme $\mathrm{Int}\, x\, (C_0)$, avec $x \in B(\overline{k})$.)

Les groupes $\mathbf{g}$ et $B(\overline{k})$ opèrent sur M. De plus, le groupe d'isotropie d'un élément $C \in M$ est le groupe $N(\overline{k})$ des $\overline{k}$-points du *normalisateur* de C dans $B \otimes \overline{k}$. Toutes les hypothèses de 4.9 sont donc vérifiées. On en conclut qu'il existe un espace homogène V de B et une bijection $V(\overline{k}) \to M$ compatible avec les opérations de $\mathbf{g}$ et de $B(\overline{k})$. En particulier, les éléments de $V(k)$ correspondent aux sous-groupes $C \in M$ qui sont définis sur k. On dit que V est la *variété des sous-groupes appartenant à* M. Si $C \in M$, et si N est le normalisateur de C, il est clair que $V \otimes \overline{k}$ est isomorphe à $(B \otimes \overline{k})/N$.

Lorsque k est un corps *fini*, et B *connexe*, alors, d'après un théorème de Lang [12], V possède un point à valeurs dans k, et il existe $C \in M$ défini sur k. Si de plus le normalisateur N de C est *connexe*, et $C' \in M$ est défini sur k, il existe $b \in B(k)$ tel que Int $(C) = C'$; en effet, le sous-schéma de B formé des éléments transformant C dans C' est un espace homogène de N, et possède donc un point à valeurs dans k, d'après le théorème de LANG cité ci-dessus.

Comme exemple d'ensemble M, on peut prendre, lorsque B est linéaire, l'ensemble des tores maximaux (resp. des sous-groupes de CARTAN, resp. des sous-groupes résolubles connexes maximaux) de $B \otimes \bar{k}$. D'après ce qui précède, si B est linéaire connexe, et si k est fini, B possède un tore maximal (resp. un sous-groupe de CARTAN, resp. un sous-groupe résoluble connexe maximal) défini sur k.

4.11. Remarque. Dans tout ce qui précède, nous nous sommes placés dans le cadre de la catégorie C_k ; on pourrait également considérer la catégorie plus vaste C_k^* formée des sommes disjointes de schémas algébriques non nécessairement réduits (ou même la catégorie de tous les schémas). Si X est une k-variété, le foncteur Aut $X : C_k \to Gr$ se prolonge en un foncteur Aut* $X : C_k^* \to Gr$. *Il se peut que* Aut X *soit représentable sans que* Aut* X *le soit*, et cela même si la caractéristique de k est zéro, et même si X est un groupe linéaire. Un exemple simple est fourni par $X = \mathbf{G}_a \times \mathbf{G}_m$. Dans ce cas, Aut X est représentable par le groupe $A = \mathbf{G}_m \times \mathbf{Z}/2\,\mathbf{Z}$ opérant de manière évidente ; si Aut* X était représentable, il le serait par le même groupe A (cela résulte du fait que, en caractéristique zéro, tout schéma en groupes localement de type fini est automatiquement réduit). Or, il est facile de voir que, si l'on prend $V = \mathrm{Spec}\ k[T]/(T^2)$, le groupe Aut* $X(V)$ est strictement plus grand que le groupe $A(V)$ [le groupe $\mathbf{G}_a \times \mathbf{G}_m$ a des automorphismes «infinitésimaux» qui ne proviennent pas de l'action d'un groupe algébrique]. Le foncteur Aut* X n'est donc pas représentable.

§ 5. Groupe d'automorphismes d'un groupe algébrique linéaire

Dans ce paragraphe, k désigne un corps *de caractéristique zéro*, et $\mathbf{g}$ le groupe de GALOIS de $\bar{k}$ sur k. Nous nous proposons de démontrer que le foncteur d'automorphismes d'un k-groupe algébrique linéaire est un groupe localement algébrique de type (ALA) (cf. n° 4.3). Pour passer des groupes réductifs ou unipotents au cas général, nous utiliserons la proposition suivante, qui précise un résultat de MOSTOW [14] :

5.1. Proposition. *Soit G un k-groupe algébrique linéaire, et soit U son radical unipotent. Il existe alors un sous-groupe algébrique H de G (défini sur k) tel que G soit produit semi-direct de H et de U. Deux tels sous-groupes sont* *conjugués par un élément de $U(k)$. Tout k-sous-groupe réductif L de G est conjugué à un sous-groupe de H par un élément de $U(k)$.*

(On rappelle que le *radical unipotent* d'un groupe algébrique linéaire G est par définition le plus grand sous-groupe algébrique distingué unipotent de G.)

Cette proposition est connue lorsque $G(k)$ est dense dans G (cf. [14], théorème 7.1), donc en particulier lorsque $k = \bar{k}$. Par conséquent, $U(\bar{k})$ opère transitivement, par automorphismes intérieurs, sur l'ensemble M des sous-groupes algébriques $\bar{H}$ de $G \otimes \bar{k}$ tels que $G \otimes \bar{k}$ soit produit semi-direct de $\bar{H}$ et de $U \otimes \bar{k}$. Les conditions de 4.10 sont vérifiées; il existe donc un espace homogène V de U tel que $M = V(\bar{k})$. Puisque U est unipotent et k parfait, un théorème de ROSENLICHT ([15], théorème 10) montre que $V(k)$ est non vide, d'où la première assertion. Si H et H' sont deux éléments de $V(k)$, soit P l'ensemble des éléments de $U(\bar{k})$ qui transforment H en H'; si U' désigne le groupe de stabilité de H dans U, il est clair que le g-ensemble P est un espace homogène principal de $U'(\bar{k})$. Comme U' est unipotent, P possède un point invariant par g, d'où la deuxième assertion. Enfin, soit $G' = L \cdot U$. Le groupe G' est produit semi-direct de U et de $L' = G' \cap H$; la dernière assertion résulte donc des deux premières, appliquées à G'.

Remarque. Soit H un k-sous-groupe de G. La proposition précédente montre que G est produit semi-direct de H et de U si et seulement si H est réductif maximal.

5.2. Notations. Dans toute la suite de ce paragraphe, G est un k-groupe algébrique *linéaire*, U son radical unipotent, H un sous-groupe algébrique réductif maximal de G. On note H^0 la composante neutre de H et S (resp. T) le groupe dérivé (resp. la composante neutre du centre) de H^0; le groupe S est *semi-simple*, le groupe T est un *tore*, on a $H^0 = S \cdot T$, et $S \cap T$ est fini.

Si M est un groupe algébrique quelconque, on note $Z(M)$ le *centre* de M. Si N est un sous-groupe algébrique de M, soit $N' = N \cap Z(M)$; le groupe N/N' opère effectivement sur M (par automorphismes intérieurs); on le note $\mathrm{Int}_M N$. En particulier, $\mathrm{Int}_H H^0$ s'identifie à $H^0/(H^0 \cap Z(H))$.

5.3. Cas élémentaires. Enumérons d'abord quelques cas où il est facile de vérifier que Aut G est, soit un groupe algébrique linéaire, soit un groupe discret de type arithmétique:

(a) Si G est *fini*, Aut G est fini.

(b) Si G est *unipotent*, l'exponentielle permet de transformer Aut G en le foncteur d'automorphismes de *l'algèbre de* LIE de G; cela montre que Aut G est un groupe algébrique linéaire.

(c) Si G est *semi-simple connexe*, Aut G est un groupe algébrique dont la composante neutre est le *groupe adjoint* $\mathrm{Int}_G\, G = G/Z(G)$ de G.

(d) Si $G = (\mathbf{G}_m)^n$, Aut G est le groupe localement algébrique de dimension zéro défini par $\mathbf{GL}_n(\mathbf{Z})$, considéré comme groupe discret sur lequel $\mathbf{g}$ opère trivialement; cela s'écrit plus simplement :

$$\mathrm{Aut}\,(\mathbf{G}_m)^n = \mathbf{GL}_n(\mathbf{Z}) \,.$$

Plus généralement, si T est un *tore* quelconque, et si Y désigne le $\mathbf{g}$-module des groupes à 1 paramètre de $T \otimes \overline{k}$ (cf. [6], exposé 11), Aut T s'identifie au $\mathbf{g}$-groupe Aut Y, cf. n° 3.5.

Nous ramènerons le cas général à ces différents cas particuliers en utilisant les décompositions $G = H \cdot U$ et $H^0 = S \cdot T$ introduites ci-dessus.

5.4. Le groupe Aut H.

Proposition. (a) *Le foncteur d'automorphismes* Aut H *est un groupe localement algébrique dont la composante neutre* B^0 *est égale à* $\mathrm{Int}_H\, H^0$.

(b) *Le noyau* B_1 *du morphisme canonique* $\mathrm{Aut}\, H \to \mathrm{Aut}\, T \times \mathrm{Aut}\,(H/H^0)$ *est un sous-groupe ouvert de* Aut H.

Pour démontrer (a), il suffit (cf. prop. 4.8) de prouver que, si V est une $\overline{k}$-variété connexe agissant sur $H \otimes \overline{k}$ par l'intermédiaire d'un morphisme ϱ, et si $\varrho(v) \in \mathrm{Int}_H\, H^0(\overline{k})$ pour un $v \in V(\overline{k})$, alors $\varrho(V(\overline{k})) \subset \mathrm{Int}_H\, H^0(\overline{k})$. Quitte à remplacer k par $\overline{k}$, on peut supposer k algébriquement clos, ce qui nous permettra d'identifier une variété algébrique à l'ensemble de ses points.

Il est clair que V agit trivialement sur T et sur H/H^0, et que V agit sur S par automorphismes intérieurs (cf. 5.3). Commençons par traiter un cas particulier :

(i) *V agit trivialement sur* H^0. Soit F le groupe fini H/H^0; le groupe F opère par automorphismes intérieurs sur le groupe abélien $Z(H^0)$. Soit L (resp. M) le groupe des 1-cocycles (resp. 1-cobords) de F à valeurs dans $Z(H^0)$. Si $n = \mathrm{Card}\,(F)$, les groupes L et M s'identifient à des sous-groupes algébriques du produit de n copies de $Z(H^0)$. De plus M est *ouvert* dans L; en effet, $M/L = H^1(F, Z(H^0))$ est un groupe algébrique dont tout élément est d'ordre divisant n (en vertu d'une propriété bien connue des groupes de cohomologie), et, puisque k est de caractéristique zéro, c'est un groupe fini. Si $z \in L$, l'application $u_z : H \to H$ définie par la formule :

$$u_z(x) = z(\overline{x}) \cdot x \quad (x \in H,\ \overline{x}\ \text{ projection de }\ x\ \text{ dans }\ F) \,,$$

est un automorphisme de H. On vérifie immédiatement que l'on obtient ainsi une bijection σ de L sur l'ensemble des automorphismes de H qui agissent trivialement sur H^0 et H/H^0; on a $\sigma(M) = \mathrm{Int}_H Z(H^0) = \sigma(L) \cap \mathrm{Int}_H H^0$. L'application $\sigma^{-1} \circ \varrho : V \to L$ est alors un morphisme qui applique v dans M, donc aussi V dans M puisque V est connexe et M ouvert dans L; cela démontre (a) dans le cas considéré.

(ii) *Cas général.* Soit V' la sous-variété de $V \times S$ formée des couples (x, s) tels que $\varrho(x) = \mathrm{Int}_{H^0}(s^{-1})$ sur H^0; on vérifie facilement que c'est un revêtement galoisien étale de V, de groupe de GALOIS le groupe $Z(S)$. L'image réciproque de v dans V' est formée de couples $(v, s_i)_{i \in I}$, avec $s_i \epsilon S$; on sait (c'est une propriété générale des revêtements étales) que toute composante connexe V'_α de V' contient l'un des (v, s_i). Faisons agir V' sur H en faisant correspondre à $(x, s) \epsilon V'$ l'automorphisme $\varrho(x) \cdot \mathrm{Int}_H(s)$. Cette action de V' sur H est triviale sur H^0 et sur H/H^0; de plus, les automorphismes correspondant aux points (v, s_i) appartiennent à $\mathrm{Int}_H H^0$ par hypothèse. En appliquant (i) aux V'_α, on en conclut que tous les $\varrho(x) \cdot \mathrm{Int}_H(s)$, avec $(x, s) \epsilon V'$, appartiennent à $\mathrm{Int}_H H^0$; il en est alors de même des $\varrho(x)$, ce qui achève la démonstration de (a).

Passons à (b). Il est clair que B_1 contient B^0, donc est un sous-groupe localement algébrique ouvert de $B = \mathrm{Aut}\, H$. Pour prouver que c'est un groupe algébrique, il faut montrer que B_1/B^0 est fini. Ici encore on peut supposer que $k = \bar{k}$. Soit B_2 le sous-groupe de B_1 formé des éléments dont la restriction à S est un automorphisme intérieur. Comme B_1/B_2 est fini, on est ramené à prouver que B_2/B^0 est fini. Les groupes B_2 et B^0 contiennent respectivement les groupes $\sigma(L)$ et $\sigma(M)$ introduits dans (i). De plus, L/M s'applique *sur* B_2/B^0; en effet, si $u \epsilon B_2$, il existe $s \epsilon S$ tel que u et $\mathrm{Int}_H(s)$ coïncident sur S, donc aussi sur H^0, et l'on peut écrire u sous la forme $u = u' \cdot \mathrm{Int}_H(s)$, avec $u' \epsilon \sigma(L)$. Comme L/M est fini (cf. (i)), notre assertion est établie.

5.5. Les groupes C_1 et C_2. La projection canonique $\pi : G \to G/U$ induit un isomorphisme de H sur G/U (cf. n° 5.1). On posera $\bar{T} = \pi(T)$. Tout automorphisme de G définit par passage au quotient un automorphisme de G/U qui laisse stable $\bar{T}$, d'où des homomorphismes

$$\mathrm{Aut}\, G \to \mathrm{Aut}\, G/U \to \mathrm{Aut}\, \bar{T}\,.$$

D'autre part, G opère par automorphismes intérieurs sur lui-même en laissant stable U, d'où un morphisme $r : G \to \mathrm{Aut}\, U$.

Soit C le sous-foncteur de $\mathrm{Aut}\, G$ formé des automorphismes qui laissent stable H et dont l'image dans $\mathrm{Aut}\, H$ appartient au groupe $B^0 = \mathrm{Int}_H H^0$ (cf. prop. 5.4). On a un morphisme injectif

$$C \to \mathrm{Aut}\, U \times B^0\,,$$

et l'on vérifie facilement que C est un sous-groupe algébrique du groupe algébrique Aut $U \times B^0$. De façon plus précise, $C(\bar{k})$ est l'ensemble des couples (u, v) ($u \in$ Aut $U(\bar{k})$, $v \in B^0(\bar{k})$) qui vérifient la condition

$$u \cdot r(h) = r(v(h)) \cdot u \quad \text{pour tout} \ \ h \in H(\bar{k}) ;$$

un tel couple opère sur $G \otimes \bar{k}$ par la formule:

$$(u, v)(x \cdot y) = u(x) \cdot v(y) \quad (x \in U(\bar{k}) , \ y \in H(\bar{k})) .$$

Comme Aut U et B^0 sont des k-groupes algébriques linéaires, il en est de même de C.

Le groupe C opère par restriction sur U, et l'on peut former le produit semi-direct $U \cdot C$. On vérifie aisément que $U \cdot C$ opère sur G par l'intermédiaire d'une opération qui associe au produit $u \cdot c$ ($u \in U(\bar{k})$, $c \in C(\bar{k})$) l'automorphisme $\mathrm{Int}_G(u) \cdot c$ de $G \otimes \bar{k}$. Cette opération n'est pas effective en général; nous noterons C_1 le quotient de $U \cdot C$ qui opère effectivement sur G; c'est encore un groupe algébrique linéaire, et il s'identifie (en tant que foncteur) à un sous-foncteur de Aut G.

Nous noterons d'autre part C_2 le noyau du morphisme canonique

$$\text{Aut } G \to \text{Aut } \bar{T} \times \text{Aut } (G/G^0) .$$

Il est clair que C_1 est contenu dans C_2.

5.6. Proposition. *Le foncteur d'automorphismes* Aut G *est un groupe localement algébrique, contenant* C_1 *et* C_2 *comme sous-groupes algébriques distingués ouverts. Un élément de* Aut $G(\bar{k})$ *appartient à* $C_1(\bar{k})$ *si et seulement si son image dans* Aut $G/U(\bar{k})$ *appartient à* $\mathrm{Int}_{G/U}(G/U)^0(\bar{k})$.

Démontrons d'abord la deuxième assertion. Il est trivial que la condition est nécessaire. Réciproquement, soit x un élément de Aut $G(\bar{k})$ la vérifiant. Le transformé $x(\bar{H})$ de $\bar{H} = H \otimes \bar{k}$ par x est un sous-groupe réductif maximal de $G \otimes \bar{k}$; d'après 5.1, il existe donc $u \in U(\bar{k})$ tel que $x(\bar{H}) = \mathrm{Int}_G(u)(\bar{H})$. Quitte à multiplier x à droite par $\mathrm{Int}_G(u^{-1})$, on peut donc supposer que $x(\bar{H}) = \bar{H}$. L'hypothèse entraîne alors que x appartient à $C(\bar{k})$, donc aussi à $C_1(\bar{k})$, ce qui achève de prouver notre assertion. On en déduit en particulier que $C_1(\bar{k})$ est un sous-groupe *distingué* de Aut $G(\bar{k})$; le même résultat est évident pour C_2.

D'autre part, l'homomorphisme canonique de Aut G dans Aut $G/U \approx$ $\approx$ Aut H induit une application de $C_2(\bar{k})/C_1(\bar{k})$ dans $B_1(\bar{k})/B^0(\bar{k})$ (les notations étant celles de 5.4); d'après ce que l'on vient de voir, cette application est injective. Comme $B_1(\overline{k})/B^0(\overline{k})$ est fini (cf. 5.4), il en est de même de $C_2(\overline{k})/C_1(\overline{k})$.

Pour terminer la démonstration, il suffit donc de prouver que la composante neutre C_1^0 de C_1 vérifie la condition (*) de la prop. 4.8. Soit donc V une $\overline{k}$-variété connexe agissant sur $G \otimes \overline{k}$, et supposons que, pour un élément v de $V(\overline{k})$, l'automorphisme correspondant appartienne à $C_1^0(\overline{k})$. D'après la prop. 5.4, les automorphismes de $G/U \otimes \overline{k}$ définis par les éléments de $V(\overline{k})$ appartiennent au groupe $\mathrm{Int}_{G/U}(G/U)^0(\overline{k})$. La deuxième assertion de 5.6, démontrée ci-dessus, entraîne alors que les automorphismes de $G \otimes \overline{k}$ définis par les éléments de $V(\overline{k})$ appartiennent à $C_1(\overline{k})$. D'après le lemme 4.7, cela signifie que V agit sur $G \otimes \overline{k}$ par l'intermédiaire d'un morphisme $V \to C_1 \otimes \overline{k}$. Comme V est connexe, ce morphisme est nécessairement à valeurs dans $C_1^0 \otimes \overline{k}$, ce qui achève la démonstration.

5.7. Le plus grand tore central de G^0. Le tore T opère sur U. On sait (cf. [6], exposé 9, lemme 2) qu'il existe une suite de composition (U_i) de $U \otimes \overline{k}$ formée de sous-groupes algébriques connexes stables par T, telle que U_i/U_{i+1} soit isomorphe au groupe additif $\mathbf{G}_a$; de plus T opère sur U_i/U_{i+1} par multiplication par un caractère χ_i et U_i est engendré par U_{i+1} et par les éléments qui sont invariants par le noyau de χ_i. Si T' désigne le plus grand tore central de G^0, on voit que $T' \otimes \overline{k}$ est la composante neutre de l'intersection des noyaux des χ_i. Chaque χ_i s'identifie à un caractère de $T/T' \otimes \overline{k}$, donc aussi de $\overline{T}/\overline{T'} \otimes \overline{k}$, avec $\overline{T} = \pi(T)$, $\overline{T'} = \pi(T')$, cf. n° 5.5. L'ensemble Ψ des caractères de $\overline{T}/\overline{T'} \otimes \overline{k}$ ainsi obtenus est indépendant du choix de T (c'est-à-dire du choix de H) ainsi que du choix de la suite (U_i); cela résulte de la prop. 5.1 combinée avec les résultats de [6], *loc. cit.* Par suite, tout automorphisme $\sigma \in \mathrm{Aut}\, G(\overline{k})$ définit un automorphisme de $\overline{T}/\overline{T'} \otimes \overline{k}$ qui laisse stable Ψ. Notons $X(\overline{T}/\overline{T'})$ le groupe des caractères de $\overline{T}/\overline{T'} \otimes \overline{k}$; comme $\overline{T'} \otimes \overline{k}$ est la composante neutre de l'intersection des noyaux des χ_i, le groupe engendré par Ψ est d'indice fini dans $X(\overline{T}/\overline{T'})$, et les automorphismes de $\overline{T}/\overline{T'} \otimes \overline{k}$ qui laissent stable Ψ forment un groupe fini. Il est clair d'autre part que le groupe C_2 opère trivialement sur $\overline{T}$. On obtient donc le résultat suivant:

5.8. Lemme. *Soit Γ le groupe $\mathrm{Aut}\, G(\overline{k})/C_2(\overline{k})$. Les éléments de Γ opèrent sur $\overline{T} \otimes \overline{k}$ en laissant stable $\overline{T'} \otimes \overline{k}$. Leurs images dans $\mathrm{Aut}(\overline{T}/\overline{T'})(\overline{k})$ forment un groupe fini Φ.*

5.9. Le groupe Γ_+ . Les automorphismes intérieurs du groupe G/U opèrent sur la composante neutre $(G/U)^0$ et en particulier sur le tore $\overline{T}$. Comme $\overline{T}$ est contenu dans le centre de $(G/U)^0$, on voit que le quotient $(G/U)/(G/U)^0 = G/G^0$ opère sur $\overline{T}$. On notera ϑ l'homomorphisme de G/G^0 dans $\mathrm{Aut}\,\overline{T}$ ainsi défini.

Posons $W = G/G^0(\overline{k})$; les groupes W et $\mathrm{Aut}\,W$ sont des **g**-*groupes finis*.

Lemme. *Soit* Γ_+ *l'ensemble des couples* $(\alpha, \beta) \in \mathrm{Aut}\,\overline{T}(\overline{k}) \times \mathrm{Aut}\,W$ *qui vérifient les deux conditions suivantes*:

(a) $\alpha \cdot \vartheta(w) \cdot \alpha^{-1} = \vartheta(\beta(w))$ *pour tout* $w \in W$,

(b) α *laisse stable* $\overline{T}' \otimes \overline{k}$ *et l'élément de* $\mathrm{Aut}\,(\overline{T}/\overline{T}')(\overline{k})$ *qu'il définit appartient au groupe* Φ (cf. 5.8).

Alors Γ_+ *est un* **g**-*sous-groupe de* $\mathrm{Aut}\,\overline{T}(\overline{k}) \times \mathrm{Aut}\,W$ *contenant* Γ.

La définition de Γ au moyen de C_2 (cf. 5.8) montre que Γ s'identifie de façon naturelle à un **g**-sous-groupe de $\mathrm{Aut}\,\overline{T}(\overline{k}) \times \mathrm{Aut}\,W$. On a déjà vu que ses éléments vérifient (b), et il est immédiat qu'ils vérifient également (a). D'autre part, chacune des conditions (a) et (b) définit un sous-groupe de $\mathrm{Aut}\,\overline{T}(\overline{k}) \times \mathrm{Aut}\,W$ stable par **g** : c'est évident pour (b), et, pour (a), cela résulte du fait que $\vartheta : W \to \mathrm{Aut}\,\overline{T}(\overline{k})$ est un morphisme de **g**-groupes.

5.10. Lemme. *Le groupe* Γ_+ *est un* **g**-*groupe de type arithmétique* (cf. 3.4), *et* Γ *est d'indice fini dans* Γ_+. *En particulier,* Γ *est un* **g**-*groupe de type arithmétique.*

Soit Y le **g**-module des *groupes à 1 paramètre* de $\overline{T} \otimes \overline{k}$, autrement dit le dual du groupe $X(\overline{T})$ des caractères de $\overline{T} \otimes \overline{k}$, et soit Y' celui de $\overline{T}'$ (cf. [6], exposé 11). Ce sont des **Z**-modules libres de type fini, et Y' est facteur direct de Y (comme groupe abélien — pas nécessairement comme **g**-module). Les **g**-groupes $\mathrm{Aut}\,\overline{T}(\overline{k})$ et $\mathrm{Aut}\,\overline{T}'(\overline{k})$ s'identifient respectivement à $\mathrm{Aut}\,Y$ et $\mathrm{Aut}\,Y'$ (cf. 5.3), et Φ s'identifie à un **g**-sous-groupe de $\mathrm{Aut}\,Y/Y'$. On peut donc considérer Γ_+ comme le sous-groupe de $\mathrm{Aut}\,Y \times \mathrm{Aut}\,W$ formé des couples (α, β) vérifiant des conditions (a') et (b') qui se déduisent immédiatement de (a) et de (b).

Pour préciser ceci, choisissons une base $(e_1, \ldots, e_n)$ de Y dont les m premiers éléments forment une base de Y' ($n = \dim \cdot T$, $m = \dim \cdot T'$). Les éléments de Φ s'identifient à des éléments de $\mathbf{GL}_{n-m}(\mathbf{Z})$. De même, les automorphismes $\vartheta(w)$ et les éléments α vérifiant (b') s'identifient à des éléments de $\mathbf{GL}_n(\mathbf{Z})$ de la forme

$$\begin{pmatrix} * & * \\ 0 & \varphi \end{pmatrix}, \quad \text{avec} \quad \varphi \in \Phi.$$

Soit $Tr_{n,m}$ le sous-groupe de $\mathbf{GL}_n$ dont les éléments sont de la forme $\left(\begin{smallmatrix} * & * \\ 0 & * \end{smallmatrix}\right)$ comme ci-dessus; c'est un groupe algébrique sur $\mathbf{Q}$ (provenant d'ailleurs d'un schéma en groupes sur $\mathbf{Z}$), et l'on a un homomorphisme canonique $Tr_{n,m} \to \mathbf{GL}_{n-m}$. L'image réciproque de Φ par cet homomorphisme est un $\mathbf{Q}$-groupe algébrique Σ. Dans le produit $\Sigma \times \operatorname{Aut} W$, les couples (α, β) vérifiant la condition (a′) sont les points d'un $\mathbf{Q}$-sous-groupe algébrique Σ', dont Γ_+ est un sous-groupe de type arithmétique. De plus, les éléments de $\mathbf{g}$ définissent des automorphismes de Σ' qui laissent stable Γ_+; cela achève de prouver que Γ_+ est un $\mathbf{g}$-groupe de type arithmétique.

Il reste à montrer que Γ est d'indice fini dans Γ_+. Nous allons plus précisément construire un sous-groupe de $\operatorname{Aut} G(\bar{k})$ qui s'applique isomorphiquement sur un sous-groupe Γ_- d'indice fini de Γ_+.

D'après le lemme 5.11 ci-après, il existe un sous-groupe fini $\bar{W}$ de $H(\bar{k})$ qui s'applique *sur* $W = H/H^0(\bar{k})$ par la projection canonique. La décomposition $H^0(\bar{k}) = T(\bar{k}) \cdot S(\bar{k})$ permet d'écrire les éléments de $\bar{W} \cap H^0(\bar{k})$ sous la forme $t_i \cdot s_i$, avec $t_i \epsilon T(\bar{k})$ et $s_i \epsilon S(\bar{k})$. Soit P le sous-ensemble de $T(\bar{k})$ réunion des t_i et de $T(\bar{k}) \cap S(\bar{k})$; c'est un ensemble fini, dont tout élément est d'ordre fini dans $T(\bar{k})$. Soit $\bar{P}$ le sous-ensemble correspondant de $\bar{T}(\bar{k})$. On définit Γ_- comme le sous-groupe de Γ_+ formé des couples (α, β) tels que:

(i) $\beta = 1$ (ce qui, vu (a), entraîne que α commute à tous les $\vartheta(w)$, $w \epsilon W$),

(ii) α laisse fixes les éléments de $\bar{P}$,

(iii) l'élément de Φ défini par α est égal à 1.

Il est clair que Γ_- est un sous-groupe d'indice fini de Γ_+. Soit d'autre part $(\alpha, 1) \epsilon \Gamma_-$. Identifions α à un automorphisme de $T \otimes \bar{k}$, et prolongeons-le en un automorphisme de $H^0 \otimes \bar{k}$ qui soit l'identité sur $S \otimes \bar{k}$; c'est possible puisque α laisse fixes les éléments de $T(\bar{k}) \cap S(\bar{k})$, en vertu de la condition (ii). De même, (i) et (ii) permettent de prolonger α en un automorphisme de $H \otimes \bar{k}$ qui soit l'identité sur $\bar{W}$. Enfin, la décomposition $G = H \cdot U$ permet de prolonger α en un automorphisme de la variété $G \otimes k$, au moyen de la formule

$$\alpha(h \cdot u) = \alpha(h) \cdot u \quad (h \epsilon H(\bar{k}),\ u \epsilon U(\bar{k})) .$$

On obtient de cette manière un automorphisme du *groupe* algébrique $G \otimes \bar{k}$; en effet, il suffit de vérifier que l'on a l'identité

$$h u h^{-1} = \alpha(h) u \alpha(h)^{-1} \quad (h \epsilon H(\bar{k}),\ u \epsilon U(\bar{k})) .$$

Or, il résulte de (iii) que $\alpha(h) = t' \cdot h$ avec $t' \epsilon T'(\bar{k})$, et l'identité ci-dessus provient du fait que T' centralise U. On a bien obtenu ainsi un relèvement du groupe Γ_- dans $\operatorname{Aut} G(\bar{k})$.

Il nous reste à démontrer le lemme suivant, utilisé en cours de démonstration:

5.11. Lemme. *Soit L un k-groupe algébrique linéaire. Il existe un sousgroupe fini $\overline{W}$ de $L(\overline{k})$ qui rencontre chaque composante connexe de $L(\overline{k})$.*

On peut supposer que $\overline{k} = k$, ce qui nous permettra d'identifier L et $L(\overline{k})$. Soit C un sous-groupe de Cartan de L^0 et soit N son normalisateur dans L. Comme les sous-groupes de Cartan de L^0 sont conjugués par automorphismes intérieurs, N rencontre chaque composante connexe de L; de plus, par définition même des sous-groupes de Cartan, on a $N^0 = C$. On est donc ramené au cas où $L^0 = C$ est *nilpotent*, puis (en utilisant la suite centrale descendante de L^0), au cas où L^0 est *commutatif*. Soit $W = L/L^0$; l'extension L de W par L^0 est alors caractérisée par une classe de cohomologie $\gamma \in H^2(W, L^0)$. Soit n l'ordre de W. L'homomorphisme $f: L^0 \to L^0$ qui applique x sur x^n est surjectif et de noyau R fini (puisque k est de caractéristique zéro). La suite exacte de cohomologie, appliquée à la suite exacte de coefficients

$$0 \to R \to L^0 \overset{f}{\to} L^0 \to 0$$

donne alors la suite exacte

$$H^2(W, R) \to H^2(W, L^0) \overset{n}{\to} H^2(W, L^0).$$

Comme W est d'ordre n, l'homomorphisme $n: H^2(W, L^0) \to H^2(W, L^0)$ est nul. Il s'ensuit que γ est l'image d'un élément $\overline{\gamma}$ de $H^2(W, R)$. Ce dernier représente une extension $\overline{W}$ de W par R, qui s'applique dans L par un homomorphisme compatible avec la projection sur W. Le groupe $\overline{W}$ répond donc à la question.[1]

5.12. Théorème. *Le foncteur des automorphismes d'un groupe algébrique linéaire sur un corps de caractéristique zéro est un groupe localement algébrique de type (ALA).*

Cela résulte de la proposition 5.4 et du lemme 5.10.

5.13. Remarque. Si G est un groupe algébrique quelconque (non nécessairement linéaire) nous ignorons si Aut G est encore un groupe localement algébrique. En tout cas, ce n'est pas toujours un groupe de type (ALA), comme le montre l'exemple du produit semi-direct d'une variété abélienne par un groupe fini (le groupe fini opérant de façon non triviale sur la variété abélienne).

---

[1]) Si L est réductif, C est un tore, et la démonstration vaut sans changement sur un corps de caractéristique quelconque. En fait, le lemme 5.11 est valable pour un groupe algébrique (non nécessairement linéaire) sur un corps parfait quelconque K, et l'on peut en outre démontrer l'existence d'un sous-groupe W défini sur K.

§ 6. Théorèmes de finitude (corps locaux)

6.1. Théorème. *Si* k *est un corps localement compact de caractéristique zéro, et si* A *est un* k*-groupe de type* (ALA), $H^1(k, A)$ *est fini.*

(En particulier, $H^1(k, G)$ est fini quand G est un groupe algébrique linéaire défini sur k.)

On sait que tout corps commutatif localement compact de caractéristique zéro est isomorphe à **R**, à **C**, où à un corps p-adique (extension finie du corps $\mathbf{Q}_p$). Il est classique qu'un tel corps n'a qu'*un nombre fini d'extensions de degré donné*. Indiquons-en brièvement une démonstration: il suffit de prouver que tout corps p-adique n'a qu'un nombre fini d'extensions totalement ramifiées de degré n donné. Or une telle extension est définie par une «équation d'EISENSTEIN» de degré n; ces équations forment un espace compact (pour la topologie définie par la convergence des coefficients), et deux équations assez voisines définissent des extensions isomorphes (cf. par exemple [1], p. 41); d'où la finitude en question.

Le théorème 6.1 est donc une conséquence du suivant:

6.2. Théorème. *Soit* k *un corps parfait n'ayant qu'un nombre fini d'extensions de degré donné. Si* A *est un* k*-groupe de type* (ALA), $H^1(k, A)$ *est fini.*

(Outre le cas p-adique, l'hypothèse faite sur k est vérifiée par les corps finis, et par les corps de séries formelles à une variable sur un corps algébriquement clos de caractéristique zéro.)

Notons $\mathbf{g}$ le groupe de GALOIS de $\bar{k}/k$. Nous allons procéder par étapes:

(a) *Finitude de* $H^1(\mathbf{g}, \Gamma)$ *lorsque* Γ *est un* $\mathbf{g}$*-groupe de type arithmétique.*

D'après la proposition 3.3, il existe un entier n tel que tout sous-groupe fini de Γ soit d'ordre $\leqslant n$. Soit $\mathbf{g}_0$ un sous-groupe ouvert distingué de $\mathbf{g}$ opérant trivialement sur Γ. Vu l'hypothèse faite sur le corps k, les sous-groupes ouverts de $\mathbf{g}_0$ d'indice $\leqslant n$ sont en nombre fini; leur intersection $\mathbf{g}_1$ est un sous-groupe ouvert distingué de $\mathbf{g}$. Tout homomorphisme continu de $\mathbf{g}_0$ dans Γ a une image finie, donc est trivial sur $\mathbf{g}_1$. *A fortiori*, l'application composée:

$$H^1(\mathbf{g}, \Gamma) \to H^1(\mathbf{g}_0, \Gamma) \to H^1(\mathbf{g}_1, \Gamma)$$

est triviale. D'après le n° 1.27, il s'ensuit que $H^1(\mathbf{g}, \Gamma)$ s'identifie à $H^1(\mathbf{g}/\mathbf{g}_1, \Gamma)$; comme Γ est un $\mathbf{g}/\mathbf{g}_1$-groupe de type arithmétique (cf. proposition 3.6), la proposition 3.8 montre que $H^1(\mathbf{g}/\mathbf{g}_1, \Gamma)$ est fini, d'où la finitude de $H^1(\mathbf{g}, \Gamma)$.

(b) *Finitude de* $H^1(k, A)$ *lorsque* A *est un groupe algébrique linéaire résoluble et connexe.*

En appliquant la prop. 1.17, on se ramène au cas où A est unipotent et au cas où A est un tore. Dans le premier cas, on a $H^1(k, A) = 0$, cf. par exemple [18], prop. 3.3.1. Supposons donc que A soit un tore. Il existe alors une extension galoisienne finie k'/k telle que $A \otimes k'$ soit isomorphe à un produit de groupes $\mathbf{G}_m$. Comme $H^1(k', \mathbf{G}_m) = 0$, la prop. 1.27 montre que $H^1(k, A)$ s'identifie à $H^1(k'/k, A)$. En particulier, si $n = [k' : k]$, on a $nx = 0$ pour tout $x \in H^1(k, A)$. Soit A_n le noyau de l'homothétie $n : A \to A$. La suite exacte:

$$0 \to A_n \to A \xrightarrow{n} A \to 0$$

donne naissance à une suite exacte de cohomologie. Cette dernière montre que $H^1(k, A_n)$ s'applique sur le noyau de $n : H^1(k, A) \to H^1(k, A)$, c'est-à-dire sur $H^1(k, A)$ tout entier. Mais $A_n(\overline{k})$ est un g-groupe fini, donc de type arithmétique, et le cas (a) montre que $H^1(k, A_n)$ est fini; il en est donc bien de même de $H^1(k, A)$.

(c) *Finitude de $H^1(k, A)$ lorsque A est un groupe algébrique linéaire.*

D'après Rosenlicht [16], il existe un sous-groupe de Cartan H de A; soit N le normalisateur de H dans A. Le quotient N/H est fini; d'après (a), $H^1(k, N/H)$ est aussi fini; en appliquant (b) et le cor. 1.20, on en déduit que $H^1(k, N)$ est fini. Comme $H^1(k, N) \to H^1(k, A)$ est surjectif (cor. 2.14), il en résulte bien que $H^1(k, A)$ est fini [2]).

(d) *Cas général.*

Puisque A est de type (ALA), on peut trouver une suite exacte

$$0 \to A^1 \to A \to \Gamma \to 0 \,,$$

où A^1 est algébrique linéaire, et où $\Gamma(\overline{k})$ est un g-groupe de type arithmétique (cf. n° 4.3). D'après (a), $H^1(k, \Gamma)$ est fini, D'autre part, pour tout $x \in Z^1(\mathfrak{g}, A)$, le groupe tordu A^1_x est linéaire, et d'après (c), $H^1(k, A^1_x)$ est fini. La finitude de $H^1(k, A)$ résulte alors du cor. 1.20.

Dans les corollaires ci-après, k désigne un corps *vérifiant les conditions du théorème* 6.2.

6.3. Corollaire. (i) *Les k-formes d'une variété abélienne définie sur k sont en nombre fini (à isomorphisme près).*

(ii) *Il en est de même des k-formes d'une algèbre de dimension finie sur k.*

(iii) *Lorsque k est de caractéristique zéro, il en est de même des k-formes d'un groupe algébrique linéaire.*

Précisons que, dans (i), nous prenons le terme de «variété abélienne» au sens de «variété abélienne munie d'une structure de groupe algébrique». Si C est une telle variété, on sait que $\operatorname{Aut} C(\overline{k})$ est un g-groupe de type arithmétique (cf. n° 3.5) et l'assertion (i) résulte alors de la correspondance entre «k-formes de C» et éléments de $H^1(k, \operatorname{Aut} C)$, cf. n° 2.6. De même, l'assertion (ii) résulte de ce que le foncteur d'automorphismes d'une k-algèbre de dimen-

---

[2]) L'idée d'utiliser le corollaire 2.14 nous a été indiquée par T. Springer. Notre démonstration initiale était plus compliquée.

sion finie est un groupe algébrique linéaire (le même argument s'applique plus généralement à la structure formée par un *espace vectoriel muni de tenseurs* de types quelconques, cf. [19], Chap. III, n° 1.1). Enfin, (iii) résulte de ce que, en caractéristique zéro, le foncteur d'automorphismes d'un groupe algébrique linéaire est de type (ALA), cf. théorème 5.12.

6.4. Corollaire. *Soit G un k-groupe algébrique, et soit V un espace homogène de G. Les orbites de $G(k)$ dans $V(k)$ sont en nombre fini.*

La variété V est réunion d'un nombre fini d'orbites de la composante neutre de G; cela permet de se ramener *au cas où G est connexe.* Si $V(k) = \varnothing$, il n'y a rien à démontrer. Sinon, soit $v \epsilon V(k)$, et soit H le groupe de stabilité de v. L'application canonique $G/H \to V$ est radicielle. Comme k est parfait, cette application définit une bijection de $(G/H)(k)$ sur $V(k)$. Or, d'après la prop. 1.12, les orbites de $G(k)$ dans $(G/H)(k)$ correspondent bijectivement aux éléments du noyau de l'application canonique

$$\alpha : H^1(k, H) \to H^1(k, G) \,.$$

Il nous suffira donc de prouver que α est *propre* (cf. n° 1.14).

Soit R le plus grand sous-groupe linéaire connexe de G, soit $S = R \cap H$, et soient $C = G/R$, $B = H/S$. Le groupe C est une variété abélienne (théorème de Chevalley), et B s'envoie injectivement dans C. On a un diagramme commutatif:

$$\begin{array}{ccc}
H^1(k, H) & \overset{\alpha}{\to} & H^1(k, G) \\
\downarrow \gamma & & \downarrow \beta \\
H^1(k, B) & \overset{\delta}{\to} & H^1(k, C) \,.
\end{array}$$

Pour tout $z \epsilon Z^1(\mathfrak{g}, H)$, le groupe tordu S_z est linéaire; d'après le théorème 6.2, $H^1(k, S_z)$ est fini. Appliquant le cor. 1.20, on en déduit que γ est propre. D'autre part, en utilisant le «théorème de complète réductibilité» de Weil (cf. [20], p. 94), on voit qu'il existe une variété abélienne B' de même dimension que B, et un homomorphisme $C \to B'$ tels que le composé $B \to C \to B'$ soit surjectif. Comme le noyau de $B \to B'$ est fini, l'argument utilisé ci-dessus pour prouver la propreté de γ montre que le composé

$$H^1(k, B) \overset{\delta}{\to} H^1(k, C) \to H^1(k, B')$$

est propre. Il s'ensuit que δ est propre, donc aussi $\delta \circ \gamma = \beta \circ \alpha$, donc aussi α, cqfd.

6.5. Corollaire. *Soit G un groupe algébrique linéaire défini sur k. Les tores maximaux (resp. les sous-groupes de Cartan) de G définis sur k forment un nombre fini de classes pour la conjugaison par les éléments de $G(k)$.*

Cela résulte du corollaire 6.4, appliqué à la «variété des tores maximaux» (resp. à la «variété des sous-groupes de Cartan») de G, cf. n° 4..10.

6.6. Corollaire. *Supposons k de caractéristique zéro. Soit G un groupe algébrique semi-simple défini sur k, et soit $\mathfrak{g}$ son algèbre de Lie. Les éléments unipotents de $G(k)$ (resp. les éléments nilpotents de $\mathfrak{g}(k)$) forment un nombre fini de classes pour la conjugaison par les éléments de $G(k)$.*

Soit U (resp. N) la sous-variété de G (resp. $\mathfrak{g}$) formée des éléments unipotents (resp. nilpotents). L'exponentielle définit une bijection de $N(k)$ sur $U(k)$; il suffit donc de prouver la finitude de $N(k)/G(k)$. D'après Kostant (cf. [11], cor. 3.7 et lemme 5.1), les orbites de $G(\bar{k})$ dans $N(\bar{k})$ sont en nombre fini (en fait, Kostant se borne au cas de $\mathbf{C}$, mais le cas général s'en déduit par application du «principe de Lefschetz»). Il suffit donc de prouver que, si X est une telle orbite, les éléments de $X(k) = X \cap N(k)$ forment un nombre fini de classes modulo les opérations de $G(k)$. C'est clair si $X(k) = \varnothing$. Sinon soit $x \in X(k)$, et soit $H \subset G$ le stabilisateur de x. On peut identifier X (resp. $X(k)$) à l'ensemble des points de G/H à valeurs dans $\bar{k}$ (resp. dans k); notre assertion résulte alors du cor. 6.4, appliqué à l'espace homogène G/H.

6.7. Le cas réel. Les résultats des n°ˢ précédents s'appliquent bien entendu au corps $\mathbf{R}$. Certains peuvent d'ailleurs s'obtenir de façon plus simple par des arguments topologiques. Ainsi par exemple le corollaire 6.4 résulte du fait (démontré par Whitney) que, si V est une $\mathbf{R}$-variété algébrique, l'espace topologique $V(\mathbf{R})$ n'a qu'un nombre fini de composantes connexes.

Nous allons voir que, pour certains groupes, on peut aller plus loin et déterminer explicitement H^1:

Partons d'un groupe de Lie compact K. Soit R l'algèbre des fonctions continues sur K qui sont combinaisons linéaires de coefficients de représentations matricielles (complexes) de K, et soit R_0 la sous-algèbre de R formée des fonctions à valeurs réelles. On a $R = R_0 \otimes_{\mathbf{R}} \mathbf{C}$. On définit de façon naturelle un homomorphisme $\delta : R_0 \to R_0 \otimes R_0$. Si l'on pose $G = \mathrm{Spec}(R_0)$, δ définit un morphisme de $G \times G$ dans G, et l'on obtient ainsi sur G une structure de *groupe algébrique sur* $\mathbf{R}$ (cf. Chevalley [5], Chap. VI, § VIII). Le groupe $G(\mathbf{R})$ des points réels de G s'identifie à K. Le groupe $G(\mathbf{C})$ s'appelle le *complexifié* de K. Le groupe de Galois $\mathfrak{g} = \mathfrak{g}(\mathbf{C}/\mathbf{R})$ opère sur $G(\mathbf{C})$.

6.8. Théorème. *L'application canonique $H^1(\mathfrak{g}, K) \to H^1(\mathfrak{g}, G(\mathbf{C}))$ est bijective.*

(Comme $\mathfrak{g}$ opère trivialement sur K, $H^1(\mathfrak{g}, K)$ est l'ensemble des classes dans K, modulo conjugaison, des éléments x tels que $x^2 = 1$.)

Le groupe g opère sur l'algèbre de L$_{IE}$ g$(\mathbf{C})$ de $G(\mathbf{C})$; les éléments invariants forment l'algèbre de L$_{IE}$ $\mathfrak{k}$ de K, et les éléments anti-invariants forment un supplémentaire p de $\mathfrak{k}$ dans g$(\mathbf{C})$. L'exponentielle définit un isomorphisme analytique réel de p sur une sous-variété fermée P de $G(\mathbf{C})$; on a $xPx^{-1} = P$ pour tout $x \epsilon K$; de plus (C$_{HEVALLEY}$, *loc. cit.*, p. 201) tout élément $z \epsilon G(\mathbf{C})$ s'écrit de façon unique sous la forme $z = xp$, avec $x \epsilon K$ et $p \epsilon P$.

Ces résultats étant rappelés, montrons que $H^1(\mathbf{g}, K) \to H^1(\mathbf{g}, G(\mathbf{C}))$ est *surjectif*. Un élément de $Z^1(\mathbf{g}, G(\mathbf{C}))$ s'identifie à un élément $z \epsilon G(\mathbf{C})$ tel que $z\bar{z} = 1$. Si l'on écrit z sous la forme xp, avec $x \epsilon K$ et $p \epsilon P$, on a $xpx\bar{p}^{-1} = 1$ (car $\bar{p} = p^{-1}$), d'où $p = x^2 \cdot x^{-1}px$. Mais $x^{-1}px$ appartient à P, et l'unicité de la décomposition $G(\mathbf{C}) = K \cdot P$ montre que $x^2 = 1$ et $x^{-1}px = p$. Si P_x est la partie de P formée des éléments commutant à x, on voit facilement que P_x est l'exponentielle d'un sous-espace vectoriel de p. On en conclut qu'on peut écrire p sous la forme $p = q^2$, avec $q \epsilon P_x$. On en tire $z = qxq$, et comme $\bar{q} = q^{-1}$, on voit que le cocycle z est cohomologue au cocycle x, qui est à valeurs dans K.

Montrons maintenant que $H^1(\mathbf{g}, K) \to H^1(\mathbf{g}, G(\mathbf{C}))$ est *injectif*. Soient $x \epsilon K$ et $x' \epsilon K$ deux éléments tels que $x^2 = 1$, $x'^2 = 1$, et supposons qu'ils soient cohomologues dans $G(\mathbf{C})$, autrement dit qu'il existe $z \epsilon G(\mathbf{C})$ tel que $x' = z^{-1}x\bar{z}$. Ecrivons z sous la forme $z = yp$, avec $y \epsilon K$ et $p \epsilon P$. On a:

$$x' = p^{-1}y^{-1}xyp^{-1}, \quad \text{d'où} \quad x' \cdot x'^{-1}px' = y^{-1}xy \cdot p^{-1}.$$

Appliquant à nouveau l'unicité de la décomposition $G(\mathbf{C}) = K \cdot P$, on en tire $x' = y^{-1}xy$, ce qui signifie que x et x' sont conjugués dans K, et achève la démonstration.

Exemples. (a) Supposons que K soit *connexe*, et soit T l'un de ses tores maximaux. Soit T_2 l'ensemble des $t \epsilon T$ tels que $t^2 = 1$; on sait que tout élément $x \epsilon K$ tel que $x^2 = 1$ est conjugué d'un élément $t \epsilon T_2$; de plus deux éléments de T_2 sont conjugués dans K si et seulement si ils sont transformés l'un en l'autre par un élément du *groupe de* W$_{EYL}$ W de K (par rapport à T). Il résulte donc du théorème 6.8 que $H^1(\mathbf{R}, G)$ (égal par définition à $H^1(\mathbf{g}, G(\mathbf{C}))$) *s'identifie à l'ensemble quotient* T_2/W.

(b) Prenons pour K le *groupe des automorphismes* d'un groupe compact semi-simple connexe S. Soit A (resp. G) le groupe algébrique associé comme ci-dessus à K (resp. à S). On sait que A *est le groupe d'automorphismes* Aut G de G. Les éléments de $H^1(\mathbf{R}, A)$ correspondent donc aux *formes réelles* du groupe G, et le théorème 6.8 redonne la classification de ces formes au moyen de classes d'«involutions» de S (résultat dû à E$_{LIE}$ C$_{ARTAN}$).

§ 7. Théorèmes de finitude (corps de nombres)

Dans tout ce paragraphe, k désigne un *corps de nombres algébriques*, c'est-à-dire une extension finie du corps $\mathbf{Q}$, et $\mathbf{g}$ désigne le groupe de Galois $\mathbf{g}(\bar{k}/k)$. On note V l'ensemble des *places* de k; rappelons qu'une place est une classe d'équivalence de valeurs absolues non discrètes (deux valeurs absolues étant dites équivalentes si elles définissent la même topologie). Si $v \in V$, on note k_v le complété de k pour la topologie définie par v; les corps k_v sont des corps localement compacts de caractéristique zéro.

Si A est un groupe localement algébrique sur k, chacun des plongements $k \to k_v$ définit une application $\omega_v : H^1(k, A) \to H^1(k_v, A)$, cf. 2.2. Pour toute partie S de V, ces applications définissent une application canonique

$$\omega_S : H^1(k, A) \to \prod_{v \in V - S} H^1(k_v, A) .$$

7.1. Théorème. *Si S est un ensemble fini de places de k, et si A est un k-groupe de type* (ALA), *l'application*

$$\omega_S : H^1(k, A) \to \prod_{v \in V - S} H^1(k_v, A)$$

est propre (cf. n° 1.14).

En particulier, on voit que les éléments de $H^1(k, A)$ qui sont «nuls localement» (c'est-à-dire dont l'image dans chacun des $H^1(k_v, A)$ est nulle) sont en nombre fini; cela généralise un résultat de [3], résultat qui va d'ailleurs être utilisé dans la démonstration ci-dessous.

Avant de donner cette démonstration, il est bon de faire deux remarques:

(i) Si le théorème est vrai pour A et pour un certain S, il est vrai pour A et pour tout autre ensemble fini S'. Cela résulte simplement de la finitude des $H^1(k_v, A)$, démontrée au § 6. En particulier, on peut toujours se ramener au cas où $S = \varnothing$.

(ii) Pour prouver le théorème 7.1, il suffit de prouver que *le noyau de* ω_S *est fini* pour le groupe A considéré et pour tous les groupes tordus A_x, avec $x \in Z^1(k, A)$. Cela résulte des propriétés élémentaires de la torsion, cf. n° 1.10.

7.2. Démonstration du théorème 7.1 lorsque A est connexe. Puisque A est de type (ALA), c'est un groupe linéaire connexe. Soit U la partie unipotente du radical de A et soit $H = A/U$. Le groupe H est réductif connexe. Comme la cohomologie d'un groupe unipotent est triviale, les applications canoniques:

$$H^1(k, A) \to H^1(k, H) \quad \text{et} \quad H^1(k_v, A) \to H^1(k_v, H)$$

sont injectives (cf. n⁰ 1.20). On est donc ramené à prouver le théorème pour le groupe H. De plus, d'après les remarques (i) et (ii) du n⁰ 7.1, il suffit de montrer que, pour tout $x \epsilon Z^1(k, H)$, le noyau de

$$H^1(k, H_x) \to \overline{\prod_{v \epsilon V}} H^1(k_v, H_x)$$

est fini. Or, c'est là un résultat connu ([3], th. 6.8).

7.3. Lemme. *Soit* A *un* **g**-*groupe discret sur lequel* **g** *opère trivialement, et soit* S *un sous-ensemble fini de* V. *Le noyau de l'application*

$$\omega_S : H^1(k, A) \to \overline{\prod_{v \epsilon V - S}} H^1(k_v, A)$$

est alors réduit à 0.

Soit x un élément de ce noyau, et choisissons une extension galoisienne k'/k telle que x appartienne à $H^1(k'/k, A)$. Soit **g**$'$ le groupe de GALOIS de k'/k; comme **g**$'$ opère trivialement sur A, l'élément x est représenté par un cocycle ξ qui est un *homomorphisme de* **g**$'$ dans A. Soit V' l'ensemble des places de k', et soit S' le sous-ensemble de V' formé des places qui prolongent les places de S. Soit $w \epsilon V' - S'$, soit v la place correspondante de $V - S$, et soit **g**$'_w$ le groupe de décomposition de w. Dire que x donne zéro dans $H^1(k_v, A)$ signifie que la restriction de ξ à **g**$'_w$ est triviale (i. e. ξ applique **g**$'_w$ dans l'élément neutre de A). Mais l'on sait que, lorsque w parcourt $V' - S'$, *la réunion des* **g**$'_w$ *est* **g**$'$ *tout entier*; de façon plus précise, tout sous-groupe cyclique de **g**$'$ est un groupe de décomposition **g**$'_w$ pour une infinité de w (cela résulte du théorème de densité d'ARTIN-ČEBOTAREV, ou même, plus simplement, du théorème de densité de FROBENIUS [7]). Il s'ensuit que ξ est trivial, d'où le lemme.

7.4. Démonstration du théorème 7.1 lorsque A est discret. En vertu de la remarque (ii) du n⁰ 7.1, il suffit de prouver que *le noyau de* ω_S est fini. Nous noterons ce noyau $H^1_S(k, A)$.

Puisque A est de type (ALA), c'est une extension d'un **g**-groupe Γ de type arithmétique par un **g**-groupe fini C; on a $A/C = \Gamma$. Comme Γ est de type fini (cf. n⁰ 3.3), il en est de même de A; comme **g** opère continûment, on en conclut qu'il existe un sous-groupe ouvert distingué **h** de **g** qui opère trivialement sur A. On notera k' le sous-corps de k correspondant à **h**, et S' l'ensemble des places de k' prolongeant l'une des places de S. Si $x \epsilon H^1_S(k, A)$, l'image x' de x dans $H^1(k', A)$ appartient à $H^1_{S'}(k', A)$: c'est immédiat. Mais, puisque **h** opère trivialement sur A, le lemme 7.3 (appliqué à k' et S') montre que $H^1_{S'}(k', A)$ est réduit à 0. D'après 1.27, il en résulte que x appartient à $H^1(k'/k, A) = H^1(\mathbf{g}/\mathbf{h}, A)$. Mais **g**/**h** est fini, donc aussi $H^1(\mathbf{g}/\mathbf{h}, \Gamma)$, cf. n⁰ 3.8; il en est de même des $H^1(\mathbf{g}/\mathbf{h}, C_x)$, où x désigne un 1-cocycle de **g**/**h** dans A; d'après 1.20, il s'ensuit que $H^1(\mathfrak{g}/\mathfrak{h}, A)$ est fini. Comme $H^1_S(k, A)$ est un sous-ensemble de $H^1(\mathfrak{g}/\mathfrak{h}, A)$, cela démontre bien le théorème 7.1 dans ce cas.

7.5. Pour aller plus loin, nous aurons besoin d'un résultat élémentaire sur les «réductions mod. p». Indiquons rapidement de quoi il s'agit (pour plus de détails, voir [9], Chap. IV):

Soit X un schéma de type fini sur le corps k, et soit $\mathfrak{o}_k$ l'anneau des entiers de k; le corps k est le corps des fractions de $\mathfrak{o}_k$, c'est-à-dire le corps des fonctions rationnelles de $\mathrm{Spec}\,(\mathfrak{o}_k)$. Il en résulte facilement qu'on peut trouver une «forme de X sur $\mathfrak{o}_k$», c'est-à-dire un schéma X_0 de type fini sur $\mathrm{Spec}\,(\mathfrak{o}_k)$ tel que $X = X_0 \otimes_{\mathfrak{o}_k} k$. Si v est un idéal maximal de $\mathfrak{o}_k$, de corps résiduel $\kappa(v)$, on peut parler du schéma $X_0 \otimes_{\mathfrak{o}_k} \kappa(v)$ obtenu à partir de X_0 par réduction en v. Ses points rationnels sur $\kappa(v)$ forment un ensemble fini $(X_0)_{\kappa(v)}$. De même, si $\mathfrak{o}_v$ désigne l'anneau des entiers du corps local k_v, X_0 définit par changement de base un schéma sur $\mathfrak{o}_v$; les points de ce schéma à valeurs dans $\mathfrak{o}_v$ seront notés $(X_0)_{\mathfrak{o}_v}$. En fait, on emploie très souvent les notations incorrectes $X_{\kappa(v)}$ et $X_{\mathfrak{o}_v}$ pour désigner $(X_0)_{\kappa(v)}$ et $(X_0)_{\mathfrak{o}_v}$. Cet abus d'écriture est sans danger si l'on ne s'intéresse qu'à des questions où n'interviennent que «presque tous les v». En effet, on démontre sans peine que, si X_0 et X_0' sont deux formes de X sur $\mathfrak{o}_k$, il existe un ouvert non vide U de $\mathrm{Spec}\,(\mathfrak{o}_k)$ au-dessus duquel X_0 et X_0' sont isomorphes; si $v \in U$ on peut alors identifier $(X_0)_{\kappa(v)}$ et $(X_0')_{\kappa(v)}$ et de même pour les points à valeurs dans $\mathfrak{o}_v$. C'est cette notation que nous utiliserons dans la proposition suivante:

7.6. Proposition. *Soit G un k-groupe algébrique connexe, et soit P un espace fibré principal pour G dont la base B est une variété algébrique de dimension zéro. Il existe alors une partie finie S de V, contenant toutes les places archimédiennes, et telle que, si $v \in V - S$, on ait $B_{\mathfrak{o}_v} = B_{k_v}$ et que $P_{\mathfrak{o}_v} \to B_{\mathfrak{o}_v}$ soit surjectif.*

(L'hypothèse faite sur P équivaut à dire que $P \otimes \bar{k}$ est réunion disjointe d'espaces homogènes principaux sur $G \otimes \bar{k}$.)

On choisit comme ci-dessus des formes de G, P, B sur $\mathfrak{o}_k$; on les notera encore G, P, B (c'est l'abus d'écriture signalé plus haut). Si l'on se place sur un ouvert $U = \mathrm{Spec}\,(\mathfrak{o}_k) - S$ assez petit de $\mathrm{Spec}\,(\mathfrak{o}_k)$, ces schémas sont *simples* sur U (cf. Grothendieck [10]), B est propre, G est un schéma en groupes à fibres connexes, et P est un espace fibré principal de base B et de groupe G (cela signifie ici que le morphisme évident de $P \times_U G$ dans $P \times_U P$ est un isomorphisme). Soit $v \in U$; comme B est simple et propre sur U, on a $B_{\kappa(v)} = B_{\mathfrak{o}_v} = B_{k_v}$. De même, la simplicité de P entraîne que $P_{\mathfrak{o}_v} \to P_{\kappa(v)}$ soit surjectif (pour ces propriétés de la simplicité – qui ne sont au fond qu'une forme du «lemme de Hensel» — voir Grothendieck [10]). D'autre part, si

$x \in B_{\kappa(v)}$, la fibre de x dans P est un espace homogène principal sur le groupe $G \otimes_{0_k} \kappa(v)$ déduit de G par réduction en v; comme $G \otimes_{0_k} \kappa(v)$ est connexe, cet espace a un point à valeurs dans $\kappa(v)$ (cf. LANG [12]). Il s'ensuit que $P_{\kappa(v)} \to B_{\kappa(v)}$ est surjectif, d'où le résultat cherché.

7.7. Corollaire. *L'application $P_{k_v} \to B_{k_v}$ est surjective pour tout $v \in V - S$.* C'est évident.

Remarque. Lorsque, dans la proposition 7.6, on ne suppose plus que B soit de dimension zéro, il est encore vrai que $P_{0_v} \to B_{0_v}$ est surjectif pour presque tout v.

7.8. Revenons maintenant à la démonstration du théorème 7.1. Soit G la composante neutre du groupe A, et soit $L = A/G$; nous identifierons (par abus de langage) L au g-groupe $L(\bar{k})$.

7.9. Lemme. *Il existe un sous-ensemble fini S de V tel que l'application $H^0(k_v, A) \to H^0(k_v, L)$ soit surjective pour tout $v \in V - S$.*

Comme A est de type (ALA), le groupe L est extension d'un g-groupe Γ de type arithmétique par un g-groupe fini C. Comme on l'a déjà remarqué, il existe un sous-groupe ouvert distingué $\mathbf{h}$ de $\mathbf{g}$ qui opère trivialement sur L. Posons $\mathbf{g}' = \mathbf{g}/\mathbf{h}$, et soit $\mathbf{h}'$ un sous-groupe de $\mathbf{g}'$. On a la suite exacte:

$$0 \to H^0(\mathbf{h}', C) \to H^0(\mathbf{h}', L) \to H^0(\mathbf{h}', \Gamma) \to H^1(\mathbf{h}', C),$$

où $H^0(\mathbf{h}', C)$ et $H^1(\mathbf{h}', C)$ sont finis. Comme d'autre part $H^0(\mathbf{h}', \Gamma)$ est un groupe de type fini (cf. 3.3 et 3.6) on en déduit facilement que $H^0(\mathbf{h}', L)$ est aussi de type fini. Puisque les $\mathbf{h}'$ sont en nombre fini, on en conclut qu'il existe un *sous-ensemble fini B de L tel que $B \cap H^0(\mathbf{h}', L)$ engendre $H^0(\mathbf{h}', L)$ pour tout $\mathbf{h}'$*; on peut en outre s'arranger pour que B soit stable par $\mathbf{g}$, ce qui permet de l'identifier à une k-sous-variété de dimension zéro de $L = A/G$. Soit P l'image réciproque de cette sous-variété dans A; c'est un espace fibré principal pour G de base B. Appliquant 7.7, on voit qu'il existe une partie finie S de V telle que $H^0(k_v, P) \to H^0(k_v, B)$ soit surjectif pour tout $v \in V - S$. *Cette propriété entraîne que*

$$H^0(k_v, A) \to H^0(k_v, L)$$

est surjectif pour tout $v \in V - S$. En effet, l'image I_v de cet homomorphisme contient l'image de $H^0(k_v, P)$, qui est égale à $H^0(k_v, B)$. Or, si $\mathbf{g}'_v$ désigne l'un des groupes de décomposition associés à v dans $\mathbf{g}'$, on peut identifier $H^0(k_v, L)$ à $H^0(\mathbf{g}'_v, L)$ et $H^0(k_v, B)$ à $H^0(\mathbf{g}'_v, B)$, c'est-à-dire à $B \cap H^0(\mathbf{g}'_v, L)$. Il en résulte (vu la définition de B) que $H^0(k_v, B)$ engendre $H^0(k_v, L)$. Comme I_v contient $H^0(k_v, B)$, on a donc bien $I_v = H^0(k_v, L)$, ce qui achève la démonstration du lemme.

7.10. Fin de la démonstration du théorème 7.1. On conserve les notations ci-dessus. On suppose en outre que $S = \varnothing$, ce qui est licite vu la remarque (i). On a un diagramme commutatif

$$
\begin{array}{ccccc}
H^1(k, G) & \overset{\alpha}{\to} & H^1(k, A) & \overset{\beta}{\to} & H^1(k, L) \\
\downarrow\omega' & & \downarrow\omega & & \downarrow\omega'' \\
\underset{v \in V}{\textstyle\prod} H^1(k_v, G) & \overset{\varphi}{\to} & \underset{v \in V}{\textstyle\prod} H^1(k_v, A) & \overset{\psi}{\to} & \underset{v \in V}{\textstyle\prod} H^1(k_v, L)\,.
\end{array}
$$

On va d'abord prouver *que l'application*

$$
(\beta, \omega) : H^1(k, A) \to H^1(k, L) \times \underset{v \in V}{\textstyle\prod} H^1(k_v, A)
$$

est propre.

Pour cela, soit $x \in Z^1(k, A)$, soit ξ l'image de x dans $H^1(k, A)$ et soit X l'ensemble des éléments de $H^1(k, A)$ dont l'image par (β, ω) est la même que celle de ξ; on doit prouver que X est *fini*. Quitte à remplacer A par A_x, on peut supposer que $x = 0$. Il en résulte que X est contenu dans l'image de $\alpha : H^1(k, G) \to H^1(k, A)$. Soit $Y = \alpha^{-1}(X)$, et soit

$$
Y' = \omega'(Y) \subset \underset{v \in V}{\textstyle\prod} H^1(k_v, G).
$$

Comme $\omega \circ \alpha = \varphi \circ \omega'$, les éléments de Y' appartiennent au noyau de φ. Mais, d'après 7.9, il existe un sous-ensemble fini S de V tel que $H^0(k_v, A) \to H^0(k_v, L)$ soit surjectif pour $v \in V - S$; vu 1.17, ceci entraîne que le noyau de $H^1(k_v, G) \to H^1(k_v, A)$ est réduit à 0 (pour $v \in V - S$). On en conclut que l'image Y'' de Y dans $\underset{v \in V - S}{\textstyle\prod} H^1(k_v, G)$ est réduite à 0, ce qui montre (cf. 7.2) que Y est fini; comme $X = \alpha(Y)$, il en est de même de X, ce qui démontre notre assertion.

Notons maintenant que, d'après 7.4, l'application ω'' est propre. Il en est donc de même de l'application $(\omega'' \times 1) \circ (\beta, \omega)$ qui applique $H^1(k, A)$ dans $\underset{v \in V}{\textstyle\prod} H^1(k_v, L) \times \underset{v \in V}{\textstyle\prod} H^1(k_v, A)$. Comme cette application se factorise par l'application ω, on en conclut que ω est propre, c q f d.

7.11. Corollaire. *Soit G une variété abélienne (resp. un groupe algébrique linéaire, resp. une algèbre de dimension finie) sur k, et soit S un ensemble fini de places de k. Les k-formes de G qui sont k_v-isomorphes à G pour tout v non dans S sont en nombre fini (à isomorphisme près).*

Cela résulte de ce que, dans chaque cas, Aut (G) est un groupe de type (ALA).

7.12. Corollaire. *Soit G un k-groupe algébrique, soit V un espace homogène de G, soit $x \in V(k)$ et soit S un ensemble fini de places de k. Soit X la partie de*

$V(k)$ formée des éléments x' tels que pour tout $v \in V - S$, il existe $g_v \in G(k_v)$ qui transforme x en x'. Supposons enfin que le groupe de stabilité H de x soit un groupe linéaire. Les orbites de $G(k)$ dans X sont alors en nombre fini.

Soit Y le quotient de X par $G(k)$; d'après 1.12, Y s'identifie à un sous-ensemble de $H^1(k, H)$. De plus, la définition même de X montre que l'image de Y dans chacun des $H^1(k_v, H), v \in V - S$, est réduite à 0. Comme H est linéaire, le théorème 7.1 s'applique, et montre que Y est fini, cqfd.

Remarques. (1) Si G est un groupe algébrique linéaire, le corollaire ci-dessus s'applique à la variété des tores maximaux (resp. des sous-groupes de Cartan, resp. etc.) de G. Nous laissons au lecteur le soin d'expliciter le résultat obtenu.

(2) Les hypothèses et notations étant celles de 7.12, soit x' un point de X. Par hypothèse, pour tout $v \in V - S$, il existe $g_v \in G_{k_v}$ qui transforme x en x'. *On peut s'arranger pour que g_v appartienne à G_{o_v} pour presque tout v.* En effet, soit P la sous-variété algébrique de G dont les points sont les éléments g transformant x en x'; c'est un espace homogène principal sur H. Si H^0 désigne la composante neutre de H, P peut aussi être considérée comme espace principal de groupe H^0 et de base finie $B = P/H^0$. D'après 7.6, l'application $P_{o_v} \to B_{k_v}$ est surjective pour presque tout v; d'autre part, l'existence d'un $g_v \in P_{k_v}$ montre que B_{k_v} est non vide. Il s'ensuit que P_{o_v} est non vide pour presque tout v, ce qui démontre notre assertion.

Ceci permet de traduire 7.12 en termes adéliques, comme dans [3].

BIBLIOGRAPHIE

[1] E. Artin. *Algebraic numbers and algebraic functions.* New York Univ., 1951 (notes polycopiées).

[2] A. Borel. *Arithmetic properties of algebraic groups.* Cong. Inter., Stockholm, 1962, p. 10–22.

[3] A. Borel, *Some finiteness properties of adele groups over number fields.* Publ. Math. Inst. Hautes Etudes Sci., n° 16, 1963, p. 5–30.

[4] A. Borel and Harish-Chandra, *Arithmetic subgroups of algebraic groups.* Annals of Maths., 75, 1962, p. 485–535.

[5] C. Chevalley, *Theory of LIE groups.* Princeton Math. Ser., n° 8, 1946.

[6] C. Chevalley. *Classification des groupes de LIE algébriques.* Séminaire Ecole Normale Sup., 1956–58.

[7] G. Frobenius. *Über Beziehungen zwischen den Primidealen eines algebraischen Körpers und den Substitutionen seiner Gruppe.* Sitzber. Berlin Akad., 32, 1896, p. 689–705.

[8] A. Grothendieck, *A general theory of fibre spaces with structure sheaf.* Univ. Kansas, Report n° 4, 1955.

[9] A. Grothendieck, *Eléments de géométrie algébrique* (rédigés avec la collaboration de J. Dieudonné). Inst. Hautes Etudes Sci., Publ. Math. 1960–61–62–63– . . .

[10] A. Grothendieck. *Séminaire de géométrie algébrique.* Inst. Hautes Etudes Sci., 1960–61.

[11] B. Kostant, *The principal three-dimensional subgroup and the BETTI numbers of a complex simple LIE group.* Amer. J. of Maths., 81, 1959, p. 973–1032.

[12] S. Lang, *Algebraic groups over finite fields*. Amer. J. of Maths., *78*, 1956, p. 555–563.

[13] S. Lang and J. Tate, *Principal homogeneous spaces over abelian varieties*. Amer. J. of Maths., *80*, 1958, p. 659–684.

[14] G. Mostow, *Fully reducible subgroups of algebraic groups*. Amer. J. of Maths., *78*, 1956, p. 200- 221.

[15] M. Rosenlicht, *Some basic theorems on algebraic groups*. Amer. J. of Maths., *78*, 1956, p. 401–443.

[16] M. Rosenlicht, *Some rationality questions on algebraic groups*. Annali di Mat. Pura Appl., *43*, 1957, p. 25–50.

[17] J.-P. Serre, *Groupes algébriques et corps de classes*. Act. Sci. Ind., n° 1264, Paris, Hermann, 1959.

[18] J.-P. Serre, *Cohomologie galoisienne des groupes algébriques linéaires*. Colloque de Bruxelles, 1962, p. 53–67.

[19] J.-P. Serre, *Cohomologie galoisienne*. Collège de France, 1963 (notes polycopiées).

[20] A. Weil. *Variétés abéliennes et courbes algébriques*. Act. Sci. Ind., n° 1064, Paris, Hermann, 1948.

[21] A. Weil, *Adeles and algebraic groups*. Inst. Adv. Study, 1961 (notes polycopiées rédigées par M. Demazure et T. Ono).

(Reçu le 16 janvier 1964)

65.

Cohomologie et rigidité d'espaces compacts localement symétriques

Séminaire Bourbaki, Exp. 265 (1963/64)

Soient X un espace riemannien symétrique à courbure négative (sans composante euclidienne), G la composante neutre du groupe des isométries de X et Γ un sous-groupe de G qui est discret et uniforme (i.e. tel que $\Gamma \backslash G$ soit compact). On s'intéressera aux espaces de cohomologie $H^p(\Gamma, F_\varrho)$ de Γ, à coefficients dans l'espace F_ϱ d'une représentation ϱ de dimension finie de G, ou en fait plus particulièrement à la nullité de $H^1(\Gamma, F_\varrho)$. Ces espaces sont intervenus jusqu'à présent principalement dans trois questions, dont deux seront discutées dans les § 3 et § 4 et la troisième mentionnée en 2.5. On suppose ϱ réelle.

Notations. On rappelle que X est homéomorphe à un espace euclidien, que G est un groupe de Lie semi-simple, de centre réduit à (e), sans facteur compact $\neq (e)$, et que $X = G/K$, où K est un sous-groupe compact maximal de G. Soient $\mathfrak{g}$, $\mathfrak{k}$ les algèbres de Lie de G et K, et $\mathfrak{m}$ le complément orthogonal de $\mathfrak{k}$ par rapport à la forme de Killing $B(x, y) = \mathrm{Tr}\,(\mathrm{ad}\,x \circ \mathrm{ad}\,y)$. On a les relations

$$(1) \qquad\qquad [\mathfrak{k}, \mathfrak{k}] \subset \mathfrak{k}, \quad [\mathfrak{k}, \mathfrak{m}] \subset \mathfrak{m}, \quad [\mathfrak{m}, \mathfrak{m}] \subset \mathfrak{k},$$

qui expriment le fait que $s\colon k + m \mapsto k - m$ ($k \in \mathfrak{k}, m \in \mathfrak{m}$) est un automorphisme de $\mathfrak{g}$ (involution de Cartan). On désignera par (X_λ) ($1 \leq \lambda \leq n = \dim \mathfrak{g}$) une base orthonormale de $\mathfrak{g}$, par rapport à la forme (positive non-dégénérée) $B^*(x, y) = -B(x, s(y))$, dont les $N = \dim \mathfrak{m}$ premiers éléments engendrent $\mathfrak{k}$, et les $n - N$ derniers engendrent $\mathfrak{k}$, et par (ω^λ) la base duale de $\mathfrak{g}^*$. On identifie $\mathfrak{g}$ et $\mathfrak{g}^*$ respectivement aux champs de vecteurs et 1-formes invariants à gauche sur G. La translation à gauche (resp. à droite) définie par un élément $g \in G$ est notée L_g (resp. R_g). Les indices i, j, k, h vont de 1 à N, et λ de 1 à n.

1. Les espaces de cohomologie $H^p(\Gamma, X, \varrho)$

1.1. Le groupe Γ opère proprement sur X, et le quotient $M = \Gamma \backslash X$ est compact. Γ opère librement si et seulement s'il est sans torsion, auquel cas M est une variété compacte localement symétrique (une forme de Clifford-Klein compacte de X). Pour tout ce qui est discuté ici, on se ramène aisément à ce cas en considérant un

416

sous-groupe distingué d'indice fini sans torsion Γ' de Γ, dont l'existence est assurée par un théorème de Selberg [10].

1.2. Soit $A^p(X, F_\varrho)$ ou simplement A^p l'espace des formes différentielles C^∞ sur X, à valeurs dans $F_\varrho(p = 0, 1, \ldots)$ et $A(X, F_\varrho)$, ou A, la somme directe des A^p. On fait opérer G sur A^p en associant à g l'application $\theta \mapsto \varrho(g^{-1}) \cdot (\theta \circ L_g)$. Ces opérations commutent à la différentielle extérieure et la somme directe des espaces $A^p(\Gamma, X, F_\varrho)$ d'éléments fixes par Γ est un sous-complexe, dont le p-ième espace de cohomologie est noté $H^p(\Gamma, X, \varrho)$ (cf. [8]).

Soit σ la projection de X sur M. En associant à tout ouvert $U \subset M$ l'espace des sections sur $\sigma^{-1}(U)$ du faisceau constant $X \times F_\varrho$, on définit sur M un faisceau $\mathscr{F}_\varrho$. On a alors pour tout p des isomorphismes canoniques

$$(2) \qquad H^p(\Gamma, F_\varrho) \cong H^p(\Gamma, X, \varrho) \cong H^p(M, \mathscr{F}_\varrho) \,.$$

Si Γ est sans torsion, $\mathscr{F}_\varrho$ est localement constant, de fibre type F_ϱ, la première égalité résulte de la suite spectrale des revêtements et de l'acyclicité de X, la deuxième du théorème de de Rham à coefficients tordus. Le cas général peut se ramener à celui-là à l'aide du groupe Γ' de 1.1. La première égalité résulte aussi du corollaire au théorème 5.3.1 de [3].

2. Formes harmoniques

2.1. Pour étudier $H^p(\Gamma, F_\varrho)$, on transporte tout sur $\Gamma\backslash G$, en utilisant les projections:

$$(3) \qquad \begin{array}{ccc} G & \xrightarrow{\ \pi\ } & X = G/K \\ \downarrow & & \downarrow{\scriptstyle \sigma} \\ \Gamma\backslash G & \longrightarrow & M = \Gamma\backslash G/K. \end{array}$$

On utilisera la même notation pour un élément de $\mathfrak{g}$ (resp. $\mathfrak{g}^*$) et le champ de vecteurs (resp. la 1-forme) qu'il définit canoniquement sur $\Gamma\backslash G$. Soit $\theta \in A^p(\Gamma, X, F_\varrho)$. Pour tout $g \in G$, posons $\bar{\theta}_g = \varrho(g^{-1}) \cdot (\theta \circ \pi)_g$. On définit ainsi une p-forme $\bar{\theta}$ sur G, visiblement invariante à gauche par Γ, d'où une p-forme θ^0 sur $\Gamma\backslash G$. On vérifie aisément que $\varphi \colon \theta \mapsto \theta^0$ est un isomorphisms de $A^p(\Gamma, X, \varrho)$ sur l'espace $B^p(\Gamma, \varrho)$ des p-formes sur $\Gamma\backslash G$, à valeurs dans F_ϱ, qui vérifient

$$(4) \qquad \Theta(X)\, \theta^0 = -\varrho(X) \cdot \theta^0, \quad i(X) \cdot \theta^0 = 0 \quad (X \in \mathfrak{k}) \,,$$

(où $\Theta(X)$ est la transformation infinitésimale associée à X, et $i(X)$ le produit intérieur par X). Une p-forme Θ sur $\Gamma\backslash G$ sera écrite sous la forme

$$\theta = \Sigma_I \theta_i \cdot \omega^I = \Sigma_{\lambda_1, \ldots, \lambda_p} \theta_{\lambda_1, \ldots, \lambda_p}\, \omega^{\lambda_1} \ldots \omega^{\lambda_p},$$

où, par conséquent,

$$\theta_I = \theta(X_{\lambda_1}, \ldots, X_{\lambda_p}), \quad (I = (\lambda_1, \ldots, \lambda_p)),$$

est une fonction sur $\Gamma \backslash G$, à valeurs dans F_ϱ. Si $\theta \in B^p(\Gamma, \varrho)$, la deuxième égalité de (4) montre que $\theta_I = 0$ si I contient un indice $> N$.

Munissons F_ϱ d'un produit scalaire $(,)$ positif non-dégénéré, tel que $\varrho(X)$ soit antisymétrique (resp. self-adjoint) si $X \in \mathfrak{k}$ (resp. $X \in \mathfrak{m}$) ce qui est toujours possible, et $\Gamma \backslash G$ de la métrique donnée par $ds^2 = \Sigma\, \omega^\lambda \cdot \omega^\lambda$. On définit alors sur l'espace des p-formes sur $\Gamma \backslash G$, à valeurs dans $F_{\acute\varrho}$, un produit scalaire par

$$\langle \theta, \theta' \rangle = \Sigma_I \int\limits_G (\theta_I, \theta'_I)\, dv \qquad (dv = \omega' \wedge \ldots \wedge \omega^n)\,.$$

On pose bien entendu $\langle \theta, \theta' \rangle = 0$ si θ et θ' sont homogènes de degrés différents.

2.2. Proposition [8]. *La différentielle $d^0 = \varphi \circ d \circ \varphi^{-1}$ sur B a un adjoint $\delta^0 \colon B \to B$ diminuant le degré de 1. Si $\theta \in B^p(\Gamma, \varrho)$, on a*

(i) $$(d^0\theta)_{i_1 \ldots i_{p+1}} = \Sigma_u (-1)^{u-1} (X_{i_u} + \varrho(X_{i_u}))\, \theta_{i_1, \ldots, \hat{\imath}_u, \ldots, i_{p+1}},$$

(ii) $$(\delta^0\theta)_{i_1 \ldots i_{p-1}} = \Sigma_k (\varrho(X_k) - X_k)\, \theta_{k i_1 \ldots i_{p-1}} \qquad (p \geqq 1);\ \delta^0\theta = 0$$

si $p = 0$.

Tout élément de $H^p(\Gamma, F_\varrho)$ est représenté par une et une seule forme harmonique.

($\theta \in B^p(\Gamma, \varrho)$ est harmonique si $d^0\theta = \delta^0\theta = 0$, donc si $\varDelta\theta = 0$ avec $\varDelta = d^0\delta^0 + \delta^0 d^0$.) Pour esquisser la démonstration, nous reprenons les notations $\theta, \theta', \theta^0$ de 2.1. La relation (i) provient de l'égalité

$$\varrho(X) = \varrho(g^{-1}) \cdot d\varrho_g(X), \qquad (X \in \mathfrak{g},\ g \in G)$$

et du fait que, dans l'expression usuelle de $d\theta(X_{i_1}, \ldots, X_{i_{p+1}})$, les termes où figurent les crochets $[X_{i_j}, X_{i_k}]$ sont nuls vu (1) et (4).

Soient $f, f' \colon G \to F_\varrho$ des fonctions différentiables. Des formules

$$d \cdot i(X) + i(X) \cdot d = \Theta(X), \qquad df = \Sigma\, X_\lambda f \cdot \omega^\lambda,$$

on déduit que $X_\lambda f \cdot dv$ est un cobord, d'où, par Stokes,

$$\langle X_\lambda f, f' \rangle = - \langle f, X_\lambda f' \rangle\,.$$

On tire alors (ii) de (i) et du fait que $\varrho(X)$ est self-adjoint si $X \in \mathfrak{m}$.

Supposons Γ sans torsion. On montre que

$$\langle \theta, \theta' \rangle = c \cdot \langle \theta, \theta' \rangle_M \qquad (\theta, \theta' \in A^p(\Gamma, X, \varrho))\,,$$

où $\langle,\rangle_M$ est le produit scalaire associé à la métrique $ds^2 = \Sigma\, \omega^i \cdot \omega^i$: par suite ϱ envoie formes harmoniques sur formes harmoniques, et la dernière assertion résulte du théorème de Hodge (à coefficients tordus). Pour le cas général, on peut s'appuyer sur [1], ou utiliser 1.1. (cf. [8]).

2.3. Pour simplifier les notations, nous supposons ici $p = 1$ (et renvoyons à [8], § 7, pour le cas général). Soient $D \colon B^1 \to B^2$ et $D' \colon B^1 \to B^0$ les opérateurs définis par

$$(D\,\theta)_{ij} = X_i \cdot \theta_j - X_j \cdot \theta_i, \quad D'\theta = -\Sigma_k X_k \cdot \theta_k \quad (\theta \in B^1(\Gamma, \varrho)) \,.$$

On vérifie sans grand mal que D et D' sont adjoints l'un de l'autre et que

$$(5) \qquad (\Delta\theta)_i = \Sigma_k (\varrho\,(X_k)^2 - X_k^2)\,\theta_i + \Sigma_k (\varrho\,([X_i, X_k]) - [X_i, X_k]) \cdot \theta_k \,,$$

$$(6) \qquad \langle \Delta\theta, \theta \rangle = \langle D\,\theta, D\,\theta \rangle + \langle D'\theta, D'\theta \rangle + \Sigma_{i,k} \langle \varrho\,(X_k)^2 \theta_i, \theta_i \rangle +$$
$$+ \Sigma_{i,j,k} \langle \varrho\,([X_i, X_k])\,\theta_j, \theta_i \rangle \,.$$

Soit $V_\varrho = \mathrm{Hom}\,(\mathfrak{m}, F_\varrho)$. On définit sur V_ϱ une forme quadratique $Q\,(u, v)$ en posant

$$(7) \qquad Q\,(u, v) = \Sigma_{i,k}(\varrho\,(X_k)^2 u\,(X_i) + \varrho\,([X_i, X_k])\,u\,(X_k), v\,(X_i)) \,,$$

qui, si l'on y remplace u et v par les valeurs de θ en un point de G, n'est autre que l'intégrant du dernier terme de droite de (6). Si maintenant θ est harmonique, le premier membre est nul, d'où la

2.4. Proposition. *Si la forme quadratique de (7) est positive non-dégénérée, alors* $H^1(\Gamma, F_\varrho) = 0$.

2.5. Généralisations. [8] considère en fait les groupes $H^p(\Gamma, X, \varrho)$ lorsque Γ est un sous-groupe discret uniforme d'une extension connexe G^* de G par un groupe compact P, et ϱ une représentation réelle ou complexe de G^*. Dans [8], on considère aussi le cas où X est un domaine borné symétrique et où l'on fait agir G sur $A^p(X, F_\varrho)$ par l'intermédiaire d'un *facteur d'automorphie* (morphisme de $G^* \times X$ dans $GL(F_\varrho)$ tel que $j\,(g \cdot g'; x) = j\,(g', x) \cdot j\,(g, g' \cdot x))$ qui est holomorphe sur X pour tout $g \in G^*$). Les espaces $H^p(\Gamma, X, j)$, $H^p(M, \mathscr{F}_\varrho)$ sont alors bigradués. Dans certains cas, $H^{N,0}(\Gamma, X, j)$ s'identifie à un espace de formes automorphes ([8], p. 413), ce qui généralise des résultats de Eichler et Shimura. Voir aussi [9] pour d'autres applications.

3. Trivialité des déformations de Γ et de M

3.1. Théorème. *Supposons que G ne contienne pas de sous-groupe distingué de dimension trois. Alors* $H^1(\Gamma, \mathfrak{g}_{\mathrm{Ad}}) = 0$.

Soit θ une forme harmonique, à valeurs dans $\mathfrak{g}$ vu comme espace de la représentation adjointe. Elle est somme de deux formes, à valeurs dans $\mathfrak{k}$ et $\mathfrak{m}$ respectivement, qui, vu (1) et (5), sont aussi harmoniques. Il suffit donc de faire voir que $\theta = 0$ lorsque θ est à valeurs soit dans $\mathfrak{k}$, soit dans $\mathfrak{m}$. Dans le premier cas, on déduit de (1), (3) que $B^*([X_i, \theta_j], X_k) = 0$ quels que soient i, j, k, ce qui montre que θ est à valeurs dans le centralisateur de $\mathfrak{m}$ dans $\mathfrak{k}$, mais ce dernier est nul puisque G est effectif. Si θ est à valeurs dans $\mathfrak{m}$, écrivons $\theta_i = \Sigma f_i^j X_j$. On voit alors que, en un point $g \in G$,

$$(8) \qquad Q\,(\theta, \theta) = (\tfrac{1}{2})\,\Sigma_{ij}(f_i^j)^2 + \Sigma_{ijhk} R_{ihkj} f_j^i f_k^h,$$

où R_{lhkj} est défini par $[[X_j, X_k], X_h] = \Sigma_i R_{ihkj} X_i$, et est donc une composante du tenseur de courbure de X. On montre assez facilement que cette forme est positive, et est non-dégénérée si $\mathfrak{g}$ n'a pas d'idéal de dimension trois (cf. [12]), et 3.1 résulte alors de 2.4.

3.2. Supposons Γ sans torsion. Alors, vu (2), $H^1(\Gamma, \mathfrak{g}_{Ad})$ s'identifie au premier espace de cohomologie de M, à coefficients dans le faisceau des germes de champs de vecteurs de Killing. D'après la théorie des déformations (cf. [2], et un article à paraître de Griffiths), la nullité de $H^1(\Gamma, \mathfrak{g}_{Ad})$ signifie alors que «M n'a pas de modules», en tant que variété localement symétrique, c'est-à-dire que toute famille différentiable de formes de Clifford-Klein compactes de X est triviale (en fait, on n'a considéré ici que des quotients $\Gamma \backslash X$ avec $\Gamma \subset G$, mais le cas où Γ est dans le groupe total des isométries de X s'y ramène aisément).

3.3. Soient P un groupe de Lie connexe, H un groupe de type fini, (h_i) $(1 \leq i \leq m)$ un système générateur de H, et R l'ensemble des homomorphismes de H dans P. L'ensemble R s'identifie à un sous-espace fermé du produit P^m de m copies de P (voir ci-dessous), d'où une topologie sur R. Soit $r \in R$. On dira que toute déformation de $r(H)$ dans P est triviale si les homomorphismes Int $p \circ r$: $H \to P$ ($p \in P$) forment un voisinage de p dans R. On a alors

3.4. Théorème (Weil [12]). *Soient G^* un groupe de Lie connexe semi-simple de centre fini, sans composante compacte ou de dimension trois, et Γ^* un sous-groupe discret uniforme de G^*. Alors toute déformation de Γ^* dans G^* est triviale.*

Le groupe adjoint Ad G^* de G^* s'identifie au groupe G de plus haut, pour un choix convenable de X, et l'image Γ de Γ^* dans Ad G^* est un sous-groupe discret, uniforme. D'autre part, par la suite spectrale des extensions de groupes, on voit tout de suite que

$$H^1(\Gamma^*, \mathfrak{g}_{Ad}) \cong H^1(\Gamma, \mathfrak{g}_{Ad}) \ .$$

Comme Γ^* est de type fini [car si C est un ouvert relativement compact de G^* tel que $G^* = C \cdot \Gamma^*$, alors Γ^* est engendré par $\Gamma^* \cap C \cdot C^{-1}$], 3.4 résulte de 3.1 et de la proposition suivante:

3.5. Proposition (Weil [13]). *Reprenons les notations de 3.3 et soit $\mathfrak{p}$ l'algèbre de Lie de P. Si $H^1(r(H), \mathfrak{p}_{Ad}) = 0$, alors toute déformation de $r(H)$ dans P est triviale.*

Esquisse de démonstration: Soient H' le groupe libre sur $h_1, \ldots, h_m$ et $w_i (i \in I)$ une famille de mots en $2m$ indéterminées $X_1, \ldots, X_{2m}$ tels que les conjugués, dans H', des éléments de H' obtenus en remplaçant, dans les w_i, X_i par h_i et X_{m+i} par $h_i^{-1} (1 \leq i \leq m)$ engendrent le noyau de l'homomorphisme naturel $u: H' \to H$. Tout mot w_i définit de façon évidente un morphisme de G^m dans G, qui sera noté W_i. Un homomorphisme $r': H' \to G$ est caractérisé par le point $x = (r'(h_i))$ de G^m, qui est arbitraire. Il se factorise par u si et seulement si x est dans l'intersection V des sous-ensembles $W_i^{-1}(e)$ ($i \in I$), donc R s'identifie à un fermé de G^m.

Soit $r: H \to P$ un homomorphisme; soient Z, Z' les espaces de 1-cocycles, à valeurs dans $\mathfrak{p}$, de H, H' opérant par Ad $\circ r$ et Ad $\circ r \circ u$, et B l'espace des 1-cobords de H dans $\mathfrak{p}$. En vertu de l'identité des cocycles:

$$z(x \cdot x') = z(x) + \varrho(x)(z(x')) \ ,$$

on voit qu'un cocycle $z' \in Z'$ est complètement déterminé par les éléments $z'(h_i)$, qui sont arbitraires, et que z' représente (i.e. provient par inflation d') un cocycle de H si et seulement si $z'(w_i) = 0$ ($i \in I$). Si l'on explicite convenablement ces conditions, on voit que Z s'identifie à l'intersection des noyaux des applications tangentes en $(r(h_i))$ aux morphismes W_i. Par ailleurs $z \in Z$ est un cobord si et seulement s'il existe $p \in \mathfrak{p}$ tel que $z(h) = p - \mathrm{Ad}\, r(h)(p)$, et il est nécessaire et suffisant pour cela que cette condition soit vérifiée pour $h = h_1, \ldots, h_m$. On en déduit facilement que B est l'image de l'espace $\mathfrak{p}$ par l'application tangente en e à $p \mapsto (\mathrm{Int}\, p\,(r(h_1)), \ldots, \mathrm{Int}\, p\,(r(h_m)))$. Le fait que $Z = B$ implique la trivialité des déformations de $r(H)$ dans P résulte alors du lemme élémentaire suivant [13]:

3.6. Lemme. *Soient U, V, $M_i (i \in I)$ des variétés différentiables, $\varphi \colon U \to V$, $\psi_i \colon V \to M_i$ des morphismes, tels que $\psi_i \circ \varphi$ soit constant; soient*

$$m_i = \psi_i \circ \varphi(U) \quad \text{et} \quad R = \bigcap_i \psi_i^{-1}(m_i)\,.$$

Soient $a \in U$, $b = \varphi(a)$, B l'image de l'espace tangent à U en a par $d\varphi$, et A l'intersection des noyaux des différentielles en b des morphismes ψ_i. Supposons que $A = B$. Alors, si V_0 est un voisinage ouvert suffisamment petit de b, $R \cap V_0$ est une sous-variété, contenue dans $\varphi(U)$.

3.7. Remarque. 3.2. a été démontré tout d'abord pour les espaces à courbure constante négative par Calabi (non publié), et 3.4 pour $\mathbf{SL}(n, \mathbf{R})$ par Selberg [10]. Le résultat principal de [12], établi par une généralisation convenable de la méthode de Calabi, est essentiellement équivalent à 3.1, qui est démontré explicitement dans [8], par la même méthode. 3.4 a été d'abord obtenu en utilisant [11] et [12]. Depuis, Weil a remarqué que l'on pouvait remplacer [11] par 3.5 (cf. [13]). Ajoutons encore que [11] et [12] démontrent 3.4 dans un cas un peu plus général: G peut avoir des facteurs non-compacts de dimension 3, pourvu que les projections de Γ^* sur ces facteurs ne soient pas discrètes.

4. Nombres de Betti de M

4.1. On prend ici pour ϱ la représentation triviale de dimension 1. Alors

$$b_p = \dim H^p(\Gamma, \mathbf{R}) = \dim H^p(M, \mathbf{R})$$

est le p-ième nombre de Betti de M. Le problème ici est de savoir si, lorsque G ne contient pas de facteur de dimension trois, on a $b_1(M) = 0$ (contrairement à ce qui se passe lorsque $G = \mathbf{PSL}(2, \mathbf{R})$ et X est le plan de Poincaré). Cela équivaut à: le groupe des commutateurs $[\Gamma \colon \Gamma]$ de Γ est d'indice fini dans Γ; en effet, si Γ est sans torsion, alors $\Gamma = \pi_1(M)$, donc $H_1(M) = \Gamma/[\Gamma, \Gamma]$, et on passe de là au cas général en utilisant 1.1.

Ce problème a été étudié d'abord dans [6]. A toute algèbre de Lie simple non compacte $\mathfrak{q}$, Matsushima associe une certaine constante $A(\mathfrak{q})$, (par exemple, $A(\mathfrak{q}) = (\dim \mathfrak{q} - \dim \mathfrak{s})/4 \cdot \dim \mathfrak{s}$, où $\mathfrak{s}$ est une sous-algèbre compacte maximale, si $\mathfrak{s}$ est simple), et considère la forme quadratique $H(\mathfrak{q})$ sur $\mathrm{Hom}\,(\mathfrak{t}, \mathfrak{t})$ (où $\mathfrak{q} = \mathfrak{s} + \mathfrak{t}$ est une décomposition de Cartan de q) obtenue à partir de la forme $Q(u, v)$ de (7) en remplaçant le facteur $1/2$ par $A(q)$. On a alors le

4.2. Théorème (Matsushima [6]). *Supposons que la forme $H(q)$ soit positive non-dégénérée pour chaque composante simple q de g. Alors $b_1(M) = 0$.*

Soit θ une forme harmonique de $B^1(\Gamma, 1)$, (notations du § 2). Pour montrer que $\theta = 0$, il suffit de voir qu'elle est invariante par G, donc que $X_\lambda \theta_i = 0$ quels que soient λ, i. En effet, une forme invariante $\neq 0$ donnerait lieu à un élément non nul de $H^1(g)$, qui est nul comme on sait. En fait, vu que $[\mathfrak{m}, \mathfrak{m}] = \mathfrak{k}$, il suffit de voir que $X_i \theta_j = 0$ $(i, j \leq N)$. Pour cela, on considère l'intégrale J sur $\Gamma \backslash G$ de la somme $\Phi(\theta)$ des fonctions $(X_i X_j \theta_k - X_j X_i \theta_k)^2$. Envisageons en chaque point de $\Gamma \backslash G$ les $f_j^i = X_i \theta_j$ comme les composantes d'un élément $u \in \mathrm{Hom}(\mathfrak{m}, \mathfrak{m})$. En transformant Φ, on montre que $J \geq 0$ entraîne $0 \leq \int_{\Gamma \backslash G} H_\mathfrak{g}(u, u)\, dv$, où $H_\mathfrak{g}$ est la somme des $H(q)$ correspondant aux idéaux simples de g, d'où le théorème.

4.3. Il reste à savoir quand $H(\mathfrak{g})$, g simple, est > 0. Cela a lieu dans les cas suivants: X est un domaine borné, non isomorphe à une boule unité [6], g provient par restriction des scalaires d'une algèbre complexe $\neq \mathfrak{s}\,\mathfrak{l}(2, \mathbf{C})$ [4], $\mathbf{g_C}$ simple, G non de type G_2 ou de type: groupe unitaire d'une forme d'indice un (sur $\mathbf{R}$, $\mathbf{C}$, $\mathbf{H}$) [5].

4.4. Soit $I_p(X)$ l'espace des p-formes sur X, invariantes par G. On sait que

$$I_p(X) = H^p(\mathfrak{g}, \mathfrak{k}) = H^p(G^u/K),$$

où G^u est un sous-groupe compact maximal de la complexification de G, et que ses éléments sont des formes harmoniques. De plus, $I_p(X)$ s'identifie à l'ensemble des éléments de $B^p(\Gamma, 1)$ invariants par G. On a donc une inclusion canonique $I_p(X) \subset H^p(\Gamma, \mathbf{R})$, et, comme $I_1(X) = 0$, le problème de 4.1 est un cas particulier du suivant étudié dans [7]: quand a-t-on $I_p(X) = H^p(M)$? On a un théorème analogue à 4.2, où $H(q)$ est remplacé par la forme $H(q)_p$ obtenue en substituant $A(q)/p$ à $A(q)$ dans $H(q)$. On trouvera dans [5] et [7] des tables de valeurs de q et p pour lequel $H(q)_p > 0$.

Bibliographie

1. Baily, W. L.: The decomposition theorem for V-manifolds, Amer. J. of Math., **78**, 1956, p. 862−888
2. Calabi, E.: On compact Riemannian manifolds with constant curvature, I., Differential geometry, p. 155−180. − Providence, American mathematical Society, 1961 (Proc. Symp. in pure Math. **3**)
3. Grothendieck, A.: Sur quelques points d'algèbre homologique, Tôhoku math. J., **9**, 1957, p. 119−121
4. Kanetuki, S., Nagano, T.: On the first Betti numbers of compact quotient spaces of complex semi-simple Lie groups by discrete subgroups, Scient. Papers Coll gen. Educ. Univ. Tokyo, **12**, 1962, p. 1−11
5. Kanetuki, S., Nagano, T.: On certain quadratic forms related to symetric Riemannian spaces, Osaka math. J., **14**, 1962, p. 241−252

6. Matsushima, Y.: On the first Betti number of compact quotient spaces of higher-dimensional symmetric spaces, Annals of Math., (2) **75,** 1962, p. 312−330
7. Matsushima, Y.: On Betti numbers of compact, locally symmetric Riemannian manifolds, Osaka math. J., **14,** 1962, p. 1−20
8. Matsushima, Y., Murakami, S.: On vector bundle valued harmonic forms and automorphic forms on symmetric spaces, Annals of Math., (2) **78,** 1963, p. 365−416
9. Matsushima, Y., Shimura, G.: On the cohomology groups attached to certain vector-valued differential forms on the product of the upper-half planes, *ibidem* p. 417−449.
10. Selberg, A.: On discontinuous groups in higher-dimensional symmetric spaces. Contributions to function theory, International Colloquium on function theory [1960, Bombay], p. 147−164. − Bombay, Tata Institute of fundamental Research, 1960
11. Weil, A.: On discrete subgroups of Lie groups, Annals of Math. (2), **72,** 1960, p. 369−384
12. Weil, A.: On discrete subgroups of Lie groups, II., Annals of Math. (2) **75,** 1962, p. 578−602
13. Weil, A.: Remarks on the cohomology of groups, [Annals of Math. (2) **80** (1964), 149−157]

1 Ce travail est consacré principalement à l'étude de propriétés d'un groupe algébrique linéaire connexe réductif G qui sont relatives à un corps de définition k de G. Il s'agit notamment d'introduire, à l'aide de k-sous-groupes (i.e. de sous-groupes fermés définis sur k) des notions qui généralisent naturellement celles de tore maximal, racines, groupe de Weyl, etc. de la théorie classique, de démontrer l'existence d'une décomposition de Bruhat pour le groupe G_k des points de G rationnels sur k, et des théorèmes de conjugaison « sur k », i.e. par des éléments de G_k. A cet effet, on utilisera essentiellement deux familles de sous-groupes de G : les tores déployés ou décomposés sur k (i.e. isomorphes sur k à un produit de groupes multiplicatifs $\mathbf{G}_m$) maximaux et les k-sous-groupes P paraboliques (i.e. tels que G/P soit une variété projective) minimaux. Ils jouent respectivement le rôle des tores maximaux et des sous-groupes résolubles connexes maximaux sur un corps algébriquement clos.

Ces résultats font l'objet des §§ 4 à 7 de I. La deuxième partie apporte divers compléments ou applications dans lesquels G n'est pas toujours réductif ou bien k est soumis à certaines restrictions.

Les §§ 0 à 2 sont de nature préliminaire. Le § 0 fixe des conventions et notations; le § 1 rassemble quelques résultats sur les tores; le § 2 rappelle les propriétés fondamentales d'un groupe réductif G, dont nous aurons besoin : structure de G lorsque G est déployé sur k (ce qui équivaut à l'existence d'un tore maximal déployé sur k), densité de G_k si k est infini, existence d'un tore maximal défini sur k et d'une extension séparable de k sur laquelle G est déployé (2.14). Le § 3 est consacré à l'étude de certains k-sous-groupes H normalisés par un tore maximal de G défini sur k. On établit en particulier l'existence et la conjugaison sur k de décompositions de Levi (3.13, 3.14), puis, dans certains cas, l'existence d'une suite de composition à quotients successifs vectoriels sur lesquels un tore donné opère par homothéties, lorsque H est unipotent (3.17, 3.18), et de k-sections locales de la fibration de G par H (3.24).

Dans le § 4, ces résultats sont appliqués à l'intersection de deux k-sous-groupes paraboliques P, P'. On montre en outre notamment que : la fibration de G par P possède des k-sections locales; si P et P' sont conjugués sur une extension K de k, ils

———

(¹) A certaines périodes durant l'élaboration de ce travail, nous avons bénéficié de l'aide financière de la N.S.F. : le premier auteur à l'Université de Chicago et à Paris (contract GP-2403), le second auteur à l'Institute for Advanced Study (contract GP-779) et à l'Université de Chicago (contract GP-574).

sont conjugués sur k ; si P et P′ sont minimaux, ils sont conjugués sur k (4.13). On établit également la conjugaison sur k des tores déployés sur k maximaux (4.21) et on relie ces sous-groupes aux k-sous-groupes paraboliques minimaux (4.16). Les démonstrations font jouer un rôle important à la notion de sous-groupes paraboliques P, P′ opposés $(P \cap R_u(P') = P' \cap R_u(P) = \{e\}$, cf. 4.8, 4.10).

Soit S un tore déployé sur k maximal. Par définition, le groupe de Weyl $_kW(G)$ de G relatif à k est le quotient $\mathcal{N}(S)/\mathcal{Z}(S)$ du normalisateur de S par le centralisateur de S, et les k-racines de G par rapport à S sont les caractères non triviaux de S dans la représentation adjointe de G. On montre (5.3) que $_kW(G)$, vu comme groupe d'automorphismes de $X^*(S) \otimes_{\mathbf{Z}} \mathbf{R}$, (où $X^*(S)$ est le groupe des caractères rationnels de S), muni d'un produit scalaire convenable (,), est le groupe engendré par les symétries par rapport aux hyperplans orthogonaux aux k-racines, et (5.8) que l'ensemble $_k\Phi(G)$ de ces dernières est un « système de racines » (i.e., outre l'invariance par le groupe de Weyl, vérifie la condition $2(a, b)(b, b)^{-1} \in \mathbf{Z}$ $(a, b \in {}_k\Phi(G))$). De plus, on a $\mathcal{N}(S) = \mathcal{N}(S)_k \cdot \mathcal{Z}(S)$ et les doubles classes de G_k suivant P_k (P un k-sous-groupe parabolique minimal) correspondent biunivoquement aux éléments de $_kW(G)$, « décomposition de Bruhat », (5.15).

Le § 6 met en relations les racines et groupes de Weyl relatifs à k et à une extension K de k, les sous-groupes paraboliques sur K et sur k, et les décompositions de Bruhat de G_K et G_k. On montre aussi que $G_K = R_u(P)_k \cdot U_{\bar{K}}^- \cdot P_K$ où P est un k-sous-groupe parabolique minimal et U^- le radical unipotent d'un K-sous-groupe opposé à P (6.25).

L'ensemble Φ_0 des éléments de $_k\Phi(G)$ dont le double n'est pas une racine est aussi un système de racines. Dans le § 7, on construit un k-sous-groupe déployé de G dont S et Φ_0 sont respectivement un tore maximal et le système de racines.

Dans le § 10 on montre (10.2) que dans un groupe algébrique quelconque G, la classe de conjugaison d'un élément semi-simple est fermée (et réciproquement si G est semi-simple) et que le centralisateur d'un k-tore, k étant un corps de définition pour G, est aussi défini sur k (10.5). Dans le § 11, on retrouve des résultats de Grothendieck et Rosenlicht sur l'existence et la conjugaison sur k de k-sous-groupes de Cartan de G lorsque G est résoluble.

Dans les paragraphes restants, on fait des hypothèses restrictives sur le corps de base k. Le § 8 le suppose parfait et prouve, par un argument de points fixes, la conjugaison sur k des sous-groupes connexes trigonalisables sur k (i.e. isomorphes sur k à un groupe de matrices triangulaires) maximaux (8.2).

Si k est localement compact de caractéristique zéro, G_k est muni naturellement d'une structure de groupe topologique. Les §§ 9, 13, 14 sont consacrés à diverses questions mettant en jeu cette topologie; en particulier : condition nécessaire et suffisante pour que G_k/H_k ou $(G/H)_k$ soit compact (9.3) ou pour que G_k soit engendré par un voisinage compact de l'élément neutre (13.4), structure du groupe des composantes connexes de G_k lorsque $k = \mathbf{R}$ et G est irréductible (14.5). Dans le § 14, on fait également le lien entre le point de vue développé ici et la théorie transcendante (décomposition d'Iwasawa, groupe de Weyl et racines d'une « paire symétrique »).

Enfin, le § 12 suppose G semi-simple connexe, k de caractéristique zéro, et discute des questions de rationalité des représentations linéaires ou projectives irréductibles de G et des propriétés de leurs poids restreints à S. On caractérise, notamment, les poids dominants des représentations projectives définies sur k (12.6) et des représentations linéaires fortement rationnelles sur k, c'est-à-dire qui sont définies sur k, et dont l'espace de représentation contient une droite définie sur k stable par un sous-groupe parabolique de G (12.10).

Les résultats de ce travail ont été en partie annoncés antérieurement [2, 34, 35], souvent en supposant k parfait, restriction qui a pu être levée grâce aux résultats de Grothendieck [11] rappelés en 2.14. Remarquons à ce propos qu'un certain nombre de complications techniques ou de détours sont dus à la nécessité de montrer que certains sous-groupes k-fermés sont définis sur k, ou que certaines classes de sous-groupes contiennent un représentant défini sur k, résultats qui sont évidents ou bien connus si l'on suppose le corps de base parfait. Sous cette dernière hypothèse, une partie des résultats des §§ 4, 5, 8 est aussi démontrée par Satake [28].

§ o. NOTATIONS ET RAPPELS

o.1. Les corps sont toujours commutatifs. Si k est un corps, k^*, k_s et $\bar{k}$ désignent respectivement le groupe multiplicatif des éléments non nuls de k, une clôture séparable de k et une clôture algébrique de k; si k' est une extension normale séparable de k, $\mathrm{Gal}(k'/k)$ désigne le groupe de Galois de k' sur k.

o.2. Soient G un groupe, H une partie de G. On notera souvent ^gH le transformé de H par l'automorphisme intérieur $\mathrm{Int}\, g : x \mapsto g.x.g^{-1}$ $(g, x \in G)$. L'image d'un sous-groupe L de G dans le groupe Int G des automorphismes intérieurs de G sera notée $\mathrm{Int}_G L$.

$\mathcal{N}_G(H)$ ou $\mathcal{N}(H)$ est le normalisateur de H dans G, et $\mathcal{Z}_G(H)$ ou $\mathcal{Z}(H)$ le centralisateur de H dans G. Le i-ième groupe dérivé (resp. le i-ième terme de la série centrale descendante) de G est noté $\mathcal{D}^i G$ (resp. $\mathcal{C}^i G$), et l'on pose $\mathcal{D}^\infty G = \bigcap_i \mathcal{D}^i G$, $\mathcal{C}^\infty G = \bigcap_i \mathcal{C}^i G$. On a donc

$$\mathcal{D}^i G = (\mathcal{D}^{i-1}G, \mathcal{D}^{i-1}G), \qquad \mathcal{C}^i G = (G, \mathcal{C}^{i-1}G) \qquad (i \geq 1)$$

en convenant que $G = \mathcal{D}^0 G = \mathcal{C}^0 G$, où (A, B) désigne le groupe engendré par les commutateurs $(a, b) = a.b.a^{-1}.b^{-1}$ $(a \in A, b \in B)$, A et B étant deux parties de G. L'élément neutre de G sera noté e ou 1. Soient A_i $(1 \leq i \leq m)$ des sous-ensembles de G. L'application $(a_i) \mapsto a_1 \ldots a_m$ de $A_1 \times \ldots \times A_m$ dans G sera appelée *l'application produit*.

o.3. Par groupe algébrique sur k, ou défini sur k, ou k-groupe algébrique [(1)], on entend dans ce travail un groupe linéaire algébrique défini sur k, au sens de [1], ou, si l'on tient à éviter le recours au corps universel, un sous-groupe algébrique sur k

---

[(1)] Sur ce point, nous nous écartons donc de la terminologie de [1].

du groupe $\mathbf{GL}_n$ des matrices inversibles d'ordre n, au sens de [3]. Nous laissons au lecteur le soin de choisir le cadre dans lequel il veut se placer, ou d'envisager les k-groupes algébriques comme des schémas en groupes affines absolument réduits sur k [11] s'il désire être plus intrinsèque.

Soit G un k-groupe algébrique. G^0 désigne la composante connexe de e dans G. Elle est aussi définie sur k. Si H est un sous-ensemble fermé de G, défini sur k, alors $\mathscr{N}_G(H)$ et $\mathscr{Z}_G(H)$ sont algébriques, k-fermés (i.e. définis sur une extension purement inséparable de k). On notera $\mathscr{N}_G(H)^0$ (resp. $\mathscr{Z}_G(H)^0$) le normalisateur (resp. centralisateur) connexe de H dans G.

Pour toute extension K de k, G_K désigne l'ensemble des points de G qui sont rationnels sur K (et de même pour une variété algébrique définie sur k). En général, on identifiera G et $G_{\bar{k}}$. Deux parties M, M' de G sont *conjuguées* (resp. *conjuguées sur k*) s'il existe $g \in G$ (resp. $g \in G_k$) tel que $M' = {}^g M$.

On écrira souvent $\mathbf{G}_m$ pour $\mathbf{GL}_1$, et $\mathbf{G}_a$ désignera le groupe additif à une dimension. $GL(\mathfrak{v})$ est le groupe des automorphismes d'un espace vectoriel $\mathfrak{v}$.

L'algèbre de Lie d'un groupe algébrique G, H, L, ... sera notée par la lettre gothique correspondante $\mathfrak{g}$, $\mathfrak{h}$, $\mathfrak{l}$, ...

0.4. Soient G, G' des k-groupes algébriques. Un *morphisme $f : G \to G'$ de groupes algébriques* est un morphisme des variétés sous-jacentes qui est un homomorphisme de groupes. On omettra souvent « de groupes algébriques » lorsque cela ne prêtera pas à confusion, et on dira *k-morphisme* si f est défini sur k.

On dira que G *opère morphiquement* sur la k-variété algébrique V s'il opère sur V par l'intermédiaire d'un morphisme $\varphi : G \times V \to V$, et qu'il opère k-morphiquement si φ est défini sur k. Si V est un k-groupe, il sera entendu que G opère comme groupe d'automorphismes.

On note Ad g l'image de $g \in G$ par la représentation adjointe $G \to GL(\mathfrak{g})$. C'est donc la différentielle en e de Int g.

0.5. Soit G un groupe algébrique. Un *caractère*, ou caractère rationnel, de G est un morphisme de G dans $\mathbf{G}_m$. L'ensemble des caractères, muni de la loi de composition définie par le produit des valeurs, est un groupe commutatif de type fini [23, § 1], libre si G est connexe, noté $X^*(G)$, et dont l'ensemble des éléments définis sur k est noté $X^*(G)_k$. Pour les caractères, on utilisera fréquemment la notation exponentielle, en désignant par g^a la valeur en g d'un caractère a; la loi de composition dans $X^*(G)$ sera alors notée additivement.

Un *groupe multiplicatif à un paramètre de* G est un morphisme $\mathbf{G}_m \to G$. Si G est commutatif, l'ensemble des groupes multiplicatifs à un paramètre, muni de la loi de composition définie par le produit des valeurs, est un groupe commutatif libre de type fini, noté $X_*(G)$, dont l'ensemble des éléments définis sur k est noté $X_*(G)_k$.

Un morphisme $f : G \to G'$ de groupes algébriques induit canoniquement des morphismes $f^* : X^*(G') \to X^*(G)$ et $f_* : X_*(G) \to X_*(G')$.

o.6. Un groupe algébrique G est le *produit direct* des sous-groupes invariants $G_1, \ldots, G_m$ fermés si l'application produit $G_1 \times \ldots \times G_m \to G$ est un isomorphisme de groupes algébriques; G est *produit presque direct* des G_i si cette application est une *isogénie* (i.e. surjective et de noyau fini).

G est le *produit semi-direct* du sous-groupe fermé H et du sous-groupe fermé distingué N si l'application produit induit un isomorphisme de variétés algébriques de $H \times N$ sur G. Il faut et il suffit pour cela que : (i) G soit le produit semi-direct de H et N en tant que groupe abstrait; (ii) les algèbres de Lie de H et N soient linéairement indépendantes. Rappelons que, en caractéristique non nulle, (ii) ne résulte pas de (i), même si G est commutatif.

o.7. Un k-groupe algébrique G est *semi-simple* (resp. *réductif*) si le radical $R(G^0)$ (resp. radical unipotent $R_u(G^0)$) de G^0 est réduit à l'élément neutre. Rappelons que le radical $R(H)$ (resp. radical unipotent $R_u(H)$) d'un k-groupe algébrique connexe H est le plus grand sous-groupe distingué fermé résoluble (resp. unipotent) connexe de H; il est k-fermé.

Un k-groupe algébrique connexe est *simple* (resp. *presque simple*) si tout sous-groupe invariant fermé propre est réduit à $\{e\}$ (resp. fini), k-*simple* (resp. *presque k-simple*) si cette condition est vérifiée par les k-sous-groupes invariants fermés. Un groupe $\neq \{e\}$ est semi-simple si et seulement s'il est produit presque direct de groupes presque simples non commutatifs [10, exp. 17, prop. 1, 2]. En particulier, si G est semi-simple, il est égal à son groupe des commutateurs.

o.8. Soit G un k-groupe algébrique. On appellera *sous-groupe de Levi* de G tout sous-groupe fermé réductif H tel que G soit le produit semi-direct de H et de son radical unipotent $U = R_u(G)$, et *décomposition de Levi* une telle présentation de G [1]. Si k est de caractéristique zéro, G possède toujours un k-sous-groupe de Levi H, et tout k-sous-groupe réductif de G est conjugué à un sous-groupe de H par un élément de U_k (voir [20] et aussi [5, prop. 5.1]). Par contre, si k est de caractéristique non nulle, G ne possède pas nécessairement de sous-groupe de Levi, d'après Chevalley (non publié), deux sous-groupes de Levi ne sont pas nécessairement conjugués (3.15), et si L est un sous-groupe réductif tel que G soit ensemblistement le produit semi-direct de L et U, les algèbres de Lie de L et U ne sont pas nécessairement transverses (3.15).

o.9. Suivant l'usage, **Z**, **N**, **N**$^+$, **Q**, **R**, **C** désignent respectivement les entiers, les entiers ≥ 0, les entiers > 0, les nombres rationnels, les nombres réels et les nombres complexes.

Dans toute la suite de ce travail, G désigne un groupe algébrique, k un corps de définition de G, et p la caractéristique de k.

---

[1] Ceci pour abréger, et bien que cette terminologie ne soit pas en complet accord avec celle qui est usuelle en théorie des algèbres de Lie, où une sous-algèbre de Levi est une sous-algèbre semi-simple, supplémentaire du radical.

I. GROUPES RÉDUCTIFS

§ 1. TORES

1.1. Un groupe algébrique T est un *tore* s'il est connexe et vérifie les conditions équivalentes suivantes : (i) T est formé d'éléments semi-simples; (ii) T est diagonalisable; (iii) T est isomorphe au produit de n copies de $\mathbf{G}_m$ ($n = \dim T$).

Un tore est *anisotrope sur le corps* K, ou K-*anisotrope*, s'il est défini sur K et si $X^*(T)_K = 1$. Il est *décomposé* ou *déployé sur* K s'il est défini sur K et K-isomorphe à un produit de groupes $\mathbf{G}_m$. Un K-tore est toujours déployé sur $\overline{\mathrm{K}}$.

1.2. Soit T un tore. Les groupes $X^*(T)$ et $X_*(T)$ sont alors commutatifs libres, de rang égal à $\dim T$. Si $a \in X_*(T)$, $b \in X^*(T)$, alors $b \circ a \in X^*(\mathbf{G}_m)$, donc $b \circ a$ est de la forme $x \mapsto x^q$ ($x \in \mathbf{G}_m$; $q \in \mathbf{Z}$). En associant q au couple (a, b), on définit sur $X_*(T) \times X^*(T)$ une forme bilinéaire à valeurs entières, notée $<, >$, qui met ces deux groupes en dualité [10, Exp. 9, § 5]. Si f est un morphisme de T dans un tore T', alors

$$<f_* a, b> = <a, f^* b> \qquad (a \in X_*(T), b \in X^*(T')).$$

Si f est surjectif, alors f^* est injectif. Comme un sous-tore d'un tore est toujours facteur direct [1, 7.4, 7.5], une suite exacte de morphismes

$$1 \to T'' \xrightarrow{f} T \xrightarrow{g} T' \to 1$$

donne lieu à des suites *exactes*

$$1 \to X_*(T'') \xrightarrow{f_*} X_*(T) \xrightarrow{g_*} X_*(T') \to 1$$
$$1 \to X^*(T') \xrightarrow{g^*} X^*(T) \xrightarrow{f^*} X^*(T'') \to 1.$$

Si ψ est une partie de $X^*(T)$, alors T_ψ dénotera la composante neutre de l'intersection des noyaux des caractères $a \in \psi$.

1.3. *Proposition.* — *Soit* T *un k-tore. Alors les conditions suivantes sont équivalentes :*
a) *L'image de* T *par tout k-morphisme* $T \to \mathbf{GL}_m$ ($m = 1, 2, \ldots$) *est diagonalisable sur k.*
b) T *est diagonalisable sur k.*
c) T *est déployé sur k.*
d) *On a* $X^*(T) = X^*(T)_k$.
e) *On a* $X_*(T) = X_*(T)_k$.

Il est clair que $a) \Rightarrow b)$. Pour $b) \Rightarrow c)$, voir [1, Prop. 7.4]. Les caractères de $\mathbf{G}_m$ étant de la forme $x \mapsto x^q$ ($q \in \mathbf{Z}$), il est clair que $c) \Rightarrow d), e)$. Si f est un morphisme de T dans $\mathbf{GL}_m$, les valeurs propres de $f(t)$ ($t \in T$) définissent des caractères de T, donc $d) \Rightarrow a)$ résulte de la Prop. 5 de [23]. Enfin, si $e)$ est vraie, T est l'image de $(\mathbf{G}_m)^n$ ($n = \dim T$) par un k-morphisme, donc $e) \Rightarrow a)$ en vertu de l'équivalence, déjà établie, des quatre premières conditions, appliquée à $(\mathbf{G}_m)^n$.

1.4. La proposition 1.3 implique que l'image d'un tore déployé sur k par un k-morphisme est un tore déployé sur k, et qu'un k-tore T possède un unique plus grand sous-tore déployé sur k, le groupe engendré par les images des k-sous-groupes à un paramètre, qui sera noté T_d. L'intersection des noyaux des éléments de $X^*(T)_k$ sera désignée par T_a. On verra ci-dessous que T_a est le plus grand sous-tore k-anisotrope de T et que ce dernier est produit presque direct de T_d et T_a.

1.5. *Proposition.* — *Soit* T *un k-tore. Alors il existe une extension galoisienne de degré fini de k sur laquelle* T *se décompose.*

Cette proposition est connue [21, Prop. 1.2.1]. Pour la commodité du lecteur, nous en indiquons une démonstration, communiquée par J. Tate.

Comme $X^*(T)$ est de type fini, il suffit, vu 1.3, de faire voir que tout élément $a \in X^*(T)$ est défini sur k_s. Or T est diagonalisable sur $\bar{k}$, donc a est en tout cas défini sur $\bar{k}$. Remplaçant k par k_s, on voit qu'il suffit de prouver que si a est défini sur une extension purement inséparable de k, alors il est défini sur k.

Il n'y a rien à démontrer si $p = 0$. Sinon, soit $q = p^s$ ($s \in \mathbf{Z}$, $s > 0$) une puissance de p assez grande pour que a soit défini sur $k^{1/q}$. On a $a(t^q) = a^q(t) \in k(t)$, donc

$$a(t^q) \subset k(t) \cap k^{1/q}(t^q) \qquad\qquad (t \in T).$$

Mais si t est générique sur k, le corps $k(t)$ est linéairement disjoint de $\bar{k}$, donc $k(t) \cap k^{1/q}(t^q) = k(t^q)$, et $a(t^q) \subset k(t^q)$. L'élément t^q est aussi générique sur k, puisque $x \mapsto x^q$ est un morphisme bijectif de T sur lui-même; l'inclusion $a(t^q) \subset k(t^q)$ montre alors que a est défini sur k.

1.6. *Corollaire.* — *Tout sous-groupe fermé de* T *est défini sur k_s, et est défini sur k si* T *est déployé sur k. Tout morphisme $f : T \to T'$ de* T *dans un k-tore est un k_s-morphisme, et un k-morphisme si* T *et* T′ *sont déployés sur k.*

La première assertion résulte du fait que T est diagonalisable sur k_s (resp. sur k) et que tout sous-groupe fermé d'un tore diagonal est défini sur le corps premier [1, Prop. 7.4]. La deuxième est conséquence de la première, appliquée au graphe de f dans $T \times T'$.

1.7. Le groupe de Galois $\Gamma = \mathrm{Gal}(k_s/k)$ opère de façon naturelle sur $X^*(T)$ et $X_*(T)$. On écrira souvent $^\gamma u$ pour le transformé par $\gamma \in \Gamma$ d'un élément u de $X^*(T)$ ou $X_*(T)$. On a en particulier

$$^\gamma u(t) = \gamma(u(\gamma^{-1}(t))), \qquad\qquad (u \in X^*(T),\ t \in T_{k_s}),$$
$$<{}^\gamma u,\ v> = <u,\ {}^\gamma v>, \qquad\qquad (u \in X_*(T),\ v \in X^*(T)).$$

$X^*(T)_k$ (resp. $X_*(T)_k$) est l'ensemble des points fixes de Γ dans $X^*(T)$ (resp. $X_*(T)$). C'est donc un facteur direct, ce qui montre que le groupe T_a introduit en 1.4 est bien connexe. Une partie de $X^*(T)$ sera dite *définie sur k* si elle est invariante par Γ.

Les opérations de Γ sur $X^*(T)$ et $X_*(T)$ sont continues (ce qui, ici, équivaut au fait que les orbites de Γ sont finies). Comme $X^*(T)$ et $X_*(T)$ sont de type fini, il existe dans Γ un sous-groupe ouvert distingué Γ' d'indice fini agissant trivialement sur ces groupes. Alors $K = (k_s)^{\Gamma'}$ est une extension galoisienne finie de k sur laquelle T se décompose. L'action considérée est donc en fait celle d'un groupe fini $\Gamma/\Gamma' \cong \mathrm{Gal}(K/k)$.

1.8. *Proposition.* — *Soient* T *un k-tore et* S *un k-sous-tore. Alors* $X^*(T)_k$ *et* $X_*(T)_k$ *sont de même rang. Il existe un k-sous-tore* S' *de* T *tel que* T *soit produit presque direct de* S *et* S'. *Si* $S = T_d$, *alors* S' *est unique, égal à* T_a, *et* T_a *est k-anisotrope. Si f est un k-morphisme de* T *dans un k-tore* T', *alors* $f(T_d) \subset T'_d$, $f(T_a) \subset T'_a$.

Pour la démonstration, voir [4, §§ 8, 13].

1.9. *Corollaire.* — a) *Soient* T, T' *deux k-tores isogènes sur k. Alors* T *est déployé sur k si et seulement si* T' *l'est.*

b) *Soit* $1 \to T' \to T \to T'' \to 1$ *une suite exacte de k-tores et k-morphismes. Alors* T *est déployé sur k si et seulement si* T' *et* T'' *le sont.*

a) L'hypothèse signifie qu'il existe un k-tore T'' et des k-isogénies de T'' sur T et T'. Il suffit donc de démontrer *a)* lorsqu'il existe une k-isogénie $f : T \to T'$. Dans ce cas, f^* applique injectivement $X^*(T')$ (resp. $X^*(T')_k$) sur un sous-groupe d'indice fini de $X^*(T)$ (resp. $X^*(T)_k$). Comme $X^*(T)_k$ et $X^*(T')_k$ sont facteurs directs dans $X^*(T)$ et $X^*(T')$ d'après 1.7, il est alors clair que les égalités $X^*(T)_k = X^*(T)$ et $X^*(T') = X^*(T')_k$ sont équivalentes, et *a)* résulte de 1.3.

b) Si T est déployé sur k, les sous-tores et tores quotients de T le sont aussi (1.4, 1.6), donc T' et T'' sont déployés sur k. Réciproquement, si T'' est déployé sur k, alors $T_a \subset T'$ vu 1.8; si de plus T' est déployé sur k, alors $T_a = \{e\}$ et $T = T_d$ d'après 1.8.

1.10. *Proposition.* — *Supposons k infini, et* G *connexe et soit* T *un k-tore opérant k-morphiquement sur* G. *Alors il existe un élément* $t \in T_k$ *dont l'ensemble des points fixes dans* G *est égal à celui de* T. *En particulier, si* T *est contenu dans* G *et opère sur lui par automorphismes intérieurs,* $\mathscr{Z}_G(t) = \mathscr{Z}_G(T)$.

Quitte à remplacer G par son produit semi-direct avec T (relatif à l'action donnée de T sur G) on peut supposer que $T \subset G$, et que T opère sur G par automorphismes intérieurs. Alors, l'ensemble V des $t \in T$ tels que $\mathscr{Z}_G(t) \neq \mathscr{Z}_G(T)$ est contenu dans la réunion des noyaux d'un nombre fini de caractères non triviaux de T [1, Prop. 7.10]; c'est donc un sous-ensemble algébrique propre, et il suffit de savoir que T_k est Zariski-dense dans T, ce qui résulte du fait que T est unirationnel sur k [23, Prop. 10].

§ 2. GROUPES DÉPLOYÉS

2.1. *Systèmes de racines.* — Soient M un module libre de type fini, N un sous-module facteur direct, Φ un sous-ensemble fini de M et P le sous-module engendré par Φ. On dira que Φ est un système de racines dans (M, N) si Φ est un système de racines au sens usuel [6, § 7] dans $P \otimes \mathbf{Q}$, si $M \otimes \mathbf{Q}$ est somme directe de $P \otimes \mathbf{Q}$ et $N \otimes \mathbf{Q}$, et si le groupe de Weyl $W(\Phi)$ de Φ est la restriction d'un groupe d'automorphismes de $M \otimes \mathbf{Q}$

laissant M stable et opérant trivialement sur N[1]. Munissons $M \otimes \mathbf{R}$ d'un produit scalaire $(,)$ positif non dégénéré et *admissible*, c'est-à-dire invariant par $W(\Phi)$. Si Φ est un système de racines dans (M, N), alors : $o \notin \Phi$, $\Phi = -\Phi$, Φ est invariant par $W(\Phi)$, les nombres $2(a, b).(b, b)^{-1} = n_{ab}$ sont entiers $(a, b \in \Phi)$, et $W(\Phi)$ est engendré par les symétries r_a par rapport aux hyperplans orthogonaux aux racines, hyperplans qui contiennent $N \otimes \mathbf{R}$. En particulier, $N \otimes \mathbf{R}$ et $P \otimes \mathbf{R}$ sont nécessairement orthogonaux.

Une partie de Φ est *connexe* si elle n'est pas réunion disjointe de deux parties non vides orthogonales l'une à l'autre.

Rappelons que si $a = \lambda.b$ $(a, b \in \Phi, \lambda \in \mathbf{R})$ alors $\lambda = \pm 1, \pm(1/2), \pm 2$ [6, § 7, n° 3, prop. 8]. Le système Φ est dit *réduit* si aucune racine n'est le double d'une autre racine. Une racine a est *indivisible* (resp. *non multipliable*) si $(1/2)a \notin \Phi$ (resp. $2a \notin \Phi$).

Les *chambres de Weyl* (composantes connexes du complémentaire de la réunion des hyperplans orthogonaux aux racines) sont ici les produits par $N \otimes \mathbf{R}$ des chambres de Weyl dans $P \otimes \mathbf{R}$. Elles sont permutées de manière simplement transitive par $W(\Phi)$. Chacune détermine un ensemble Φ^+ de racines positives. On note Φ^- (resp. Δ) l'ensemble des racines négatives (resp. simples). Rappelons qu'une racine est simple si elle est positive et non somme de deux racines positives, et que toute racine est de façon unique combinaison linéaire de racines simples à coefficients entiers tous de même signe. On a $r_a(b) = b - n_{ba}.a$ $(a, b \in \Phi)$. Le groupe $W(\Phi)$ est engendré par les réflexions r_a $(a \in \Delta)$, appelées *réflexions fondamentales*.

Le *diagramme* de Φ est la donnée pour tout $a \in \Delta$ du plus grand entier n_a tel que $n_a.a \in \Phi$ et pour toute paire a, b de racines simples des entiers n_{ab}, n_{ba}. Il existe un et un seul élément $w \in W(\Phi)$ qui applique Φ^+ sur Φ^-, donc aussi un et un seul automorphisme i de Φ vérifiant $w(i(a)) = -a$ $(a \in \Phi)$. L'automorphisme i est involutif, laisse Φ^+ et Δ stables, et est appelé *l'involution d'opposition*. Deux parties de Φ^+ transformées l'une de l'autre par i sont dites *opposées*.

La proposition suivante est bien connue. Ne pouvant fournir de référence, nous en donnons une démonstration pour la commodité du lecteur.

2.2. *Proposition.* — *Supposons* G *connexe. Alors les conditions suivantes sont équivalentes :*

(i) G *est réductif (i.e.* R(G) *est un tore).*

(ii) G *est produit presque direct d'un tore et de son groupe dérivé* $\mathscr{D}$G, *qui est semi-simple.*

(iii) G *possède une représentation linéaire de noyau fini complètement réductible.*

(ii) $\Rightarrow$ (iii). Le groupe G est produit presque direct de groupes presque simples $G_1, \ldots, G_n$. Il suffit donc de prendre une somme directe de représentations irréductibles non triviales des quotients $G/(G_1 \ldots G_{i-1}.G_{i+1} \ldots G_n)$.

(iii) $\Rightarrow$ (i). Soit $\rho : G \to GL(V)$ une représentation complètement réductible et soit $U = R_u(G)$. L'espace V_0 des points fixes par U est $\neq o$ d'après le théorème de Lie-

---

(1) Cette variante de la définition de [6] est adoptée ici pour pouvoir parler commodément du système de racines d'un groupe réductif non nécessairement semi-simple. Bien entendu, les propriétés générales des systèmes de racines démontrées dans [6] restent valables, et nous nous y référerons sans autre commentaire.

Kolchin, et est évidemment stable par G. Il existe donc un supplémentaire V_1 de V_0 invariant par G, dans lequel la représentation de G est encore complètement réductible. En raisonnant par récurrence sur dim V, on voit que U opère trivialement sur V, d'où l'implication annoncée.

(i) $\Rightarrow$ (ii). Soit $S = R(G)$. C'est un tore invariant dans G, donc central. Le groupe G/S est semi-simple, donc égal à son groupe des commutateurs, ce qui entraîne que $G = S . \mathscr{D}G$. Il reste à faire voir que $S \cap \mathscr{D}G$ est fini, ce qui résulte du lemme suivant :

Lemme. — *Soient* H *un groupe algébrique connexe, et* S *un tore central de* H. *Alors* $S \cap \mathscr{D}H$ *est fini.*

Identifions H à un groupe d'automorphismes d'un espace vectoriel V. Ce dernier est somme directe de sous-espaces V_i stables par G et sur lesquels S agit par homothéties. Le lemme résulte alors du fait que tout morphisme $H \to \mathbf{GL}_m$ applique $\mathscr{D}H$ dans $\mathbf{SL}_m$.

2.3. *Groupes réductifs.* — Nous rappelons ici quelques propriétés fondamentales des groupes réductifs. La référence générale pour ces résultats est [10].

Soient G réductif connexe, T un tore maximal de G, et Z le centre connexe de G. Le groupe G est donc produit presque direct de Z et de $\mathscr{D}G$; par conséquent, $\mathscr{D}G$ est le plus grand sous-groupe semi-simple de G et contient tout sous-groupe unipotent de G. L'ensemble des *racines* de G par rapport à T, i.e. des caractères non triviaux de T dans la représentation adjointe de G, est noté $\Phi(T, G)$ ou simplement $\Phi(G)$, ou Φ, si cela ne prête pas à confusion. On note U_b le groupe radiciel à un paramètre associé à $b \in \Phi$. Il est caractérisé par l'existence d'un isomorphisme $\theta_b : \mathbf{G}_a \to U_b$ tel que :

$$t . \theta_b(x) . t^{-1} = \theta_b(t^b . x) \qquad\qquad (t \in T,\ x \in U_b).$$

Soient T_b la composante neutre de ker b, $Z_b = \mathscr{Z}_G(T_b)$ et $Z_b' = \mathscr{D}Z_b$. On sait que Z_b est connexe, réductif, de centre connexe T_b, que Z_b' est presque simple, de dimension trois, engendré par U_b et U_{-b}, et que $H_b = Z_b' \cap T$ est un tore maximal de Z_b'. On notera aussi H_b le groupe multiplicatif à un paramètre de T dont l'image est $Z_b' \cap T$, et tel que $<H_b, b> = 2$.

Soit $W = W(G) = \mathscr{N}(T)/\mathscr{Z}(T)$. C'est un groupe fini, qui opère fidèlement sur $X^*(T)$ et $X_*(T)$. L'ensemble des H_b $(b \in \Phi)$ (resp. l'ensemble Φ) est un système de racines dans $(X_*(T), X_*(Z))$ (resp. $(X^*(T), N)$, où N est l'ensemble des caractères nuls sur $(T \cap \mathscr{D}G)^0$), de groupe de Weyl W. La réflexion r_b provient d'un élément de Z_b'. Elle applique U_b sur U_{-b}, H_b sur H_{-b} et opère trivialement sur T_b.

Fixons un ordre sur Φ; soient $U = U^+$ (resp. U^-) le sous-groupe engendré par les U_a $(a \in \Phi^+)$ (resp. $a \in \Phi^-$), et $B = B^+ = T . U$, $B^- = T . U^-$. Le groupe U est unipotent.

Pour tout ordre sur Φ^+, l'application produit de $\prod U_a$ $(a \in \Phi^+)$ dans U est un isomorphisme de variétés algébriques. Le sous-groupe B est fermé résoluble connexe maximal, ou encore, est un sous-groupe de Borel de G $(^1)$; il est égal à son normalisateur et au normalisateur de U, son groupe dérivé est U et tout sous-groupe de Borel lui est conjugué. De même pour B^- et U^-. Les sous-groupes de Borel de G contenant T correspondent biunivoquement aux ensembles d'éléments positifs pour les différents ordres possibles sur Φ, et sont permutés de manière simplement transitive par W(G). L'application produit induit un isomorphisme de variétés de $U^- \times B$ sur un ouvert de G. Le groupe G est engendré par $(\mathscr{Z}(G))^0$ et les groupes U_a, U_{-a} $(a \in \Delta)$. Tout sous-groupe de U normalisé par T est fermé, connexe, et produit semi-direct multiple des groupes U_a qu'il contient [10, Exp. 13, th. 1 d].

Remarque. — Soient H un groupe algébrique connexe et S un tore maximal de H. Alors il existe deux sous-groupes unipotents connexes fermés U, U′ normalisés par S tels que H soit engendré par S et U, U′, et que tout sous-groupe unipotent connexe de H fasse partie du sous-groupe engendré par U et U′. En effet, cela est vrai si H est réductif d'après ce qui précède, et on se ramène tout de suite à ce cas en considérant le quotient $H/R_u(H)$, qui est réductif, et en remarquant que l'image de S dans $H/R_u(H)$ en est un tore maximal.

Il s'ensuit en particulier que tout sous-groupe H connexe fermé de G normalisé par T est engendré par $(T \cap H)^0$ et les sous-groupes U_a qu'il contient, et est réductif si et seulement si $U_a \subset H$ entraîne $U_{-a} \subset H$ quel que soit $a \in \Phi$. Par exemple, le centralisateur d'une partie d'un tore de G est réductif.

2.4. *Décomposition de Bruhat.* — Pour tout élément $w \in W(G)$, choisissons un représentant n_w dans $\mathscr{N}(T)$. Soit d'autre part U_w' (resp. U_w'') le sous-groupe de U engendré par les U_a $(a \in \Phi^+, w^{-1}(a) \in \Phi^-)$ (resp. $a, w^{-1}(a) \in \Phi^+$). Alors l'application produit définit un isomorphisme de variétés de $U_w' \times U_w''$ sur U et de $U_w' \times n_w \times B$ sur $B.n_w.B = U.n_w.B = U_w'.n_w.B$. Le groupe G est réunion disjointe des doubles classes $B.n_w.B$ $(w \in W(G))$. Deux doubles classes $U.n.U$ et $U.n'.U$ $(n, n' \in \mathscr{N}(T))$ ne sont égales que si $n = n'$ $(^2)$.

Cela implique en particulier que *si B′ est un sous-groupe de Borel de* G, *alors* $B' \cap B$ *contient un tore maximal de* G. En effet, puisque $G = B.\mathscr{N}(T).B$ on peut trouver $x \in B$ et $n \in \mathscr{N}(T)$ tels que $^xB' = {}^nB$, d'où

$$^x(B' \cap B) = {}^xB' \cap B = {}^nB \cap B \supset T,$$

et $B' \cap B \supset x^{-1}.T.x$.

La classe latérale $n_w.T$ de T dans $\mathscr{N}(T)$ ne dépend évidemment que de w. Dans la suite on conviendra souvent de la désigner par $w.T$. Plus généralement, si A est une partie de $\mathscr{N}(T)$, l'ensemble A.T ne dépend que de l'image C de A dans W et

---

$(^1)$ L'un des auteurs insistant pour que l'on adopte cette terminologie, aujourd'hui généralement admise, l'autre auteur s'y résigne.
$(^2)$ En fait, cette dernière propriété ne semble pas figurer explicitement dans [10]. Elle se démontre à partir des précédentes exactement comme l'assertion correspondante dans 5.15, dont elle est du reste un cas particulier.

sera souvent noté C.T. De même, si F est une partie de G normalisée par T, l'ensemble ^{n_w}F sera aussi noté ^{w}F.

2.5. *Proposition.* — *On conserve les notations précédentes. Soient* $a, b \in \Phi(G)$, *non proportionnelles. Soit* $U_{(a,b)}$ *le sous-groupe de G engendré les groupes* U_c, *où c parcourt l'ensemble* (a, b) *des racines de la forme* $i.a + j.b$ $(i, j > 0)$. *Alors* $U_{(a,b)}$ *est unipotent, ne contient aucun autre groupe radiciel à un paramètre, et l'on a* $(U_a, U_b) \subseteq U_{(a,b)}$.

Cette proposition résulte de faits établis dans [9, 10], et nous nous bornons à en esquisser la démonstration. On peut tout d'abord fixer un ordre sur Φ tel que $a, b \in \Phi^+$ [9, lemme 1]. Supposons $a < b$. Il résulte de [10, Exp. 17, Cor. au Th. 1] que U_a, U_b normalisent $U_{(a,b)}$ et que si l'on range les U_c $(c \in (a, b))$ par ordre croissant, l'application produit de $U_a \times U_b \times \prod_{c \in (a,b)} U_c$ dans U est un isomorphisme de variétés de ce produit sur le groupe engendré par U_a, U_b et $U_{(a,b)}$. De plus, le lemme 1 de [10, Exp. 17] montre que si $x \in U_b$ et $y \in U_a$,

$$x.y \in U_a . U_b . U_{(a,b)},$$

et que les images des composantes de $x.y$ dans U_a et U_b par θ_a^{-1} et θ_b^{-1} sont de la forme $c_a . \theta_a^{-1}(x)$ et $c_b . \theta_b^{-1}(y)$ où c_a, c_b sont des constantes. Faisant $y = 1$ ou $x = 1$, on voit que $c_a = c_b = 1$, d'où

$$x.y \in y.x. U_{(a,b)},$$

ce qui termine la démonstration.

Remarque. — On trouvera dans [9, p. 27] une formule beaucoup plus précise, établie en caractéristique zéro, mais valable en général vu les théorèmes de classification de [10]. Si G est presque simple, cette formule montre que (U_a, U_b) peut être un sous-groupe propre de $U_{(a,b)}$ seulement dans les cas suivants :

$$p = 2, \ G = \mathbf{B}_n, \ \mathbf{C}_n, \ \mathbf{G}_2, \ \mathbf{F}_4; \qquad p = 3, \ G = \mathbf{G}_2.$$

Pour la commodité des références, nous insérons ici un lemme élémentaire connu.

2.6. *Lemme.* — *Supposons G connexe et soit B un k-sous-groupe fermé de G, tel que la fibration de G par B admette une section locale définie sur k. Si k est fini ou G_k Zariski dense dans G, alors la projection canonique* $G_k \to (G/B)_k$ *est surjective.*

(Rappelons qu'une section locale sur k de la fibration d'un k-groupe algébrique H par un k-sous-groupe B est un k-morphisme σ d'un ouvert V défini sur k de H/B dans H tel que $\pi \circ \sigma = 1$, où $\pi : H \to H/B$ désigne la projection canonique. L'application produit définit alors un k-isomorphisme de $\sigma(V) \times B$ sur un ouvert de H. En particulier B et V sont irréductibles si H l'est.)

Si G_k est Zariski dense dans G, les translatés du domaine de définition V d'une section locale sur k par les éléments de G_k forment un recouvrement de G/B, d'où la surjectivité de $G_k \to (G/B)_k$. Si k est fini, cette dernière résulte de [16], puisque B est connexe.

2.7. *Groupes résolubles déployés.* — Un k-groupe algébrique résoluble H est dit *déployé sur k* ou *k-déployé* ou *k-résoluble* s'il est connexe et possède une suite de composition $H = H_0 \supset H_1 \supset \ldots \supset H_l = \{e\}$ formée de k-sous-groupes connexes dont les quotients successifs sont k-isomorphes à $\mathbf{G}_a$ ou $\mathbf{G}_m$ (c'est le « k-solvable » de [26]). Un k-groupe résoluble connexe est toujours déployé sur $\bar{k}$. Pour un tore, cette notion coïncide bien avec celle introduite en 1.1, vu 1.9 b). Toute extension sur k d'un groupe k-résoluble par un groupe k-résoluble est k-résoluble. Toute image d'un groupe k-résoluble par un k-morphisme est k-résoluble [23, Prop. 6]. Rappelons encore deux propriétés fondamentales d'un groupe k-résoluble G :

(i) Si G opère k-morphiquement sur une variété complète V et si $V_k \neq \emptyset$, alors V_k contient un point fixe par G [23, p. 35].

(ii) Tout k-espace homogène de G possède un point rationnel sur k [22, p. 425]. En particulier, si G est un k-sous-groupe d'un k-groupe algébrique L, alors la projection $L_k \to (L/G)_k$ est surjective.

2.8. *Groupes réductifs déployés.* — Supposons G réductif, et soient T, θ_a, U_a comme en 2.3. Alors $\{T, \theta_a\}$ ou, par abus de langage, $\{T, U_a\}$ $(a \in \Phi)$ est une *donnée de déploiement sur k* si T est déployé sur k et si les θ_a sont définis sur k. Le groupe G est *déployé sur k* s'il possède une donnée de déploiement sur k; il est toujours déployé sur $\bar{k}$. D'après un résultat (non publié) de Cartier (qui ne sera pas utilisé ici), tout tore maximal déployé sur k de G fait partie d'une donnée de déploiement (voir aussi [11]). Il s'ensuit en particulier que G est déployé sur k si et seulement s'il possède un sous-groupe de Borel déployé sur k. On pourrait donc prendre cette dernière condition comme définition de la notion de groupe algébrique déployé sur k. Cependant, nous n'en aurons pas besoin en dehors des cas réductif et résoluble.

Dans les lemmes 2.9, 2.10, et 2.11, le groupe G est réductif, déployé sur k, et les notations de 2.3 se rapportent à une donnée de déploiement sur k.

2.9. *Lemme.* — *Soit V un sous-groupe de U normalisé par T. Soient $t \in T_k$ et $v \in V$ tels que $v.t.v^{-1} \in B_k$. Alors $v.t.v^{-1}$ est conjugué à t par un élément de V_k.*

Soient $a_1, \ldots, a_m$ les racines positives telles que $U_{a_i} \subset V$ rangées par ordre croissant, et V_i le sous-groupe engendré par les groupes U_{a_j} $(1 \leq i \leq j \leq m)$. Le théorème 1 d) de [10, Exp. 13] montre que l'application produit est un k-isomorphisme de variétés du produit des U_{a_j} $(j \geq i)$ sur V_i et, joint à 2.5, prouve que V_i est distingué dans V et est le produit semi-direct de U_{a_i} avec V_{i+1} $(i = 1, \ldots, m$; on pose $V_{m+1} = \{e\}$).

La démonstration procède par récurrence descendante sur le plus grand entier $r \leq m+1$ tel que $v \in V_r$, notre assertion étant évidente si $r = m+1$. Posons

$$v = v_r.v' = v''.v_r \qquad (v', v'' \in V_{r+1}; v_r \in U_{a_r}),$$

et soit $x = \theta_{a_r}^{-1}(v_r) \in \mathbf{G}_a$. Si $t^{a_r} = 1$, alors t centralise v_r, donc $v.t.v^{-1} = v''.t.v''^{-1}$ et l'on peut appliquer l'hypothèse de récurrence. Soit donc $t^{a_r} = c^{-1} \neq 1$. On a

$$t^{-1}.v.t.v^{-1} \in \theta_{a_r}((c-1).x).\mathrm{V}_{r+1},$$

d'où $(c-1).x \in k$, donc $x \in k$, et $v_r \in \mathrm{U}_k$. Par conséquent,

$$v'.t.v'^{-1} = v_r^{-1}.(v.t.v^{-1}).v_r \in \mathrm{G}_k.$$

En vertu de l'hypothèse d'induction t et $v'.t.v'^{-1}$ sont conjugués par un élément de V_k; il en est alors de même pour t et $v.t.v^{-1} = v_r.(v'.t.v'^{-1}).v_r^{-1}$.

Les deux lemmes suivants sont des cas particuliers de propositions qui seront établies ultérieurement (n°s 4.21, 5.15, 11.4).

2.10. *Lemme.* — *Soient* V *comme en* 2.9 *et* C = T.V. *Tout tore* S *défini sur* k *de* C (*resp. déployé sur* k *de* G) *est conjugué par un élément de* C_k (*resp.* G_k) *à un sous-tore de* T.

Le tore S admet un point fixe dans $(\mathrm{G}/\mathrm{B})_k$ d'après 2.7 (i). Vu 2.3, 2.6, S est donc conjugué sur k à un sous-tore de B, et il suffit de démontrer l'assertion relative à C. Supposons k infini. Il existe alors $s \in \mathrm{S}_k$ tel que $\mathscr{Z}_{\mathrm{C}}(s) = \mathscr{Z}_{\mathrm{C}}(\mathrm{S})$ (1.10). On a la décomposition $s = t.v$ ($t \in \mathrm{T}$, $v \in \mathrm{V}$) et, puisque $s \in \mathrm{C}_k$ et que C est produit semi-direct sur k de T et V, on a aussi $t, v \in \mathrm{C}_k$. Les éléments t et s sont conjugués dans $\mathrm{V}_{\bar{k}}$ [1, § 12] donc (2.9) aussi dans V_k, et l'on peut se ramener au cas où $s = t$; mais alors $\mathscr{Z}_{\mathrm{C}}(\mathrm{S}) = \mathscr{Z}_{\mathrm{C}}(s) \supset \mathrm{T}$, donc $\mathrm{S} \subset \mathrm{T}$. Soit maintenant k fini. Le centralisateur de S est défini sur k, donc [23, footnote p. 45] contient un tore maximal T' défini sur k. Comme $\mathrm{T} = \mathscr{Z}_{\mathrm{C}}(\mathrm{T}) = \mathscr{N}_{\mathrm{C}}(\mathrm{T})$, l'ensemble des éléments $v \in \mathrm{V}$ tels que $v.\mathrm{T}.v^{-1} = \mathrm{T}'$ est un espace homogène de T, évidemment défini sur k, et non vide [1, § 12]. Il contient alors un point rationnel sur k (2.7 (ii) ou [16]).

2.11. *Lemme.* — *Posons* $\mathrm{N} = \mathscr{N}_{\mathrm{G}}(\mathrm{T})$. *Alors* $\mathrm{N} = \mathrm{N}_k.\mathrm{T}$ *et* $\mathrm{G}_k = \mathrm{U}_k.\mathrm{N}_k.\mathrm{U}_k$; *deux doubles classes* $\mathrm{U}_k.n.\mathrm{U}_k$, $\mathrm{U}_k.n'.\mathrm{U}_k$ *correspondant à des éléments distincts* n, n' *de* N_k *sont distinctes.*

Ce lemme est connu lorsque k est algébriquement clos (2.4). Il suffit donc de démontrer les deux premières égalités.

Soit $n \in \mathrm{N}$. Le groupe $^n\mathrm{B}$ contient T, donc (2.3) est déployé sur k et (2.7 (i)) possède un point fixe dans $(\mathrm{G}/\mathrm{B})_k$. Vu 2.6, 2.10, il existe alors $g \in \mathrm{G}_k$ tel que $^{gn}\mathrm{B} = \mathrm{B}$ et $^g\mathrm{T} = \mathrm{T}$, d'où $g \in \mathrm{N}_k$ et $g.n \in \mathrm{B} \cap \mathrm{N} = \mathrm{T}$.

Soit $g \in \mathrm{G}$. D'après le lemme sur $\bar{k}$, il existe $n \in \mathrm{N}$ tel que $g \in \mathrm{U}.n.\mathrm{U}$. Soient w l'image de n dans W(G), n' un représentant de w dans N_k, et V (resp. V') le produit des groupes U_a où $a \in \Phi^+$, $w(a) > 0$, (resp. $w(a) < 0$), pris dans un ordre quelconque et soit $\mathrm{Y} = {}^n\mathrm{V}' \subset \mathrm{U}^-$. On a

$$g.n'^{-1} \in \mathrm{U}.\mathrm{T}.n'.\mathrm{U}.n'^{-1} = \mathrm{U}.\mathrm{T}.n'.\mathrm{V}.\mathrm{V}'.n'^{-1} = \mathrm{U}.\mathrm{T}.\mathrm{Y},$$

d'où, vu 2.3, et l'égalité $n.\mathrm{V}'.n^{-1} = n'.\mathrm{V}'.n'^{-1}$,

$$g.n'^{-1} \in (\mathrm{U}.\mathrm{T}.\mathrm{Y})_k = \mathrm{U}_k.\mathrm{T}_k.\mathrm{Y}_k,$$
$$g \in \mathrm{U}_k.\mathrm{T}_k.n'.\mathrm{U}_k \subset \mathrm{U}_k.\mathrm{N}_k.\mathrm{U}_k.$$

2.12. *Lemme.* — *Soient* G, G′ *deux groupes réductifs connexes isomorphes,* T, T′ *des tores maximaux respectifs de* G *et* G′, α *un isomorphisme de* T *sur* T′ *qui applique* $\Phi(\mathrm{G})$ *sur* $\Phi(\mathrm{G}')$. *Soient* ψ *un ensemble de racines de* G *linéairement indépendantes et* $\beta_b : \mathrm{U}_b \to \mathrm{U}_{b'}$ $(b \in \psi,\ b' = {}^t\alpha^{-1}(b))$ *un isomorphisme. Alors il existe un isomorphisme* γ *de* G *sur* G′ *dont les restrictions à* T *et à* U_b *sont respectivement* α *et* β_b $(b \in \psi)$.

Vu [10, Exp. 24, Cor. 1] on peut, sans nuire à la généralité, identifier G à G′ et supposer que α est l'identité. Soit $\mathrm{K} \supset k$ un corps algébriquement clos de définition pour G′ et les groupes U_b. Un K-isomorphisme de U_b sur $\mathbf{G}_a$ transforme β_b en un automorphisme de $\mathbf{G}_a$, c'est-à-dire en la multiplication par un élément $s_b \in \mathrm{K}^*$. Puisque les éléments de ψ sont indépendants, il existe $t \in \mathrm{T}_\mathrm{K}$ tel que $t^b = s_b$ $(b \in \psi)$. Il suffit alors de prendre $\gamma = \mathrm{Int}\ t$.

2.13. *Théorème.* — *Deux groupes* G, G′ *réductifs connexes déployés sur* k *et isomorphes sont isomorphes sur* k.

Soient $(\mathrm{T}, \mathrm{U}_a)$ et $(\mathrm{T}', \mathrm{U}'_{a'})$ des données de déploiement sur k de G et G′, et K une extension algébriquement close de k sur laquelle G et G′ sont isomorphes. Vu 2.3 et la conjugaison des tores maximaux, il existe un K-isomorphisme η de G sur G′ qui applique $(\mathrm{T}, \mathrm{U}_a)$ sur $(\mathrm{T}, \mathrm{U}'_{a'})$. Fixons dans $\Phi(\mathrm{G})$ et $\Phi(\mathrm{G}')$ des ordres qui se correspondent par ${}^t\eta_{|\mathrm{T}}$. Soient U et U′ (resp. U^- et U'^-) les sous-groupes unipotents de G et G′ engendrés par les groupes U_a et $\mathrm{U}'_{a'}$ correspondant aux racines positives (resp. négatives). Pour toute racine simple $a \in \Delta$, soient $\beta_a : \mathrm{U}_a \to \mathrm{U}'_{a'}$ un k-isomorphisme $(a' = {}^t\eta_{|\mathrm{T}}^{-1}(a))$ et u_a un élément différent de 1 de $\mathrm{U}_{a,k}$. D'après 2.11, il existe un unique élément $n_a \in \mathscr{N}(\mathrm{T})$ tel que $u_a \in \mathrm{U}^-.n_a.\mathrm{U}^-$, et cet élément fait partie de $\mathscr{N}(\mathrm{T})_k$. En appliquant ce même lemme et 2.4 au centralisateur Z_a de la composante neutre du noyau de a (cf. 2.3), on voit que l'image de n_a dans $\mathrm{W}(\mathrm{G})$ est la réflexion fondamentale r_a; par suite, le sous-groupe M engendré par les n_a est un sous-groupe de $\mathscr{N}(\mathrm{T})_k$ qui s'applique sur $\mathrm{W}(\mathrm{G})$ par la projection canonique.

D'après 2.12, il existe un K-isomorphisme $\gamma : \mathrm{G} \to \mathrm{G}'$ qui prolonge $\eta_{|\mathrm{T}}$ et les β_a $(a \in \Delta)$. On a $\gamma(u_a) = \beta_a(u_a) \in \mathrm{U}'^-.\gamma(n_a).\mathrm{U}'^-$. Comme $\beta_a(u_a) \in \mathrm{G}'_k$ par hypothèse, 2.11 implique que $\gamma(u_a)$ est aussi rationnel sur k, donc $\gamma(\mathrm{M}) = \mathrm{M}' \subset \mathrm{G}'_k$. Soit maintenant $c \in \Phi(\mathrm{G})$. Puisque M contient au moins un représentant de chaque élément de $\mathrm{W}(\mathrm{G})$, il existe $m \in \mathrm{M}$ et $a \in \Delta$ tels que $\mathrm{U}_c = m.\mathrm{U}_a.m^{-1}$. La restriction de γ à U_c est donc égale à $\mathrm{Int}\ \gamma(m).\beta_a.\mathrm{Int}\ m^{-1}$, donc est définie sur k puisque β_a l'est et que m, $\gamma(m)$ sont rationnels sur k. Comme la restriction de η à T est définie sur k (1.6), il s'ensuit que la restriction de γ à l'ouvert $\mathrm{U}^-.\mathrm{T}.\mathrm{U}$ est définie sur k, d'où le théorème.

Remarque. — Le théorème 2.13 est un cas très particulier d'un résultat de Demazure [11].

2.14. *Théorème* (Grothendieck [11]). — *Supposons* G *connexe. Alors :*

a) G *possède un tore maximal défini sur* k [11, Exp. XIV, Théor. 1.1].

b) *Si* G *est réductif,* G_k *se déploie sur une extension séparable de degré fini de* k.

c) *Si* G *est réductif et* k *infini, alors* G_k *est Zariski-dense dans* G [11, Exp. XIV, 6.5, 6.7].

Lorsque k est parfait, c), pour un groupe non nécessairement réductif, et a) sont dus à Rosenlicht [23]. Il en est de même pour a) si G est résoluble [26, Theorem 4] (*voir* aussi 11.4).

2.15. Nous indiquons ici quelques conséquences immédiates de 2.14.

a) *Si* G *est réductif, son centre est défini sur* k.

En effet, $\mathscr{Z}(G)$ est k-fermé, contenu dans tout tore maximal de G, donc dans un tore maximal T défini sur k vu 2.14 a), et est par conséquent aussi défini sur k_s (1.6).

b) *Supposons* G *semi-simple et soit* N *un* k-*sous-groupe distingué connexe de* G. *Alors* N *est produit presque direct des* k-*sous-groupes distingués de* G *presque simples sur* k *qu'il contient, et* G *est produit presque direct de* N *et d'un* k-*sous-groupe distingué connexe* N′.

En effet, d'après 2.14 b), les sous-groupes presque simples sur $\overline{k}$ de G sont définis sur k_s; ils sont donc permutés par le groupe de Galois Γ de k_s/k, et un sous-groupe distingué connexe de G est presque simple sur k si et seulement s'il est produit presque direct de facteurs presque simples sur $\overline{k}$ de G formant une orbite de Γ.

c) *Supposons* G *semi-simple. Il existe un* k-*groupe* G_1 *et une* k-*isogénie* $\nu : G_1 \rightarrow G$ *déterminés à un isomorphisme près par les conditions suivantes :* G_1 *est produit direct de ses sous-groupes distingués presque simples sur* $\overline{k}$ *et* ν *induit des* $\overline{k}$-*isomorphismes de ceux-ci sur les facteurs presque simples de* G.

Les facteurs presque simples de G étant définis sur k_s, vu 2.14 b), notre assertion est immédiate sur k_s. Que G_1 et ν soient alors définis sur k résulte immédiatement de critères de descente du corps de base [37].

d) *Supposons* G *réductif. Soient* S *un* k-*tore de* G *et* A *une partie de* S_k. *Alors* $\mathscr{Z}(A)^0$ *et* $\mathscr{Z}(S)$ *sont réductifs, définis sur* k *et* S *est contenu dans un tore maximal de* G *défini sur* k ([1]).

$\mathscr{Z}(A)^0$ et $\mathscr{Z}(S)$ sont k-fermés. Il suffit donc de faire la démonstration lorsque k est séparablement clos. Vu 2.14 et 2.10 on peut alors supposer que $S \subset T$, où T fait partie d'une donnée de déploiement de G sur k. Comme $\mathscr{Z}(S)$ est connexe [1, prop. 18.4], il résulte alors de 2.3 que $\mathscr{Z}(S)$ et $\mathscr{Z}(A)^0$ sont engendrés par T et les groupes U_a, où a parcourt les racines de G relatives à T égales à 1 sur S et A respectivement. Ils sont définis sur k, et aussi réductifs (2.3, remarque).

e) *Si* G *possède une décomposition de Levi sur* k (0.8), *alors son radical est défini sur* k.

Soit $G = H.R_u(G)$ une décomposition de Levi sur k de G, et soit Z le centre connexe de H. Il est défini sur k vu a), et est le radical de H, donc $R(G) = Z.R_u(H)$ est défini sur k.

([1]) Pour une démonstration plus directe d'un résultat plus général, *voir* 10.3, 10.5.

§ 3. RACINES ET SOUS-GROUPES NORMALISÉS
PAR UN TORE MAXIMAL

Dans tout ce paragraphe, G *désigne un groupe réductif connexe.*

3.0. Soient M un **Z**-module libre de type fini, Φ un ensemble fini d'éléments non nuls de M, et ψ, η des parties de Φ. On dit que ψ *normalise* η (dans Φ) si quels que soient $m, n \in \mathbf{N}^+$, $a_1, \ldots, a_m \in \psi$ et $b \in \eta$:

(∗) $a = a_1 + a_2 + \ldots + a_m + nb \in \Phi$ entraîne $a \in \eta$.

Si de plus $\psi \supset \eta$ (resp. $\psi = \eta$), on dit que η est un *idéal* de ψ (resp. est *clos* dans Φ). L'ensemble ψ est *symétrique* si $\psi = -\psi$; *convexe* s'il contient toute combinaison linéaire à coefficients rationnels positifs de ses éléments qui fait partie de Φ. Une partie convexe de Φ est évidemment close dans Φ.

Rappelons que si Φ est un système de racines (2.1), la condition (∗) équivaut à cette même condition où l'on fait $m = n = 1$. Cela résulte aisément du lemme 19 de [6, § 7, n° 6]. Supposons Φ symétrique. On posera $\psi_s = \psi \cap (-\psi)$, $\psi_u = \psi \cap \complement(-\psi)$. Ce sont des parties disjointes de ψ dont ψ est la réunion. Si $\Phi = -\Phi$ et si ψ est clos dans Φ, ψ_s et ψ_u sont clos et ψ_u est un idéal de ψ. C'est clair pour ψ_s. Si ψ_u n'était pas un idéal on pourrait trouver $n \in \mathbf{N}^+$, $a_1, \ldots, a_m \in \psi$ et $b \in \psi_u$ tels que $a = a_1 + \ldots + a_m + nb \in \psi_s$, mais alors $-b = -a + a_1 + \ldots + a_m + (n-1)b \in \psi$, d'où $b \in \psi_s$. Une intersection d'ensembles clos dans Φ est close dans Φ, ce qui permet de parler du plus petit ensemble clos dans Φ contenant η, que l'on appellera la *clôture* de η dans Φ.

3.1. *Proposition.* — *Soient* A *un sous-anneau de* **R** *et* $0 \to M' \to M \xrightarrow{\pi} M'' \to 0$ *une suite exacte de* A-*modules libres de type fini. Alors :*

(i) *Il existe sur* M *un et un seul ordre total compatible avec un ordre total donné sur* M'' *et induisant un ordre total donné sur* M'.

(ii) *Étant donné un ensemble fini* Φ *d'éléments de* M *dont toutes les combinaisons* A-*linéaires à coefficients* >0 *sont* $\neq 0$, *il existe un ordre total sur* M *par rapport auquel les éléments de* Φ *sont* >0.

(i) On pose $a > b$ si, soit $\pi(a) > \pi(b)$, soit $\pi(a) = \pi(b)$ et $a - b > 0$ pour l'ordre donné sur M'.

(ii) Quitte à remplacer A par son corps des fractions, on peut supposer que A est un corps. On procède par récurrence sur la dimension de M. On peut se borner au cas où aucun élément de Φ n'est combinaison linéaire à coefficients >0 d'autres éléments de Φ. Soient M' le sous-espace engendré par un élément $a \in \Phi$ et π la projection de M sur $M'' = M/M'$. Soit Φ'' l'ensemble des éléments non nuls de $\pi(\Phi)$. Vu les hypothèses sur Φ, il vérifie aussi la condition imposée à Φ dans (ii), d'où l'existence d'un ordre total sur M'' tel que Φ'' soit formé d'éléments positifs. On prend alors sur M l'ordre compatible avec l'ordre construit sur M'' et pour lequel $a > 0$.

3.2. *Racines par rapport à un tore.* — Soit S un tore opérant morphiquement sur G. Les *poids de S dans G* sont les caractères intervenant dans la représentation induite de S dans $\mathfrak{g}$. Si S est un sous-tore de G, opérant par automorphismes intérieurs, les *racines de G relativement à S* sont les poids non nuls de S dans G. Leur ensemble est noté $\Phi(S, G)$, ou simplement Φ. Si S′ est un sous-tore de S, alors $\Phi(S', G)$ est l'ensemble des éléments non nuls de l'image de $\Phi(S, G)$ par restriction. Si S est un sous-tore de G, et H un sous-groupe fermé de G normalisé par S, les *poids de S dans H* sont les poids associés à l'action de S par automorphismes intérieurs. Les poids non nuls de S dans H sont donc les racines de G, relatives à S, dont l'espace propre a une intersection non nulle avec $\mathfrak{h}$.

Si S *est un tore maximal, les poids non nuls de S dans* H *sont aussi les racines* $a \in \Phi(S, G)$ *telles que* $U_a \subset H$. En effet, si $U_a \subset H$, l'algèbre de Lie $\mathfrak{u}_a$ de U_a fait partie de l'algèbre de Lie $\mathfrak{h}$ de H, donc a est un poids non nul de S dans H. La réciproque résulte de 2.3.

Dans le lemme suivant, qui généralise la Proposition 1, p. 13-02 de [10], on se place dans la catégorie des groupes à opérateurs avec domaine d'opérateurs P fixé. En particulier, tous les sous-groupes considérés sont stables par P, et les homomorphismes commutent à P. Rappelons que l'on appelle sous-quotient d'un groupe, toute image homomorphe d'un sous-groupe de celui-ci.

3.3. *Lemme.* — *Soit* H *un groupe nilpotent (à opérateurs); soient* $H_1, \ldots, H_m$ *des sous-groupes de* H *sans sous-quotients isomorphes. Alors les conditions suivantes sont équivalentes :*

(i) *Pour tout sous-groupe* $A \subset H$, *l'application produit est une bijection de* $(A \cap H_1) \times \ldots \times (A \cap H_m)$ *sur* A.

(ii) H *possède une suite de composition à quotients commutatifs dont chaque terme est engendré par ses intersections avec les groupes* H_i.

(iii) H *possède une suite de composition* $H = L_0 \supset L_1 \supset \ldots \supset L_n = \{e\}$ *ayant la propriété :* a) *pour tout* $i \leq n$, *il existe* $j(i) \leq m$ *tel que* $L_{i-1} = L_i \cdot (L_{i-1} \cap H_{j(i)})$.

(iv) H *possède une suite distinguée* (L_i) *centrale (i.e.* H/L_{i-1} *est dans le centre de* H/L_i *pour* $i = 1, \ldots, n$*) ayant la propriété* a).

Avant de passer à la démonstration remarquons tout d'abord que :

b) Toute suite de composition (L_i) à quotients commutatifs, dont les termes sont engendrés par leurs intersections avec les H_i, possède un raffinement vérifiant a).

Pour obtenir un tel raffinement, il suffit en effet d'intercaler entre L_{i-1} et L_i les sous-groupes $L_i \cdot (L_{i-1} \cap H_1) \ldots (L_{i-1} \cap H_j)$ $(j = 1, 2, \ldots, m-1)$. (Ce sont bien des sous-groupes invariants dans L_{i-1} puisque L_{i-1}/L_i est commutatif, et ils vérifient a) puisque, vu l'hypothèse, L_{i-1}/L_i est le produit des images canoniques des groupes $L_{i-1} \cap H_j$).

(iv) $\Rightarrow$ (i). Posons $L_{n-1} = L$. Si $n = 1$, il n'y a rien à démontrer. Si $n = 2$, on peut évidemment se borner au cas où H_1 et H_2 sont non triviaux. Alors a) implique que $H = H_1 \times H_2$ et que H_1, H_2 sont commutatifs. Quitte à diviser H par $(H_1 \cap A) \times (H_2 \cap A)$, on peut supposer que $H_1 \cap A = H_2 \cap A = \{e\}$. Mais alors les projections de A sur H_1 et H_2 sont isomorphes, donc réduites à l'élément neutre.

Dans le cas général, on procède par induction sur n. En appliquant l'hypothèse de récurrence à l'image de A dans H/L par la projection canonique, on voit que

$$(1) \qquad A \subset (A \cap (H_1 . L)) . (A \cap (H_2 . L)) \dots \dots (A \cap (H_m . L)).$$

Mais, en vertu de $a)$, L est contenu dans l'un des groupes H_i, disons dans H_s, et l'on a $H_i . L = H_i \times L$ ou $H_i . L = H_i$ suivant que $i \neq s$ ou $i = s$, d'où encore, compte tenu de ce qui a déjà été démontré :

$$A \cap (H_i . L) \subset (A \cap H_i) . (A \cap L).$$

Comme $A \cap L \subset A \cap H_s$, l'inclusion (1) devient

$$(2) \qquad A \subset (A \cap H_1) . (A \cap H_2) \dots \dots (A \cap H_m).$$

Enfin, soit

$$a_1 . a_2 \dots \dots a_m = b_1 . b_2 \dots \dots b_m \qquad (a_i, b_i \in A \cap H_i ; \quad i = 1, \dots, m).$$

L'hypothèse de récurrence, appliquée à l'image de A dans H/L, donne $a_i \in b_i . L$, donc $a_i \in b_i . H_s$ pour tout i, d'où $a_i = b_i$ si $i \neq s$, et finalement aussi $a_s = b_s$.

(i) $\Rightarrow$ (ii) $\Rightarrow$ (iii). Cela résulte du fait que H possède au moins une suite de composition à quotients commutatifs, et de $b)$.

(iii) $\Rightarrow$ (iv). On procède par récurrence sur n, le cas $n = 1$ étant trivial. Vu $b)$ il suffit de prouver que les termes de la suite centrale ascendante sont engendrés par leurs intersections avec les groupes H_i. Par une récurrence évidente sur la longueur de cette suite, on voit qu'il suffit de montrer que le centre Z de H est engendré par les groupes $H_i \cap Z$.

D'après $a)$, il existe s tel que $H = L_1 . H_s$. Soit $B = Z . H_s$. L'hypothèse de récurrence, appliquée à L_1 et aux groupes $L_1 \cap H_i$, et l'implication (iv) $\Rightarrow$ (i) déjà établie, montrent que $B \cap L_1$ est engendré par ses intersections avec les H_i. Il en est donc de même pour $B = H_s . (B \cap L_1)$. Une suite centrale de B constituée par B, H_s et une suite centrale de H_s, est alors formée de groupes qui sont engendrés par leurs intersections avec les H_i, donc B vérifie (iv) vu $b)$. Il suit alors de l'implication (iv) $\Rightarrow$ (i) que tout sous-groupe de B, en particulier Z, est engendré par ses intersections avec les H_i.

Remarques. — 1) les conditions du lemme précédent sont aussi équivalentes à :

(v) *Toute suite de composition de* G *possède un raffinement vérifiant* a).

En effet (v) $\Rightarrow$ (iii) et, compte tenu de $b)$, (i) $\Rightarrow$ (v).

2) Le groupe des transformations affines $x \mapsto a . x + b$ de la droite (sans opérateurs), et les sous-groupes H_1 des homothéties et H_2 des translations donnent un exemple de groupe non nilpotent et de sous-groupes vérifiant (iii), mais non (i).

3.4. *Proposition*. — *Soient* T *un tore maximal de* G, H *un sous-groupe connexe fermé normalisé par* T, *et* ψ *l'ensemble des racines* $a \in \Phi(T, G)$ *telles que* $U_a \subset H$. *Soit* H* *le groupe engendré par les* U_a $(a \in \psi)$. *Alors :*

a) $H = H^* . (T \cap H)^0$.

b) *Si* η *est une partie close de* Φ *contenue dans* ψ *et si* H* *est engendré par les groupes* U_a *avec* $a \in \eta$, *alors* $\eta = \psi$.

a) Soient B un sous-groupe de Borel de H, S un tore maximal de B et U le plus grand sous-groupe unipotent de B. Le groupe B fait partie d'un sous-groupe de Borel B′ de T.H et, par conjugaison dans T.H, ce qui laisse H stable, on peut supposer que $S \subset T$ [1, Cor. 16.6]. On a d'autre part $B' = T.U$. Il s'ensuit en particulier que H et T.H ont les mêmes sous-groupes unipotents. D'après la remarque à 2.3, le sous-groupe engendré par les sous-groupes unipotents de T.H est déjà engendré par deux sous-groupes unipotents maximaux normalisés par T. Vu 2.3 ces derniers sont engendrés par les sous-groupes U_a qu'ils contiennent, donc le sous-groupe engendré par les sous-groupes unipotents de H n'est autre que H*. Cela entraîne que $H^* . (T \cap H)^0$ est un sous-groupe de H qui contient un tore maximal S de H et tous les sous-groupes unipotents de H; il est donc égal à H (2.3, rem.).

b) Soient $\eta_s = \eta \cap (-\eta)$, $\eta_u = \eta \cap \complement(-\eta)$. Ce sont des parties closes de Φ (3.0). On sait [6, Prop. 22] que l'on peut fixer un ordre sur Φ tel que $\eta_u \subset \Phi^+$. Soient $\eta^{\pm} = \eta \cap \Phi^{\pm}$ et $V^{\pm}$ le groupe engendré par les U_a $(a \in \eta^{\pm})$. L'ensemble $\eta^{\pm}$ est aussi clos dans Φ, et 2.5 montre que $V^{\pm}$ ne contient pas de groupe U_b avec $b \notin \eta^{\pm}$.

L'ensemble η_s, étant clos et symétrique, est aussi un système de racines. Posons $W' = W(\eta_s)$ et montrons que $M = V^+ . W . T . V^-$ est un groupe (on utilise la convention du n° 2.4). Vu 2.3, il est engendré par T et les groupes U_a, où a parcourt η^+ et les opposés des racines simples de η_s, pour l'ordre fixé. Il suffit donc de faire voir que M est invariant par translation à gauche par T et par les U_a sus-mentionnés. C'est évident pour T, et pour U_a si $a \in \eta^+$. Soit donc a une racine simple de η_s. La Proposition 2.5 montre que le sous-groupe Y de V^+ engendré par les U_b $(b \in \eta^+, b \neq a)$ est normalisé par U_a, U_{-a}, donc en particulier que $V^+ = U_a . Y = Y . U_a$, d'où

$$U_{-a} . M \subset Y . U_{-a} . U_a . W' . T . V^-.$$

La décomposition de Bruhat (2.4) de Z_a donne

$$(1) \qquad U_{-a} . U_a \subset T . U_a \cup U_a . r_a . T . U_a,$$

r_a étant la réflexion fondamentale associée à a, d'où

$$(2) \qquad U_{-a} . V^+ . w' . T . V^- \subset M \cup V^+ . U_{-a} . w . T . V^- \qquad (w' \in W' ; w = r_a . w').$$

Si $w^{-1}(a) > 0$, alors $w^{-1}(-a) \in \eta^-$, donc $U_{-a} . w . T . V^- \subset w . T . V^-$ et le dernier terme de (2) fait partie de M. Soit donc $b = w^{-1}(-a) > 0$. On a

$$(3) \qquad U_{-a} . w . T . V^- = w . T . U_b . V^-.$$

La formule (1), avec a remplacé par $-b$, donne

$$(4) \qquad \mathrm{T.U}_b.\mathrm{V}^- \subset \mathrm{T.V}^- \cup \mathrm{T.U}_{-b}.r_b.\mathrm{V}^-,$$

d'où, vu (3),

$$\mathrm{U}_{-a}.w.\mathrm{T.V}^- \subset w.\mathrm{T.V}^- \cup \mathrm{U}_a.w.r_b.\mathrm{T.V}^-,$$

et, vu (2), $\mathrm{U}_{-a}\mathrm{M} \subset \mathrm{M}$.

L'ensemble M est donc bien un groupe. La décomposition de Bruhat du quotient de M par son radical unipotent montre alors que ce quotient a η_s comme système de racines. Comme V^+ et V^- ne contiennent aucun U_b ($b \notin \eta$) il en est alors de même de M, donc $\eta = \psi$.

3.5. *Lemme.* — *Soient* S *un tore de* G, T *un tore maximal de* G *contenant* S, *et* H *un sous-groupe fermé de* G *contenant* $\mathscr{Z}(\mathrm{S})$. *Alors le sous-groupe* H^* *de* H *engendré par les* U_a ($a \in \Phi(\mathrm{T}, \mathrm{G})$) *contenus dans* H, *mais non dans* $\mathscr{Z}(\mathrm{S})$, *est normalisé par* $\mathscr{Z}(\mathrm{S})$.

Soient a, b deux éléments de $\Phi(\mathrm{T}, \mathrm{G})$ tels que $\mathrm{U}_a \in \mathscr{Z}(\mathrm{S})$, $\mathrm{U}_b \notin \mathscr{Z}(\mathrm{S})$, $\mathrm{U}_b \subset \mathrm{H}$. Le groupe $(\mathrm{U}_a, \mathrm{U}_b)$ est contenu dans le sous-groupe engendré par les U_c où c parcourt les racines de la forme $ma + nb$ ($m, n > 0$) (2.5), et engendré par ceux de ces sous-groupes qu'il contient (2.3). Ces derniers ne peuvent faire partie de $\mathscr{Z}(\mathrm{S})$ (puisque $n \neq 0$), mais sont dans H, donc dans H^*, ce qui prouve que $(\mathrm{U}_a, \mathrm{U}_b) \subset \mathrm{H}^*$. Par conséquent, le normalisateur de H^* contient tout sous-groupe U_a faisant partie de $\mathscr{Z}(\mathrm{S})$; comme il contient aussi T, le lemme est démontré.

3.6. *Proposition.* — *Soient* H *un k-sous-groupe réductif connexe de rang maximum de* G *et* U *un k-sous-groupe unipotent connexe de* G *normalisé par* H. *Soit* S *le plus grand sous-tore déployé sur k du centre de* H. *Alors il existe un ordre sur* $\mathrm{X}^*(\mathrm{S})$ *tel que les poids de* S *dans* U *soient* > 0.

Remarquons tout d'abord que le centre de H est défini sur k (2.15), ce qui légitime la définition de S. Soit T un tore maximal de H, donc de G, défini sur k, qui existe vu 2.14, et soit ψ l'ensemble des racines a de G par rapport à T telles que $\mathrm{U}_a \subset \mathrm{U}$. C'est aussi (3.2) l'ensemble des poids de T dans U. Vu 3.1, il suffit donc, pour établir la proposition, de faire voir que toute combinaison linéaire à coefficients entiers > 0 d'une famille non vide d'éléments de ψ a une restriction non nulle à S. Supposons le contraire, et soit c une telle combinaison linéaire qui s'annule sur S. Le groupe de Galois Γ de k_s/k opère sur $\mathrm{X}^*(\mathrm{T})$, en permutant les éléments de ψ, et sur $\mathrm{X}_*(\mathrm{T})$, en laissant $\mathrm{X}_*(\mathrm{S})$ fixe point par point; quitte à remplacer c par la somme de ses transformés par Γ, on peut donc supposer c invariant par Γ, donc défini sur k. Il est alors trivial sur la composante neutre Z du centre de H (1.8). D'autre part, le groupe de Weyl $\mathrm{W}(\mathrm{H})$ de H opère aussi sur ψ, donc la somme c' des transformés de c par $\mathrm{W}(\mathrm{H})$ est aussi une combinaison linéaire à coefficients entiers > 0 d'éléments de ψ, qui s'annule sur Z. Ce dernier fait montre que c' est combinaison linéaire, à coefficients rationnels, de racines de H par rapport à T. Comme elle est invariante par $\mathrm{W}(\mathrm{H})$, elle doit alors être nulle. Mais T.U fait partie d'un groupe de Borel de G, donc ψ est contenu dans l'ensemble des racines positives de G pour un ordre convenable, et $c' \neq 0$, d'où une contradiction.

3.7. *Corollaire. — Soient* U *un k-sous-groupe unipotent connexe de* G *et* S *un k-tore de* G *dont le centralisateur* $\mathscr{Z}$(S) *normalise* U. *Alors les poids de* S *dans* U *sont tous non nuls.*

Le groupe $H = \mathscr{Z}(S)$ est réductif, connexe, défini sur k (2.15 d)) et de rang maximum. Soient T et ψ comme dans la démonstration précédente. Les poids de S dans U sont les restrictions à S des éléments de ψ. Si $a \in \psi$ a une restriction à S nulle, c'est une racine de H par rapport à T, et il est forcément nul sur le plus grand tore central de H, en contradiction avec 3.6.

3.8. *Ensembles quasi-clos de racines et sous-groupes associés.* — (i) Soient S un tore de G et T un tore maximal de G contenant S. Une partie μ de $\Phi(T, G) = \Phi$ est *quasi-close* si le groupe engendré par les U_a $(a \in \mu)$ ne contient aucun autre groupe U_b $(b \in \Phi)$. D'après 3.4, un ensemble clos est aussi quasi-clos. (Remarquons en passant que la réciproque ne peut être en défaut que dans les cas mentionnés dans la remarque à 2.5, où (U_a, U_b) est strictement plus petit que $U_{(a, b)}$ $(a, b \in \Phi, a + b \neq 0)$.)

Une partie ψ de $\Phi(S, G)$ est quasi-close si l'ensemble des racines de G par rapport à T dont la restriction à S est contenue dans $\psi \cup \{o\}$ est quasi-close. Comme deux tores maximaux de G contenant S sont conjugués dans $\mathscr{Z}(S)$, il est clair que cette condition sera alors vérifiée pour tout tore maximal de $\mathscr{Z}(S)$. Toute intersection de parties quasi-closes est quasi-close. Il y a donc un plus petit ensemble quasi-clos contenant une partie σ de $\Phi(S, G)$, qui sera appelé la *quasi-clôture* de σ. Si $\psi \subset \Phi(S, G)$ est quasi-close, $-\psi$ l'est aussi ; pour le montrer, il suffit évidemment de considérer le cas où $S = T$, et cela résulte alors de l'existence d'un automorphisme de G conservant T et induisant sur T l'automorphisme $t \mapsto t^{-1}$ ([10, exp. 24, cor. 1]).

(ii) Soient ψ une partie quasi-close de $\Phi(S, G)$ et μ (resp. ν) l'ensemble des éléments de $\Phi(T, G)$ dont la restriction à S est contenue dans $\psi \cup \{o\}$ (resp. ψ). On note $G_\psi^{(S)}$ ou G_ψ le groupe engendré par T et les U_a $(a \in \mu)$, et $G_\psi^{*(S)}$ ou G_ψ^* le groupe engendré par les U_a $(a \in \nu)$. Par définition, G_ψ ne contient aucun autre groupe U_b $(b \notin \mu)$, donc est engendré par $\mathscr{Z}(S)$ et G_ψ^*. De plus G_ψ^* est normalisé par $\mathscr{Z}(S)$ d'après 3.5, donc $G_\psi = \mathscr{Z}(S) . G_\psi^*$. Cela montre aussi que G_ψ et G_ψ^* ne changent pas si l'on remplace dans la définition précédente T par un autre tore maximal de G contenant S.

Par définition, on a

$$(1) \qquad G_\mu^{(T)} = G_\psi^{(S)}, \qquad G_{\nu'}^{*(T)} = G_\psi^{*(S)}$$

où ν' est la quasi-clôture de ν (il se peut que $\nu' \neq \nu$).

Plus généralement soit S' un tore contenant S et soient σ (resp. τ) la quasi-clôture de l'ensemble des racines relatives à S' dont la restriction à S appartient à $\psi \cup \{o\}$ (resp. ψ). On a alors visiblement

$$(2) \qquad G_\sigma^{(S')} = G_\psi^{(S)}, \qquad G_\tau^{*(S')} = G_\psi^{*(S)}.$$

Les poids de S dans G_ψ (resp. G_ψ^*) sont les éléments de $\psi \cup \{o\}$ (resp. de ψ ou $\psi \cup \{o\}$). Remarquons encore que si $\psi = \Phi(S, G)$, alors $G_\psi = G$ et G_ψ^* est le produit des facteurs presque simples de G non centralisés par S.

(iii) Soient ψ, η des parties quasi-closes de $\Phi(S, G)$. On dit que η est un *quasi-idéal* de ψ si G_η^* est un sous-groupe distingué de G_ψ^*. On verra (3.9) qu'un idéal est un quasi-idéal.

(iv) Si G_ψ^* est unipotent, on le notera souvent U_ψ, et l'on dira que ψ est *unipotent*. Si ψ est quasi-clos et formé de racines >0 pour un ordre convenable sur $X^*(S)$, alors il est unipotent, et ν est quasi-clos unipotent. En effet, soit Φ^+ l'ensemble des racines relatives à T positives pour un ordre sur $X^*(T)$ compatible avec l'ordre donné sur $X^*(S)$ (3.1). On a $\nu \subset \Phi^+$, donc aussi $\nu' \subset \Phi^+$, et $G_{\nu'}^{*(T)}$ fait partie du groupe unipotent maximal U engendré par les U_a $(a \in \Phi^+)$. De plus, si $a, b \in \nu$, $m, n \in \mathbf{N}^+$ et $m.a + n.b \in \nu'$, alors la restriction à S de $m.a + n.b$ ne peut être nulle, donc fait partie de ψ et $m.a + n.b \in \nu$. Il résulte alors de 2.5 que ν est quasi-clos et que S n'a pas le poids zéro dans U_ψ.

Si $S = T$, alors réciproquement tout ensemble quasi-clos unipotent est formé de racines positives, puisque $T.U_\psi$ est contenu dans un sous-groupe de Borel de G. Cela vaut aussi si S est défini sur k et contient le tore déployé sur k maximal du centre de $\mathscr{Z}(S)$, vu 3.6. Mais des exemples simples montrent que cette réciproque n'est pas vraie en général.

3.9. *Proposition.* — *On conserve les notations de 3.8. Soient ψ, η des parties quasi-closes de $\Phi(S, G)$ telles que ψ normalise η. Alors G_ψ normalise G_η^*.*

Soient T un tore maximal de G contenant S et $j : X^*(T) \to X^*(S)$ l'homomorphisme de restriction. Comme $\mathscr{Z}(S)$ normalise G_η^*, il suffit de montrer que si $a \in j^{-1}(\psi) \cap \Phi(T, G)$ et $b \in j^{-1}(\eta) \cap \Phi(T, G)$, alors $(U_a, U_b) \subset G_\eta^*$. Supposons d'abord que $m.j(a) + n.j(b) \neq 0$ quels que soient $m, n \in \mathbf{N}^+$; comme ψ normalise η, on a alors $m.a + n.b \in \eta$ toutes les fois que $m.a + n.b \in \Phi$, et notre assertion résulte de 2.5. Si, par contre, il existe $m, n \in \mathbf{N}^+$ tels que $m.j(a) + n.j(b) = 0$, alors $j(a) = (m+1).j(a) + n.j(b) \in \eta$, donc $U_a \in G_\eta^*$, d'où évidemment $(U_a, U_b) \subset G_\eta^*$.

3.10. *Proposition.* — *Soient S un tore et ψ, η deux parties quasi-closes de $\Phi(S, G)$ positives pour un ordre donné sur $X^*(S)$. Soit σ l'ensemble des éléments de $\Phi(S, G)$ qui sont somme d'une combinaison linéaire non nulle à coefficients entiers ≥ 0 d'éléments de ψ et d'une combinaison linéaire non nulle à coefficients entiers ≥ 0 d'éléments de η. Alors σ est clos unipotent, et $(U_\psi, U_\eta) \subset U_\sigma$.*

Il est clair que σ est clos. Soient T un tore maximal de G contenant S, et Φ^+ l'ensemble des racines de G par rapport à T, positives pour un ordre sur $X^*(T)$ compatible avec l'ordre donné sur $X^*(S)$. Soient ψ', η', σ' les images réciproques de ψ, η, σ dans Φ. Vu 3.8 (iv), on a $U_\psi = U_\psi^{(T)}$ et de même pour η, σ. D'autre part σ' est obtenu à partir de ψ', η' comme σ l'est à partir de ψ et η. La proposition est donc conséquence de 2.3, 2.5 et des identités $(x, yz) = (x, y).^y(x, z)$ et $(xy, z) = {}^x(y, z).(x, z)$.

3.11. *Proposition.* — *Soient S un tore de G, ψ un ensemble quasi-clos unipotent de racines relatives à S, $\psi_1, \ldots, \psi_n$ des parties quasi-closes unipotentes de ψ dont ψ est la réunion disjointe, et H un sous-groupe unipotent fermé connexe de U_ψ (resp. G_ψ) normalisé par S. Alors*

l'application produit définit un isomorphisme de variétés de $(H \cap U_{\psi_1}) \times \ldots \times (H \cap U_{\psi_n})$ *(resp.*
$(H \cap \mathscr{Z}(S)) \times (H \cap U_{\psi_1}) \times \ldots \times (H \cap U_{\psi_n}))$, *sur* H.

D'après 3.5, U_ψ est normalisé par $\mathscr{Z}(S)$, donc est le radical unipotent de G_ψ,
et $S.H.U_\psi$ est un sous-groupe fermé résoluble de G_ψ. Soient B un sous-groupe de Borel
de G contenant $S.H.U_\psi$ et T un tore maximal de B qui contient S, donc fait aussi partie
de $\mathscr{Z}(S)$. Les poids de T dans le radical unipotent U de B sont les éléments positifs de
$\Phi = \Phi(T, G)$ pour un ordre convenable. Soient $j : X^*(T) \to X^*(S)$ l'homomorphisme de
restriction et $\eta_0 = j^{-1}(o) \cap \Phi^+$, $\eta_i = j^{-1}(\psi_i) \cap \Phi$ ($1 \leq i \leq n$), $\eta = j^{-1}(\psi) \cap \Phi$ et $\sigma = \eta \cup \eta_0$. Évidemment η_0 est clos et, vu 2.3, $U_{\eta_0} = \mathscr{Z}(S) \cap U$; vu 3.8 (iv), $\eta_1, \ldots, \eta_n$, η sont des
parties quasi-closes unipotentes de Φ^+, η est la réunion disjointe de $\eta_1, \ldots, \eta_n$, et l'on a
$U_\psi = U_\eta^{(T)}$, $U_{\psi_i} = U_{\eta_i}^{(T)}$ ($1 \leq i \leq n$). Il résulte aussi de 2.3 que σ est l'ensemble des poids
de T dans $V = G_\psi \cap U$, donc que σ est quasi-clos et que $V = U_\sigma$. Soient $a_1, \ldots, a_q$
(resp. $b_1, \ldots, b_r$) les éléments de η (resp. σ) rangés par ordre croissant. D'après 2.5
l'application produit définit un isomorphisme de variétés de $U_{a_j} \times \ldots \times U_{a_q}$ (resp.
$U_{b_j} \times \ldots \times U_{b_r}$) sur un sous-groupe distingué V_j (resp. Y_j) de U_η (resp. U_σ) pour tout
$j \leq q$ (resp. $j \leq r$). Les groupes V_j (resp. Y_j) forment donc une suite de composition
de U_η (resp. U_σ) dont les éléments sont engendrés par leurs intersections avec les groupes
U_{η_i} ($i = 1, \ldots, n$) (resp. $i = 0, \ldots, n$). La proposition résulte alors de 3.3 (où l'on prend S
comme groupe d'opérateurs).

3.12. *Proposition.* — *Soient* S *un tore,* H *un sous-groupe fermé connexe de* G *normalisé
par* S *et* ψ *un ensemble convexe de racines de* G *relatives à* S. *Supposons que tout poids de* S *dans* H
appartienne à ψ *(resp.* $\psi \cup \{o\}$*). Alors* H *est contenu dans* G_ψ^* *(resp.* G_ψ*).*

Soit S' un tore maximal de $S.H$ contenant S. Le groupe $S.H$ est engendré par S'
et deux sous-groupes unipotents maximaux normalisés par S' (2.3, remarque), donc H
est engendré par $(S' \cap H)^0$ et deux sous-groupes unipotents normalisés par S. Il suffit
par conséquent de considérer le cas où H est unipotent.

Soient B un sous-groupe de Borel de G contenant $S.H$, T un tore maximal de B
contenant S, et U le radical unipotent de B. Soient Φ^+ l'ensemble des racines de U
relatives à T, et φ_0 l'ensemble des éléments de Φ^+ dont la restriction à S' est nulle.
Soient $\varphi_0, \varphi_1, \ldots, \varphi_n$ les classes d'équivalence de la relation obtenue dans Φ^+ en
considérant comme équivalentes deux racines dont les restrictions à S ne diffèrent que par
un coefficient strictement positif; $\varphi_0, \varphi_1, \ldots, \varphi_n$ sont convexes et (2.3) $\mathscr{Z}(S) \cap U = U_{\varphi_0}$.
Soient ψ_i l'ensemble des restrictions à S des éléments de φ_i, et

$$U_i = U_{\psi_i} = U_{\varphi_i}^{(T)} \quad (1 \leq i \leq n), \qquad U_0 = U_{\varphi_0}^{(T)} = \mathscr{Z}(S) \cap U.$$

Il résulte de 3.11 que $H = (H \cap \mathscr{Z}(S)).(H \cap U_1) \ldots (H \cap U_n)$, et que $H \cap \mathscr{Z}(S) \neq \{e\}$
seulement si S a le poids zéro dans H. Soit $i \geq 1$. Alors $H \cap U_i \neq \{e\}$ entraîne $\psi \cap \psi_i \neq \emptyset$,
donc, puisque ψ est convexe, $\psi_i \subset \psi$, et $U_i \subset U_\psi$, d'où la proposition.

3.13. *Théorème.* — *Soient* S *un* k*-tore de* G, H *un sous-groupe connexe* k*-fermé de* G *normalisé
par* $\mathscr{Z}(S)$ *et* ψ *une partie quasi-close de* $\Phi(S, G)$. *Alors :*

a) H *est défini sur k, et possède un unique sous-groupe de Levi* L *normalisé par* $\mathscr{Z}(\mathrm{S})$. *Les groupes* L *et* $\mathrm{R}_u(\mathrm{H})$ *sont définis sur k.*

b) *Les groupes* G_ψ *et* G_ψ^* (3.8) *sont définis sur k si et seulement si* ψ *l'est* (1.7). *Les ensembles* $\psi_s = \psi \cap (-\psi)$ *et* $\psi_u = \psi \cap \complement(-\psi)$ *sont quasi-clos, et* ψ_u *est unipotent. Si* $\mathrm{H} = \mathrm{G}_\psi$ (*resp.* $\mathrm{H} = \mathrm{G}_\psi^*$), *alors* $\mathrm{L} = \mathrm{G}_{\psi_s}$ (*resp.* $\mathrm{L} = \mathrm{G}_{\psi_s}^*$) *et* $\mathrm{R}_u(\mathrm{H}) = \mathrm{U}_{\psi_u}$.

a) Il suffit de faire la démonstration lorsque k est séparablement clos. D'après 2.10 et 2.14, S est contenu dans un tore maximal T de G faisant partie d'une donnée de déploiement sur k. Le groupe $\mathrm{R}_u(\mathrm{H})$ (resp. H) est alors engendré par les groupes radiciels U_a qu'il contient (resp. et $(\mathrm{T} \cap \mathrm{H})^0$) et est donc défini sur k.

Soient η l'ensemble des $a \in \Phi(\mathrm{T}, \mathrm{G})$ tels que $\mathrm{U}_a \subset \mathrm{H}$, et $\eta_s = \eta \cap (-\eta)$, $\eta_u = \eta \cap \complement(-\eta)$. Montrons que l'ensemble η_u est égal à l'ensemble μ des $a \in \Phi(\mathrm{T}, \mathrm{G})$ tels que $\mathrm{U}_a \subset \mathrm{R}_u(\mathrm{H})$. Soit $a \in \mu$. Si $-a$ faisait partie de η, alors le groupe Z_a' engendré par U_a et U_{-a} normaliserait $\mathrm{R}_u(\mathrm{H})$, son radical unipotent contiendrait $\mathrm{R}_u(\mathrm{H}) \cap \mathrm{Z}_a'$, donc U_a, et serait $\neq \{e\}$ ce qui est absurde puisque Z_a' est presque simple de dimension trois (2.3). Ainsi, $-a \notin \eta$, donc $a \in \eta_u$ et $\mu \subset \eta_u$. D'autre part, le quotient de $\mathrm{T}.\mathrm{H}$ par $\mathrm{R}_u(\mathrm{H})$ est réductif, donc a un ensemble de racines symétrique, d'où $\mu \supset \eta_u$, et $\mu = \eta_u$, ce qui montre en particulier que η_u est quasi-clos unipotent.

L'ensemble η_s est quasi-clos puisque η et $-\eta$ le sont (3.8(i)) et le groupe L engendré par les U_a ($a \in \eta_s$) et $\mathrm{T}' = (\mathrm{T} \cap \mathrm{H})^0$ est réductif. Vu 2.3, 3.4, il contient tout sous-groupe de H réductif normalisé par T, et l'on a $\mathrm{H} = \mathrm{L}.\mathrm{R}_u(\mathrm{H})$. Les résultats rappelés en 2.3 montrent aussi que les algèbres de Lie de L et $\mathrm{R}_u(\mathrm{H})$ sont transverses et que $\mathrm{L} \cap \mathrm{R}_u(\mathrm{H})$ est connexe, donc réduit à $\{e\}$. Le groupe H est par suite produit semi-direct de L et $\mathrm{R}_u(\mathrm{H})$, et L est l'unique sous-groupe réductif normalisé par T ayant cette propriété. Il reste à voir que L est stable par $\mathscr{Z}(\mathrm{S})$. Le sous-groupe L' de H engendré par les ${}^x\mathrm{L}$ ($x \in \mathscr{Z}(\mathrm{S})$) est réductif, car il est stable par un automorphisme de G qui induit $t \mapsto t^{-1}$ sur T (et dont l'existence résulte du cor. 1 de [10, exp. 24]), donc possède un ensemble de racines v par rapport à T qui est symétrique. On a évidemment $\eta_s \subset v \subset \eta$, d'où $v = \eta_s$ et $\mathrm{L}' = \mathrm{L}$, ce qui termine la démonstration de *a*). Remarquons que l'on a montré en outre que

$$(\mathrm{1}) \qquad \mathrm{L} = (\mathrm{T} \cap \mathrm{H})^0.\mathrm{G}_{\eta_s}^{*(\mathrm{T})}, \qquad \mathrm{R}_u(\mathrm{H}) = \mathrm{G}_{\eta_u}^{*(\mathrm{T})}.$$

b) Si G_ψ (resp. G_ψ^*) est défini sur k, il en est évidemment de même de ψ; réciproquement, si ψ est défini sur k alors G_ψ et G_ψ^* sont k-fermés, normalisés par $\mathscr{Z}(\mathrm{S})$, donc définis sur k vu *a*).

Soit $j : \mathrm{X}^*(\mathrm{T}) \to \mathrm{X}^*(\mathrm{S})$ l'homomorphisme de restriction, et soit η comme ci-dessus. Si $\mathrm{H} = \mathrm{G}_\psi$ (resp. $\mathrm{H} = \mathrm{G}_\psi^*$), alors on a $j^{-1}(\psi) \cap \Phi(\mathrm{T}, \mathrm{G}) \subset \eta \subset j^{-1}(\psi \cup \{o\}) \cap \Phi(\mathrm{T}, \mathrm{G})$, d'où en particulier $j^{-1}(\psi_s) \cap \Phi(\mathrm{T}, \mathrm{G}) \subset \eta_s$ et $j^{-1}(\psi_u) \cap \Phi(\mathrm{T}, \mathrm{G}) \subset \eta_u$. Compte tenu de (1), il s'ensuit que

$$(\mathrm{2}) \qquad \mathrm{G}_{\psi_s} \ (\text{resp. } \mathrm{G}_{\psi_s}^*) \subset \mathrm{L} \quad \text{et} \quad \mathrm{G}_{\psi_u}^* \subset \mathrm{R}_u(\mathrm{H}).$$

Mais $G^*_{\psi_u}$ est normalisé par G_{ψ_s} (3.0, 3.9), donc $H = G_{\psi_s}.G^*_{\psi_u}$ (resp. $G^*_{\psi_s}.G^*_{\psi_u}$), et les inclusions (1) sont en fait des égalités.

3.14. Corollaire. — *Soit* H *un* k-*sous-groupe connexe de* G *de rang maximum. Alors* H *possède une décomposition de Levi* (0.8) *sur* k, *et deux* k-*sous-groupes de Levi de* H *sont conjugués par un unique élément de* $R_u(H)_k$. *Les* k-*sous-groupes de Levi de* H *sont les centralisateurs dans* H *des tores maximaux de* R(H) *qui sont définis sur* k.

H contient un tore maximal défini sur k (2.14 *a*)), donc a une décomposition de Levi sur k vu 3.13. Soient $H = L.R_u(H)$ une telle décomposition et T un tore maximal de L. Le radical de H est défini sur k et égal à $Z.R_u(H)$, où Z est le centre connexe de L et aussi un tore maximal de R(H) défini sur k (2.15 *e*)). On a $L \subset \mathscr{Z}_H(Z)$ par construction. D'autre part, $\mathscr{Z}_H(Z) \cap R_u(H)$ est stable par T, donc engendré par les sous-groupes U_a ($a \in \Phi(T, G)$) qu'il contient (2.3), ce qui, compte tenu de 3.6, montre que $\mathscr{Z}_H(Z) \cap R_u(H) = \{e\}$, d'où $L = \mathscr{Z}_H(Z)$, et notre dernière assertion.

Soient L′ un deuxième k-sous-groupe de Levi et Z′ son centre connexe. Les tores Z et Z′ sont maximaux dans R(H), donc conjugués par un élément $u \in R_u(H)_{\bar{k}}$ [1, Th. 12.2]. Cet élément est unique, car le normalisateur et le centralisateur de Z dans R(H) coïncident [1, 10.2] et on a vu ci-dessus que ce dernier se réduit à Z. L'élément u est donc rationnel sur une extension inséparable de k, et, pour démontrer qu'il est rationnel sur k (1), on peut supposer k séparablement clos, donc G déployé sur k. Vu 2.10, appliqué à G et à $T.R_u(H)$, on peut supposer que T fait partie d'une donnée de déploiement et trouver $v \in R_u(H)_k$ tel que $v.Z′.v^{-1} \subset T$. On a alors $v.Z′.v^{-1} \subset L \cap R(H)$, donc $v.Z′.v^{-1} = Z$ et $v.L′.v^{-1} = L$.

3.15. *Contre-exemples.* — Nous supposons ici la caractéristique p de k non nulle. Soient V un espace vectoriel sur k de dimension deux, V_p sa p-ième puissance symétrique, et x, y une base de V_k. On identifie V_p à $\bar{k}^{p+1}$ en prenant comme base les monômes x^p, y^p, $x^{p-1}.y$, ..., $x.y^{p-1}$. Le groupe GL(V) opère canoniquement sur V_p, et l'on dénote L l'image dans $\mathbf{GL}_{p+1}$ du groupe SL(V) des éléments de déterminant un de GL(V) par cette représentation.

Soit P le sous-groupe de $\mathbf{GL}_{p+1}$ laissant stable le sous-espace W engendré par x^p et y^p, et soit $U = R_u(P)$. Il est immédiat que $U \cong (\mathbf{G}_a)^{2(p-1)}$ et que l'on a une décomposition de Levi $P = M.U$ avec $M \cong \mathbf{GL}_2 \times \mathbf{GL}_{p-1}$. La projection $\pi : P \to M′ = P/U$ induit un isomorphisme ψ de M sur M′. D'autre part, il est élémentaire que L laisse W stable, donc que $L \subset P$. Soit $H = L.U$ le groupe engendré par L et U. On a $L \cap U = \{e\}$ donc H est au point de vue ensembliste le produit semi-direct de L et U. On peut aussi écrire $H = L′.U$, où $L′ = H \cap M = \psi^{-1} \circ \pi(L)$ est k-isomorphe à L. Les algèbres de Lie de L′ et U sont transverses (puisque celles de M et de U le sont), donc L′ est un sous-groupe de Levi de H.

---

(1) Ce qui résulte aussi d'un théorème de Rosenlicht (*voir* 11.4).

Les sous-groupes L et L′ de H ne sont pas conjugués dans H, car s'ils l'étaient, on pourrait trouver un sous-espace W′ supplémentaire de W dans V_p stable par L, et un calcul élémentaire (ou bien [10, Exp. 20, p. 12]) montre qu'il n'en est rien.

On peut prendre comme base de l'algèbre de Lie $\mathfrak{l}$ de L les vecteurs H, X, Y tangents aux images des groupes de matrices diagonales, triangulaires unipotentes supérieures et triangulaires unipotentes inférieures respectivement. Un calcul élémentaire laissé au lecteur montre que H est diagonale, et que

 a) si $p=2$, X et Y forment une base de l'algèbre de Lie $\mathfrak{u}$ de U;

 b) si $p \geqslant 3$, $\mathfrak{l}$ et $\mathfrak{u}$ sont transverses.

(On vérifie par exemple que X, Y ont leurs termes diagonaux nuls et que si $p \geqslant 3$, $X_{p,p+1} \neq 0$, $X_{p+1,p} = 0$, $Y_{p,p+1} = 0$, $Y_{p+1,p} \neq 0$.) Si $p \geq 3$, H est donc un exemple de groupe possédant deux sous-groupes de Levi L, L′ non conjugués. Si $p=2$, H est un sous-groupe de $\mathbf{GL}_3$, normalisé par le tore maximal des matrices diagonales de $\mathbf{GL}_3$, donc auquel 3.13 s'applique, mais qui possède un sous-groupe réductif maximal L tel que $H = L.U$ et dont l'algèbre de Lie a une intersection non nulle avec celle de U.

3.16. *Lemme.* — *Soient* U *un k-groupe unipotent connexe,* S *un k-tore opérant k-morphiquement sur* U, *et f un caractère non trivial de* S. *Supposons que sur k_s,* U *possède une structure d'espace vectoriel telle que l'opération d'un élément $s \in S_{\bar{k}}$ soit l'homothétie de rapport s^f. Alors il en est déjà ainsi sur k, et f est défini sur k* [1].

L'ensemble des poids de S dans U se réduit à *f*, et est d'autre part stable par le groupe de Galois Γ de k_s/k, donc *f* est défini sur *k*.

Les hypothèses entraînent que l'on a $^\gamma(c.u) = {}^\gamma c.{}^\gamma u$, quels que soient $\gamma \in \Gamma$, $c \in k_s$ et $u \in U_{k_s}$. Il suffit donc de faire voir que U possède une base formée d'éléments rationnels sur *k*. Il est clair que U possède une base formée d'éléments rationnels sur une extension galoisienne convenable K de *k* de degré fini. Soit $\Gamma' = \mathrm{Gal}(K/k)$ et soit $(u_1, \ldots, u_n)$ une partie de U_k libre dans U_{k_s} (vu comme espace vectoriel sur k_s) maximale. Alors, pour tout $u \in U_K$ et $c \in K$, l'élément $\sum_{\gamma \in \Gamma'} {}^\gamma c.{}^\gamma u$ est rationnel sur *k*, donc combinaison linéaire des $u_i (1 \leq i \leq n)$. Il en est alors de même de *u* d'après le théorème de la base normale, donc $(u_1, \ldots, u_n)$ est une base de U_{k_s}.

3.17. *Théorème.* — *Soient* S *un k-tore de* G, ψ *un ensemble quasi-clos unipotent d'éléments de* $\Phi(S, G)$ *défini sur k, et f un élément défini sur k de* ψ *tel que* $\eta = \psi - \{f\}$ *soit un quasi-idéal de* ψ *(3.8). Soit* H *un k-sous-groupe de* U_ψ, *normalisé par un tore maximal T de* G *défini sur k et contenant* S. *Alors le quotient* $H/(H \cap U_\eta)$ *a une structure d'espace vectoriel définie sur k telle que l'automorphisme induit par* Int *s* $(s \in S_{\bar{k}})$ *soit l'homothétie de rapport s^f.*

Vu 3.16, on peut supposer que $k = k_s$, donc que G est déployé sur *k* et que T

---

[1] Ce lemme est aussi une conséquence facile du lemme p. 109 de [26]. Il est aussi vrai lorsque $f = 0$ [26, th. 3, p. 107].

fait partie d'une donnée de déploiement. Il existe un sous-groupe de Borel de G qui contient $T.U_\psi$, et ce sous-groupe est engendré par les groupes radiciels U_a, où a parcourt l'ensemble Φ^+ des racines >0 pour un ordre convenable, donc est défini sur k. Soit μ (resp. ν) l'ensemble des éléments de Φ^+ dont la restriction à S est contenue dans ψ (resp. η). Alors (2.3, 3.11) l'application produit induit un k-isomorphisme α de variétés du produit des U_a, $a\in\mu$ (resp. U_b, $b\in\nu$), sur U_ψ (resp. U_η) et un morphisme bijectif du produit des $U_a\cap H$, $a\in\mu$ (resp. $U_b\cap H$, $b\in\nu$) sur $U_\psi\cap H$ (resp. $U_\eta\cap H$), où les éléments de μ (resp. ν) sont rangés dans un ordre quelconque. Mais $U_a\cap H$ est égal à $\{e\}$ ou à U_a (2.3) et α commute à l'action de T. Cette application induit donc un k-isomorphisme de variétés du produit des U_a ($a\in\mu$, $a\notin\nu$, $U_a\subset H$) sur $H/(H\cap U_\eta)$ qui commute à T, d'où le théorème.

3.18. *Corollaire.* — *Tout k-sous-groupe unipotent connexe* H *de* G *normalisé par un tore maximal* T *de* G *défini sur* k *est déployé sur* k. *Si* σ *est une partie quasi-close unipotente de* $\Phi(S, G)$ *définie sur* k, *alors* U_σ *est un k-groupe unipotent déployé sur* k.

Vu 3.6, il existe un ordre sur $X^*(T_d)$ tel que les poids de T_d dans H soient >0. Soient $a_1, \ldots, a_m$ les éléments de $\Phi^+(T_d, G)$, rangés par ordre croissant. Pour tout i compris entre 1 et m, $\{a_i, \ldots, a_m\}$ est un quasi-idéal de $\{a_1, \ldots, a_m\}$. La première partie du corollaire résulte alors du théorème par récurrence descendante sur le plus petit entier i tel que $H\subset U_{\{a_i, \ldots, a_m\}}$.

Le groupe U_σ est défini sur k (3.13), normalisé par $\mathscr{Z}(S)$ (3.8 (ii)), donc aussi par un tore maximal de G défini sur k (2.15 d)). La deuxième assertion du corollaire est donc un cas particulier de la première.

Remarque. — Supposons S déployé sur k, et σ formé d'éléments positifs de $\Phi(S, G)$ pour un ordre convenable sur $X^*(S)$. Soient $b_1, \ldots, b_m$ les éléments de σ rangés par ordre croissant. Pour tout $i\geq 1$, l'ensemble $\sigma_i=\{b_i, \ldots, b_m\}$ est une partie quasi-close de $\Phi(S, G)$, car c'est l'intersection de σ avec l'ensemble des racines $\geq b_i$, qui est visiblement clos. Comme S est déployé sur k, tout élément de $\Phi(S, G)$ est défini sur k, donc $U_i=U_{\sigma_i}$ est un k-groupe unipotent. Vu 3.10, U_{i+1} est un sous-groupe invariant dans U_i, et vu 3.17, U_i/U_{i+1} admet une structure d'espace vectoriel sur k, telle que l'automorphisme intérieur par $s\in S_{\bar k}$ y induise une homothétie de rapport s^{b_i}. Il existe donc dans ce cas une suite de composition sur k, à quotients successifs vectoriels sur k, sur lesquels S agit par homothéties, ce qui précise 3.18.

L'hypothèse de positivité faite sur σ est vérifiée d'elle-même si $\sigma=(c)$ est l'ensemble des multiples entiers positifs d'un élément $c\in\Phi(S, G)$, ou, vu 3.6, si S est le plus grand tore déployé sur k du centre de $\mathscr{Z}(S)$, donc, plus particulièrement encore, si S est un tore déployé sur k maximal de G.

3.19. *Corollaire.* — *Soient* S *un tore de* G, f *un caractère non trivial de* S *et* α *un isomorphisme de* $\mathbf{G}_a$ *sur un sous-groupe* H *de* G *normalisé par* S *vérifiant* $s.\alpha(t).s^{-1}=\alpha(s^f.t)$ ($s\in S$, $t\in\mathbf{G}_a$). *Alors* H *est contenu dans le sous-groupe* $U_{(f)}^{(S)}$, *où* (f) *désigne l'ensemble des éléments de* $\Phi(S, G)$ *de la forme* $n.f$ ($n\in\mathbf{N}^+$).

Soit ψ l'ensemble des racines relatives à S de la forme $c.f$ avec $c > 0$. Cet ensemble est convexe, donc $H \subset U_\psi$ d'après 3.12.

Soit η la plus petite partie quasi-close de ψ telle que $H \subset U_\eta$.

Supposons que $\eta \not\subset (f)$, et soit c le plus petit non-entier tel que $c.f \in \eta$. L'ensemble $\mu = \eta - \{c.f\}$ est un quasi-idéal de η. Soit π la projection canonique de U_η sur l'espace vectoriel $U_\eta/U_\mu = V$, et soit β le morphisme de variétés de $\mathbf{G}_m$ dans V qui applique $t \in \mathbf{G}_m$ sur $\pi(\alpha(t))$. D'après 3.17, on a

$$(1) \qquad \beta(s^f.t) = s^{cf}.\beta(t) \qquad\qquad (s \in S,\ t \in \mathbf{G}_m).$$

Comme $H \subset U_\mu$, $\beta_{/H}$ n'est pas nul, mais alors (1) et le fait que β est un morphisme entraînent que c est entier, d'où une contradiction.

3.20. *Corollaire.* — *Soit* H *un* k-*sous-groupe connexe de* G, *normalisé par un tore maximal* T *de* G *défini sur* k. *Si* k *est infini, alors* H_k *est* $\mathcal{Z}$*ariski-dense dans* H.

Le groupe H admet une décomposition de Levi $H = L.R_u(H)$ sur k d'après 3.13. Le groupe $R_u(H)$ est normalisé par T, donc déployé sur k vu 3.18, donc isomorphe sur k à un espace affine [26, p. 101], ce qui implique que $R_u(H)_k$ est Zariski-dense dans $R_u(H)$. Enfin, L_k est Zariski-dense dans L d'après 2.14.

3.21. *Lemme.* — *Soient* S *un tore de* G, ψ *une partie convexe de* $\Phi(S, G)$ *et* A *la composante neutre du noyau des caractères appartenant à* ψ_s.

a) ψ_s *contient tout élément de* $\Phi(S, G)$ *qui s'annule sur* A.

b) *Si* S *est un tore maximal de* G, *tout élément* $w \in W(S, G)$ *qui opère trivialement sur* $X_*(A)$ *est produit de réflexions* r_b $(b \in \psi_s)$.

a) Soit $c \in \Phi(S, G)$ s'annulant sur A. Il existe donc des éléments de ψ_s dont c est combinaison linéaire à coefficients rationnels. Comme ψ_s est symétrique, on peut supposer ces coefficients > 0; d'autre part, ψ_s est convexe, puisque ψ l'est, donc $c \in \psi_s$.

b) D'après [6], w est produit de réflexions r_b, où b est une racine nulle sur $X_*(A)$, donc un caractère de S nul sur A, donc un élément de ψ_s vu *a)*.

3.22. *Proposition.* — *Soient* S *un tore de* G *et* ψ, η *deux parties quasi-closes de* $\Phi(S, G)$. *Alors :*

a) $\mathfrak{g}_\psi \cap \mathfrak{g}_\eta = \mathfrak{g}_{\psi \cap \eta}$ *et* $(G_\psi \cap G_\eta)^0 = G_{\psi \cap \eta}$.

b) *Si* $\psi \cap \eta$ *est convexe*, $G_\psi \cap G_\eta = G_{\psi \cap \eta}$.

c) *Si* ψ *est unipotent*, $U_\psi \cap G_\eta = U_\psi \cap G_\eta^* = U_{\psi \cap \eta}$.

Soient T un tore maximal de G contenant S et $j : X^*(T) \to X^*(S)$ l'homomorphisme de restriction.

a) Un sous-groupe radiciel U_a $(a \in \Phi(T, G))$ par rapport à T fait partie de G_ψ si et seulement si a est un poids de T dans G_ψ (3.8), donc si et seulement si $j(a) \subset \psi \cup \{0\}$. Cela vaut aussi pour η ou $\psi \cap \eta$, d'où *a)*.

b) Il suffit de considérer le cas où k est algébriquement clos, donc où T fait partie d'une donnée de déploiement sur k de G. D'autre part, si $\mu \subset \Phi(S, G)$ est convexe, alors $\Phi(T, G) \cap j^{-1}(\mu \cup \{o\})$ est aussi convexe, et évidemment

$$j^{-1}((\psi \cap \eta) \cup \{o\}) = j^{-1}(\psi \cup \{o\}) \cap j^{-1}(\eta \cup \{o\}).$$

On peut donc supposer que $S = T$.

Soit $x \in G_\psi \cap G_\eta$. Nous devons montrer que $x \in G_{\psi \cap \eta}$. Le tore $x.T.x^{-1}$ étant conjugué à T dans $(G_\psi \cap G_\eta)^0$, on peut, vu $a)$, se borner au cas où $x \in \mathcal{N}(T)$. Soient $A = T_{\psi_s}$ et $B = T_{\eta_s}$ (1.2). D'après 3.13, 3.14, $G_{\psi_s} = \mathscr{Z}(A)$ et $G_{\eta_s} = \mathscr{Z}(B)$ sont des sous-groupes de Levi de G_ψ et G_η respectivement, donc l'élément w du groupe de Weyl $W(T, G)$ de G défini par x fait partie de $W(T, G_{\psi_s}) \cap W(T, G_{\eta_s})$; il centralise $X_*(A)$, $X_*(B)$ et par conséquent aussi $X_*(C)$, où C est la composante neutre de l'intersection des noyaux des éléments de $\psi_s \cap \eta_s = (\psi \cap \eta)_s$. Notre assertion résulte alors de 3.21 $b)$.

$c)$ Ici aussi, on peut supposer k algébriquement clos et $S = T$, et $c)$ résulte alors des propriétés de U rappelées en 2.3.

3.23. *Corollaire.* — *Supposons S défini sur k et soient ψ, η deux parties quasi-closes de $\Phi(S, G)$ définies sur k.*

(i) $G^*_{\psi, k} \subset G_{\eta, k}$ *si et seulement si* $\psi \subset \eta$.

(ii) $G^*_{\psi, k} = G^*_{\eta, k}$ *(resp.* $G_{\psi, k} = G_{\eta, k}$*) si et seulement si* $\psi = \eta$.

Les groupes G_ψ, G_η, G^*_ψ, G^*_η sont définis sur k d'après 3.13. L'assertion (ii) est une conséquence immédiate de (i) et il est clair que $\psi \subset \eta$ entraîne $G^*_{\psi, k} \subset G_{\eta, k}$. Il reste à faire voir que $G^*_{\psi, k} \subset G_{\eta, k}$ implique $\psi \subset \eta$.

Soient $a \in \Phi(S, G)$ et (a) l'ensemble des multiples entiers positifs de a contenus dans $\Phi(S, G)$. Il est clos, unipotent (3.8 (iv)), donc $\mu = (a) \cap \psi$ est quasi-clos et unipotent. Comme ψ est réunion de parties de ce type, on peut supposer qu'il est unipotent. Dans ce cas, $U_\psi \cap G_\eta = U_{\psi \cap \eta}$ d'après 3.22. Les groupes $U_{\psi \cap \eta}$ et U_ψ sont unipotents déployés sur k (3.18), donc la variété sous-jacente à U_ψ est k-isomorphe au produit de $U_{\psi \cap \eta}$ par un espace vectoriel sur k [26, cor. 1, 2 to th. 1]. On ne peut alors avoir $U_{\psi, k} \subset U_{\psi \cap \eta, k}$ que si $U_\psi = U_{\psi \cap \eta}$, donc si $\psi \subset \eta$.

3.24. *Proposition.* — *Soient S un k-tore déployé et ψ une partie convexe de $\Phi(S, G)$. Alors, il existe des parties unipotentes convexes $\psi_1, \ldots, \psi_m$ de $\Phi(S, G)$ telles que $\Phi(S, G)$ soit la réunion disjointe de $\psi, \psi_1, \ldots, \psi_m$ et que l'application produit définisse un k-isomorphisme de variétés de $G_\psi \times U_{\psi_1} \times \ldots \times U_{\psi_m}$ sur un ouvert de G. Si $\psi \cup (-\psi) = \Phi(S, G)$, on peut faire $m = 1$ (d'où $\psi_1 = \Phi(S, G) - \psi$).*

Posons $X = X^*(S) \otimes \mathbf{Q}$, et soit Y le sous-espace vectoriel de X engendré par ψ_s. Si une combinaison linéaire à coefficients rationnels positifs d'éléments de ψ_u appartient à Y, elle est identiquement nulle; en effet, de la relation

$$\sum_{i=1}^{r} c_i a_i \in Y \qquad (a_i \in \psi_u,\, c_i \in \mathbf{Q},\, c_i \geqslant o)$$

on tire

$$-c_1 a_1 = \sum_{i=2}^{r} c_i a_i + \sum_{j=1}^{t} d_j b_j \qquad (b_j \in \psi_s,\, d_j \geqslant o)$$

et si c_1 n'était pas nul, $-a_1$ devrait appartenir à ψ, ce qui est absurde. En vertu de 3.1, il existe dans X/Y un ordre tel que les images canoniques dans X/Y des éléments de ψ_u soient positives. Soit ψ' l'ensemble de tous les éléments de $\Phi(S, G)$ dont l'image canonique dans X/Y est positive pour l'ordre en question. Il est clair que ψ' est convexe, que $\psi'_s = \psi_s$, et que $\psi_u \subset \psi'_u$, ces deux ensembles étant égaux si $\psi \cup (-\psi) = \Phi(S, G)$. Soient $\psi_1, \ldots, \psi_{m-1}$ les parties convexes minimales de ψ'_u non contenues dans ψ_u. L'ensemble ψ'_u est la réunion disjointe de $\psi_u, \psi_1, \ldots, \psi_{m-1}$; par conséquent (3.11) l'application produit est un isomorphisme de variété de $U_{\psi_u} \times U_{\psi_1} \times \ldots \times U_{\psi_{m-1}}$ sur $U_{\psi'_u}$, donc (3.13) de $G_\psi \times U_{\psi_1} \times \ldots \times U_{\psi_{m-1}}$ sur $G_{\psi'}$. Enfin, si on pose $\psi_m = -\psi'_u$, l'application produit est un isomorphisme de variétés de $G_{\psi'} \times U_{\psi_m}$ sur un ouvert de G. En effet, $G_{\psi'}$ et U_{ψ_m} sont normalisés par tout tore maximal contenu dans $\mathscr{Z}(S)$ et il résulte de 2.3 que leurs algèbres de Lie sont linéairement indépendantes et de somme égale à $\mathfrak{g}$; d'autre part, $G_{\psi'} \cap U_{\psi_m} = \{e\}$ en vertu de 3.22 $c)$.

3.25. *Corollaire.* — *La variété* G/G_ψ *est rationnelle sur* k. *La fibration de* G *par* G_ψ *possède une k-section locale, et la projection* $G_k \to (G/G_\psi)_k$ *est surjective.*

Le composé de l'application produit et de la projection $\pi : G \to G/G_\psi$ induit un k-isomorphisme de variétés de $U_{\psi_1} \times \ldots \times U_{\psi_m}$ sur un ouvert V de G/G_ψ, et la variété sous-jacente à U_{ψ_i} est k-isomorphe à un espace affine [26, p. 101], d'où la première assertion. La k-section locale est définie sur V comme l'inverse de $\pi : U_{\psi_1} \ldots U_{\psi_m} \to V$, et la dernière assertion résulte de 2.6, 2.14.

§ 4. SOUS-GROUPES PARABOLIQUES

Dans tout ce paragraphe, G est réductif, connexe.

4.1. Un sous-groupe fermé P d'un groupe algébrique H est *parabolique* s'il contient un sous-groupe de Borel de H. Vu [1, § 16], il revient au même d'exiger que G/P soit une variété complète ou une variété projective.

4.2. *Sous-groupes paraboliques standards.* — (i) On fixe un tore maximal T de G et un ordre sur l'ensemble Φ des racines de G par rapport à T, et on utilise les notations de 2.3. Pour toute partie θ de Δ, on note $[\theta]$ l'ensemble des racines qui sont combinaisons linéaires à coefficients entiers d'éléments de θ, et l'on pose

$$(1) \qquad \pi_\theta = [\theta] \cup \Phi^+, \ \pi_\theta^- = [\theta] \cup \Phi^-, \ \alpha_\theta = \complement[\theta] \cap \Phi^+, \ \alpha_\theta^- = \complement[\theta] \cap \Phi^-.$$

Évidemment

$$(2) \qquad \pi_\theta^- = -\pi_\theta, \ [\theta] = \pi_\theta \cap \pi_\theta^- = (\pi_\theta)_s, \ (\pi_\theta)_u = \alpha_\theta, \ (\pi_\theta^-)_u = \alpha_\theta^- = -\alpha_\theta.$$

Les ensembles π_θ et π_θ^- sont les seules parties fermées de Φ contenant respectivement Φ^+ et Φ^- [6, § 7, n° 7, Prop. 20].

(ii) On écrira P_θ, P_θ^-, Z_θ, V_θ, V_θ^- respectivement pour G_{π_θ}, $G_{\pi_{\bar\theta}}$, $\mathscr{Z}(T_\theta)$, U_{α_θ}, $U_{\alpha_{\bar\theta}}$ (cf. 3.8; suivant la convention de 1.2, T_θ désigne la composante neutre de l'intersection des noyaux des caractères $a \in \theta$). On a les décompositions de Levi

$$(3) \qquad P_\theta = Z_\theta . V_\theta, \qquad P_\theta^- = Z_\theta . V_\theta^-.$$

Les sous-groupes P_θ $(\theta \subset \Delta)$ sont les *sous-groupes paraboliques standards* (associés à T et Φ^+). Évidemment $P_\varnothing = B$ et $P_\Delta = G$.

(iii) Posons $W_\theta = W(Z_\theta)$. C'est le sous-groupe de $W(G)$ engendré par les réflexions fondamentales r_b $(b \in \theta)$. D'après 2.4, et avec la convention d'écriture qui y est faite, on a

$$Z_\theta = (B \cap Z_\theta) . W_\theta . (B \cap Z_\theta),$$

d'où

$$(4) \qquad P_\theta = B . W_\theta . B, \qquad P_\theta^- = B^- . W_\theta . B^-.$$

Le groupe $V_\theta^- \cap P_\theta$ est stable par T, ne contient aucun groupe U_a $(a \in \Phi)$, donc est réduit à l'élément neutre (2.3). D'autre part, l'algèbre de Lie de $\mathfrak{g}$ est somme directe des algèbres de V_θ^- et P_θ. Par suite, *l'application produit induit un isomorphisme de variétés de $V_\theta^- \times P_\theta$ sur un ouvert de G, et la fibration de G par P_θ possède une section locale* (cf. aussi 3.24, 3.25).

4.3. De 4.2 et [33, Th. 3], il résulte que les sous-groupes P_θ $(\theta \subset \Delta)$ sont les seuls sous-groupes de G contenant B, et que tout sous-groupe parabolique de G est connexe, égal à son normalisateur, et conjugué à un et un seul sous-groupe parabolique standard. (Le fait que les axiomes de [33] sont vérifiés résulte immédiatement des calculs de [9], qu'on retrouvera d'ailleurs plus loin (5.16).) Il s'ensuit que si deux sous-groupes paraboliques P et P' de G sont conjugués, et si $P \cap P'$ est parabolique, alors $P = P'$.

4.4. *Proposition. — Soient P, P' deux sous-groupes paraboliques de G et L un sous-groupe de Levi de P.*

a) *Si $P \subset P'$, le transporteur $T(P, P') = \{x \in G \,|\, {}^x P \subset P'\}$ de P dans P' est égal à P'.*

b) *$(P \cap P') . R_u(P)$ est un sous-groupe parabolique de G. Il est égal à P si et seulement si P' contient un sous-groupe de Levi de P. C'est un sous-groupe de Borel de G si P' en est un.*

c) *Les sous-groupes paraboliques de G contenus dans P sont les produits semi-directs des sous-groupes paraboliques de L par $R_u(P)$. Deux sous-groupes paraboliques de G contenus dans P sont conjugués dans G si et seulement si leurs intersections avec L sont conjuguées dans L.*

a) *Si ${}^x P \subset P'$, alors ${}^x P' \cap P' \supset {}^x P$, donc ${}^x P' = P'$, et $x \in P'$ vu 4.3.*

b) *L'intersection $P \cap P'$ contient un tore maximal de G (2.4). Après conjugaison par un élément convenable de G, on peut donc supposer, dans les notations de 4.2, que $P = P_\theta$ et $P \cap P' \supset T$. Soit ψ l'ensemble des racines de P' par rapport à T. Il contient au moins un élément de chaque paire $(a, -a)$ $(a \in \Phi)$; il en est donc de même pour $\eta = (\pi_\theta \cap \psi) \cup \alpha_\theta$. Mais cet ensemble est clos dans Φ, car $\pi_\theta \cap \psi$ l'est et α_θ est un idéal dans π_θ. Vu la Prop. 20 de [6, § 7, n° 7], η contient l'ensemble des racines positives* de G pour un ordre convenable. Comme η est aussi l'ensemble des racines de $(P \cap P') . R_u(P)$ par rapport à T, il s'ensuit que $(P \cap P') . R_u(P)$ contient un sous-groupe de Borel de G (2.3), d'où la première assertion de *b)*.

Si P' contient un sous-groupe de Levi de P, il en est de même de $P \cap P'$, donc $(P \cap P') . R_u(P) = P$. Réciproquement, cette égalité entraîne que les éléments $a \in [\theta]$ sont tous des racines de $P \cap P'$ par rapport à T, donc que $P \cap P' \supset Z_\theta$. Supposons enfin que P' soit un sous-groupe de Borel de G. Alors l'ensemble ψ des racines de P' ne contient qu'un élément de chaque paire $(a, -a)$ $(a \in \Phi)$. Il en est alors de même de l'ensemble η considéré plus haut, qui est donc l'ensemble des racines positives pour un ordre convenable [6, *loc. cit.*], et $(P \cap P') . R_u(P)$ est alors le sous-groupe de Borel correspondant.

c) Tout sous-groupe de Borel de P contient $R_u(P)$ et se projette sur un sous-groupe de Borel de $P/R_u(P)$ vu [1, th. 22.1]. Par suite, les sous-groupes de Borel de P sont les produits semi-directs des sous-groupes de Borel de L avec $R_u(P)$, d'où la première partie de *c)*. Si deux sous-groupes paraboliques R, R' contenus dans P sont conjugués dans G, ils le sont dans P par *a)*, donc leurs images dans $P/R_u(P)$ sont aussi conjuguées, ce qui équivaut à dire que $L \cap R$ et $L \cap R'$ sont conjugués dans L. La réciproque est évidente.

4.5. *Corollaire.* — $P \cap P'$ *est connexe. Si* P' *est conjugué à* P *et contient* $R_u(P)$, *alors* $P = P'$.

Le groupe $(P \cap P') . R_u(P)$ est parabolique, donc connexe, donc égal à $(P \cap P')^0 . R_u(P)$, ce qui montre que $P \cap P'$ est engendré par $(P \cap P')^0$ et $H = (P \cap P') \cap R_u(P)$. Mais H est conjugué à un sous-groupe de U stable par T; il est donc connexe (2.3), d'où la première assertion.

Si $P' \supset R_u(P)$, alors $P \cap P'$ contient $(P \cap P') . R_u(P)$, qui est parabolique (4.4), et la deuxième assertion résulte de 4.3.

4.6. *Lemme.* — *Supposons* G *déployé sur* k. *Soient* P, P' *deux* k-*sous-groupes paraboliques de* G, *et* (T, U_a) *une donnée de déploiement sur* k *de* G. *Alors il existe* $g \in G_k$ *et* $\theta \subset \Delta$ *tels que* $^gP = P_\theta$ *et* $^gP' \supset T$. *En particulier,* $P \cap P'$ *contient un tore maximal conjugué à* T *sur* k, *et* $R(P)$, $R_u(P)$ *sont déployés sur* k.

Il existe un unique θ tel que P soit conjugué à P_θ sur $\overline{k}$ (4.3). Nous montrerons en premier lieu que P est conjugué à P_θ sur k. Le radical unipotent V_θ de P_θ est contenu dans un seul conjugué de P_θ (4.5), donc admet dans G/P un unique point fixe M, qui est alors rationnel sur k vu 2.7 (i), puisque V_θ est déployé sur k. L'application $g \mapsto g . M$ est alors un k-morphisme séparable de G sur G/P, constant sur les classes $x . P_\theta$; elle induit par suite un k-isomorphisme de variétés φ de G/P_θ sur G/P. Il est immédiat que $\varphi^{-1}(P)$ est le point fixe M' de P dans G/P_θ, qui est donc rationnel sur k. Mais la fibration de G par P_θ possède une section locale d'image V_θ^- (4.2), donc définie sur k, et la projection $G_k \to (G/P_\theta)_k$ est surjective (2.6, 2.14). Si $x \in G_k$ s'envoie sur M', alors $^xP_\theta = P$.

Cela montre que l'on peut se ramener au cas où $P = P_\theta$, et qu'il existe $x \in G_k$

tel que $P'' = {}^x P' \supset B$. Mais (2.11) on peut écrire $x = a \cdot n^{-1} \cdot b$ avec a, $b \in B_k$ et $n \in \mathcal{N}(T)_k$. On a alors ${}^b P = P$ et ${}^b P' = {}^n P'' \supset T$, ce qui démontre le lemme.

4.7. Proposition. — *Soient P, P′ deux k-sous-groupes paraboliques de G. Alors $P \cap P'$ contient un tore maximal de G défini sur k. Tout sous-groupe connexe k-fermé de G normalisé par $P \cap P'$ est défini sur k. En particulier, $P \cap P'$, $R(P \cap P')$, $R_u(P \cap P')$ sont définis sur k. Le groupe $P \cap P'$ possède des k-sous-groupes de Levi et deux quelconques d'entre eux sont conjugués par un unique élément de $R_u(P \cap P')_k$.*

G est déployé sur k_s (2.14), donc $P \cap P'$ est défini sur k_s (4.5, 4.6). Comme il est k-fermé, il est aussi défini sur k. D'autre part, il est de rang maximum (4.6), donc contient un tore maximal de G défini sur k (2.14). Les autres assertions sont alors conséquences de 3.13 et 3.14.

4.8. *Sous-groupes paraboliques opposés.* — Deux sous-groupes paraboliques de G sont *opposés* si leur intersection est un sous-groupe de Levi de chacun d'eux.

Il est clair que P_θ^- est le seul sous-groupe parabolique opposé à P_θ contenant Z_θ. Par conséquent, vu 4.3 et la conjugaison des décompositions de Levi dans un sous-groupe parabolique (4.7), étant donné un sous-groupe parabolique P, l'application $\alpha : P' \mapsto P \cap P'$ est une bijection de l'ensemble des sous-groupes paraboliques opposés à P sur l'ensemble des sous-groupes de Levi de P. Il résulte alors de 4.7 que *si P est un k-sous-groupe parabolique, G contient un k-sous-groupe parabolique opposé à* P, et que *deux tels sous-groupes sont conjugués par un unique élément de* $R_u(P)_k$. En particulier, P et P′ sont opposés si et seulement s'il existe $g \in G$ et $\theta \subset \Delta$ tels que ${}^g P = P_\theta$ et ${}^g P' = P_\theta^-$.

4.9. Deux classes de conjugaison $\mathscr{P}$, $\mathscr{P}'$ de sous-groupes paraboliques sont dites opposées s'il existe des sous-groupes paraboliques $P \in \mathscr{P}$, $P' \in \mathscr{P}'$ qui sont opposés. Vu 4.8, il existe une et une seule classe opposée à une classe donnée, et tout sous-groupe parabolique opposé à un élément de $\mathscr{P}$ (resp. $\mathscr{P}'$) fait partie de $\mathscr{P}'$ (resp. $\mathscr{P}$). Si $\mathscr{P} = \mathscr{P}'$, on dit que $\mathscr{P}$ est *auto-opposée*.

Fixons un ordre sur $\Phi(G)$ et soit i l'involution d'opposition de Φ (2.1). Si $n \in \mathcal{N}(T)$ représente l'élément de W qui applique Φ^+ sur Φ^-, alors

$$ {}^n P_\theta = P_{i(\theta)}^-, \qquad\qquad (\theta \subset \Delta); $$

par conséquent, si $P_\theta \subset \mathscr{P}$, alors la classe opposée contient $P_{i(\theta)}$. Si la symétrie $a \mapsto -a$ $(a \in \Phi)$ fait partie de $W(G)$, alors i est l'identité, donc toute classe de conjugaison de sous-groupes paraboliques est auto-opposée.

4.10. Proposition. — *Soient P, P′ deux sous-groupes paraboliques de G et $\mathscr{P}$, $\mathscr{P}'$ leurs classes de conjugaison. Alors les conditions suivantes sont équivalentes :*

a) P et P′ sont opposés.

b) $\mathscr{P}$ et $\mathscr{P}'$ sont opposées et P, P′ contiennent deux sous-groupes de Borel opposés de G.

c) $(P \cap P') \cdot R_u(P) = P$, $(P \cap P') \cdot R_u(P') = P'$ et P, P′ contiennent deux sous-groupes de Borel opposés de G.

d) *L'application produit est un isomorphisme de variétés de* $P \times R_u(P')$ *sur un ouvert de* G.

e) $P \cap R_u(P') = P' \cap R_u(P) = \{e\}$.

Si P et P' sont opposés, alors (4.8) il existe $g \in G$ et $\theta \subset \Delta$ tels que ${}^g P = P_\theta$ et ${}^g P' = P_\theta^-$, donc $a) \Rightarrow b)$, $c)$, $e)$ et, compte tenu de 4.2, $a) \Rightarrow d)$.

Supposons maintenant que P et P' contiennent deux sous-groupes de Borel opposés. Vu 4.8, on peut, après conjugaison par un élément convenable de G, supposer que $P \supset B$ et $P' \supset B^-$, donc (4.3) que $P = P_\theta$ et $P' = P_\eta^-$ $(\theta, \eta \subset \Delta)$. Mais P' est le seul élément de $\mathscr{P}'$ contenant B^- (4.3); si $\mathscr{P}'$ est opposée à $\mathscr{P}$, on doit donc avoir $\eta = \theta$, ce qui prouve que $b) \Rightarrow a)$. D'autre part, si $c)$ est vraie, alors $P \cap P' \supset Z_\theta$, Z_η et Z_θ, Z_η sont des sous-groupes de Levi de $P \cap P'$. Comme ils contiennent T, ils sont égaux (3.13), donc $c) \Rightarrow a)$.

Montrons maintenant que $d) \Rightarrow a)$. On peut supposer que $P = P_\theta$ et $P' \supset T$ (4.6). Alors $R_u(P') \subset V_\theta^-$ et $\dim R_u(P') = \dim G - \dim P = \dim V_\theta^-$, donc $R_u(P') = V_\theta^-$ et $P' = P_\theta^-$.

Il reste à prouver que $e) \Rightarrow a)$. On peut de nouveau admettre que $P = P_\theta$ et $P' \supset T$. L'égalité $P \cap R_u(P') = \{e\}$ entraîne $R_u(P') \subset V_\theta^-$. D'autre part, l'égalité $P' \cap V_\theta = \{e\}$ montre que le sous-groupe de Levi L de P' contenant T fait partie de Z_θ. L'ensemble $\Phi(T, L)$ est évidemment symétrique; s'il était strictement plus petit que $[\theta]$, ou bien si $R_u(P') \neq V_\theta^-$, alors on pourrait trouver une racine a telle que $U_a \notin P'$ et $U_{-a} \notin P'$, ce qui est absurde, donc $P' = P_\theta^-$.

4.11. Proposition. — *Soient* P, P' *des sous-groupes paraboliques opposés de* G. *Alors le sous-groupe engendré par* $R_u(P)$ *et* $R_u(P')$ *est distingué dans* G.

On peut supposer que $P = P_\theta$ et $P' = P_\theta^-$ (4.8). Le sous-groupe M engendré par $R_u(P) = V_\theta$ et $R_u(P') = V_\theta^-$ est fermé connexe [10, Exp. 3, th. 2] normalisé par V_θ, V_θ^- et Z_θ, donc par P_θ et P_θ^-. Par suite (2.3), le normalisateur de M est ouvert dans G, donc égal à G.

4.12. Lemme. — *Soient* P, P' *deux sous-groupes paraboliques de* G. *Alors l'ensemble* M *des éléments* $g \in G$ *tels que* ${}^g P$ *et* P' *contiennent des sous-groupes de Borel opposés est un ouvert de* G *de la forme* $P'.x.P$ $(x \in G)$; *il possède un point rationnel sur* k *si* k *est infini ou si* P *et* P' *sont définis sur* k.

Nous montrons tout d'abord que M est un ouvert de la forme $P'.x.P$ $(x \in G)$. Pour cela on peut remplacer P et P' par des sous-groupes conjugués, donc (4.3) supposer que P, P' contiennent respectivement B et B^-. D'après 4.8, $x \in M$ si et seulement s'il existe $g \in G$ tel que ${}^g B \subset {}^x P$ et ${}^g B^- \subset P'$, ce qui signifie (4.4 a)) que $g^{-1}.x \in P$ et $g \in P'$, d'où $x \in P'.P$. L'ensemble $P'.P$ est ouvert puisque $B^-.B$ l'est (2.3).

Si k est infini, alors G_k est Zariski dense dans G (2.14), donc $M_k \neq \varnothing$. Supposons maintenant k fini et P, P' définis sur k. Alors P et P' contiennent des sous-groupes de Borel définis sur k et deux sous-groupes de Borel définis sur k de G sont conjugués sur k ([23, p. 45], [5, 4.10]). Il suffit donc de faire voir que si B est un sous-groupe de Borel de G défini sur k, alors il existe un sous-groupe de Borel de G défini sur k et opposé à B, mais cela résulte de 4.8.

4.13. *Théorème.* — *Soient* P, P′ *des k-sous-groupes paraboliques de* G.

a) *La fibration de* G *par* P *possède une section locale définie sur* k; *la projection* $G_k \to (G/P)_k$ *est surjective.*

b) *Si* P *et* P′ *sont minimaux (parmi les k-sous-groupes paraboliques), ils sont conjugués sur* k.

c) *Si* P *et* P′ *sont conjugués sur une extension de* k, *ils sont aussi conjugués sur* k.

a) D'après 4.8, il existe un k-sous-groupe parabolique P″ opposé à P′; son radical unipotent est l'image d'une k-section locale de la fibration de G par P, vu 4.10; la deuxième partie de *a)* résulte alors de 2.6 et 2.14.

b) Si k est fini, P et P′ sont des k-sous-groupes de Borel de G [23, p. 45], donc sont conjugués sur k [5, 4.10].

Supposons donc k infini, et remarquons d'abord que si Q et Q′ sont des k-sous-groupes paraboliques minimaux, contenant des sous-groupes de Borel opposés, alors Q et Q′ sont opposés. En effet $(Q \cap Q') . R_u(Q)$ et $(Q \cap Q') . R_u(Q')$ sont des sous-groupes paraboliques (4.4) définis sur k (4.7), donc égaux à Q et Q′ respectivement, et la condition 4.10 *c)* est vérifiée.

Soit M (resp. M′) l'ensemble des éléments $g \in G$ tels que ${}^gP$ (resp. ${}^gP'$) et P contiennent des sous-groupes de Borel opposés; M et M′ sont des ouverts non vides de G (4.12). Il en est donc de même pour $M \cap M'$, et comme G_k est dense dans G, il existe $x \in M_k \cap M'_k$. D'après la remarque faite précédemment, ${}^xP$ et ${}^xP'$ sont tous deux opposés à P. Étant définis sur k, ils sont alors conjugués par un élément de $R_u(P)_k$ d'après 4.8, d'où *b)*.

c) P′ est le seul sous-groupe conjugué à P contenant $R_u(P')$ vu 4.5, donc P′ admet un seul point fixe A sur G/P, qui est nécessairement k-fermé. Mais (2.14) G est déployé sur k_s; donc (4.6) le radical unipotent de P′ est aussi déployé sur k_s, et (2.7 (i)) possède un point fixe dans G/P rationnel sur k_s. Par conséquent, A est aussi rationnel sur k_s, donc rationnel sur k. D'après *a)*, on peut trouver $x \in G_k$ qui est appliqué sur A par la projection naturelle. On a alors ${}^xP = P'$.

Remarques. — 1) La démonstration précédente montre que si k est infini, deux k-sous-groupes paraboliques minimaux possèdent un opposé commun défini sur k. On verra plus loin que cela est encore vrai sur un corps quelconque pour un k-groupe parabolique P et un élément arbitraire de la classe de conjugaison de P (6.27).

2) Vu 4.15, 4.13 *a)* est un cas particulier de 3.25.

4.14. *Corollaire.* — *La classe de conjugaison* $\mathscr{P}$ *de sous-groupes paraboliques contenant les k-sous-groupes paraboliques minimaux est auto-opposée.*

En effet la classe opposée contient un élément défini sur k (4.8), de même dimension que les k-sous-groupes paraboliques minimaux, donc minimal, et conjugué à un élément de $\mathscr{P}$ vu 4.13.

4.15. *Théorème.* — a) *Soient* S *un tore déployé sur* k *de* G *et* Φ^+ *(resp.* Φ^-*) l'ensemble des racines de* G *par rapport à* S *qui sont* > 0 *(resp.* < 0*) pour un ordre donné sur* $X^*(S)$. *Alors* G_{Φ^+}

et G_{Φ^-} *sont deux k-sous-groupes paraboliques opposés ayant* $\mathscr{Z}(S)$ *comme k-sous-groupe de Levi commun.*

b) *Soient* P *un k-sous-groupe parabolique de* G *et* S *un tore déployé sur k maximal de* R(P). *Alors il existe un ordre dans* $\Phi(S, G)$ *tel que* $P = G_{\Phi^+}$. *Les k-sous-groupes de Levi de* G *sont les centralisateurs des tores déployés sur k maximaux de* R(P).

a) Les groupes G_{Φ^+} et G_{Φ^-} sont définis sur k puisque Φ^+ et Φ^- le sont. Soit T un tore maximal de G défini sur k et contenant S (2.15 *d*)). Fixons sur $X^*(T)$ un ordre compatible avec l'ordre donné sur $X^*(S)$ (3.1). Alors $G_{\Phi^+} = G_\eta^{(T)} = G_\eta$, $G_{\Phi^-} = G_{-\eta}^{(T)}$, où η est l'ensemble des éléments de $\Phi(T, G)$ dont la restriction à S fait partie de $\Phi^+ \cup \{o\}$. Évidemment $\eta \supset \Phi(T, G)^+$, donc $G_\eta, G_{-\eta}$ sont paraboliques, et (4.2) il existe une partie θ de l'ensemble des racines simples de G par rapport à T telle que $G_\eta = P_\theta$, donc $G_{-\eta} = P_\theta^-$.

b) Le groupe $(\mathscr{Z}(S) \cap P)^0$ est défini sur k. Cela résulte de 10.5, ou peut se voir ainsi : P contient un tore maximal T′ de G défini sur k (2.14), qui fait donc partie d'une donnée de déploiement de G sur k_s (2.14, 2.10); S est conjugué dans P sur k_s à un sous-tore de T′ (2.10), donc $(\mathscr{Z}(S) \cap P)^0$ est conjugué sur k_s à un groupe normalisé par T′, k_s-fermé, donc défini sur k_s (3.13). Par suite, $(\mathscr{Z}(S) \cap P)^0$ est k-fermé et défini sur k_s, donc défini sur k.

Grâce à 2.14 *a)*, on peut trouver un tore maximal T de G défini sur k tel que $S \subset T \subset \mathscr{Z}(S) \cap P$. Soit $S' = R(P) \cap T$. C'est un tore maximal de R(P), et un sous-groupe de T qui est k-fermé, donc (1.6) défini sur k. Son centralisateur $\mathscr{Z}(S')$ est alors défini sur k (2.15 *d*)), et est un sous-groupe de Levi de P dont S′ est le centre connexe (4.2, 3.14). Par construction, S est le plus grand tore déployé sur k de S′, donc (3.6) les poids de S dans $R_u(P)$ sont $> o$ pour un ordre convenable sur $X^*(S)$. Mettons sur $X^*(T)$ un ordre compatible avec celui qui vient d'être fixé sur $X^*(S)$. Vu la structure des groupes paraboliques (4.2, 4.3), on a $\Phi(T, G) = \mu \cup \nu \cup (-\nu)$ où μ est l'ensemble des racines nulles sur S′ et ν l'ensemble des poids de T dans $R_u(P)$. Par conséquent, une racine est nulle sur S′ si et seulement si elle est nulle sur S, et $\mu \cup \nu$ est l'ensemble des racines dont la restriction à S est $\geq o$. On a alors $P = G_{\Phi^+}$ par définition (3.8) et $\mathscr{Z}(S) = \mathscr{Z}(S')$. Ainsi, le centralisateur d'un tore déployé sur k maximal de R(P) est un k-sous-groupe de Levi de P. Comme les k-sous-groupes de Levi de P sont conjugués sur k (3.14), cela termine la démonstration du théorème.

4.16. *Corollaire.* — *Soit* P *un k-sous-groupe parabolique de* G. *Alors les conditions suivantes sont équivalentes :*

(i) P *est minimal.*

(ii) R(P) *contient un tore déployé sur k maximal de* G.

(iii) *Les k-sous-groupes de Levi de* P *sont les centralisateurs des tores déployés sur k maximaux de* G *contenus dans* R(P).

(iv) *Tout tore déployé sur k maximal de* G *contenu dans* P *est contenu dans* R(P).

(i) $\Rightarrow$ (ii). Soient S un tore déployé sur k maximal de R(P) et S′ un tore déployé sur k maximal de G contenant S. Choisissons dans $\Phi = \Phi(S, G)$ un ordre tel que $P = G_{\Phi^+}$, ce qui est possible d'après 4.15 *b)*, et dans $\Psi = \Phi(S', G)$ un ordre compatible avec celui-ci. Le groupe $P' = G_{\Psi^+}^{(S')}$ est parabolique, admet $\mathscr{Z}(S')$ comme sous-groupe de Levi, est défini sur k (4.15 *a)*) et contenu dans P. Ces groupes sont donc égaux, et $\mathscr{Z}(S')$ est un sous-groupe de Levi de P, donc $S' \subset R(P)$ et finalement $S' = S$.

(ii) $\Rightarrow$ (iii). Cela résulte de 4.15 *b)* et de 4.7.

(iii) $\Rightarrow$ (iv) Soit S (resp. S') un tore déployé sur k maximal de G contenu dans P (resp. R(P)). On a $P = \mathscr{Z}(S') . R_u(P)$. La projection S'' de S sur le facteur $\mathscr{Z}(S')$ de cette décomposition est un tore déployé sur k (1.9) ; il en est donc de même de S'.S'' (1.4). Puisque S' est maximal, cela signifie que $S'' \subset S'$, d'où $S \subset S' . R_u(P) = R(P)$.

(iv) $\Rightarrow$ (i). Soit P' un k-sous-groupe parabolique minimal contenu dans P, et S un tore déployé sur k maximal de G contenu dans R(P'); en vertu des implications (i) $\Rightarrow$ (ii) $\Rightarrow$ (iii) déjà établies, un tel tore S existe et $\mathscr{Z}(S) \subset P'$. D'autre part, $S \subset R(P)$ et on a (4.15) $P = \mathscr{Z}(S) . R_u(P) \subset P'$, d'où $P = P'$.

4.17. *Corollaire.* — *Le groupe* G *possède un k-sous-groupe parabolique propre si et seulement s'il contient un tore déployé sur k non central.*

C'est une conséquence évidente de 4.15 *a)* et 4.16.

4.18. *Corollaire.* — *L'intersection de deux k-sous-groupes paraboliques* P, P' *de* G *contient le centralisateur d'un tore déployé sur k maximal de* G.

On peut supposer P et P' minimaux. D'après 4.7, $P \cap P'$ contient un tore maximal T de G défini sur k. Le groupe $T \cap R(P)$ est un tore maximal de R(P), défini sur k vu 1.6. En vertu de 4.2, 4.3 et 2.15 *d)*, $\mathscr{Z}(T \cap R(P))$ est un k-sous-groupe de Levi de P, donc (4.6) $T \cap R(P)$ contient un tore déployé sur k maximal de G. Ce dernier est alors nécessairement égal au plus grand tore T_d déployé sur k de T. De même, $T_d \subset R(P')$, et l'inclusion $\mathscr{Z}(T_d) \subset P \cap P'$ est conséquence de 4.16.

4.19. *Corollaire.* — *Soient* $P \subset P'$ *des k-sous-groupes paraboliques de* G. *Alors les tores déployés sur k maximaux de* R(P') *sont les intersections de* R(P) *avec les tores déployés sur k maximaux de* R(P).

Comme $R(P) \supset R(P')$, cela est vrai pour au moins un k-tore déployé sur k maximal de R(P'), donc pour tous vu 4.15, la conjugaison sur k des k-sous-groupes de Levi (3.14), et le fait que R(P') est distingué dans R(P).

4.20 *Corollaire.* — *Soient* P, P' *des k-sous-groupes paraboliques de* G. *Alors* $P = P'$ *si et seulement si* $P_k = P'_k$.

Si k est infini, cela résulte du fait que P_k et P'_k sont denses dans P et P' (2.14, 3.20), mais la démonstration ci-dessous vaut sans hypothèse sur k.

Soient Q, Q' des k-sous-groupes paraboliques minimaux de G contenus dans P et P' respectivement. Vu 4.16, 4.18, on peut trouver un tore S déployé sur k maximal de G tel que $\mathscr{Z}(S)$ soit un k-sous-groupe de Levi commun à Q et Q'. Soit $V = R(P) \cap S$. D'après 4.19 et 4.15 il existe une partie α quasi-close de $\Phi(V, G)$ telle que $P = G_\alpha^{(V)}$. Si μ est l'ensemble des éléments de $\Phi(S, G)$ dont la restriction à V fait partie de $\alpha \cup \{o\}$, alors μ est quasi-clos et $P = G_\mu^{(S)}$ (3.8). De même, il existe une partie quasi-close ν de $\Phi(S, G)$ telle que $P' = G_\nu^{(S)}$. Notre assertion est maintenant conséquence de 3.23.

4.21. *Théorème. — Les tores déployés sur k maximaux de G sont conjugués sur k.*

Soient S, S′ deux tores déployés sur k maximaux de G. D'après 4.15 et 4.16, il existe deux k-sous-groupes paraboliques minimaux P et P′ tels que $\mathscr{Z}(S)$ et $\mathscr{Z}(S')$ soient des k-sous-groupes de Levi de P et P′ respectivement. Vu 4.13, P et P′ sont conjugués sur k et d'après 4.7, les k-sous-groupes de Levi de P sont conjugués sur k. Par conséquent il existe $g \in G_k$ tel que ${}^g\mathscr{Z}(S) = \mathscr{Z}(S')$. Mais alors ${}^gS$ et S′ coïncident avec le plus grand tore décomposé sur k du centre Z de $\mathscr{Z}(S)$ (rappelons que Z^0 est défini sur k vu 2.15 *a)*).

Remarque. — Le théorème vaut plus généralement pour un k-groupe algébrique connexe dont le radical unipotent est déployé sur k (11.6).

4.22. *Corollaire. — Soient S, S′ des tores déployés sur k maximaux de G et A, A′ des parties de S_k et S'_k ou des sous-tores de S et S′ respectivement. Soit $x \in G_k$ tel que ${}^xA = A'$. Alors il existe $y \in G_k$ tel que ${}^yS = S'$ et ${}^ya = {}^xa$ ($a \in A$).*

Soit $S'' = {}^xS$. Les groupes S′ et S″ sont des tores déployés sur k maximaux du groupe $\mathscr{Z}(A')^0$, qui est défini sur k (1.6, 2.15 *d)*). D'après le théorème, on peut trouver $z \in \mathscr{Z}(A')_k^0$ tel que ${}^zS'' = S'$. Alors $y = z.x$ vérifie nos conditions.

4.23. *Définitions.* — La dimension (commune vu 4.21) des tores déployés sur k maximaux de G est le *k-rang de* G et est notée $r_k(G)$. Le *k-rang semi-simple* de G est le k-rang du groupe dérivé de G. Si k est algébriquement clos on omet en général le préfixe ou indice k. Le groupe G est *anisotrope sur k* si son k-rang est nul ou, ce qui revient au même d'après 4.17, s'il ne contient pas de k-sous-groupe parabolique propre et si son centre connexe est anisotrope sur k.

4.24. Supposons que G soit la composante neutre du groupe orthogonal $\mathbf{O}(Q)$ d'une forme quadratique Q sur un k-espace vectoriel V. Alors G *est anisotrope sur k si et seulement si V_k ne contient pas de vecteur isotrope non nul*, autrement dit si Q ne représente pas zéro sur k.

En effet, si V_k contient un vecteur isotrope non nul, on peut classiquement choisir des coordonnées dans V_k telles que $Q = c.x_1.x_2 + Q'(x_3, \ldots, x_n)$, $(c \in k^*)$, et G contient les transformations $(x_1 \mapsto \lambda.x_1, x_2 \mapsto \lambda^{-1}.x_2, x_i \mapsto x_i \ (i \geq 3))(\lambda \in \bar{k}^*)$ qui définissent un tore déployé sur k de G.

Réciproquement, si G contient un tore $S \neq \{e\}$ déployé sur k, alors, en diagonalisant S sur k, on voit qu'il existe $x \in V_k - \{o\}$, et un entier $m \geq 1$ tels que $Q(x) = Q(\lambda^m.x) = \lambda^{2m}.Q(x)$ quel que soit $\lambda \in \bar{k}$, d'où $Q(x) = o$.

De manière similaire, on voit que si Q est non-dégénérée, le k-rang de G est égal à l'indice de Q, c'est-à-dire à la dimension des espaces isotropes maximaux (rationnels sur k bien entendu) de Q.

4.25. *Proposition.* — *Soient* G′ *un k-groupe réductif et* $f : \mathrm{G} \to \mathrm{G}'$ *une k-isogénie séparable. Alors les tores déployés sur k maximaux de* G′ *(resp. de* G*) sont les images par f (resp. les composantes neutres des images réciproques) des tores déployés sur k maximaux de* G *(resp.* G′*). En particulier,* G *et* G′ *ont le même k-rang et le même k-rang semi-simple.*

(La notion de morphisme séparable est rappelée en 10.3; ici, cela revient à exiger que f induise un isomorphisme des algèbres de Lie.)

Comme l'image d'un k-tore déployé sur k par un k-morphisme est un tore déployé sur k (1.3), il suffit, vu 4.21, de montrer que si S′ est un tore déployé sur k maximal de G′, alors $\mathrm{S} = (f^{-1}(\mathrm{S}'))^0$ est un tore déployé sur k de G.

Soit T′ un tore maximal de G′ défini sur k et contenant S′ ($2.15\,d)$). Alors $\mathrm{T} = (f^{-1}(\mathrm{T}'))^0$ est un tore maximal de G [1, th. 22.1]. Il contient par conséquent le noyau de f, donc T est toute l'image réciproque de T′. Comme f est séparable, T est un cycle rationnel sur k [22, Prop. 1, Cor.] irréductible, donc T est un k-sous-groupe [39, Prop. 1, p. 208]; 1.8 montre alors que S est le plus grand tore déployé de T.

4.26. *Remarque.* — La proposition précédente n'est pas vraie sans restriction sur f, comme le montre l'exemple suivant.

Supposons k non parfait de caractéristique deux, et soient t un élément de k qui ne soit pas un carré dans k, Q la forme quadratique à trois variables donnée par $\mathrm{Q}(x, y, z) = x^2 + t.y.z$, et $\mathrm{G} = \mathbf{O}(\mathrm{Q})$ le groupe orthogonal de Q. La forme Q ne représente pas zéro sur k, donc G est anisotrope sur k (4.24). D'autre part, la forme bilinéaire associée à Q admet la droite $y = z = 0$ comme noyau. On en déduit immédiatement que les éléments de $\mathrm{G}_{\bar{k}}$ sont les matrices de la forme

$$M = \begin{pmatrix} 1 & (t.a.b)^{1/2} & (t.c.d)^{1/2} \\ 0 & a & c \\ 0 & b & d \end{pmatrix}, \qquad (a.d - b.c = 1),$$

donc que $M \mapsto \begin{pmatrix} a & c \\ b & d \end{pmatrix}$ est une k-isogénie de G sur $\mathbf{SL}_2 = \mathrm{G}'$. Cependant $r_k(\mathrm{G}) = 0$, et $r_k(\mathrm{G}') = 1$.

4.27. *Proposition.* — *Soient* $\mathrm{G}_1, \ldots, \mathrm{G}_n$ *des k-sous-groupes distingués connexes dont* G *soit le produit presque direct* (0.6), S *un tore déployé sur k maximal de* G *et* $\mathrm{S}_i = (\mathrm{S} \cap \mathrm{G}_i)^0$ $(1 \leqq i \leqq n)$. *Alors* S *est le produit presque direct des* S_i *et* S_i *est un tore déployé sur k maximal de* G_i $(1 \leqq i \leqq n)$. *En particulier* $r_k(\mathrm{G}) = \sum_i r_k(\mathrm{G}_i)$.

Grâce à 4.25, on peut se ramener au cas où G est le produit direct des G_i. La projection S'_i de S sur G_i $(1 \leqq i \leqq n)$ est un tore déployé sur k (1.3), donc $\prod \mathrm{S}'_i = \mathrm{S}$, et $\mathrm{S}'_i = \mathrm{S}_i$ $(1 \leqq i \leqq n)$. D'autre part, si S''_i est un tore déployé sur k de G_i contenant S_i, alors $\mathrm{S}.\mathrm{S}''_i$ est un tore déployé sur k (1.3, 1.9), donc égal à S, d'où $\mathrm{S}''_i = \mathrm{S}_i$, et S_i est bien maximal dans G_i.

4.28. *Corollaire.* — G *possède un plus grand k-sous-groupe connexe* H *distingué anisotrope sur k. Le groupe* G *est le produit presque direct de ce sous-groupe, de ses k-sous-groupes distingués presque simples sur k de k-rang* >0 *et de son plus grand tore déployé sur k central.*

Le sous-groupe en question est engendré par les k-sous-groupes distingués connexes presque simples sur k et anisotropes sur k de $\mathscr{D}G$ (voir 2.15 b)) et par le plus grand sous-tore anisotrope sur k de $\mathscr{Z}(G)$ (2.15 a), 1.8).

Nous concluons ce paragraphe par deux résultats qui seront utilisés au § 8.

4.29. *Proposition.* — *Soient* $\{H_i\}_{i \in I}$ *un ensemble de k-sous-groupes connexes de* G, P *un k-sous-groupe parabolique minimal de* G *et* H *le sous-groupe engendré par* $R_u(P)$ *et les* H_i $(i \in I)$. *Supposons que pour tout* $i \in I$, *le groupe* H_i *soit normalisé par un k-sous-groupe parabolique* P_i. *Alors* $\mathscr{N}(H) = P'$ *est un k-sous-groupe parabolique contenant* P, *et* $R_u(P') = R_u(H)$.

Le groupe $(P \cap P_i).R_u(P)$ est un k-sous-groupe parabolique contenu dans P, donc égal à P, et P_i contient un k-sous-groupe de Levi L_i de P (4.4, 4.7). Le groupe F_i engendré par H_i et $R_u(P)$ est normalisé par L_i, donc par $P = L_i.R_u(P)$ et par conséquent $\mathscr{N}(H) \supset P$. Le groupe $\mathscr{N}(H)$, étant k-fermé, est alors défini sur k (4.7). Puisque $P' \supset P$, on a $R_u(P') \subset R_u(P) \subset H$; comme en outre $R_u(P')$ est normalisé par $H \subset P'$, on a aussi $R_u(P') \subset R_u(H)$. D'autre part $R_u(H)$ est normalisé par P', donc $R_u(H) \subset R_u(P')$.

4.30. *Corollaire.* — *Soit* $\{P_i\}_{i \in I}$ *un ensemble de k-sous-groupes paraboliques minimaux. Alors le sous-groupe* H *engendré par les groupes* $R_u(P_i)$ *est distingué dans le groupe engendré par les* P_i $(i \in I)$.

En effet, 4.29 montre que P_i normalise H quel que soit $i \in I$.

§ 5. RACINES, GROUPES DE WEYL
ET DÉCOMPOSITIONS CELLULAIRES RELATIFS

Dans tout ce paragraphe, G *est réductif connexe, et* S *est un tore déployé sur k maximal de* G.

5.1. *Groupe de Weyl relatif et k-racines.* — Les racines de G par rapport à S seront appelées *k-racines* ou racines relatives à k de G (par rapport à S si cette précision paraît nécessaire). On écrira ${}_k\Phi(G)$ ou ${}_k\Phi$ pour $\Phi(S, G)$. Le *groupe de Weyl relatif à* k de G est le quotient $\mathscr{N}(S)/\mathscr{Z}(S)$ et est noté ${}_kW$, ou ${}_kW(G)$, ${}_kW(S, G)$. Comme $\mathscr{Z}(S)$ est la composante neutre de $\mathscr{N}(S)$, ce groupe est fini. Il opère canoniquement sur $X^*(S)$, (en laissant ${}_k\Phi$ stable), sur $X_*(S)$ (en permutant les noyaux des k-racines), donc aussi sur $X^*(S) \otimes A$ et $X_*(S) \otimes A$, où A est un anneau quelconque. Un produit scalaire sur $X^*(S) \otimes \mathbf{R}$ ou $X_*(S) \otimes \mathbf{R}$ invariant par ${}_kW$ est dit *admissible*, et sera noté en général (,). Comme ${}_kW$ est fini, il en existe toujours. On identifiera souvent $X^*(S) \otimes \mathbf{R}$ et $X_*(S) \otimes \mathbf{R}$ au moyen d'un produit scalaire admissible.

Si k est algébriquement clos, ou bien si G est déployé sur k, alors ${}_kW$ et ${}_k\Phi$ s'identifient au groupe de Weyl et aux racines usuels, et l'on omettra la mention de k.

Vu 4.21, les ensembles de racines $\Phi(S, G)$ et les groupes de Weyl relatifs associés à deux tores déployés sur k maximaux sont isomorphes.

5.2. Étant donné $a \in \Phi(S, G)$, on désignera par $_kU_a$ ou $U_{(a)}$ le groupe noté $G^{*(S)}_{(a)}$ dans 3.8, où (a) est l'ensemble des k-racines qui sont multiples entiers positifs de a. Ce groupe est unipotent (3.8 (iv)), défini sur k (3.13).

Soient $w \in {}_kW$, n un représentant de w dans $\mathcal{N}(S)$, et ψ une partie quasi-close de $_k\Phi$. Dans les notations de 3.8, il est clair que

$$(1) \qquad\qquad {}^nG_\psi = G_{w(\psi)}, \qquad {}^nG^*_\psi = G^*_{w(\psi)}.$$

Les groupes ${}^nG_\psi$ et ${}^nG^*_\psi$ ne dépendent donc que de l'image w de n dans $_kW$ et seront aussi notés ${}^wG_\psi$ et ${}^wG^*_\psi$. Ces groupes sont définis sur k (3.13).

Soient A une partie de $\mathcal{N}(S)$ (resp. $\mathcal{N}(S)_k$) et A′ son image canonique dans $_kW(G)$. Des classes d'éléments de G (resp. G_k) telles que P.A, P.A.P (resp. $P_k.A$, $P_k.A.P_k$) qui ne dépendent que de A′ seront aussi notées P.A′, P.A′.P (resp. $P_k.A'$, $P_k.A'.P_k$).

5.3. *Théorème.* — *Le rang de $_k\Phi(G)$ est égal au k-rang semi-simple de G (4.23). Le groupe $_kW(G)$, vu comme groupe d'automorphismes de $X_*(S) \otimes \mathbf{R}$, muni d'un produit scalaire admissible, est engendré par les symétries par rapport aux hyperplans annulant une k-racine. Chaque composante connexe de $\mathcal{N}(S)$ rencontre G_k.*

Soit S_a $(a \in {}_k\Phi)$ la composante neutre du noyau de a, et soit Z l'intersection des groupes S_a. Le groupe Z^0 est la composante neutre de $\mathcal{Z}(G) \cap S$, car la représentation adjointe de G a $\mathcal{Z}(G)$ comme noyau [23, lemma 1, p. 39]. (Cela résulte aussi du fait que si $z \in Z$, alors toutes les racines de G par rapport à un tore maximal de G contenant S sont nulles sur z, et de 2.3.) Comme $r_k(G) = r_k(\mathcal{Z}(G)) + r_k(\mathcal{D}G)$ vu 2.2 (ii) et 4.27, cela établit la première assertion.

Montrons maintenant que si $r_k(\mathcal{D}G) \neq 0$, alors $\mathcal{N}(S)_k \neq \mathcal{Z}(S)_k$. D'après 4.15, il existe deux k-sous-groupes paraboliques opposés P et P′ ayant $\mathcal{Z}(S)$ comme sous-groupe de Levi commun. Vu 4.16, ces groupes sont minimaux parmi les k-sous-groupes paraboliques, donc on peut trouver $x \in G_k$ tel que $^xP = P'$ (4.13). Les groupes $^x(\mathcal{Z}(S))$ et $\mathcal{Z}(S)$ sont deux k-sous-groupes de Levi de P′, donc (4.7), quitte à multiplier x par un élément de $R_u(P')_k$, on peut supposer que x normalise $\mathcal{Z}(S)$. Il doit alors aussi normaliser S, qui est le plus grand tore déployé sur k du centre de $\mathcal{Z}(S)$; d'autre part, il n'agit pas trivialement sur $_k\Phi$ vu 4.15, donc $x \in \mathcal{N}(S)_k$ et $x \notin \mathcal{Z}(S)_k$.

Prouvons maintenant que $_kW$ contient le groupe W′ engendré par les symétries r_a par rapport aux hyperplans $X_*(S_a) \otimes \mathbf{R}$ de $X_*(S) \otimes \mathbf{R}$ $(a \in {}_k\Phi)$.

Soit $M = \mathcal{Z}(S_a)$. C'est un k-groupe réductif connexe (2.15 d)). Il contient le groupe unipotent $U_{(a)}$ (5.2), donc $\mathcal{Z}(S) \neq M$, et S n'est pas central dans M. Vu 4.27, et le fait que S_a est de codimension un dans S, cela entraîne que $r_k(\mathcal{D}M) = 1$ et que $V = (S \cap \mathcal{D}M)^0$ est un tore déployé sur k maximal de $\mathcal{D}M$. D'après ce qui a déjà été démontré, on peut trouver dans $(\mathcal{D}M)_k$ un élément x normalisant, mais ne centralisant pas V. Comme dim V $= 1$, on a nécessairement $(\mathrm{Int}\, x)\,(t) = t^{-1}$ $(t \in V)$. D'autre part, x centralise S_a, donc normalise S et induit un automorphisme involutif non trivial de $X_*(S) \otimes \mathbf{R}$ laissant $X_*(S_a) \otimes \mathbf{R}$ fixe point par point. C'est nécessairement une symétrie par rapport à cet hyperplan pour tout produit scalaire admissible, donc $r_a \in {}_kW$ et $W' \subset {}_kW$; de plus, toute composante connexe de $\mathcal{N}(S)$ se projetant dans W' contient un élément de G_k.

Il reste à faire voir que ${}_kW = W'$. Vu les propriétés classiques des groupes engendrés par des réflexions [6], il suffit pour cela de prouver que si $w \in {}_kW$ laisse stable une chambre de Weyl C de W' dans $X_*(S) \otimes \mathbf{R}$, alors w est l'identité. Il existe un point $D \in C$ fixe par w. On peut donc trouver un ordre sur $X^*(S)$ tel que si $a \in X^*(S)$ n'est pas nul sur D, alors $a > o$ si et seulement si $a(D) > o$. Comme les racines ne sont pas nulles sur D, l'élément w permute l'ensemble ψ des k-racines $> o$, donc tout $n \in \mathcal{N}(S)$ représentant w normalise le groupe $G_\psi = P$, qui est un k-groupe parabolique minimal vu 4.15, 4.16. Comme P est son propre normalisateur (4.3), on a $n \in P$ et il suffit de montrer que

$$\mathcal{N}(S) \cap P = \mathcal{Z}(S),$$

ce qui est immédiat : en effet, on peut écrire $n = u.v$ $(u \in \mathcal{Z}(S), v \in R_u(P))$ (4.16), d'où $v \in R_u(P) \cap \mathcal{N}(S)$, et l'on sait que dans un groupe résoluble connexe, le normalisateur d'un tore est égal à son centralisateur [1, prop. 10.2].

5.4. Corollaire. — *Le plus grand k-sous-groupe M invariant connexe anisotrope sur k de $\mathcal{Z}(S)$ est distingué dans $\mathcal{N}(S)$.*

Le groupe M est évidemment normalisé par $\mathcal{N}(S)_k$. D'autre part (4.28), $\mathcal{Z}(S)$ est le produit presque direct de S et de M, donc M est invariant dans $\mathcal{Z}(S)$, et par suite dans $\mathcal{N}(S) = \mathcal{N}(S)_k . \mathcal{Z}(S)$.

Remarque. — Un peu plus généralement, cela montre que tout k-sous-groupe distingué de $\mathcal{Z}(S)$ invariant par tout k-automorphisme de $\mathcal{Z}(S)$ est distingué dans $\mathcal{N}(S)$.

5.5. Corollaire. — *Soient K une extension de k, T un tore déployé sur K maximal de G contenant S. Soient $\mathcal{N}(S, T) = \mathcal{N}(S) \cap \mathcal{N}(T)$ et ${}_KW_k = \mathcal{N}(S, T)/\mathcal{Z}(T)$. Alors $\mathcal{N}(S) = \mathcal{N}(S, T) . \mathcal{Z}(S)$, et ${}_kW$ s'identifie à la restriction de ${}_KW_k$ à S.*

Pour démontrer la première égalité, il suffit, vu le théorème, de faire voir que $\mathcal{N}(S)_k \subset \mathcal{N}(S, T) . \mathcal{Z}(S)$, ce qui résulte de 4.22. La deuxième assertion n'est qu'une autre manière d'énoncer la première.

5.6. Corollaire. — *Soient $f : G \to G'$ une k-isogénie séparable de G sur un k-groupe réductif connexe G', et $S' = f(S)$. Alors f induit une isogénie de $\mathcal{N}(S)$ sur $\mathcal{N}(S')$, un isomorphisme de ${}_kW(G)$ sur ${}_kW(G')$ et $f^* : X^*(S') \to X^*(S)$ induit une bijection de ${}_k\Phi(G')$ sur ${}_k\Phi(G)$.*

Le groupe S' est un tore déployé sur k maximal de G' d'après 4.25, et f induit un isomorphisme des algèbres de Lie, d'où la dernière partie du corollaire. Les autres assertions résultent alors du théorème.

5.7. Proposition. — *Soient ψ une partie de ${}_k\Phi(G)$ qui contient tous les multiples rationnels de ses éléments faisant partie de ${}_k\Phi(G)$, w un élément de ${}_kW(G)$ qui soit produit de réflexions r_a $(a \in \psi)$, et $c \in X^*(S)$. Alors $w(c) - c$ est combinaison linéaire à coefficients entiers d'éléments de ψ.*

Soient T un tore maximal de G contenant S et $j : X^*(T) \to X^*(S)$ l'homomorphisme de restriction. Soit $\eta = j^{-1}(\psi) \cap \Phi(T, G)$. D'après 5.5, $W = W(T, G)$ contient au moins un élément w_a dont la restriction à S est r_a $(a \in {}_k\Phi)$. Un tel élément agit trivialement sur l'hyperplan V_a de $X_*(S) \otimes \mathbf{R}$ annulant a, donc [6] est produit de réflexions $r_b \in W$, où b appartient à l'ensemble des racines qui s'annulent sur V_a. Ces dernières font partie de η, vu l'hypothèse faite sur ψ, donc W contient un élément w' dont la restriction à S est w, et qui est produit de réflexions r_b $(b \in \eta)$. Comme d'autre part S est facteur direct dans T [1, § 7], j est surjectif. On est donc ramené au cas où S = T est un tore maximal de G et où ${}_kW = W$, ${}_k\Phi = \Phi(G)$, pour lequel notre assertion est connue [6, § 7, no. 9, prop. 27].

5.8. *Corollaire. — Soient (,) un produit scalaire admissible sur $X^*(S) \otimes \mathbf{R}$, $a \in \Phi_k(G)$ et $c \in X^*(S)$. Alors $2(a, c)/(a, a)$ est un entier. En particulier, ${}_k\Phi(G)$ est un système de racines au sens de 2.1 dans $(X^*(S), N)$ où N désigne l'ensemble des caractères de S qui sont triviaux sur $S \cap \mathscr{D}G$.*

La restriction $X^*(S) \to X^*(S')$, où $S' = (S \cap \mathscr{D}G)^0$, est un isomorphisme de ${}_k\Phi(G)$ sur ${}_k\Phi(\mathscr{D}G)$, et le rang de ${}_k\Phi(\mathscr{D}G)$ est égal à $r_k(\mathscr{D}G)$ d'après 5.3, donc ${}_k\Phi(G)$ engendre dans $X^*(S) \otimes \mathbf{Q}$ un sous-espace supplémentaire de $N \otimes \mathbf{Q}$. D'autre part, on a

$$r_a(c) = c - 2 \cdot (a, c) \cdot (a, a)^{-1} \cdot a,$$

donc $2(a, c) \cdot (a, a)^{-1} \in \mathbf{Z}$ d'après 5.7. Les autres conditions imposées à un système de racines sont vérifiées vu 5.3.

5.9. *Corollaire. — Si P est un k-sous-groupe parabolique minimal contenant S, alors $P = G_\psi^{(S)}$, où ψ est l'ensemble des k-racines positives pour un ordre convenable, et réciproquement. Le groupe ${\rm Int}_G \mathscr{N}(S)$ induit un groupe de permutations simplement transitif, isomorphe à ${}_kW$, des k-sous-groupes paraboliques minimaux contenant S.*

La première assertion résulte de 4.15, 4.16, la deuxième de 5.3, 5.8 et de la simple transitivité du groupe de Weyl sur les chambres de Weyl.

5.10. *Proposition. — Soient $G_1, \ldots, G_n$ des k-sous-groupes distingués connexes de G dont G est le produit presque direct et $S_i = (S \cap G_i)^0$ $(1 \leqq i \leqq n)$. Alors, pour tout $a \in {}_k\Phi(G)$, il y a une et une seule valeur de l'indice i telle que la restriction de a à S_i ne soit pas nulle ; le groupe ${}_kU_a$ est contenu dans le groupe G_i correspondant. Si Φ_i est l'ensemble des k-racines non nulles sur S_i, alors Φ est somme directe des Φ_i et l'homomorphisme de restriction $X^*(S) \to X^*(S_i)$ induit un isomorphisme de Φ_i sur ${}_k\Phi(G_i)$.*

Soient T un tore maximal de G contenant S et $b \in \Phi(T, G)$ une racine dont la restriction à S soit a. Pour tout i, on a les implications suivantes

$$a|_{S_i} \neq 0 \Rightarrow b|_{S_i} \neq 0 \Rightarrow U_b \subset G_i \Rightarrow b|_{T \cap G_j} = 0 \ (j \neq i) \Rightarrow a|_{S_j} = 0 \ (j \neq i) \Rightarrow a|_{S_i} \neq 0$$

dont la deuxième et la troisième résultent de 2.3 et les autres sont évidentes. Ces assertions sont donc toutes équivalentes, d'où la première partie de la conclusion. La seconde en découle immédiatement.

5.11. *Corollaire. — Les k-sous-groupes distingués connexes presque simples sur k non anisotropes sur k de* G *sont les groupes* G_ψ^*, *où* ψ *parcourt les facteurs directs irréductibles de* $_k\Phi(G)$. *En particulier, si* G *est presque simple sur k,* $_k\Phi$ *est un système de racines irréductible. Tout k-sous-groupe distingué connexe de* G *est produit d'un groupe* G_ψ^*, *où* ψ *est un idéal de* Φ, *d'un tore déployé sur k et d'un groupe anisotrope sur k déterminés de façon unique.*

Soit ψ un facteur direct irréductible de $_k\Phi$. C'est un idéal, donc G_ψ^* est un k-sous-groupe distingué connexe de G; la prop. 5.10 montre que G_ψ^* est presque simple sur k et que tout k-groupe distingué connexe presque simple sur k, de k-rang non nul, s'obtient ainsi. La dernière assertion résulte alors de 4.28.

5.12. *k-sous-groupes paraboliques standards.* — On fixe un ordre sur $_k\Phi$ et on note $_k\Delta$ l'ensemble des k-racines simples correspondantes. A toute partie θ de $_k\Delta$, on peut associer comme en 4.2 des parties π_θ, π_θ^-, etc. de $_k\Phi$ et des sous-groupes P_θ, P_θ^-, etc., qui sont définis sur k vu 3.13; les paragraphes 4.2 (i), (ii) restent valables, une fois T et Φ remplacés par S et $_k\Phi$. Pour éviter des confusions avec les notions correspondantes sur $\bar{k}$, on écrira souvent $_kP_\theta$, $_kP_\theta^-$, $_kZ_\theta$, $_kV_\theta$, $_kV_\theta^-$ pour les groupes qui, dans les notations de 3.8, seraient désignés par $G_{\pi_\theta}^{(S)}$, $G_{-\pi_\theta}^{(S)}$, $\mathscr{Z}(S_\theta)$, $U_{\alpha_\theta}^{(S)}$ et $U_{-\alpha_\theta}^{(S)}$. Les groupes $_kP_\theta$ $(\theta \subset {}_k\Delta)$ sont les k-*sous-groupes paraboliques standards de* G. Le groupe $_kP_\theta$ est minimal si $\theta = \varnothing$, égal à G si $\theta = {}_k\Delta$. Le radical unipotent $_kV_\varnothing$ de $_kP_\varnothing$ sera aussi noté $_kU$. Jusqu'au n° 5.20 inclus, on écrira P pour $_kP_\varnothing$.

5.13. Soit $w \in {}_kW$ et soient $\mu_w = {}_k\Phi^+ \cap w(_k\Phi^-)$ et $\nu_w = {}_k\Phi^+ \cap w(_k\Phi^+)$. Ce sont des ensembles de k-racines convexes dont $_k\Phi^+$ est réunion disjointe. On posera

$$(1) \qquad\qquad _kU_w' = U_w' = U_{\mu_w}^{(S)}, \qquad _kU_w'' = U_w'' = U_{\nu_w}^{(S)}.$$

D'après 3.11, l'application produit est un k-isomorphisme de variétés de $U_w' \times U_w''$ sur $_kU$.

Plus généralement, soient $s \in {}_kW$,

$$\mu_{s,w} = s(_k\Phi^+) \cap w(_k\Phi^-), \qquad \nu_{s,w} = s(_k\Phi^+) \cap w(_k\Phi^+).$$

Ce sont des ensembles convexes de k-racines, dont $s(_k\Phi^+)$ est la réunion disjointe. On a

$$(2) \qquad\qquad U_{\mu_{s,w}}^{(S)} = {}^s(_kU) \cap {}^w(_kU^-) \qquad U_{\nu_{s,w}}^{(S)} = {}^s(_kU) \cap {}^w(_kU),$$

et l'application produit est un isomorphisme de variétés de $U_{\mu_{s,w}}^{(S)} \times U_{\nu_{s,w}}^{(S)}$ sur $^s(_kU)$.

5.14. *Proposition. — Les k-sous-groupes paraboliques standards sont les seuls k-sous-groupes contenant* $P = {}_kP_\varnothing$. *Tout k-sous-groupe parabolique de* G *est conjugué sur k à un seul k-sous-groupe parabolique standard;* G *est engendré par* P *et les sous-groupes* $U_{(-a)}$ $(a \in {}_k\Delta)$.

Soit P′ un k-sous-groupe parabolique contenant P. Vu 4.16, 4.19, l'intersection $S' = R(P') \cap S$ est un tore déployé sur k maximal de $R(P')$, et, vu 4.15, on peut écrire $P' = G_\psi^{(S')}$, où ψ est l'ensemble des éléments de $\Phi(S', G)$ qui sont >0 pour un ordre convenable. On a donc aussi $P' = G_\eta^{(S)}$, où η est l'ensemble des k-racines dont la restriction à S′ est soit nulle soit contenue dans ψ. Comme ψ est clos dans $\Phi(S', G)$, l'ensemble η

est clos dans $_k\Phi$, contient $_k\Phi^+$ puisque $P' \supset P$, et l'existence de $\theta \subset {}_k\Delta$ tel que $\eta = \pi_\theta$ résulte alors de [6, § 7, n° 7, prop. 20].

Soit Q un k-sous-groupe parabolique de G. Vu 4.13 et ce qui vient d'être démontré, Q est conjugué sur k à au moins un k-sous-groupe parabolique standard. L'unicité de ce dernier est conséquence de 4.3. Enfin, ce qui précède montre que le sous-groupe Q engendré par P et les groupes $U_{(-a)}$ $(a \in {}_k\Delta)$ est le groupe P_θ $(\theta = {}_k\Delta)$, donc est égal à G.

5.15. *Théorème. — Posons* $N = \mathcal{N}(S)$ *et* $U = {}_kU$. *Soient* $s \in {}_kW$ *et* $n, n' \in N$. *Alors* $G_k = {}^sU_k \cdot N_k \cdot U_k$. *On a* $n = n'$ *si et seulement si* ${}^sU \cdot n \cdot U = {}^sU \cdot n' \cdot U$. *L'ensemble* ${}^sU \cdot n \cdot U$ *est localement fermé dans* G *et l'application produit est un isomorphisme de variétés de* $({}^sU \cap {}^nU^-) \times \{n\} \times U$ *sur* ${}^sU \cdot n \cdot U$, *qui est défini sur* k *si* $n \in N_k$. *Le groupe* G_k *est réunion disjointe des doubles classes* $P_k \cdot w \cdot P_k$ $(w \in {}_kW)$.

En utilisant le fait que N_k contient un élément dont l'image canonique dans $_kW$ est s, on se ramène immédiatement au cas où $s = 1$.

Soit $g \in G_k$. D'après 4.18, le groupe $P \cap g^{-1} \cdot P \cdot g$ contient le centralisateur $\mathcal{Z}(S')$ d'un tore déployé sur k maximal S' de G. Les groupes $\mathcal{Z}(S)$ et $\mathcal{Z}(S')$ sont deux k-sous-groupes de Levi de P, donc (3.14) sont conjugués par un élément de $R_u(P)_k$, ce qui montre l'existence de $u \in P_k$ tel que $u \cdot g^{-1} \cdot P \cdot g \cdot u^{-1} \supset \mathcal{Z}(S)$. En utilisant 5.9, on voit que l'on peut alors trouver $v \in N_k$ tel que $v \cdot u \cdot g^{-1} \in \mathcal{N}(P)_k$; comme P est égal à son normalisateur (4.3), cela prouve que $g \in P_k \cdot N_k \cdot P_k$, donc aussi que $g \in U_k \cdot N_k \cdot U_k$, vu l'existence de la décomposition de Levi $P = \mathcal{Z}(S) \cdot U$.

L'ensemble $U \cdot n \cdot U$ est une orbite de $U \times U$ opérant sur G par translations à gauche et à droite, donc est ouvert dans son adhérence [1, § 15.2], i.e. est localement fermé. Soit w l'image canonique de n dans $_kW$. On a, dans les notations de 5.13,

$$(1) \qquad\qquad U \cdot n \cdot U = U'_w \cdot U''_w \cdot n \cdot U = U'_w \cdot n \cdot U,$$

puisque $n^{-1} \cdot U''_w \cdot n \subset U$, donc l'application produit définit un morphisme surjectif $\varphi : U'_w \times \{n\} \times U \to U \cdot n \cdot U$. Comme $n^{-1} \cdot U'_w \cdot n \subset U^-$, le composé de φ et de la translation à gauche par n^{-1} est la restriction à une partie fermée de l'application produit $U^- \times U \to U^- \cdot U$. Comme cette dernière est un isomorphisme (4.10), il s'ensuit que φ est un isomorphisme; de plus, les groupes U'_w et U étant définis sur k, φ l'est aussi si $n \in N_k$. Prouvons maintenant que $U \cdot n \cdot U' = U \cdot n' \cdot U$ entraîne $n = n'$, autrement dit que $N \cap U \cdot n \cdot U = \{n\}$. Vu (1) cela équivaut encore à $N \cap n^{-1} \cdot U'_w \cdot n \cdot U = \{e\}$, ce qui est conséquence de

$$(2) \qquad\qquad N \cap U^- \cdot U = \{e\}.$$

Il suffit donc d'établir (2). Soit $n = v \cdot u$ $(n \in N, v \in U^-, u \in U)$. Quel que soit $s \in S$, on a évidemment

$$(v^{-1} \cdot {}^ns \cdot v \cdot {}^ns^{-1}) \cdot ({}^ns \cdot s^{-1}) \cdot (s \cdot u \cdot s^{-1} \cdot u^{-1}) = e.$$

Mais les facteurs définis par les parenthèses appartiennent respectivement à U^-, S, U, et doivent donc être tous égaux à 1 vu 4.10. Cela entraîne que $u, v \in \mathcal{Z}(S)$, d'où $u = v = 1$ et $n = 1$.

La première partie du théorème et la décomposition $P_k = \mathscr{Z}(S)_k . U_k$ montrent que G_k est réunion des doubles classes $P_k . w . P_k$. Il reste à voir que ces dernières sont disjointes. Soit $w' \in P_k . w . P_k$ et soient n', n des représentants de w' et w dans N_k. Il existe alors $u, v \in U_k$ et $a, b \in \mathscr{Z}(S)_k$ tels que $n' = u.a.n.b.v$, d'où $n' = a.n.b$ et $w' = w$.

5.16. Proposition. — *Soient* $w \in {}_k W$ *et* r *une réflexion fondamentale. On a*

(i) $r . P_k . \{w, r.w\} . P_k = P_k . \{w, r.w\} . P_k.$

(ii) $r . P_k . r \neq P_k.$

Soit a la k-racine simple telle que $r = r_a$. Quitte à remplacer w par $r.w$, on peut supposer que $w^{-1}(a) \in {}_k\Phi^+$. Le groupe U est produit semi-direct sur k du radical unipotent V_a du k-groupe parabolique standard $P_{\{a\}}$ et du groupe $U_{(a)}$, et ce dernier est l'intersection de U avec $\mathscr{Z}(S_a)$. On a par conséquent

$$r . P_k . \{w, r.w\} . P_k = \mathscr{Z}(S)_k . r . U_k . \{w, r.w\} . P_k = \mathscr{Z}(S)_k . r . V_{a,k} . U_{(a),k} . \{w, r.w\} . P_k,$$
$$r . P_k . \{w, r.w\} . P_k = \mathscr{Z}(S)_k . V_{a,k} . r . U_{(a),k} . \{w, r.w\} . P_k.$$

En appliquant 5.15 à $\mathscr{Z}(S_a)$, on en déduit

$$r . P_k . \{w, r.w\} . P_k \subset P_k . \mathscr{Z}(S_a)_k . w . P_k = P_k . U_{(a),k} . \{\mathbf{1}, r\} . U_{(a),k} . w . P_k$$

donc, puisque $w^{-1}(a) > 0$,

$$r . P_k . \{w, r.w\} . P_k \subset P_k . \{r.w, w\} . P_k.$$

En multipliant à gauche les deux membres par r, on en tire aussi l'inclusion contraire, d'où (i).

(ii) est une conséquence de 3.23 et du fait que $r({}_k\Phi^+) \neq {}_k\Phi^+$.

5.17. Proposition. — *Soient* θ *une partie de* ${}_k\Delta$ *et* W_θ *le sous-groupe de* ${}_k W$ *engendré par les réflexions fondamentales correspondant aux éléments de* θ. *Alors* $P_{\theta,k} = P_k . W_\theta . P_k.$

Le groupe de Weyl relatif de Z_θ (cf. 5.12) est W_θ. En vertu de 5.15, on a donc

$$P_{\theta,k} = Z_{\theta,k} . V_{\theta,k} = (U \cap Z_\theta)_k . W_\theta . (U \cap Z_\theta)_k . V_{\theta,k},$$

d'où

$$P_{\theta,k} = P_k . P_{\theta,k} . P_k = P_k . W_\theta . P_k.$$

5.18. Corollaire. — (i) *Les groupes* $P_{\theta,k}$ *sont les seuls sous-groupes de* G_k *contenant* P_k.

(ii) *Soient* $\theta, \theta' \subset {}_k\Delta$ *et* $g \in G_k$. *Alors* ${}^g P_{\theta',k} \subset P_{\theta,k}$ *si et seulement si* $\theta' \subset \theta$ *et* $g \in P_{\theta,k}$. *En particulier,* $P_{\theta,k}$ *est son propre normalisateur dans* G_k.

5.15 et 5.16 montrent que les axiomes de [33] sont vérifiés ; 5.18 résulte alors de 5.17 et des th. 2, 3, 4 de [33].

5.19. Corollaire. — *Soit* Q *un* k-*sous-groupe parabolique de* G. *Alors* Q_k *est le normalisateur dans* G_k *de* $R_u(Q)_k$.

Vu 5.14, on peut supposer que $Q = P_\theta \,(\theta \subset {}_k\Delta)$. Le normalisateur M de $R_u(Q)_k$ contient alors $P_{\theta,k}$, donc (5.18) est égal à un groupe $P_{\theta',k} \,(\theta \subset \theta' \subset {}_k\Delta)$. Supposons que $\theta' \neq \theta$. Soient a un élément de θ' non contenu dans θ, et n un représentant dans N_k de la réflexion fondamentale r_a. On a $U_{(a), k} \subset R_u(P_\theta)_k$, d'où aussi

$$U_{(-a), k} = {}^n U_{(a), k} \subset R_u(P_\theta)_k \subset P_k,$$

ce qui est absurde, donc $\theta' = \theta$ et $M = Q_k$.

5.20. Corollaire. — *Soient θ, θ' deux parties de $_k\Delta$. Alors il existe entre les doubles classes $P_{\theta, k} \cdot g \cdot P_{\theta', k}$ $(g \in G_k)$ de G_k et les doubles classes $W_\theta \cdot w \cdot W_{\theta'}$ de $_kW$ $(w \in {}_kW)$ une correspondance biunivoque caractérisée par la relation*

$$P_{\theta, k} \cdot g \cdot P_{\theta', k} = P_k \cdot W_\theta \cdot w \cdot W_{\theta'} \cdot P_k.$$

Soit $g \in G_k$. Soient n l'élément de N_k tel que $g \in U_k \cdot n \cdot U_k$ (5.15), et w l'image canonique de n dans $_kW$. En utilisant 5.16 (i) et 5.17, on a

$$P_{\theta, k} \cdot g \cdot P_{\theta', k} = P_k \cdot W_\theta \cdot P_k \cdot w \cdot P_k \cdot W_{\theta'} \cdot P_k \subset P_k \cdot W_\theta \cdot w \cdot W_{\theta'} \cdot P_k \subset P_{\theta, k} \cdot w \cdot P_{\theta', k} = P_{\theta, k} \cdot g \cdot P_{\theta', k}.$$

Il résulte alors de la dernière assertion de 5.15 que deux doubles classes distinctes $W_\theta \cdot w \cdot W_{\theta'}$, $W_\theta \cdot w' \cdot W_{\theta'}$ correspondent à des doubles classes distinctes de G_k modulo $P_{\theta, k}$, $P_{\theta', k}$.

Nous terminerons ce paragraphe par quelques remarques simples sur les classes de conjugaison de sous-groupes paraboliques.

5.21. Proposition. — *Soient $\mathscr{P}$, $\mathscr{Q}$ deux classes de conjugaison de sous-groupes paraboliques de G et $P, P' \in \mathscr{P}$, $Q, Q' \in \mathscr{Q}$ tels que $P \cap Q$ et $P' \cap Q'$ soient des sous-groupes paraboliques. Alors :*

(i) *Si $P \subset Q$, tout élément de $\mathscr{P}$ est contenu dans un seul élément de $\mathscr{Q}$.*

(ii) *Les groupes $P \cap Q$ et $P' \cap Q'$ sont conjugués.*

(iii) *Les sous-groupes paraboliques R, R' engendrés par P, Q et P', Q' respectivement sont conjugués.*

L'assertion (i) est contenue dans 4.3.

Reprenons les notations de 4.2. On peut trouver $g, g' \in G$ tels que

$$^g(P \cap Q) \supset B, \qquad ^{g'}(P' \cap Q') \supset B;$$

d'après 4.3, il existe alors $\theta, \psi \subset \Delta$ tels que

$$^g P = {}^{g'} P' = P_\theta, \qquad ^g Q = {}^{g'} Q' = P_\psi.$$

Il est alors clair que

$$^g(P \cap Q) = {}^{g'}(P' \cap Q') = P_{\theta \cap \psi},$$

$$^g R = {}^{g'} R' = P_{\theta \cup \psi},$$

d'où la proposition.

5.22. Définition. — Soient $\mathscr{P}$, $\mathscr{Q}$ deux classes de conjugaison de sous-groupes paraboliques de G. On écrit $\mathscr{P} \prec \mathscr{Q}$ si tout élément de $\mathscr{P}$ est contenu dans un élément de $\mathscr{Q}$. La proposition précédente montre qu'il suffit pour cela qu'il existe un élément de $\mathscr{P}$ contenu dans un élément de $\mathscr{Q}$.

D'après 5.21, les sous-groupes paraboliques de la forme $P \cap Q$ ($P \in \mathscr{P}$, $Q \in \mathscr{Q}$) (resp. engendrés par deux sous-groupes $P \in \mathscr{P}$, $Q \in \mathscr{Q}$ tels que $P \cap Q$ soit parabolique) forment une seule classe de conjugaison, qui sera désignée par $\mathscr{P} \curlywedge \mathscr{Q}$ (resp. $\mathscr{P} \curlyvee \mathscr{Q}$).

5.23. On notera ${}_k\mathscr{P}_\theta$ la classe de conjugaison du k-groupe parabolique standard ${}_kP_\theta$ ($\theta \subset {}_k\Delta$) (5.12). On a évidemment

$$ {}_k\mathscr{P}_\theta \subset {}_k\mathscr{P}_\psi \Leftrightarrow \theta \subset \psi \qquad\qquad (\theta, \psi \subset {}_k\Delta). $$

Si $\mathscr{P}$ et $\mathscr{Q}$ contiennent chacune un élément défini sur k, il en est de même pour $\mathscr{P} \curlywedge \mathscr{Q}$ et $\mathscr{P} \curlyvee \mathscr{Q}$ vu 4.7, 4.13. D'après 5.14, il existe $\theta, \psi \subset {}_k\Delta$ tels que $\mathscr{P} = {}_k\mathscr{P}_\theta$ et $\mathscr{Q} = {}_k\mathscr{P}_\psi$. On a évidemment :

$$ {}_k\mathscr{P}_\theta \curlywedge {}_k\mathscr{P}_\psi = {}_k\mathscr{P}_{\theta \cap \psi}, \qquad {}_k\mathscr{P}_\theta \curlyvee {}_k\mathscr{P}_\psi = {}_k\mathscr{P}_{\theta \cup \psi} \qquad (\psi, \theta \subset {}_k\Delta). $$

Enfin, on voit exactement comme en 4.9 que si i est l'involution d'opposition de ${}_k\Phi$, alors la classe opposée à ${}_k\mathscr{P}_\theta$ est ${}_k\mathscr{P}_{i(\theta)}$.

5.24. Le groupe $\Gamma = \mathrm{Gal}(k_s/k)$ opère à la manière usuelle sur l'ensemble des k_s-sous-variétés de G et laisse visiblement invariant l'ensemble des k_s-sous-groupes paraboliques de G. Dans la suite, nous aurons à considérer la condition suivante, imposée à une classe de conjugaison $\mathscr{P}$ de sous-groupes paraboliques :

(i) *L'ensemble $\mathscr{P}_{k_s}$ des éléments de $\mathscr{P}$ définis sur k_s est stable par Γ,*
et l'on dira que $\mathscr{P}$ est défini sur k si (i) est vérifiée. Bien que cela ne soit pas nécessaire dans ce travail, nous en donnerons ici une autre interprétation.

Un groupe parabolique est égal à son normalisateur, donc étant donné $P_0 \in \mathscr{P}$, on peut identifier $\mathscr{P}$ à G/P_0 par l'application φ_0 qui associe à $P \in \mathscr{P}$ l'unique point fixe de P dans G/P_0. L'application φ_0 commute à G, opérant sur $\mathscr{P}$ par automorphismes intérieurs et sur G/P_0 par translations à gauche, d'où une structure d'espace homogène (de groupe algébrique) sur $\mathscr{P}$, qui ne dépend visiblement pas du choix de l'origine P_0. C'est toujours d'elle qu'il sera question ci-dessous.

Si $\mathscr{P}$ contient un élément P défini sur une extension K de k, alors K est un corps de définition de l'espace homogène $\mathscr{P}$ (i.e. de $\mathscr{P}$ et du morphisme $G \times \mathscr{P} \to \mathscr{P}$ définissant l'action de G) comme on le voit en prenant P comme origine. En particulier (2.14, 4.3), l'espace homogène $\mathscr{P}$ est défini sur k_s. Cela étant, la condition (i) équivaut à :

(ii) *L'espace homogène $\mathscr{P}$ admet k comme corps de définition.*
En effet, (i) $\Rightarrow$ (ii) d'après les critères de descente du corps de base (voir [37], ou aussi [5, 2.12, 4.10]). La réciproque est immédiate.

Remarquons qu'il peut arriver que $\mathscr{P}$ soit défini sur k, mais ne possède pas d'élément défini sur k. Par exemple la classe de conjugaison des sous-groupes de Borel de G est toujours définie sur k.

§ 6. DESCENTE DU CORPS DE BASE

6.1. *Notations.* — Dans ce paragraphe, K désigne toujours une extension de k, $\Gamma = \mathrm{Aut}(K/k)$ le groupe des k-automorphismes de K, G un groupe réductif connexe, S un tore déployé sur k maximal de G et T un tore déployé sur K maximal de G contenant S. Les groupes $X^*(S)$ et $X^*(T)$ sont dotés d'ordres compatibles (i.e. la restriction à S d'un caractère positif de T est positive), et ${}_k\Delta$, ${}_K\Delta$ désignent les ensembles de racines relatives simples qui y correspondent. Notons toutefois que l'hypothèse $S \subset T$ (et *a fortiori* la compatibilité des ordres) n'intervient pas avant le n° 6.6.

6.2. *Une action de Γ sur $X^*(T)$.* — Soit γ un élément de Γ. En vertu de 4.21, 5.9, il existe un automorphisme intérieur α par un élément de G_K transformant le tore ${}^\gamma T$ en T et l'ensemble de racines simples ${}^\gamma({}_K\Delta)$ relatives à ${}^\gamma T$ en ${}_K\Delta$. L'automorphisme de $X^*(T)$ induit par le composé $\alpha \circ \gamma$ est indépendant de l'automorphisme intérieur α choisi (deux tels α différant à gauche par un automorphisme intérieur centralisant T) et sera noté ${}_K\Delta\gamma$, ou, par abus de notation, ${}_\Delta\gamma$. Cet automorphisme laisse évidemment invariants ${}_K\Delta$ et ${}_K\Phi$, donc aussi le diagramme de ${}_K\Delta$ (2.1). On a la formule

$$(\mathrm{1}) \qquad\qquad ({}_\Delta\gamma(c))(\alpha({}^\gamma t)) = \gamma(c(t)) \qquad\qquad (c \in X^*(T),\ t \in T_K,\ \gamma \in \Gamma).$$

Soit P un K-sous-groupe parabolique de G contenant T. Il est immédiat, à partir de 5.12, 5.14, que l'homomorphisme de restriction $X^*(P) \to X^*(T)$ identifie $X^*(P)_K$ à un sous-groupe de $X^*(T)$, qui est d'indice fini si P est minimal. On peut donc considérer ${}_\Delta\gamma$ comme opérant sur $X^*(P)_K$. Si β est un automorphisme intérieur de G transformant ${}^\gamma P$ sur P, cette action vérifie

$$(\mathrm{2}) \qquad\qquad ({}_\Delta\gamma(c))(\beta({}^\gamma p)) = \gamma(c(p)) \qquad\qquad (c \in X^*(P)_K,\ p \in P_K,\ \gamma \in \Gamma),$$

comme on le voit à partir de (1), en remarquant que $P = \mathcal{N}(P)$ et que P agit trivialement sur ses caractères par automorphismes intérieurs.

Soit θ une partie de ${}_K\Delta$. On a la relation évidente

$$(\mathrm{3}) \qquad\qquad \gamma({}_K P_\theta) = {}_K P_{\theta'} \qquad\qquad (\theta' = {}_\Delta\gamma(\theta)).$$

Celle-ci constitue une nouvelle définition de ${}_\Delta\gamma$, si on ajoute le fait que, pour tout caractère $c \in X^*(T)$ qui s'annule sur $(T \cap \mathscr{D}G)^0$, on a ${}_\Delta\gamma(c) = {}^\gamma c$.

Le groupe Γ opère sur $X^*(T)$ par l'intermédiaire des ${}_\Delta\gamma$. Lorsque T est invariant par Γ (en particulier, lorsque T est défini sur k) il convient évidemment de distinguer cette action de celle envisagée habituellement, notamment ci-dessus (1.7), et que nous appellerons parfois l'action *usuelle* de Γ sur $X^*(T)$. Le lien entre ces deux actions peut être caractérisé comme suit : pour tout $\gamma \in \Gamma$, il existe un unique élément w du groupe de Weyl ${}_K W$ tel que $w(\gamma({}_K\Delta)) = {}_K\Delta$, et on a alors ${}_\Delta\gamma = w \circ \gamma$. Remarquons encore que si P est défini sur k, on a, d'après (2),

$$(4) \qquad\qquad {}_\Delta\gamma(c) = {}^\gamma c \qquad\qquad (c \in X^*(P)_K).$$

En particulier, c est alors défini sur k si et seulement s'il est fixe par tous les ${}_\Delta\gamma$ ($\gamma \in \Gamma$).

Il résulte de 4.21 et 5.9 que le Γ-module obtenu en faisant opérer Γ sur $X^*(T)$ par l'intermédiaire de ${}_\Delta\gamma$ ne dépend pas, à isomorphisme près, de T ni de ${}_k\Delta$. C'est donc un invariant du groupe G (relatif aux deux corps k, K). Par contre, si T et T′ sont deux tores déployés sur K maximaux, tous deux invariants par Γ, les Γ-modules usuels $X^*(T)$ et $X^*(T')$ ne sont pas nécessairement isomorphes ([1]).

6.3. *Descente et sous-groupes paraboliques.* — Soit θ une partie de ${}_k\Delta$. La relation ${}_k\mathscr{P}_\theta = {}_K\mathscr{P}_{\theta'}$ définit une partie θ' de ${}_K\Delta$ qui sera notée ${}_K\eta_k(\theta)$. Il résulte immédiatement de 4.9 et 5.23 que ${}_K\eta_k$, qui est donc une application de l'ensemble des parties de ${}_k\Delta$ dans l'ensemble des parties de ${}_K\Delta$, est compatible avec l'intersection, la réunion et l'opposition. Cela étant, il existe une application ${}_K\rho_k : {}_K\Delta \to {}_k\Delta \cup \{o\}$, caractérisée par la relation

$$ {}_K\eta_k(\theta) = {}_K\rho_k^{-1}(o) \cup \bigcup_{a \in \theta} {}_K\rho_k^{-1}(a). $$

La signification de cette application apparaîtra dans la suite (6.8). Nous noterons encore

$$ {}_K\Delta_k^0 = {}_K\eta_k(\varnothing) = {}_K\rho_k^{-1}(o), $$

la partie de ${}_K\Delta$ correspondant à la classe de conjugaison des sous-groupes paraboliques définis sur k minimaux. Dans toutes les notations introduites ci-dessus, le double indice K, k sera éventuellement omis, si aucune confusion ne risque d'en résulter.

Une partie θ de ${}_K\Delta$ sera dite *apparente* sur k si elle est l'image par η d'une partie de ${}_k\Delta$, c'est-à-dire si ${}_K P_\theta$ est conjugué à un sous-groupe parabolique défini sur k, et *définie* sur k si la classe de conjugaison de ${}_K P_\theta$ est définie sur k (cf. 5.24). Il résulte immédiatement de 4.7 que

(1) *Une partie de ${}_K\Delta$ est apparente sur k si et seulement si elle est définie sur k et contient* ${}_K\Delta_k^0$.

L'ensemble des parties apparentes sur k et l'ensemble des parties définies sur k sont invariants par Γ (pour l'action définie au n° 6.2) et par l'involution d'opposition.

Lorsque T est défini sur k, la notion d'ensemble de caractères défini sur k, introduite au n° 1.7, ne coïncide pas avec celle introduite ici dans le cas où les caractères en question sont des racines simples. Dans ce paragraphe, l'expression « défini sur k », en parlant d'un ensemble de racines simples, sera toujours comprise au sens qui lui a été donné ici.

6.4. *Caractérisation galoisienne des parties de ${}_K\Delta$ définies sur k.* — Soit Δ le système des racines simples de G. On a les identifications canoniques $\Delta = {}_{k_s}\Delta = {}_{K_s}\Delta$. Le groupe $\Gamma(k) = \mathrm{Gal}(k_s/k)$ opère sur ${}_{k_s}\Delta$, donc sur Δ. Notons η_K l'application ${}_{K_s}\eta_K$, considérée comme application de l'ensemble des parties de ${}_K\Delta$ dans l'ensemble des parties de Δ. Par une descente galoisienne immédiate, on voit que les parties de ${}_{k_s}\Delta = \Delta$ définies sur k sont les parties invariantes par $\Gamma(k)$. Par conséquent

---

[1] Un exemple de J.-P. Serre montre du reste qu'il peut n'y avoir aucun tore déployé sur K maximal stable par Γ.

(1) *Une partie de $_{\mathrm{K}}\Delta$ est définie sur k si et seulement si sa transformée par η_{K} est invariante par $\Gamma(k)$.*

Lorsque K est une extension normale de k, $\Gamma \cong \Gamma(k)/\Gamma(\mathrm{K})$, donc $\Gamma(k)$ opère sur $_{\mathrm{K}}\Delta$ par l'intermédiaire de Γ. Il est clair que l'action de $\Gamma(k)$ permute avec l'application η_{K}, donc

(2) *Lorsque K est une extension normale de k, une partie de $_{\mathrm{K}}\Delta$ est définie sur k si et seulement si elle est invariante par Γ. En particulier (d'après* 6.3, (1)*), les images par $_{\mathrm{K}}\rho_k^{-1}$ des éléments de $_k\Delta$ sont les orbites de Γ dans $_{\mathrm{K}}\Delta$ non contenues dans $_{\mathrm{K}}\Delta_k^0$.*

Lorsque K est une extension radicielle de k, toute partie de $_{\mathrm{K}}\Delta$ est définie sur k. Plus généralement, toute classe de conjugaison de sous-groupes paraboliques définie sur K l'est aussi sur k (5.24). Par contre, il n'est pas toujours vrai que toute partie de $_{\mathrm{K}}\Delta$ soit apparente sur k (c'est-à-dire que les tores déployés sur k maximaux soient aussi déployés sur K maximaux) comme le montre l'exemple du groupe $\mathbf{O}_3^+(\mathrm{Q})$ correspondant à une forme quadratique Q anisotrope, non dégénérée et de défaut 1 sur un corps k de caractéristique 2, K étant une extension radicielle de k sur laquelle Q acquiert des zéros non triviaux.

6.5. *k-formes, indice, classification.* — Soit $\overline{\mathrm{G}}$ un groupe algébrique défini sur K. Le groupe G est appelé une *k-forme* de $\overline{\mathrm{G}}$ s'il est isomorphe à $\overline{\mathrm{G}}$ sur K. Identifions $\overline{\mathrm{G}}$ à G, sur K, de façon à pouvoir considérer T et $_{\mathrm{K}}\Delta$ comme un tore déployé et un ensemble de K-racines simples de $\overline{\mathrm{G}}$. On appelle *indice* de la k-forme G la donnée constituée par $_{\mathrm{K}}\Delta_k^0$ et par l'ensemble des parties de $_{\mathrm{K}}\Delta$ définies sur k (d'où on déduit aussi, par 6.3, (1), l'ensemble des parties apparentes). Si l'extension K/k est normale, l'*indice galoisien* est constitué par la donnée de $_{\mathrm{K}}\Delta_k^0$ et de l'action de Γ sur $_{\mathrm{K}}\Delta$.

De la seule donnée du diagramme de $_{\mathrm{K}}\Delta$, on peut déduire, indépendamment du groupe $\overline{\mathrm{G}}$ et des corps K et k, des conditions auxquelles doivent satisfaire un ensemble de parties de $_{\mathrm{K}}\Delta$, fermé par intersection et réunion, et un élément Δ^0 de cet ensemble, pour pouvoir être l'indice d'une forme. Certaines conditions se déduisent déjà immédiatement de 6.3 et 6.4. On sait notamment que l'indice doit être invariant par l'involution d'opposition. Notons que cette dernière condition est, à elle seule, très restrictive (cf. par ex. [32], n° 5), si on tient compte du fait que la restriction de l'indice à une partie Δ' de $_{\mathrm{K}}\Delta$ apparente sur k est aussi un indice de k-forme (c'est l'indice des k-sous-groupes de Levi d'un k-sous-groupe parabolique appartenant à $_{\mathrm{K}}\mathscr{P}_{\Delta'}$). D'autres conditions imposées aux indices possibles résultent du théorème 6.13 ci-après. On trouvera des éléments de classification dans [34, 35, 36].

Signalons encore l'existence d'un « théorème de Witt », permettant de caractériser une k-forme d'un groupe donné sur $\overline{k}$ par son indice galoisien et par la k-structure (avec spécification de l'isomorphisme sur $\overline{k}$) du centralisateur $\mathscr{Z}(\mathrm{S})$ d'un tore déployé sur k maximal (cf. [28], [32]; la démonstration de [28] s'étend à un corps non parfait si on fait usage de 2.14).

6.6. *Lemme.* — *Soient* T *et* T′ *deux tores déployés sur* K *maximaux contenant* S, *soient donnés dans* $X^*(T)$ *et* $X^*(T')$ *des ordres compatibles avec un même ordre dans* $X^*(S)$, *et soient* $_K\Delta$ *et* $_K\Delta'$ *les ensembles de racines simples correspondants. Alors, tout élément* $g \in G$ *tel que* Int g *transforme* T *en* T′ *et* $_K\Delta$ *en* $_K\Delta'$ *appartient à* $\mathscr{Z}(S)$.

Soit P (resp. Q ; resp. Q′) le sous-groupe parabolique défini sur k (resp. K ; resp. K) minimal standard, correspondant à l'ordre choisi dans $X^*(S)$ (resp. $X^*(T)$; resp. $X^*(T')$). On a P⊃Q, P⊃Q′ et $^gQ = Q'$. En vertu de 4.3, il s'ensuit que $^gP = P$, donc $g \in P$ et $g = g'.g''$ avec $g' \in R_u(P)$ et $g'' \in \mathscr{Z}(S)$. Pour tout $s \in S$, on a $(g, s) = (g', s) \in R_u(P)$, et $(g, s) = {}^gs.s^{-1} \in T'.s^{-1} \subset \mathscr{Z}(S)$, donc $(g, s) = e$ et $g \in \mathscr{Z}(S)$.

6.7. *Proposition.* — *Supposons* T *invariant par* Γ. *Alors, pour tout* $c \in X^*(T)$ *et tout* $\gamma \in \Gamma$, *la différence* $_\Delta\gamma(c) - \gamma(c)$ *est une combinaison linéaire à coefficients entiers de racines s'annulant sur* S.

Soit w l'élément de $_KW$ tel que $w(\gamma(_K\Delta)) = {}_K\Delta$. Le transformé par γ de l'ordre donné dans $X^*(T)$ est compatible avec l'ordre donné dans $X^*(S)$, et l'ensemble de racines simples qui lui correspond est $\gamma(_K\Delta)$. En vertu du lemme précédent, w centralise donc S, et [6] est un produit de réflexions relatives à des racines s'annulant sur S. La proposition est alors une conséquence de 5.7 et du fait que $_\Delta\gamma = w \circ \gamma$.

6.8. *Proposition.* — *Soit* $j = X^*(T) \to X^*(S)$ *l'homomorphisme de restriction. Alors* $j(a) = {}_K\rho_k(a) \in {}_k\Delta \cup \{0\}$ $(a \in {}_K\Delta)$.

Soient $k_1 \subset k_2 \subset k_3$ trois corps. On a la relation de transitivité évidente $_{k_2}\rho_{k_1} \circ {}_{k_3}\rho_{k_2} = {}_{k_3}\rho_{k_1}$ (en posant $\rho(0) = 0$). Il s'ensuit que

(1) si la proposition est vraie pour $k = k_1$, $K = k_3$ et pour $k = k_2$, $K = k_3$, elle l'est aussi pour $k = k_1$, $K = k_2$;

(2) si la proposition est vraie pour $k = k_1$, $K = k_2$, et si un tore déployé maximal sur k_2 est aussi déployé maximal sur k_3, la proposition est vraie pour $k = k_1$, $K = k_3$.

En appliquant (1) avec k_3 algébriquement clos, on se ramène au cas où K est algébriquement clos et ensuite, en appliquant (2) avec $k_2 = k_s$, à celui où l'on a $K = k_s$, ce que nous supposerons dorénavant. Soient $b \in {}_K\Delta$, $a = \rho(b)$, $a' = b|_S$, (x) l'ensemble des racines de la forme $n.x$ avec $n \in \mathbf{N}$, et $_kP_\theta$ $(\theta \subset {}_k\Delta)$ les sous-groupes paraboliques standards par rapport à S et $_k\Delta$ (5.12). Il résulte de la définition de ρ que $a = 0$ si et seulement si $U_{(-b)} \subset {}_kP_\emptyset$, et que si $a \neq 0$, $U_{(-b)} \subset {}_kP_a$. D'autre part, si $a' \neq 0$, $U_{(-b)} \subset U_{(-a')}$. Compte tenu de 3.22 $c)$, on a donc les implications

$$a' \neq 0 \Leftrightarrow U_{(-b)} \not\subset {}_kP_\emptyset \Leftrightarrow a \neq 0 \Rightarrow U_{(-b)} \subset {}_kP_a \Rightarrow a' \in \{a, 2a, 0\}.$$

Il s'ensuit que, dans tous les cas, $a' = a$ ou $a' = 2a$.

Du lemme 6.6, il résulte que deux éléments d'une même orbite de Γ dans $_K\Delta$ (par rapport à l'action définie au n° 6.2) ont même restriction à S. En vertu de 6.4, (2), on a donc, pour tout $a \in {}_k\Delta$, $\rho^{-1}(a)|_S = \{a\}$ ou $\{2a\}$. Mais toute K-racine est combinaison linéaire à coefficients entiers d'éléments de $_K\Delta$, donc toute k-racine est combinaison linéaire à coefficients entiers d'éléments de $_K\Delta|_S$, par conséquent, $\rho^{-1}(a)|_S=\{a\}$ pour tout a.

6.9. *Corollaire.* — *Soit* S′ *la composante connexe de l'élément neutre du sous-groupe de* T *défini par les équations* $a=1$ *pour tout* $a\in\Delta^0$, *et* $b=c$ *pour tous* $b,c\in{_K\Delta}$ *tels que* $\rho(b)=\rho(c)$ (*c'est-à-dire, du noyau de l'ensemble de caractères* $\Delta^0\cup\{b-c\,|\,b,c\in{_K\Delta},\ \rho(b)=\rho(c)\}$). *Alors* $(S\cap\mathscr{D}G)^0=(S'\cap\mathscr{D}G)^0$.

Cela résulte de la proposition précédente et d'un calcul de dimension évident.

6.10. *Groupes de Weyl.* — Posons

$$\mathscr{N}(S,T)=\mathscr{N}(S)\cap\mathscr{N}(T),\qquad \mathscr{N}^0(S,T)=\mathscr{Z}(S)\cap\mathscr{N}(T).$$

On désignera par $_KW_k=\mathscr{N}(S,T)/\mathscr{Z}(T)$ et $_KW_k^0=\mathscr{N}^0(S,T)/\mathscr{Z}(T)$ les images canoniques respectives de $\mathscr{N}(S,T)$ et $\mathscr{N}^0(S,T)$ dans $_KW$. En particulier, $_KW_k^0$ est le k-groupe de Weyl de $\mathscr{Z}(S)$, engendré par les réflexions fondamentales correspondant aux éléments de Δ^0. C'est un sous-groupe distingué de $_KW_k$, et on a (cf. 5.5) un isomorphisme canonique $_kW={_KW_k}/{_KW_k^0}$.

On notera R (resp. $_kR$) le sous-réseau de $X^*(T)$ (resp. $X^*(S)$) engendré par les K-racines (resp. k-racines) et on posera $X=R\otimes\mathbf{R}$ et $Y={_kR}\otimes\mathbf{R}$, les groupes R et $_kR$ étant identifiés à des sous-réseaux de X et Y. L'application de restriction $j:R\to{_kR}$ s'étend en une application linéaire $X\to Y$, qui coïncide avec ρ sur $_K\Delta$, et dont le noyau N est engendré par Δ^0 et par les différences $b-c$ de racines $b,c\in{_K\Delta}$ telles que $\rho(b)=\rho(c)$ (cf. 6.9). Ayant choisi dans X un produit scalaire admissible (5.1), on peut identifier Y à $N^\perp$, ce que nous ferons. Alors, $_KW$ étant vu comme groupe de transformations linéaires de X, $_KW_k$ (resp. $_KW_k^0$) est le normalisateur (resp. le centralisateur) de Y dans $_KW$, et $_kW$ est la restriction de $_KW_k$ à Y. En particulier, la restriction à Y du produit scalaire choisi dans X est admissible. Des produits scalaires admissibles dans X et Y qui sont entre eux dans la relation ainsi décrite seront dits *compatibles*.

Si les éléments de $_K\Delta$ sont tous définis sur k au sens de 6.3 (par exemple, si l'extension K$/k$ est normale et si Γ opère trivialement sur $_K\Delta$), $_KW_k$ est le normalisateur de $_KW_k^0$ dans $_KW$; en effet, Y est alors l'espace de tous les points fixes de $_KW_k^0$ dans X.

6.11. Supposons que T soit défini sur k et que l'extension K$/k$ soit normale. Alors, $N^\perp$ est l'espace des points fixes de l'action *usuelle* de Γ sur X. Il s'ensuit que dans l'identification de Y avec $N^\perp$, l'image canonique $j(a)$ dans Y d'un élément $a\in X$ (par exemple, la restriction à S d'un caractère de T nul sur $\mathscr{Z}(G)$) est identifiée avec l'orbite

$$a^0=(1/|\Gamma|).\sum_{\gamma\in\Gamma}{}^\gamma a,$$

de a sous l'action de Γ (lorsque Γ est infini, il faut le remplacer, dans la formule précédente, par un quotient fini par l'intermédiaire duquel il opère sur X et dont l'ordre est noté $|\Gamma|$). En particulier, on a, pour des produits scalaires compatibles, et quels que soient $a,b\in X$,

$$(1)\qquad\qquad (j(a),j(b))=(a^0,b^0)=(a,b^0)=(a^0,b).$$

Enfin, $_{\mathrm{K}}W_k$ est alors le groupe de tous les $w\in{}_{\mathrm{K}}W$ tels que $^{\gamma}w\equiv w$ (mod. $_{\mathrm{K}}W_k^0$) pour tout $\gamma\in\Gamma$, où l'exposant à gauche est relatif à l'action de Γ sur $_{\mathrm{K}}W$ induite par l'action usuelle de Γ sur X.

6.12. Le théorème suivant permet de déterminer le diagramme de $_k\Delta$ à partir du diagramme de $_{\mathrm{K}}\Delta$ et de l'ensemble des $\rho^{-1}(a)$. Pour en simplifier l'énoncé, on utilisera la notation suivante. Soit $M=((m_{ab}))$ $(a,b\in{}_{\mathrm{K}}\Delta)$ une matrice dont les lignes et les colonnes sont indexées par $_{\mathrm{K}}\Delta$. Alors, $\rho(M)$ désignera la matrice $((m'_{cd}))$ $(c,d\in{}_k\Delta)$, dont les lignes et les colonnes sont indexées par $_k\Delta$ et dont les coefficients sont donnés par

$$m'_{cd}=\sum_{a\in\rho^{-1}(c)}\ \sum_{b\in\rho^{-1}(d)}m_{ab}.$$

6.13. *Théorème.* — (i) *Les espaces* $\mathrm{X}=\mathrm{R}\otimes\mathbf{R}$ *et* $\mathrm{Y}={}_k\mathrm{R}\otimes\mathbf{R}$ *étant dotés de produits scalaires compatibles, si* $_{\mathrm{K}}M$ *(resp.* $_kM$*) désigne la matrice des produits scalaires des K-racines simples (resp. des k-racines simples), on a* $_kM=(\rho(_{\mathrm{K}}M^{-1}))^{-1}$.

(ii) *Soit* $a\in{}_k\Delta$. *Alors, le plus grand entier* n_a $(n_a=1,2)$ *tel que* $n_a.a$ *soit une k-racine est le maximum de la somme des coefficients des éléments de* $\rho^{-1}(a)$ *dans les K-racines qui sont combinaisons linéaires d'éléments de* $\rho^{-1}(a)\cup\Delta^0$.

Soit $\varkappa=\mathrm{K}$ ou k, et soient $\bar{m}_{xy}$ les coefficients de la matrice $_{\varkappa}M^{-1}$. Pour tout $x\in{}_{\varkappa}\Delta$, posons $x^*=\sum_y\bar{m}_{xy}.y$. On a $(x^*,y)=\delta_{xy}$ c'est-à-dire que $\{x^*\}$ est la base duale de $_{\varkappa}\Delta$ (dans X ou dans Y, selon que $\varkappa=\mathrm{K}$ ou k).

Comme au n° 6.10, on identifiera Y à $\mathrm{N}^{\perp}$, où N désigne le noyau de l'application canonique de X dans Y. Alors, pour tout $c\in{}_k\Delta$, c^* est défini par les relations

$$(1)\qquad\qquad (c^*,a)=\delta_{c,\rho(a)}\qquad\qquad\text{pour tout}\quad a\in{}_{\mathrm{K}}\Delta.$$

En effet, l'élément de X ainsi défini, que nous désignerons provisoirement par c', est orthogonal à toutes les différences $a-b$ de K-racines simples telles que $\rho(a)=\rho(b)$, ainsi qu'à tous les éléments de Δ^0, donc aussi à N. Par conséquent, il appartient à Y et on a, pour tout $a\in{}_{\mathrm{K}}\Delta$, $(c',\rho(a))=(c',a)=\delta_{c,\rho(a)}$, d'où $c'=c^*$. De (1), il résulte que $c^*=\sum_{a\in\rho^{-1}(c)}a^*$, c'est-à-dire

$$\sum_{d\in{}_k\Delta}{}_k\bar{m}_{cd}.d=\sum_{a\in\rho^{-1}(c)}\ \sum_{b\in{}_{\mathrm{K}}\Delta}{}_{\mathrm{K}}\bar{m}_{ab}.b.$$

En formant le produit scalaire des deux membres avec d^*, il vient, toujours en tenant compte de (1),

$$_k\bar{m}_{cd}=\sum_{a\in\rho^{-1}(c)}\ \sum_{b\in\rho^{-1}(d)}{}_{\mathrm{K}}\bar{m}_{ab},$$

ce qui démontre (i).

(ii) est une conséquence immédiate de 6.8.

Le théorème précédent permet notamment de calculer, à partir de $_{\mathrm{K}}\Delta$ et des $\rho^{-1}(a)$, l'ordre du produit de deux réflexions fondamentales dans $_kW$ (ou, ce qui revient au même, l'angle de deux k-racines simples). Cependant, celui-ci peut aussi être obtenu à l'aide d'une formule plus simple, que nous établirons à présent.

6.14. *Proposition.* — *Soient a, b deux k-racines simples et soit φ_0 (resp. φ_a, resp. φ_b, resp. φ_{ab}) le cardinal du système de racines réduit ayant pour ensemble de racines simples Δ^0 (resp. $\eta(a) = \Delta^0 \cup \rho^{-1}(a)$; resp. $\eta(b)$; resp. $\eta(a, b)$). Alors, l'ordre m_{ab} du produit des réflexions fondamentales correspondant à a et b dans $_k W$ est donné par*

$$m_{ab} = 2(\varphi_{ab} - \varphi_0)/(\varphi_a + \varphi_b - 2\varphi_0).$$

Dans cette démonstration, les racines dont il est question ne sont jamais données qu'à un facteur de proportionnalité strictement positif près. En particulier, l'égalité $a = b$ prend le sens de $a = r \cdot b$ $(r \in \mathbf{R}, r > 0)$.

Soient ψ le système des k-racines qui sont combinaisons linéaires de a et b, $W(\psi)$ son groupe de Weyl et $2m_{ab}$ l'ordre de $W(\psi)$; pour toute k-racine x, soit $f(x)$ le cardinal de l'ensemble de K-racines dont la restriction à S est x. Pour que la restriction à S d'une K-racine c soit nulle (resp. proportionnelle à a; resp. à b; resp. soit combinaison linéaire de a et b), il faut et il suffit (6.8) que c soit combinaison linéaire de racines simples appartenant à Δ^0 (resp. $\eta(a)$; resp. $\eta(b)$; resp. $\eta(a, b)$). On a donc

$$(1) \qquad f(a) = \frac{1}{2}(\varphi_a - \varphi_0), \qquad f(b) = \frac{1}{2}(\varphi_b - \varphi_0);$$

$$(2) \qquad \sum_{x \in \psi} f(x) = \varphi_{ab} - \varphi_0.$$

D'autre part, il est clair que pour tout $w \in {}_k W, f(w(x)) = f(x)$. Or on sait que tout élément de ψ est exactement de deux façons le transformé d'une racine simple (a ou b) par un élément de $W(\psi)$. Par conséquent,

$$(3) \qquad 2 \sum_{x \in \psi} f(x) = 2m_{ab} \cdot f(a) + 2m_{ab} \cdot f(b).$$

La formule de l'énoncé est une conséquence immédiate de (1), (2), (3).

6.15. *Proposition.* — *Soient a, b deux k-racines simples distinctes et m_{ab} l'ordre du produit des réflexions fondamentales correspondantes. Alors, les trois conditions suivantes sont équivalentes :*

(i)   *La paire a, b est connexe ; autrement dit, $m_{ab} > 2$.*

(ii)   *Pour toute K-racine $a' \in \rho^{-1}(a)$, il existe une K-racine $b' \in \rho^{-1}(b)$ et une partie θ de Δ^0 telles que $\theta \cup \{a', b'\}$ soit connexe.*

(iii)   *Il existe des K-racines $a' \in \rho^{-1}(a)$ et $b' \in \rho^{-1}(b)$, et une partie θ de Δ^0, telles que $\theta \cup \{a', b'\}$ soit connexe.*

On peut, sans nuire à la généralité, supposer que $_k\Delta = \{a, b\}$: il suffit de remplacer G par le sous-groupe $G_{\Phi'}$, où Φ' est le système des k-racines qui sont combinaisons linéaires de a et b.

Comme précédemment, on notera (x) l'ensemble des racines de la forme $n \cdot x$ $(n \in \mathbf{N})$. Soit ψ (resp. ψ') le plus petit idéal du système $_k\Phi$ (resp. du système $_K\Phi$) contenant b (resp. $\rho^{-1}(b)$). Le plus petit sous-groupe distingué de G contenant tous les $U_{(b')}$ avec $b' \in \rho^{-1}(b)$, est aussi le plus petit sous-groupe distingué contenant $U_{(b)}$. Ce sous-groupe, que nous noterons H, est défini sur k, puisqu'il est k-fermé et défini sur k_s (2.14 a)).

Il résulte alors de 5.11 que $H = G_\psi^* = G_{\psi'}^*$. Cela étant, on a les implications immédiates suivantes :

$$(i) \Rightarrow a \in \psi \Rightarrow H \supset U_{(a)} \Rightarrow \psi' \supset \rho^{-1}(a) \Rightarrow (ii) \Rightarrow (iii) \Rightarrow \psi' \cap \rho^{-1}(a) \neq \emptyset \Rightarrow H \cap U_{(a)} \neq \{e\} \Rightarrow (i).$$

6.16. *Corollaire.* — *Soient* $\theta = \{a_1, \ldots, a_t\}$ *une partie connexe de* $_k\Delta$ *et* $a_1' \in \rho^{-1}(a_1)$. *Alors il existe* $a_i' \in \rho^{-1}(a_i)$ $(2 \leq i \leq t)$ *et une partie* ψ *de* Δ^0 *tels que* $\{a_1', \ldots, a_t'\} \cup \psi$ *soit connexe.*
Cela résulte de 6.15 par récurrence sur t.

6.17. *Le foncteur* $R_{K/k}$. — Dans ce numéro et les quatre suivants, K est une extension séparable de degré fini d de k, contenue dans k_s, et $\sigma_1 = \mathrm{id}$, $\sigma_2, \ldots, \sigma_d$ sont les différents monomorphismes de K dans k_s. On renvoie à [38, Chap. I], (voir aussi [5, 2.8]), pour la définition du foncteur $R_{K/k}$, de restriction des scalaires, qui va de la catégorie des K-groupes (ou K-variétés quasi-projectives) à celle des k-groupes (ou k-variétés quasi-projectives). Si H est un K-groupe connexe et $H' = R_{K/k}H$, alors il existe un K-morphisme $\mu : H' \to H$ tel que

$$(1) \qquad\qquad \mu^0 = (^{\sigma_1}\mu, \ldots, {}^{\sigma_d}\mu) : H' \xrightarrow{\sim} {}^{\sigma_1}H \times \ldots \times {}^{\sigma_d}H$$

soit un k_s-isomorphisme, et cela caractérise la paire (H', μ) à un k-isomorphisme près. Tout K-morphisme $\alpha : G \to H$ s'écrit d'une et d'une seule manière sous la forme $\mu \circ \beta$, où $\beta : G \to H'$ est un k-morphisme.

$(\mu^0)^{-1}$ applique $H = {}^{\sigma_1}H$ sur un sous-groupe de H' qui est défini sur k_s et visiblement stable par tout élément de $\mathrm{Gal}(k_s/K)$, donc en fait défini sur K, d'où l'existence d'un K-morphisme $\nu : H \to H'$ tel que $\nu^0 = (\nu, {}^{\sigma_2}\nu, \ldots, {}^{\sigma_d}\nu)$ soit l'inverse de μ^0 [3, § 1].

6.18. *Proposition.* — *On conserve les notations de* 6.17. *Alors* $R_{K/k}$ *définit une bijection des* K-*sous-groupes connexes de* H *sur l'ensemble des* k-*sous-groupes connexes* L' *de* H' *tels que* $\mu^0(L')$ *soit le produit de ses intersections avec les* $^{\sigma_i}H$. *Si* $L' = R_{K/k}(L)$, *et si* $\mathscr{Z}(L)^0$ *(resp.* $\mathscr{N}(L)^0$*) est défini sur* K, *alors* $R_{K/k}\mathscr{Z}(L)^0 = \mathscr{Z}(L')^0$ *(resp.* $R_{K/k}\mathscr{N}(L)^0 = \mathscr{N}(L')^0$*).*

Il est clair que si L est un K-sous-groupe connexe de H, alors $L' = R_{K/k}L$ a la propriété indiquée. Soit réciproquement L' un k-sous-groupe connexe de H' tel que $\mu^0(L')$ soit produit de ses projections sur les $^{\sigma_i}H$. Ce qu'on a dit précédemment montre que $\mu^0(L') \cap {}^{\sigma_1}H = L$ et la projection η de L' sur L sont définis sur K, ce qui entraîne immédiatement que (L', η) vérifie la propriété caractéristique (1) de 6.17.

La deuxième assertion est une conséquence immédiate de la première.

6.19. *Corollaire.* — *Le foncteur* $R_{K/k}$ *définit une bijection des sous-groupes de Cartan définis sur* K *(resp. tores maximaux définis sur* K, *resp.* K-*sous-groupes paraboliques) de* H *sur les sous-groupes de Cartan définis sur* k *(resp. tores maximaux définis sur* k, *resp.* k-*sous-groupes paraboliques) de* H'.

Remarquons tout d'abord que le foncteur $R_{K/k}$ préserve la propriété d'être un tore ou d'être commutatif, ou nilpotent. On utilisera aussi le fait que les sous-groupes paraboliques d'un produit direct sont les produits directs de sous-groupes paraboliques de facteurs; cela résulte par exemple de 4.2, 4.3, après une réduction évidente au cas semi-simple.

Soit L un K-sous-groupe de H. Si L est un sous-groupe de Cartan (resp. un tore maximal; resp. un sous-groupe parabolique), il en est de même du sous-groupe $^{\sigma_i}L$ de $^{\sigma_i}H$, donc aussi du sous-groupe $R_{K/k}L$ de H'. Réciproquement, si L' est un sous-groupe de Cartan (resp. un tore maximal; resp. un sous-groupe parabolique) de H', alors $\mu^0(L')$ est produit direct de ses intersections avec les $^{\sigma_i}H$; si de plus L' est défini sur k, il existe (vu 6.18) un K-sous-groupe L de H tel que $L'=R_{K/k}L$, et L est un sous-groupe de Cartan de H, car il est nilpotent et égal à son normalisateur connexe (resp. est un tore maximal de H pour des raisons de dimension; resp. est un sous-groupe parabolique de H parce que $R_{K/k}(H/L)=H'/L'$ est une variété projective, et qu'il en est donc de même de H/L).

6.20. Soient T un tore défini sur K et $T'=R_{K/k}T$. Alors l'application $b\mapsto\sum_i{}^{\sigma_i}(b\circ\mu)$ définit un isomorphisme α de $X^*(T)_K$ sur $X^*(T')_k$ [21, § 1.4]; en particulier (1.8), T_d et T'_d ont même dimension. On a un homomorphisme naturel $X^*(T)_K\to X^*(T'_d)$, le composé de α et de la restriction $X^*(T')\to X^*(T'_d)$, qui est injectif, de conoyau fini, mais n'est pas bijectif en général.

Supposons T déployé sur K. Alors la projection de $\mu^0(T')$ sur $^{\sigma_i}T$ induit un isomorphisme de T'_d sur $^{\sigma_i}T$ pour tout i, d'où un isomorphisme $\beta_i : X^*(^{\sigma_i}T)\to X^*(T'_d)$; il suffit de le vérifier pour $T=\mathbf{G}_m$, pour lequel c'est élémentaire. En particulier, β_1, qu'on notera aussi β, est un isomorphisme de $X^*(T)=X^*(T)_K$ sur $X^*(T'_d)=X^*(T'_d)_k$.

6.21. (i) Supposons maintenant H réductif. De 6.19, 6.20, on tire : $r_K(H)=r_k(H')$; si A est un tore déployé sur K maximal de H, alors le plus grand tore A'_d déployé sur k de $A'=R_{K/k}A$ est un tore déployé sur k maximal de H'; l'isomorphisme canonique $\beta : X^*(A)\to X^*(A'_d)$ de 6.20 induit une bijection de $_K\Phi(H)$ sur $_k\Phi(H')$, et $\mathscr{Z}(A')=\mathscr{Z}(A'_d)$.

Soit ψ une partie quasi-close de $_K\Phi(H)$ (3.8). Le groupe H_ψ peut être caractérisé comme le sous-groupe de H contenant $\mathscr{Z}(A)$ et dont l'algèbre de Lie est somme de celle de $\mathscr{Z}(S)$ et des espaces propres maximaux de A dans $\mathfrak{h}$ correspondant aux éléments de ψ. Vu 6.20 et l'égalité $\mathscr{Z}(A')=\mathscr{Z}(A'_d)$, il s'ensuit que

$$(1)\qquad\qquad R_{K/k}H_\psi=H'_{\beta(\psi)}.$$

On voit de manière analogue que si η est quasi-clos, formé de racines positives pour un ordre convenable, alors $R_{K/k}H^*_\eta=H'^*_{\beta(\eta)}$. Comme H^*_ψ est engendré par des groupes de ce type, on a aussi

$$(2)\qquad\qquad R_{K/k}H^*_\psi=H'^*_{\beta(\psi)}.$$

Il résulte en particulier de (1) que si l'on fixe sur $X^*(A)$ et $X^*(A'_d)$ des ordres se correspondant par β, alors, dans les notations de 5.12,

$$(3)\qquad\qquad R_{K/k}{}_KP_\theta={}_kP_{\beta(\theta)},$$

θ étant une partie de l'ensemble des k-racines simples de H.

(ii) Si H est semi-simple et presque simple sur K, alors H′ est semi-simple et presque simple sur k. En effet les groupes $^{\sigma_i}$H sont alors les facteurs presque simples de H′ et ils sont permutés transitivement par $\Gamma = \mathrm{Gal}(k_s/k)$. Tout groupe G semi-simple et presque simple sur k est k-isogène à un groupe $\mathrm{R}_{k'/k}\mathrm{H}$ où k' est une extension de k convenablement choisie, et où H est un k'-groupe, presque simple sur $\bar{k}$. En effet, quitte à le remplacer par un groupe k-isogène, on peut supposer que G est $\bar{k}$-isomorphe à un produit direct de sous-groupes presque simples. Soit H l'un d'eux et soit $k' \subset \bar{k}$ son plus petit corps de définition contenant k. C'est une extension séparable de degré fini de k (2.14), que l'on peut donc supposer contenue dans k_s. Les facteurs simples de G sont permutés par Γ, nécessairement de façon transitive. On peut donc les écrire sous la forme $^{\nu_1}$H, …, $^{\nu_d}$H où ν_1, …, ν_d forment un système de représentants des classes à gauche de $\mathrm{Gal}(k_s/k)$ modulo $\mathrm{Gal}(k_s/k')$. Les restrictions des ν_i à k' sont évidemment les isomorphismes de k' dans k_s. D'autre part le produit des $^{\nu_i}$H différents de H est défini sur k_s, stable par $\mathrm{Gal}(k_s/k')$, donc aussi défini sur k', et ainsi la projection π de G sur H est définie sur k'. Par conséquent, la paire (G, π) vérifie la propriété (1) de 6.17 et $\mathrm{G} \cong \mathrm{R}_{k'/k}\mathrm{H}$.

6.22. *Proposition.* — *Soient* $\theta_i\ (i = 1, 2)$ *deux parties de* $_k\Delta$, $\mathrm{P}_i = {}_k\mathrm{P}_{\theta_i}$ *les k-sous-groupes paraboliques standards correspondants* (5.12), $_k\mathrm{W}_i$ *(resp.* $_\mathrm{K}\mathrm{W}_i$*) le sous-groupe de* $_k\mathrm{W}$ *(resp.* $_\mathrm{K}\mathrm{W}$*) engendré par les réflexions fondamentales correspondant aux éléments de* θ_i *(resp.* $_\mathrm{K}\eta_k(\theta_i)$ *(6.3)) et* $w \in {}_\mathrm{K}\mathrm{W}$. *Alors, la double classe* $\mathrm{P}_{1,\mathrm{K}}.w.\mathrm{P}_{2,\mathrm{K}} \subset \mathrm{G}_\mathrm{K}$ (5.20) *possède des points rationnels sur k si et seulement si la double classe correspondante* $_\mathrm{K}\mathrm{W}_1.w._\mathrm{K}\mathrm{W}_2$ *a une intersection non vide avec* $_\mathrm{K}\mathrm{W}_k$ (6.10). *Lorsqu'il en est ainsi, cette intersection est l'image réciproque par l'application canonique* $_\mathrm{K}\mathrm{W}_k \to {}_k\mathrm{W}$ *d'une double classe* $_k\mathrm{W}_1.w'._k\mathrm{W}_2$, *avec* $w' \in {}_k\mathrm{W}$, *et la double classe* $\mathrm{P}_{1,k}.w'.\mathrm{P}_{2,k}$ *est l'ensemble des points rationnels sur k de* $\mathrm{P}_{1,\mathrm{K}}.w.\mathrm{P}_{2,\mathrm{K}}$.

Soient x' un élément quelconque de $_k\mathrm{W}$, x un représentant de x' dans $_\mathrm{K}\mathrm{W}_k$ et n un représentant de x' dans $\mathscr{N}(\mathrm{S})_k$. Puisque T et ^{n}T sont deux tores déployés sur K maximaux de $\mathscr{Z}(\mathrm{S})$, il existe un $z \in \mathscr{Z}(\mathrm{S})_\mathrm{K}$ tel que $^z\mathrm{T} = {}^n\mathrm{T}$, c'est-à-dire tel que $n' = n^{-1}.z \in \mathscr{N}(\mathrm{S}, \mathrm{T})_\mathrm{K}$. L'image canonique x'' de n' dans $_\mathrm{K}\mathrm{W}_k$ est congrue à x (mod. $_\mathrm{K}\mathrm{W}_k^0$). On a donc, puisque $\mathscr{Z}(\mathrm{T})_\mathrm{K} \subset \mathrm{P}_{i,\mathrm{K}}$,

$$\mathrm{P}_{1,k}.x'.\mathrm{P}_{2,k} = \mathrm{P}_{1,k}.n.\mathrm{P}_{2,k} \subset \mathrm{P}_{1,\mathrm{K}}.n.\mathrm{P}_{2,\mathrm{K}} = \mathrm{P}_{1,\mathrm{K}}.n'.\mathrm{P}_{2,\mathrm{K}} = \mathrm{P}_{1,\mathrm{K}}.x''.\mathrm{P}_{2,\mathrm{K}} = \mathrm{P}_{1,\mathrm{K}}.x.\mathrm{P}_{2,\mathrm{K}}.$$

La première partie de l'énoncé en résulte, compte tenu de 5.20.

Pour établir la seconde, il suffit à présent de montrer que si $\bar{x}' \in {}_k\mathrm{W}$ est tel que $\mathrm{P}_1.\bar{x}'.\mathrm{P}_2 = \mathrm{P}_1.x'.\mathrm{P}_2$, alors $\bar{x}' \in {}_k\mathrm{W}_1.x'._k\mathrm{W}_2$. Posons $\mathrm{P} = {}_k\mathrm{P}_\theta$, pour tout parabolique $\mathrm{Q} \supset \mathscr{Z}(\mathrm{S})$, notons $^u\mathrm{Q}$ son unique opposé contenant $\mathscr{Z}(\mathrm{S})$, et soient P′ un k-sous-groupe parabolique tel que $\mathscr{Z}(\mathrm{S}) \subset \mathrm{P}' \subset (^{ux'}\mathrm{P}_2 \cap \mathrm{P}_1).\mathrm{R}_u(\mathrm{P}_1)$, et $\mathrm{P}'' = (^{x^{-1}u}\mathrm{P}' \cap \mathrm{P}_2).\mathrm{R}_u(\mathrm{P}_2)$. D'après 4.4, 5.9, il existe $y' \in {}_k\mathrm{W}_1$ et $y'' \in {}_k\mathrm{W}_2$ tels que $^{y'}\mathrm{P}' = \mathrm{P}''$ et $^{y''}\mathrm{P} = \mathrm{P}''$. Il résulte de la construction de P′, P″ que $\mathrm{P}'.^x\mathrm{P}''$ est dense dans $\mathrm{P}_1.^x\mathrm{P}_2$, donc que $\mathrm{P}.y'.x'.y''.\mathrm{P}$ est dense dans $\mathrm{P}_1.x'.\mathrm{P}_2$. De même, il existe $\bar{y}' \in {}_k\mathrm{W}_1$ et $\bar{y}'' \in {}_k\mathrm{W}_2$ tels que $\mathrm{P}.\bar{y}'.\bar{x}'.\bar{y}''.\mathrm{P}$ soit dense dans $\mathrm{P}_1.\bar{x}'.\mathrm{P}_2$. Mais alors $y'.x'.y'' = \bar{y}'.\bar{x}'.\bar{y}''$, en vertu de 5.15, et $\bar{x}' \in {}_k\mathrm{W}_1.x'._k\mathrm{W}_2$.

6.23. *Corollaire.* — *Soient* P *et* P′ *deux sous-groupes paraboliques de* G *définis sur* k. *Alors, deux doubles classes distinctes de* P_k, P'_k *dans* G_k *sont contenues dans des doubles classes distinctes de* P_K, P'_K *dans* G_K.

6.24. *Remarques.* — *a)* La proposition 6.22 prend une forme particulièrement simple lorsque $P_1 = P_2 = {}_kP_\emptyset$, k-sous-groupe parabolique minimal, que nous noterons encore P. Dans ce cas, les doubles classes de P_K dans G_K correspondent aux doubles classes de ${}_KW_k^0$ dans ${}_KW$, et la proposition affirme que les doubles classes de P_K qui possèdent un point rationnel dans k sont celles qui correspondent aux doubles classes de ${}_KW_k^0$ dans ${}_KW_k$; ces dernières sont en fait des classes latérales simples, puisque ${}_KW_k^0$ est distingué dans ${}_KW_k$. L'ensemble des points rationnels de la double classe de P_K associée à une classe latérale donnée de ${}_KW_k^0$ dans ${}_KW_k$ est alors la double classe latérale de P_k dans G_k qui correspond à l'élément de ${}_kW$ représentant celle-ci.

b) Lorsque $K = \bar{k}$ (ou, si on préfère, lorsque K est le domaine universel), la proposition 6.22 donne la condition pour qu'une double classe « géométrique » $P_1.g.P_2$ possède des points rationnels sur k. Notons qu'une telle double classe peut être définie sur k sans posséder de points rationnels.

6.25. *Proposition.* — *Soient* P *et* P⁻ *deux* k-*sous-groupes paraboliques minimaux opposés, et soient* U *et* U⁻ *leurs radicaux unipotents. On a alors* $G_K = U_k.U_K^-.P_K$. *En particulier,* $G = U_k.U^-.P$.

Dans la suite nous supposons, ce qui est loisible (4.8, 4.13, 4.16) que $P \cap P^- = \mathscr{Z}(S)$. Avant de démontrer cette proposition, nous établirons deux corollaires.

6.26. *Corollaire.* — *Tout élément de* ${}_kW$ *possède un représentant dans l'ensemble* $U_k.U_k^-.U_k$. En effet, il résulte de 6.25 que $G_k = U_k.U_k^-.P_k = U_k.U_k^-.U_k.\mathscr{Z}(S)_k$.

6.27. *Corollaire.* — *Étant donnés deux sous-groupes paraboliques* Q, Q′ *conjugués donc l'un est défini sur* k, *il existe un sous-groupe parabolique défini sur* k *qui leur est opposé à tous deux.*

On ne nuit pas à la généralité en supposant, dans les notations de 5.12 et 6.25, que $Q = {}_kP_\theta \supset P$ et que ${}_kP_\theta^- = P^-$ (4.8, 5.14). Soit $g \in G$ tel que $Q' = {}^gQ$. Vu 6.25, on peut poser $g = u.u'.p$ avec $u \in U_k$, $u' \in U^-$ et $p \in P$. Alors, le sous-groupe parabolique ${}^u({}_kP_\theta^-) = {}^{u.u'}({}_kP_\theta^-)$ est opposé simultanément à ${}^u({}_kP_\theta) = Q$ et à ${}^{u.u'}({}_kP_\theta) = Q'$.

On aurait pu tout aussi facilement déduire la proposition 6.25 du corollaire 6.27. C'est d'ailleurs essentiellement ce qu'exprime le lemme suivant, première étape de la démonstration de 6.25.

6.28. *Lemme.* — *Les notations étant celles de* 6.25, *supposons que, pour tout sous-groupe parabolique* P′ *conjugué à* P, *il existe un sous-groupe parabolique défini sur* k *et opposé simultanément à* P *et à* P′. *Alors,* $G_K = U_k.U_K^-.P_K$ *pour toute extension* K *de* k.

Soit $g \in G_K$ et soit Q un sous-groupe parabolique défini sur k opposé à P et à ${}^gP$. Alors (4.13) il existe $u \in U_k$ tel que ${}^uQ = P^-$. Le groupe ${}^{u.g}P$ est opposé à P⁻ et défini sur K, donc il existe un $u' \in U_K^-$ tel que ${}^{u'.u.g}P = P$. Mais alors,

$$u'.u.g \in P_K \qquad \text{et} \qquad g \in u^{-1}.u'^{-1}.P_K \subset U_k.U_K^-.P_K.$$

6.29. *Lemme.* — *Supposons que k soit égal au corps fini $\mathbf{F}_q$ à q éléments, et que G soit k-isogène à $\mathbf{SU}_3$. Alors, G possède $(q^3 + 1)$ k-sous-groupes de Borel et le nombre des k-sous-groupes de Borel non opposés à un sous-groupe de Borel donné quelconque est inférieur à $2(q + 1)$.*

Soit $K = \mathbf{F}_{q^2}$ et soit γ l'élément de $\mathrm{Gal}(K/k)$ distinct de l'identité. Le groupe G est déployé sur K et la variété des sous-groupes de Borel de G (5.24) est définie sur K et peut être identifiée avec la variété des drapeaux (paires formées d'un point et d'une droite le contenant) d'un plan projectif M défini sur K (au sens de la géométrie algébrique), et dont l'ensemble M_K des points rationnels sur K est donc un plan projectif sur K (au sens usuel). Pour tout point $x \in M_K$, le transformé par γ du groupe de stabilité de x laisse invariant une et une seule droite que nous noterons D_x. La correspondance $x \mapsto D_x$ est une « polarité hermitienne » dans M_K (si $G = SU_3(f)$, c'est la polarité associée à la forme hermitienne f). L'identification mentionnée plus haut associe à tout sous-groupe de Borel B le drapeau F qu'il laisse stable, et F est défini sur k si et seulement si B l'est. Il est immédiat que les drapeaux définis sur k sont les drapeaux de la forme (x, D_x), avec $x \in M_K$. Leur nombre est égal à celui des points $x \in M_K$ qui sont isotropes (i.e. qui appartiennent à leur droite polaire D_x), donc à $q^3 + 1$.

Soit (y, D) un drapeau de M, et soit B son groupe de stabilité. Pour que le groupe d'isotropie d'un drapeau (x, D_x) rationnel sur k ne soit pas opposé à B, il faut et il suffit que l'on ait soit $x \in D$, soit $y \in D_x$. Comme une droite quelconque contient 0, 1 ou $q + 1$ points isotropes rationnels sur K, le nombre de drapeaux (x, D_x) vérifiant la première condition est égal à 0, 1 ou $q + 1$. Il en est de même du nombre de drapeaux vérifiant la deuxième condition ; en effet, si $y \in M_K$, cette dernière peut s'écrire $x \in D_y$, et sinon, y est contenu dans au plus une droite définie sur K.

6.30. *Démonstration de la proposition* 6.25.

a) *k est infini.* — La classe de conjugaison $\mathscr{P}$ des k-sous-groupes paraboliques minimaux est auto-opposée (4.14), donc deux éléments de cette classe sont opposés si et seulement s'ils contiennent des sous-groupes de Borel opposés (4.10). Il suit alors de 4.12 que les hypothèses de 6.28 sont vérifiées.

b) *Réduction au cas du rang semi-simple relatif* 1. — Nous nous proposons de montrer que la proposition est vraie en général si elle l'est pour les groupes de rang semi-simple relatif 1. Dans les calculs qui suivent, les indices K sont omis (on peut d'ailleurs, si on veut, se ramener immédiatement au cas où K est le domaine universel, soit en utilisant la relation $(U^-.P)_K = U_K^-.P_K$, soit en combinant la démonstration du corollaire 6.27 et le lemme 6.28). Posons $G^* = U_k.U^-.P$; soit a une k-racine simple quelconque, et désignons par (a) (resp. $(a)^*$) l'ensemble des racines positives proportionnelles (resp. non proportionnelles) à a. La proposition étant supposée vraie pour le groupe $G' = G_{(a) \cup (-a)}$, on a

$$G^*.U_{-(a)} = U_k.U^-.\mathscr{L}(S).U_{(a)}.U_{(a)^*}.U_{-(a)} = U_k.U_{-(a)^*}.U_{-(a)}.\mathscr{L}(S).U_{(a)}.U_{-(a)}.U_{(a)^*} =$$
$$= U_k.U_{-(a)^*}.G'.U_{(a)^*} = U_k.U_{-(a)^*}.U_{(a),k}.U_{-(a)}.\mathscr{L}(S).U_{(a)}.U_{(a)^*} =$$
$$= U_k.U_{(a),k}.U_{-(a)^*}.U_{-(a)}.\mathscr{L}(S).U_{(a)}.U_{(a)^*} = U_k.U^-.P = G^*.$$

Or $G^* . P = G^*$. Mais P et les groupes $U_{-(a)}$ correspondant à toutes les racines simples engendrent G (5.14). Par conséquent, $G^* . G = G^*$, et $G^* = G$.

c) *k est fini et* $r_k(\mathscr{D}G) = 1$.

Soit $k = \mathbf{F}_q$. Nous voulons montrer que G vérifie l'hypothèse du lemme 6.28. Celle-ci ne change pas si on remplace G par $\mathscr{D}G$. Nous supposerons donc G semi-simple et $r_k(G) = 1$, il est alors presque simple sur k puisqu'il n'existe pas de groupe semi-simple anisotrope non trivial sur un corps fini. Soient $G_0, \ldots, G_{n-1}$ les sous-groupes invariants quasi-simples de G et soit k' le plus petit corps de définition commun des G_i. Le groupe $\mathrm{Gal}(k'/k)$ permute transitivement les G_i, et puisqu'il est cyclique, il est d'ordre n et permute cycliquement les G_i. En particulier, on a $k' = \mathbf{F}_{q^n}$. Soit γ un générateur de $\mathrm{Gal}(k'/k)$. Nous supposerons que G_i est le transformé de G_0, que nous noterons encore G', par γ^i. Le groupe G est k-isogène à $R_{k'/k}(G')$ (6.21 (ii)) donc le k'-rang de G' est égal à 1 (6.21 (i)).

Tout sous-groupe de Borel B de G est un produit $B_0 . B_1 \ldots B_{n-1}$ de sous-groupes de Borel de $G_0, G_1, \ldots, G_{n-1}$ respectivement. D'après 6.19, le nombre de k-sous-groupes de Borel de G est égal à celui des k'-sous-groupes de Borel de G'.

Soit $B = B_0 \ldots B_{n-1}$ un sous-groupe de Borel de G. Un autre sous-groupe de Borel $B' = B_0' . B_1' \ldots B_{n-1}'$ lui est opposé si et seulement si B_i' est opposé à B_i pour tout i. D'autre part, on tire de 4.18, 5.9 que dans un k-groupe réductif, de k-rang semi-simple égal à un, deux k-sous-groupes paraboliques minimaux distincts sont opposés. Par conséquent, pour établir notre assertion il suffira de montrer que si N_i désigne le nombre de sous-groupes de Borel de G_i définis sur k' et non opposés à B_i, on a $\sum_i N_i < N - 1$ quel que soit B. Or il résulte de la classification que G' est, à une k'-isogénie près, le groupe $\mathbf{SL}_2$ ou le groupe $\mathbf{SU}_3$. Dans le premier cas, on a $N = q^n + 1$ et $0 \leq N_i \leq 1$, d'où $\sum_i N_i \leq n < N - 1$. Dans le second cas, il suit du lemme 6.29 que $N = q^{3n} + 1$ et $N_i \leq 2(q^n + 1)$, d'où $\sum_i N_i \leq 2n . (q^n + 1) < N - 1$, ce qui achève la démonstration.

Remarque. — La prop. 6.25 répond à une question de Grothendieck et Demazure. Elle a aussi été démontrée par Steinberg et Demazure.

§ 7. UN SOUS-GROUPE DÉPLOYÉ MAXIMAL

7.1. Lemme. — *Soient donnés un système de racines de rang 2, et un ordre dans celui-ci. Soient a, b les racines simples, r_a, r_b les réflexions fondamentales correspondantes et W le groupe de Weyl. Pour tout entier $i \geq 1$, posons $c_i = a$ ou b selon que i est pair ou impair et soit $w_i \in W$ défini inductivement par les relations $w_1 = r_b$ et $w_i = r_{c_i} . w_{i-1}$ pour $i \geq 2$. Soit m (resp. m'; resp. m'') la plus petite valeur de l'entier $i \geq 2$ telle que $(r_a . r_b)^i = e$ (resp. telle que $w_{i-1}(a)$ soit une racine simple; resp. telle que $w_i(a)$ soit une racine négative). Alors, $m = m' = m''$ et $w_{m-1}(a) = c_m$.*

La racine $w_{m''-1}(a)$ est positive et sa transformée par $r_{c_{m''}}$ est négative. Par conséquent, $w_{m''-1}(a)$ ne peut être que la racine simple $c_{m''}$ [6, § 7, n° 6, Prop. 17]. En particulier, $m' \leq m''$.

Posons $w_{m'-1}(a) = c$ où c est donc égal à a ou b. On a $w_{m'-1} \cdot r_a \cdot w_{m'-1}^{-1} = r_c$. Selon que $c = c_{m'-1}$ ou $c_{m'}$, cette dernière relation peut encore s'écrire $(r_a \cdot r_b)^{m'-1} = e$ ou $(r_a \cdot r_b)^{m'} = e$. Il en résulte que $m \leq m'$.

Enfin, la relation $(r_a \cdot r_b)^m = e$ peut s'écrire $w_{m-1} \cdot r_a \cdot w_{m-1}^{-1} = r_{c_m}$, ou encore $w_{m-1}(a) = \pm c_m$. Puisque $m \leq m''$, c'est le signe $+$ qui doit être retenu. Mais alors, $w_m(a) = r_{c_m}(c_m) = -c_m$, donc $m'' \leq m$, ce qui achève la démonstration.

Avant d'énoncer le théorème de ce paragraphe, rappelons que dans un système de racines Φ, l'ensemble (d) des multiples entiers positifs d'une racine est égal à $\{d\}$ ou $\{d, 2d\}$ et que d est dite non multipliable dans le premier cas. L'ensemble Φ' des racines non multipliables est un système de racines réduit.

7.2. Théorème. — *Supposons G réductif connexe. Soient S un tore déployé sur k maximal de G, $\Phi = \Phi(S, G)$ le système des k-racines de G, Φ' celui des racines non multipliables de Φ, et Δ l'ensemble des racines simples de Φ' pour un ordre donné sur $X^*(S)$. Pour tout $a \in \Delta$, soit E_a un k-sous-groupe de dimension un de $U_{(a)}$ normalisé par S et soit V le groupe engendré par les E_a. Alors G possède un unique sous-groupe réductif F connexe déployé sur k contenant S.V. Le tore S est un tore maximal de F et $\Phi' = \Phi(S, F)$. En particulier, $W(F) \cong_k W(G)$. Pour tout $d \in \Phi$, l'intersection $U_{(d)} \cap F$ est le groupe radiciel $U_{d'}$ de F correspondant à la racine d' de Φ' contenue dans (d).*

Pour tout élément $b \in \Phi'$, on a $(b) = \{b\}$, donc le groupe $U_{(b)}$ admet une structure d'espace vectoriel sur k telle que $\operatorname{Int} s \ (s \in S)$ soit l'homothétie de rapport s^b (3.18, rem.). Par suite, tout k-sous-groupe de dimension un de $U_{(b)}$ stable par S est k-isomorphe au groupe additif $\mathbf{G}_a$. Cela vaut en particulier pour les groupes E_a de l'énoncé.

Soit v un élément quelconque de $E_{a,k}$, distinct de l'élément neutre. D'après le théorème 5.15, appliqué au groupe réductif $G_{\{a,-a\}}$ (3.4, 3.8), v possède une décomposition unique de la forme

$$(1) \qquad v = v' \cdot n_a \cdot v'', \qquad (v', v'' \in U_{(-a)} ; n_a \in \mathcal{N}(S), \quad \text{et} \quad v', v'', n_a \in G_k).$$

Soit s un élément de S tel que $s^a = -1$. On a

$$v = {}^s(v^{-1}) = {}^s(v''^{-1}) \cdot {}^s(n_a^{-1}) \cdot {}^s(v'^{-1}) = v'' \cdot {}^s(n_a^{-1}) \cdot v',$$

d'où, en vertu de l'unicité de (1), $v' = v''$, ${}^s(n_a^{-1}) = n_a$, et

$$(2) \qquad n_a^2 = (n_a, s) \in S_k.$$

Posons $E_{-a} = \{{}^s v' \mid s \in S\} \cup \{e\}$. On a vu que $U_{(-a)}$ a une structure d'espace vectoriel sur k telle que la conjugaison par $s \in S$ soit la multiplication scalaire par s^{-a}. Donc E_{-a} est un k-groupe k-isomorphe au groupe additif. On a, pour tout $s \in S$,

$$S \cdot E_{-a} \cdot {}^s v = S \cdot E_{-a} \cdot {}^s n_a \cdot {}^s v';$$

comme $^s n_a \in n_a . S = S . n_a$, on en tire, par réunion étendue à S,

$$(3) \qquad S . E_{-a} . (E_a - \{e\}) = S . E_{-a} . n_a . (E_{-a} - \{e\}).$$

Multiplions membres à membres cette relation et celle qu'on en déduit en remplaçant chaque membre par son inverse; il vient, compte tenu de (2),

$$S . E_{-a} . E_a . E_{-a} = S . E_{-a} . n_a . E_{-a} . n_a . E_{-a}.$$

Posons

$$(4) \qquad Y_a = S . E_{-a} . \{e, n_a\} . E_{-a}.$$

De (3) multipliée à droite par E_{-a}, il résulte que Y_a est aussi égal à $S . E_{-a} . E_a . E_{-a}$. Donc

$$S . E_{-a} . n_a . E_{-a} . n_a . E_{-a} = Y_a.$$

Il s'ensuit que $Y_a . Y_a \subset Y_a$. D'autre part, $Y_a^{-1} = Y_a$. Par conséquent Y_a est un groupe qui, étant engendré par E_a, E_{-a} et S, est connexe et défini sur k. Du théorème 5.15 (où il faut remplacer l'ensemble des racines positives par celui des racines négatives) il résulte que $Y_a \cap U_{(-a)} = E_{-a}$. En transformant cette relation par n_a, on voit que $n_a . E_{-a} . n_a^{-1}$ contient E_a; pour raison de dimension, il s'ensuit que

$$(5) \qquad n_a . E_{-a} . n_a^{-1} = E_a.$$

Transformons à présent la relation (4) par n_a; il vient

$$(6) \qquad Y_a = S . E_a . \{e, n_a\} . E_a.$$

Soient $a, b \in \Delta$ deux racines simples quelconques, (a, b) l'ensemble de toutes les racines de la forme $m.a + n.b$ avec m et n entiers strictement positifs, E_{ab} le groupe engendré par E_a et E_b, et $E'_{ab} = (E_a, E_b)$ le groupe engendré par les commutateurs (x, y) avec $x \in E_a$ et $y \in E_b$. On a $E'_{ab} = E'_{ba}$, et d'après 3.10,

$$(7) \qquad E'_{ab} \subset U_{(a, b)}.$$

En vertu de l'identité

$$^v(x, y) = (v.x, y) . (v, y)^{-1},$$

E'_{ab} est normalisé par E_a, donc le produit $E_a . E'_{ab}$ est un groupe que nous noterons X_{ab}. Celui-ci est normalisé par E_b, car E_b normalise E'_{ab} et $(E_b, E_a) = E'_{ab} \subset X_{ab}$. Par conséquent

$$(8) \qquad E_{ab} = X_{ab} . E_b = E_a . E'_{ab} . E_b.$$

Il résulte alors de (7) et 3.11 qu'on a $E_{ab} \cap U_{(a)} = E_a$ et $E_{ab} \cap U_{(b)} = E_b$, ce qui nous permet, sans risque de contradiction dans les notations, de poser pour tout $c \in (a, b) \cup \{a, b\}$,

$$E_c = E_{ab} \cap U_{(c)}.$$

L'ensemble $A = \{^x y \mid x \in E_b - \{e\}, y \in E_a\}$ est un système générateur de X_{ab}. En effet, cet ensemble étant normalisé par S, il en est de même du groupe qu'il engendre, et il résulte alors de 3.11 qu'en même temps que $^x y = (x, y) . y^{-1}$, ce groupe contient (x, y)

et y^{-1}. Aucune combinaison linéaire $m.a-n.b$, avec m et n entiers strictement positifs, n'étant une racine, les groupes E_a et E_{-b} commutent comme on le voit à partir de 3.10, en utilisant un ordre sur $X^*(S)$ tel que $a, -b > 0$. Par conséquent, si on pose

$$Y_b^* = (E_b - \{e\}).S.E_{-b} = Y_b - \{e, n_b\}.S.E_{-b},$$

on a $A = \{{}^x y \mid x \in Y_b^*, y \in E_a\}$. Or, d'après (2) et la deuxième expression donnée pour l'ensemble Y_b^*, celui-ci est invariant par multiplication à gauche par n_b. Il s'ensuit que A est normalisé par n_b et il en est donc de même de X_{ab} :

$$(9) \qquad n_b.X_{ab}.n_b^{-1} = X_{ab}.$$

De (5), (8), (9) et 3.11, il résulte que, si r_a désigne (pour tout $a \in \Delta$) la réflexion fondamentale correspondant à a, on a, pour tout $c \in (a, b) \cup \{a, b\}$,

$$(10) \qquad n_b.E_c.n_b^{-1} = E_{r_b(c)}.$$

Soient $m(a, b)$ l'ordre du produit $r_a.r_b$ dans le groupe de Weyl relatif ${}_k W$, $n_{ab} = \ldots n_a.n_b$ le produit de $m(a, b) - 1$ facteurs alternativement égaux à n_b et à n_a, en commençant par la droite, et $c(a, b)$ la racine simple égale à a ou b selon que $m(a, b)$ est pair ou impair. On a alors, en vertu de (10) et du lemme 7.1,

$$n_{ab}.E_a.n_{ab}^{-1} = E_{c(a, b)},$$

et, par ailleurs

$$n_{ab}.U_{(-a)}.n_{ab}^{-1} = U_{(-c(a, b))}.$$

Transformons à présent la relation (1) par n_{ab}; on voit, compte tenu des deux relations précédentes et du théorème 5.15, que

$$n_{ab}.n_a.n_{ab}^{-1} \in n_{c(a, b)}.S_k,$$

ou encore, en utilisant (2),

$$(11) \qquad (n_a.n_b)^{m(a, b)} \in S_k.$$

Les relations (2) et (11) et la définition classique des groupes de Weyl par générateurs et relations montrent que, si M désigne le groupe engendré par S et les n_a $(a \in \Delta)$, on a

$$(12) \qquad M \cap \mathcal{Z}(S) = S,$$

et que l'application canonique $M_k \to {}_k W$ est surjective.

Soit V_a^* le groupe engendré par les X_{ba} $(b \in \Delta - \{a\})$; puisque E_a normalise les X_{ba}, il normalise aussi V_a^*, donc

$$V = E_a.V_a^* = V_a^*.E_a.$$

Soit $(a)'$ (resp. $(a)^*$) l'ensemble des racines positives proportionnelles (resp. non proportionnelles) à a. En vertu de (7), V_a^* est contenu dans $U_{(a)^*}$, donc on a, d'après 3.11,

$$(13) \qquad V \cap U_{(a)'} = E_a,$$

et

$$(14) \qquad V \cap U_{(a)^*} = V_a^*.$$

D'autre part, il résulte de (9) que

$$(15) \qquad n_a . V_a^* . n_a^{-1} = V_a^* .$$

Nous nous proposons à présent de montrer que

(16) si $m \in M$ et si l'image canonique w de m dans $_k W$ transforme a en une racine positive, alors $m . E_a . m^{-1} \subset V$.

La démonstration se fera par induction sur le nombre de racines positives transformées en racines négatives par w^{-1}. Si ce nombre est nul, $w = e$, et l'assertion est évidente. Soit $w \neq e$, soit $b \in \Delta$ tel que $w^{-1}(b) < 0$, posons $w' = r_b . w$ et $m' = n_b . m$. En vertu de l'hypothèse d'induction, on a $m' . E_a . m'^{-1} \subset V = E_b . V_b^*$. Mais $w'(a)$ n'est pas proportionnelle à b (puisque $w(a) > 0$), donc, d'après (14), $m' . E_a . m'^{-1} \subset V_b^*$. En transformant les deux membres de cette relation par n_b, il vient, tenant compte de (15), $m . E_a . m^{-1} \subset V_b^* \subset V$, ce qui établit (16).

Faisant usage de (4), (15), (16), et du fait que Y_a est un groupe, on peut reproduire ici les calculs de 5.16, d'où il résulte que, pour tout $m \in M$ et tout $a \in \Delta$, on a

$$n_a . V . \{n_a, n_a . m\} . V . S = V . \{n_a, n_a . m\} . V . S.$$

Il en résulte immédiatement (cf. par ex. [33]) que le produit $F = V . M . V$ est un groupe, qui, étant engendré par S, les E_a et les E_{-a}, est connexe et défini sur k. Soit d une racine quelconque, soient (d) et d' définis comme dans l'énoncé, et soit m un élément de M_k dont l'image canonique w dans $_k W$ transforme d' en un élément de Δ que nous désignerons par a, de sorte que $w((d)) = (a)$ ou $(a)'$. Il résulte alors du théorème 5.15 et de (13) que

$$m . (F \cap U_{(d)}) . m^{-1} = E_a .$$

Par conséquent, $F \cap U_{(d)} = m^{-1} . E_a . m$ est un sous-groupe de $U_{(d')}$ isomorphe sur k au groupe additif, et que nous noterons $E_{d'}$; il est clair que cette notation est cohérente avec celles introduites plus haut. Si d' est positive, $E_{d'}$ est contenu dans V (par exemple en vertu de (16)), donc on a, d'après 3.11,

$$(17) \qquad V = \prod_{d' \in \Phi'^+} E_{d'} .$$

Le groupe Y_a n'est pas résoluble, car s'il l'était, son radical unipotent contiendrait E_a et E_{-a}, donc n_a, donc aussi tout commutateur (n_a, s), avec $s \in S$, ce qui est absurde. Tout sous-groupe Q de F contenant le produit S . V et distinct de lui contient au moins un élément de M dont l'image canonique w dans $_k W$ est distincte de e. Il existe alors au moins une racine simple a dont la transformée par w est négative. Mais alors Q contient E_{-a}, donc Y_a, et n'est pas résoluble. Le sous-groupe S . V étant résoluble (puisque contenu dans S . U^+), c'est un sous-groupe de Borel de F, et V en est un sous-groupe unipotent maximal. Soit m un élément de M dont l'image canonique dans $_k W$ transforme les racines positives en les racines négatives. Le transformé $V^- = {}^m V$ est aussi un sous-groupe unipotent maximal de F et on a $R_u(F) \subset V \cap V^- = \{e\}$ (3.22), donc F est réductif. En vertu du théorème 5.15, $F \cap \mathscr{Z}(S) = \mathscr{Z}_F(S) = S$, par conséquent S est un tore maximal de F. Les $E_{d'}(d' \in \Phi')$ sont des sous-groupes radiciels à un paramètre de F relativement à S, et ce sont les seuls, en vertu de la relation (17) et de sa transformée par m. Le groupe F est donc déployé et Φ' est le système des racines de F par rapport à S.

Enfin, soit $\overline{F}$ un sous-groupe réductif déployé de G contenant S.V. Le tore S est un tore maximal de $\overline{F}$. En vertu de (17), les $E_{d'}$, avec $d' \in \Phi'^+$ sont des sous-groupes radiciels à un paramètre de $\overline{F}$ relatifs à S. En particulier, le système des racines de $\overline{F}$ contient Φ'^+. Comme, d'autre part, il est nécessairement réduit, et contenu dans Φ, il ne peut être que Φ'. Soient a une racine simple, $v \in E_{a,k}$ et $\overline{E}_{-a}$ le sous-groupe radiciel à un paramètre de $\overline{F}$ relatif à $-a$. En appliquant le théorème 5.15 au groupe engendré par S, E_a et $\overline{E}_{-a}$, on voit qu'il doit exister des éléments $\overline{n}_a \in \mathscr{N}(S)_k$ et $\overline{v}', \overline{v}'' \in \overline{E}_{-a,k}$ tels que $v = \overline{v}'.\overline{n}_a.\overline{v}''$. Comme $\overline{E}_{-a}$ est contenu dans $U_{(-a)}$ (3.19), cette décomposition de v doit coïncider avec la décomposition (1). En particulier, $\overline{v}' = v'$, donc $\overline{E}_{-a} = E_{-a}$ et $\overline{F} = F$, ce qui achève la démonstration du théorème.

7.3. *Remarques.* — Le théorème 7.2 suggère les questions suivantes :

(i) Les sous-groupes F correspondant à divers choix des E_a sont-ils conjugués entre eux ?

(ii) Lorsqu'ils sont conjugués, deux sous-groupes F le sont-ils sur k ?

(iii) Le groupe F est-il le seul sous-groupe réductif déployé contenant V ?

(iv) Les sous-groupes F du théorème 7.2 sont-ils les seuls sous-groupes réductifs déployés maximaux de G ?

La réponse aux quatre questions est négative, comme le montrent les exemples suivants.

Soient $f = x.y + f_1(z_1, z_2, z_3)$ une forme quadratique à cinq variables d'indice 1 (i.e. f_1 est anisotrope) et de défaut 1 si la caractéristique est 2, et $G = \mathbf{O}^+(f)$. Le groupe G n'a que deux racines relatives opposées, $\pm a$, et on voit immédiatement que (i) (resp. (ii)) a une réponse affirmative si et seulement si deux sous-espaces vectoriels à une dimension de $U_{(a)}$, définis sur k, sont conjugués par un élément de $\mathscr{Z}(S)$ (resp. $\mathscr{Z}(S)_k$). Dans le cas présent, $\mathscr{Z}(S) = S \times \mathbf{O}^+(f_1)$, et le second facteur opère sur $U_{(a)}$ (qui a 3 dimensions) par la représentation usuelle. On obtient donc un contre-exemple à (i) en supposant la caractéristique égale à 2 (k est alors nécessairement non parfait) et un contre-exemple à (ii) en supposant la caractéristique $\neq 2$, et en prenant pour f_1 une forme dont les valeurs ne forment pas une seule classe de restes quadratiques.

Soit à présent $G = SU_3(f)$ un groupe spécial unitaire, correspondant à une forme hermitienne f d'indice 1 (groupe quasi-déployé : cf. [35]). Ce groupe, de k-rang égal à 1, possède quatre k-racines $\pm d$ et $\pm 2d$. Le groupe $U_{(2d)}$ a une dimension. On obtient un contre-exemple à (iii) en conjuguant F par un élément de $U_{(d),k}$ n'appartenant pas à $U_{(2d)}$. Supposons à présent k de caractéristique 0, et soit $\overline{V}' \neq U_{(2d)}$ un sous-groupe de dimension 1 de $U_{(d)}$ défini sur k. Il résulte du théorème de Jacobson-Morozov [14, th. 17, p. 100] que V' est contenu dans un sous-groupe de G isogène sur k à $\mathbf{SL}_2$, lequel constitue un contre-exemple à (iv), car on vérifie aisément que V' et $U_{(2d)}$ ne sont pas conjugués (par exemple, le normalisateur de V' est S.V' tandis que celui de $U_{(2d)}$ est $\mathscr{Z}(S).U_{(d)}$).

II. — COMPLÉMENTS ET APPLICATIONS

§ 8. SOUS-GROUPES TRIGONALISABLES (k PARFAIT)

8.1. Un sous-groupe fermé de $\mathbf{GL}_n$ est dit *trigonalisable sur un corps* K s'il est défini sur K et conjugué sur K à un sous-groupe du groupe $\mathbf{T}_n$ des matrices triangulaires supérieures. Il est alors nécessairement résoluble.

Supposons G connexe résoluble. La prop. 5 de [23] et l'existence d'un tore maximal défini sur k (2.14, ou [26], ou 11.4) montrent que les propriétés suivantes sont équivalentes :

(i) G est k-isomorphe à un groupe matriciel trigonalisable sur k.

(ii) G possède un tore maximal déployé sur k, et son radical unipotent G_u est défini sur k.

(iii) Toute image de G par un k-morphisme $G \to \mathbf{GL}_m$ est trigonalisable sur k.

On dira que G est *trigonalisable sur k* s'il vérifie ces propriétés. Tout groupe résoluble connexe déployé sur k est trigonalisable sur k [23, prop. 7]. La réciproque n'est pas vraie en général puisque un k-groupe unipotent est toujours trigonalisable sur k [23, cor. 1, p. 33] mais n'est pas nécessairement déployé sur k [23, p. 35]. Elle l'est cependant sur un corps parfait [23, cor. 2, p. 34]. Vu (ii), un tore est déployé sur k si et seulement s'il est trigonalisable sur k.

8.2. *Théorème. — Supposons k parfait. Alors les sous-groupes résolubles déployés sur k maximaux (resp. les tores déployés sur k maximaux, resp. les k-sous-groupes unipotents connexes maximaux) de G sont conjugués sur k. Si H est un sous-groupe résoluble déployé sur k maximal, alors $G_k/H_k = (G/H)_k$ s'identifie à l'ensemble des points rationnels sur k d'une k-variété projective sur laquelle G opère.*

Soit H un sous-groupe résoluble déployé sur k de dimension maximum de G. On choisit une réalisation linéaire de G comme k-groupe d'automorphismes d'un espace vectoriel V sur k contenant une droite W définie sur k telle que $H = \{g \in G \mid g(W) = W\}$, ce qui est toujours possible [25, p. 222]. Comme H est résoluble, déployé sur k, son image par le morphisme naturel dans GL(V/W) est trigonalisable sur k (cf. 8.1). Il existe donc un drapeau F_0 de V, défini sur k, stable par H, dont la droite est W. Le groupe H est alors *a fortiori* le groupe de stabilité de F_0 dans G. Soit B le groupe de stabilité de F_0 dans GL(V). C'est un groupe de Borel de GL(V), la variété des drapeaux $\mathscr{F}(V)$ s'identifie au quotient $GL(V)/B$, et la projection canonique $\pi : GL(V) \to \mathscr{F}(V)$ est surjective pour les points rationnels sur k (4.13).

Montrons que

$$(1) \qquad G(F_0)_k = \overline{(G(F_0))}_k.$$

Soient $P \in \overline{(G(F_0))}_k$ et G_P le groupe d'isotropie de P dans G. Comme P est un drapeau défini sur k, son groupe de stabilité est trigonalisable sur k, donc la composante neutre de G_P est un groupe résoluble déployé sur k; par conséquent, $\dim G_P \leq \dim H$ et $\dim G(P) \geq \dim G(F_0)$. Mais l'adhérence de $G(F_0)$ est réunion de $G(F_0)$ et d'orbites de dimensions strictement plus petites [1, § 16], donc $P \in G(F_0)_k$.

Soit maintenant L un sous-groupe résoluble déployé sur k de G. Il opère sur $\mathscr{F}(V)$ en laissant la variété complète $\overline{G(F_0)}$ stable. Comme cette dernière est définie sur k et possède au moins un point rationnel sur k, le groupe L admet un point fixe $P \in \overline{(G(F_0))}_k$ (2.7 (i)), qui, vu (1), est en fait dans $G(F_0)_k$. Mais, H étant déployé sur k, la projection canonique $G_k \to G(F_0)_k$ est surjective (2.7 (ii)). Il existe alors $g \in G_k$ tel que $g.F_0 = P$, donc tel que $g^{-1}.L.g \subset H$, ce qui montre que tout sous-groupe résoluble déployé sur k de G est conjugué sur k à un sous-groupe de H. L'assertion de conjugaison des sous-groupes unipotents résulte alors du fait que tout k-groupe unipotent connexe est déployé sur k, celle des tores maximaux du fait que, dans H, les tores maximaux définis sur k sont conjugués sur k (voir [26], ou aussi (11.4); on peut du reste donner une démonstration de ce fait plus simple dans le cas des groupes déployés sur un corps parfait).

8.3. *Lemme. — Supposons G connexe et k de caractéristique zéro. Alors tout k-sous-groupe unipotent V de G est contenu dans un k-sous-groupe parabolique minimal P de G, et l'on a $P \neq G$ si $V \not\subset R_u(G)$.*

Soit $\pi : G \to G' = G/R_u(G)$ la projection canonique. Les k-sous-groupes de G contenant $R_u(G)$ sont les images réciproques des k-sous-groupes de G' et si H est l'un d'eux, $G/H \simeq G'/\pi(H)$. L'application $H \mapsto \pi(H)$ est donc une bijection pour les k-sous-groupes paraboliques minimaux, et l'on peut se borner au cas où G est réductif.

Soit V un k-sous-groupe unipotent $\neq \{e\}$ de G. Il contient un k-sous-groupe unipotent V' de dimension un. D'après le théorème de Jacobson-Morosow [14, Chap. III, th. 17, p. 100], il existe un homomorphisme de $\mathfrak{sl}_{2,k}$ dans $\mathfrak{g}_k$ dont l'image contient $\mathfrak{v}'_k$. Comme, en caractéristique zéro, toute représentation de l'algèbre de Lie $\mathfrak{h}$ d'un groupe semi-simple connexe et simplement connexe H, est la différentielle d'une représentation (rationnelle) de H, on voit qu'il existe aussi un k-morphisme $f : \mathbf{SL}_2 \to G$ dont l'image contient V'. Ce morphisme est de noyau fini, donc G contient un tore $S \neq \{e\}$ déployé sur k (par exemple l'image par f du groupe des matrices diagonales de $\mathbf{SL}_2$), et par conséquent un k-sous-groupe parabolique propre P (4.17). Le groupe V est déployé sur k, donc possède un point fixe dans $(G/P)_k$ vu 2.7, et est alors conjugué sur k à un sous-groupe de P d'après 4.13 $a)$, d'où le lemme.

8.4. *Proposition.* — *Supposons* G *connexe et* k *de caractéristique zéro. Alors les* k*-sous-groupes unipotents maximaux de* G *sont les radicaux unipotents des* k*-sous-groupes paraboliques minimaux de* G.

Comme précédemment, on voit en considérant la projection de G sur $G/R_u(G)$ que l'on peut se borner au cas où G est réductif.

Soient P un k-sous-groupe parabolique minimal de G, S un tore déployé sur k maximal de G contenu dans P. On a la décomposition de Levi $P = \mathscr{Z}(S).R_u(P)$ et $R(P) \supset S.R_u(P)$ (4.16). Soit V un k-sous-groupe unipotent non trivial de G. Vu 8.3, il suffit de considérer le cas où $V \subset P$, et nous avons alors à montrer que $V \subset R_u(P)$, ou, ce qui revient au même, que $V \subset R(P)$. Si ce n'était pas le cas, le groupe réductif $P/(S.R_u(P)) \cong \mathscr{Z}(S)/S$ contiendrait un k-sous-groupe unipotent non trivial, donc un tore déployé sur k non trivial (8.3, 4.17), ce qui contredirait l'hypothèse faite sur S, vu 1.9 b).

8.5. *Corollaire.* — *Supposons* G *réductif et* k *de caractéristique zéro. Alors les conditions suivantes sont équivalentes :*

    (i)   G *est anisotrope sur* k.

    (ii) G *ne contient aucun* k*-sous-groupe unipotent non trivial, et* $X^*(G^0)_k = \{o\}$.

    (iii) G_k *est formé d'éléments semi-simples, et* $X^*(G^0)_k = \{o\}$.

Le groupe G est produit presque direct de son groupe dérivé, qui est semi-simple, par son centre connexe Z (2.2), donc $X^*(G^0)_k$ s'identifie à un sous-groupe d'indice fini de $X^*(Z)_k$ et, vu 1.7, 1.8, on a

$$(1) \qquad\qquad X^*(G^0)_k = \{o\} \Leftrightarrow Z_d = \{1\}.$$

En caractéristique zéro, tout groupe algébrique unipotent est connexe, donc chacune des conditions de 8.5 équivaut à cette même condition pour G^0, et l'on peut supposer G connexe. Alors, (i) $\Rightarrow$ (ii) d'après 8.3, 4.17 et (1). L'implication (ii) $\Rightarrow$ (iii) résulte de l'existence d'une décomposition de Jordan pour les éléments de G_k [1, th. 8.4], et (iii) entraîne (i) en vertu de (1) et 4.17.

8.6. *Proposition.* — *Supposons* G *connexe et* k *de caractéristique zéro. Soient P un* k*-sousgroupe parabolique minimal et* H *un* k*-sous-groupe connexe contenant* $R_u(P)$. *Alors* $\mathscr{N}(R_u(H)) = P'$ *est un* k*-sous-groupe parabolique contenant* P, *et* $R_u(P') = R_u(H)$. *Le groupe* H *contient le plus petit* k*-sous-groupe distingué de* P' *contenant* $R_u(P)$.

On se ramène comme plus haut au cas où G est réductif. Soit $\pi : H \to H' = H/R_u(H)$ la projection canonique. Vu 8.4, $\pi(R_u(P)) = U'$ est un k-sous-groupe unipotent maximal de H'. Soit V' le radical unipotent d'un k-sous-groupe parabolique minimal de H' opposé à $\mathscr{N}(U')$ (4.9). D'après 4.11, U' et V' engendrent un sous-groupe distingué de H', donc $U = R_u(P) = \pi^{-1}(U')$ et $V = \pi^{-1}(V')$ engendrent un k-sous-groupe distingué H_1 de H. Comme H_1 contient $R_u(H)$, on a $R_u(H) = R_u(H_1)$. Le groupe $\mathscr{N}(V)$ étant un k-sous-groupe parabolique (8.4), 4.29 montre que $P'' = \mathscr{N}(H_1)$ est un k-sous-groupe parabolique contenant P, dont le radical unipotent est égal à celui de H_1. On a donc aussi $P'' = \mathcal{N}(R_u(H_1)) = \mathcal{N}(R_u(H)) = P'$ et $R_u(P') = R_u(H_1) = R_u(H)$. Par construction $H \supset H_1 \supset R_u(P)$ et H_1 est engendré par $R_u(P)$ et par un sous-groupe V qui est aussi conjugué à U dans P' vu 8.4, donc H_1 est le plus petit sous-groupe distingué de P' contenant $R_u(P)$.

8.7. *Proposition.* — *Supposons* G *réductif et* k *de caractéristique* o. *Soient* H *et* H' *deux* k-*sous-groupes de* G *conjugués. Alors si* H *contient un* k-*sous-groupe unipotent (resp. un sous-groupe résoluble déployé sur* k) *maximal de* G, *il en est de même pour* H'.

On peut supposer H connexe. Dans les deux cas envisagés, il existe, vu 8.4, un k-sous-groupe parabolique minimal P tel que $H \supset U = R_u(P)$. Soient $Q = \mathcal{N}(R_u(H))$ et $Q' = \mathcal{N}(R_u(H'))$. D'après 8.6, Q est un k-sous-groupe parabolique et $R_u(Q) = R_u(H)$. Comme H et H' sont conjugués sur une extension convenable K de k, les groupes Q et Q' sont conjugués sur K, et Q' est aussi un sous-groupe parabolique. Il est évidemment défini sur k, donc conjugué sur k à Q (4.13). On peut par conséquent se borner au cas où $Q = Q'$. Soit H_1 le plus petit sous-groupe distingué de Q contenant U. Vu 8.6, il fait partie de H. D'autre part, un élément $x \in G_K$ tel que $^x H = H'$ doit normaliser Q, donc faire partie de Q (4.3), donc normaliser H_1, d'où $H' = {}^x H \supset H_1 \supset U$.

Supposons maintenant que H contienne un sous-groupe résoluble déployé sur k maximal de G. Vu 8.2, 8.4, on peut supposer que $H \supset S . R_u(P)$, où S est un tore déployé sur k maximal de G contenu dans $R(P)$, et il reste à prouver que H' contient un tore déployé sur k maximal de G. En utilisant ce qui a déjà été obtenu, et 8.6, on se ramène au cas où $Q = \mathcal{N}(H) = \mathcal{N}(H')$ et $R_u(Q) = R_u(H) = R_u(H')$, et où il existe $x \in Q$ tel que $^x H = H'$.

Le groupe Q admet une décomposition de Levi $Q = \mathcal{Z}(S') . R_u(Q)$, où $S' = S \cap R(Q)$ est un tore déployé sur k maximal de $R(Q)$, donc est aussi le plus grand tore déployé sur k central de $\mathcal{Z}(S')$ (4.15, 4.19). Le groupe $S' . R_u(Q)$ est invariant dans Q, contenu dans H, donc aussi dans H'. Le groupe $\mathcal{Z}(S')$ est produit presque direct sur k de S' et d'un k-groupe M, qui ne possède pas de tore déployé sur k central non trivial, et l'on a $H = (H \cap M) . S' . R_u(Q)$, $H' = (H' \cap M) . S' . R_u(Q)$. D'après 4.27, il suffit de faire voir que $H' \cap M$ contient un tore déployé sur k maximal de M. Comme les tores déployés sur k maximaux se correspondent par isogénie séparable (4.25), on voit, en considérant les images de H et H' dans $Q/(S' . R_u(Q))$, que l'on est ramené au cas où G n'a pas de tore central non trivial déployé sur k et ou H possède un k-sous-groupe H_1 réductif, qui est invariant dans G et contient $R_u(P)$. Le groupe G est alors produit presque direct de H_1 et d'un k-groupe réductif H_2. Ce dernier n'a pas de k-sous-groupe non trivial unipotent (sinon $R_u(P)$ ne serait pas un k-groupe unipotent maximal) et ne possède pas de tore déployé sur k central non trivial, puisque G n'en a pas; par suite H_2 est anisotrope sur k (8.5), et (4.27) H_1 contient tout tore déployé sur k maximal de G. Comme $H' \supset H_1$ cela termine la démonstration.

Remarque. — On renvoie à un article en préparation de l'un de nous pour des extensions de 8.3 à 8.7 à des corps de base plus généraux.

§ 9. SOUS-GROUPES TRIGONALISABLES
(CORPS LOCALEMENT COMPACT DE CARACTÉRISTIQUE ZÉRO)

9.1. Soit K un corps localement compact de caractéristique zéro. (K est donc isomorphe à **R**, **C** ou à un corps p-adique, *i.e.* à une extension finie d'un corps $\mathbf{Q}_p$ (p premier).) Soit V une variété algébrique sur K. Alors V_K est munie canoniquement d'une topologie d'espace localement compact séparé. Si $f : V \to W$ est un K-morphisme de V dans une variété algébrique W sur K, alors $f_K : V_K \to W_K$ est continue. Enfin, si V est irréductible, non singulière, de dimension n, alors V_K est une K-variété analytique de dimension m [39, App. III].

Rappelons encore que si V est un espace homogène sur K d'un K-groupe algébrique H, alors V_K est réunion d'un nombre fini d'orbites de H_K [5, § 6.4], qui sont des sous-variétés analytiques ouvertes et fermées de V_K.

9.2. *Lemme. — Supposons k localement compact de caractéristique zéro et G résoluble, déployé sur k. Soit V une k-variété algébrique sur laquelle G opère k-morphiquement. Soit M un sous-ensemble compact non vide de V_k stable par G_k. Alors M contient un point fixe par G_k* (¹).

La démonstration procède par récurrence sur dim G. Supposons tout d'abord G de dimension un, donc isomorphe à $\mathbf{G}_a$ ou $\mathbf{G}_m$. Soient $v \in M$ et Y l'adhérence de $G.v$. Admettons tout d'abord que $Y \cap M = G.v \cap M$. Alors $G.v \cap M$ est fermé donc compact. Mais $(G.v)_k$ est réunion d'un nombre fini d'orbites de G_k, qui sont ouvertes et fermées dans $(G.v)_k$. Par suite, $G.v \cap M$ est réunion d'un nombre fini d'orbites de G_k, qui sont compactes. Chacune est de la forme G_k/H_k, où H est un k-sous-groupe fermé de G, et comme $G = \mathbf{G}_a$ ou $\mathbf{G}_m$, cela entraîne $G = H$; donc v est fixe par G. Supposons maintenant $Y \cap M \neq G.v \cap M$, et soit $y \in M \cap Y$, $y \notin G.v$. Son orbite $G.y$ est de dimension strictement plus petite que celle de v [1, § 16], donc est de dimension zéro, et y est un point fixe.

Si dim $G > 1$, le groupe G contient un k-sous-groupe invariant connexe N de codimension un déployé sur k. L'ensemble W des points fixes par N est une k-sous-variété algébrique de V, et $W \cap M$ est non vide par hypothèse d'induction. Le groupe G/N opère k-morphiquement sur W [25, prop. 2] et $(G/N)_k$ laisse $W \cap M$ stable. Comme G est déployé sur k, G/N l'est aussi, et l'on est ramené au cas précédent.

9.3. *Proposition. — Supposons k localement compact de caractéristique zéro, et soit H un k-sous-groupe fermé de G. Alors les conditions suivantes sont équivalentes :*

(i) *H contient un sous-groupe connexe trigonalisable sur k maximal de G^0.*

(ii) G_k/H_k *est compact.*

(iii) $(G/H)_k$ *est compact.*

Le groupe $(G^0)_k$ est ouvert, d'indice fini dans G_k. On peut donc se borner au cas où G est connexe.

(¹) L'énoncé original supposait V complète. Indépendamment, J. I. Hano a mentionné dans une lettre à l'un de nous ce lemme lorsque $k = \mathbf{R}$, ce qui nous a conduit à lever la restriction faite d'abord sur V.

(i) $\Rightarrow$ (ii) Soit L un sous-groupe connexe trigonalisable sur k maximal de G contenu dans H. Alors G_k/L_k s'identifie à l'ensemble des points rationnels sur k d'une k-variété projective (8.2), donc est compact. *A fortiori*, G_k/H_k est compact.

(ii) $\Rightarrow$ (i) En appliquant 9.2 au cas où $V=G/H$ et $M=G_k/H_k$, on voit que tout sous-groupe résoluble connexe déployé sur k de G admet un point fixe dans G_k/H_k, donc est conjugué sur k à un sous-groupe de H.

$(G/H)_k$ est réunion d'un nombre fini d'orbites de G_k (9.1). On peut donc trouver un nombre fini de k-sous-groupes H_i $(1\leq i\leq m,\ H_1=H)$ de G tels que $(G/H)_k=\bigcup_i G_k/H_{i,k}$, et chaque orbite est ouverte et fermée dans $(G/H)_k$. Cela prouve que (iii) $\Rightarrow$ (ii).

Il reste à montrer que (i) $\Rightarrow$ (iii). Les groupes H_i sont évidemment conjugués de $H=H_1$ sur une extension convenable K de k (puisque ce sont des groupes de stabilité dans une même orbite de G). Il résulte donc de 8.7 que pour tout i, H_i contient un sous-groupe résoluble déployé sur k maximal de G. Vu l'implication (i) $\Rightarrow$ (ii) déjà établie, il s'ensuit que $G_k/H_{i,k}$ est compact pour tout i, donc que $(G/H)_k$ est compact.

Remarque. — L'équivalence de (i) et (ii) a été aussi obtenue par I. Satake (*Publ. Math.*, I.H.E.S., Paris, n° 18 (1964), prop. 1, § 1), et pour $k=\mathbf{R}$ par R. Hermann (*Proc. Nat. Acad. Sci.*, U.S.A., *51* (1964), 456-461) et J. I. Hano.

9.4. *Corollaire.* — *Le groupe G_k est compact si et seulement si G est réductif et anisotrope sur k.*

Si G_k est compact, alors tout k-sous-groupe connexe trigonalisable sur k est réduit à l'élément neutre, donc G est réductif, et, vu 8.5, anisotrope sur k. La réciproque est conséquence de l'implication (i) $\Rightarrow$ (ii) de 9.3.

§ 10. CLASSES DE CONJUGAISON FERMÉES ET CENTRALISATEURS DE TORES

10.0. *Notations.* — Soit V une variété algébrique. L'espace tangent à V en un point simple v sera noté $T(V)_v$. Si $f:V\to W$ est un morphisme de V dans une variété algébrique, on notera df sa différentielle. Si $v\in V$ et $f(v)=w$ sont des points simples, alors df induit une application linéaire de $T(V)_v$ dans $T(W)_w$ qui sera désignée en général par df_v.

10.1. *Lemme.* — *Soient H un sous-groupe fermé de G et s un élément semi-simple de G normalisant H. Alors l'espace $\mathfrak{z}(s)\cap\mathfrak{h}$ des points fixes de Ad s dans l'algèbre de Lie $\mathfrak{h}$ de H est l'algèbre de Lie du groupe $\mathscr{Z}(s)\cap H$.*

On peut évidemment supposer que $G=\mathbf{GL}_n$. On démontrera le lemme tout d'abord lorsque $H=\mathbf{GL}_n$. Pour cela on considère une décomposition de Bruhat de $\mathbf{GL}_n$, et l'on reprend les notations de 2.3, en supposant $s\in T$. Soit U^- le groupe unipotent engendré par les U_a $(a<0)$. Alors $(x,t,y)\mapsto x.t.y$ est une immersion ouverte de $U^-\times T\times U$ dans G, dont l'image sera notée V. Vu 2.3, la restriction de Int s à U_a, identifié à $\mathbf{G}_a$, est la multiplication par $a(s)$, donc la restriction de la différentielle Ad s de Int s à l'algèbre de $\mathfrak{u}_a$ de U_a est aussi la multiplication par $a(s)$. Par conséquent $\mathscr{Z}(s) \cap V$ est engendré par T et par les groupes U_a, où a parcourt les racines égales à un sur s, et

$$\mathfrak{z}(s) = \mathfrak{t} \oplus \sum_{a(s)=1} \mathfrak{u}_a, \qquad (\mathfrak{t} \text{ algèbre de Lie de T}),$$

donc $\mathfrak{z}(s)$ est l'algèbre de Lie de $\mathscr{Z}(s)$.

Soit c_s l'application de G dans lui-même qui envoie g sur le commutateur (s, g). Il est clair que $(dc_s)_e = \operatorname{Ad} s - 1$, donc

$$\mathfrak{m} = \operatorname{Im}((dc_s)_e) = \sum_{a(s) \neq 1} \mathfrak{u}_a,$$

$$(1) \qquad \mathfrak{gl}_n = \mathfrak{g} = \mathfrak{z}(s) \oplus \mathfrak{m}.$$

Soit $M = c_s(G)$. C'est la translatée par s d'une orbite de G, opérant sur lui-même par automorphismes intérieurs, donc M est une variété non singulière. D'autre part, une fois U_a identifié à $\mathbf{G}_a$, on a $c_s(u) = (a(s) - 1) \cdot u \ (u \in U_a)$, donc $\mathfrak{m}$ fait partie de l'espace tangent en e à M. Mais $s^{-1} \cdot M$ est la classe de conjugaison de s^{-1}; donc sa dimension est égale à celle de $G/\mathscr{Z}(s)$, c'est-à-dire à celle de $\mathfrak{m}$, vu ce qui a été démontré plus haut. Il s'ensuit que $\mathfrak{m} = T(M)_e$.

Soit maintenant H quelconque. Vu (1) et le fait que $^s H = H$, on a

$$(2) \qquad \mathfrak{h} = \mathfrak{z}(s) \cap \mathfrak{h} + \mathfrak{m} \cap \mathfrak{h}.$$

Notons aussi, en vue de 10.3, que la restriction de $(dc_s)_e$ à $\mathfrak{m}$ étant un isomorphisme,

$$(3) \qquad \mathfrak{m} \cap \mathfrak{h} = (dc_s)_e \mathfrak{h}.$$

L'espace tangent $\mathfrak{c}$ à $\mathscr{Z}(s) \cap H$ en e est contenu dans $\mathfrak{z}(s) \cap \mathfrak{h}$ et l'espace tangent $\mathfrak{m}'$ en e à $M' = c_s(H)$ fait partie de $\mathfrak{m} \cap \mathfrak{h}$, car $M' \subset M \cap H$. Comme c_s est un morphisme de H sur M' dont les fibres sont les classes à gauche $h \cdot (\mathscr{Z}(s) \cap H) \ (h \in H)$, on doit avoir $\dim \mathfrak{m}' + \dim \mathfrak{c} = \dim \mathfrak{h}$, d'où, vu (2),

$$(4) \qquad \mathfrak{c} = \mathfrak{z}(s) \cap \mathfrak{h}, \qquad \mathfrak{m}' = \mathfrak{m} \cap \mathfrak{h}.$$

10.2. *Théorème.* — *Soient g un élément de G, et $\mathscr{C}(g)$ sa classe de conjugaison.*

(i) *Si g est semi-simple, et H un sous-groupe fermé de G normalisé par g, alors l'ensemble* $\operatorname{Int}_G H(g)$ *des conjugués de g par H est fermé. En particulier, $\mathscr{C}(g)$ est fermé.*

(ii) *Si G est semi-simple et connexe, l'adhérence $\overline{\mathscr{C}(g)}$ de $\mathscr{C}(g)$ contient la partie semi-simple g_s de g; en particulier $\mathscr{C}(g)$ n'est pas fermé lorsque g n'est pas semi-simple.*

On désignera par $C(X, \lambda)$ et $M(X, \lambda)$ le polynôme caractéristique et le polynôme minimal d'une matrice carrée X, où λ est une indéterminée.

(i) $\operatorname{Int}_G H(g)$ est la classe de conjugaison de g dans le groupe algébrique engendré par H et le groupe $\mathscr{A}(g)$ adhérent à g (qui normalise H). On peut donc supposer $H = G$. On identifie G avec une de ses réalisations linéaires. Soit

$$L = \{x \in G \mid M(g, x) = 0, \ C(\operatorname{Ad} x, \lambda) = C(\operatorname{Ad} g, \lambda)\}.$$

C'est un sous-ensemble algébrique de G, stable par automorphismes intérieurs. On considère L comme espace de transformations de G, opérant par automorphismes intérieurs. L'orbite de $x \in$ L est la classe de conjugaison de x dans G et sa dimension est par suite égale à celle de $G/\mathscr{Z}(x)$. Or, si $x \in$ L, il annule le polynôme minimal de g, donc $M(x, \lambda)$ divise $M(g, \lambda)$ et n'a que des facteurs simples. Par conséquent, x est aussi semi-simple, et 10.1 entraîne que dim $\mathscr{Z}(x)$ est égale à la multiplicité de la valeur propre un de Ad x. Mais Ad x et Ad g ont même polynôme caractéristique, donc dim $\mathscr{Z}(x)$ est aussi égale à la multiplicité de la valeur propre un pour Ad g, et par conséquent (10.1) à dim $\mathscr{Z}(g)$. Les orbites de G dans L ont ainsi toutes la même dimension. Elles sont donc fermées [1, § 16].

(ii) Soit $g = g_s \cdot g_u$ la décomposition multiplicative de Jordan de g. On suppose $k = \bar{k}$ et on reprend les notations de 2.3, en admettant que $g_s \in$ T, $g_u \in$ U. On peut trouver un groupe à un paramètre $\mu : \mathbf{G}_m \to$ T tel que $<\mu, a> = m_a$ soit $>$o pour toute racine simple, donc pour toute racine positive. Notons u_a la composante de g_u dans U_a, identifié à $\mathbf{G}_a$, et soit φ le morphisme de variétés algébriques de $\mathbf{G}_m$ dans U qui applique x sur $\operatorname{Int} \mu(x)(g_u)$. La composante dans U_a de $\varphi(x)$ est égale à $x^{m_a} \cdot u_a$ $(a \in \Phi^+)$. Comme les m_a sont $>$o, cela montre que φ est la restriction à $\mathbf{G}_m$ d'un morphisme de la droite affine qui applique l'origine sur l'élément neutre. Ce dernier fait donc partie de l'adhérence de $\varphi(\mathbf{G}_m)$. Comme $\mu(\mathbf{G}_m)$ centralise g_s, il s'ensuit que g_s fait partie de l'adhérence de $\operatorname{Int}_G \mu(\mathbf{G}_m)(g)$, donc de $\overline{\mathscr{C}(g)}$.

10.3. *Proposition. — Soient* H *un sous-groupe fermé de* G, s *un élément semi-simple du normalisateur de* H *dans* G, *et* F $= \mathscr{Z}(s) \cap$ H. *Alors l'application* $c_s : h \mapsto (s, h)$ *est un morphisme séparable de* H *sur* M $= c_s($H$)$, *qui induit par passage au quotient un isomorphisme de* H$/$F *sur* M. *Si* H *est connexe, et si* H, s *sont définis sur* k, *alors* F^0 *est défini sur* k.

(Rappelons qu'un morphisme $f :$ V $\to$ W de k-variétés algébriques irréductibles qui est surjectif (ou génériquement surjectif, *i.e.* tel que $f($V$)$ contienne un ouvert non vide de W) est *séparable* si le corps $\bar{k}($V$)$ des fonctions rationnelles sur V est une extension séparable de l'image de $\bar{k}($W$)$ par le comorphisme f^0 associé à f. Il suffit pour cela que $df_v :$ T$($V$)_v \to$ T$($W$)_{f(w)}$ soit surjectif lorsque v et $f(w)$ sont des points simples. En effet, l'espace tangent en un point générique étant identifié à l'espace des dérivations du corps des fonctions rationnelles (*voir* [39], prop. 15, p. 12 et chap. IV, n° 6), cette condition signifie que toute dérivation de $f^0(\bar{k}($W$))$ sur $\bar{k}$ se prolonge en une dérivation de $\bar{k}($V$)$ sur $\bar{k}$, donc que $\bar{k}($V$)$ est une extension séparable de $f^0(\bar{k}($W$))$ d'après un critère connu (*voir* Samuel-Zariski, *Commutative Algebra*, I, Van Nostrand, chap. II, th. 42).)

Il est immédiat que $c_s(x) = c_s(y)$ $(x, y \in$ H$)$ si et seulement si $x \in y \cdot$ F, donc c_s induit un morphisme bijectif de H$/$F sur M. Pour établir la première assertion, il suffit de faire voir que $(dc_s)_h :$ T$($H$)_h \to$ T$($M$)_m$ $(m = c_s(h))$ est surjectif quel que soit $h \in$ H. Identifiant l'algèbre de Lie $\mathfrak{h}$ de H à T$($H$)_e$, on a T$($H$)_h = h \cdot \mathfrak{h}$, et

$$dc_s(h \cdot \mathrm{X}) = s \cdot h \cdot \mathrm{X} \cdot s^{-1} \cdot h^{-1} - s \cdot h \cdot s^{-1} \cdot \mathrm{X} \cdot h^{-1} \qquad (\mathrm{X} \in \mathfrak{h}),$$

ou encore
$$dc_s(h \cdot \mathrm{X}) = c_s(h) \cdot \operatorname{Ad} s \cdot h \cdot (\operatorname{Ad} s - \mathrm{I})(\mathrm{X}),$$

d'où

(1) $$dc_s(\mathrm{T(H)}_h) = c_s(h).\mathrm{Ad}\, s.h.(\mathfrak{m}),$$

avec

(2) $$\mathfrak{m} = (\mathrm{Ad}\, s - 1)(\mathfrak{h}) = (dc_s)_e(\mathfrak{h}).$$

Mais, vu 10.1, (3) et (4), $\mathfrak{m} = \mathrm{T(M)}_e$; donc $(dc_s)_h$ est surjective.

Supposons maintenant H connexe. Soit $\Gamma \subset \mathrm{H} \times \mathrm{M}$ le graphe de l'application c_s. On a $\Gamma \cap (\mathrm{H} \times \{e\}) = \mathrm{F} \times \{e\}$ au point de vue ensembliste. Mais la surjectivité de dc_s entraîne que $\mathrm{H} \times \{e\}$ et Γ se coupent transversalement; d'après le critère de multiplicité un [39, VI, § 2, th. 6], on a alors

$$\mathrm{F} = \mathrm{pr_H}(\mathrm{F} \times \{e\}) = \mathrm{pr_H}(\Gamma \cdot (\mathrm{H} \times \{e\})),$$

où le point désigne ici une intersection de cycles. Cela signifie que le cycle somme des composantes irréductibles de F, chacune affectée du coefficient 1, est rationnel sur k. Comme F^0 est stable par $\mathrm{Gal}(k_s/k)$, c'est une composante rationnelle sur k de F. Étant irréductible, F^0 est alors aussi défini sur k [39, prop. 1, p. 208].

10.4. *Corollaire.* — *L'application* $\varphi : h \mapsto h.s.h^{-1}$ *induit un isomorphisme d'espaces homogènes de* $\mathrm{H/F}$ *sur* $\mathrm{Int_G}\, \mathrm{H}(s)$.

En effet, φ est le composé de $c_{s^{-1}}$ et de la translation à gauche par s.

10.5. *Corollaire.* — *Supposons* G *connexe et soit* S *un* k-*tore de* G. *Alors* $\mathscr{Z}(\mathrm{S})$ *est défini sur* k.

Le groupe $\mathscr{Z}(\mathrm{S})$ est connexe [1, prop. 18.4] et k-fermé. Il reste à montrer qu'il est défini sur k_s. On peut supposer que $k = k_s$, donc que k est infini. D'après 1.10, il existe alors $s \in \mathrm{S}_k$ tel que $\mathscr{Z}(s) = \mathscr{Z}(\mathrm{S})$, et notre assertion résulte de 10.3.

10.6. *Proposition.* — *Supposons* G *connexe et soit* $f : \mathrm{G} \to \mathrm{G}'$ *un* k-*morphisme séparable de* G *sur un* k-*groupe* G'. *Alors les tores maximaux définis sur* k *(resp. les tores déployés sur* k *maximaux) de* G' *sont les images par* f *des tores maximaux définis sur* k *(resp. des tores déployés sur* k *maximaux) de* G.

On sait déjà que l'image d'un tore maximal de G est un tore maximal de G' [1, th. 22.1]. Soit T' un tore maximal de G' défini sur k et soit $\mathrm{H} = f^{-1}(\mathrm{T}')^0$. Ce groupe est défini sur k. En effet, d'après [22, prop. 1] et [39, Chap. VIII, th. 4], le cycle $f^{-1}(\mathrm{T}')$, dont chaque composante est affectée du coefficient un, est rationnel sur k. Comme H est k-fermé, il est alors aussi défini sur k [39, chap. VIII, prop. 1].

Le groupe H contient un tore maximal de G [1, *loc. cit.*] donc aussi (2.14) un tore maximal T de G défini sur k. Alors $\mathrm{T}' = f(\mathrm{T})$.

Si S est un tore déployé sur k de G, alors $f(\mathrm{S})$ est aussi déployé sur k (1.8). Pour terminer la démonstration, il suffit de prouver que si S est un tore déployé sur k de G et si S' est un tore déployé sur k maximal de G' contenant $f(\mathrm{S})$, alors S fait partie d'un tore S'' déployé sur k tel que $f(\mathrm{S}'') = \mathrm{S}'$.

Le centralisateur de S′ est défini sur k (10.5), donc contient un tore maximal T′ de G′ défini sur k (2.14). Le groupe $H = f^{-1}(T')^0$ considéré ci-dessus est défini sur k, contient S et est de rang maximum, donc $(\mathscr{Z}(S) \cap H)^0$ est défini sur k (10.5) et est de rang maximum. Vu 2.14, ce groupe contient par conséquent un tore maximal T de G défini sur k contenant S. On a alors $S \subset T_d$, et $f(T) = T'$, d'où aussi (1.8) $f(T_d) = T'_d = S'$.

§ 11. SOUS-GROUPES DE CARTAN DES GROUPES RÉSOLUBLES

Ce paragraphe donne de nouvelles démonstrations de théorèmes d'existence et de conjugaison dans les groupes résolubles, dus à Rosenlicht [26] et Grothendieck [11]. Elles utilisent le lemme 11.1, qui généralise et précise le lemme 9.6 de [1].

11.1. *Lemme.* — *Soit* H *un k-groupe et supposons que* G *soit un sous-groupe unipotent connexe de* H. *Soit s un élément semi-simple de* H_k *normalisant* G. *Soient* F *le centralisateur de s dans* G, c_s *l'application* $g \mapsto (s, g)$ *de* G *dans* G *et* $M = c_s(G)$. *Alors* $M \cap F = \{e\}$; *le groupe* F *est connexe, défini sur k*; M *est une k-sous-variété irréductible fermée; l'application produit* $\alpha : M \times F \to G$ *et la restriction de c_s à* M *sont des isomorphismes de variétés.*

Remarquons tout d'abord que la restriction de c_s au centre $\mathscr{Z}(G)$ de G est un homomorphisme de $\mathscr{Z}(G)$ dans lui-même et que, plus généralement :

$$(1) \qquad c_s(x.y) = c_s(x).c_s(y) \qquad\qquad (x \in G,\ y \in \mathscr{Z}(G)).$$

Soit $v = (s, g) \in F$. Dans l'égalité $g.s^{-1}.g^{-1} = s^{-1}.v$, le membre de droite est produit d'un élément semi-simple et d'un élément unipotent commutant entre eux, tandis que le membre de gauche est semi-simple, d'où $v = e$, $g \in F$ et $M \cap F = \{e\}$.

M est le translaté par s de l'ensemble des conjugués par G de l'élément semi-simple s^{-1}. C'est donc une sous-variété irréductible fermée (10.2), évidemment définie sur k.

Nous voulons montrer maintenant que

(*)   F est connexe et α est bijective.

Si $\dim G = 1$, ou plus généralement si G est commutatif, (*) résulte de [1, lemme 9.6]. Nous procédons donc par récurrence sur $\dim$ G et supposons G non commutatif. Comme (*) est de nature géométrique, nous pouvons aussi admettre que $k = \bar{k}$. Soit $V \neq \{e\}$ un sous-groupe fermé connexe du centre de G, stable par Int s. Soit π la projection du normalisateur N de V dans H, sur $N' = N/V$. Le groupe $G' = \pi(G)$ est connexe unipotent, $s' = \pi(s)$ est semi-simple, normalise G′, et l'on peut appliquer l'hypothèse d'induction à $c_{s'}$, G′, $M' = c_{s'}(G')$ et $F' = c_{s'}^{-1}(e)$. On a évidemment $c_{s'} \circ \pi = \pi \circ c_s$, donc $M' = \pi(M)$ et $\pi(F) \subset F'$.

Soient $g' \in F'$ et $g \in \pi^{-1}(g')$. On a $s.g.s^{-1} = g.z$ $(z \in V)$ et, compte tenu de l'hypothèse d'induction, on peut écrire

$$z = s.x.s^{-1}.x^{-1}.y \qquad\qquad (x \in V,\ y \in F \cap V),$$

d'où, en posant $h = x^{-1}.g = g.x^{-1}$, l'égalité $s.h.s^{-1} = h.y$, qui montre que $y \in M \cap F$, donc que $y = e$, et que $h \in \pi^{-1}(g') \cap F$. Par conséquent, la restriction de π à F est un morphisme de F sur F', de noyau $F \cap V$. Comme F' et $F \cap V$ sont connexes par hypothèse d'induction, F est aussi connexe.

De l'égalité $G' = M'.F'$, on tire alors $G = M.F.V$; mais $V = c_s(V).(F \cap V)$ d'où, vu (1)

$$G = c_s(G).F.c_s(V).(F \cap V) = c_s(G).c_s(V).F = c_s(G.V).F = M.F,$$

ce qui montre que α est surjective. Il reste à prouver que α est injective. Comme F est un groupe, cela revient à faire voir que si $a, b \in M$ et $a = b.f$ ($f \in F$), alors $f = e$. On a $\pi(a) = \pi(b).\pi(f)$, donc $\pi(f) = e$, et $f \in V$. Mais si $x, y \in G$ sont tels que $c_s(x) = a$, $c_s(y) = b$, on tire immédiatement de l'égalité $a = b.f$ et du fait que $f \in V$ est central que $f = c_s(y^{-1}.x)$, d'où $f \in M \cap F$ et $f = e$, ce qui termine la démonstration de (*).

On a $c_s(x) = c_s(y)$ $(x, y \in G)$ si et seulement si $x \in y.F$, ce qui, vu (*), montre que la restriction β de c_s à M est un k-morphisme bijectif de M sur M. On a déjà vu que α est aussi bijective. Comme les variétés G, M, F sont non singulières, donc normales, il suffit, pour terminer la démonstration du lemme (en vertu du Main Theorem de Zariski [17, chap. V, § 2]), de faire voir que α et β sont birationnelles, donc que leurs différentielles sont bijectives sur les espaces tangents.

Soient $\mathfrak{g}$, $\mathfrak{f}$ les algèbres de Lie de G et F respectivement. D'après 10.3 (1), (2), on a

$$(2) \qquad\qquad dc_s(T(G)_g) = c_s(g).\operatorname{Ad} g(\mathfrak{m}) \qquad (\mathfrak{m} = (\operatorname{Ad} s - 1)(\mathfrak{g}); \, g \in G),$$

$$(3) \qquad\qquad dc_s(T(G)_g) = T(M)_m \qquad\qquad\qquad (m = c_s(g)).$$

Montrons que $T(M)_m.f$ et $m.T(F)_f$ sont linéairement indépendants quel que soit $f \in F$. Vu (2), cela équivaut à

$$m.\operatorname{Ad} g(\mathfrak{m}).f \cap m.\mathfrak{f}.f = \{0\}$$

donc à

$$(4) \qquad\qquad\qquad \operatorname{Ad} g(\mathfrak{m}) \cap \mathfrak{f} = \{0\}.$$

Le groupe Ad G est un sous-groupe unipotent connexe de $\operatorname{GL}(\mathfrak{g})$, et $\operatorname{Ad}_{\mathfrak{g}} s$ est un élément semi-simple de $\operatorname{GL}(\mathfrak{g})$ normalisant Ad G. On peut trouver une base (X_i) $(1 \leq i \leq \dim \mathfrak{g})$ de $\mathfrak{g}$ formée de vecteurs propres de $\operatorname{Ad}_{\mathfrak{g}} s$ par rapport à laquelle Ad G est représenté par des matrices triangulaires supérieures. Pour le voir, on raisonne par récurrence sur dim V. Soit W l'espace des points fixes de Ad G et soit W' un supplémentaire de W stable par s. On a $W \neq 0$, donc on peut admettre que V/W possède une base (Y_j) ayant les propriétés requises. On prend pour (X_i) une base formée de vecteurs propres de s dont les premiers engendrent W et dont les suivants sous-tendent W' et s'appliquent sur les vecteurs Y_j par la projection canonique. Soit $w \in \mathfrak{m} - \{0\}$. On a donc $w = \sum_i m_i.X_i$ avec $m_i \in \bar{k}$, $m_i = 0$ si $X_i \in \mathfrak{f}$. Si j est le plus grand indice tel que $m_j \neq 0$, alors $\operatorname{Ad} g(w) = m_j X_j$ modulo une combinaison linéaire de $X_1, \ldots, X_{j-1}$. Comme $m_j \neq 0$, $X_j \notin \mathfrak{f}$, donc $\operatorname{Ad} g(w) \notin \mathfrak{f}$, ce qui prouve (4).

Pour des raisons de dimension, $T(M)_m \cdot f$ et $m \cdot T(F)_f$ engendrent $T(G)_{m \cdot f}$. Comme ils font évidemment partie de l'image de $d\alpha_{(m.f)}$, il s'ensuit que $d\alpha_{(m.f)}$ est bijectif.

Soit $z = c_s(m.f) = c_s(m)$. D'après (3), $(dc_s)_{m.f}$ applique $T(G)_{m.f}$ sur $T(M)_z$. Son noyau, qui contient $m \cdot T(F)_f$, doit donc être égal à $m \cdot T(F)_f$ pour des raisons de dimension. Mais $m \cdot T(F)_f$ n'a que zéro en commun avec $T(M)_m \cdot f$ d'après ce qui a déjà été démontré, donc la restriction de dc_s à $T(M)_m \cdot f$ est injective, et $d\beta(T(M)_m) = T(M)_z$.

11.2. Proposition. — *Supposons G connexe résoluble. Soit L un sous-groupe de G_k formé d'éléments semi-simples. Alors $\mathscr{Z}_G(L)$ est connexe, défini sur k, et L est contenu dans un tore maximal de G.*

Le groupe L étant commutatif [1, Prop. 10.2], une récurrence immédiate permet de se ramener au cas où L est engendré par un élément x. Ce dernier fait partie d'un tore maximal T de G [1, Théorème 12.6], et G est le produit semi-direct de T par son radical unipotent G_u [1, Théorème 12.2]. Par suite, $\mathscr{Z}_G(x) = T \cdot (\mathscr{Z}_G(x) \cap G_u)$. Comme $\mathscr{Z}_G(x) \cap G_u$ est connexe (11.1), le groupe $\mathscr{Z}_G(x)$ l'est aussi. Il est alors défini sur k d'après 10.3.

Le lemme suivant est conséquence immédiate de quelques résultats de Rosenlicht.

11.3. Lemme. — *Supposons G connexe unipotent commutatif, et tel que $G^p = \{e\}$ si la caractéristique p de k est non nulle. Soit T un k-tore qui opère k-morphiquement sur G, et n'a que l'élément neutre de G comme point fixe. Alors G est déployé sur k.*

D'après [24, Prop. 1.2], G est $\bar{k}$-isomorphe à un produit de groupes $\mathbf{G}_a$. Le groupe T étant décomposé sur k_s (1.5), le lemme de la p. 109 de [26] (dans lequel on a $G_0 = \{e\}$ vu notre hypothèse), montre que G est k_s-isomorphe à un produit de groupes $\mathbf{G}_a$. Ainsi, G est déployé sur k_s, donc sur k [26, Théorème 3].

11.4. Théorème. — *Soit G résoluble connexe. Alors G possède un sous-groupe de Cartan (resp. un tore maximal) défini sur k. Deux tels sous-groupes sont conjugués par un élément de $(\mathscr{C}^\infty G)_k$.*

Un sous-groupe de Cartan C possède un unique tore maximal T dont il est le centralisateur [1, § 20]. Si C (resp. T) est défini sur k, il en est de même de T (resp. C) d'après [23, Prop. 9] (resp. (10.3)). Pour prouver 11.4, on peut donc se borner à considérer les sous-groupes de Cartan.

(i) Nous donnerons tout d'abord une démonstration, différente de celle de [11], de l'existence d'un k-sous-groupe de Cartan lorsque k est infini. (Si k est fini, nous renvoyons à [23, Note p. 45].) On procède par récurrence sur dim G.

Admettons tout d'abord que $\mathscr{C}^\infty G$ possède un k-sous-groupe connexe propre $N \neq \{e\}$ normal dans G, et soit $\pi : G \to G' = G/N$ la projection canonique. Soit C' un k-sous-groupe de Cartan de G'. Comme $\mathscr{C}^\infty G' = \pi(\mathscr{C}^\infty G) \neq \{e\}$, le groupe G' n'est pas nilpotent, donc $C' \neq G'$ et $H = \pi^{-1}(C')$ est un k-sous-groupe connexe propre de G. Il contient un sous-groupe de Cartan de G [1, Théor. 22.2], donc tout sous-groupe de Cartan de H est un sous-groupe de Cartan de G; par l'hypothèse de récurrence, il en existe au moins un qui est défini sur k.

Si $\mathscr{C}^\infty G$ ne contient pas de sous-groupe N vérifiant les conditions ci-dessus, alors $\mathscr{C}^\infty G$ est commutatif, et $(\mathscr{C}^\infty G)^p = \{e\}$ lorsque la caractéristique p de k est non nulle. Montrons l'existence d'un élément semi-simple $a \in G_k$, tel que $\mathscr{Z}(a) \cap \mathscr{C}^\infty G = \{e\}$.

Soit $\pi : G \to G' = G/\mathscr{C}^\infty G$ la projection canonique. Le quotient G' est un k-groupe nilpotent, son unique tore maximal T' est défini sur k [23, Prop. 9], et $H = \pi^{-1}(T')$ est un k-groupe connexe. Il opère par automorphismes intérieurs sur $\mathscr{C}^\infty G$; comme $\mathscr{C}^\infty G$ est commutatif, il opère trivialement sur lui-même, donc, par passage au quotient, G' opère k-morphiquement sur $\mathscr{C}^\infty G$ [25, Prop. 2]. L'intersection de $\mathscr{C}^\infty G$ avec le centralisateur d'un tore maximal de H est réduite à $\{e\}$ [1, Théorème 13.4], donc le seul point fixe de T' dans $\mathscr{C}^\infty G$ est l'élément neutre. Par suite, $\mathscr{C}^\infty G$ est déployé sur k (11.3), et la fibration de G sur G' possède une section régulière s définie sur k [26, Theorem 1].

Soit V l'ensemble des éléments de T' dont l'ensemble des points fixes dans $\mathscr{C}^\infty G$ est différent de l'élément neutre. C'est la réunion d'un nombre fini de sous-groupes propres fermés, et l'on peut trouver $x \in T'_k - V_k$ (1.10). Soit $y = s(x)$. Si k est parfait, alors la partie semi-simple y_s de y est rationnelle sur k, et $\pi(y_s) = x$, donc y_s est l'élément cherché. Si $p \neq 0$, alors il existe une puissance $q = p^m$ de p telle que $z = y^q$ soit semi-simple, et évidemment rationnel sur k. On a alors $\pi(z) = \pi(y)^q = x^q$. Mais $t \mapsto t^q$ est une bijection de T' sur lui-même laissant stable tout sous-groupe; on a donc encore $x^q \in T'_k - V_k$, d'où $\mathscr{Z}(z) \cap \mathscr{C}^\infty G = \{e\}$ ce qui établit notre assertion.

Le groupe $\mathscr{Z}(a)$ est connexe, défini sur k (10.3) et contient un sous-groupe de Cartan C de G. On a $G = C . \mathscr{C}^\infty G$ et $\mathscr{Z}(a) \cap \mathscr{C}^\infty G = \{e\}$, donc $C = \mathscr{Z}(a)$ est défini sur k.

(ii) Soient T un tore maximal de G défini sur k et g un élément semi-simple de G_k. Nous voulons montrer l'existence de $x \in (\mathscr{C}^\infty G)_k$ tel que $x . g . x^{-1} \in T$.

La projection $\pi : G \to G' = G/\mathscr{C}^\infty G$ induit un isomorphisme de T sur le k-tore maximal T' de G'. En effet, $\pi : T \to T'$ est bijectif, $\pi(T)$ est un tore maximal de G' [1, Théor. 22.2], $\pi : G \to G'$ est séparable, et le noyau de $d\pi : \mathfrak{g} \to \mathfrak{g}'$ est l'algèbre de Lie de $\mathscr{C}^\infty G$; mais G est, en tant que variété, le produit de T par G_u [1, Théor. 12.2], donc $d\pi$ induit un isomorphisme de $\mathfrak{t}$ sur $\mathfrak{t}'$, d'où notre assertion. Il existe donc un unique élément $t \in T_k$ tel que $\pi(t) = \pi(g)$. On a $t^{-1} = g^{-1} . u$, avec $u \in (\mathscr{C}^\infty G)_k$. Le théorème de conjugaison sur k des tores maximaux montre l'existence de $v \in \mathscr{C}^\infty G$ tel que $v . g^{-1} . v^{-1} = t^{-1}$, d'où

$$u = (g, v) \in c_g(\mathscr{C}^\infty G),$$

et 11.1 entraîne l'existence de $x \in (\mathscr{C}^\infty G)_k$ tel que $u = (g, x)$. On a alors

$$t^{-1} = g^{-1} . g . x . g^{-1} . x^{-1} = x . g^{-1} . x^{-1},$$

d'où (ii).

(iii) Soient C, C′ deux k-sous-groupes de Cartan de G, et T, T′ leurs tores maximaux respectifs. Supposons tout d'abord k infini. Vu 1.10, on peut trouver $a' \in T'_k$ tel que $\mathscr{Z}(a') = \mathscr{Z}(T') = C'$. Vu (ii), il existe $x \in (\mathscr{C}^\infty G)_k$ tel que $x . a' . x^{-1} = a \in T$. Évidemment, $x . \mathscr{Z}(a') . x^{-1} = \mathscr{Z}(a)$. Ainsi, $\mathscr{Z}(a)$ est un sous-groupe de Cartan contenant C, donc égal à C, d'où $x . C' . x^{-1} = C$.

Supposons maintenant k fini, ou plus généralement parfait, et soit

$$V = \{ x \in \mathscr{C}^\infty G \,|\, x . C . x^{-1} = C' \}.$$

C'est un ensemble algébrique [1, 2.5] défini sur k puisque C et C′ le sont (et k est parfait), non vide vu le théorème de conjugaison des sous-groupes de Cartan sur $\bar{k}$, et qui est un espace homogène pour le groupe $H = \mathscr{N}_G(T) \cap \mathscr{C}^\infty G$. Mais $\mathscr{N}_G(T) = \mathscr{Z}_G(T) = C$ [1, Prop. 10.2, Théor. 12.2], et 11.1 montre que H est connexe. Étant unipotent, H est alors déployé sur k [23, Cor. 2, p. 34], donc V possède un point rationnel sur k [22, Theor. 10].

11.5. *Corollaire. — Soient* T *un tore maximal défini sur k de* G, S *un k-tore de* G *et* L *un sous-groupe de* G_k *formé d'éléments semi-simples. Alors* S (*resp.* L) *est conjugué à un sous-groupe de* T *par un élément de* $(\mathscr{C}^\infty G)_k$, *et le groupe* $\mathscr{A}(L)$ *adhérent à* L [1, § 3] *est défini sur k.*

Les groupes $\mathscr{Z}(S)$ et $\mathscr{Z}(L)$ sont connexes et définis sur k d'après 10.5 et 11.2. Soit T′ un tore maximal de $\mathscr{Z}(S)$ (resp. $\mathscr{Z}(L)$) défini sur k, ce qui existe vu 11.4. Il contient S (resp. L) et est un tore maximal de G (11.2). La première assertion résulte donc de 11.4 appliqué à T et T′. Pour la deuxième, on peut par suite supposer $L \subset T$. Alors $\mathscr{A}(L)$ est un sous-groupe de T qui est k-fermé, donc défini sur k (1.6).

Remarque. — Le théorème 11.4 pour les tores maximaux est dû à Rosenlicht [26, Theor. 4]. L'existence de k-sous-groupes de Cartan définis sur k dans un k-groupe algébrique connexe quelconque a été démontrée par Grothendieck [11, Exp. XII, Théor. 1.1]. Dans [11, Exp. XIV, § 6] il est également montré que l'espace homogène G/C (G résoluble connexe, C un sous-groupe de Cartan défini sur k) est k-isomorphe à un espace affine; cela résulte ici du fait que $\mathscr{C}^\infty G$ est déployé sur k [26, Théor. 4, Cor. 2] et de ce qu'un k-espace homogène d'un groupe unipotent déployé sur k qui possède un point rationnel sur k est k-isomorphe à un espace affine [11, Exp. XIV, lemme 6.5]. Remarquons encore que si un tore maximal T de G défini sur k possède un point s rationnel sur k tel que $\mathscr{Z}(s) = \mathscr{Z}(T)$, ce qui est toujours le cas lorsque k est infini (1.10), alors la variété $M = c_s(\mathscr{C}^\infty G) = \{ s . x . s^{-1} . x^{-1}, (x \in \mathscr{C}^\infty G) \}$ est k-isomorphe à G/C. En effet, $\mathscr{C}^\infty G$ est k-isomorphe, en tant que variété, au produit $(\mathscr{Z}(s) \cap \mathscr{C}^\infty G) \times M = (C \cap \mathscr{C}^\infty G) \times M$, d'après 11.1, donc la projection de G sur G/C est un isomorphisme de M sur G/C.

Nous terminons ce paragraphe par une application à des groupes non nécessairement résolubles.

11.6. *Proposition. — Supposons que* $R_u(G^0)$ *soit déployé sur k. Alors les tores déployés sur k maximaux de* G *sont conjugués par des éléments de* G^0_k.

(Si k est parfait, l'hypothèse faite sur $R_u(G^0)$ est vérifiée d'elle-même [23, Cor. 1, 2 à la Prop. 5]), donc 11.6 généralise l'assertion de conjugaison de tores déployés faite dans 8.2.)

Pour la démonstration, on peut supposer G connexe. Soit π la projection canonique de G sur $G'=G/R_u(G)$, et soient S, S' deux tores déployés sur k maximaux de G. Vu 10.6, $\pi(S)$ et $\pi(S')$ sont des tores déployés sur k maximaux de G', donc sont conjugués sur k (4.21). Comme $R_u(G)$ est déployé sur k, l'application $\pi : G_k \to G'_k$ est surjective; il existe donc $g \in G_k$ tel que ${}^g S \subset \pi^{-1}(\pi(S')) = S'.R_u(G)$, et l'on est ramené à 11.4.

§ 12. QUESTIONS DE RATIONALITÉ POUR LES REPRÉSENTATIONS LINÉAIRES

Dans ce paragraphe, la caractéristique de k est nulle, G est connexe, semi-simple, et $\widetilde{G}$ désigne le revêtement universel de G. On fixe un tore maximal T de G et un sous-groupe de Borel B de G contenant T, et on désigne par Δ l'ensemble correspondant des racines simples. On écrit Γ pour $\mathrm{Gal}(\overline{k}/k)$.

Nous rappelons tout d'abord quelques propriétés classiques des représentations linéaires de G (*voir* [10, 14, 29, 31]).

12.1. Soit $\rho : G \to GL(V)$ une représentation irréductible de G dans l'espace vectoriel V. Il existe dans V une et une seule droite D_ρ stable par B. L'orbite C_ρ de D_ρ par G sera appelée le *cône de la représentation* ρ. Les groupes de stabilité des droites de C_ρ forment une classe de conjugaison de sous-groupes paraboliques de G, notée $\mathscr{P}_\rho$, la *classe de sous-groupes paraboliques de* ρ.

Soit V' l'espace projectif des droites de V, et Aut V' le groupe des automorphismes de V'. La représentation ρ définit une représentation projective ρ' de G, i.e. un morphisme $\rho' : G \to \mathrm{Aut}\,V'$, qui est irréductible, c'est-à-dire ne laisse invariant aucun sous-espace projectif propre de V'. A C_ρ est associée une sous-variété fermée C'_ρ de V', qui s'identifie à l'espace homogène G/P, $P \in \mathscr{P}_\rho$, et est la seule orbite fermée de G dans V'. Un élément $P' \in \mathscr{P}_\rho$ n'a qu'un seul point fixe dans V', et ce point fixe fait partie de C'_ρ.

12.2. La représentation de T dans D_ρ induite par ρ définit un caractère l_ρ de T, le *poids dominant* de ρ, qui caractérise ρ à une équivalence près. Si (,) désigne un produit scalaire admissible sur $X^*(T) \otimes \mathbf{R}$, on a

$$(1) \qquad\qquad 2(l_\rho, a) = (a, a).m_a \qquad\qquad (m_a \in \mathbf{N},\ a \in \Delta),$$

et réciproquement tout élément de $X^*(T)$ vérifiant les conditions (1) est poids dominant d'une représentation irréductible.

Supposons $G = \widetilde{G}$. Il existe alors pour tout ensemble d'entiers $m_a \geq 0$ $(a \in \Delta)$ un poids dominant vérifiant (1). Les poids dominants l_a définis par

$$(2) \qquad\qquad 2(l_a, a) = (a, a), \qquad (l_a, b) = 0\ (a \neq b) \qquad\qquad (a, b \in \Delta),$$

sont les *poids dominants fondamentaux*, et les poids dominants de G sont les combinaisons linéaires à coefficients entiers $\geqq 0$ des poids fondamentaux. Même si G n'est pas simplement connexe, on exprimera tout poids dominant l_ρ comme combinaison linéaire des l_a, ce qui se justifie en envisageant soit l_ρ comme poids dominant de $\widetilde{G}$, soit les l_a comme des éléments de $X^*(T) \otimes \mathbf{Q}$. Le coefficient de l_a sera quelquefois noté $c_a(l_\rho)$. On a donc

$$(3) \qquad l_\rho = \sum_{a \in \Delta} c_a(l_\rho) . l_a, \qquad c_a(l_\rho) = 2(l_\rho, a) . (a, a)^{-1} \in \mathbf{N}.$$

Rappelons encore que $\mathscr{P}_\rho = \mathscr{P}_\theta$, où θ est l'ensemble des $a \in \Delta$ orthogonaux à l_ρ.

12.3. On notera $\mathscr{R}(G)$ ou $\mathscr{R}$ (resp. $\mathscr{R}'(G)$ ou $\mathscr{R}'$) l'ensemble des classes d'équivalence de représentations linéaires (resp. projectives) irréductibles de G. Les notions introduites précédemment qui ne dépendent que de la classe d'équivalence de ρ ou de ρ' seront aussi indexées par la classe de ρ ou ρ', et $d(\xi)$ ($\xi \in \mathscr{R}$) (resp. $d(\xi')$, $\xi' \in \mathscr{R}'$) désignera la dimension de l'espace vectoriel (resp. projectif) de toute représentation contenue dans ξ (resp. ξ').

Il est clair que $\mathscr{R}'(\widetilde{G}) = \mathscr{R}'(G)$; on a une application naturelle de $\mathscr{R}(G)$ dans $\mathscr{R}'(G)$, qui fait correspondre à toute représentation linéaire la représentation projective associée, et qui est bijective lorsque G est simplement connexe.

Un élément $\xi \in \mathscr{R}$ est *rationnel sur k* s'il contient un k-morphisme $G \to \mathbf{GL}_d$ ($d = d(\xi)$). Comme deux espaces vectoriels de même dimension définis sur k sont k-isomorphes, il revient au même d'exiger l'existence dans ξ d'un k-morphisme $G \to GL(V)$, où V est un espace vectoriel défini sur k, de dimension $d(\xi)$. Un tel morphisme est déterminé à une équivalence sur k près. En effet, si $\rho : G \to GL(V)$ et $\rho' : G \to GL(V')$ sont deux k-morphismes appartenant à ξ, il existe un $\bar{k}$-isomorphisme $A : V \to V'$ tel que

$$A . \rho(x) = \rho'(x) . A, \qquad\qquad (x \in G_{\bar{k}})$$

d'où, pour tout $\gamma \in \Gamma$ et $x \in G_{\bar{k}}$,

$$^\gamma A . \rho(^\gamma x) = \rho'(^\gamma x) . {}^\gamma A,$$

et le lemme de Schur entraîne l'existence de $a_\gamma \in \bar{k}^*$ tel que $A = a_\gamma . {}^\gamma A$. Il est immédiat que $\gamma \mapsto a_\gamma$ est un 1-cocycle de Γ à valeurs dans $\bar{k}^*$, donc un cobord [30, prop. 2, p. 158], ce qui signifie qu'il existe $b \in \bar{k}^*$ tel que $a_\gamma = b^{-1} . {}^\gamma b$ quel que soit $\gamma \in \Gamma$. On a alors $^\gamma(b . A) = b . A$, donc B est un k-isomorphisme de V sur V' qui réalise une équivalence de ρ avec ρ'.

Il est clair que si ξ est rationnelle sur k, la classe de sous-groupes paraboliques $\mathscr{P}_\xi$ associée à ξ est stable par Γ, c'est-à-dire est définie sur k (5.24), mais la réciproque est inexacte, comme on le voit déjà en considérant une forme unitaire de $\mathbf{SL}_2$.

Un élément $\xi \in \mathscr{R}$ est *fortement rationnel sur k* s'il est rationnel sur k et si $\mathscr{P}_\xi$ contient un k-groupe.

Un élément $\eta \in \mathscr{R}'$ est *rationnel sur k* s'il existe une k-forme V' de l'espace projectif

$\mathbf{P}_d$ $(d=d(\eta))$ et un k-morphisme $\sigma : G \to \text{Aut } V'$ appartenant à η. Comme $\mathbf{P}_d$ a en général des k-formes non triviales, les « variétés de Severi-Brauer » [30, chap. X, 6], cette condition est plus faible que l'existence d'un k-morphisme $G \to \mathbf{PGL}_{d+1} \cong \text{Aut } \mathbf{P}_d$ appartenant à η. Si η provient de $\xi \in \mathscr{R}$, cette dernière condition équivaut au fait que ξ est rationnelle sur k. En effet, si $\pi : \mathbf{GL}_d \to \mathbf{PGL}_d$ est la projection naturelle et si $\rho : G \to \mathbf{GL}_d$ est tel que $\rho' = \pi \circ \rho$ soit un k-morphisme, alors ρ est défini sur k, car c'est l'unique morphisme de G dans $\mathbf{GL}_d$ tel que $\rho' = \pi \circ \rho$. La réciproque est évidente.

Dans la suite, on s'intéressera aux trois conditions suivantes portant sur un élément $\xi \in \mathscr{R}(G)$ et sur son image $\xi' \in \mathscr{R}'(G)$.

  a) ξ' est rationnel sur k,

  b) ξ est rationnel sur k,

  c) ξ est fortement rationnel sur k.

Il est clair que *c) ⇒ b) ⇒ a)*. Si G est déployé sur k, alors $\widetilde{G}$ l'est aussi, et tout élément de $\mathscr{R}(G)$ (resp. $\mathscr{R}'(G)$) est fortement rationnel (resp. rationnel) sur k. Cela résulte du fait que les représentations irréductibles de $\mathfrak{g}$ peuvent être construites rationnellement sur le corps de base [14, chap. VII]. On peut aussi le voir en remarquant que tous les systèmes linéaires de diviseurs sur G/B qui interviennent dans la construction des représentations irréductibles de G (voir [29, 31] en caractéristique zéro, [10] dans le cas général) sont rationnels sur k. Il s'ensuit en particulier que tout élément de $\mathscr{R}(G)$ est fortement rationnel sur $\bar{k}$.

12.4. Le groupe Γ opère de façon naturelle sur les représentations linéaires ou projectives, en respectant l'équivalence, donc aussi sur $\mathscr{R}$ et $\mathscr{R}'$. Si $\rho : G \to \mathbf{GL}_d$ est un $\bar{k}$-morphisme, on a

$$(1) \qquad \gamma(\rho)(g) = \gamma(\rho(\gamma^{-1}(g))) \qquad (\gamma \in \Gamma, g \in G_{\bar{k}}),$$

autrement dit

$$(2) \qquad \gamma(\rho) = \gamma \circ \rho \circ \gamma^{-1}.$$

12.5. *Lemme.* — *On conserve les notations précédentes. Soient $\xi \in \mathscr{R}(G)$ et $\gamma \in \Gamma$. Alors le poids dominant de $\gamma(\xi)$ est $_{\Delta}\gamma(l_\xi)$, dans les notations de 6.2.*

Soient $\rho : G \to \text{GL}(V)$ un élément de ξ, D la droite de V stable par B, u un élément de $G_{\bar{k}}$ tel que $u.{}^{\gamma}B.u^{-1} = B$ et $E = {}^{\gamma}\rho(u).{}^{\gamma}D \subset V$. Soient encore $b \in B$, $x \in D - \{o\}$ et $y = {}^{\gamma}\rho(u)({}^{\gamma}x)$. Le poids dominant $l_\rho \in X^*(T)$ possède une extension unique à B ; on la notera aussi l_ρ. On a alors

$${}^{\gamma}\rho(u.{}^{\gamma}b.u^{-1}).y = {}^{\gamma}\rho(u).\gamma(\rho(b)).{}^{\gamma}\rho(u^{-1}).y = {}^{\gamma}\rho(u).\gamma(\rho(b).x) = \gamma(l_\rho(b)).y,$$

donc E est stable par B, agissant dans ${}^{\gamma}V$ par l'intermédiaire de ${}^{\gamma}\rho$, et le caractère l' de B dans E est donné par

$$\gamma(l_\rho(b)) = l'(u.{}^{\gamma}b.u^{-1}),$$

ce qui (6.2 (2)) signifie précisément que $l' = {}_{\Delta}\gamma(l_\rho)$.

12.6. *Proposition. — Soit* $\xi \in \mathscr{R}(G)$. *Alors les conditions suivantes sont équivalentes :*

(i) $\ _{\Delta}\gamma(l_\xi) = l_\xi$ *quel que soit* $\gamma \in \Gamma$.

(ii) ξ *est stable par* Γ.

(iii) *La classe de représentations projectives* ξ' *est rationnelle sur* k.

(iv) *Il existe une algèbre à division centrale* C_k *sur* k *et un* k-morphisme

$$G \to \mathbf{GL}_m(C) \qquad\qquad (d(\xi) = m \cdot c, \quad c^2 = [C : k]).$$

L'équivalence de (i) et (ii) résulte de 12.2 et 12.5.

(ii) $\Rightarrow$ (iii), (iv). Soit $\rho : G \to \mathbf{GL}_n$ $(n = d(\xi))$ un élément de ξ défini sur $\bar{k}$. Vu l'hypothèse (ii), il existe pour tout $\gamma \in \Gamma$ une matrice $A_\gamma \in \mathbf{GL}_{n,\bar{k}}$ telle que

$$^\gamma\rho(g) = A_\gamma^{-1} \cdot \rho(g) \cdot A_\gamma, \qquad\qquad (g \in G_{\bar{k}}).$$

D'après le lemme de Schur, A_γ est déterminée à la multiplication par un élément de $\bar{k}^*$ près, donc l'image $A_\gamma' = \pi(A_\gamma)$ de A_γ dans $\mathbf{PGL}_n$ par la projection canonique est univoquement déterminée. Si $\delta \in \Gamma$, on a

$$^{\delta \cdot \gamma}\rho(g) = \delta(^\gamma\rho(\delta^{-1}(g))) = \delta(A_\gamma^{-1} \cdot \rho(\delta^{-1}(g)) \cdot A_\gamma) = \delta(A_\gamma^{-1}) \cdot {}^\delta\rho(g) \cdot \delta(A_\gamma),$$

d'où l'on déduit que $A_\delta \cdot \delta(A_\gamma)$ est le produit de $A_{\delta\gamma}$ par un scalaire, donc que $A' : \gamma \mapsto A_\gamma'$ est un 1-cocycle de Γ, à valeurs dans $\mathbf{PGL}_n$, au sens de la cohomologie galoisienne [5, 30]. Comme $\mathbf{PGL}_n \cong \operatorname{Aut} \mathbf{P}_{n-1} \cong \operatorname{Aut} \mathbf{M}_n$, ce cocycle définit une k-forme V' de $\mathbf{P}_{n-1}$ et une k-forme M' de $\mathbf{M}_n$, (voir [5, § 2.6] ou [30, Chap. X, § 6]). Il existe alors des $\bar{k}$-isomorphismes $\varphi : \mathbf{P}_{d-1} \to V'$ et $\psi : \mathbf{M}_n \to M'$ tels que $^\gamma\varphi = \varphi \cdot A_\gamma'$, $^\gamma\psi = \psi \cdot A_\gamma'$ $(\gamma \in \Gamma)$. Soit alors $\sigma = \varphi \circ \rho' \circ \varphi^{-1}$. C'est un $\bar{k}$-morphisme de G dans Aut V'. Pour tout $\gamma \in \Gamma$, on a

$$^\gamma\sigma = {}^\gamma\varphi \cdot {}^\gamma\rho' \cdot {}^\gamma\varphi^{-1} = \varphi \cdot A_\gamma' \cdot A_\gamma'^{-1} \cdot \rho' \cdot A_\gamma' \cdot A_\gamma'^{-1} \cdot \varphi^{-1} = \sigma,$$

donc σ est défini sur k, ce qui montre que (ii) $\Rightarrow$ (iii). On voit de même que $\psi \circ \rho \circ \psi^{-1}$ est un k-morphisme de G dans le groupe M'^* des éléments inversibles de M'. Mais M_k' est une algèbre centrale simple de degré n^2 sur k, donc est de la forme $\mathbf{M}_m(C_k)$ où C_k est une algèbre à division centrale sur k, d'où (ii) $\Rightarrow$ (iv).

Réciproquement, soient V' une k-forme de $\mathbf{P}_{n-1}$ (resp. M' une k-forme de $\mathbf{M}_n$), $\sigma : G \to \operatorname{Aut} V'$ (resp. $\sigma : G \to M'^*$) un k-morphisme de ξ et $\varphi : \mathbf{P}_{n-1} \to V'$ (resp. $\varphi : \mathbf{M}_n \to M'$) un $\bar{k}$-isomorphisme. Alors $A' : \gamma \mapsto A_\gamma' = \varphi^{-1} \cdot {}^\gamma\varphi$ est un 1-cocycle de Γ à valeurs dans $\mathbf{PGL}_n$. Si l'on pose $\rho = \varphi^{-1} \circ \sigma \circ \varphi$, le calcul précédent montre que $^\gamma\rho = A_\gamma'^{-1} \cdot \rho \cdot A_\gamma'$, d'où les implications (iv) $\Rightarrow$ (ii) et (iii) $\Rightarrow$ (ii).

12.7. A toute classe ξ de représentations vérifiant les conditions de 12.6, correspond un élément β_ξ du groupe de Brauer $H^2(k^*)$: l'image par l'application cobord $\delta : H^1(\Gamma, \mathbf{PGL}_n) \to H^2(k^*)$ de la classe de cohomologie du cocycle A' considéré plus haut. Rappelons que δ fait partie de la suite exacte de cohomologie non commutative associée à la suite exacte

$$1 \to \mathbf{G}_m \to \mathbf{GL}_n \to \mathbf{PGL}_n \to 1,$$

et que, comme $H^1(\Gamma, \mathbf{GL}_n) = 0$, on a en fait une suite exacte

$$0 \to H^1(\Gamma, \mathbf{PGL}_n) \overset{\delta}{\to} H^2(k^*).$$

Bien entendu, l'image β_ξ de la classe de cohomologie du cocycle A' de 12.6 par δ correspond à l'algèbre à division C_k (voir [30, Chap. X] pour plus de détails).

12.8. *Proposition.* — *Soient $\xi \in \mathcal{R}(G)$ et ξ' son image dans $\mathcal{R}'(G)$. Les deux conditions suivantes sont équivalentes :*

(i) *ξ est rationnelle sur k.*

(ii) *ξ' est rationnelle sur k, et l'élément β_ξ du groupe de Brauer associé à ξ' dans 12.7 est nul.*

Si ξ est rationnelle sur k, alors ξ' l'est aussi, et le cocycle A' de 12.6 est trivial, donc (i) $\Rightarrow$ (ii). Réciproquement, si $\beta_\xi = 0$, alors, vu 12.7, A' est un cobord, donc il existe $B \in \mathbf{PGL}_{n,\bar{k}}$ tel que $A'_\gamma = B^{-1} \cdot {}^\gamma B$ ($\gamma \in \Gamma$). Si l'on pose $\sigma'(g) = B \cdot \rho'(g) \cdot B^{-1}$, on a

$${}^\gamma\sigma'(g) = {}^\gamma B \cdot {}^\gamma\rho'(g) \cdot {}^\gamma B^{-1} = {}^\gamma B \cdot A'^{-1}_\gamma \cdot \rho'(g) \cdot A'_\gamma \cdot {}^\gamma B^{-1} = B \cdot \rho'(g) \cdot B^{-1},$$

donc ${}^\gamma\sigma' = \sigma'$, et σ' est défini sur k, ce qui montre que ξ' est représenté par un k-morphisme de G dans $\mathbf{PGL}_n$. Donc ξ est rationnelle sur k (12.3).

12.9. Dans la suite, nous supposerons que T contient un tore déployé sur k maximal S et que B est contenu dans un k-sous-groupe parabolique minimal P. On a donc $P = P_\theta$ ($\theta = \{a \in \Delta \mid a|_S = 0\}$).

12.10. *Proposition.* — *Soit $\xi \in \mathcal{R}(G)$. Alors les deux conditions suivantes sont équivalentes :*

(i) *l_ξ est stable par Γ, et le coefficient $c_a(l_\xi)$ est nul si $a|_S = 0$.*

(ii) *ξ est fortement rationnelle sur k.*

Supposons (ii) vérifiée. Alors ξ est stable par Γ vu 12.6. D'autre part, $\mathscr{P}_\xi$ contient un élément défini sur k, donc un élément $P' \supset P$. On a alors $P' = P_{\theta'}$ avec $\theta \subset \theta' \subset \Delta$ (5.14), et l_ξ est orthogonal à θ' (12.2) donc à θ, d'où la deuxième partie de (i).

Supposons maintenant que ξ vérifie (i). Vu 12.6, il existe une k-forme V' de $\mathbf{P}_{d-1}$ ($d = d(\xi)$) et un k-morphisme $\rho' : G \to \mathrm{Aut}\, V'$ appartenant à ξ'. De plus, $\mathscr{P}_\xi \succ \mathscr{P}_\theta$, donc $\mathscr{P}_\xi$ possède un élément défini sur k (4.7).

Soit P' un k-groupe de $\mathscr{P}_\xi$. D'après ce qui a été rappelé en 12.1, P' a un seul point fixe Q dans V', qui est alors nécessairement rationnel sur k. Un résultat de F. Châtelet (*Annales E.N.S.*, *61* (1944), 249-300) implique alors que V' est k-isomorphe à $\mathbf{P}_{d-1}$; par suite, ξ' contient un k-morphisme de G dans $\mathbf{PGL}_d$ et ξ est rationnelle sur k (12.3).

12.11. *Corollaire.* — *ξ est fortement rationnelle sur k si et seulement si l_ξ est la restriction à T d'un élément de $X^*(P)_k$.*

Supposons ξ fortement rationnelle sur k. Soit $\rho : G \to \mathbf{GL}_d$ un k-morphisme contenu dans ξ et soit D la droite invariante par B. La classe $\mathscr{P}_\xi$ contient un élément défini sur k. L'unique élément P' de $\mathscr{P}_\xi$ qui contient P est donc défini sur k (4.7), et il en est de même de la droite D, unique droite stable par P'. Le poids dominant peut s'envisager comme le caractère de P′ dans D ; c'est par conséquent la restriction à T d'un élément de $X^*(P')_k$, donc *a fortiori* d'un élément de $X^*(P)_k$.

Supposons maintenant que l_ξ soit dans l'image de $X^*(P)_k \to X^*(T)_k$. Il est fixe par Γ d'après 6.2 (3). D'autre part, l_ξ est trivial sur le groupe dérivé de $\mathscr{Z}(S)$, donc orthogonal aux racines simples qui sont égales à 1 sur S. On retrouve ainsi la condition (i) de 12.10.

12.12. Soit π une représentation irréductible de G. Les *k-poids* de π sont les restrictions à S des poids de π. Si G est déployé sur k, il n'y a donc pas de distinction à faire entre poids et k-poids. Le k-poids dominant (restriction du poids dominant) sera noté m_π ou m_ξ, ξ étant la classe de π.

Dans la suite, on supposera, outre les conditions de 12.9, que T est défini sur k, et que l'on a choisi sur $X^*(T)$ et $X^*(S)$ des ordres compatibles, dont les ensembles positifs de racines correspondent à P et B. On note j la restriction $X^*(T) \to X^*(S)$ et (,) des produits scalaires admissibles et compatibles sur $X^*(T) \otimes \mathbf{R}$ et $X^*(S) \otimes \mathbf{R}$ (6.10).

Pour tout $b \in_k \Delta$, soit $l_{(b)}$ la somme des l_a $(a \in \rho^{-1}(b))$. Il résulte de 12.10 que le poids dominant d'une représentation fortement rationnelle sur k est combinaison linéaire à coefficients entiers $\geqq 0$ des $l_{(b)}$ (et réciproquement si G est simplement connexe). Si m_b désigne la restriction de $l_{(b)}$ à S, on a, d'après 6.11 :

$$(1) \qquad\qquad 2(m_b, b) = c_b . (b, b), \qquad (m_b, c) = 0 \qquad\qquad (b, c \in_k \Delta ; b \neq c),$$

avec c_b rationnel > 0, (en fait $c_b = (a, a)/(b, b)$, où $a \in \Delta$ est un élément quelconque de $\rho^{-1}(b)$), d'où la proposition suivante :

12.13. *Proposition. — Soit $\xi \in \mathscr{R}(G)$ fortement rationnelle sur k. Alors m_ξ est combinaison linéaire à coefficients entiers $\geqq 0$ des poids m_b $(b \in_k \Delta)$. Il existe un entier $d_b \geqq 1$ tel que $z . d_b . m_b$ soit le k-poids dominant d'une représentation fortement rationnelle sur k pour tout $z \in \mathbf{N}$.*

Appelons représentation fortement rationnelle sur k *fondamentale*, toute représentation fortement rationnelle sur k dont le k-poids dominant soit le plus petit multiple possible d'un m_b. Il existe donc $r = r_k(G)$ représentations fortement rationnelles sur k fondamentales, et leurs poids dominants forment une base de $X^*(S) \otimes \mathbf{Q}$. De plus, toute combinaison linéaire à coefficients entiers $\geqq 0$ de ces derniers est le k-poids dominant d'une représentation fortement rationnelle sur k, mais la réciproque est inexacte en général. Elle l'est toutefois si G est simplement connexe, et dans ce cas les k-poids dominants fondamentaux sont les m_b $(b \in_k \Delta)$.

12.14. Nous terminerons ce paragraphe par quelques propriétés des k-poids d'une représentation irréductible π (non nécessairement rationnelle sur k). On sait que tout poids de π est de la forme $l = l_\pi - \Sigma c_a(l) . a$ $(a \in \Delta, c_a(l) \in \mathbf{N})$, où l_π est le poids dominant de π. Comme $\rho(\Delta) \subset_k \Delta \cup \{o\}$, il s'ensuit qu'un k-poids q de π peut s'écrire

$$(1) \qquad\qquad q = m_\pi - \sum_{b \in_k \Delta} d_b(q) . b, \qquad\qquad (d_b(q) \in \mathbf{N}),$$

où m_π est le k-poids dominant de π. On notera $\theta(l)$ (resp. $\theta(q)$) l'ensemble des $a \in \Delta$ (resp. $b \in {}_k\Delta$) tels que $d_a(l) \neq 0$ (resp. $d_b(q) \neq 0$), et $|l|$ (resp. $|q|$) la somme des $d_a(l)$ (resp. $d_b(q)$).

L'ensemble des k-poids de π est évidemment invariant par le groupe de Weyl ${}_k W(G)$ relatif à k (5.1).

12.15. Lemme. — *Soit q un k-poids de π, différent du k-poids dominant. Alors il existe $c \in \theta(q)$ tel que $q + c$ soit un k-poids de π.*

Soit V l'espace de la représentation π. Il est somme directe des espaces propres V_r de S correspondant aux différents k-poids r de π, espaces qui sont stables par $\mathscr{Z}(S)$. D'après un raisonnement élémentaire et bien connu, on a

$$(1) \qquad \rho(X)(V_r) \subset V_{r+c} \qquad\qquad (c \in {}_k\Phi,\ X \in \mathfrak{g}_c),$$

où l'on a posé $\qquad \mathfrak{g}_c = \{ X \in \mathfrak{g} \mid \operatorname{Ad} s(X) = s^c . X\ (s \in S) \} \qquad\qquad (c \in {}_k\Phi).$

Le radical unipotent U de B est engendré par les groupes radiciels U_a ($a \in \Delta$) (cf. 2.3), comme cela résulte par exemple de 2.5 et de la remarque à 2.5 (c'est d'ailleurs vrai en toute caractéristique). Si $q + c$ n'était pas un k-poids quel que soit $c \in {}_k\Delta$, V_q serait stable par U, vu (1), donc par B (puisque $B \subset \mathscr{Z}(S) . U$), et contiendrait par conséquent une droite fixe par B, ce qui est absurde, vu 12.1. Il existe donc $c \in {}_k\Delta$ tel que $q + c$ soit un k-poids de π. Mais, si $c \notin \theta(q)$, alors $q + c$ n'est pas un k-poids vu 12.14, d'où le lemme.

La proposition suivante est due à Satake lorsque $k = \mathbf{R}, \mathbf{C}$ [27, lemmes 5, 7].

12.16. Proposition. — *On conserve les notations précédentes. Une partie θ de ${}_k\Delta$ est de la forme $\theta(q)$, où q est un k-poids de la représentation irréductible π, si et seulement si $\{m_\pi\} \cup \theta$ est connexe.*

Prouvons tout d'abord que $\theta(q) \cup \{m_\pi\}$ est connexe. On procède par récurrence sur le nombre d'éléments de $\theta(q)$, l'assertion étant vide si $\theta(q)$ l'est. Par application répétée de 12.15, on voit qu'il existe un k-poids l de π et un élément $c \in \theta(q)$ tels que $\theta(q) = \theta(l) \cup \{c\}$, $c \notin \theta(l)$, et que $l' = l - c$ soit un k-poids de π. Par hypothèse d'induction, $\{m_\pi\} \cup \theta(l)$ est connexe. Il reste donc à montrer que c n'est pas orthogonal à $\{m_\pi\} \cup \theta(l)$. S'il l'était, alors $(l', c) = -(c, c)$, donc le transformé

$$r_c(l') = l' - 2(l', c) . (c . c)^{-1} . c = l' + 2c = l + c,$$

de l' par la réflexion r_c serait un k-poids, ce qui est absurde vu 12.14 et le fait que $c \notin \theta(l)$.

Réciproquement, soit $\theta \subset {}_k\Delta$ tel que $\{m_\pi\} \cup \theta$ soit connexe. On peut numéroter les éléments b_i de θ de manière à ce que $(m_\pi, b_1) > 0$, et que pour tout i ($2 \leqq i \leqq m = \operatorname{card}(\theta)$) il existe un j ($1 \leqq j < i$) tel que $(b_j, b_i) \neq 0$ (donc < 0). On vérifie alors immédiatement, par récurrence sur i, que

$$q_i = r_{b_i} \ldots r_{b_1}(m_\pi)$$

est un k-poids tel que $\theta(q_i) = \{b_1, \ldots, b_i\}$ ($1 \leqq i \leqq m$) et que $(q_i, b_{i+1}) > 0$ si $i < m$.

12.17. *Proposition.* — *On conserve les notations précédentes. Soient* $w \in {}_k W(G)$ *et* $q = w(m_\pi)$. *Alors il existe* $w_1, w_2 \in {}_k W(G)$ *tels que* $w_2(m_\pi) = m_\pi$, *que* w_1 *soit un produit de réflexions* r_b $(b \in \theta(q))$, *et que* $w = w_1 . w_2$.

On procède par récurrence sur $|q|$, la proposition étant évidente si $|q| = 0$.

On sait que l'adhérence d'une chambre de Weyl est un domaine fondamental pour ${}_k W(G)$ [6]. Par conséquent, tout élément de l'orbite de m_π par ${}_k W(G)$, différent de m_π, a un produit scalaire < 0 avec au moins un élément de ${}_k \Delta$. Il existe donc $c \in {}_k \Delta$ tel que $(q, c) < 0$. Soit alors

$$q' = r_c(q) = q - 2(q, c) . (c, c)^{-1} . c.$$

C'est un k-poids. Vu 12.14 et l'inégalité $(q, c) < 0$, on doit avoir $c \in \theta(q)$, donc $\theta(q') \subset \theta(q)$, et $|q'| < |q|$. Il suffit alors d'appliquer l'hypothèse d'induction à q' et de remarquer que l'on a aussi $q = r_c(q')$.

§ 13. GROUPES p-ADIQUES A ENGENDREMENT COMPACT

13.1. Un groupe topologique est à *engendrement compact* s'il est égal au sous-groupe engendré par un sous-ensemble compact convenable. On sait que tout groupe de Lie réel ou complexe, n'ayant qu'un nombre fini de composantes connexes, est à engendrement compact. Il n'en n'est pas toujours ainsi du groupe des points rationnels d'un groupe algébrique H sur un corps p-adique K. Par exemple, si H est le groupe additif $\mathbf{G}_a$, alors H_K est non compact, mais réunion d'une suite croissante de sous-groupes compacts ouverts. Le but de ce paragraphe est de donner une condition nécessaire et suffisante pour que H_K soit à engendrement compact. La remarque suivante sera utile :

(i) Soient L un groupe topologique localement compact, M un sous-groupe fermé à engendrement compact. Alors si L/M est compact, L est à engendrement compact.

Cela résulte du fait que tout compact de L/M est image d'un compact de L par la projection canonique.

13.2. *Lemme.* — *Soient* G *unipotent et* k *de caractéristique zéro. Alors* $(\mathcal{D}G)_k = \mathcal{D}(G_k)$.

Le lemme est évident si G est commutatif. Sinon, soit V un k-sous-groupe central non trivial de G contenu dans $\mathcal{D}G$, par exemple le dernier sous-groupe non trivial de la série centrale descendante. Si x, y sont des éléments de l'algèbre de Lie $\mathfrak{g}$ de G dont le crochet fait partie de l'algèbre de Lie $\mathfrak{v}$ de V, alors, la formule de Campbell-Hausdorff entraîne immédiatement que $(\exp x, \exp y) = \exp[x, y]$. Par conséquent,

$$(1) \qquad\qquad\qquad V_k \subset \mathcal{D}(G_k).$$

Soit $\pi : G \to G' = G/V$ la projection canonique. Raisonnant par récurrence sur la dimension, on peut supposer que $\mathcal{D}(G'_k) = (\mathcal{D}G')_k$. Comme $G'_k = G_k/V_k$ (2.7), cela donne $(\mathcal{D}G)_k \subset \mathcal{D}(G_k) . V_k$, et le lemme résulte alors de (1).

13.3. *Lemme. — Supposons k de caractéristique zéro. Soient U un k-sous-groupe unipotent distingué de G, $\pi : G \to G' = G/\mathscr{D}U$ la projection canonique et L une partie de G_k. Si $\pi(L)$ engendre G'_k, alors L engendre G_k.*

Démonstration par récurrence sur $\dim G$. Si U est commutatif, il n'y a rien à prouver. Sinon, soit V le dernier sous-groupe non-trivial de la série centrale descendante de U. Il est défini sur k, et invariant dans G. On a évidemment $G' = (G/V)/\mathscr{D}(U/V)$.

Soit L^* le sous-groupe engendré par L. L'hypothèse de récurrence, et l'égalité $G_k/V_k = (G/V)_k$ (2.7) montrent que $G_k = L^* . V_k$, donc aussi que $U_k = (L^* \cap U_k) . V_k$. Or, si

$$x = a.v, \qquad y = b.w \qquad\qquad (a, b \in L^* \cap U_k; v, w \in V_k),$$

on a, puisque v, w sont dans le centre de U, $(x, y) = (a, b)$, d'où $\mathscr{D}(U_k) \subset L^* \cap U_k$ et aussi, vu 13.2, $V_k \subset L^*$, donc $G_k = L^*$.

13.4. *Théorème. — Supposons que k soit un corps p-adique de caractéristique zéro. Soient U le radical unipotent de G, $V = U/\mathscr{D}U$, et ρ la représentation de G dans l'algèbre de Lie $\mathfrak{v}$ de V définie à partir de la représentation adjointe de G. Alors les deux conditions suivantes sont équivalentes :*

(i) G_k est à engendrement compact.

(ii) $\mathfrak{v}$ ne contient aucun sous-espace $\mathfrak{w} \neq 0$ défini sur k, stable par G, tel que l'image de G_k dans $GL(\mathfrak{w})$ par ρ soit compacte.

En particulier, tout k-groupe réductif est à engendrement compact [1].

Le groupe $(G^0)_k$ étant d'indice fini dans G_k [5, 6.4], on peut supposer G connexe. Nous prouverons tout d'abord que (ii) $\Rightarrow$ (i), en distinguant plusieurs cas.

a) $G = \mathbf{G}_m$. Alors $G_k = k^*$ est engendré par le groupe M des unités de k, qui est compact, et par t, où t est une uniformisante de k (élément dont l'image dans k^*/M, qui est cyclique infini, engendre ce groupe).

b) $G = T$ *est un tore.* Alors T contient un tore décomposé T_d tel que $T' = T/T_d$ soit anisotrope sur k (1.8). Le groupe T'_k est compact (9.4) et l'on a $T_k/T_{d,k} = T'_k$ (2.7). Notre assertion résulte alors de *a)*, et 13.1 (i).

c) $G = T.U$ *est le produit semi-direct d'un k-tore T et d'un k-sous-groupe invariant unipotent U de dimension un.*

L'hypothèse (ii) entraîne que T opère non trivialement par automorphismes intérieurs sur U. Il existe donc $\chi \in X^*(T)_k$, $\chi \neq 0$, et un k-isomorphisme $\theta : \mathbf{G}_a \to U$, tels que

$$\theta(\chi(t).x) = t.\theta(x).t^{-1} \qquad\qquad (x \in \mathbf{G}_a, t \in T).$$

On peut alors trouver un k-morphisme $\lambda : \mathbf{G}_m \to T$ tel que $\chi \circ \lambda$ soit de la forme $s \mapsto s^m$ où m est un entier > 0 (cf. § 1). Par conséquent $\chi(T_k) \supset (k^*)^m$. Si L est un système générateur compact de T_k, qui existe d'après *b)*, alors G_k est engendré par L et par les $\theta(x^i.y)$, où x est une uniformisante de k, $i \in \mathbf{N}$, $i < m$ et y parcourt les entiers de k, d'où *c)*.

---

[1] Ce théorème répond à une question de M. Kneser, et intervient dans [15]. Nous lui devons en outre plusieurs suggestions pour le passage du cas réductif au cas général.

d) G est réductif. Soit P un k-sous-groupe parabolique minimal de G. Écrivons-le sous la forme $P = M.S.U$ où U est le radical unipotent de P, S un tore déployé sur k maximal de G, et M la partie anisotrope sur k de $\mathscr{Z}(S)$ (cf. 4.28). Le groupe U admet une suite de composition sur k dont les quotients successifs sont des espaces vectoriels sur lesquels S opère par des homothéties non triviales (3.18, remarque). Il résulte alors immédiatement de *c)* que $(S.U)_k = S_k.U_k$ est à engendrement compact. Le groupe $\mathscr{Z}(S)/S$ est isogène à M, donc anisotrope sur k, et $(\mathscr{Z}(S)/S)_k$ est compact (9.4). Comme S est déployé, on a $(\mathscr{Z}(S)/S)_k = \mathscr{Z}(S)_k/S_k$, donc vu *b)* et 13.1 (i), $\mathscr{Z}(S)_k$ est à engendrement compact. Alors $P_k = \mathscr{Z}(S)_k.U_k = \mathscr{Z}(S)_k.(S.U)_k$ est aussi à engendrement compact, et il en est de même de G_k, qui est réunion d'un nombre fini de doubles classes modulo P_k (5.15).

e) G n'est pas réductif. Il est le produit semi-direct de son radical unipotent U et d'un k-sous-groupe réductif maximal H (cf. 0.8). Soit q le plus petit sous-espace vectoriel de $\mathfrak{v}$ contenant toute droite définie sur k qui est stable par un tore décomposé sur k n'opérant pas trivialement sur elle. Ce sous-espace est défini sur k, stable par H_k, donc aussi par H, puisque H_k est Zariski dense dans H [23]. Les représentations rationnelles de H étant complètement réductibles, on peut trouver un supplémentaire $\mathfrak{r}$ de q dans $\mathfrak{v}$, défini sur k et stable par H. Tout tore décomposé sur k de H opère trivialement sur $\mathfrak{r}$, donc l'image de l'homomorphisme $\sigma : H \to GL(\mathfrak{r})$ défini par ρ est anisotrope sur k, et $\sigma(H_k)$ est compact (9.4). Vu l'hypothèse (ii), on a donc $\mathfrak{r} = \{o\}$. On déduit alors immédiatement de la définition de q et de *c)* l'existence d'un sous-ensemble compact M de $(G/\mathscr{D}U)_k$ tel que V_k soit contenu dans le sous-groupe M^* engendré par M. Vu *d)*, il s'ensuit que $(G/\mathscr{D}U)_k = H_k.V_k$ est à engendrement compact. Puisque $(G/\mathscr{D}U)_k = G_k/(\mathscr{D}U)_k$, on voit qu'il existe un compact $L \subset G_k$ dont l'image dans $(G/\mathscr{D}U)_k$ engendre ce groupe ; L engendre alors G_k d'après 13.3, ce qui termine la démonstration de l'implication (ii) $\Rightarrow$ (i).

(i) $\Rightarrow$ (ii). Le groupe $(G/\mathscr{D}U)_k = G_k/(\mathscr{D}U)_k$ est aussi à engendrement compact ; on peut donc supposer U commutatif, $U = V$. Soit q un k-sous-espace de $\mathfrak{v}$ stable par G, et soit $\mathfrak{r}$ un supplémentaire sur k de q dans $\mathfrak{v}$, stable par le sous-groupe de Levi H de G, donc aussi par G. Soient $Q = \exp q$ et $R = \exp \mathfrak{r}$ les sous-groupes correspondants. V est donc le produit direct de Q et R, et G le produit semi-direct de H, Q, R. Soit encore L un ensemble compact engendrant G_k, que l'on suppose de la forme $L = A.B.C$ ($A \subset H_k$, $B \subset Q_k$, $C \subset R_k$), ce qui est évidemment loisible. Supposons que l'image de G_k dans $GL(q)$ soit compacte. Alors, quitte à remplacer B par $\underset{g \in G_k}{\bigcup} {}^g B$, qui est aussi compact, on peut supposer que B est invariant par les automorphismes intérieurs de G_k. Pour tout entier $n \geq 1$, on a alors

$$H_k.B^n.R_k.A.B.C \subset H_k.B^{n+1}.R_k,$$

ce qui montre que G_k est la réunion des sous-ensembles $H_k.B^n.R_k$ ($n = 1, 2, \ldots$), donc que Q_k est engendré par le compact B. Comme Q_k est le groupe additif d'un espace vectoriel sur k, cela entraîne que $Q = \{e\}$, d'où (ii).

§ 14. GROUPES RÉELS

14.1. Suivant l'usage, $\pi_0(X)$ désigne l'ensemble des composantes connexes par arcs de l'espace topologique X. C'est un groupe si X est un groupe topologique. On sait que si X est une **R**-variété algébrique, alors $\pi_0(X_\mathbf{R})$ est fini [40]. (Ici et dans la suite on considère bien entendu $X_\mathbf{R}$ comme muni de la topologie usuelle, définie à partir de celle de **R**, cf. 9.1.) La dimension topologique de $X_\mathbf{R}$ est au plus égale à dim X, et si X est une variété non singulière, chaque composante connexe de $X_\mathbf{R}$ est une variété analytique de dimension dim X [40]. Lorsque X est un groupe, on notera également $(X_\mathbf{R})^0$ la composante neutre de $X_\mathbf{R}$.

14.2. *Proposition.* — *Supposons* G *connexe, défini sur* **R**, *et soit* P *un* **R**-*sous-groupe parabolique de* G. *Alors* $(G/P)_\mathbf{R}$ *est une variété compacte connexe.*

$(G/P)_\mathbf{C}$ est une variété projective complexe non singulière, donc $(G/P)_\mathbf{R}$ est une variété réelle (9.1), projective, donc compacte.

Soit $G = H.U$ une décomposition de Levi de G. Le groupe P contient U et $G/P = H/(H \cap P)$. Le groupe $H \cap P$ est donc un **R**-sous-groupe parabolique de H, et l'on peut supposer que G est réductif.

Soit V le radical unipotent d'un **R**-sous-groupe parabolique opposé à P. On a vu (4.10) que la projection de G sur G/P identifie V à un ouvert de G/P, donc (9.1) $V_\mathbf{R}$ à un ouvert de $(G/P)_\mathbf{R}$. Le complémentaire Y de V dans G/P est un **R**-sous-ensemble algébrique de dimension strictement plus petite que $m = \dim G/P$. La dimension topologique de $Y_\mathbf{R}$ est donc $< m$ (14.1), et $V_\mathbf{R}$ est dense dans $(G/P)_\mathbf{R}$. Comme $V_\mathbf{R}$ est connexe, la proposition est démontrée.

14.3. Dans [8, Chap. VI, § 5, n° 2, prop. 2] il est montré que « tout groupe linéaire compact est algébrique ». Dans la terminologie suivie ici, cela signifie qu'un sous-groupe compact H de $\mathbf{GL}(n, \mathbf{R})$ est l'ensemble des points réels d'un sous-groupe algébrique de $\mathbf{GL}(n, \mathbf{C})$ défini sur **R**, que l'on peut prendre égal à l'adhérence de Zariski $\mathscr{A}(H)$ de H. Cela implique en particulier que *si* $G \subset \mathbf{GL}(n, \mathbf{C})$ *est connexe, défini sur* **R**, *et si* $G_\mathbf{R}$ *est compact, alors* $G_\mathbf{R}$ *est connexe.* En effet, $G_\mathbf{R}^0 = \mathscr{A}((G_\mathbf{R})^0)_\mathbf{R}$ donc si $G_\mathbf{R} \neq (G_\mathbf{R})^0$, le groupe $\mathscr{A}((G_\mathbf{R})^0)$ est un sous-groupe propre de même dimension de G, et G n'est alors pas connexe.

14.4. *Théorème* (Matsumoto [18]). — *Supposons* G *connexe, défini sur* **R**, *et soit* S *un tore déployé sur* R *maximal de* G. *Alors* $G_\mathbf{R} = (G_\mathbf{R})^0 . S_\mathbf{R}$.

Nous avons à montrer que l'homomorphisme naturel $\pi_0(S_\mathbf{R}) \to \pi_0(G_\mathbf{R})$ est surjectif. Soient H un **R**-sous-groupe réductif maximal de G contenant S et U le radical unipotent de G. Le groupe $U_\mathbf{R}$ est connexe et $G_\mathbf{R}$ est homéomorphe au produit $H_\mathbf{R} \times U_\mathbf{R}$. L'homomorphisme $\pi_0(H_\mathbf{R}) \to \pi_0(G_\mathbf{R})$ est donc bijectif, et il suffit de faire la démonstration lorsque G est réductif.

Soit P un **R**-sous-groupe parabolique minimal de H contenant S. C'est le produit semi-direct de son radical unipotent par $\mathscr{Z}(S)$, donc $\pi_0(\mathscr{Z}(S)_\mathbf{R}) \to \pi_0(P_\mathbf{R})$ est bijectif. Mais $(G/P)_\mathbf{R}$ est égal à $G_\mathbf{R}/P_\mathbf{R}$ (4.13) et est connexe (14.2). La suite exacte d'homotopie

$$(1) \qquad \pi_0(P_\mathbf{R}) \overset{\iota}{\to} \pi_0(G_\mathbf{R}) \to \pi_0(G_\mathbf{R}/P_\mathbf{R})$$

montre donc que ι est surjectif. Il reste à prouver que $\pi_0(S_\mathbf{R}) \to \pi_0(\mathscr{Z}(S)_\mathbf{R})$ est surjectif.

Le groupe $M' = \mathscr{Z}(S)/S$ est anisotrope sur $\mathbf{R}$, connexe, donc $M'_\mathbf{R}$ est compact (9.4) et connexe (14.3). Comme S est déployé sur $\mathbf{R}$, on a $M'_\mathbf{R} = \mathscr{Z}(S)_\mathbf{R}/S_\mathbf{R}$ et notre assertion résulte de la suite d'homotopie (1), appliquée à la fibration de $\mathscr{Z}(S)$ par S.

14.5. *Corollaire.* — *On a* $\pi_0(G_\mathbf{R}) \cong (\mathbf{Z}/2\mathbf{Z})^d$, *où* $d \leq r_\mathbf{R}(G)$.

En effet, $S_\mathbf{R}$ est un produit de $r = r_\mathbf{R}(G)$ facteurs $\mathbf{R}^*$, donc $\pi_0(S_\mathbf{R}) \cong (\mathbf{Z}/2\mathbf{Z})^r$, et $\pi_0(G_\mathbf{R})$ est un quotient de $\pi_0(S_\mathbf{R})$ d'après le théorème.

14.6. *Corollaire.* — *Supposons* G *réductif. Alors tout élément du groupe de Weyl relatif* $_\mathbf{R}W(G)$ (*cf.* 5.1) *admet un représentant dans* $(G_\mathbf{R})^0$.

En effet, si $x \in \mathscr{N}(S)_\mathbf{R}$, il existe, vu 14.4, $y \in S_\mathbf{R}$ tel que $y.x \in (G_\mathbf{R})^0$, et $y.x$ représente évidemment le même élément de $_\mathbf{R}W(G)$ que x.

14.7. Nous terminons par quelques remarques pour faire le lien entre la théorie des groupes et algèbres de Lie semi-simples réels et certaines notions introduites dans ce travail. On suppose G semi-simple, défini sur $\mathbf{R}$. Il s'agit principalement de voir que le groupe A.N de la décomposition d'Iwasawa est la composante neutre de l'ensemble des points réels d'un $\mathbf{R}$-groupe trigonalisable maximal, et que $_\mathbf{R}W(G)$ et $_\mathbf{R}\Phi$ s'identifient au groupe de Weyl et au système de racines de la paire symétrique $(G_\mathbf{R}, K)$, où K est un sous-groupe compact maximal de $G_\mathbf{R}$.

Nous rappelons tout d'abord les décompositions de Cartan et d'Iwasawa. (Pour plus de détails, et les démonstrations, voir par exemple [13], ainsi que [4, § 1] pour divers compléments, concernant en particulier les groupes réels non connexes.)

Soient K un sous-groupe compact maximal, $\mathfrak{k}$ son algèbre de Lie, $\mathfrak{p}$ le complément orthogonal de $\mathfrak{k}$ dans l'algèbre de Lie $\mathfrak{g}_\mathbf{R}$ de $G_\mathbf{R}$ par rapport à la forme de Killing, $\mathfrak{a}$ une sous-algèbre maximale de $\mathfrak{p}$, nécessairement commutative, et $P = \exp \mathfrak{p}$, $A = \exp \mathfrak{a}$. On sait que $(k, p) \mapsto k.\exp p$ est un homéomorphisme de $K \times \mathfrak{p}$ sur $G_\mathbf{R}$ (décomposition de Cartan) et que $s : k.p \mapsto k.p^{-1}$ $(k \in K, p \in P)$ est un automorphisme involutif de $G_\mathbf{R}$ (involution de Cartan), dont K est l'ensemble des points fixes. Les endomorphismes ad a $(a \in \mathfrak{a})$ sont simultanément diagonalisables sur $\mathbf{R}$, et $\mathfrak{g}_\mathbf{R}$ est somme directe du centralisateur $\mathfrak{z}(\mathfrak{a})$ de $\mathfrak{a}$ et d'espaces propres $\mathfrak{v}_a$, où a parcourt un ensemble ψ fini de formes linéaires non nulles sur $\mathfrak{a}$, les *racines de* $\mathfrak{g}_\mathbf{R}$ par rapport à $\mathfrak{a}$. Choisissons un ordre sur $\mathfrak{a}^*$ et soit $\mathfrak{n} = \sum_{a>0} \mathfrak{v}_a$. Alors $\mathfrak{n}$ est une algèbre de Lie formée d'éléments x tels que ad x soit nilpotent, donc nilpotente, qui est normalisée par $\mathfrak{a}$. Soient $A = \exp \mathfrak{a}$ et $N = \exp \mathfrak{n}$ les sous-groupes analytiques de $G_\mathbf{R}$ engendrés par $\mathfrak{a}$ et $\mathfrak{n}$. Ils sont fermés, et $(k, a, n) \mapsto k.a.n$ est un homéomorphisme de $K \times A \times N$ sur $G_\mathbf{R}$ (décomposition d'Iwasawa). De plus N est unipotent, A diagonalisable sur $\mathbf{R}$, et A.N trigonalisable sur $\mathbf{R}$. Soit $L = \mathscr{A}(A.N)$

l'adhérence de Zariski de A.N dans G. Alors L est un **R**-sous-groupe trigonalisable sur **R** maximal de G, et A.N $=(L_\mathbf{R})^0$. En effet, L est trigonalisable sur **R**, puisque A.N l'est. D'autre part $G_\mathbf{R}/A.N$ est compact (il est homéomorphe à K), donc $G_\mathbf{R}/L_\mathbf{R}$ l'est aussi, et 9.3 implique que L contient un **R**-sous-groupe trigonalisable maximal de G.

Cela entraîne aussi que $A=(S_\mathbf{R})^0$ où $S=\mathscr{A}(A)$ est un tore déployé sur **R** maximal, donc $\mathscr{N}(S)=\mathscr{N}(A)$. L'involution de Cartan s laisse stable A, donc aussi $\mathscr{N}(S)_\mathbf{R}$. On a $\mathscr{N}(S)_\mathbf{R}=(K\cap\mathscr{N}(S)_\mathbf{R}).\exp\mathfrak{m}$, où $\mathfrak{m}$ est l'intersection de $\mathfrak{p}$ avec l'algèbre de Lie de $\mathscr{N}(S)_\mathbf{R}$ (cf. [4, 1.10]). Mais la relation $[\mathfrak{p},\mathfrak{p}]\subset\mathfrak{k}$ montre alors que $[\mathfrak{m},\mathfrak{a}]=o$, d'où $\mathfrak{m}=\mathfrak{a}$ et

$$\mathscr{N}(S)_\mathbf{R}=(K\cap\mathscr{N}(S)_\mathbf{R}).A,$$

d'où aussi, puisque A.N est connexe

$$(G_\mathbf{R})^0\cap\mathscr{N}(S)=(K^0\cap\mathscr{N}(S)).A.$$

Tenant compte de 14.6, on voit que tout élément de $_\mathbf{R}W(G)$ possède un représentant dans K^0, d'où aussi

$$_\mathbf{R}W(G)=(K^0\cap\mathscr{N}(A))/(K^0\cap\mathscr{Z}(A)).$$

Le groupe $_\mathbf{R}W(G)$ coïncide donc avec le groupe introduit dans la théorie des espaces symétriques par E. Cartan; de plus, comme $\mathfrak{a}$ est la partie réelle de l'algèbre de Lie de S, les racines de $\mathfrak{g}_\mathbf{R}$ par rapport à $\mathfrak{a}$ s'identifient aux **R**-racines de G (ou plus précisément à leurs différentielles). On retrouve ainsi les définitions originales de ces notions, posées sans référence aux groupes algébriques.

The Institute for advanced study, Princeton N. J.,
Universität Bonn.

BIBLIOGRAPHIE

[1] BOREL (A.), Groupes linéaires algébriques, *Annals of Math.* (2), *64*, 1956, 20-80.

[2] — Ensembles fondamentaux pour les groupes arithmétiques, *Coll. sur la théorie des groupes algébriques*, Bruxelles, 1962, 23-40.

[3] — Some finiteness properties of adele groups over number fields, *Publ. Math.*, I.H.E.S., n° *16* (1963), 101-126.

[4] BOREL (A.) et HARISH-CHANDRA, Arithmetic subgroups of algebraic groups, *Annals of Math.* (2), *75* (1962), 485-535.

[5] BOREL (A.) et SERRE (J.-P.), Théorèmes de finitude en cohomologie galoisienne, *Comm. Math. Helv.*, *39* (1964), 111-164.

[6] BOURBAKI (N.), *Groupes et algèbres de Lie*, chap. II : « Systèmes de racines » *(à paraître)*.

[7] CHEVALLEY (C.), *Theory of Lie groups*, Princeton University Press, Princeton 1946.

[8] — *Théorie des groupes de Lie*, II : « *Groupes algébriques* », Act. Sci. Ind., n° 1151, Hermann, Paris, 1951; III : « *Théorèmes généraux sur les algèbres de Lie* », Act. Sci. Ind., n° 1226, Hermann, Paris, 1955.

[9] — Sur certains groupes simples, *Tohoku Math. Jour.* (2), *7* (1955), 14-66.

[10] — *Séminaire sur la classification des groupes de Lie algébriques*, 2 vol., Paris, 1958 (notes polycopiées).

[11] DEMAZURE (M.) et GROTHENDIECK (A.), *Schémas en groupes*, I.H.E.S., 1964 (notes polycopiées).

[12] GODEMENT (R.), Groupes linéaires algébriques sur un corps parfait, *Sém. Bourbaki*, n° 206, Paris, 1960.

[13] HELGASON (S.), *Differential Geometry and Symmetric Spaces*, Acad. Press, New York, 1962.

[14] JACOBSON (N.), *Lie algebras*, Intersc. tracts in pure math., *10*, Interscience publ., New York, 1962.

[15] KNESER (M.), Erzeugende und Relationen verallgemeinerter Einheitsgruppen, *Jour. f. reine u. ang. Mat.*, *214/15* (1964), 345-349.

[16] LANG (S.), Algebraic groups over finite fields, *Amer. J. Math.*, *78* (1956), 555-563.

[17] — *Introduction to algebraic geometry*, Intersc. tracts in pure math., *5*, Interscience publ., New York, 1958.

[18] MATSUMOTO (H.), Quelques remarques sur les groupes algébriques réels, *Proc. Japan. Ac.*, *40* (1964), 4-7.

[19] — Self-adjoint group, *Ann. of Math.* (2), *62* (1955), 44-45.

[20] MOSTOW (G. D.), Fully reducible subgroups of algebraic groups, *Amer. Jour. Math.*, *78* (1956), 200-221.

[21] ONO (T.), Arithmetic of algebraic tori, *Ann. of Math.*, (2), *74* (1961), 101-139.

[22] ROSENLICHT (M.), Some basic theorems on algebraic groups, *Amer. Jour. Math.*, *78* (1956), 401-443.

[23] — Some rationality questions on algebraic groups, *Annali di Mat.* (IV), *43* (1957), 25-50.

[24] — Extensions of vector groups by abelian varieties, *Amer. Jour. Math.*, *80* (1958), 685-714.

[25] — On quotient varieties and the affine embedding of certain homogeneous spaces, *Trans. Amer. Math. Soc.*, *101* (1961), 211-223.

[26] — Questions of rationality for solvable algebraic groups over nonperfect fields, *Annali di Math.* (IV), *61* (1963), 97-120.

[27] SATAKE (I.), On representations and compactifications of symmetric Riemannian spaces, *Annals of Math.* (2), *71* (1960), 77-110.

[28] — On the theory of reductive algebraic groups over a perfect field, *Jour. Math. Soc. Japan*, *15* (1963), 210-235.

[29] SERRE (J.-P.), Représentations linéaires et espaces homogènes kähleriens des groupes de Lie compacts, *Sém. Bourbaki*, n° 100, Paris, 1954.

[30] — *Corps locaux*, Act. Sci. Ind., n° 1296, Paris, Hermann, 1962.

[31] TITS (J.), Sur certaines classes d'espaces homogènes de groupes de Lie, *Mém. Acad. Royale Belgique*, *29* (3) (1955).

[32] — Sur la classification des groupes algébriques semi-simples, *C.R. Acad. Sci. Paris*, *249* (1959), 1438-1440.

[33] — Théorème de Bruhat et sous-groupes paraboliques, *C.R. Acad. Sci. Paris*, *254* (1962), 2910-2912.

[34] — Groupes algébriques semi-simples et géométries associées, *Coll. alg. top. found. geom. Utrecht* (1959), 175-192, Pergamon Press, Oxford, 1962.

[35] — Groupes semi-simples isotropes, *Coll. sur la théorie des groupes algébriques*, Bruxelles, 1962, 137-147.

[36] — Groupes simples et géométries associées, *Proc. Int. Congr. Math. Stockholm*, 1962, 197-221.

[37] WEIL (A.), The field of definition of a variety, *Amer. Jour. Math.*, *78* (1956), 509-524.

[38] — *Adeles and algebraic groups* (Notes by M. DEMAZURE and T. ONO), The Institute for Advanced Study, Princeton, 1961.

[39] — *Foundations of algebraic geometry*, Amer. Math. Soc. Colloquium Publ., n° 29, 2e éd., 1962.

[40] WHITNEY (H.), Elementary structure of real algebraic varieties, *Annals of Math.* (2), *66* (1957), 545-556.

Reçu le 1er novembre 1964.

67.

Statement of the index theorem. Outline of proof

Chap. I in Seminar on the Atiyah-Singer Index Theorem by R. Palais et al.,
Ann. Math. Stud. **57** (1965) 1–11

As an introduction to the subject matter of this seminar, this lecture gives the statement and a rough description of the proof of the index theorem. More details on the proofs of the results stated and on the concepts discussed here will be found in the subsequent lectures.

Manifolds are compact, smooth (i.e., C^∞), orientable and *oriented*, consist of connected components of the same dimension, and, unless otherwise stated, have no boundary. Complex vector bundles will usually be denoted by the same letter as their total spaces. E_x denotes the fibre at a point x of a bundle E and $C^\infty(E)$ the space of smooth cross sections of a smooth complex vector bundle E over a manifold.

§1. The index theorem

1. Let X be a manifold, E, F smooth complex vector bundles on X. A differential operator d from E to F is a linear map $d: C^\infty(E) \to C^\infty(F)$ which is given locally by a matrix of ordinary (i.e., scalar) differential operators. More precisely, let U be the domain of definition of a local chart, $x_1, \ldots, x_n$ ($n = \dim X$) local coordinates, $C^\infty(U)$ the space of complex valued C^∞-functions on U, and choose sections of E and F over U which at each point $x \in U$ form a basis of E_x and F_x. Then d defines a map $C^\infty(U)^p \to C^\infty(U)^q$ ($p = \dim E_x$, $q = \dim F_x$ ($x \in U$)) given by a matrix $P(x, d) = (P(x, \partial/\partial x_1, \ldots, \partial/\partial x_n)_{ij})$ ($1 \leq i \leq p;\ 1 \leq j \leq q$), where $P_{ij} \in C^\infty(U)[\partial/\partial x_1, \ldots, \partial/\partial x_n]$. The order r of

521

d is the maximum of the degrees of the P_{ij} in the partial derivatives.
The characteristic matrix $\mathbf{Q}(x, \xi)$ of d at x is obtained from P by
taking the terms of degree r, substituting indeterminates ξ_i for the
partial derivatives $\partial/\partial x_i$, and multiplying by $(-1)^{r/2}$. To each cotangent
vector $\xi_1 dx + \ldots + \xi_n dx_n$, there is then associated a $p \times q$ matrix $\mathbf{Q}(x, \xi)$
with complex coefficients. The system is elliptic if p = q and if this
matrix is invertible for all $\xi \neq 0$ and for all $x \in X$. Under those cir-
cumstances, it is known that ker d and coker $d = C^{\infty}(F)/\mathrm{Im}\ d$ are finite
dimensional; the difference of their dimensions is the *index of* d. In
order to distinguish it from another index, to be defined below, we shall
call this the *analytical index* of d and denote it by $i_a(d)$. Thus

(1) $i_a(d) = \dim \ker d - \dim \mathrm{coker}\ d$.

It is known that $i_a(d)$ is invariant under deformations of d, and this led
Gelfand to ask whether it could be expressed in terms of topological data.
The index theorem provides a positive answer to that question. In order to
formulate it, a mixed rational cohomology class $\mathrm{ch}\ d \in H^*(X; \mathbf{Q})$, depending
on the symbol of d (see section 2, below), is introduced. Let moreover
$\mathscr{T}(X)$ be the Todd class or the complexified tangent bundle of X. It is
obtained by considering the product of formal power series

$$-x_i^2(1-e^{-x_i})^{-1} \cdot (1-e^{x_i})^{-1} ,$$

where the x_i (i = 1, ..., s $\geq$ dim X/2) are indeterminates, expressing it
as a formal sum of homogeneous polynomials, which are then symmetric in the
x_i^2, writing these as polynomials in the elementary symmetric functions in
the x_i^2 's, and then replacing the j-th symmetric function by the j-th
Pontrjagin class of X (j = 1, 2, ...). The *topological index* $i_t(d)$ is
then defined as

(2) $i_t(d) = (\mathrm{ch}\ d \cdot \mathscr{T}(X))[X]$ (n = dim X)   ,

where the right hand side stands for the value of the n-dimensional component
of ch d. $\mathscr{T}(X)$ on the fundamental cycle of X. We have then the

> THEOREM (Atiyah-Singer). Let d be an elliptic operator
> on the manifold X. Then $i_a(d) = i_t(d)$.

This is not the most general form of the theorem. Atiyah and Singer have extended it: (a) to elliptic complexes, i.e., sequences of operators whose symbols form an exact sequence, but this can easily be reduced to the case of one operator; (b) to a wider class of operators; this is quite important for the proof, and will be dealt with at length in this seminar; (c) to boundary value problems; this will probably not be touched upon here, for lack of time and material. [See however, Appendix I, by M. Atiyah.] A fourth generalization, which would include Grothendieck's version of the Riemann-Roch theorem over **C**, is contemplated but, as far as I know, has not yet been carried out.

The proof falls naturally into two parts: a topological one, which investigates the properties of $i_t(d)$, and an analytical one, concerned with $i_a(d)$. The former one also gives the motivation for the latter one, and here we shall summarize it first. Before doing that, however, we give some more details on ch d.

2. *The definition of* ch d. Let X be a finite CW-complex. The Grothendieck group K(X) of X is the quotient of the free commutative group generated by the isomorphism classes of complex vector bundles on X by the subgroup generated by the elements E - E' - E'' where $0 \to E'' \to E \to E' \to 0$ is an exact sequence. If a base point $x \in X$ has been chosen, then $\tilde{K}(X)$ is the kernel of the homomorphism $K(X) \to$ **Z** which assigns to each bundle E the dimension of E_x. If Y is a closed subcomplex of X, the relative Grothendieck group K(X, Y) is by definition $\tilde{K}(X/Y)$, where X/Y means X with Y pinched to a point, which is then taken as base point.

Let now X be again a manifold. Let B(X) (resp. S(X)) be the unit ball (resp. unit sphere) bundle of the cotangent bundle T*(X), with respect to some Riemannian metric, and $\pi: B(X) \to X$ the natural projection. It is known that, via cup product, $H^*(B(X), S(X))$ (any ring of coefficients) is a free module over $H^*(X)$, with a canonical generator U of degree n = dim X, whence the existence of an isomorphism $\varphi_*: H^*(B(X),$ $\varphi_*: H^*(B(X), S(X)) \to H^*(X)$, the Thom isomorphism, which decreases dimensions by n, the inverse of $a \mapsto a \cup U(a \in H^*(X))$.

Let $d: C^\infty(E) \to C^\infty(F)$ be a differential operator of order r
on X. It is well-known that the matrix $Q(x, \xi)$, which was defined above
using local coordinates, has in fact an intrinsic meaning and associates to
each $\xi \in T^*(X)_x$ a linear map of E_x into F_x. This linear map depends
of course smoothly on x, ξ, whence a homomorphism $\sigma_r(d): \pi^*E \to \pi^*F$ of
the bundles on $B(X)$ lifted from E and F via π, to be called the
symbol of d. Ellipticity means that $\sigma_r(d)$, restricted to $S(X)$, is an
isomorphism. In that case, one can associate to $(\pi^*E, \pi^*F, \sigma_r(d)|S(X))$ a
difference element $[\sigma_r(d)] \in K(B(X), S(X))$ (see [2]). Then ch d is defined by

$$(3) \qquad\qquad \mathrm{ch}\ d = (-1)^{n(n+1)/2}\ \varphi_*\mathrm{ch}[\sigma_r(d)] \quad ,$$

where $\mathrm{ch}[\sigma_r(d)] \in H^*(B(X), S(X); \mathbf{Q})$ is the Chern character of $[\sigma_r(d)]$.

§2. The topological index

The right hand side of (3) makes sense for any element $a \in$
$K(B(X), S(X))$ regardless of whether it comes from an elliptic operator or
not, and therefore so does i_t. The topological index may thus be viewed
as a function on $K(B(X), S(X))$; it is a homomorphism of the latter into
$\mathbf{Q}$, whose main properties are the following:

3. *Multiplicativity under* $\otimes$. Let X, X' be two manifolds.
There is then a pairing

$$K(B(X), S(X)) \times K(B(X'), S(X')) \to K(B(X \times X'), S(X \times X'))$$

to be called a tensor product, and we have

$$(4) \qquad i_t(a \otimes b) = i_t(a) \cdot i_t(b) \qquad (a \in K(B(X), S(X)),\ b \in K(B(X'), S(X')))$$

Briefly, $a \otimes b$ is defined as follows. Choose bundles E, F
on $B(X)$, E', F' on $B(X')$ and isomorphisms $\sigma: E \to F$, $\sigma': E' \to F'$ on
$S(X)$ and $S(X')$ respectively, such that a and b are the corresponding
difference elements, which is always possible. On $X \times Y$, the tensor product

$$0 \to E \otimes E' \to E \otimes F' + E' \otimes F \to E' \otimes F' \to 0$$

of the complexes $0 \to E \xrightarrow{\sigma} F \to 0$, $0 \to E' \xrightarrow{\sigma'} F' \to 0$, is easily seen to be an exact sequence on $S(X \times Y)$. The corresponding difference element [2] is then $a \otimes b$.

4. Let X be even dimensional. As we shall see later, there exists on X an elliptic differential system D_0, whose topological index is the L-genus of X, and whose analytical index is zero if $\dim X \equiv 2 \bmod 4$, equal to the index of X if $\dim X \equiv 0 \bmod 4$. Let $\varkappa_0 = [\sigma(D_0)]$.

On the other hand, $K(B(X), S(X))$ is a module over $K(X)$. We may therefore define i_t also on $K(X)$ by putting

$$(5) \qquad i_t(X, V) = i_t(V \cdot \varkappa_0), \qquad (V \in K(X)) \quad .$$

This is a homomorphism of $K(X)$ into $\mathbf{Q}$. The subgroup $K(X) \; \varkappa_0$ is of finite index in $K(B(X), S(X))$, and therefore i_t is completely determined by its values on this subgroup. The function i_t on $K(X)$ has the following properties:

(a) $i_t(X + Y, a + b) = i_t(X, a) + i_t(Y, b)$,

$\qquad i_t(X, a \oplus b) = i_t(X, a) + i_t(X, b)$

where $+$ is disjoint sum, $\oplus$ Whitney sum.

(b) $i_t(X \times Y, a \otimes b) = i_t(X, a) \cdot i_t(Y, b)$

(c) $i_t(X, a) = 0$ if $(X, a) \sim 0$, i.e., if there exists a manifold Y with boundary $X = \partial Y$, and an element $a' \in K(Y)$ whose restriction to X is equal to a.

(d) $i_t(P_{2n}(\mathbf{C})), 1) = 1$, $i_t(S^{2m}, V_m) = 2^m$, where $V_m, 1$ form a system of generators of $K(S^{2m})$.

5. Finally, a *uniqueness theorem* asserts that there is only one real valued function on the groups $K(X)$, X running through the even-dimensional manifolds, which satisfies (a) to (d).

This is in fact a consequence of a cobordism theorem. Let Σ be the set of isomorphism classes of objects $[X, a]$, where X is an even-dimensional manifold and $a \in K(X)$. let A be the set of equivalence classes in Σ under the relation: $[X, a] \sim [X', a']$ if there exists a manifold Y with $\partial Y = X - X'$ and an element $b \in K(Y)$ whose restriction

to X (resp. X') is equal to a (resp. a'). The set A may be given a
ring structure, with addition and multiplication defined by means of disjoint sum and tensor product, respectively. Moreover, the set A_1 of classes of elements [X, a], with a of virtual dimension one, is a subring.
It may be shown that $A_1 \otimes \mathbf{Q}$ is a ring of polynomials over the classes of
the elements which occur in (d) above. Now a real valued function on Σ
satisfying (a), (b), (c) may be thought of as a homomorphism of $A \otimes \mathbf{Q}$
into $\mathbf{R}$. It follows rather directly from the structure theorem on $A_1 \otimes \mathbf{Q}$
just mentioned that such a function is completely determined by its values
on the above elements.

§3. The analytical index

        The uniqueness theorem above suggests naturally to try to con-
sider i_a as a function on K(B(X), S(X)) and prove that it has the same
properties as i_t. This will be possible if the two following assertions
are true

        (i)   if d, d' are elliptic operators whose symbols define
              the same element of K(B(X), S(X)), then $i_a(d) = i_a(d')$;
        (ii) every element of K(B(X), S(X)) is the class of a symbol
              of an elliptic operator.

        In (i), the assumption means that the symbols are homotopic.
On the other hand, homotopic operators have the same analytical index, therefore (i) will be a consequence of

        (iii) a homotopy between symbols can be "raised" to a homotopy
               between elliptic operators.

        It appears that (ii), (iii) are not true in general for differ-
ential elliptic operators. This has led one to consider a wider class of
elliptic operators, for which both i_a and i_t are still defined, in which
there are enough elements to fulfill (ii), (iii), and to prove the index
theorem for these operators. This brings us to the first topic of the
analytical part of the proof:

        6. *The Seeley algebra* $\mathrm{Int}_r(E, F)$ *of operators of order* $\leq r$
[5]. In order to give a rough idea of what these operators are, we recall

first some facts on operators on $\mathbf{R}^n$, using the standard multiplicative notation: $\alpha = (\alpha_1, \ldots, \alpha_n)$, $(\alpha_i \geq 0, \alpha_i \in \mathbf{Z})$, $|\alpha| = \Sigma \, \alpha_i$, $D^\alpha = \partial^{|\alpha|}/\partial x_1^{\alpha_1} \ldots \partial x_n^{\alpha_n}$, $\xi^\alpha = \xi_1^{\alpha_1} \ldots \xi_n^{\alpha_n}$. We put also $|\xi|^2 = \Sigma \, \xi_i^2$, and denote by $\mathfrak{F}$ the Fourier transform $f \to \hat{f}$ where $\hat{f}(\xi) = (2\pi)^{-n/2} \int f(x) e^{i(x,\xi)} dx$.

The Riesz operator R^α is $\mathfrak{F}^{-1} \circ (\xi/|\xi|)^\alpha \circ \mathfrak{F}$, where, of course, the middle factor stands for the multiplication by $(\xi/|\xi|)^\alpha$. To a finite linear combination $A = \Sigma_\alpha \, a_\alpha(x) R^\alpha$ of Riesz operators, with coefficients in the space $B^\infty = B^\infty(\mathbf{R}^n)$ of C^∞-functions all of whose partial derivatives are bounded, we associate the symbol $\sigma(A) = \Sigma \, a_\alpha(x)(\xi/|\xi|)^\alpha$, which may be viewed as a function on $\mathbf{R}^n \times S^{n-1}$. The symbol map $A \to \sigma(A)$ is injective, and extends by continuity (in suitable topologies) to a 1-1 map of a certain class of operators of $H^\infty(\mathbf{R}^n)$ to $H^\infty(\mathbf{R}^n)$, (where $H^\infty(\mathbf{R}^n)$ is the space of functions all of whose partial derivatives are square integrable), to be called B^∞ singular integral operators of order zero, onto the space of functions on $\mathbf{R}^n \times S^{n-1}$, all of whose derivatives are bounded. A B^∞ operator of order r is then a product $B = A\Lambda^r$, where A is of order zero, and $\Lambda = \mathfrak{F}^{-1} \circ (1 + |\xi|^2)^{1/2} \circ \mathfrak{F}$ is the square root of $1 + \Delta$, (Δ being the Laplacian). The symbol $\sigma_r(B)$ is then $\sigma(A) \cdot |\xi|^r$.

These operators, modulo operators of lower order, include the differential operators $D = \Sigma_{|\alpha|=r} \, a_\alpha(x) D^\alpha$ of order r, with coefficients in B^∞, because, since $D^\alpha = \mathfrak{F}^{-1} \circ (-i\xi)^\alpha \circ \mathfrak{F}$, we may write $D = (\Sigma \, a_\alpha(x) R^\alpha) \cdot (-i\Lambda_0)^r$, where $\Lambda_0 = \mathfrak{F}^{-1} \circ |\xi| \circ \mathfrak{F}$ is the "square root" of Δ, and it can be shown that $\Lambda^r - \Lambda_0^r$ is of order $\leq r-2$.

Let now X be a manifold, E, F smooth complex vector bundles on X. A linear map $A: C^\infty(E) \to C^\infty(F)$ belongs to $\mathrm{Int}_r(E, F)$ if:

   (1) for $\varphi, \psi \in C^\infty(X)$ with disjoint supports, $\varphi A \psi$ is of
       order $\leq r-1$;

   (2) for $\varphi, \psi \in C^\infty(X)$ with supports in some coordinate neighborhood U, over which sections of E, F are chosen as in
       section 1, $\varphi A \psi$ is given, modulo operators of order $\leq r-1$,
       by a matrix (A_{ij}) of B^∞ operators of order r.

The matrix $(\sigma_r(A_{ij}))$ of the symbols is shown to have an intrinsic meaning, and associates to every $\xi \in T^*(X)_x - 0$ $(x \in U)$ a linear transformation of E_x into F_x, whence a homomorphism $\sigma_r(A): \pi^*E \to \pi^*F$, defined on $S(X)$, the *symbol of* A. The operator A is elliptic of order r if it is in $\mathrm{Int}_r(E, F)$ and if $\sigma_r(A)$ is an isomorphism. The set of such operators is denoted $\mathrm{Ell}_r(E, F)$. If $A \in \mathrm{Ell}_r(E, F)$, then, using $\sigma_r(A)$, the cohomology class $\mathrm{ch}\, A$ and the topological index $i_t(A)$ are defined as in the case of differential operators. Moreover, it can be shown that $\ker A$, $\mathrm{coker}\, A$ are finite dimensional, so that $i_a(A)$ is again defined, and that (ii), (iii) are true. Consequently i_a may now be viewed as a function on $K(B(X), S(X))$ and the next steps of the proof consist in showing that it has the same properties as i_t, namely:

7. *Multiplicativity*. One wants to know: if A, B, C, are elliptic operators on X, Y and $X \times Y$ such that $[\sigma(A)] \cdot [\sigma(B)] = [\sigma(C)]$, in the sense of the pairing of §2, section 3, then $i_a(A) \cdot i_a(B) = i_a(C)$. This will be proved by making use of a pairing of operators: $(A,B) \mapsto A \# B$ which is suggested by the construction underlying section 3 above. This, however, meets with some technical difficulties since $A \# B$ is not in general an integro-differential operator. It will nevertheless be a limit of such operators and this will allow one to keep track of symbols and indices.

8. If X is even-dimensional, and V a bundle on X, we define

$$i_a(X, V) = i_a(V \cdot x_0) \quad ,$$

where the right hand side stands for the analytical index of any operator with symbol belonging to $V \cdot x_0$. In fact, using a connection on V and the operator D_0 mentioned in 4, we may find a differential operator with that symbol. One then has to prove that (a) to (d) in section 4 are true with i_t replaced by i_a. This is fairly standard, except, however, for (c), which is one of the main parts of the whole proof.

The index theorem for even-dimensional manifolds then follows from 4, 5 and 8. In order to extend it to odd-dimensional manifolds, we need the following statement:

9. *There exists on the circle* S^1 *an elliptic operator* E_0, *from the trivial bundle to the trivial bundle, with* $i_a(E_0) = i_t(E_0) \neq 0$, which, together with the multiplicativity properties 3, 7, allows one to reduce the odd-dimensional case to the even-dimensional one by multiplying with E_0.

10. The analytical index is an integer by definition. Since every element of $K(B(X), S(X))$ is the class of a symbol of an elliptic operator in the Seeley algebra, the main theorem for these operators implies the

COROLLARY. Let X be a manifold, and $\alpha \in K(B(X), S(X))$. Then the topological index of α is an integer.

As we shall see, this yields all the integrality theorems pertaining to the Todd genus or the $\hat{A}$-genus.

§4. Appendix

The order of exposition in the sequel does not coincide with the one adopted in the previous outline. In order to orient the reader, we make here some comments on the contents of the different chapters.

Chapters II to X give some background material both for the topological and the analytical parts. In conformity with the purpose of this seminar, the treatment of the latter one is practically self-contained, omitting only proofs of some quite standard facts, while much is taken for granted on the topological side. Chapter II reviews briefly K-theory, Chern characters, and shows that $K(X, Y)$ may be defined as the set of equivalence classes of suitable sequences of vector bundles. Chapter III describes a method to compute the topological index when the bundles underlying the differential operator and the tangent bundle to the base manifold are associated in a suitable way to a given principal bundle.

Chapter IV introduces some basic material on differential operators: definition, symbols, the jet bundle exact sequence, adjoints, Green operators, some classical differential operators.

Chapter V defines the differential operator leading to the index of a manifold and checks the index theorem in some special cases.

Chapters VI to X review some notions and results in functional analysis: in particular bounded operators on Banach spaces with finite dimensional kernel and cokernel (to be called Fredholm operators) the Sobolev spaces H^k, the Sobolev inequality, and Rellich's theorem.

The Seeley algebra is introduced axiomatically in XI, by means of five conditions, the existence proof being postponed to XVI. Chapters XII, XIII, XIV, and part of XV are devoted to the main properties of the Seeley algebra: regularity properties of elliptic operators (XI), homotopy invariance of i_a (XII), (this is (i) of §3; property (ii) is essentially built in the axioms, so that its validity is really part of the existence proof), behaviour under Whitney sums (XIII); and the multiplicativity property 7 above, which is dealt with with the help of the sixth condition imposed on the Seeley algebra (XIV).

Chapter XV proves that both i_t and i_a satisfy the condition conditions (a), (b), (d) of 4, and that i_t also verifies (c). The proof that (c) also is true for i_a is much harder and is given in XVII.

Chapter XVI proves the existence of the Seeley algebra of integro-differential operators, with the properties postulated in XI and XIV. The method finally adopted in these Notes is different from that of [5], after which the above summary was patterned. This turned out to be more convenient for a self-contained exposition. Chapter XVI also contains the construction of an elliptic scalar operator on the circle with both indices equal to -1, comments on the sign conventions made in defining i_a and i_t, and historical remarks on singular integral operators.

Chapter XVIII is topological, and gives the proof of the uniqueness theorem in Section 5 of 1, using the main results of Thom's cobordism theory.

How the results of Chapters XI to XVIII yield a proof of the index theorem is briefly recapitulated in Chapter XIX, which also contains an extension to the non-orientable case, further remarks on the theorem, and some of the main applications (the Riemann-Roch theorem, the Hirzebruch index theorem, various integrality theorems of algebraic topology).

Appendix I is devoted to the index theorem on manifolds with boundary. It gives the precise statement, and discusses the notions underlying it; but the proof is only briefly sketched. Finally, there is an appendix by Weishu Shih, pertaining to Chapter III, which treats in a more general setting, characteristic classes and the topological index of differential operators associated to G-structures.

REFERENCES

[1] M. F. Atiyah, "The index of elliptic operators on compact manifolds," Sem. Bourbaki, Mai 1963, Exp. 253.

[2] M. F. Atiyah and F. Hirzebruch, "Analytic cycles on complex manifolds, Topology 1 (1962), pp. 25-46.

[3] M. F. Atiyah and I. M. Singer, "The index of elliptic operators on compact manifolds," Bull. A. M. S. 69 (1963), pp. 422-433.

[4] R. T. Seeley, "Singular integrals on compact manifolds," Amer. J. M. 81 (1959), pp. 658-690.

[5] R. T. Seeley, Integro-differential operators on vector bundles, Trans. A.M.S. 117 (1965), pp. 167-204.

68.

A spectral sequence for complex analytic bundles

Appendix Two in: F. Hirzebruch, Topological Methods in Algebraic Geometry,
3rd edition, 202–217, Springer 1966

The spectral sequence to be discussed here relates the $\bar{\partial}$-cohomology of the total space, the base space and the typical fibre of a complex fibre bundle with compact connected fibres. In addition to the usual fibre- and base-degrees, it carries a bigrading stemming from the type of differential forms. The precise statement is given in 2.1, the proof in Sections 3 to 6. The latter proceeds along more or less expected lines, albeit in a rather cumbersome notation, one point of interest however being the exactness of the sequence 3.7 (4), which is essentially a consequence of smoothness properties of a GREEN operator. The main applications of 2.1 given here concern the multiplicative behaviour of the χ_y-genus (8.1) and the $\bar{\partial}$-cohomology of the CALABI-ECKMANN manifolds (9.5).

Familiarity with spectral sequences of fibre bundles is assumed. As to the rest, we follow the notation and conventions of this book, with some minor deviations to be mentioned explicitly. References to sections of this paper are in ordinary type; those to other sections of this book in boldface.

This is a revised version of a paper written in 1953, quoted in the bibliography of the first edition of this book, but not published.

§ 1. Preliminaries

1.1. Manifolds are HAUSDORFF and paracompact; smooth means differentiable of class C^∞. The sheaves on a manifold M are always $\mathbf{C}_b(M)$ modules, [where $\mathbf{C}_b(M)$ is the sheaf of germs of smooth complex valued functions on M], and tensor products of sheaves are over $\mathbf{C}_b(M)$.

1.2. Let M be a complex manifold, W a complex vector bundle over M, and $\mathfrak{W}$ the sheaf of germs of smooth sections of W. Then $A_M^{p,q}(W)$ denotes the space of smooth exterior differential forms on M, of type (p, q), with coefficients in W (see **15.4**), and $\mathfrak{A}_M^{p,q}(W)$ is the sheaf of germs of such forms. If $W = 1$ is the trivial bundle $M \times \mathbf{C}$, then we omit (W) in the preceding notation. We have

$$\mathfrak{A}_M^{p,q}(W) \cong \mathfrak{W} \otimes \mathfrak{A}_M^{p,q}, \quad A_M^{p,q}(W) \cong \Gamma(\mathfrak{A}_M^{p,q}(W)). \tag{1}$$

$A_M^i(W)$ denotes the sum of the $A_M^{p,q}(W)$, where $p + q = i$, $A_M(W)$ the sum of the $A_M^i(W)$, and similarly for the corresponding sheaves.

532

Let U be an open subset of M over which W may be (and has been) identified with the trivial bundle $U \times \mathbf{C}_d$. We recall that $A_U^{p,q}(W|_U)$ is canonically identified to the set of d-ples of ordinary exterior differential (p, q)-forms on U. If $\omega \in A_U^{p,q}(W|_U)$ corresponds to $(\omega_1, \ldots, \omega_d)$, then $\bar{\partial}\omega$ corresponds to $(\bar{\partial}\omega_1, \ldots, \bar{\partial}\omega_d)$. Assume moreover that U is a co-ordinate neighbourhood, with local coordinates $z_1, \ldots, z_n$. For a subset $I = \{i_1, \ldots, i_k\}$ of $\{1, \ldots, n\}$, we put

$$dz_I = dz_{i_1} \wedge \cdots \wedge dz_{i_k}, \quad d\bar{z}_I = d\bar{z}_{i_1} \wedge \cdots \wedge d\bar{z}_{i_k}.$$

Then the above form ω_i may be written uniquely

$$\omega_i = \sum_{I,J} f_{i,I,J} \cdot dz_I \wedge d\bar{z}_J, \tag{2}$$

where I (resp. J) runs through the subsets of p (resp. q) elements of $\{1, \ldots, n\}$, and $f_{i,I,J}$ is a smooth complex valued function on U.

1.3. The direct sum of the spaces $H^{p,q}(M)$ [resp. $H^{p,q}(M, W)$, see **15.4**] is denoted $H_{\bar{\partial}}(M)$ [resp. $H_{\bar{\partial}}(M, W)$], and $h^{p,q}$ or $h^{p,q}(M)$ [resp. $h^{p,q}(W)$ or $h^{p,q}(M, W)$] is the dimension of $H^{p,q}(M)$ [resp. $H^{p,q}(M, W)$]. The space $H_{\bar{\partial}}(M)$ is in a natural way an anticommutative bigraded algebra. If $W = M \times F$ is a trivial bundle, then $H_{\bar{\partial}}(M, W) \cong H_{\bar{\partial}}(M) \otimes F$, as follows directly from the definitions.

1.4. Let now M be compact. The spaces $H^{p,q}(M, W)$ are then finite dimensional (**15.4.2**). Let further G be a LIE group operating continuously on M, by means of bi-holomorphic transformations, and let $\varphi : G \to \mathrm{Aut}M$ be the map defined by this action. Then φ induces a continuous representa-tion φ^0 of G into $H^{p,q}(M)$. *If M is kählerian, then φ^0 is constant on each connected component of G*; in fact, in this case, $H_{\bar{\partial}}(M)$ may be canonically identified with the usual cohomology algebra $H^*(M, \mathbf{C})$ of M (see e. g. WEIL [2], Chap. IV), by an isomorphism which clearly commutes with the natural representations of G in $H_{\bar{\partial}}(M)$ and $H^*(M,\mathbf{C})$; our assertion is then a consequence of the homotopy axiom. In the non-kählerian case however, this need not be true, as is shown by an example of KODAIRA [cf. GUGENHEIM and SPENCER, Proc. A. M. S. 7 (1956), 144—152].

1.5. Let $\xi = (E, B, F, \pi)$ be a complex analytic bundle (**3.2**), where E is the total space, B the base space, F the standard fibre and $\pi : E \to B$ the projection map. We assume F to be compact connected. By defini-tion (loc. cit.) the structure group G of ξ is a complex LIE group, acting on F by means of a holomorphic map $\psi : G \times F \to F$. Let ξ be defined by means of the transition functions $f_{\alpha\beta} : U_\alpha \cap U_\beta \to G$, where $(U_\alpha)_{\alpha \in \mathscr{A}}$ is a suitable covering of B. It is clear that $\bigcup_{b \in B} H^{p,q}(F_b)$, $(F_b = \pi^{-1}(b)$, $b \in B)$, is in a natural way the total space of a smooth vector bundle over B, whose transition functions $f_{\alpha\beta}^0$ are obtained by composing the

$f_{\alpha\beta}$ with the given representation φ^0 of G in $\mathbf{GL}(H^{p,q}(F))$. This bundle is denoted $\mathbf{H}^{p,q}(F)$, and $\mathbf{H}_{\bar\partial}(F)$ is the direct sum of the $\mathbf{H}^{p,q}(F)$.

If φ^0 is constant on the connected components of G, in particular if F is kählerian, then $\mathbf{H}_{\bar\partial}(F)$ is a holomorphic complex vector bundle over B (with locally constant transition functions).

In fact, the $f^0_{\alpha\beta}$ are then locally constant functions, hence may be viewed as holomorphic maps of $U_\alpha \cap U_\beta$ into $\mathbf{GL}(H_{\bar\partial}(F))$.

§ 2. The spectral sequence

2.1. Theorem. *Let $\xi = (E, B, F, \pi)$ be a complex analytic fibre bundle, where E, B, F are connected and F is compact. Let W be a complex vector bundle on B, and $\hat{W} = \pi^* W$ its inverse image on E. Assume that every connected component of the structure group G of ξ acts trivially on $H_{\bar\partial}(F)$. Then there exists a spectral sequence (E_r, d_r), $(r \geq 0)$, with the following properties:*

(i) *E_r is 4-graded, by the fibre-degree, the base-degree and the type. Let $^{p,q}E_r^{s,t}$ be the subspace of elements of E_r of type (p, q), fibre-degree s, base degree t. We have $^{p,q}E_r^{s,t} = 0$ if $p + q \neq s + t$, or if one of p, q, s, t is < 0. The differential d_r maps $^{p,q}E_r^{s,t}$ into $^{p,q+1}E_r^{s+r,t-r+1}$.*

(ii) *If $p + q = s + t$, we have*

$$^{p,q}E_2^{s,t} \cong \sum_{i \geq 0} H^{i,s-i}(B, W \otimes \mathbf{H}^{p-i,q-s+1}(F)) .$$

(iii) *The spectral sequence converges to $H_{\bar\partial}(E, \hat{W})$. For all $p, q \geq 0$, we have*

$$\operatorname{Gr} H^{p,q}(E, \hat{W}) = \sum_{s+t=p+q} {}^{p,q}E_\infty^{s,t} ,$$

for a suitable filtration of $H^{p,q}(E, \hat{W})$.

(iv) *If $W = 1$, then (E_r, d_r) consists of differential anticommutative algebras, and the isomorphism of (iii) is compatible with the products.*

2.2. Remarks. (1) Under our assumption on G, the bundle $\mathbf{H}_{\bar\partial}(F)$ is holomorphic so that (ii) makes sense. This condition is automatically fulfilled if F is kählerian (1.4).

(2) 2.1 (ii) shows that E_2 has a 4-grading which is finer than the one mentioned in 2.1 (i), namely the 4-grading given by the type of differential forms on B and on F. The proof will show that this 4-grading is also present in E_0, E_1.

Since $^{p,q}E_r^{s,t} = 0$ unless $p + q = s + t$, the superscript t is in fact redundant and it would be more correct to say that the spectral sequence is trigraded by the type (p, q) and s, where s will turn out to be the degree associated to the filtration underlying the spectral sequence. The total degree is of course $p + q$. The degree t has been added however to bring closer the analogy with the usual spectral sequence of fibre bundles, but it will be omitted in §§ 4, 5, 6.

§ 3. Auxiliary sheaves and exact sequences

3.1. Until § 6 inclusive, ξ, W, $\hat{W}$, G are as in 2.1, $\mathfrak{W}$ is the sheaf of germs of smooth sections of W, and $\mathbf{C}_b$ stands for $\mathbf{C}_b(B)$. We remark however that no assumption about the action of G on $H_{\bar{\partial}}(F)$ is needed before 6.1.

$\mathscr{U} = (U_\alpha)_{\alpha \in \mathscr{A}}$ is a locally finite open covering of B by coordinate neighbourhoods over which W and ξ are trivial. We let

$$\varphi_\alpha : W|_{U_\alpha} \to U_\alpha \times \mathbf{C}_m \quad \text{and} \quad \psi_\alpha : \pi^{-1}(U_\alpha) \to U_\alpha \times F, \quad (\alpha \in \mathscr{A})$$

be allowable trivialisations and

$$\varphi_{\alpha\beta} : U_\alpha \cap U_\beta \to \mathbf{GL}(m, \mathbf{C}) \quad \text{and} \quad \psi_{\alpha,\beta} : U_\alpha \cap U_\beta \to G, \quad (\alpha, \beta \in \mathscr{A}),$$

be the corresponding transition functions.

For every $z \in U_\alpha \cap U_\beta$, the map $\psi_{\alpha\beta}(z)$ induces an automorphism of A_F, to be denoted sometimes by $\psi'_{\alpha\beta}(z)$.

$(z_1^\alpha, \ldots, z_n^\alpha)$ is a set of local coordinates on U_α, and $\eta_{\alpha\beta}$ is the change of local coordinates in $U_\alpha \cap U_\beta (\alpha, \beta \in \mathscr{A})$.

3.2. Let φ be the complex tangent vector bundle along the fibres of ξ (BOREL-HIRZEBRUCH [1], § 7.4). We let $\varphi^{a,b}$ be the bundle of (a, b)-forms associated to φ. Thus $\varphi^{a,b} = (\lambda^a \varphi) \wedge (\lambda^b \bar{\varphi})$, where $\bar{\varphi}$ is the conjugate bundle to φ. Let $\mathscr{F}^{a,b}$ be the space of smooth sections of $\varphi^{a,b}$. For every $z \in B$, the restriction x_z of an element $x \in \mathscr{F}^{a,b}$ to the fibre $F_z = \pi^{-1}(z)$ is an (a, b)-form on F_z, and thus x may be viewed as a family of (a, b)-forms on the fibres, parametrized by B, and smooth in an obvious sense. x is called a *fibre (a, b)-form* (on B). There is a $\mathbf{C}_b$-linear map $\bar{\partial}_F : \mathscr{F}^{a,b} \to \mathscr{F}^{a,b+1}$ characterised by $r_z(\bar{\partial}_F x) = \bar{\partial}(r_z x), (z \in B, x \in \mathscr{F}^{p,q})$ (for all this, see KODAIRA-SPENCER [5], I, § 2).

Let $\mathfrak{F}^{a,b}$ be the sheaf of germs on B of fibre (a, b)-forms. We have $\Gamma(\mathfrak{F}^{a,b}) = \mathscr{F}^{a,b}$, and $\bar{\partial}_F$ is the map of sections induced by a homomorphism of $\mathbf{C}_b$-modules of $\mathfrak{F}^{a,b}$ into $\mathfrak{F}^{a,b+1}$, also denoted by $\bar{\partial}_F$. Let $\mathfrak{Z}^{a,b} \subset \mathfrak{F}^{a,b}$ be its kernel. By definition, the *sequence*

$$0 \to \mathfrak{Z}^{a,b} \xrightarrow{\;i\;} \mathfrak{F}^{a,b} \xrightarrow{\;\bar{\partial}_F\;} \bar{\partial}_F(\mathfrak{F}^{a,b}) \to 0, \tag{1}$$

where i is the inclusion map, *is exact.*

3.3. More generally we shall consider the fibre $\hat{W}$-(a, b)-forms on B. They may be defined first as the smooth sections of $W \otimes \varphi^{a,b}$ (see KODAIRA-SPENCER [5], I, § 2); for this, $\hat{W}$ could of course be any complex vector bundle on E. If x is such a form, then $r_z(x)$ is a (a, b)-form on F_z, with coefficients in the trivial bundle $V_z \times F_z$, where V_z is the fibre over z of W. Clearly, we may identify these forms with the sections of the sheaf $\mathfrak{W} \otimes \mathfrak{F}^{a,b}$.

3.4. Let

$$M^{a,b,c,d} = \Gamma(\mathfrak{W} \otimes \mathfrak{F}^{a,b} \otimes \mathfrak{A}_B^{c,d}) \,, \quad (a,\, b,\, c,\, d \in \mathbf{Z}; \, a,\, b,\, c,\, d \geqq 0) \,.$$

Its elements are to be thought of as "$(c,\, d)$-forms on B with coefficients in the fibre $\hat{W}$-$(a,\, b)$-forms". In the notation of 3.1, an element $h \in M^{a,b,c,d}$ is given by its restrictions h_α to the open subsets U_α, and h_α is an array of differential m-forms $h_{\alpha, i}$ which may be written

$$h_{\alpha, i} = \sum_{I, J} h_{\alpha, i, I, J} \, dz_I^\alpha \wedge d\bar{z}_J^\alpha \,,$$

where I and J run respectively through the subsets of c and d elements of $\{1, \ldots, n\}$, and where $h_{\alpha, i, I, J} \in \mathscr{F}^{a,b}(U_\alpha)$, is a fibre $(a,\, b)$-form on U_α.

Thus h_α is identified with a $\hat{W}$-$(a + c,\, b + d)$ differential form on $\pi^{-1}(U_\alpha)$. But of course, this identification depends essentially on the local trivialisations, and h itself cannot be viewed as a differential form. To be more precise, h_α and h_β are related by the transformations defined by $\varphi_{\alpha\beta}$, $\psi_{\alpha\beta}$, and $\eta_{\alpha\beta}$. However, if we want to describe the differential form h_α in $U_\beta \cap U_\alpha$ by means of the local coordinates (z_i^β) and the local trivialisations over U_β, then we have also to take into account the derivatives of the $\psi_{\alpha\beta}$ with respect to z. This implies that in these new coordinates, h_α will be equal to the sum of h_β and of differential forms of base-degree $> c + d$.

3.5. Although this is not needed in the sequel, we remark here, without going into details, that if we allow vector bundles to have infinite dimensional fibres, we may also view the elements of $M^{a,b,c,d}$ as $(c,\, d)$-forms on B with coefficients in a vector bundle.

In fact, $A_F^{a,b}$ is a FRECHET space in a natural way (SERRE [3]), and any automorphism of F induces a homeomorphism of $A_F^{a,b}$. Thus the transition functions $\psi'_{\alpha, \beta} : U_\alpha \cap U_\beta \to \operatorname{Aut} A_F^{a,b}$ allow one to define over B an associated bundle $\mu^{a,b}$, with standard fibre $A_F^{a,b}$. Furthermore, the transition functions are smooth in the sense that if $\varrho : U_\alpha \cap U_\beta \to A_F^{a,b}$ is smooth, then $\psi_{\alpha\beta} \circ \varrho$ is also smooth. Thus it makes sense to speak of the smooth sections of $\mu^{a,b}$. It may then be seen that the elements of $M^{a,b,c,d}$ are just the $(c,\, d)$-forms on B, with coefficients in $W \otimes \mu^{a,b}$.

3.6. The sheaf $\mathfrak{W} \otimes \mathfrak{A}_B^{c,d}$ is locally free over $\mathbf{C}_b$, therefore the sequence

$$0 \to \mathfrak{W} \otimes \mathfrak{Z}^{a,b} \otimes \mathfrak{A}_B^{c,d} \to \mathfrak{W} \otimes \mathfrak{F}^{a,b} \otimes \mathfrak{A}_B^{c,d} \to \mathfrak{W} \otimes \bar{\partial}_F(\mathfrak{F}^{a,b}) \otimes \mathfrak{A}_B^{c,d} \to 0 \,, \quad (2)$$

obtained by tensoring 3.2 (1) by $\mathfrak{W} \otimes \mathfrak{A}_B^{c,d}$ is also exact. Moreover, since $\mathfrak{A}_B^{c,d}$ is fine (**3.5**), the sequence

$$\begin{aligned} 0 \to \Gamma(\mathfrak{W} \otimes \mathfrak{Z}^{a,b} \otimes \mathfrak{A}_B^{c,d}) &\to \Gamma(\mathfrak{W} \otimes \mathfrak{F}^{a,b} \otimes \mathfrak{A}_B^{c,d}) \to \\ &\to \Gamma(\mathfrak{W} \otimes \bar{\partial}_F(\mathfrak{F}^{a,b}) \otimes \mathfrak{A}_B^{c,d}) \to 0 \,, \end{aligned} \quad (3)$$

derived from (2), is exact (**2.10.1, 2.11.1**).

3.7. Let σ be the map which sends a $\bar\partial$-closed form on F into its $\bar\partial$-cohomology class. This map defines a $\mathbf{C}_b$-homomorphism, also to be denoted σ, of $\mathfrak{Z}^{a,b}$ into the sheaf $\mathfrak{H}^{a,b}(F)$ of germs of smooth sections of the bundle $\mathbf{H}^{a,b}(F)$ defined in 1.5. We claim that *the sequence*

$$0 \to \bar\partial_F(\mathfrak{F}^{a,b-1}) \overset{i}{\longrightarrow} \mathfrak{Z}^{a,b} \overset{\sigma}{\longrightarrow} \mathfrak{H}^{a,b}(F) \to 0, \quad (a \geqq 0, b \geqq 1), \qquad (4)$$

is exact.

That $\sigma \circ i = 0$ is clear. Furthermore, since $H^{a,b}(F)$ is finite dimensional, it is readily seen that σ is surjective. There remains to prove that $\operatorname{im} i \supset \ker \sigma$. This amounts to the following assertion:

Let $z \in B$ and U an open neighbourhood of z in B, ω a fibre (a, b)-form over U, *i. e.* a map assigning to $x \in U$ an (a, b)-form $\omega(x)$ on F depending smoothly on x. Assume that for each x, there exists an $(a, b-1)$-form ν_x on F such that $\omega(x) = \bar\partial \nu_x$. Then there exists a neighbourhood V of z in U and a fibre $(a, b-1)$-form τ on V such that $\omega(x) = \bar\partial \tau(x)$ for all $x \in V$.

In other words, we may choose ν_x so as to depend smoothly on x. This assertion is contained in Theorems 7, 8 of KODAIRA-SPENCER [7].

3.8. In the same way as the exactness of (3) was deduced from that of (1), it follows from 3.7 that the sequence

$$0 \to \Gamma(\mathfrak{W} \otimes \bar\partial_F(\mathfrak{F}^{a,b-1}) \otimes \mathfrak{A}_B^{c,d}) \to$$

$$\to \Gamma(\mathfrak{W} \otimes \mathfrak{Z}^{a,b} \otimes \mathfrak{A}_B^{c,d}) \overset{\sigma}{\longrightarrow} \Gamma(\mathfrak{W} \otimes \mathfrak{H}^{a,b}(F) \otimes \mathfrak{A}_B^{c,d}) \to 0 \qquad (5)$$

is exact. On the other hand, there is a natural isomorphism

$$\mathfrak{W} \otimes \mathfrak{H}^{a,b}(F) = \mathfrak{S}(W \otimes \mathbf{H}^{a,b}(F)), \qquad (6)$$

where $\mathfrak{S}(W \otimes \mathbf{H}^{a,b}(F))$ is the sheaf of germs of smooth sections of the tensor product bundle $W \otimes \mathbf{H}^{a,b}(F)$. Therefore, we also have

$$\Gamma(\mathfrak{W} \otimes \mathfrak{H}^{a,b}(F) \otimes \mathfrak{A}_B^{c,d}) \simeq A_B^{c,d}(W \otimes \mathbf{H}^{a,b}(F)) \simeq \Gamma(\mathfrak{S}(W \otimes \mathbf{H}^{a,b}(F)) \otimes \mathfrak{A}_B^{c,d}).$$
$$(7)$$

§ 4. The filtration. Proof of 2.1 (i), (iii), (iv)

4.1. Let us say that an open subset $U \subset E$ is *small* if ξ and W are trivial over $\pi(U)$ and if an allowable trivialisation of ξ over $\pi(U)$ carries U into the product of coordinate neighbourhoods of B and F. For every small open set U and positive integer k, let $L_k(U)$ be the set of elements of $A_U(\hat W|_U)$ which, when expressed in terms of local coordinates (z_i) on B and (y_j) on F, are sums of monomials $dz_I \wedge d\bar z_J \wedge dy_{I'} \wedge d\bar y_{J'}$, in which $|I| + |J| \geqq k$, where $|A|$ denotes the number of elements in a finite set A. It is clear that $L_k(U)$ is invariant under change of coordinates (but the set of elements for which $|I| + |J| = k$ is not, and consequently the filtration introduced below is not associated to a grading). Let

$$L_k = \{\omega \in A_E(\hat{W}); \; \omega|_U \in L_k(U) \text{ for every small open subset } U \text{ of } E\}. \quad (1)$$

It is of course enough to check this condition when U runs through the elements of one open covering of E by small sets. We have

$$L_0 = A_E(\hat{W}), \; L_k = 0 \, (k > \dim_{\mathbf{R}} B); \; L_k \supset L_{k+1}, \; \bar{\partial}(L_k) \subset L_k \; (k \geq 0), \quad (2)$$

which shows that the L_k define a bounded decreasing filtration of the differential $\mathbf{C}$-module $(A_E(\hat{W}), \bar{\partial})$ by submodules stable under $\bar{\partial}$. The corresponding spectral sequence is by definition the spectral sequence (E_r, d_r) of 2.1. We have clearly

$$L_k = \sum_{p,q} {}^{p,q}L_k, \quad ({}^{p,q}L_k = L_k \cap A_E^{p,q}(\hat{W})),$$

which means that the filtration is compatible with the bigrading provided by the type, hence also with the total degree. Moreover, $\bar{\partial}$ is homogeneous of degree 1 in q, of degree 0 in p, hence this bigrading is also present in the spectral sequence. We denote by ${}^{p,q}E_r^{s,t}$ or ${}^{p,q}E_r^s$ [see 2.2 (2)] the space of elements of E_r of type (p, q), total degree $s + t$, and degree s in the grading defined by the filtration. As is usual, s and t will be called respectively base-degree and fibre-degree. Of course, ${}^{p,q}E_r^{s,t} = 0$ if $p + q \neq s + t$.

The assertions 2.1 (i), (iii) then follow from standard general facts about convergent spectral sequences of filtered-graded differential modules.

If now $\hat{W} = 1$ is the trivial bundle with fibre $\mathbf{C}$, then $A_E(\hat{W})$ is an anticommutative differential algebra. Again from general principles, this product shows up in the spectral sequence, and we have 2.1 (iv). There remains to prove 2.1 (ii).

4.2. We give here a slight reformulation of the definition of the filtration which will be useful below.

Let $V_\alpha = \pi^{-1}(U_\alpha)$, and identify V_α to $U_\alpha \times F$ by means of ψ_α. We denote by $M_\alpha^{a,b,c,d}$ the space of $W\text{-}(c, d)$-forms on B with coefficients in the (a, b)-forms of the fibre (DE RHAM [1], Chap. II, § 7). Using ψ_α, we see that

$$M_\alpha^{a,b,c,d} \cong \Gamma_{U_\alpha}(\mathfrak{W} \otimes \mathfrak{F}^{a,b} \otimes \mathfrak{A}_B^{c,d}), \quad (3)$$

$$A_{V_\alpha}(\hat{W}|_{U_\alpha}) = \sum_{a,b,c,d} M_\alpha^{a,b,c,d}. \quad (4)$$

Then

$$L_s = \{\omega \in A_E(\hat{W}); \; \omega|_{V_\alpha} \in L_{s,\alpha}(\alpha \in \mathscr{A})\}. \quad (5)$$

where

$$L_{s,\alpha} = \sum_{c+d \geq s} M_\alpha^{a,b,c,d}. \quad (6)$$

We remark that the isomorphism (3) and the direct sum decomposition (4) depend on the trivialisation ψ_α but, as before, the condition $\omega|_{V_\alpha} \in L_{s,\alpha}$ does not.

§ 5. The terms E_0, E_1

5.1. Lemma. *There exists a canonical isomorphism*

$$^{p,q}k_0^s : {}^{p,q}E_0^s \xrightarrow{\sim} \sum_i \Gamma(\mathfrak{W} \otimes \mathfrak{F}^{p-i,\,q-s+i} \otimes \mathfrak{A}_B^{i,\,s-i}) \quad (p, q, s \geq 0) .$$

The sum k_0 of the maps $^{p,q}k_0^s$ carries d_0 onto δ_F.

We keep the previous notation. Let $\omega \in {}^{p,q}L_s$ and ω_α be its restriction to $\pi^{-1}(U_\alpha)$ $(\alpha \in \mathscr{A})$. We may write (4.2):

$$\omega_\alpha = \sum_i \omega_\alpha^{p-i,\,q-s+i,\,i,\,s-i} \bmod L_{s+1,\alpha} , \tag{1}$$

where

$$\omega_\alpha^{a,\,b,\,c,\,d} \in M_\alpha^{a,\,b,\,c,\,d} = \Gamma_{U_\alpha}(\mathfrak{W} \otimes \mathfrak{F}^{a,\,b} \otimes \mathfrak{A}_B^{c,\,d}) . \tag{2}$$

We claim that, for each i, the forms $\omega_\alpha^{p-i,\,q-s+i,\,i,\,s-i}$ $(\alpha \in \mathscr{A})$ match so as to define a section $\omega^{p-i,\,q-s+i,\,i,\,s-i}$ of $\mathfrak{W} \otimes \mathfrak{F}^{p-i,\,q-s+i} \otimes \mathfrak{A}_B^{i,\,s-i}$. In fact, let α, $\beta \in \mathscr{A}$ be such that $U_\alpha \cap U_\beta \neq \emptyset$. The elements ω_α and ω_β represent the same differential form on $U_\alpha \cap U_\beta$, hence are related by a transformation $f_{\alpha\beta}$ associated to the coordinate transformations $\psi_{\alpha,\beta}$, $\varphi_{\alpha,\beta}$, $\eta_{\alpha,\beta}$. Now $f_{\alpha\beta}$ also involves the derivatives of the $\psi_{\alpha,\beta}$ with respect to the local coordinates on B. However, as was already pointed out (3.4), each term in which such a derivative occurs has a strictly bigger total base-degree, hence belongs to $L_{s+1,\alpha}$. Thus to go from $\omega_\alpha^{p-i,\,q-s+i,\,i,\,s-i}$ to $\omega_\beta^{p-i,\,q-s+i,\,i,\,s-i}$, one may neglect these derivatives and just apply the transformation defined by $\psi_{\alpha,\beta}$, $\varphi_{\alpha,\beta}$, $\eta_{\alpha,\beta}$; but this is precisely how sections of the sheaf $\mathfrak{W} \otimes \mathfrak{F}^{p-i,\,q-s+i} \otimes \mathfrak{A}_B^{i,\,s-i}$ over U_α and U_β have to match in order to define a section over $U_\alpha \cup U_\beta$.

We now associate to ω the sum of the $\omega^{p-i,\,q-s+i,\,i,\,s-i}$. This defines a map

$$^{p,q}k^s : {}^{p,q}L_s \to \sum_i \Gamma(\mathfrak{W} \otimes \mathfrak{F}^{p-i,\,q-s+i} \otimes \mathfrak{A}_B^{i,\,s-i}) ,$$

which is obviously linear, with kernel $^{p,q}L_{s+1}$, whence an injective linear map

$$^{p,q}k_0^s : {}^{p,q}E_0^s \cong {}^{p,q}L_s/{}^{p,q}L_{s+1} \to \sum_i \Gamma(\mathfrak{W} \otimes \mathfrak{F}^{p-i,\,q-s+i} \otimes \mathfrak{A}_B^{i,\,s-i}) .$$

To compute $d_0(\bar\omega)$, $(\bar\omega \in {}^{p,q}E_0^s)$, we have to apply δ to a representative ω of $\bar\omega$ in L_s, and reduce $\bmod L_{s+1}$. In local coordinates, this means that we may disregard differentiation with respect to local coordinates on B, and take into account only the coordinates on the fibres. But this is how δ_F is defined, whence

$$^{p,q}k_0^s(d_0\,\bar\omega) = \delta_F({}^{p,q}k_0^s(\bar\omega)) .$$

There remains to show that $^{p,q}k_0^s$ is surjective. Let $u \in M^{a,b,c,d}$. Put $p = a + c$, $q = b + d$, $s = c + d$. We have to find $\omega \in {}^{p,q}L_s$ such that $^{p,q}k^s(\omega) = u$.

There exists a countable locally finite covering $\mathscr{V} = (V_j)$ $(j = 1, 2, \ldots)$ of E by small open subsets (see 4.1) such that for each j there exists $\alpha = \alpha(j) \in \mathscr{A}$ for which $\pi(V_j) \subset U_\alpha$. Since E is paracompact, we may further find a sequence of open coverings $\mathscr{V}^{(l)} = (V_j^{(l)})$ $(l = 1, 2, \ldots)$ such that

$$V_j^{(1)} = V_j, \quad \overline{V}_j^{(l)} \subset V_j^{(l-1)} \quad (j, l \geq 1) .$$

Let us put

$$V_j^{(\infty)} = \bigcap_{l \geq 1} V_j^{(l)} \quad (j \geq 1) .$$

Since $\mathscr{V}$ is locally finite, it is clear that the $V_j^{(\infty)}$ also form a covering of E (not necessarily open of course). Therefore, if (n_j) is a sequence of strictly positive integers, the union of the $V_j^{(n_j)}$ is also an open covering. The form ω will be defined by means of its restrictions to the elements of such a covering.

In each V_j we choose local coordinates once and for all. The restriction u_j of u to V_j may then be identified with a differential form, with coefficients in the typical fibre of W, also denoted u_j. By definition $\omega^{(1)} = u_1$ on V_1. If $V_1 \cap V_2^{(2)} = \emptyset$, we put $\omega^{(2)} = u_1$ on V_1, $\omega^{(2)} = u_2$ on $V_2^{(2)}$. Suppose now $V_1 \cap V_2^{(2)} \neq \emptyset$. In that intersection, we have $\omega^{(1)} = u_2 + \sigma$, where σ is a form whose base-degree (i. e. degree in the differentials of local coordinates on B) is $> c + d$. We can find a form τ on $V_2^{(1)}$ which coincides with σ on $\overline{V}_1^{(2)} \cap \overline{V}_2^{(2)}$ (this is a trivial extension problem, since σ is already defined in an open neighbourhood of $V_1^{(2)} \cap V_2^{(2)}$). We then let $\omega^{(2)}$ be the differential form on $V_1^{(2)} \cup V_2^{(2)}$ which is equal to u_1 on $V_1^{(2)}$, to $u_2 + \tau$ on $V_2^{(2)}$.

Let now $l \geq 2$. Assume that there is a sequence of l strictly positive integers $n_{j,l}(j = 1, \ldots, l)$ and a differential form $\omega^{(l)}$ defined on

$$V_{(l)} = \bigcup_{1 \leq j \leq l} V_j^{(n_{j,l})} ,$$

such that

$$(\omega^{(l)} - u_j)\big|_{V_j^{(n_{j,l})}} \in L_{s+1}(V_j^{(n_{j,l})}) \quad (1 \leq j \leq l) . \tag{3}$$

Let now I be the set of integers j between 1 and l for which

$$V_{l+1} \cap V_j^{(n_{j,l})} \neq \emptyset .$$

In the intersection

$$V_{l+1} \cap V_{(l)} = V_{l+1} \cap \Big(\bigcup_{j \in I} V_j^{(n_{j,l})} \Big) ,$$

the difference $\sigma = \omega^{(l)} - u_{l+1}$ belongs to L_{s+1}. As before, we may find a form τ on V_{l+1} which coincides with σ on

$$\overline{V}_{l+1}^{(2)} \cap \Big(\bigcup_{j \in I} \overline{V}_j^{(n_{j,l+1})} \Big) .$$

Let us define a sequence $(n_{j,\,l+1})$ of $l+1$ integers by

$$n_{j,\,l+1} = n_{j,\,l} + 1, \ (j \in I); \ n_{j,\,l+1} = n_{j,\,l}, \ (1 \le j \le l, j \notin I); \ n_{l+1,\,l+1} = 2 \,.$$

We let then $\omega^{(l+1)}$ be the form on

$$V_{(l+1)} = \bigcup_{1 \le j \le l+1} V_j^{(n_{j,\,l+1})} \,,$$

which is equal to $\omega^{(l)}$ on $V_j^{(n_{j,\,l+1})}$ for $j \le l$ and to $u_{l+1} + \tau$ on $V_{l+1}^{(2)}$. Then it satisfies the condition (3) with l replaced by $l+1$.

In order to go from the domain of definition of $\omega^{(l)}$ to that of $\omega^{(l+1)}$ we may have to shrink some of the $V_j^{(n_{j,\,l})}$, but not if $V_j \cap V_{l+1} = \emptyset$. Our covering being locally finite, given $m \ge 1$, there exists $l(m)$ such that $V_m \cap V_l = \emptyset$ for all $l \ge l(m)$. As a consequence, for fixed j, the sequence $n_{j,\,l}$ becomes stationary and there exists an integer n_j such that $V_j^{(n_j)}$ belongs to the domain of definition of $\omega^{(l)}$ for all $l \ge 1$. By construction, we have then $\omega^{(l)} = \omega^{(l')}$ on $V_j^{(n_j)}$ for $l, l' \ge n_j$. There exists therefore a differential form ω on E such that $\omega = \omega^{(n_j)}$ on $V_n^{(n_j)}$ for all j. It follows then from (3) that $^{p,\,q}k^s(\omega) = u$.

5.2. Lemma. *The map* $^{p,\,q}k_0^s$ *of 5.1 induces an isomorphism* $^{p,\,q}k_1^s$ *of* $^{p,\,q}E_1^s$ *onto* $\sum_i A_B^{i,\,s-i}(W \otimes \mathbf{H}^{p-i,\,q-s+i}(F))$.

This follows from 5.1, from the exactness of the sequences (3), (5) of § 3, and from the isomorphisms 3.8 (7).

§ 6. The term E_2. Proof of 2.1 (ii)

We let k_0 (resp. k_1) be the direct sum of the maps $^{p,\,q}k_0^s$ (resp. $^{p,\,q}k_1^s$). In view of our assumption on the structure group of ξ, the image space of k_1 is the space of forms on B with coefficients in a holomorphic vector bundle (1.5), therefore it is a differential module under $\bar\delta$. The assertion 2.1 (ii) will then be a consequence of the

6.1. Lemma. *The map k_1 carries d_1 onto $\bar\delta$. For all p, q, s, it induces an isomorphism*

$$^{p,\,q}k_2^s : {}^{p,\,q}E_2^s \xrightarrow{\sim} \sum_i H^{i,\,s-i}(B, W \otimes \mathbf{H}^{p-i,\,q-s+i}(F)) \,.$$

The second assertion follows directly from the first one, which we prove now.

Let $\varkappa_1^0$ be the map of the space $Z(E_0)$ of d_0-cocycles of E_0 onto E_1. In view of 5.1, 5.2, we have the following commutative diagram

$$
\begin{array}{ccc}
\sum\limits_{c+d \ge s} \Gamma(\mathfrak{W} \otimes \mathfrak{Z} \otimes \mathfrak{A}_B^{c,\,d}) & \xrightarrow{\ \sigma\ } & \sum\limits_{c+d \ge s} \Gamma(\mathfrak{W} \otimes \mathfrak{H}(F) \otimes \mathfrak{A}_B^{c,\,d}) \\[2ex]
\Big\uparrow{\scriptstyle k_0} & & \Big\uparrow{\scriptstyle k_1} \\[2ex]
Z(E_0^s) & \xrightarrow{\ \varkappa_1^0\ } & E_1^s
\end{array}
\qquad (1)
$$

where

$$\mathfrak{B} = \sum \mathfrak{B}^{a,b}, \quad E_i^s = \sum_{p,q} {}^{p,q}E_i^s \quad (i = 0, 1, \ldots), \tag{2}$$

and σ is as in 3.8 (5). We denote μ_u the projection of L_u onto $E_0^u = L_u/L_{u+1}$ $(u = 0, 1, \ldots)$.

Let a, b, c, d be positive integers, and set $p = a + c$, $q = b + d$, $s = c + d$. Let $u \in \Gamma(\mathfrak{W} \otimes \mathfrak{H}^{a,b}(F) \otimes \mathfrak{A}_B^{c,d})$ and u' an element of $\Gamma(\mathfrak{W} \otimes \mathfrak{B}^{a,b} \otimes \mathfrak{A}_B^{c,d})$ such that $\sigma(u') = u$, which exists by 3.8. Let $v = k_1^{-1}(u)$ and $v' = k_0^{-1}(u')$. We have to prove:

$$k_1(d_1 v) = \bar{\partial} u. \tag{3}$$

By definition, $v' \in Z(E_0^s)$. There exists therefore $v'' \in L_s$ such that $\bar{\partial}(v'') \in L_{s+1}$ and $\mu_s(v'') = v'$. By the above, we have then

$$u = k_1 \cdot \varkappa_1^0 \cdot \mu_s(v'') = \sigma \cdot k_0 \cdot \mu_s(v''). \tag{4}$$

On the other hand, the definition of d_1 gives $d_1 v = \varkappa_1^0 \cdot \mu_{s+1}(\bar{\partial} v'')$ hence, also, by (1),

$$k_1(d_1 v) = \sigma \cdot k_0 \cdot \mu_{s+1}(\bar{\partial} v''). \tag{5}$$

Therefore, (3) is equivalent to

$$\sigma \cdot k_0 \cdot \mu_{s+1}(\bar{\partial} v'') = \bar{\partial} u. \tag{6}$$

It is enough to prove this for the restriction of v'' to $\mu^{-1}(U_\alpha)$, for all $\alpha \in \mathscr{A}$. We may write (4.2):

$$v'' = v^{a,b,c,d} + v^{a-1,b,c+1,d} + v^{a,b-1,c,d+1} \bmod L_{s+2,\alpha},$$

where $v^{e,f,j,k} \in \Gamma_{U_\alpha}(\mathfrak{W} \otimes \mathfrak{F}^{e,f} \otimes \mathfrak{A}_B^{j,k})$; by construction, $v^{a,b,c,d}$ may be identified with u'. We have then

$$\bar{\partial} v'' = \bar{\partial} u' + \bar{\partial}(v^{a-1,b,c+1,d} + v^{a,b-1,c,d+1}) \bmod L_{s+2,\alpha}.$$

Since we compute $\bmod L_{s+2,\alpha}$, we may neglect all terms of base degree $> c + d + 1$; this means that we also have:

$$\bar{\partial} v'' = \bar{\partial} u' + \bar{\partial}_F(v^{a-1,b,c+1,d} + v^{a,b-1,c,d+1}) \bmod L_{s+2,\alpha},$$

$$k_0 \mu_{s+1} \bar{\partial}(v'') = \bar{\partial} u' + \bar{\partial}_F(v^{a-1,b,c+1,d} + v^{a,b-1,c,d+1}).$$

The second term on the right hand side, being a $\bar{\partial}_F$-coboundary, is annihilated by σ, hence

$$\sigma k_0 \mu_{s+1} \bar{\partial}(v'') = \sigma \bar{\partial} u'.$$

But it is clear that

$$\sigma \cdot \bar{\partial}(u') = \bar{\partial}(\sigma(u')) = \bar{\partial} u,$$

whence the equality (6).

Remark. A similar proof yields a construction by means of differential forms of the spectral sequence in real cohomology of a differentiable fibre bundle. The differential algebra is the space of real valued differential forms on E, filtered by the degree in base coordinates, as in 4.1. The proof is basically the same, simpler in notation since we dispense with W and the type. If F is compact, the exactness of the sequence corresponding to 3.7 (4) follows again from the smoothness properties of the GREEN operator (DE RHAM [1], p. 157). In the general case, it is a consequence of a result of VAN EST [Proc. Konikl. Neder. Ak. van Wet. Series A, **61** (1958), 399–413, Cor. 1 to Thm. 1].

§ 7. Elementary properties and applications of the spectral sequence

We keep the notation and assumptions of 2.1.

7.1. *If the bundle* $\mathbf{H}_{\bar{\mathfrak{z}}}(F)$ *is trivial, in particular if the structure group of ξ is connected, then*

$$^{p,q}E_2^{s,t} \cong \sum_i H^{i,s-i}(B, W) \otimes H^{p-i,q-s+i}(F) \,.$$

This follows from 2.1 (ii) and 1.3.

7.2. *The space* $^{p,q}E_r^{p+q,0}$ *is a quotient of* $^{p,q}E_{r-1}^{p+q,0}$ $(r \geq 3)$. *The composition of the natural maps*

$$H^{p,q}(B, W) \cong {}^{p,q}E_2^{p+q,0} \to {}^{p,q}E_\infty^{p+q,0} \subset H^{p,q}(E, \hat{W}) \,,$$

is π^*. It is injective if $q = 0$.

The first assertion follows in the usual way from the construction of the spectral sequence and from standard facts about "edge homomorphisms". Since no element of type $(p, 0)$ can be a d_r-coboundary $(r \geq 0)$, the second one is then obvious.

7.3. By our assumption on G, the bundle $\mathbf{H}_{\bar{\mathfrak{z}}}(F)$ has the discrete structure group G/G^0, where G^0 is the identity component of G. There is then, in the usual manner, a homomorphism of the fundamental group $\pi_1(B)$ of B into $\mathrm{Aut}\, H_{\bar{\mathfrak{z}}}(F)$, and $\mathbf{H}_{\bar{\mathfrak{z}}}(F)$ may be viewed as a local system of coefficients. From this it is easily seen that if B is compact, then $H^{0,0}(B, \mathbf{H}^{p,q}(F))$ is isomorphic to the space $H^{p,q}(F)^\pi$ of fixed points of $\pi_1(B)$ under the above action. Thus

$$^{p,q}E_2^{0,p+q} \cong H^{p,q}(F)^\pi \,.$$

7.4. *The space* $^{p,q}E_r^{0,p+q}$ *may be identified with the space of d_{r-1}-cocycles of* $^{p,q}E_{r-1}^{0,p+q}$ $(r \geq 3)$. *If $W = 1$ and B is compact, the composition of the natural maps*

$$H^{p,q}(E) \to {}^{p,q}E_\infty^{0,p+q} \subset {}^{p,q}E_2^{0,p+q} = H^{p,q}(F)^\pi \subset H^{p,q}(F) \,,$$

is induced by the homomorphism associated to the inclusion map of a fibre.

This follows again from elementary facts about spectral sequences.

7.5. *If the structure group of ξ is connected, then*

$$h^{p,q}(E, \hat{W}) \leq \sum_{\substack{a+c=p \\ b+d=q}} h^{c,d}(B, W) \cdot h^{a,b}(F) \,.$$

This is a consequence of 7.1 and of the relations

$$h^{p,q}(E, \hat{W}) = \dim {}^{p,q}E_\infty \leq \dim {}^{p,q}E_2 \,,$$

where we have put

$$^{p,q}E_r = \sum_{s,t \geq 0} {}^{p,q}E_r^{s,t} \,.$$

7.6. Finally we note that *if G is connected, and if $i^* : H_{\bar{\partial}}(E) \to H_{\bar{\partial}}(F)$ is surjective, then $H_{\bar{\partial}}(E)$ is additively isomorphic to $H_{\bar{\partial}}(B) \otimes H_{\bar{\partial}}(F)$.*

In fact, E_2 may then be identified as an algebra with the tensor product of the algebras $H_{\bar{\partial}}(F) \otimes 1$ and $1 \otimes H_{\bar{\partial}}(B)$, which consist of permanent cocycles.

§ 8. The multiplicative property of the χ_y-genus

8.1. Theorem. *Let $\xi = (E, B, F, \pi)$ be a complex analytic fibre bundle with connected structure group, where E, B, F are compact, connected, and F is kählerian. Let W be a complex analytic vector bundle on B. Then $\chi_y(E, \pi^*W) = \chi_y(B, W) \cdot \chi_y(F)$.*

For the notation χ_y and χ^p, see **15.5.** Since G is connected and F is kählerian, G acts trivially on the $\bar{\partial}$-cohomology of the standard fibre (1.4), therefore we may apply 2.1; we have moreover (7.1):

$$E_2 \cong H_{\bar{\partial}}(B, W) \otimes H_{\bar{\partial}}(F) \,. \tag{1}$$

Let us put, in the notation of 7.4,

$$\chi^p(E_r) = \sum_q (-1)^q \dim {}^{p,q}E_r \,,$$

$$\chi_y(E_r) = \sum_p \chi^p(E_r) \cdot y^p \,.$$

It follows from 2.1 (iii) that

$$\chi_y(E, \pi^*W) = \chi_y(E_\infty) \,. \tag{2}$$

A simple calculation, using (1), yields

$$\chi_y(E_2) = \chi_y(B, W) \cdot \chi_y(F) \,. \tag{3}$$

Let $^{(p)}E_r = \sum_q {}^{p,q}E_r$ $(r \geq 2)$. This is a graded space, whose EULER characteristic $\chi(^{(p)}E_r)$ is equal to $\chi^p(E_r)$; it is stable under d_r, and its derived group is $^{(p)}E_{r+1}$. By a well-known and elementary fact, we have then $\chi(^{(p)}E_r) = \chi(^{(p)}E_{r+1})$, hence $\chi^p(E_r) = \chi^p(E_{r+1})$, $r \geq 2$, which, together with (2) and (3), ends the proof.

§ 9. The $\bar{\partial}$-cohomology of the CALABI-ECKMANN manifolds

9.1. We shall denote by $A^0(X)$ the identity component of the group $A(X)$ of complex analytic homeomorphisms of a compact connected complex manifold X. Although this is not really needed below, we recall that, by a well-known result of BOCHNER-MONTGOMERY, $A(X)$ is a complex LIE group. If X is the total space of a complex analytic fibering (X, Y, F, π), then every element of $A^0(X)$ commutes with π, whence a natural homomorphism $\pi^0 : A^0(X) \to A^0(Y)$ (BLANCHARD [3], Prop. I. 1, p. 160). In particular, if M and N are connected complex analytic compact manifolds, then $A^0(M \times N) = A^0(M) \times A^0(N)$ (BLANCHARD [3], p. 161).

9.2. We let $\mathbf{M}_{u,v}$, $(u, v \in \mathbf{Z}; u, v \geq 0)$ be the product $\mathbf{S}^{2u+1} \times \mathbf{S}^{2v+1}$ endowed with one of the complex structures of CALABI-ECKMANN [1]. It is the total space of a principal complex analytic fibre bundle $\xi_{u,v}$ over $\mathbf{B}_{u,v} = \mathbf{P}_u(\mathbf{C}) \times \mathbf{P}_v(\mathbf{C})$, with standard fibre and structure group a complex torus $\mathbf{T}$ of complex dimension one. We have

$$A^0(\mathbf{M}_{u,v}) = (\mathbf{GL}(u + 1, \mathbf{C}) \times \mathbf{GL}(v + 1, \mathbf{C}))/\Gamma,$$

where Γ is an infinite cyclic discrete central subgroup (BLANCHARD [1]), and the map

$$\nu_{u,v} : A^0(\mathbf{M}_{u,v}) \to A^0(\mathbf{B}_{u,v}) = \mathbf{PGL}(u + 1, \mathbf{C}) \times \mathbf{PGL}(v + 1, \mathbf{C}),$$

associated to the projection $\pi_{u,v}$ of $\mathbf{M}_{u,v}$ onto $\mathbf{B}_{u,v}$ (9.1) is the obvious homomorphism.

For $u = 0$ or $v = 0$, $\mathbf{M}_{u,v}$ is a HOPF manifold. Let $\sigma_{u,v}$ be the projection $\pi_{u,v}$ followed by the projection of $\mathbf{B}_{u,v}$ on its first factor. Then $\sigma_{u,v}$ is the projection of a complex analytic fibering $\eta_{u,v}$ with typical fibre $\mathbf{M}_{0,v}$. To see this, one may for instance use the fact (BLANCHARD [1]) that $\mathbf{M}_{u,v}$ is the base space of a complex analytic principal bundle with total space $\mathbf{M}_{u,0} \times \mathbf{M}_{0,v}$, structure group a complex 1-dimensional torus, and projection map ν, such that

$$\pi_{u,v} \circ \nu = \pi_{u,0} \times \pi_{0,v} : \mathbf{M}_{u,0} \times \mathbf{M}_{0,v} \to \mathbf{B}_{u,v}.$$

9.3. Lemma. *The group* $A^0(\mathbf{M}_{u,v})$ *acts trivially on* $H_{\bar{\partial}}(\mathbf{M}_{u,v})$.

The fibre bundle has a connected structure group and a kählerian fibre, namely $\mathbf{T}$, and 2.1 applies. By 9.1, the group $A^0(\mathbf{M}_{u,v})$ is an automorphism group of the fibred structure, hence it operates on the spectral sequence. This action is trivial on $E_2 = H_{\bar{\partial}}(\mathbf{B}_{u,v}) \otimes H_{\bar{\partial}}(\mathbf{T})$, since both $\mathbf{B}_{u,v}$ and $\mathbf{T}$ are kählerian (1.4), hence also on E_∞. But $E_\infty = \mathrm{Gr}(H_{\bar{\partial}}(\mathbf{M}_{u,v}))$. By full reducibility, any compact subgroup of $A^0(\mathbf{M}_{u,v})$ acts trivially on $H_{\bar{\partial}}(\mathbf{M}_{u,v})$. The kernel of the action of $A^0(\mathbf{M}_{u,v})$ on $H_{\bar{\partial}}(\mathbf{M}_{u,v})$ is then a normal subgroup which contains all compact subgroups, hence is equal to the whole group.

9.4. Lemma. *We have* $H^{1,0}(\mathbf{M}_{0,v}) = 0$, $(v \geq 1)$.

This lemma is known. We recall a proof for the sake of completeness. $\mathbf{M}_{0,v}$ may be defined as the quotient of $\mathbf{C}_{v+1} - \{0\}$ by the discrete group generated by a homothetic transformation $\gamma : z \to c \cdot z$ $(c \neq 1)$. Let ω be a holomorphic differential on $\mathbf{M}_{0,v}$. Its inverse image ω^* in $\mathbf{C}_{v+1} - \{0\}$ may be written as $\omega^* = g_1 \cdot dz_1 + \cdots + g_{v+1} \cdot dz_{v+1}$ where the z_i's are coordinates and the g_i's are holomorphic in $\mathbf{C}_{v+1} - \{0\}$. The form ω^* is invariant under γ; this implies

$$g_i(c^n \cdot z) = c^{-n} \cdot g_i(z) \quad (n \in \mathbf{Z}, i = 1, \ldots, v + 1) ,$$

and shows that if $g_i \not\equiv 0$, then g_i is not bounded near the origin, in contradiction with HARTOG's theorem.

The $\bar{\partial}$-cohomology of $\mathbf{M}_{u,v}$ will be generated by pure elements, and subscripts will indicate the type.

9.5. Theorem. *Let* $u \leq v$. *Then*

$$H_{\bar{\partial}}(\mathbf{M}_{u,v}) \cong \mathbf{C}[x_{1,1}]/(x_{1,1}^{u+1}) \otimes \Lambda(x_{v+1,v}, x_{0,1}) .$$

We consider first the case where $u = 0$. In the spectral sequence of the fibering $\xi_{0,v}$ we have

$$E_2 \cong \mathbf{C}[x_{1,1}]/(x_{1,1}^{v+1}) \otimes \Lambda(x_{1,0}, x_{0,1}) ,$$

where the first factor on the right hand side represents the cohomology of the base $\mathbf{P}_v(\mathbf{C})$, and the second one the cohomology of the fibre $\mathbf{T}$. The element $x_{0,1}$ generates $^{0,1}E_2^{0,1}$ and is mapped by d_2 into $^{0,2}E_2^{2,0}$, which is zero, hence $d_2(x_{0,1}) = 0$. If $d_2(x_{1,0}) = 0$, then $x_{1,0}$ would be a permanent cocycle and would show up in E_∞ (see 7.4), which would contradict 9.4. We may therefore assume that $d_2(x_{1,0}) = x_{1,1}$. A routine computation then yields:

$$E_3 \cong \Lambda(y_{v+1,v}, x_{0,1}) ,$$

where

$$y_{v+1,v} = \varkappa_3^2(x_{1,1}^{v+1} \otimes x_{1,0}) \in {}^{v+1,v}E_3^{2v,1} .$$

$y_{v+1,v}$ and $x_{0,1}$ have fibre degree one, hence are d_r-cocycles for all $r \geq 3$, whence $E_3 \cong E_\infty$. Since E_∞ is a free anticommutative graded algebra, we have $E_\infty \cong H_{\bar{\partial}}(\mathbf{M}_{u,v})$ also multiplicatively.

If now $0 < u \leq v$, consider the fibering $\eta_{u,v}$ of $\mathbf{M}_{u,v}$ over $\mathbf{P}_u(\mathbf{C})$, with fibre $\mathbf{M}_{0,v}$ (9.2). Its structure group is connected, since the base is simply connected, and it acts trivially on the $\bar{\partial}$-cohomology of the standard fibre (9.3). We may therefore apply 2.1, and we have

$$E_2 = \mathbf{C}[x_{1,1}]/(x_{1,1}^{u+1}) \otimes \Lambda(x_{v+1,v}, x_{0,1}) .$$

As before, it is seen that $x_{0,1}$ is a d_2-cocycle, hence a permanent cocycle. Since $u \leq v$, there is no element of type $(v + 1, v + 1)$ with a strictly positive base degree in the spectral sequence, hence $x_{v+1,v}$ is a permanent cocycle too. The base terms being always permanent cocycles, it follows that $d_r = 0$, $(r \geq 2)$, and that $E_2 \cong E_\infty$. We have therefore $E_\infty \cong H_{\bar{\partial}}(\mathbf{M}_{u,v})$ at least additively. But representatives of $x_{v+1,v}$, $x_{0,1}$ in $H_{\bar{\partial}}(\mathbf{M}_{u,v})$ are always of square zero, and there is a representative $y_{1,1}$ of $x_{1,1}$, namely $\pi_{u,v}^*(x_{1,1})$, such that $y_{1,1}^{u+1} = 0$. From this it follows immediately that E_∞ and $H_{\bar{\partial}}(\mathbf{M}_{u,v})$ are also isomorphic as algebras, which proves the theorem.

Remark. The $\bar{\partial}$-cohomology of the HOPF manifold $M_{0,v}$ is computed in KODAIRA-SPENCER [5], § 15 for $v = 1$, in ISE [1] for any v. Theorem 4 of ISE [1] also describes the $\bar{\partial}$-cohomology of a HOPF manifold with coefficients in a line bundle.

69.

(with W. L. Baily Jr.)

**Compactification of arithmetic quotients
of bounded symmetric domains**

Ann. Math. (2) **84** (1966) 442–528

TABLE OF CONTENTS

Introduction

This paper is chiefly concerned with a bounded symmetric domain X and an arithmetically defined discontinuous group Γ of automorphisms of X. Its main goals are to construct a compactification V^* of the quotient space $V = X/\Gamma$, in which V is open and everywhere dense, to show that V^* may be endowed with a structure of normal analytic space which extends the natural one on V, and to establish, using automorphic forms, an isomorphism of V^* onto a normally projective variety, which maps V onto a Zariski-open subset of the latter.

We now proceed to a synopsis of the contents and methods of this paper, making, for convenience in this introduction, the following assumptions, which are no essential loss in generality: $X = K\backslash G_{\mathbf{R}}$ is the quotient by a maximal compact subgroup K of the group $G_{\mathbf{R}}$ of real points of a connected algebraic matric group defined over $\mathbf{Q}$, *simple over* $\mathbf{Q}$, and Γ is an arithmetic subgroup of G (i.e., Γ is commensurable with the group $G_{\mathbf{Z}}$ of integral matrices in G,

---

* Partial support by N.S.F. grants GP-91 and GP-3903 for the first-named author, by N.S.F. grant GP-2403 for the second-named author.

548

see 3.1). Our program may be roughly divided into three parts:

I. Construction and properties of the compactification V^* of V as a topological space.

II. Study of certain automorphic forms, and of their behavior under a Φ-operator.

III. Analytic structure on V^*, projective embedding.

Part I is covered in §§ 1–4. The first paragraph deals with the *natural compactification* of X; i.e., the closure $\bar{D}$ of the Harish-Chandra realization of X as a bounded domain D [22]. We recall that $\bar{D} - D$ is the union of locally closed analytic subsets of the ambient vector space, which are themselves (equivalent to) bounded symmetric domains in a smaller number of dimensions, called the *boundary components of* $\bar{D}$. The *normalizer* $N(F) = \{g \in G_{\mathbf{R}}^0 \mid F \cdot g = F\}$ of the boundary component F is a maximal parabolic subgroup of the topological identity component $G_{\mathbf{R}}^0$ of $G_{\mathbf{R}}$, and conversely. It contains as a normal subgroup the *centralizer* $Z(F) = \{g \in G_{\mathbf{R}}^0 \mid x \cdot g = x (x \in F)\}$ of F. To F there is associated a(n essentially) canonical unbounded realization S_F of X, and a complex analytic mapping σ_F of X onto F, whose fibres are affine subspaces of the ambient vector space, and are the orbits of $Z(F)^0$. These results, due to Pyateckii-Shapiro for the classical domains [30], were extended to the general case by Korányi-Wolf [27]. In §1 we review those facts which are needed later, establish some properties of functional determinants, and some technical lemmas for later use.

Our case of interest is when X/Γ is not compact. This implies that G has a non-trivial maximal $\mathbf{Q}$-split torus, and a non-trivial system $_{\mathbf{Q}}\Phi$ of $\mathbf{Q}$-roots (see 2.1). Section 2 is mainly devoted to the study of the natural restriction map from $\mathbf{R}$-roots to $\mathbf{Q}$-roots. This will show notably that $_{\mathbf{Q}}\Phi$ is of one of the two types occurring for the systems of $\mathbf{R}$-roots of irreducible bounded symmetric domains (2.9).

Section 3 introduces the notion of rational boundary component F by means of two conditions:

(i) $U(F)/(U(F) \cap \Gamma)$ is compact, if $U(F)$ is the unipotent radical of $N(F)$,

(ii) $\Gamma(F) = (N(F) \cap \Gamma)/(Z(F) \cap \Gamma)$ is discontinuous on F.

The condition (i) is equivalent to $N(F)_{\mathbf{C}}$ being defined over $\mathbf{Q}$. The main result of §3 shows that, in our case, this in fact implies (ii), or rather more precisely implies that $\Gamma(F)$ is of arithmetic type. The map $F \mapsto N(F)_{\mathbf{C}}$ induces then a bijection of the set of rational boundary components onto the set of proper maximal parabolic $\mathbf{Q}$-groups (3.7). If X is not the unit disc, then $\dim_{\mathbf{C}} F \leqq \dim_{\mathbf{C}} X - 2$, (3.15).

Section 4 is devoted to the construction of V^*, following the pattern of

Satake's paper [33]: the union X^* of X and its rational boundary components is endowed with a topology, defined by means of a suitable fundamental set in X, such that each $g \in G_{\mathbf{Q}} \cap G_{\mathbf{R}}^0$ operates continuously on X^* and such that X^*/Γ, supplied with the quotient topology, is a compact Hausdorff space. This is the sought-for compactification V^* of V. It is the union of V and of the quotients $F_i/\Gamma(F_i) = V_i$, where F_i runs through a set of representatives of the different Γ-orbits of rational boundary components (4.9, 4.11). It is shown that every $x \in V^*$ has a basis of open neighborhoods $\{U_\alpha\}$ such that each $U_\alpha \cap V$ is connected (4.15).

Sections 5–8 are devoted to automorphic forms, and in particular to those which are called here Poincaré-Eisenstein series (P-E series for short); they generalize simultaneously Poincaré series and Eisenstein series. They are first introduced in §6 in a general setting, suggested by results of Harish-Chandra and Godement on Poincaré series and Eisenstein series, proved or stated in §§5, 6. In §7 we turn to the more special P-E series which play a central role in our paper. (On the generalized upper half-plane, they are different from, although related to, the series introduced by Maass [28], under the name of Poincaré series.) They are defined as follows: let F be a rational boundary component and σ_F be the canonical projection of X onto F (see *supra*). A P-E series adapted to F, of weight m, is a series of the form

$$E(x) = \sum_{\Gamma/\Gamma_0} \varphi(\sigma_F(x \cdot \gamma)) \cdot J(x, \gamma)^m ,$$

where φ is a polynomial on F, (in the coordinates of the canonical bounded realization of F), J the functional determinant in the unbounded realization of X associated to F, and Γ_0 a suitable subgroup of Γ. The convergence of these series follows from the results of §6. Their behavior at rational boundary components is studied in §§7, 8, where an operator similar to the Φ-operator of Maass is developed, at least for P-E series. The main idea is to prove the existence of normal (absolute) majorants of the above series in certain sufficiently big sets, which are parts of Siegel domains, so that it becomes possible to deal with such series termwise in such sets. This majorant is constructed by means of a suitable rational representation of G, and to discuss the behavior of an individual term, we use mainly the Bruhat decomposition of $G_{\mathbf{Q}}$ and some properties of weights of representations (7.6, 7.8). Our main result is that a P-E series (adapted to F) has, in a suitable sense, a holomorphic limit $\Phi_{F'}E$ as we approach any rational boundary component F'; the image of E under Φ is by definition the collection of the limits $\Phi_{F'}E$; if $\dim F' \leq \dim F$ and $F' \not\subset F \cdot \Gamma$, then $\Phi_{F'}E = 0$; moreover, the image of Φ_F contains the module of all Poincaré series of F with respect to $\Gamma(F)$ for infinitely many weights (8.5).

Part III consists of §§ 9, 10. In the latter, we endow V^* with the sheaf $\mathcal{A}$ of germs of continuous functions whose restrictions to the V_i's are analytic. Section 9 proves a prolongation theorem of analytic structure, similar in spirit to those of [2, 18, 35], which, combined with the results of § 8 on P-E series, and known facts on Poincaré series [19], allows us to prove that $(V^*, \mathcal{A})$ is an irreducible normal analytic space (10.4). The existence of a projective embedding of V^* by means of automorphic forms, whose image is projectively normal, follows then in the usual manner (10.11).

Let $\dim G > 3$. Then we have $\dim_{\mathrm{C}}(V^* - V) \leqq \dim_{\mathrm{C}} V - 2$. Standard facts about normal spaces imply therefore that every Γ-automorphic function, i.e., every meromorphic function on V, extends to a meromorphic function on V^*; hence the field of Γ-automorphic functions is an algebraic function field, each element of which is the quotient of two automorphic forms of the same weight (10.12). Also, an extension theorem of Serre [36] shows then that every automorphic form of the classical type extends to a holomorphic cross section of an algebraic coherent sheaf on V^* (10.14); this generalizes Koecher's principle.

Finally, an appendix (§ 11) contains some remarks on the full groups of isometries and of automorphisms of X.

The main results of this paper were announced in [7], and are also described in [5]. Similar theorems have been stated independently, with sketches of some proofs, by Pyateckii-Shapiro [31]. Earlier special cases may be found notably in [2, 3, 30, 35]. These are mostly connected with families of abelian varieties, and the construction of the compactification gives a concrete realization, in many cases, of the variety of moduli of such varieties. In this paper, we leave untouched the question of the minimal field of definition for a projective model of V^*, and of the possible connection of V^* with moduli of algebraic structures. For the known results in that direction, we refer to [37] where other references to related work are also given.

0. Notation and conventions

In this paragraph, we collect some notation to be used frequently in this paper without further reference.

0.1. As is usual, $\mathbf{Z}$, $\mathbf{Q}$, $\mathbf{R}$, and C denote respectively the ring of integers, and the fields of rational, real, and complex numbers. If A is a commutative ring, $\mathbf{GL}\,(n, A)$ or $\mathbf{GL}_{n,A}$ is the group of $n \times n$ matrices with coefficients in A whose determinant is a unit of A, and $\mathbf{SL}\,(n, A)$ or $\mathbf{SL}_{n,A}$, the group of elements of determinant one in $\mathbf{GL}_{n,A}$. The group of units of a ring B is denoted by B^*.

0.2. If G is a group, and M a non-empty subset of G, then $N(M)$ or $N_G(M)$

(resp. $Z(M)$ or $Z_G(M)$) is the normalizer (resp. centralizer) of M in G. Thus

$$N(M) = \{g \in G \mid g \cdot M \cdot g^{-1} = M\} \,,$$
$$Z(M) = \{g \in G \mid g \cdot m \cdot g^{-1} = m \, (m \in M)\} = \bigcap_{m \in M} N(m) \,.$$

The inner automorphism $h \mapsto g \cdot h \cdot g^{-1} (h \in G)$ is denoted Int g. Often, we write ^gM for Int $g(M)$, and M^g for Int $g^{-1}(M)$.

0.3. As regards algebraic groups, we follow in general the notation of [14]. However, our *universal field* is C, and so, in this paper algebraic group stands for complex linear algebraic group. An algebraic group here may always be (and will tacitly be whenever convenient) identified with an algebraic subgroup of $\mathbf{GL}(n, \mathbf{C})$. Algebraic group defined over k and k-group will be used synonymously. For a subring A of C, we put $G_A = G \cap \mathbf{GL}(n, A)$. The algebraic group G will be identified with $G_\mathbf{C}$.

The Lie algebra of an algebraic group, or of a Lie group, $G, H, \cdots$ will usually be denoted by the corresponding lower case German letter. If G is algebraic, defined over k, then $\mathfrak{g} = \mathfrak{g}_k \otimes \mathbf{C}$, where $\mathfrak{g}_k$ is a uniquely determined Lie algebra over k. If k' is an overfield of k, then $\mathfrak{g}_{k'} = \mathfrak{g}_k \otimes_k k'$.

In both the algebraic and Lie group cases, Ad denotes the adjoint representation of G into $\mathfrak{g}$, where Ad $g(g \in G)$ is the differential of Int g at e. The restriction of Ad g to a subspace $\mathfrak{v}$ is denoted $\mathrm{Ad}_\mathfrak{v} \, g$.

0.4. Let G be a k-group. Unless otherwise said, a *character* of G is a rational character, i.e., a morphism of algebraic groups of G into $\mathbf{GL}(1, \mathbf{C})$. The characters of G form a finitely generated commutative group, denoted $X(G)$, which is free if G is connected. The subgroup of elements of $X(G)$ which are defined over k is denoted by $X(G)_k$.

The value of $a \in X(G)$ on $g \in G$ will be written $a(g)$, or more often g^a. In the latter case, it is implied that the group operation in $X(G)$ is written additively, and that usually no notational distinction is made between a and its differential, which is a linear form on $\mathfrak{g}$. In particular, we have, by convention, $g^a = \exp a(X) \, (X \in \mathfrak{g}, \, g = \exp X)$.

0.5. An algebraic group G is a *torus* if it is isomorphic to a product of groups $\mathbf{C}^*$; a torus *splits over k*, or is k-trivial, if it is moreover defined over k and isomorphic over k to a product of groups $\mathbf{C}^*$.

Let G be a k-group. Its *radical $R(G)$* (resp. *unipotent radical $R_u(G)$*, resp. *split radical*) is the greatest connected normal solvable subgroup of G (resp. normal unipotent subgroup of G, resp. the normal subgroup generated by $R_u(G)$ and the k-split tori of $R(G)$). G is *reductive* (resp. *semi-simple*) if $R_u(G) = \{e\}$ (resp. $R(G) = \{e\}$). G is *simple over k* (resp. *almost simple* *over* k) if it has no (resp. connected) proper normal k-subgroup. G is an *almost direct product* of normal subgroups G_i if it is the quotient by a finite group of the product of the G_i's.

0.6. The identity component of a topological group H is denoted by H^0. We recall that if G is algebraic, then $G_{\mathbf{C}}$ is connected as a topological group if and only if it is connected as an algebraic group (i.e. the underlying algebraic variety is irreducible). However, if G is connected, defined over $\mathbf{R}$, the group $G_{\mathbf{R}}$, viewed as a real Lie group may have more than one connected component, but will always have only finitely many connected components.

0.7. Let A be a set. A function on A, with values in a locally compact space, is *bounded* if its range is relatively compact. A function with values in the space $\mathbf{R}^+$ of *strictly* positive real numbers is *multiplicatively bounded* if there are strictly positive constants c, c' such that $c \leq f(a) \leq c'(a \in A)$.

Let u, v be functions on A with values in the set of positive real numbers. We write $u \prec v$ if there exists a strictly positive constant c such that

$$u(a) \leq c \cdot v(a) , \qquad\qquad (a \in A)$$

and $u \succ v$ (resp. $u \asymp v$) if $v \prec u$ (resp. $u \prec v$ and $v \prec u$).

I. THE COMPACTIFICATION V^* AS A TOPOLOGICAL SPACE

1. Natural compactification and Cayley transforms of a bounded symmetric domain.

1.1. The following notation will be used in this section.

G is a connected reductive algebraic group defined over $\mathbf{R}$ which has no non-trivial character defined over $\mathbf{R}$. Thus $G_{\mathbf{R}}^0$ is a connected Lie group with reductive Lie algebra and compact center.[1] We denote by $\mathfrak{g}$ the Lie algebra of $G_{\mathbf{R}}^0$.

K is a maximal compact subgroup of $G_{\mathbf{R}}^0$. The symmetric space $X = K \backslash G_{\mathbf{R}}^0$ is assumed to carry an invariant complex structure. It is then equivalent to a bounded symmetric domain [22, 24], and is hermitian symmetric.

$\mathfrak{p}$ is the orthogonal complement of the Lie algebra $\mathfrak{k}$ of K in $\mathfrak{g}$ with respect to the Killing form, hence $\mathfrak{g} = \mathfrak{k} + \mathfrak{p}$ is a Cartan decomposition of $\mathfrak{g}$. Since X is hermitian symmetric, we have the direct sum decomposition

$$\mathfrak{g}_{\mathbf{C}} = \mathfrak{k}_{\mathbf{C}} \oplus \mathfrak{p}^+ \oplus \mathfrak{p}^- \qquad\qquad (\mathfrak{p}^+ \oplus \mathfrak{p}^- = \mathfrak{p}_{\mathbf{C}}) ,$$

where $\mathfrak{p}^\pm$ is a commutative subalgebra normalized by $\mathfrak{k}_{\mathbf{C}}$.

---

[1] Essentially, it would suffice to consider the case where G is semi-simple, without compact factors. However, it is more convenient for future references in this paper to start from a slightly more general assumption.

$\mathfrak{h}$ denotes a Cartan subalgebra of $\mathfrak{k}$, and therefore also of $\mathfrak{g}$, in view of our assumption on X, and $\Phi = \Phi(\mathfrak{h}_C, \mathfrak{g}_C)$ is the set of roots of $\mathfrak{g}_C$ with respect to $\mathfrak{h}_C$. We let $E_\mu (\mu \in \Phi)$ be root vectors, and H_μ be elements of $\mathfrak{h}_C$ verifying

$$[E_\mu, E_{-\mu}] = H_\mu, \qquad \nu(H_\mu) = 2(\nu, \mu) \cdot (\mu, \mu)^{-1} \qquad (\mu, \nu \in \Phi),$$

where $(\ ,\ )$ is the restriction of the Killing form to $\mathfrak{h}_C$, and such that the complex conjugation of $\mathfrak{g}_C$ with respect to $\mathfrak{g}$ permutes E_μ and $E_{-\mu}$ whenever $E_\mu \in \mathfrak{p}^\pm$. Let $\pi^\pm = \{\mu \in \Phi \mid E_\mu \in \mathfrak{p}^\pm\}$. The elements $E_\mu (\mu \in \pi^\pm)$ form a basis of $\mathfrak{p}^\pm$, and the elements

$$X_\mu = E_\mu + E_{-\mu}, \qquad Y_\mu = i(E_\mu - E_{-\mu}) \qquad (\mu \in \pi^+)$$

form a basis (over $\mathbf{R}$) of $\mathfrak{p}$.

Two independent roots μ, ν are said to be strongly orthogonal if neither $\mu + \nu$ nor $\mu - \nu$ are roots. We fix once for all a maximal set $(\mu_1, \cdots, \mu_t)$ of strongly orthogonal roots in π^+, as in [24], and write $H_i, E_i, E_{-i}, X_i, Y_i$ for $H_{\mu_i}, E_{\mu_i}, E_{-\mu_i}, X_{\mu_i}, Y_{\mu_i}$.

1.2. *The system of* $\mathbf{R}$-*roots.* We let $\mathfrak{a}$ be the subalgebra of $\mathfrak{p}$ spanned by $X_1, \cdots, X_t$, and $_{\mathbf{R}}\Phi = {}_{\mathbf{R}}\Phi(\mathfrak{a}, \mathfrak{g})$, the set of roots of $\mathfrak{g}$ with respect to $\mathfrak{a}$, to be called the $\mathbf{R}$-roots of $\mathfrak{g}$. The algebra $\mathfrak{a}$ is a maximal commutative subalgebra of $\mathfrak{p}$, and is maximal among the subalgebras of $\mathfrak{g}$ which can be diagonalized in the adjoint representation. $\mathfrak{g}$ is the direct sum of the centralizer $\mathfrak{z}(\mathfrak{a})$ of $\mathfrak{a}$ and of the root spaces

$$\mathfrak{g}_\alpha = \{X \in \mathfrak{g} \mid [a, x] = \alpha(a) \cdot x, a \in \mathfrak{a}\} \qquad (\alpha \in {}_{\mathbf{R}}\Phi).$$

Assume X to be irreducible. Then $_{\mathbf{R}}\Phi$ is known to be of one of two types, to be denoted by C_t and BC_t. If (γ_i) are coordinates with respect to the basis $((1/2)X_i)$, then C_t consists of the roots $\pm(\gamma_i \pm \gamma_j)/2, (1 \leqq i < j \leqq t), \pm\gamma_i (1 \leqq i \leqq t)$ and BC_t is the union of C_t and of the set of elements $\pm\gamma_i/2\,(1 \leqq i \leqq t)$. In both cases we always take as ordering the lexicographic ordering defined by the basis (X_i). The set $_{\mathbf{R}}\Delta$ of simple $\mathbf{R}$-roots consists then of

$$\alpha_i = (\gamma_i - \gamma_{i+1})/2 \qquad (1 \leqq i < t),$$

and of $\alpha_t = \gamma_t$ (resp. $\alpha_t = \gamma_t/2$) if $_{\mathbf{R}}\Phi$ is of type C_t (resp. BC_t).

The numbering of the simple $\mathbf{R}$-roots thus defined will be referred to as the *canonical numbering*.

1.3. *Maximal parabolic subgroups.* A *parabolic subgroup* of $G_{\mathbf{R}}^0$ is the intersection of $G_{\mathbf{R}}^0$ with an algebraic subgroup P of G which is parabolic, (i.e., such that G/P is a projective variety) and defined over $\mathbf{R}$. The description of the parabolic subgroups of an algebraic group is recalled, in a more general setting, in 2.2. Here we introduce the minimal ones and the maximal ones, which will play a fundamental role in this paper.

Let $\mathfrak{n}$ be the sum of the $\mathfrak{g}_\alpha$ ($\alpha > 0$), and $A = \exp \mathfrak{a}$, $N = \exp \mathfrak{n}$. These are closed subgroups, with N unipotent, normalized by A, and $A \cdot N$ is maximal among the connected subgroups of $G_\mathbf{R}^0$ which can be put in triangular form over $\mathbf{R}$. The normalizer P of N is equal to the semi-direct product $P = Z(A) \cdot N$ and $Z(A) = M \times A$ with $M = Z(A) \cap K$. The group P is generated by P^0 and a commutative subgroup of type $(2, 2, \cdots, 2)$ of M, which can be described as $K \cap \exp i \cdot \mathfrak{a}$, as follows from [14, 14.4]. Every minimal parabolic subgroup of $G_\mathbf{R}^0$ is conjugate to P.

Assume, for convenience, X to be irreducible. We let $\mathfrak{a}_b$ ($1 \leq b \leq t$) be the one-dimensional subspace on which all simple $\mathbf{R}$-roots but α_b are zero, and $A_b = \exp \mathfrak{a}_b$. The space $\mathfrak{a}_b$ is spanned by $X_1 + \cdots + X_b$. We let P_b be the subgroup generated by $Z(A_b)$ and N, and V_b be its unipotent radical. The group P_b is the semi-direct product of V_b by $Z(A_b)$. The Lie algebra $\mathfrak{v}_b$ is the sum of the root spaces $\mathfrak{g}_\alpha$ where α is >0 and not zero on $\mathfrak{a}_b$. Therefore, α runs through the roots

$$(\gamma_i \pm \gamma_j)/2, (1 \leq i \leq b < j \leq t), \qquad (\gamma_i + \gamma_j)/2, (1 \leq i \leq j \leq b),$$

and the roots $\gamma_i/2$ ($1 \leq i \leq b$) in the case BC_t. Let $\mathfrak{l}_b$ (resp. $\mathfrak{l}_b'$) be the sum of the subspaces $\mathfrak{g}_\alpha + [\mathfrak{g}_\alpha, \mathfrak{g}_{-\alpha}]$, where α runs through the $\mathbf{R}$-roots which are linear combinations of $\alpha_{b+1}, \cdots, \alpha_t$ (resp. $\alpha_1, \cdots, \alpha_{b-1}$). These are two simple ideals of $\mathfrak{z}(\mathfrak{a}_b)$, clearly normalized by the Lie algebra $\mathfrak{m}$ of M, and $\mathfrak{z}(\mathfrak{a}_b)$ is the direct sum of $\mathfrak{l}_b$, $\mathfrak{l}_b'$, $\mathfrak{a}_b$ and of an ideal $\mathfrak{m}_b$ of $\mathfrak{n}$. The group $Z(A_b)$ is generated by the analytic groups L_b, L_b', A_b, with Lie algebras $\mathfrak{l}_b$, $\mathfrak{l}_b'$, $\mathfrak{a}_b$, and by M. Let $\mathfrak{z}_b = \mathfrak{m}_b \oplus \mathfrak{a}_b \oplus \mathfrak{l}_b' \oplus \mathfrak{v}_b$. It is an ideal of $\mathfrak{p}_b$ such that $\mathfrak{p}_b = \mathfrak{l}_b + \mathfrak{z}_b$. Denote by Z_b^0 the analytic subgroup of $G_\mathbf{R}^0$ with Lie algebra $\mathfrak{z}_b$. We let Z_b be the inverse image in P_b of the centralizer of $(P_b/Z_b^0)^0$ in P_b/Z_b^0. It is a closed normal subgroup of P_b, with Lie algebra $\mathfrak{z}_b$, whose intersection with L_b is the center of L_b. It contains every normal subgroup of P_b with Lie algebra $\mathfrak{z}_b$: in fact, the image in P_b/Z_b^0 of such a subgroup is a finite normal subgroup, and therefore centralizes $(P_b/Z_b^0)^0$. In particular, Z_b contains $(Z_b^0)_\mathbf{C} \cap G_\mathbf{R}^0$, whence

$$(1) \qquad\qquad Z_b = (Z_b)_\mathbf{C} \cap G_\mathbf{R}^0,$$

where $(Z_b)_\mathbf{C}$ and $(Z_b^0)_\mathbf{C}$ denote the smallest algebraic subgroups of G containing Z_b and Z_b^0 respectively.

By the general conjugacy theorems on parabolic groups (2.2 below), every maximal proper parabolic subgroup of $G_\mathbf{R}^0$ is conjugate to one and only one of the groups P_b. It will sometimes be convenient to extend the definition of P_b to $b = 0$, by putting $P_0 = L_0 = G_\mathbf{R}^0$, $\mathfrak{l}_0' = 0$; then $\mathfrak{z}_0 = \mathfrak{v}_0 = 0$ and Z_0 is the center of $G_\mathbf{R}^0$.

1.4. *The natural compactification.* Let $P^\pm = \exp \mathfrak{p}^\pm$ and K_{C} be the analytic subgroup of G_{C} with Lie algebra $\mathfrak{k}_{\mathrm{C}}$. These are closed subgroups, and the semi-direct product $K_{\mathrm{C}} \cdot P^\pm$ is a parabolic subgroup of G_{C}. The map $(x, k, y) \mapsto e^x \cdot k \cdot e^y$ is a biregular map of $\mathfrak{p}^- \times K_{\mathrm{C}} \times \mathfrak{p}^+$ onto a Zariski-open subset $\Omega = P^- \cdot K_{\mathrm{C}} \cdot P^+$ of G_{C}, which contains G_{R}^0 [24]. An element $g \in \Omega$ will often be written

$$g = g_- \cdot g_0 \cdot g_+ \qquad\qquad (g_0 \in K_{\mathrm{C}};\ g_\pm \in P^\pm)\ ,$$

and the map $g \mapsto \log g_+$ of Ω onto $\mathfrak{p}^+$ will be denoted ζ. It is known [22], [24] that ζ induces an isomorphism of $X = K \backslash G_{\mathrm{R}}^0$ onto $\zeta(G) = D$, and that D is a bounded domain in $\mathfrak{p}^+$. This is the Harish-Chandra realization of X as a bounded domain. Its closure $\bar{D}$ is therefore compact, and will be called the *natural compactification* of X. The action of G_{R}^0 on D is defined by right translations; i.e., by

$$(1) \qquad\qquad p \cdot g = \zeta(e^p \cdot g) \qquad\qquad (p \in D, g \in G_{\mathrm{R}}^0)$$

and is known to extend to a continuous action on $\bar{D}$. Then (1) is true with $p \in \bar{D}$.

1.5. *Boundary components* (see [27], [29], [30]). (i) Assume first X to be irreducible. We use the notation of **1.3**. We have the direct sum decomposition

$$\mathfrak{l}_{b,\mathrm{C}} = \mathfrak{k}_{b,\mathrm{C}} \oplus \mathfrak{p}_b^+ \oplus \mathfrak{p}_b^- \qquad (\mathfrak{p}_b^\pm = \mathfrak{l}_{b,\mathrm{C}} \cap \mathfrak{p}^\pm)\ ,$$

the space $X_b = K_b \backslash L_b$ is hermitian symmetric, and the restriction of ζ to L_b yields the Harish-Chandra realization D_b of X_b as a bounded domain.

Let $o_b = -(E_1 + \cdots + E_b)$ $(1 \leq b \leq t)$, and put $o_0 = o$. Then

$$\bar{D} = \bigcup_{0 \leq b \leq t} o_b \cdot G_{\mathrm{R}}^0\ .$$

Moreover, if $g \in L_b$, then $o_b \cdot g = o_b + \zeta(g)$. Therefore, the orbit F_b of o_b under L_b is just $o_b + D_b$, and is contained in an affine subspace of $\mathfrak{p}^+$. The transforms of the F_b's by elements of G are the *boundary components* of $\bar{D}$. We allow here b to be equal to zero, and view D itself as a boundary component (sometimes called the *improper boundary component* of $\bar{D}$).

If X is not irreducible, then it is a product of irreducible hermitian symmetric spaces X_i corresponding to the different semi-simple, simple, non-compact ideals of $\mathfrak{g}$, D is the product of the Harish-Chandra realizations D_i of the X_i, and $\bar{D}$ the product of the $\bar{D}_i$. A boundary component is a product of boundary components of the different factors. The F_b's or, if X is not irreducible, the products of components F_b's corresponding to the different irreducible factors of X, are the *standard boundary components*.

The above construction is *hereditary*: if F is a boundary component then its closure $\bar{F}$ in $\bar{D}$ may be identified with the natural compactification of F, and its boundary components are also boundary components of X. More specifically, if X is irreducible and $F = F_b$, then $\bar{F}_b = \bar{D}_b + o_b$ and the standard boundary components of $\bar{D}_b$ may be identified with the F_c's ($b \leq c \leq t$); in fact F_c would have $c - b$ as index in the canonical numbering for X_b. The groups L_c and $P_c \cap L_b$ are in the same relationship to L_b as L_b and P_b are to $G_{\mathbf{R}}^0$. This is clear from the construction.

We refer to [30] for various more geometric definitions of the boundary components in the natural compactification and to [29] for a proof of their equivalence.

For every boundary component F, we put

$$N(F) = \{g \in G_{\mathbf{R}}^0 \mid F \cdot g = F\} ,$$
$$Z(F) = \{g \in G_{\mathbf{R}}^0 \mid f \cdot g = f(f \in F)\} ,$$
$$G(F) = N(F)/Z(F) ,$$

and let $U(F)$ be the unipotent radical of $N(F)$. The group $N(F)$ is the *normalizer*, and $Z(F)$ the *centralizer*, of F. If X is irreducible, we have, in the notation of 1.3,

$$(1) \qquad N(F_b) = P_b , \qquad U(F_b) = V_b .$$

Moreover

$$(2) \qquad Z(F_b) = Z_b .$$

In fact, $Z(F_b)$ is a normal subgroup of P_b with Lie algebra $\mathfrak{z}_b$, hence $Z(F_b) \subset Z_b$ by 1.3; on the other hand, the image in $G(F_b)$ of an element $z \in Z_b$ centralizes the image L_b'' of L_b, and therefore the maximal compact subgroups of L_b'', hence it acts trivially on F_b, and $Z_b \subset Z(F_b)$.

Returning to the general case, we see, by applying 1.3 (1) and 1.5 (2) to each irreducible factor of X, that $Z(F)$ is the intersection of $G_{\mathbf{R}}^0$ with an **R**-subgroup of G. Furthermore, by 11.2, each element of $N(F)$ induces a complex analytic homeomorphism of F, hence (11.6), $G(F)$ is connected, with center reduced to $\{e\}$; equivalently, if X is irreducible, we have $N(F_b) = L_b \cdot Z_b$.

(ii) If F and F' are two boundary components such that $F' \subset \bar{F}$, then there exists $g \in G_{\mathbf{R}}^0$ such that $F \cdot g$ and $F' \cdot g$ are both standard boundary components. To see this, we may assume X to be irreducible. Let then b, c be the indices such that $F \subset F_b \cdot G_{\mathbf{R}}^0$, $F' \subset F_c \cdot G_{\mathbf{R}}^0$, and let $u \in G_{\mathbf{R}}^0$ be such that $F \cdot u = F_b$. Then $F' \cdot u$ and F_c are both boundary components of F_b, of the same dimension. Consequently, there exists $v \in L(F_b)$ such that $F' \cdot u \cdot v$ is standard. $F' \cdot u \cdot v$ is then equal to F_c; hence, $g = u \cdot v$ verifies our condition.

(iii) *If X is irreducible, and $\dim_C X \geq 2$, then $\dim_C X \geq \dim_C F + 2$ for every proper boundary component F of X.*

To see this, we may assume that $F = F_b$ $(1 \leq b \leq t)$. If $b = t$, F_b is a point, and there is nothing to prove. So assume $b \neq t$. Then $t \geq 2$, and $\mathfrak{n}$ contains at least three root spaces $\mathfrak{g}_\alpha$, whose sum intersects $\mathfrak{n} \cap \mathfrak{l}_b$ only at the origin, namely those corresponding to $\alpha = (\gamma_b \pm \gamma_t)/2$, γ_b, hence,

$$\dim \mathfrak{n} - \dim (\mathfrak{l}_b \cap \mathfrak{n}) \geq 3 \ .$$

On the other hand $\dim_R X = \dim \mathfrak{a} + \dim \mathfrak{n}$, $\dim_R F_b = \dim (\mathfrak{a} \cap \mathfrak{l}_b) + \dim (\mathfrak{l}_b \cap \mathfrak{n})$, and $\dim \mathfrak{a} - \dim (\mathfrak{a} \cap \mathfrak{l}_b) \geq 1$, whence our assertion.

1.6. *The Cayley transforms of X.* The space X also admits certain unbounded realizations, introduced by Pyateckii-Shapiro [40] in the classical cases under the name of Siegel domains of type I, II or III, and discussed in general by Korányi and Wolf [27]. In this and the next section, we summarize only the results which are used in the sequel. We assume again X to be irreducible.

The *Cayley transform* c_b is, by definition,

$$c_b = \prod_{1 \leq i \leq b} \exp{(\pi/4)} \cdot (E_{-i} - E_i), (1 \leq b \leq t); \qquad c_0 = e \ .$$

It verifies

$$(1) \qquad \operatorname{Ad} c_b(H_i) = X_i \ , \qquad \operatorname{Ad} c_b(X_i) = -H_i \ , \qquad (1 \leq i \leq b)$$

$$(2) \qquad \operatorname{Ad} c_b(H_i) = H_i \ , \qquad \operatorname{Ad} c_b(X_i) = X_i \ , \qquad (b < i \leq t)$$

and is transformed into its inverse by the complex conjugation of G_C with respect to G_R. Moreover

$$(3) \qquad \qquad c_b \cdot g = g \cdot c_b \ , \qquad\qquad\qquad (g \in L_b) \ .$$

$$(4) \qquad \qquad G \cdot c_b \subset P^- \cdot K_C \cdot P^+ \ .$$

We put then $S_b = \zeta(G \cdot c_b)$, and let G act on S_b by

$$(5) \qquad\qquad s \cdot g = \zeta(e^s \cdot c_b^{-1} \cdot g \cdot c_b) \ .$$

The map $g \mapsto g \cdot c_b$ induces then an isomorphism ν_b of X onto S_b; by definition $S_0 = D$, and S_0 is just the bounded realization. Often, we shall denote also by o the fixed point of K in S_b.

In the next proposition, we denote by $\mathfrak{q}_b$ the subspace of $\mathfrak{p}^+$ spanned by the vectors $E_\mu(\mu \in \pi^+, \mu(H_i) \neq 0$ for at least one $i \leq b)$. Thus $\mathfrak{p}^+ = \mathfrak{p}_b^+ \oplus \mathfrak{q}_b$.

1.7. PROPOSITION. *We keep the preceding notation. We have $c_b^{-1} \cdot Z_b \cdot c_b \subset K_C \cdot P^+$, and the action of Z_b (resp. V_b) on S_b extends to an action of Z_b on $\mathfrak{p}^+$ by means of affine transformations (resp. affine transformations with unipotent linear homogeneous parts) which leave $\mathfrak{q}_b$ stable and induce the* *identity on* $\mathfrak{p}^+/\mathfrak{q}_b$. *The projection* σ_b *of* $\mathfrak{p}^+$ *onto* $\mathfrak{p}_b^+$ *with kernel* $\mathfrak{q}_b$ *maps* S_b *onto* D_b, *and its fibres in* S_b *are the orbits of* Z_b^0 $(0 \leq b \leq t)$. *Its restriction to* S_b *commutes with* $N(F_b)$.

This is contained in the more precise results of [27, §7]. Let $z \in Z_b$. We have then $z' = c_b^{-1} \cdot z \cdot c_b = z_0' \cdot z_+'$ $(z_0' \in K_C, z_+' \in P^+)$. The action of z on S_b or $\mathfrak{p}^+$ is therefore given by

$$(1) \qquad\qquad s \cdot z = \mathrm{Ad}\, z_0'^{-1}(s) + \log z_+' \qquad\qquad (s \in \mathfrak{p}^+) .$$

Since $\mathfrak{p}^+$ is commutative, we can replace z_0' by z', whence

$$(1') \qquad s \cdot z = \mathrm{Ad}\, c_b^{-1} \cdot z \cdot c_b(s) + \log (c_b^{-1} \cdot z \cdot c_b)_+ , \qquad (s \in \mathfrak{p}^+, z \in Z_b) .$$

If $g \in L_b$, then it commutes with c_b, therefore 1.6 (5) becomes

$$(2) \qquad\qquad s \cdot g = \zeta(e^s \cdot g) \qquad\qquad (s \in S_b, g \in L_b) .$$

In particular

$$(3) \qquad\qquad s \cdot g = \mathrm{Ad}\, g^{-1}(s) \qquad\qquad (s \in S_b, g \in K_b) .$$

REMARK. Let F be a boundary component, and $g \in G_R^0$ be such that $F \cdot g = F_b$. Then $x \mapsto \sigma_b(x \cdot g) \cdot g^{-1}$ is a holomorphic map of X onto F. If g' is such that $F \cdot g' = F_b$, then $g' = g \cdot n$ $(n \in N(F_b))$; since translation by n commutes with σ_b, we get $\sigma_b(x \cdot g) \cdot g^{-1} = \sigma_b(x \cdot g') \cdot g'^{-1}$ $(x \in X)$. We have thus defined a canonical holomorphic projection of X onto F, equivariant with respect to $N(F)$, to be denoted σ_F. If $F' \subset \bar{F}$, then we have a factorization

$$\sigma_{F'} = \sigma_{F'F} \circ \sigma_F ,$$

where $\sigma_{F'F}$ is the canonical projection of F on its boundary component F'. In fact, there exists by 1.5 an element $g \in G_R^0$ such that $F \cdot g = F_b$ and $F' \cdot g = F_c$ $(b \leq c)$, and it is clear from Proposition 1.7 that $\sigma_c = \sigma_{c,b} \circ \sigma_b$ where $\sigma_{c,b}$ is the canonical projection of F_b onto the standard boundary component F_c.

The remark extends obviously to non-irreducible bounded symmetric domains.

1.8. *Automorphy factors, functional determinants.* Let M be a complex manifold, H a group of automorphisms of M, and Q a complex Lie group. We recall that a (holomorphic) automorphy factor for H on M, with values in Q, is a map $\mu: M \times H \to Q$ which, for fixed $h \in H$, is holomorphic in $x \in M$, and which verifies the identity

$$(1) \qquad \mu(x, h \cdot h') = \mu(x, h) \cdot \mu(x \cdot h, h') \qquad (x \in M; h, h' \in H) ,$$

to be referred to as the *cocycle formula*; it implies

$$(2) \qquad \mu(x, h \cdot h' \cdot h'') = \mu(x, h) \cdot \mu(x \cdot h, h') \cdot \mu(x \cdot h \cdot h', h'')$$
$$(x \in M; h, h', h'' \in H) .$$

It follows immediately from (1) that the set R of elements $h \in H$ for which $\mu(x, h) = \rho(h)$ is independent of x is a subgroup, and that

$$(3) \qquad \mu(x, h \cdot r) = \mu(x, h) \cdot \rho(r) \qquad (x \in M; \, h \in H; \, r \in R).$$

If M is a domain in $\mathbf{C}^n$, then the jacobian Jac (x, h) which associates to $h \in H$ and $x \in M$ the differential of h at x, is an automorphy factor with values in $\mathbf{GL}(n, \mathbf{C})$, and $J(x, h) = \det \mathrm{Jac}\,(x, h)$ is an automorphy factor with values in $\mathbf{C}^*$.

Let $M = X$ be an irreducible bounded symmetric domain. We shall denote by $\mathrm{Jac}_b\,(x, g)$ the jacobian of $g \in G_\mathbf{R}^0$ at $x \in S_b$, in the unbounded realization associated to F_b, by $J_b(x, g)$ its determinant, and by $j_b(x, g)$ the functional determinant of $g \in P_b$ at $x \in D_b$ $(0 \leq b \leq t)$. Our next aim is to obtain some information on $J_b(x, g)$ when $g \in P_b$, which will be used in studying Poincaré-Eisenstein series.

It is immediate that $\mu_b(x, g) = (e^x \cdot c_b \cdot g \cdot c_b^{-1})_0$ $(x \in S_b, \, g \in G_\mathbf{R}^0)$ is a holomorphic automorphy factor, with values in $K_\mathbf{C}$. It is called the *canonical automorphy factor* for the unbounded realization S_b. The automorphy factors usually considered in the theory of holomorphic automorphic forms are of the form $\rho(\mu_b(x, g))$ where $\rho \colon K_\mathbf{C} \mapsto \mathbf{GL}(m, \mathbf{C})$ is a holomorphic representation. The following lemma asserts that Jac_b is of this type. It is well-known for the bounded realization; the proof is essentially the same in the general case, and is included for the sake of completeness.

1.9. Lemma. *We keep the notation of* **1.8**, *and identify the tangent space to a point* $x \in \mathfrak{p}^+$ *with* $\mathfrak{p}^+$ *by translation. Then*

$$\mathrm{Jac}_b\,(x, g) = \mathrm{Ad}_{\mathfrak{p}^+}\, h_0^{-1} \qquad \left(x \in S_b, \, g \in G_\mathbf{R}^0; \, h = (e^x \cdot c_b \cdot g \cdot c_b^{-1})\right).$$

Let $X \in \mathfrak{p}^+$. Then $x + X$ goes under the differential dg of the automorphism of S_b defined by g onto an element $Y + x \cdot g$; we have to prove that

$$(1) \qquad Y = \mathrm{Ad}\, h_0^{-1}(X).$$

Write g' for $c_b \cdot g \cdot c_b^{-1}$. By definition

$$(2) \qquad x \cdot g = \zeta(e^x \cdot g') = \log\,(e^x \cdot g')_+ = \log h_+ ,$$

and

$$Y + x \cdot g = d\zeta \left\{ \frac{d}{dt}(e^{tX} \cdot e^x \cdot g') \Big|_{t=0} \right\}.$$

Clearly

$$e^{tX} \cdot e^x \cdot g' = h_- \cdot e^{tU} \cdot h_0 \cdot h_+ \qquad (U = \mathrm{Ad}\, h_-^{-1}(X)).$$

But the bracket relations

(3) $$[\mathfrak{k}_\mathbb{C}, \mathfrak{p}^\pm] \subset \mathfrak{p}^\pm , \qquad [\mathfrak{p}^+, \mathfrak{p}^-] \subset \mathfrak{k}_\mathbb{C} ,$$

imply readily that if $p \in \mathfrak{p}^-$ and $X \in \mathfrak{p}^+$, then $\mathrm{Ad}\, \exp p(X) \equiv X$, modulo $\mathfrak{k}_\mathbb{C} \oplus \mathfrak{p}^-$. We may therefore write $U = X + Z$ $(Z \in \mathfrak{k}_\mathbb{C} \oplus p^-)$, whence

$$e^{tX} \cdot h = h_- \cdot h_0 \cdot e^{t(X'+Z')} \cdot h_+ \qquad \left(X' = \mathrm{Ad}\, h_0^{-1}(X),\ Z' = \mathrm{Ad}\, h_0^{-1}(Z)\right) .$$

We have then

$$\frac{d}{dt}(e^{tX} \cdot h)\Big|_{t=0} = h_- \cdot h_0 \cdot (X' + Z') \cdot h_+ .$$

Since Z' also belongs to $\mathfrak{k}_\mathbb{C} + \mathfrak{p}^-$, the image of the right-hand side under $d\zeta$ is $X' + \log h_+$, which, in view of (2), proves our contention.

1.10. Lemma. *Let β_b $(0 \leq b \leq t)$ be the sum of roots $\mu \in \pi^+$ such that $E_\mu \in \mathfrak{p}_b^+$. Then $m_b = \beta_b(H_i)$ is a strictly positive integer independent of i $(b < i \leq t)$, and $m_b > m_c$ if $0 \leq b < c \leq t$.*

In view of the "hereditary" character of the natural compactification (1.5), it is enough to prove this when $b = 0$, $G_\mathbb{R}^0 = L_b$.

Let $\mathfrak{h}_t = \mathrm{Ad}\, c_t^{-1}(\mathfrak{a}_\mathbb{C})$. It is the subalgebra of $\mathfrak{h}$ spanned by the vectors H_i $(1 \leq i \leq t)$. We denote the coordinates with respect to the basis $(H_i/2)$ also by γ_i. We choose an ordering on Φ verifying the following conditions:

The elements of π^+ are positive, the restrictions to $\mathfrak{h}_t$ of the elements of π^+ are the linear forms $(\gamma_i + \gamma_j)/2$ $(1 \leq i \leq j \leq t)$, and also the forms $\gamma_i/2$ in the case BC_t; the positive roots of $\mathfrak{k}_\mathbb{C}$ restrict to the differences $(\gamma_i - \gamma_j)/2$ $(1 \leq i < j \leq t)$, and also to $\gamma_i/2$ in the case BC_t.

This is always possible [22, § 6]. Let $\Delta = \{\nu_1, \cdots, \nu_l\}$ be the corresponding set of simple roots. It is known that we may assume $\theta = \{\nu_1, \cdots, \nu_{l-1}\}$ to be the set of simple roots of $\mathfrak{k}_\mathbb{C}$, and that the elements of π^+ are the roots which are congruent to ν_l modulo a linear combination of elements in θ, [22]. Moreover, since $\mathfrak{k}_\mathbb{C}$ normalizes $\mathfrak{p}^+$, its Weyl group permutes the elements of π^+ and leaves β_0 invariant. In particular, β_0 is left fixed by the fundamental reflections r_ν $(\nu \in \theta)$, whence

(1) $$(\beta_0, \nu_i) = 0 \qquad\qquad (1 \leq i < l) .$$

The sum of two elements in π^+ is never a root, hence

(2) $$(\mu, \nu) \geq 0 \qquad\qquad (\mu, \nu \in \pi^+) .$$

We have therefore

(3) $$(\beta_0, \mu) = (\beta_0, \nu_l) \geq (\nu_l, \nu_l) > 0 \qquad\qquad (\mu \in \pi^+) .$$

But $\nu(H_i) = 2(\nu, \gamma_i) \cdot (\gamma_i, \gamma_i)^{-1}$ is an integer for every $\nu \in \Phi$, and each i; therefore,

(4) $$\beta_0(H_i) = 2 \cdot (\beta_0, \gamma_i) \cdot (\gamma_i, \gamma_i)^{-1} \in \mathbf{Z},\ \beta_0(H_i) > 0 .$$

The relative Weyl group of $\operatorname{Ad} c_i^{-1}(\mathfrak{g})$ with respect to $\operatorname{Ad} c_i^{-1}(\mathfrak{a})$ contains the permutations of the γ_i. But every such transformation is induced by an element of the Weyl group of $\mathfrak{g}_{\mathbf{C}}$ with respect to $\mathfrak{h}_{\mathbf{C}}$ (see e.g. [14, 5.5]). It follows then that (γ_i, γ_i) is independent of i, whence our first assertion.

The difference $m_0 - m_c$ $(c \geq 1)$ is the sum of the numbers $\mu(H_i)$ where μ runs through the elements of π^+ such that $E_\mu \notin \mathfrak{p}_c^+$; these numbers are all ≥ 0 by the above. But there is at least one such μ, for instance one which restricts to $(\gamma_1 + \gamma_i)/2$, for which $\mu(H_i) \neq 0$, which ends the proof.

1.11. PROPOSITION. *Let X be irreducible. Let J_b be the functional determinant function for $G_{\mathbf{R}}^0$ acting on S_b, and j_b the functional determinant function for L_b acting on D_b. Then*

(i) *The function $J_b(x, g)$ is constant along the fibres of the projection $\sigma_b \colon S_b \to D_b$ of 1.7 if $g \in P_b$, is independent of x if $g \in Z_b$, and is equal to one if g is a unipotent element of Z_b. The restriction η_b of J_b to Z_b is a rational character.*

(ii) *If $g \in L_b$, we have $J_b(x, g)^{m_b} = j_b(\sigma_b(x), g)^{m_0}$, with m_0, m_b as in 1.10.*

PROOF OF (i). If $g \in Z_b$ (resp. $g \in Z_b$ and is unipotent), then g acts on S_b by means of an affine transformation (resp. with unipotent linear part) in $\mathfrak{p}^+$; therefore, $J_b(x, g) = \eta_b(g)$ is independent of x (resp. is equal to one); then η_b is a rational character by 1.7 (1').

Write $g = l \cdot u$ $(l \in L_b, u \in Z_b)$, and let $z \in Z_b^0$. Using the cocycle formula, we have

$$J_b(x \cdot z, l \cdot u) = J_b(x \cdot z, l) \cdot \eta_b(u) = J_b(x, z \cdot l) \cdot J_b(x, z)^{-1} \cdot \eta_b(u) \,,$$
$$J_b(x \cdot z, l \cdot u) = J_b(x, l) \cdot \eta_b(z') \cdot \eta_b(z)^{-1} \cdot \eta_b(u) \,,$$

where $z' = l^{-1} \cdot z \cdot l$. But Z_b^0 is the semi-direct product of V_b by a reductive group which centralizes L_b (see 1.3); therefore, $\eta_b(z) = \eta_b(z')$ and

$$J_b(x \cdot z, l \cdot u) = J_b(x, l \cdot u) \,.$$

Since the fibres of σ_b are the orbits of Z_b^0, this ends the proof of (i).

PROOF OF (ii). For every element $g \in K_{b,\mathbf{C}}$, let us put

$$\Psi(g) = \det\left(\operatorname{Ad}_{\mathfrak{p}^+} g^{-1}\right) \,, \qquad \psi(g) = \det\left(\operatorname{Ad}_{\mathfrak{p}_b^+} g^{-1}\right) \,.$$

We want to prove

$$(4) \qquad\qquad\qquad \Psi(g)^{m_b} = \psi(g)^{m_0} \qquad\qquad\qquad (g \in K_{b,\mathbf{C}}) \,.$$

Assume first that $g = \exp(\lambda_{b+1} H_{b+1} + \cdots + \lambda_t H_t)$. In this case, $\Psi(g)$ (resp. $\psi(g)$) is the product of the numbers $\exp \mu(-\log g)$ where μ runs through the roots μ such that $E_\mu \subset \mathfrak{p}^+$ (resp. $E_\mu \subset \mathfrak{p}_b^+$). Using 1.10, we get

$$(5) \qquad \Psi(g) = \prod_{b < i \leq t} \exp -\lambda_i \cdot m_0 \,, \qquad \psi(g) = \prod_{b < i \leq t} \exp -\lambda_i \cdot m_b \,,$$

which proves our contention in this case. It is also clear from (5) that Ψ and ψ are not identically equal to one on the subgroup just considered. The group $K_{b,C}$ is generated by its derived group, on which both Ψ and ψ are equal to one, and by its one-dimensional center. It is therefore also generated by its derived group and the group of elements considered in (5), which proves (4).

Any element $g \in L_b$ commutes with the Cayley transform c_b; therefore, we have by 1.9, applied to L_b operating on S_b and on D_b:

$$(6) \qquad J_b(x, g) = \Psi(g) , \qquad j_b(y, g) = \psi(g) , \qquad (x \in S_b; y \in D_b; g \in K_b)$$

$$(7) \qquad J_b(o, g) = \Psi(g_0) , \qquad j_b(o_b, g) = \psi(g_0) , \qquad (g \in L_b) .$$

Given $x \in S_b$, there exists $l \in L_b$ such that $\sigma_b(x) = o_b \cdot l$. The points x and $o_b \cdot l$ belong to the same fibre of σ_b, hence $J_b(x, g) = J_b(o_b \cdot l, g)$ by (i), and the latter functional determinant has to be compared with $j_b(o_b \cdot l, g)$. The desired relationship then follows from the cocycle formula and what has already been proved.

1.12. PROPOSITION. *Let* $a = \exp(\lambda_1 X_1 + \cdots + \lambda_t X_t)$ *be an element of* A. *Then*

$$J_b(o, a) = \prod_{1 \leq i \leq b} e^{-\lambda_i \cdot m_0} \cdot \prod_{b < i \leq t} (\cosh \lambda_i)^{-m_0} .$$

Write $a = u \cdot v$ $(u = \exp(\lambda_1 X_1 + \cdots + \lambda_b X_b); v = \exp(\lambda_{b+1} X_{b+1} + \cdots + \lambda_t X_t))$. We have then $u \in Z_b$, $v \in L_b$, and therefore, by 1.11,

$$J_b(o, a) = J_b(o, u) \cdot J_b(o, v) .$$

Since $c_b^{-1} \cdot u \cdot c_b = u' = \exp(\lambda_1 H_1 + \cdots + \lambda_b H_b) \in K_C$, the action of u on S_b is given by

$$s \cdot u = \operatorname{Ad} u'^{-1}(s) , \qquad (s \in S_b) ,$$

hence

$$(1) \qquad J_b(x, u) = \Psi(c_b^{-1} \cdot u \cdot c_b) = \prod_{1 \leq i \leq b} e^{-\lambda_i \cdot m_0} .$$

On the other hand, a standard computation on the three-dimensional simple group (see e.g. [24, p. 316]) shows that the K_C-component v_0 of v is

$$v_0 = \prod_{i \leq t} \exp \log \cosh \lambda_i \cdot H_i ,$$

and our assertion now follows from 1.11 (5), (6).

1.13. COROLLARY. *Let* $h_c(\lambda) = \exp \lambda(X_1 + \cdots + X_c)$ $(1 \leq c \leq t, \lambda \in \mathbf{R})$. *Then* $J_c(o, h_c(\lambda)) = J_b(o, h_c(\lambda))$ *if* $c \leq b$, *and* $J_b(o, h_c(\lambda)) \cdot J_c(o, h_c(\lambda))^{-1}$ *tends monotonically to zero as* $\lambda \to -\infty$ *if* $c > b$.

By 1.11, we have

$$J_c(o, h_c(\lambda)) = \exp -\lambda \cdot m_0 \cdot c ,$$

and

$$J_b\big(o,\, h_c(\lambda)\big) = \exp -\lambda\cdot m_0\cdot c\;, \qquad\qquad (c \leqq b)\;,$$
$$J_b\big(o,\, h_c(\lambda)\big) = \exp -\lambda\cdot m_0\cdot b(\cosh \lambda)^{-m_0(c-b)}\;, \qquad\qquad (c > b)\;,$$

whence our assertion.

1.14. *Remark on the Bergman kernel function.* Let $K_b(z,\, w)$ be the Bergman kernel function in S_b. We have therefore

$$K_b(z\cdot g,\, w\cdot g) = K_b(z,\, w)\cdot |\, J_b(z,\, g)\,|\cdot |\, J_b(w,\, g)\,|\;, \qquad\qquad (z,\, w \in S_b;\, g \in G_{\mathbf{R}}^0)\;.$$
$$K_b(z\cdot k,\, w\cdot k) = K_b(z,\, w) \qquad\qquad (z,\, w \in S_b;\, k \in K)\;.$$

Since $G = K\cdot A\cdot K$ and o is fixed under K, this shows that $K_b(z,\, z)$ is completely determined by $K_b(o\cdot a,\, o\cdot a),\, (a \in A)$, which is given by

$$K_b(o\cdot a,\, o\cdot a) = K_b(o,\, o)\cdot |\, J_b(o,\, a)\,|^2\;.$$

We may then insert the expression of $J_b(o,\, a)$ given by 1.12; the formula thus obtained in the two extreme cases $b = 0$, $S_b = D$, and $b = t$ have been given, in a slightly different form, and with the value of the constant $K_b(o,\, o)$, by Bott-Korányi [27, 5.7] and Korányi [27, 5.5] respectively.

Our next aim is to relate $J_b(x,\, g)$ to the determinant of $\operatorname{Ad} g$ in $\mathfrak{v}_b$ (cf. 1.3) when $g \in Z_b$. For this, we need the following lemma:

1.15. LEMMA. *Let X be irreducible. Let u (resp. v) be the multiplicity of the $\mathbf{R}$-roots $(\gamma_i \pm \gamma_j)/2\ (i \neq j)$ (resp. $\gamma_i/2$ in the case BC_t). Let ν_b be the restriction of α_b to $\mathfrak{a}_b\ (1 \leqq b \leqq t)$. Then the weights of $\mathfrak{a}_b$ in $\mathfrak{g}$, for the adjoint representation are 0, $\pm\nu_b$, and possibly $\pm 2\cdot\nu_b$. Let p_b (resp. q_b) be the multiplicity of ν_b (resp. $2\cdot\nu_b$):*

(i) *if $_{\mathbf{R}}\Phi$ is of type C_t and $b = t$, then $p_b = t + u\cdot\binom{t}{2}$, $q_b = 0$.*

(ii) *if $_{\mathbf{R}}\Phi$ is of type C_t and $b \neq t$, then $p_b = 2\cdot u\cdot b\cdot(t - b)$, $q_b = b + u\cdot\binom{b}{2}$.*

(iii) *if $_{\mathbf{R}}\Phi$ is of type BC_t, then $p_b = v\cdot b + 2\cdot u\cdot b\cdot(t - b)$, $q_b = b + u\binom{b}{2}$.*

The $\mathbf{R}$-roots are linear combinations of the simple ones with coefficients 0, ± 1, ± 2. Since $\mathfrak{a}_b$ annihilates all the simple $\mathbf{R}$-roots except α_b, this proves the first assertion.

We have $\nu_b = \gamma_t$ in the case (i), and $\nu_b = \gamma_b/2$ in the other cases. In the case (i), p_b is the sum of the multiplicities of the $\mathbf{R}$-roots $\gamma_i\ (i \leqq t)$, which are all equal to one, and of the $\mathbf{R}$-roots $(\gamma_i + \gamma_j)/2\ (1 \leqq i < j \leqq t)$, while $q_b = 0$. In cases (ii) and (iii), q_b is the sum of the multiplicities of the $\mathbf{R}$-roots $\gamma_i\ (i \leqq b)$ and $(\gamma_i + \gamma_j)/2\ (1 \leqq i < j \leqq b)$. In case (ii), p_b is the sum of the multiplicities of the roots $(\gamma_i \pm \gamma_j)/2\ (1 \leqq i \leqq b < j \leqq t)$, and in case (iii), we have to add also the multiplicities of the roots $\gamma_i/2\ (1 \leqq i \leqq b)$, whence the lemma.

In the next proposition, the important point is not the explicit value of

n_b, but rather the fact that it is > 0 and completely determined by p_b and q_b. This will play an important role in our discussion of Eisenstein series.

1.16. PROPOSITION. *Let η_b be the restriction to Z_b of the functional determinant J_b, and let $\chi_b(g) = \det \mathrm{Ad}_{v_b} g \ (g \in P_b)$. Then $\eta_b(g) = \chi_b(g)^{-n_b}$ if $g \in A_b \cdot V_b$, and $|\eta_b(g)| = |\chi_b(g)|^{-n_b}$ if $g \in Z_b$, where $n_b = 1$ in case* (i) *of* 1.15, *and $n_b = (p_b + 4q_b) \cdot (2p_b + 4q_b)^{-1}$ in the cases* (ii), (iii) *of* 1.15.

By 1.7 (1'), we have

$$
(1) \qquad
\begin{aligned}
J_b(x, g) &= \eta_b(g) \\
&= \det (\mathrm{Ad}_{\mathfrak{p}^+} c_b^{-1} \cdot g^{-1} \cdot c_b) = \Psi_b(c_b^{-1} \cdot g \cdot c_b) \qquad (x \in S_b; g \in Z_b) \, .
\end{aligned}
$$

Both χ_b and η_b are rational characters of Z_b. They are therefore equal to one on V_b and on the derived group of Z_b. On the compact subgroup $K \cap Z_b$ they are both of modulus one. Since Z_b is generated by its intersection with K, a semi-simple subgroup L_b', its unipotent radical V_b, and A_b (see 1.3), it remains to check 1.16 on A_b.

The group A_b belongs to the center of the maximal reductive subgroup $Z(A_b)$ of the parabolic subgroup P_b; hence, the weights of A_b in $\mathfrak{v}_b$ are the restrictions of the positive **R**-roots which are not equal to one on A_b and therefore

$$
(2) \qquad \chi_b(g) = \det \mathrm{Ad}_{v_b} g = \nu_b(g)^{p_b + 2q_b} \qquad (g \in A_b) \, .
$$

We may write $a \in A_b$ in the form $a = \exp \lambda(X_1 + \cdots + X_b)$. Therefore (1.6) we have $c_b^{-1} \cdot a \cdot c_b = \exp \lambda(H_1 + \cdots + H_b)$. Let ν_b' be the image of ν_b under Int c_b^{-1}. Its value on $c_b^{-1} \cdot a \cdot c_b$ is again equal to the restriction of γ_b in case (i), of $\gamma_b/2$ in cases (ii), (iii), where γ_i are now coordinates in $\mathfrak{h}_t$ with respect to the base $(H_i/2)$. By definition $\eta_b(a^{-1})$ is equal to the product of the values on $c_b^{-1} \cdot a \cdot c_b$ of the roots $\mu \in \pi^+$. Therefore $\eta_b(a^{-1}) = \nu_b'(a)^{r_b}$ where the exponent r_b is in case (i),

the number of elements of π^+, which is equal to p_b;
in case (ii),

the number of elements of π^+ restricting to one of $(\gamma_i + \gamma_j)/2$, $(1 \leq i \leq b < j \leq t)$, plus twice the number of elements in π^+ restricting to one of $\gamma_i \ (1 \leq i \leq b)$, or of $(\gamma_i + \gamma_j)/2 \ (1 \leq i < j \leq c)$, which gives

$$
(3) \qquad r_b = u \cdot b(t - b) + 2b + u \cdot b(b - 1) = p_b/2 + 2q_b \, ;
$$

in case (iii),

it is the same as in case (ii) plus the number of $\mu \in \pi^+$ restricting to one of $\gamma_i/2 \ (1 \leq i \leq b)$. According to Lemma 14 in [22], this last number is *half* the multiplicity of the **R**-root $\gamma_i/2$, which gives

$$(4) \qquad r_b = u \cdot b(t - b) + 2b + u \cdot b(b - 1) + v \cdot b/2 = p_b/2 + 2q_b \,,$$

and our assertion follows from (2), (3) and (4).

1.17. PROPOSITION. *We keep the notation of* 1.11, 1.16, *and let* $q_b = m_0/m_b$. *Then*

$$| J_b(x, g) | = | j_b(\sigma_b(x), g) |^{q_b} \cdot | \chi_b(g) |^{-n_b} , \qquad (x \in S_b; \ g \in N(F_b)) \,.$$

In view of 1.3, we may write $g = l \cdot z$ $(l \in L(F_b), z \in Z(F_b))$. By the cocycle formula

$$J_b(x, g) = J_b(x, l) \cdot J_b(x \cdot l, z) \,.$$

Since z acts trivially on F_b, we have

$$j_b(y, l) = j_b(y, l \cdot z) \,, \qquad (y \in F_b) \,.$$

The proposition follows then from 1.11, 1.16.

1.18. PROPOSITION. *Let* $1 \leqq b < d \leqq t$, *and* $\nu_{b,d} = \nu_d \circ \nu_b^{-1}\colon S_b \to S_d$, *where* $\nu_b\colon X \to S_b$ *is as in* 1.6. *Then*

(i) $J_b(x, g) = J_d(\nu_{b,d}(x), g)$, $(g \in Z(F_b)^0; x \in S_b)$.

(ii) *the functional determinant* $j(x, \nu_{b,d})$ *of* $\nu_{b,d}$ *is constant along the fibres of the canonical projection* $\sigma_b\colon S_b \to F_b$.

The group $Z(F_b)^0$ is the semi-direct product of a reductive group R_b with Lie algebra $\mathfrak{m}_b + \mathfrak{a}_b + \mathfrak{l}_b'$, in the notation of 1.3, by the unipotent radical V_b of $P_b = N(F_b)$, and it is contained in $Z(F_d)$. Both $J_b(x,\ )$ and $J_d(x,\ )$ are equal to one if $z \in V_b$ by 1.11. Therefore, it suffices to prove (i) when $g \in R_b$. By the definition of the Cayley transform, $c_b^{-1} \cdot c_d \in L_{b,\mathbb{C}}$; therefore, $c_b^{-1} \cdot c_d$ centralizes R_b, and (i) follows from 1.7 (1').

By the composition rule for functional determinants, we have

$$j(x, \nu_{b,d}) \cdot J_d(\nu_{b,d}(x), g) = J_b(x, g) \cdot j(x \cdot g, \nu_{b,d}) \qquad (x \in S_b, g \in G_{\mathbb{R}}^0) \,;$$

therefore, (ii) follows from (i) and from the transitivity of $Z(F_b)^0$ on the fibres of σ_b, (1.7).

We end this section with a result which will be used in discussing rational boundary components. The following lemma will be needed.

1.19. LEMMA. *Let* X *be irreducible. Then*

(i) $[x, \mathfrak{g}_{(\gamma_i \mp \gamma_j)/2}] \neq \{0\}$ $(x \in \mathfrak{g}_{(\gamma_i \pm \gamma_j)/2} - \{0\}; 1 \leqq i < j \leqq t)$,

(ii) *if* $_{\mathbb{R}}\Phi$ *is of type* BC_t, $[x, \mathfrak{g}_{\gamma_i/2}] \neq \{0\}$ $(x \in \mathfrak{g}_{\gamma_i/2} - \{0\}; 1 \leqq i \leqq t)$.

PROOF OF (i). By Lemmas 13, 15 of [22], the roots $\mu \in \Phi$ restricting on $\mathfrak{h}_t$ to $(-\gamma_i + \gamma_j)/2$ $(i \neq j)$ are compact, those restricting to $(\gamma_i + \gamma_j)/2$ are in π^+, and $\mu \mapsto \mu + \gamma_i$ is a bijective map of the first set C_{ij} onto the second one P_{ij}. Let $\mathfrak{c}_{ij}$ (resp. $\mathfrak{p}_{ij}$) be the $\mathbb{C}$-subspace of $\mathfrak{g}_{\mathbb{C}}$ spanned by the vectors E_μ, $(\mu \in C_{ij},$

resp. $\mu \in P_{ij}$). With c_t being as in 1.6, we have

$$(1) \qquad \operatorname{Ad} c_t(\mathfrak{c}_{ij}) \cap \mathfrak{g} = \mathfrak{g}_{(\gamma_j - \gamma_i)/2}, \qquad \operatorname{Ad} c_t(\mathfrak{p}_{ij}) \cap \mathfrak{g} = \mathfrak{g}_{(\gamma_i + \gamma_j)/2}.$$

Furthermore, $\mathfrak{g}_{\gamma_i, \mathbf{C}}$ is one-dimensional and spanned by $\operatorname{Ad} c_t(E_{\gamma_i})$. From the result just quoted, and the standard fact $[E_\mu, E_\nu] \neq 0$ if $\mu + \nu$ is a root, we get then

$$(2) \qquad [\mathfrak{g}_{\gamma_i}, \mathfrak{g}_{(-\gamma_i \pm \gamma_j)/2}] = \mathfrak{g}_{(\gamma_i \pm \gamma_j)/2},$$

in particular, we may write

$$x = [y, u] \qquad\qquad (y \in \mathfrak{g}_{(-\gamma_i \pm \gamma_j)/2}, \, u \in \mathfrak{g}_{\gamma_i}).$$

It is well-known that, given $v \in \mathfrak{g}_\alpha$ ($\alpha \in {}_{\mathbf{R}}\Phi$), there exists $v' \in \mathfrak{g}_{-\alpha}$ such that $[v, v']$ is a non-zero multiple of the element $h_\alpha \in \mathfrak{a}$ such that $\beta(h_\alpha) = (\beta, \alpha)$ ($\beta \in {}_{\mathbf{R}}\Phi$). (One may take, for instance, for v' the transform of v under a suitable Cartan involution.) There exists therefore $z \in \mathfrak{g}_{(\gamma_i \mp \gamma_j)/2}$ such that $[z, y] = c \cdot h_{(\gamma_i \mp \gamma_j)/2}$ ($c \neq 0$). We have then

$$[[z, y], u] = c \cdot \gamma_i(h_{(\gamma_i \mp \gamma_j)/2}) \cdot u \neq 0.$$

Since $[z, u] = 0$, because $3\gamma_i/2 \pm \gamma_j/2$ is not an $\mathbf{R}$-root, the Jacobi identity shows that $[x, z] = [[y, u], z] \neq 0$, which proves (i).

PROOF OF (ii). Let C_i' (resp. P_i) be the set of compact (resp. non-compact) roots which restrict on $\mathfrak{h}_t$ to $\gamma_i/2$ and $\mathfrak{c}_i'$ (resp. $\mathfrak{p}_i$) the space spanned by the vectors E_μ ($\mu \in C_i'$, resp. $\mu \in P_i$). We have

$$\operatorname{Ad} c_t(\mathfrak{c}_i' + \mathfrak{p}_i) \cap \mathfrak{g} = \mathfrak{g}_{\gamma_i/2},$$

therefore (ii) is equivalent to

$$(3) \qquad \mathfrak{z}(\mathfrak{c}_i' + \mathfrak{p}_i) \cap (\mathfrak{c}_i' + \mathfrak{p}_i) = \{0\}.$$

By Lemma 14 of [22], the map $\alpha \mapsto \alpha + \gamma_i$ is a bijection of $C_i = -C_i'$ onto P_i. Hence, given $\mu \in C_i'$ (resp. $\mu \in P_i$), there exists $\nu \in P_i$ (resp. $\nu \in C_i'$) such that $\mu + \nu = \gamma_i$. Then $[E_\mu, E_\nu] \neq 0$. Since the left hand side of (3), being stable under $\mathfrak{h}_{\mathbf{C}}$, is spanned by root vectors, this proves (3).

1.20. PROPOSITION. *Let X be irreducible. Let W_b be the center of V_b ($1 \leq b \leq t$), and C be the connected centralizer of W_b in P_b. Then $\mathfrak{c}$ is the direct sum of $\mathfrak{l}_b + \mathfrak{v}_b$, which is an ideal of $\mathfrak{c}$, and of its intersection with $\mathfrak{m} = \mathfrak{z}(\mathfrak{a}) \cap \mathfrak{k}$. In particular, $C/(L_b \cdot V_b)$ is compact.*

The ideal $\mathfrak{v}_b$ is the sum of the root spaces $\mathfrak{g}_\alpha$ where α runs through the $\mathbf{R}$-roots of the form

$$(\gamma \pm \gamma_j)/2 \, (1 \leq i \leq b < j \leq t), \qquad (\gamma_i + \gamma_j)/2 \, (1 \leq i < j \leq b),$$
$$\gamma_i \, (1 \leq i \leq b),$$

together with $\gamma_i/2$ $(1 \leq i \leq b)$ in the case BC_t. We want to prove that $\mathfrak{w}_b$ is the sum of the root spaces $\mathfrak{g}_\alpha$, where α runs through the roots

$$(\gamma_i + \gamma_j)/2 , \qquad\qquad (1 \leq i \leq j \leq b) .$$

The relation $[\mathfrak{g}_\alpha, \mathfrak{g}_\beta] \subset \mathfrak{g}_{\alpha+\beta}$, and the structure of $_R\Phi$, show that $\mathfrak{w}_b$ contains the root spaces just listed. Of course $\mathfrak{w}_b$ is stable under $\mathfrak{a}$, hence is the sum of its intersections with the $\mathfrak{g}_\alpha$. In order to prove our assertion, it is therefore enough to show that

$$\mathfrak{w}_b \cap \mathfrak{g}_{(\gamma_i \pm \gamma_j)/2} = 0 \qquad\qquad (1 \leq i \leq b < j \leq t) ,$$

and that

$$\mathfrak{w}_b \cap \mathfrak{g}_{\gamma_i/2} = \{0\} , \qquad\qquad (1 \leq i \leq b) ,$$

if $_R\Phi$ is of type BC_t, but this follows from 1.19.

The Lie algebra $\mathfrak{c}$ is also stable under $\mathfrak{a}$. It is obvious that it contains the $\mathfrak{g}_\alpha \subset \mathfrak{l}_b$ and that $\mathfrak{c} \cap \mathfrak{a} \cap \mathfrak{z}_b = 0$. Lemma 1.19 shows moreover that, if $\mathfrak{g}_\alpha \subset \mathfrak{l}'_b$, then $\mathfrak{g}_\alpha \cap \mathfrak{c} = 0$. The proposition follows then from the facts about $\mathfrak{p}_b$ and $\mathfrak{z}_b$ recalled in 1.3.

REMARK. Proposition 1.20 was suggested by a statement in [31, § 3.3] which becomes essentially equivalent to 1.20, if the word *normalizer* there is replaced by *centralizer*.

2. Relative root systems

For most of the facts recalled below, we refer to [14]. As was already pointed out in 0.3, the ground fields may be assumed to be contained in **C**, which is then our *universal field*, although the results of 2.1, 2.2 are valid in greater generality.

2.1. *Relative roots.* Let G be a connected reductive k-group. Its maximal k-split tori are conjugate over k and their common dimension is the k-rank $\mathrm{rk}_k(G)$ of G. Let S be a maximal k-split torus of G. The k-roots, or *roots relative to k*, or *restricted roots* are the non-trivial characters of S in the adjoint representation of G, and the relative Weyl group $_kW = {}_kW(G)$ is the quotient $N(S)/Z(S)$. We denote by $_k\Phi$ or $_k\Phi(G)$ the set $\Phi(S, G)$ of k-roots. It is a *root system* in $X^*(T) \otimes \mathbf{R} = V$. This means in particular that, with respect to a scalar product $(\ ,\ )$ on V invariant under $_kW$, the group $_kW$ is generated by the reflections in the hyperplanes orthogonal to the k-roots, leaves $_k\Phi$ stable, and that $2(\alpha, \beta) \cdot (\beta, \beta)^{-1} \in \mathbf{Z}$ for all $\alpha, \beta \in {}_k\Phi$. For every $\alpha \in {}_k\Phi$ we put

$$\mathfrak{g}_\alpha = \{x \in \mathfrak{g} \mid \mathrm{Ad}\ s(x) = s^\alpha \cdot x (s \in S)\} .$$

Then $\mathfrak{g}$ is the direct sum of the $\mathfrak{g}_\alpha$ ($\alpha \in {}_k\Phi$) and of the Lie algebra $\mathfrak{z}(S)$ of the centralizer of S.

Given an ordering in $X^*(S)$, we denote by ${}_k\Delta$ the set of simple k-roots. A subset of ${}_k\Delta$ is *connected* if it is not the union of two non-empty disjoint subsets which are mutually orthogonal.

2.2. *Parabolic k-subgroups.* An algebraic subgroup P of G is *parabolic* if the quotient space G/P is a projective variety. P is then connected, equal to its normalizer, and is the normalizer of its unipotent radical.

Let U be the subgroup normalized by S whose Lie algebra $\mathfrak{u}$ is the sum of the $\mathfrak{g}_\alpha$, where α runs through the positive k-roots (for some fixed ordering). Then U is a unipotent k-subgroup, normalized by $Z(S)$.

For every subset θ of ${}_k\Delta$, let $S_\theta = (\bigcap_{\alpha \in \theta} \ker \alpha)^0$. We let ${}_kP_\theta$ be the subgroup generated by $Z(S_\theta)$ and U; it is the semi-direct product of $Z(S_\theta)$ and of its unipotent radical $U_\theta \subset U$. The split radical (0.5) of ${}_kP_\theta$ is the semi-direct product $S_\theta \cdot U_\theta$. Every parabolic k-subgroup of G is conjugate over k to one and only one ${}_kP_\theta$. Moreover, two parabolic k-subgroups are conjugate in G if and only if they are conjugate over k. The groups ${}_kP_\theta$ are the *standard parabolic k-subgroups* (for a given choice of S and U). If $\theta = \varnothing$, then ${}_kP_\theta = {}_kP = Z(S) \cdot U$ is the minimal standard parabolic k-subgroup. We can write uniquely

$$Z(S) = M \cdot S \qquad (M \text{ normal } k\text{-subgroup, } M \cap S \text{ finite}),$$

and M is anisotropic over k, i.e., $\mathrm{rk}_k(M) = 0$.

For $\theta \subset {}_k\Delta$, we let $[\theta]$ be the set of k-roots which are linear combinations of elements in θ, and let ${}_kL_\theta$ be the smallest connected k-subgroup normalized by $Z(S)$ whose Lie algebra ${}_k\mathfrak{l}_\theta$ contains the subspaces $\mathfrak{g}_\alpha$ ($\alpha \in [\theta]$). It is easily seen that $[\theta] = {}_k\Phi({}_kL_\theta)$, that $S \cap {}_kL_\theta$ is a maximal k-split torus of ${}_kL_\theta$, and that ${}_kL_\theta$ is semi-simple with Lie algebra

$${}_k\mathfrak{l}_\theta = \sum_{\alpha \in [\theta]} \mathfrak{g}_\alpha + [\mathfrak{g}_\alpha, \mathfrak{g}_{-\alpha}] \ .$$

Moreover, we have ${}_kP_\theta = M_\theta \cdot {}_kL_\theta \cdot S_\theta \cdot U_\theta$, where M_θ is the identity component of $M \cap Z({}_kL_\theta)$. *If θ is connected, then ${}_kL_\theta$ is almost k-simple*, since otherwise, by [14, 5.11, 8.5] ${}_kL_\theta$ would be the almost direct product of a k-*group* without k-rational unipotent elements $\neq e$, and of a k-group containing all unipotent elements rational over k of ${}_kL_\theta$, and ${}_kL_\theta$ could not be generated by unipotent k-subgroups.

We shall sometimes denote by ${}_kT$ a maximal k-split torus and by ${}_kU$ the unipotent radical of a minimal parabolic k-subgroup.

We recall finally the Bruhat decomposition $G_k = P_k \cdot N(S)_k \cdot P_k$ [14, 5.15].

More precisely, let n_w be a representative in $N(S)_k$ of $w \in {}_kW$. Then G_k is the disjoint union of the double classes

$$P_k \cdot n_w \cdot P_k = U_k \cdot n_w \cdot Z(S)_k \cdot U_k \ .$$

If G is connected, but not reductive, it is the semi-direct product of its unipotent radical $R_u(G)$ by a reductive k-subgroup (we are in characteristic zero), and $R_u(G)$ is contained in every parabolic subgroup. Since $N(S) \cap R_u(G) = Z(S) \cap R_u(G)$, it follows that the above decomposition is still valid, with P a minimal parabolic k-subgroup, and $N(S)$ being the normalizer of S either in G or in a maximal reductive k-subgroup containing S.

2.3. *Fundamental highest weights relative to k.* There is a basis of $X(S) \otimes \mathbf{Q}$ over $\mathbf{Q}$ consisting of elements $d_\alpha \in X(P)(\alpha \in {}_k\Delta)$ such that $(d_\alpha, \beta) = c_\alpha \delta_{\alpha\beta}(\alpha, \beta \in {}_k\Delta)$ where c_α are positive integers. The restriction to S_θ of the elements $d_\alpha(\alpha \in \theta' = {}_\mathbf{Q}\Delta - \theta)$ form a basis of $X(S_\theta) \otimes \mathbf{Q}$. Let $d \in X(S)$ be a linear combination of elements $d_\alpha(\alpha \in \theta')$ with strictly positive integral coefficients. Then there exists an absolutely irreducible representation $\rho \colon G \to \mathbf{GL}(V)$ defined over k, and a unique one-dimensional subspace $V' \subset V$ stable under P_θ and on which $g \in P_\theta$ acts *via* multiplication by $d(g)$ [14, 12.2, 12.13]. The characters $d_\alpha(\alpha \in \theta')$ will be called *fundamental highest weights* for P_θ. Let

$$\chi_0(p) = \det(\mathrm{Ad}_{\mathfrak{u}_\theta} p) \qquad\qquad (p \in P_\theta) \ .$$

We have then

$$\chi_0 = \sum_{\alpha \in \theta'} e_\alpha d_\alpha \qquad\qquad (e_\alpha \in \mathbf{Q}; e_\alpha > 0) \ .$$

In fact, by definition, χ_0 is the sum of the weights of S in $\mathfrak{u}_\theta$. These are the positive roots which involve at least one of the elements of θ', each root being of course counted with its multiplicity. χ_0 is stable under $N(S) \cap P_\theta$, hence under the fundamental symmetries $s_\beta(\beta \in \theta)$; therefore it is orthogonal to θ, and is a linear combination of the elements $d_\alpha(\alpha \in \theta')$. The coefficient e_α of d_α is equal to $(\chi_0, \alpha) \cdot c_\alpha^{-1}$. Let χ_1 be the sum of the positive elements in $[\theta]$, counted with their multiplicities, and $\chi = \chi_0 + \chi_1$. Then $(\chi_1, \alpha) \leqq 0$ for $\alpha \in \theta'$ and, by a standard argument $(\chi, \alpha) > 0$, (in fact it is equal to $(c + 2d) \cdot (\alpha, \alpha)$, where c and d are the multiplicities of α and 2α), hence $(\chi_0, \alpha) > 0$ and $e_\alpha > 0$.

2.4. *Restriction of relative roots.* Let K be an overfield of k, T a maximal K-split torus of G containing S and $r \colon X(T) \to X(S)$ the restriction homomorphism. Two orderings of $X(T)$ and $X(S)$ are *compatible* if $\alpha > 0$, $r(\alpha) \neq 0$ imply $r(\alpha) > 0$ $(\alpha \in X(T))$.

The existence of an ordering on $X(T)$ compatible with a given ordering on $X(S)$ is immediate [14, 3.1]. Let ${}_K\Delta$ and ${}_k\Delta$ be the sets of simple roots for compatible orderings. Then we have

(i) $_k\Delta \subset r(_K\Delta) \subset {}_k\Delta \cup \{0\}$.

(ii) *Let* $\theta \subset {}_k\Delta$, $\psi \subset {}_K\Delta$ *and* $\alpha \in {}_K\Delta \cap r^{-1}(\theta)$. *If* ψ *is connected, then* $r(\psi) \cap {}_k\Delta$ *is connected. If* θ *is connected, there exists a connected subset* θ' *of* $_K\Delta$ *containing* α *such that*

$$\theta \subset r(\theta') \subset \theta \cup \{0\} \ .$$

For the proofs, see [14, 6.8, 6.15, 6.16]. A simple K-root will be said to be *critical* if it restricts onto a simple k-root. Thus a simple K-root either is critical or restricts to zero.

(iii) *Let* θ *be a connected subset of* $_k\Delta$, *and assume there exists a unique greatest connected subset* ψ *of* $_K\Delta$ *such that* $r(\psi) \cap {}_k\Delta = \theta$. *Then* $_kL_\theta = {}_KL_\psi$. *In particular,* $_KL_\psi$ *is defined over* k.

PROOF. Let $\beta \in {}_k\Phi$. Then the space $\mathfrak{g}_\beta$ is the direct sum of the eigenspaces $\mathfrak{g}_\alpha$ of T, where α runs through the K-roots whose restriction to S is equal to β. It is a standard fact about root systems that, if a root α is expressed as linear combination of simple roots, then the set of simple roots which occur in α with a non-zero coefficient is connected. Therefore if $\beta \in [\theta]$ and $\alpha \in {}_K\Phi$ restricts to β, then $\alpha \in [\psi]$. This implies that $_kL_\theta \subset {}_KL_\psi$. Moreover, $Z(S)$ normalizes $_kL_\theta$, and it is clear from the definitions that $_KL_\psi \subset Z(S) \cdot {}_kL_\theta$. Consequently, $_kL_\theta$ is a normal subgroup of $_KL_\psi$. However the latter is almost K-simple (2.2), whence our assertion.

2.5. PROPOSITION. *Let* k *be an algebraic number field,* k_v *its completion with respect to an archimedean valuation* v, *and* G *a connected reductive* k-*group. Then every maximal torus defined over* k_v *of* G *is conjugate over* k_v *to a maximal torus defined over* k.

(i) We show first that, if L is a connected k-group, then L_k is dense in $(L_{k_v})^0$ in the usual topology. By [32, p. 41], there exists a generically surjective rational map of an affine space into L which is defined over k. In other words, we may find a Zariski k-open subset U of an affine space, and a k-morphism $f: U \to L$ whose image contains a non-empty Zariski k-open subset V of L. Since we are in characteristic zero, f is separable, and there exists a non-empty Zariski k-open subset U' of U such that $f: U'_{k_v} \to L_{k_v}$ is open. Of course $f(U'_k) \subset L_k$ and U'_k is dense in U'_{k_v}. Thus, L_k is dense in a non-empty open subset of L_{k_v}, hence in $(L_{k_v})^0$.

(ii) If $k_v = \mathbf{C}$, then all maximal tori of G_{k_v} are conjugate, and 2.5 amounts to the existence of a maximal torus defined over k, which is known [32, p. 45]. Assume now that $k_v = \mathbf{R}$. Let T be a maximal torus of G defined over $\mathbf{R}$ and H' the set of regular elements of $(T_\mathbf{R})^0$. It is well known that the set C of conjugates of elements of H' by elements of $(G_{\mathbf{R}})^0$ is an open subset of $(G_{\mathbf{R}})^0$. By (i), it contains an element x rational over k. Then $Z(x)^0$ is the desired maximal torus.

2.6. COROLLARY. *Let k be a subfield of $\mathbf{R}$. Then G has a maximal torus defined over k which contains a maximal $\mathbf{R}$-split torus of G.*

2.7. REMARKS. (1) Proposition 2.5 is also valid for an arbitrary connected k-group G. In fact, let U be the unipotent radical of G and $\pi: G \to G' = G/U$ the canonical projection. Let T be a maximal torus defined over k_v of G. Then $\pi(T)$ is a maximal torus of G' [9, § 22], obviously defined over k_v. By 2.5, it is conjugate over k_v to a maximal torus T' defined over k of G'. Since U is unipotent, and k is perfect, the map $G_{k_v} \to G'_{k_v}$ is surjective by Rosenlicht's cross-section theorem, whence the existence of $x \in G_{k_v}$ such that $^x T \subset \pi^{-1}(T') = Q$. The group Q is a connected solvable k-group, hence its maximal tori defined over k_v, are conjugate over k_v and one of them is defined over k (see e.g. [14, 11.4]; or, in characteristic zero, Borel-Mostow, Annals of Math. 61 (1955), 389–405).

Proposition 2.5 is then of course also true if maximal tori are replaced by Cartan subgroups, since the latter are the centralizers of the former.

(2) Although this will not be needed in this paper, we point out that, if G is a connected k-group, G_k is dense in G_{k_v}, not only in $(G_{k_v})^0$, as was shown in (i). If $k_v = \mathbf{C}$, then G_{k_v} is connected, and there is nothing new to prove. If $k_v = \mathbf{R}$, there remains to show that G_k meets every connected component of $G_{\mathbf{R}}$. By 2.6, applied to a maximal connected reductive k-subgroup of G, (or by remark (1) above), there exists a maximal torus T of G defined over k and containing a maximal $\mathbf{R}$-split torus of G. By [14, 14.4], each connected component of $G_{\mathbf{R}}$ contains one of $T_{\mathbf{R}}$, so that we are reduced to the case of a torus, where our assertion follows from a result of Serre quoted in [25, 5.1].

In the terminology of [25], this means essentially that G has the weak approximation property for archimedean valuations. As a matter of fact, it has been checked here only for one such valuation, but the case of several is easily reduced to that of one by considering the group $R_{k/\mathbf{Q}}G$.

2.8. Let k be a subfield of $\mathbf{R}$ and G a connected semi-simple and absolutely simple k-group. Let $_{\mathbf{R}}T$ be a maximal $\mathbf{R}$-split torus of G containing a maximal k-split torus $_k T$. We let $r: X(_{\mathbf{R}}T) \to X(_k T)$ be the restriction map, endow $X(_{\mathbf{R}}T)$ and $X(_k T)$ with compatible orderings, and denote by $_k\Delta$, $_{\mathbf{R}}\Delta$ the corresponding sets of simple relative roots.

We assume further that the riemannian symmetric space $K\backslash G_{\mathbf{R}}$, where K is a maximal compact subgroup of $G_{\mathbf{R}}$, is a bounded symmetric domain, necessarily irreducible since $G_{\mathbf{R}}$ is simple. Then $_{\mathbf{R}}\Phi$ is either of type C_t or of type BC_t (1.2). In fact, the following proposition, and its proof are valid under that last assumption. On $_R\Phi = \{\alpha_1, \cdots, \alpha_t\}$, we use the canonical numbering of 1.2. For each $\beta \in {}_k\Delta$, let $m(\beta)$ be the greatest value of the index i such that $r(\alpha_i) = \beta$. We number the elements $\beta_1, \cdots, \beta_s$ of $_k\Delta$ in such a way that $i < j$ if and only if $m(\beta_i) < m(\beta_j)$, and then write $m(j)$ for $m(\beta_j)$.

2.9. PROPOSITION. *We keep the notation and assumptions of 2.8. We assume that* $\dim {}_k T > 0$. *Then*

(a) $_k\Phi$ *is of type* BC_s *if either* $_R\Phi$ *is of type* BC_t *or* $_R\Phi$ *is of type* C_t *and* $r(\alpha_t) = 0$, *and is of type* C_s *otherwise. The numbering of* $_k\Delta$ *defined in* 2.8 *is the canonical one.*

(b) *Each* $\beta \in {}_k\Delta$ *is the restriction of one and only one simple* **R**-*root.*

By our choice of the numberings, any final segment $(\beta_i, \cdots, \beta_s)$ in $_k\Delta$ (possibly with zero added), is the restriction of a final segment of $_R\Delta$, hence is connected (2.4). Conversely, any connected subset θ of $_k\Delta$ containing β_a, β_b $(a < b)$ contains β_i for every i between a and b. In fact, there exists by 2.4 (ii) a connected subset θ' of $_R\Delta$ containing $\alpha_{m(b)}$ such that $\theta \subset r(\theta') \subset \theta \cup \{0\}$. The set θ' contains then at least one simple **R**-root α_c with $c \leq m(a)$. In view of the structure of $_R\Delta$, the set θ' must then contain all simple **R**-roots with index i between c and $m(b)$, hence in particular all **R**-roots $\alpha_{m(j)}$ $(a \leq j \leq b)$, whence our contention.

This shows that the graph of $_k\Delta$ is a chain (no branch point). $_k\Delta$ is therefore of one of the types $A_s, B_s, C_s, G_2, F_4, BC_s$, where the first five symbols refer to the standard Cartan-Killing classification. We now distinguish some cases.

(i) $_R\Phi$ *is of type* BC_t. In this case, the set of $\alpha \in {}_R\Delta$ whose double is a root spans $X(_R T) \otimes \mathbf{R}$, hence contains at least one element γ whose restriction is not zero. Thus $r(\gamma)$ and $2 \cdot r(\gamma)$ are k-roots, and $_k\Delta$ is of type BC_s. Moreover, the highest root in $_R\Delta$ is $\gamma_1 = 2(\alpha_1 + \cdots + \alpha_t)$; therefore, if $\beta \in {}_k\Delta$ is the restriction of at least two simple **R**-roots, then β has a coefficient ≥ 4 in the highest k-root, which is impossible in type BC_s. Also, $\alpha_{m(s)} + \cdots + \alpha_t = \gamma_{m(s)}/2$ and its double are roots, hence β_s and $2\beta_s$ are k-roots; and, by the above, the connected subsets of $_k\Delta$ containing β_s are the final segments. This shows that our numbering is the canonical one, and ends the proof of the proposition in this case.

From now on, $_R\Delta$ is of type C_t. Its highest root is then

$$2(\alpha_1 + \cdots + \alpha_{t-1}) + \alpha_t \, .$$

Let c_i denote the number of simple **R**-root restricting onto β_i.

(ii) Assume that $r(\alpha_t) \neq 0$. Then the highest k-root is

$$\delta = 2c_1 \cdot \beta_1 + \cdots + 2 \cdot c_{s-1} \cdot \beta_{s-1} + (2(c_s - 1) + 1)\beta_s \, .$$

If $c_i \geqq 2$ for some $i < s$, then there is a coefficient $\geqq 4$ in δ, and $_k\Phi$ is of type F_4. There is in this case in δ a coefficient 3, which must then be the coefficient of β_s. However, in F_4 the simple root with coefficient 3 in the highest root is not an end point of the graph. Therefore $c_i = 1$ for $i < s$.

Assume now that $\beta_s = r(\alpha_t) = r(\alpha_j)$ for some $j < t$. (There is at most one such j since no simple root has a coefficient $\geqq 5$ in δ in the systems under consideration here). Then β_s has a coefficient $\geqq 3$ and all other simple k-roots have coefficient 2. This occurs only in G_2. If $m(1) < j$, then

$$3 \cdot \beta_2 = r\big(2(\alpha_j + \cdots + \alpha_{t-1}) + \alpha_t\big)$$

is a k-root, which is absurd. If $j < m(1)$, then

$$r\left(\frac{1}{2}(\gamma_{m(t)} + \gamma_j)\right) = 2(\beta_1 + \beta_2)$$

is a k-root, which is absurd because $\beta_1 + \beta_2$ is a k-root, and G_2 has no root whose double is a root. This proves that $c_i = 1 \ (1 \leqq i \leqq s)$, hence that

$$\delta = 2(\beta_1 + \cdots + \beta_{s-1}) + \beta_s \, .$$

Therefore $_k\Phi$ is of type C_t, and the numbering is the canonical one.

(iii) Assume that $r(\alpha_t) = 0$. In this case the highest k-root is

$$\delta = 2 \cdot c_1 \cdot \beta_1 + \cdots + 2 \cdot c_s \cdot \beta_s \, .$$

By the classification of root systems, this implies that $_k\Phi$ is of type BC_s and that $c_i = 1 \ (1 \leqq i \leqq s)$. Furthermore, $2\beta_s = r(\gamma_{m(s)})$ is a k-root, so that again the numbering is the canonical one. This completes the proof of the proposition.

2.10. COROLLARY. (a) *The proper maximal parabolic k-subgroups of G are also proper maximal among parabolic* **R**-*subgroups.*

(b) *Let ψ be an initial (resp. a final) segment of $_R\Delta$ consisting of all roots which come before (resp. after) a critical root and $\theta = r(\psi) \cap {}_k\Delta$. Then $_RL_\psi = {}_kL_\theta$ and is defined over k.*

PROOF OF (a). Let P be a proper maximal parabolic k-subgroup of G. It follows from 2.2 that there exists a simple k-root β such that P is conjugate over k to the group $_kP_\theta \ (\theta = {}_k\Delta - \{\beta\})$. Let α be the unique critical simple **R**-root which restricts onto β and $\psi = {}_R\Delta - \{\alpha\}$. Then $T_\psi = S_\theta$ by 2.9. Furthermore, since the given orderings are compatible, we have $_RU \subset Z(_kT) \cdot {}_kU$, and therefore $_RP_\psi \subset {}_kP_\theta$. Since $_RP_\psi$ is a proper maximal parabolic **R**-group, we have $_RP_\psi = {}_kP_\theta$, which proves (a).

PROOF OF (b). By 2.9, θ is an initial or final segment of $_k\Delta$, and ψ is the greatest connected subset of $_K\Delta$ such that $\theta = {}_k\Delta \cap r(\psi)$. Therefore (b) follows from 2.4 (iii).

3. Rational boundary components

3.1. Let G be an algebraic group defined over $\mathbf{Q}$. A subgroup Γ of $G_{\mathbf{Q}}$ is *arithmetic* if for one (and hence for every [13, 6.3]) faithful $\mathbf{Q}$-morphism $\rho: G \to \mathbf{GL}_m$, the group $\rho(\Gamma)$ is commensurable with $\rho(G)_{\mathbf{Z}}$.

We recall that, if $f: G \to G'$ is a surjective $\mathbf{Q}$-morphism of G onto a $\mathbf{Q}$-group G', and Γ is an arithmetic subgroup of G, then $f(\Gamma)$ is also arithmetic. (See [13, 6.11] for isogenies, [11, Th. 6] for the generalization to surjective morphisms.) Since we are interested in automorphism groups of symmetric spaces, we may, without restricting generality, limit ourselves to centerless groups whenever convenient. Moreover, it follows from [13, 6.11] that, if G is an almost direct (or a semi-direct) product of two $\mathbf{Q}$-subgroups G_1, G_2 and Γ is an arithmetic subgroup of G, then $(\Gamma \cap G_i)$ is an arithmetic subgroup of G_i ($i = 1, 2$) and $(\Gamma \cap G_1) \cdot (\Gamma \cap G_2)$ is commensurable with Γ.

Let G be simple over $\mathbf{Q}$. Then there exists an algebraic number field k and an absolutely simple k-group G' such that $G = R_{k/\mathbf{Q}}G'$ [14, 6.21 (ii)], where $R_{k/\mathbf{Q}}$ is the functor of restriction of the ground field [38, Ch. I], from k to $\mathbf{Q}$.

3.2. LEMMA. *Let k be an algebraic number field. G' a connected semi-simple and absolutely simple k-group, and $G = R_{k/\mathbf{Q}}G'$. Let K be a maximal compact subgroup of $G_{\mathbf{R}}$, $X = K\backslash G_{\mathbf{R}}$, and Γ be an arithmetic subgroup of G.*

(a) If K has the same rank as G, in particular if X is a bounded domain, then k is totally real.

(b) If X/Γ is not compact, then $G_{\mathbf{R}}^0$ has no compact factor $\neq \{e\}$, and $\mathrm{rk}_k(G') \neq 0$.

Let V be the set of normalized archimedean valuations of k, and k_v the completion of k with respect to $v \in V$. Then

$$G_{\mathbf{R}} \cong \prod_{v \in V} G'_{k_v}, \qquad\qquad [38, 1.3.2]$$

hence

$$X = \prod_{v \in V} X_v \qquad\qquad (X_v = (K \cap G'_{k_v})\backslash G'_{k_v})$$

and $K_{(v)} = K \cap G'_{k_v}$ is a maximal compact subgroup of G'_{k_v}. If $k_v = \mathbf{C}$, then G'_{k_v} is a complex Lie group, viewed as real Lie group, and its rank as such is twice the rank of $K_{(v)}$, whence (a).

The groups G'_{k_v} are the simple factors of $G_{\mathbf{R}}$. If one of them is compact, then $G'_k \cong G_{\mathbf{Q}}$ consists of semi-simple elements, hence $G_{\mathbf{R}}/\Gamma$ is compact [13, 11.6], which proves (b), and G has no proper $\mathbf{Q}$-parabolic subgroup.

3.3. (i) Let X be a bounded symmetric domain, $H(X)$ its group of holomorphic automorphisms and Is (X) its group of isometries with respect to the underlying riemannian structure. Let $\mathfrak{h}$ be the Lie algebra of $H(X)$. It is known that $\operatorname{Ad}\mathfrak{h} = H(X)^0 \subset \operatorname{Is} X = \operatorname{Aut}\mathfrak{h}$. Thus $H(X)$ is identified with a group of finite index in the group of real points of an algebraic $\mathbf{R}$-group, namely $\operatorname{Aut}\mathfrak{h}_{\mathbf{C}}$. Assume that we have put on $\operatorname{Aut}\mathfrak{h}_{\mathbf{C}}$ a structure of $\mathbf{Q}$-group subordinated to its natural $\mathbf{R}$-structure. This is equivalent to putting a $\mathbf{Q}$-structure on $\mathfrak{h}$; i.e., fixing a Lie subalgebra $\mathfrak{h}_{\mathbf{Q}}$ over $\mathbf{Q}$ of $\mathfrak{h}$ such that $\mathfrak{h} = \mathfrak{h}_{\mathbf{Q}} \otimes {}_{\mathbf{Q}}\mathbf{R}$. An *arithmetic subgroup* Γ of $H(X)$ is then an arithmetic subgroup of $\operatorname{Aut}\mathfrak{h}_{\mathbf{C}}$, viewed as a $\mathbf{Q}$-group. More correctly, one should say that Γ is arithmetically definable, since this definition presupposes the determination of a $\mathbf{Q}$-structure, for which there is usually a wide choice. However, we shall just say arithmetic for the sake of brevity. It is then understood that $\operatorname{Ad}\mathfrak{h}_{\mathbf{C}}$ has been identified with a semi-simple $\mathbf{Q}$-group G which has no center; i.e., with $\operatorname{Ad}\mathfrak{g}_{\mathbf{C}}$, and $H(X)$ with a subgroup of $\operatorname{Aut}\mathfrak{g}_{\mathbf{R}}$. The space X is then the quotient of $\operatorname{Aut}\mathfrak{g}_{\mathbf{R}}$, or $H(X)$, or $\operatorname{Ad}\mathfrak{g}_{\mathbf{R}}$, by a maximal compact subgroup.

The group G is the direct product of its normal simple $\mathbf{Q}$-groups G_i ($1 \leq i \leq m$), and X the product of the symmetric spaces $(K \cap G_{i\mathbf{R}})\backslash G_{i\mathbf{R}}$, which are then also bounded symmetric domains. Let $\Gamma_i = \Gamma \cap G_{i\mathbf{R}}^0$ ($1 \leq i \leq m$) and Γ' be the subgroup generated by the Γ_i. It is arithmetic, normal, of finite index, in Γ.

Our problem is the compactification of X/Γ. It turns out that the passage from X/Γ' to X/Γ offers no difficulty (cf. 8.9). Since X/Γ' is the product of the X_i/Γ_i, the essential case to consider is when G is simple over $\mathbf{Q}$, and $\Gamma \subset G_{\mathbf{R}}^0$.

(ii) We introduce some notation pertaining to our main case of interest. We keep the assumption of (i), and assume moreover G to be simple over $\mathbf{Q}$, and $\Gamma \subset G_{\mathbf{R}}^0$. Then $G = R_{k/\mathbf{Q}}G'$, where G' is an absolutely simple k-group, and k a totally real number field. Let Σ be the set of distinct isomorphisms of k into $\mathbf{R}$. There is a 1–1 correspondence between elements of Σ and normalized archimedean valuations of k given by $|a|_v = |\sigma(a)|$ ($a \in k$), and we have $G_{k_v} \cong ({}^\sigma G')_{\mathbf{R}}$, [38, Ch. I]. We may then also write

$$X = \prod_{\sigma \in \Sigma} X_\sigma, \qquad (X_\sigma = K_{(\sigma)}\backslash {}^\sigma G_{\mathbf{R}} = K_{(\sigma)}^0 \backslash {}^\sigma G_{\mathbf{R}}^0) ,$$

where X_σ is an irreducible symmetric bounded domain. For simplicity, we shall also write G_σ for ${}^\sigma G_{\mathbf{R}}^0$.

We assume further that if Γ is an arithmetic subgroup of G, then X/Γ is not compact. This implies (3.2) that no X_σ is reduced to a point and that G' has a non-trivial maximal k-split torus, say S'. Then ${}^\sigma S'$ is a maximal $\sigma(k)$-

split torus of $^\sigma G'$; there is a canonical isomorphism $\varphi_\sigma: S' \to {^\sigma}S'$ which induces an isomorphism of $_k\Phi$ onto $_{\sigma(k)}\Phi(^\sigma G') = {_k}\Phi_\sigma$. Furthermore the maximal $\mathbf{Q}$-split subtorus S of $R_{k/\mathbf{Q}}S'$ is a maximal $\mathbf{Q}$-split torus of G. It is canonically isomorphic to S' and is diagonally embedded in $R_{k/\mathbf{Q}}S'$. This means more precisely that the projection pr_σ of S into $^\sigma G'$ is the composition of the canonical isomorphisms $\varphi: S \to S'$ and $\varphi_\sigma: S' \to {^\sigma}S'$. The isomorphism also induces an isomorphism of $_\mathbf{Q}\Phi(G)$ onto $_k\Phi(G')$. We shall identify $_k\Phi(G')$, $_{\sigma(k)}\Phi(^\sigma G')$, and $_\mathbf{Q}\Phi(G)$ by means of these isomorphisms.

In each group $^\sigma G'$, we choose a maximal $\mathbf{R}$-split torus $T_\sigma \supset {^\sigma}S'$, contained in a maximal torus defined over $\sigma(k)$ (apply 2.6 to $Z(^\sigma S')$). We fix an ordering on $X(S')$, hence, using φ and φ_σ, also an ordering on $X(^\sigma S')$ and $X(S)$. For each σ, choose an ordering on $X(T_\sigma)$ compatible with the given one on $X(^\sigma S')$, and let $r: X(T_\sigma) \to X(^\sigma S') \cong X(S)$ be the restriction homomorphism. By 2.9, the canonical numbering on the set $_\mathbf{R}\Delta_\sigma$ of simple $\mathbf{R}$-roots of G with respect to T_σ is compatible by restriction with the canonical numbering of $_\mathbf{Q}\Delta$.

Let $_k\Delta = \{\beta_1, \cdots, \beta_s\}$. For i between 1 and s, we let $c(i, \sigma)$ be the index of the critical simple $\mathbf{R}$-root of $^\sigma G$ restricting on β_i. Then, the remark just made shows that $i < j$ implies $c(i, \sigma) < c(j, \sigma)$ for all $\sigma \in \Sigma$.

A sequence of elements indexed by Σ will often be denoted in boldface and used as a multi-index or a multi-exponent. In particular, let b be between 1 and s. Then

$$F_\mathbf{b} = \prod_{\sigma \in \Sigma} F_{c(b,\sigma)} ,$$

is the product of the standard boundary components $F_{c(b,\sigma)}$ of X_σ, where *standard* refers to the choice of T_σ and $_\mathbf{R}\Delta_\sigma$. It is also understood that the Lie algebra of $K_{(\sigma)}$ is orthogonal to that of T_σ. Since $c(j, \sigma)$ is an increasing function of j, for each σ, we have $\bar F_\mathbf{j} \subset \bar F_\mathbf{i}$ $(1 \leq i \leq j \leq s)$.

Let $F = \prod_\sigma F_{i(\sigma)}$ be a product of standard boundary components. We let $S_F = \prod S_{i(\sigma)}$ be the product of the unbounded realizations associated to the $F_{i(\sigma)}$, (1.6), J_F be the functional determinant in S_F, and j_F be the functional determinant in the Harish-Chandra bounded realization of F. If $F = F_\mathbf{b}$ we also write $S_\mathbf{b}, J_\mathbf{b}, j_\mathbf{b}$ for S_F, J_F and j_F. In the notation of 1.8, we have therefore

$$J_F(x, g) = \prod_\sigma J_{i(\sigma)}(x_\sigma, g_\sigma) \qquad (x = (x_\sigma), g = (g_\sigma); x_\sigma \in X_\sigma, g_\sigma \in G_\sigma) ,$$

$$j_F(x, g) = \prod_\sigma j_{i(\sigma)}(x_\sigma, g_\sigma) \qquad (x = (x_\sigma), g = (g_\sigma); x_\sigma \in F_{i(\sigma)}, g_\sigma \in L(F_\sigma)) .$$

By 1.11, applied to each irreducible factor of X, the functional determinant $J_F(x, g)$, $(g \in N(F))$, is constant along the fibres of the canonical projection σ_F of X onto F, (defined in 1.7, remark).

The natural compactification $\bar{X}$ of X is the product of the natural compactifications $\bar{X}_\sigma$ of the X_σ. We shall also write o_b for the point with components $o_{c(b,\sigma)} \in \bar{X}_\sigma$, in the notation of 1.5. Thus

$$F_b = o_b \cdot N(F_b) = o_b \cdot L(F_b) = o_b \cdot G(F_b) , \qquad (1 \leqq b \leqq s) .$$

We recall that $G(F_b) = N(F_b)/Z(F_b)$. We shall denote by ϖ_b the natural projection of $N(F_b)$ onto $G(F_b)$. Applying 1.3 and 1.5 to each irreducible factor of X, we see that $G(F_b)$ is connected, that $N(F_b) = L(F_b) \cdot Z(F_b)$, and that $Z(F_b)$ is the greatest normal subgroup of $N(F_b)$ with identity component $Z(F_b)^0$.

3.4. It will be sometimes convenient to use the following variation on the notion of arithmetic group.

Let H be a connected *real* Lie group. A subgroup Γ is of *arithmetic type*, or *arithmetically definable*, if there exists a connected $\mathbf{Q}$-group G, a continuous surjective homomorphism $f \colon G_\mathbf{R}^0 \to H$ with *compact* kernel N and an arithmetic subgroup Γ' of G such that $f(\Gamma') = \Gamma$. Since N is compact, the group Γ is then obviously discrete.

Let H be semi-simple. It is easily seen that, without restricting generality, G may be assumed to be semi-simple, and to be almost simple over $\mathbf{Q}$ if H is simple. If G is simple over $\mathbf{Q}$, and $\dim N > 0$, then H/Γ is compact. In fact, we have in this case $G = R_{k/\mathbf{Q}}G'$, where k is an algebraic number field, and G' an absolutely simple k-group. The group $G_\mathbf{R}$ is the product of the groups G'_{k_v} where k_v runs through the archimedean completions of k, and these are the simple normal subgroups of $G_\mathbf{R}$. Therefore N contains at least one of them, G'_k consists of semi-simple elements, and $G_\mathbf{R}/\Gamma'$ is compact [13, 11.6]. This implies of course the compactness of H/Γ and of $K\backslash H/\Gamma$, where K is a compact subgroup of H.

3.5. *Rational boundary components.* Let G be a connected semi-simple $\mathbf{Q}$-group, whose symmetric space of non-compact type, $X = K\backslash G_\mathbf{R}$, where K is a maximal compact subgroup of $G_\mathbf{R}$, is a bounded symmetric domain. For a discrete subgroup Γ of $G_\mathbf{R}$ and a boundary component F of X, (1.2), we let $\Gamma(F)$ be the image of $\Gamma \cap N(F)$ in $G(F) = N(F)/Z(F)$ by the natural projection. The component F is said to be Γ-rational if

    (i) the quotient $U(F)/(U(F) \cap \Gamma)$ is compact,

    (ii) the group $\Gamma(F)$ is discrete.

Clearly a Γ-rational component is Γ'-rational for any group Γ' commensurable with Γ. In particular, if F is Γ-rational for one arithmetic group Γ, it is so for all arithmetic groups; in that case, we shall drop the prefix Γ- and speak of *rational boundary components*.

Let now Γ be arithmetic. We remark first that (i) is equivalent to

(i)′ $N(F)_{\mathbf{C}}$ is defined over $\mathbf{Q}$.

The implication (i)′ $\Rightarrow$ (i) follows from the standard fact that, if U is a unipotent $\mathbf{Q}$-group and Γ an arithmetic subgroup of U, then $U_{\mathbf{R}}/\Gamma$ is compact. Assume now (i) to hold. Let V be the smallest algebraic subgroup of $U(F)_{\mathbf{C}}$ containing $U(F) \cap \Gamma$. It is invariant under all automorphisms of $\mathbf{C}$, since $\Gamma \subset G_{\mathbf{Q}}$, hence it is defined over $\mathbf{Q}$. Since $U(F)/\Gamma$ is compact, the quotient $U_{\mathbf{R}}/V_{\mathbf{R}}$ is compact, too. But $U_{\mathbf{R}}$ and $V_{\mathbf{R}}$ are homeomorphic to euclidean spaces, hence $U(F)_{\mathbf{C}} = V$, which shows that $U(F)_{\mathbf{C}}$ is defined over $\mathbf{Q}$. The group $N(F)_{\mathbf{C}}$, being equal to the normalizer of its unipotent radical, is then also defined over $\mathbf{Q}$.

It will turn out that (i)′ $\Rightarrow$ (ii) in our case; however we have preferred to start from a definition which makes sense for any symmetric space and any Satake compactification.

3.6. Clearly, (ii) is implied by

(ii)′ the group $\Gamma(F)$ is of arithmetic type.

Assuming (i), we now prove that (ii)′ is implied by either of the two following equivalent conditions:

(iii) There exists a normal connected $\mathbf{Q}$-subgroup C of $N(F)_{\mathbf{C}}$ containing $U(F)_{\mathbf{C}}$ and $L(F)_{\mathbf{C}}$ and such that $C_{\mathbf{R}}/L(F) \cdot U(F)$ is compact.

(iv) There exists a connected normal $\mathbf{Q}$-subgroup B of $N(F)_{\mathbf{C}}$, contained in $Z(F)_{\mathbf{C}}^{0}$, containing $U(F)_{\mathbf{C}}$, such that $Z(F)/(B_{\mathbf{R}} \cap G_{\mathbf{R}}^{0})$ is compact.[2]

We show, to begin with, that (iii) and (iv) are equivalent. First assume (iii). Let H be a maximal connected reductive $\mathbf{Q}$-subgroup of C, and L a maximal connected reductive $\mathbf{Q}$-subgroup of $N(F)_{\mathbf{C}}$ containing H. We may write $L = H \cdot H'$, with H' normal, defined over Q, and $H \cap H'$ finite. Then $D = H' \cdot U(F)_{\mathbf{C}}$ is a normal $\mathbf{Q}$-subgroup such that $Z(F)/D_{\mathbf{R}}^{0}$ is compact. Moreover, $D_{\mathbf{R}} \cap G_{\mathbf{R}}^{0} \subset Z(F)$ by (1) of 1.3, whence (iv). The other implication is proved in the same way.

Assume (iv) holds. The projection $N(F) \to G(F)$ is then the composition of the restriction to $N(F)$ of the $\mathbf{Q}$-morphism $N(F)_{\mathbf{C}} \to N(F)_{\mathbf{C}}/B$ with the projection of $N(F)/(B_{\mathbf{R}} \cap G_{\mathbf{R}}^{0})$ onto $G(F)$, which has a compact kernel, whence (ii)′.

We note finally that if $G = G_1 \times \cdots \times G_m$ is a direct product of normal $\mathbf{Q}$-subgroups, then a boundary component $F = F_1 \times \cdots \times F_m$ of X is rational if and only if F_i is a rational boundary component of $X_i = (K \cap G_i)\backslash G_{i\mathbf{R}}$ for all i. This follows immediately from the two following facts: the parabolic subgroups of G are the products of the parabolic subgroups of the G_i; the

---

[2] These conditions are of course fulfilled if $Z(F)_{\mathbf{C}}$ is defined over $\mathbf{Q}$. In fact, this is the requirement made in [10]; however, it has turned out to be too restrictive.

group Γ is commensurable with the product of the groups $\Gamma \cap G_i$, which are arithmetic (3.1).

3.7. Theorem. *We keep the assumption of 3.3 (i). A boundary component F of X is rational if and only if $N(F)_{\mathbf{C}}$ is defined over $\mathbf{Q}$. If F is rational, $\Gamma(F)$ is of arithmetic type. The map $F \mapsto N(F)_{\mathbf{C}}$ defines a bijection of the set of proper rational boundary components onto the set of proper maximal parabolic $\mathbf{Q}$-subgroups of $G_{\mathbf{C}}$.*

By the last remark of 3.6, we may assume G to be simple over $\mathbf{Q}$. If X/Γ is compact, where Γ is an arithmetic group, then $\mathrm{rk}_{\mathbf{Q}}(G) = 0$ [13, 11.4, 11.6], G has no proper parabolic $\mathbf{Q}$-subgroup [14, 8.3-5], and there is no proper rational boundary component.

Assume now X/Γ to be non-compact. In the notation of 3.3 (ii), we have $G = R_{k/\mathbf{Q}}G'$ with G' absolutely simple and k totally real. Let F be a boundary component of X. If F is rational, then $N(F)_{\mathbf{C}}$ is a $\mathbf{Q}$-subgroup by 3.5 (i)$'$. Assume conversely that $N(F)_{\mathbf{C}}$ is defined over $\mathbf{Q}$. We have then $N(F)_{\mathbf{C}} = R_{k/\mathbf{Q}}P$, where P is a parabolic k-subgroup of G', [14, 6.19], hence

$$F = \prod_{\sigma \in \Sigma} F_\sigma , \qquad N(F) = \prod N(F_\sigma) ,$$
$$\left(N(F_\sigma) = ({}^\sigma P)_{\mathbf{R}} \cap G_\sigma, \sigma \in \Sigma \right) ,$$

where F_σ is a boundary component of X_σ. Let V'_0 be the center of the unipotent radical $V' = R_u(P)$ of P, and C' the connected centralizer of V'_0 in P. The groups V', V'_0 and C' are clearly defined over k. It follows immediately from the properties of the functor $R_{k/\mathbf{Q}}$ that $C = R_{k/\mathbf{Q}}C'$ is the centralizer in $N(F)_{\mathbf{C}}$ of the center $V_0 = R_{k/\mathbf{Q}}V'_0$ of the unipotent radical of $N(F)_{\mathbf{C}}$, and that $C^0_{\mathbf{R}} = \prod_\sigma C^0_\sigma$, where C_σ is the connected centralizer in $N(F_\sigma)$ of the center of $U(F_\sigma)$. By 1.20, C_σ contains $L(F_\sigma) \cdot U(F_\sigma)$ and the quotient $C_\sigma/L(F_\sigma) \cdot U(F_\sigma)$ is compact. Therefore condition (iii) of 3.6 is fulfilled; since, together with (i), it implies (ii), (ii)$'$, by 3.6, our first two assertions are proved.

Let F be rational. Then $N(F)_{\mathbf{C}} = \prod_\sigma N(F_\sigma)_{\mathbf{C}}$ and $N(F_\sigma)_{\mathbf{C}}$ is a proper maximal parabolic $\mathbf{R}$-subgroup of ${}^\sigma G'$, hence $N(F)_{\mathbf{C}}$ is a proper maximal parabolic $\mathbf{Q}$-subgroup of G.

Conversely, let P be a proper maximal parabolic $\mathbf{Q}$-subgroup of G. We have $P = R_{k/\mathbf{Q}}P'$, where P' is a proper maximal parabolic k-subgroup of G', and therefore $P = \prod_\sigma {}^\sigma P'$, and ${}^\sigma P'$ is a proper maximal parabolic $\sigma(k)$-subgroup of ${}^\sigma G'$. By 2.10, ${}^\sigma P'$ is also a proper maximal parabolic $\mathbf{R}$-subgroup of ${}^\sigma G'$; consequently (1.5), ${}^\sigma P'_{\mathbf{R}} \cap G^0_{\mathbf{R}} = N(F_\sigma)$ where F_σ is a boundary component of X, and $P_{\mathbf{R}} \cap G_{\mathbf{R}} = N(F)$, $(F = \prod F_\sigma)$. The boundary component F is then rational by the first part of the theorem. Since two boundary components with the same normalizer are identical, the proof of the theorem is complete.

3.8. THEOREM. *We keep the notation and conventions of 3.3 (ii). Let[3]* $\theta(b) = \{\beta_{b+1}, \cdots, \beta_s\}$, $L'_b = {}_kL_{\theta(b)}$ *(cf. 2.2) and* $L_b = R_{k/Q}L'_b$, $(1 \le j \le s)$. *If* $b \ne s$, *then* $(L_{b,\mathbf{R}})^0 \cong L(F_b)$. *In particular* $L(F_b)_\mathbf{C}$ *is defined over* **Q**, *almost* **Q**-*simple, and its* **Q**-*rank is equal to* $s - b$. *For any arithmetic subgroup* Γ *of* G, *the quotient* $F_s/\Gamma(F_s)$ *is compact. Given a rational boundary component* F, *there exist one and only one index* b $(1 \le b \le s)$ *and an element* $x \in G_\mathbf{Q}$ *such that* $F = F_b \cdot x$.

By 2.10 (b), we have $L'_b = L(F_{c(b,\sigma)})_\mathbf{C}$ for all $\sigma \in \Sigma$. Since $L_b = \prod_\sigma {}^\sigma L'_b$, this proves the first assertion. The set $\theta(b)$ being connected, L'_b is almost k-simple, (2.2), hence L_b is almost **Q**-simple [14, 6.21 (ii)].

For each σ, the index $c(s, \sigma)$ is the last critical index, therefore (1.3, 1.5) $Z(F_s)$ contains $S'_\mathbf{R}$. Let C be the connected centralizer in $N(F_s)_\mathbf{C}$ of the center of the unipotent radical of $N(F_s)_\mathbf{C}$. By 1.20, the intersection of ${}^\sigma G' \cap C$ with ${}^\sigma S'$ is finite for every $\sigma \in \Sigma$. In particular, S normalizes C and $S \cap C$ is finite. This implies that the **Q**-rank of C is zero, for otherwise there would exist a maximal **Q**-split subtorus T of $N(F)_\mathbf{C}$ with $\dim (T \cap C) \ne 0$, and it could not be conjugate to S (since C is normal), contradicting the conjugacy theorem for maximal split tori. It follows then from [14, 8.5] that the unipotent elements of $C_\mathbf{Q}$ all belong to the unipotent radical of C, and that $X(C)_\mathbf{Q} = 0$. The quotient $C_\mathbf{R}/(\Gamma \cap C)$ is then compact [13, 11.8]. Since the projection $N(F_s) \to G(F_s)$ maps $C_\mathbf{R}^0$ onto $G(F_s)$ and $\Gamma \cap C_\mathbf{R}^0$ into $\Gamma(F_s)$, we see that $G(F_s)/\Gamma(F_s)$ is compact, whence our second assertion.

Let F be a rational boundary component. Then $N(F)_\mathbf{C}$ is a proper maximal parabolic **Q**-subgroup of G. On the other hand, the group $N(F_b)_\mathbf{C}$ is, in the notation of 2.2, the standard parabolic group ${}_\mathbf{Q}P_{\psi(b)}$, $(\psi(b) = {}_\mathbf{Q}\Delta - \{\beta_b\}$; $b = 1, \cdots, s)$. The groups ${}_\mathbf{Q}P_{\psi(j)}$ $(j = 1, \cdots, s)$ are all the proper standard maximal parabolic **Q**-subgroups (2.2); there exists therefore one and only one b for which we may find $x \in G_\mathbf{Q}$ such that $x \cdot N(F)_\mathbf{C} \cdot x^{-1} = N(F_b)_\mathbf{C}$. We have then $F = F_b \cdot x$, which ends the proof.

3.9. COROLLARY. *Let* F, F' *be rational boundary components of* X *such that* $F' \subset \bar{F}$. *Let* b, c *be the integers such that* $F \subset F_b \cdot G_\mathbf{Q}$ *and* $F' \subset F_c \cdot G_\mathbf{Q}$. *Then there exists* $g \in G_\mathbf{Q}$ *such that* $F \cdot g = F_b$, $F' \cdot g = F_c$.

There is nothing to prove unless $b \ne c$; in particular, we may assume $b \ne s$. Therefore (3.8), $L(F_b)_\mathbf{C}$ is defined over **Q**, almost **Q**-simple, and we may apply 3.8 to $X = F_b$. The proof of 3.9 is then the same as that of the

---

[3] We hope the reader will not be unduly confused by the occasionally similar (or, by chance, even coinciding) notation used for real Lie groups in §1, and for their complex forms in §§2, 3. Also, L_b has a different meaning here than in §1.3.

similar remark made in 1.5, using 3.8 to insure that u, v may be taken in $G_\mathbf{Q}$ and $L(F_\mathrm{b})_\mathbf{Q}$ respectively.

3.10. REMARKS. (1) The group $G(F_\mathrm{b})$ is the quotient of $L(F_\mathrm{b})$ by its center, which is finite; it may be identified with the adjoint group of $L(F_\mathrm{b})$. Thus, if $b \neq s$, the group $G(F_\mathrm{b})_\mathrm{C}$ may be viewed in a canonical way as a group defined over $\mathbf{Q}$, in such a way that ϖ_b induces a $\mathbf{Q}$-morphism of $N(F_\mathrm{b})_\mathrm{C}$ onto $G(F_\mathrm{b})_\mathrm{C}$, and that $\Gamma(F_\mathrm{b})$ is arithmetic. The images of $P \cap L(F_\mathrm{b})_\mathrm{C}$ and $S \cap L(F_\mathrm{b})_\mathrm{C}$ under ϖ_b are a minimal parabolic $\mathbf{Q}$-subgroup and a maximal $\mathbf{Q}$-split torus respectively.

(2) Theorem 3.8 shows, independently of 3.7, that F is rational if $N(F)_\mathrm{C}$ is defined over $\mathbf{Q}$ and conjugate to one of the groups $N(F_\mathrm{b})_\mathrm{C}$ with $b \neq s$. Our original proof of 3.7 consisted of 3.8 and of a separate discussion of the case $b = s$. The proof of 3.7 given here, which is based on 1.20, was suggested by [31].

3.11. PROPOSITION. *We keep the notation and assumptions of 3.3* (ii). *Let* $\chi_\mathrm{b}(g) = \det \mathrm{Ad}_\mathfrak{u}\, g$ $(g \in N(F_\mathrm{b}))$, *where* $\mathfrak{u} = \mathfrak{u}(F_\mathrm{b})$ *is the Lie algebra of* $U(F_\mathrm{b})$, *and let* η_b *be the restriction to* $Z(F_\mathrm{b})$ *of the functional determinant* J_b. *Then there exists a rational number* $n_b > 0$ *such that*

$$\eta_\mathrm{b}(g) = \chi_\mathrm{b}(g)^{-n_b} \qquad\qquad \big(g \in A_b \cdot U(F_\mathrm{b})\big) ,$$
$$|\,\eta_\mathrm{b}(g)\,| = |\,\chi_\mathrm{b}(g)\,|^{-n_b} \qquad\qquad \big(g \in Z(F_\mathrm{b})\big) .$$

Using the notation of 1.16, we may write

(1)
$$\chi_\mathrm{b}(g) = \textstyle\prod_{\sigma \in \Sigma} \det \cdot \mathrm{Ad}_{\mathfrak{u}_\sigma}\, g_\sigma$$
$$= \textstyle\prod_{\sigma \in \Sigma} \chi_{c(b,\sigma)}(g_\sigma) \qquad\qquad \big(g = (g_\sigma),\ \mathfrak{u}_\sigma = \mathfrak{u}(F_{c(b,\sigma)})\big) ,$$

and similarly

(2)
$$\eta_\mathrm{b}(g) = \textstyle\prod_{\sigma \in \Sigma} \eta_{c(b,\sigma)}(g_\sigma) .$$

By 1.16, there exists a rational number $n_{c(b,\sigma)} > 0$ such that

(3)
$$|\,\eta_{c(b,\sigma)}(g)\,| = |\,\chi_{c(b,\sigma)}(g)\,|^{-n_{c(b,\sigma)}} , \qquad\qquad \big(g \in Z(F_{c(b,\sigma)})\big)$$

(4)
$$\eta_{c(b,\sigma)}(g) = \chi_{c(b,\sigma)}(g)^{-n_{c(b,\sigma)}} \qquad\qquad \big(g \in A_{c(b,\sigma)}\big) .$$

Let $g = (g_\sigma) \in {}_\mathbf{Q}A$. We have already remarked that $g_\sigma = \varphi_\sigma \circ \varphi(g)$, in the notation of 3.3; by the properties of the functor $R_{k/\mathbf{Q}}$, this implies

(5)
$$\det \mathrm{Ad}_{\mathfrak{u}_\sigma}\, g = \chi_{c(b,\sigma)}(g_\sigma) \qquad\qquad (\sigma \in \Sigma;\ g \in {}_\mathbf{Q}A) .$$

The proposition will then be a consequence of (1) to (5) once it is shown that $n_{c(b,\sigma)}$ is independent of σ. By 2.9, $_k\Phi$ is of type C_s if and only if $_\mathrm{R}\Phi(^\sigma G')$ is of type C_{t_σ} and $c(s, \sigma) = t_\sigma$ for *every* $\sigma \in \Sigma$. Thus if $_k\Phi$ is of type C_s and $b = s$, we are in case (i) of 1.15 for every σ, hence $n(b, \sigma) = 1$ $(\sigma \in \Sigma)$. If either

$b \neq s$ or $_k\Phi$ is of type BC_s, then, for every $\sigma \in \Sigma$, we are in one of the cases (ii), (iii) of 1.15. Then $n_{c(b,\sigma)}$ is given by an expression which depends only on the multiplicities of the restrictions to $A_{c(b,\sigma)}$ of $\alpha_{\sigma,c(b,\sigma)}$ and $2\alpha_{\sigma,c(b,\sigma)}$. In view of the different identifications made in 3.3, these are also the multiplicities of the restrictions of β_b and $2 \cdot \beta_b$ to A_b, and they are consequently independent of σ. This proves our assertion; and, in view of 1.16, shows that

$$(6) \qquad n_b = 1 \qquad\qquad\qquad (_k\Phi \text{ of type } C_s \text{ and } b = s)$$

$$(7) \qquad n_b = (p_b + 4q_b) \cdot (2p_b + 4q_b)^{-1} \qquad \text{otherwise,}$$

where p_b and q_b are the multiplicities of the restrictions of β_b and $2 \cdot \beta_b$ to the intersection $_QA_b$ of the kernels of the simple $\mathbf{Q}$-roots β_i $(i \neq b)$.

3.12. *The functional determinant on F_b.* Let $\mathbf{q}_b = (q_{c(b,\sigma)})_{\sigma \in \Sigma}$, where $q_{c(b,\sigma)}$ is the rational number ≥ 1 attached by 1.17 to X_σ and the boundary component $F_{c(b,\sigma)}$. Put

$$|j_F(\sigma_b(x), g)|^{\mathbf{q}_b} = \prod_{\sigma \in \Sigma} |j_{c(b,\sigma)}(\sigma_{c(b,\sigma)}(x_\sigma), g_\sigma)|^{q_{c(b,\sigma)}} \qquad (g = (g_\sigma) \in N(F_b)) \ .$$

If we apply 1.17 to each component $F_{c(b,\sigma)}$ and use 3.11, we see that

$$(1) \qquad |J_b(x, g)| = |j_F(\sigma_b(x), g)|^{\mathbf{q}_b} |\chi_b(g)|^{-n_b}, \qquad (x \in S_b; g \in N(F_b)) \ .$$

3.13. Let B be a normal connected $\mathbf{Q}$-subgroup of $N(F_b)_{\mathbb{C}}$ satisfying (iv) of 3.6. We may write B as a semi-direct product over $\mathbf{Q}$ of $S_b \cdot U(F_b)_{\mathbb{C}}$ by a reductive $\mathbf{Q}$-group H. Write further $H = H' \cdot \mathfrak{D}H$ as an almost direct product of its connected center H' by its derived group $\mathfrak{D}H$. We know that $Z(F_b)/U(F_b)$ is the almost direct product of $B_{\mathbf{R}}/U(F_b)$ by a compact group. Furthermore, $Z(F_b)/_QA_b \cdot U(F_b)$ modulo its derived group is compact. This follows from 1.3, 1.5, applied to each factor $N(F_{c(b,\sigma)})$. It follows then that $B_{\mathbf{R}}/U(F_b)$ modulo its derived group is compact, therefore $H'_{\mathbf{R}}$ is compact. Since J_b and χ_b are both characters on $Z(F_b)$, they are equal to one on $\mathfrak{D}H$, whence the equality

$$(1) \qquad \chi_b(g)^{-n_b} = J_b(x, g) \qquad (g \in (\mathfrak{D}H) \cdot _QA_b \cdot U(F_b)) \ .$$

Let now Γ_0 be an arithmetic subgroup of $N(F_b)_{\mathbb{C}}$. The group $\Gamma_0 \cap H$ is commensurable with $(\Gamma_0 \cap H') \cdot (\Gamma_0 \cap \mathfrak{D}H)$, where both factors are arithmetic [13; 6.4, 6.11]. Since $H'_{\mathbf{R}}$ is compact, the group $\Gamma_0 \cap H'$ is finite. Of course, χ_b takes the values ± 1 on Γ_0. It follows then that the image of $\Gamma_0 \cap B$ under J_b is a finite group (of roots of unity). This proves therefore the

3.14. PROPOSITION. *We keep the previous notation. Let Γ_0 be an arithmetic subgroup of $N(F_b)_{\mathbb{C}}$. Then there exists a positive integer d such that*

$$J_b(x, g \cdot \gamma)^d = J_b(x, g)^d \qquad (x \in X, g \in N(F_b), \gamma \in \Gamma_0 \cap B) \ .$$

3.15. PROPOSITION. *Let G be as in 3.3 (i). Assume that G has no normal*

Q-*subgroup of dimension* 3. *Then every proper rational boundary component has complex codimension* $\geqq 2$.

It suffices to prove this when G is simple over **Q**. If it is absolutely simple, then our assertion follows from 1.5 (iii). Let now G be not absolutely simple. Then the set Σ of 3.3 (ii) has at least two elements. By 3.7, a rational boundary component F is a product $\prod_\sigma F_\sigma$, where F_σ is a *proper* boundary component of X_σ. We have then $\dim_{\mathbf{C}} X_\sigma - \dim_{\mathbf{C}} F_\sigma \geqq 1$ for each σ; since card $\Sigma \geqq 2$, we are done.

4. Fundamental sets and compactification

4.1. Let G be a connected semi-simple **Q**-group, and P a minimal parabolic **Q**-subgroup. We write $P = M \cdot S \cdot U$ as in 2.2 and let $_{\mathbf{Q}}\Delta$ be the set of simple **Q**-roots for the ordering associated to U. Since M centralizes S, we also have $P = S \cdot V$ where $V = M \cdot U$ is the semi-direct product of M and U.

We shall often write $_{\mathbf{Q}}A$ for $S_{\mathbf{R}}^0$. For $t > 0$, let

$$(1) \qquad\qquad {}_{\mathbf{Q}}A_t = \{ a \in {}_{\mathbf{Q}}A \mid a^\beta \leqq t \, (\beta \in {}_{\mathbf{Q}}\Delta) \} \, .$$

A fundamental property of this subset is given by the:

LEMMA. *Let ω be a compact subset of $(M \cdot U)_{\mathbf{R}}$. Then the union of the sets $a \cdot \omega \cdot a^{-1}$, $(a \in {}_{\mathbf{Q}}A_t)$ is relatively compact.*

Since S centralizes M and normalizes U, it is enough to prove this when $\omega \subset U_{\mathbf{R}}$. But then $_{\mathbf{Q}}A_t$ is a bounded set of operators on $\mathfrak{u}_{\mathbf{R}}$ by $a \mapsto \mathrm{Ad}\, a$, in view of the very definition of $_{\mathbf{Q}}A_t$. Since the exponential is a homeomorphism of $\mathfrak{u}_{\mathbf{R}}$ onto $U_{\mathbf{R}}$, our assertion follows (see [13, Prop. 4.2]).

4.2. Let K be a maximal compact subgroup of $G_{\mathbf{R}}$ whose Lie algebra is orthogonal to that of $S_{\mathbf{R}}$, with respect to the Killing form, π the natural projection of $G_{\mathbf{R}}$ onto $X = K \backslash G_{\mathbf{R}}$, and $o = \pi(K)$. A *Siegel domain* $\mathfrak{S}'' = \mathfrak{S}''_{t,\omega}$ in $G_{\mathbf{R}}$ (with respect to K, S, and U) is a set of the form

$$\mathfrak{S}'' = \mathfrak{S}''_{t,\omega} = K \cdot {}_{\mathbf{Q}}A_t \cdot \omega \qquad\qquad (\omega \text{ compact in } V_{\mathbf{R}}) \, .$$

In this paper, we shall in fact be more concerned with the intersection $\mathfrak{S}' = (K \cap P) \cdot {}_{\mathbf{Q}}A_t \cdot \omega$ of $\mathfrak{S}''$ with $P_{\mathbf{R}}$, to be called a *Siegel domain in P*, and with

$$\mathfrak{S} = \mathfrak{S}_{t,\omega} = o \cdot \mathfrak{S}'' = o \cdot \mathfrak{S}' \, ,$$

to be called a *Siegel domain in X*. If we replace the sign $\leqq$ by $<$ in 4.1 (1), and ω by an open relatively compact subset of $V_{\mathbf{R}}$, then we get *open Siegel domains* in $G_{\mathbf{R}}$, $P_{\mathbf{R}}$ and X. We note that $_{\mathbf{Q}}A$ centralizes M and that $K \cap P \subset M$, so that we may also write $\mathfrak{S}' = {}_{\mathbf{Q}}A_t \cdot \omega'$ with $\omega' = (K \cap P) \cdot \omega$.

The **Q**-roots are, in a natural way, positive-valued functions on $P_\mathbf{R}^0$ which are equal to one on $V_\mathbf{R}^0$. Thus we also have

$$s^\beta \leqq t \qquad\qquad (s \in \mathfrak{S}_t'; \beta \in {}_\mathbf{Q}\Delta) .$$

4.3. Let Γ be an arithmetic subgroup of G. It is known that there exists a Siegel domain $\mathfrak{S}$ and a finite subset $C \subset G_\mathbf{Q}$ such that $\Omega = \mathfrak{S} \cdot C$ is a *fundamental set* for Γ in X; i.e., verifies

(F1) $\Omega \cdot \Gamma = X$, and

(F2) given $x \in G_\mathbf{Q}$, the set of $\gamma \in \Gamma$ for which $\Omega x \cap \Omega \gamma \neq \varnothing$ is finite (the *Siegel property*).

(See [20]. The result is stated there for $\Omega' = \pi^{-1}(\Omega)$ in $G_\mathbf{R}$, but this is clearly equivalent.)

If $S = \{e\}$, then X/Γ is compact, and conversely [13, 11.6, 11.4]. Of course any compact set has the Siegel property (for any discontinuous group Γ, and any $x \in G_\mathbf{R}$ in fact), so that in this case, any compact set verifying (F1) is a fundamental set. Also, in this case, a(n open) Siegel domain is just a(n open relatively) compact set.

REMARK. In considering later arithmetic subgroups of $H(X)$, we shall implicitly use a slight extension of these results to non-connected semi-simple groups, which is not stated in the literature, but reduces readily to the connected case. Namely, if G' is a **Q**-group whose connected component is G, we replace P and K by their normalizers in G' and $G_\mathbf{R}'$ respectively. Let Γ be an arithmetic subgroup of G'. The only non-obvious point is to see that a Siegel domain has the Siegel property in G'. To see this, one uses the corresponding result in G and the following facts:

(i) $G_\mathbf{Q}'$ is generated by $G_\mathbf{Q}$ and $N(P)_\mathbf{Q}$, which follows from [14, § 4.13];

(ii) if $h \in N(P)_\mathbf{R}$, then $\mathfrak{S} \cdot h$ is contained in a Siegel domain (with respect to K, S, P).

4.4. We now assume that G is simple over **Q**, X is a bounded symmetric domain, $\Gamma \subset G_\mathbf{R}^0$ and X/Γ is not compact. We take the notation and conventions of 3.3 (ii). In each group G_σ, we shall use the notation of §1. Putting

$$(1) \qquad\qquad {}_\mathbf{R}\Delta_\sigma = \{\alpha_{\sigma,1}, \cdots, \alpha_{\sigma,t_\sigma}\} ,$$

we may write,

$$(2) \qquad\qquad \alpha_{\sigma,i} = (\gamma_{\sigma,i} - \gamma_{\sigma,i+1})/2 , \qquad\qquad (1 \leqq i \leqq t_\sigma)$$

$$(3) \qquad \begin{aligned} \alpha_{\sigma,t_\sigma} &= \gamma_{t_\sigma} , & &({}_\mathbf{R}\Phi_\sigma \text{ of type } C_{t_\sigma}) , \\ \alpha_{\sigma,t_\sigma} &= \gamma_{t_\sigma}/2 . & &({}_\mathbf{R}\Phi_\sigma \text{ of type } BC_{t_\sigma}) . \end{aligned}$$

In the product of the groups $T_{\sigma\mathbf{R}}$, the torus ${}_\mathbf{Q}A$ is contained in the identity component of the intersections of the kernels of the roots $\alpha_{\sigma,i}$ ($\alpha_{\sigma,i} \in {}_R\Delta_\sigma$, $1 \leqq i \leqq t_\sigma$, i not critical). An element $a = (a_\sigma) \in {}_QA$ satisfies therefore the relations

$$(4) \qquad \gamma_{\sigma,i}(\log a_\sigma) = \gamma_{\sigma,i+1}(\log a_\sigma) \qquad (\sigma \in \Sigma, 1 \leqq i \leqq t_\sigma, i \text{ not critical})$$

with the convention that $\gamma_{\sigma,i} = 0$ if $i > t_\sigma$. Moreover, S_R being *diagonally* imbedded in G_R (see 3.3), we have

$$(5) \qquad a^{\beta_j} = a_\sigma^{\alpha_{\sigma,i}} \qquad (i = c(j, \sigma), j = 1, \cdots, s; \sigma \in \Sigma) \; .$$

Therefore, $a \in {}_QA_t$ if and only if it verifies (4), (5), and

$$(6) \qquad \gamma_{\sigma,i}(\log a_\sigma) - \gamma_{\sigma,i+1}(\log a_\sigma) \leqq 2 \log t \qquad (i = c(j, \sigma))$$

if $c(j, \sigma) \neq t_\sigma$, and

$$\gamma_{\sigma,t_\sigma}(\log a_\sigma) \leqq \log t \; \big(\text{resp. } \gamma_{\sigma,t_\sigma}(\log a_\sigma) \leqq 2 \log t\big),$$

if t_σ is critical; i.e., if $t_\sigma = c(s, \sigma)$, and if ${}_R\Phi_\sigma$ is of type C (resp. BC).

The subset ${}_QA_t \cdot \omega' = \mathfrak{S}' \cap P$ is contained in every standard parabolic subgroup, in particular in the groups $N(F_b)$. From 3.8 and the construction of the groups $L(F_b)$, we deduce immediately the following assertion: For $b \neq s$, $\mathfrak{S}' \cap L(F_b)$ is a Siegel domain of $L(F_b)$ with respect to $K \cap L(F_b)$, $S \cap L(F_b)$, and $U \cap L(F_b)$; its image under ϖ_b is a Siegel domain of $G(F_b)$; and

$$o_b \cdot \varpi_b(\mathfrak{S}'' \cap N(F_b)) = o_b \cdot \varpi_b(\mathfrak{S}')$$

is a Siegel domain of F_b.

If $b = s$, then $F_b/\Gamma(F_b)$ is compact, and $o_b \cdot \varpi_b(\mathfrak{S}'' \cap N(F_b))$ is compact.

4.5. LEMMA. *We keep the assumptions and notation of 4.4. The closure $\bar{\mathfrak{S}}$ of a Siegel domain $\mathfrak{S}$, in the natural compactification of X, is contained in the union of the standard rational boundary components F_b ($0 \leqq b \leqq s$). The intersection $F_b \cap \bar{\mathfrak{S}}$ is equal to $o_b \cdot \varpi_b(\mathfrak{S}')$ and is a Siegel domain in F_b ($0 \leqq b \leqq s$); moreover, any Siegel domain in F_b is contained in a Siegel domain obtained in this way by taking $\mathfrak{S}$ itself sufficiently large.*

Let a_ν and v_ν ($\nu = 1, 2, \cdots$) be sequences of elements in ${}_QA_t$ and ω respectively. We may assume that $v_\nu \to v$ and that $\lim a_\nu^{\beta_j} = d_j$ exists for every $j \leqq s$. If all the d_j are > 0, then $o \cdot a_\nu \cdot v_\nu$ tends to a point of $\mathfrak{S}$ itself. Otherwise let b be the greatest index such that $d_b = 0$. It follows from 4.4 that

$$(1) \qquad \gamma_{\sigma,i}(\log a_{\nu,\sigma}) \to -\infty \qquad (i \leqq c(b, \sigma), \sigma \in \Sigma)$$

and that $\gamma_{\sigma,i}(\log a_{\nu,\sigma})$ has finite limits if $i > c(b, \sigma)$. We have then

$$\lim o \cdot a_\nu = o_b \cdot a \; ,$$

where a is the element of $L(F_b) \cap \mathfrak{S}$ such that $a^{\beta_j} = d_j$ ($j > b$) if $b \neq s$, and is the identity if $b = s$. This implies

$$\lim_{\nu \to \infty} o \cdot a_\nu \cdot v_\nu = o_\mathfrak{b} \cdot a \cdot v \in o_\mathfrak{b} \cdot \varpi_\mathfrak{b}({}_\mathbf{Q} A_t \cdot \omega) \ .$$

Conversely, given $x \in o_\mathfrak{b} \cdot \pi({}_\mathbf{Q} A_t \cdot \omega)$, it is clear that we can find a sequence $a_\nu \cdot v_\nu$ as above such that $x = \lim o \cdot a_\nu \cdot v_\nu$. Taking into account the next to last paragraph of 4.4, this proves the lemma.

The following lemma is the analogue, in the present context, of Lemma 1 in [33].

4.6. Lemma. *We keep the assumptions of 4.4. Let $\mathfrak{S}$ be a Siegel domain in X, C a finite subset of $G_\mathbf{Q}$, and $\Omega = \mathfrak{S} \cdot C$. Then $\bar{\Omega}$ is contained in the union of finitely many rational boundary components. There exist finitely many elements $\gamma_i \in \Gamma$ $(1 \leq i \leq q)$ having the following property: for every $\gamma \in \Gamma$ such that $\bar{\Omega}\gamma \cap \bar{\Omega} \neq \varnothing$, there exists γ_j verifying the condition $a \cdot \gamma_j = a \cdot \gamma$ for every $a \in \bar{\Omega}\gamma^{-1} \cap \bar{\Omega}$.*

By 4.5, $\bar{\Omega}$ is contained in the union of the boundary components $F_\mathfrak{b} \cdot c (c \in C)$ which are all rational. Moreover, if we put $\mathfrak{X}_\mathfrak{b} = \mathfrak{S} \cap F_\mathfrak{b}$, then we may write $\bar{\Omega} = \bigcup_\lambda \mathfrak{X}_\lambda \cdot c_\lambda$ where λ runs through a finite set Λ, $c_\lambda \in C$, and $\mathfrak{X}_\lambda = \mathfrak{X}_{b(\lambda)}$ for a suitable $b(\lambda)$. Thus $\mathfrak{X}_\lambda$ is a Siegel domain of $F_{b(\lambda)}$ and has the Siegel property (4.4, 4.5).

For each pair $\lambda, \mu \in \Lambda$ for which there exists $\gamma \in \Gamma$ verifying

$$(1) \qquad\qquad \mathfrak{X}_\lambda \cdot c_\lambda \cdot \gamma \cap \mathfrak{X}_\mu \cdot c_\mu \neq \varnothing \ ,$$

choose one element $\gamma_{\lambda\mu} \in \Gamma$ fulfilling that condition and let $d_{\lambda\mu} = c_\lambda \cdot \gamma_{\lambda\mu} \cdot c_\mu^{-1}$. We have $b(\lambda) = b(\mu)$, $\mathfrak{X}_\lambda = \mathfrak{X}_\mu \subset F_{b(\lambda)}$, and $d_{\lambda\mu} \in N(F_{b(\lambda)})_\mathbf{Q}$. Let γ be an element of Γ satisfying (1). We define $e_{\lambda\mu}(\gamma) = c_\mu \cdot \gamma_{\lambda\mu}^{-1} \cdot \gamma \cdot c_\mu^{-1}$. We have $e_{\lambda\mu}(\gamma) \in N(F_{b(\lambda)})_\mathbf{Q}$. Condition (1) implies

$$(2) \qquad\qquad \mathfrak{X}_\lambda \cdot \varpi_\mathfrak{b}(d_{\lambda\mu}) \cdot \varpi_\mathfrak{b}\big(e_{\lambda\mu}(\gamma)\big) \cap \mathfrak{X}_\lambda \neq \varnothing \ .$$

If $b = s$, then $\mathfrak{X}_\lambda$ is relatively compact, and has trivially the Siegel property for any discontinuous group. If $b \neq s$, then $\varpi_\mathfrak{b}$ is canonically a $\mathbf{Q}$-morphism (3.10 (1)). Since $c_\mu \cdot \Gamma \cdot c_\mu^{-1}$ is arithmetic [13, §6.4] and $\mathfrak{X}_\lambda$ is a Siegel domain (4.4), the Siegel property obtains. In both cases, it shows that there are only finitely many possibilities for $\varpi_\mathfrak{b}(e_{\lambda\mu}(\gamma))$ when γ varies through the elements of Γ verifying (1). We have thus shown the existence of a finite subset D_λ of $c_\lambda^{-1}(N(F_{b(\lambda)})_\mathbf{Q})c_\mu$ such that (1) implies the existence of $\tau(\gamma) \in D_\lambda$ verifying

$$x \cdot \tau(\gamma) = x \cdot \gamma \ , \qquad\qquad\qquad (x \in \mathfrak{X}_\lambda \cdot c_\lambda) \ .$$

This means that, if $\bar{\Omega} \cdot \gamma \cap \bar{\Omega} \neq \varnothing$, there are only finitely many possibilities for the action of γ on $\bar{\Omega} \cdot \gamma^{-1} \cap \bar{\Omega}$, whence the lemma.

4.7. Lemma. *We keep the assumptions of 4.4. Let X^* be the union of the rational boundary components of X. Then there exists a fundamental* set $\Omega = \mathfrak{S}\cdot C$ for Γ such that $X^* = \bar{\Omega}\cdot\Gamma = \bar{\Omega}\cdot G_{\mathbf{Q}}$.

By 4.6, we have $\bar{\Omega} \subset X^*$, and hence $\bar{\Omega}\cdot G_{\mathbf{Q}} \subset X^*$, for every fundamental set of the type considered in 4.2.

By 4.2, 4.3, and 4.5, there exists a fundamental set Ω such that $\bar{\Omega} \cap F_{\mathbf{b}}$ is a fundamental set for $\Gamma(F_{\mathbf{b}})$ $(0 \leq b \leq s)$. For this set we have then $F_{\mathbf{b}} \subset \bar{\Omega}(\Gamma \cap N(F_{\mathbf{b}}))$. Let now F be a rational boundary component. There exists $c \in G_{\mathbf{Q}}$ and b such that $F = F_{\mathbf{b}}\cdot c$ (3.8). Let $\gamma \in N(F_{\mathbf{b}}) \cap \Gamma$. By the Siegel property, there exists a finite subset D of Γ (depending on γ), such that

$$\Omega\cdot\gamma\cdot c \subset \Omega\cdot D .$$

This implies

$$\bar{\Omega}\cdot\gamma\cdot c \subset \bar{\Omega}\cdot D \subset \bar{\Omega}\cdot\Gamma ;$$

therefore,

$$F = F_{\mathbf{b}}\cdot c \subset \bar{\Omega}\cdot\Gamma ,$$

and $X^* \subset \bar{\Omega}\cdot\Gamma$, which completes the proof.

4.8. We now turn to the general case and let Γ be an arithmetic subgroup of $H(X)$ as in 3.3 (i). We take the notation of 3.3 (i). The union X^* of the rational boundary components of X is the product of the similarly defined spaces X_i^* corresponding to the different simple $\mathbf{Q}$-factors G_i of G. The group $H(X)$ operates continuously on the natural compactification of X (11.2), and hence $H(X)_{\mathbf{Q}}$ operates on X^*. Furthermore, (11.2), the restriction of each element $h \in H(X)$ to each boundary component is holomorphic. By 4.7, applied to X_i^* and $\Gamma_i = \Gamma \cap G_{i\mathbf{R}}^0$, and 4.3, there exists then a fundamental set Ω for Γ in X such that $\bar{\Omega}\cdot\Gamma = X^*$. For $x \in X^*$, we let $\Gamma_x = \{\gamma \in \Gamma, x\cdot\gamma = x\}$. The space X^* will be endowed with the topology $\mathfrak{S}(\Omega, \Gamma)$ in which a fundamental set of neighborhoods of $x \in X^*$ is the set of all subsets U of X^* containing x and having the following properties: $U\cdot\gamma = U$ if $\gamma \in \Gamma_x$ and $U\cdot\gamma \cap \bar{\Omega}$ is a neighborhood of $x\cdot\gamma$ in $\bar{\Omega}$, in the natural topology of $\bar{\Omega}$ whenever $x\cdot\gamma \in \bar{\Omega}$. This topology was introduced, in a similar setting, in [33], and will be called the *Satake topology* for X^*.

4.9. Theorem. *We keep the assumptions of 4.8. The Satake topology is the unique topology on X^* having the following properties:*

(i) *It induces the natural topology on X and on the closure of any fundamental set Ω for any arithmetic group $\Gamma \subset H(X)$.*

(ii) *The elements of the group $H(X)_{\mathbf{Q}}$ operate continuously on X^*.*

(iii) *If x and x' are not equivalent with respect to Γ, then there exist neighborhoods U of x and U' of x' such that $U\cdot\Gamma \cap U' = \varnothing$.*

(iv) *For each $x \in X^*$, there exists a fundamental set of neighborhoods $\{U\}$ of x such that $U \cdot \gamma = U$ if $\gamma \in \Gamma_x$, and $U \cdot \gamma \cap U = \varnothing$ if $\gamma \notin \Gamma_x$.*

PROOF. We wish to appeal to Theorem 1' of [33]. We first consider a fundamental set as in 4.8, used to define the topology. As was remarked in 4.8, every element of $H(X)_\mathbf{Q}$ operates continuously on X^*, viewed as a subspace of $\bar{X}$ in the natural topology, and this implies condition (2) of Theorem 1', loc. cit. As to the condition (3), it is just Lemma 4.6. Thus, Theorem 1' applies. Also, by [33, Remark 2, p. 563], the topology induced by $\mathbb{S}(\Omega, \Gamma)$ on X is the natural one. This proves 4.9, with $H(X)_\mathbf{Q}$ replaced by Γ in (ii).

Let now $c \in H(X)_\mathbf{Q}$, and let Ω be a fundamental set for $c \cdot \Gamma \cdot c^{-1} \cap \Gamma$. It is also one for Γ and $c \cdot \Gamma \cdot c^{-1}$ in view of the Siegel property. It follows directly from the definitions, and from the fact that c acts continuously on $\bar{X}$, that c carries $\mathbb{S}(c \cdot \Gamma \cdot c^{-1}, \Omega)$ onto $\mathbb{S}(\Gamma, \Omega \cdot c)$. However, as pointed out in [33, Remark 3, p. 563], these two topologies of X^* are identical, whence (ii).

In the sequel we shall also write $\mathbb{S}$ for $\mathbb{S}(\Omega, \Gamma)$.

4.10. PROPOSITION. *Let F be a rational boundary component of X, and $x \in F$. Then x has a neighborhood U in X^* verifying 4.9 (iv), and such that a rational boundary component F' intersects U if and only if $F \subset \bar{F}'$. The closure of F in X^* is the union of the rational boundary components contained in the closure of F in the natural compactification $\bar{X}$ of X.*

We let $\mathbb{S}$ and $\mathcal{T}$ denote respectively the Satake topology of X^* and the topology of the natural compactification $\bar{X}$. Let $\bar{\Omega}$ be as in 4.8. Then $\mathbb{S}$ and $\mathcal{T}$ coincide on $\bar{\Omega}$. We may (and shall) assume that $x \cdot \Gamma$ contains an interior point of $\bar{\Omega} \cap F$. It follows from 4.6 that we may find finitely many elements $\gamma_1, \cdots, \gamma_m \in \Gamma$ such that $\bar{\Omega} \cap x \cdot \Gamma = \{x \cdot \gamma_1, \cdots, x \cdot \gamma_m\}$. Let U_i $(1 \le i \le m)$ be a subset of X^* such that $U_i \cdot \gamma_i$ is a neighborhood of $x \cdot \gamma_i$ in $\bar{\Omega}$. Then $U = \bigcup_i U_i \cdot \Gamma_x$ is a neighborhood of x, which satisfies 4.9 (iv) if the U_i are sufficiently small, and we get in this way a fundamental system of neighborhoods of x in $\mathbb{S}$ [33, p. 562]. By 4.6, $\bar{\Omega}$ is the union of its intersections with finitely many rational boundary components. We may therefore assume U_i to be such that if the rational boundary component F' intersects $U_i \cdot \gamma_i$, then $x \cdot \gamma_i$ belongs to the closure of $F' \cap \bar{\Omega}$.

Let now $F' \cap U \ne \varnothing$. There exist then $\gamma \in \Gamma_x$ and an index i such that $F' \cap U_i \cdot \gamma \ne \varnothing$, whence $F' \cdot \gamma^{-1} \cdot \gamma_i \cap U_i \cdot \gamma_i \ne \varnothing$; by the condition just imposed on U_i, this implies that $x \cdot \gamma_i$ belongs to the closure of $F' \cdot \gamma^{-1} \cdot \gamma_i \cap \bar{\Omega}$, either in $\mathbb{S}$ or $\mathcal{T}$ since both coincide on $\bar{\Omega}$.

Consequently x belongs to the closure of $F' \cdot \gamma^{-1}$, and hence to that of F', in both $\mathbb{S}$ and $\mathcal{T}$. This proves our first assertion and shows that if x belongs to the closure of F' in $\mathfrak{S}$, then it does so in $\mathfrak{T}$, too. To prove the converse, we may assume G to be $\mathbf{Q}$-simple, and $F = F_\mathfrak{b}$ to be standard. Then $F_\mathfrak{c}\,(c \geqq b)$ belongs to the closure of $F_\mathfrak{b}$ in both the $\mathfrak{S}$ and the $\mathfrak{T}$ topologies.

Let F' be a rational boundary component contained in the $\mathfrak{T}$-closure of F. There exist $c \geqq b$ and $g \in N(F)_\mathbf{Q}$ such that $F' \cdot g = F_\mathfrak{c}$ (3.8, 3.9). Since g defines a homeomorphism in the $\mathfrak{S}$-topology (4.9 (ii)), it follows that F' is also in the $\mathfrak{S}$-closure of F.

4.11. COROLLARY. *The quotient $V^* = X^*/\Gamma$, endowed with the quotient topology, is a compact Hausdorff space. $V = X/\Gamma$ is an open everywhere dense subset. V^* is the finite union of subspaces $V_i = F_i/\Gamma(F_i)$, where F_i runs through a set of representatives of equivalence classes modulo Γ of rational boundary components. The closure of V_i is the union of V_i and of subspaces V_j of strictly smaller dimension.*

The first three assertions are obvious consequences of 4.9 and 4.10, once it is shown that, when G is $\mathbf{Q}$-simple, $F_\mathfrak{b} \cdot H(X)_\mathbf{Q}/\Gamma$ is covered by the images of finitely many rational boundary components. Let $\Omega = \mathfrak{S} \cdot C$ be the fundamental set of 4.7, where $\mathfrak{S}$ is a Siegel domain and C a finite subset of $G_\mathbf{Q}$. Let F be a rational boundary component of type $F_\mathfrak{b}$. There exist $c \in C$ and $\gamma \in \Gamma$ such that $\bar{\mathfrak{S}} \cdot c \cdot \gamma \cap F \neq \varnothing$. But the intersection of $\bar{\mathfrak{S}}$ with the orbit of $F_\mathfrak{b}$ under $G_\mathbf{R}^0$ is contained in $F_\mathfrak{b}$ by 4.5; therefore, $F_\mathfrak{b} \cdot c \cdot \gamma \cap F \neq \varnothing$, and $F_\mathfrak{b} \cdot c \cdot \gamma = F$, so that $F_\mathfrak{b} \cdot H(X)_\mathbf{Q}/\Gamma$ is a quotient of $F_\mathfrak{b} \cdot C$.

The last assertion of the corollary follows from 4.10, because in $\bar{X}$, the closure of a boundary component F is the union of F and of boundary components of strictly smaller dimension (1.5).

REMARK. The proper rational boundary components correspond to the proper maximal parabolic $\mathbf{Q}$-subgroups (3.7). Since parabolic $\mathbf{Q}$-subgroups are conjugate if and only if they are conjugate by an element of $G_\mathbf{Q}$ (2.2), the V_i's of the corollary are in one-to-one correspondence with the orbits of Γ in $G_\mathbf{Q}/P_\mathbf{Q}$, where P runs through the standard maximal parabolic $\mathbf{Q}$-subgroups.

Let G be the quotient of the symplectic group $\mathbf{Sp}(2n, \mathbf{C})$ by its center, and $\Gamma = \mathbf{Sp}(2n, \mathbf{Z})$ be Siegel's modular group. The proper maximal parabolic $\mathbf{Q}$-subgroups are the stability groups of the rational isotropic subspaces. It is elementary that two rational isotropic subspaces of the same dimension q are transforms of each other by an element of Γ, so that we have $G_\mathbf{Q} = \Gamma \cdot P_\mathbf{Q}$ for every maximal parabolic $\mathbf{Q}$-subgroup. The boundary component corresponding to such a subspace is isomorphic to Siegel's upper half-plane of degree q, and thus $V^* = \bigcap_{0 \leqq q \leqq n} V_q$, where V_q is the quotient of the Siegel upper half-plane of degree q by the corresponding modular group. We get therefore again the compactification introduced by Satake and later considered in [35].

4.12. *Truncated Siegel domains.* In order to give a more precise description of a fundamental set of neighborhoods of a point in V^*, we need to introduce certain subsets of a Siegel domain. To define them, we assume again, for convenience, that G is simple over $\mathbf{Q}$.

Let $\mathfrak{S}' = {}_{\mathbf{Q}}A_t \cdot \omega$ be a Siegel domain in $P_{\mathbf{R}}$ and $\mathfrak{S} = o \cdot \mathfrak{S}'$ the corresponding Siegel domain in X (4.2). Fix an index $b \leqq s$. For a positive number u and a subset $E \subset F_{\mathbf{b}}$, we let

$$(1) \qquad \mathfrak{S}'_b(u, E) = \{s \in \mathfrak{S}' \mid s^{\beta_b} \leqq u, \, o_{\mathbf{b}} \cdot s \in E\} \,,$$

and

$$(2) \qquad \mathfrak{S}_b(u, E) = o \cdot \mathfrak{S}'_b(u, E) \,.$$

The sets described by (1), (2) will be called $F_{\mathbf{b}}$-*adapted truncated Siegel domains*. The subscript b will sometimes be omitted, if it is clear from the context, or replaced by F.

An element $s \in \mathfrak{S}'$ can be written uniquely in the form

$$(3) \qquad s = a_1 \cdot h_b \cdot a_2 \cdot w \qquad (a_1 \in \mathfrak{D}Z(F_{\mathbf{b}}) \cap {}_{\mathbf{Q}}A, \, a_2 \in L(F_{\mathbf{b}}) \cap {}_{\mathbf{Q}}A, \, h_b \in {}_{\mathbf{Q}}A_b, \, w \in \omega) \,,$$

where ${}_{\mathbf{Q}}A_b$ is the kernel of all simple $\mathbf{Q}$-roots β_i $(i \neq b)$. Of course, all elements on the right hand side depend (continuously) on s. However, for simplicity, we shall not make this explicit in the notation as long as no confusion arises. We collect a few remarks about this decomposition.

(i) The element h_b annihilates all simple $\mathbf{Q}$-roots except β_b, and a_1 (resp. a_2) annihilates β_i for $i \geqq b + 1$ (resp. $i < b$); hence,

$$(4) \qquad s^{\beta_i} = a_1^{\beta_i} \,, \qquad\qquad (i \leqq b - 1)$$

$$(5) \qquad s^{\beta_i} = a_2^{\beta_i} \,, \qquad\qquad (i \geqq b + 1) \,.$$

If the roots β_i $(i \geqq b + 1)$ are multiplicatively bounded on a_2, then a_2 is bounded; hence, there exists t' such that the set of products $h \cdot a_1$ is contained in ${}_{\mathbf{Q}}A_{t'}$. In view of 4.5, it is clear that the β_i's $(i \geqq b + 1)$ are multiplicatively bounded if E is relatively compact, as will be the case unless the contrary is stated. If so, the discussion of 4.5 implies

$$(6) \qquad \bigcap_{u>0} \bar{\mathfrak{S}}'_b(u, E) = o_{\mathbf{b}} \cdot \mathfrak{S}'_b(u_0, E) \qquad\qquad (\text{some } u_0 > 0) \,.$$

(ii) The group ${}_{\mathbf{Q}}A_b$ centralizes $K' = K \cap P_{\mathbf{R}}$, and is contained in the isotropy group of $o_{\mathbf{b}}$. Since $\mathfrak{S}' = K' \cdot {}_{\mathbf{Q}}A_t \cdot \omega$, $(\omega \subset V_{\mathbf{R}})$, we have then

$$(7) \qquad h \cdot \mathfrak{S}'_b(u, E) \subset \mathfrak{S}'_b(v, E) \,, \qquad\qquad (v = u \cdot h^{\beta_b}, \, h \in {}_{\mathbf{Q}}A_b) \,,$$

and hence also

$$o_{\mathrm{b}} \cdot \mathfrak{S}' = o_{\mathrm{b}} \cdot \mathfrak{S}'_b(u, F_{\mathrm{b}}) \,, \qquad\qquad (u > 0) \,,$$

$$o_{\mathrm{b}} \cdot \mathfrak{S}'(u, E) = o_{\mathrm{b}} \cdot \mathfrak{S}'(v, E) \qquad\qquad (u, v > 0) \,.$$

(iii) Let (X_i) be the basis of $_{\mathrm{Q}}\mathfrak{a}$ with respect to which the simple **Q**-roots are written as in 1.2. Then the Lie algebra of $_{\mathrm{Q}}A \cap \mathfrak{D}Z(F_{\mathrm{b}})$ (resp. $_{\mathrm{Q}}A \cap L(F_{\mathrm{b}})$, resp. $_{\mathrm{Q}}A_b$) is generated by $X_i - X_{i+1}$ $(i = 1, \cdots, b-1)$ (resp. X_i $(i = b+1, \cdots, s)$, resp. $X_1 + \cdots + X_b$).

4.13. LEMMA. *We keep the assumptions of* 4.4. *Let* $x \in F_{\mathrm{b}}$, *and let* x^* *be the image of* x *in* V^*. *Let* $\Omega = \mathfrak{S} \cdot C$ *be a fundamental set verifying* 4.7, *where* $\mathfrak{S}$ *is an open Siegel domain whose closure contains* x. *Then there exist finitely many elements* $e_i \in N(F_{\mathrm{b}})_{\mathrm{Q}}$ *with the following property: the image* U *in* V^* *of* $U' = \bigcup_i \bar{\mathfrak{S}}(u, E_i) \cdot e_i$, *where* E_i *is a relatively compact neighborhood of* $x \cdot e_i^{-1}$ *in* F_{b}, *is a neighborhood of* x^*; *the set* U *(resp.* $U' \cdot \Gamma_x$) *describes a fundamental set of neighborhoods of* x^* *in* V^* *(resp.* $x \in X^*$) *if* $u \to 0$ *and, independently,* E_i *runs through a fundamental set of neighborhoods of* $x \cdot e_i^{-1}$ *in* F_{b}. *Furthermore, we get an equivalent set of neighborhoods of* x^* *if we replace* $\mathfrak{S}$ *by any open Siegel domain* $\mathfrak{S}' \supset \mathfrak{S}$.

It is clear from 4.5 that $\bar{\mathfrak{S}}_b(u, E)$ contains an open neighborhood of x in $\bar{\mathfrak{S}}_b$, and 4.12 shows that we get in this way a fundamental set of neighborhoods of x in $\bar{\mathfrak{S}}$. We let D be the finite set of elements $c \in C$ such that $x \in \bar{\mathfrak{S}} \cdot c$. If $d \in D$, then we get similarly a fundamental set of neighborhoods of x in $\bar{\mathfrak{S}} \cdot d$ by taking subsets $\bar{\mathfrak{S}}_b(u, E) \cdot d$, where E is a neighborhood of $x \cdot d^{-1}$ in F_{b}. Since $\bar{\Omega}$ is the finite union of the closed subsets $\bar{\mathfrak{S}} \cdot c$, $(c \in C)$, it follows that x has a fundamental set of neighborhoods in $\bar{\Omega}$ of the form $\bigcup_d \bar{\mathfrak{S}}_b(u, E_d) \cdot d$, where d runs over D and E_d is a neighborhood of $x \cdot d^{-1}$.

By 4.6, $x \cdot \Gamma \cap \bar{\Omega}$ is finite. Let γ_j $(1 \leq j \leq q)$ be elements of Γ such that $x \cdot \Gamma \cap \bar{\Omega} = \{x \cdot \gamma_1, \cdots, x \cdot \gamma_q\}$, and put $x \cdot \gamma_j = x_j$. Let Γ_x be the isotropy group of x in Γ. From 4.10, it follows that if $U_j \cdot \gamma_j$ is a neighborhood of x_j in $\bar{\Omega}$, then $\bigcup_j U_j \cdot \Gamma_x$ is a neighborhood of x in X^*, and that we get in this way a fundamental set of neighborhoods. For U_j, we may by the above take a finite union of sets $\bar{\mathfrak{S}}_b(u, E) \cdot c$, where $c \in C$ is such that $x_j \in \bar{\mathfrak{S}} \cdot c$, and E is a neighborhood of $x_j \cdot c^{-1}$ in F_{b}. From this, our first assertion follows, with e_i running over the set of products $c \cdot \gamma_j^{-1}$ for which $x \cdot \gamma_j \cdot c^{-1} \in \bar{\mathfrak{S}}$. Such elements are rational over **Q**, and they belong to $N(F_{\mathrm{b}})$, since they transform x into a point of F_{b}.

As for the second assertion, the images of the sets $\bigcup_j U'_j$, which are constructed simply by replacing $\mathfrak{S}$ by $\mathfrak{S}'$ while allowing c and γ_j to run over the same sets of objects as in the definition of U_j, are relatively compact neighborhoods of x^*, and for $u \to 0$ and $E_{j,c}$ decreasing through a basis of compact neighborhoods of $x_j \cdot c^{-1}$ for each j and c, we obtain a decreasing family of *compact* neighborhoods of x^*; it follows from 4.12 (6) that the intersection of all these is just x^*. Since V^* is compact and Hausdorff, and since all the spaces we consider are second countable, it follows that these neighborhoods also give a basis of neighborhoods of x^* in V^*, which proves the last assertion.

REMARK. Let $E_n\,(n = 1, \cdots)$ be a decreasing sequence of neighborhoods of x in F_b, whose intersection is x, and let u_n be a sequence of real numbers tending to 0. Then the images in V^* of the sets $U_n = \bigcup_i \mathfrak{S}(u_n, E_n \cdot e_i^{-1}) \cdot e_i$ form a fundamental set of neighborhoods of x^*.

4.14. LEMMA. *Let $b\,(1 \leq b \leq s)$ and a Siegel domain $\mathfrak{S}'_0$ in P_R be given. Then there exists a Siegel domain $\mathfrak{S}' \supset \mathfrak{S}'_0$ in P_R such that $\mathfrak{S}_b(u, E) = o \cdot \mathfrak{S}'(u, E)$ is connected for every $u > 0$ and every connected open subset E of $o_b \cdot \mathfrak{S}'$.*

We choose an Iwasawa decomposition $P_R = K' \cdot A \cdot N$ of P_R where

$$K' = K \cap P, \qquad A = (M \cap A) \times {}_Q A, \qquad N = (M \cap N) \cdot U_R.$$

We may assume as before that ${}_Q T \subset {}_R T \subset T$, and choose compatible orderings on the root systems. The groups $Z(F_b) \cap A \cdot N$ and $L(F_b) \cap A \cdot N$ are connected (for otherwise they would have infinitely many components, which is impossible since $Z(F_b)_C$, $(A \cdot N)_C$, and $L(F_b)_C$ are algebraic); they are also contractible, and the product mapping yields a homeomorphism of

$$(L(F_b) \cap A \cdot N) \times (Z(F_b) \cap A \cdot N)$$

onto $A \cdot N$. Furthermore $(L(F_b) \cap A)$ and $(L(F_b) \cap N)$ are the A and N part of an Iwasawa decomposition of $L(F_b)$. It follows that $(a, n) \mapsto o \cdot a \cdot n$ and $(a, n) \mapsto o_b \cdot a \cdot n$ yield homeomorphisms of $A \times N$ onto X and of

$$(L(F_b) \cap A) \times (L(F_b) \cap N)$$

onto F_b. We also have ($\approx$ meaning homeomorphism)

$$L(F_b) \cap A \cdot N \approx \left({}_Q A \cap L(F_b)\right) \times (L(F_b) \cap V_R \cap A \cdot N),$$
$$Z(F_b) \cap A \cdot N \approx \left({}_Q A \cap Z(F_b)\right) \times (Z(F_b) \cap V_R \cap A \cdot N),$$

where the homeomorphisms are given by the product mapping.

Let us write $\mathfrak{S}'_0 = K' \cdot {}_Q A_{t_0} \cdot \omega_0$, $(\omega_0 \subset V_R)$. We take $t > t_0$, and choose ω_2, ω_1 open, relatively compact, and connected in $L(F_b) \cap V_R \cap A \cdot N$ and $Z(F_b) \cap V_R \cap A \cdot N$ such that $K' \cdot \omega_1 \cdot \omega_2 \supset K' \cdot \omega_0$, (note that K' meets every connected component of P_R). We claim that $\mathfrak{S}' = \mathfrak{S}'_{t,\omega}$, $(\omega = K' \cdot \omega_1 \cdot \omega_2)$, verifies our conditions. Since K' is contained in the isotropy group of o_b in P_R, and is the isotropy group of o in P_R, an elementary argument shows that it suffices to prove that $K' \backslash S'_b(u, x)$ is connected for every $u > 0$ and every

$x \in o_{\mathrm{b}} \cdot \mathfrak{S}'$. We may write

$$x = o_{\mathrm{b}} \cdot a' \cdot w' \qquad\qquad \left(a' \in L(F_{\mathrm{b}}) \cap {}_{\mathrm{Q}}A, \ w' \in \omega_2 \right) .$$

An arbitrary element $s \in \mathfrak{S}'$ can be written as a product

$$s = k' \cdot a_1 \cdot h_b \cdot a_2 \cdot w_1 \cdot w_2 \, ,$$

where a_1, h_b, a_2 are as in 4.12, and $k' \in K'$, $w_i \in \omega_i$ $(i = 1, 2)$. The remarks made above imply that $o_{\mathrm{b}} \cdot s = x$ if and only if $a_2 = a'$, $w_2 = w'$. Thus w_1 runs through ω_1, which is connected by assumption, k' through K', and $a_1 \cdot h_b$ runs through the elements $y \in Z(F_{\mathrm{b}}) \cap {}_{\mathrm{Q}}A$ satisfying the conditions

$$y^{\beta_i} \leqq t \ (i < b) \, , \qquad y^{\beta_b} \leqq u \cdot t \, ,$$

which define clearly a connected set; whence our contention.

4.15. PROPOSITION. *We keep the notation of 4.11. Every point $v^* \in V^*$ has a fundamental set of open neighborhoods U such that $U \cap V$ is connected.*

Let Γ' be a normal subgroup of finite index of Γ. Then $V^* = V^*(\Gamma) = X^*/\Gamma$ is the quotient of X^*/Γ' by Γ/Γ'. Let G_i $(1 \leq i \leq m)$ be the simple **Q**-factors of G and, as in 3.3 (i), let Γ' be the group generated by the $\Gamma'_i = G^0_{i\mathrm{R}} \cap \Gamma$. Then $X^*/\Gamma' = \prod_i X^*_i/\Gamma_i$. It suffices therefore to consider the case where G is simple over **Q** and $\Gamma \subset G^0_{\mathrm{R}}$. We may furthermore assume $x \in F_{\mathrm{b}}$. We use the notation of 4.13, 4.14, take $\mathfrak{S}'_0$ with $\mathfrak{S}_0$ verifying 4.7, and $\bar{\mathfrak{S}}_0$ containing x. We choose the elements $e_i \in N(F_{\mathrm{b}})_{\mathrm{Q}}$ as in the proof of 4.13; in particular, $x \cdot e_i^{-1} \in o_{\mathrm{b}} \cdot \mathfrak{S}'_0$. Moreover, we may so arrange things that each $x \cdot e_i^{-1}$ is contained in the (relative) interior of $F_{\mathrm{b}} \cap \bar{\mathfrak{S}}_0$; this is an easy consequence of the two following facts: the fixed set C of 4.13 is finite, and the number of elements in the intersection of any orbit of Γ in X^* with the closure of any Siegel domain (constructed with respect to the given torus, ordering, etc.) is finite.

Let E be a connected open neighborhood of x in F, small enough so that $E \cdot e_i^{-1} \subset o_{\mathrm{b}} \cdot \mathfrak{S}'_0$ for all i. Given $u > 0$, there exists $g_0 \in \mathfrak{S}'_{0,b}(u, E)$ such that $x = o_{\mathrm{b}} \cdot g_0$ (see 4.12 (ii)). By construction, $o_{\mathrm{b}} \cdot g_0 \cdot e_i^{-1} \in o_{\mathrm{b}} \cdot \mathfrak{S}'_0$; hence (loc. cit.), $o_{\mathrm{b}} \cdot g_0 \cdot e_i^{-1} \in o_{\mathrm{b}} \cdot \mathfrak{S}'_{0,b}(u, E \cdot e_i^{-1})$; there exists therefore u_i in the isotropy subgroup of o_{b} in $N(F_{\mathrm{b}})$ such that $u_i \cdot g_0 \cdot e_i^{-1} \in \mathfrak{S}'_{0,b}(u, E \cdot e_i^{-1})$.

The groups $K \cap L(F_{\mathrm{b}})$ and $K \cap Z(F_{\mathrm{b}})$ are maximal compact in $L(F_{\mathrm{b}})$ and $Z(F_{\mathrm{b}})$, and the first one is the isotropy group of o_{b} in $L(F_{\mathrm{b}})$. Moreover,

$$Z(F_{\mathrm{b}}) = \left(K \cap Z(F_{\mathrm{b}}) \right) \cdot \left(P \cap Z(F_{\mathrm{b}}) \right) .$$

Since $N(F_{\mathrm{b}}) = L(F_{\mathrm{b}}) \cdot Z(F_{\mathrm{b}})$ by 3.3, we may write (not uniquely):

$$u_i = k'_i \cdot k''_i \cdot s_i \qquad \left(k'_i \in K \cap L(F_{\mathrm{b}}); \ k''_i \in K \cap Z(F_{\mathrm{b}}); \ s_i \in Z(F_{\mathrm{b}}) \cap P \right) .$$

Let $\mathfrak{S}'$ be an open Siegel domain in $P_{\mathbf{R}}$ satisfying 4.14 which contains $\mathfrak{S}'_0$ and the elements $s_i \cdot g_0$. The images in V^* of the sets $U = \bigcup_i \bar{\mathfrak{S}}_b(u, E' \cdot e_i^{-1}) \cdot e_i$, as u runs over all strictly positive numbers, and E' over a basis of connected open neighborhoods of x contained in E, form a fundamental set of neighborhoods of v^*, and their interiors form a fundamental set of open neighborhoods. It is then enough to show that the interior of $U \cap X$ is connected. This latter set is the union over i, of the open sets $o \cdot \mathfrak{S}'_b(u, E \cdot e_i^{-1}) \cdot e_i$. We are therefore reduced (4.14) to proving that

$$(1) \qquad o \cdot \mathfrak{S}'_b(u, E' \cdot e_i^{-1}) \cdot e_i \cap o \cdot \mathfrak{S}'_b(u, E') \neq \varnothing \; ,$$

for all $u > 0$, all i, and all E' subject to our previous conditions.

We let h be an element of $_{\mathbf{Q}}A_b$ such that $h^{8b} \leqq t^{-1} \cdot u$.

By construction, $o_b \cdot s_i \cdot g_0 = x$ and $s_i \cdot g_0 \in \mathfrak{S}'$; hence, $s_i \cdot g_0 \in \mathfrak{S}'_b(t, E')$, and therefore (4.12 (ii)):

$$(2) \qquad h \cdot s_i \cdot g_0 \in \mathfrak{S}'_b(u, E') \; .$$

By 4.12 (ii), we have

$$o \cdot h \cdot u_i \cdot g_0 \cdot e_i^{-1} \in o \cdot \mathfrak{S}'_b(u, E' \cdot e_i^{-1}) \; .$$

But $_{\mathbf{Q}}A_b$ centralizes $L(F_b)$ and $K \cap Z(F_b)$; therefore,

$$o \cdot h \cdot u_i \cdot g_0 \cdot e_i^{-1} = o \cdot h \cdot k_i' \cdot k_i'' \cdot s_i \cdot g_0 \cdot e_i^{-1} = o \cdot h \cdot s_i \cdot g_0 \cdot e_i^{-1} \; ,$$

whence

$$(3) \qquad o \cdot h \cdot s_i \cdot g_0 \in o \cdot \mathfrak{S}'_b(u, E' \cdot e_i^{-1}) \cdot e_i \; ,$$

which, together with (2), proves (1).

4.16. PROPOSITION. *We keep the previous notation. Let $J(x, g)$ be the functional determinant function in the unbounded realization associated to the rational boundary component $F = F_b$ (3.3) and $g \in N(F)_{\mathbf{Q}}$. Then $J(x, g)$ is multiplicatively bounded on $\mathfrak{S}_F$, where $\mathfrak{S}_F = \mathfrak{S}_b(u, E)$ (E compact in F) is an F-adapted truncated Siegel domain.*

We have $\mathfrak{S}_F = o \cdot \mathfrak{S}'_F$, and

$$(1) \qquad J(x, g) = J(o \cdot s, g) = J(o, s)^{-1} \cdot J(o, s \cdot g) \qquad\qquad (s \in \mathfrak{S}'_F)$$

We want to estimate both factors on the right hand side. We write as in 4.12,

$$(2) \qquad s = a_1 \cdot h \cdot a_2 \cdot q \qquad (a_1 \in \mathfrak{D}Z(F) \cap {}_{\mathbf{Q}}A, \, a_2 \in L(F) \cap {}_{\mathbf{Q}}A, \, h \in {}_{\mathbf{Q}}A_b, \, q \in \omega) \; .$$

As pointed out in 4.12 (i), since $s \in \mathfrak{S}'_F$, the element a_2 is bounded on $\mathfrak{S}'_F$, and there exists $t' \geqq t$ such that

$$(3) \qquad h \cdot a_1 \in {}_{\mathbf{Q}}A_{t'} \; .$$

From (2), we get

$$s = a_2 \cdot h \cdot a_1 \cdot q \cdot (h \cdot a_1)^{-1} \cdot h \cdot a_1 = c \cdot h \cdot a_1 \qquad \left(c = a_2 \cdot h \cdot a_1 \cdot q \cdot (h \cdot a_1)^{-1} \right) ;$$

hence, by 1.11,

$$(4) \qquad J(o, s) = J(o, c) \cdot J(o, h) .$$

By 4.1, and the above remark on a_2, the element c varies in a compact set when $s \in \mathfrak{S}'_F$; hence,

$$(5) \qquad J(o, s) \asymp J(o, h) \qquad (s \in \mathfrak{S}'_F) .$$

The Bruhat decomposition gives

$$g = u \cdot t \cdot w \cdot v , \qquad (u, v \in U_{\mathbf{Q}}, t \in Z(S)_{\mathbf{Q}}, w \in N_{N(F)}(S)_{\mathbf{Q}})$$

whence

$$s \cdot g = d \cdot f \cdot v \qquad (d = c \cdot h \cdot a_1 \cdot u \cdot (h \cdot a_1)^{-1} \cdot t \cdot w; f = w^{-1} \cdot h \cdot a_1 \cdot w) .$$

Since c varies in a compact set, 4.1 and (2) show that so does d. We may write

$$v = v' \cdot v'' \qquad (v' \in L(F) \cap U, v'' \in Z(F) \cap U) .$$

The element v' commutes with the subtorus $Z(F) \cap S = S_\theta$ ($\theta = \{\beta_{b+1}, \cdots, \beta_s\}$). Since θ is a connected component of the set $_{\mathbf{Q}}\Delta - \{\beta_b\}$ of simple $\mathbf{Q}$-roots of $N(F)$, the Weyl group relative to $\mathbf{Q}$ of $N(F)$ also leaves S_θ stable. Therefore v' commutes with f, we have $s \cdot g = d \cdot v' \cdot f \cdot v''$, and, by (1.11),

$$(6) \qquad J(o, s \cdot g) = J(o, d \cdot v') \cdot J(o, f) \cdot J(o, v'') .$$

But v' and v'' are fixed and d varies in a compact set. Consequently

$$(7) \qquad J(o, s \cdot g) \asymp J(o, f) \qquad (s \in \mathfrak{S}'_F) .$$

The restriction of $J(o, g)$ to $Z(F)$ is a character (1.8, 1.11), which is of course equal to one on the derived group. Furthermore, the relative Weyl group of $N(F)$ is generated by the symmetries with respect to the hyperplanes annihilating the simple $\mathbf{Q}$-roots $\beta_i \neq \beta_b$, and hence it acts trivially on $_{\mathbf{Q}}A_b$. We have therefore

$$(8) \qquad J(o, f) = J(o, h) ,$$

and the proposition follows now from (1), (5), (7), and (8).

II. Automorphic Forms

5. Poincaré series

5.1. In this section, G is a complex connected reductive group defined over $\mathbf{R}$, whose group of real points has a compact center, H, a subgroup of finite index of $G_{\mathbf{R}}$, K, a maximal compact subgroup of H, V, a finite dimensional complex Hilbert space, and $\rho \colon K \to \mathbf{GL}(V)$ a unitary representation.

A function $f\colon H \to V$ is of *type* ρ on the left (resp. right) if $f(k\cdot g) = \rho(k)\cdot f(g)$ (resp. $f(g\cdot k) = \rho(k^{-1})\cdot f(g)$) $(g \in H,\ k \in K)$. It is of *finite type*, with respect to K, or K-finite, on the right (resp. left) if the set of right translates $r_k f$ (resp. left translates $l_k f$) of f by elements of K spans a finite dimensional vector space over C of V-valued functions. This is in particular the case if f is of type ρ, and the general case can be subsumed into that one, at the cost of changing V, as is easily seen.

5.2. The universal enveloping algebra $\mathfrak{U}(\mathfrak{g})$ of $\mathfrak{g}$ is identified in the customary manner with the algebra of right invariant differential operators on H. In particular, if $X \in \mathfrak{g}$ and f is a differentiable V-valued function, then

$$Xf(g) = \left(\frac{d}{dt} f(e^{t\cdot x}\cdot g)\right)\Big|_{t=0}.$$

The center $\mathfrak{Z}(\mathfrak{g})$ of $\mathfrak{U}(\mathfrak{g})$ is then identified with the algebra of left and right invariant differential operators on H. A smooth function (or a distribution) is called $\mathfrak{Z}(\mathfrak{g})$-finite if it is annihilated by an ideal $\mathfrak{U}$ of finite codimension of $\mathfrak{Z}(\mathfrak{g})$. If $\mathfrak{U}$ has codimension one, such a function is an eigenfunction of $\mathfrak{Z}(\mathfrak{g})$, which is our main case of interest. It is known that if f is $\mathfrak{Z}(\mathfrak{g})$-finite, and is K-finite on the right (resp. left), then it is annihilated by an elliptic right (resp. left) invariant, hence analytic, differential operator, and is consequently necessarily analytic. However, since analyticity is obvious for the functions to be considered later, we omit the proof. The convergence proofs for the Poincaré series could be based on that fact (and indeed are in [35, Exp. 10]). Here, we shall use instead the following result (in which, in fact, the assumptions on H may be slightly relaxed) of Harish-Chandra [23, Th. 1].

5.3. LEMMA. *Let U be a neighborhood of the identity in H. Let $f\colon H \to V$ be a C^∞-function which is of finite type on the right (resp. left) with respect to K, and is $\mathfrak{Z}(\mathfrak{g})$-finite. Then there exists $\alpha \in C_c^\infty(U)$ such that $\alpha(k\cdot g\cdot k^{-1}) = \alpha(g)$ $(g \in H,\ k \in K)$ and that $f = f * \alpha$ (resp. $f = \alpha * f$).*

As is usual, C_c^∞ refers to C^∞-functions with compact support, and $*$ stands for the convolution. Thus in particular

$$f * \alpha(g) = \int_H f(g\cdot h^{-1})\cdot\alpha(h)dh = \int_H f(h)\cdot\alpha(h^{-1}\cdot g)dh\ ,$$

where dh is a Haar measure on H, chosen once and for all.

5.4. THEOREM. *Let Γ be a discrete subgroup of H. Let $f\colon H \to V$ be a function which belongs to $L^1(H) \otimes V$, is $\mathfrak{Z}(\mathfrak{g})$-finite, and is of finite type on the right, with respect to K. Then the series*

$$p_f(g) = \sum_{\gamma\in\Gamma} f(g\cdot\gamma)\ , \qquad p_{\|f\|}(g) = \sum_{\gamma\in\Gamma} \|f(g\cdot\gamma)\|\ ,$$

are absolutely and uniformly convergent on compact subsets, and are bounded on H.

It is enough to prove this for $p_{\|f\|}$. Since Γ is discrete, we may find a symmetric compact neighborhood U of e such that $U^2 \cap \Gamma = \{e\}$. By 5.3, there exists $\alpha \in C_c^\infty(U)$ such that

$$f(g \cdot \gamma) = \int_H f(g \cdot \gamma \cdot h^{-1}) \cdot \alpha(h)dh \; ; \qquad (g \in H, \gamma \in \Gamma)$$

this can also be written

$$f(g \cdot \gamma) = \int_H f(g \cdot h^{-1})\alpha(h \cdot \gamma)dh \; .$$

Let M be the maximum of $|\alpha|$ on H. Since α has its support in $U = U^{-1}$, we have then

$$(1) \qquad \|f(g \cdot \gamma)\| \leq M \cdot \int_{U \cdot \gamma^{-1}} \|f(g \cdot h^{-1})\| \cdot dh \; .$$

But $U \cdot \gamma \cap U \cdot \gamma' = \varnothing$ if $\gamma \neq \gamma'$, hence

$$\sum_{\gamma \in \Sigma} \|f(g \cdot \gamma)\| \leq M \int_H \|f(g \cdot h^{-1})\| \, dh \leq M \cdot \|f\|_{L^1} \; ,$$

which proves that $p_{\|f\|}$ is bounded on H.

Let now C be a compact subset of H. Given $\varepsilon > 0$, there exists a compact subset D of H such that

$$(2) \qquad \int_{H-D} \|f(h)\| \, dh \leq \varepsilon \; .$$

The inequality (1) can also be written

$$(3) \qquad \|f(g \cdot \gamma)\| \leq M \int_{g \cdot \gamma \cdot U} \|f(h)\| \, dh \; .$$

Let ψ be the set of elements $\gamma \in \Gamma$ for which $C \cdot \gamma \cdot U \cap D \neq \varnothing$; it is finite. Given $g \in C$, the translates $g \cdot \gamma \cdot U(\gamma \in \Gamma, \gamma \notin \psi)$ are pairwise disjoint subsets of $H - D$, hence

$$\sum_{\gamma \in \Gamma - \psi} \|f(g \cdot \gamma)\| \leq M \cdot \int_{H-D} \|f(h)\| \, dh \leq M \cdot \varepsilon \qquad (g \in C) \; ,$$

from which the uniform convergence of $p_{\|f\|}$ on C follows.

5.5. REMARKS. (1) The above proof is due to Harish-Chandra. Our original argument was longer, and was a variation on one of Godement's [35, Exp. 10].

(2) If f is of finite type on the *left*, and satisfies the other assumptions of the theorem, then a similar argument, or the one of Godement, shows readily that p_f is absolutely and uniformly convergent on compact sets. However, it does not seem necessarily true then that p_f is bounded.

5.6. The series p_f, where f satisfies the assumptions of 5.4, will be called a *Poincaré series*. Our next aim is to show that the usual Poincaré series on bounded symmetric domains are associated in a simple way to Poincaré series in the above sense.

Up to the end of this section, we assume that $X = K \backslash H$ is a bounded symmetric domain, let $H = G_{\mathbf{R}}^0$, and use the notation of §1. Let, further,

$$\mu(x, g) = (e^{tx} \cdot g)_0 \in K_{\mathbf{C}}\,, \qquad\qquad (g \in H, x \in D)$$

be the canonical automorphy factor of the bounded realization D of X (1.8), and let

$$(1) \qquad\qquad \mu_\rho(x, g) = \rho\big(\mu(x, g)\big)\,,$$

where ρ also denotes the natural extension to $K_{\mathbf{C}}$ of ρ. Since $\mu(x, k) = k$ $(k \in K)$, we have in particular,

$$(2) \qquad\qquad \mu_\rho(o, k) = \rho(k) \qquad\qquad (k \in K)\,.$$

To a function $F\colon D \to V$, we associate $f\colon H \to V$ by

$$(3) \qquad\qquad f(g) = \mu_\rho(o, g) \cdot F\big(\zeta(g)\big) \qquad\qquad (g \in H)\,.$$

From (2), and the cocycle identity (1.8), it follows that

$$(4) \qquad\qquad f(k \cdot g) = \rho(k) \cdot f(g) \qquad\qquad (k \in K, g \in H)\,.$$

It is easily seen that $F \mapsto f$ is in fact a bijection of the set of V-valued functions on D onto the set of V-valued functions on H which satisfy (4).

5.7. Lemma. *We keep the notation of* 5.6. *Then*

(a) *the function F on D is holomorphic if and only if $Y \cdot f = 0$ for all $Y \in \mathfrak{p}^-$;*

(b) *Let $f\colon H \to V$ be a function which satisfies* 5.6 (4) *and $Y \cdot f = 0$, $(Y \in \mathfrak{p}^-)$. Then f is $\mathfrak{Z}(\mathfrak{g})$-finite.*

The second assertion is proved in [35, Exp. 10, pp. 6–8]. As a matter of fact, it is only explicitly stated there that f is an eigenfunction of $\mathfrak{Z}(\mathfrak{g})$ if ρ is irreducible, but the proof also yields (b).

Part (a) is also known, and mentioned in [35, Exp. 10, p. 6]. For the sake of completeness, we indicate a proof. If we view $\mathfrak{g}$ as a Lie algebra of differential operators on D, *via* the action of $G_{\mathbf{R}}^0$, then the elements of $\mathfrak{p}^-$ are the linear combinations with constant coefficients of the partial derivatives $\partial/\partial \bar{z}_i$, where the z_i are coordinates in $\mathfrak{p}^+$. Therefore it is enough to show

$$(1) \qquad\qquad Yf(g) = \mu_\rho(o, g) \cdot YF\big(\zeta(g)\big) \qquad\qquad (Y \in \mathfrak{p}^-; g \in H)\,.$$

Let first $Y \in \mathfrak{g}_\mathbf{R}$. We may write

$$(2) \qquad\qquad Y = Y_- + Y_0 + Y_+ \qquad\qquad (Y_\pm \in \mathfrak{p}_\pm;\; Y_0 \in \mathfrak{k}_\mathbf{C})$$

where, obviously

$$(3) \qquad\qquad Y_0 = \left(\frac{d}{dt}\,(e^{tY})_0\right)_{t=0}.$$

We have

$$Yf(e) = \frac{d}{dt}\left(\mu_\rho(o,\, e^{tY})\cdot F(o\cdot e^{tY})\right)\Big|_{t=0},$$

hence, using (3), and denoting by $d\rho\colon \mathfrak{g}_\mathbf{C} \to \mathfrak{gl}(V)$ the differential of ρ,

$$(4) \qquad\qquad Yf(e) = d\rho(Y_0)\cdot F(o) + Y\cdot F(o).$$

By linearity this formula extends to all $Y \in \mathfrak{g}$. If $Y \in \mathfrak{p}^-$, then $Y_0 = 0$, and (4) yields (1) for $g = e$. Now the correspondence 5.6 (3) associates to the right translate $f' = r_g\cdot f(g \in H)$ the function F' given by $F'(x) = \mu_\rho(x,\, g)\cdot F(x\cdot g)$. Since μ_ρ is holomorphic in x, we have $YF'(o) = \mu_\rho(o, g)\cdot YF(o\cdot g)$; on the other hand $Y(r_g f)(e) = Yf(g)$, hence (4), applied to f' and F', yields (1).

5.8. **Lemma.** *Let $J(x,\, g)$ be the functional determinant function in the bounded realization of X. Then $g \mapsto J(o,\, g)^a$ is in $L^1(H)$ for every integer $a \geq 2$.*

We know that $J(x,\, k) = \det \mathrm{Ad}_{\mathfrak{p}^+}\, k^{-1}$ is a scalar independent of x, of modulus one $(k \in K)$. By the cocycle formula, $g \mapsto |\,J(o,\, g)\,|^a$ is therefore left and right invariant under K, and in particular may be viewed as a function on D. We have

$$\int_{G_\mathbf{R}} |\,J(o,\, g)\,|^a\, dg = \int_D |\,J(o,\, x)\,|^a\, dx\,,$$

where dx is a suitable invariant volume element. Up to a positive factor,

$$dx = |\,J(o,\, x)\,|^{-2}\, \omega\,,$$

where ω is the euclidean volume element in $\mathfrak{p}^+$. The domain of integration being bounded, it is then enough to show that $|\,J(o,\, x)\,|$ is bounded on D. Since $G_\mathbf{R} = K\cdot A\cdot K$, it suffices to check this on $o\cdot A$, where it follows from 1.12.

5.9. Let $X = X_1 \times \cdots \times X_q$ be the decomposition of X into irreducible bounded symmetric domains. The X_i's correspond canonically to the almost simple, non-compact, almost direct factors of the derived group of $G_\mathbf{R}$, and are therefore stable under $G_\mathbf{R}$. Each $g \in G_\mathbf{R}^0$ induces a complex analytic homeomorphism g_i of X_i such that

$$(x_1,\, \cdots,\, x_q)\cdot g = (x_1\cdot g_1,\, \cdots,\, x_q\cdot g_q) \qquad\qquad (x_i \in X_i)\,.$$

Letting J_i be the functional determinant function in the canonical bounded realization D_i of X_i, we have

$$J(x, g) = \prod_i J_i(x_i, g_i) .$$

If $\mathbf{a} = (a_1, \cdots, a_q)$ is a sequence of integers, define $J^{\mathbf{a}}$ by

$$J(x, g)^{\mathbf{a}} = \prod_i J_i(x_i, g_i)^{a_i} .$$

Let Γ be a discrete subgroup of H. Let $\varphi: D \to V$ be a polynomial mapping. Put

$$(1) \qquad P_\varphi(x) = P(x) = \sum_{\gamma \in \Gamma} J(x, \gamma)^{\mathbf{a}} \cdot \varphi(x \cdot \gamma) .$$

Up to the fact that (for later use) we allow a multi-exponent $\mathbf{a}$, this is just a Poincaré series in the usual sense. If it converges, it represents an automorphic form of weight $\mathbf{a}$, i.e., it verifies

$$(2) \qquad P(x) = J(x, \gamma)^{\mathbf{a}} \cdot P(x \cdot \gamma) \qquad\qquad (x \in X; \gamma \in \Gamma) ,$$

as follows from the cocycle formula. To φ we associate, as in 5.6, the function $f: H \to V$ defined by

$$f(g) = J(o, g)^{\mathbf{a}} \cdot \varphi(\zeta(g)) \qquad\qquad (g \in H) .$$

Then by the cocycle formula, we have formally

$$(3) \qquad p_f(g) = \sum_{\gamma \in \Gamma} f(g \cdot \gamma) = J(o, g)^{\mathbf{a}} \cdot P(\zeta(g)) .$$

As before, let

$$(4) \qquad p_{\|f\|}(g) = \sum_{\gamma \in \Gamma} \| f(g \cdot \gamma) \| .$$

5.10. THEOREM. *Assume $a_i \geq 2$ $(i = 1, \cdots, q)$. Then the series p_f is a Poincaré series in the sense of 5.6. Consequently P_φ, p_f and $p_{\|f\|}$ converge absolutely and uniformly on compact sets, and $p_{\|f\|}$ is bounded on H.*

By 5.6 (4), f is of finite type on the *left*. Together with 5.7, this shows that f is $\mathfrak{Z}(\mathfrak{g})$-finite. Since φ is a polynomial mapping on D, it is bounded; hence, f is in $L^1(H) \otimes V$ by 5.8. We have

$$J(o, g \cdot k) = J(o, g) \cdot \prod_i \cdot (\det \mathrm{Ad}\, k_i^{-1})^{a_i} \qquad\qquad \left(k = (k_1, \cdots, k_q),\, k_i \in K_i \right) .$$

On the other hand, since k acts on D by means of a linear transformation, namely $\mathrm{Ad}_{\mathfrak{p}+}\, k^{-1}$, it transforms φ into a polynomial mapping of the same degree, hence the set of transforms of φ under K is contained in a finite dimensional vector space. It follows then that f is K-finite *on the right*, too. It satisfies therefore all the assumptions of 5.4.

5.11. We conclude this paragraph with some remarks to be used in § 10, in the application of our main embedding theorem. Let ν be the one-dimensional representation $k \mapsto \det \mathrm{Ad}_{\mathfrak{p}^+} k^{-1}$ of $K_{\mathbf{C}}$. Then, in the notation of 5.6, the automorphy factor μ_ν is just the functional determinant J in the bounded realization of X. Letting $\rho(m)$ ($m \in \mathbf{Z}$) stand for the tensor product $\rho \otimes \nu^m$, we have

$$(1) \qquad \mu_{\rho(m)}(x, g) = J(x, g)^m \cdot \mu_\rho(x, g) , \qquad\qquad (x \in X, g \in H) .$$

For a linear transformation A of a finite dimensional Hilbert space we let $\| A \|^2 = \mathrm{Tr}\,(A^* \cdot A)$, where A^* is the adjoint of A. We claim that, given ρ, there exists m_0 such that the function $\beta \colon g \mapsto \| \mu_{\rho(m)}(0, g) \|$ belongs to $L^1(H)$ for all $m \geqq m_0$.

PROOF. We have

$$(2) \qquad \mu(o, k \cdot g \cdot k') = k \cdot \mu(o, g) \cdot k' \qquad\qquad (g \in H, k, k' \in K \cap H) ,$$

so that β is right and left invariant under K. Clearly, $\mu_{\rho(n)} = J^{n-m} \cdot \mu_{\rho(m)}$, $(n, m \in \mathbf{Z})$; therefore, as in 5.8, it is enough to show that β is bounded on A for m big enough. For $a \in A$, we have

$$(3) \qquad \begin{aligned} \mu(o, a) &= a_0 \\ &= \prod_i \exp\,(\log \cosh \gamma_i \cdot H_i) , \qquad (a = \exp\,(\gamma_1 X_1 + \cdots + \gamma_t X_t)) , \end{aligned}$$

as was recalled in 1.12. The transformations $\rho(m)(a_0)$ are simultaneously diagonalizable, and their eigenvalues are of the form a_0^δ, where δ runs through the weights of $\rho(m)$. These are the sums $m \cdot \nu + \lambda$, where λ runs through the weights of ρ. From 1.10, applied to each irreducible factor of X, we see that $\nu(H_i) < 0$ for all i. Consequently, there exists m_0 such that $\delta(H_i) < 0$ for all i and all weights δ of $\rho(m)$ if $m \geqq m_0$ By (3), β is then bounded on A.

As in 5.10, it follows that if $\varphi \colon X \to V$ is a polynomial mapping, the series

$$\textstyle\sum_{\gamma \in \Gamma} J(x, \gamma)^m \cdot \mu_\rho(x, g) \cdot \varphi(x \cdot \gamma) ,$$

is a Poincaré series for $m \geqq m_0$, and any discrete subgroup Γ of H, to which the conclusion of 5.10 applies.

6. Poincaré-Eisenstein series

6.1. In this section, G is a connected semi-simple $\mathbf{Q}$-group, H a subgroup of finite index of $G_{\mathbf{R}}$, K a maximal compact subgroup of H, V a finite dimensional complex Hilbert space, P a parabolic $\mathbf{Q}$-subgroup of G, and $\chi_0 \colon p \mapsto \det \mathrm{Ad}_{\mathrm{u}} p$ ($p \in P$) the determinant of $\mathrm{Ad}\,p$ in the Lie algebra of the unipotent radical $U = R_{\mathrm{u}}(P)$ of P.

Furthermore we assume P to be in the standard form (2.2). Thus $P = {}_{\mathbf{Q}}P_\theta$ where $\theta \subset {}_{\mathbf{Q}}\Delta$. Let $\theta' = {}_{\mathbf{Q}}\Delta - \theta$. We have then (2.3):

$$\chi_0 = \textstyle\sum_{\alpha \in \theta'} e_\alpha d_\alpha , \qquad\qquad (e_\alpha \in \mathbf{Q}, e_\alpha > 0) ,$$

where the d_α are fundamental highest weights for P. For every set $s =$

$(s_\alpha)_{\alpha \in \theta'}$ of complex numbers, we let $\Delta(p, s)$ be the complex valued function on $P_{\mathbf{R}}$ defined by

$$(1) \qquad \Delta(p, s) = \prod_{\alpha \in \theta'} |d_\alpha(p)|^{-s_\alpha} .$$

Let $f: H \to V$ be a continuous function which satisfies

$$(2) \qquad f(g \cdot p) = f(g) \cdot \Delta(p, s) \qquad\qquad (g \in H, p \in P \cap H) .$$

Let Γ be an arithmetic subgroup of G, contained in H, and Γ_∞ a subgroup of finite index of $\Gamma \cap P$. The series

$$(3) \qquad E_f(g) = \sum_{\gamma \in \Gamma/\Gamma_\infty} f(g \cdot \gamma) \qquad\qquad (g \in H)$$

is called an *Eisenstein series*. It follows from a theorem of Godement (unpublished; for a sketch of the proof, see [12]) that this series converges absolutely and uniformly on compact sets if

$$(4) \qquad \mathfrak{R}s_\alpha > e_\alpha \qquad\qquad (\alpha \in \theta') .$$

We note that since a rational character, defined over $\mathbf{Q}$, takes only the values ± 1 on an arithmetic subgroup, (2) implies that f is right invariant under Γ_∞, so that the summation in (3) makes sense, and E_f is right invariant under Γ. We shall need the following generalization of this result.

6.2. THEOREM. *We keep the notation of 6.1. Let $f': H \to V$ be a continuous function which is right invariant under Γ_∞ and such that*

$$m(g) = \sup_p \| f'(g \cdot p) \| \cdot | \Delta(p, s) |^{-1} \qquad\qquad (p \in P \cap H)$$

is finite for every $g \in H$, and is bounded on compact sets. If s verifies 6.1(4), then

$$(1) \qquad E_{f'} = \sum_{\gamma \in \Gamma/\Gamma_\infty} f'(x \cdot \gamma) ,$$

converges absolutely and uniformly on compact sets.

We only show how this reduces to the Godement theorem. Replacing f' by $\|f'\|$ we may assume the s_α to be real, and f' to be a real-valued positive function. Let f'' be a strictly positive continuous function on H such that

$$f''(g \cdot p) = f''(g) \cdot \Delta(p, s) \qquad\qquad (g \in H, p \in P_{\mathbf{R}} \cap H) .$$

Such functions obviously exist; we may for instance simply put $f''(k \cdot p) = \Delta(p, s)$. Since Δ is trivial on $P \cap K$, this is legitimate, and defines the required function on $H = K \cdot (P \cap H)$. Writing $g = k \cdot p$ ($k \in K, p \in P \cap H$), we have

$$f'(g) \cdot \big(f''(g) \big)^{-1} = f'(k \cdot p) \cdot \Delta(p, s)^{-1} \cdot f''(k)^{-1} .$$

f'' has a strictly positive minimum on K, and $f'(k \cdot p)\Delta(p, s)^{-1}$ remains bounded when k runs through K and p through $P \cap H$ by assumption. There exists therefore a strictly positive constant c such that

$$f'(g) \leqq c \cdot f''(g) \qquad\qquad (g \in H),$$

whence the reduction to the case of 6.1.

We now introduce a generalization of Eisenstein series and Poincaré series, to be called Poincaré-Eisenstein series.

6.3. Let B be a normal connected **Q**-subgroup of P which contains the split radical $S \cdot U$ of P, and let $C = P/B$. This is a reductive connected **Q**-group which has no non-trivial rational character defined over **Q**. The natural projection maps $H \cap P$ and $P_{\mathbf{R}}$ onto subgroups of finite index of $C_{\mathbf{R}}$. Furthermore, if $f: H \to V$ verifies

$$(1) \qquad\qquad \| f(h \cdot c) \| = \| f(h) \cdot \Delta(c, s) \|, \qquad\qquad (h \in H, c \in B_{\mathbf{R}} \cap H),$$

then the restriction of $\| f \cdot \Delta(\ , s)^{-1} \|$ to $H' \cap P$ is right invariant under $B \cap H$, and may be identified with a function on the open subgroup $(H \cap P)/(H \cap B) = C_1$ of $C_{\mathbf{R}}$.

Let now Γ and Γ_∞ be as in 6.1, and $\Gamma_0 = \Gamma_\infty \cap B$. We define the Poincaré-Eisenstein series E_f by:

$$(2) \qquad\qquad E_f(h) = \sum_{\gamma \in \Gamma/\Gamma_0} f(h \cdot \gamma) \qquad\qquad (h \in H).$$

6.4. THEOREM. *We keep the notation of 6.3, and assume that $C_{\mathbf{R}}^0$ has a compact center. Let $f: H \to V$ and $f': C_1 \to V$ be continuous functions which verify*

(i) $\| f(k \cdot h \cdot b) \| = \| f(h) \| \cdot \Delta(b, s)$ $(k \in K, h \in H, b \in B \cap H)$, *where* $s = (s_\alpha)$ *is real, and*

(ii) *the function f' belongs to $L^1(C_1) \otimes V$, is $\mathfrak{Z}(\mathfrak{c}_1)$-finite (cf. 5.1), is of finite type on the right with respect to some maximal compact subgroup, and is equal in norm to $f \cdot \Delta(\ , s)^{-1}$.*

Then, if s satisfies 6.1 (4), the series 6.3 (2) converges absolutely and uniformly on compact sets.

PROOF. Note first that

$$(1) \qquad \| f(h \cdot b \cdot p) \| = \| f(h \cdot p) \| \cdot \Delta(b, s) \qquad (h \in H, b \in B \cap H, p \in P \cap H).$$

In fact, we have $b \cdot p = p \cdot b'$ $(b' = p^{-1} \cdot b \cdot p)$, hence

$$\| f(h \cdot b \cdot p) \| = \| f(h \cdot p \cdot b') \| = \| f(h \cdot p) \| \cdot \Delta(b', s).$$

But $\Delta(b', s) = \Delta(b, s)$ since P acts trivially by inner automorphisms on its character group, whence (1). The function Δ, being equal to one on Γ_0, and Γ_0 being normal in Γ_∞, this implies in particular

$$(2) \qquad\qquad f(h \cdot \tau \cdot \sigma) = f(h \cdot \sigma) \qquad\qquad (h \in H, \tau \in \Gamma_0, \sigma \in \Gamma_\infty).$$

Let

$$(3) \qquad \begin{aligned} p_f(h) &= \textstyle\sum_{\sigma \in \Gamma_\infty / \Gamma_0} f(h \cdot \sigma) \ , \\ p_{\|f\|}(h) &= \textstyle\sum_{\sigma \in \Gamma_\infty / \Gamma_0} \| f(h \cdot \sigma) \| \end{aligned} \qquad\qquad (h \in H) \ .$$

Write $h = k_h \cdot p_h$, where $k_h \in K$ and $p_h \in P$ are determined up to an element of $K \cap P$. In view of (i), (ii), we have, π denoting the projection $H \cap P \to C_1$,

$$(4) \qquad p_{\|f\|}(h \cdot p) \cdot \Delta(p, s)^{-1} = \Delta(p_h, s) \cdot p_{\|f'\|}\bigl(\pi(p_h \cdot p)\bigr) \ ,$$

where

$$(5) \qquad p_{\|f'\|}\bigl(\pi(q)\bigr) = \textstyle\sum_{\sigma \in \Gamma_\infty / \Gamma_0} \| f'\bigl(\pi(q \cdot \sigma)\bigr) \| \qquad\qquad (q \in P \cap H) \ .$$

By (ii) and 5.4, the series in (5) is uniformly bounded; therefore, the left hand side of (4) is bounded when p varies in $P \cap H$ and p_h runs over a compact set. Moreover, (2) and (3) show that p_f is right invariant under Γ_∞. We may write

$$(6) \qquad E_f = \textstyle\sum_{\gamma \in \Gamma / \Gamma_\infty} p_f(h \cdot \gamma) \ ,$$

$$(7) \qquad E_{\|f\|} = \textstyle\sum_{\gamma \in \Gamma / \Gamma_\infty} p_{\|f\|}(h \cdot \gamma) \ ,$$

so that the theorem now follows from 6.2.

7. Poincaré-Eisenstein series on bounded symmetric domains

In this section, G is a simple, connected $\mathbf{Q}$-group satisfying the assumptions of 3.3 (ii); the notation of 3.3 is used.

7.1. Let $F = F_b$ $(1 \leq b \leq s)$ be a standard rational boundary component, $\sigma_b \colon X \to F$ the canonical projection. The group $N(F)_C$ is defined over $\mathbf{Q}$, and has a connected normal $\mathbf{Q}$-subgroup B, containing the split radical $S_b \cdot U_b$ of $N(F)_C$, such that $B_R^0 \subset Z(F)$ and $Z(F)/B_R^0$ is compact (3.6, 3.7).

Let J_F or J_b, or simply J, be the functional determinant in the unbounded realization S_b associated to F, and φ a polynomial on F, in the coordinates of the canonical bounded realization of F. Let Γ be an arithmetic subgroup of G, contained in G_R^0, $\Gamma_\infty = N(F) \cap \Gamma$, and $\Gamma_0 = B_R^0 \cap \Gamma$; let l be a positive integer. We shall consider the series

$$(1) \qquad E(x) = E_{\varphi, l, \Gamma}(x) = \textstyle\sum_{\gamma \in \Gamma / \Gamma_0} \varphi\bigl(\sigma_b(x \cdot \gamma)\bigr) \cdot J_F(x, \gamma)^l \qquad\qquad (x \in X) \ .$$

For this to make good sense, each term of the right hand side should be right invariant under Γ_0. More generally, we wish to know that

$$(2) \qquad \varphi\bigl(\sigma_b(x \cdot g \cdot \lambda)\bigr) \cdot J_F(x, g \cdot \lambda)^l = \varphi\bigl(\sigma_b(x \cdot g)\bigr) \cdot J_F(x, g)^l \quad (g \in G_R^0, x \in X, \lambda \in \Gamma_0) \ .$$

Since $B_R^0 \subset Z(F)$ acts trivially on F, the invariance of the first factor is clear. For the second one, it is enough that l be a multiple of the integer d of 3.14, as we shall assume.

In what follows, E may be supplied with a subscript consisting of any subset of $\{\varphi, l, \Gamma\}$ sufficient to characterize it in a given context. It will be called a Poincaré-Eisenstein series (P-E series for short) *adapted to F*. More generally, for any $g \in G_\mathbf{Q}$, we shall also consider the transform $E \circ g$ of E by g, defined by

$$(3) \qquad E \circ g(x) = J_F(x, g^{-1})^l \cdot E(x \cdot g^{-1}) \qquad (x \in X) .$$

By the cocycle identity:

$$(4) \qquad E \circ g(x) = \sum_{\gamma \in \Gamma/\Gamma_0} \varphi\big(\sigma_\mathrm{b}(x \cdot g^{-1} \cdot \gamma)\big) \cdot J_F(x, g^{-1} \cdot \gamma)^l .$$

We shall see shortly that this series converges absolutely for suitable l. It represents then an automorphic form of weight l for the group $\Gamma^g = g^{-1} \cdot \Gamma \cdot g$.

7.2. Theorem. *There exists a positive integer l_0 such that, if l is a positive multiple of l_0, then the series $E_{\varphi, l, \Gamma} \circ g$ converges absolutely and uniformly on compact sets.*

It suffices to prove this for E.

The group $N(F)_\mathbf{C}$ is the standard parabolic group ${}_\mathbf{Q}P_\vartheta$ where $\theta = {}_\mathbf{Q}\Delta - \{\beta_b\}$, in the notation of 2.2. Therefore the set $\theta' = {}_\mathbf{Q}\Delta - \theta$ reduces to $\{\beta_b\}$, and $\chi_b(p) = \det \mathrm{Ad}_\mathrm{u}\, p (p \in N(F)_\mathbf{C})$ may be taken as the fundamental highest weight relative to $\mathbf{Q}$ (2.3). By 3.12 (1), we have then

$$(1) \qquad | J_F(x, g) | = | j_F\big(\sigma_\mathrm{b}(x), g\big) |^{q_\mathrm{b}} \cdot \Delta(g, n_b) , \qquad (g \in N(F))$$

where $\Delta(g, n_b) = \chi_\mathrm{b}(g)^{-n_b}$.

We claim that 7.2 holds if we take for l_0 the smallest positive integer l verifying 7.1 (2) and:

$$(2) \qquad l \cdot n_b \in \mathbf{Z} , \qquad l \cdot n_b > 1 ; \qquad l \cdot q_{o(b, \sigma)} \geqq 2 \qquad (\sigma \in \Sigma) .$$

Let f be the complex valued function on $G_\mathbf{R}^0$ defined by

$$(3) \qquad f(g) = \varphi\big(\sigma_\mathrm{b}(\zeta(g))\big) \cdot J_F(o, g)^l .$$

Since the maximal compact subgroup K leaves o fixed, and $| J_F(o, k) | = 1$, if $k \in K$, we have

$$(4) \qquad | f(k \cdot g) | = | f(g) | \qquad (k \in K \cap G_\mathbf{R}^0, g \in G_\mathbf{R}^0) .$$

The cocycle formula and 3.11 imply

$$(5) \qquad | f(g \cdot b) | = | f(g) | \cdot \Delta(b, l \cdot n_b) , \qquad (g \in G_\mathbf{R}^0, b \in B_\mathbf{R} \cap G_\mathbf{R}^0) .$$

From (1), we get

$$(6) \qquad | f(g) | \cdot \Delta(g, l \cdot n_b)^{-1} = | \varphi\big(\sigma_\mathrm{b} \cdot \zeta(g)\big) | \cdot | j_F(o_\mathrm{b}, g) |^{l \cdot q_\mathrm{b}} \qquad (g \in N(F) \cap G_\mathbf{R}^0) .$$

Taking (4), (5) and 5.7, 5.8 into account, we see that all conditions of 6.4 are fulfilled by $| f |$, if l is a positive multiple of l_0; therefore

$$(7) \qquad E_f(g) = \sum_{\gamma \in \Gamma/\Gamma_0} f(g\cdot\gamma) , \qquad\qquad (g \in G_{\mathbf{R}}^0)$$

converges absolutely and uniformly on compact subsets. Since

$$(8) \qquad E_{\varphi,l}(o\cdot g) = J(o, g)^{-l}\cdot E_f(g) , \qquad\qquad (g \in G_{\mathbf{R}}^0)$$

by the cocycle identity, the theorem is proved.

As in 6.4, we may write E in the form

$$(9) \qquad E(o\cdot g) = J(o, g)^{-l}\cdot\sum_{\gamma \in \Gamma/\Gamma_\infty} p_f(g\cdot\gamma) ,$$

where

$$(10) \qquad p_f = \sum_{\lambda \in \Gamma_\infty/\Gamma_0} f(g\cdot\lambda) = \sum_{\lambda \in \Gamma_\infty/\Gamma_0} \varphi\big(\sigma_b \circ \zeta(g\cdot\lambda)\big)\cdot J_F(o, g\cdot\lambda)^l .$$

The argument used in proving 6.4 shows that

$$|f(g\cdot b\cdot p)| = |f(g\cdot p)|\,\Delta(b, l\cdot n_b) \qquad \big(g \in G_{\mathbf{R}}^0,\, b \in B \cap G_{\mathbf{R}}^0,\, p \in N(F)\big) ;$$

therefore, if we put again

$$(11) \qquad p_{\|f\|} = \sum_{\lambda \in \Gamma_\infty/\Gamma_0} |f(g\cdot\lambda)| ,$$

we have

$$(12) \qquad p_{\|f\|}(k\cdot g\cdot b) = p_{\|f\|}(g)\cdot\Delta(b, l\cdot n_b) \qquad (g \in G_{\mathbf{R}}^0,\, b \in G_{\mathbf{R}}^0 \cap B,\, k \in K) .$$

As was proved in 6.4, $p_{\|f\|}(g\cdot p)\Delta(p, l\cdot n_b)^{-1}$ is bounded when g varies in a compact set of $G_{\mathbf{R}}^0$ and p in $P \cap G_{\mathbf{R}}^0$.

We henceforth assume l to be divisible by the integer l_0 defined above.

7.3. From the geometric point of view, an automorphic form ω on X of weight l for Γ is a holomorphic cross section, invariant under Γ, of the complex line bundle $(\Lambda^n\tau)^{-l}$ ($n = \dim_{\mathbf{C}} X$), where $\Lambda^n\tau$ is the nth exterior power of the tangent bundle τ to X. If X is realized as a domain in $\mathbf{C}^n$, then τ is canonically trivialized, and ω is represented by a holomorphic function ω' on X which verifies

$$(1) \qquad \omega'(x) = j(x, \gamma)^l\cdot\omega'(x\cdot\gamma) , \qquad\qquad (x \in X, \gamma \in \Gamma)$$

where j is the functional determinant in the coordinates of the ambient vector space.

By a slight abuse of language, we shall say that ω *is* a P-E series adapted to F_b, if it is represented by such a series in the unbounded realization S_b associated to $F = F_b$. Let $F^* = F_{b^*}$ be another standard rational boundary component. We let $\nu = \nu_{F^*F} = \nu_{b^*}\circ\nu_b^{-1}$, where $\nu_c \colon D \to S_c$ ($1 \leq c \leq s$) is the canonical isomorphism (cf. §1). The P-E series E adapted to F is then represented on S_{b^*} by the function E^* given by

$$(2) \qquad E^*\big(\nu(x)\big) = j(x, \nu)^{-l}\cdot E(x) \qquad\qquad (x \in S_b) ,$$

where $j(x, \nu)$ is the functional determinant of ν at x. In studying E^*, and more generally $E^* \circ g$ $(g \in G_{\mathbf{Q}})$, we shall use the following notation, where J, J_* stand for J_F, J_{F^*}:

$$(3) \qquad \alpha(s) = J(o, s)^l ,$$

$$(4) \qquad \alpha^*(s) = J_*(o, s)^l ,$$

$$(5) \qquad \beta(s) = \sum_{\lambda \in \Gamma_\infty / \Gamma_0} \varphi(\sigma_{\mathrm{b}} \cdot \zeta(s \cdot \lambda)) \alpha(s \cdot \lambda) , \qquad\qquad (s \in G_{\mathbf{R}}^0) .$$

Thus, β is the function p_f of 7.2 (10), and we have, by 7.1 (4) and 7.2 (9):

$$(6) \qquad (E \circ g)(o \cdot s) = \alpha(s)^{-1} \cdot \sum_{\gamma \in \Gamma / \Gamma_\infty} \beta(s \cdot g^{-1} \cdot \gamma) \qquad\qquad (s, g \in G_{\mathbf{R}}^0) .$$

More generally

$$(7) \qquad \begin{aligned} & (E^* \circ g)(o, s) \\ & \quad = d^{-1} \cdot \alpha^*(s)^{-1} \cdot \sum_{\gamma \in \Gamma / \Gamma_\infty} \beta(s \cdot g^{-1} \cdot \gamma) \qquad (s, g \in G_{\mathbf{R}}^0; d = j(o, \nu)^l) . \end{aligned}$$

In fact, the obvious equality

$$(8) \qquad j(x, \nu) \cdot J_*(\nu(x), s) = J(x, s) \cdot j(x \cdot s, \nu) , \qquad\qquad (s \in G_{\mathbf{R}}^0, x \in S_{\mathrm{b}})$$

shows that

$$(9) \qquad d \cdot \alpha^*(s) = \alpha(s) \cdot j(o \cdot s, \nu)^l .$$

On the other hand

$$\begin{aligned} (E^* \circ g)(o \cdot s) &= J_*(o \cdot s, g^{-1})^l \cdot E^*(o \cdot s \cdot g^{-1}) \\ &= J_*(o \cdot s, g^{-1})^l \cdot j(o \cdot s \cdot g^{-1}, \nu)^{-l} \cdot E(o \cdot s \cdot g^{-1}) \\ &= J_*(o, s)^{-l} \cdot J_*(o, s \cdot g^{-1})^l \cdot j(o \cdot s \cdot g^{-1}, \nu)^{-l} \cdot E(o \cdot s \cdot g^{-1}) . \end{aligned}$$

Together with (9), this gives

$$(E^* \circ g)(o \cdot s) = \alpha^*(s)^{-1} \cdot d^{-1} \cdot \alpha(s \cdot g^{-1}) \cdot E(o \cdot s \cdot g^{-1}) ,$$

so that (7) follows from (6), and from the definition 7.1 (3) of $E \circ g$.

7.4. LEMMA[4]. *We keep the notation of 7.3, fix b, b^*, l, put $\Lambda = l n_{\mathrm{b}} \cdot \chi_{\mathrm{b}}$ if $b \geqq 1$, $\Lambda_* = l n_{\mathrm{b}^*} \cdot \chi_{\mathrm{b}^*}$ if $b^* \geqq 1$, and $\Lambda = 0$, $\Lambda_* = 0$ otherwise. Let $\mathfrak{S}'_{\mathrm{b}^*} = \mathfrak{S}^*$ be an F_{b^*}-adapted truncated Siegel domain (4.12) in $P_{\mathbf{R}}^0$. For $s \in \mathfrak{S}^*$, we write $s = a(s) \cdot v(s)$ $(a(s) \in {}_{\mathbf{Q}}A, v(s) \in V_{\mathbf{R}})$. Then*

   (i) $\alpha^*(s) \asymp \alpha^*(a(s)) \asymp a(s)^{-\Lambda_*}, (s \in \mathfrak{S}^*)$

   (ii) $\alpha^*(a(s)) \cdot a(s)^\Lambda \asymp 1, (b^* \leqq b; s \in \mathfrak{S}^*)$

   (iii) $\alpha^*(a(s)) \cdot a(s)^\Lambda \asymp \prod_{b < i \leqq b^*} a(s)^{-l_i \beta_i} \; (b < b^*; l_i \in \mathbf{Z}, l_i > 0, (b < i \leqq b^*)).$

By 3.12 (1), we have

$$|\alpha^*(s)| = |j_{F^*}(\sigma_{\mathrm{b}^*}(o_*), \varpi_{\mathrm{b}^*}(s))|^{l \cdot q_{\mathrm{b}}} \cdot |\chi_{\mathrm{b}^*}(s)|^{-n_{\mathrm{b}^*} \cdot l}$$

$$|\alpha^*(a(s))| = |j_{F^*}(\sigma_{\mathrm{b}^*}(o_*), \varpi_{\mathrm{b}^*}(a(s)))|^{l \cdot q_{\mathrm{b}}} \cdot |\chi_{\mathrm{b}^*}(a(s))|^{-n_{\mathrm{b}^*} \cdot l} .$$

---

[4] In this proof, s occurs in two capacities: as the **Q**-rank of G and as an element of a truncated Siegel domain. We trust this will not cause any confusion.

By the definition of a truncated Siegel domain, $\varpi_{b^*}(\mathfrak{S}^*)$ is a set of elements in $G(F^*)$ which bring the origin into a compact set, and so is relatively compact. Since $v(s)$ varies in a compact set, the set of elements $\varpi_{b^*}(v(s))$ is then also relatively compact. Thus the factors j_{F^*} in the two above equalities are multiplicatively bounded on $\mathfrak{S}^*$. Since $\chi_{b^*}(s) = \chi_{b^*}(a(s))$ for any $s \in N(F^*)$, this proves the first assertion.

If $b^* = 0$, then $\mathfrak{S}^*$ is compact, and the remaining assertions are obvious. So we assume $b^* \geqq 1$.

As in 4.12, we may write, with reference to the index b, an element $a \in {}_{Q}A$ as $a = a_1 \cdot h \cdot a_2$; the set of all a_1 (resp. h, resp. a_2) which occur in this way form a subgroup A_1 (resp. H, resp. A_2) of ${}_{Q}A$, we have

$$(1) \qquad \begin{aligned} H &= {}_{Q}A_b\,, & A_1 \cdot H &= Z(F_b) \cap {}_{Q}A\,, \\ A_2 &= L(F_b) \cap {}_{Q}A\,, & A_1 &= \mathfrak{D}Z(F_b) \cap {}_{Q}A\,, \end{aligned}$$

and the rational character Λ is trivial on $A_1 \cdot A_2$. We have $\beta_i = (\gamma_i - \gamma_{i+1})/2$, $(1 \leqq i < s)$, and $\beta_s = \gamma_s$ (resp. $\beta_s = \gamma_s/2$) in case C_s (resp. BC_s). It is clear that any character χ of ${}_{Q}A$ trivial on $A_1 \cdot A_2$ is of the form $\chi = m(\gamma_1 + \cdots + \gamma_b)$ with some $m \in \mathbf{Q}$. In particular $\Lambda = m_b(\gamma_1 + \cdots + \gamma_b)$, and $m_b > 0$, if $b \geqq 1$, because the simple $\mathbf{Q}$-roots appear with positive coefficients in $\det \mathrm{Ad}_{\mathfrak{u}} = \chi_b$, where $\mathfrak{u}$ is the Lie algebra of $U(F)$. We want to prove that m_b is *independent of b for $b \geqq 1$*. We have

$$(2) \qquad \Lambda = 2m_b\big(\beta_1 + 2 \cdot \beta_2 + \cdots + b(\beta_b + \cdots + \beta_{s-1}) + \nu \cdot b \cdot \beta_s\big)\,,$$

where $\nu = 1/2$ (resp. 1) in case C_s (resp. BC_s). There is an analogous formula for the restriction of Λ to each irreducible factor ${}^{\sigma}G'$ of G (notation of 3.3 (ii)). Moreover (3.3 (ii)) the first simple $\sigma(k)$-root of ${}^{\sigma}G'$ is the restriction of only one simple $\mathbf{R}$-root, with index $c(1, \sigma)$. Using the fact that ${}_{Q}A$ is diagonally embedded in $(R_{k/Q}S')_{\mathbf{R}}$ and applying 1.12 to $G_{\sigma} = ({}^{\sigma}G')_{\mathbf{R}}^{0}$ for each σ, we see that

$$m_b = l \cdot \sum_{\sigma} m_{0,\sigma} \cdot c(1, \sigma)\,,$$

where $m_{0,\sigma}$ is the positive integer associated to G_{σ} by 1.10, and denoted by m_0 there. This expression is indeed independent of $b \geqq 1$. From now on, we write m for m_b. From (1), we get

$$(3) \qquad a^{\Lambda} = a^{2m\nu b\beta_s} \cdot \prod_{1 \leqq i \leqq b} a^{2mi\beta_i} \cdot \prod_{b < i < s} a^{2mb\beta_i} \qquad (b \geqq 1, m > 0, a \in {}_{Q}A)\,.$$

The roots β_i $(i > b^*)$ are multiplicatively bounded on $a(s)$ $(s \in \mathfrak{S}^*)$. Therefore if $\chi = c_1\beta_1 + \cdots + c_s\beta_s$ $(c_i \in \mathbf{Q})$ is a character on ${}_{Q}A$, we have

$$(4) \qquad a(s)^{\chi} \asymp \prod_{1 \leqq i \leqq b^*} a(s)^{c_i \cdot \beta_i}\,, \qquad\qquad (s \in \mathfrak{S}^*)\,.$$

In particular, taking (i) into account:

(5) $\alpha^*\big(a(s)\big) \asymp \prod_{1 \leq i \leq b^*} a(s)^{-2m \cdot i \cdot \beta_i}$ $(1 \leq b^* < s)$,

(6) $\alpha^*\big(a(s)\big) \asymp a^{-2ms\nu\beta_s} \cdot \prod_{1 \leq i < s} a^{-2mi\beta_i}$, $(s \in \mathfrak{S}^*; \ b^* = s)$.

These relations and (3) yield (ii) and

(7) $\alpha^*\big(a(s)\big) \cdot a(s)^\Lambda \asymp \prod_{b < i \leq b^*} a(s)^{-2m(i-b)\beta_i}$, $(s \in \mathfrak{S}^*; \ b < b^* < s)$.

(8) $\alpha^*\big(a(s)\big) \cdot a(s)^\Lambda \asymp a^{-2m(s-b)\nu\beta_s} \cdot \prod_{b < i \leq s} a(s)^{-2m(i-b)\beta_i}$ $(b < b^* = s)$,

which proves (iii).

7.5. Our main aim in this section is to study the behavior of P-E series near boundary components in X^*. For this purpose, we need to construct a function that will help us to majorize these series in a certain way and to study them termwise.

Let $\rho \colon G \to \mathbf{GL}(V)$ be an irreducible representation defined over $\mathbf{Q}$, with highest weight $\Lambda = l \cdot n_b \cdot \chi_b$, where n_b is as in 3.11, such that $V_\mathbf{Q}$ contains an element $e_0 \neq 0$ which spans a line stable under $N(F_b)_\mathbf{C}$, and on which the latter group acts *via* its one dimensional representation Λ. This always exists (2.3). We endow V with a hermitian structure such that K and S are represented by unitary and self-adjoint operators respectively, and such that e_0 has norm one. The function we shall use is defined by

(1) $c(g) = \| \rho(g) \cdot e_0 \|^{-1}$ $(g \in G_\mathbf{R})$.

Obviously

(2) $c(k \cdot g \cdot p) = c(g) \cdot |\, p^{-\Lambda} \,|$ $\big(p \in N(F)\big)$.

It is also clear that if h varies in a compact set C and $g \in G_\mathbf{R}$, then $c(g) \asymp c(h \cdot g)$. In particular, if $\mathfrak{S}'$ is a Siegel domain in the minimal standard parabolic $\mathbf{Q}$-group P, then, using 4.1, we see that

(3) $c\big(a(s) \cdot g\big) \asymp c(s \cdot g)$, $(s \in \mathfrak{S}', g \in G_\mathbf{R})$

where, as in 7.4, $a(s)$ denotes the component in $_\mathbf{Q}A$ of s. The main properties of interest to us of the function c are given by the following lemma:

7.6. LEMMA. *We keep the notation of 7.4, 7.5, choose $t > 0$, and let a_0 be the element of $_\mathbf{Q}A$ on which all simple $\mathbf{Q}$-roots β_i take the value t. Then,*

(i) $a^\Lambda \cdot c(a \cdot h) \leq a_0^\Lambda \cdot c(a_0 \cdot h) \ (h \in G_\mathbf{R}; \ a \in {}_\mathbf{Q}A_t)$.

(ii) $\lim_{a^{\beta_{b^*}} \to 0} a^\Lambda \cdot c(a \cdot g) = 0$ *if* $b^* \geq b$ *and* $g \notin N(F_{b^*}) \cdot N(F_b)$, $(g \in G_\mathbf{Q}, a \in {}_\mathbf{Q}A_t)$.

We refer to the situation in 7.5. Since $_\mathbf{Q}A$ is represented by self-adjoint operators in V, the space V is the direct sum of the mutually orthogonal subspaces

$$V_\mu = \{ v \in V, \ \rho(a) \cdot v = a^\mu \cdot v, \ (a \in {}_\mathbf{Q}A) \},$$

corresponding to the different **Q**-weights μ of ρ. The space V_Λ being one-dimensional, spanned by the unit vector e_0, we may write

$$\rho(h)\cdot e_0 = d(h)\cdot e_0 + \sum_\mu f_\mu(h) , \qquad \big(f_\mu(h) \in V_\mu; d(h) \in \mathbf{R}; h \in G_\mathbf{R}\big) ,$$

whence

$$c(h)^{-2} = d(h)^2 + \sum_\mu \| f_\mu(h) \|^2 ,$$

and

$$\rho(a\cdot h)\cdot e_0 = a^\Lambda \cdot d(h)\cdot e_0 + \sum_\mu a^\mu \cdot f_\mu(h) , \qquad (a \in {}_\mathbf{Q}A, h \in G_\mathbf{R}) .$$

It is known [14, §12.14] that every **Q**-weight is of the form

$$\mu = \Lambda - \sum_{1 \le i \le s} m_i(\mu)\cdot \beta_i \qquad (m_i(\mu) \in \mathbf{Z}, m_i(\mu) \ge 0) ;$$

hence

$$(1) \qquad a^{-\Lambda}\cdot\rho(a\cdot h)\cdot e_0 = d(h)\cdot e_0 + \sum_\mu \Big(\prod_i a^{-m_i(\mu)\cdot\beta_i}\Big)\cdot f_\mu(h) \qquad (a \in {}_\mathbf{Q}A, h \in G_\mathbf{R}) .$$

Using (1) and the definition (7.5) of c, we have

$$(2) \qquad | c(a\cdot h)\cdot a^\Lambda |^{-2} = d(h)^2 + \sum_\mu \Big(\prod_i a^{-2m_i(\mu)\beta_i}\Big) \| f_\mu(h) \|^2 .$$

Since the $m_i(\mu)$ are ≥ 0, we have

$$a^{-2m_i(\mu)\beta_i} \ge a_0^{-2m_i(\mu_i)\beta_i} = t^{-2m_i(\mu)} ,$$

for all μ and all i, whence (i).

Let $P = M\cdot S\cdot U$ be the minimal standard parabolic **Q**-group (2.2) which underlies the definition of the standard boundary components. As recalled in 2.2, the element $g \in G_\mathbf{Q}$ may be written as

$$(3) \qquad g = u\cdot n_g\cdot u' \qquad (u, u' \in U_\mathbf{Q}; n_g \in N(S)_\mathbf{Q}) ,$$

where n_g is uniquely determined by g. Let w_g be the image of n_g in the relative Weyl group ${}_\mathbf{Q}W(G) = N(S)/Z(S)$.

We have $a\cdot g = a\cdot u\cdot a^{-1}\cdot n_g\cdot n_g^{-1}\cdot a\cdot n_g\cdot u'$, and consequently

$$c(a\cdot g) = c(a\cdot u\cdot a^{-1}\cdot n_g)\cdot (n_g^{-1}\cdot a\cdot n_g)^{-\Lambda} .$$

But (4.1), $a\cdot u\cdot a^{-1}$ remains in a relatively compact set, since $a \in {}_\mathbf{Q}A_t$, and so the first factor is multiplicatively bounded. Therefore, we are solely concerned with the behavior of

$$a^\Lambda \cdot (n_g^{-1}\cdot a\cdot n_g)^{-\Lambda} = a^\Lambda \cdot a^{-w_g(\Lambda)} .$$

The transform $\nu = w_g(\Lambda)$ of Λ by w_g is a weight of ρ, and therefore has the form

$$\nu = \Lambda - \sum_i m_i(\nu)\beta_i , \qquad \big(m_i(\nu) \ge 0, m_i(\nu) \in \mathbf{Z}\big) .$$

Thus we are reduced to studying the product

$$\prod_i a^{m_i(\nu)\cdot\beta_i} .$$

Since $a \in {}_{\mathbf{Q}}A_t$, each factor is bounded above. Let $\theta(\nu) = \{\beta_i \mid m_i(\nu) > 0\}$. It follows from [14, §12.16] that $\theta(\nu) \cup \Lambda$ is connected; by construction, Λ is orthogonal to all simple **Q**-roots except β_b. Now, if $g \notin N(F_b)$, then $\nu \neq \Lambda$; therefore $\theta(\nu)$ is non-empty, and we must have $m_b(\nu) \neq 0$. It follows that

$$\lim\nolimits_{a^{\beta_b} \to 0} \prod\nolimits_i a^{m_i(\nu) \cdot \beta_i} = 0 \qquad\qquad (a \in {}_{\mathbf{Q}}A_t) ,$$

which proves (ii) if $b^* = b$. Let now $b^* > b$, and assume that the limit is not zero. Then $m_{b^*}(\nu) = 0$, and $\theta(\nu)$ is contained in the set ${}_{\mathbf{Q}}\Delta - \{\beta_{b^*}\}$ of **Q**-simple roots of $N(F_{b^*})_{\mathbf{C}}/U(F_{b^*})_{\mathbf{C}}$. Then, by [14, 12.17], there exist

$$n_1 \in N(S)_{\mathbf{Q}} \cap N(F_{b^*}) , \qquad n_2 \in N(S)_{\mathbf{Q}} \cap N(F_b) ,$$

such that $n_g = n_1 \cdot n_2$. Consequently

$$g = g_1 \cdot g_2 , \qquad g_1 = u \cdot n_1 \in N(F_{b^*})_{\mathbf{Q}} , \qquad g_2 = n_2 \cdot u' \in N(F_b)_{\mathbf{Q}} ,$$

which completes the proof of (ii).

7.7. Lemma. *We keep the notation of 7.4, 7.5. Let $g \in G_{\mathbf{Q}}$.*

(i) *There exists a convergent series of positive constant terms which majorizes termwise, in the truncated Siegel domain $\mathfrak{S}^*$, the series (see 7.3 (7)):*

$$(E^* \circ g)(o \cdot s) = \sum\nolimits_{\gamma \in \Gamma/\Gamma_\infty} \alpha^*(s)^{-1} \cdot \beta(s \cdot g^{-1} \cdot \gamma) ;$$

(ii) *if $b^* \leqq b$ and $g \notin N(F_{b^*}) \cdot N(F_b)$, then $\lim_{s^{\beta_{b^*}} \to 0} \alpha^*(s)^{-1} \cdot \beta(s \cdot g) = 0$, $(s \in \mathfrak{S}^*)$;*

(iii) *if $b < b^*$, then $\lim_{s^{\beta_{b^*}} \to 0} \alpha^*(s)^{-1} \cdot \beta(s \cdot g) = 0$, $(s \in \mathfrak{S}^*)$.*

The function β is equal to p_f, in the notation of 7.2(10), and is majorized by $p_{\|f\|}$. We have already remarked (7.2(12)) that

$$p_{\|f\|}(h \cdot p) \cdot \Delta(p, n_b \cdot l)^{-1} = p_{\|f\|}(h \cdot p) \cdot | p^{-\Lambda} |$$

is bounded when h varies in a compact set of $G_{\mathbf{R}}$ and $p \in P \cap G_{\mathbf{R}}^0$. Since $c(h \cdot p) = c(h) \cdot | p^{-\Lambda} |$ by 7.5 (2), it follows then, as in 6.2, that $p_{\|f\|} \cdot c^{-1}$ is bounded on $G_{\mathbf{R}}^0$. There exists then a constant $\delta > 0$ such that

$$(1) \qquad\qquad | \beta(h) | \leqq \delta \cdot c(h) , \qquad\qquad (h \in G_{\mathbf{R}}^0) ,$$

and therefore such that

$$(2) \qquad\qquad \delta \cdot \sum\nolimits_{\gamma \in \Gamma/\Gamma_\infty} \alpha^*(s)^{-1} \cdot c(s \cdot h \cdot \gamma) ,$$

is a normal majorant of $(E^* \circ h)(s)$ for all s, $h \in G_{\mathbf{R}}^0$, which converges in view of 6.1.

We have $c(s \cdot h) \asymp c(a(s) \cdot h)$ for s in a Siegel domain and $h \in G_{\mathbf{R}}^0$ by 7.5 (3), and $\alpha^*(a(s)) \asymp \alpha^*(s)$ for $s \in \mathfrak{S}^*$, $(s = a(s) \cdot v, a(s) \in {}_{\mathbf{Q}}A, v \in V_{\mathbf{R}})$ by 7.4. Together with (1), this yields

(3) $\qquad |\alpha^*(s)^{-1}\cdot\beta(s\cdot h)| \prec \cdot \alpha^*(a(s))^{-1}\cdot c(a(s)\cdot h) \qquad (s \in \mathfrak{S}^*, h \in G^0_{\mathbf{R}})$.

Using 7.4, we get

(4) $\qquad |\alpha^*(s)^{-1}\cdot\beta(s\cdot h)| \prec (\prod_{i=b+1}^{b^*} a(s)^{l_i\beta_i})\cdot(c(a(s)\cdot h)\cdot a(s)^\Lambda)$,

where the first factor on the right hand side stands for the constant one if $b^* \leqq b$, and the l_i are strictly positive integers if $b^* > b$. Let, as in 7.6, a_0 be the element of $_{\mathbf{Q}}A$ on which the simple $\mathbf{Q}$-roots take the value t. Since the first factor on the right hand side of (4) is bounded on $\mathfrak{S}^*$, Lemma 7.6 (i) and (1) imply the existence of a constant $\delta' > 0$ such that

$$c'(h) = \delta'\cdot c(a_0\cdot h)\cdot a_0^\Lambda \geqq |\alpha^*(s)^{-1}\cdot\beta(s\cdot h)| \qquad (s \in \mathfrak{S}^*; h \in G^0_{\mathbf{R}})$$.

As a consequence, $E^*\circ g$ is majorized in $\mathfrak{S}^*$ by the series

$$\sum c'(g^{-1}\cdot\gamma) ,$$

which converges, as was noted above (6.1 and 7.5 (2)), whence (i).

In view of (1), (2) it suffices to prove the statements (ii), (iii) with $\alpha^*(s)^{-1}\cdot\beta(s\cdot g)$ replaced by

$$(\prod_{i=b+1}^{b^*} a^{l_i\beta_i})\cdot c(a\cdot g)\cdot a^\Lambda \qquad (a \in {}_{\mathbf{Q}}A_t) .$$

Let $b^* \leqq b$ and $g \notin N(F_{b^*})\cdot N(F_b)$. Then the first factor is one, and the second one tends to zero as $a^{\beta_{b^*}} \to 0$ by 7.6 (ii), which proves (ii).

Let $b^* > b$. Then $l_{b^*} > 0$ by 7.4; hence, the first factor tends to zero. The second one remains bounded by 7.6 (i), whence (iii).

7.8. THEOREM. *Let E and E^* be as in 7.1, 7.4. Let $\mathfrak{S}'_{F^*}$ be a truncated Siegel domain adapted to F^*; let $\varepsilon > 0$ and $g \in G_{\mathbf{Q}}$ be given.*

(i) *Assume $b < b^*$. Then there exists $u_0 > 0$ such that*

$$|E^*\circ g(o\cdot s)| < \varepsilon \qquad (s \in \mathfrak{S}'_{F^*}, a(s)^{\beta_{b^*}} < u_0) .$$

(ii) *Assume $b \geqq b^*$. Then there exists $u_0 > 0$ such that*

$$\left|E^*\circ g(o\cdot s) - j(o, \nu)^l \sum_{\gamma\in g\cdot N(F^*)N(F)\cap\Gamma/\Gamma_\infty} \alpha^*(s)^{-1}\cdot\beta(s\cdot g^{-1}\cdot\gamma)\right| < \varepsilon$$

for all $s \in \mathfrak{S}'_{F^*}$ satisfying $a(s)^{\beta_{b^*}} < u_0$.

By 7.7 (i) the series $E^*\circ g$ has a constant majorant series in $\mathfrak{S}^*$, so that we may investigate its behavior termwise; 7.7 (ii) and 7.7 (iii) allow us to do this, with the theorem as an immediate consequence.

7.9. PROPOSITION. *We keep the previous notation and assume $b^* \leqq b$. Then the series*

(1) $\qquad E^*\circ g(o\cdot s) = \alpha^*(s)^{-1}\cdot\sum_{\gamma\in(gN(F^*)N(F)\cap\Gamma)/\Gamma_\infty} \beta(s\cdot g^{-1}\cdot\gamma) \qquad (g \in G_{\mathbf{Q}})$

is an automorphic form on F^*, for the group

$$\Gamma^g(F^*) = (Z(F^*) \cap g^{-1}\Gamma g)\backslash(N(F^*) \cap g^-\Gamma g) ,$$

lifted to X by means of the canonical projection $\sigma_{b^*}\colon X \to F^*$, of type $J_*(x,\ )^l$.

Since $\mathbf{b}^* \leqq \mathbf{b}$ we have a canonical factorization $\sigma = \sigma_b = \tau \circ \sigma_*$ where $\tau\colon F^* \to F$ is a holomorphic map. In fact τ may be identified with the canonical projection of F^* onto F, the latter being viewed as a standard rational boundary component of F^*.

We show first that $E^* \circ g$ is holomorphic, constant along the fibres of σ_*. By definition

$$E^* \circ g(o \cdot s)$$
$$= \alpha^*(s)^{-1} \sum_{\gamma \in \Gamma / \Gamma_\infty} \sum_{\lambda \in \Gamma_\infty / \Gamma_0} \varphi\big(\sigma(o \cdot s \cdot g^{-1} \gamma \cdot \lambda)\big) \cdot J(o,\, s \cdot g^{-1} \cdot \gamma \cdot \lambda)^l\ ,$$

the range of γ being as in (1). For fixed γ, the product of the series on the right hand side by $\alpha(s \cdot g^{-1} \cdot \gamma)^{-1}$ is a Poincaré series for Γ_∞ / Γ_0 on F, lifted to X by σ. It is therefore *a fortiori* lifted from a holomorphic function on F^*. In order to finish the proof of our assertion, it suffices to show that

$$\alpha^*(s)^{-1} \cdot \alpha(s \cdot u \cdot v)\ , \qquad\qquad \big(u \in N(F^*),\, v \in N(F)\big)$$

is holomorphic, as a function of $o \cdot s \in X$, and constant along fibres of σ_*. The equality 7.3 (8) yields

$$\alpha^*(s)^{-1} \cdot \alpha(s \cdot u \cdot v) = J_*(o,\, s)^{-l} \cdot J(o,\, s \cdot u \cdot v)^l$$
$$= j(o,\, \nu)^l \cdot j(o \cdot s,\, \nu)^{-1} \cdot J(o,\, s)^{-l} \cdot J(o,\, s \cdot u \cdot v)^l$$
$$= d \cdot j(o \cdot s,\, \nu)^{-l} \cdot J(o \cdot s,\, u \cdot v)^l\ .$$

The factors on the right hand side are of course holomorphic in $o \cdot s$; by 3.3 (ii), $j(o \cdot s,\, \nu)^l$ is constant along the fibres of σ_*. By 1.7, $Z(F^*)^0$ is transitive along the fibers of σ_*. In view of the cocycle identity, it suffices to prove

$$(2) \qquad\qquad J(x,\, u) = J(xz,\, u) \qquad\qquad \big(z \in Z(F^*),\, u \in N(F^*)\big)\ ,$$

$$(3) \qquad\qquad J(xu,\, v) = J(xzu,\, v)\ .$$

We have $zu = uz'$ $(z' \in Z(F^*))$. Since $\mathbf{b}^* \leqq \mathbf{b}$, $F \subset \bar{F}$, whence $Z(F^*) \subset Z(F)$, so that (2) and (3) follow from 3.3 (ii).

Let now $\gamma_0 \in N(F^*) \cap g^{-1} \cdot \Gamma \cdot g$. Thus $\gamma_0 = g^{-1} \cdot \gamma' \cdot g$ $(\gamma' \in \Gamma)$, and it is clear that the series on the right hand side of (1) remains unaltered if s is replaced by $s \cdot \gamma_0$. Moreover

$$\alpha^*(s \cdot \gamma_0)^{-1} = J_*(o,\, s \cdot \gamma_0)^{-l} = J_*(o,\, s)^{-l} \cdot J_*(o \cdot s,\, \gamma_0)^{-l}\ ;$$

hence,

$$E^* \circ g(o \cdot s) = J_*(o \cdot s,\, \gamma_0)^l \cdot E^* \circ g(o \cdot s \cdot \gamma_0)\ ,$$

which ends the proof of the proposition.

8. The operator Φ

8.1. Up to 8.7, we keep the notation and assumptions of 3.3 (ii). As in 4.8, X^* denotes the union of the rational boundary components of X, endowed with the Satake topology. A *good neighborhood* of $x \in X^*$ is one which verifies 4.9 (iv) and 4.10. We let F denote a rational boundary component, and ϖ_F be the canonical epimorphism of $N(F)$ onto $G(F)$.

An automorphic form ω of weight l, for the arithmetic group Γ, will be said to be a P-E series adapted to F if its transform $\omega \circ g$ under some element $g \in G_\mathbf{Q}$ which carries F onto the rational boundary component F_b is a P-E series adapted to F_b for Γ^g, (7.6).

8.2. The functional determinant $J_\mathrm{b}(x, g)$ $(g \in N(F_\mathrm{b}), x \in X)$ is by 3.3 (ii) constant along the fibres of the projection $\sigma_\mathrm{b}\colon X \to F_\mathrm{b}$. It defines therefore an automorphy factor on F_b for $N(F_\mathrm{b})$, hence an action of $N(F_\mathrm{b})$ on the trivial line bundle $F_\mathrm{b} \times \mathbf{C}$ given by

$$(x, c) \cdot n = \big(x \cdot n, J_\mathrm{b}(x, n) \cdot c\big) \qquad \big(x \in F_\mathrm{b}, c \in \mathbf{C}, n \in N(F_\mathrm{b})\big) \, .$$

The $N(F_\mathrm{b})$-bundle thus defined will be denoted by ξ_b. If $F = F_\mathrm{b} \cdot g^{-1}$ is as above, the translation by g^{-1} carries ξ_b over onto an $N(F)$-bundle denoted by ξ_F. The isomorphism class of ξ_F, viewed as an $N(F)$-bundle, depends only on F.

Let Λ be a discrete subgroup of $N(F)$ whose image Λ' under ϖ_F is discrete, and l an integer. An automorphic form ω for Λ', of type ξ_F^l, is a Λ'-invariant holomorphic cross section of ξ_F^l. The transform $\omega \circ g$ is then an automorphic form of type ξ_b^l for $\varpi_\mathrm{b}(\Lambda^g)$. It is therefore represented in the canonical bounded realization of F_b by a function f which satisfies

$$(1) \qquad f(x \cdot \lambda) = J_\mathrm{b}(x, \lambda)^{-l} \cdot f(x) \, , \qquad (x \in F_\mathrm{b}, \lambda \in \Lambda^g)$$

where, by abuse of notation, we write $J_\mathrm{b}(x, g)$ for $J_\mathrm{b}(x', g)$ if $x \in F_\mathrm{b}$, $x' \in \sigma_\mathrm{b}^{-1}(x)$, $g \in N(F_\mathrm{b})$.

Let $c \leq b$ and $\nu_{\mathrm{b},\mathrm{c}}\colon S_\mathrm{c} \to S_\mathrm{b}$ be the canonical isomorphism. We have

$$(2) \qquad j(x, \nu_{\mathrm{b},\mathrm{c}}) \cdot J_\mathrm{b}\big(\nu_{\mathrm{b},\mathrm{c}}(x), g\big) = J_\mathrm{c}(x, g) \cdot j(x \cdot g, \nu_{\mathrm{b},\mathrm{c}}) \qquad (x \in S_\mathrm{c}, g \in G_\mathrm{R}^0) \, .$$

Let $g \in N(F_\mathrm{c})$. Then, using 1.11, we see that $J_\mathrm{b}(x, g)$ is constant along the fibres of the projection $\sigma_\mathrm{c}\colon S_\mathrm{c} \to F_\mathrm{c}$, and hence defines an automorphy factor on F_c, and an action of $N(F_\mathrm{c})$ on the trivial line bundle. However, (2) shows that the automorphy factors given by J_b and J_c are equivalent, and hence the $N(F_\mathrm{c})$-bundle just defined is isomorphic, as an $N(F_\mathrm{c})$-bundle, to ξ_c.

8.3. Let U be an open subset of X^* which intersects F, and Λ a discrete subgroup of $N(F)$ leaving U invariant. Let ω be an automorphic form of weight l for Λ in U. Let $g \in G_Q$ be such that $F \cdot g = F_b$ for some $\mathbf{b}$. We say that ω extends to $F \cap U$ if $\omega \circ g$ is represented, in the unbounded realization S_b, by a function f which extends by continuity to a holomorphic function f' on $(U \cap F) \cdot g$. The extension f' clearly represents an automorphic form σ on $(U \cap F) \cdot g$ of type ξ_b^l, for Λ^g, or rather for the image $\varpi_b(\Lambda^g)$ of Λ^g in $G(F_b)$. Its transform ω' under g^{-1} is then an automorphic form of type ξ_F^l for $\varpi_F(\Lambda)$, to be called sometimes the extension of ω. This form depends only on ω and F. In fact if $F \cdot g' = F_b (g' \in G_Q)$, then $g' = g \cdot n$ $(n \in N(F_b)_Q)$; hence, $\omega \circ g'$ is represented by $f^*(x) = f(x, n^{-1}) \cdot J(x, n^{-1})^l$. The function f^* extends by continuity to $f^{*\prime} = x \mapsto f'(x, n^{-1}) J_b(x, n^{-1})^l$, which represents $\sigma \circ n$. Its transform under g'^{-1} is then again ω' (cf. Remark of 1.7).

Let $d \geqq b$, and f_d be the function which represents $\omega \circ g$ in the unbounded realisation S_d associated to F_d. Then ω extends to $F \cap U$ (where still $F \cdot g = F_b$) if and only if f_d extends by continuity to a holomorphic function f'_d on $(U \cap F) \cdot g$. This follows from the equality $f_d = f \cdot J(\ , \nu_{d,b})^l$ and the constancy of $J(\ , \nu_{d,b})$ along the fibres of σ_b. Thus, in order to check that ω extends to $U \cap F$, we may use any unbounded realization S_d $(\mathbf{d} \geqq \mathbf{b})$. Furthermore, the last remark of 8.2 implies easily that $\omega' \circ g$ is represented by f'_d in the trivialization of ξ_d^l which is given by J_d.

8.4. *Local integral automorphic forms.* Let Γ be an arithmetic subgroup of G, and $x \in X^*$. We suppose $x \in F$. An automorphic form ω of weight l for Γ_x on $X \cap N(x)$, where $N(x)$ is a good neighborhood of x, is *integral* if it extends to an automorphic form for $\varpi_F(N(F') \cap \Gamma)$ on $F' \cap N(x)$ for every rational boundary component F' which meets $N(x)$.

If $y \in N(x)$, then $\Gamma_y \subset \Gamma_x$ by 4.9; therefore, the restriction to a good neighborhood $N(y) \subset N(x)$ of y of an integral automorphic form on $N(x)$ is an integral automorphic form, whose extension to $N(y) \cap F'$ is the restriction of the extension of ω to $F' \cap N(x)$.

The form ω is integral on $N(x)$ if and only if for every $g \in G_Q$ such that $F \cdot g = F_b$, the transform $\omega \circ g$ is represented in S_b by a function which extends by continuity to a holomorphic function on $F_c \cap N(x)$ for all $c \leqq b$. In fact, as before, this condition is insensitive to a change of g; moreover, by 3.9, if $F' \cap N(x) \neq \varnothing$, there exists $g \in G_Q$ such that $F' \cdot g = F_c$, $F \cdot g = F_b$ $(c \leqq b)$.

From 3.9 we also deduce that ω is integral if and only if for one $g \in G_Q$ which maps F onto F_b, the transform $\omega \circ g$ extends by continuity to $N(x) \cdot g$, and is holomorphic on $(F' \cap N(x)) \cdot g$ for every rational boundary component F'.

8.5. *Integral automorphic forms. The operator* Φ. An automorphic form ω on X of weight l, for Γ, is *integral* if its restriction to $X \cap N(x)$ is integral for every $x \in X^*$ and every good neighborhood $N(x)$ of x. This is the case if and only if for every F and $g \in G_Q$ such that $F \cdot g = F_b$ is standard, the transform $\omega \circ g$ is represented on S_b by a function which extends by continuity to a holomorphic function on F_b. Then this function also extends by continuity to a holomorphic function on F_c for every $c \leq b$.

Let ω be integral. It then has an extension to any rational boundary component F, which is an automorphic form for $\Gamma(F)$ of type ξ_F^l, and which will often be denoted by $\Phi_F\omega$. The operator Φ is, by definition, the operator which associates to ω the collection of automorphic forms $\Phi_F\omega$. The definition of the extension of ω given in 8.3 implies that for any rational boundary component F and $g \in G_Q$:

$$(1) \qquad\qquad (\Phi_{Fg^{-1}}\omega) \circ g = \Phi_F(\omega \circ g) \, .$$

8.6. THEOREM. *Let E be a* P-E *series adapted to the rational boundary component F, for the arithmetic group Γ, of weight l. Then E is an integral automorphic form. Let F^* be a rational boundary component. Then $\Phi_{F^*}E = 0$ if $\dim F^* \leq \dim F$ and $F^* \not\subset F \cdot \Gamma$. The operator Φ_F maps the module of* P-E *series adapted to F, of weight l, onto the module of Poincaré series for $\Gamma(F)$, of type ξ_F^l.*

If the statement is true for E, F, Γ, then it is also true for $E \circ g, F \cdot g$, and Γ^g ($g \in G_Q$). We may therefore, without loss of generality, assume that $F = F_b$ is a standard rational boundary component.

In order to prove the first assertion, it is enough to show that, for any $x \in X^*$ and $g \in G_Q$ such that $x \cdot g \in F_c$ ($1 \leq c \leq s$), the transform $E \circ g$ is represented on S_c by a function which extends by continuity to a holomorphic function around $x \cdot g$ on F_c.

Let E^* be the function which represents E on S_c. Then $E \circ g$ is represented by the function defined by $(E^* \circ g)(x) = J_c(x, g^{-1})^l \cdot E^*(x \cdot g^{-1})$. Let $\mathfrak{S}'$ be a Siegel domain in P such that $F_c \cap \bar{\mathfrak{S}}$, where $\mathfrak{S} = o \cdot \mathfrak{S}'$, contains $x \cdot g$ in its interior. This is possible by 4.5. There exists a finite subset C of $N(F_c)_Q$, containing the identity, such that $\bar{\Omega} \cdot \Gamma_{x \cdot g}$, where $\Omega = \bigcup_{a \in C} \mathfrak{S}(u, V_a) \cdot a$, runs through a fundamental set of neighborhoods of x in X^* when $u \to 0$ and V_a runs through a fundamental set of neighborhoods of $x \cdot g \cdot a^{-1}$ (see 4.13). We have

$$(E^* \circ g)(s \cdot h) = J_c(s, h)^{-l} \cdot (E^* \circ gh^{-1})(s) \qquad (s \in \mathfrak{S}(u, V_a), h \in a\Gamma_{x \cdot g}, a \in C) \, .$$

The function $J_c(\ , h)$ is constant along the fibres of $\sigma_c \colon S_c \to F_c$ (3.3 (ii)), and is multiplicatively bounded on $\mathfrak{S}(u, V_a)$ by 4.16. Consequently, we are reduced to showing that $E^* \circ g \cdot h^{-1}$ extends by continuity to a holomorphic function on the interior of $\bar{\mathfrak{S}}(u, V_a) \cap F_c$. Since this series has a normal majorant in $\mathfrak{S}(u, V_a)$ by 7.7, this follows from 7.8, 7.9.

In order to study the limit of $E^* \circ g$ around $x \cdot g$ on F_c, it is enough to consider its behavior on $\mathfrak{S}(u, V)$, where V is a neighborhood of $x \cdot g$. If $c > b$, i.e., if $\dim F^* < \dim F$, this limit is zero by 7.8 (i). Let now $c = b$. Theorem 7.8 (ii) implies that the limit is zero unless $g \cdot N(F_b) \cap \Gamma \neq \varnothing$, i.e., unless $F_b \cdot g^{-1} \subset F_b \cdot \Gamma$. Since

$$\Phi_{F_b}(E \circ g) = (\Phi_{F_b \cdot g^{-1}} E) \circ g \ ,$$

this ends the proof of the second assertion.

Let now $g = e$. Then 7.8 shows that the limit of E on F_b is the limit of the "constant" term $a(s)^{-1} \cdot b(s)$, which is by 7.9 the Poincaré series of type ξ_b^l:

$$\sum\nolimits_{\lambda \in \Gamma_\infty / \Gamma_0} \varphi \big(\sigma_b(x \cdot \lambda) \big) J_b(x, \lambda)^l \ ,$$

(see 8.2). Furthermore, it is clear that every such Poincaré series can be obtained in this way by suitable choice of φ in the definition of E.

8.7. The automorphy factor for $N(F_b)$ defined on F_b by J_b is equal in absolute value to $|\, j_b(\sigma_b(x), g)\,|^{q_b}$ (cf. 3.12). Therefore, (5.10), it satisfies the condition imposed by H. Cartan [35 Exp. 10 bis; 19 p. 170] so that we may apply Theorems 2, 3 of loc. cit. to the Poincaré series formed by means of J_b on F_b. Since every rational boundary component is the transform of some F_b by an element of G_Q, this implies:

(1) Let F be a rational boundary component of X, $a_1, \cdots, a_q$, q points of F, no two of which are equivalent under $\Gamma(F)$, and t a positive integer. Then there is a positive integer l_0 with the property that, for any multiple l of l_0, there exist a Poincaré series P of type ξ_F^l (see 8.2) which has pre-assigned admissible (i.e., locally invariant, cf. [19, pp. 170-171]) Taylor developments of order t (in suitable local coordinates), at $a_1, \cdots, a_q$.

This means in particular that, for suitable l, we may find P which is not zero at a_1 and zero at a_i ($i \geq 2$).

8.8. PROPOSITION. *Let F, F' be two rational boundary components such that $\dim F \geq \dim F'$, and that either $F = F'$ or $F \not\subset F' \cdot \Gamma$. Let $x \in F$, $y \in F'$ be not equivalent under Γ. There is an integer l_0, such that if l is a multiple of l_0, there exists an integral automorphic form E which verifies $\Phi_F E(x) \neq 0$, $\Phi_{F'} E(y) = 0$.*

This follows from 8.6 and the result of Cartan mentioned in 8.7 (1).

8.9. We now want to extend 8.8 to the case where G is not necessarily

Q-simple, and where Γ is an arithmetic group of holomorphic automorphisms of X. The notation of 3.3 (i) is used. In particular, X is the product of the spaces $X_i = (K \cap G_{i\mathbf{R}})\backslash G_{i\mathbf{R}}$, the space X^* is the product of the X_i^*, where G_i runs through the **Q**-simple factors of G, and Γ' is the product of the groups $\Gamma_i = \Gamma \cap G_{i\mathbf{R}}^0$. We do not exclude the possibility that X_i/Γ_i is compact for some i, in which case $X_i = X_i^*$. Let $\mathrm{pr}_i \colon X^* \to X_i^*$ be the natural projection. The notion of an integral automorphic form on X for Γ is defined as in 8.5. If E_i is an integral automorphic form on X_i for Γ_i, of weight l, then the product of the forms $E_i \circ \mathrm{pr}_i$ is an integral automorphic form of weight l on X for Γ'. This shows first that 8.6, 8.8 are valid for Γ'.

Let F, F' be rational boundary components of X, such that $F' \not\subset F \cdot \Gamma$, or $F = F'$, $\dim F' \leq \dim F$, and let $x \in F$, $y \in F'$, $y \notin x \cdot \Gamma$. Let $\Gamma(F)_x$ be the isotropy group of x in $\Gamma(F) = (N(F) \cap \Gamma)/(Z(F) \cap \Gamma)$, where $N(F)$ and $Z(F)$ are respectively the normalizer and the centralizer of F in $H(X)$. The orbit of x under $N(F) \cap \Gamma$ is the union of finitely many orbits of $N(F) \cap \Gamma'$. By 8.6, 8.8 for Γ', there exists l_0' such that for any multiple l of l_0' we may find a P-E series E' adapted to F of weight l for Γ' verifying

$$(1) \quad \begin{aligned} &\Phi_F E'(x) \neq 0 , \quad \Phi_F E'(x') = 0 \quad (x' \in x \cdot \Gamma \cap F;\ x' \notin x \cdot \Gamma' \cap F) , \\ &\Phi_{F'} E'(y') = 0 . \quad\quad\quad\quad\quad\quad\ (y' \in y \cdot \Gamma \cap F') . \end{aligned}$$

Assume now that l is also a multiple of the order of $\Gamma(F)_x$, and put

$$E = \sum_{\gamma \in \Gamma'\backslash \Gamma} E' \circ \gamma .$$

Each summand on the right-hand side is an integral automorphic form of weight l for Γ', hence E is an integral automorphic form of weight l for Γ. We claim that

$$(2) \quad\quad \Phi_F E(x) \neq 0 , \quad\quad \Phi_{F'} E(y) = 0 ,$$

which will prove our contention, if we take for l_0 some multiple of l_0' and of the order of $\Gamma(F)_x$. We have, for any rational boundary component F^*

$$(3) \quad\quad (\Phi_{F^* \cdot \gamma^{-1}} E') \circ \gamma = \Phi_{F^*}(E' \circ \gamma) ,$$

(see 8.5). Let $F' \not\subset F \cdot \Gamma$ and $F^* = F'$. Then $F^* \cdot \gamma^{-1} \not\subset F \cdot \Gamma'$, the left-hand side of (3) is zero by 8.6, whence the second part of (2) if $F \neq F'$. Let now $F^* = F = F'$. If $\gamma \notin (N(F) \cap \Gamma) \cdot \Gamma'$, then $F \cdot \gamma^{-1} \not\subset F \cdot \Gamma'$, the left-hand side of (3) is zero by 8.6, whence

$$\Phi_F E(z) = \sum_{\Gamma'\backslash(N(F)\cap\Gamma)\cdot\Gamma'} \big((\Phi_F E') \circ \gamma\big)(z) \quad\quad (z \in F)$$

which implies $\Phi_F E(y) = 0$ and, because of the conditions imposed in (1).

$$\Phi_F E(x) = \sum_{(N(F)\cap\Gamma')\backslash(N(F)\cap\Gamma)} j_F(x, \gamma)^{-1} \cdot \Phi_F E'(x \cdot \gamma) .$$

If $x \cdot \gamma \notin x \cdot \Gamma'$, then the corresponding summand is zero by construction of E'. There remains to consider those terms for which $x \cdot \gamma \in x \cdot \Gamma'$. In that case $\gamma \in (N(F) \cap \Gamma') \cdot \Gamma(F)_x$. Since we sum modulo $N(F) \cap \Gamma'$, we may assume that $\gamma \in \Gamma(F)_x$. Then $j_F(x, \gamma)^l = 1$, since l is a multiple of the order of $\Gamma(F)_x$, and the corresponding summand is equal to $\Phi_F E'(x)$. As a result, $\Phi_F E(x)$ is a non-zero multiple of $\Phi_F E'(x)$, which ends the proof.

8.10. Assume again for convenience, G to be **Q**-simple. An automorphic form of weight l is a *cusp form* if it belongs to the kernel of Φ.

It is known [35; Exp. 10, §4] that every cusp form whose weight is a multiple of some suitable fixed integer l_0 is a linear combination of Poincaré series; in particular, if X/Γ is compact, every automorphic form of weight $m l_0$ is a linear combination of Poincaré series. It follows therefore from 8.6, by an obvious induction procedure on $\dim F$, that there exists an integer l_0 with the following property: every automorphic form for Γ, of weight l divisible by l_0, is a linear combination of transforms under elements of $G_\mathbf{Q}$ of P-E series for conjugates of Γ under $G_\mathbf{Q}$.

III. The Compactification as an Analytic Space

9. An analyticity criterion

9.1. In this section, V is a locally compact Hausdorff space, satisfying the second axiom of countability, which is the union of a locally finite countable family of disjoint subspaces $V_0, V_1, \cdots$, each of which is provided with the structure of an irreducible normal analytic space.

An $\mathcal{C}$-function on an open subset U of V is a complex valued continuous function on U whose restriction to $U \cap V_i$ is analytic ($0 \leq i \leq m$). If we associate to U the C-module of $\mathcal{C}$-functions defined on U, we get a presheaf which is easily seen to be a sheaf, the *sheaf $\mathcal{C}$ of germs of $\mathcal{C}$-functions*. The continuous sections of $\mathcal{C}$ over an open subset U of V are the $\mathcal{C}$-functions defined on U. We let $\mathcal{C}_v$ be the stalk of $\mathcal{C}$ at $v \in V$.

9.2. THEOREM. *We keep the notation of 9.1 and make the following assumptions:*

(i) *For each positive integer d, the union $V_{(d)}$ of the V_i's whose dimension is $\leq d$ is closed. $\dim V_0 = \dim V$, $\dim V_i < \dim V_0$ if $i \neq 0$ and V_0 is dense in V.*

(ii) *Each point $v \in V$ has a fundamental set of open neighborhoods (U_α) such that $U_\alpha \cap V_0$ is connected for every α.*

(iii) *The restrictions to V_i of local $\mathcal{C}$-functions define the structural sheaf of V_i.*

(iv) *Each point $v \in V$ has a neighborhood U_v whose points are separated by the $\mathfrak{A}$-functions defined on U.*

Then $(V, \mathfrak{A})$ is an irreducible normal analytic space and for each $d \leq \dim V_0$, $V_{(d)}$ is an analytic subspace of $(V, \mathfrak{A})$ with dimension equal to $\max_{\dim V_i \leq d} (\dim V_i)$.

The proof of 9.2, will be broken up into several lemmas, and will be concluded at the end of 9.7. We note first that, in view of (i), the subspace V_i is locally closed in $V_{(d)}$ ($d = \dim V_i$), hence is locally closed in V.

We shall use the following remark on normal analytic spaces.

9.3. Lemma. *Let Y be a normal analytic space. Then the ring of analytic functions on Y is integrally closed in the ring of complex-valued continuous functions on Y.*

Being normal, Y is the disjoint union of its irreducible components, which are open in Y. We may therefore assume Y to be irreducible.

Let h be a continuous, complex-valued function on Y which satisfies a relation

$$(1) \qquad h^n(y) + \sum_{0 < i \leq n} a_i(y) \cdot h^{n-i}(y) = 0 \qquad (y \in Y),$$

where the a_i are analytic functions on Y.

Let $a \in Y$, $\mathcal{O}_a$ be the local ring of Y at a, and K_a be the field of quotients of $\mathcal{O}_a$. Let $P = P(T) = T^n + a_1 \cdot T^{n-1} + \cdots + a_n \in \mathcal{O}_a[T]$, where a_i also denotes the germ defined at a by a_i. We assume a to be a regular point. Using the Gauss factorization lemma, and (1), we can find a factor

$$Q = T^m + b_1 \cdot T^{m-1} + \cdots + b^m \in \mathcal{O}_a[T]$$

of P which is irreducible in $K_a[T]$, and such that

$$(2) \qquad Q_y(h(y)) = h^m(y) + b_1(y) \cdot h^{m-1}(y) + \cdots + b_m(y) = 0 \qquad (y \in U),$$

where U is a sufficiently small neighborhood of a. Here $Q_y \in C[T]$ denotes the polynomial obtained from Q by replacing b_i by $b_i(y)$, ($y \in U$). By consideration of the resultant of Q and dQ/dT, we see that the set of points $y \in U$, for which Q_y and dQ_y/dT have a common root, is a proper analytic subset Z of U. If $y \notin Z$, then the implicit function theorem and (2) show that h is analytic around y. Hence, h is analytic at a set of points defined locally as the complement of a proper, local analytic subset of U. It is then analytic in U by Riemann's extension theorem [1, 44.42, p. 420]. Since the set of singular points of Y is a proper analytic subset, a further application of the Riemann extension theorem shows that h is analytic on Y.

9.4. Lemma. *We keep the notation of 9.1 and assume (i), (ii) of 9.2. Then $\mathfrak{A}_v$ is integrally closed for every $v \in V$.*

Except for the use of 9.3, the proof is the same as that of the corresponding assertion in [35, Exp. 11, p. 7], and we describe it briefly.

By (ii), v has a fundamental system of neighborhoods U such that $U \cap V_0$ is an irreducible analytic space. Therefore, if f, g are $\mathcal{Q}$-functions on U whose product is identically zero, then one of them must be identically zero on $U \cap V_0$, hence on U by continuity. This shows that $\mathcal{Q}_v$ is integral.

Let now $f, g \in \mathcal{Q}_v$ with g not identically zero and f/g in the integral closure of $\mathcal{Q}_v$. There exists then a relation of the form

$$(1) \qquad (f/g)^n + \sum_{0 < i \leqq n} a_i \cdot (f/g)^{n-i} = 0 \qquad (a_i \in \mathcal{Q}_v; i = 0, \cdots, n - 1) .$$

If U is a sufficiently small neighborhood of v, then f, g may be viewed as $\mathcal{Q}$-functions on U, and g is not identically zero on $U \cap V_0$. Since $U \cap V_0$ is normal, there exists then an analytic function h on $U \cap V_0$ such that

$$h(x) \cdot g(x) = f(x) \qquad\qquad (x \in U \cap V_0) .$$

As in [35, loc. cit.] it follows from (1) and (i) that h extends by continuity to a continuous function on U, which will then verify

$$(2) \qquad\qquad h^n(x) + \sum_{0 < i \leqq n} a_i(x) \cdot h^{n-i}(x) = 0 ,$$

for all $x \in U$. By 9.3, the restriction of h to $V_i \cap U$ is then analytic; hence, h is an $\mathcal{Q}$-function on U, and $f/g \in \mathcal{Q}_v$.

9.5. LEMMA. *We keep the notation of* 9.1 *and the assumptions* (i), (iv) *of* 9.2. *Let* $v \in V$, U' *be an open relatively compact neighborhood of* v *whose points are separated by* $\mathcal{Q}$-*functions, and* U *be a neighborhood of* v *whose closure is contained in* U'. *Then there exist finitely many* $\mathcal{Q}$-*functions on* U *which separate the points of* U.

Let $f_1, \cdots, f_s$ be a finite set of $\mathcal{Q}$-functions on U'. Define a holomorphic map $f: U' \to \mathbf{C}^s$ by $f(u) = (f_1(u), \cdots, f_s(u))$, and let

$$\varphi = f \times f: U' \times U' \to \mathbf{C}^s \times \mathbf{C}^s .$$

Let Δ and D be the diagonals of $U' \times U'$ and $\mathbf{C}^s \times \mathbf{C}^s$ respectively. Clearly $\varphi^{-1}(D) \supset \Delta$, and we have $\varphi^{-1}(D) = \Delta$ if and only if f is injective.

$U' \times U'$ is the disjoint union of the locally closed analytic spaces $(U' \cap V_i) \times (U' \cap V_j)$. Similarly $(U' \times U') - \Delta$ is a disjoint union of locally closed subspaces, each endowed with the structure of a separable normal analytic space, namely the complements of the diagonal in the subspaces $(U' \cap V_i) \times (U' \cap V_j)$. Therefore $U' \times U' - \Delta$ may be written as disjoint union of countably many subspaces M_j, each of which is an irreducible analytic space, locally analytically embedded in some V_i. The restriction of φ to M_j is analytic, hence $U \cap \varphi^{-1}(D) \cap M_j$ is an analytic subspace. Let M_{jk}

be its irreducible components, and let

$$\delta_U(f_1, \cdots, f_s) = \max_{j,k} \dim M_{jk}$$

Put $\delta_U(f_1, \cdots, f_s) = -1$ if all the M_{jk} are empty, i.e., if f is injective on U. It is clearly enough to show that if, $\delta_U(f_1, \cdots, f_s) \geqq 0$, then there exists an open neighborhood U'' of U, and finitely many $\mathcal{C}$-functions $f_1', \cdots, f_t'$ on U'' such that $\delta_U(f_1', \cdots, f_t') < \delta_U(f_1, \cdots, f_s)$.

Let us enumerate the M_{jk} as $Y_1, Y_2, \cdots$, and let $y_i = (u_i, v_i) \in Y_i$ ($i = 1, 2, \cdots$). Then $u_i \neq v_i$, so there exists an $\mathcal{C}$-function g_i on U' such that $g_i(u_i) \neq g_i(v_i)$ ($i = 1, \cdots$). Define g_i^* on $U' \times U'$ by $g_i^*(x, y) = g_i(x) - g_i(y)$. We may, and shall, assume that $|g_i| \leqq 1/2$ on U', hence that $|g_i^*| \leqq 1$ on $U' \times U'$. Let U'' be an open neighborhood of $\bar{U}$ whose closure is contained in U'. We claim that we may choose constants c_i such that the sequence $g_{(m)}^* = \sum_{1 \leqq i \leqq m} c_i g_i^*$ converges uniformly on $\bar{U}'' \times \bar{U}''$ to a function g^* such that $g^*(u_i, v_i) \neq 0$ ($i = 1, 2, \cdots$). In fact, supposing $c_1, \cdots, c_{m-1}$ chosen in such a way that $g_{(i)}^*(u_i, v_i) \neq 0$ for $i = 1, \cdots, m - 1$, we select c_m verifying the following conditions

$$|c_m| \leqq 4^{-m} ; \qquad g_{(m)}^*(u_m, v_m) \neq 0 ,$$
$$|c_m \cdot g_m^*(u_i, v_i)| \leqq 4^{-m} \cdot \mathrm{Min}_{1 \leqq j < m} |g_{(j)}^*(u_i, v_i)| \qquad (1 \leqq i < m) .$$

Then the constants c_i are easily proven to satisfy our condition. In this case, $g = \sum c_i \cdot g_i$ converges uniformly on U'' and is an $\mathcal{C}$-function on U'' such that $g^*(x, y) = g(x) - g(y)$, $(x, y \in U'')$. This implies that $g(u_i) - g(v_i) = g^*(u_i, v_i) \neq 0$ ($i = 1, \cdots$), hence that

$$\delta_U(f_1, \cdots, f_s, g) < \delta_U(f_1, \cdots, f_s) .$$

9.6. **Lemma.** *We keep the notation of* 9.1 *and the assumptions* 9.2 (i), 9.2 (ii). *Let U be a relatively compact open neighborhood of $v \in V$, $f_1, \cdots, f_s$ a finite set of $\mathcal{C}$-functions on U which separate the points of U, and $f: u \mapsto (f_1(u), \cdots, f_s(u))$ the associated mapping of U into $\mathbf{C}^s$. Then there exists a relatively compact neighborhood U' of v in U such that f induces a homeomorphism of U' (resp. $U' \cap V_i$, $i = 0, 1, \cdots$) onto an analytic (resp. locally analytic) set in some open domain N of $\mathbf{C}^s$, and that $f(U')$ is locally analytically irreducible at each of its points.*

Let U_1 be an open neighborhood of v such that $\bar{U}_1$ is contained in U. Since f is injective on U, there is an open neighborhood N of $f(v)$ such that $f(\bar{U}_1 - U_1) \cap N$ is empty. Put $U' = f^{-1}(N) \cap U_1$. Let C be compact in N, and $C' = f^{-1}(C) \cap U'$. The set C' is contained in $f^{-1}(C) \cap \bar{U}_1$, which is compact. Let b belong to the closure of C' in $\bar{U}_1$. Then $f(b) \in C \subset N$, so $b \in U_1 \cap f^{-1}(C) \subset U'$; thus $b \in f^{-1}(C) \cap U' = C'$, so C' is compact. Consequently. f is proper on U', and therefore is a homeomorphism of U' onto $f(U') \subset N$. Now, let $V_{i_1}, \cdots, V_{i_r}$ be those V_i of smallest dimension d_0 which meet U'. By 9.2(i), the intersection of each with U' is closed in U' and since f is proper on U', it follows that $f((V_{i_1} \cap \cdots \cap V_{i_r}) \cap U')$ is a closed analytic subset of dimension d_0 of N. Assume now that for some integer $d \geqq d_0$ we have proved that $S = f((V_{(a)}) \cap U')$ is a closed analytic set in N of dimension $\leqq d$. Let V_j be of dimension $d + 1$. By [21, Ch. V, C5, p. 162] $f(V_j \cap U')$ is analytic of dimension $d + 1$ in $N - S$. Then, by a theorem of Remmert-Stein [21, Ch. V, D5, p. 169] the closure of $f(V_j \cap U)$ in N is an analytic set in N. The fact that $f(U')$ is an analytic set now follows by induction on d. Since f is bijective on U' and since each of its coordinates is an $\mathcal{C}$-function, it follows that for each $x \in U'$, f induces an injection of the local ring of $f(U')$ at $f(x)$ into $\mathcal{C}_x$; the latter being an integral domain, we see that $f(U')$ is irreducible at every point.

9.7. Lemma. *We keep the notation of* 9.1 *and the assumptions of* 9.2. *Let* U' *be as in* 9.6, *and put* $Y = f(U')$. *Let* $\tilde{Y}$ *be the normalization of* Y. *Then* f *induces an isomorphism of ringed spaces of* $(U', \mathcal{C}|_{U'})$ *onto* $\tilde{Y}$.

Since Y is analytically irreducible (of dimension $d = \dim V_0$) at each point, the canonical projection of $\tilde{Y}$ onto Y is a homeomorphism, and we may identify $\tilde{Y}$ with Y, endowed with the structural sheaf $\tilde{\mathcal{O}}$ whose stalk at y is the integral closure $\tilde{\mathcal{O}}_y$ of the local ring $\mathcal{O}_y$ of Y at y. We have to prove that f induces an isomorphism of $\mathcal{C}_u$ onto $\tilde{\mathcal{O}}_{f(u)}$ for every $u \in U'$.

Let first $g \in \tilde{\mathcal{O}}_{f(u)}$. There is a neighborhood of $f(u)$ in which g defines a continuous function which satisfies an integral dependence relation

$$(1) \qquad g^n(x) + \sum_{0 < i \leqq n} b_i(x) \cdot g^{n-i}(x) = 0 \, ,$$

where the b_i are analytic on Y around $f(u)$. The function $g \circ f$ is then continuous around u, and satisfies there a relation similar to (1), with b_i replaced by $a_i = b_i \circ f$. The a_i's are continuous around u. By 9.6, they are $\mathcal{C}$-functions; hence (9.3), the restriction of $g \circ f$ to V_i around u is analytic. Therefore $g \circ f$ is an $\mathcal{C}$-function.

It is well-known [8, p. 179] that an analytic homeomorphism of one complex manifold onto another is an isomorphism. Let a be an $\mathcal{C}$-function at $u \in U'$, i.e., $a \in \mathcal{C}_u$, and let N be a neighborhood of $f(u)$. If N is chosen small enough, then $a \circ f^{-1}$ is continuous in N and analytic on $f(V_0) \cap N$ (viewed as a subset of $\tilde{Y}$), except possibly at the image points of the singularities of V_0. Hence [1, 44.42, p. 420], $a \circ f^{-1}$ is analytic on all of $f(V_0) \cap N$. By 9.6, $(f(U') - f(U' \cap V_0)) \cap N$ is a proper analytic subset of $\tilde{Y} \cap N$. Hence, $a \circ f^{-1}$ is analytic on $\tilde{Y} \cap N$, so that, finally, f^* is an isomorphism of $\tilde{\mathcal{O}}_{f(u)}$ onto $\mathcal{C}_u$.

By 9.5, each point $v \in V$ has a neighborhood U' as in 9.6. Therefore, the first assertion of the theorem follows from 9.7. The assertion about $V_{(a)}$ follows at once from the induction procedure indicated in 9.6.

9.8. COROLLARY. *Let U be an open subset of V, and f a continuous function on U which is analytic on $V_0 \cap U$. Then f is an $\mathcal{C}$-function.*

This follows from the theorem and the Riemann extension theorem.

10. Analytic structure and projective embeddings of the compactification

10.1. We now revert to the set up of 3.3 (i). In particular, G is a connected semi-simple **Q**-group, with center reduced to $\{e\}$, whose symmetric space $X = K\backslash G_\mathbf{R}$ of non-compact type is a bounded domain and $H(X)$ is the group of all holomorphic automorphisms of X, in which $G_\mathbf{R}^0$ is of finite index. Moreover, X^* is the union of the rational boundary components of X, endowed with the Satake topology (4.8), Γ an arithmetic group of automorphisms of X, $V^* = X^*/\Gamma$ the compactification of $V = X/\Gamma$ introduced in § 4, and $\pi: X^* \to V^*$ the canonical projection. There are finitely many rational boundary components F_i ($0 \leq i \leq m$, $F_0 = X$) such that V^* is the disjoint union of the quotients $V_i = F_i/\Gamma(F_i)$. Since $V_i \neq V_j$ if $i \neq j$, we have $F_i \not\subset F_j \cdot \Gamma (i \neq j)$.

10.2. The group $\Gamma(F_i)$ acts in a properly discontinuous fashion on F_i; hence, V_i is canonically endowed with the structure of an irreducible normal analytic space [17]. We are thus in the situation of 9.1 and introduce the sheaf $\mathcal{C}$ of germs of $\mathcal{C}$-functions on V^*. An $\mathcal{C}$-function on an open subset U of V^* is a continuous complex-valued function whose restriction to $V_i \cap U$ is analytic ($0 \leq i \leq m$).

Let $x \in X^*$ and $v = \pi(x)$. Let U be a good neighborhood of x in X^* (8.1). Then $U' = \pi(U)$ may be identified with U/Γ_x, hence $V_i \cap U'$ with $(F_i \cap U)/(\Gamma_x \cap N(F_i))$. The definitions of the analytic structure on V_i and of $\mathcal{C}$ imply that $f: U' \to \mathbf{C}$ is an $\mathcal{C}$-function if and only if $f \circ \pi$ is a continuous function on U, which is invariant under Γ_x, and whose restriction to $F \cap U$ is analytic for every rational boundary component F. In particular, the quotient ω/ω' of two integral automorphic forms ω, ω' for Γ_x on U, of the same weight, where ω' does not take the value zero in U, may be identified with an $\mathcal{C}$-function of U'.

10.3. Let i be the index such that $v \in V_i$. The canonical projection $\sigma_i: X \to F_i$ (1.7, remark) induces an analytic map σ_i' of $(X \cap U)/\Gamma_x = V \cap U'$ onto a neighborhood of v in $V_i \cap U' = (U \cap F_i)/\Gamma_x$. Let j be such that $v \in \bar{V}_j$, and let $w \in V_j \cap U'$. There exists $\gamma \in \Gamma$ such that $F_j \cdot \gamma \cap U \neq \varnothing$ and $\bar{F}_j \cdot \gamma \ni x$.

The canonical projection $\sigma_{F_i, F_j \cdot \gamma}$ induces a holomorphic map of

$$\pi(F_j \cdot \gamma \cap U) = (F_j \cdot \gamma \cap U)/(N(F_j \cdot \gamma) \cap \Gamma_x)$$

onto a neighborhood of v in V_i. We have the factorization (1.7, remark)

$$\sigma_{F_i} = \sigma_{F_i, F_j \cdot \gamma} \circ \sigma_{F_j \cdot \gamma} \, .$$

Let now f be a holomorphic function around v on V_i. The above remarks imply that $f \circ \sigma'_{F_i}$ extends by continuity to an $\mathcal{O}$-function near v in V, whose restriction to V_i around v is equal to f, and whose restriction to V_j near w, lifted to $F_j \cdot \gamma$, is equal to

$$f \circ \pi \circ \sigma_{F_i, F_j \cdot \gamma} \, .$$

10.4. THEOREM. *We keep the assumptions and notation of 10.1, 10.2. Then $(V^*, \mathcal{O})$ is an irreducible normal analytic space, in which each V_i is embedded as a locally closed analytic space.*

To prove the theorem, it is enough to check that the conditions (i) to (iv) of 9.2, with V and V_0 replaced by V^* and V respectively, hold true in the present situation.

Conditions (i), (ii) and (iii) are consequences of 4.11, 4.15 and 10.3, respectively.

It remains to check the separation of points by $\mathcal{O}$-functions. Let $v \in V_i$ and $x \in F_i$ be such that $\pi(x) = v$. By 8.8, 8.9 there exists an integral automorphic form E, of some weight l, such that $\Phi_{F_i} E(x) \neq 0$. Let U be a good neighborhood of x in X^*, on which the extension of E does not take the value zero, and let $U' = \pi(U)$. Let $p', q' \in U'$ and j, k be the indices such that $p' \in V_j$, $q' \in V_k$. Let $p \in F_j \cap \pi^{-1}(p')$, $q \in F_k \cap \pi^{-1}(q')$. Assume $\dim F_j \geqq \dim F_k$. Since, by construction we have either $j = k$, or $F_j \not\subset F_k \cdot \Gamma$, there exists (8.8, 8.9) a multiple $l' = l \cdot m$ of l, and an integral automorphic form E' of weight l' for Γ, such that

$$\Phi_{F_j} E'(p) \neq 0 \, , \qquad \Phi_{F_k} E'(q) = 0 \, .$$

The quotient E'/E^m is then an $\mathcal{O}$-function on U' which separates p' from q'. Thus 9.2 (iv) also holds true in V^*.

10.5. COROLLARY. *Assume that G has no normal **Q**-subgroup of dimension 3. Let U be open in V^*. Then every meromorphic function on $U \cap V$ is the restriction of a meromorphic function on U. In particular, the restriction to V yields an isomorphism of the field of meromorphic functions on V^* onto the field of meromorphic functions on V.*

The assumption on G implies, by 3.15, that $\dim (V^* - V) \leqq \dim V^* - 2$.

Therefore 10.5 follows from 10.4 and a well-known extension theorem on normal analytic spaces.

We recall the map $f \mapsto \pi \circ f$ identifies the meromorphic functions on V with the Γ-invariant meromorphic functions on X, i.e., with the automorphic functions for Γ.

10.6. LEMMA. *We keep the notation of* 10.1, 10.2. *There exist a weight* l *and finitely many integral automorphic forms* $E_0, \cdots, E_N$ *of weight* l *such that the forms* $\Phi_{F_i} \cdot E_j$ *are nowhere simultaneously zero* $(0 \leq i \leq m)$.

Given $x \in F_i$, there exists a weight l_x, and an integral automorphic form E_x of weight l such that $\Phi_F E_x(x) \neq 0$, (8.8, 8.9). There is then a good neighborhood $N(x)$ of x such that the extension of E to $N(x)$ is nowhere zero. By compactness, V is covered by the images of finitely many such neighborhoods $N(x_j)$. The lemma follows then by taking for l the l.c.m. of the l_{x_j} and for E_i's suitable powers of the E_{x_j}.

10.7. Let E_j $(0 \leq j \leq N)$ be as in 10.6. If we trivialize the bundle $\xi^l_{F_i}$, the forms $\Phi_{F_i} E_j$ are identified with holomorphic functions which are nowhere simultaneously zero; their values at $x \in F_j$ are the coordinates of a point in $\mathbf{C}^{N+1} - 0$. If we change the trivialization, these coordinates are all multiplied by the same non-zero constant, hence define the same point in the associated projective space $\mathbf{P}(N, \mathbf{C})$. Thus, to $x \in F$ there is associated a well-defined point in $\mathbf{P}(N, \mathbf{C})$, whose homogeneous coordinates will be denoted by $\Phi_{F_i} E_j(x)$. Since the E_j are automorphic forms of the same weight, two points x and $x \cdot \gamma (\gamma \in \Gamma)$ will have the same image in $\mathbf{P}(N, \mathbf{C})$, whence a map $f \colon V^* \to \mathbf{P}(N, \mathbf{C})$ defined by

$$f\big(\pi(x)\big) = \big(\Phi_{F_i} E_0(x), \cdots, \Phi_{F_i} E_N(x)\big) \qquad (x \in F_i, \ i = 0, \cdots, m) .$$

Since the quotient of two integral automorphic forms is an $\mathcal{C}$-function outside the set of zeros of the denominator, f is a holomorphic mapping.

10.8. LEMMA. *We keep the notation of* 10.1, 10.2. *There exist a weight* l *and finitely many integral automorphic forms* $E_0, \cdots, E_N$ *of weight* l *for* Γ, *satisfying* 10.5, *such that the map* $f \colon V^* \to \mathbf{P}(N, \mathbf{C})$ *associated to the* E_i*'s is a homeomorphism of* V^* *onto* $f(V^*)$.

The proof is essentially the same as in the symplectic case [2], and will be described briefly. It is enough to show that for suitable E_i's the map f is injective.

Let D and Δ be the diagonals in $V^* \times V^*$ and $\mathbf{P}(N, \mathbf{C}) \times \mathbf{P}(N, \mathbf{C})$ respectively, and $S = (f \times f)^{-1}(\Delta)$. Then S is an analytic subset of $V^* \times V^*$ containing D, which is equal to D if and only if f is injective. Since, in a compact analytic space, a decreasing sequence of analytic subsets is stationary, it is enough to show that if $S \neq D$, then there exists a similar map $f' : V^* \to \mathbf{P}(N', \mathbf{C})$ associated to integral automorphic forms of some weight l' for which $S' = (f' \times f')^{-1}(\Delta) \subsetneqq S$.

Let $x \in F_i$, $y \in F_j$ ($\dim F_i \geqq \dim F_j$), $x' = \pi(x)$, $y' = \pi(y)$, be such that $(x', y') \in S - D$. Then $x \notin y \cdot \Gamma$; by 8.8, 8.9 there exists a multiple l' of l and an integral automorphic form E of weight l such that $\Phi_{F_i} E(x) \neq 0$, $\Phi_{F_j} E(y) = 0$. We then take as E_j'''s all the monomials of degree l'/l in the E_j's, and E. We have $S' \subset S$, and $(x, y) \notin S'$, hence $S' \neq S$.

10.9. Let $\mathcal{Q}(\Gamma)$ be the graded ring of automorphic forms of positive weight for Γ on X. It may be identified with the set of invariants of Γ in a ring $B = \sum_{i \geqq 0} B_i$ of holomorphic functions, on which $H(X)$ operates by

$$(f \circ g^{-1})(x) = J(x, g)^i \cdot f(x \cdot g) \qquad (f \in B_i; \, x \in X) \, ,$$

where J is the functional determinant in some realization of X as a domain in euclidean space. Since X is connected, it follows that $\mathcal{Q}(\Gamma)$ is integrally closed [35, Exp. 17, No. 5]. We claim that the subring $\mathcal{Q}'(\Gamma)$ of integral automorphic forms is also integrally closed. Since $\mathcal{Q}(\Gamma)$ is, this amounts to showing that, if h is an automorphic form of weight l which verifies an integral dependence relation

$$(1) \qquad\qquad h^n + \sum_{0 < i \leqq n} a_i(x) \cdot h^{n-i} = 0 \, ,$$

where a_i is an integral automorphic form of weight $l \cdot i$, then h is integral. Let $x \in X^*$ and U be a good neighborhood of x. We may identify h and the a_i's with holomorphic functions on $X \cap U$. Moreover the a_i's extend by continuity to continuous functions whose restrictions to $F \cap U$ are holomorphic for any rational boundary component F. The relation (1) and the condition 9.2 (ii) imply again, as in 9.4, that h extends by continuity to a continuous function on U. If follows then from 9.3 that h is analytic on $F \cap U$ for every F. Thus h is integral.

10.10. Let $(E_i)_{0 \leqq i \leqq N}$ be a set of integral automorphic forms verifying 10.8, A be the subring of $A'(\Gamma)$ generated by the E_i's, and $\tilde{A}$ its integral closure. The latter is a finitely generated algebra over $\mathbf{C}$ [15, Ch. 5, § 3, No. 2] and is contained in $A'(\Gamma)$ by 10.9. It is elementary that there exists an integer d such that the subring $\tilde{A}^{(d)}$ of elements in $\tilde{A}$ whose degree is a multiple of d is generated by $\tilde{A}_d$ [15, Ch. 3, § 1, No. 3, Prop. 3]. Moreover, $\tilde{A}^{(d)}$ is also integrally closed [15, Ch. 5, § 1, No. 8, Cor. 3]. Therefore $\tilde{A}^{(d)}$ is a normally projective algebra over $\mathbf{C}$, in the sense of [35, Exp. 17]. Let E_i ($0 \leq i \leq M$) be a basis of $\tilde{A}_d$. Then $\tilde{A}^{(d)} \cong \mathbf{C}[T_0, \cdots, T_M]/I$, where I is the ideal of the relations between the E_i. The projective variety $V(\widetilde{A}^{(d)}) \subset \mathbf{P}(M, \mathbf{C})$ defined by I is then normally projective. The map $f\colon V^* \to \mathbf{P}(M, \mathbf{C})$ associated to the E_i's is well-defined, injective; its image is an analytic, hence algebraic, variety, contained in $V(\widetilde{A}^{(d)})$. It is in fact equal to $V(\widetilde{A}^{(d)})$ since otherwise there would exist a polynomial $P \in \mathbf{C}[T_0, \cdots, T_M]$, not contained in I, such that $P(E_0, \cdots, E_M)$ would be identically zero on X, in contradiction with the definition of I. Thus f is a bijective holomorphic map of V^* onto $V(\widetilde{A}^{(d)})$. Since both V^* and $V(\widetilde{A}^{(d)})$ are normal analytic spaces, f is an isomorphism of analytic spaces. Thus we have proved the following:

10.11. Theorem. *We keep the notation of 10.1, 10.2. There exist a weight l and finitely many integral automorphic forms E_i of weight l whose extensions to X^* are nowhere simultaneously zero, such that the associated map $f\colon V^* \to \mathbf{P}(N, \mathbf{C})$ is an isomorphism of V^* onto a normally projective subvariety of $\mathbf{P}(N, \mathbf{C})$.*

10.12. Corollary. *Assume that G has no normal $\mathbf{Q}$-subgroup of dimension 3. Then the field of automorphic functions for Γ is canonically isomorphic with the field of rational functions on $f(V^*)$. In particular, it is an algebraic function field of transcendence degree equal to $\dim_{\mathbf{C}} X$. Every automorphic function is the quotient of two integral automorphic forms of the same weight.*

This follows from 10.5, and from the fact that a meromorphic function on a projective variety is rational by Chow's theorem.

10.13. Let $\rho\colon K^0 \to \mathbf{GL}(E)$ be a finite dimensional unitary representation of K^0. It defines on X a complex vector bundle ξ_ρ, the bundle associated by ρ to $G_{\mathbf{R}}^0$, viewed as principal K^0-bundle by left translations. The total space is therefore the quotient $G_{\mathbf{R}}^0 \times_{K^0} E$ of $G_{\mathbf{R}}^0 \times E$ by the equivalence relation

$$(g, v) \approx (k \cdot g, \rho(k) \cdot v) \qquad\qquad (k \in K^0,\, g \in G_{\mathbf{R}}^0,\, v \in E).$$

It can also be written as $P^- \cdot K_{\mathbf{C}}^0 \cdot G_{\mathbf{R}}^0 \times_{P^- \cdot K_{\mathbf{C}}^0} E$, where ρ is extended in the obvious fashion to a representation of $P^- \cdot K_{\mathbf{C}}^0$ which is trivial on P^-; hence, it is a holomorphic vector bundle. An automorphic form of type ξ_ρ is a Γ-invariant, holomorphic cross-section of ξ_ρ. These forms correspond in a canonical fashion to the holomorphic V-valued functions on X which satisfy the relation

$$f(x \cdot \gamma) = \mu_\rho(x, \gamma)^{-1} \cdot f(x) \qquad\qquad (x \in X,\, \gamma \in \Gamma)\,,$$

where μ_ρ is the automorphy factor introduced in 5.6. We let $\mathcal{A}_\rho$ be the sheaf of germs of automorphic functions of type ξ_ρ, for Γ on X/Γ. It is reflexive, torsionless, and is known to be an analytic coherent sheaf [34, Exp. XX].

REMARK. We have tacitly assumed that Γ operates on ξ_ρ. This is certainly the case if $\Gamma \subset G_R^0$. Otherwise we assume that ρ extends to a subgroup K' of finite index of $K \cap H(X)$ such that $\Gamma \subset K' \cdot G_R^0$. Replacing K^0 and K_C^0 by K' and K_C' respectively in the above construction, we see easily that the action of G_R^0 extends to one of $K'G_R^0$.

10.14. THEOREM. *Assume that G has no normal* **Q**-*subgroup of dimension 3, and let $\mathcal{C}_\rho$ be as in 10.13. Then the direct image $i_*\mathcal{C}_\rho$ in V^* of the sheaf of germs of automorphic forms of type ξ_ρ is an algebraic coherent sheaf. In particular, if U is an open subset of V^*, every holomorphic section of $\mathcal{C}_\rho$ over $U \cap V$ extends to a holomorphic section over U. The space of automorphic forms of type ξ_ρ is canonically isomorphic to the space of holomorphic cross-sections of $i_*\mathcal{C}_\rho$ over V^*, and is finite dimensional. The ring of automorphic forms of positive weight is finitely generated.*

We identify V^* with its image under the map of 10.11. Then the restriction to V of the line bundle $\mathcal{O}$ of $\mathbf{P}(N, \mathbf{C})$ attached to the divisor of a hyperplane is the sheaf $\mathcal{C}_d$ of germs of automorphic forms of weight d. We know (5.11) that if m is large enough, the product $J^{d \cdot m} \cdot \mu_\rho$ is an automorphy factor which satisfies the condition allowing one to construct Poincaré series. Therefore, Theorem 3 of [19] applies. It shows that given $x \in X/\Gamma$, there exist finitely many analytic cross-sections of the sheaf $\mathcal{C}_\rho \otimes \mathcal{O}^n$ which generate the fibre of $\mathcal{C}_\rho$ at x. Since $V^* - V$ has codimension $\geqq 2$ (3.15), Serre's extension theorem [36] applies, and yields the theorem, except for the last assertion. We now know that $A(\Gamma) = A'(\Gamma)$. Let l and E_i be as in 10.11. The automorphic forms of weight $m \cdot l$ (m a positive integer) may be identified with the holomorphic cross-sections of $\mathcal{O}^m$. They are therefore the polynomials of degree m in the E_i's. This means that the algebra $A(\Gamma)^{(l)}$ is generated by the E_i's. Each space $A(\Gamma)_i$ is finite dimensional. Therefore, in order to establish the second assertion, it suffices to show the existence of an integer n_0 such that

$$(1) \qquad\qquad A(\Gamma)_l \cdot A(\Gamma)_{p+n \cdot l} = A(\Gamma)_{p+(n+1) \cdot l} \qquad\qquad (n \geqq n_0, p \geqq n_0) \, .$$

Since the sheaves involved extend to algebraic coherent sheaves on a projective variety, the proof of (1) given by Serre [34, XX, nos. 9, 10] when X/Γ is compact applies without change to our case.

APPENDIX

11. Connected components of automorphism groups

In this section, we collect some partly known remarks on connected components, whose use in the preceding sections has allowed for some slight

technical simplification. X is a bounded symmetric domain, $H(X)$ the group
of complex analytic homeomorphisms (i.e., of automorphisms) of X, and $\mathrm{Is}(X)$
the group of isometries of X with respect to the riemannian structure defined
by the Bergman metric.

11.1. As is well-known, $\mathrm{Is}(X)$ and $H(X)$ are semi-simple Lie groups,
with finitely many connected components, X is the quotient of $\mathrm{Is}(X)$ by a
maximal compact subgroup K, and $\mathrm{Is}(X)^0 = H(X)^0$ is a non-compact semi-
simple Lie group with center reduced to $\{e\}$. Thus $\mathrm{Is}(X)^0 = \mathrm{Ad}\,\mathfrak{g}$ where $\mathfrak{g}$ is
the Lie algebra of $\mathrm{Is}(X)$. Furthermore, it is known that $\mathrm{Is}(X) \cong \mathrm{Aut}\,\mathfrak{g}$.

Assume now X to be irreducible. Then *its isometries are either holomorphic or anti-holomorphic, and $H(X)$ has index two in* $\mathrm{Is}(X)$. In fact,
let $\mathfrak{g} = \mathfrak{k} + \mathfrak{p}$ be the Cartan decomposition of $\mathfrak{g}$ associated to the Lie algebra $\mathfrak{k}$
of K. Then $\mathrm{Is}(X) = K \cdot P$ $(P = \exp \mathfrak{p})$ and $H(X) = (K \cap H(X)) \cdot P$. The
identity component S of the center of the identity component of K is one-
dimensional, and an automorphism $\mathrm{Int}\,k$ $(k \in K)$ is either the identity or the
inversion $s \mapsto s^{-1}$ on S. On the other hand, the multiplication by $\sqrt{-1}$ in the
tangent space X_0 of X at K is induced by $\mathrm{Ad}\,s_0$, where s_0 is an element of
order 4 in S. Therefore $\mathrm{Int}\,k(s_0)$ is equal either to s_0 or to s_0^{-1}. The transformation k is holomorphic in the first case, anti-holomorphic in the second
one. In particular

$$(1) \qquad\qquad\qquad H(X) \cap K = Z(S) \,,$$

and $H(X)$ has index ≤ 2 in $\mathrm{Is}(X)$. On the other hand, there is clearly a linear
orthogonal transformation A on X_0 which carries the given complex structure
onto its conjugate. By standard facts on simply connected riemannian symmetric spaces, A extends to an isometry of X, which is then anti-holomorphic,
hence $\mathrm{Is}(X) \neq H(X)$.

If X is the product of r irreducible components, it is clear that $\mathrm{Is}(X)$
(resp. $H(X)$) is generated by products of isometries (resp. automorphisms) of
the different factors, and permutations of isomorphic factors.

These remarks have already been made by E. Cartan, who has also given
the structure of $H(X)$ in all irreducible cases; it is connected, except in the
cases mentioned in 11.4, where $H(X)^0$ has index two in $H(X)$, [16, p. 152].

11.2. PROPOSITION. *Let $\bar{D}$ be the natural compactification of X (1.4).
The action of $H(X)$ on X extends by continuity to a continuous action on $\bar{D}$,
and the restriction of $h \in H(X)$ to any boundary component is holomorphic.*

It suffices to prove this for $K \cap H(X)$ since $H(X)$ is generated by this
group and by $H(X)^0$. But, if $k \in K \cap H(X)$, then k commutes with the ele-

ment s_0 considered in 11.1, and the extension to $\mathfrak{p}_C$ of $\mathrm{Ad}_{\mathfrak{p}}\,k$ leaves the two subspaces $\mathfrak{p}^+$, $\mathfrak{p}^-$ stable; hence, the action of k on X extends to a linear transformation of $\mathfrak{p}^+$. Therefore k operates continuously on $\bar{D}$. Furthermore, the boundary components are open subsets of complex affine subspaces of $\mathfrak{p}^+$; hence, the restriction of k to such a component is holomorphic.

11.3. PROPOSITION. *Let G be a connected simple algebraic group defined over* **R** *such that the symmetric space of non-compact type X of $G_{\mathbf{R}}$ is a bounded symmetric domain. Then $H(X) \cap G_{\mathbf{R}} = G_{\mathbf{R}}^0$. The group $G_{\mathbf{R}}$ has either one or two connected components.*

Since G is simple, X is irreducible, and we may identify $G_{\mathbf{R}}$ with a subgroup of $\mathrm{Aut}\,\mathfrak{g}_{\mathbf{R}}$, namely $\mathrm{Ad}\,\mathfrak{g}_{\mathbf{C}} \cap \mathrm{Aut}\,\mathfrak{g}_{\mathbf{R}}$. We keep the notation of 11.1; in particular, $K \cap G_{\mathbf{R}}$ is a maximal compact subgroup of $G_{\mathbf{R}}$, and $G_{\mathbf{R}} = (K \cap G_{\mathbf{R}}) \cdot P$.

The second assertion follows from the first one and 11.1; in view of 11.1 (1), the first assertion is equivalent to: $G_{\mathbf{R}} \cap Z(S)$ is connected, which we now prove.

Being the centralizer of a torus in $G_{\mathbf{C}}$, the group $Z(S)_{\mathbf{C}}$ is connected, since G is (cf. [9, § 18]). It is defined over **R**, and its Lie algebra is $\mathfrak{k}_{\mathbf{C}}$. Therefore $\mathfrak{k}$ is a compact real form of $\mathfrak{k}_{\mathbf{C}}$, and K^0 is the identity component of a maximal compact subgroup L of $Z(S)_{\mathbf{C}}$. Since $Z(S)_{\mathbf{C}}$ is connected, so must be L, whence $Z(S) \cap K = K^0$.

11.4. REMARK. In the type IV_n (IV refers to Siegel's notation; it is III in [16]) of bounded symmetric domains, $G = \mathbf{PSO}(n + 2, \mathbf{C})$ is the quotient of the special orthogonal group in $n + 2$ variables by its center, $G_{\mathbf{R}} = \mathbf{PSO}(n, 2)$, and $K \cap G_{\mathbf{R}}$ is the group of elements of determinant one in $\mathbf{O}(n) \times \mathbf{O}(2)$ (divided by $\{\pm 1\}$ if n is even). From this we see readily that

(a) if n is odd, $G_{\mathbf{R}} = \mathrm{Is}\,(X)$, $G_{\mathbf{R}}^0 = H(X)$, and

(b) if n is even, $\mathrm{Is}\,(X)/\mathrm{Is}\,(X)^0 = \mathbf{Z}_2 + \mathbf{Z}_2$, $G_{\mathbf{R}}/G_{\mathbf{R}}^0 = \mathbf{Z}_2$, $H(X)/H(X)^0 = \mathbf{Z}_2$. The situation (b) also occurs for type $I_{n,n}$ $(n \geq 2)$.

11.5. LEMMA. *Let G be a connected semi-simple group defined over* **R**. *Assume that $G_{\mathbf{R}}^0$ has a center reduced to $\{e\}$, and has the same rank as its maximal compact subgroups. Then the center of $G_{\mathbf{C}}$ is reduced to $\{e\}$.*

Let $\mathfrak{t}$ be a Cartan subalgebra of a maximal compact subgroup K of $G_{\mathbf{R}}$. The assumption implies that $\mathfrak{t}_{\mathbf{C}}$ is a Cartan subalgebra of $\mathfrak{g}$. It is the Lie algebra of a maximal torus $T_{\mathbf{C}}$ of $G_{\mathbf{C}}$, which is defined over **R**, and whose subgroup of real points is compact. The latter is then necessarily a maximal compact subgroup of $T_{\mathbf{C}}$, and is connected and equal to $\exp\mathfrak{t}$. Let now z be in the center of $G_{\mathbf{C}}$. It belongs to $T_{\mathbf{C}}$ and is of finite order; hence, z and $\exp\mathfrak{t}$ generate a compact subgroup of $T_{\mathbf{C}}$. Thus $z \in \exp\mathfrak{t} \cap G_{\mathbf{R}}^0$, and $z = e$.

REMARK. If we drop the assumption on the rank of K, the lemma becomes false as is shown by the case where $G_\mathbf{R} = \mathbf{SO}(p, q)$, $(p, q$ odd).

11.6. We now revert to the notation of 1.3, 1.5, and prove that $G(F)$ is connected, as asserted in 1.5. An obvious reduction shows that it suffices to do this when X is irreducible. In view of 1.5 (1), (2) this amounts to proving that the group P_b/Z_b of 1.3 is connected.

The group $Q_\mathbf{C} = P_\mathbf{C}/Z_{b,\mathbf{C}}$ is almost simple, connected, defined over $\mathbf{R}$, and $Q_\mathbf{R}^0 = \mathrm{Ad}\, l_b$. On the other hand, the symmetric space F_b of non-compact type of L_b is a bounded symmetric domain; hence, L_b has the same rank as its maximal compact subgroups. By 11.5, we have then $Q_\mathbf{C} = \mathrm{Ad}\, \mathbf{q}_\mathbf{C}$, which implies that $Q_\mathbf{R}$ is a *subgroup* of $\mathrm{Is}\,(F_b)$. By 11.4, all elements of P_b induce complex analytic homeomorphisms of F_b; therefore (11.3), the image of P_b in $Q_\mathbf{R}$ is connected, equal to $\mathrm{Ad}\, l_b$. The kernel of the homomorphism $P_b \to Q_\mathbf{R}$ is $P_b \cap Z_{b,\mathbf{C}}$. This is a normal subgroup of P_b, with Lie algebra $\mathfrak{z}_b$, which contains Z_b. It is therefore equal to Z_b (see 1.3), whence the result.

UNIVERSITY OF CHICAGO
INSTITUTE FOR ADVANCED STUDY

REFERENCES

1. S. ABHYANKAR, Local Analytic Geometry, Academic Press, New York, 1964.
2. W. L. BAILY, JR., *On Satake's compactification of V_n*, Amer. J. Math., 80 (1958), 348–364.
3. ———, *On the Hilbert-Siegel modular space*, ibid., 81 (1959), 846–874.
4. ———, *On the theory of automorphic functions and the problem of moduli*, Bull. Amer. Math. Soc., 69 (1963), 727–732.
5. ———, "On compactification of orbit spaces of arithmetic discontinuous groups acting on bounded symmetric domains", in Proc. Sympos. Pure Math., vol. 9, Amer. Math. Soc., Providence, R. I., 1966, pp.   .
6. ———, Fourier-Jacobi series, ibid.
7. ———, and A. BOREL, *On the compactification of arithmetically defined quotients of bounded symmetric domains*, Bull. Amer. Math. Soc., 70 (1964), 588–593.
8. S. BOCHNER and W. T. MARTIN, Several Complex Variables, Princeton University Press, Princeton, 1948.
9. A. BOREL, *Groupes linéaires algébriques*, Ann. of Math. (2), 64 (1956), 20–82.
10. ———, "Ensembles fondamentaux pour les groupes arithmétiques", Coll. s. l. théorie d. groupes algébriques, Bruxelles 1962, 23–40.
11. ———, *Density and maximality of arithmetic subgroups*, J. Reine. Angew. Math. (to appear).
12. ———, "Introduction to automorphic forms", in Proc. Sympos. Pure Math., vol. 9, Amer. Math. Soc., Providence, R. I., 1966.
13. A. BOREL, and HARISH-CHANDRA, *Arithmetic subgroups of algebraic groups*, Ann. of Math. (2), 75 (1962), 485–535.
14. A. BOREL and J. TITS, "Groupes réductifs", in Publ. I.H.E.S. 27, 1965, 55–150.
15. N. BOURBAKI, Algèbre commutative, Chapters 3, 4, Act. Sci. Ind. 1293, Chapters 5, 6, Act. Sci. Ind. 1308, Hermann, Paris.
16. E. CARTAN, *Sur les domaines bornés homogènes de l'espace de n variables*, Abh. Math. Sem. Hamburg, 11 (1935), 116–162.

17. ——, "Quotient d'un espace analytique par un groupe d'automorphismes". in Symposium in honor of S. Lefschetz, Princeton University Press, Princeton, 1957, 90-102.

18. ——, *Prolongement des espaces analytiques normaux*, Math. Ann., 136 (1958), 97-110.

19. ——, *Fonctions automorphes et séries de Poincareé*, J. Analyse Math., 6 (1958), 169-175.

20. R. Godement, *Domaines fondamentaux des groupes arithmétiques*, Sém. Bourbaki, 15, 1962-63, Exp. 257.

21. R. C. Gunning and H. Rossi, Analytic Functions of Several Complex Variables, Prentice-Hall, 1965.

22. Harish-Chandra, *Representations of semi-simple Lie groups*: VI, Amer. J. Math., 78 (1956), 564-628.

23. ——, *Discrete series for semi-simple Lie groups*: II, Acta Math. (to appear).

24. S. Helgason, Differential Geometry and Symmetric Spaces, Academic Press, New York, 1962.

25. M. Kneser, "Schwache Approximation in algebraischen Gruppen", in Coll. s. l. théorie d. groupes algébriques, Bruxelles 1962, 41-52.

26. A. Korányi, *The Poisson integral for generalized half-planes and bounded symmetric domains*, Ann. of Math. (2), 82 (1965), 332-350.

27. ——, and J. Wolf, *Generalized Cayley transforms of bounded symmetric domains*, Amer. J. Math., 87 (1965), 899-939.

28. H. Maass, *Ueber die Darstellung der Modulformen n-ten Grades durch Poincarésche Reihen*, Math. Ann., 123 (1951), 125-151.

29. C. C. Moore, *Compactifications of symmetric spaces* II: *The Cartan domains*, Amer. J. Math., 86 (1964), 358-378.

30. I. I. Pyateckii-Shapiro, Geometry of Classical Domains and Automorphic Functions (Russian), Fizmatgiz, Moscow, 1961.

31. ——, *Arithmetic groups on complex domains*, Uspehi Mat. Nauk., XIX (6) (1964), 93-121, translated in Russian Mathematical Surveys XIX (6), pp. 83, ff.

32. M. Rosenlicht, *Some rationality questions on algebraic groups*, Annali di Mat. (IV), 63 (1957), 25-50.

33. I. Satake, *On compactifications of the quotient spaces for arithmetically defined discontinuous groups*, Ann. of Math. (2), 72 (1960), 555-580.

34. Séminaire E. N. S. Paris, 6ième année, 1953-54.

35. ——, 10ième année, 1957-58.

36. J. P. Serre, *Prolongement de faisceaux cohérents* [Ann. Inst. Fourier 16 (1966), 363-374].

37. G. Shimura, *On the field of definition for a field of automorphic functions*, Ann. of Math. (2), 80 (1964), 160-189.

38. A. Weil, Adèles and Algebraic Groups, Notes by M. Demazure and T. Ono, Institute for Advanced Study, Princeton, 1961.

(Received February 21, 1965)

70.

Density and maximality of arithmetic subgroups

J. Reine Angew. Math. **224** (1966) 78–89

Let G be a connected semi-simple group defined over the field Q of rational numbers and H an *arithmetic subgroup* of G. By definition, then, H is a subgroup of G_Q and if $\varrho : G \to GL_n$ is an injective Q-morphism, then $\varrho(H)$ is commensurable with $\varrho(G)_Z$ (see [4]). We show that if G fulfills an obviously necessary condition, H is Zariski-dense in G (Thm 1), and is contained in only finitely many discrete subgroups of G_R (Cor. to Thm 4). Theorem 2 describes the group $C(H)$ of elements $g \in G_C$ such that $g \cdot H \cdot g^{-1}$ is commensurable with H. In particular, $C(H)$ is equal to G_Q if G has no normal Q-subgroup $N \neq \{e\}$ whose group N_R of real points is compact. Similar results hold for groups defined over a number field k of finite degree over Q (Thms 3, 5). In § 7 we consider the case where G "splits over k" and H is the group of units of certain lattices, and prove that H is then discrete maximal if either $k = Q$ or k has class number one and G has no center (Thm 7).

Theorem 1 is closely related to the density theorem of [1]; as a matter of fact, it can easily be derived from it and from the finiteness of the volume of G_R/H. Theorem 2 was stated in [2]. For both, the proofs given here are different from the original ones and were obtained jointly with J.-P. Serre. Theorem 4 generalizes a result of Ramanathan [11] and the proof is in part patterned after his. Theorem 7 applies in particular to $Sp(2\mathrm{n}, Z)$, $SL(n, Z)$ or the group of units of the standard quadratic form of maximal index in $2n$ variables; in those cases, it is proved in [11] (and goes back to Hecke for $SL(2, Z)$). Theorem 7 was suggested by these results.

For algebraic groups, we follow the notation and conventions of [5]. In particular: k-group or "algebraic group defined over a field k" are used synonymously; the identity component of a k-group G is denoted G^0; a connected semi-simple k-group is *almost simple over k* if it has no normal *connected* k-subgroup $N \neq \{e\}$. Often, ^gA stands for $g \cdot A \cdot g^{-1}$.

Throughout this paper, k is a number field of finite degree over Q, $\bar{k}$ an algebraic closure of k, $\mathfrak{o}$ the ring of integers of k, G a connected semi-simple k-group, and H an arithmetic subgroup of G.

§ 1. Density

Theorem 1. *Let $k = Q$. Assume that G contains no connected normal Q-subgroup $N \neq \{e\}$ such that N_R is compact. Then H is Zariski-dense in G.*

For any subgroup L of G_C, let $\mathscr{A}(L)$ be the smallest algebraic group containing L. As is obvious and well-known, $\mathscr{A}(L)$ is also the Zariski closure of L in G. If L' is a sub-

635

group of finite index of L, then $L \cdot \mathscr{A}(L')$ consists of finitely many cosets of $\mathscr{A}(L')$, hence is an algebraic set, and contains $\mathscr{A}(L)$. Therefore $\mathscr{A}(L)^0 = \mathscr{A}(L')^0$, and it follows that if L and M are commensurable subgroups of G_C, then $\mathscr{A}(L)^0 = \mathscr{A}(M)^0$.

Let $g \in G_Q$. Then ${}^gH$ is commensurable with H, hence

$$g \cdot \mathscr{A}(H)^0 \cdot g^{-1} = \mathscr{A}({}^gH)^0 = \mathscr{A}(H)^0,$$

which shows that G_Q normalizes $\mathscr{A}(H)^0$. But G_Q is Zariski-dense in G [12], hence $\mathscr{A}(H)^0$ is a normal subgroup of G; it is defined over Q since $H < G_Q$. Let C be a normal connected Q-subgroup of G which is almost simple over Q and different from $\{e\}$. Since G is semi-simple, there exists a normal Q-subgroup C' of G such that G is isogeneous to $C \times C'$. It follows then (from [4], 6. 11) that $H \cap C$ is an arithmetic subgroup of C. By assumption C_R is not compact, and by ([4], 7. 8), $C_R/(H \cap C)$ has finite invariant measure. Therefore $H \cap C$ is infinite, and $\mathscr{A}(H)^0 \cap C$ has strictly positive dimension, hence is equal to C. This being true for any such factor C, we have $\mathscr{A}(H)^0 = G$.

§ 2. The commensurability group

Definition. Let U be a group and V a subgroup of U. The *commensurability group* $C(V)$ of V (in U) is the set of elements $u \in U$ such that ${}^uV$ is commensurable with V.

Since "to be commensurable with each other" is an equivalence relation between subgroups of a group, $C(V)$ is indeed a group. Clearly, if $V' < U$ is commensurable with V, then $C(V) = C(V')$. If $f : U \to U'$ is a homomorphism, then $f(C(V)) < C(f(V))$.

In classical cases (in particular when U is the group of conformal transformations of the upper half plane and V is discrete), $C(V)$ is traditionally called the "group of transformations" of G/V.

Theorem 2. *Let $k = Q$. Let N be the greatest normal Q-subgroup of G such that N_R is compact, and $\pi : G \to G' = G/N$ the canonical projection. Then the commensurability group of H in G_C is equal to $\pi^{-1}(G'_Q)$.*

We first prove

$$(1) \qquad\qquad C(H) = \pi^{-1}(C(\pi(H))).$$

Let N' be a connected normal Q-subgroup of G such that $G = N \cdot N'$ and $N \cap N'$ is finite. The group H is commensurable with $(H \cap N) \cdot (H \cap N')$, by [4, 6. 4, 6. 11], hence with $H \cap N'$, since $H \cap N$ is finite.

We have already remarked that $\pi(C(H)) < C(\pi(H))$. Let now $g \in \pi^{-1}(C(\pi(H)))$. There exists then a subgroup of finite index L of $H \cap N'$ such that ${}^gH \cdot (N \cap N')_C$ contains L. Since $H \cap N'$ is commensurable with H and $(N \cap N')_C$ is finite, this proves the other inclusion.

There remains to prove that $C(\pi(H)) = G'_Q$. The group $\pi(H)$ being also arithmetic (see Theorem 6 below), this means that we are reduced to the case where $N = \{e\}$, and have then to prove that $C(H) = G_Q$.

For any subfield K of C, let $K[G]$ denote the K-algebra of K-morphic functions on G. We have $K[G] \cong K \otimes_Q Q[G]$. If we identify G, over Q, with a subgroup of SL_n, then $K[G]$ is generated, as a K-algebra, by the matrix coefficients g_{ij} of g. We let G_C

operate on $C[G]$ by $f \mapsto {}^g f$ where

$$^g f(x) = f(g^{-1} \cdot x \cdot g) \qquad (f \in C[G];\ g, x \in G_C).$$

We get in this way a rational representation σ, defined over Q, of G into the vector space V spanned in $C[G]$ by the g_{ij}'s. Since $C[G] = C \otimes Q[G]$, it is clear that V_Q is spanned over Q by the g_{ij}'s. But $N = \{e\}$, hence G has no center, σ is faithful and is a Q-isomorphism of G onto $\sigma(G)$. Consequently, $g \in G_C$ belongs to G_Q if and only if $^g f \in Q[G]$ whenever $f \in Q[G]$.

Let now $g \in C(H)$ and $f \in Q[G]$. We may write

$$^g f = f_0 + c_1 \cdot f_1 + \cdots + c_m \cdot f_m \quad (f_i \in Q[G],\ (0 \leq i \leq m);\ c_i \in C,\ (1 \leq i \leq m))$$

with $(1, c_1, \ldots, c_m)$ linearly independent over Q. Let $x \in H$ be such that $g^{-1} \cdot x \cdot g = y \in H$. Then

$$f_0(x) + c_1 \cdot f_1(x) + \cdots + c_m \cdot f_m(x) = f(y) \in Q,$$

whence

$$(2) \qquad\qquad f_i(x) = 0 \ (i \geq 1), \quad f_0(x) = {}^g f(x).$$

But $^g H$ is commensurable with H by assumption, hence $^g H \cap H$ is Zariski-dense in G by Theorem 1. Since (2) is true for every $x \in {}^g H \cap H$, we have then $f_i \equiv 0$ $(i \geq 1)$, and $^g f = f_0 \in Q[G]$.

§ 3. Groups over number fields

We recall that a subgroup L of G is *arithmetic* if it belongs to G_k and if for any faithful k-morphism ϱ of G into some linear group GL_n the group $\varrho(L)$ is commensurable with $\varrho(G)_0$. It suffices that this condition be fulfilled for one such morphism, as follows from [4, 6. 3] and the properties of the functor $R_{k/Q}$ recalled below.

As is well-known, this notion is not more general than the one considered so far in this paper. In fact, if $G' = R_{k/Q}G$, where $R_{k/Q}$ is the functor of restriction of the ground-field from k to Q ([13], Chap. I), then there is a canonical isomorphism $\varphi_{k/Q} : G_k \to G'_Q$ which induces a bijection between arithmetic subgroups.

More generally, if k' is a subfield of k, then $R_{k/k'}$ is a functor from k-groups and k-morphisms to k'-groups and k'-morphisms; if P is a k-group and $P' = R_{k/k'}P$, there is a canonical isomorphism $\varphi_{k/k'} : P_k \to P'_{k'}$ which induces a bijection between arithmetic subgroups. Moreover ([5], 6. 22 (ii)), P is semi-simple and almost simple over k if and only if P' so is over k'.

These remarks and Theorem 1 imply the first assertion of the following

Theorem 3. (a) *If G contains no connected normal k-subgroup $N \neq \{e\}$ such that $(R_{k/Q}N)_R$ is compact, then H is Zariski-dense in G.*

(b) *If G has no normal k-subgroup $N \neq \{e\}$ such that $(R_{k/Q}N)_R$ is compact, then the commensurability group $C(H)$ of H in $G_{\bar{k}}$ is equal to G_k.*

The assertion (b) is proved in exactly the same way as the equality $C(H) = G_Q$ in § 2. One has merely to replace Q and C by k and $\bar{k}$ in the above proof.

§ 4. Almost maximality

Theorem 4. *Let $k = \boldsymbol{Q}$. Assume that G has no connected normal $\boldsymbol{Q}$-subgroup $N \neq \{e\}$ such that N_R is compact. Then there are in G_C only finitely many subgroups which contain H as a subgroup of finite index.*

The quotient G_R/H has finite invariant measure ([4], 7. 8). As a consequence, if a discrete subgroup L of G_R contains H, then H must have finite index in L. The theorem implies therefore the following

Corollary. *Under the assumption of Theorem 4, there are only finitely many discrete subgroups of G_R which contain H.*

We now prove Theorem 4. Let Z be the center of G and $\pi: G \to G' = G/Z$ the canonical projection. Since Z is finite, it is of course sufficient to show that there are only finitely many subgroups of G'_C which contain $\pi(H)$ as a subgroup of finite index. But $\pi(H)$ is arithmetic ([4], 6. 11), and G' clearly verifies the condition imposed on G. It suffices therefore to prove the theorem when $Z = \{e\}$. In this case, G verifies the assumption of Theorem 2, hence $C(H) = G_Q$. Of course, any subgroup of G_C in which H has finite index belongs to $C(H)$. Our assertion is then a consequence of the following theorem.

Theorem 5. *Assume that G contains no normal k-subgroup $N \neq \{e\}$ such that $(R_{k/Q}N)_R$ is compact. Then there are only finitely many subgroups of G_k which contain H as a subgroup of finite index.*

We first reduce the proof to the case where G is absolutely simple. Assume to begin with that $G = G_1 \times G_2$ where G_1 and G_2 are normal k-subgroups. We may of course replace H by a subgroup of finite index, in particular by $(H \cap G_1) \times (H \cap G_2)$, which implies that we may assume G to be simple over k. If now G is not absolutely simple, there exists an extension K of k of finite degree, an absolutely simple K-group G' such that $G = R_{K/k}G'$ ([5], 6. 22 (ii)). In view of the remarks made in § 3, it is enough to prove the theorem for the subgroup $\varphi_{K/k}^{-1}(H)$ of G'_K, which is arithmetic, whence our reduction.

From now on, G is absolutely simple. We identify it with its image under a faithful absolutely irreducible representation defined over k, for instance the adjoint representation, whose degree is denoted by n. Since we may replace H by a subgroup of finite index, we may also assume that $H < G_0 < \boldsymbol{GL}(n, \mathfrak{o})$. From now on, the proof is essentially the same as Ramanathan's [11].

Let $\mathscr{H}$ be the enveloping algebra of H over $\mathfrak{o}$, in other words the $\mathfrak{o}$-subalgebra of $\boldsymbol{M}(n, \mathfrak{o})$ generated by the elements of H. Since it belongs to $\boldsymbol{M}(n, \mathfrak{o})$, and $\mathfrak{o}$ is noetherian, it is a finitely generated $\mathfrak{o}$-module. Let us prove that it contains a vector space basis of $\boldsymbol{M}(n, k)$ over k. If not, there would exist a proper subspace V of $\boldsymbol{M}(n, k)$ containing $\mathscr{H}$. It would be stable under H, acting on $\boldsymbol{M}(n, k)$ by left translations, say. But H is Zariski-dense in G (Thm 3), therefore, $V \otimes \bar{k}$ would be stable under $G_{\bar{k}}$, acting by left translations, hence also under the enveloping $\bar{k}$-algebra $\mathscr{G}$ of $G_{\bar{k}}$. But $G_{\bar{k}}$ is absolutely irreducible, as a linear group, hence $\mathscr{G} = \boldsymbol{M}(n, \bar{k})$ by the Burnside theorem. Since $1 \in V$, this is a contradiction.

It follows that $\mathscr{H}$ is a lattice in $\boldsymbol{M}(n, \mathfrak{o})$. There exists then an ideal $\mathfrak{a}$ of $\mathfrak{o}$ such that $\mathfrak{a} \cdot M(n, \mathfrak{o}) < \mathscr{H}$, and hence also a strictly positive rational integer d such that $d \cdot \boldsymbol{M}(n, \mathfrak{o}) < \mathscr{H}$.

Let now L be a subgroup of G_k containing H as a subgroup of finite index. We want to prove:

$$(3) \qquad\qquad d \cdot x \in M(n, \mathfrak{o}) \qquad\qquad (x \in L).$$

Given $x \in L$, there exists a strictly positive rational integer r such that $x^r \in H < \boldsymbol{GL}(n, \mathfrak{o})$. As a consequence, the eigenvalues of x, and its trace $tr\ x$, belong to the ring $\mathfrak{O}$ of algebraic integers of $\bar{k}$. We have therefore $tr(x \cdot y) \in \mathfrak{O}$ for every $y \in H$, hence also for every $y \in \mathcal{H}$. Since $d \cdot M(n, \mathfrak{o}) < \mathcal{H}$, we see that $d \cdot tr(x \cdot z) \in \mathfrak{O}$ for every $z \in M(n, \mathfrak{O})$. This implies immediately that $d \cdot x \in M(n, \mathfrak{O})$. But $d \cdot x \in M(n, k)$, whence (3).

The transforms of the lattice $\mathfrak{o}^n$ under L are finite in number (since H has finite index in L). By (3), they all belong to the lattice $d^{-1} \cdot \mathfrak{o}^n$. Their sum Γ is thus a lattice invariant under L verifying

$$(4) \qquad\qquad \mathfrak{o}^n < \Gamma < d^{-1} \cdot \mathfrak{o}^n.$$

But it is standard that there are only finitely many lattices Γ verifying (4). There exists consequently finitely many arithmetic subgroups of G, namely the stability groups of these lattices, such that any subgroup of G_k containing H as a subgroup of finite index belongs to one of them; this proves the theorem.

§ 5. A remark on surjective morphisms

We have used the fact that the image of an arithmetic group under an isogeny is an arithmetic group. We take this opportunity to point out an easy generalization of this fact.

Theorem 6. *Let L, L' be k-groups, $f: L \to L'$ a surjective k-morphism, and M an arithmetic subgroup of L. Then $f(M)$ is an arithmetic subgroup of L'.*

It is enough to prove this when L is connected. Moreover, in the notation of § 3, we have $R_{k/\boldsymbol{Q}}(f) \circ \varphi_{k/Q} = \varphi_{k/Q} \circ f : L_k \to L'_{\boldsymbol{Q}}$, which shows that it suffices to prove the theorem when $k = \boldsymbol{Q}$.

Let $N = \ker f$. The morphism f is the composition of the projection $L \to L/N^0$ and of the isogeny $L/N^0 \to L'$. Since the theorem is known for isogenies ([4], 6. 11), we may assume N to be connected. This group is the semi-direct product over $\boldsymbol{Q}$ of a reductive $\boldsymbol{Q}$-subgroup S and of its unipotent radical $U = R_u(N)$ [10]. We have the factorisation $L \to L/U \to L' \cong (L/U)/(N/U)$, which reduces us to the cases where N is unipotent or reductive.

Write $L = B \cdot R_u(L)$, with B reductive. If N is unipotent, it belongs to $R_u(L)$; if it is reductive, it belongs to B, because it is conjugate to a subgroup of B by an element of $R_u(L)_k$ [10], and is normal in L. On the other hand, the semi-direct product of an arithmetic subgroup of B by an arithmetic subgroup of $R_u(L)$ is an arithmetic subgroup of L ([4], 6. 4). As a consequence, it is enough to consider the cases where L is unipotent or reductive.

(i) *L unipotent.* By ([4], 6. 3), $f(M)$ is contained in an arithmetic subgroup M' of L', hence is discrete in $L'_{\boldsymbol{R}}$. On the other hand, it is well-known that $L_{\boldsymbol{R}}/M$ is compact, hence $L'_{\boldsymbol{R}}/f(M)$ is also compact, and $f(M)$ has finite index in M'.

(ii) *L reductive*. There exists then a normal connected Q-subgroup N' of L such that $L = N \cdot N'$ and $N \cap N'$ is finite. By ([4], 6. 4, 6. 11), the subgroup $(M \cap N) \cdot (M \cap N')$ has finite index in M, and $M \cap N'$ is arithmetic in N'. Therefore $f(M)$ is commensurable with $f(M \cap N')$; but the latter is arithmetic by ([4, 6. 11]) applied to the isogeny $f : N' \to L' \cong N'/(N' \cap N)$.

§ 6. Maximality

1. The group G is said to *split over k* if it contains a maximal torus which splits over k (i. e. whose image under any k-morphism into a linear group is diagonalisable over k, see [5, § 1]). Such groups are called "déployés" in [5], and are sometimes referred to as group of "Chevalley type" or "Tohoku type". It is known that if G' is a k-group which splits over k and is isomorphic to G over $\bar{k}$, then G' is already isomorphic to G over k (see [5], § 2 for instance).

Assume that G splits over k. Let T be a maximal torus which splits over k, $\Phi = \Phi(T, G)$ the system of roots of G with respect to T, and Δ (resp. Φ^+) the set of simple (resp. positive) roots with respect to some ordering. For each $b \in \Phi$, there is a unique faithful k-morphism Θ_b of the additive group G_a onto a subgroup U_b of G normalized by T such that $\Theta_b(b(t) \cdot x) = t \cdot \Theta_b(x) \cdot t^{-1}$, $(x \in G_a, t \in T)$. We let U be the subgroup generated by the U_b ($b \in \Phi^+$) and $B = T \cdot U$.

Let $\mathfrak{g}$, $\mathfrak{t}$ be the Lie algebras of G and T. We denote by (h_a, u_b) $(a \in \Delta, b \in \Phi)$ a *Chevalley basis* of $\mathfrak{g}$. We have therefore in particular

$$b(h_a) = 2 \cdot (b, a) \cdot (a, a)^{-1}, \quad [h, u_a] = a(h) \cdot u_a, \quad [u_a, u_{-a}] = h_a \qquad (a, b \in \Phi)$$

where $(,)$ denotes the scalar product on $\mathfrak{t}^*$ defined by the Killing form.

For $b \in \Phi$, let T_b be the identity component of ker b and Z_b the derived group of the centralizer of T_b. It is known that Z_b is almost simple, of dimension three, that $U \cap Z_b = U_b$ and that there is a unique k-morphism

$$\tau_b : SL_2 \to Z_b,$$

whose differential $d\tau_b$ maps

$$h = \begin{pmatrix} 1 & 0 \\ 0 & -1 \end{pmatrix}, \quad e = \begin{pmatrix} 0 & 1 \\ 0 & 0 \end{pmatrix}, \quad e_- = \begin{pmatrix} 0 & 0 \\ 1 & 0 \end{pmatrix}$$

onto h_b, u_b and u_{-b} respectively. We shall denote by E the subgroup of upper triangular unipotent matrices of SL_2. Thus, τ_b induces an isomorphism of E onto U_b.

Definition. We keep the previous notation. A k-morphism $\varrho : G \to GL_m$ is *special* if: (i) $\mathfrak{v}^m$ has a basis formed by eigenvectors of $\varrho(T)$; (ii) for every $b \in \Delta$, the coefficients of $\varrho(SL_2)$, viewed as morphic functions on SL_2, belong to the ring $Z[SL_2]$ generated by the coefficients in the identity representation of SL_2. The arithmetic subgroup H of G is *special* if there exists a faithful special k-morphism such that $\varrho(H) = \varrho(G)_0$.

Let V be a vector space defined over k and $\varrho : G \to GL(V)$ a k-morphism. A lattice Γ in V is *special* if it is free over $\mathfrak{v}$ and if ϱ becomes special once $GL(V)$ has been identified to GL_m, ($m = \dim V$) by means of basis of Γ.

Remark. Let $k = \boldsymbol{Q}$. A lattice $\Gamma \subset V$ is *admissible* in the sense of Chevalley [8] if it verifies (i) and

(iii) For every $b \in \Delta$ and every positive integer j, the linear transformation $(j!)^{-1} \cdot d\varrho(u_b)^j$ leaves Γ stable.

It is easily seen that (iii) $\Rightarrow$ (ii), and hence that an admissible lattice is always special. That admissible lattices always exist is stated in [8], and proved in [7] for the adjoint representation. Furthermore the uniqueness theorem of [8] of the "structure over $\boldsymbol{Z}$" defined by an admissible lattice implies that the groups of units of two admissible lattices (in faithful representation spaces) are identical. According to Cartier (unpublished) this uniqueness is also true for groups of units of special lattices. However, we shall not need these facts.

2. Let W be the Weyl group of G. By definition, W is the quotient of the normalizer $\mathcal{N}(T)$ by the centralizer $\mathcal{Z}(T)$ of T. If H is special, then it meets every connected component of $\mathcal{N}(T)$, i. e. every element of W is represented by some element of H. In fact, by condition (ii), H contains representatives of the fundamental reflections r_b ($b \in \Delta$), and these are well-known to generate W. This shows in particular that condition (ii) for special morphisms is then satisfied for every root $b \in \Phi$.

Lemma 1. *We keep the previous notation, assume G to split over k and to be identified to a subgroup of $\boldsymbol{GL}_n$ by means of a special faithful k-morphism. Let k_v be the completion of k with respect to a non-archimedean valuation v, and $\mathfrak{o}_v$ the ring of integers of k_v. Then*

(a) (Bruhat), $\qquad\qquad G_{k_v} = G_{\mathfrak{o}_v} \cdot B_{k_v}.$

(b) *If k has class number one, $G_k = G_{\mathfrak{o}} \cdot B_k$.*

The assertion (a) is proved in ([6], Prop. 15) for the adjoint group, and a lattice in $\mathfrak{g}$ spanned by a Chevalley basis, but the proof depends only on property (ii) above, and goes over without change to the case considered here.

Proof of (b). We use the notation of [3]. In particular, G_v stands for G_{k_v}, G_A is the adele group of G, and G_A^∞ the product of the groups $G_{\mathfrak{o}_w}$, where w runs through the normalized non-archimedean valuations of k, by the groups G_v, where v runs through the normalized archimedean valuations of k. It follows from (a) that

$$(1) \qquad\qquad G_A = G_A^\infty \cdot B_A.$$

We have moreover

$$T_A = T_A^\infty \cdot T_k; \quad U_A = U_A^\infty \cdot U_k,$$

since T splits over k and k has class number one ([3], 2.3) and U is unipotent ([3], 2.5), hence

$$(2) \qquad\qquad B_A = T_A \cdot U_A = B_A^\infty \cdot B_k.$$

The equalities (1) and (2) yield $G_A = G_A^\infty \cdot B_k$, hence $G_k = (G_k \cap G_A^\infty) \cdot B_k = G_{\mathfrak{o}} \cdot B_k$ ([3], 7.5).

In order to get a unified treatment for the two cases considered in Theorem 7 below, we cast the main part of its proof in the form of the following lemma:

Lemma 2. *Assume that k has class-number one and that G splits over k. Let H be a special arithmetic subgroup of G and L a subgroup of G_k containing H as a subgroup of finite index. Then $H = L$.*

We identify G to a subgroup of $\boldsymbol{GL}_n$ by a special faithful k-morphism, and therefore H to $G_\mathfrak{o}$.

Assume first that $G = \boldsymbol{SL}_2$, and $H = \boldsymbol{SL}_{2,\mathfrak{o}}$. There exists a lattice $\Gamma \supset \mathfrak{o}^2$ invariant under L, e. g. the sum of the transforms of $\mathfrak{o}^2$ under L. Since k has class number one, we may find $u, v \in \mathfrak{o}$ and $g \in \boldsymbol{SL}_{2,\mathfrak{o}}$ such that Γ is spanned by $u^{-1} \cdot g(e_1)$ and $v^{-1} \cdot g(e_2)$, where e_1, e_2 is the standard basis of k^2. But $\boldsymbol{SL}_{2,\mathfrak{o}}$ contains an element which sends e_1 and e_2 onto e_2 and $-e_1$ respectively, hence also an element which maps $g(e_1)$ and $g(e_2)$ onto $g(e_2)$ and $-g(e_1)$ respectively; this implies that $u \cdot v^{-1}$ is a unit of k; therefore L is homothetic to $\mathfrak{o}^2$ and has the same group of units as $\mathfrak{o}^2$, which proves our statement in this case.

In the general case, let $b_1, \ldots, b_m$ be the positive roots ranged in increasing order, and $U_i = Z_{b_i} \cap U$. The one-parameter groups U_j $(j \geq i)$ generate a normal k-subgroup V_i of B, and V_i is the semi-direct product of U_i and V_{i+1} $(i = 1, \ldots, m; V_{m+1} = \{e\})$. An element $u \in U$ can be written in one and only one way as a product $u_1 \circ \cdots \circ u_m$ $(u_i \in U_i)$ and, for every i, the map $u \mapsto u_i$ is a k-morphism of varieties (see e. g. [5], § 2).

By Lemma 1, $G_k = H \cdot B_k$, therefore

$$(3) \qquad\qquad L = H \cdot (L \cap B_k).$$

We prove first

$$(4) \qquad\qquad L = H \cdot (L \cap U_k).$$

Let $\pi_i \colon V_i \to V_i/V_{i+1} = U_i'$ be the canonical projection. Then $\pi_i \circ \tau_{b_i}$ induces an isomorphism of E onto U_i'. The image of $V_i \cap L$ in U_i' is an arithmetic subgroup, stable under $L \cap B$ acting by inner automorphisms. Let $x \in L \cap B$. We can write it uniquely as

$$x = t \cdot u \qquad\qquad (t \in T_k; \; u \in U_k);$$

it follows from the facts recalled at the beginning of this paragraph that if we identify U_i' to the additive group G_a, then x transforms $y \in U_i'$ onto $t^{b_i} \cdot y$. This transformation leaves the subgroup $L_i = \pi(L \cap V_i)$ of G_a invariant. The group L_i is arithmetic, hence is generated, as an additive group, by a vector space basis over $\boldsymbol{Q}$ of k. There exist then strictly positive rational integers a, b such that $a \cdot L_i \subset \mathfrak{o}$ and $b \in a \cdot L_i$. We have therefore $(t^{b_i}) \cdot b \in \mathfrak{o}$. But, for every $m \in \boldsymbol{Z}$, the T-component of x^m is t^m, therefore we also get

$$(t^{b_i})^m \cdot L_i = L_i, \; (t^{b_i})^m \cdot b \in \mathfrak{o}, \qquad\qquad (m \in \boldsymbol{Z})$$

which implies that t^{b_i} is a unit of k. The diagonal coefficients of t are of the form t^c, where c runs through the weights of the identity representation of G, hence through a generating set of $X^*(T)$. The roots generate a subgroup of finite index of $X^*(T)$. We have therefore shown that each diagonal coefficient t_{ii} has a power lying in $\mathfrak{o}$. Since $t_{ii} \in k$ by assumption, t_{ii} is itself in $\mathfrak{o}$, and $t \in H$, whence (4). In order to finish the proof of the lemma, it is then enough to show:

(*) Let i be an integer between 1 and m, and $x \in V_i \cap L$. Write $x = x_i \cdot y$ $(x_i \in U_i, v \in V_{i+1})$. Then $x_i \in H \cap U_i$.

We proceed by induction on i, and distinguish two cases, the first of which contains the starting point $i = 1$ of the induction.

(a) b_i *is simple.* Let V' be the subgroup of U generated by the U_b's with $b \neq b_i$. The group U can also be written as a semi-direct product $U = U_i \cdot V'$ with V' normal, containing V_{i+1}. The corresponding unique decomposition of x is still $x = x_i \cdot y$ $(x_i \in U_i,\ y \in V')$.

As is well-known and simple to check, Z_{b_i} normalizes V', and the group $Q = Z_{b_i} \cdot V'$ generated by Z_{b_i} and V' is isomorphic to the semi-direct product of Z_{b_i} and V'. Let σ be the natural projection of Q onto $Q' = Q/V'$. The map $\sigma \circ \tau_{b_i}$ is then a k-isogeny of SL_2 onto Q'. Let $L' = \sigma(L \cap Q)$ and $L'' = \sigma \circ \tau_{b_i}(SL_{2,0})$. Both L' and L'' are arithmetic (Theorem 6), and L' contains L'' (by property (ii)), necessarily as a subgroup of finite index. We want to prove

$$(5) \qquad\qquad \sigma(L \cap U) < L''.$$

Assume this is false. Then $L'' \cap \sigma(U_i)$ is a proper subgroup of $\sigma(L \cap U)$. Since $\sigma \circ \tau_{b_i}$ is a k-isomorphism of E onto $\sigma(U_i)$, it follows that

$$(\sigma \circ \tau_{b_i})^{-1}(L') \cap SL_{2,k}$$

is a subgroup of $SL_{2,k}$ containing $SL_{2,0}$ properly as a subgroup of finite index, in contradiction with the first part of the proof of this lemma.

There exists therefore $h \in SL_{2,0}$ such that $\sigma(x) = \sigma \circ \tau_{b_i}(h)$, whence $x_i = \tau_{b_i}(h) \cdot v'$ with $v' \in V'_k$. We have then $\tau_{b_i}(h) \in U \cap \tau_{b_i}(SL_2) = U_{b_i}$, and therefore

$$x_i = \tau_{b_i}(h), \quad v' = y,$$

which proves (*) in the case under consideration.

(b) *The root b_i is not simple.* We first recall a "commutation rule" in U (see e. g. [5], 2. 5):

Let $b,\ c \in \Phi^+$ with $b > c$ and $u \in U_b$, $v \in U_c$. Then the commutator $(u, v) = u \cdot v \cdot u^{-1} \cdot v^{-1}$ is contained in the subgroup generated by the U_a's, where a runs through the roots of the form $a = m \cdot b + n \cdot c\ (m, n \in \mathbf{Z},\ m, n \geq 1)$.

This means that the component $(u \cdot v)_j$ of $u \cdot v$ in U_j is equal to v if $b_j = c$, to u if $b_j = b$, and is otherwise zero unless $b_j = m \cdot b + n \cdot c\ (m, n \in \mathbf{Z},\ m, n \geq 1)$. Repeated use of this rule allows one to prove in particular the following assertion:

(**) Let $(i_1, \ldots, i_q)$ be distinct integers between 1 and m. Let $u(i_j)$ be an element of U_{i_j} and $u = u(i_1) \cdot u(i_2) \cdots u(i_q)$. Write $u = u_1 \cdots u_m\ (u_i \in U_i)$, and let i_s be the smallest of the i_j's. Then $u_i = e$ for $i < i_s$, i. e. $u \in V_{i_s}$, and moreover $u_{i_s} = u(i_s)$.

We now go over to the proof of (*) in the case (b). Since $(b_i, b_i) > 0$, there exists a simple root a such that $(b_i, a) > 0$. We let

$$b_{i'} = r_a(b_i) = b_i - 2 \cdot (a, b_i) \cdot (a, a)^{-1} \cdot a.$$

From $b_i \neq a$, and $(b_i, a) > 0$, we get

$$0 < b_{i'} < b_i, \quad a < b_i.$$

Thus for any $j \geq i$, $b_j \neq a$, hence $r_a(b_j) = b_{j'} > 0$, and r_a maps the set $(b_i, \ldots, b_m)$ into Φ^+.

For any element $y \in V_i$, let $\Theta(y)$ be the set of b_j's such that $y_j \neq e$, and let $n_a(y)$ be the index of the smallest element in $r_a(\Theta(y))$. If $y \notin V_{i+1}$, then, clearly, $n_a(y) \leq i' < i$.

Let $n \in \tau_a(SL_{2,v})$ be an element which induces the Weyl reflection r_a. We have then

$$^n U_j = U_{j'}, \qquad\qquad (j \geq i),$$

$$^n z = {}^n z_i \cdot {}^n z_{i+1} \cdots {}^n z_m, \qquad\qquad (z \in V_i,\ z_j \in U_j;\ j \geq i)$$

with

$$^n z_j \in U_{j'}. \qquad\qquad (j \geq i).$$

The proof of (*) will be divided into two steps. We assume $x \notin V_{i+1}$, since otherwise there is nothing to prove.

(i) *Assume that* $n_a(x) = i'$. We have then $^n x_j \in V_{i'+1}$ for all $j > i$, hence

$$^n x = {}^n x_i \cdot {}^n y \qquad\qquad (^u y \in V_{i'+1}).$$

Since $n \in H$, the element $^n x$ belongs to $L \cap V_{i'}$. By our induction assumption on i, we have then $^n x_i \in H$, hence also $x_i \in H$.

(ii) In the general case we prove (*) by descending induction on $n_a(x)$. If $n_a(x) = i-1$, we have necessarily $i' = i - 1$, and we are back to case (i). Thus we may assume (*) to be proven for all $x' \in L \cap V_i$ such that $n_a(x') > n_a(x)$. Let s be the index such that $s' = n_a(x)$. In view of (i), there remains the case where $s \neq i$. By (**) we have

$$^n x \in V_{n_a(x)},\ {}^n x = {}^n x_s \cdot y' \qquad\qquad (y' \in V_{n_a(x)+1}).$$

By the induction assumption on i, this implies $^n x_s \in H$, hence $x_s \in H$.

Let

$$z = x_s^{-1} \cdot x = \prod_{i \leq j < s} (x_s^{-1} \cdot x_j \cdot x_s) \cdot \prod_{j > s} x_j.$$

Since $x_s \in V_{i+1} \cap H$, we have of course

$$z \in V_i \cap L,\ z \notin V_{i+1}.$$

Furthermore the commutation rule in U recalled above implies

$$(1) \qquad\qquad z_j = x_j, \qquad\qquad (i \leq j < s)$$

$$(2) \qquad\qquad z_s = e,$$

and shows that if $b \in \Theta(z)$ does not belong to $\Theta(x)$, then b is of the form

$$b = r \cdot b_s + r_1 \cdot b_{i_1} + \cdots + r_q \cdot b_{i_q},$$

where $r, r_1, \ldots, r_q$ are strictly positive integers, $q \geq 1$, and $i_j \geq i$ for all j's. Thus, if $b \in \Theta(z)$, $b \notin \Theta(x)$, then

$$r_a(b) > r_a(b_s).$$

On the other hand, if $b \in \Theta(z) \cap \Theta(x)$, then $b \neq b_s$ by (2), hence $r_a(b) > r_a(b_s)$ too. Therefore,

$$n_a(z) > n_a(x).$$

By the induction hypothesis on $n_a(x)$, we have then $z_i \in H$. Since $x_i = z_i$ by (1), this concludes the proof of (*), and of the lemma.

Theorem 7. *Let G be split over k and H a special arithmetic subgroup of G. Then*

(a) *if $k = Q$, the group H is discrete maximal in G_R,*

(b) *if the class number of k is one and the center of G is reduced to $\{e\}$, the group $R_{k/Q} H$ is discrete maximal in $(R_{k/Q}G)_R$.*

In both cases, let L be a discrete subgroup of $(R_{k/Q}G)_R$ containing H. We have to prove that $L = H$.

Proof of (a). Let Z be the center of G and $\pi : G \to G/Z = G'$ the natural projection. The group G' is the direct product of normal Q-subgroups which are simple over Q and split over Q. The groups of real points of these factors are therefore not compact, and we may apply Theorem 2, which shows that

$$(1) \qquad\qquad \pi(L) < G'_Q.$$

Let $x \in L$. Then $\pi({}^xB)$ is defined over Q. Since π is Q-isogeny, ${}^xB$ is also defined over Q; there exists therefore $y \in G_Q$ such that ${}^xB = {}^yB$ ([5], 4. 13); since B is equal to its normalizer, we have then $x \in y \cdot B$. But $y \in H \cdot B_Q$ by Lemma 1, whence $x \in H \cdot B$, and

$$(2) \qquad\qquad L = H \cdot (B \cap L).$$

Let $x \in B \cap L$ and write $x = t \cdot u$ ($t \in T, u \in U$). Let σ be the composition of the restriction of π to B and of the canonical projection $\pi(B) \to \pi(B)/\pi(U) = T'$. Since $\sigma(L \cap B)$ belongs to T'_Q by (1) and contains $\sigma(H \cap B)$, it is an arithmetic subgroup of T'; but $\sigma : T \to T'$ is a Q-isogeny, hence $\sigma(T)$ is also a split torus ([5], 1. 3). Therefore $\sigma(L \cap B)$ is a finite group, and so is $T \cap \sigma^{-1}(\sigma(L \cap B))$. Thus t is a real diagonal matrix of finite order. Its diagonal coefficients are then equal to ± 1, whence $t \in H$, and

$$(3) \qquad\qquad L = H \cdot (U \cap L).$$

The kernel of π belongs to T, and π induces a Q-isomorphism of U onto $\pi(U)$. In view of (1), we have then $L \cap U < U_Q$, and, by (3),

$$L = H \cdot (L \cap U_Q) < G_Q.$$

The equality $L = H$ follows now from Lemma 2.

Proof of (b). A normal k-subgroup $N \neq \{e\}$ of G which is simple over k is also split over k, therefore $R_{k/Q}N$ contains a non-trivial torus which splits over k ([5]), 6. 22, and $(R_{k/Q}N)_R$ is not compact. By Theorem 3, we have $L < G_k$, and we are reduced to Lemma 2.

Remark. Let $G = SL_2$ and $k \neq Q$. Then $SL_{2,0}$ does not contain the center of $(R_{k/Q}G)_R$, which shows that the assumption on the center in the second case of the theorem is necessary in general. We do not know to what extent the hypothesis on the class-number of k is needed. In this connection we note that there is some overlap, but apparently no inclusion relation, between the results of [9] and Theorem 7. More precisely, let G be the quotient of SL_2 by its center and H the group of units of the lattice spanned in the Lie algebra $\mathfrak{g}$ of G by the elements h, e, e_- introduced earlier in this paragraph. The Lie group $M = (R_{k/Q}G)_R$ has 2^r connected components, where r is the number of real places of k. Then main result of [9] implies that $H \cap M^0$ is maximal in the identity component M^0 of M, if k is totally real. If k has class number one, this follows also from Theorem 7.

Bibliography

[1] *A. Borel*, Density properties for certain subgroups of semi-simple groups without compact components, Annals of Math. (2) **72** (1960), 179—188.

[2] *A. Borel*, Ensembles fondamentaux pour les groupes algébriques, Colloque sur la théorie des groupes algébriques, Bruxelles 1962, 23—40.

[3] *A. Borel*, Some finiteness properties of adele groups over number fields, Journal Inst. Hautes Etudes Sci. **16** (1963), 101—126.

[4] *A. Borel* and *Harish-Chandra*, Arithmetic subgroups of algebraic groups, Annals of Math. (2) **75** (1962), 485—535.

[5] *A. Borel* and *J. Tits*, Groupes réductifs, Journal Inst. Hautes Etudes Sci. **27** (1965).

[6] *F. Bruhat*, Distributions sur un groupe localement compact et applications à l'étude des représentations des groupes p-adiques, Bull. Soc. Math. France **89** (1961), 43—75.

[7] *C. Chevalley*, Sur certains groupes simples, Tohoku Math. J. (2) **7** (1955), 14—66.

[8] *C. Chevalley*, Certains schémas de groupes semi-simples, Sém. Bourbaki **219**, Paris 1961.

[9] *H. Maass*, Über die Erweiterungsfähigkeit der Hilbertschen Modulgruppe, Math. Annalen **51** (1948), 255—261.

[10] *G. D. Mostow*, Fully reducible subgroups of algebraic groups, Amer. J. Math. **78** (1956), 200—221.

[11] *K. G. Ramanathan*, Discontinuous groups II, Nachr. Akad. Wiss. Göttingen Math.-Phys. Klasse **22** (1964), 145—164.

[12] *M. Rosenlicht*, Some rationality questions on algebraic groups, Annali di Mat. pura e appl. IV, **63** (1957), 25—50.

[13] *A. Weil*, Adeles and algebraic groups (Notes by M. Demazure and T. Ono), Inst. for Advanced Study, Princeton, N. J., 1961.

Eingegangen 22. Mai 1965

71.

Opérateurs de Hecke et fonctions zêta

Séminaire Bourbaki, Exp. 307 (1965/66)

Soit Γ un groupe d'automorphismes proprement discontinu du demi-plan de Poincaré H. Les travaux dont il est question dans cet exposé ont pour but d'exprimer, pour certains groupes arithmétiques Γ, la fonction zêta globale $Z(s; C/k)$ d'un modèle convenable C sur un corps de nombres k du quotient H/Γ (ou de certaines variétés fibrées sur H/Γ) à l'aide de séries de Dirichlet attachées à des opérateurs de Hecke agissant sur des espaces de formes automorphes. De ce rapprochement on déduit notamment des renseignements, d'une part sur la nature de $Z(s; C/k)$, et d'autre part sur les valeurs propres d'opérateurs de Hecke. En particulier, cela démontre (ou ramène aux conjectures de Weil) des analogues de la conjecture de Ramanujan-Petersson.

Ces résultats ont été tout d'abord obtenus par Eichler [1] pour certains sous-groupes de congruence de $\mathbf{SL}(2, \mathbf{Z})$ puis, dans des cas de généralité croissante, par Shimura [7, 11] et Kuga-Shimura [6]. Cependant, la conjecture de Ramanujan proprement dite, qui est en somme à l'origine de ces recherches n'a pu jusqu'à présent être insérée dans ce cadre, (sinon heuristiquement).

Cet exposé est consacré principalement au cas le plus simple considéré dans [11]. Le dernier paragraphe donne quelques indications sur les cas plus généraux de [11, 6].

§ 1. Corps de quaternions. Anneau de Hecke

1.1. L désignera toujours un corps de quaternions sur $\mathbf{Q}$, indéfini sur $\mathbf{R}$. Il existe donc deux entiers $q, d > 0$ tels que L puisse se représenter comme l'ensemble des matrices

$$a = \begin{pmatrix} x & y \\ -q \cdot \bar{y} & \bar{x} \end{pmatrix}, \quad (x, y \in K = \mathbf{Q}(\sqrt{d})),$$

où $x \to \bar{x}$ est l'automorphisme $u + v \cdot \sqrt{d} \to u - v \cdot \sqrt{d}$, $(u, v \in \mathbf{Q})$, de K. L'involution fondamentale $a \to a'$ de L est induite par l'involution

$$\begin{pmatrix} \alpha & \beta \\ \gamma & \delta \end{pmatrix} \mapsto \begin{pmatrix} \delta & -\beta \\ -\gamma & \alpha \end{pmatrix}, \quad (\alpha, \beta, \gamma, \delta \in \mathbf{R}),$$

647

de $\mathbf{M}(2, \mathbf{R})$. La trace $\mathrm{tr}\,(x)$ et la norme $n(x)$ de $x \in L$ sont données par

$$\mathrm{tr}\,(x) = x + x', \quad n(x) = x \cdot x' = \det x \,,$$

et $n(x) \neq 0$ si $x \neq 0$.

Un *ordre* de L est un sous-anneau qui est un $\mathbf{Z}$-module de type fini contenant une $\mathbf{Q}$-base de L (et 1). Les éléments x d'un ordre sont *entiers* (i.e. $n(x)$, $\mathrm{tr}\,(x) \in \mathbf{Z}$). On fixe une fois pour toutes un ordre maximal, qui sera noté o. Par idéal, on entendra ici, sauf mention expresse du contraire, «o-idéal entier à gauche» (sous-$\mathbf{Z}$-module de type fini de o, stable par multiplication à gauche par o). Ces idéaux sont toujours principaux (Eichler). La norme d'un idéal $\mathrm{o} \cdot a$ est égale à $|n(a)|$.

Il existe un nombre fini ≥ 1 de nombres premiers p, dont le produit sera noté $d(L)$, tels que $L_p = L \otimes \mathbf{Q}_p$ est une algèbre à division si et seulement si $p \,|\, d(L)$. Si $p \nmid d(L)$, il existe un isomorphisme de L_p sur $M(2, \mathbf{Q}_p)$ qui applique o sur $\mathbf{M}\,(2, \mathbf{Z}_p)$ et induit un isomorphisme de $\mathrm{o}/p \cdot \mathrm{o}$ sur $\mathbf{M}(2, \mathbf{Z}/p \cdot \mathbf{Z})$.

1.2. Soient

$$\Gamma = \{a \in \mathrm{o}, \det a = 1\}, \quad \Delta = \{a \in \mathrm{o}, \det a > 0\} \,.$$

Pour tout $a \in \Delta$, les groupes $a \cdot \Gamma \cdot a^{-1}$ et Γ sont commensurables (i.e. leur intersection est d'indice fini dans chacun d'eux). On notera $R(\Gamma, \Delta)$ ou R l'*anneau de Hecke* associé à Γ et Δ. C'est le groupe abélien libre sur les doubles classes $\Gamma \cdot a \cdot \Gamma \, (a \in \Delta)$ muni d'un produit défini par la règle suivante: soient $u = \Gamma \cdot a \cdot \Gamma$, $v = \Gamma \cdot b \cdot \Gamma$ deux doubles classes, et $u = \bigcup \Gamma \cdot a_i$, $v = \bigcup \Gamma \cdot v_j$ des décompositions de u et v en classes à droite disjointes. Pour $c \in \Delta$, soit $d(u, v; c)$ le nombre de paires (i, j) telles que $a_i \cdot b_j \subset \Gamma \cdot c$. Alors

$$(\Gamma \cdot a \cdot \Gamma) \cdot (\Gamma \cdot b \cdot \Gamma) = \sum d(u, v; c)\, \Gamma \cdot c \cdot \Gamma \,,$$

la somme étant étendue aux doubles classes $\Gamma \cdot c \cdot \Gamma \subset \Gamma \cdot a \cdot \Gamma \cdot b \cdot \Gamma$. On montre que R est un anneau associatif, et commutatif, car $\Gamma \cdot a \cdot \Gamma = \Gamma \cdot a' \cdot \Gamma$ pour tout $a \in \Delta$ ([11], p. 281).

[Soit M le $\mathbf{Z}$-module libre engendré par les éléments de $\Gamma \backslash \Delta$. Associons à u un endomorphisme s_u de M défini par $s_u(\Gamma \cdot c) = \sum \Gamma \, a_i \cdot c$. Alors le produit précédent est défini de manière à ce que l'on ait $s_u \cdot s_v = s_{u \cdot v}$.]

1.3. Pour tout entier $n \geq 1$ on note $T(n)$ la somme des doubles classes $\Gamma \cdot a \cdot \Gamma$ $(a \in \Delta, \det a = n)$, (somme qui est finie, et se réduit à un terme si n est premier). Les opérateurs $T(n)$ ont des propriétés formelles analogues à celles des opérateurs de Hecke dans le cas classique $(\Gamma = \mathbf{SL}(2, \mathbf{Z}))$ dont on déduit que la série de Dirichlet formelle à coefficients dans R,

$$D(s) = \sum (\Gamma \cdot a \cdot \Gamma) \cdot (\det a)^{-s} = \sum_{n \geq 1} T(n) \cdot n^{-s},$$

admet une décomposition en produit eulérien

$$D(s) = \prod_{p\ \text{premier}} H(p^{-s}; p)^{-1}$$

avec

$$H(u; p) = 1 - T(p) \cdot u \qquad\qquad (p\,v \mid d(L)),$$
$$H(u; p) = 1 - T(p) \cdot u + p \cdot T(p, p) \cdot u^2 \quad (p \nmid d(L); T(p, p) = \Gamma \cdot p \cdot \Gamma)\,.$$

1.4. On note $d(\Gamma \cdot a \cdot \Gamma)$ le nombre de classes à droite $\Gamma \cdot b$ contenues dans $\Gamma \cdot a \cdot \Gamma$, (qui est ici égal au nombre de classes à gauche dans $\Gamma \cdot a \cdot \Gamma$, vu $\Gamma \cdot a' \cdot \Gamma = \Gamma \cdot a \cdot \Gamma$). Soit $\Gamma(a) = a^{-1} \cdot \Gamma \cdot a \bigcap \Gamma$. On voit immédiatement que

$$(1) \qquad\qquad \Gamma = \cup\, \Gamma(a) \cdot c_i \Rightarrow \Gamma \cdot a \cdot \Gamma = \cup\, \Gamma \cdot a \cdot c_i,$$

les deux unions étant simultanément disjointes ou non. En particulier $d(\Gamma \cdot a \cdot \Gamma) = [\Gamma : \Gamma(a)]$. On a $d(T(p, p)) = 1$ et, si $p \nmid d(L)$, $d(T(p)) = p + 1$. Notons encore que les classes à droite contenues dans $T(p)$ sont les intersections de $T(p)$ avec les idéaux de norme p.

§ 2. Formes automorphes. Opérateurs de Hecke

Le groupe $G = \{g \in \mathbf{GL}(2, \mathbf{R}), \det g > 0\}$ opèra à la manière usuelle sur H, l'automorphisme associé à $g = \begin{pmatrix} a & b \\ c & d \end{pmatrix}$ étant $z \mapsto (az + b) \cdot (cz + d)^{-1}$. On pose $j(g, z) = cz + d$. Le groupe discret Γ opère proprement, et H/Γ est compact. Soit $S_k(\Gamma)$ ou S_k l'espace des formes automorphes pour Γ de poids k. Ses éléments sont les fonctions holomorphes sur H vérifiant

$$f(\gamma \cdot z) = j(\gamma, z)^k \cdot f(z) \quad (\gamma \in \Gamma; z \in H)\,,$$

A une double classe $\Gamma \cdot a \cdot \Gamma = \cup\, \Gamma \cdot a_i$ ($a \in \Delta$) on associe un endomorphisme $(\Gamma \cdot a \cdot \Gamma)_k$ de S_k défini par

$$(f \mid (\Gamma \cdot a \cdot \Gamma)_k(z) = (\det a)^{k-1} \sum_i f(a_i \cdot z) \cdot j(a_i, z)^{-k}.$$

On voit facilement que le membre de droite ne dépend que de f et de $\Gamma \cdot a \cdot \Gamma$, et que l'on obtient ainsi une représentation de $R(\Gamma, \Delta)$ dans $S_k(\Gamma)$. Les opérateurs $(\Gamma \cdot a \cdot \Gamma)_k$ sont des *opérateurs de Hecke*. Ils sont self-adjoints par rapport à la métrique de Petersson, donc simultanément diagonalisables, à valeurs propres réelles. Une majoration aisée [6, p. 491] fait voir que la série de Dirichlet, à valeurs dans $\mathrm{End}\,(S_k(\Gamma))$:

$$(1) \qquad D(s, k) = \sum (\Gamma \cdot a \cdot \Gamma)_k(\det a)^{-s} = \sum T(n)_k\, n^{-s} = \prod_p H(p^{-s}; p, k)^{-1},$$

où

$$H(u; p, k) = 1 - T(p)_k \qquad\qquad\qquad\quad \text{si } p \mid d(L)$$
$$H(u; p, k) = 1 - T(p)_k \cdot u + p \cdot T(p, p)_k \cdot u^2 \quad \text{si } p \nmid d(L)$$

converge pour $Rs > (k/2) + 1$. Shimura [11, 1.6] a montré que $D(s, k)$ se prolonge analytiquement en une fonction entière, avec équation fonctionnelle reliant $D(s, k)$ et $D(k - s, k)$. Comme $-1 \in \Gamma$, on a $S_k(\Gamma) = 0$ si k est impair. Si k est pair, les opérateurs de Hecke laissent invariant un réseau de S_k, donc leurs valeurs propres sont des entiers algébriques [12, Prop. 9.1; 17, 5.2.5].

§ 3. Résultats

3.1. Soient X une variété projective irréductible sur le corps $\mathbf{F}_p$ à p éléments (p premier) et $Z = Z(u; X)$ la fonction zêta de X sur $\mathbf{F}_p$. On a donc

$$(1) \qquad \frac{d}{du}(\log Z) = \sum_{m \geq 1} N_m \cdot u^{m-1}, \quad Z(0) = 1 ,$$

où N_m désigne le nombre de points de X rationnels sur l'extension de degré m de $\mathbf{F}_p$. On sait (Dwork) que Z est une fonction rationnelle de u. Si X est une courbe lisse, alors

$$(2) \qquad Z(u; X) = (1 - u)^{-1} \cdot (1 - p \cdot u)^{-1} \cdot \det(1 - M_l(\pi) \cdot u) ,$$

où $M_l(\pi)$ est une représentation l-adique (l premier, $l \neq p$) de l'endomorphisme de Frobenius de la jacobienne $J(X)$ de X, et les valeurs propres de $M_l(\pi)$ sont des entiers algébriques de valeur absolue $p^{1/2}$ [18]. Si X est lisse de dimension n, alors [2]:

$$(3) \qquad Z(u; X) = \prod_{0 \leq i \leq 2n} F_i(u)^{(-1)^{i+1}},$$

où F_i est un polynôme dont on espère (conjectures de Weil) que les racines inverses sont des entiers algébriques de valeur absolue $p^{i/2}$. On a en particulier

$$F_0 = 1 - u, \quad F_{2n} = 1 - p^n \cdot u .$$

3.2. Soit maintenant $X \subset \mathbf{P}(n, \mathbf{C})$ une variété projective irréductible lisse sur $\mathbf{Q}$. Le plongement projectif définit une structure sur $\mathbf{Z}$, donc, pour tout nombre premier, un cycle $p(X)$ sur $\mathbf{F}_p$, la réduction mod p de X. Pour presque tout p, (les «bons» p), $p(X)$ est une variété irréductible lisse, avec multiplicité un. La fonction zêta globale de X sur $\mathbf{Q}$ est, par définition pour certains, au produit par une fonction élémentaire de s près pour d'autres, le produit

$$(1) \qquad Z(s; X/\mathbf{Q}) = \prod_p Z(p^{-s}; p(X)) ,$$

étendu aux bons p. Il converge pour $R s$ assez grand et, suivant Hasse-Weil, on conjecture qu'il se prolonge en une fonction méromorphe sur le plan complexe.

3.3. Théorème. *On reprend les notations des §§ 1, 2. Il existe un modèle C projectif lisse sur $\mathbf{Q}$ de H/Γ tel que l'on ait*

$$(i) \qquad Z(u; p(C)) = (1 - u)^{-1} \cdot (1 - p \cdot u)^{-1} \cdot \det H(u, p, 2)$$

pour presque tout nombre premier p ne divisant pas $d(L)$;

(ii) $$Z(s, C/\mathbf{Q}) = f(s) \cdot \zeta(s) \cdot \zeta(s-1) \cdot (\det D(s, 2))^{-1},$$

où f est une fonction élémentaire et ζ la fonction zêta de Riemann.

Il est clair que (ii) résulte de (i) et des définitions. La démonstration de (i) est l'objet des §§ 4, 5. Le point central en est la formule de congruence (4.4), qui relie la caractéristique zéro et la caractéristique p. Elle permet de transporter en caractéristique zéro des calculs relatifs à l'application de Frobenius, et d'utiliser l'isomorphisme de $S_2(\Gamma)$ avec l'espace des différentielles holomorphes de degré un sur C.

3.4. *Conséquences.* (1) Joint au résultat mentionné à la fin du § 2, le théorème montre que $Z(s; C/\mathbf{Q})$ est une fonction méromorphe sur le plan complexe, conformément à la conjecture de Hasse-Weil. Jusqu'à présent, cette dernière a été vérifiée principalement dans les cas mentionnés dans cet exposé, dont 3.3 est un exemple typique [1, 6, 7, 11], et pour des variétés abéliennes à multiplication complexe [16, Chap. IV].

(2) Les valeurs propres de $M_l(\pi)$, (cf. 3.1), étant égales à $p^{1/2}$ en valeur absolue, 3.3 entraîne que les valeurs propres de $T(p)_2$ sont $\leq 2p^{1/2}$ en valeur absolue, ce qui établit l'analogue de la conjecture de Ramanujan-Petersson pour le groupe Γ considéré ici, $k = 2$, et presque tout p.

§ 4. Correspondances modulaires. Démonstration du théorème à partir de la formule de congruence

Dans tout ce paragraphe, H/Γ est identifiée au modèle C sur lequel porte 3.3.

4.0. Soient X, Y, u des indéterminées, et $H(u; X, Y) = 1 - X \cdot u + Y \cdot u^2 \in \mathbf{Z}[X, Y][u]$. Il existe évidemment des polynômes $f_m(X, Y) \in \mathbf{Z}[X, Y]$ tels que

$$\frac{d}{du}(\log H(u; X, Y)^{\pm 1}) = \pm (1 - X + 2Yu) \cdot H(u; X, Y)^{-1} = \mp \sum_{m \geq 1} f_m(X, Y) \cdot u^{m-1}.$$

(1)

Soient V_i des espaces vectoriels de dimension finie sur $\mathbf{C}$ et a_i des entiers $(1 \leq i \leq q)$. Si X_i, Y_i sont des endomorphismes de V_i commutant entre eux, on déduit immédiatement de (1) que l'on a, dans $\mathbf{C}[[u]]$:

$$(2) \qquad \frac{d}{du}\left(\log \prod_i (\det H(u; X_i, Y_i))^{-a_i}\right) = \sum_{i, m} a_i \cdot \operatorname{tr}(f_m(X_i, Y_i)) \cdot u^{m-1}.$$

4.1. Une *correspondance* (propre) d'une courbe algébrique lisse V est un diviseur de $V \times V$ (sans composante de la forme $v \times V$ ou $V \times v$ ($v \in V$)). Les correspondances propres sur V forment un anneau: l'addition est définie par celle des diviseurs, et le produit $X \circ Y$ de X, Y est la projection pr_{13}, sur le produit du premier et du troisième facteur de $V \times V \times V$, du cycle $(V \times X) \cdot (Y \times V)$. La permutation des deux facteurs de $V \times V$ induit un antiautomorphisme (de Rosati) $X \mapsto {}^tX$ de l'anneau des correspondances propres.

On note $d(X)$, $d'(X)$ les entiers tels que $pr_1(X) = d(X) \cdot V$, $pr_2(X) = D'(X) \cdot V$. Evidemment $d'(X) = d({}^tX)$. L'entier $d(X)$ est aussi le degré du cycle

$$X(u) = X \cdot (u \times V) \qquad (u \in V).$$

4.2. Soit $a \in \Delta$. Posons $\Gamma(a) = \Gamma \cap a^{-1} \cdot \Gamma \cdot a$, et soit f_1 la projection canonique de $H/\Gamma(a)$ sur C. On a visiblement $a \cdot \Gamma(a) \cdot z \subset \Gamma \cdot a \cdot z$ $(z \in H)$, d'où une application $f_2 : H/\Gamma(a) \to C$. Alors l'image de $H/\Gamma(a)$ dans $C \times C$ par (f_1, f_2) est une correspondance propre, ne dépendant que de $\Gamma \cdot a \cdot \Gamma$, appelée *correspondance modulaire*, qui sera notée $X(\Gamma \cdot a \cdot \Gamma)$. En fait, on ne s'intéressera qu'à $X_p = X(T(p))$, en notant que $X(T(p, p))$ est l'identité (p premier). L'application

$$\Gamma \cdot a \cdot \Gamma \mapsto X(\Gamma \cdot a \cdot \Gamma)$$

induit un homomorphisme de $R(\Gamma, \Delta)$ dans l'anneau des correspondances propres, rationnelles sur $\mathbf{Q}$, de C. Si $\Gamma \cdot a \cdot \Gamma = \bigcup \Gamma \cdot a_i$, alors, en utilisant 1.4 (1), on voit immédiatement que:

$$X(v(z)) = \sum v(a_i(z)),$$

v désignant la projection canonique de H sur C.

4.3. Le ième groupe de cohomologie à coefficients complexes $H^i(C)$ de C est de façon naturelle un espace de représentation pour $R(\Gamma, \Delta)$. Si l'on utilise l'isomorphisme canonique $H^i(H/\Gamma', \mathbf{C}) \cong H^i(\Gamma', \mathbf{C})$, où Γ' est un sous-groupe discret de G, et $H^i(\Gamma', \mathbf{C})$ le ième groupe de ième groupe de cohomologie de Γ' à coefficients dans le module trivial $\mathbf{C}$, on peut définir l'endomorphisme $X(\Gamma \cdot a \cdot \Gamma)^{(i)}$ de $H^i(C)$ associé à $\Gamma \cdot a \cdot \Gamma$ comme le composé des homomorphismes

$$(1) \qquad H^i(\Gamma; \mathbf{C}) \to H^i(a^{-1} \cdot \Gamma \cdot a; \mathbf{C}) \xrightarrow{\text{res}} H^i(\Gamma(a); \mathbf{C}) \xrightarrow{\text{cores}} H^i(\Gamma, \mathbf{C}),$$

la première flèche étant l'isomorphisme associé à $\gamma \mapsto a^{-1} \cdot \gamma \cdot a\, (\gamma \in \Gamma)$.

Pour $i = 0, 2$, $X(\Gamma \cdot a \cdot \Gamma)^i$ est l'homothétie de rapport $d(\Gamma \cdot a \cdot \Gamma)$. En effet, si $i = 0$ les espaces vectoriels de (1) s'identifient canoniquement à $\mathbf{C}$, les deux premiers homomorphismes sont l'identité, et le troisième est la multiplication par $[\Gamma : \Gamma(a)] = d(\Gamma \cdot a \cdot \Gamma)$, (cf. 1.4). Si $i = 2$, ces espaces s'identifient de nouveau à $\mathbf{C}$ via l'isomorphisme de $H^2(H/\Gamma', \mathbf{C})$ sur $\mathbf{C}$ qui applique la classe fondamentale de H/Γ' sur 1 ($\Gamma' = \Gamma, \Gamma(a), a^{-1} \cdot \Gamma \cdot a$). On voit alors que le premier et le troisième homomorphismes sont l'identité, tandis que le deuxième est la multiplication par $[a^{-1} \cdot \Gamma \cdot a : \Gamma(a)]$. Or il est immédiat que cet indice est le nombre de classes à gauche contenues dans $\Gamma \cdot a \cdot \Gamma$; il est donc aussi égal à $d(\Gamma \cdot a \cdot \Gamma)$ (cf. 1.4).

4.4. *La formule de congruence.* Elle s'écrit

$$p(X_p) = \pi + {}^t\pi,$$

où p est un bon nombre premier ne divisant pas $d(L)$, $p(X_p)$ la réduction mod p du cycle X_p, vue comme correspondance sur $p(C)$, et π la correspondance de Frobenius de $p(C)$, (autrement dit le graphe de l'application qui associe au point x de coordonnées homogènes (x_i) le point x^p de coordonnées homogènes (x_i^p)).

Le principe de démonstration de cette égalité sera donné au § 5. Nous indiquons ici comment on en déduit 3.3 (i).

4.5. *Démonstration de* 3.3 (i). Nous en donnons tout d'abord une version plus longue que celle de [11], mais qui est utilisée dans le cas des variétés fibrées [6]. Etant donné une correspondance propre X de C (ou de $p(C)$), on note $I_0(X)$ le degré de $X \cdot D$, où D est la diagonale. Si X est une correspondance sur $\mathbf{Q}$ de C et si p est bon, alors

$$(1) \qquad I_0(X) = I_0(p(X)) \,.$$

N_m étant comme dans 3.1, on a

$$I_0(\pi^m) = I_0({}^t\pi^m) = N_m \quad (m \in \mathbf{Z}, m \geqq 1) \,.$$

D'autre part, $\pi \cdot {}^t\pi = p \cdot \mathrm{Id}$, donc si l'on pose $X = \pi + {}^t\pi$ et $Y = p \cdot \mathrm{Id}$ dans 4.0 (1), on obtient

$$\pi^m + {}^t\pi^m = f_m(\pi + {}^t\pi, p \cdot \mathrm{Id}) \,, \quad (m \geqq 1) \,,$$

d'où

$$2 \cdot \frac{d}{du} \log Z(u; p(C)) = \sum_{m \geqq 1} I_0(f_m(\pi + {}^t\pi, p \cdot \mathrm{Id})) \cdot u^{m+1} \,.$$

En vertu de 4.4 et de (1), cela s'écrit

$$2 \cdot \frac{d}{du} \log Z(u; p(C)) = \sum_{m \geqq 1} I_0(f_m(X_p, p \cdot \mathrm{Id})) \cdot u^{m-1} \,.$$

Mais la formule des points fixes de Lefschetz entraîne

$$I_0(f_m(X_p, p \cdot \mathrm{Id})) = \sum_{i = 0, 1, 2} (-1)^i \operatorname{tr} f_m(X_p^{(i)}, p \cdot \mathrm{Id}) \,,$$

d'où, vu 4.0 (2):

$$(2) \qquad Z(u; p(C))^2 = \prod_{0 \leqq i \leqq 2} \det(1 - X_p^{(i)} u + p \cdot u^2)^{(-1)^{i+1}} \,.$$

Comme $d(X_p) = p + 1$, les endomorphismes $X_p^{(0)}$ et $X_p^{(2)}$ sont la multiplication par $p + 1$ (cf. 4.3), donc

$$(3) \qquad \det(1 - X_p^{(i)} \cdot u + p \cdot u^2) = (1 - p \cdot u)(1 - u) \quad (i = 0, 2) \,.$$

On a $H^1(C) = H^{1,0}(C) + H^{0,1}(C)$. L'espace $H^{1,0}(C)$ des différentielles holomorphes de degré 1 sur C s'identifie canoniquement à $S_2(\Gamma)$, et la conjugaison complexe induit un **R**-isomorphisme de $H^{1,0}$ sur $H^{0,1}$. On vérifie que ces isomorphismes commutent à $R(\Gamma, \varDelta)$, opérant dans S_2 par les opérateurs de Hecke (cf. § 2) et dans $H^1(C)$ par les $X(\Gamma \cdot a \cdot \Gamma)^{(1)}$. Comme les valeurs propres des opérateurs de Hecke sont réelles, et que $T(p, p)_2$ est l'identité, on en déduit

$$(4) \qquad \det(1 - X_p^{(1)} \cdot u + p \cdot u^2) = (\det(1 - T_{p,2} \cdot u + p \cdot u^2))^2 = \det H(u; p, 2))^2,$$

ce qui, joint à (2), (3), démontre 3.3 (i).

4.6. *Démonstration de 3.3* (i), *(2ème version).* Il suffit de prouver:

$$(1) \qquad\qquad \det(1 - T_{p,2} \cdot u + p \cdot u^2) = \det(1 - M_l(\hat{\pi}) \cdot u),$$

où $\hat{\pi}$ est l'endomorphisme de Frobenius de la jacobienne $J(p(C))$ de $p(C)$, (cf. 3.1). Le membre de droite est aussi égal à $\det(1 - M_l({}^t\hat{\pi}) \cdot u)$, donc vu $\hat{\pi} \cdot {}^t\hat{\pi} = p \cdot \mathrm{Id}$,

$$(2) \qquad\qquad (\det(1 - M_l(\hat{\pi}) \cdot u))^2 = \det(1 - M_l(\hat{\pi} + {}^t\hat{\pi}) \cdot u + p \cdot u^2).$$

Soit $p(X_p)\hat{}$ la correspondance de $J(p(C))$ canoniquement associée à $p(X_p)$. La formule de congruence implique

$$p(X_p)\hat{} = \hat{\pi} + {}^t\hat{\pi}$$

d'autre part $J(p(C)) = p(J(C))$ et $p(X_p)\hat{}$ est la réduction $\mathrm{mod}\, p$ de l'endomorphisme $\hat{X}_p$ de $J(C)$ associée à X_p. On sait que l'on a alors

$$M_l(X_p) = M_l(p(X_p)\hat{})$$

pour un choix convenable de coordonnées l-adiques, d'où

$$(3) \qquad\qquad (\det(1 - M_l(\pi) \cdot u))^2 = \det(1 - M_l(X_p) \cdot u + p \cdot u^2).$$

Il est bien connu que $M_l(X_p)$ est équivalente à la somme $M^d(X_p) + \bar{M}^d(X_p)$, où $M^d(X_p)$ est l'endomorphisme de l'espace des différentielles de première espèce sur C induit par X_p. En utilisant l'isomorphisme canonique de ce dernier espace sur $S_2(\Gamma)$ on voit que le membre de droite de (3) est égal à $\det(1 - T_{p,2} \cdot u + p \cdot u^2)^2$, d'où le résultat.

§ 5. Familles de variétés abéliennes. Formule de congruence

5.1. (Pour le contenu de ce n°, cf. [10]). Pour $z \in C$, on note $e(z)$ le vecteur $(z, 1)$ de $\mathbf{C}^2$. Etant donné $z \in H$, l'ensemble $\mathfrak{o} \cdot e(z) = D_z$ est un réseau de $\mathbf{C}^2$. Soit $c \in \mathfrak{o}$ tel que c^2 soit rationnel et < 0. On montre qu'un multiple de la forme **R**-bilinéaire E sur $\mathbf{C}^2$ définie par

$$E(a \cdot e(z), b \cdot e(z)) = \mathrm{tr}(c \cdot a \cdot b'), \qquad (a, b \in L \otimes \mathbf{R}),$$

est une forme de Riemann sur D_z, d'où une structure de variété abélienne et une polarisation C_z sur le quotient $A_z = \mathbf{C}^2/D_z$. La transformation linéaire de $\mathbf{C}^2$ définie par $a \in \mathfrak{o}$ laisse D_z stable, donc définit un endomorphisme de A_z, d'où un monomorphisme θ_z de $\mathfrak{o}$ dans l'anneau $A(A_z)$ des endomorphismes de A_z, (compatible avec la polarisation en ce sens que

$$E(\theta_z(a) \cdot x, y) = E(x, \theta_z(c^{-1} \cdot a \cdot c) \cdot y) \qquad (a \in \mathfrak{o},\, x, y \in \mathbf{C}^2) .$$

On a ainsi obtenu une famille analytique de variétés abéliennes polarisées $P_z = (A_z, C_z, \theta_z)$ de type $\mathfrak{o}$, paramétrée par H. Un homomorphisme de P_x sur $P_y (x, y \in H)$ est une isogénie $f : A_x \to A_y$ commutant à $\mathfrak{o}$ et telle que $f^{-1}(C_y) = C_x$. On montre que P_x est isomorphe à P_y si et seulement si $x \in \Gamma \cdot y$. La courbe C paramètre donc les classes d'isomorphisme de tels systèmes. On peut préciser cela en construisant dans un espace projectif une famille F de sous-variétés $F_z (z \in H)$ ayant notamment les propriétés suivantes: $F(x) = F(y)$ si et seulement si $x \in \Gamma \cdot y (x, y \in H)$. Le point de Chow $c(F_z)$ de F_z décrit une courbe C' définie sur $\mathbf{Q}$ et $\mathbf{Q}(c(F_z))$ est le «corps des modules» de P_z (i.e. le plus petit sous-corps k de $\mathbf{C}$ ayant la propriété suivante: si K est un corps de définition pour $A_z, C_z, \theta_z(a)$ $(a \in \mathfrak{o})$ et σ est un monomorphisme de K dans C, alors $K \supset k$ et P_z est isomorphe à $(A_z^\sigma, C_z^\sigma, \theta_z^\sigma)$ si et seulement si σ est l'identité sur k). Il existe un nombre fini de fonctions f_i sur H automorphes pour Γ, définies en dehors de la réunion W d'un nombre fini d'orbites de Γ, qui engendrent sur $\mathbf{C}$ le corps $K(\Gamma)$ des fonctions automorphes pour Γ, telles que pour $z \in H - W$, les coordonnées de $c(F_z)$ soient $(1, f_1(z), \ldots, f_m(z))$.

Le corps $L = \mathbf{Q}(u)$, (u générique sur $\mathbf{Q}$) s'identifie donc à un sous-corps de $K(\Gamma)$ tel que $K(\Gamma) = \mathbf{C} \cdot L$. La courbe C du théorème est un modèle lisse sur $\mathbf{Q}$ de L. Elle est analytiquement isomorphe à H/Γ. Dans la suite, on n'aura à considérer que des points génériques, aussi ne distinguerons-nous pas entre C et C'.

[Pour donner une idée de la construction de F, indiquons comment on obtient un système de représentants des classes d'isomorphisme de variétés abéliennes polarisées (A_z, C_z): on part d'une famille de plongements projectifs $\mu_z : A_z \to P = \mathbf{P}(n, \mathbf{C})$ dépendant holomorphiquement de $z \in H$, tels que C_z soit induite par les sections hyperplanes (ce qui se construit à l'aide de fonctions thêta, cf. [8]). En associant à $\varphi \in \mathrm{Aut}\, P$ le point de Chow $c(\varphi(\mu_z(A_z)))$ de $\varphi(\mu_z(A_z))$ on définit un morphisme v_z de $\mathrm{Aut}\, P$ dans un espace projectif. Alors $F(A_z, C_z)$ est l'adhérence de Zariski de l'image de v_z.]

5.2. La correspondance X_p définie dans 4.2 est de degré $p + 1$. Ecrivons, pour $u \in C$ générique

$$X_p(u) = X_p \cdot (u \times C) = u_0 + \ldots + u_p.$$

Pour démontrer la formule de congruence (4.3), il suffit de faire voir que si k est un corps sur lequel u et les u_i sont rationnels et $\mathfrak{p}$ est une place de k prolongeant p, alors on a, $^-$ dénotant la réduction mod $\mathfrak{p}$:

$$(1) \qquad\qquad\qquad \bar{u}_0 = \bar{u}^p, \qquad \bar{u}_i^p = \bar{u}_0.$$

En effet, on a $\pi(\bar{u}) = \bar{u}^p$ et $'\pi(\bar{u}) = p \cdot \bar{u}^{1/p}$; la réduction commutant à l'intersection de cycles positifs, (1) entraîne que $p(X_p) - (\pi + {}'\pi)$ est un diviseur de la forme $\mathfrak{a} \times C$, où a est un diviseur de C; comme $\pi + {}'\pi$ n'a pas de composante de ce type, a doit être positif, mais d'autre part il est de degré zéro, puisque $d(p(X_p)) = d(\pi + {}'\pi) = p + 1$.

La démonstration de (1) se fera en établissant des relations similaires pour les réductions mod $\mathfrak{p}$ des variétés abéliennes polarisées de type $\mathfrak{o}$ représentées par les points de C.

5.3. Soit v la projection de H sur C. On fixe un point $u \in C$ «suffisamment général», et $y \in v^{-1}(u)$. Soit p un bon nombre premier ne divisant pas $d(L)$. Soient $\mathfrak{o} \cdot a_i (0 \leqq i \leqq p)$ les idéaux de norme p et $y_i = a_i(y)$. On a donc $T(p) = \bigcup \Gamma \cdot a_i$ et le cycle $X_p(u)$ est la somme des $v(y_i)$. En accord avec 5.2, on pose $u_i = v(y_i)$. On écrira A_i, C_i, θ_i, P_i, F_i, D_i pour A_{y_i}, C_{y_i}, θ_{y_i}, P_{y_i}, F_{y_i}, D_{y_i}. Admettons que P_i et F_i soient rationnels sur k et se réduisent bien mod $\mathfrak{p}$. Alors (1) équivaut à

$$(2) \qquad \overline{c(F_0)} = \overline{c(F_u)}^p, \; \overline{c(F_i)^p} = \overline{c(F_u)} \qquad (1 \leqq i \leqq p) \, .$$

Vu les propriétés de la famille F, cela revient à

$$(3) \qquad \bar{A}_0 = (\bar{A}_y)^p, \; (\bar{A}_i)^p = \bar{A}_y \qquad (1 \leqq i \leqq p) \, ,$$

et à des relations semblables pour les C_i et θ_i. On se bornera ici à démontrer (3), le reste se prouvant de la même manière.

5.4. Le groupe $g(p, A_y)$ des éléments d'ordre p de A_y est d'ordre p^4 et s'identifie canoniquement, comme $\mathfrak{o}$-module, à $\mathfrak{o}/p \cdot \mathfrak{o} = \mathbf{M}(2, \mathbf{Z}/2\,\mathbf{Z})$. C'est le noyau de l'isogénie $p \cdot \mathrm{Id}$. Il possède $p + 1$ sous-$\mathfrak{o}$-modules g_i d'ordre p^2, dont les images inverses sont, dans $\mathfrak{o}$ les idéaux $\mathfrak{o} \cdot a_i$ et dans $\mathbf{C}^2$ les réseaux $p^{-1} \cdot \mathfrak{o} \cdot a_i(e(y))$, $(0 \leqq i \leqq p)$. On a évidemment, dans les notations du § 2,

$$a_i \cdot e(y) = j(a_i, y) \cdot e(y_i) \, ,$$

ce qui entraîne que l'homothétie de $\mathbf{C}^2$ de rapport $p \cdot j(a_i, y)^{-1}$ envoie $p^{-1} \cdot \mathfrak{o} a_i(e(y))$ sur D_i. Elle induit donc une isogénie $\lambda_i : A_y \to A_i$ de noyau g_i, qui visiblement, commute à $\mathfrak{o}$ et est compatible avec les polarisations; c'est donc en fait une isogénie de P_y sur P_i. Comme $\mathrm{Ker}\, \lambda_i \subset g(p, A_y)$, on a une factorisation $p \cdot \mathrm{Id} = \mu_i \cdot \lambda_i$, où μ_i est une isogénie de P_i sur P_y. Quitte à passer à une extension de k, on peut supposer les éléments de $g(p, A_y)$ rationnels sur k, donc λ_i, μ_i définis sur k. Supposons que $P_y, P_i, F_y, F_i (0 \leqq i \leqq p)$ se réduisent bien mod $\mathfrak{p}$. La réduction mod $\mathfrak{p}$ définit un homomorphisme r de $g(p, A_y)$ dans le groupe $g(\bar{A}_y)$ des éléments d'ordre p de $\bar{A}_y$, qui est surjectif [16, Prop. 16, p. 98]. On montre d'autre part que $g(p, \bar{A}_y)$ est d'ordre p^2 [6, 5.11, p. 516]. Le noyau de r est donc l'un des g_i, que l'on supposera être g_0. Sous les hypothèses faites $r(\ker \bar{\lambda}_i) = \ker \bar{\lambda}_i (0 \leqq i \leqq p)$ [16, Prop. 13, p. 96]. D'autre part, g_i et g_j sont supplémentaires si $i \neq j$, donc

$$r(g_i) = g(p, \bar{A}_y) \qquad (1 \leqq i \leqq p) \, .$$

On a donc

$$\ker \bar{\lambda}_0 = (0)$$

$$\ker \bar{\lambda}_i = g\,(p, \bar{A}_y) \qquad (1 \leqq i \leqq p)\,.$$

Mais $\mu_i \cdot \lambda_i = p \cdot \mathrm{Id}$ entraîne $\bar{\mu}_i \cdot \bar{\lambda}_i = p \cdot \mathrm{Id}$. Comme le degré d'une isogénie se conserve par réduction et que celui de $p \cdot \mathrm{Id}$ est p^4, on a un diagramme d'extensions de corps (voir page suivante); dans ce diagramme, $k\,(A)$ est le corps des fonctions rationnelles sur k de la k-variété A, f^0 le morphisme de fonctions associé à un k-morphisme f de variétés, et $(a, b) = $ (degré séparable, degré inséparable).

$$(4) \qquad
\begin{array}{ccc}
 & k\,(\bar{A}_y) & \\
{\scriptstyle(1,\,p^2)}\nearrow & & \nwarrow{\scriptstyle(1,\,p^2)} \\
k\,(\bar{A}_y^p) & & \bar{\lambda}_0^0\,(k\,(\bar{A}_0)) \\
{\scriptstyle(p^2,\,1)}\searrow & & \swarrow{\scriptstyle(p^2,\,1)} \\
 & (p \cdot \mathrm{Id})^0\,(k\,(\bar{A}_y)) &
\end{array}$$

Cela entraîne que $k\,(\bar{A}_y^p) = \bar{\lambda}_0^0\,(k\,(\bar{A}_0)) \cong k\,(\bar{A}_0)$, et par suite que $\bar{A}_y^p$ et $\bar{A}_0$ sont birationnellement k-isomorphes, donc k-isomorphes. C'est la première égalité de (3). En remplaçant dans ce qui précède $\bar{A}_y$, $\bar{A}_0$ et $\bar{\lambda}_0$ par $\bar{A}_i$, $\bar{A}_y$ et $\bar{\mu}_i (i \geqq 1)$, on démontre exactement de la même manière la deuxième partie de (3). (Pour cette démonstration, dans le cas plus général de 6.1, *voir* [11, § 5].)

5.5. *Remarque.* Il y a quelques précautions à prendre avec les conditions de bonne réduction mod $\mathfrak{p}$, car d'une part tout doit marcher en dehors d'un ensemble fini de nombres premiers, et d'autre part les P_i, y_i *dépendent de* p. Ayant fixé u, on commence par choisir une extension k_0 de type fini de $\mathbf{Q}$, sur laquelle P_y est défini et, pour tout bon p, une place $\mathbf{p}_0$ de k_0 prolongeant p. Pour presque tout p, P_y et F_y se réduisent bien mod $\mathbf{p}_0$. Choisissons un tel p. Soient k une extension algébrique de k_0 sur laquelle les éléments de $g\,(p, A_y)$ sont rationnels, et $\mathbf{p}$ une extension de $\mathbf{p}_0$ à k. Comme P_i est l'image de P_y par une k-isogénie, il résulte d'un théorème de Koizumi-Shimura [3] que l'on peut trouver un modèle $P_i^* = (A_i^*, C_i^*, \theta_i^*)$ de P_i tel que P_i^* et les isogénies λ_i^*, μ_i^* correspondant à λ_i, μ_i se réduisent bien mod $\mathfrak{p}$. Au changement de modèle des P_i près, qui ne modifie pas substantiellement la démonstration, on est donc bien parvenu à la situation de 5.4 pour presque tout p.

§ 6. Généralisations

Le théorème 3.3 a été généralisé dans trois directions, que nous allons passer brièvement en revue.

Dans ce paragraphe, $\mathfrak{b} = \mathfrak{o} \cdot b$ est un $\mathfrak{o}$ idéal entier bilatère de L, on note b_0 l'entier rationnel > 0 tel que $b \cap \mathbf{Z} = b_0 \cdot \mathbf{Z}$, et l'on pose

$$\Delta_{\mathfrak{b}} = \{x \in \Delta, (\det x, b_0) = 1\}\,,$$

$$\Gamma_{\mathfrak{b}} = \{\gamma \in \Gamma, \gamma \equiv 1 \bmod \mathfrak{b}\}\,.$$

L'application $\Gamma_\mathfrak{b} \cdot a \cdot \Gamma_\mathfrak{b} \mapsto \Gamma \cdot a \cdot \Gamma$ induit un homomorphisme de l'anneau de Hecke $R(\Gamma_\mathfrak{b}, \varDelta_\mathfrak{b})$, qui n'est pas toujours commutatif, dans $R(\Gamma, \varDelta)$.

6.1. *Sous-groupes de congruence* [11]. On suppose ici $\mathfrak{b}$ premier à $\mathfrak{o} \cdot d(L)$. On peut alors prendre $b = b_0$ et il existe un isomorphisme $h_\mathfrak{b}$ de $\mathfrak{o}$-modules de $\mathfrak{o}/\mathfrak{b}$ sur $\mathbf{M}(2, \mathbf{Z}/b\,\mathbf{Z})$. On note $\varDelta_\mathfrak{b}^*$ l'ensemble des éléments de $\varDelta_\mathfrak{b}$ dont l'image par $h_\mathfrak{b}$ est de la forme $\left(\begin{smallmatrix} 1 & 0 \\ 0 & d \end{smallmatrix}\right)$. L'homomorphisme $R(\Gamma_\mathfrak{b}, \varDelta_\mathfrak{b}) \to R(\Gamma, \varDelta)$ induit un isomorphisme de $R_\mathfrak{b} = R(\Gamma_\mathfrak{b}, \varDelta_\mathfrak{b}^*)$ sur $R(\Gamma, \varDelta_\mathfrak{b})$ [11, Prop. 1.15]. On désigne par $T_\mathfrak{b}(p)$ et $T_\mathfrak{b}(p, p)$ les éléments correspondant par cet isomorphisme à $T(p)$ pour p premier à b et à $T(p, p)$ pour p premier à $b \cdot d(L)$. On fait opérer $R_\mathfrak{b}$ sur l'espace $S_k(\Gamma_\mathfrak{b})$ des formes automorphes de poids k comme au § 2, et on en déduit une série de Dirichlet

$$D_\mathfrak{b}(s, k) = \sum (\Gamma_\mathfrak{b} \cdot a \cdot \Gamma_\mathfrak{b})_k \cdot (\det a)^{-s} = \prod_{\substack{p \text{ premier} \\ (p, \mathfrak{b}) = 1}} H_\mathfrak{b}(p^{-s}; p, k)$$

où $H_\mathfrak{b}(u; p, k)$ est le polynôme obtenu à partir du polynôme $H(u; p, k)$ du § 2 en remplaçant $T(p)_k$ et $T(p, p)_k$ par les endomorphismes $T_\mathfrak{b}(p)_k$ et $T_\mathfrak{b}(p, p)_k$ de $S_k(\Gamma_\mathfrak{b})$ définis par $T_\mathfrak{b}(p)$ et $T_\mathfrak{b}(p, p)$; cette série se prolonge en une fonction entière avec équation fonctionnelle [11, § 1].

On reprend les notations du § 5. Pour $z \in H$, soit t_z l'image dans A_z du point $b^{-1} \cdot e(z)$. On vérifie que $x \mapsto \theta_z(x) \cdot t_z$ induit un isomorphisme de $\mathfrak{o}$-modules de $\mathfrak{o}/\mathfrak{b}$ sur le groupe $g(\mathfrak{b}, A_z) = \{t \in A_z, \theta_z(\mathfrak{b}) \cdot t = 0\}$ des points d'ordre $\mathfrak{b}$ de A_z. On associe alors à z le système $Q_z = (P_z, t_z)$ formé de P_z, avec les points d'ordre $\mathfrak{b}$ de A_z marqués (dans un ordre déterminé). On montre que $Q_x \cong Q_y (x, y \in H)$ si et seulement si $x \in \Gamma_\mathfrak{b} \cdot y$ [14, Prop. 4.4]. Le système Q_z admet un corps de modules; les résultats de 5.1, s'étendent et conduisent à un modèle projectif lisse $C_\mathfrak{b}$ sur $\mathbf{Q}$ de $H/\Gamma_\mathfrak{b}$ pour lequel 3.3 est valable, pour presque tout p ne divisant pas $b \cdot d(L)$, une fois $H(u; p, 2)$ et $D(s, 2)$ remplacés par $H_\mathfrak{b}(u; p, 2)$ et $D_\mathfrak{b}(s, 2)$.

La démonstration est semblable en principe à celle de 3.3 (i), mais présente quelques complications techniques, en particulier à propos de la formule de congruence. En effet, l'isogénie $\lambda_i : P_y \to P_{y_i}$ de 5.4 applique t_y sur $\pm p \cdot t_{y_i}$ pour $i \geqq 1$, et n'est donc pas une isogénie de Q_y sur Q_{y_i}. En introduisant un automorphisme convenable $Y_\mathfrak{b}$ de $C_\mathfrak{b}$, laissant stables les fibres de $C_\mathfrak{b} \to C$, et induisant la bijection

$$\{\pm t\} \mapsto \{\pm p \cdot t\}$$

de $g(\mathfrak{b}, A_y)/\{\pm 1\}$, on est conduit à une formule de congruence de la forme

$$\text{(1)} \qquad\qquad \bar{X}_p = \pi + {}^t\pi \circ \bar{Y}_p$$

où $\bar{}$ est la réduction $\bmod\, p$, qui se complète par

$$\text{(2)} \qquad\qquad {}^t\pi \circ \bar{Y}_p = {}^t\bar{Z} \circ {}^t\pi \circ \bar{Z} \,,$$

où Z est un automorphisme défini sur $\mathbf{Q}$ de $C_\mathfrak{b}$ [11, Theorems 4,5]. La partie de la démonstration résumée au § 4 subsiste avec peu de changements.

6.2. *Variétés fibrées* [6]. On suppose jusqu'à la fin de cet exposé que $\Gamma_\mathfrak{b}$ opère librement sur H, (ce qui a lieu notamment si $b_0 \geqq 3$). Soit $W_\mathfrak{b}$ le quotient de

$H \times (L_R/\mathfrak{o})$ par $\Gamma_\mathfrak{b}$, opérant via $\gamma(z, u) = (\gamma \cdot z, \gamma \cdot u)$. C'est une variété fibrée sur $C_\mathfrak{b}$, de fibre type le tore $T = L_\mathbf{R}/\mathfrak{o}$, de groupe structural $\Gamma_\mathfrak{b}$. On munit $W_\mathfrak{b}$ d'une structure de variété analytique complexe telle que la projection $\varphi_\mathfrak{b} : W_\mathfrak{b} \to C_\mathfrak{b}$ soit un morphisme, que les zéros des fibres forment une section holomorphe $\psi_\mathfrak{b}$, et que la fibre sur un point $v_\mathfrak{b}(z)$ soit isomorphe à $A_z (z \in H; v_\mathfrak{b}$ projection de H sur $H/\Gamma_\mathfrak{b})$. Soit $W_{m,\mathfrak{b}}$ ou W_m le produit fibré sur $C_\mathfrak{b}$ de m copies de $W_\mathfrak{b}$, $\varphi_{m\mathfrak{b}}$ la projection de $W_{m,\mathfrak{b}}$ sur $C_\mathfrak{b}$ et $\psi_{m\mathfrak{b}}$ la section zéro de $W_{m\mathfrak{b}}$. Il existe des modèles projectifs lisses de W_m, $C_\mathfrak{b}$ définis sur le corps cyclotomique $K_\mathfrak{b} = \mathbf{Q} (\exp. 2\pi i \cdot k_0^{-1})$ tels que $\psi_{m\mathfrak{b}}$ et $\varphi_{m\mathfrak{b}}$ soient définis sur $K_\mathfrak{b}$ [14, §§ 5, 6].

Dans la suite de ce n°, on suppose $\mathfrak{b}$ premier à $d(L)$. On peut alors remplacer $K_\mathfrak{b}$ par $\mathbf{Q}$ dans l'assertion précédente [6, Prop. 7.4]. Le théorème 3.3(i) se généralise de la manière suivante [6, 7.7]:

6.3. Théorème. *On conserve les notations précédentes. Il existe des entiers* $a(m, i, q) \geqq 0$ $(0 \leqq i \leqq 4m; 0 \leqq q \leqq i)$ *tels que l'on ait,*

$$
(1) \qquad
\begin{aligned}
Z(u, p(W_{m,\mathfrak{b}})) &= \prod_{0 \leqq j \leqq 2m} ((1 - p^j \cdot u) \cdot (1 - p^{j+1} \cdot u))^{-a(m, 2j, 0)} \times \\
&\times \prod_{0 \leqq i \leqq 4m} \prod_{0 \leqq q \leqq i} \det(H_\mathfrak{b}(p^{(i-q)/2} \cdot u; p, q+2)^{(-1)^i a(m,i,q)}
\end{aligned}
$$

pour presque tout nombre premier p ne divisant pas $\mathfrak{b} \cdot d(L)$.

Il en résulte en particulier que la fonction zêta globale $Z(s, W_{m,\mathfrak{b}}/\mathbf{Q})$ est le produit d'une fonction élémentaire de s, de translatées de $\zeta(s)$, et de

$$
(2) \qquad \prod_{0 \leqq i \leqq 4m} \cdot \prod_{0 \leqq q \leqq i} \det D_\mathfrak{b}(s - (i-q)/2, q+2)^{(-1)^{i+1} a(m,i,q)},
$$

donc que $Z(s, W_{m,\mathfrak{b}})$ se prolonge en une fonction méromorphe sur le plan complexe.

Pour la démonstration de 6.3 on définit un homomorphisme $\Gamma \cdot a \cdot \Gamma \mapsto X_m(\Gamma \cdot a \cdot \Gamma)$ de $R_\mathfrak{b}$ dans l'anneau des correspondances propres de W_m rationnelles sur $\mathbf{Q}$, commutant à $\varphi_{m\mathfrak{b}}$, et on montre [6, 7.5] que la réduction mod p de $X_m(T_\mathfrak{b}(p))$ satisfait á des formules de congruence qui s'écrivent comme 6.1 (1), (2), sauf que ${}^t\pi$ est remplacé par une correspondance π^* qui induit ${}^t\pi$ sur $C_\mathfrak{b}$ et l'isogénie $u \mapsto p \cdot u$ sur chaque fibre, (i.e. dont le diviseur est le lieu des points $(x^p, p \cdot x)$). Des calculs presque identiques à ceux de 4.5 mènent alors à

$$
(3) \qquad Z(u, p(W_m))^2 = \prod_{0 \leqq i \leqq 4m+2} \det(1 - X_{mp}^{(i)} \cdot u + p \cdot U_{mp}^{(i)} \cdot u^2)^{(-1)^{i+1}},
$$

où $X_{mp}^{(i)}$ et $U_{mp}^{(i)}$ désignent les endomorphismes de $H^i(W_m; \mathbf{C})$ induits par les correspondances $X_m(T_\mathfrak{b}(p))$ et $X_m(T_\mathfrak{b}(p,p))$. La détermination de ceux-ci est faite dans [4]. On établit tout d'abord un isomorphisme d'espaces vectoriels, où la cohomologie est à coefficients complexes:

$$
(4) \qquad H^i(W_m) = \sum_{c+d=i} H^c(\Gamma_\mathfrak{b}; H^d(T^m)), \qquad (T = L_\mathbf{R}/\mathfrak{o}),
$$

les termes de droite étant nuls pour $c \geqq 3$; on montre que $H^d(T^m)$ s'identifie, comme $\Gamma_\mathfrak{b}$-module, au dual de $\Lambda^d(L_\mathbf{C} + \ldots + L_\mathbf{C})$, ($m$ copies), sur lequel $\Gamma_\mathfrak{b}$ agit par la représentation congragrédiente de $\Lambda^d \circ \varrho^{(m)}$, où $\varrho^{(m)}$ est la somme directe de m copies de la représentation de G dans $L_\mathbf{C}$ définie par multiplication à gauche [4, Chap. I–II]. On décompose ensuite ces espaces en sommes directes de G-modules irréductibles. On parvient ainsi à une somme de termes de la forme $H^c(\Gamma_\mathfrak{b}; M_q)$, où M_q est l'espace des polynômes homogènes de degré q sur $\mathbf{C}^2$, muni de sa structure usuelle de $\Gamma_\mathfrak{b}$-module. Pour $c = 0,2$, on montre que $X_m(\Gamma \cdot a \cdot \Gamma)^{(c+d)}$ opère sur $H^c(\Gamma_\mathfrak{b}; H^d(T^m))$ par homothétie de rapport $(\det a)^{d/2} \cdot d(\Gamma \cdot a \cdot \Gamma)$, [4, Thm IV-2-3]. Pour $i = 1$, on utilise *l'isomorphisme de Eichler-Shimura* [12, 17]:

$$(5) \qquad\qquad H^1(\Gamma_\mathfrak{b}; M_q) \cong S_{q+2}(\Gamma_\mathfrak{b}) + \overline{S_{q+2}(\Gamma_\mathfrak{b})}\,.$$

On en déduit alors une décomposition de $X_{mp}^{(i)}$ et $U_{mp}^{(i)}$ en somme d'opérateurs de Hecke et d'opérateurs scalaires [4, Chap. IV, 2], qui conduit au résultat.

L'opérateur $T_\mathfrak{b}(p, p)_k$ est la multiplication par p^{k-2}. On tire alors de 6.3 et des conjectures de Weil (cf. 3.1) que les valeurs propres de $T_\mathfrak{b}(p)_k$ sont $\leqq 2 \cdot p^{(k-1)/2}$ en valeur absolue (cf. [6], 6.12, 6.14).

6.4. *Variétés fibrées et fonctions L* [6]. Soient $G_\mathfrak{b}$ le groupe des éléments inversibles de $\mathfrak{o}/\mathfrak{b}$ et $S_\mathfrak{b}$ le sous-groupe de $G_\mathfrak{b}$ formé des classes de restes contenant un élément a de déterminant $\equiv 1 \bmod \mathfrak{b}$. La projection de $\mathfrak{o}$ sur $\mathfrak{o}/\mathfrak{b}$ induit un isomorphisme de $\Gamma/\Gamma_\mathfrak{b}$ sur $S_\mathfrak{b}$ [6, Prop. 1.8]. On définit d'autre part un homomorphisme $s \mapsto Y(s)$ de $S_\mathfrak{b}$ sur un groupe d'automorphismes de W_m définis sur $K_\mathfrak{b}$, injectif si $m \geqq 1$, de noyau $\{\pm 1\}$ si $m = 0$ [6, 6.7].

Soit π une représentation de $G_\mathfrak{b}$ dans un espace vectoriel complexe U de dimension finie. On note $S_k(\Gamma, \pi)$ l'espace vectoriel des applications holomorphes $f : H \to U$ vérifiant $f(\gamma(z)) = j(\gamma, z)^k \cdot \pi(\gamma) \cdot f(z)$ ($z \in H$; $\gamma \in \Gamma$).

Les résultats mentionnés plus haut sont en fait des cas particuliers des théorèmes principaux de [6]. Ces derniers établissent des relations semblables entre opérateurs de Hecke opérant sur certains espaces $S_k(\Gamma, \pi)$ et séries L de $\mathfrak{p}(W_{m,b})$, ($\mathfrak{p}$ idéal premier de $K_\mathfrak{b}$), définies par rapport au groupe d'automorphismes $\mathfrak{p}(Y(s))$ ($s \in S_\mathfrak{b}$) et à certains caractères de $S_\mathfrak{b}$. On renvoie à [6], pour les énoncés (Thms 6.8, 6.11) et les démonstrations.

Bibliographie

Outre les mémoires cités dans le texte, elle comprend quelques articles d'exposition ([5], [9], [13], [15]).

1. Eichler, M.: Quaternäre quadratische Formen und die Riemannsche Vermutung für die Kongruenzzetafunktion. Archiv der Math. **5** (1954), 355–366
2. Grothendieck, A.: Formule de Lefschetz et rationalité des fonctions L. Sém. Bourbaki, 17e année, 1964–65, Exp. 279
3. Koizumi, S., Shimura, G.: On specializations of abelian varieties. Sci. Papers Coll. Gen. Ed. Univ. Tokyo **9** (1959), 187–211
4. Kuga, M.: Fibre varieties over a symmetric space whose fibres are abelian varieties. Lecture Notes, Univ. of Chicago, 1963–64
5. Kuga, M.: Fibre varieties over a symmetric space whose fibres are abelian varieties. Proc. Symp. Pure Math. **9**, A.M.S., Providence R.I., 1966

6. Kuga, M., Shimura, G.: On the zeta function of a fibre variety whose fibres are abelian varieties. Annals of Math. (2) **82** (1965), 478−539
7. Shimura, G.: Correspondances modulaires et les fonctions ζ des courbes algébriques. J. Math. Soc. Japan **10** (1958), 1−28
8. Shimura, G.: Modules de variétés abéliennes polarisées et fonctions modulaires III. Sém. E.N.S. 1957−58, Exp. 20
9. Shimura, G.: Fonctions automorphes et correspondances modulaires. Proc. Int. Congress Math. Edinburgh 1958, Cambridge Univ. Press, p. 330−338
10. Shimura, G.: On the theory of automorphic functions. Annals of Math. (2) **70** (1959), 101−144
11. Shimura, G.: On the zeta-functions of curves uniformized by certain automorphic functions. J. Math. Soc. Japan **13** (1961), 275−331
12. Shimura, G.: On Dirichlet series and abelian varieties attached to automorphic forms. Annals of Math. (2) **76** (1962), 237−294
13. Shimura, G.: The zeta-function of an algebraic variety and automorphic functions. Summer Institute on algebraic geometry, Woodshole 1964, (Notes)
14. Shimura, G.: On the field of definition for a field of automorphic functions II. Annals of Math. (2) **81** (1965), 124−165
15. Shimura, G.: Moduli of abelian varieties and number theory. Proc. Symp. pur. Math. **9,** A.M.S. Providence R.I. 1966
16. Shimura, G., Taniyama, Y.: Complex multiplication of abelian varieties and its applications to number theory. Publ. Math. Soc. Japan **6** (1961)
17. Verdier, J.-L.: Sur les intégrales attachées aux formes automorphes. Sém. Bourbaki 13e année, 1960−1961, Exp. 216
18. Weil, A.: Variétés abéliennes et courbes algébriques. Act. Sci. Ind. 1064, Hermann éd., Paris 1948

73.

Linear algebraic groups

Proc. Symp. Pure Math. **9**, Amer. Math. Soc. (1966) 3–19

This is a review of some of the notions and facts pertaining to linear algebraic groups. From §2 on, the word linear will usually be dropped, since more general algebraic groups will not be considered here.

1. **The notion of linear algebraic group.** According to one's taste about naturality and algebraic geometry, it is possible to give several definitions of linear algebraic groups. The first one is not intrinsic at all but suffices for what follows.

1.1. *Algebraic matrix group.* Let Ω be an algebraically closed field. We shall denote by $M(n, \Omega)$ the group of all $n \times n$ matrices with entries in Ω and by $GL(n, \Omega)$ the group of all $n \times n$ invertible matrices. $GL(n, \Omega)$ is an affine subvariety of Ω^{n^2+1} through the identification

$$g = (g_{ij}) \mapsto (g_{11}, g_{12}, \cdots, g_{nn}, (\det g)^{-1}).$$

The set $M(n, \Omega)$ carries a topology—the Zariski topology—the closed sets being the algebraic subsets of $M(n, \Omega) = \Omega^{n^2}$. The ring $GL(n, \Omega)$ is an open subset of $M(n, \Omega)$ and carries the induced topology.

A subgroup of G of $GL(n, \Omega)$ is called an algebraic matrix group if G is a closed subset of $GL(n, \Omega)$, i.e., if there exist polynomials $p_\alpha \in \Omega[X_{11}, X_{12}, \cdots, X_{nn}]$ $(\alpha \in J)$ such that

$$G = \{g = (g_{ij}) \in GL(n, \Omega) | p_\alpha(g_{ij}) = 0, \ (\alpha \in J)\}.$$

The coordinate ring $\Omega[G]$ of G, i.e., the ring of all regular functions on G, is the Ω-algebra generated by the coefficients g_{ij} and $(\det g)^{-1}$. It is the quotient ring $\Omega[X_{ij}, Z]/I$, where I is the ideal of polynomials in the $n^2 + 1$ letters X_{ij}, Z vanishing on G, considered as a subset of Ω^{n^2+1}, via the above imbedding of $GL(n, \Omega)$ in Ω^{n^2+1}.

When B is a subring of Ω, we shall denote by $GL(n, B)$ the set of $n \times n$ matrices g with entries in B, such that $\det g$ is a unit in B, and by G_B the intersection $G \cap GL(n, B)$.

Let k be a subfield of Ω. The algebraic matrix group G is *defined over k* or is a k-*group* if the ideal I of polynomials annihilated by G has a set of generators in $k[X_{ij}, Z]$. If I_k denotes the ideal of all polynomials with coefficients in k vanishing on G, the quotient ring $k[X_{ij}, Z]/I_k = k[G]$ is the coordinate ring of G over k.

REMARK. If the field k is not perfect, it is not enough to assume that G is k-closed (i.e., that G is defined by a set of equations with coefficients in k) to

662

conclude that G is defined over k; one can only infer that G is defined over a purely inseparable extension k' of k.

The following variant of the definition eliminates the choice of a basis.

1.2. *Algebraic groups of automorphisms of a vector space.* Let V be an n-dimensional vector space over Ω, and $GL(V)$ the group of all automorphisms of V. Every base of V defines an isomorphism of $GL(V)$ with $GL(n, \Omega)$. A subgroup G of $GL(V)$ is called an algebraic group of automorphisms of V if any such isomorphism maps G onto an algebraic matrix group.

Let k be a subfield of Ω. Assume that V has a k-structure, i.e., that we are given a vector subspace V_k over k of V such that $V = V_k \otimes_k \Omega$. The subgroup of G of $GL(V)$ is then defined over k if there exists a basis of V_k such that the corresponding isomorphism $\beta: GL(V) \to GL(n, \Omega)$ maps G onto a k-group in the previous sense.

1.3. *Affine algebraic group.* Let G be an affine algebraic set. It is an affine algebraic group if there are given morphisms

$$\mu: G \times G \to G, \qquad \mu(a, b) = ab,$$
$$\rho: G \to G, \qquad \rho(a) = a^{-1},$$

of affine sets, with the usual properties. G is an affine algebraic group defined over k if G, μ and ρ are defined over k. One can prove that every affine algebraic group defined over k is isomorphic to an algebraic matrix group defined over k.

1.4. *Functorial definition of affine algebraic groups.* Sometimes one would like not to emphasize a particular algebraically closed extension of the field k. For instance in the case of adèle groups, an algebraically closed field containing every p-adic completion of a number field k is a cumbersome object. Let G be a k-group in the sense of §1.1. Then for any k-algebra A, we may consider the set G_A of elements of $GL(n, A)$ whose coefficients annihilate the polynomials in I_k. It is a group, which may be identified to the group $\mathrm{Hom}_k(k[G], A)$ of k-homomorphisms of $k[G]$ into A. Furthermore, to any homomorphism $\rho: A \to B$ of k-algebras corresponds canonically a homomorphism $G_A \to G_B$. Thus we may say that a k-group is a functor from k-algebras to groups, which is representable by a k-algebra $k[G]$ of finite type, such that $\bar{k} \otimes k[G]$ has no nilpotent element, $\bar{k}$ being an algebraic closure of k. (The last requirement stands for the condition that $I_k \otimes \Omega$ is the ideal of all polynomials vanishing on G, it would be left out in a more general context.) This definition was introduced by Cartier as a short cut to the notion of (absolutely reduced) "affine scheme of groups over k." The functors corresponding to the general linear group and the special linear group, will be denoted GL_n and SL_n, and $(GL_n)_A$ by $GL_n(A)$ or $GL(n, A)$.

Usually the more down to earth point of view of algebraic matrix groups will be sufficient.

1.5. *Connected component of the identity.* An algebraic set is reducible if it is the union of two proper closed subsets; it is nonconnected if it is the union of two proper disjoint closed subsets. An algebraic group is irreducible if and only if it is connected. To avoid confusion with the irreducibility of a linear group, we shall usually speak of connected algebraic groups. The connected component of the identity of G will be denoted by G^0. The index of G^0 in G is finite.

If $\Omega = C$, every affine algebraic group G can be viewed as a complex Lie group; then G is connected as an algebraic group, if and only if G is connected as a Lie group. When G is defined over R, G_R is a closed subgroup of $GL(n, R)$ and hence a real Lie group. It is not true that for a connected algebraic R-group, the Lie group G_R is also connected, but in any case it has only finitely many connected components. The connected component of the identity for the usual topology will be denoted G_R^0.

Let G be connected. Then $\Omega[G]$ is an integral domain. Its field of fractions $\Omega(G)$ is the field of rational functions on G. The quotient field of $k[G]$ is a subfield of $\Omega(G)$, consisting of those rational functions which are defined over k.

1.6. *The Lie algebra of an algebraic group.* A group variety is nonsingular, so that the tangent space at every point is well defined. The tangent space $\mathfrak{g}$ at e can be identified with the set of Ω-derivations of $\Omega[G]$ which commute with right translations. $\mathfrak{g}$, endowed with the Lie algebra structure defined by the usual vector space structure and bracket operations on derivations, is the *Lie algebra of* $\mathfrak{g}$. Of course, G and G^0 have the same Lie algebra. If G is connected, $\mathfrak{g}$ could alternatively be defined as the Lie algebra of Ω-derivations of the field $\Omega(G)$, which commute with right translations (and the definition would then be valid for any algebraic group linear or not). If G is defined over k, we have $\mathfrak{g} = \mathfrak{g}_k \otimes_k \Omega$, where $\mathfrak{g}_k$ is the set of derivations which leave $k[G]$ stable. If the characteristic p of k is $\neq 0$, then $\mathfrak{g}$ and $\mathfrak{g}_k$ are restricted Lie algebras, in the sense of Jacobson. However the connection between an algebraic group and its Lie algebra in characteristic $p \neq 0$ is weaker than for a Lie group; for instance, there does not correspond a subgroup to every restricted Lie subalgebra of $\mathfrak{g}$, and it may happen that several algebraic subgroups have the same Lie algebra.

The group G operates on itself by inner automorphisms. The differential of $\mathrm{Int}\, g : x \mapsto g \cdot x \cdot g^{-1}$ $(x, g \in G)$ at e is denoted $\mathrm{Ad}_{\mathfrak{g}} g$. The map $g \mapsto \mathrm{Ad}_{\mathfrak{g}} g$ is a k-morphism (in the sense of 2.1) of G into $GL(\mathfrak{g})$, called the *adjoint representation of G*.

1.7. *Algebraic transformation group.* If G is an algebraic group and V is an algebraic set, G operates morphically on V (or G is an algebraic transformation group) when there is given a morphism $\tau : G \times V \to V$ with the usual properties of transformation groups. It operates k-morphically if G, V and τ are defined over k.

An elementary, but basic, property of algebraic transformation groups is the existence of at least one closed orbit (e.g. an orbit of smallest possible dimension [1, §16]).

2. Homomorphism, characters, subgroups and quotient groups of algebraic groups.

2.1. *Homomorphisms of algebraic groups.* Let ρ, G, G' be algebraic groups and $\rho : G_\Omega \to G'_\Omega$ be a map. It is a morphism of algebraic groups if:

(1) ρ is a group homomorphism from G_Ω to G'_Ω;

(2) the transposed map ρ^0 of ρ is a homomorphism of $\Omega[G']$ into $\Omega[G]$ (if $f \in \Omega[G']$, f is a map from G'_Ω to Ω and $\rho^0(f) = f \circ \rho$). In case G and G' are defined over k, the map ρ is a k-morphism if moreover ρ^0 maps $k[G']$ into $k[G]$. The differential $d\rho$ at the identity element of the morphism $\rho: G \to G'$ defines a homomorphism $d\rho: \mathfrak{g} \to \mathfrak{g}'$ of the corresponding Lie algebras.

A rational representation of G is a morphism $\rho: G \to GL_m$. Let G be considered as a matrix group so that $\Omega[G] = \Omega[g_{11}, \cdots, g_{nn}, (\det g)^{-1}]$. Each coefficient of the matrix $\rho(g)$, $g \in G$, is then a polynomial in $g_{11}, g_{12}, \cdots, g_{nn}, (\det g)^{-1}$.

2.2. *Characters.* A *character* of G is a rational representation of degree 1; $\chi: G \to GL_1$. The set of characters of G is a commutative group, denoted by $X(G)$ or $\hat{G}$. The group $\hat{G}$ is finitely generated; it is free if G is connected [**8**]. If one wants to write the composition-law in $\hat{G}$ multiplicatively, the value at $g \in G$ of $\chi \in G$ should be noted $\chi(g)$. But since one is accustomed to add roots of Lie algebras, it is also natural to write the composition in $\hat{G}$ additively. The value of χ at g will then be denoted by g^χ. To see the similarity between roots and characters take $\Omega = C$; if $X \in \mathfrak{g}$, the Lie algebra of G, $(e^X)^\chi = e^{d\chi(X)}$, where $d\chi$ is the differential of $d\chi$ at e; $d\chi$ is a linear form over $\mathfrak{g}$. In the sequel, we shall often not make any notational distinction between a character and its differential at e.

2.3. *Subgroups, quotients* [**4, 7**]. Let G be an algebraic group defined over k, H a closed subgroup of G; it is a k-subgroup of G if it is defined over k as an algebraic group. H is in particular k-closed. The converse need not be true. The homogeneous space G/H can be given in a natural way a structure of quasi-projective algebraic set defined over k. (A quasi-projective algebraic set is an algebraic set isomorphic to an open subset of a projective set.) The projection $\pi: G \to G/H$ is a k-morphism of algebraic sets which is "separable" ($d\pi$ is surjective everywhere) such that every morphism $\phi: G \to V$, constant along the cosets of H, can be factored through π. Moreover, G acts on G/H as an algebraic group of transformation; if H is a normal k-subgroup of G, then G/H is an algebraic group defined over k.

Assume that in G there exist a subgroup H and a normal subgroup N such that
(1) G is the semidirect product of H and N as abstract group,
(2) the map $\mu: H \times N \to G$, with $\mu(h, n) = hn$, is an isomorphism of algebraic varieties.

Then G is called the semidirect product of the algebraic groups H and N.

In characteristic zero the condition (2) follows from (1). In characteristic $p > 0$, it is equivalent with the transversality of the Lie algebras of H and N or with the regularity of $d\mu$ at the origin, but does not follow from (1).

2.4. *Jordan decomposition of an element of an algebraic group* [**1, 4**]. Let

$$g \in GL(n, \Omega),$$

g can be written uniquely as the product $g = g_s \cdot g_n$, where g_s is a semisimple matrix (i.e., g_s can be made diagonal) and g_n is a unipotent matrix (i.e., the only eigenvalue of g_n is 1, or equivalently $g_n - I$ is nilpotent) and $g_s \cdot g_n = g_n \cdot g_s$. If G is an algebraic matrix group and $g \in G$, one proves that g_n and g_s belong also to G and that the decomposition of g in a semisimple and an unipotent part does not depend on the representation of G as a matrix group. More generally, if $\phi: G \to G'$ is a morphism of algebraic groups and $g \in G$, then $\phi(g_s) = [\phi(g)]_s$ and $\phi(g_n) = [\phi(g)]_n$. If $g \in G_k$, g_s and g_n are rational over a purely inseparable extension of k.

3. **Algebraic tori** [1, 3, 4]. An algebraic group G is an *algebraic torus* if G is isomorphic to a product of d copies of $\mathbf{GL}_1$ (where $d = \dim G$).

If $\Omega = \mathbf{C}$, an algebraic torus is isomorphic to $(\mathbf{C}^*)^d$, and so is not an ordinary torus. However, the algebraic tori have many properties analogous to those of usual tori in compact real Lie groups. Since in what follows the tori in the topological sense will occur rarely, the adjective "algebraic" will be dropped.

3.1. THEOREM. *For a connected algebraic group G the following conditions are equivalent*:

(1) *G is a torus*;

(2) *G consists only of semisimple elements*;

(3) *G, considered as matrix group, can be made diagonal.*

Property (3) means that there always exists a basis of Ω^n such that G is represented by diagonal matrices with respect to that basis. Each diagonal element of the matrix, considered as a function on G, is then a character.

Let T be a torus of dimension d. Every element $x \in T$ can be represented by $(x_1, \cdots, x_d)$, with $x_i \in \Omega^*$. A character χ of T can then be written

$$\chi(x) = x_1^{n_1} x_2^{n_2} \cdots x_d^{n_d}$$

with $n_i \in \mathbf{Z}$ hence $\hat{T} \cong \mathbf{Z}^d$.

3.2. THEOREM. *Let T be a torus defined over k. The following conditions are equivalent*:

(1) *All characters of T are defined over k: $\hat{T} = \hat{T}_k$.*

(2) *T has a diagonal realization over k.*

(3) *For every representation $\rho: T \to \mathbf{GL}_m$, defined over k, the group $\rho(T)$ is diagonalizable over k.*

DEFINITION. If T satisfies these three equivalent conditions, T is called a *split k-torus*, and is said to *split over k*.

If T splits over k, so does every subtorus and quotient of T. There always exists a finite separable Galois-extension k'/k such that T splits over k'. The Galois-group operates on $\hat{T}$. This action determines completely the k-structure of T. The subgroup $\hat{T}_k$ is the set of characters left fixed by the Galois group.

DEFINITION. A torus T is called *anisotropic over k* if $\hat{T}_k = \{1\}$. The anisotropic tori are very close to the usual compact tori. Let $k = \mathbf{R}$. If $\dim T = 1$, there are two possibilities; either T splits over k, and then $T_{\mathbf{R}} \cong \mathbf{R}^*$, or T is anisotropic over k; then T is isomorphic over k with $\mathbf{SO}_2$, and $T_{\mathbf{R}} = SO(2, \mathbf{R})$ is the circle group. In the general case $T_{\mathbf{R}}$ is compact if and only if T is anisotropic over $\mathbf{R}$

(this is also true if R is replaced by a p-adic field). In this case, T_R is a topological torus (product of circle groups).

3.3. THEOREM. *Let T be a k-torus. There exist two uniquely defined k-subtori T_d and T_a, such that*

(1) T_d *splits over k,*

(2) T_a *is anisotropic over k,*

(3) $T_d \cap T_a$ *is finite and $T = T_d \cdot T_a$.*

This decomposition is compatible with morphisms of algebraic groups. (Property 3 will be abbreviated by saying that T is the *almost direct product of T_d and T_a*.)

If S is a k-subtorus of T, then there exists a k-subtorus S' such that T is almost direct product of S and S'.

EXAMPLE. If $k = R$, $T = T_1 \cdot T_2 \cdot \cdots \cdot T_d$ where every T_i is one dimensional. The product is an almost direct product.

4. Solvable, nilpotent and unipotent groups.

4.1. DEFINITION. The algebraic group G is *unipotent* if every element of G is unipotent.

EXAMPLE. If $\dim G = 1$ and G is connected unipotent then G is isomorphic to the additive group of the field;

$$G \cong G_a = \left\{ g \in GL_2 \,\big|\, g = \begin{pmatrix} 1 & x \\ 0 & 1 \end{pmatrix} \right\}.$$

A connected and unipotent matrix group is conjugate to a group of upper-triangular matrices with ones in the diagonal. Hence it is nilpotent; more precisely there exists a central series

$$G = G_0 \supset G_1 \supset \cdots \supset G_i \supset G_{i+1} \supset \cdots \supset G_n = \{e\}$$

such that $G_i/G_{i+1} \cong G_a$. Conversely, if there exists a normal series ending with $\{e\}$ such that $G_i/G_{i+1} \cong G_a$, where G_i is an algebraic subgroup of G, then G is unipotent.

In characteristic 0, a unipotent algebraic group is connected, and the exponential is a bijective polynomial mapping from the Lie algebra $\mathfrak{g}$ to G; the inverse map is the logarithm. In characteristic $p > 0$, this is no more true; in that case, G is a p-group.

DEFINITION. G is a *solvable* (resp. *nilpotent*) algebraic group, if it is solvable (resp. nilpotent) as an abstract group.

4.2. We now state some basic properties of a connected solvable group G.

(1) (Theorem of Lie–Kolchin): If G is represented as a matrix group, it is conjugate (over Ω) to a group of triangular matrices [1].

(2) If G operates on a complete algebraic variety (in particular on a projective variety) then G has a fixed point [1].

(3) The set of unipotent elements in G is a normal connected subgroup U. If G is defined over k, it has a maximal torus defined over k; G is the semidirect

product, as algebraic group, of T and U; any two maximal tori defined over k are conjugate by an element of G_k (Rosenlicht, Annali di Mat. (iv), **61** (1963), 97–120; see also [**3**, §11]).

(4) G has a composition series

$$G = G_0 \supset G_1 \supset \cdots \supset G_i \supset G_{i+1} \supset \cdots \supset G_n = \{e\}$$

where the G_i are algebraic subgroups of G such that G_i/G_{i+1} is isomorphic with G_a or GL_1.

(5) The group G is nilpotent if and only if it is the direct product of a maximal torus T and of a unipotent subgroup U. In this case T (resp. U) consists of all semisimple (resp. unipotent) elements of G.

Properties (1) and (2) are closely connected. In fact (2) implies (1): take the manifold of full flags (see §5.3) of the ambient vector space V, on which G acts in a natural way. Let F be a flag fixed by G; if one chooses a basis of V adapted to F, then G is triangular. On the other hand, for projective varieties, (2) follows immediately from (1). Property (4) is an immediate consequence of (1). In (3), one has to take care that, in contradistinction to the existence of a maximal torus defined over k, the normal subgroup U need not be defined over k, although it is k-closed [**8**].

4.3. DEFINITION. Let G be a connected solvable group defined over k. G *splits over* k if there exists a composition series

$$G = G_0 \supset G_1 \supset \cdots \supset G_i \supset G_{i+1} \supset \cdots \supset G_m = \{e\}$$

consisting of connected k-subgroups of G such that G_i/G_{i+1} is isomorphic over k with G_a or GL_1.

In particular, every torus T of G splits then over k. Conversely, when k is perfect, if the maximal tori of G which are defined over k split over k, then so does G.

Let G be a connected solvable k-group which splits over k, and V a k-variety on which G operates k-morphically. Then: (a) if V is complete and V_k is not empty, G has a fixed point in V_k [**8**]; (b) if G is transitive on V, the set V_k is not empty [**7**].

5. Radical. Parabolic subgroups. Reductive groups.

5.1. DEFINITIONS. Let G be an algebraic group. The *radical* $R(G)$ of G is the greatest connected normal subgroup of G; the *unipotent radical* $R_u(G)$ is the greatest connected unipotent normal subgroup of G. The group G is *semisimple* (resp. *reductive*) if $R(G) = \{e\}$ (resp. $R_u(G) = \{e\}$).

The definitions of $R(G)$ and $R_u(G)$ make sense, because if H, H' are connected normal and solvable (resp. unipotent) subgroups, then so is $H \cdot H'$. Both radicals are k-closed if G is a k-group. Clearly, $R(G) = R(G^0)$ and $R_u(G) = R_u(G^0)$.

The quotient $G/R(G)$ is semisimple, and $G/R_u(G)$ is reductive. In characteristic zero, the unipotent radical has a complement; more precisely: Let G be defined over k. There exists a maximal reductive k-subgroup H of G such that

$$G = H \cdot R_u(G),$$

the product being a semidirect product of algebraic groups. If H' is a reductive subgroup of G defined over k, then H' is conjugate over k to a subgroup of H. In characteristic $p > 0$, this theorem is false (not just for questions of inseparability): according to Chevalley, there does not always exist a complement to the unipotent radical, moreover there are easy counter-examples to the conjugacy property [3].

5.2. THEOREM. *Let G be an algebraic group. The following conditions are equivalent*:

(1) G^0 *is reductive*,

(2) $G^0 = S \cdot G'$, *where S is a central torus and G' is semisimple*,

(3) G^0 *has a locally faithful fully reducible rational representation*,

(4) *If moreover the characteristic of Ω is 0, all rational representations of G are fully reducible*.

If G is a k-group, and $k = \mathbf{R}$, these conditions are also equivalent to the existence of a matrix realization of G such that $G_{\mathbf{R}}$ is "self-adjoint"

$$(g \in G_{\mathbf{R}} \Rightarrow {}^t g \in G_{\mathbf{R}}).$$

In property (2) G' is the commutator subgroup $\mathscr{D}(G)$ of G; it contains every semisimple subgroup of G. The group G is separably isogenous to $S \times G'$ and every torus T of G is separably isogenous to $(T \cap S) \times (T \cap G')$.

5.3. THEOREM [1]. *Let G be a connected algebraic group.*

(1) *All maximal tori of G are conjugate. Every semisimple element is contained in a torus. The centralizer of any subtorus is connected.*

(2) *All maximal connected solvable subgroups are conjugate. Every element of G belongs to one such group.*

(3) *If P is a closed subgroup of G, then G/P is a projective variety if and only if P contains a maximal connected solvable subgroup.*

The *rank* of G is the common dimension of the maximal tori, (notation $rk(G)$).

A closed subgroup P of G is called parabolic, if G/P is a projective variety. Following a rather usual practice, the speaker will sometimes allow himself to abbreviate "maximal connected closed solvable subgroup" by "Borel subgroup."

EXAMPLE. $G = \mathbf{GL}_n$. A *flag* in a vector space V is a properly increasing sequence of subspaces $0 \neq V_1 \subset \cdots \subset V_t \subset V_{t+1} = V$. The sequence (d_i)

$$(d_i = \dim V_i, i = 1, \cdots t))$$

describes the type of the flag. If $d_i = i$ and $t = \dim V - 1$, we speak of a *full flag*.

A parabolic subgroup of GL_n is the stability group of a flag F in Ω^n. G/P is the manifold of flags of the same type as F, and is well known to be a projective variety. A Borel subgroup is the stability group of a full flag. In a suitable basis, it is the group of all upper triangular matrices.

5.4. With respect to rationality question one can state that if G is a connected algebraic group defined over k, then

(1) G has a maximal torus defined over k (Grothendieck [5], see also [2]). The centralizer of any k-subtorus is defined over k ([5], [3, §10]).

(2) If k is infinite and G is reductive, G_k is Zariski dense in G (Grothendieck [5], see also [2]).

(3) If k is infinite and perfect, G_k is Zariski dense in G (Rosenlicht [8]).

Rosenlicht has constructed an example of a one dimensional unipotent group defined over a field k of characteristic 2 such that G is not isomorphic to G_a over k and G_k is not dense in G [8]. An analogous example exists for every positive characteristic (Cartier).

6. **Structure theorems for reductive groups.** The results stated below are proved in [3]. Over perfect fields, some of them are established in [6], [9].

6.1. *Root systems.* Let V be a finite dimensional real vector space endowed with a positive nondegenerate scalar product. A subset Φ of V is a root system when

(1) Φ consists of a finite number of nonzero vectors that generate V, and is symmetric ($\Phi = -\Phi$).

(2) for every $\alpha \in \Phi$, $s_\alpha(\Phi) = \Phi$, where s_α denotes reflection with respect to the hyperplane perpendicular to α.

(3) if $\alpha, \beta \in \Phi$, then $2(\alpha, \beta)/(\alpha, \alpha) \in Z$. The group generated by the symmetrics $s_\alpha (\alpha \in \Phi)$ is called the Weyl group of Φ (notation $W(\Phi)$). It is finite. The integers $2(\alpha, \beta)/(\alpha, \alpha)$ are called the Cartan integers of Φ. Condition (3) means that for every α and β of Φ, $(s_\alpha(\beta) - \beta)$ is an integral multiple of α, since

$$s_\alpha(\beta) = \beta - 2\alpha(\alpha, \beta)/(\alpha, \alpha).$$

For the theory of reductive groups we shall have to enlarge slightly the notion of root system: if M is a subspace of V, we say that Φ is a root system in (N, M) if it generates a subspace P supplementary to M, and is a root system in P. The Weyl group $W(\Phi)$ is then understood to act trivially on M.

A root system Φ in V is the direct sum of $\Phi' \subset V'$ and $\Phi'' \subset V''$, if $V = V' \oplus V''$ and $\Phi = \Phi' \cup \Phi''$. The root system is called *irreducible* if it is not the direct sum of two subsystems.

6.2. *Properties of root systems.*

(1) Every root system is direct sum of irreducible root systems.

(2) If α and $\lambda\alpha \in \Phi$, then $\lambda = \pm 1, \pm\frac{1}{2}$, or ± 2.

The root system Φ is called *reduced* when for every $\alpha \in \Phi$, the only multiples of α belonging to Φ are $\pm\alpha$. To every root system Φ, there belongs two natural reduced systems by removing for every $\alpha \in \Phi$ the longer (or the shorter) multiple of α:

$$\Phi_s = \{\alpha \in \Phi | \tfrac{1}{2}\alpha \notin \Phi\},$$

$$\Phi_e = \{\alpha \in \Phi | 2\alpha \notin \Phi\}.$$

(3) The only reduced irreducible root systems are the usual ones:

$$A_n \quad (n \geq 1), \quad B_n \quad (n \geq 2), \quad C_n \quad (n \geq 3), \quad D_n \quad (n \geq 4),$$

$$G_2, \quad F_4, \quad E_6, \quad E_7, \quad E_8.$$

(4) For each dimension n, there exists one irreducible nonreduced system, denoted by BC_n (see below).

EXAMPLES. B_n: Take R^n with the standard metric and basis $\{x_1, \cdots, x_n\}$.

$$B_n = \{\pm(x_i \pm x_j) \quad (i < j) \text{ and } \pm x_i \quad (1 \leq i \leq n)\}.$$

$W(B_n) = \{s \in GL(n, R) | s$ a product of a permutation matrix
with a symmetry with respect to a coordinate subspace$\}$

$$C_n = \{\pm(x_i \pm x_j) \quad (i < j) \text{ and } \pm 2x_i \quad (1 \leq i \leq n)\},$$

$W(C_n) = W(B_n),$

$$BC_n = \{\pm(x_i \pm x_j) \quad (i < j), \pm x_i \text{ and } \pm 2x_i \quad (1 \leq i \leq n)\},$$

$W(BC_n) = W(B_n).$

DEFINITION. A hyperplane of V is called *singular* if it is orthogonal to a root $\alpha \in \Phi$. A Weyl-chamber C^0 is a connected component of the complement of the union of the singular hyperplanes.

To a Weyl-chamber, is associated an ordering of the roots defined by:

$$\alpha > 0, \text{ if } (\alpha, v) > 0 \text{ for every } v \text{ in } C^0.$$

The root α is *simple* (relative to the given ordering) if it is not the sum of two positive roots. The set of simple roots is denoted by Δ. Δ is connected if it cannot be written as the union of $\Delta' \cup \Delta''$ where Δ' is orthogonal to Δ''.

(5) The Weyl group acts simply transitively on the Weyl-chambers (i.e., there is exactly one element of the Weyl group mapping a given Weyl-chamber on to another one).

(6) Every root of Φ is the sum of simple roots with integral coefficients of the same sign.

(7) The root system Φ is irreducible if and only if Δ is connected.

6.3. *Roots of a reductive group, with reference to a torus.* Let G be a reductive group, and S a torus of G. It operates on the Lie-algebra $\mathfrak{g}$ of G by the adjoint representation. Since S consists of semisimple elements, $\mathrm{Ad}_\mathfrak{g} S$ is diagonalizable

$$\mathfrak{g} = \mathfrak{g}_0^{(S)} \oplus \amalg_\alpha \mathfrak{g}_\alpha^{(S)}$$

where

$$g_\alpha^{(S)} = \{X \in \mathfrak{g} \mid \text{Ad } s(X) = s^\alpha \cdot X\} \qquad (\alpha \in \hat{S}; \ \alpha \neq 0).$$

The set $\Phi(G, S)$ of roots of G relative to the torus S is the set of nontrivial characters of S appearing in the above decomposition of the adjoint representation. If $T \supset S$, every root of G relative to T that is not trivial on S defines a root relative to S. If T is maximal $\Phi(G, T) = \Phi(G)$ is the set of roots of G in the usual sense.

6.4. *Anisotropic reductive groups.* A connected reductive group G defined over k is called *anisotropic* over k, if it has no k-split torus $S \neq \{e\}$.

EXAMPLES. (1) Let F be a nondegenerate quadratic form on a k-vector space V with coefficients in k. Let $G = O(F)$ be the orthogonal group of F. The group G is anisotropic over k if and only if F does not represent 0 over k, i.e., if V_k has no nonzero isotropic vector.

PROOF. Assume v is an isotropic vector. Then there exists a hyperbolic plane through v and in a suitable basis the quadratic form becomes

$$F(x_1, \cdots, x_n) = x_1 x_2 + F'(x_3, \cdots, x_n).$$

If $\lambda \in \Omega^*$, the set of transformations

$$x_1' = \lambda x_1, \qquad x_2' = \lambda^{-1} \cdot x_2, \qquad x_i' = x_i \qquad (i \geq 3)$$

is a torus of G split over k. Conversely if there exists a torus S of G which splits over k, diagonalize S. There is a vector $v \in V_k - \{0\}$ and a nontrivial character $\chi \in \hat{S}$, such that $s(v) \equiv s^\chi v$. Since $s^\chi \neq \pm 1$ for some s, and $F(v) = F(s(v))$, one has $F(v) = 0$ and v is isotropic.

(2) If $k = \mathbf{R}$ or is a p-adic field, G is anisotropic over k if and only if G_k is compact. If k is an arbitrary field of characteristic 0, G is anisotropic over k if and only if G_k has no unipotent element $\neq e$ and $\hat{G}_k = \{1\}$.

6.5. *Properties of reductive k-groups.* Let G be a connected reductive group defined over k.

(1) The maximal k-split tori of G are conjugate over k (i.e., by elements of G_k). If S is such a maximal k-split torus, the dimension of S is called the k-rank of G (notation: $rk_k(G)$). $Z(S)$ is the connected component of $N(S)$. The finite group $N(S)/Z(S)$ is called the Weyl group of G relative to k (notation: $_kW(G)$). Every coset of $N(S)/Z(S)$ is represented by an element rational over k: $N(S) = N(S)_k Z(S)$.

(2) The elements of $\Phi(G, S)$, where S is a maximal k-split torus are called the k-roots, or roots relative to k. We write $_k\Phi$ or $_k\Phi(G)$ for $\Phi(G, S)$. This is a root system in $(\hat{S} \otimes \mathbf{R}, M)$ where M is the vector space over $\mathbf{R}$ generated by the characters which are trivial on $S \cap \mathscr{D}(G)$. The Weyl group of G relative to k and the Weyl group of $_k\Phi$ are isomorphic:

$$W(_k\Phi) \cong {}_kW(G).$$

If G is simple over k, $_k\Phi$ is irreducible.

(3) The minimal parabolic k-subgroups P of G are conjugate over k. Furthermore there exists a k-split torus S such that

$$P = Z(S) \cdot R_u(P)$$

where the semidirect product is algebraic and everything is defined over k. If P and P' are minimal parabolic k-subgroups containing a maximal k-split torus S, then $P \cap P'$ contains the centralizer of S. The minimal parabolic k-subgroups containing $Z(S)$ are in $(1, 1)$ correspondence with the Weyl chambers: P corresponds to the Weyl chamber C if the Lie algebra of $R_u(P)$ is $\sum_{\alpha > 0} \mathfrak{g}_\alpha^{(S)}$, where the ordering of the roots is associated to the Weyl-chamber C. The Weyl group $_kW(G)$ permutes in a simply transitive way the minimal parabolic k-subgroups containing $Z(S)$. The unipotent radical of a minimal parabolic k-subgroup is a maximal unipotent k-subgroup, at least for a field of characteristic 0.

(4) Bruhat decomposition of G_k. Put $V = R_u(P)$, where P is a minimal parabolic k-subgroup. Then

$$G_k = U_k \cdot N(S)_k \cdot U_k,$$

and different elements of $N(S)_k$ define different double cosets; more generally if $n, n' \in N(S)$: $UnU = Un'U \Leftrightarrow n = n'$. Choose for every $w \in {}_kW$ a representative $n_w \in N(S)_k$; then the above equality can be written as

$$G_k = \bigcup_{w \in {}_kW} U_k \cdot n_w \cdot P_k,$$

the union being disjoint.

One can phrase this decomposition in a more precise way. If we fix $w \in {}_kW$, then there exist two k-subgroups U'_w and U''_w of U, such that $U = U'_w \times U''_w$ as an algebraic variety and such that the map of $U'_w \times P$ onto Un_wP sending (x, y) onto xn_wy is a biregular map defined over k. This decomposition gives rise to a cellular decomposition of G_k/P_k. Let π be the projection of G onto G/P. Then

$$(G/P)_k = G_k/P_k = \bigcup_{w \in {}_kW} \pi(U'_{w,k}).$$

If k is algebraically closed, $U'_{w''}$ as a unipotent group is isomorphic to an affine space. So one gets a cellular decomposition of G/P.

(5) *Standard parabolic k-subgroups* (with respect to a choice of S and P). Let $_k\Phi$ be the root system of G relative to k defined by the torus S. The choice of the minimal parabolic k-subgroup P determines a Weyl chamber of $_k\Phi$ and so a set of positive roots. Let $_k\Delta$ be the set of simple k-roots for this ordering. If Θ is a subset of $_k\Delta$, denote by S_Θ the identity component of $\bigcap_{\alpha \in \Theta} \ker \alpha$. S_Θ is a k-split torus, the dimension of which is dim $S_\Theta = rk_k(G) -$ card Θ. The standard parabolic k-subgroup defined by Θ is then the subgroup $_kP_\Theta$ generated by $Z(S_\Theta)$ and U. That subgroup can be written as the semidirect product $Z(S_\Theta) \cdot U_\Theta$, where $U_\Theta = R_u(P_\Theta)$. The Lie algebra of U_Θ is $\sum \mathfrak{g}_\alpha$, the sum going over all positive roots that are not linear combination of elements in Θ.

(6) Every parabolic k-subgroup is conjugate over k to one and only one standard parabolic k-subgroup. In particular, if two parabolic k-subgroups are conjugate over Ω, they are already conjugate over k.

(7) Let W_Θ be the subgroup of the Weyl group $_kW$ generated by the reflections s_α for $\alpha \in \Theta$. Then if Θ and Θ' are two subsets of $_k\Delta$,

$$_kP_{\Theta,k}\backslash G_k/_kP_{\Theta',k} \cong W_\Theta\backslash_kW/W_{\Theta'}.$$

6.6. EXAMPLES. (1) $G = \mathbf{GL}(n)$,

$$S = \text{group of diagonal matrices} = \left\{ \begin{pmatrix} s^{\lambda_1} & & & 0 \\ & s^{\lambda_2} & & \\ & & \ddots & \\ 0 & & & s^{\lambda_n} \end{pmatrix} \right\}$$

where $\lambda_i \in \hat{S}$ is such that $s^{\lambda_i} = s_{ii}$. S is obviously a split torus and is maximal. A minimal parabolic k-subgroup P is given by the upper triangular matrices, which is in this case a Borel subgroup. The unipotent radical U of P is given by the group of upper triangular matrices with ones in the diagonal. If e_{ij} is the matrix having all components zero except that with index (i,j) equal to 1, $\mathrm{Ad}_G\, s(e_{ij}) = (s^{\lambda_i}/s^{\lambda_j})e_{ij}$. So the positive roots are $\lambda_i - \lambda_j$ $(i < j)$ since the Lie algebra of U is generated by e_{ij} $(i < j)$. The simple roots are $(\lambda_1 - \lambda_2, \lambda_2 - \lambda_3, \cdots, \lambda_{n-1} - \lambda_n)$. The Weyl group is generated by s_α, where α is a positive root; since for $\alpha = \lambda_i - \lambda_j$, s_α permutes the i and j axis, $_kW \cong \mathfrak{S}_n$, the group of permutations of the basis elements. The parabolic subgroups are the stability groups of flags.

(2) G "splits over k" (i.e., G has a maximal torus which splits over k). Example (1) enters in this category. The k-roots are just the usual roots. A minimal parabolic k-subgroup is a maximal connected solvable subgroup. If k is algebraically closed G always splits over k and this gives just the usual properties of semisimple or reductive linear groups.

(3) G is the orthogonal group $SO(F)$ of a nondegenerate quadratic form F on a vector space V_k (where, to be safe, one takes char $k \neq 2$). In a suitable basis

$$F(x_1, \cdots, x_n) = x_1x_n + x_2x_{n-1} + \cdots + x_qx_{n-q+1} + F_0(x_{q+1}, \cdots, x_{n-q})$$

where F_0 does not represent zero rationally. The index of F, the dimension of the maximal isotropic subspaces in V_k, is equal to q. A maximal k-split torus S is given by the set of following diagonal matrices:

$$\begin{pmatrix} s^{\lambda_1} & & & & & & & & \\ & s^{\lambda_2} & & & & & & & \\ & & \ddots & & & & & & \\ & & & s^{\lambda_q} & & & & & \\ & & & & 1 & & & & \\ & & & & & \ddots & & & \\ & & & & & & 1 & & \\ & & & & & & & s^{-\lambda_q} & \\ & & 0 & & & & & & s^{-\lambda_{q-1}} \\ & & & & & & & & \ddots \\ & & & & & & & & & s^{-\lambda_1} \end{pmatrix}$$

Let $SO(F_0)$ denote the proper orthogonal group of the quadratic form F_0, imbedded in $SO(F)$ by acting trivially on $x_1, \cdots, x_q,\ x_{n-q+1}, \ldots, x_n$. Then $Z(S) = S \times SO(F_0)$. The minimal parabolic k-subgroups are the stability groups of the full isotropic flags. For the above choice of S, and ordering of the coordinates, the standard full isotropic flag is

$$[e_1] \subset [e_1, e_2] \subset \cdots \subset [e_1, \cdots, e_q].$$

The corresponding minimal parabolic k-subgroup takes then the form

$$P = \left\{ \begin{pmatrix} A_0 & A_1 & A_2 \\ 0 & B & A_3 \\ 0 & 0 & A_4 \end{pmatrix} \right\}$$

where A_0 and A_4 are upper triangular $q \times q$ matrices, $B \in SO(F_0)$, with additional relations that insure that $P \subset SO(F)$. The unipotent radical U of P is the set of matrices in P, where $B = I$, A_0, A_4 are unipotent, and

$$A_4 = {}^{\sigma}A_0^{-1}; \qquad Q \cdot A_3 + {}^{t}A_1 \cdot J \cdot A_4 = 0,$$

$$ {}^{t}A_4 \cdot J \cdot A_2 + {}^{t}A_3 \cdot Q \cdot A_3 + {}^{t}A_2 \cdot J \cdot A_4 = 0,$$

where Q is the matrix of the quadratic form F_0, J is the $q \times q$ matrix with one's in the nonprincipal diagonal and zeros elsewhere, and σ is the transposition with respect to the same diagonal, $({}^{\sigma}M = J^{t}MJ)$. To determine the positive roots, one has to let S operate on U. To compute the root spaces it is easier to diagonalize $Q: q_{ij} = d_i \cdot \delta_{ij}$. Three cases are to be considered.

$i < j \leqq q$; $\lambda_i - \lambda_j$ is a root; the corresponding root space is generated by $e_{ij} - e_{n-j+1, n-i+1}$; the multiplicity of the root is 1.

$i \leqq q < j \leqq n - q$; λ_i is a root with multiplicity $n - 2q$; the corresponding root space is generated by

$$e_{ij} - d_j^{-1} e_{j, n-i+1} \qquad (q + 1 \leqq j < n - q).$$

$i < j \leqq q$; $\lambda_i + \lambda_j$ is a root with multiplicity one; the corresponding root space is generated by $e_{i, n-j+1} - e_{j, n-i+1}$. The simple roots are

$$\lambda_1 - \lambda_2, \lambda_2 - \lambda_3, \cdots, \lambda_{q-1} - \lambda_q,$$

and λ_q if $n \neq 2q$, $\lambda_{q-1} + \lambda_q$ if $n = 2q$. The Weyl group consists of all products of permutation matrices with symmetries with respect to a coordinate subspace (of any dimension if $n \neq 2q$ of even dimension if $n = 2q$). The group $SO(F)$ splits if and only if $q = [n/2]$. If it does not split, there exist roots with multiplicity > 1. The parabolic k-subgroups are the stability subgroups of rational isotropic flags. The parabolic k-subgroups are conjugate over k if and only if there exists an element of G_k mapping one flag onto the other; by Witt's theorem this is possible if and only if the two flags have the same type.

(4) When one starts with a hermitian form, the same considerations apply, except that one gets a root system of type BC_q.

(5) For real Lie groups, this theory is closely connected with the Iwasawa and Cartan decompositions. If $\mathfrak{g}$ is the real Lie algebra of G_R, G being a connected algebraic reductive group, then $\mathfrak{g} = \mathfrak{k} + \mathfrak{p}$, where $\mathfrak{g}$ is the Lie algebra of a maximal compact subgroup of G_R. Then $G = K \cdot (\exp \mathfrak{n})$. Let $\mathfrak{a}$ be a maximal commutative subalgebra of $\mathfrak{n}$. Then $A = \exp \mathfrak{a}$ is the topological connected component of the groups of real points in a torus S which is maximal among R-split tori. (On the Riemannian symmetric space G_R/K, it represents a maximal totally geodesic flat subspace.)

$$N(A)_R = N(S)_R = [K \cap N(A)] \cdot A,$$

and

$$Z(S)_R = (K \cap Z(A)) \cdot A;$$

the group $K \cap Z(A)$ is usually denoted by M. The Weyl group $_RW(G, S)$ is isomorphic to $(K \cap N(A))/M$, i.e., to the Weyl group of the symmetric space G/K as introduced by E. Cartan. Similarly $_R\Phi$ may be identified to the set of roots of the symmetric pair (G, K). Let $\mathfrak{n}$ be the Lie subalgebra generated by the root spaces corresponding to positive roots $\mathfrak{n} = \sum_{\alpha > 0} \mathfrak{n}_\alpha^{(S)}$, $\alpha \in {}_R\Phi(G, S)$, for some ordering. Let $N = \exp \mathfrak{n}$. Then $G = K \cdot A \cdot N$ is an Iwasawa decomposition and $M \cdot A \cdot N$ is the group of real points of a minimal parabolic R-group. Assume G_R simple and G/K to be a bounded symmetric domain. Then there are two possibilities for the root system $_R\Phi$:

$$G_R/K \text{ is a tube domain} \Leftrightarrow {}_R\Phi \text{ is of type } C_t,$$

$$G_R/K \text{ is not a tube domain} \Leftrightarrow {}_R\Phi \text{ is of type } BC_t.$$

7. **Representations in characteristic zero [3].** We assume here the ground field to be of characteristic zero, and G to be semisimple, connected. Let $P = Z(S) \cdot U$ be a minimal parabolic k-group, where $U = R_u(P)$, and S is a maximal k-split torus. We put on $X(S)$ an ordering such that u is the sum of the positive k-root spaces.

Assume first k to be algebraically closed. Let $\rho : G \to \mathrm{GL}(V)$ be an irreducible representation. It is well known that there is one and only one line $D_\rho \subset V$ which is stable under P. The character defined by the 1-dimensional representation of P in V is the highest weight λ_ρ of ρ. The orbit $G(D_\rho) = \mathscr{C}_\rho$ is a closed homogeneous cone (minus the origin). The stability group of D_ρ is a standard parabolic group $P_\rho \supset P$. The stability groups of the lines in $\mathscr{C}_\rho$ are conjugate to P_ρ, and these lines are the only ones to be stable under some parabolic subgroup of G. Every highest weight λ_ρ is a sum $\lambda_\rho = \sum_{\alpha \in \Delta} c_\alpha \cdot \Lambda_\alpha (c_\alpha \geq 0, c_\alpha \in Z)$ of the fundamental highest weights Λ_α (and conversely if G is simply connected), where Λ_α is defined by $2(\Lambda_\alpha, \beta) \cdot (\beta, \beta)^{-1} = \delta_{\alpha\beta} (\alpha, \beta \in \Delta)$.

We want to indicate here a "relativization" of these facts for a nonnecessarily algebraically closed k.

Let T be a maximal torus of G, defined over k, containing S. We choose an ordering on $X(T)$ compatible with the given one on $X(S)$ (i.e., if $\alpha > 0$, and $r(\alpha) \neq 0$, then $r(\alpha) > 0$ where $r: X(T) \to X(S)$ is the restriction homomorphism. The k-weights of ρ are the restrictions to S of the weights of ρ with respect to T; the highest k-weight μ_ρ is the restriction of λ_ρ. It follows from standard facts that every k-weight μ is of the form

$$\mu = \mu_\rho - \sum m_\alpha(\mu)_\alpha, \qquad (\alpha \in \Delta),$$

with

$$m_\alpha(\mu) \in \mathbf{Z}, \qquad m_\alpha(\mu) \geq 0.$$

Let

$$\Theta(\mu) = \{\alpha \in {}_k\Delta \,|\, m_\alpha(\mu) \neq 0\}.$$

Then $\Theta \subset {}_k\Delta$ is a $\Theta(\mu)$, for some k-weight μ, if and only if $\Theta(\mu) \cup \mu_\rho$ is connected.

Let us say that ρ is *strongly rational* over k if it is defined over k and if the cone $G(D_\rho)$ has a rational point over k. This is the case if and only if the above coefficients $\subset_\alpha$ satisfy the following conditions:

$$\subset_\alpha = 0 \text{ if } r(\alpha) = 0, \qquad \subset_\alpha = \subset_\beta \text{ if } r(\alpha) = r(\beta) \qquad (\alpha, \beta \in \Delta).$$

The highest weight of a strongly rational representation is a sum, with positive integral coefficients, of fundamental highest weights $M_\beta(\beta \in {}_k\Delta)$ where

$$M_\beta = \sum_{\alpha \in \Delta, r(\alpha) = \beta} r(\Lambda_\alpha)$$

(and conversely if G is simply connected). The $M_\beta(\beta \in {}_k\Delta)$ satisfy relations of the form $(M_\beta, \gamma) = d_\beta \cdot \delta_{\beta, \gamma}$, with $d_\beta > 0$. They will be called the *fundamental highest k-weights*.

Assume $k \subset \mathbf{C}$. Let ρ be strongly rational over k. Put on the representation space a Hilbert structure. Let $v \in D_\rho - 0$. Then the function $\phi: G \to \mathbf{R}^+$ defined by

$$\phi(g) = \|\rho(g) \cdot v\|,$$

satisfies

$$\phi(g \cdot p) = \phi(g)|p^{\mu_\rho}| \qquad (g \in G, \quad p \in P_\rho).$$

If in particular $k = \mathbf{Q}$, such functions appear in the discussion of fundamental sets and of Eisenstein series for arithmetic groups.

References

1. A. Borel, *Groupes linéaires algébriques*, Ann. of Math. (2) **64** (1956), 20–80.

2. A. Borel and T. A. Springer, *Rationality properties of linear algebraic groups*, Proc. Sympos. Pure Math., vol. 9, Amer. Math. Soc., Providence, R.I., 1966, pp. 26–32.

3. A. Borel and J. Tits, *Groupes réductifs*, Publ. I.H.E.S. **27** (1965), 55–150.

4. C. Chevalley, *Séminaire sur la classification des groupes de Lie algébriques*, 2 vols., Inst. H. Poincaré, Paris, 1958. (Mimeographed notes)

5. M. Demazure and A. Grothendieck, *Schémas en groupes*, I.H.E.S. Bures-sur-Yvette, France, 1964. (Mimeographed notes)

6. R. Godement, *Groupes linéaires algébriques sur un corps parfait*, Sém. Bourbaki, Exposé 206, Paris, 1960.

7. M. Rosenlicht, *Some basic theorems on algebraic groups*, Amer. J. Math. **78** (1956), 401–443.

8. ———, *Some rationality questions on algebraic groups*, Annali di Mat. (IV) **43** (1957), 25–50.

9. I. Satake, *On the theory of reductive groups over a perfect field*, J. Math. Soc. Japan **15** (1963), 210–235.

(From Notes by F. Bingen)

74.

Reduction theory for arithmetic groups

Proc. Symp. Pure Math. **9**, Amer. Math. Soc. (1966) 20–25

This lecture is devoted to the statement of results concerning fundamental sets for arithmetic groups. The notation of [3] is used.

1.1. **Arithmetic groups.** We recall that two subgroups A, B of a group C are commensurable if $A \cap B$ has finite index in A and in B.

DEFINITION. Let G be an algebraic Q-group. A subgroup Γ of G_Q is called an *arithmetic group* (or an arithmetic subgroup of G) if there exists a faithful rational representation $\rho : G \to GL_n$ defined over Q such that $\rho(\Gamma)$ is commensurable with $\rho(G) \cap GL(n, Z)$. (The same condition is then automatically fulfilled for every faithful Q-representation of G.)

The arithmetic group Γ is a discrete subgroup of G_R. It will act on G_R by right translations. If $X = K \backslash G_R$, where K is a maximal compact subgroup of G_R, then Γ operates also on X as a properly discontinuous group of translations.

One could apparently generalize the definition of arithmetic groups by starting from a number field k, an algebraic group G defined over k, a faithful k-representation $\rho : G \to GL_n$ and taking as arithmetic group a subgroup of G_k commensurable with $G_{\mathfrak{o}_k} = GL_n(n, \mathfrak{o}_k) \cap \rho(G)$, where $\mathfrak{o}_k$ is the ring of integers of k. But this class of arithmetic groups is the same as the one which was first defined. Indeed if $G' = R_{k/Q}G$ is obtained from G by restriction of the ground field from k to Q (see [8]), using a basis of $\mathfrak{o}_k$ over Z, it is easy to see that $G'_Z \cong G_{\mathfrak{o}_k}$. On the other hand Γ will usually not be discrete in G_C.

1.2. THEOREM. *Let $\rho : G \to G'$ be a surjective Q-morphism of algebraic groups. If Γ is an arithmetic subgroup of G, then $\rho(\Gamma)$ is an arithmetic subgroup of G'.*

This is proved for isogenies in [4, §6], for general Q-morphisms in [2]. Two simple consequences are:

(1) Let $G = H \cdot N$ be a semidirect product defined over Q. Then $\Gamma_1 \cdot \Gamma_2$ is an arithmetic subgroup of G if Γ_1 is arithmetic in H and Γ_2 in N.

(2) Let G be the almost direct product of two normal Q-subgroups G_1 and G_2 (i.e., G is Q-isogeneous to $G_1 \times G_2$). If Γ is an arithmetic subgroup of G, then $\Gamma_i = \Gamma \cap G_i$ is arithmetic in G_i and $\Gamma_1 \cdot \Gamma_2$ is commensurable with Γ.

DEFINITION. Let Γ be an arithmetic group in G. The subgroup

$$C(\Gamma) = \{g \in G_R | g \cdot \Gamma \cdot g^{-1} \text{ commensurable with } \Gamma\},$$

is called the *commensurability subgroup* of Γ. One has always that $G_Q \subset C(\Gamma)$.

If G_R is compact, Γ is finite, every conjugate of Γ is commensurable to Γ, and so $C(\Gamma) = G_R$.

1.3. THEOREM [2]. *Let N be the greatest normal Q-subgroup of a semisimple algebraic Q-group G, such that N_R is compact. If π is the projection of G onto $G' = G/N$, then $C(\Gamma) = \pi^{-1}(G'_Q) \cap G_R$.*

For instance, if G is Q-simple, with center reduced to $\{e\}$, then $C(\Gamma) = G_Q$. However, if G_C has a nontrivial center, this need not be so, as is already seen in the case where

$$G = SL_n, \qquad \Gamma = SL_{n,Z}.$$

1.4. **Fundamental sets for arithmetic groups.** Let G be an algebraic Q-group and Γ an arithmetic subgroup of G.

DEFINITION. A subset Ω of G_R is called a *fundamental set* for Γ if:

(F0) $K\Omega = \Omega$, where K is some maximal compact subgroup of G_R;

(F1) $\Omega \cdot \Gamma = G_R$;

(F2) for any g in $C(\Gamma)$, the set of translates $\Omega\gamma$, $\gamma \in \Gamma$, that meet $\Omega \cdot g$ is finite.

One could replace (F2) by the weaker condition:

(F2') $\{\gamma \in \Gamma | \Omega \cap \Omega \cdot \gamma \neq \varnothing \}$ is finite.

But the stronger condition ensures that when one has a fundamental set Ω for Γ one can construct a fundamental set Ω' for a commensurable subgroup Γ' by taking $\Omega' = \bigcup_{\xi \in \Gamma'/\Gamma \cap \Gamma'} \Omega\xi$. The condition (F2) then goes over to Γ'. This would apparently not be the case with the weaker condition (F2').

Due to condition (F0), the projection $\Omega' = \pi(\Omega)$ of a fundamental set Ω in G_R into $X = K\backslash G_R$ satisfies the conditions

$$(\text{F1})_X \qquad\qquad \Omega' \cdot \Gamma = X,$$

and (F2). A subset of X verifying $(\text{F1})_X$ and (F2) will be called a fundamental set for Γ in X. Thus Ω is a fundamental set for Γ in G_R if and only $\Omega' = \pi(\Omega)$ is one in X, and then $\Omega = \pi^{-1}(\Omega')$.

If G is unipotent, then G_R/Γ is compact. Since G is the semidirect product $G = H \cdot R_u(G)$ of a reductive Q-group and of its unipotent radical, it follows from (1.2) that the discussion of fundamental sets is easily reduced to the case of reductive groups, or of semisimple groups and tori.

When G_R/Γ is compact, there is often no need to have more information about the shape of a fundamental set. The purpose of the reduction theory outlined here is (a) to give a criterion for compactness, (b) in the noncompact case, to describe fundamental sets in which the complement of big compact subsets is a union of subsets which have properties similar to those of the cusps of fundamental domains for fuchsian groups.

1.5. THEOREM [4, 7]. *Let G be a Q-group. G_R/Γ is compact if and only if $\hat{G}_Q^0 = 0$ and every unipotent element of G_Q belongs to $R_u(G)$.*

Let in particular G be reductive. Then $G_\mathbf{R}/\Gamma$ is compact if and only if G is anisotropic over $\mathbf{Q}$.

EXAMPLES. (1) $SO(F)$ where F is an anisotropic quadratic form with rational coefficients.

(2) Let G' be the multiplicative group of a finite extension field k of $\mathbf{Q}$. $G = R_{k/\mathbf{Q}}G'$ is an algebraic group defined over $\mathbf{Q}$. It contains a $\mathbf{Q}$-subgroup N consisting of all elements of k of norm 1. The group N is anisotropic and so N/Γ, where Γ is an arithmetic subgroup of N, is compact. This is equivalent with the main part of Dirichlet's unit theorem.

1.6. **Siegel domains.** Siegel domains are defined in the cases where $G_\mathbf{R}/\Gamma$ is not compact. The reductive group G contains then a nontrivial maximal $\mathbf{Q}$-split torus S. Let P be a minimal parabolic subgroup containing S. The group P is different from G and $P = Z(S) \cdot U = M \cdot S \cdot U$, where U is the unipotent radical of P, $M \cap S$ is finite, M is reductive and anisotropic over $\mathbf{Q}$ (in particular every $\mathbf{Q}$-character of P or $Z(S)$ is trivial on M). Denote by ${}_\mathbf{Q}\Phi$ the sets of roots of G with respect to S. The choice of P orders this set; let ${}_\mathbf{Q}\Delta$ be the set of simple roots of ${}_\mathbf{Q}\Phi$ with respect to this ordering. For every $t \in \mathbf{R}^+$, define in $A = S_\mathbf{R}^0$ the subset

$$A_t = \{a \in A \,|\, a^\alpha \leqq t \text{ for every } \alpha \in {}_\mathbf{Q}\Delta\}.$$

Since every positive root is a positive linear combination of simple roots, there exists a $C > 0$ such that also $a^\alpha \leqq C$ for every positive root α in ${}_\mathbf{Q}\Phi$. A fundamental property of A_t is expressed by the following:

1.7. LEMMA. *If w is a compact set in $(M \cdot U)_\mathbf{R}$, then the set*

$$\{awa^{-1} \,|\, a \in A_t\}$$

is relatively compact.

PROOF. w is contained in a product $w_1 \cdot w_2$ where w_1 is compact in M and w_2 is compact in U. Since M centralizes S, $awa^{-1} \subset w_1 a w_2 a^{-1}$ and it is enough to prove the lemma for w compact in U. Since U is unipotent over a field of characteristic zero, the logarithm is a bijection of U onto its Lie algebra $\mathfrak{u}$. If $u \in U$, $\log u = \sum_{\alpha > 0} c_\alpha \cdot X_\alpha$ and $\log(a \cdot u \cdot a^{-1}) = \sum_{a > 0} a^\alpha \cdot c_\alpha \cdot X_\alpha$. But a^α for $\alpha > 0$ stays bounded in A_t. So $\log(a \cdot u \cdot a^{-1})$ stays bounded as $a \in A_t$ and $u \in w$. The exponential being continuous, this proves the lemma.

1.8. **Siegel domains.** We keep the same notation as above. Let K be a maximal compact subgroup of $G_\mathbf{R}$ such that the Lie algebra of K is orthogonal to that of $S_\mathbf{R}$. Let w be a compact neighborhood of e in $(M \cdot U)_\mathbf{R} = M_\mathbf{R} \cdot U_\mathbf{R}$. The subset $\mathfrak{S} = K \cdot A_t \cdot w$ is called a Siegel domain for G. If $\pi: G \to X = K\backslash G_\mathbf{R}$, then $\pi(\mathfrak{S}) = \mathfrak{S}' = \sigma \cdot \mathfrak{S}$ (where o is the coset K) is called a Siegel domain in X. The set $(K \cap P) \cdot A_t \cdot w$ is a Siegel domain in $P_\mathbf{R}$. Since $G_\mathbf{R}$ is generated by K and $P_\mathbf{R}$, one has then $\mathfrak{S}' = o \cdot \mathfrak{S} = o \cdot A_t \cdot w$.

EXAMPLES. (1) $G = SL(2, \mathbf{R})$, $\Gamma = SL(2, \mathbf{Z})$, and $X = SO(2, \mathbf{R})\backslash SL(2, \mathbf{R})$ is the upper half plane. To be in agreement with the rest of this lecture, we let G act on X on the right by putting

$$z \cdot g = (a \cdot z + c) \cdot (b \cdot z + d)^{-1}, \qquad \left(g = \begin{pmatrix} a & b \\ c & d \end{pmatrix} \in \mathbf{SL}(2, \mathbf{R}) \right).$$

Let

$$P = \left\{ \begin{pmatrix} a & 0 \\ c & a^{-1} \end{pmatrix} \right\}, \qquad S = \left\{ \begin{pmatrix} a & 0 \\ 0 & a^{-1} \end{pmatrix} \right\}, \qquad U = \left\{ \begin{pmatrix} 1 & 0 \\ c & 1 \end{pmatrix} \right\}.$$

The positive root corresponding to this choice of U is $\alpha(a) = a^{-2}$. The fixed point of K is i. Let w be the set of elements in U for which $|c| \leq C$. Then $\mathfrak{S}'$ is the rectangular domain:

$$\mathfrak{S}' = i \cdot A_t \cdot w = \{ z \in X, |\operatorname{Re} z| \leq C, \operatorname{Im} z \geq t^{-1} \}.$$

It contains the classical fundamental domain $|z| \geq 1$, $\operatorname{Re} z \leq \frac{1}{2}$ if $C \geq \frac{1}{2}$ and $t^2 \geq \frac{4}{3}$.

(2) $G = \mathbf{GL}(n, \mathbf{R})$, $K = \mathbf{O}(n, \mathbf{R})$, S (resp. U, resp. P) group of diagonal (resp. unipotent upper triangular, resp. upper triangular) matrices. Then X is the space of positive nondegenerate quadratic forms in n variables. A_t is the set of diagonal matrices with positive entries a_i verifying $a_i/a_{i+1} \leq t$. The natural projection of G onto $X = K \backslash G$ is the map $g \mapsto {}^t g \cdot g$. The image of the Siegel domain $K \cdot A_t \cdot w$ in X is then the set of matrices

$$ {}^t u \cdot a \cdot u \qquad (u \in w; a \in A_{t^2}).$$

It is well known to be a fundamental set if $t^2 \geq \frac{4}{3}$, and if w contains all matrices $u = (u_{ij}) \in U$ such that $|u_{ij}| \leq \frac{1}{2}$ $(i \neq j)$.

1.9. LEMMA. *A Siegel domain has finite Haar measure.*

PROOF. We have

$$G_\mathbf{R} = K \cdot P_\mathbf{R} = K \cdot A \cdot (M \cdot U)_\mathbf{R}.$$

The second decomposition is not unique but determined up to an element of the compact group $K \cap M$. By standard facts on Haar measures, we have:

$$\int_{\mathfrak{S}} dg = \int_{K \cdot A \cdot (M \cdot U)_\mathbf{R}} \phi \, dg$$

$$= c \cdot \int_K dk \cdot \int_{A_t \cdot w} a^\chi \, da \, dv$$

where ϕ is the characteristic function of $\mathfrak{S}$, dv is the Haar measure of $M \cdot U$ and $\chi = \det Ad_u a$. We have $\chi = \sum_{\alpha \in \Delta} c_\alpha \alpha$ and $c_\alpha > 0$. The only integral one has to evaluate to prove the finiteness is that extended over A_t; up to a constant factor, it is a product over $\alpha \in {}_\mathbf{Q}\Delta$ of integrals of the form

$$\int_{-\infty}^t \exp(c_\alpha x) \, dx,$$

which are finite since $c_\alpha > 0$.

1.10. THEOREM. *Let G be a connected semisimple algebraic group defined over Q, and Γ an arithmetic subgroup of G.*

(1) *There exists a finite subset C of G_Q and a Siegel domain $\mathfrak{S}$ such that $\Omega = \mathfrak{S}C$ is a fundamental set in G_Q for Γ. The set C contains then at least one representative for every double coset $\Gamma\backslash G_Q/P_Q$. (In particular the number of such double cosets is finite.)*

(2) *Conversely if C is a finite subset of G_Q containing a representative for every double coset of $\Gamma\backslash G_Q/P_Q$, then there exists a Siegel domain $\mathfrak{S}$ such that $\Omega = \mathfrak{S}\cdot C$ is a fundamental set in G_R for Γ.*

It follows immediately from the lemma and the theorem that G_R/Γ has finite invariant volume.

1.11. THEOREM. *Let G be an algebraic group defined over Q, Γ an arithmetic subgroup of G_Q. Then G_R/Γ has finite volume if and only if G_Q^0 has no characters defined over Q. $(\hat{G}_Q^0 = \{0\}.)$*

Writing $G^0 = H \cdot Z \cdot U$, where U is the unipotent radical and Z the central torus of a maximal reductive subgroup, the condition states that Z is anisotropic over Q.

Theorem 1.11 is proved in [4] and Theorem 1.10 is announced in [1]. For proofs which are different from those of [1, 4], see [6].

In Ω, the complement of a compact set is the union of sets of the form $KA_r \cdot w \cdot c$ ($c \in G_Q$, $r > 0$, r sufficiently small). These are to be viewed as the analogues of the "cusps" for fuchsian groups. The minimum number of cusps is then the number of elements in $\Gamma\backslash G_Q/P_Q$ or, equivalently, the number of conjugacy classes of minimal parabolic Q-subgroups under Γ.

EXAMPLES. If F is a quadratic form defined over Q and $G = SO(F)$, P_Q is the stability group of a full isotropic flag. Then the minimal number of cusps for a fundamental set Ω is the number of transitivity classes of full isotropic flags under Γ. The same is true for $Sp(n)$. Since $Sp(n, Z)$ is transitive on the full isotropic flags, the minimum number of cusps for the modular group is 1, as is well known by Siegel's construction of a fundamental domain in this case. More generally, [2, Lemma 1], we have $G_Q = G_Z \cdot P_Q$ if G splits over Q, and G_Z is the group of integral points for the canonical Z-structure on G introduced by Chevalley, and described in [5]. In this case, there is only one cusp.

1.12. **Minimum principles connected with fundamental domains.**

EXAMPLE. 1. Let $X = \{Z \in M(n, C) | {}^t Z = Z, \operatorname{Im} Z > 0\}$ be the Siegel upper half plane, $\mathfrak{S}$ a Siegel domain which is a fundamental domain for the modular group Γ. Consider for fixed $Z \in X$ the function $f(Z\gamma) = \det(\operatorname{Im} Z \cdot \gamma)^{-1}$ defined on Γ. It is well known that this function has a minimum on $Z \cdot \Gamma$ and that this minimum is taken in a point of $Z \cdot \Gamma \cap \mathfrak{S}$.

2. Let $G = GL(n)$, and $\mathfrak{S}$ a suitable big Siegel domain in G_R. It is known from the Hermite or the Minkowski reduction theory that, if $\|x\|$ denotes the length of the vector $x \in R^n$, then for every $g \in R$, the function $f_g(\gamma) = \|g\gamma(e_1)\|$ from $GL(n, Z)$ to R^+ attains a minimum in $\mathfrak{S}$.

Such minimum principles hold for every semisimple algebraic group defined over Q. Let P be a minimal parabolic Q-subgroup of G; $_Q\Delta$ the set of simple roots of G relative to this choice of a minimal parabolic subgroup. Take in $\hat{P}$ a set of fundamental weights Λ_α such that $(\Lambda_\alpha, \beta) = d_\alpha\delta_{\alpha\beta}$, $(\alpha, \beta \in {}_Q\Delta)$, and $d_\alpha > 0$.

1.13. THEOREM. *Let* $\chi \in \hat{P}$ *with* $\chi = \sum c_\alpha\Lambda_\alpha$ $(c_\alpha \geq 0)$ *and* f *a function from* G_R *to* R^+ *satisfying* $f(x, p) = f(x)|p^\chi|$ $(p \in P_R)$. *Take for* C *a set of representatives of the double cosets of* $\Gamma \backslash G_Q/P_Q$. *Then there exists a Siegel domain* $\mathfrak{S}$ *in* G_R *such that for any* $x \in G_R$ *the function* $f_x(c, \gamma) = f(x \cdot c \cdot \gamma)$ *attains a minimum in* $C \cdot \Gamma \cap x^{-1} \cdot \mathfrak{S}$; *in other words, there exist* $c_0 \in C$, $\gamma_0 \in \Gamma$ *such that* $xc_0\gamma_0 \in \mathfrak{S}$ *and* $f(xc_0\gamma_0) \leq f(xc\gamma)$ $(c \in G', \gamma \in \Gamma)$.

The Minkowski reduction theory in GL_n makes use of $n - 1$ successive minima. This approach was generalized to adèle groups of arbitrary semisimple groups in [6]. The number of successive minima is equal to $rk_Q(G)$, and the functions which are minimized are associated to fundamental strongly Q-rational representations, in the sense of [3, §7]. There is an analogue of this for G_R and Γ, which also generalizes §1.13, where, given an integer r in $(1 \leq r \leq rk_Q G)$, one takes successive minima of r functions. The two cases just mentioned then correspond to $r = 1$, $r = rk_Q G$. The formulation of this result is however more complicated than in the adèle case, because fundamental sets for Γ in G_R have in general more than one cusp. Details will be given in a future publication of the speaker.

REFERENCES

1. A. Borel, *Ensembles fondamentaux pour les groupes arithmétiques*, Coll. Théorie des groupes algébriques (Bruxelles 1962), pp. 23–40. Librairie Universitaire, Louvaine; Gauthier-Villars, Paris.

2. ——, *Density and maximality of arithmetic subgroups*, (to appear).

3. ——, *Linear algebraic groups*, Proc. Sympos. Pure Math., vol. 9, Amer. Math. Soc., Providence, R.I., 1966, pp. 3–19.

4. A. Borel and Harish-Chandra, *Arithmetic subgroups of algebraic groups*, Ann. of Math. (2) **75** (1962), 485–535.

5. P. Cartier, *Groups over Z*. Proc. Sympos. Pure Math., vol. 9, Amer. Math. Soc., Providence, R.I., 1966, pp. 90–98.

6. R. Godement, *Domaines fondamentaux des groupes arithmétiques*, Séminaire Bourbaki, vol. 15, (1962–1963), Exposé 257, Paris.

7. G. D. Mostow and T. Tamagawa, *On the compactness or arithmetically defined homogeneous spaces*, Ann. of Math. (2) **76** (1962), 446–463.

8. T. Tamagawa, *Adèles*, Proc. Sympos. Pure Math., vol. 9, Amer. Math. Soc., Providence, R.I., 1966, pp. 113–121.

(From Notes by F. Bingen)

<h1 style="text-align:center">75.</h1>

<h1 style="text-align:center">Introduction to automorphic forms</h1>

Proc. Symp. Pure Math. **9,** Amer. Math. Soc. (1966) 199–210

The classical notion of automorphic form in one complex variable is well known. Let Γ be a Fuchsian group: the function f, holomorphic in the upper half plane H, is called an automorphic form of weight $2k$ (k an integer) if for every $\gamma \in \Gamma$

$$f(z) = J_\gamma(z)^k \cdot f(\gamma \cdot z),$$

where

$$J_\gamma(z) = (cz + d)^{-2}, \qquad \gamma = \begin{pmatrix} a & b \\ c & d \end{pmatrix}, \qquad ad - bc = 1;$$

(and if it satisfies certain regularity conditions at the cusps of Γ, when H/Γ is not compact). This notion was generalized in two directions: on the one hand to holomorphic functions of several complex variables (Poincaré, Hilbert–Blumenthal, Siegel, etc.), on the other hand to nonholomorphic functions, for instance Eisenstein series of the form $\sum_{(c,d)=1} |cz + d|^{-s}$ where s is not necessarily an integer (Maass, Selberg, etc.). Our first aim here is to define a notion of automorphic form on a semisimple Lie group, which encompasses the two types just mentioned, introduced by Harish-Chandra [6].

1. **Definition of automorphic forms.** Let G be a real semisimple Lie group. To simplify matters, we assume that it is of finite index in the set of real points of a semisimple algebraic **R**-group. The universal enveloping algebra $U(\mathfrak{g})$ of the Lie algebra $\mathfrak{g}$ (with complex coefficients) can be identified with the algebra $D(G)$ of right invariant differential operators on G: to $y \in \mathfrak{g}$ is associated a differential operator such that

$$Yf(g) = \frac{d}{dt}f(\exp tY \cdot g)\bigg|_{t=0};$$

this linear map of $\mathfrak{g}$ into $D(G)$ extends to an isomorphism of $U(\mathfrak{g})$ onto $D(G)$. On G^0, the center $Z(\mathfrak{g})$ of $U(\mathfrak{g})$ corresponds then to the left and right invariant differential operators and is isomorphic with a polynomial ring in l letters, where l is the rank of G.

A vector valued function $f: G \to V$ is called $Z(\mathfrak{g})$-finite if $Z(\mathfrak{g}) \cdot f$ is a finite-dimensional vector space, or equivalently, if f is annihilated by an ideal I of $Z(\mathfrak{g})$

of finite codimension. The most important case is when I has codimension one, i.e. when f is an eigenfunction of every operator in $Z(\mathfrak{g})$.

DEFINITION. Let Γ be a discrete subgroup of G, K a maximal compact subgroup of G, ρ a representation of K in $GL(V)$, where V is a finite-dimensional complex vector space. The smooth vector valued function $f: G \to V$ is called an *automorphic form for* Γ if

(1) $f(k \cdot g \cdot \gamma) = \rho(k) \cdot f(g)$, f is left equivariant for K and right invariant for Γ.

(2) f is $Z(\mathfrak{g})$-finite.

(3) f satisfies a certain growth condition, to be specified later.

If I is an ideal of finite codimension in $Z(\mathfrak{g})$ which annihilates f, then f is called an automorphic form of type (ρ, I).

2. **Remark on the smoothness conditions.** The above definition makes sense for distributions. However, the $Z(\mathfrak{g})$-finiteness and the K-equivariance imply that f is a real analytic function:

PROOF: Let $\mathfrak{g} = \mathfrak{k} + \mathfrak{n}$ be a Cartan decomposition of $\mathfrak{g}$. Choosing orthonormal basis X_i of $\mathfrak{n}$ and Y_j of $\mathfrak{k}$ allows one to construct the Casimir operator $C = \sum X_i^2 - \sum Y_j^2$, which belongs to $Z(\mathfrak{g})$. The universal algebra $U(\mathfrak{k})$ is a subalgebra of $U(\mathfrak{g})$. For $D \in U(\mathfrak{k})$, $Df = d\rho(D)f$, due to the K-equivariance of f. Since ρ is a finite-dimensional representation, the function f is annihilated by an ideal of $U(\mathfrak{k})$ of finite codimension. Moreover, f being $Z(\mathfrak{g})$-finite, the vector space $Z(\mathfrak{g}) \cdot U(\mathfrak{k})f = W$ is finite dimensional. The operator $w = C + 2\sum Y_j^2 = \sum X_i^2 + \sum Y_j^2$ is elliptic and belongs to $Z(\mathfrak{g}) \cdot U(\mathfrak{k})$. Hence it keeps W invariant. Since W is finite dimensional, there exists a polynomial P such that $P(w)f = 0$. The operator $P(w)$ is an analytic elliptic operator, hence f is analytic.

3. **Geometric interpretation of condition** (1). G is a principal bundle with basis the Riemannian symmetric space $X = K \backslash G$ and structural group K. Consider the associated bundle $G \times_K V$ corresponding to the representation ρ of K in $GL(V)$: it is the quotient of $G \times V$ by the equivalence relation $(g, v) \approx (k \cdot g, \rho(k) \cdot v)$, and is a bundle with basis X and typical fibre V. Denote by $[G, V]_k$ the set of maps f from G to V that are K-equivariant: $f(kg) = \rho(k) \cdot f(g)$. The group G acts by right translations on $[G, V]_k$. The map f defines then a map $s: G \to G \times V$, such that $s(g) = (g, f(g))$, compatible with the equivalence relation defined by K. Going over to the quotient, s defines a cross section $\sigma: X \to G \times_k V$. This map is an isomorphism of $[G, V]_k$ onto the vector space $S[G \times_k V]$ of sections of $G \times_k V$. The group G operates on $S[G \times_k V]$ by right translations, and this isomorphism is compatible with the action of G. Condition (1) says then that f defines a Γ-invariant cross-section of $G \times_k V$.

4. **Automorphy factors and condition** (1). We shall start with a space X in a certain category (e.g. complex analytic, real analytic, C^∞ manifold $\cdots$), a group Γ of morphisms of X, and a group H acting on a space V. The group Γ will operate on the right on X, but not necessarily as a properly discontinuous transformation group. Denote by $[X, H]$ the set of morphisms of X into H

(i.e., holomorphic, real analytic, C^∞ maps $\cdots$). The group Γ acts on $[X, H]$ through $(\gamma \cdot \phi)(x) = \phi(x \cdot \gamma)$.

DEFINITIONS. (i). An *automorphy factor* of Γ is a 1-cocycle μ of F with value in $[X, H]$, i.e., a map $\mu : X \times \Gamma \to H$ such that

$$(1) \qquad \mu(x, \gamma \cdot \gamma') = \mu(x, \gamma) \cdot \mu(x\gamma, \gamma'), \qquad (\dot{x} \in X; \gamma, \gamma' \in \Gamma).$$

(ii). An *automorphic form of type* μ is a morphism $f : X \to V$ satisfying

$$(2) \qquad f(x) = \mu(x, \gamma) \cdot f(x \cdot \gamma), \qquad (x \in X, \gamma \in \Gamma).$$

The automorphy factor μ allows one to define an action of Γ on $X \times V$ by:

$$(3) \qquad (x, v) \cdot \gamma = (x \cdot \gamma, \mu(x, \gamma) \cdot v).$$

Due to the cocycle condition, Γ is indeed a transformation group on $X \times V$. If Γ operates freely on X as a properly discontinuous transformation group, $(X \times V)/\Gamma$ is a fibre bundle with fibre V and basis X/Γ. The automorphic forms are the cross sections of this bundle, lifted to X via the natural projection.

Assume now that $X = K\backslash G$ as above. To express the condition (2) above, and connect the automorphic forms with geometric objects on X/Γ, we need only to have an automorphy factor on Γ. However, in order to relate such an automorphic form to a function on G satisfying condition (1) of §1, we assume the automorphy factor to be defined on G.

Let 0 be the fixed point of K in X. Then for $k, k' \in K$

$$\mu(0, kk') = \mu(0, k) \cdot \mu(0, k').$$

Thus $\rho : k \mapsto \mu(0, k) = \rho(k)$, is a homomorphism of K into H. Let $\alpha : G \to H$ be the map defined by $\alpha(g) = \mu(0, g)$. Due to the cocycle condition, α is left equivariant for the representation ρ of $K : \alpha \in [G, H]_K$. Hence, it defines a cross-section of $P = G \times_K H$. The space P is obtained by identifying in $G \times H$ the pairs (g, h) and $(k \cdot g, \rho(k) \cdot h)$ $(k \in K)$. The space P is a principal bundle with basis $X = K\backslash G$ and fibre H. The cross-section α makes it possible to identify $G \times_K H$ with $X \times H$. Assuming V to be a vector space, $E = P \times_K V$ is the vector bundle over X associated to $P : E = P \times_H V$. The identification of P with $X \times H$, by means of α, allows one to identify E with $X \times V$ and every section of E with a map from X to V:

$$[G, V]_K \xrightarrow{\sim} [X, V].$$

If $\dot{g} = 0 \cdot g = K \cdot g$ denotes the image of the left K-coset of g in X, the inverse identification is given by $f \mapsto F$, where $F(g) = \alpha(g) \cdot f(\dot{g})$. This map is obviously a G-homomorphism, where the action of G is defined on $[G, V]_K$ by right translations, and on $[X, V]$ by §4 (3). The functions that appear in condition (1) of the definition of automorphic forms are then identified with the functions from X to V which satisfy (ii).

Conversely a cross-section σ of the principal bundle $P = G \times_K H$ defines, in the usual way, a left K-equivariant map $\alpha : G \to H$, and an automorphy factor

$\mu: X \times G \to H$, given by $\mu(\dot{g}, g') = \alpha(g)^{-1}$, $\alpha(g \cdot g') = \alpha(k \cdot g)^{-1} \cdot \alpha(k \cdot g \cdot g')$ where $g, g' \in G$, $k \in K$ and $\dot{g} = 0 \cdot g \in X$. It is again possible to identify $[G, V]_K$ with $[X, V]$, by means of μ.

5. **Example.** Let X be a bounded symmetric domain, $X = K \backslash G$. If $\mathfrak{g}$ and $\mathfrak{k}$ are the Lie algebras of G and K respectively, $\mathfrak{g}_C$ and $\mathfrak{k}_C$ their complexifications, then as vector space $\mathfrak{g}_C = \mathfrak{k}_C \oplus \mathfrak{p}_C = \mathfrak{k}_C \oplus \mathfrak{n}^+ \oplus \mathfrak{n}^-$, where $\mathfrak{n}^+$ and $\mathfrak{n}^-$ are commutative subalgebras of $\mathfrak{g}_C$ contained in $\mathfrak{p}_C$. Putting $P^{\pm} = \exp(\mathfrak{n}^{\pm})$, then $G \subset P^- \cdot K_C \cdot P^+$, and the map of $P^- \times K_C \times P^+$ into G_C given by $(x, y, z) \mapsto x \cdot y \cdot z$ is biholomorphic onto a Zariski-open subset of G_C. Let $g = g_- g_0 g_+$ be the decomposition of $g \in G$ with respect to P^-, K_C, P^+. A theorem of Harish-Chandra asserts that the map $G \to \mathfrak{p}^+$, which sends g onto $\log g_+$, identifies $X = K \backslash G$ with a bounded domain D of $\mathfrak{n}^+$. The subspace $P^- \cdot K_C \cdot G$ is open in G_C. Hence

$$P^- \backslash P^- K_C G \cong K_C \times_K G,$$

is a complex analytic principal bundle over $X = K \backslash G$ with K_C as fibre. This fibre space has a natural cross-section v defined by P^+, to which corresponds an automorphy factor μ. If $x \in D \in \mathfrak{n}^+$ and $g \in G$, then

$$\exp x \cdot g = (\exp x \cdot g)_- \cdot (\exp x \cdot g)_0 \cdot (\exp x \cdot g)_+,$$

and

$$\mu(x, g) = (\exp x \cdot g)_0 \in K_C.$$

This is the "canonical automorphy factor" on X, considered in [9]. To recover completely the situation of §§1, 4, one gives a representation $\rho: K_C \to GL(V)$. Then $\mu_\rho(x, g) = \rho((\exp x \cdot g)_0)$. (To get the Jacobian determinant of the bounded realization as an automorphy factor, take $\rho: K_C \to \boldsymbol{GL}_1$ defined by $\rho(k) = \det \cdot \mathrm{Ad}_{y^+} k^{-1}$.)

It is not necessary to take P^+ as cross section for the bundle $K_C \times_K G$. Let P' be conjugate to P^+, such that $G \subset P^- \cdot K_C \cdot P'$. Every element $g \in G$ can again be decomposed as $g = g^- \cdot g'_0 g'_+$. Then $\mu'(x, g) = (\exp x \cdot g)'_0$ is again an automorphy factor. There exist a number of canonical choices of P', defined by the so-called partial Cayley transforms, that give rise to unbounded realizations of X (see [1]).

EXAMPLE. $G = \boldsymbol{Sp}(n, \boldsymbol{R})$, $K = \boldsymbol{U}(n)$, $K_C = \boldsymbol{GL}(n, \boldsymbol{C})$.
If

$$g = \begin{pmatrix} A & B \\ C & D \end{pmatrix}$$

and $K \backslash G$ is realized as the Siegel upper half plane, then the corresponding canonical automorphy factor is

$$\mu(Z, g) = CZ + D.$$

6. **Connection between holomorphy and condition** (2). Let $f: X \to V$ be a holomorphic automorphic form defined on X, with automorphy factor μ_ρ; then the corresponding automorphic form F on G is $Z(\mathfrak{g})$-finite, where $F(g) = \mu(0, g) \cdot f(\dot{g})$. This was pointed out in [**11**, Exp. 10]. One way to see it is to prove first that,

$$(4) \qquad Y \cdot F(g) = \mu_\rho(0, g)(\tilde{Y} \cdot f)(\dot{g}) \qquad (Y \in \mathfrak{n}^-),$$

where, due to the definition of the complex structure of X, $\tilde{Y}$ is the derivative with respect to the conjugate of some coordinate-variable in X (see [**2**, §5]). The function f is holomorphic if and only if $\tilde{Y}f = 0$ $(Y \in \mathfrak{n}^-)$ by (4); this is equivalent to $Y \cdot F = 0$ $(Y \in \mathfrak{n}^-)$. The fact that F is then $Z(\mathfrak{g})$-finite depends now on some properties of $Z(\mathfrak{g})$. As a vector space $\mathfrak{g}_C = \mathfrak{n}^+ \oplus \mathfrak{k}_C \oplus \mathfrak{n}^-$. Hence as a vector space $U(\mathfrak{g}) = U(\mathfrak{n}^+) \otimes U(\mathfrak{k}) \otimes U(\mathfrak{n}^-)$: every $x \in U(\mathfrak{g})$ can accordingly be written as $x = \sum p_i^+ \cdot k_i \cdot p_i^-$. One shows then that if $x \in Z(\mathfrak{g})$, the only terms occurring in this sum are such that p_i^+ and p_i^- are simultaneously zero or different from zero. Hence there exists a linear map $v: Z(\mathfrak{g}) \to U(\mathfrak{k}_C)$ such that

$$z - v(z) \in U(\mathfrak{g}_C) \cdot \mathfrak{n}^-, \qquad \text{for } z \in Z(\mathfrak{g}).$$

Since F is annihilated by $\mathfrak{n}^-$, $z \cdot F = v(z)F$. The equivariance of F with respect to K insures that F is annihilated by an ideal in $U(\mathfrak{k})$ of finite codimension. The same will then be true in $Z(\mathfrak{g})$.

7. **Growth condition.** Let W be a finite-dimensional complex vector space and σ a locally faithful representation of G in $GL(W)$. Put on W a structure of Hilbert space invariant by $\sigma(K)$. Let $\|g\| = \operatorname{tr}(g^* \cdot g)$.

Then

$$\|g \cdot h\| \leq \|g\| \, \|h\| \qquad (g, h \in G),$$

and

$$\|k \cdot g \cdot k'\| = \|g\| \qquad (k, k' \in K, g \in G).$$

If A is the connected component of a maximal $\boldsymbol{R}$-split torus, then $G = K \cdot A^+ \cdot K$ (where A^+ is a positive Weyl chamber, for some ordering of the $\boldsymbol{R}$-roots), and the second relation shows that what matters is the behavior of the seminorm on A: for $\alpha \in A$, $\|a\| < c \cdot a^\Lambda$, where c is a constant and Λ is some weight of A^+, dominating the weights of σ ($\Lambda = \sum c_\alpha \Lambda_\alpha$ with $c_\alpha \geq 0$, Λ_α being fundamental weights). As an example, let σ be the adjoint representation, $\mathfrak{g} = \mathfrak{k} + \mathfrak{n}$ a Cartan decomposition of the Lie algebra of G, and s the corresponding Cartan involution. Then we may put

$$(5) \qquad \|g\| = \operatorname{Tr}(\operatorname{Ad} s(g)^{-1} \cdot \operatorname{Ad} g).$$

If $| \ |$ is a norm in the vector space V, and $f: G \to V$, the growth condition imposed on f is:

$$\exists c > 0 \qquad \text{and} \qquad m \in \boldsymbol{Z}, m \geq 0,$$

such that

$$|f(g)| \leq c \|g\|^m.$$

This growth condition really does not depend on the chosen representation σ, since it is easy to see, due to the behavior of the seminorm on A, that if τ is another locally faithful representation of G, there exists a positive integer n, $C_1 > 0$ such that

$$\|g\|_\sigma \leqq C_1 \|g\|_\tau^n, \qquad (g \in G).$$

8. **Construction of automorphic forms.** Let $X = K\backslash G$, and $\mu : X \times G \to \mathrm{GL}(V)$ be an automorphy factor. For $\phi : X \to V$, consider the series

$$(6) \qquad \sum_{\gamma \in \Gamma} \mu(x, \gamma) \cdot \phi(x \cdot \gamma).$$

If it converges, it will be equal to an automorphic form under suitable assumptions for ϕ (for instance, holomorphy if X is a bounded domain, and μ is itself holomorphic). One can work analogously on G instead of X. Starting from $f : G \to V$, assumed to be $Z(\mathfrak{g})$-finite and K-equivariant, then the series $\sum_\Gamma f(g \cdot \gamma)$, if properly convergent, will represent an automorphic form. It may happen that f is already invariant under some subgroup Γ_∞ of Γ. If Γ_∞ is infinite, the summation will of course be taken over Γ/Γ_∞. The analogous situation for (6) is when ϕ is invariant under Γ_∞, and $\mu(x, \gamma) = 1$ $(x \in X, \gamma \in \Gamma_\infty)$. Two standard examples are the Poincaré series and the Eisenstein series. The former are obtained by imposing on f conditions strong enough so that $\sum_\Gamma \rho(g\gamma)$ converges for any discrete group Γ. For the Eisenstein series, Γ_∞ is in general infinite, and conditions are imposed on Γ and ϕ. We now describe natural generalizations of these notions in the present context.

9. **Poincaré series.** A vector valued function f on G is said to be K-finite on the left (resp. right) if the set of right (resp. left) translates of f under elements of K is a finite-dimensional vector space.

9.1. THEOREM. *Let V be a finite-dimensional vector space, and f a function from G to V. Assume*:
(a) *$f \in L^1(G) \otimes V$,*
(b) *f is $Z(\mathfrak{g})$-finite,*
(c) *f is K-finite on the left (respectively on the right). Then the series*

$$P_f(g) = \sum_{\gamma \in \Gamma} f(g \cdot \gamma)$$

converges absolutely and uniformly on compact sets (respectively, and moreover $\sum_\Gamma |f(g \cdot \gamma)|$ is bounded on G).

Proof of the first assertion (Godement [**11**, Exp. 10]):

$$\int_G |f(g)| \, dg = \int_{G/\Gamma} \sum_\Gamma |f(g \cdot \gamma)| \, dg < \infty.$$

$\sum_\Gamma |f(g) \cdot \gamma)|$ converges in $L^1(G/\Gamma)$, and so converges almost everywhere and also converges in the distribution sense. Now, by assumptions (b) and (c), f is annihilated by an elliptic operator (see (2)). By a general principle (essentially an application of the closed graph theorem) the series converges then in the C^∞-topology; in particular, it converges uniformly on compact sets of G.

To prove the uniform boundedness we use the following lemma of Harish-Chandra [7, Theorem 1]:

9.2. LEMMA. *Let $f: G \to V$ be $Z(\mathfrak{g})$-finite and K-finite on the right (resp. on the left) and U be a neighborhood of e in G. Then there exists an $\alpha \in C_c^\infty(U)$ invariant by inner automorphisms of K such that $f = f * \alpha$ (resp. $f = \alpha * f$).*

(C_c^∞ refers to C^∞-functions with compact support, and $*$ to convolution. Thus

$$f * \alpha(g) = \int_G f(g \cdot u^{-1})\alpha(u)\, du).$$

Proof of uniform boundedness when f is K-finite on the right:

$$f(g \cdot \gamma) = \int_G f(g \cdot \gamma \cdot u^{-1})\alpha(u)\, du = \int_G f(g \cdot v^{-1})\alpha(v \cdot \gamma)dv,$$

$$|f(g \cdot \gamma)| \leq M \int_{U \cdot \gamma} |f(gv^{-1})|\, dv,$$

if U is small enough $U \cdot \gamma \cap U \cdot \gamma' = \phi$ when $\gamma \neq \gamma'$. So

$$\sum_{\gamma \in \Gamma} |f(g \cdot \gamma)| \leq M \sum_{\gamma} \int_{U \cdot \gamma} |f(gv^{-1})|\, dv$$

$$\leq M \int_G |f(v)|\, dv = M\|f\|_{L_1}.$$

REMARK. The above proof for the uniform boundedness is due to Harish-Chandra. A slight variation of it also yields the first assertion. That $\sum f(x \cdot \gamma)$ is uniformly bounded (when f is K-finite on the *right*, and verifies (a), (b)) was proved in the holomorphic case first by Godement [11, Exp. 10]. His argument may be extended to the general case.

10. **Example of the classical Poincaré series.** Let $X = K\backslash G$ be a bounded symmetric domain. Take as automorphy factor $J(x, g)$ the determinant of the Jacobian. If f is a polynomial in $C^N \supset X$, then

$$P(x) = \sum_{\gamma \in \Gamma} J(x, \gamma)^l \cdot f(x\gamma),$$

converges absolutely and uniformly on compact sets for $l \geq 2$, and

$$\tilde{P}(g) = J(o, g)^l \cdot P(\dot{g})$$

is uniformly bounded. More strongly:

$$\sum_{\Gamma} |J(o, g\gamma)|^{|l|} |f(0 \cdot g\gamma)| < C < \infty$$

where o denotes the coset K in X.

PROOF. In the identification of X with a bounded domain $\mathfrak{y}^{+} = C^N$ the coset o is mapped into the origin. Hence K is linear and the polynomial f is K-finite on the right. The function $F(g) = J(o, g)^l \cdot f(g)$ is $Z(\mathfrak{g})$-finite (cf. §6). To apply the previous theorem, it is enough to show that F belongs to $L^1(G) \otimes V$ for $l \geq 2$. X being a bounded domain, f is bounded on X. So

$$\int_G |F(g)| \, dg \leqq C \cdot \int_G |J(o, g)|^l \, dg.$$

It is easily seen that

$$|J(o, kgk')| = |J(o, g)| \qquad (k, k' \in K; g \in G);$$

hence $|J(o, g)|$ may be viewed as a K-invariant function on X. Therefore,

$$\int |J(o, g)|^l \, dg \leqq C' \cdot \int_X |J(o, x)|^l \, dw,$$

where dw is an invariant measure on X. But $dw = c \cdot |J(o, x)|^{-2} \, dx \, (c > 0)$, where dx is the usual Lebesgue measure on R^{2n}. Consequently, we are reduced to showing that $J(o, x)$ is bounded on X. Since $X = o \cdot A \cdot K$, it is enough to check this on $o \cdot A$, which is easy (see for instance [2, §1]).

If $\tau : K \to GL(V)$ and $\mu_\tau(x, g)$ is the automorphy factor of §5, then there exists an l_0 such that

$$\sum_{\gamma \in \Gamma} J(x, \gamma)^l \mu_\tau(x, \gamma) \cdot f(x \cdot \gamma)$$

converges for $l \geq l_0$ (see [2, §5] for more details).

11. **Eisenstein series.** Let G be a connected semisimple algebraic group defined over $k \subset R$, P a standard (not necessarily minimal) parabolic k subgroup of G, for a maximal k-split torus S. If P_0 is a minimal parabolic k-subgroup between S and P, let θ be the subset $_k\Delta$ of simple roots of $\Phi(G, S)$, for the ordering defined by P_0, such that $P = {}_kP_\theta$ (see [3, §6.5]). Then

$$P_0 = Z(S_0) \cdot U_0,$$

$$P = Z(S_\theta) \cdot U_0 = Z(S_\theta) \cdot U,$$

where $U = R_u(P)$ and $S_\theta = (\bigcap_{\alpha \in \theta} \mathrm{Ker}\, \alpha)^0$. Put $\theta' = {}_k\Delta - \theta$ and $\chi = \det \mathrm{Ad}_u$, where $\mathfrak{u}$ is the Lie algebra of U: for $p \in P_R$ $p^\chi = \det \mathrm{Ad}_u p$. The character χ is then a positive linear combination of the fundamental highest k-weights Λ_α of $G : \chi = \sum_{\alpha \in \theta'} e_\alpha \Lambda_\alpha$, $(e_\alpha > 0)$, where the Λ_α verify $(\Lambda_\alpha, \beta) = d_\alpha \delta_{\alpha\beta}$ $(\alpha, \beta \in {}_k\Delta, d > 0)$. Let $s = (s_\alpha)_{\alpha \in \theta'}$ be a set of complex numbers. Put $\Lambda_s = \sum_{\alpha \in \theta'} s_\alpha \Lambda_\alpha$ and $p^{\Lambda_s} = \Pi |p^{\Lambda_\alpha}|^{s_\alpha}$:

If $A = S_R^0$, $G_R = K \cdot P_R = K \cdot M_R \cdot A \cdot U$. If $g = kmau$ is the corresponding decomposition of $g \in G$, then $k \cdot m$, a and u are uniquely determined; we let $a(g)$ denote the A-component of $g \in G$.

11.1. LEMMA (GODEMENT). *Let* Γ *be a discrete subgroup of* G_R *and* Γ_∞ *a subgroup of* $\Gamma \cap (M \cdot U)_R$. *Assume that*

(1) $a(\gamma)^{\Lambda_\alpha} \geq d > 0$ *for all* $\gamma \in \Gamma$ *and* $\alpha \in \theta'$,

(2) $(M \cdot U)_R / \Gamma_\infty$ *has finite-invariant measure,*

(3) $Rs_\alpha > e_\alpha$, *for all* $\alpha \in \theta'$.

Then $E(g, s) = \sum_{\gamma \in \Gamma/\Gamma_\infty} a(g \cdot \gamma)^{-\Lambda_s}$ *converges uniformly on any compact set of* G.

(Note that, $M \cdot U$ being contained in the kernel of every k-character of P, $a(g \cdot \gamma)^{\Lambda_\alpha} = a(g)^{\Lambda_\alpha}$ for $\gamma \in \Gamma_\infty$ and $M \cdot U$ is unimodular.)

SKETCH OF THE PROOF. We may assume the s_α to be real. Let C be a compact subset of G. Then there exist $d, d' > 0$ such that

$$d \cdot a(g)^{-\Lambda_s} \leq a(c \cdot g)^{-\Lambda_s} \leq d' \cdot a(g)^{-\Lambda_s} \qquad (g \in G, c \in C).$$

This implies readily that the uniform convergence on compact sets is equivalent to the convergence at one point, say e. Furthermore, if C is a neighborhood of e, the convergence at e is equivalent to the convergence of the series

$$\sum_\gamma \int_C a(c \cdot \gamma)^{-\Lambda_s} \, dg.$$

Take C small enough so that $C \cdot C^{-1} \cap \Gamma = \{e\}$. Then the series is majorized by

$$I = \int_{C \cdot \Gamma / \Gamma_\infty} a(g)^{-\Lambda_s} \, dg.$$

Let

$$A(t) = \{a \in A \,|\, a^{\Lambda_\alpha} \geq t \qquad (\alpha \in \theta')\}.$$

By assumption $\Gamma \subset K \cdot M \cdot A(t') \cdot U$, for some $t' > 0$, whence the existence of $t > 0$ such that $C \cdot \Gamma \subset K \cdot M \cdot A(t) \cdot U$.

We have $K \cdot M \cdot A(t) \cdot U = K \cdot A(t) \cdot M \cdot U$. By assumption there exists $w \subset M \cdot U$, of finite measure, such that $M \cdot U = w \cdot \Gamma_\infty$. From this we deduce without difficulty that

$$I \leq \delta \cdot \int_{KA(t)w} a^{-\Lambda_s} \, dg, \qquad (\delta > 0).$$

By standard facts about Haar measures, the last integral is, up to a constant, equal to

$$\int_K dk \int_w dv \int_{A(t)} a^{-\Lambda_s + \chi} \, da = d \int_{A(t)} a^{-\Lambda_s + \chi} \, da,$$

where $\chi = \sum e_\alpha \cdot \Lambda_\alpha = \det \mathrm{Ad}_u$ (see beginning of this section), and dk, dv, da, are

Haar measures on K, $M \cdot U$ and A respectively. The last integral is a product, (over $\alpha \in \theta'$) of integrals of the form

$$\int_t^\infty \exp[(-s_\alpha + e_\alpha) \cdot t]\, dt,$$

hence converges, since we assume $s_\alpha > e_\alpha$.

11.2. THEOREM. *Let $f: G_R \to V$, where V is a finite-dimensional vector space. Keeping the same notations as above, suppose that*

(i) *the assumptions (1), (2) and (3) of the lemma are satisfied,*

(ii) *$f(g \cdot \gamma) = f(g)$ for $\gamma \in \Gamma_\infty$,*

(iii) *$|f(g \cdot p)| p^{\Lambda_s}$ is bounded if g stays in a compact set of G and $p \in P_R$.*

Then the series $E_f(g) = \sum_{\Gamma/\Gamma_\infty} f(g \cdot \gamma)$ converges absolutely and uniformly on any compact set of G.

PROOF. We have $G = K \cdot P_R$, hence

$$f(g) \cdot a(g)^{\Lambda_s} = f(k \cdot p) \cdot a(k \cdot p)^{\Lambda_s}$$

$$= f(k \cdot p) \cdot a(p)^{\Lambda_s} = f(k \cdot p) \cdot p^{\Lambda_s},$$

$$|f(g) \cdot a(g)^{\Lambda_s}| = |f(k \cdot p)| p^{\Lambda_s} \leqq C.$$

Hence

$$|f(g)| \leqq C |a(g)|^{-\Lambda_s},$$

$$\sum_{\Gamma/\Gamma_\infty} |f(g \cdot \gamma)| \leqq C \cdot \sum_{\Gamma/\Gamma_\infty} |a(g)|^{-\Lambda_s}.$$

The lemma ensures that the last series converges. This proves the theorem. The series E_f will be called an Eisenstein series.

12. Special cases of Eisenstein series.

12.1. THEOREM. *We keep the notation of the previous section and assume moreover that $k = Q$, that Γ is an arithmetic group, and Γ_∞ a subgroup of finite index in $\Gamma \cap P$. Then the assumptions of (1) and (2) of the lemma in §11.1 are fulfilled.*

PROOF. (1) $(M \cdot U)_R$ has no nontrivial character defined over Q, and Γ_∞ is an arithmetic subgroup of $(M \cdot U)_R$. So $(M \cdot U)_R/\Gamma_\infty$ has finite volume by reduction theory [4, Theorem 9.1].

(2) Let ρ_α be an irreducible representation of G, which is strongly rational over Q, having Λ_α ($\alpha \in \theta'$) as highest weight [3, §7]. If e_α is a corresponding weight vector in the representation space F_α of ρ_α one has $\rho(p)(e_\alpha) = p^{\Lambda_\alpha} \cdot e_\alpha$ ($p \in P$) because $\alpha \in \theta'$. There exists a lattice L_α in $F_{\alpha Q}$ invariant under Γ. We may assume that $e_\alpha \in L_\alpha - \{0\}$. So $\rho_\alpha(\Gamma) \cdot e_\alpha \in L - \{0\}$. There exists a $d > 0$ such that $|\rho_\alpha(\gamma) \cdot e_\alpha| > d$ for every $\gamma \in \Gamma$. Write $\gamma = k \cdot m \cdot a(\gamma)u$. Since M and U lie in P and

have no Q-character, one has

$$d < |\rho_\alpha(\gamma)e_\alpha| = |\rho_\alpha(a(\gamma))\cdot e_\alpha| = a(\gamma)^{\Lambda_\alpha}.$$

This proves condition (2) of the lemma.

If $f: G_R \to V$ verifies the identity $f(g\cdot p) = f(g)p^{-\Lambda_s}$, then f satisfies the hypotheses (iii) and (ii) of the convergence theorem ((ii) because Γ_∞ lies in the kernel of all Q-characters of P). One gets with such functions a straightforward generalization of the classical Eisenstein series.

If moreover f is ρ-equivariant on the left with respect to K, then it follows from standard facts about the universal enveloping algebra $U(\mathfrak{g})$ that E_f is $Z(\mathfrak{g})$-finite. Moreover, the growth condition can also be checked using reduction theory. Thus we get automorphic forms in the sense of §1.

A typical function satisfying $f(g\cdot p) = f(g)\cdot p^{-\Lambda_\alpha}$ $(p\in P)$ is given by $f(g) = |\rho_\alpha(g)\cdot e_\alpha|$. These functions are also connected with minimum properties on Siegel-domains.

EXAMPLES 1. Let $G = Sp(n, C)$, and P be the maximal parabolic group, consisting of the matrices

$$\begin{pmatrix} A & B \\ 0 & D \end{pmatrix}$$

in the standard notation. Then, as function $f(g)$, we may take $\det(C\cdot i + D)$.

2. Let $G = SL_n$, $K = SO(n)$, $= SL(n, Z)$. The fundamental representation of G are the exterior powers of the identity representation of G. For an ordering of the roots associated to the group of upper triangular matrices, the highest weight of the ith exterior power is $e_1 \wedge \cdots \wedge e_i$, where (e_i) is the canonical basis of R^n $(1 \leq i \leq n - 1)$. The corresponding function is then

$$\Phi_i(g) = \|g\cdot e_1 \wedge \cdots \wedge g\cdot e_i\|.$$

Let P_i be the stability group of the flag

$$[e_1] \subset [e_1, e_2] \subset \cdots \subset [e_1, \cdots, e_i].$$

Then for Rs_j sufficiently big, $(1 \leq j \leq i)$,

$$\sum_{\gamma\in\Gamma/\Gamma\cap P_i} \Phi_1(g\cdot\gamma)^{-s_1}\cdots\Phi_i(g\cdot\gamma)^{-s_i},$$

is an Eisenstein series. Such series have been considered by Selberg [10].

3. We return to the situation of the theorem in §12.1. Let $\rho: K \to GL(V)$ be a finite-dimensional representation of K. Let $\Phi: G \to V$ be a continuous function which is ρ-equivariant, with respect to K, is right invariant with respect to $A\cdot U_R$, and is a cusp form on M_R for $\Gamma_\infty \cap M$ (for the notion of cusp form, see [5]). We may define a function $f: G \to V$ by $f(k\cdot m\cdot a\cdot u) = \Phi(k\cdot m)\cdot a^{\Lambda_s}$. Since a cusp form is bounded, it is easily seen that f satisfies the condition of the theorem.

The corresponding Eisenstein series is then

$$E_f = \sum_{\gamma \in \Gamma/\Gamma_\infty} f(g \cdot \gamma) = \sum_{\gamma \in \Gamma/\Gamma_\infty} \Phi(g \cdot \gamma) \cdot a(g \cdot \gamma)^{-\Lambda_s}.$$

Such series occur in the work of Selberg, and of Langlands [8].

4. Finally, we note that the Poincaré–Eisenstein series which are used in [1; 2] to study the compactification of X/Γ, when X is a bounded symmetric domain are special cases of the series considered in §11.2. In this case, M is the almost direct product of two normal Q-subgroups M_1, M_2 and, roughly speaking, the function f, restricted to M_R is the product of a constant function on M_R by a Poincaré series (hence a cusp form) on M_{2R}.

REFERENCES

1. W. Baily, *On compactifications of orbit spaces of arithmetic discontinuous groups acting on bounded symmetric domains.* Proc. Sympos. Pure Math., vol. 9, Amer. Math. Soc., Providence, R.I., 1966, pp. 281–295.

2. W. Baily and A. Borel, *Compactification of arithmetic quotients of bounded symmetric domains,* (to appear).

3. A. Borel, *Linear algebraic groups,* Proc. Sympos. Pure Math., vol. 9, Amer. Math. Soc., Providence, R.I., 1966, pp. 3–19.

4. A. Borel and Harish-Chandra, *Arithmetic subgroups of algebraic groups,* Ann. of Math. (2) **75** (1962), 485–535.

5. R. Godement, *The spectral decomposition of cusp-forms,* Proc. Sympos. Pure Math., vol. 9, Amer. Math. Soc., Providence, R.I., 1966, pp. 225–234.

6. Harish-Chandra, *Automorphic forms on a semi-simple Lie group,* Proc. Nat. Acad. Sci. U.S.A. **45,** (1959), 570–573.

7. ———, *Discrete series for semi-simple Lie groups.* II [Acta Math. **166** (1966), 1–111].

8. R. Langlands, *Eisenstein series,* Proc. Sympos. Pure Math., vol. 9, Amer. Math. Soc., Providence, R.I., 1966, pp. 235–252.

9. S. Murakami, *Cohomologies of vector-valued forms on compact, locally symmetric Riemann manifolds,* Proc. Sympos. Pure Math., vol. 9, Amer. Math. Soc., Providence, R.I., 1966, pp. 387–399.

10. A. Selberg, *Harmonic analysis and discontinuous groups,* J. Indian Math. Soc. **20** (1956), 47–87.

11. Séminaire H. Cartan, *Fonctions automorphes,* 2 vols., 10ème année (1957/1958), Paris, 1958 (Mimeographed notes).

(From Notes by F. Bingen)

76.

(with T. A. Springer)

Rationality properties of linear algebraic groups

Proc. Symp. Pure Math. **9,** Amer. Math. Soc. (1966) 26–32

In this lecture, G is a connected linear algebraic group, $\mathfrak{g}$ its Lie algebra, k a field of definition for G and p the characteristic of k. Our purpose is to sketch, for infinite k, a proof of

THEOREM A. (i) *G contains a maximal torus and a Cartan subgroup defined over k.*

(ii) *Let G be reductive. Then G is unirational over k. In particular, if k is infinite, G_k is Zariski dense in G.*

(G unirational over k means that there exists a k-morphism of a k-open subset of an affine space into G, whose image contains a nonempty open subset of G.)

This theorem is due to Rosenlicht [6] over perfect fields, to Grothendieck [5] in the general case. In contrast with [5], we shall give a proof which does not use schemes; however, it is in part based on similar ideas. We presuppose the theory of algebraic groups over an algebraically closed field [1], [4].

Notation. The Lie algebra of an algebraic group H, $M, \cdots$ is denoted by the corresponding German letter. Ad refers to the adjoint representation of G in $\mathfrak{g}$, and ad to the representation of $\mathfrak{g}$ into itself defined by $\operatorname{ad} X(Y) = [X, Y]$. As is well known, ad is the differential of Ad. ^{g}A stands for $g \cdot A \cdot g^{-1}$ $(g \in G, A \subset G)$. For $X \in \mathfrak{g}$, we let $Z_G(X) = \{g \in G, \operatorname{Ad} g(X) = X\}$ and $z_G(X) = \{Y \in \mathfrak{g}, [Y, X] = 0\}$. Clearly, $Z_G(X)$ is a closed subgroup, whose Lie algebra is contained in $z_G(X)$. If no confusion can arise, we drop the subscript G.

Let $p \neq 0$. We denote by $[p]: X \to X^{[p]}$ the pth power operation in $\mathfrak{g}$. It is defined over k: if $\mathfrak{g} \subset \mathfrak{gg}_n$, then $X^{[p]} = X^p$. In particular, if $[X, Y] = 0$, then

$$(X + Y)^{[p]} = X^{[p]} + Y^{[p]}.$$

If $q = p^s$ is a power of p, we write $[q]$ for $[p]^s$. The pth power operation commutes with differentials of morphisms.

1. Jordan decomposition in $\mathfrak{g}$. The centralizer of a semisimple element.

1.1. DEFINITION. An element $X \in \mathfrak{g}$ is semisimple (resp. nilpotent) if it belongs to the Lie algebra of a sub-torus (resp. unipotent subgroup) of G.

1.2. LEMMA. *Let $G = T \cdot U$ be solvable, with one-dimensional unipotent radical U, where T is a maximal torus of G. Let $X \in \mathfrak{t}$. Then there are two possibilities:*

(1) $[X, \mathfrak{u}] = 0$, $Z(X) = G$.

(2) $[X, \mathfrak{u}] \neq 0$, $Z(X) = T$; *the map $u \mapsto \operatorname{Ad} u(X) - X$ is an isomorphism of U onto the additive algebraic group $\operatorname{Add}(\mathfrak{u})$ of $\mathfrak{u}$. Every element $X + Z$ $(Z \in \mathfrak{u})$ is semisimple.*

697

Let $\pi: G \to G' = G/U$ be the canonical projection. Then $d\pi$ maps $\mathfrak{t}$ isomorphically onto $\mathfrak{g}'$, and G' is commutative. Therefore, $\mathrm{Ad}\, g(X) - X \in \mathfrak{u}$ $(g \in G)$.

Assume $[X, \mathfrak{u}] = 0$, and identify G to a matrix group. If $Z(X) \neq G$, there exists $u \in U$ such that $\mathrm{Ad}\, u(X) = X + U$ $(U \in \mathfrak{u} - 0)$. But then $\mathrm{Ad}\, u(X)$ is on the one hand a semisimple matrix (being tangent to a torus), and on the other hand is the sum of a semisimple and a nonzero nilpotent matrix, commuting with each other, a contradiction.

Let now $[X, \mathfrak{u}] \neq 0$. Since ad is the differential of Ad, this implies that $Z(X) \neq G$. Since $Z(X) \supset T$, it follows, for dimensional reasons, that $Z(X)^0 = T$. However, in a connected solvable group, the centralizer and the normalizer of a torus are connected, and coincide [1]. Since $[\mathfrak{t}, \mathfrak{u}] \neq 0$, T is not central, hence $\mathrm{Norm}(T) = T$ and $Z(X) = T$. Let $Y \in \mathfrak{u} - 0$. Then $\mathrm{Ad}\, u(X) - X = c(u) \cdot Y$. Obviously c is a morphism of U into the additive group of $\mathfrak{u}$, injective since $Z(X) = T$, hence bijective. Thus every element $X + Z$ $(Z \in \mathfrak{u})$ is conjugate to X, hence is semisimple. The relation $[X, \mathfrak{u}] \neq 0$ implies that the differential of c is an isomorphism, therefore c is birational, biregular.

1.3. PROPOSITION. *Every $X \in \mathfrak{g}$ can be written uniquely as $X = X_s + X_n$ with X_s semisimple, X_n nilpotent, $[X_s, X_n] = 0$. For any morphism $\pi: G \to \mathbf{GL}_n$, $d\pi(X) = d\pi(X_s) + d\pi(X_n)$ is the Jordan decomposition of $d\pi(X)$.*

Let $X = X_s + X_n$ be one decomposition of X with X_s semisimple, X_n nilpotent and $[X_s, X_n] = 0$. Then $d\pi(X_s)$ (resp. $d\pi(X_n)$) is tangent to a torus (resp. a unipotent group), hence is a semisimple (resp. nilpotent) matrix, whence the second assertion. Using a matrix realization of G, this implies the uniqueness of this decomposition. There remains to show its existence. We assume first G to be solvable, and proceed by induction on $\dim G$. Let $U = R_u(G)$ be the unipotent radical of G. There exists a connected one-dimensional subgroup N of the center of U which is normal in G. Let $\pi: G \to G' = G/N$ be the canonical projection. Let $X \in \mathfrak{g}$. By induction assumption,

$$d\pi(X) = A' + B' \qquad (A', B' \in \mathfrak{g}';\ A' \text{ semisimple, } B' \text{ nilpotent, } [A', B'] = 0).$$

Since $R_u(G') = \pi(R_u(G))$, and every torus of G' is the image of a torus of G, this yields

$$X = A + B \qquad (A \text{ semisimple, } B \text{ nilpotent, } d\pi(A) = A', d\pi(B) = B').$$

Let T be a maximal torus whose Lie algebra contains A. Since every rational representation of T is fully reducible, we may write $\mathfrak{u} = \mathfrak{r} \oplus \mathfrak{n}$, with $\mathfrak{r}$ stable under $\mathrm{Ad}\, T$. Writing

$$B = B_1 + B_2 \qquad (B_1 \in \mathfrak{r}, B_2 \in \mathfrak{n}),$$

we have then

$$[A, B_1] = 0, \qquad [A, B_2] = C \in \mathfrak{n}, \qquad [B_1, B_2] = 0.$$

If $C = 0$, we are done, so assume $C \neq 0$. By Lemma 1.2, applied to $T \cdot N$, the sum $A + B_2$ is semisimple, hence the decomposition $X = X_s + X_n$ with $X_s = A + B_2, X_n = B_1$ has the required properties.

This proves Proposition 1.3 for solvable groups. The general case follows by using the following Lemma [**5**, XIV, Théorème 4.11]:

1.4. LEMMA. *Let B be a Borel subgroup of G. Then* $\mathfrak{g} = \bigcup_{g \in G} \mathrm{Ad}\, g(\mathfrak{b})$.

The proof is similar to that given in [**3**, VI] for the corresponding statement in G; it uses the two following facts, which follow from properties of reductive groups: B is the normalizer of $\mathfrak{b}$ in G, and there exists an element $X \in \mathfrak{b}$ which is not contained in any conjugate of $\mathfrak{b}$ distinct from $\mathfrak{b}$.

1.5. PROPOSITION. *Let $X \in \mathfrak{g}$ be semisimple. Then $z(X)$ is the Lie algebra of $Z(X)$.*

Since $z(X)$ contains the Lie algebra of $Z(X)$, it is enough to show that $\dim z(X) \leqq \dim Z(X)$.

(a) *G solvable.* We let $U, N, \pi: G \to G' = G/N$ be as in the previous proof, and let T be a maximal torus whose Lie algebra contains X. Let $X' = d\pi(X)$. The Lie algebra $\mathfrak{g}$ is the direct sum of $\mathfrak{n}$ and of a subspace $\mathfrak{r}$ stable under $\mathrm{Ad}\, T$, whence immediately,

$$\dim z(X) = z(X) \cap \mathfrak{n} + \dim z(X').$$

By Lemma 1.2, $\dim z(X) \cap \mathfrak{n} = \dim Z(X) \cap N$; by induction $\dim Z_{G'}(X') = \dim z(X')$. It suffices therefore to show that $\pi(Z(X)) = Z_{G'}(X')$.

Let $g' \in Z_{G'}(X')$. There exists $g \in G$ such that $\pi(g) = g'$, hence such that $\mathrm{Ad}\, g(X) = X + Y$ $(Y \in \mathfrak{n})$. There is something to prove only if $Y \neq 0$. Since $\mathrm{Ad}\, g(X)$ is semisimple, and Y is nilpotent, we must have $[X, Y] \neq 0$ by the uniqueness of the Jordan decomposition. By Lemma 1.2, there exists then $n \in N$ such that $\mathrm{Ad}\, n(X) = X - Y$. We have then $\mathrm{Ad}\, n \cdot g(X) = X$, and $\pi(n \cdot g) = g'$.

(b) *G not solvable.* Let T be a maximal torus whose Lie algebra contains X. It follows from known facts about reductive groups that G has two Borel subgroups B, B', normalized by T, which generate G, such that $\mathfrak{g} = \mathfrak{b} + \mathfrak{b}', \mathfrak{b} \cap \mathfrak{b}' = \mathfrak{t} + \mathfrak{u}$ ($\mathfrak{u}$ Lie algebra of $U = R_u(G)$) (see [**3**, §2]). Let $\mathfrak{c}, \mathfrak{c}'$ be supplementary subspaces of $\mathfrak{t} + \mathfrak{u}$ in $\mathfrak{b}$ and $\mathfrak{b}'$ respectively, stable under $\mathrm{Ad}\, T$. The Lie algebra $z(X)$ is the direct sum of its intersections with $\mathfrak{c}, \mathfrak{c}'$ and $\mathfrak{t} + \mathfrak{u}$, say $\mathfrak{p}, \mathfrak{q}, \mathfrak{r}$. By (a) $\mathfrak{p} + \mathfrak{r}$ and $\mathfrak{q} + \mathfrak{r}$ belong to the Lie algebra of $Z(X)$, hence $\dim Z(X) \geqq \dim z(X)$.

REMARK. Using induction and the relation $\pi(Z(X)) = Z'_{G'}(X')$ proved above, one sees easily that $Z(X)$ is connected if G is solvable. Examples show that this need not be the case when G is semisimple.

1.6. PROPOSITION. *Let $X \in \mathfrak{g}$ be semisimple. Then* $\mathrm{Ad}\, G(X)$ *is closed.*

Identify G with a matrix group. Let L be the set of $Y \in \mathfrak{g}$ which annihilate the minimal polynomial of X, and such that $\mathrm{ad}\, Y$ has the same characteristic polynomial as $\mathrm{ad}\, X$. This is an algebraic subset of $\mathfrak{g}$, stable under $\mathrm{Ad}\, G$. Let $Y \in L$. Its minimal polynomial divides that of X, hence has only simple factors, and Y is a semisimple matrix. By Proposition 1.3, it is a semisimple element of $\mathfrak{g}$, in the sense of Definition 1.1, consequently, (Proposition 1.5), dim $Z(Y)$ is equal to the multiplicity of the eigenvalue zero of ad Y. By the definition of L, and Proposition 1.5, this implies dim $Z(Y) = $ dim $Z(X)$. As a consequence, the orbits of G in L are all of the same dimension, hence are closed [**1**, §16].

1.7. COROLLARY. *Assume that every semisimple element of* $\mathfrak{g}$ *is central in* $\mathfrak{g}$. *Then the set of all semisimple elements of* $\mathfrak{g}$ *is a subspace* $\mathfrak{s}$ *defined over k, which is the Lie algebra of every maximal torus of G.*

Let T be a maximal torus of G. Then $\mathfrak{t} \subset \mathfrak{s}$. Let $X \in \mathfrak{g}$ be semisimple. By the conjugacy of maximal tori in G, there exists $g \in G$ such that Ad $g(X) \subset \mathfrak{t}$. By assumption $[X, \mathfrak{g}] = 0$, hence (Proposition 1.5), $Z(X) = G$ and Ad $g(X) = X$. Thus $\mathfrak{s} \subset \mathfrak{t}$. There remains to see that $\mathfrak{s}$ is defined over k, in other words, that it is generated over $\bar{k}$ by elements of $\mathfrak{g}_k$. This is obvious in char. 0 (or in fact over a perfect field, since in this case the Jordan decomposition is over the groundfield), so we assume $p \neq 0$. There exists a power q of p such that $X^{[q]} = 0$, whenever X is nilpotent. For arbitrary X we have $X^{[q]} = (X_s + X_n)^{[q]} = X_s^{[q]} + X_n^{[q]}$ $= X_s^{[q]} \subset \mathfrak{t}$ (notation of §1.3). Thus $[q]: X \mapsto X^{[q]}$ maps $\mathfrak{g}$ into $\mathfrak{t}$. On the other hand, $\mathfrak{t}$ can be diagonalized, hence the restriction of $[q]$ to $\mathfrak{t}$ is bijective. Thus, Im$[q] = \mathfrak{t}$. But $[q]$ is a morphism of $\mathfrak{g}$, viewed as an algebraic set, into itself, obviously defined over k. Therefore, Im$[q]$ is also defined over k.

2. **Inseparable isogenies.** The main tools which will allow us to get hold of fields of definition in char. $p \neq 0$ are the criterion of multiplicity one of intersection theory, and the following result of Barsotti and Serre ([**7**, Théorème 1], [**3**, §7]).

2.1. PROPOSITION. *Let m be an ideal of* $\mathfrak{g}$ *which is stable under the pth power operation and under* Ad G. *Then there exists one and, up to k-isomorphism, only one k-group G′ with the following properties:* (i) *there exists a purely inseparable k-isogeny* $\pi: G \to G'$ *such that* ker $d\pi = \mathfrak{m}$; (ii) *every purely inseparable k-isogeny* $\theta: G \to G''$ *such that* ker $d\theta \supset \mathfrak{m}$ *can be k-factored through* π.

We also write $G/\mathfrak{m}$ for G'. Note that, since π is purely inseparable, π is a bijective morphism of G onto G'.

2.2. PROPOSITION. *Let G be not nilpotent, and assume that every semisimple element of* $\mathfrak{g}$ *is central. Let T be a maximal torus of G. Then there exists a k-group G′, such that not all semisimple elements of* $\mathfrak{g}'$ *are central, and a purely inseparable k-isogeny* $\pi: G \to G'$, *whose differential* $d\pi$ *had* $\mathfrak{t}$ *as kernel. The Lie algebra* $\mathfrak{g}'$ *is the direct sum of* $d\pi(\mathfrak{g})$ *and of the Lie algebra of any maximal torus.*

Let $\Phi = \Phi(G, T)$ be the set of roots of G with respect to T (nontrivial characters of T in $\mathfrak{g}$, under the adjoint representation). Since G is not nilpotent, T is not central, and Φ is not empty. (To see this, use the fact that G is generated by two solvable subgroups containing T, and the "lemme de dévissage" of [**4**], Exp. 9, Lemme 2.)

$\mathfrak{g}$ is the direct sum of the Lie algebra of $Z(T)$ and of the root spaces

$$\mathfrak{g}_a = \{X \in \mathfrak{g} | \mathrm{Ad}\, t(X) = t^a \cdot X, (t \in T)\} \qquad (a \in \Phi).$$

Let da be the differential of $a: T \to \mathbf{GL}_1$. It is a linear form on $\mathfrak{t}$, and we have

$$[Y, X] = da(Y) \cdot X \qquad (Y \in \mathfrak{t}, X \in \mathfrak{g}_a).$$

It is easily seen that the differential of a character x of T is zero if and only if x is divisible by p, in the character group $X(T)$ of T. Let c be the smallest positive integer such that $\Phi \not\subset p^{c+1} \cdot X(T)$. The elements of $\mathfrak{t}$ are central in $\mathfrak{g}$ if and only if $c \geq 1$. We prove Proposition 2.2 by induction on c.

By Corollary 1.7, $\mathfrak{t}$ is the set of all semisimple elements of $\mathfrak{g}$ and is defined over k. By Proposition 1.5, every $X \in \mathfrak{t}$ is centralized by G. Since $\mathfrak{t}$ is the Lie algebra of an algebraic group, it is stable under pth power operation therefore (Proposition 2.1), there exists a purely inseparable isogeny $\pi_1: G \to G_1 = G/\mathfrak{t}$.

It follows from Proposition 1.3 that $d\pi_1(\mathfrak{g})$ consists of nilpotent elements; therefore if T' is a maximal torus of G, then $\mathfrak{t}' \cap d\pi_1(\mathfrak{g}) = 0$, hence, for dimensional reasons, $\mathfrak{g}' = \mathfrak{t}' \oplus d\pi_1(\mathfrak{g})$ (direct sum of vector spaces). This applies in particular to the Lie algebra $\mathfrak{t}_1$ of $T_1 = \pi(T)$, which implies readily that ${}^t\pi_1$ maps $\Phi(G_1, T_1)$ onto $\Phi(G, T)$. On the other hand, it follows from $d\pi(\mathfrak{t}) = 0$ that ${}^t\pi_1$ maps $X(T')$ into $p \cdot X(T)$. Thus, if d is the smallest positive integer such that

$$\Phi(G_1, T_1) \not\subset p^{d+1} \cdot X(T_1),$$

we have $d < c$. If $d = 0$, then $\mathfrak{t}'$ is not central, and we take $G' = G_1$. If $d \geq 1$, we apply the induction assumption and get $\pi': G_1 \to G'$, with $\ker d\pi' = \mathfrak{t}_1$, satisfying our conditions for G_1. Then $\pi = \pi' \circ \pi_1: G \to G'$ has all the required properties.

3. **Proof of Theorem** A.

3.1. *Regular elements.* Let nil X be the multiplicity of the eigenvalue zero of ad X ($X \in \mathfrak{g}$), and let $n(\mathfrak{g}) = \min_{X \in \mathfrak{g}} \mathrm{nil}(X)$. An element X is *regular* if $\mathrm{nil}(X) = n(\mathfrak{g})$, *singular* otherwise. Clearly nil $X = \mathrm{nil}\, X_s$ (notation of §1.3), hence X is regular if and only if X_s is so. If $p \neq 0$, then X and $X^{[p]}$ are simultaneously regular or singular.

Let k be *infinite*. Then there exists a semisimple regular element $Y \in \mathfrak{g}_k$. In fact, the set S of singular elements is a proper algebraic subset (the set of zeros of the last nonidentically vanishing coefficient of the Killing equation), hence we may find $X \in \mathfrak{g}_k$ which is regular. Let $X = X_s + X_n$ be its Jordan decomposition. If $p = 0$, take $Y = Y_s$; if not, there exists a power q of p such that $X_n^{[q]} = 0$. Then put $Y = X^{[q]}$.

In the sequel, k is infinite.

3.2. *Proof of Theorem* A (i). A Cartan subgroup is the centralizer of a maximal torus [1], and the centralizer of a k-torus is defined over k [2, §10]. It suffices therefore to prove the existence of a maximal torus defined over k. We use induction on dim G.

Let first G be nilpotent, then $G = T \times U$ (T unique maximal torus, U unipotent radical). If k is perfect, T is defined over k. If not, we let q be a power of p such that $U^q = \{e\}$. Then $G^q = T^q = T$, hence $g \mapsto g^q$ is a morphism of G onto T, clearly defined over k, and T is defined over k. (This argument, phrased slightly differently, is due to Rosenlicht [6].) Let now G be not nilpotent. We let $\pi : G \mapsto G'$ be the identity if not all of the semisimple elements of $\mathfrak{g}$ are central, and be as in Proposition 2.2 otherwise.

By §3.1, there exists a semisimple regular element $Y \in \mathfrak{g}'_k$. By construction of G', we have $n(\mathfrak{g}') \neq \dim G$, hence $z(Y) \neq \mathfrak{g}'$.

We let G operate on $\mathfrak{g}'$ by Ad $\circ \pi$. Clearly, $G(Y) = \text{Ad } G'(Y)$, hence $G(Y)$ is closed (Proposition 1.6). Let $\tilde{Z}_G(Y)$ be the isotropy group of Y. Then $\pi(\tilde{Z}_G(Y)) = Z_{G'}(Y)$, hence (Proposition 1.5), $\dim \tilde{Z}_G(Y) = \dim \mathfrak{g} - n(\mathfrak{g}')$ and $\tilde{Z}_G(Y)^0$ is a *proper* closed subgroup of G. We want to prove that it is defined over k. This is clear in char. 0 (or if k is perfect) so let $p \neq 0$. Let f be the map $g \mapsto \text{Ad } \pi(g)(Y)$ of G onto $G(Y)$. It is defined over k. We claim that it is separable. To show this, it suffices to prove that $df_e : X \mapsto Y + [d\pi(X), Y]$ maps $\mathfrak{g}$ onto the tangent space to $G(Y)$ at Y. By 1.5, this tangent space is equal to $Y + [\mathfrak{g}', Y]$, so that our assertion is clear if π is the identity. If $\pi \neq \text{Id}$, we are in the situation of Proposition 2.2. Then $d\pi(\mathfrak{g})$ is an ideal of $\mathfrak{g}'$, obviously stable under Ad G', which is a supplementary subspace of the Lie algebra of any maximal torus of G'. Since Y is contained in such an algebra, we see that $[Y, \mathfrak{g}'] = [Y, d\pi(\mathfrak{g})]$, whence our assertion. It follows that the graph Γ of f in $G \times G(Y)$, and $G \times \{Y\}$, cut each other transversally. By the criterion of multiplicity one [8, VI, §2, Theorem 6] the cycle, sum of the irreducible components of $\Gamma \cap (G \times \{Y\}) = Z_G(Y)$, each with coefficient one, is rational over k, hence $(\tilde{Z}_G(Y))^0$ is defined over k [8, Proposition 1, p. 208]. By induction assumption $\tilde{Z}_G(Y)^0$ has a maximal torus T defined over k. But $\tilde{Z}_G(Y)$ contains at least one maximal torus of G, hence T is a maximal torus of G.

3.3. *Proof of Theorem* A (ii). Let now G be reductive. The result is known if G is a torus [6], so we again use induction on dim G. The group $\tilde{Z}_G(Y)^0$ considered above is reductive; in fact, it contains a maximal torus T, and it is clear that $\Phi(\tilde{Z}_G(Y)^0, T)$ is a symmetric subset of $\Phi(G, T)$ (see [2, §2.3, Remark]). We want to prove that the groups $\tilde{Z}_G(Y)^0$, where Y varies over the regular semisimple elements $Y \in \mathfrak{g}'_k$ generate G. Let H be the group they generate and $H' = \pi(H)$; assume $H' \neq G$. There exists then a regular element $X \in \mathfrak{g}'_k$ not in $\mathfrak{h}'$. We have then $z(X_s) \not\subset \mathfrak{h}'$, hence $z(X_s^{[q]}) = z(X_s) \not\subset \mathfrak{h}'$ for every power q of p. We can therefore find a regular semisimple element $Y \in \mathfrak{g}'_k$ such that nil $z(Y) \not\subset \mathfrak{h}'$. Since $z(Y)$ is the Lie algebra of $Z(Y) = \pi(\tilde{Z}_G(Y))$ by Proposition 1.5, it follows that $\tilde{Z}_G(Y)^0 \not\subset H$, contradicting the definition of H. Thus $H = G$. There exist consequently finitely many connected reductive proper k-subgroups H_i ($i = 1, \cdots, t$), such that the product mapping $(h_1, \cdots, h_t) \mapsto h_1 \cdot \cdots \cdot h_t$ of $H_1 \times \cdots \times H_t$ into G is a surjective

k-morphism of the underlying algebraic varieties. By induction assumption, each H_i is unirational over k, hence so is G.

REMARKS. (1) Let X be a regular element in $\mathfrak{g}$. Its nilspace $\mathfrak{n}(X)$, i.e. the set of Y in $\mathfrak{g}$ which are annihilated by some power of ad Y, is by definition in [5] a Cartan subalgebra of $\mathfrak{g}$. We have $\mathfrak{n}(X) = \mathfrak{n}(X_s) = z(X_s)$, and, by Proposition 1.5, $z(X_s)$ is the Lie algebra of $Z(X_s)$. It follows that the subgroups of type (C) of [5] are the centralizers in G of the regular semisimple elements of $\mathfrak{g}$.

(2) In the same context, one can also prove that a reductive k-group splits over a finite separable extension of the groundfield, that the variety of Cartan subgroups of G is rational over k [5], give alternate proofs of some structure theorems of Rosenlicht's about unipotent groups acted upon by tori, and of the conjugacy over k of maximal k-tori in a solvable k-group. Details will be given elsewhere.

REFERENCES

1. A. Borel, *Groupes linéaires algébriques*, Ann. of Math. (2) **64** (1956), 20–80.
2. A. Borel and J. Tits, *Groupes réductifs*, Publ. Math. I.H.E.S. n° **27** (1965), 55–150.
3. P. Cartier, *Isogénies de variétés de groupes*, Bull. S.M.F. **87** (1959), 191–220.
4. C. Chevalley, *Séminaire sur la classification des groupes de Lie algébriques*, 2 vols. Paris, 1958. (Mimeographed)
5. M. Demazure and A. Grothendieck, *Schémas en groupes*, I.H.E.S. 1964. (Mimeographed)
6. M. Rosenlicht, *Some rationality questions on algebraic groups*, Annali di Mat. (IV) **43** (1957), 25–50.
7. J-P. Serre, *Quelques propriétés des variétés abéliennes en caractéristique p*, Amer. J. Math. **80** 1958, 715–739.
8. A. Weil, *Foundations of algebraic geometry*, Amer. Math. Soc. Colloq. Publ. vol. **29**, 2nd ed., Amer. Math. Soc., Providence, R.I. 1962.

77.

(with R. Narasimhan)

Uniqueness conditions for certain holomorphic mappings

Invent. Math. **2** (1967) 247–255

The main purpose of this paper is to show that a morphism $u\colon X \to M$ of connected complex spaces is completely determined by its value at one point $a \in X$, and its effect on the fundamental group of X based at a, if M is covered by a bounded domain, and X is either the complement of a nowhere dense analytic subset of a compact complex space or is pseudo-concave at each point (see Theorem 3.6, where the assumptions on X and M are in fact more general). In particular, under those conditions on X and M, two morphisms of X into M which coincide at one point, and are continuously homotopic, are identical.

This theorem is a direct consequence of Theorem 3.2, which implies that if the above space X is the quotient of a connected complex space Y by a complex Lie group of transformations G, then any orbit of G on Y is a uniqueness set for bounded holomorphic functions. The proof of this uses mainly some properties of pseudo-convex functions which are stated or proved in §§1, 2.

Theorem 3.6 is in particular true if X is a projective variety and $M = D/L$ is the quotient of Siegel's upper half-plane D, of some degree n, by a congruence subgroup L of Siegel's modular group which acts freely. In that case, it was obtained first by Grothendieck, as a corollary to a theorem on abelian schemes in characteristic zero [5]. Furthermore, Grothendieck conjectured in [5] that this should remain true if M were the quotient of a bounded symmetric domain by an arithmetic group acting freely. This conjecture is at the origin of the present paper.

§ 1. Pseudo-Convex Functions

In this paper, all complex spaces are assumed to be reduced and countable at infinity. Morphism is meant in the category of complex spaces and is therefore synonymous with holomorphic mapping.

1.1. Definition. *Let X be a complex space. A function s on X is pseudo-convex if it takes its values in $[-\infty, \infty)$, is upper semicontinuous, and if for any morphism f of a domain W of the complex plane $\mathbb{C}$ into X, the function $s \circ f$ is subharmonic in W.*

---

* Second-named author supported by the Air Force Office of Scientific Research grant AF-AFOSR-1071-66.

704

These functions are called plurisubharmonic in [4]. However, it has become customary to reserve that name for an apparently different notion, which is however equivalent to the above one on complex manifolds. Whether they are in fact different does not seem to be known even on normal complex spaces.

1.2. We first recall some standard properties of pseudo-convex functions, which follow directly from the definition and known facts about subharmonic functions. X is a complex space.

(a) A pseudo-convex function on a connected complex space which has a maximum is constant. In particular, any pseudo-convex function on a compact connected complex space is constant.

(b) If f is holomorphic on X, then $|f|$, $|f|^2$ and $\log |f|$ are pseudo-convex.

(c) A function f on X is pseudo-convex if and only if each point of X has a neighborhood on which f is pseudo-convex.

(d) Let S be a family of pseudo-convex functions on X. If the function

$$x \mapsto \hat{s}(x) = \sup_{s \in S} s(x),$$

has its values in $[-\infty, \infty)$, and is upper semi-continuous, then it is pseudo-convex.

(e) Let Y be a complex space, $f: X \to Y$ a morphism, and s a pseudo-convex function on Y. Then $s \circ f$ is pseudo-convex on X.

We shall need a partial converse to (e):

1.3. Proposition. *Let X, Y be complex spaces, and $f: X \to Y$ a surjective morphism. Let s be a real valued continuous function on Y such that $s \circ f$ is pseudo-convex on X. Then s is pseudo-convex.*

Let $g: W \to Y$ be a morphism of a plane domain W into Y. We have to show that $s \circ g$ is subharmonic. Let

$$Z = \{(w, x) \in W \times X, g(w) = f(x)\},$$

be the "pull-back" of X by g. It is the inverse image of the diagonal in $Y \times Y$ under the morphism $g \times f: W \times X \to Y \times Y$, hence has a natural structure of complex space, such that the projections of $W \times X$ onto the two factors induce morphisms $g': Z \to X$ and $f': Z \to W$ verifying $f \circ g' = g \circ f'$. The map f' is surjective since f is. Moreover, our assumptions and 1.2 (e) imply that $(s \circ g) \circ f'$ is pseudo-convex. Since $s \circ g$ is continuous, we are reduced to proving that if Y is a domain in $\mathbb{C}$ then s is subharmonic.

Let $A \subset Y$ be the union of the images of the irreducible components of X on which f is not constant. It is open. Its complement B is the image of a set of irreducible components of X on which f is constant, hence is countable. By a well-known fact (see remark below), it suffices to prove that s is subharmonic on A. Let $a \in A$ and Z be an irreducible component of X such that $f(Z)$ is an open neighborhood of a, and choose $z \in Z \cap f^{-1}(a)$. The fibres of f on Z have all codimension one. Therefore (see proof of Prop. 3, Chap. VII in [8]) there exists an analytic set T of dimension one in the neighborhood of z in Z such that $f|_T$ is a proper map of T onto a neighborhood N of a. The map $f|_T$ is then a local isomorphism outside a discrete set E whose image $f(E)$ is discrete in N. The function s is then clearly subharmonic on $N - f(E)$. Since $f(E)$ is discrete and s is continuous, it is then subharmonic in N. Thus s is subharmonic in A.

1.4. We have used the known fact that a continuous function u on a domain $U \subset \mathbb{C}$, which is subharmonic in the complement of a closed countable set B of U, is subharmonic in U. For the sake of completeness, we recall briefly a proof.

We may assume U to be bounded. There exists a subharmonic function p on U which is <0, and is equal to $-\infty$ exactly at the points of B. This is true for any set B of capacity zero, but is readily checked in our case. Namely, we can take

$$p(x) = \sum_{b \in B} e_b \cdot \log |x - b| / r,$$

where

$$\sum_{b \in B} e_b$$

is a convergent series of strictly positive numbers, and r a positive constant $> |x - y|$ $(x, y \in U)$.

It follows then that $u_\varepsilon = u + \varepsilon \cdot p$ is subharmonic on U for every strictly positive ε. Let K be a compact subset of U and h a continuous function which majorizes u on the boundary ∂K of K and is harmonic in the interior Int K of K. Then h majorizes a fortiori u_ε on ∂K hence $u_\varepsilon(x) \leqq h(x)$ for $x \in K$ and every $\varepsilon > 0$. Letting $\varepsilon \to 0$, we see that $u(x) \leqq h(x)$ for every $x \in K - (K \cap B)$, hence, by continuity, for every $x \in K$, which shows that u is subharmonic in U. We note that this (known) argument depends only on the properties of p stated above, hence is valid whenever B has capacity zero.

§ 2. Non-Existence of Pseudo-Convex Functions

2.1. Proposition. *Let X be a connected complex space, Y the complement of a nowhere dense analytic subset in a compact complex space C, and $f: Y \to X$ a surjective morphism. Then every pseudo-convex function on X which is bounded above is constant.*

Let (C_n, π) be the normalization of C. It is compact, and $f \circ \pi$ is a surjective morphism of $\pi^{-1}(Y)$ onto X. Thus it suffices to consider the case where C is normal.

Let s be a pseudo-convex function on X which is bounded above. Then $s \circ f$ is pseudo-convex on Y, and bounded above. By an extension theorem of Grauert and Remmert [4, Satz 3], $s \circ f$ extends to a pseudo-convex function on C, hence is locally constant on Y. Since X is connected, s is then constant.

Remark. The same argument, using Satz 4 of [4] instead of Satz 3, shows that every pseudo-convex function on X is constant if $C - Y$ has complex codimension ≥ 2 in C.

2.2. A complex space X is called *pseudo-concave with respect to a point $a \in X$* if there is a connected relatively compact open set U containing a with the following property:

(PCV) for every $x \in \partial U$ there exists a neighborhood V of x in X and an irreducible analytic set S of dimension > 0 in V such that

$$S \subset \overline{U}, \qquad S \cap \partial U = \{x\}.$$

A complex space X is *pseudo-concave* if it is pseudo-concave with respect to any of its points.

2.3. Proposition. *Let X be a connected pseudo-concave complex space. Then any pseudo-convex function on X is constant.*

Let $a \in X$ and let U be a neighborhood of a having property (PCV) of 2.2. We assert that every pseudo-convex function s on X is constant on U. In fact, let x be a point of ∂U where $s(x) = \sup s(y)$ $(y \in \overline{U})$. Let V and S be as in (PCV) above. Clearly, if V is small enough, $S \cap \partial V \cap \partial U = \emptyset$, and $S \cap \partial V \subset U$. But

$$s(x) \leq \sup_{z \in S \cap V} s(z) = \sup_{y \in S \cap \partial V} s(y),$$

(since the restriction of s to S is pseudo-convex), hence there exists a point $y_0 \in U$ such that $s(y_0) = \sup s(y)$ $(y \in U)$, which implies our contention (1.2a).

This shows that s is locally constant; since X is connected, s is constant.

2.4. Remark. If X is a complex manifold, the above proposition remains valid if the condition (PCV) in 2.2. is replaced by the following weaker one:

$(PCV)'$: every $x \in \partial U$ has a fundamental set of neighborhoods V in X such that the set of points $y \in V$ for which

$$|g(y)| \leq \sup_{z \in U \cap V} |g(z)|$$

for every holomorphic function g on V is a neighborhood of x.

§ 3. Uniqueness Theorems

3.1. Lemma. *Let X be a complex space and F a family of holomorphic functions on X bounded in modulus by a constant M. Then F is equicontinuous. The function*

$$x \mapsto g(x) = \sup_{f \in F} |f(x)|$$

is a continuous pseudo-convex function.

If g is continuous, then it is pseudo-convex by 1.2b, d. On the other hand, it is elementary that g is continuous if F is equicontinuous. It suffices therefore to prove the first assertion.

This is a local question, so we may assume X to be an analytic set in some polydisc D. Let D' be a polydisc whose closure is contained in D. Then there exists a constant M' such that every holomorphic function f on X which is bounded in modulus by M is on $D' \cap X$ the restriction of a holomorphic function on D' which is bounded in modulus by M'. This is a known fact: the spaces $H^0(D, \mathcal{O}_D)$ and $H^0(X, \mathcal{O}_X)$, endowed with the topology of uniform convergence on compact sets, are Fréchet spaces. The restriction map $H^0(D, \mathcal{O}_D) \to H^0(X, \mathcal{O}_X)$ is continuous, and is surjective by Theorem A, hence is open by the closed graph theorem, which implies our assertion. (See e.g. [3, p. 240], where however this happens to be stated only for X normal. In fact, by lifting our functions to the normalization of X, we could anyway reduce the proof of 3.1 to that case.)

We are thus reduced to the case where X is a domain in $\mathbb{C}^n$, for which the equicontinuity is well known, and follows from CAUCHY's formula.

3.2. Theorem. *Let Y be a connected complex space and G a complex transformation group of Y. Assume that the quotient ringed space $X = Y/G$ is a complex space on which every pseudo-convex function bounded above is constant. Let $a \in Y$, and f be a bounded holomorphic function on Y which takes only finitely many values on $G \cdot a$. Then f is constant.*

(A complex transformation group of Y is a complex Lie group G of automorphisms of Y whose action on Y is defined by a holomorphic mapping of $G \times Y$ into Y.)

Let $c_1, \ldots, c_m$ be the values of f on $G \cdot a$. Then

$$\prod_i (f - c_i)$$

is zero on $G \cdot a$. It suffices therefore to show that f vanishes identically on Y if it is zero on $G. a$.

For every $g \in G$, let us denote by $g \cdot f$ the function on Y defined by $g \cdot f(y) = f(g \cdot y)$, $(y \in Y)$ and let

$$s(y) = \sup_{g \in G} |g \cdot f(y)|, \quad (y \in Y).$$

The functions $g \cdot f$ are bounded in modulus by any upper bound for $|f|$. Hence (3.1) s is a continuous pseudo-convex function which is bounded above. It is clearly invariant under G, therefore (2.4) it is of the form $s' \circ \pi$, where $\pi: Y \to X$ is the canonical projection, and s' is a continuous pseudo-convex function on X, which is then bounded above. By our assumption, s', hence s, is constant. Since $f(g \cdot a) = 0$ for all $g \in G$, the function s is zero at a, hence everywhere, which proves our contention.

3.3. Corollary. *If G has a compact orbit on Y, then any bounded holomorphic function on Y is constant.*

This follows from the theorem since any holomorphic function is constant on every connected component of a compact orbit.

3.4. We make a few remarks on our assumptions. (a) The second condition imposed on X is fulfilled in particular if X is the image under a morphism of the complement of a nowhere dense analytic subset A of a compact complex space C (2.1), or if X is pseudo-concave (2.3). In fact, in the latter case, or in the former if $\dim_{\mathbb{C}} C - \dim_{\mathbb{C}} A \geq 2$, no global boundedness from above is needed, so that the conclusion remains true if the assumption "$|f|$ bounded" is replaced by

$$\sup_{g \in G} |f(g \cdot y)|$$

is finite and locally bounded on Y.

(b) The quotient of a complex space by a complex transformation group carries a natural structure of ringed space, but is not necessarily a complex space. When it is, one often says that "the quotient Y/G exists". If Y is connected, a necessary condition for this is that all orbits of G be closed, of the same dimension [6, Satz 7, p. 337]. However, by a theorem of Cartan [2], the quotient exists if G is a discrete group which is properly discontinuous (i.e. given a compact set $K \subset Y$, the set of $g \in G$ such that $g \cdot K \cap K \neq \emptyset$ is finite). More generally Holmann has shown that the quotient Y/G exists either if Y/G is Hausdorff and Y

satisfies the "weak Riemann extension theorem", [1, Satz 15, p. 350], or if G is almost proper [1, Satz 19, p. 356]. This last condition means the following: given a compact subset $K \subset Y$, there exists a compact subset $L \subset G$ such that for any $x \in K$, the set of $g \in G$ for which $g \cdot x \in K$ is contained in $L \cdot G_x$, where $G_x = \{g \in G, g \cdot x = x\}$.

3.5. As is usual, we denote the fundamental group of a topological space X, based at a point $x \in X$, by $\pi_1(X, x)$. A continuous map $f: X \to Y$ induces a homomorphism $\pi_1(X, x) \to \pi_1(Y, f(x))$ to be denoted $\pi_{1,x}(f)$.

Let X be a connected complex space. It is arcwise connected and locally simply connected. Let Y be its universal covering, and $\pi: Y \to X$ the canonical projection. There is on Y a unique structure of complex space such that π is a local isomorphism at every point. The group L of covering transformations is then a properly discontinuous group of automorphisms of Y, acting freely (i.e. only the identity has a fixed point), and $X \cong Y/L$ (as complex spaces). Furthermore, for any $y \in Y$ there is a canonical isomorphism of L onto $\pi_1(X, \pi(y))$.

By a *bounded covering* of a complex space X we mean an unramified covering of X which is isomorphic to an analytic subset of a bounded domain in $\mathbb{C}^n$.

3.6. Theorem. *Let X be a connected complex space which carries no non-constant pseudo-convex function which is bounded above. Let $a \in X$ and M be a complex space admitting a bounded covering. Let u, v be two morphisms of X into M which verify $u(a) = v(a)$ and $\pi_{1,a}(u) = \pi_{1,a}(v)$. Then $u = v$.*

Let $X = Y/G$ and $M = D_0/L$, where Y and D_0 are the universal coverings of X and M respectively and G, L the corresponding groups of covering transformations. Let D be a covering of M which can be embedded in a bounded domain in $\mathbb{C}^n$. Choose points $a' \in Y$ lying over a and $b' \in D$, $b_0 \in D_0$ lying over $b = u(a) = v(a)$. Let $r: G \to L$ be the homomorphism defined by means of $\pi_{1,a}(u) = \pi_{1,a}(v)$ via the canonical isomorphisms $G \cong \pi_1(X, a)$, $L \cong \pi_1(M, b)$ associated to a', b_0. Let $p: D_0 \to D$ be the canonical projection verifying $p(b_0) = b'$. Then u, v can be lifted uniquely to morphisms $u_0, v_0: Y \to D_0$ which verify

$$u_0(a') = v_0(a') = b_0; \quad u_0(g \cdot a') = r(g) \cdot u_0(a'),$$

$$v_0(g \cdot a') = r(g) \cdot v_0(a'), \quad (g \in G),$$

so that u_0, v_0 coincide on $G \cdot a'$. Then $u' = p \circ u_0$ and $v' = p \circ v_0$ are liftings of u, v to morphisms $Y \to D$ which map a' onto b' and clearly coincide on $G \cdot a'$. By assumption, D may be embedded in a bounded domain in $\mathbb{C}^n$. The coordinates of $u' - v'$ are then bounded holomorphic functions on Y which vanish on $G \cdot a'$. By 3.2, they vanish identically.

We remark that this proof remains valid if M is assumed to have an unramified covering whose points are separated by bounded holomorphic functions.

3.7. Theorem 3.6 shows that morphisms into M are to a large extent governed by their effect on the fundamental groups. As another instance of this, we would like to point out a consequence of an extension theorem of [1], which answers a question raised by Grothendieck, also à propos of [5]. To state it, we recall a definition of [1]: an open subset U of a complex space X is *weakly pseudo-convex at a boundary point* $a \in \partial U$ if there is no injective morphism of the open unit disc into X whose image contains a and belongs to $U \cup \{a\}$. The set U is weakly pseudo-convex if it is so at every point.

3.8. Theorem. *Let W be the complement of a nowhere dense analytic set A in a normal connected complex space X. Let $f: W \to M$ be a morphism of W into a complex space M having an unramified covering D which is isomorphic to a relatively compact weakly pseudo-convex open subset of a Stein space. Assume that for some $w \in W$, there exists a homomorphism $\sigma: \pi_1(X, w) \to \pi_1(M, f(w))$ such that*

$$\pi_{1,w}(f) = \sigma \circ \pi_{1,w}(j),$$

where j is the inclusion map of W in X. Then f extends to a morphism of X into M.

In view of the way the fundamental group of an arcwise connected space depends on the base point, it is clear that our assumption is verified for any $w \in W$. Let X' be an open connected subset of X and $W' = X' \cap W$. We have a commutative diagram of homomorphisms

$$\begin{array}{ccc}
\pi_1(W', x) & \to & \pi_1(X', x) \\
\downarrow & & \downarrow \\
\pi_1(W, x) & \to & \pi_1(X, x)
\end{array} \qquad (x \in W'),$$

induced by inclusion maps, which shows that our assumptions are still fulfilled by X', W' and $f|_{W'}$. Since the extension problem is local, and since each point of X has an open, connected and simply connected neighborhood, we see that it suffices to consider the case where $\pi_{1,w}(f)$ is the trivial map. In this case, f can be lifted to a morphism of W into the universal covering of M, hence also to a morphism f' of W into D. By Theorem 5 of [1], f' extends then to X, whence the result.

Remarks. As to the assumption on D, we recall that a bounded symmetric domain is weakly pseudo-convex [7, Theorem 7]. More generally, if D' is relatively compact and Runge in a Stein space, then it is weakly pseudo-convex.

The same argument, using Theorem 2 of [*1*], shows that the conclusion of 3.8 is valid if A has complex codimension $\geqq 2$ and M is covered by a Stein space.

Thus 3.8 applies in particular if $M = D/L$, where D and L are as in the introduction. In this case, it is the transcendental analogue of Theorem 4.1 in [5].

References

[*1*] ANDREOTTI, A., and W. STOLL: Extension of holomorphic maps, Annals of Math. (2) **72**, 312—349 (1960).

[2] CARTAN, H.: Quotient d'un espace analytique par un groupe d'automorphismes, Algebraic Geometry and Algebraic Topology, a Symposium in honor of S. Lefschetz, 90—102 Princeton: Princeton University Press 1957.

[3] GRAUERT, H.: Charakterisierung der holomorph vollständigen komplexen Räume. Math. Annalen **129**, 233—259 (1955).

[4] GRAUERT, H., and R. REMMERT: Plurisubharmonische Funktionen in komplexen Räumen. Math. Z. **65**, 175—194 (1956).

[5] GROTHENDIECK, A.: Un théorème sur les homomorphismes de schémas abéliens. Inventiones math. **2**, 59—78 (1966).

[6] HOLMANN, H.: Komplexe Räume mit komplexen Transformationsgruppen. Math. Annalen **150**, 327—360 (1963).

[7] MOORE, C. C.: Compactification of symmetric spaces II: The Cartan domains. Amer. J. Math. **86**, 358—378 (1964).

[8] NARASIMHAN, R.: Introduction to the theory of analytic spaces. Lecture Notes 25, Berlin-Heidelberg-New York: Springer 1966.

The Institute for Advanced Study
Princeton, New Jersey

(*Received December 17, 1966*)

78.

Sur une généralisation de la formule de Gauss-Bonnet

An. Acad. Bras. Cienc. **39** (1967) 31–37

La formule de Gauss-Bonnet-Allendoerfer-Weil-Fenchel-Chern... exprime la caracté-ristique d'Euler-Poincaré d'une variété V^{2n} riemannienne compacte orientée comme l'inté-grale d'une 2n-forme qui, sur l'espace des repères orthonormaux, est un polynome universel en les composantes de la forme de courbure (*voir* par exemple [4, p. 38], l'énoncé est rappelé au §3). Récemment, AVEZ [1] et CHERN [5] ont montré que cette formule est encore valable, au signe près, sur une variété pseudo-riemannienne. Nous nous proposons ici de déduire ce fait de remarques générales simples sur les connexions. Il s'agit surtout de relier les homo-morphismes caractéristiques, au sens de la géométrie différentielle, d'un espace fibré principal différentiable ξ de groupe structural un groupe semi-simple linéaire réel G et de l'espace fibré principal associé dont le groupe structural est une forme compacte de la complexification de G, (*voir* Théorème 2.2.). L'application à la classe d'Euler-Poincaré (d'un fibré qui n'est pas nécessairement un fibré tangent) est faite dans le §3, où l'on examine aussi, brièvement, de ce point de vue, les classes de Pontrjagin, et les classes de Chern dans le cas pseudo-her-mitien.

§1. CONNEXIONS

Le but de ce paragraphe est le lemme 1.5, qui est élémentaire et bien naturel. Si le paragraphe est plus long que ce lemme ne le mérite, c'est que, tout en fixant les notations, nous avons rappelé au passage des définitions et faits connus.

1.1. Soit G un groupe de Lie réel. On note **g** son algèbre de Lie, Ad **g** la repré-sentation adjointe de G dans **g**, S (G) (resp. I (G)) l'algèbre des polynomes (resp. des poly-nomes invariants par G), à coefficients réels, sur **g**.

Si $\lambda : G \longrightarrow H$ est un homomorphisme de G dans un groupe de Lie H, la différentielle $\delta\lambda$ de λ induit un homormorphisme de **g** dans **h**, d'où un homomorphisme de S (H) dans S (G), qui sera noté λ°, et qui applique visiblement I (H) dans I (G).

1.2. Soit ξ une fibration principale différentiable de groupe structural $G = G_\xi$ de Lie (pour les espaces fibrés, nous utilisons en général les notations et définitions de [12, 3]). Le groupe G opère librement, à droite, sur l'espace total E_ξ de ξ, d'où, pour tout $x \in E_\xi$, un isomorphisme canonique α_x de **g** sur l'espace tangent V_x à la fibre de x. Rappelons qu'une *connexion* sur ξ peut être définie comme la donnée d'une 1-forme différentielle sur E_ξ, à valeurs dans **g** telle que

$$(1) \qquad \operatorname{Ad}(g^{-1}) \cdot \omega = \omega \cdot g \,;\; \omega(\alpha_x(z)) = z, \; (g \in G; \, z \in \mathbf{g}; \, x \in E_\xi).$$

713

Le noyau de l'application $\omega_x : E_{\xi,x} \rightarrow \mathbf{g}$, où $E_{\xi,x}$ est l'espace tangent à E_ξ en x, définie par ω est un sous-espace H_x supplémentaire de V_x, l'espace des *vecteurs horizontaux* en x; la famille des plans (H_x) $(x \in E_\xi)$ dépend différentiablement de x et est invariant par G. Réciproquement une telle famille de plans détermine univoquement une connexion [4, Chap. III; 8, Chap. II, §1].

1.3. Soit Ω la forme de courbure de la connexion ω. C'est une 2-forme, à valeurs dans $\mathbf{g}$, définie par

$$(1) \qquad \Omega = d\omega + (1/2)\,[\omega, \omega] \quad,$$

qui vérifie

$$(2) \qquad \mathrm{Ad}\,(g^{-1}) \cdot \Omega = \Omega \cdot g \qquad (g \in G).$$

Ses composantes (Ω_i) relativement à une base (e_i) de $\mathbf{g}$ sont donc des 2-formes différentielles scalaires.

Soit $P = P(X_1, \ldots, X_m) \in S(G)$ et soit k le degré de P. Comme les Ω_i sont dans le centre de l'algèbre des formes différentielles sur E_ξ, on peut former $P(\Omega) = P(\Omega_1, \ldots, \Omega_m)$, qui est une forme différentielle extérieure de degré 2k sur E_ξ. Si de plus $P \in I(G)$, alors $P(\Omega)$ est une forme de base fermée, ce qui signifie qu'il existe une forme fermée μ_P sur la base B_ξ de ξ, de degré 2k, telle que $\pi_\xi^* \mu_P = P(\Omega)$, ($\pi_\xi$ étant la projection canonique de E_ξ sur B_ξ). En associant à $P \in I(G)$ la classe de cohomologie de μ_P dans l'algèbre de cohomologie réelle $H^*(B_\xi; \mathbf{R})$ de B_ξ, on définit un homorphisme $\sigma_\xi : I_G \rightarrow H^*(B_\xi; \mathbf{R})$, dit *caractéristique*, qui, d'après un théorème de WEIL [4, pp. 57–58], ne dépend que de ξ, et non de la connexion. Pour abréger, on dira que la classe de cohomologie $\sigma_\xi(P)$ est *représentée par* la forme différentielle $P(\Omega)$.

1.4. Soient $Z : G \rightarrow H$ un homomorphisme de G dans un groupe de Lie H, et $\eta = \lambda(\xi)$ la fibration principale obtenue par extension du groupe structural à partir de ξ [3, §6]. On a donc

$$(1) \qquad G_\eta = H, \; B_\xi = B_\eta, \; E_\eta = E_\xi \times_G H,$$

et l'application $x \longmapsto (x, e)$, où e est l'élément neutre de H, induit un morphisme $f : E_\xi \rightarrow E_\eta$ qui vérifie

$$(2) \qquad f(x \cdot g) = f(x) \cdot \lambda(g), \qquad \pi_\eta \circ f = \pi_\xi \;(x \in E_\xi \,;\, g \in G).$$

LEMME. *Étant donnée une connexion sur ξ, de forme ω et de forme de courbure Ω, il existe une et une seule connexion sur η dont la forme ω' vérifie*

$$(3) \qquad f^* \omega' = \delta\lambda \circ \omega.$$

La forme de courbure Ω' de cette connexion satisfait à la relation

$$(4) \qquad f^* \Omega' = \delta\lambda \circ \Omega.$$

La relation (2) montre immédiatement que la différentielle de f en $x \in E_\xi$ est un isomorphisme de H_x sur un sous-espace H'_y de $E_{\eta,y}$ $(y = f(x))$ qui est un supplémentaire de l'espace tangent à la fibre de y. De plus, si $y' = f(x')$ est dans la même fibre que y, alors, d'après (2), il existe $g \in G$ tel que $x' = x \cdot g$, d'où $H_{y'} = H_y \cdot \lambda(g)$. Il s'ensuit que l'ensemble des translatés par H des plans H'_y $(y \in f(E_\xi))$ est l'ensemble des plans horizontaux d'une connexion dont la forme ω' vérifie (3) par construction; cette connexion est unique puisque $f(E_\xi)$ rencontre chaque fibre de η.

L'égalité (4) résulte alors de (3), 1.3(1) et des relations évidentes

$$\mathrm{d}\,(\delta\lambda \circ \omega) = \delta\lambda \circ \mathrm{d}\omega\,, \quad [\delta\lambda \circ \omega,\, \delta\lambda \circ \omega] = \delta\lambda \circ [\omega, \omega]\,.$$

On dira que ω' est *l'extension de ω à* E_η. Si λ est l'inclusion d'un sous-groupe fermé, le seul cas intéressant pour la suite de cet article, alors E_ξ s'identifie à une sous-variété de E_η, et ω n'est autre que la restriction de ω' à E_ξ.

LEMME **1.5**. *Soient* G, H *des groupes de Lie,* $\lambda : G \to H$ *un homomorphisme,* ξ *une fibration principale de groupe structural* G, *base* B, *et* η *la fibration principale de groupe structural* H *obtenue par extension de* ξ *suivant* λ. *Soient* $\sigma_\eta : \mathrm{I}\,(\mathrm{H}) \to \mathrm{H}^*\,(\mathrm{B}; \mathbf{R})$ *et* $\sigma_\xi : \mathrm{I}\,(\mathrm{G}) \to \mathrm{H}^*\,(\mathrm{B}; \mathbf{R})$ *les homomorphismes caractéristiques de* ξ, η *et* $\lambda^\circ : \mathrm{I}\,(\mathrm{H}) \to \mathrm{I}\,(\mathrm{G})$ *l'homomorphisme induit par* λ. *Alors* $\sigma_\eta = \sigma_\xi \circ \lambda^\circ$.

Soient ω la forme d'une connexion sur ξ, et ω' l'extension de ω à E_η (cf. 1.4). Soient $\mathrm{Q} \in \mathrm{I}\,(\mathrm{H})$ et $\mathrm{P} = \lambda^\circ(\mathrm{Q})$. Soient Ω, Ω' les formes de courbure de ω et ω', et α_P, α_Q les formes différentielles sur B telles que

$$\pi_\xi^*(\alpha_\mathrm{P}) \;=\; \mathrm{P}\,(\Omega) \qquad \pi_\eta^*(\alpha_\mathrm{Q}) \;=\; \mathrm{Q}\,(\Omega')\,.$$

Par définition, $\sigma_\xi(\mathrm{P})$ (resp. $\sigma_\eta(\mathrm{Q})$) est la classe de cohomologie de α_P, (resp. α_Q). Il suffit donc de faire voir que $\alpha_\mathrm{P} = \alpha_\mathrm{Q}$ ou encore, puisque π_ξ^* est injectif sur les formes différentielles, que $\pi_\xi^* \alpha_\mathrm{P} = \pi_\xi^* \alpha_\mathrm{Q}$. Or, cela est immédiat car, si $\mathrm{f} : E_\xi \to E_\eta$ est le morphisme canonique, on a (cf. 1.4)

$$\pi_\xi^* \alpha_\mathrm{Q} \;=\; \mathrm{f}^* \cdot \pi_\eta^*(\alpha_\mathrm{Q}) \;=\; \mathrm{f}^*\,(\mathrm{Q}\,(\Omega')) \;=\; \mathrm{Q}\,(\mathrm{f}^*(\Omega'))\,, \quad ,$$

d'où, vu 1.4(4),

$$\pi_\xi^* \alpha_\mathrm{Q} \;=\; \mathrm{Q}\,(\delta\lambda \circ \Omega) \;=\; \mathrm{P}\,(\Omega) \;=\; \pi_\xi^* \alpha_\mathrm{P} \quad .$$

§2. GROUPES SEMI-SIMPLES

2.1. Soient $\mathrm{G}, \mathrm{G}' \subset \mathbf{GL}\,(\mathrm{n}, \mathbf{R})$ des groupes linéaires réels fermés ayant un nombre fini de composants connexes et dont l'intersection K est un sous-groupe compact maximal de G.

Soit ξ une fibration principale de groupe structural G. Comme G/K est homéomorphe à un espace euclidien, d'après le théorème de Cartan-Malcev, (qui vaut aussi lorsque G a un nombre fini de composantes connexes [9]) on peut restreindre le groupe structural de ξ à K, et la fibration principale θ ainsi obtenue est unique à un isomorphisme près [12, Th. 12.7]. Soit ξ' la fibration de groupe structural G' obtenue à partir de θ par extension du groupe structural de K à G'. Les fibrations θ et ξ' peuvent être supposées différentiables si ξ l'est. On dira que ξ' *est associée à* ξ; cette fibration est déterminée à un isomorphisme près.

THÉORÈME **2.2**. *Soient* $\mathrm{G}, \mathrm{G}', \mathrm{K}, \xi, \xi'$ *comme en* 2.1 *et soient* $\mathrm{P} \in \mathrm{I}\,(\mathrm{G})$, $\mathrm{Q} \in \mathrm{I}\,(\mathrm{G}')$. *Si les restrictions de* P *et* Q *à l'algèbre de Lie* k *de* K *sont égales, alors* $\sigma_\xi(\mathrm{P}) = \sigma_{\xi'}(\mathrm{Q})$.

Cela résulte du lemme 1.5, appliqué d'une part à θ et ξ, d'autre part à θ et ξ'.

2.3. Le théorème précédent montre que pour exprimer une classe caractéristique de ξ', correspondant à $\mathrm{Q} \in \mathrm{I}\,(\mathrm{G}')$, à l'aide de la forme de courbure d'une connexion sur ξ, il suffit de connaître un polynome $\mathrm{P} \in \mathrm{I}\,(\mathrm{G})$ dont la restriction à $\mathbf{k}$ est égale à celle de Q. Pour pouvoir déterminer P dans certains cas particuliers, nous intercalons ici quelques remarques sur les invariants dans le cas où G, G' sont semi-simples et où il existe un élément $\mathrm{A} \in \mathbf{GL}\,(\mathrm{n}, \mathbf{C})$, centralisant K, tel que $\mathrm{G}_\mathrm{u} = \mathrm{A}^{-1} \cdot \mathrm{G}' \cdot \mathrm{A}$ soit un sous-groupe compact maximal de la *complexification* G_c de G, (c'est à dire du plus petit sous-groupe analytique complexe G_c de $\mathbf{GL}\,(\mathrm{n}, \mathbf{C})$ contenant G).

Soit $\mathbf{g} = \mathbf{k} + \mathbf{p}$ une *décomposition de Cartan* de $\mathbf{g}$. L'espace $\mathbf{p}$ est donc le complément orthogonal de $\mathbf{k}$ par rapport à la forme de Killing et $s : x + y \longmapsto x - y$ ($x \in \mathbf{k}$, $y \in \mathbf{p}$) est un automorphisme involutif de $\mathbf{g}$, une *involution de Cartan* de $\mathbf{g}$. On sait que l'on peut supposer que l'algèbre de Lie $\mathbf{g}_u$ de G_u est $\mathbf{g}_u = \mathbf{k} + \sqrt{-1} \cdot \mathbf{p}$, et que s induit un automorphisme involutif de $\mathbf{g}_u$, d'où aussi un automorphisme involutif s' de l'algèbre de Lie $\mathbf{g}'$ de G', dont $\mathbf{k}$ est l'ensemble des points fixes (*voir* p. ex. [6, Chap. III]).

On désignera par $(x_1, \ldots, x_a, y_1, \ldots, y_b)$ des coordonnées sur $\mathbf{g}$ par rapport à une base $(e_1, \ldots, e_a, f_1, \ldots, f_b)$ dont les a premiers éléments sous-tendent $\mathbf{k}$ et les b derniers sous-tendent $\mathbf{p}$, et par $(x'_1, \ldots, x'_a, y'_1, \ldots, y'_b)$ les coordonnées sur $\mathbf{g}'$ par rapport à la base $(A \cdot e_i \cdot A^{-1}, -\sqrt{-1} \cdot A \cdot f_j \cdot A^{-1})$ $(1 \leq i \leq a; 1 \leq j \leq b)$. Pour les monomes, on adopte la notation multiplicative usuelle:

$$x^u = x_1^{u_1} \cdot \ldots \cdot x_a^{u_a} \qquad |u| = u_1 + \ldots + u_a \ ,$$

etc.

LEMME **2.4**. *Nous conservons les notations et conventions de 2.3. Soient* $i^* : I(G) \longrightarrow I(K)$ *et* $i'^* : I(G') \longrightarrow I(K)$ *les homomorphismes de restriction, et soit* $Q = \sum c_{u,v} \cdot x'^u \cdot y'^v \in I(G')$. *Alors*

$$(1) \qquad P_Q = \sum c_{u,v} \cdot (\sqrt{-1})^{|v|} \cdot x^u \cdot y^v \ ,$$

est un élément de $I(H) \otimes \mathbf{C}$ *dont la restriction à* $\mathbf{k}$ *est égale à celle de Q. Si* $\mathbf{g}$ *et* $\mathbf{k}$ *ont même rang alors* i^* *et* i'^* *sont injectifs, chaque monome d'un polynome* $I(G)$ (*resp.* $I(G')$) *est de degré pair en les variables* y_i (*resp.* y'_i), *et* $P_Q \in I(G)$.

On peut envisager Q comme un polynome sur l'algèbre de Lie $\mathbf{g}_c$ de G_c, qui est alors nécessairement invariant par Ad G_c. Sa restriction à $\mathbf{g}$ est donc un élément de $I(G) \otimes C$ qui, vu les relations entre G' et G_u et les conventions de 2.3, est donnée par (1).

Supposons maintenant $\mathbf{g}$ et $\mathbf{k}$ de même rang. Cela signifie qu'une sous-algèbre de Cartan $\mathbf{c}$ de $\mathbf{k}$ est aussi une sous-algèbre de Cartan de $\mathbf{g}$, donc de $\mathbf{g}_u$ et de $\mathbf{g}'$. L'ensemble de ses conjugués par G' (resp. G) est $\mathbf{g}'$ (resp. un ouvert de $\mathbf{g}$) donc un polynome invariant est complètement déterminé par sa restriction à $\mathbf{c}$.

On sait d'autre part qui si $\mathbf{g}$ et $\mathbf{k}$ ont même rang, les involutions s, s' sont des automorphismes intérieures définis par un élément de K. Par suite, tout polynome de $I(G)$ (resp. $I(G')$) est invariant par s (resp. s'), d'où la dernière assertion.

§3. CLASSES D'EULER-POINCARÉ ET DE PONTRJAGIN

3.1. On note $\mathbf{O}(p,q)$ (resp. $\mathbf{O}(p,q)^+$) le groupe orthogonal réel (resp. propre) de la forme quadratique

$$I_{p,q} = x_1^2 + \ldots + x_p^2 - x_{p+1}^2 - \ldots - x_{p+q}^2 \ ,$$

sur $\mathbf{R}^n$ ($n = p + q$).

Soient $G = \mathbf{O}(p,q)$ (resp. $\mathbf{O}(p,q)^+$) et $G' = \mathbf{O}(n)$ (resp. $\mathbf{SO}(n)$). Alors $K = G \cap G'$ est égal à $\mathbf{O}(p) \times \mathbf{O}(q)$ (resp. $(\mathbf{O}(p) \times \mathbf{O}(q)) \cap \mathbf{SO}(n)$) et est un sous-groupe compact maximal de G. Le groupe G_c est le groupe orthogonal complexe (resp. propre) de $I_{p,q}$. Soit $(e_i)_{1 \leq i \leq n}$ la base canonique de $\mathbf{C}^n$. Il est clair que la transformation linéaire A de $\mathbf{C}^n$ définie par

$$A(e_i) = e_i, \ (1 \leq i \leq p), \ A(e_i) = \sqrt{-1} \cdot e_i, \ (p < i \leq n),$$

centralise K et applique G_c sur $\mathbf{O}(n,\mathbf{C})$ (resp. $\mathbf{SO}(n,\mathbf{C})$). Il en résulte que $G_u = A^{-1} \cdot \mathbf{O}(n) \cdot A$ (resp. $G_u = A^{-1} \cdot \mathbf{SO}(n).A$) est un sous-groupe compact maximal de G_c. On est bien dans le cas envisagé dans **2.3**.

Soit E_j^i la matrice dont le coefficient à l'intersection de la i-ème ligne et de la j-ième colonne est égal à 1 et les autres coefficients sont nuls. Alors $\mathbf{g} = \mathbf{k} + \mathbf{p}$, où $\mathbf{k}$ est sous-tendue par les matrices

$$e_j^i = E_j^i - E_i^j \,, \quad (1 \le i < j \le p, \quad \text{ou} \quad p < i < j \le n) \,,$$

et $\mathbf{p}$ par les matrices

$$f_j^i = E_j^i + E_i^j \,, \quad (1 \le i \le p < j \le n) \,.$$

L'algèbre de Lie $so(n)$ de $\mathbf{so}(n)$ est engendrée par

$$e'^i_j = E_j^i - E_i^j \,, \quad (1 \le i < j \le p \,, \quad \text{ou} \quad p < i < j \le n) \,,$$

$$f'^i_j = E_j^i - E_i^j \,, \quad (1 \le i \le p < j \le n) \,,$$

et l'on a

$$A^{-1} \cdot e'^i_j \cdot A = e_j^i \,, \quad A^{-1} \cdot f'^i_j \cdot A = \sqrt{-1} \cdot f_j^i \,.$$

On notera z_j^i (resp. z'^i_j) les coordonnées correspondantes sur $\mathbf{g}$ (resp. $\mathbf{g}'$). On a donc, dans les notations de **2.3**:

$$z_j^i = x_j^i \ (1 \le i < j \le p \ \text{ou} \ p < i < j \le n) \,, \quad z_j^i = y_j^i \,, \quad (1 \le i \le p \le j \le n) \,,$$

et des formules similaires pour $\mathbf{g}'$.

3.2. Soient Ω'^i_j (resp. Ω_j^i) les composantes du tenseur de courbure d'une connexion sur un fibré différentiable, de groupe structural G (resp. G'), par rapport à la base ci-dessus de $\mathbf{g}$ (resp. $\mathbf{g}'$), et Ω_{ij} (resp. Ω'_{ij}) les composantes obtenues par abaissement de l'indice contravariant. On a

(1) $$\Omega'^i_j = \Omega'_{ij} \,, \quad \Omega'_{ij} = -\Omega'_{ji} \,, \quad (1 \le i, j \le n),$$

(2) $$\Omega_j^i = \Omega_{ij}, \ (1 \le i \le p); \ \Omega_j^i = (p < i \le n); \ \Omega_{ij} = -\Omega_{ji} \ (1 \le i, j \le n) \,.$$

THÉORÈME **3.3.** *Soient ξ une fibration principale différentiable de groupe structural* $G = \mathbf{O}(p,q)^+$, $(p + q = 2m = n)$, *ξ' la fibration principale associée de groupe structural* $\mathbf{SO}(2m)$. *Soient Ω_{ij} $(1 \le i,j \le 2m)$ les composantes covariantes de la forme de courbure d'une connexion sur ξ. Alors la classe d'Euler-Poincaré réelle $W_m(\xi)$ de ξ est représentée par la forme différentielle*

$$\Omega_n = (-1)^{[p/2]} \cdot (4\pi)^{-m} \cdot (m!)^{-1} \sum_{\pi \in S_n} \varepsilon_\pi \cdot \Omega_{\pi(1)\pi(2)} \cdot \ \cdots \ \cdot \Omega_{\pi(n-1)\pi(n)} \,,$$

et est nulle si p ou q est impair.

(S_n désigne le groupe des permutations de n objets, et ϵ_π est la signature de $\pi \in S_n$).

On notera $\mathbf{T}_n$ l'ensemble des permutations $\pi \in S_n$ pour lesquelles

$$\pi(2j - 1) < \pi(2j) \ (j = 1, \ldots, m), \ \pi(1) < \pi(3) < \ldots < \pi(n - 1) \,.$$

Comme Ω_{ij} est antisymétrique en i, j, on a évidemment

(1) $$\Omega_n = (-1)^{[p/2]} \cdot (2\pi)^{-m} \cdot \sum_{\pi \in \mathbf{T}_n} \varepsilon_\pi \cdot \Omega_{\pi(1)\pi(2)} \cdot \ \cdots \ \cdot \Omega_{\pi(n-1)\pi(n)} \cdot$$

Soient (Ω'_{ij}) les composantes covariantes de la forme de courbure d'une connexion sur ξ'. On sait [4, p. 38] que la classe d'Euler-Poincaré $W_m(\xi')$ de ξ' est représentée par la forme

$$(2) \qquad \Omega'_n = (-2\pi)^{-m} \sum_{\pi \epsilon T_n} \varepsilon_\pi \, \Omega'_{\pi(1)\pi(2)} \cdot \, \cdots \, \cdot \Omega'_{\pi(n-1)\pi(n)} \, ,$$

et que cette forme provient, par l'homomorphisme caractéristique, du polynome invariant

$$(3) \qquad Q = (-2\pi)^{-m} \sum_{\pi \epsilon T_n} \varepsilon_\pi \cdot z'^{\pi(1)}_{\pi(2)} \cdot \, \cdots \, \cdot z'^{\pi(n-1)}_{\pi(n)} \, ,$$

(voir [10, §4]). Si $a(\pi)$ ($\pi \epsilon T_n$) désigne le nombre de paires $(\pi(2j-1), \pi(2j))$ pour lesquelles $\pi(2j-1) \leq p$, et $\pi(2j) > p$, on a alors (cf. 2.4(1)).

$$(4) \qquad P_Q = (-2\pi)^{-m} \sum_{\pi \epsilon T_n} (\sqrt{-1})^{a(\pi)} \cdot \varepsilon_\pi \cdot z^{\pi(1)}_{\pi(2)} \cdot \, \cdots \, \cdot z^{\pi(n-1)}_{\pi(n)} \, ,$$

Notant $b(\pi)$ le nombre de paires $\pi(2j-1), \pi(2j)$ dans lesquelle $\pi(2j-1) > p$ on en tire, compte tenu de 3.2(2),

$$(5) \qquad P_Q(\Omega) = (-2\pi)^{-m} \sum_{\pi \epsilon T_n} (\sqrt{-1})^{a(\pi)} \cdot (-1)^{b(\pi)} \cdot \varepsilon_\pi \cdot \Omega_{\pi(1)\pi(2)} \cdots \Omega_{\pi(n-1)\pi(n)} \cdot$$

Il est clair que

$$(6) \qquad a(\pi) + 2.b(\pi) = q.$$

Supposons maintenant p et q *pairs*. Alors G,G',K sont de même rang m, et $a(\pi) = 2.a'(\pi)$ est pair. Vu (6), $P_Q(\Omega) = \Omega_n$ donc, vu 2.2, 2.4 Ω_n représente la classe d'Euler-Ponicaré $W_m(\xi')$.

Soient p et q impairs. Alors $a(\pi) = 2.a'(\pi) + 1$ est impair. Par suite

$$(\sqrt{-1})^{a(\pi)} \cdot (-1)^{b(\pi)} = (\sqrt{-1}) \cdot (-1)^{[q/2]}$$

et $(\sqrt{-1}) \cdot P_Q(\Omega)$ est égal au membre de droite de (1). Comme p et q sont impairs, il est clair que chaque monome de Q (resp. de P_Q) contient au moins une variable y'^i_j (resp. y^i_j), donc la restriction de Q et P_Q à $\mathbf{k}$ est nulle. Le lemme 1.5 et ce qui précède montrent alors que Ω_n et Ω'_n sont cohomologues à zéro.

3.4. REMARQUE. Soit X une variété compacte orientée de dimension 2m. On sait que X possède une métrique pseudo-riemann'enne de signature (p,q) si et seulement si elle admet un champ continu de p-plans tangents [12, §40.11]. On retrouve donc le fait connu [11] que la caractéristique d'Euler-Poincaré de X est nulle si on peut construire sur X un champ continu de p plans, avec p impair.

THÉORÈME **3.5.** *Soient ξ une fibration principale de groupe structural* $G = \mathbf{O}(p,q)$, *(Ω_{ij}) les composantes covariantes de la forme de courbure d'une connexion sur ξ, et ξ' la fibration principale associée à ξ de groupe structural* $G' = \mathbf{O}(p+q)$. *Alors la k-ième classe de Pontrjagin réelle P_k de ξ' est représentée par la forme différentielle*

$$(1) \qquad \psi_{4k} = (2\pi)^{-2k} \cdot \sum_{1 \leq a_1 < \ldots < a_{2k} \leq n} \sum_{\pi \epsilon S_{2k}} (-1)^{c(a,\pi)} \cdot \varepsilon_\pi \cdot \Omega_{a_1 a_{\pi(1)}} \cdots \Omega_{a_{2k} a_{\pi(2k)}} \, ,$$

où $n = p + q$ *et où $c(a,\pi)$ désigne le nombre de paires $(a_i, a_{\pi(i)})$ dont les deux éléments sont $> p$.*

L'entier k étant fixé, on écrira $\sum_{a,\pi}$ pour la double sommation de (1), a parcourant donc les parties de $2k$ éléments de $(1,\ldots,n)$ et π le groupe des permutations de $2k$ objets. Soient $d(a,\pi)$ le nombre de paires $(a_i, a_{\pi(i)})$ dont exactement un élément est $\leq p$ et $e(a)$ le nombre d'éléments a_i de a qui sont $> p$. Il est immédiat que $d(a,\pi)$ est pair et que

$$(2) \qquad (d(a,\pi)/2 + c(a,\pi) = e(a).$$

D'après [4, Thm. 2, p. 82], P_k est représenté par la forme

$$(3) \qquad \psi'_{4k} = (2\pi)^{-2k} \cdot \sum_{a,\pi} \varepsilon_\pi \cdot \Omega'^{a_1}_{a_{\pi(1)}} \cdot \, \cdots \, \cdot \Omega'^{a_{2k}}_{a_{\pi(2k)}}$$

où (Ω'^i_j) est la forme de courbure d'une connexion sur ξ', qui, [10, §2], provient par l'homomorphisme caractéristique du polynome invariant

$$Q_{2k} = (2\pi)^{-2k} \cdot \sum_{a,\pi} \varepsilon_\pi \cdot z'^{a_1}_{a_{\pi(1)}} \cdot \ \cdots \ \cdot z'^{a_{2k}}_{a_{\pi(2k)}}$$

(notations de 3.1, 3.2). On a alors (cf. 2.4).

$$P_{Q_{2k}} = (2\pi)^{-2k} \cdot \sum_{a,\pi} \varepsilon_\pi (\sqrt{-1})^{d(a,\pi)} z^{a_1}_{a_{\pi(1)}} \cdot \ \cdots \ \cdot z^{a_{2k}}_{\pi(2k)}$$

donc, compte tenu de 3.2 (2),

$$P_{Q_{2k}}(\Omega) = (2\pi)^{-2k} \cdot \sum_{a,\pi} \varepsilon_\pi (\sqrt{-1})^{d(a,\pi)} (-1)^{e(a)} \Omega_{a_1 a_{\pi(1)}} \cdot \ \cdots \ \cdot \Omega_{a_{2k} a_{\pi(2k)}} \ ,$$

et le théorème résulte maintenant de (2).

3.6. REMARQUE. La signature de la métrique pseudo-riemannienne n'intervient dans 3.5(1) que par l'intermédiaire du signe $(-1)^{e(a,\pi)}$. Comme d'autre par Ω_{ij} est antisymmétrique en i,j, on voit en particulier que le membre de droite de 3.5(1) est indépendant de la signature lorsque $q \leq 2$.

Supposons V orientée, de dimension n=4m multiple de quatre. Alors, d'après HIRZEBRUCH [7, Satz 8.2.2], son index est égal à la valeur prise sur le cycle fondamental par un polynome universel $L(P_1, \ldots P_m)$ en les classes de Pontrjagin. Remplaçant les P_k par les formes différentielles ψ_{4k}, on voit que l'index de V est intégrale sur V d'une forme différentielle en les Ω_{ij} qui ne dépend pas de p,q si $q \leq 2$. Pour n=4, on retrouve ainsi un résultat de AVEZ [2].

3.7. Soit $\mathbf{U}(p, q)$ le groupe unitaire de la forme hermitienne

$$H = x_1 \cdot \bar{x}_1 + \ldots + x_p \cdot \bar{x}_p - x_{p+1} \cdot \bar{x}_{p+1} - \ldots - x_m \cdot \bar{x}_m \ ,$$

$(m=p+q)$. En utilisant le plongement standard de $\mathbf{GL}(m,\mathbf{C})$ dans $\mathbf{GL}(2m,\mathbf{R})$, on voit que $G = \mathbf{U}(p,q)$ et $G' = \mathbf{U}(p+q)$ sont dans la situation décrite dans 2.3, 2.4. Si ξ est une fibration principale de groupe structural G et ξ' la fibration principale associée, on peut aussi représenter la classe de Chern $C_i(\xi')$ par une forme différentielle $\Phi(\Omega)$ en les composantes (Ω_{ij}) du tenseur de courbure d'une connexion sur ξ. On montre exactement comme en 3.5 que Φ_i s'obtient à partir du polynome de Chern [4, p. 77; 10, §3] donnant $C_k(\xi)$ exactement comme ψ_i s'obtient à partir de ψ'_i dans 3.3. En particulier, il n'y a pas de changement de signe si $q = 1$.

BIBLIOGRAPHIE

[1] Avez, A., (1962), *C. R. Acad. Sci., Paris,* **255**, 2049–2051.

[2] Avez, A., (1964), *C. R. Acad. Sci. Paris,* **259**, 1934–1937.

[3] Borel, A. and F. Hirzebruch, (1958), *Amer. Jour. Math.,* **80**, 458–538.

[4] Chern, S. S., (1951), *Topics in Differential Geometry,* The Institute for Advanced Study, Princeton, N. J., mimeographed notes.

[5] Chern, S. S., (1963), *Pseudo-Riemannian geometry and the Gauss-Bonnet formula, An. Acad. brasil. Ci.,* **35**, 17–26.

[6] Helgason, S., (1962), *Differential Geometry and Symmetric Spaces,* Academic Press, New York.

[7] Hirzebruch, F., (1956), *Ergebnisse d. M. u. i. Grenzgeb. N. F.,* **9**, Springer.

[8] Kobayashi, S. and K. Nomizu, (1963), Interscience Tracts in Pure and Applied Mathematics, **15**, Interscience Publishers, New York.

[9] Mostow, G. D., (1955), *Annals of Math.,* **62**, 44–55.

[10] Murakami, S., (1956), *Osaka Math. Jour.,* **8**, 187–224.

[11] Samelson, H., (1951), *Portugaliae Math.,* **10**, 129–133.

[12] Steenrod, N., (1951), *Princeton Math. Series,* **14**, Princeton University Press, Princeton, N. J.

80.

(with T. A. Springer)

Rationality properties of linear algebraic groups II

Tôhoku Math. J., (2) **20** (1968) 443–497

This paper is a development of [4], and gives a more detailed treatment of the topic named in the title. It includes in particular the birational equivalence with affine space, over the groundfield, of the variety of Cartan subgroups of a k-group G, the splitting of G over a separable extension of k if G is reductive, some results on unipotent groups operated upon by tori, and on the existence of subgroups of G whose Lie algebra contains a given nilpotent element of the Lie algebra $\mathfrak{g}$ of G.

Discussing as it does a number of known results (due mostly to Rosenlicht and Grothendieck), this paper is to be viewed as partly expository. In fact, besides proving some new results, our main goal is to provide a rather comprehensive, albeit not exhaustive, account of our topic, from the point of view sketched in [4].

Our basic tools are some rationality properties of transversal intersections and of separable mappings, the Jordan decomposition in $\mathfrak{g}$, and purely inseparable isogenies of height one. They are reviewed or discussed in section 1.13, §3 and §5 respectively. Thus Lie algebras of algebraic groups play an important role in this paper and, for the sake of completeness, we have collected in §1 a number of definitions and facts pertaining to them.

§2 reproves a result of Grothendieck ([12], Exp. XIV) stating that $\mathfrak{g}$ is the union of the subalgebras of its Borel subgroups. Its main use for us is to reduce to Lie algebras of solvable groups the existence proof of the Jordan decomposition.

§4 discusses subalgebras $\mathfrak{s}$ of $\mathfrak{g}$ consisting of semi-simple elements, to be called "toral subalgebras" of $\mathfrak{g}$. They are tangent to maximal tori, and have several properties similar to that of tori in G, in particular: the centralizer $Z(\mathfrak{s}) = \{g \in G,\ \mathrm{Ad}\,g(X) = X (X \in \mathfrak{s})\}$ of $\mathfrak{s}$ in G is defined over k if $\mathfrak{s}$ is, (see 4.3 for $Z(\mathfrak{s})^0$, 6.14 for $Z(\mathfrak{s})$), its Lie algebra is $\mathfrak{z}(\mathfrak{s}) = \{X \in \mathfrak{g}, [\mathfrak{s}, X] = 0\}$. If $\mathfrak{s}$ is spanned by one element X, the conjugacy class of X is isomorphic to $G/Z(\mathfrak{s})$. This paragraph also gives some conditions under which a subalgebra of $\mathfrak{g}$ is algebraic, and reproves some results of Chevalley [8] in characteristic zero.

§6 introduces regular elements, Cartan subalgebras in $\mathfrak{g}$, and the subgroups of type (C) of ([12], Exp. XIII) in G. By definition here, a Cartan subalgebra

720

of $\mathfrak{g}$ is the centralizer in $\mathfrak{g}$, and a subgroup of type (C) is the identity component of the centralizer in G, of a maximal toral subalgebra $\mathfrak{t}$ of $\mathfrak{g}$; subgroups of type (C) always contain Cartan subgroups of G, but may be bigger. The map which associates to a closed subgroup of G its Lie algebra yields a one-one correspondence between subgroups of type (C) and Cartan subalgebras, preserving fields of definition (6.6). The existence of Cartan subalgebras defined over k, which is an easy fact, yields then, as a first step to rationality properties, the existence of subgroups of type (C) defined over k. In fact, G is separably generated by such subgroups (6.10). Moreover, if k is separably closed, two Cartan subgroups or two Cartan subalgebras defined over k are conjugate under $G(k)$, (6.13).

§7 gives first some conditions under which the field of definition of a homogeneous space of G can be brought down to k (7.6). They apply in particular to the set of Cartan subgroups or of Cartan subalgebras of G (7.7). We then show that both are rational varieties over k (7.9). The proof is to a large extent an adaptation in our framework of the one given by Grothendieck ([12], Exp. XIV) for the former variety. Some consequences are derived. §8 is devoted to the splitting of a reductive k-group over a separable extension of k.

In characteristic $p \neq 0$, a nilpotent element $X \in \mathfrak{g}(k)$ need not be tangent to a one-dimensional subgroup, even if $X^{[p]} = 0$ (9.2). §9 is mainly concerned with the finding of subgroups to which X is tangent, assuming X to be an eigenvector of a subtorus T of G, corresponding to a non-trivial character b of T. If G is reductive or solvable, X is tangent to a unipotent k-subgroup stable under T, in which the weights of T are certain multiples of b. Sufficient conditions under which H may be chosen to be commutative, or isomorphic to G_a, are given (9.8, 9.16). As an application, some properties of unipotent groups operated upon by tori are derived.

0. Notations and conventions. *Throughout the paper, k is a commutative field, p its characteristic, K an algebraically closed extension of k, $\bar{k}$ the algebraic closure of k in K, and k_s the separable closure of k in $\bar{k}$. G is an affine k-group, $\mathfrak{g}$ its Lie algebra.*

The notation is basically that of [3], with which familiarity is assumed, with the following additions or modifications.

0.1. Our varieties are those of "classical" algebraic geometry, although not necessarily irreducible, and are all affine or quasi-projective. Algebraic variety defined over k and k-variety will be used synonymously. In order to avoid any ambiguity, let us briefly define that notion in the quasi-projective case. An algebraic subset V of affine (resp. projective) space is defined over

k if the ideal of polynomials (resp. homogeneous polynomials) vanishing on V, with coefficients in $\bar{k}$, is generated by polynomials with coefficients in k. A k-variety is, in this paper, either an affine or projective algebraic set defined over k, or the complement of a k-closed subset in a projective k-variety. For a quasi-projective variety V, the following conditions are equivalent: (i) V is defined over k; (ii) the irreducible components of V are defined over k_s and are permuted by the Galois group of k_s over k, acting by conjugation; (iii) the cycle sum of the irreducible components of V, with coefficients one, is rational over k in the sense of ([27], Chap. VIII, §1). This follows immediately from ([27], Lemma 2, p. 209). Our notion of k-variety is also equivalent to that of absolutely reduced quasi-projective scheme over k. If V is defined over k, the points of V rational over k_s are dense in V (see e.g. S. Lang, Introduction to Algebraic Geometry, Interscience Publ., Prop. 10, p. 76).

0.2. Let V be a k-variety. Then $k[V]$ denotes the k-algebra of regular functions defined over k on V, or, as we shall say, of k-morphic functions on V. The fact that V is defined over k implies that for any extension field k' of k, we have $k'[V] = k[V] \otimes_k k'$. If V is irreducible, $k(V)$ is the field of rational functions on V, defined over k, on V.

The set of points of V rational over an extension field k' of k will be denoted $V(k')$, (and not $V_{k'}$, as in [3]). Often, we let V stand for $V(\bar{k})$, i.e., we identify V with its set of points over $\bar{k}$. The tangent space to V at a simple point x is denoted $T(V)_x$. If $x \in V(k)$, the space $T(V)_x$ carries a canonical k-structure. Let W be a k-variety and $f: W \to V$ a k-morphism. If $x \in W(k)$ and $f(x)$ are simple points, f induces a linear map $T(W)_x \to T(V)_{f(x)}$, defined over k, to be denoted df_x and to be called the *differential* of f at x. The map f induces a homomorphism $f^0: k[V] \to k[W]$, the *"comorphism"* associated to f. Similarly, if V, W are irreducible, and $f(W)$ is dominant in V, then f induces an injective homomorphism of $k(V)$ into $k(W)$, also to be denoted f^0. We recall that $f: W \to V$ is *dominant* if $f(W)$ is dense in V. In that case it contains an open everywhere dense subset of V.

0.3. If A, B are two subsets of a group H, we let $\operatorname{Tr}(A, B)$ or $\operatorname{Tr}_H(A,B)$ be the set of $h \in H$ such that $\operatorname{Int} h(A) \subset B$. If $A = B$, then $\operatorname{Tr}(A, B)$ is the normalizer $N_H(A)$ or $N(A)$ of A in H.

0.4. Let $\mathfrak{h}$ be a Lie algebra over k. We let $\operatorname{ad} X$ denote the map $Y \mapsto [X, Y]$ of $\mathfrak{h}$ into itself. Let $p \neq 0$. We recall that $\mathfrak{h}$ is *restricted* if it is endowed with a map $[p]: X \mapsto X^{[p]}$ into itself, having the following properties:

 (a) $(\operatorname{ad} X)^p = \operatorname{ad} X^{[p]}$

(b) $(a \cdot X)^{[p]} = a^p \cdot X^{[p]}$ $(X \in \mathfrak{h},\ a \in k)$

(c) $(X_1 + X_2)^{[p]} = X_1^{[p]} + X_2^{[p]} + \Sigma'_{0<i<p} i^{-1} \cdot s_i(X_1, X_2)$

where $s_i(X_1, X_2)$ is the coefficient of λ^{i-1} in $(\mathrm{ad}(\lambda X_1 + X_2))^{p-1}(X_1)$. If s is a positive integer and $q = p^s$, we write $[q]: X \mapsto X^{[q]}$ for the s-th power of $[p]$.

By convention, if $p=0$, any Lie algebra over k is restricted.

0.5.   Let $\mathfrak{a}, \mathfrak{b}$ be subsets of $\mathfrak{g}$. We put

$$\mathfrak{z}(\mathfrak{a}) = \mathfrak{z}_\mathfrak{g}(\mathfrak{a}) = \{X \in \mathfrak{g}, [X, \mathfrak{a}] = 0\},$$

$$Z_G(\mathfrak{a}) = Z(\mathfrak{a}) = \{g \in G, \mathrm{Ad}\, g(X) = X (X \in \mathfrak{a})\},$$

$$\mathfrak{n}(\mathfrak{a}) = \mathfrak{n}_\mathfrak{g}(\mathfrak{a}) = \{X \in \mathfrak{g}, [X, \mathfrak{a}] \subset \mathfrak{a}\}$$

$$N_G(\mathfrak{a}) = N(\mathfrak{a}) = \{g \in G, \mathrm{Ad}\, g(\mathfrak{a}) = \mathfrak{a}\}$$

$$\mathfrak{tr}(\mathfrak{a}, \mathfrak{b}) = \{X \in \mathfrak{g}, [X, \mathfrak{a}] \subset \mathfrak{b}\}$$

$$\mathrm{Tr}_G(\mathfrak{a}, \mathfrak{b}) = \mathrm{Tr}(\mathfrak{a}, \mathfrak{b}) = \{g \in G, \mathrm{Ad}\, g(\mathfrak{a}) \subset \mathfrak{b}\}.$$

$Z(\mathfrak{a})$ and $N(\mathfrak{a})$ are the *centralizer* and the *normalizer* of $\mathfrak{a}$ in G. (The definition of $\mathfrak{g}$ and of the adjoint representation will be recalled in §1.)

0.6.   Let H (resp. H') be a group and A (resp. A') a set on which it operates. Let $f: H \to H'$ be a homomorphism. A map $u: A \to A'$ is *f-equivariant* if $u(h \cdot a) = f(h) \cdot u(a)(h \in H, a \in A)$. If $H=H'$ and $f = \mathrm{id.}$, then u is said to be equivariant or H-equivariant.

1. The Lie algebra of a linear group.

1.1.   In this paragraph, we let A stand for the algebra $k[G]$ of k-morphic functions on G. We have $k[G \times G] = A \otimes_k A$, hence the comorphism associated to the product map $v: G \times G \to G$ is a homomorphism $\mu: A \to A \otimes A$. We let ι be the inverse map $x \mapsto x^{-1}$ of G and ε the homomorphism $a \mapsto a(e)$ of A into k defined by the neutral element e of G. We have

$$(1) \qquad (\varepsilon \otimes \mathrm{id}) \circ \mu = \mathrm{id} = (\mathrm{id} \otimes \varepsilon) \circ \mu.$$

The existence of the group law on G is equivalent to a number of properties of μ, ε, ι which can be found, e.g., in [14, 0.8.2].

1.2. The Lie algebra of G.   The group G operates by left and right translations on $\bar{k}[G] = A \otimes_k \bar{k}$. For $g \in G$, let $\lambda_g, \rho_g : \bar{k}[G] \to \bar{k}[G]$ be defined by

$$\lambda_g a(x) = a(g^{-1} \cdot x), \qquad \rho_g a(x) = a(x \cdot g).$$

Then $\lambda : g \mapsto \lambda_g$ (resp. $\rho : g \mapsto \rho_g$) is a representation of G into $\bar{k}[G]$, called the *left* (resp. *right*) *regular representation of G in $\bar{k}[G]$*. The latter space is the union of finite dimensional subspaces, defined over k, stable under λ and ρ. The restriction of λ (resp. ρ) to such a subspace is a rational representation defined over k. The representations λ and ρ commute with each other.

A *k-derivation X of A is left-* (resp. *right-*) *invariant* if its natural extension to $A \otimes \bar{k}$ commutes with the left (resp. right) translations. Left-invariance can also be described by the relation

$$(1) \qquad\qquad \mu \circ X = (\mathrm{id} \otimes X) \circ \mu.$$

We denote by $L(G)(k)$ the k-vector space of all left invariant k-derivations of A. Endowed with the bracket operation $[X, Y] = X \cdot Y - Y \cdot X$ and the p-th power operation $X^{[p]} = X^p$, it is readily seen to be a restricted Lie algebra over k. If k' is an extension of k, then $L(G)(k') = L(G)(k) \otimes_k k'$. We shall also write $L(G)$ for $L(G)(\bar{k})$ and call $L(G)$ the *Lie algebra of G*. We shall soon identify $L(G)$ and $T(G)_e$. Until then, we let $\mathfrak{g}$ stand for $T(G)_e$. Since e is the identity, it is clear that

$$d\nu_{(e,e)}(X, 0) = d\nu_{(e,e)}(0, X) = X, \quad (X \in \mathfrak{g}),$$

whence

$$(2) \qquad\qquad d\nu_{(e,e)}(X, Y) = X + Y, \qquad (X, Y \in \mathfrak{g}).$$

Since the composition of the map $G \to G \times G$ defined by $x \mapsto (x^{-1}, x)$, with ν is the constant map, it follows that

$$(3) \qquad\qquad (d\iota)_e(X) = -X \quad (X \in \mathfrak{g}).$$

Furthermore, using induction and the composition law of differentials, we deduce from (2)

1.3. PROPOSITION. *Let $V_1, \cdots, V_m$ be k-varieties, $u_i \in V_i$, and $f_i : V_i \to G$ a k-morphism which maps u_i onto e. Let $u = (u_1, \cdots, u_m)$ and let $f : V_1 \times \cdots \times V_m \to G$ be the k-morphism defined by $f(v_1, \cdots, v_m) = f_1(v_1) \cdot \cdots \cdot f_m(v_m)$. Then $df_u(X_1, \cdots, X_m) = df_1(X_1) + \cdots + df_m(X_m)$ $(X_i \in T(V_i)_{u_i}, i = 1, \cdots, m)$.*

1.4. In this section and the following, $\mathfrak{g}$ stands for $T(G)_e$. We recall that $\mathfrak{g}(k)$ may be viewed as the k-vector space of k-derivations of A into k, where k is made into an A-module by means of ε. Let now $X \in L(G)(k)$. We associate to X a k-linear map $\sigma(X)$ of A into k by

$$(1) \qquad \sigma(X)(a) = \varepsilon \cdot Xa = Xa(e), \qquad (a \in A).$$

X is in fact a k-derivation of A into k, hence σ is a k-linear map of $L(G)(k)$ into $\mathfrak{g}(k)$, obviously compatible with field extensions.

On the other hand,

$$(2) \qquad \tau(Y) = (\mathrm{id} \otimes Y) \circ \mu, \qquad (Y \in \mathfrak{g}(k)),$$

is a derivation of A into A. The obvious equality

$$(3) \qquad \mu \circ \lambda_g = (\lambda_g \otimes \mathrm{id}) \circ \mu,$$

shows at once that $\mathrm{id} \otimes \tau(Y)$ and $\lambda_g (g \in G)$ commute. Therefore τ is a k-linear map of $\mathfrak{g}(k)$ into $L(G)(k)$.

1.5. PROPOSITION. *We keep the previous notation. σ and τ are isomorphisms, inverse of each other. In particular, $\dim_k L(G)(k) = \dim G$, and $L(G) = L(G^0)$.*

To prove 1.5, we may assume $k = \bar{k}$. It suffices to show that σ is injective and that $\sigma \circ \tau = \mathrm{id}$. Let $\sigma(X) = 0$. Then, by 1.4(1), applied to $\lambda_g a$, $(g \in G, a \in A)$, we get $Xa(g) = 0$, whence $X = 0$. Let $X \in \mathfrak{g}(k)$. Then, in view of 1.1(1), 1.2(1):

$$\sigma \circ \tau(X) = (\mathrm{id} \otimes \varepsilon) \circ (\mathrm{id} \otimes X) \circ \mu = (\mathrm{id} \otimes \varepsilon) \circ \mu \circ X = X.$$

From now on, we identify $L(G)(k)$ and $\mathfrak{g}(k)$ via σ. Thus $\mathfrak{g}(k)$ is canonically endowed with a structure of restricted Lie algebra over k. An explicit description of this structure is as follows: for a positive integer s, let $\mu^{(s)} : A \to \otimes^s A$ be the comorphism defined by the product map $G^s \to G$. For $a \in A$, let us write

$$(1) \qquad \mu^{(s)} a = \Sigma_i a_{i,1} \otimes \cdots \otimes a_{i,s}.$$

Then, for $s = 3$, we have

$$(2) \qquad [X, Y]a = \Sigma_i (Xa_{i1} \cdot Ya_{i2} - Ya_{i1} \cdot Xa_{i2}) a_{i3}(e), \qquad (X, Y \in \mathfrak{g}(k)),$$

and, if $p > 0$, for $s = p + 1$

$$(3) \qquad X^{[p]}a = \Sigma_i (Xa_{i1} \cdot \cdots \cdot Xa_{ip}) \cdot a_{i,p+1}(e), \qquad (X \in \mathfrak{g}(k)),$$

as follows from 1.4(2).

1.6. Let H be a k-group and $f: H \to G$ a k-morphism. Then $df_e: \mathfrak{h}(k) \to \mathfrak{g}(k)$ is a homomorphism of restricted Lie algebras over k, as follows from 1.5(2),(3), and it is then clear that $G \mapsto L(G)(k)$ is a functor from the category of k-groups and k-morphisms to the category of restricted Lie algebras over k. We write also $L(f)$ or df for df_e. If f is the inclusion of a subgroup, then $L(f)$ identifies $L(H)$ with a restricted subalgebra of $L(G)(k)$.

1.7. Examples. (a) $G = \boldsymbol{G}_a$, the additive group of the one-dimensional vector space. We have $A = k[T]$ and $\mu(T) = T \otimes 1 + 1 \otimes T$, $\mathcal{E}(T) = 0$. An easy computation, based on 1.5(2),(3), shows that $\mathfrak{g}$ is one-dimensional and $\mathfrak{g}^{[p]} = 0$.

(b) $G = \boldsymbol{GL}_n$. Then $A = k[T_{11}, T_{12}, \cdots, T_{nn}, D^{-1}]$, where $D = \det(T_{ij})$, $(1 \leqq i, j \leqq n)$. The comorphism $\mu: A \to A \otimes A$ is given by

$$(1) \qquad \mu(T_{ij}) = \Sigma_k T_{ik} \otimes T_{kj},$$

and

$$(2) \qquad \mathcal{E}(T_{ij}) = \delta_{ij}, \quad (1 \leqq i, j \leqq n).$$

$X \in \mathfrak{g}(k)$ is determined by the elements $a_{ij} = XT_{ij}$ of k, hence by a matrix $(a_{ij}) \in \boldsymbol{M}_n(k)$. In the notation of 1.4, we have, by 1.4(2):

$$(3) \qquad \tau(X)(T_{ij}) = \Sigma_k T_{ik} \cdot a_{kj}.$$

It follows from (3) that $X \mapsto (XT_{ij})$ identifies $\mathfrak{g}(k)$ with $\boldsymbol{M}_n(k)$, endowed with the usual commutator $[X, Y] = XY - YX$, and the *ordinary* p-th power as $[p]$-operation.

(c) $G = T$ is a torus and $p > 0$. Then $\mathfrak{g}$ is commutative, $[p]$ is a bijective F_p-linear map of $\mathfrak{g}(\bar{k})$ onto itself, where F_p is the Frobenius automorphism $x \mapsto x^p$ of $\bar{k}$. The set $\mathfrak{g}_0$ of fixed elements of $[p]$ is a vector space over the prime field $\boldsymbol{F}_p$ with p elements, such that $\mathfrak{g}(\bar{k}) = \mathfrak{g}_0 \otimes \bar{k}$. If T splits over k, then $\mathfrak{g}_0 \subset \mathfrak{g}(k)$ and $\mathfrak{g}(k) = \mathfrak{g}_0 \otimes k$ (the tensor products being over $\boldsymbol{F}_p$).

Let X_* be the group of morphisms of $\boldsymbol{GL}_1$ into T. It is a free abelian group of rank equal to $\dim T$. Let us associate to $x \in X_*$ the element $dx(1) \in \mathfrak{g}(\bar{k})$. This is a homomorphism, which induces a surjective homomorphism $f: X_* \otimes \bar{k} \to \mathfrak{g}(\bar{k})$. The image of X_* is pointwise fixed under $[p]$, and contains a basis of $\mathfrak{g}$ over k. If T splits over k, then $f(X_* \otimes 1) = \mathfrak{g}_0$. All these facts follow readily from (b), since T splits over $\bar{k}$, and a k-split torus is a direct product over k of some copies of $\boldsymbol{GL}_1$.

1.8. The adjoint representation. Let $X \in L(G)$. Since left translations commute with right translations

$$\operatorname{Ad} g(X) = \lambda_g \circ X \circ \lambda_g^{-1} \quad (g \in G)$$

is again an element of $L(G)$. From the definition of the bracket and p-th power operations in $L(G)$, it is clear that $\operatorname{Ad} g$ is an automorphism of $L(G)$. The map $g \mapsto \operatorname{Ad} g$ is a representation of G into $L(G)$, to be called the *adjoint representation* of G. If we identify $\mathfrak{g}(\bar{k})$ and $L(G)$ by τ, we see immediately that $\operatorname{Ad} g$ *is the differential at e of the inner automorphism* Int $g : x \mapsto g \cdot x \cdot g^{-1}$. If $f : H \to G$ is a k-morphism, then

$$(1) \qquad L(f)(\operatorname{Ad} h(X)) = \operatorname{Ad} f(h)(L(f)(X)) \quad (X \in L(H), \, h \in H).$$

1.9. PROPOSITION. *The adjoint representation of G is a k-morphism of G into* $GL(\mathfrak{g})$. *The differential of* Ad *is the homomorphism* ad : $X \mapsto \operatorname{ad} X$ *defined by* $\operatorname{ad} X(Y) = [X, Y]$ $(X, Y \in \mathfrak{g})$.

In view of 1.8(1), it suffices to prove that if H is a k-subgroup of $\boldsymbol{GL}_n$ and $\mathfrak{m}$ a subspace defined over k of $\mathfrak{gl}_n$ stable under $\operatorname{Ad} g$ $(g \in H)$, (where Ad is the adjoint representation of $\boldsymbol{GL}_n$), then the representation $h \mapsto \operatorname{Ad} h|_{\mathfrak{m}}$ of h into $\mathfrak{m}$ is defined over k, and its differential is $\operatorname{ad} X|_{\mathfrak{m}}$. This follows readily from the formula

$$(1) \qquad \operatorname{Ad} g(X) = g \cdot X \cdot g^{-1} \quad (g \in \boldsymbol{GL}_n, \, X \in \mathfrak{gl}_n)$$

which in turn is a simple consequence of 1.7(3).

1.10. Applications. (a) We identify $T(G)_g$ to $\mathfrak{g}$ by means of the left translation $l_g : x \mapsto g \cdot x$. Then, by definition $(dl_g)_e = \operatorname{id}$. The right translation $r_g : x \mapsto x \cdot g$ can be written as $l_g \circ \operatorname{Int} g^{-1}$. Therefore $(dr_g)_e = \operatorname{Ad} g^{-1}$. As a consequence the differential at e of the map $g \mapsto g \cdot a \cdot g^{-1}$ is $(\operatorname{Ad} a^{-1} - \operatorname{Id})$.

(b) Let $X \in \mathfrak{g}$. Identify the tangent space to $\mathfrak{g}$ at X in the usual manner to $\mathfrak{g}$. Then the differential at e of $g \mapsto \operatorname{Ad} g(X)$ is $-\operatorname{ad} X$. This follows, e.g., from 1.9(1).

1.11. PROPOSITION. *Let L, M be two connected k-subgroups of G and $Q = (L, M)$ be the group generated by the commutators $(x, y) = x \cdot y \cdot x^{-1} \cdot y^{-1}$ $(x \in L, y \in M)$. Then the Lie algebra $\mathfrak{q}$ of Q contains all elements of the form* $\operatorname{Ad} x(X) - X$ $(x \in L, X \in \mathfrak{m})$ *or* $(x \in M, X \in \mathfrak{l})$ *and all commutators* $[X, Y], (X \in \mathfrak{l}, Y \in \mathfrak{m})$.

(We recall that Q is a connected k-group, see, e.g., ([1], Lemme 4.3).) Let $x \in L$. Then $y \mapsto x \cdot y \cdot x^{-1} \cdot y^{-1}$ $(y \in M)$, may be viewed as the composition of the two morphisms (of varieties) $y \mapsto x \cdot y \cdot x^{-1}$ and $y \mapsto y^{-1}$ of M into Q, with the product in Q. By 1.2, its differential at (e, e) is the sum of the differential of $\mathrm{Int}\, x$, which is $\mathrm{Ad}\, x$, and of the differential $-\mathrm{Id}$ of $y \mapsto y^{-1}$ (see 1.2(3)). Thus $\mathrm{Ad}\, x(\dot X) - X \in \mathfrak{q}$ if $X \in \mathfrak{m}$. Similarly, $\mathrm{Ad}\, x(X) - X \in \mathfrak{q}$ if $x \in M$, $X \in \mathfrak{l}$.

Let $X \in \mathfrak{m}$. Then $x \mapsto \mathrm{Ad}\, x(X) - X$ is a morphism of L into $\mathfrak{q}$, whose differential at e is $Y \mapsto \mathrm{ad}\, Y(X) = [Y, X]$, $(Y \in \mathfrak{l})$, whence the second assertion.

1.12. COROLLARY. (i) *If G is solvable (resp. nilpotent, resp. commutative), then $\mathfrak{g}$ is solvable (resp. nilpotent, resp. commutative).* (ii) *The Lie algebra $\mathfrak{z}$ of the center Z of G belongs to the center of $\mathfrak{g}$.*

REMARK. If $p > 0$, the converse to (i) or (ii) is well known to be false. To mention one example, $\mathfrak{sl}_2$ is solvable with one-dimensional center if $p = 2$, while $\boldsymbol{SL_2}$ is simple.

1.13. Separable morphisms. Let X, Y be irreducible k-varieties and $f : X \to Y$ a dominant k-morphism. The comorphism f^0 defines a monomorphism of $k(Y)$ into $k(X)$. f is said to be *separable* if $k(X)$ is a separable extension of $k(Y)$. It is then so for every extension k' of k. We recall that the following conditions on f are equivalent:

(i) f is dominant, separable.

(ii) f is dominant. There exists a simple point $a \in X$ whose image b is simple and such that $df_a : T(X)_a \to T(Y)_b$ is surjective.

(iii) There exists an open dense subset U of X, consisting of simple points, whose image consists of simple points, and such that df_x is surjective for all $x \in U$.

Let $F_x = f^{-1}(f(x))$ be the fibre through $x \in X$. Let m be a positive integer. It is well kown that the set of points for which $\dim F_x \geq m$ is closed, and that $\min \dim F_x = \dim X - \dim f(X)$, and so $\dim F_x = \dim X - \dim f(X)$ for x in an open dense subset of X. Also the set of points for which $\dim \ker df_x \geq m$ is closed. If x is simple on X and F_x, then $T(F_x)_x \subset \ker df_x$. It follows then that the minimum of $\dim \ker df_x$ is attained on a non-empty open set, and is $\geq \dim X - \dim f(X)$.

(ii) $\Longrightarrow$ (iii). We have $\dim \ker df_a = \dim X - \dim f(X)$, hence by the above, there is an open set U in X such that $\dim \ker df_x = \dim \ker df_a$ for $x \in U$, whence (iii).

(iii) $\Longrightarrow$ (ii). We have $\dim \mathrm{Im}\, df_x \leq \dim f(X)$ for all simple $x \in X$, hence df_x cannot be surjective if $\dim f(X) < \dim Y$, which implies that f is dominant. For the equivalence (i) $\Longleftrightarrow$ (iii), see, e.g., ([27], Prop. 15, p. 12, and Chap. IV, §6).

We shall be mainly concerned with the case where X, Y are smooth over k, X is affine, f is surjective, and df_x is surjective for every $x \in X$. In this case, if Z is a closed subvariety defined over k of Y, then $f^{-1}(Z)$ is defined over k. In fact, $f^{-1}(Z)$ is obviously k-closed. By (0.1), it suffices to show that its irreducible components are defined over k_s. We may therefore assume Z to be irreducible. The assumptions imply that $X \times Z \subset X \times Y$ cuts the graph of f transversally, therefore the cycle sum of the irreducible components of $f^{-1}(Z)$, each affected with coefficient one, is rational over k ([27], Thm. 6, p. 200, Thm. 4, p. 223). But (0.1) this is equivalent to $f^{-1}(Z)$ being defined over k as an algebraic set. In particular $f^{-1}(Z)(k_s) \neq \emptyset$. If we apply this to the k_s-points of Y, we see that $f(X(k_s)) = Y(k_s)$.

1.14. Separable actions. Let G act k-morphically on the k-variety X. We say that G acts separably if, for each $x \in X$, the map $\varphi_x \colon g \mapsto g \cdot x$ is a separable map of G onto the orbit $G(x)$ of x. In view of 1.13, this is the case if and only if the kernel of $(d\varphi_x)_e$ is the Lie algebra of the stability group G_x of x. It suffices of course to check this condition for one point of each orbit.

1.15. Generation by subgroups. Assume G to be connected. Let $\mathcal{M} = (H_i)_{1 \leq i \leq m}$ be a family of connected k-subgroups of G. We say that $\mathcal{M}$ *spans* G (separably) if the product morphism $f \colon H_1 \times \cdots \times H_m \to G$ is surjective (and separable). Assume now that $\mathcal{M}$ *generates* G, i.e., that no proper closed subgroup of G contains all the H_i's. It is then known that there exists a family $\mathcal{M}' = (H_j')_{1 \leq j \leq n}$, such that each H_j' is one of the H_i's, which spans G. We say that $\mathcal{M}$ *generates* G *separably* if there exists such an $\mathcal{M}'$ which spans G separably.

1.16. PROPOSITION. *Let G be connected, and $(H_i)_{1 \leq i \leq m}$ a family of connected k-subgroups of G.*

(i) *If the Lie algebras $\mathfrak{h}_i$ span $\mathfrak{g}$, then (H_i) generates G separably.*

(ii) *If (H_i) generates G separably, then $\mathfrak{g}$ is spanned by the subalgebras* $\operatorname{Ad} g(\mathfrak{h}_i)$, $(i = 1, \cdots, m,\ g \in G)$.

(i) Let H be the subgroup generated by the H_i. It is a connected k-subgroup whose Lie algebra contains the $\mathfrak{h}_i$, hence is equal to $\mathfrak{g}$. Therefore (1.5), $\dim H = \dim G$, and $H = G$. Thus the H_i generate G. If (H_j') is a family consisting of elements of (H_i), which spans G, then the differential of the product morphism at e is surjective, whence the separability (1.13).

(ii) Let $(H_j')_{1 \leq j \leq n}$ span G separably, where each H_j' is one of the H_i's. There exists then a point $h = (h_1, \cdots, h_n)$ $(h_j \in H_j')$ such that df_h is surjective,

After having made a translation by $f(h)^{-1}$ we may assume that $f(h) = e$. Put now

$$v_i = h_1 \cdots h_i, \quad H_i'' = \operatorname{Int} v_i(H_i') \qquad (1 \leqq i \leqq n).$$

Let q be the isomorphism

$$\operatorname{Int} v_1 \times \cdots \times \operatorname{Int} v_n: \quad H_1' \times \cdots \times H_n' \to H_1'' \times \cdots \times H_n'',$$

and r the product morphism: $H_1'' \times \cdots \times H_n'' \to G$. Then $r \circ q$ is the map $(x_1, \cdots, x_n) \mapsto h_1 \cdot x_1 \cdots h_n \cdot x_n \cdot v_n^{-1}$. Therefore, $d(r \circ q)_e$ is surjective. But (1.3) the image of dr_e is $\mathfrak{h}_1'' + \cdots + \mathfrak{h}_m''$. Since $\mathfrak{h}_i'' = \operatorname{Ad} v_i(\mathfrak{h}_i')$, this proves our contention.

1.17. Example. Let $G = \boldsymbol{SL}_2$. Then $\mathfrak{g} = \mathfrak{sl}_2$ is the Lie algebra of 2×2 matrices with trace zero. Let T be the torus of G consisting of the diagonal matrices of determinant one. Its Lie algebra $\mathfrak{t}$ is the Lie algebra of diagonal 2×2 matrices with trace zero. Let now $p = 2$. Then $\mathfrak{t}$ consists of scalar matrices and is pointwise fixed under the adjoint representation. Since the 1-dimensional tori of G are conjugate by inner automorphisms it follows that $\mathfrak{t}$ is the Lie algebra of any such torus. Thus $\mathfrak{g}$ is not generated by the Lie algebras of the tori of G, although G is generated by its tori, in fact, by two suitably chosen tori.

1.18. PROPOSITION. *Let G' be a k-group and $f: G \to G'$ a surjective k-morphism. Let H' be a k-subgroup of G' and $H = f^{-1}(H')$. Assume that $\mathfrak{g}' = \mathfrak{h}' + df(\mathfrak{g})$. Then G acts separably on G'/H' via f, the group H is defined over k, and f induces an f-equivariant k-isomorphism of G/H onto G'/H'.*

Let $\pi: G' \to G'/H'$ be the canonical projection. Then $d\pi_e$ is surjective with kernel $\mathfrak{h}'$, therefore $d(\pi \circ f)_e$ is surjective, and $\pi \circ f$ is surjective, separable. Then, $H = f^{-1}(\pi(H'))$ is defined over k (1.13), and f induces a bijective f-equivariant k-morphism $\bar{f}$ of G/H onto G'/H'. In view of the assumption $d\bar{f}$ is surjective at the origin, hence $\bar{f}$ is an isomorphism.

2. Solvable subgroups. The following lemma is a Lie algebra analogue of [1, Prop. 17.1]. More general results may be found in [12, Exp. XIII, §1].

2.1. LEMMA. *Let H be a closed subgroup of G. Assume that there exists $X \in \mathfrak{h}$ such that the set $\operatorname{Tr}(X, \mathfrak{h})$ of $g \in G$ for which $\operatorname{Ad} g(X) \in \mathfrak{h}$ consists of finitely many left classes $\bmod H$. Then $N_G(\mathfrak{h})^0 = H^0$, and*

$V = \bigcup_{g \in G} \mathrm{Ad}\, g(\mathfrak{h})$ *contains an open non-empty subset of* $\mathfrak{g}$. *If* G/H *is complete, then* $V = \mathfrak{g}$.

$N_G(\mathfrak{h})$ is a closed subgroup of G contained in $\mathrm{Tr}(X, \mathfrak{h})$, hence its identity component is equal to H^0.

Let M be the set of pairs (gH, Y) in $G/H \times \mathfrak{g}$ such that $\mathrm{Ad}\, g^{-1}(Y) \in \mathfrak{h}$. We claim that M is a closed subvariety of $G/H \times \mathfrak{g}$. To see this, we consider the morphisms

$$G \times \mathfrak{h} \xrightarrow{\; a \;} G \times \mathfrak{g} \xrightarrow{\; b \;} G/H \times \mathfrak{g},$$

where $a(g, Y) = (g, \mathrm{Ad}\, g(Y))$ and b is the canonical projection on the first factor, the identity on the second factor. The morphism a is a closed immersion, hence $\mathrm{Im}\, a$ is closed. Clearly, $(g, Y) \in \mathrm{Im}\, a$ implies $(g \cdot H, Y) \in \mathrm{Im}\, a$, hence $\mathrm{Im}\, a = b^{-1}(b(\mathrm{Im}\, a))$ and $\mathrm{Im}(b \circ a)$ is closed. But $\mathrm{Im}(b \circ a) = M$. The morphism $p_1 : M \to G/H$ induced by the projection of $G/H \times \mathfrak{g}$ onto its first factor is surjective, and its fibres are isomorphic to $\mathfrak{h}$, hence $\dim M = \dim G/H + \dim \mathfrak{h} = \dim G$. Let now $p_2 : M \to \mathfrak{g}$ be the morphism induced by the projection on the second factor. By definition $V = \mathrm{Im}\, p_2$. By assumption, at least one fibre of p_2 consists of finitely many points, hence $\dim V = \dim M = \dim \mathfrak{g}$, which implies our first assertion. If G/H is moreover complete, then p_2 is a closed morphism, and V is also closed, which yields the last part of the lemma.

2.2. A *Borel subalgebra* $\mathfrak{b}$ of $\mathfrak{g}$ is a subalgebra which is the Lie algebra of a Borel subgroup of G. The conjugacy of Borel subgroups by inner automorphisms of G^0 implies the conjugacy of Borel subalgebras under $\mathrm{Ad}\, G^0$. The main result on Borel subalgebras is 2.3, due to A. Grothendieck ([12], Exp. XIV, Thm. 4.11, p. 33). The proof presented here is somewhat different from that of Grothendieck's. It was sketched in [4].

2.3. PROPOSITION. *The Lie algebra* $\mathfrak{g}$ *is the union of its Borel subalgebras.*

Let R be the radical of G (i.e., the greatest connected solvable normal subgroup of G). It is contained in every Borel subgroup and its Lie algebra is contained in all Borel subalgebras of $\mathfrak{g}$. We may therefore replace G by G/R and assume G to be semi-simple.

Let B be a Borel subgroup of G, T a maximal torus of B and Φ the set of roots of G with respect to T. We let Δ denote the set of simple roots for the ordering defined by B. We have

$$(1) \qquad \mathfrak{g} = \mathfrak{t} \oplus \Sigma_{a \in \Phi} \mathfrak{g}_a \qquad \mathfrak{b} = \mathfrak{t} + \Sigma_{a>0} \mathfrak{g}_a$$

where

$$(2) \qquad \mathfrak{g}_a = \{X \in \mathfrak{g}, \ \mathrm{Ad}\, t(X) = t^a \cdot X\}\,.$$

Furthermore, given $b \in \dot{\Phi}$, there is an isomorphism θ_b of $\boldsymbol{G}_a$ onto a unipotent subgroup U_b of G such that

$$(3) \qquad \theta_b(t^b \cdot x) = t \cdot \theta_b(x) \cdot t^{-1} \qquad (x \in \boldsymbol{G}_a, \ t \in T),$$

and $\mathfrak{g}_b$ is the Lie algebra of U_b. Since G/B is complete, 2.3 will follow from 2.1 and the following lemma

2.4. LEMMA. *We keep the notation of* 2.3. *Let* X_a *be a non-zero element of* $\mathfrak{g}_a$ $(a \in \Delta)$ *and* $X = \Sigma_{a \in \Delta} X_a$. *Then* $\{g \in G, \ \mathrm{Ad}\, g(X) \subset \mathfrak{b}\} = B$.

Let U be the unipotent radical of B, and W the Weyl group $N(T)/T$. For $w \in W$, let n_w be a representative of w in $N(T)$. Let $g \in \mathrm{Tr}(X, \mathfrak{b})$. By the Bruhat decomposition [10, Exp. 13], we have $g = b' \cdot n_w \cdot b$ $(b, b' \in B)$, with $b \in U$, $b' \in B$. Since $\mathrm{Ad}_G B(\mathfrak{b}) = \mathfrak{b}$, we may assume $b' = e$. By 1.11, $\mathrm{Ad}\, b(X) - X$ lies in the Lie algebra of the commutator subgroup (U, U) of U. But (U, U) is contained in the subgroup of U generated by the $U_c(c > 0, \ c \notin \Delta)$ (see, e.g., [3], Prop. 2.5, p. 66), hence

$$\mathrm{Ad}\, b(X) - X \in \Sigma_{c>0, c \notin \Delta}\, \mathfrak{g}_c\,.$$

We conclude that

$$\mathrm{Ad}\, g(X) = \Sigma_{a \in \Phi} \xi_a X_{w(a)}$$

with $\xi_a \neq 0$ if $a \in \Delta$. Since the $w(a)$ are distinct roots, the $X_{w(a)}$ are linearly independent, and $\mathrm{Ad}\, g(X) \in \mathfrak{b}$ implies that $w(a) > 0$ for $a \in \Delta$. As is well known, we have then $w = e$, hence $g \in B$.

2.5. PROPOSITION. *Assume* G *to be connected. Let* U *be the unipotent radical and* T *a maximal torus of* G. *There exist two Borel subgroups* B, B' *of* G, *which generate* G *separably, such that* $B \cap B' = T \cdot U$ *and* $\mathfrak{g} = \mathfrak{b} + \mathfrak{b}'$, $\mathfrak{b} \cap \mathfrak{b}' = \mathfrak{t} + \mathfrak{u}$.

It suffices to prove this for the reductive group G/U. For the latter take two opposed Borel subgroups containing T (see [3], 2.3, p. 64).

2.6. PROPOSITION. *Assume* G *to be connected. Let* P *be a parabolic subgroup of* G. *Then* $P = N_G(\mathfrak{p})$.

Let B be a Borel subgroup of G contained in P and let $Q = N_G(\mathfrak{p})$. The Borel subalgebras of $\mathfrak{q}$ are the conjugates of $\mathfrak{b}$, hence are contained in $\mathfrak{p}$. We have then $\mathfrak{p} = \mathfrak{q}$ by either 2.3 or 2.5. Since parabolic subgroups of a connected group are connected, it follows that $P = Q$.

2.7. For the sake of reference we recall here some facts about groups over finite fields.

Let k be finite. Then G has a Cartan subgroup (resp. maximal torus, resp. Borel subgroup) defined over k. Any two Borel subgroups defined over k are conjugate by an element of $G(k)$. Let H be a k-group and f: $G \to H$ a surjective k-morphism. Then a Cartan subgroup (resp. maximal torus, resp. Borel subgroup) defined over k of H is the image of such a subgroup of G.

For the existence, see [17, p. 45], or apply Lang's theorem [16] to the variety of maximal tori or of Borel subgroups (see §7). The conjugacy follows from Lang's theorem, too.

2.8. Let k be finite, G be connected reductive. Let B and $T \subset B$ be a Borel subgroup and a maximal torus of G defined over k. Then the Borel subgroup B' opposed to B and containing T is defined over k. Since the unipotent radical of a k-group is defined over k when k is perfect, it follows from 2.7 that if k is finite, we may choose B, B' and T in 2.5 to be defined over k. We want to use this remark to prove 2.9, which answers a question raised in ([12], Exp. XIV, p. 46). A proof has also been given by Steinberg (unpublished).

2.9. PROPOSITION. *Let k be finite and G be connected. Then G is generated by its Cartan subgroups which are defined over k.*

Let B, B', T as in 2.8. Since the Cartan subgroups defined over k of a k-group are the centralizers of its maximal tori defined over k, we see that the Cartan subgroups defined over k of B or B' are contained in Cartan subgroups defined over k of G. Consequently, it is enough to consider the case where G is solvable, where we proceed by induction on $\dim G$. If G is nilpotent, in particular if $\dim G = 1$, our assertion is obvious. Let U be the unipotent part of G and N a non-trivial connected normal k-subgroup of G contained in the center of U, of minimal dimension. Let H be the subgroup of G generated by the Cartan subgroups of G defined over k. If we apply the induction assumption to G/N, and use 2.7, we see that $G = H \cdot N$. Let T' be a maximal torus of G/N defined over k and L its inverse image in G. By 2.7, the maximal tori of L are maximal tori in G, hence the Cartan subgroups defined over k of L are contained in Cartan subgroups defined over

k of G. If $L \neq G$, then, by induction, L is generated by Cartan subgroups defined over k. Since $N \subset L$, this proves our assertion in this case.

There remains to consider the case where $G/N = T'$. Then (2.7), $G = T \cdot N$ where T is a maximal torus defined over k of G. Furthermore, the assumptions made on N imply that N has no proper non-trivial connected subgroup normalized by T. In particular, if $H \cap N \neq N$, then $(H \cap N)^0 = \{e\}$ and $H = T$, and since $Z(T)$ is connected [1, §13], we have either $Z(T) \cap N = N$, and then $G = Z(T)$ is its own Cartan subgroup, or $Z(T) \cap N = \{e\}$. So assume $Z(T) \cap N = \{e\}$. In view of the minimality assumption on N, we have $N^p = \{e\}$, hence ([18], Prop. 1,2, p. 688), N is isomorphic over k to a product of groups $\mathbf{G}_a$. In particular $N(k)$ has at least two elements. Let $x \in N(k)$, $x \neq e$. Then $x \cdot T \cdot x^{-1}$ is defined over k. Since $Z(T) \cap N = \{e\}$ and since in a connected solvable group, the centralizer and the normalizer of a torus coincide ([1], Prop. 10.2 p. 52), we have $x \cdot T \cdot x^{-1} \neq T$, hence $H \neq T$, $H \cap N \neq \{e\}$ and finally $H \supset N$, $G = H$.

2.10. Let H be a k-group. As in [3, 0.4], we say that it acts k-morphically on G if it acts k-morphically on the underlying variety of G and if, for each $h \in H$, the map $\varphi_h : g \mapsto h \cdot g$ is an automorphism of G. Then $h \mapsto (d\varphi_h)_e$ is a k-morphism of H into the group $\mathrm{Aut}(\mathfrak{g})$ of automorphisms of the restricted Lie algebra $\mathfrak{g}$.

2.11. PROPOSITION. *Assume G to be connected. Let S be a k-torus which acts k-morphically on G. Then S acts trivially on G if and only if it acts trivially on $\mathfrak{g}$.*

The necessity of the condition is obvious.

The set of $s \in S$ whose centralizer in $\mathfrak{g}$ is equal to that of S is a non-empty open subset A (consisting of the elements which do not annihilate the non-trivial weights of S in $\mathfrak{g}$). Similarly the set B of $s \in S$ whose centralizer in G is equal to $Z(S)$ contains an open non-empty set [3, 1.10, p. 62]. The sufficiency then follows from [3, 10.1] applied to an element of $A \cap B$.

3. Jordan decomposition in the Lie algebra of an algebraic group.

3.1. An element $X \in \mathfrak{g}$ is *semi-simple* (resp. *nilpotent*) if it belongs to the Lie algebra of a subtorus (resp. unipotent subgroup) of G.

(a) *Any element $X \in \mathfrak{g}$ can be written uniquely as $X = X_s + X_n$ where X_s is semi-simple, X_n is nilpotent and $[X_s, X_n] = 0$. If $f : G \to H$ is a morphism then $df(X_s) = (df(X))_s$, $df(X_n) = (df(X))_n$. If $G = \mathbf{GL}_n$ then $X = X_s + X_n$ is the Jordan decomposition of the linear transformation X.*

For the proof see [4]. The decomposition $X = X_s + X_n$ is called the

Jordan decomposition of X, and X_s (resp. X_n) is the *semi-simple* (resp. *nilpotent*) *part* of X. We state some consequences of (a).

(b) *If k is perfect, and $X \in \mathfrak{g}(k)$, then $X_s, X_n \in \mathfrak{g}(k)$.*

To see this we may identify G to a k-subgroup of $\boldsymbol{GL}_n$, and then we use the corresponding fact for the Jordan decomposition of a linear transformation over a perfect field.

(c) *Let G be connected, solvable. Then $X \in \mathfrak{g}$ is nilpotent if and only if it is tangent to the unipotent part of G.*

This follows from the fact that the unipotent radical of G contains all unipotent elements of G [1, Prop. 10.1, p. 52].

(d) *Let G be connected. It is a torus (resp. a unipotent group) if and only if its Lie algebra consists of semi-simple (resp. nilpotent) elements.*

The necessity of the condition is obvious. Let $g \in G$. It belongs to a Borel subgroup B of G, which, in view of (c) is a torus (resp. unipotent group). Thus g is semi-simple (resp. unipotent). We then use the fact that a connected group consisting of semi-simple (resp. unipotent) elements is a torus (resp. unipotent group) [1, §19].

3.2. PROPOSITION. *Let $p > 0$. Then the p-th power operation $[p]$ is a k-morphism of $\mathfrak{g}$ into itself, which maps the set of semi-simple (resp. nilpotent) elements onto (resp. into) itself. There exists a power q of p such that $X^{[q]} = X_s^{[q]}$ for all $X \in \mathfrak{g}$. The image of $[q]$ is the set of semi-simple elements of $\mathfrak{g}$.*

We may assume $\mathfrak{g} \subset \mathfrak{gl}_n$. Then (1.7(b)), $X^{[p]} = X^p$, which proves that $[p]$ is a k-morphism of $\mathfrak{g}$ into itself leaving the set of semi-simple (resp. nilpotent) elements stable. Moreover, $X^{[q]} = 0$ if $q > n$ and X is nilpotent. For such a q we have then, since $[X_s, X_n] = 0$,

$$X^{[q]} = (X_s + X_n)^{[q]} = X_s^{[q]} + X_n^{[q]} = X_s^{[q]},$$

which shows that $\operatorname{Im}[q]$ consists of semi-simple elements. On the other hand, $[p]$ is surjective on $\mathfrak{g}$ if G is a torus (1.7(c)), hence $\operatorname{Im}[p^s]$ contains the set of semi-simple elements of $\mathfrak{g}$ for every $s \geq 1$.

3.3. Let X be a non-zero nilpotent element of $\mathfrak{g}(k)$. If $p = 0$, then it is well known that X is tangent to a unique one-dimensional unipotent k-subgroup U of G. If $G \subset \boldsymbol{GL}_n$, then U is the set of elements

$$\exp s \cdot X = \Sigma_{n \geq 0} (n!)^{-1} \cdot s^n \cdot X^n$$

(see, e.g., [8], Prop. 1, p. 159).

However, if $p>0$, there is not always a one-dimensional unipotent subgroup U of G whose Lie algebra is spanned by X. We shall come back to this question in §9.

4. Toral subalgebras.

4.1. PROPOSITION. *Let $X \in \mathfrak{g}(k)$ be semi-simple. Then $\mathfrak{z}(X)$ is the Lie algebra of $Z(X)$, and $Z(X)$ is defined over k. The conjugacy class $\operatorname{Ad} G(X)$ of X in $\mathfrak{g}$ is closed, and the map $g \mapsto \operatorname{Ad} g(X)$ induces a k-isomorphism of $G/Z(X)$ onto $\operatorname{Ad} G(X)$.*

By ([4], 1.5, 1.6, p. 28), $\operatorname{Ad} G(X)$ is closed in $\mathfrak{g}$, and $\mathfrak{z}(X)$ is the Lie algebra of $Z(X)$. Since $\operatorname{ad} X$ is semi-simple, we have $\mathfrak{g}=\mathfrak{z}(X)+[X,\mathfrak{g}]$. On the other hand (1.10), the differential of $f: g \mapsto \operatorname{Ad} g(X)$ at e is $-\operatorname{ad} X$. Therefore f is a separable map of G onto $\operatorname{Ad} G(X)$, which implies the other assertions of the proposition.

4.2. DEFINITION. A subalgebra $\mathfrak{s}$ of $\mathfrak{g}$ is called a *toral subalgebra* if it consists of semi-simple elements.

4.3. PROPOSITION. *Let $\mathfrak{s}$ be a toral subalgebra of $\mathfrak{g}$ which is defined over k. Then $\mathfrak{z}(\mathfrak{s})$ is the Lie algebra of $Z(\mathfrak{s})$. The group $Z(\mathfrak{s})$ is defined over k, contains maximal tori of G, and $\mathfrak{s}$ belongs to the Lie algebra of every maximal torus of $Z(\mathfrak{s})$. The group $Z(\mathfrak{s})^0$ is equal to $N(\mathfrak{s})^0$.*

We show first that $\mathfrak{s}$ is commutative. Assume it is not. Since $\operatorname{ad} X$ $(X \in \mathfrak{s})$ is a semi-simple linear transformation of $\mathfrak{g}$, there exist then $X, Y \in \mathfrak{s}$, not zero, such that $[X,Y]=Y$. But then $\operatorname{ad} Y$ leaves the two-dimensional space $\mathfrak{m}$ spanned by X, Y stable, and its restriction to $\mathfrak{m}$ is non-zero, nilpotent, a contradiction.

Let $X \in \mathfrak{s}(k)$. Assume first that $[X,\mathfrak{g}]=0$. Let T be a maximal torus of G whose Lie algebra contains X. Since $G^0 \subset Z(X)$, by 4.1, X belongs to the Lie algebra of the group $g \cdot T \cdot g^{-1}$ $(g \in G^0)$, which runs through all maximal tori of G as g runs through G^0. Thus, if $\mathfrak{s}$ is central in $\mathfrak{g}$, then $\mathfrak{s}$ belongs to the Lie algebra of all maximal tori of G, and $Z(\mathfrak{s}) \supset G^0$. Since $\mathfrak{s}$ is defined over k, $Z(\mathfrak{s})$ is defined over a purely inseparable extension of k. But any subgroup of G containing G^0 is defined over a separable extension of k, hence $Z(\mathfrak{s})$ is defined over k if $[\mathfrak{s},\mathfrak{g}]=0$.

Let now $\mathfrak{s}$ not be central in $\mathfrak{g}$. There exists then $X \in \mathfrak{s}(k)$ such that $[X,\mathfrak{g}]\neq 0$. We have then $\dim Z(X)\neq \dim G$ by 4.1. The group $Z(X)$ is defined over k, contains a maximal torus T whose Lie algebra contains X. We may then prove the second assertion by induction on $\dim G$.

Let T be a maximal torus of $Z(\mathfrak{s})^0$. Then $Z(T) \subset Z(\mathfrak{s})^0$. Using the conjugacy of maximal tori in $Z(\mathfrak{s})^0$, we see that $N(\mathfrak{s}) = (N(T) \cap N(\mathfrak{s})) \cdot Z(\mathfrak{s})^0$. Since $Z(T) \subset Z(\mathfrak{s})^0$ and $N(T)^0 = Z(T)$, it follows that $N(\mathfrak{s})^0 = Z(T) \cdot Z(\mathfrak{s})^0 = Z(\mathfrak{s})^0$.

4.4. COROLLARY. *Let G' be a k-group and $f\colon G \to G'$ a separable surjective morphism. Then $f(Z(\mathfrak{s})^0) = Z(df(\mathfrak{s}))^0$.*

We have $f(Z(\mathfrak{s})^0) \subset Z(df(\mathfrak{s}))^0$. To prove surjectivity, it is enough, in view of 4.3, to show that $\mathfrak{z}(df(\mathfrak{s})) = df(\mathfrak{z}(\mathfrak{s}))$, which follows from the full reducibility of $\mathrm{ad}_\mathfrak{g}\,\mathfrak{s}$.

4.5. COROLLARY. *The maximal toral subalgebras are the Lie algebras of the maximal tori of G, and are conjugate under G^0.*

REMARK. Toral subalgebras have also been introduced by Humphreys in [15, §13]. 4.5 is also proved there.

We shall have to use frequently the known global analogues of the previous results. For the ease of references, we collect them, and sharpen them slightly, in the following proposition.

4.6. PROPOSITION. *Let H be a k-subgroup of G, S a k-torus of G and $s \in G(k)$ a semi-simple element which normalizes H. Then $Z(S) \cap H$ and $Z(s) \cap H$ are defined over k, the orbit $M = \mathrm{Int}_G H(s)$ is closed in G, and the map $h \mapsto h \cdot s \cdot h^{-1}$ induces a k-isomorphism of $H/(Z(s) \cap H)$ onto M. The Lie algebra of $Z(S) \cap H$ (resp. $Z(s) \cap H$) is $\mathfrak{z}(S) \cap \mathfrak{h}$ (resp. $\mathfrak{z}(s) \cap \mathfrak{h}$). If G is connected, G' is a k-group and $f\colon G \to G'$ a surjective k-morphism, then $f(Z(S)) = Z(f(S))$ and $f(Z(s)^0) = f(Z(s))^0$.*

For the part pertaining to s of the first assertion, see ([3], 10.2, p. 128 and 10.3, p. 129), taking into account that the connectedness restrictions made in 10.3 are superfluous, in view of the facts recalled in 0.1. For the second assertion about s, see ([3], 10.1, p. 128).

The group $S(k_s)$ contains an element t such that $Z(t) = Z(S)$ ([3], 1.10, p. 62); consequently, $Z(S) \cap H$ is defined over k_s. Since it is purely inseparable over k, it is then defined over k. If G is connected, $Z(S)$ is connected by ([1], Prop. 18.4, p. 72).

We now prove the last assertion. Clearly $f(Z(S)) \subset Z(f(S))$. Since both groups are connected, it suffices to show that they have the same dimension.

Let N be the kernel of f. By the above, the Lie algebras of $Z(S)$ and $Z(S) \cap N$ are $\mathfrak{z}(S)$ and $\mathfrak{z}(S) \cap \mathfrak{n}$, hence

$$\dim f(Z(S)) = \dim Z(S) - \dim (Z(S) \cap N) = \dim df(\mathfrak{z}(S)).$$

But $\mathrm{Ad}_G S$ is fully reducible, whence $df(\mathfrak{z}(S)) = \mathfrak{z}(f(S))$ and $\dim df(\mathfrak{z}(S))$ $= \dim \mathfrak{z}(f(S)) = \dim Z(f(S))$. The proof of the equality $f(Z(s)^0) = Z(f(s))^0$ is quite similar.

4.7. PROPOSITION. *Let G be connected, solvable, and $\mathfrak{s}$ a toral subalgebra of $\mathfrak{g}$. Then $N(\mathfrak{s})$ and $Z(\mathfrak{s})$ are equal, and connected.*

Let T be a maximal torus of $Z(\mathfrak{s})$. Then $\mathfrak{s} \subset \mathfrak{t}$, hence $Z(T) \subset Z(\mathfrak{s})$. Let $x \in N(\mathfrak{s})$. Then $x \cdot T \cdot x^{-1}$ is a maximal torus of $Z(\mathfrak{s})^0$, and there exists $y \in Z(\mathfrak{s})^0$ such that $y \cdot x \in N(T)$, which shows that $N(\mathfrak{s}) \subset N(T) \cdot Z(\mathfrak{s})^0$. But $Z(T)$ is equal to $N(T)$ ([1], Prop. 10.2, p. 52), and is connected ([1], Thm. 13.2, p. 60), whence $N(T) \subset Z(\mathfrak{s})^0$, and $N(\mathfrak{s}) \subset Z(\mathfrak{s})^0$.

4.8. A Lie subalgebra of $\mathfrak{g}$ is *algebraic* if it is the Lie algebra of an algebraic subgroup. In the remaining part of this paragraph, we reprove some known results on algebraic Lie algebras.

Let $p = 0$. Then the Lie algebra of the intersection of two algebraic subgroups is the intersection of the Lie algebras of the two subgroups. Therefore, if M is a subset of $\mathfrak{g}$, there is a unique smallest algebraic subgroup, to be denoted $\mathcal{A}(M)$, whose Lie algebra contains M, namely the intersection of all algebraic subgroups whose Lie algebras contain M. If M is a subspace of $\mathfrak{g}$, defined over k, then $\mathcal{A}(M)$ is defined over k. If X is a non-zero nilpotent element of $\mathfrak{g}$, then $\mathcal{A}(X)$ is the one-dimensional group constructed in 3.3.

If $p \neq 0$, then the opening statement of the preceding paragraph may be false, as is already shown by the example (1.17) of the tori in $\boldsymbol{SL_2}$ in characteristic two.

4.9. PROPOSITION. *Let $\mathfrak{s}$ be a toral subalgebra of $\mathfrak{g}$.*
(i) *If $p = 0$, then $\mathcal{A}(\mathfrak{s})$ is a torus.*
(ii) *([13], Prop. 2, p. 5) If $p > 0$, then $\mathfrak{s}$ is algebraic if and only if* $\mathfrak{s}^{[p]} \subset \mathfrak{s}$.

$\mathfrak{s}$ belongs to the Lie algebra of a torus, whence (i).

To prove (ii), we may assume that G is a torus (4.3), and, by going over to an extension of k, that G splits over k.

If $\mathfrak{s}$ is algebraic, then $\mathfrak{s}^{[p]} = \mathfrak{s}$. Assume conversely that $\mathfrak{s}^{[p]} \subset \mathfrak{s}$. We have then $\mathfrak{s}^{[p]} = \mathfrak{s}$. It is elementary that this implies

$$\mathfrak{s} = \mathfrak{s}_0 \otimes_{F_p} k , \qquad (\mathfrak{s}_0 = \mathfrak{s} \cap \mathfrak{g}_0),$$

where $\mathfrak{g}_0$ is the fixed point set of $[p]$ (see 1.7(c)). There exists then a direct summand Y of $X_* = X_*(G)$, of rank equal to dim $\mathfrak{s}$, such that $f(Y \otimes k) = \mathfrak{s}$, in the notation of 1.7(c). But then the images in G of the elements of Y generate a subtorus of G whose Lie algebra is $\mathfrak{s}$.

4.10. LEMMA. *Let $p = 0$. Let $X \in \mathfrak{g}$ be either semi-simple or nilpotent. Let V be a finite dimensional vector space over k and $f: G \to GL(V)$ be a morphism. Let W be a subspace of V stable under $df(X)$. Then $f(g)(W) \subset W$ for all $g \in \mathcal{A}(X)$.*

Let X be nilpotent. Then it follows from 3.3 that $f(g)$ $(g \in \mathcal{A}(X))$ is a polynomial in $df(X)$, hence $f(g)$ leaves W stable.

Let now X be semi-simple. We may assume k to be algebraically closed. Then $df(X)$ is diagonalizable over k, and it suffices to consider the case where W is spanned by one element, say Y. Since $\mathcal{A}(X)$ is a torus, $f(\mathcal{A}(X))$ is also diagonalizable. Let then (X_i) $(1 \leq i \leq n)$ be a basis of V and a_i be characters of $\mathcal{A}(X)$ such that $f(t) \cdot X_i = t^{a_i} \cdot X_i$ $(i = 1, \cdots, n)$. We have then $df(U) \cdot X_i = da_i(U) \cdot X_i$ for U in the Lie algebra of $\mathcal{A}(X)$ $(i = 1, \cdots, n)$. Write $df(X) \cdot Y = a \cdot Y$ and $Y = \Sigma c_i X_i$; let J be the set of indices for which $c_i \neq 0$. Then $c_i = da_i(X)$ $(i \in J)$. Since $\mathcal{A}(X)$ is the smallest torus whose Lie algebra contains X, this implies that the a_i's $(i \in J)$ are equal to one another, hence Y is an eigenvector of $\mathcal{A}(X)$.

4.11. PROPOSITION. *Let $p = 0$. A subalgebra $\mathfrak{h}$ of $\mathfrak{g}$ is algebraic if and only if it satisfies the following conditions:*
(i) *$X \in \mathfrak{h}$ implies $X_s, X_n \in \mathfrak{h}$*
(ii) *if $X \in \mathfrak{h}$ is semi-simple, then the Lie algebra of $\mathcal{A}(X)$ belongs to $\mathfrak{h}$.*

The necessity of these conditions is obvious. Let us assume conversely that they are fulfilled. Let L be the algebraic subgroup of G generated by the groups $\mathcal{A}(X)$, $(X \in \mathfrak{h}(k)$, X semi-simple or nilpotent). Then, by 4.10, $\mathfrak{h}$ is stable under $\mathrm{Ad}_\mathfrak{g} L$. It follows then from 1.16 that the Lie algebra $\mathfrak{l}$ of L is spanned by the elements $\mathrm{Ad}\, x(X)$ $(x \in L; X \in \mathfrak{h}; X$ semi-simple or nilpotent). Therefore $\mathfrak{l} \subset \mathfrak{h}$. But $\mathfrak{l} \supset \mathfrak{h}$ in view of (i); hence $\mathfrak{l} = \mathfrak{h}$, and $\mathfrak{h}$ is algebraic.

4.12. COROLLARY. *Let V be a finite dimensional vector space over $\bar{k}$ and $G \to GL(V)$ a morphism. Let $\mathfrak{h}$ be a subalgebra of $\mathfrak{g}$, and W a subspace of V such that $df(\mathfrak{h})(W) \subset W$. Then $f(\mathcal{A}(\mathfrak{h})) \cdot (W) \subset W$.*

Let $\mathfrak{l} = \{X \in \mathfrak{g}, df(X)(W) \subset W\}$. Since the semi-simple and the nilpotent parts of a linear transformation Y are polynomials in Y, it follows from 3.1(a) that $\mathfrak{l}$ verifies the condition (i) of 4.9. By 4.10, it also fulfills condition (ii) of 4.11, hence $\mathfrak{l}$ is algebraic. The connected algebraic group L with Lie algebra $\mathfrak{l}$ is generated by the subgroups $\mathcal{A}(X)$ ($X \in \mathfrak{l}$, X semi-simple or nilpotent), hence L leaves W stable (4.10). Moreover, since $\mathfrak{l} \supset \mathfrak{h}$, we have $L \supset A(\mathfrak{h})$.

4.13. COROLLARY. *Let $\mathfrak{h}$ be a subalgebra of $\mathfrak{g}$. Assume $\mathfrak{h}$ to be spanned by algebraic subalgebras. Then $\mathfrak{h}$ is algebraic.*

We may write $\mathfrak{h} = \mathfrak{h}_1 + \cdots + \mathfrak{h}_q$, with $\mathfrak{h}_i$ algebraic ($1 \leq i \leq q$). Let H_i be the connected group with Lie algebra $\mathfrak{h}_i$, and L the group generated by the H_i's. We have $[\mathfrak{h}_i, \mathfrak{h}] \subset \mathfrak{h}$ whence $\mathrm{Ad}_G H_i(\mathfrak{h}) = \mathfrak{h}$ by 4.12, and therefore $\mathrm{Ad}_G L(\mathfrak{h}) = \mathfrak{h}$. But $\mathfrak{l}$ is generated by subalgebras of the form $\mathrm{Ad}\, g(\mathfrak{h}_i)$ ($g \in L$, $1 \leq i \leq q$) by 1.16, therefore $\mathfrak{l} \subset \mathfrak{h}$, and finally $\mathfrak{l} = \mathfrak{h}$.

REMARK. 4.11 to 4.13 are known results of Chevalley [8].

5. Inseparable isogenies.

5.1. In this paragraph, we assume $p \neq 0$. An *inseparable k-isogeny of height 1* $\pi: G \to H$ of k-groups is a k-isogeny such that the image of $k[H]$ under the comorphism π^0 contains $(k[G])^p$. The next result is known (see [21], Theorem 1 or [7], §3). It is a very special result about the existence of quotients in group schemes (for which we refer to [12], exp. V). For the convenience of the reader a sketch of a proof is given below.

5.2. PROPOSITION. *Let $\mathfrak{m}$ be an ideal in $\mathfrak{g}$, which is stable under the p-th power operation and under $\mathrm{Ad}(G)$, and which is defined over k. Then there exists a k-group $G/\mathfrak{m}$ and a purely inseparable k-isogeny π of height 1, which has the following properties:*
  (i) $\mathrm{Ker}\, d\pi = \mathfrak{m}$,
  (ii) *if $\pi': G \to G'$ is a purely inseparable k-isogeny with $\mathrm{Ker}\, \pi' \supset \mathfrak{m}$, then there exists a unique k-isogeny $\theta: G/\mathfrak{m} \to G'$ such that $\pi' = \theta \circ \pi$. The pair $(G/\mathfrak{m}, \pi)$ is unique up to k-isomorphism.*

The proof of uniqueness is standard and will be left to the reader. Let $A = k[G]$. By §1, $\mathfrak{g}(k)$ is an algebra of derivations of A and so is $\mathfrak{m}(k)$. Let B be the ring of invariants of $\mathfrak{m}$ in A:

$$B = \{a \in A, Xa = 0 \text{ for all } X \in \mathfrak{m}(k)\}\,.$$

Since $B \supset k[A^p]$, a well-known argument shows that B is finitely generated over k (see, e.g., [22], p. 50, Lemma 10). Moreover if l is an extension of k, then $B \otimes_k l$ is canonically imbedded in $A \otimes_k l$, which is a reduced ring (i.e., a ring without nilpotents $\neq 0$), because G is defined over k. It follows that $B \otimes_k l$ is reduced.

We conclude that B is the k-algebra of a variety $G/\mathfrak{m}$ which is defined over k. The injection $B \to A$ induces a morphism of k-varieties $\pi \colon G \to G/\mathfrak{m}$ which is bijective on $G(\bar{k})$ (since $B \supset k[A^p]$). We next prove that $G/\mathfrak{m}$ is an algebraic group and that π is a homomorphism. It will follow that π is a k-isogeny.

Let $\mu \colon A \to A \otimes_k A$ define the group law on G. It suffices to prove that μB is contained in the subring $B \otimes B$ of $A \otimes A$. Let $X \in \mathfrak{m}(k)$. Then the right invariance of X implies by 1.2(1) that $(X \otimes \mathrm{id})\mu b = 0$ for $b \in B$. The invariance of $\mathfrak{m}$ under $\mathrm{Ad}(G)$ implies that $(\mathrm{id} \otimes X)\mu b = 0$ for $b \in B$. From these two facts it follows that $\mu B \subset B \otimes B$. To finish the existence proof it remains to be shown that $\mathrm{Ker}\, d\pi = \mathfrak{m}$. Let $X \in \mathfrak{g}(k)$. The canonical image Y of X in the Lie algebra of $G/\mathfrak{m}$ is obtained as follows (see 1.5). Let $b \in B$, $\mu b = \Sigma_i b_i \otimes b'_i$, then $Yb = \Sigma_i Xb_i(e)b'_i$. From this it is clear that $\mathfrak{m} \subset \mathrm{Ker}\, d\pi$. Next let $X \in \mathrm{Ker}\, d\pi$. Then $\Sigma_i Xb_i(e) b'_i = 0$. Taking (as we may) the b'_i to be linearly independent, we get $Xb_i(e) = 0$. Clearly $\mathrm{Ker}\, d\pi$ is invariant under $\mathrm{Ad}(G)$, which implies that $Xb_i = 0$. Since $b = \Sigma_i b'_i(e)b_i$, we see, finally, that $Xb = 0$ for $b \in B$, $X \in \mathrm{Ker}\, d\pi \cap \mathfrak{g}(k)$. The proof will be finished if we show: if k is algebraically closed and if $X \in \mathfrak{g}$, $Xb = 0$ for all $b \in B$, then $X \in \mathfrak{m}$.

We may then take G to be connected. Let L (resp. M) be the quotient field of the integral domain A (resp. B). L is a purely inseparable extension of M of height 1, $\mathfrak{n} = \mathfrak{m} \otimes_k M$ is an algebra of deriviations of L/M and M is exactly the field of invariants of $\mathfrak{n}$. We can now apply the Galois theory of purely inseparable extensions of height 1, due to Jacobson, to get the desired result (see [6], Prop. 6, p. 194). To prove (ii), let $f \colon C \to A$ be the L-algebra homomorphism defining π'. Then $\mathrm{Ker}\, d\pi' \supset \mathfrak{m}$ implies that $f(C) \subset B$, whence the result.

5.3. PROPOSITION. *Let S be a subtorus of G which is not contained in the center of G and whose Lie algebra $\mathfrak{s}$ is central in $\mathfrak{g}$. Then there exists an algebraic group G' and a purely inseparable isogeny $\pi \colon G \to G'$, whose differential $d\pi$ has kernel $\mathfrak{s}$ and such that the Lie algebra $\mathfrak{s}'$ of $S' = \pi(S)$ is non-central in $\mathfrak{g}'$. The algebra $\mathfrak{g}'$ is the direct sum of $d\pi(\mathfrak{g})$ and $\mathfrak{s}'$.*

Let $\Phi = \Phi(S, G)$ be the set of roots of G with respect to S. By 2.11, $\Phi(S, G) \neq \emptyset$. $\mathfrak{g}$ is the direct sum of the Lie algebra $\mathfrak{z}(S)$ of $Z(S)$ (see (4.6)) and of the root spaces

$$\mathfrak{g}_a = \{X \in \mathfrak{g} \,|\, \mathrm{Ad}(s)X = s^a X \text{ for } s \in S\},$$

where $a \in \Phi$.

The differential of $a \colon S \to \boldsymbol{GL}_1$ can be identified with a linear form on $\mathfrak{s}$ and we have $[Y, X] = da(Y)X$ for $Y \in \mathfrak{s}$, $X \in \mathfrak{g}_a$. The form da is zero if and only if a is divisible by p in the character group $X^*(S)$ of S.

Let c be the smallest positive integer such that $\Phi \not\subset p^{c+1} X^*(S)$. $\mathfrak{s}$ is central in $\mathfrak{g}$ if and only if $c \geq 1$. We prove the first statement of 5.3 by induction on c.

From 4.3 it follows that 5.2 is applicable with $\mathfrak{m} = \mathfrak{s}$. Let $G_1 = G/\mathfrak{s}$, let $\pi_1 \colon G \to G_1$ be the isogeny of 5.2. Let $S_1 = \pi_1(S)$. From $d\pi_1(\mathfrak{s}) = 0$ it follows that the transposed homomorphism ${}^t\pi_1 \colon X^*(S_1) \to X^*(S)$ maps $X^*(S_1)$ into $pX^*(S)$. If c_1 has the same meaning for S_1 as c for S we have $c_1 < c$. If $c_1 = 0$, we can take $G' = G_1$, $\pi = \pi_1$, if $c_1 > 0$ we can apply induction to get the required pair (G', π).

To prove the last assertion, it suffices, by dimensions, to show that $d\pi(\mathfrak{g}) \cap \mathfrak{s}' = 0$. Let $X \in \mathfrak{g}$, $d\pi(X) \in \mathfrak{s}'$. Write $X = X_s + X_n$ according to 3.1(a). Then $d\pi(X_s) + d\pi(X_n)$ is in $\mathfrak{s}'$, hence semi-simple. By 3.1 this means that $X_n \in \mathrm{Ker}\, d\pi = \mathfrak{s}$. But $\mathfrak{s}$ consists of semi-simple elements, hence $X_n = 0$. So X is semi-simple. By 4.2, X and $\mathfrak{s}$ are contained in the Lie algebra $\mathfrak{t}$ of a maximal torus T of G. Replacing G by T, we are reduced to prove the assertion for the case that G is a *torus*. 4.9 (ii) then shows that $\mathfrak{s}$ is the Lie algebra of a subtorus of G. We may then assume $S = (\boldsymbol{GL}_1)^m$, $G = (\boldsymbol{GL}_1)^n$, with the imbedding $(x_1, \cdots, x_m) \mapsto (x_1, \cdots, x_m, 1, \cdots, 1)$. The isogeny $\pi \colon G \to G'$ can then be described as follows, identifying G' with $(\boldsymbol{GL}_1)^n$:

$$\pi(x_1, \cdots, x_n) = (x_1^{a_1}, \cdots, x_m^{a_m}, x_{m+1}, \cdots, x_n),$$

where the a_i are p-powers > 1. Identifying $\mathfrak{g}$ and $\mathfrak{g}'$ with n-dimensional space, we have

$$d\pi(x_1, \cdots, x_n) = (0, \cdots, 0, x_{m+1}, \cdots, x_n),$$

and $\mathfrak{s}'$ consists of the elements $(x_1, \cdots, x_n, 0, \cdots, 0)$ of $\mathfrak{g}'$. This shows that $d\pi(\mathfrak{g}) \cap \mathfrak{s}' = 0$, as had to be proved.

6. Regular elements, Cartan subalgebras, subgroups of type (C).

6.1. Regular elements. For $X \in \mathfrak{g}$, the nilspace of X is the space of elements in $\mathfrak{g}$ annihilated by some power of $\mathrm{ad}\, X$, and nil (X) denotes the dimension of the nilspace of X, i.e., the multiplicity of the eigenvalue zero of $\mathrm{ad}\, X$. If X is semi-simple, the nilspace of X is then $\mathfrak{z}(X)$.

We may write

$$\det(\operatorname{ad} X - T) = T^m \cdot (P_0(X) + P_1(X) \cdot T + \cdots + P_{n-m}(X) \cdot T^{n-m}),$$

($n = \dim \mathfrak{g}$), where T is an indeterminate, and the P_i's are non-identically vanishing polynomial functions on $\mathfrak{g}$, defined over k. Then X is regular if and only if $P_0(X) \neq 0$. The set $R(\mathfrak{g})$ of regular elements of $\mathfrak{g}$ is therefore open, non-empty.

6.2. LEMMA. *Let $X \in \mathfrak{g}$ and $X = X_s + X_n$ its Jordan decomposition.*
(i) *X is regular if and only if X_s is regular.*
(ii) *Let $p > 0$. Then X is regular if and only if $X^{[p]}$ is regular.*
(iii) *Let k be infinite. Then $R(\mathfrak{g}) \cap \mathfrak{g}(k)$ is Zariski-dense in $\mathfrak{g}$. The space $\mathfrak{g}(k)$ contains regular semi-simple elements.*

(i) Since $\operatorname{ad} X = \operatorname{ad} X_s + \operatorname{ad} X_n$ is the Jordan decomposition of $\operatorname{ad} X$ (3.1), $\operatorname{ad} X$ has the same eigenvalues as $\operatorname{ad} X_s$, hence $\operatorname{nil}(X) = \operatorname{nil}(X_s)$.

(ii) Since $\mathfrak{g}$ is restricted, $\operatorname{ad} X^{[p]} = (\operatorname{ad} X)^p$, hence $\operatorname{nil}(X) = \operatorname{nil}(X^{[p]})$.

(iii) The set of regular elements is open non-empty, and $\mathfrak{g}(k)$ is Zariski-dense in $\mathfrak{g}$, whence the first part of (iii). Let $X \in \mathfrak{g}(k)$ be regular. If $p = 0$, then $X_s \in \mathfrak{g}(k)$, and is regular by (i). If $p > 0$, there is a power q of p such that $X^{[q]} = X_s^{[q]}$ (3.2). But then $X^{[q]}$ is regular (by (ii)), semi-simple, and rational over k.

6.3. LEMMA. *Let S be a torus in G. Let $\Phi = \Phi(S, G)$ be the set of roots of G with respect to S, and Φ' the set of $a \in \Phi$ whose differential da is zero. The set $R_{\mathfrak{s}}$ of $X \in \mathfrak{s}$ such that $da(X) \neq 0$ $(a \in \Phi - \Phi')$ is the set of elements $X \in \mathfrak{s}$ for which $\mathfrak{z}(X) = \mathfrak{z}(\mathfrak{s})$, and is non-empty open in $\mathfrak{s}$. If $X \in R_{\mathfrak{s}}$, then $Z(X)^0 = Z(\mathfrak{s})^0$. If S is a maximal torus, then $R_{\mathfrak{s}}$ consists of regular elements of $\mathfrak{g}$. If k is infinite and $\mathfrak{s}$ is defined over k, then $\mathfrak{s}(k) \cap R_{\mathfrak{s}} \neq \emptyset$.*

(In this statement, the differential db of a morphism $b: S \to \boldsymbol{GL}_1$ is identified to a linear form on $\mathfrak{s}$.)

For a character b of S, let

$$(1) \qquad \mathfrak{g}_b^{(S)} = \mathfrak{g}_b = \{X \in \mathfrak{g}, \operatorname{Ad} s \cdot X = s^b \cdot X, \quad (s \in S)\}.$$

We have then

$$(2) \qquad \mathfrak{g} = \mathfrak{g}_0 \oplus \Sigma_{a \in \Phi} \mathfrak{g}_a.$$

Moreover,

$$(3) \qquad [X, Y] = db(X) \cdot Y, \qquad (X \in \mathfrak{s}, Y \in \mathfrak{g}_b);$$

consequently:

$$(4) \qquad \mathfrak{z}(\mathfrak{s}) = \mathfrak{g}_o \oplus \Sigma_{a \in \Phi'} \mathfrak{g}_a$$

$$(5) \qquad \mathfrak{z}(X) = \mathfrak{g}_o \oplus \Sigma_{a \in \Phi,\, da(x)=0} \mathfrak{g}_a$$

from which our first assertion follows. The second one is then a consequence of 4.3. Since $R_\mathfrak{s}$ is open and non-empty, its intersection with $\mathfrak{s}(k)$ is not empty if k is infinite. Let finally S be a maximal torus. By 6.2, $\mathfrak{g}$ has a regular semi-simple element Y, which belongs necessarily to a maximal toral subalgebra of $\mathfrak{g}$. By conjugacy, $\mathfrak{s}$ then contains regular semi-simple elements of $\mathfrak{g}$. But, clearly, $\mathrm{nil}(X) = \min \mathrm{nil}(Y)$ $(Y \in \mathfrak{s})$, if $X \in R_\mathfrak{s}$, which implies our last assertion.

If k is finite, $\mathfrak{g}(k)$ does not always contain regular elements. It was shown however by Chevalley that this is the case if G is the adjoint group of a semi-simple group (see [12] Exp. XIV, Appendix by J.-P. Serre). We outline briefly how the same method can be used for any reductive group.

6.4. PROPOSITION. *Let k be finite and G be reductive. Then* $\mathfrak{g}(k)$ *contains regular elements.*

One shows first (loc. cit. lemme 1) that it suffices to consider the case where G is quasi-simple over $\bar{k}$. Let T be a maximal torus of G which is defined over k, and $\Phi=\Phi(T,G)$. Let Φ' be the set of $a \in \Phi$ whose differential is identically zero. By 6.3, $X \in \mathfrak{g}$ is regular if and only if $da(X) \neq 0$ for all $b \in \Phi-\Phi'$.

The Galois group $\Gamma = \mathrm{Gal}(\bar{k}/k)$ operates on $T(\bar{k})$, on Φ, and leaves Φ' stable. It is proved in loc. cit. Lemme 3, that one can choose T in such a way that there are $r=\dim T$ roots $a_1, \cdots, a_r$, which form a basis of the lattice spanned by the roots, and such that $\Phi = \cup_i \Gamma \cdot a_i$. It suffices then to exhibit $X \in \mathfrak{t}(k)$ such that $da_i(X) \neq 0$ whenever $a_i \in \Phi-\Phi'$, $i = 1, \cdots, r$. Such an element exists by lemme 4, of loc. cit.

6.5. DEFINITION. A *Cartan subalgebra* of $\mathfrak{g}$ is the centralizer in $\mathfrak{g}$ of the Lie algebra of a maximal torus of G. A *subgroup of type* (C) of G is the identity component of the centralizer in G of the Lie algebra of a maximal torus of G.

In view of 6.6, 6.7 below, these notions are equivalent to those introduced in ([12], Exp. XIII) under the same terminology. A subgroup of type (C) always contains a Cartan subgroup, but it may be bigger; for example, in characteristic two, SL_2 is its own subgroup of type (C), while its Cartan subgroups are its maximal tori.

6.6. PROPOSITION. (i) *Two Cartan subalgebras of $\mathfrak{g}$ (resp. subgroups of type (C) of G) are conjugate.*

(ii) *The transform of a subgroup of type (C) of G (resp. Cartan subalgebra of $\mathfrak{g}$) by an automorphism of G (resp. the differential of an automorphism of G) is a subgroup of type (C) (resp. a Cartan subalgebra).*

(iii) *The Cartan subalgebras defined over k of $\mathfrak{g}$ are the Lie algebras of the subgroups of type (C) defined over k of G.*

(i) and (ii) follow from the definitions and the conjugacy of maximal tori, (iii) from the definitions and 4.3.

6.7. PROPOSITION. *Let $\mathfrak{h}$ be a subalgebra (not necessarily restricted) of $\mathfrak{g}$. The following conditions are equivalent:*

  (i) $\mathfrak{h}$ *is a Cartan subalgebra of $\mathfrak{g}$;*

  (ii) $\mathfrak{h}$ *is the nilspace of a regular element;*

  (iii) $\mathfrak{h}$ *is the centralizer of a regular semi-simple element;*

  (iv) $\mathfrak{h}$ *is nilpotent, equal to its normalizer.*

If those conditions are fulfilled, $\mathfrak{h}$ is maximal nilpotent, contains one and only one maximal toral subalgebra $\mathfrak{t}$ of $\mathfrak{g}$, and $\mathfrak{t}$ is the set of all semi-simple elements of $\mathfrak{h}$.

The equivalence of (i), (ii), (iii) follows from 6.2, 6.3 and from the fact that the nilspace of X is equal to the centralizer of X_s.

We now prove that (i) implies (iv) and the last assertion. By assumption $\mathfrak{h} = \mathfrak{z}(\mathfrak{t})$, where $\mathfrak{t}$ is a maximal toral subalgebra of $\mathfrak{g}$. Let $X \in \mathfrak{t}$ be such that $\mathfrak{h} = \mathfrak{z}(X)$; such an X exists and is regular by 6.3. Let $\mathfrak{n}$ be the normalizer of $\mathfrak{h}$. It is stable under $\mathrm{Ad}_G T$, hence, in the notation of 6.3:

$$\mathfrak{n} = \mathfrak{g}_0 + \Sigma_{a \in \Phi} \mathfrak{n} \cap \mathfrak{g}_a .$$

On the other hand

$$\mathfrak{h} = \mathfrak{g}_0 + \Sigma_{a \in \Phi, da(X)=0} \mathfrak{g}_a .$$

Let $Y \in \mathfrak{n} \cap \mathfrak{g}_a$. Then $[X, Y] = da(X) \cdot Y \in \mathfrak{h}$, hence $da(X) = 0$, and $Y \in \mathfrak{h}$. Thus $\mathfrak{h} = \mathfrak{n}$. The algebra $\mathfrak{h}$ is algebraic (6.6), therefore $Z \in \mathfrak{h}$ implies $Z_s, Z_n \in \mathfrak{h}$, and, by 4.3, $Z_s \in \mathfrak{t}$, which shows that $\mathfrak{t}$ contains all semi-simple elements of $\mathfrak{h}$. This also proves that $[Z, U] = [Z_n, U]$ ($U \in \mathfrak{h}$). Consequently, ad Z, restricted to $\mathfrak{h}$ is nilpotent. $\mathfrak{h}$ is then nilpotent by Engel's theorem. Let finally $\mathfrak{m}$ be a nilpotent algebra containing $\mathfrak{h}$. If $\mathfrak{h} \neq \mathfrak{m}$, then the normalizer of $\mathfrak{h}$ in $\mathfrak{m}$ is not equal to $\mathfrak{h}$, a contradiction with what has already been proved.

(iv)$\Longrightarrow$(i). Let $X \in \mathfrak{h}$. By 3.1 and standard facts on linear transformations, $\operatorname{ad} X_s$ and $\operatorname{ad} X_n$ are polynomials in $\operatorname{ad} X$, hence X_s, X_n normalize $\mathfrak{h}$, and therefore belong to $\mathfrak{h}$. But the restriction of $\operatorname{ad} X_s$ to $\mathfrak{h}$ is nilpotent, since $\mathfrak{h}$ is nilpotent, and semi-simple since X_s is semi-simple. Therefore X_s is in the center of $\mathfrak{h}$, and the set $\mathfrak{s}$ of all semi-simple elements of $\mathfrak{h}$ is a toral subalgebra, central in $\mathfrak{h}$. We want to prove that $\mathfrak{s}$ is maximal. Let $\mathfrak{t}$ be a maximal toral subalgebra containing $\mathfrak{s}$. If $\mathfrak{s} \neq \mathfrak{t}$, then the fixed point set V of $\mathfrak{s}$ in $\mathfrak{g}/\mathfrak{h}$ is not zero. The image of $\mathfrak{h}$ in $\mathfrak{gl}(V)$ consists of nilpotent transformations, hence, by Engel's theorem, $\mathfrak{h}$ has a non-trivial fixed point set in V; but this contradicts the assumption $\mathfrak{h} = \mathfrak{n}(\mathfrak{h})$. Thus $\mathfrak{s} = \mathfrak{t}$, and $\mathfrak{h} \subset \mathfrak{z}(\mathfrak{t})$. By the first part of the proof, $\mathfrak{z}(\mathfrak{t})$ is nilpotent, and therefore $\mathfrak{h} = \mathfrak{z}(\mathfrak{t})$.

6.8. PROPOSITION. (a) *Let C be a subgroup of type* (C) *of G, and $\mathfrak{t}$ the maximal toral subalgebra of its Lie algebra $\mathfrak{c}$. The following conditions are equivalent*: (i) C *is defined over k,* (ii) $\mathfrak{c}$ *is defined over k,* (iii) $\mathfrak{t}$ *is defined over k.*

(b) *Let k be infinite. A subgroup H of G (resp. subalgebra $\mathfrak{h}$ of $\mathfrak{g}$) is of type* (C), *defined over k (resp. a Cartan subalgebra defined over k) if and only if it is the identity component of the centralizer in G (resp. the centralizer in $\mathfrak{g}$) of a regular semi-simple element of $\mathfrak{g}$ contained in $\mathfrak{g}(k)$.*

(a) Clearly (i)$\Longrightarrow$(ii). Let $\mathfrak{c}$ be defined over k. By 6.7, $\mathfrak{t}$ is the set of all semi-simple elements of $\mathfrak{c}$, hence is defined over k (3.2). This proves that (ii)$\Longrightarrow$(iii). We have $C = Z(\mathfrak{t})^0$ by definition, hence (iii)$\Longrightarrow$(i) follows from 4.3.

(b) follows from (a), 4.3 and 6.3.

6.9. LEMMA. *Let H be a closed connected subgroup of G whose Lie algebra contains a Cartan subalgebra $\mathfrak{c}$ of $\mathfrak{g}$. Then $H = N(\mathfrak{h})^0$, has finite index in its normalizer, and is the only closed connected subgroup of G with Lie algebra $\mathfrak{h}$.*

Obviously, $H \subset N(\mathfrak{h})^0$. Let now $g \in N(\mathfrak{h})$. Then, (6.6) $\mathfrak{c}$ and $\operatorname{Ad} g(\mathfrak{c})$ are both Cartan subalgebras of $\mathfrak{h}$, and we may find therefore $h \in H$ such that $h^{-1} \cdot g \in N(\mathfrak{c})$, whence $N(\mathfrak{h}) \subset H \cdot N(\mathfrak{c})$. By 6.7, $\mathfrak{c}$ is equal to its normalizer in $\mathfrak{g}$; therefore, if C is a subgroup of type (C) of H with Lie algebra $\mathfrak{c}$, we have $C = N(\mathfrak{c})^0$. From this we get $(H \cdot N(\mathfrak{c}))^0 = H \cdot C = H$, and $N(\mathfrak{h})^0 \subset H$. Since $N(H) \subset N(\mathfrak{h})$, it follows that H has finite index in its normalizer. Finally, the equality $H = N(\mathfrak{h})^0$ shows the uniqueness of H.

6.10. PROPOSITION. (a) *Let H be a closed connected subgroup of G, whose Lie algebra $\mathfrak{h}$ is defined over k, and contains a Cartan subalgebra* *of* $\mathfrak{g}$. *Then H is defined over k and is separably generated by its subgroups of type* (C) *defined over k.*

(b) *Let G' be a k-group and $f\colon G^0 \to G'$ a surjective separable k-morphism. Then f (resp. df) maps the set of k-subgroups of type* (C) *of G (resp. of Cartan subalgebras of $\mathfrak{g}$) defined over k onto the set of subgroups of type* (C) *of G' (resp. of Cartan subalgebras of $\mathfrak{g}'$) defined over k.*

(i) We prove first (a) when k is *infinite*. In view of 6.6 (iii), 1.16, it suffices to show that $\mathfrak{g}$ is spanned by its Cartan subalgebras which are defined over k. Let $X \in \mathfrak{g}(k)$. We let Y be equal to X_s if $p = 0$, to $X^{[q]}$ if $p \neq 0$ where q is a power of p big enough so that $Z^{[q]} = 0$ for all nilpotent $Z \in \mathfrak{g}$ (3.2). We have $[X, Y] = 0$ and $Y \in \mathfrak{g}(k)$. If X is regular, so is Y (6.2) and $\mathfrak{z}(Y)$ is a Cartan subalgebra of $\mathfrak{g}$, defined over k (6.8). Thus X belongs to a Cartan subalgebra defined over k. Since $R(\mathfrak{g}) \cap \mathfrak{g}(k)$ is dense in $\mathfrak{g}$ (6.2), this proves our assertion.

(ii) We now prove (b). In view of 6.6, it suffices to prove the part of (b) pertaining to df. Let $\mathfrak{n} = \ker df$. Let $\mathfrak{c}$ be a Cartan subalgebra defined over k and $\mathfrak{t}$ its maximal toral subalgebra. $\mathfrak{t}$ is defined over k (6.8). Let T be a maximal torus of G whose Lie algebra is $\mathfrak{t}$. Since f is separable, df is surjective, and $df(\mathfrak{t}) = \mathfrak{t}'$ is the Lie algebra of $T' = f(T)$, which is a maximal torus of G' ([1], §22). $\mathrm{ad}_{\mathfrak{g}}\,\mathfrak{t}$ is a commutative algebra of semi-simple transformations, which leaves $\mathfrak{n}$ stable. There exists then a subspace $\mathfrak{a} \subset \mathfrak{g}$ stable under $\mathrm{ad}_{\mathfrak{g}}\,\mathfrak{t}$ and such that $\mathfrak{g} = \mathfrak{a} \oplus \mathfrak{n}$. We have $\mathfrak{z}(\mathfrak{t}) = \mathfrak{z}(\mathfrak{t}) \cap \mathfrak{a} + \mathfrak{z}(\mathfrak{t}) \cap \mathfrak{n}$. This implies immediately that

$$df(\mathfrak{c}) = df(\mathfrak{z}(\mathfrak{t})) = \mathfrak{z}(df(\mathfrak{t})) = \mathfrak{z}(\mathfrak{t}'),$$

hence $df(\mathfrak{c})$ is a Cartan subalgebra defined over k.

Let now C' be a subgroup of type (C) of G' defined over k and $H = f^{-1}(C')^0$. The group H is defined over k (since f is separable), and contains a maximal torus of G ([1], §22). Let us show that a subgroup C of type (C) of H is also of type (C) in G. Let T be a maximal torus of C. It is maximal in H, hence in G, and maps onto a maximal torus T' of C'. Then $\mathfrak{c}' = \mathfrak{z}(\mathfrak{t}')$. Let $\mathfrak{b}$ be a subspace of $\mathfrak{g}$, supplementary to $\mathfrak{h}$, and stable under $\mathrm{Ad}_{\mathfrak{g}}\,T$. Then df maps $\mathfrak{b}$ isomorphically onto a supplement of $\mathfrak{c}'$ in $\mathfrak{g}'$, hence $\mathfrak{z}(\mathfrak{t}') \cap df(\mathfrak{b}) = 0$. It follows that $\mathfrak{z}(\mathfrak{t}) \cap \mathfrak{b} = 0$. Since $\mathfrak{z}(\mathfrak{t})$ is sum of its intersections with $\mathfrak{h}$ and $\mathfrak{b}$, we see that $\mathfrak{c}$ is the centralizer of $\mathfrak{t}$ in $\mathfrak{g}$.

Let now k be infinite. Then, by (i), $\mathfrak{h}$ has a Cartan subalgebra defined over k. By what has already been proved, $\mathfrak{c}$ is a Cartan subalgebra of $\mathfrak{g}$ and $df(\mathfrak{c})$ is a Cartan subalgebra of $\mathfrak{c}'$, hence is equal to $\mathfrak{c}'$.

Let k be finite. Then H has a maximal torus T defined over k (2.7), $\mathfrak{c} = \mathfrak{z}(\mathfrak{t})$ is a Cartan subalgebra of $\mathfrak{g}$ defined over k, contained in $\mathfrak{h}$, and $df(\mathfrak{c}) = \mathfrak{c}'$.

(iii) There remains to prove (a) for k finite. Let B, B', T be as in 2.8. Since the maximal tori of B, B' are maximal in G, the Cartan subalgebras defined over k of $\mathfrak{b}$ or $\mathfrak{b}'$ are contained in Cartan subalgebras defined over k of $\mathfrak{g}$. In view of 2.8, we are reduced to the case where G is solvable. We proceed by induction on $\dim G$. There is nothing to prove if G is nilpotent, in particular if $\dim G = 1$. Let N be a non-trivial normal closed connected subgroup of G, contained in the center of the unipotent radical of G, minimal for these properties. Let $f: G \to G' = G/N$ be the canonical projection. Let $\mathfrak{a}$ be the subspace of $\mathfrak{g}$ spanned by the Cartan subalgebras defined over k of $\mathfrak{g}$. In view of (ii) and of the induction assumption, applied to G', we have $\mathfrak{g} = \mathfrak{a} + \mathfrak{n}$. Let T' be a maximal torus defined over k of G' and $H = f^{-1}(T')$. Let T be a maximal torus defined over k of H. We have $H = T \cdot N$ (semi-direct). If $T' \neq G'$, then $H \neq G$, and, by induction assumption, $\mathfrak{h}$ is spanned by its Cartan subalgebras defined over k. But T is a maximal torus of G, too, hence the Cartan subalgebras defined over k of $\mathfrak{h}$ are contained in Cartan subalgebras defined over k of $\mathfrak{g}$. Since $\mathfrak{n} \subset \mathfrak{h}$, we have $\mathfrak{g} = \mathfrak{a}$ in this case. So assume $G = H = T \cdot N$. The subgroup $C = Z(\mathfrak{t})^0$ of G is of type (C), defined over k, and contains T. Then $C \cap N$ is normal connected in G. By the minimality assumption on N, we have either $C \cap N = N$, and $G = C$, in which case our assertion is proved, or $C \cap N = \{e\}$. So assume $T = C$. Then $\mathfrak{t}$ is not central in $\mathfrak{g}$, and there exists $X \in \mathfrak{t}(k)$ not central in $\mathfrak{g}$. Then (4.3), $Z(X)^0 \neq G$ whence again $Z(X)^0 = T$ and $\mathfrak{z}(X) = \mathfrak{t}$. We claim that $Z = X + Y$ is semisimple for any $Y \in \mathfrak{n}$. In fact, since $Z_n \in \mathfrak{n}$, we see, by considering the projection of G onto $G/N \cong T$, and using 3.1, that $Z_s = X + U$ $(U \in \mathfrak{n})$. Since $\mathfrak{n}$ is commutative, we have then

$$0 = [Z_s, Z_n] = [X, Z_n],$$

hence $Z_n = 0$, $U = Y$. We see also that $\mathfrak{z}(X + Y) \cap \mathfrak{n} = 0$. Thus $X + Y$ is regular, and, if $Y \in \mathfrak{n}(k)$, $\mathfrak{z}(X + Y)$ is a Cartan subalgebra defined over k. It follows that the space $\mathfrak{a}$ spanned by the Cartan subalgebras defined over k of $\mathfrak{g}$ contains $\mathfrak{t}$ and $X + \mathfrak{n}(k)$. It is then elementary that $\mathfrak{a} = \mathfrak{g}$.

6.11. COROLLARY. *Let H, K be closed subgroups of G. Assume that H is connected and that $\mathfrak{h}$ contains a Cartan subalgebra of $\mathfrak{g}$. Then $H \subset K$ if and only if $\mathfrak{h} \subset \mathfrak{k}$.*

This follows from 6.10, with k standing for a common field of definition for G, H, K.

6.12. LEMMA. *Let G be connected, and T be a maximal torus of G. Then T is defined over k if and only if $Z(T)$ is.*

If T is defined over k, then so is $Z(T)$ by (4.6). The group $Z(T)$ being connected, nilpotent, the converse follows from the fact that if a connected nilpotent group H is defined over k, then the unique maximal torus S of H is defined over k. This was noticed first in ([17], Prop. 9, p. 37), and follows from the fact that, for a suitable power q of p, the map $x \mapsto x^q$ is k-morphism of varieties of H onto S.

6.13. PROPOSITION. *Let H, K be two subgroups of type (C) (resp. Cartan subgroups, resp. maximal tori) of G and $\mathfrak{a}, \mathfrak{b}$ two Cartan subalgebras (resp. maximal toral subalgebras) of $\mathfrak{g}$. If H, K (resp. $\mathfrak{a}, \mathfrak{b}$) are defined over k, then $\mathrm{Tr}(H, K)$ (resp. $\mathrm{Tr}(\mathfrak{a}, \mathfrak{b})$) is defined over k, and H, K (resp. $\mathfrak{a}, \mathfrak{b}$) are conjugate under $G(k_s)$. The group $N(H)$ (resp. $N(\mathfrak{a})$) is defined over k if and only if H (resp. $\mathfrak{a}$) is so.*

The Lie algebra of $N(\mathfrak{a})$ is $\mathfrak{a}$ if $\mathfrak{a}$ is a Cartan subalgebra, $\mathfrak{z}(\mathfrak{a})$ if $\mathfrak{a}$ is a maximal toral subalgebra (4.3). Taking 6.8 into account, it follows that $\mathfrak{a}$ is defined over k if $N(\mathfrak{a})$ is so. If H is of type (C) or is a Cartan subgroup, it is the identity component of $N(H)$. If H is a maximal torus, then it is the unique maximal torus of the Cartan subgroups $Z(H)^0 = N(H)^0$ of G^0. Using 6.12 in the latter case, we see that H is defined over k if $N(H)$ is so.

To prove the remaining part of the proposition, we may assume $H, K, \mathfrak{a}, \mathfrak{b}$ to be defined over k. Then $\mathrm{Tr}(H, K)$ and $\mathrm{Tr}(\mathfrak{a}, \mathfrak{b})$ are defined over purely inseparable extensions of k, and it suffices to prove that they are defined over k_s. For the rest of the proof we may (and shall) assume k to be *separably closed*.

Let $\mathfrak{a}$ be a Cartan subalgebra, $\mathfrak{t}$ its maximal toral subalgebra. Since k is infinite, there exists $X \in \mathfrak{a}(k)$, regular, semi-simple, and we have $Z(X)^0 = Z(\mathfrak{t})^0$ by 6.3. Let $M = \mathrm{Ad}\, G(X)$. Then M is closed and $f: g \mapsto \mathrm{Ad}\, g(X)$ is a surjective separable k-morphism of G onto M (4.1). Let us show that M and $\mathfrak{a}$ intersect transversally. Let $g \in G$ be such that $Y = \mathrm{Ad}\, g(X) \in \mathfrak{a}$. Then $Y \in \mathfrak{t}$ and $\mathfrak{z}(Y) = \mathfrak{a}$, hence $g \in N(\mathfrak{a})$. But $N(\mathfrak{a})^0 = Z(X)^0$ by 6.9; therefore $M \cap \mathfrak{a}$ is zero-dimensional and the intersection is proper. Since $M \cap \mathfrak{a}$ is an orbit of $N(\mathfrak{a})$, it suffices to check transversality at X. The space $T(M)_x$ is the translate by X of $[X, \mathfrak{g}]$ (see 1.10). Since X is semi-simple, $\mathfrak{g} = \mathfrak{z}(X) \oplus [X, \mathfrak{g}] = \mathfrak{a} + [X, \mathfrak{g}]$, which yields our assertion. By conjugacy, M is transversal to any Cartan subalgebra. In particular $M \cap \mathfrak{b}$ is defined over k (1.13) and its points are rational over k (recall that $k = k_s$ here). A simple application of 1.13 shows then that $\mathrm{Tr}(\mathfrak{a}, \mathfrak{b}) = f^{-1}(M \cap \mathfrak{b})$ is defined over k and has points rational over k. Also, $N(\mathfrak{a}) = \mathrm{Tr}(\mathfrak{a}, \mathfrak{a})$ is defined over k. This settles the proposition for Cartan subalgebras, hence also, in view of 6.8 and 4.3, for maximal toral subalgebras and for subgroups of type (C).

The argument for Cartan subgroups is quite similar, but proceeds in the group rather than in the Lie algebra, and we outline it briefly. Let T be the unique maximal torus of H. It is defined over k (6.12) and we may find $x \in T(k)$ such that $Z(x)^0 = H$. The conjugacy class $M = \operatorname{Int} G(x)$ is closed and $g \mapsto g \cdot x \cdot g^{-1}$ is a surjective separable k-morphism of G onto M (4.6). Using 1.13 and the conjugacy of Cartan subgroups we see exactly as above that it suffices to check that M and H intersect transversally. If $g \cdot x \cdot g^{-1} \in H$, then x is a regular semi-simple element of H, $Z(g \cdot x \cdot g^{-1})^0 = H$, and $g \in N(H)$. Thus $M \cap H$ is zero dimensional, hence the intersection is proper. Since $M \cap H$ is an orbit of $N(H)$, there remains to check transversality at x. By 1.10, $T(M)_x$ is the translate by x of $(1 - \operatorname{Ad} x)(\mathfrak{g})$. Since x is semi-simple, $\mathfrak{g} = \mathfrak{z}(x) + (1 - \operatorname{Ad} x)(\mathfrak{g}) = \mathfrak{h} + (1 - \operatorname{Ad} x)(\mathfrak{g})$, whence the result. By 6.12 this also implies the corresponding assertion for maximal tori.

6.14. Corollary. *Let $\mathfrak{s}$ be a toral subalgebra of G defined over k. Then $Z(\mathfrak{s})$ is defined over k.*

In fact, $Z(\mathfrak{s})$ consists of finitely many components of $N(\mathfrak{s})$, hence is defined over k_s, and on the other hand, $Z(\mathfrak{s})$ is purely inseparable over k.

7. **Varieties of subgroups.** *In §§7, 8, for an extension k' of k in K, $C_{k'}$ denotes the category of extension fields of k' in K, where the morphisms are the inclusion homomorphisms. For $k' \in C_k$, $\bar{k'}$ is the algebraic closure of k' in K, k'_s the separable closure of k' in $\bar{k'}$, and $\Gamma(k')$ the Galois group of k'_s over k'.*

7.1. A (G, k)-set M is a set with the following properties:

(i) $G(K)$ operates on M. For each $m \in M$, the isotropy group of m is the set of K-points of an algebraic subgroup of G, to be denoted G_m.

(ii) There is given an inclusion preserving map $k' \mapsto M(k')$ from C_k to subsets of M, such that $M = M(K)$ and $M(k')$ is stable under $G(k')$.

(iii) For $k' \in C_k$, $\Gamma(k')$ operates on $M(k'_s)$ continuously, ($\Gamma(k')$ being endowed with the usual pro-finite topology, and $M(k'_s)$ with the discrete topology, (see [2])). We have

$$(1) \qquad {}^s(g \cdot m) = {}^s g \cdot {}^s m, \qquad (g \in G(k'_s),\ m \in M(k'_s),\ s \in \Gamma(k')),$$

and for $k \subset k' \subset k'' \subset K$, the following diagram

$$\begin{array}{ccc}
\Gamma(k') \times M(k'_s) & \longrightarrow & M(k'_s) \\
\uparrow{\scriptstyle j} \quad \downarrow{\scriptstyle i} & & \downarrow{\scriptstyle i} \\
\Gamma(k'') \times M(k''_s) & \longrightarrow & M(k''_s)
\end{array}$$

where i is the inclusion map, and j the natural restriction homomorphism, is commutative.

The (G, k)-set is *homogeneous* if $G(K)$ acts transitively on M.

We can view M as a functor from C_k to sets. Our conditions mean that the functors $G: k' \mapsto G(k')$ and $\Gamma: k' \mapsto \Gamma(k')$ operate on M, and verify a compatibility relation (1).

A (G, k)-set defines in an obvious way a (G, k')- set for every $k' \in C_k$.

If G' is a k-group, and $f: G' \to G$ is a k-morphism, then $G'(K)$ operates on M, and M is a (G', k)-set. It is homogeneous if M is so and f is surjective.

7.2. Examples. (a) Let X be a k-variety on which G acts k-morphically, transitively and separably. Then $M = X(K)$, with $M(k') = X(k')$, $(k' \in C_k)$, is a homogeneous (G, k)-set.

(b) Let V be a k-variety. We recall that to a k'_s-subvariety W of V $(k' \in C_k)$, and to $s \in \Gamma(k')$ there is associated a conjugate k'_s-subvariety, to be denoted ^sW. In this way $\Gamma(k')$ operates on the set of k'_s-subvarieties of V. Assume that G operates k-morphically on V. Then $G(K)$ operates on the set M of K-subvarieties of V, which is easily seen to be a (G, k)-set, $M(k')$ being by definition the set of elements of $M(K)$ which are defined over k' $(k' \in C_k)$. An orbit Q of $G(K)$ in M such that $Q(k'_s)$ is stable under $\Gamma(k')$ for every $k' \in C_k$ is then a homogeneous (G, k)-set. In this case, $Q(k')$ is the fixed point set of $\Gamma(k')$ in $M(k'_s)$.

In this paper, we shall be interested only in two special cases of this situation, namely:

(i) $V = G$, operated upon by inner automorphisms, M is a conjugacy class of closed subgroups;

(ii) $V = \mathfrak{g}$, operated upon by the adjoint representation, and M is a conjugacy class of subalgebras.

In both cases G_m is the normalizer of m in G.

7.3. Let M be a homogeneous (G, k)-set. Then $M = G(K)/G_m(K)$, which allows one to identify M with the set of K-points of a K-variety X on which G operates separably and transitively. We want to give conditions under which the field of definition of X can be brought down to k. More precisely, let X be a k-variety and φ a map from $X(K)$ to $M(K)$. We shall say that (X, φ), or simply X, *is the (G, k)-variety of elements of M*, or that M *is defined over k*, or that the functor M *is representable*, and *represented* by X, if it satisfies the following conditions:

(A) *G operates k-morphically, transitively and separably on X, and φ is a $G(K)$-equivariant bijection of $X(K)$ onto M which induces an isomorphism of (G, k)-sets.*

Clearly (X, φ) is also the (G, k')-variety of elements in M for $k' \in C_k$. It is immediate, and will follow from 7.5 (i), that (X, φ) is determined up to a unique k-isomorphism by (A), which justifies calling X "the" k-variety of elements in M. Note that in order to show that φ is compatible with 7.1 (ii), (iii) it is enough to check:

(B) $M(k')$ *is the set of fixed points of* $\Gamma(k')$ *in* $M(k'_s)$. *The map* φ *is a* $G(K)$-*equivariant bijection of* $X(K)$ *onto* $M(K)$ *which, for every* $k' \in C_k$, *induces a* $\Gamma(k')$-*equivariant bijection of* $X(k'_s)$ *onto* $M(k'_s)$.

In fact, since the fixed point set of $\Gamma(k')$ in $X(k'_s)$ is $X(k')$, φ induces then a $G(k')$-equivariant bijection of $X(k')$ onto $M(k')$, for every $k' \in C_k$, and all our conditions are fulfilled.

7.4. LEMMA. *Let* M *be a homogeneous* (G, k)-*set and* (X, φ) *the* (G, k)-*variety of elements in* M. *Let* G' *be a* k-*group and* $f: G' \to G$ *a surjective* k-*morphism. Assume that* $\mathfrak{g}$ *is spanned by* $df(\mathfrak{g}')$ *and by the Lie algebra of some isotropy group* $G_m (m \in M)$. *Then* (X, φ), *acted upon by* G' *via* f, *is the* (G', k)-*variety of elements in* M, *viewed as a* (G', k)-*set.*

Let $a \in X(K)$. By assumption $T(X)_a$ is the image of $\mathfrak{g}$ under the differential of the morphism $g \mapsto g \cdot a$. By the condition imposed on df, it follows that $T(X)_a$ is the image of $\mathfrak{g}'$ under the differential of the morphism $g' \mapsto f(g') \cdot a$ of G' onto X, hence G' acts on X transitively and *separably*. It is then clear that the other conditions in 7.3 are fulfilled for G'.

7.5. LEMMA. *Let* G' *be a* k-*group, and* $f: G' \to G$ *a surjective* k-*morphism. Let* M' *(resp.* M*) be a homogeneous* (G', k)-*set (resp.* (G, k)-*set). Assume that* $\mathfrak{g}$ *is spanned by* $df(\mathfrak{g}')$ *and the Lie algebra of* G_m *(* $m \in M$*). Let* (X', φ') *and* (X, φ) *be* (G', k)- *and* (G, k)-*variety structures on* M' *and* M *respectively. Let* $\lambda: M' \to M$ *be an* f-*equivariant map which, for each extension* k' *of* k *in* K, *induces a* $\Gamma(k')$-*equivariant map of* $M'(k'_s)$ *into* $M(k'_s)$.

(i) *There exists a unique* f-*equivariant* k-*morphism* $\psi: X' \to X$ *which coincides with* $\varphi^{-1} \circ \lambda \circ \varphi'$ *on* $X(K)$. *The map* ψ *is surjective, separable.*

(ii) *Let* k' *be an extension of* k *in* K. *Let* $m \in M(k')$, *and* $a = \varphi^{-1}(m)$. *Then* $\psi^{-1}(a)$ *is the* k'-*variety of elements in* $\lambda^{-1}(m)$, *acted upon in the natural way by* $f^{-1}(G_a)(K)$.

In view of 7.4, we may assume that $G = G'$ and f is the identity.

(i) Let $a' \in X'(k_s)$, $m' = \varphi'(a')$, $m = \lambda(m')$, $a = \varphi^{-1}(m)$. The groups $G_a = G_m$ and $G_{a'} = G_{m'}$ are defined over k_s; moreover, $G_{a'} \subset G_a$, since λ is a G-morphism.

Let $\mu: G/G_{a'} \to G/G_a$ be the canonical projection. It is a surjective separable k_s-morphism. The maps $g \mapsto g \cdot a$ and $g \mapsto g \cdot a'$ define k_s-isomorphisms $\alpha: G/G_a \xrightarrow{\sim} X$ and $\alpha': G/G_{a'} \xrightarrow{\sim} X'$ such that

$$(1) \qquad \alpha \circ \mu \circ \alpha'^{-1} = \varphi^{-1} \circ \lambda \circ \varphi' ,$$

on $X'(K)$. Let $\psi = \alpha \circ \mu \circ \alpha'^{-1}$. It is a separable surjective k_s-morphism of X' onto X. The assumption implies that it commutes with $\Gamma(k)$, hence ψ is defined over k. Since ψ is given on $X'(K)$ it is clearly unique.

(ii) The variety $Y = \psi^{-1}(a)$ is defined over k', since ψ is separable. The group G_a is also defined over k', being the inverse image of a under the k'-morphism $g \mapsto g \cdot a$ of G onto X, which is separable. It is then readily checked that 7.3 (A), with k replaced by k' is fulfilled by Y, $\varphi | Y$ and $\lambda^{-1}(m)$.

The following proposition extends, for homogeneous M, 4.9 of [2] to non-necessarily perfect groundfields. More general results on representability of such functors may be found in ([12], Exp. XIII).

7.6. PROPOSITION. *Let M be a homogeneous (G, k)-set. Assume:*

(a) *for every extension k' of k in $\bar{k}$, the set $M(k'_s)$ is a non-empty orbit of $G(k'_s)$ stable under $\Gamma(k')$, the set of fixed points of $\Gamma(k')$ in $M(k'_s)$ is $M(k')$, and $m \in M$ belongs to $M(k'_s)$ if and only if G_m is defined over k'_s.*

Then M is representable by a (G, k)-variety (7.3).

We first show the existence of (X, φ) verifying 7.3 (B) for $k' = k$. The proof is the same as that of an analogous result in [2, 4.10], where k was assumed to be perfect, and we sketch it briefly.

Let k_1 be a finite Galois extension of k in k_s such that $M(k_1) \neq \emptyset$ and let $m \in M(k_1)$. Let $Y = G/G_m$. The map $g \mapsto g \cdot m$ induces a $G(\bar{k})$-equivariant bijection α of $Y(\bar{k})$ onto M and a $G(k_s)$-equivariant bijection of $Y(k_s)$ onto $M(k_s)$. The latter allows one to define an action of $\Gamma(k)$ on $Y(k_s)$ by

$$(1) \qquad y \mapsto s(y) = \alpha^{-1}(^s(\alpha(y))), \qquad (y \in Y(k_s),\ s \in \Gamma(k)).$$

It follows from 7.1 (1) that

$$(2) \qquad s(g \cdot y) = {}^s g \cdot s(y), \qquad (y \in Y(k_s),\ g \in G(k_s)).$$

Since $G(k_s)$ is transitive on $M(k_s)$ we can, given $s \in \Gamma(k)$, choose $u_s \in G(k_s)$ such that $^s m = u_s \cdot m$. We have then

$$(3) \qquad \alpha \cdot s(y) = {}^s g \cdot \alpha(u_s \cdot m) \qquad (y = g \cdot \alpha^{-1}(m)\,;\ g \in G(k_s)).$$

It follows from 7.1 (1) that ${}^sG_m = G_{s_m}$, whence ${}^sY = G/{}^sG_m = G/G_{s_m}$, and

$$(4) \qquad\qquad u_s \cdot G_m \cdot u_s^{-1} = G_{s_m}.$$

The latter relation shows that $g \mapsto g \cdot u_s$ induces a k_s-isomorphism F_s of ${}^sY$ onto Y. Let $f_s \colon {}^sY(k_s) \to Y(k_s)$ be the bijection characterized by

$$f_s({}^sy) = s(y) \qquad (y \in {}^sY(k_s)).$$

It follows immediately from (3) that f_s is the restriction of F_s to ${}^sY(k_s)$. Thus condition (a) of ([2], lemma 2.12, p. 132) is fulfilled. Conditions (b), (c) of that lemma are checked to hold true exactly in the same way as in ([2], p. 143). The lemma yields then the existence of a k-variety X and of a k_1-isomorphism $\beta \colon X \to Y$ which commutes with $\Gamma(k)$, where $\Gamma(k)$ acts on X in the natural way, on Y by (1). We let G act on X in the obvious way via β, i.e.,

$$g \cdot x = \beta^{-1}(g(\beta(x))) \qquad (g \in G,\ x \in X).$$

This action is transitive, separable, defined over k_1. For $s \in \Gamma(k)$, $g \in G(k_s)$, $x \in X(k_s)$, we have, using (2)

$${}^sg \cdot {}^sx = \beta^{-1}({}^sg(\beta({}^sx))) = \beta^{-1}({}^sg \cdot s(\beta(x))) = \beta^{-1}(s(g \cdot \beta(x))) = {}^s(\beta^{-1}(g \cdot \beta(x)))$$

hence

$$({}^sg \cdot {}^sx = {}^s(g \cdot x),$$

which implies that the k_1-morphism $G \times X \to X$ defining the action of G on X commutes with $\Gamma(k)$, hence is defined over k. It is then clear that (X, φ), where $\varphi = \alpha \circ \beta$, verifies the condition 7.3 (B) for $k' = k$.

We now prove that (X, φ) verifies 7.3 (B) in general. Given k' $(k \subset k' \subset K)$ we have to show

$$(5) \qquad\qquad \varphi(X(k_s')) = M(k_s'),$$

$$(6) \qquad\qquad \varphi({}^sb) = {}^s(\varphi(b)), \qquad (b \in X(k_s'),\ s \in \Gamma(k')).$$

Let $b \in X(k_s')$. Since G operates separably on X, the isotropy group G_b of b is defined over k_s'. Since $G_b = G_{\varphi(b)}$, the condition imposed on M shows that $\varphi(b) \in M(k_s')$. Let now $m \in M(k_s')$. There exists $g \in G(k_s')$ such that $m = g \cdot \varphi(b)$. We have therefore $m = g \cdot \varphi(b) = \varphi(g \cdot b) \in \varphi(X(k_s'))$, whence (5).

Choose $a \in X(k_s)$, and let again $b \in X(k_s')$. The k_s-morphism $g \mapsto g \cdot a$

being separable, there exists $g \in G(k'_s)$ such that $b = g \cdot a$. We have then $^s b = {}^s g \cdot {}^s a$, hence

$$(7) \qquad \varphi(^s b) = {}^s g \cdot \varphi(^s a) \qquad (s \in \Gamma(k')).$$

Let $s' \in \Gamma(k)$ be the restriction of s to k_s. Then s and s' have the same action on $X(k_s)$ and on $M(k_s)$, (see 7.1), therefore

$$\varphi(^s a) = \varphi(^{s'} a) = {}^{s'}(\varphi(a)) = {}^s(\varphi(a)),$$

since (X, φ) verifies (B) for $k' = k$; taking (7) and 7.1(1) into account, we get

$$\varphi(^s b) = {}^s g \cdot {}^s(\varphi(a)) = {}^s(g \cdot \varphi(a)) = {}^s(\varphi(g \cdot a)) = {}^s(\varphi(b)).$$

7.7. PROPOSITION. (i) *The varieties of maximal tori and of Cartan subgroups of G are defined over k and are canonically isomorphic over k.*

(ii) *The varieties of subgroups of type* (C), *of Cartan subalgebras, and of maximal toral subalgebras of G are defined over k, and are canonically isomorphic over k.*

(iii) *Let $\mathscr{T}$ and C be the varieties of maximal tori and of subgroups of type* (C) *of G. Then the map $T \mapsto Z(\mathfrak{t})^0$ induces a surjective separable k-morphism $\tau = \tau_G : \mathscr{T} \to C$ of G-spaces. If $k' \in C_k$, and $H \in C(k')$, then $\tau^{-1}(H)$ is the k'-variety of maximal tori of H.*

Let M be one of the sets of subgroups or subalgebras listed in the statement. Then M is a conjugacy class of G (see 6.6 and [1], [10]) hence is an orbit of G in the set of irreducible subvarieties of G or $\mathfrak{g}$, operated upon by inner automorphisms or by the adjoint representation. 7.6 (a) is fulfilled in view of 6.12, 6.13, which yields the existence of the (G, k)-variety structures listed in 7.7.

As was remarked earlier (6.12), the Cartan subgroups of G defined over an extension k' of k are the centralizers of the maximal tori defined over k' of G. Consequently, the map which assigns to a maximal torus its centralizer is a bijection of the set A of maximal tori on the set B of Cartan subgroups, which induces a bijection of $A(k'_s)$ onto $B(k'_s)$ commuting with $\Gamma(k')$, whence the isomorphism of (i). The isomorphisms of (ii) follow similarly, using 6.6, 6.8.

(iii) is an application of 7.5 (ii).

7.8. PROPOSITION. *Let G' be a k-group and $f : G' \to G$ a surjective k-morphism such that $\mathfrak{g}$ is spanned by $df(\mathfrak{g}')$ and some maximal toral subalgebra. Let $\mathscr{T}'$ and C' (resp. $\mathscr{T}$ and C) be the k-varieties of maximal tori and of subgroups of type* (C) *of G' (resp. G).*

(i) *The map f induces a surjective separable k-morphism α of $\mathscr{T}'$ onto $\mathscr{T}$, which is an isomorphism if f is an isogeny.*

(ii) *If f is separable, it induces a surjective separable k-morphism $\beta: C' \to C$ such that $\tau_{G'} \circ \alpha = \beta \circ \tau_G$.*

The map f induces a surjective map of the set of maximal tori of G' onto the set of maximal tori of G ([1], §22), and is bijective if f is an isogeny. If f is separable, then f induces a surjective map of the sets of subgroups of type (C), by 6.10. The proposition follows therefore from 7.4.

The following theorem, for $\mathscr{T}$, is Grothendieck's main rationality result ([12], Exp. XIV, Th. 6.2, p. 39):

7.9. THEOREM. *The (G, k)-varieties $\mathscr{T}$ of maximal tori and C of subgroups of type (C) of G are rational varieties over k.*

We may assume G to be connected. The proof proceeds by induction on $\dim G$. There is nothing to prove if G is nilpotent, in particular if $\dim G = 1$. If G is its own subgroup of type (C), then C is reduced to a point rational over k. Using 7.8, 5.3 we see that it suffices to consider the case where the subgroups of type (C) of G are *proper*. Let $\tau: \mathscr{T} \to C$ be the canonical morphism (7.7). Let $k' \in C_k$. By 7.7 and 7.5, if $H \in C(k')$ then $\tau^{-1}(H)$ is the k'-variety of maximal tori of H. By our induction assumption, it is a rational variety over k'. In particular, $\tau^{-1}(H)(k') \neq \emptyset$, if k' is infinite. If we apply this to a generic point over k, we see that the field $k(\mathscr{T})$ is a purely transcendental extension of $k(C)$. It suffices therefore to prove that C is a rational variety over k. In the sequel, we view C as the k-variety of Cartan subalgebras of $\mathfrak{g}$, which is possible (7.7).

We assume first that $\mathfrak{g}(k)$ contains a regular semi-simple element X. This is the case notably if k is infinite (6.2) or if G is reductive (6.4). Let $\mathfrak{h} = \mathfrak{z}(X)$ and $\mathfrak{m}$ be a supplementary subspace to $\mathfrak{h}$ in $\mathfrak{g}$, defined over k. Then $N(\mathfrak{h})$ is defined over k (6.13), and C may be identified, over k, with $G/N(\mathfrak{h})$. To prove the assertion, we shall set up a k-isomorphism between non-empty open subsets of C and $\mathfrak{m}$. The set U of $Z \in \mathfrak{m}$ such that $X + Z$ is regular is non-empty, since $0 \in U$, and open, since the set of regular elements in $\mathfrak{g}$ is open (6.1). Such an element is contained in a unique Cartan subalgebra $\mathfrak{h}_z$, namely its nilspace. Let V be the set of $Z \in U$ such that $\mathfrak{h}_z \cap \mathfrak{m} = 0$. It is open, non-empty. The set of Cartan subalgebras $\mathfrak{h}'$ such that $\mathfrak{h}' \cap \mathfrak{m} = 0$ is open, not empty in C, therefore the set W of Cartan subalgebras $\mathfrak{h}_z(Z \in V)$ is open, non-empty, in C. To prove our assertion, it is then enough to show that $\alpha: Z \mapsto \mathfrak{h}_z$ is a k-isomorphism of V onto W. The nilspace of $\mathrm{ad}\,(X + Z)$ is also the kernel of $(\mathrm{ad}\,(X + Z))^n$ $(n = \dim \mathfrak{g})$ which implies immediately that

α is a k-morphism. Any subspace $\mathfrak{q}$ of $\mathfrak{g}$ supplementary to $\mathfrak{m}$ intersects $X+\mathfrak{m}$ at only one point, which is rational over the smallest field of definition of $\mathfrak{q}$ containing k, hence α is bijective, birational over k, and therefore is a k-isomorphism.

There remains to consider the case where k is *finite* and G is not reductive. Its unipotent radical U is then defined over k, (since k is perfect), and $\neq \{e\}$. Let N be a non-trivial central connected k-subgroup of exponent p, normal in G (e.g., a suitable p-th power of the last non-trivial term in the descending central series of U). Let $\pi: G \to G' = G/N$ be the canonical projection. By 7.5, 7.7, it induces a separable surjective k-morphism β of C onto the k-variety C' of subgroups of type (C) of G', and, for $H' \in C'(k')$, $\beta^{-1}(H')$ is the k'-variety of subgroups of type (C) of $\pi^{-1}(N(H'))$, $(k' \in C_k)$. If $H' \neq G'$, then induction, and the argument used earlier in the reduction to varieties of Cartan subalgebras, show that C is a rational variety over k. So assume $H' = G'$. Let A be a subgroup of type (C) of G defined over k, (6.10). Then, by 6.10 (b), we have $G = A \cdot N$, and also, since π is separable, $\mathfrak{g} = \mathfrak{a} + \mathfrak{n}$. Since $G/N = A/(A \cap N)$, it follows that $\dim(A \cap N) = \dim(\mathfrak{a} \cap \mathfrak{n})$, hence N and A intersect transversally. Then so do N and $N(A)$. Consequently, the natural bijective morphism of varieties of $N' = N/(N \cap N(A))$ onto $G/N(A) \cong C$ is a k-isomorphism. But N' is a connected commutative k-group of exponent p. Since k is perfect, N' is k-solvable, in the sense of [20], and is k-isomorphic to a vector group by Prop. 1.2, p. 688 of [18]. In particular, it is a rational variety over k.

REMARK. If G is solvable, we have the stronger result that C and $\mathcal{G}$ are k-isomorphic to affine spaces (see 9.14).

We mention some consequences of 7.9. For more results, see ([12], Exp. XIV).

7.10. PROPOSITION. *G has a maximal torus defined over k. If G is connected, it is generated by its Cartan subgroups defined over k.*

It suffices to prove the second assertion (6.12). If k is finite, see 2.9. Let k be infinite. The k-variety M of Cartan subgroups of G is a rational variety over k (7.7, 7.9), hence $M(k)$ is Zariski-dense, and in particular not empty. There exists therefore a Cartan subgroup C defined over k. Its normalizer $N(C)$ is also defined over k (6.13), and M is k-isomorphic to $G/N(C)$. Let H be the subgroup of G generated by the Cartan subgroups defined over k of G. Assume $H \neq G$. Since $C \subset H$ and C has finite index in $N(C)$, the image of H in $G/N(C)$ has a strictly smaller dimension than $G/N(C)$. But it contains $(G/N(C))(k)$, which is Zariski-dense, whence a contradiction.

7.11. COROLLARY. *Let G be connected. Assume that the Cartan subgroups of G defined over k are unirational over k. Then G is unirational over k. If k is infinite, $G(k)$ is dense in G.*

(We recall that an irreducible k-variety V is unirational over k if there exists a dominant k-morphism of an open subset of an affine space into V, or, equivalently, if the field $k(V)$ is contained in a purely transcendental extension of k. If k is infinite, this implies that $V(k)$ is Zariski-dense in V.)

In view of 7.10 we can find finitely many Cartan subgroups $C_1, \cdots, C_t$ of G, defined over k, such that the product map: $C_1 \times \cdots \times C_t \to G$ is surjective, whence our assertion.

7.12. COROLLARY. *Let G be connected. Then G is unirational over k if either k is perfect or G is reductive. In particular if k is infinite and either k is perfect or G is reductive, $G(k)$ is dense in G.*

If G is reductive, its Cartan subgroups are maximal tori, hence are unirational over their fields of definition ([17], Prop. 6, p. 39). If k is perfect, any connected nilpotent k-group is unirational over k.

REMARK. For perfect infinite fields, 7.10 to 7.12 are due to Rosenlicht [17]. For solvable groups and infinite fields the existence of a maximal torus defined over k is also proved in ([20], Thm. 4). See also ([3], §11).

8. Some results on reductive groups.

8.1. Let G be connected, reductive, T a maximal torus of G, and Φ the set of roots of G with respect to T. Given $b \in \Phi$, there exists an isomorphism θ_b of the additive group $\boldsymbol{G_a}$ onto a unipotent subgroup U_b of G, uniquely determined by b, such that

$$(1) \qquad t \cdot \theta_b(x) \cdot t^{-1} = \theta_b(t^b \cdot x), \qquad (x \in G_a;\ t \in T).$$

The group G is said to *split over k*, or to be *k-split*, if we can choose T to be k-split, and the θ_b's to be defined over k.

The theorem below is due to P. Cartier (unpublished). More general results can be found in ([12], Exp. XXII):

8.2. THEOREM. *Assume that G is connected, reductive, and contains a maximal torus which is defined over k and splits over k. Then G splits* *over k.*

We keep the notation of 8.1 and assume T to be k-split. Then b is defined over k ([3], 1.3, p. 60). Let $S = (\ker b)^0$. It is a subtorus of T, of codimension one, defined over k ([3], 1.6, p. 61). By 8.1 (1), $Z(S) \supset U_b$, U_{-b}. The group $Z(S)$ is defined over k (4.6), is reductive and contains T. It is known that its commutator subgroup H' is three-dimensional, generated by U_b, U_{-b}, ([10], Exp. 13) and defined over k. Moreover $L_b = (H' \cap T)^0$ is one-dimensional, defined over k, and split over k. The roots of H' with respect to L_b are $\pm b'$, where b' is the restriction of b to L_b. Since $\ker b' = (L_b \cap \ker b)$, it is clear that a monomorphism $\theta_{b'} : \boldsymbol{G}_a \to H'$ verifying $t \cdot \theta_{b'}(x) t^{-1} = \theta_{b'}(t^b \cdot x)$ for $x \in G_a$, $t \in L_b$ may also be viewed as a homomorphism θ_b verifying 8.1 (1). We are therefore reduced to the case where G is of type A_1; we denote by $\pm b$ the roots of G. We have $\mathfrak{g} = \mathfrak{t} + \mathfrak{g}_b + \mathfrak{g}_{-b}$. Since T splits over k, $\mathrm{Ad}_G T$ is diagonalizable over k, and $\mathfrak{g}_b$, $\mathfrak{g}_{-b}$ are defined over k. Let $\mu : G \to G'$ be as in 5.3 if $\mathfrak{t}$ is central in $\mathfrak{g}$, the identity otherwise. We want to prove that $U_{\pm b}$ is defined over k and is k-isomorphic to $\boldsymbol{G}_a$. It suffices to show that $B^{\pm} = T \cdot U_{\pm b}$ is defined over k, because $U_{\pm b}$ is its derived group, and then μ is a k-isomorphism of $U_{\pm b}$ onto its image (since $\ker d\mu = \mathfrak{t}$), and $\mu(U_{\pm b})$ is k-isomorphic to $\boldsymbol{G}_a$ by [4, 1.2]. Let $B' = \mu(B^+)$. It follows readily from 2.6 that B is the stability group of $\mathfrak{b}'$ in G, acting on $\mathfrak{g}'$ via $\mathrm{Ad} \circ \mu$. Since $L(\mu(T))$ is not central in $\mathfrak{g}'$, it is clear that $\mathfrak{b}'$ is its own normalizer in $\mathfrak{g}'$. Since $\ker d\mu = \mathfrak{t}$, it follows then that the orbit map $f : g \mapsto \mathrm{Ad}(g) \cdot (\mathfrak{b}')$ in the Grassmannian of two-planes in $\mathfrak{g}'$ is separable. Then $B = f^{-1}(\mathfrak{b}')$ is defined over k by 1.13. Similarly for B^-.

Let $\alpha : \boldsymbol{G}_a \to U_b$ be a k-isomorphism. Then $\alpha^{-1} \circ \theta_b$ is an automorphism of G_a, hence is of the form $x \mapsto c \cdot x$ for some $c \in \bar{k}^*$. But then it is obvious that 8.1 (1) holds for α too, i.e., we may choose θ_b to be defined over k.

8.3. COROLLARY. *The group G splits over a finite separable extension of k.*

This follows from 8.2 and from the fact that a k-torus splits over a separable extension of k [3, 1.5, p. 61].

8.4. PROPOSITION. *Assume that G is connected and has a unipotent radical U defined over k. Let $\mathscr{P}$ be a conjugacy class of parabolic subgroups, such that the set $\mathscr{P}(k_s)$ of elements of $\mathscr{P}$ defined over k_s is stable under $\Gamma(k)$. Then the variety of elements in $\mathscr{P}$ is defined over k. In particular, the variety of Borel subgroups of G is defined over k.*

A parabolic subgroup is equal to its normalizer (cf. [10], Exp. 9 for Borel subgroups, the general case follows immediately) therefore the last condition of 7.6 (a) is automatically fulfilled. The second one holds true, since we deal with subvarieties of a k-variety (7.2 (b)).

Let $\pi\colon G \to G/U$ be the canonical projection. Since U is assumed to be defined over k, π is a separable k'-morphism for every extension k' of k, hence defines a surjective morphism of $G(k'_s)$ onto $G/U(k'_s)$. Moreover π maps $\mathscr{P}$ onto a conjugacy class of parabolic subgroups of G/U. To check the first condition of 7.6 (a), we may therefore assume G to be reductive. Since then two parabolic subgroups of G defined over $k' \in C_k$, and conjugate over $\bar{k}$, are conjugate over k' ([3], Th. 4.13, p. 90), it suffices to show that $\mathscr{P}(k'_s)$ is stable under $\Gamma(k')$, $(k' \in C_k)$. This is clear for Borel subgroups of course.

Fix a maximal k-torus T of G, and an ordering of the roots of G with respect to T. Let Δ be the set of simple roots. Since G splits over k_s, $\mathscr{P}$ contains one and only one standard parabolic k_s-subgroup $P_\theta(\theta \subset \Delta)$, (see [3], 4.3, p. 86). Moreover, $\mathscr{P}(k_s)$ is stable under $\Gamma(k_s)$ if and only if θ is stable under the action of $\Gamma(k)$ on Δ defined in ([3], 6.2, p. 104). Since T splits over k_s, the analogous action of $\Gamma(k')$ on Δ goes through the restriction homomorphism $\Gamma(k') \to \Gamma(k)$, hence θ is also stable under $\Gamma(k')$, which implies our contention.

8.5. COROLLARY. *The variety of Borel subgroups of a k-group H is defined over k if either H is reductive or if k is perfect.*

In the first case, $U = \{e\}$, and in the second one, U is defined over k.

REMARK. Similar results in the framework of schemes are to be found in ([12], Exp. XXII; in particular Cor. 5.8.3, p. 74). If U is not defined over k, the (G, k)-set of Borel subgroups is not always representable by a (G, k)-variety ([12], Exp. XIV, Rem. 4.6, p. 30).

8.6. Using 8.2, one can extend some results of Steinberg [23] to non-perfect ground fields. The following result is proved there for perfect k (Th. 1.7, p. 51):

Let G be a connected, semi-simple simply connected k-group, which is quasi-split (i.e., contains a Borel subgroup which is defined over k). Then any regular semi-simple conjugacy class of G which is defined over k contains an element of $G(k)$.

Using 8.2, the proof given in loc. cit. carries over to arbitrary ground fields, if one observes that, with the notations of ([23], p. 305), any regular semi-simple conjugacy class intersects N in exactly one point, with multiplicity 1 (as follows from loc. cit., 8.1).

The arguments of ([23], §§10, 11) then show:

If k is a field of dimension ≤ 1 and G a connected reductive k-group, then $H^1(k, G) = 0$.

This answers a question of Serre ([23], p. 58, Remarques).

9. Nilpotent elements in Lie algebras.

9.1. Let X be a non-zero nilpotent element in $\mathfrak{g}(k)$. If $p = 0$, we have seen (3.3) that X is tangent to a unique one-dimensional unipotent k-subgroup of G. This is no longer true if $p \neq 0$. A necessary (but not sufficient) condition for this is that $X^{[p]} = 0$, a condition which easily is seen not to be always fulfilled (e.g., in $\mathfrak{gl}_n$ for $n > 2$). In this paragraph, we give a few positive results in that direction. The example below shows that the sufficient conditions of 9.8, 9.16 (ii), (iii) are also necessary. We assume $p > 0$ throughout.

9.2. EXAMPLE. Let q, r be two distinct powers of p, with strictly positive exponents. We let G be the connected unipotent k-group which is k-isomorphic, as a variety, to the 2-dimensional affine space, with the product given by

$$(1) \qquad (x, y) \cdot (z, t) = (x + z,\ y + t + x^q \cdot z^r).$$

Since $q \neq r$, G is non-commutative and so yields a (known) example of a 2-dimensional non-commutative nilpotent group in non-zero characteristic. The k-algebra of morphic k-functions on G is $A = k[T, U]$, with the coproduct $\mu : A \to A \otimes_k A$ given by

$$\mu(T) = T \otimes 1 + 1 \otimes T, \quad \mu(U) = U \otimes 1 + 1 \otimes U + T^q \otimes T^r.$$

One checks that $\mathfrak{g}$ is commutative, spanned by $X = \partial/\partial T$, $Y = \partial/\partial U$ and that $\mathfrak{g}^{[p]} = \{0\}$.

Let H be the k-subgroup of G whose elements are the pairs $(0, y)$. It is k-isomorphic to $\boldsymbol{G}_a$. We claim that any k-morphism $f \colon \boldsymbol{G}_a \to G$ has its image in H. We have $f(x) = (g(x),\ h(x))$ where g, h are polynomials which verify

$$(2) \qquad g(x + y) = g(x) + g(y), \quad g(0) = 0,$$

$$(3) \qquad h(x + y) - h(x) - h(y) = g(x)^q\, g(y)^r, \quad h(0) = 0 \quad (x, y \in \boldsymbol{G}_a).$$

If $g \not\equiv 0$, the left hand side of (3) is a symmetric polynomial in x and y, whereas the right hand side is not. This is impossible, hence $g \equiv 0$, which proves our contention. Thus X is not tangent to any one-dimensional subgroup of G and G has elements which are not contained in a connected one-dimensional subgroup. Moreover, as $q \neq r$, X is not tangent to any commutative subgroup of G, and G has elements which do not belong to any connected commutative subgroup.

Let $T = \boldsymbol{GL}_1$ and b a non-trivial character of T. We let T act on G by means of the map $\{t,(x,y)\} \mapsto (t^b x, t^{(q+r)b} y)$. It is readily seen that in this way T acts k-morphically on G, and X, Y are eigen-vectors of T with weights b and $(q+r) \cdot b$.

9.3. Let T be a k-torus, F a finite dimensional vector space defined over k, and $r: T \to GL(F)$ a k-morphism. We let $H^1(T, F)$ be the first cohomology group of T with coefficients in the T-module F, based on cochains $T \to F$ which are k-morphisms of varieties. Thus a cocycle is a k-morphism $f: T \to F$ such that $f(s \cdot t) = f(s) + r(s) \cdot f(t)$ $(s, t \in T)$ and a coboundary is a cocycle of the form $t \mapsto (1 - r(t))u$ $(u \in F(k))$.

LEMMA. *We keep the previous notation. Then* $H^1(T, F) = 0$.

This is a special case of Th. 3.1, p. 10 of [12, Exp. IX]. For the sake of completeness, we sketch a proof. It is first easily seen from the definitions that if k' is an extension of k, then

$$H^1(T \times_k k', F \otimes_k k') = H^1(T, F) \otimes_k k'.$$

It suffices therefore to consider the case where T splits over k, and consequently where F is one-dimensional. Let $a \in X^*(T)$ be such that $r(t) \cdot x = t^a \cdot x$ $(x \in F, t \in T)$. The regular k-functions on T are the finite linear combinations with coefficients in k of characters of T. Let f be a 1-cocycle. We have then

$$f(t) = \Sigma_{b \in X^*(T)} c_b \cdot t^b, \qquad (c_b \in k, \ t \in T),$$

with $c_b \neq 0$ for at most finitely many b's. The cocycle condition and the independence of the characters over $\bar{k}$ imply immediately that $f(t) = c_0(1 - t^a)$ $(t \in T)$, which yields the lemma.

9.4. LEMMA. *Let T be a k-torus, U a commutative unipotent connected k-group on which T acts k-morphically, and which verifies $U^p = \{e\}$. Let V be a connected k-subgroup of U stable under T. Assume that V and U/V are T-equivariantly k-isomorphic to vector groups over k on which T acts linearly. Then U is the direct product of V and of a* *k-subgroup stable under T.*

Let $\pi: U \to U' = U/V$ be the canonical projection. By ([18], Prop. 1, 2, p. 688) V is a direct factor over k in U. There exists therefore a k-morphism $s: U' \to U$ such that $\pi \circ s = \mathrm{id}$. We have to show the existence of such an s which is T-equivariant. Let F be the set of morphisms of U' into V. Identifying U' and V over k with vector groups, we see that F is the vector space of p-polynomial mappings of U' into V. It has a k-structure, where $F(k)$ consists of the k-morphisms of U' into V. We define a representation r of T in F by

$$r(t) \cdot f(x) = t \cdot f(t^{-1} \cdot x) \qquad (f \in F,\ t \in T,\ x \in U').$$

Then F is the union of finite dimensional subspaces defined over k, stable under T, on which this representation is rational.

To $t \in T$ let us associate the map:

$$f(t): x \longmapsto (t \cdot s(t^{-1} \cdot x)) \cdot s(x)^{-1}, \qquad (x \in U').$$

This is readily checked to be a k-morphism of U' into V, which is a cocycle of T with coefficients in a finite dimensional subspace of F, stable under T, and defined over k. By 9.3, there exists $c \in F(k)$ such that $f(t) = c - r(t) \cdot c$. It then follows that $s': x \longmapsto s(x) \cdot c(x)$ is the desired cross-section.

9.5. REMARK. The existence of an equivariant cross-section, which is not necessarily a subgroup however, can be proved similarly in a more general case, namely if U is unipotent, V is central and T-equivariantly k-isomorphic to the additive group of a vector space W over k on which T acts via a k-morphism $T \to GL(W)$.

In fact, Cor. 1, p. 100 of [20] still gives the existence of a cross-section $s: U/V \to U$ defined over k. To get an equivariant one, we argue as before with F replaced by the space of k-morphisms of varieties of U/V into V.

9.6. Under the previous conditions, U is T- and V-equivariantly k-isomorphic to $U/V \times V$. More generally let U be a unipotent k-group on which T acts k-morphically, and V a connected k-subgroup stable under T. Assume that there is a T-equivariant cross-section $s: U/V \to U$ defined over k. Then, obviously, the map $\psi: (U/V) \times V \to U$ defined by $(x, v) \mapsto s(x) \cdot v$ is a k-isomorphism of varieties which is T-equivariant and commutes with V, acting by right translations. We note also that if A is a k-variety on which T acts k-morphically, such that U is T- and V-equivariantly k-isomorphic to $A \times V$, then the projection of U on U/V induces a T-equivariant k-isomorphism of A onto U/V.

9.7. LEMMA. *Let G be solvable, connected, U the unipotent part of G and T a k-torus of G. Then G has a normal k-subgroup M contained in U, such that $G = Z(T) \cdot M$ and $U = (Z(T) \cap U) \cdot M$. The Lie algebra of M contains all eigenvectors of T not fixed by T. If U is commutative, $M = (T, G)$ and G (resp. U) is the semi-direct (resp. direct) product of $Z(T)$ (resp. $Z(T) \cap U$) and M.*

Let first U be commutative. Let $t \in T(\overline{k})$ be such that $Z(t) = Z(T)$, and M be the image of the commutator map $x \mapsto (t, x)$ $(x \in U)$. Then ([3], 11.1, p. 131), U is the direct product of $N = Z(t) \cap U$ and M. The group M is normalized by $Z(t)$, and also by U since U is commutative, hence M is normal in G. It follows that M containts (s, U) for any $s \in T$ and also, that $(s, U) = (s, G)$ for any $s \in T$, hence $M = (T, U) = (T, G)$. The Lie algebra of N is the fixed point set of T in $\mathfrak{u}$ (4.6), hence $\mathfrak{m}$ is the unique supplementary subspace to $\mathfrak{n}$ in $\mathfrak{u}$ which is stable under T. It contains then all non-zero eigenvectors of T corresponding to non-zero weights.

If V is a normal k-subgroup of G contained in U and $\pi : G \to G/V$ is the canonical projection, then $\pi(Z(T)) = Z(\pi(T))$ ([1], 13.1, p. 60). The proposition in the general case follows then by induction, dividing out by (G, G).

9.8. THEOREM. *Assume G to be connected, solvable and to have a commutative unipotent radical U. Let T be a k-torus of G. Let X be a non-zero nilpotent element of $\mathfrak{g}(k)$, and b a non-trivial character of T such that $\operatorname{Ad} t(X) = t^b \cdot X (t \in T)$. Assume that either (a): $U^p = \{e\}$, or (b): no p-power multiple of b is a root of G with respect to T. Then there exists a closed k-subgroup V of G contained in U, stable under T, whose Lie algebra is spanned by X, and a k-isomorphism $\theta : \mathbf{G}_a \xrightarrow{\sim} V$, such that $t \cdot \theta(x) \cdot t^{-1} = \theta(t^b \cdot x)$ $(t \in T,\ x \in \mathbf{G}_a)$.*

(a) We show first that the proof can be reduced to the case where T is a maximal torus, is one-dimensional, k-split, U is defined over k, $Z(T) \cap U = \{e\}$, and $Z(t)^0 \neq G$.

Replacing G by $T \cdot (T, G)$ we may, in view of 9.7, assume T to be maximal, U to be defined over k, and $Z(T) = T$. Since the representation of T in $\overline{k} \cdot X$ is defined over k, the character b is also defined over k. Let $S = (\ker b)^0$. It is a subtorus of T, defined over k, (see [3], 1.6, p. 61). The groups $Z(S)$ and $Z(S) \cap U$ are connected, defined over k (4.6), and X is in the Lie algebra of $Z(S)$. The canonical projection of $Z(S)$ onto $Z(S)/S$ induces a T-equivariant k-isomorphism of U onto the unipotent radical of $Z(S)/S$. This reduces us to $Z(S)/S$, hence we may assume that T has dimension one. It has a non-trivial character b defined over k, hence is k-split ([3], 1.3, p. 60).

If t is central in $\mathfrak{g}$, we perform an isogeny $\pi: G \to G'$ of the type of 5.3 onto a group G' whose subgroups of type (C) are proper. We have $G' = \pi(T) \cdot \pi(U)$, and π is a T-equivariant k-isomorphism of U onto $\pi(U)$. It suffices then to consider G', which yields the desired reduction.

(b) Assume $Z(t)^0 = T$. In this case, we shall exhibit a T-equivariant k-isomorphism of U onto $\mathfrak{u}$, the latter being viewed as a vector group. Let Φ be the set of roots of G with respect to T. Our assumption is equivalent to $dc \neq 0$ for all $c \in \Phi$ (1.9, 6.3 (5)). Let A be a non-zero element of $t(k)$. Then, since T is one-dimensional, $dc(A) \neq 0$ for all $c \in \Phi$. Thus A is regular, semi-simple and we have $Z(A) \subset N(T)$. Since the normalizer and the centralizer of a torus in a connected solvable group coincide and are connected ([1], 10.2, p. 52, 13.2, p. 60) we have $Z(A) = T$. For $u \in U$, we may write

$$\mathrm{Ad}\, u(A) = A + f(u) \qquad (f(u) \in \mathfrak{u})$$

and, clearly, f is a T-equivariant k-morphism. Since $Z(A) = T$, the map f is injective, hence bijective. We have $df(\mathfrak{u}) = [A, \mathfrak{u}]$ by 1.10, and therefore $df(\mathfrak{u}) = \mathfrak{u}$, which shows that f is an isomorphism. It has then the required properties.

(c) We now proceed by induction on $\dim U$. If $\dim U = 1$, then, by the condition of (a), $Z(t)^0 = T$, and we are in the case (b). So let $\dim U \geqq 2$, and assume the proposition to hold true for groups with unipotent radical of dimension $< \dim U$.

If $db = 0$, then X is in the Lie algebra of $Z(t)^0$, which is a k-subgroup (4.3), proper by (a), and induction applies. So let $db \neq 0$. By (b), we have to consider only the case where $C = Z(t)^0 \neq T$. Since $Z(T) = T$, there exists in $\mathfrak{c}(k)$ a non-zero eigenvector Y of T, with non-trivial weight a, no multiple of which is a weight of T in $\mathfrak{c}$. By induction, $Z(t) \cap U$ contains then a k-subgroup W, k-isomorphic to $\boldsymbol{G}_a$, tangent to Y, and stable under T. Since U is commutative, W is then normal in G. Let $\pi: G \to G/W$ be the canonical projection. Then $d\pi(X) \neq 0$. The induction assumption yields a one-dimensional k-subgroup V' of G', tangent to $d\pi(X)$, stable under $\pi(T)$, and T-equivariantly isomorphic to $\bar{k} \cdot X$.

Replacing G by $T \cdot \pi^{-1}(V')$, we are reduced to the case where $\dim U = 2$ and $\mathfrak{u}$ is spanned by X, Y. Let us show now that $U^p = \{e\}$, also under the assumption (b) of the proposition. The p-th power $g \mapsto g^p$ is a T-equivariant k-morphism of U or U/W, trivial on W and U/W, hence defines a T-equivariant k-morphism $\mu: U/W \to W$. Let $\alpha: \boldsymbol{G}_a \to U/W$ and $\beta: \boldsymbol{G}_a \to W$ be the k-isomorphisms given by the induction assumption. Then $f = \beta^{-1} \circ \mu \circ \alpha$ is a k-morphism of $\boldsymbol{G}_a$ into itself which verifies $f(t^b \cdot x) = t^c \cdot f(x)$ $(x \in \boldsymbol{G}_a)$. But f is a p-power polynomial. Therefore if $f \neq 0$, then $c = p^j \cdot b$ for some $j \neq 0$, a contradiction. Consequently $f = 0$ and $U^p = \{e\}$. Our assertion now follows from 9.4. As an application. we deduce a slight extension and sharpening of a result of Rosenlicht's ([20], lemma on p. 109).

9.9. COROLLARY. *Let G be connected solvable, with a commutative unipotent radical U of exponent p, and T a k-torus of G. Then U is the direct product of $Z(T) \cap U$ and of a k-subgroup M of G contained in U, stable under T, and T-equivariantly k-isomorphic to its Lie algebra, viewed as a vector group. In particular, if T splits over k, M is k-isomorphic to a product of G_a's stable under T.*

We may write $\mathfrak{u}$ as the direct sum of the fixed point set $\mathfrak{u}_0$ of T and of subspaces $\mathfrak{m}_i$ ($1 \leq i \leq r$) defined over k, stable under T, and irreducible over k as T-modules. We have $\mathfrak{u}(k_s) = \mathfrak{u}(k) \otimes_k k_s$, which allows us to view in the usual manner $\mathfrak{u}(k_s)$ as a module for the Galois group $\Gamma(k)$ of k_s over k. Since T splits over k_s, it is elementary that we may find a basis X_{ij} ($1 \leq j \leq m_i = \dim \mathfrak{m}_i$) of $\mathfrak{m}_i(k_s)$ formed by eigenvectors of T permuted transitively by $\Gamma(k)$. Let k_i be the smallest overfield of k in k_s such that $X_{i1} \in \mathfrak{m}_i(k_i)$. Choose a one-parameter k_i-subgrpup V_{i1} of U stable under T and tangent to X_{i1} (9.8). It is then stable by $\Gamma(k_i)$. It follows that the transforms ${}^sV_{i1}$, where s runs through a set of representatives of $\Gamma(k)/\Gamma(k_i)$ are m_i distinct k_s-subgroups tangent to the transforms ${}^sX_{i1}$ of X_{i1}. Let $M_i = \Pi_s\, {}^sV_{i1}$. Then M_i is defined over k, with Lie algebra $\mathfrak{m}_i$, and is T-equivariantly k-isomorphic to $\mathfrak{m}_i$. We claim that $M = M_1 \cdots M_r$ fulfills our conditions. The Lie algebra of M contains the sum of the $\mathfrak{m}_i$'s, but is not bigger for dimensional reasons. Let $f\colon M_1 \times \cdots \times M_r \to M$ be the product map. It is surjective, T-equivariant and separable. Its kernel is finite, stable under T, hence consists of fixed points of T, and therefore is reduced to $\{e\}$. Thus f is an isomorphism. Similarly, since $\mathfrak{u}_0$ is the Lie algebra of $Z(T) \cap U$ (see 4.6), it follows that $U = (Z(T) \cap U) \times M$.

9.10. PROPOSITION. *Assume G to be connected, solvable, with a commutative unipotent radical U. Let T be a k-split subtorus of G. Then U is the direct product of $Z(T) \cap U$ and of k-subgroups $U_1, \cdots, U_m$ of G, contained in U, stable under T, such that the weights of T in U_i are p-power multiples of one of them, and not zero.*

The proof is by induction on $\dim U$. In view of 9.9, it is enough to consider the case where $T = Z(T)$. Then U is defined over k (9.7). Let Φ be the set of roots of G with respect to T. Write Φ as a disjoint union of sets Φ_i where Φ_i consists of the p-power multiples of one of its elements, and is maximal among subsets having that property. Let a_i be an element of Φ_i such that no p-power multiple of a_i belongs to Φ and let $X_i \in \mathfrak{u}(k)$ be a non-zero eigenvector of T with weight a_i. By 9.8, there exists a k-subgroup N_i of U which is stable under T, tangent to X_i. Let $G_i = G/N_i$. The induction assumption yields a direct k-factor V_i of the unipotent part of G_i stable under T, such that the weights of T in $\mathfrak{v}_i$ are elements of Φ_i. Let U_i be the inverse image of V_i in U. The set of weights of T in U_i is Φ_i, and it is obvious that U is the direct product of the U_i's.

9.11. COROLLARY. *Assume that $Z(T)=T$ and that no root of G with respect to T is a p-power multiple of another one. Then U is a vector group over k, on which T acts linearly.*

We now turn to the case where U is not necessarily commutative.

9.12. COROLLARY. *Let G be connected, solvable, U its unipotent radical, T a k-torus of G, and assume that either $Z(T) \cap U = \{e\}$ or k is perfect. There exists a series $U = U_0 \supset U_1 \supset \cdots \supset U_m = \{e\}$ of connected normal k-subgroups of G stable under T with the following properties: U_i/U_{i+1} is central in U/U_{i+1} and is T-equivariantly k-isomorphic to its Lie algebra $\mathfrak{u}_i/\mathfrak{u}_{i+1}$, on which the representation of T is irreducible over k, $(i=0, \cdots, m-1)$. There is a T-equivariant k-isomorphism of varieties of U onto $\mathfrak{u}$ which maps U_i onto $\mathfrak{u}_i$ $(i=0, \cdots, m-1)$.*

U is defined over k (9.7). We may find a non-trivial connected central k-subgroup V of U, stable under T, of exponent p, (e.g., a suitable p-th power of the last non-trivial term in the descending central series of U). Our assertion is true for V: if $Z(T) \cap U = \{e\}$, it follows from 9.9; if k is perfect, it follows from 9.9 and the fact that $Z(T) \cap V$ is k-isomorphic to a vector group ([18], Prop. 1,2, p. 688). The assertion follows then using induction on $\dim U$ and 9.5.

REMARK. This implies in particular that U is k-solvable, in the sense of [20], as follows also from ([20], Cor. to Thm. 3, p. 108). If T is k-split, the groups U_i/U_{i+1} are k-isomorphic to $\boldsymbol{G}_a$. In the general case, U_i/U_{i+1} splits over k_s into a sum of one-dimensional T-modules which are inequivalent, since otherwise U_i/U_{i+1} would not be k-irreducible. Consequently, every closed subgroup of U_i/U_{i+1} stable under T is defined over k_s.

9.13. PROPOSITION. *Let G be solvable, connected, with a unipotent radical U defined over k, and T a k-torus of G. Let V be a connected k-subgroup of U stable under T. Assume that $Z(T) \cap U \subset V$ or that k is perfect. Then $M = U/V$ is T-equivariantly k-isomorphic, as a variety, to its tangent space $T(M)_0$ at the origin, and U is T- and V-equivariantly* *isomorphic, as a variety, to $U/V \times V$. In particular M is k-isomorphic to an affine space.*

In this proof, $\cong$ stands for T-equivariant k-isomorphism of varieties. Moreover, if B is a k-subgroup of U stable under T, and A a k-variety on which T operates k-morphically, then $U \cong_{(B)} A \times B$ means that U is T- and B-equivariantly k-isomorphic to $A \times B$. This is equivalent to the existence of a T-equivariant k-cross-section $s : U/B \to U$, and furthermore, in this case, $A \cong U/B$ (see 9.6).

The proof is by induction on $\dim U$ and, for fixed U, by induction on $\dim U/V$. We distinguish several cases.

(a)   There exists a connected k-subgroup N of U, stable under T, containing V, and distinct from U and V.

By induction

$$(1) \qquad U \cong_{(N)} U/N \times N, \quad N \cong_{(V)} N/V \times V$$

hence

$$(2) \qquad U \cong_{(V)} U/N \times N/V \times V.$$

By 9.6, this implies

$$(3) \qquad U/V \cong U/N \times N/V$$

hence

$$(4) \qquad U \cong_{(V)} U/V \times V.$$

By induction

$$(5) \qquad T(U/N)_0 \cong U/N, \quad T(N/V)_0 \cong N/V.$$

By full reducibility

$$(6) \qquad T(U/V)_0 \cong T(U/N)_0 \times T(N/V)_0.$$

The equality

$$T(U/V)_0 \cong U/V$$

follows then from (3), (5), (6).

From now on, we assume that we are not in case (a). Let W be a non-trivial connected central k-subgroup of U, of exponent p, stable under T, minimal for these properties.

(b)   Assume $W \subset V$, and W to be T-equivariantly k-isomorphic to a vector group on which T acts linearly. There exist then T-equivariant k-cross-sections

$$s_1 \colon U/V \to U/W, \quad s_2 \colon U/W \to U$$

as follows from the induction assumption for s_1, from 9.5 for s_2. But then, $s_2 \circ s_1$ is a T-equivariant k-cross-section for the fibration of U by V, hence (9.6):

$$U \cong_{(T)} U/V \times V.$$

Since

$$U/V \cong (U/W)/(V/W),$$

the equality

$$T(U/V)_0 \cong U/V,$$

also follows by induction.

    (c) $W \subset V$, $W \not\subset Z(T)$. By the minimality assumption on W, and 9.12, we have $W \cong \mathfrak{W}$, hence we are back to case (b).

    (d) $W \subset Z(T)$, k is perfect. Then W is k-isomorphic to a vector group, and we are back to case (b).

    (e) $W \subset Z(T)$, k is infinite. Let $Z = Z(T) \cap U$. This group is contained in V. By [3, 11.1, p. 131], it is defined over k and

$$(7) \qquad\qquad U \cong_{(Z)} U/Z \times Z, \quad V \cong_{(Z)} V/Z \times Z.$$

(The T-equivariance is clear from the construction.) Let then $s_1 \colon U/Z \to U$ be a T-equivariant k-cross-section, and $t_1 \colon U/W \to U/Z$ the natural projection. By induction, there exists a T-equivariant k-cross-secion

$$s_2 \colon U/V \cong (U/W)/(V/W) \to U/W.$$

It is then readily checked that

$$(8) \qquad\qquad s_1 \circ t_1 \circ s_2 \colon U/V \to U$$

is a T-equivariant k-cross-section, whence

$$U \cong_{(T)} U/V \times V.$$

The equality

$$T(U/V)_0 \cong U/V$$

follows from $U/V \cong (U/W)/(V/W)$ and induction.

(f) $W \not\subset V$. Since case (a) is excluded, we have then $U = W \cdot V$. In particular, V is normal in U, and the last non-trivial term V of the descending central series of V is also central in U. Let W' be a non-trivial connected k-subgroup of V', stable under T, minimal for these properties. Then, we are reduced to one of the cases (b), (c), (d), (e) (with W' playing the role of W there).

As an application of 9.13, we deduce a sharpening of 7.9 for solvable groups, due to Grothendieck in the case of $\mathscr{T}$ ([12], Exp. XVI, Cor. 6.2, p. 41):

9.14. COROLLARY. *Let G be solvable. Then the (G, k)-varieties of maximal tori $\mathscr{T}$ and of subgroups of type* (C) *C of G are k-isomorphic to affine spaces.*

We may assume G to be connected. Let T be a maximal torus of G defined over k (7.10), and $\mathfrak{t}$ its Lie algebra. We know that $N(T) = Z(T)$ and $N(\mathfrak{t}) = Z(\mathfrak{t})$ are connected (4.8, and [1], Prop. 10.1, 13.2), and defined over k (4.3, 4.6). Therefore $\mathscr{T}$ and C are k-isomorphic to $G/Z(T)$ and $G/Z(\mathfrak{t})$, respectively. By 9.7, there is a connected normal k-subgroup M of G, contained in the unipotent radical U of G, such that $G = Z(T) \cdot M$. *A fortiori*, we have $G = Z(\mathfrak{t}) \cdot M$. We claim that M intersects $Z(T)$ and $Z(\mathfrak{t})$ transversally. We have

$$\dim(Z(T) \cap M) + \dim(G/M) = \dim Z(T),$$

hence

$$\dim(Z(T) \cap M) = \dim M + \dim Z(T) - \dim G,$$

and similarly

$$\dim(Z(\mathfrak{t}) \cap M) = \dim Z(\mathfrak{t}) + \dim M - \dim G,$$

which shows that these intersections are proper. It suffices to check that the Lie algebra $\mathfrak{m}$ of M is transversal to those of $Z(T)$ and of $Z(\mathfrak{t})$. The latter are the centralizer of T and of $\mathfrak{t}$ in $\mathfrak{g}$ (4.3, 4.6). The transversality follows then from the full reducibility of $\mathrm{Ad}_{\mathfrak{g}} T$ and the fact that $\mathfrak{m}$ contains all eigenvectors of T corresponding to non-trivial characters (9.7). The natural bijective k-morphisms $M/(Z(T) \cap M) \to G/Z(T)$ and $M/(Z(\mathfrak{t}) \cap M) \to G/Z(\mathfrak{t})$ are then separable, and therefore are k-isomorphisms. Our assertion now follows from 9.13.

9.15. LEMMA. *Let U be a connected unipotent k-group, T a k-torus which acts k-morphically on U, and N a connected normal k-subgroup of*

U, stable under T, such that T has no fixed point $\neq e$ in $U'=U/N$. Let Φ and Ψ be the sets of weights of T in N and U', respectively.

(i) *Assume Φ not to contain any linear combination with strictly positive integral coefficients of elements of Ψ. Then any T-equivariant k-cross-section $s: U' \to U$ is a group homomorphism. The extension of U' by N splits T-equivariantly over k.*

1   (ii) *If Φ does not contain any element of the form $(p^i + p^j) \cdot a$ $(a \in \Psi;$ $i, j \geqq 0)$, N is central and U/N is commutative, then U is commutative.*

(i) Let $f: U' \times U' \to N$ be defined by

$$f(x, y) = s(x \cdot y) \cdot s(y)^{-1} \cdot s(x)^{-1}, \qquad (x, y \in U').$$

We have to prove that $f(x, y) = e$ for all $x, y \in U'$. The map f is clearly T-equivariant. Over $\bar{k}$, U' and N are T-equivariantly isomorphic to their Lie algebras (9.12). Hence, f yields a T-equivariant morphism of varieties:

$$F: \mathfrak{u}' \times \mathfrak{u}' \to \mathfrak{n}.$$

It suffices to show that any such F is the zero map. Take coordinates in $\mathfrak{u}'$ and $\mathfrak{n}$ with respect to bases formed by eigenvectors of T. Then the coordinates F_i of $F(X, Y)$, $(X, Y \in \mathfrak{u}')$ are polynomials in the coordinates (X_i) and (Y_i) of X and Y which satisfy relations of the form

$$F_i(t^{a_1} \cdot x_1, \cdots, t^{a_m} X_m; \; t^{a_1} \cdot Y_1, \cdots, t^{a_m} Y_m) = t^{b_i} \cdot F_i(X_1, \cdots, X_m; \; Y_1, \cdots, Y_m).$$

$(t \in T, \; \Psi = \{a_1, \cdots, a_m\}, \; b_i \in \Phi, \; X_1, \cdots, X_m, \; Y_1, \cdots, Y_m \in \bar{k})$. If $F_i \not\equiv 0$, this implies that b_i is a linear combination with positive, not all zero, integral coefficients, which contradicts the assumpton made on Φ and Ψ. Hence $F \equiv 0$, which proves the first part of (i). The second part then follows from the existence of a T-equivariant k-cross-section (9.6).

(ii) By ([10], Exp. 9, lemme 2, p. 1), we may find composition series

$$U = U_m \supset U_{m-1} \supset \cdots \supset U_0 = N, \quad N = N_0 \supset N_1 \supset \cdots N_q = \{e\},$$

of connected subgroups stable under T, such that the successive quotients are isomorphic to $\boldsymbol{G}_a$. The set of weights of T in U_i/N is then a subset of Ψ.

We prove by induction on i that U_i is central in U. This being true for $i = 0$, we may assume U_{m-1} to be in the center of U. The commutator map $(x, y) \mapsto x \cdot y \cdot x^{-1} \cdot y^{-1}$ induces then a T-equivariant morphism α of varieties of $U/U_{m-1} \times U/U_{m-1}$ into N. Using elementary facts on commutators, we check immediately that the restriction of α to $U/U_{m-1} \times \{y\}$ or to

$\{x\} \times U/U_{m-1}$ is a group homomorphism. Assume that $\operatorname{Im} \alpha \subset N_j$. We want to prove that $\operatorname{Im} \alpha \subset N_{j+1}$. The map α defines a T-equivariant morphism of varieties $\beta : U/U_{m-1} \times U/U_{m-1} \to N_j/N_{j+1}$ which is a group homomorphism if one of the two variables is left fixed. Identify U/U_{m-1} and N_j/N_{j+1} to $\boldsymbol{G}_a$. Then β is given by a polynomial $F \in \bar{k}[X, Y]$ which verifies the conditions

$$(1) \qquad F(X+X', Y) = F(X, Y) + F(X', Y), \qquad (X, Y', Y \in \boldsymbol{G}_a),$$

$$(2) \qquad F(X, Y+Y') = F(X, Y) + F(X, Y'), \qquad (X, Y, Y' \in \boldsymbol{G}_a),$$

and

$$(3) \qquad F(t^b X, t^b Y) = t^a F(X, Y), \qquad (X, Y \in \boldsymbol{G}_a,\ t \in T)$$

for some $a \in \Phi$, $b \in \Psi$. (1) and (2) show that F is a linear combination of monomials $X^{p^i} \cdot Y^{p^j}$ $(i, j \geqq 0)$. Then, if $F \neq 0$, (3) implies that a is of the from $(p^i + p^j) \cdot b$ $(i, j \geqq 0)$ which contradicts the assumption made in (ii). Hence $F = 0$. By induction on j, this shows that $(U, U) = \{e\}$.

9.16. THEOREM. *Assume that the unipotent radical U of G is defined over k or that G is solvable. Let S be a k-torus in G, b a non-trivial character of S, and X a nilpotent non-zero element of $\mathfrak{g}(k)$ such that $\operatorname{Ad} s(X) = s^b \cdot X$ $(s \in S)$.*

(i) X is tangent to a unipotent k-subgroup V of G, stable under S, in which the weights of S are non-zero multiples of b;

(ii) if $\Phi(S, G)$ dose not contain any element of the form $(p^i + p^j) \cdot b$ $(i, j \geqq 0)$, V may be chosen to be commutative and so that the weights of S in V are p-power multiples of b;

(iii) if $\Phi(S, G)$ does not contain any element of the form $p^i \cdot b$ $(i \geqq 1)$ or $(p^i + p^j) \cdot b$ $(i, j \geqq 0)$, V may be chosen to be one-dimensional, k-isomorphic to $\boldsymbol{G}_a$.

The proof proceeds by induction on $\dim G$. We distinguish three cases:

(a) G is solvable. Exactly as in 9.8 (a), we first reduce the proof to the case where S is one-dimensional, maximal, k-split, and U is defined over k. Let N be a non-trivial connected central k-subgroup of U, stable under S, and minimal for these properties. If $X \in \mathfrak{n}$, our assertion follows from 9.8. If not, applying induction to G/N, we may assume that the weights of S in U/N are non-zero multiples of b in case (i), p-power multiples of b in case (ii) and that U/N is k-isomorphic to $\boldsymbol{G}_a$ in case (iii). By 9.7, $Z(S) \cap N$ is a direct factor over k, in N. In view of the minimality assumption on N, we have either $Z(S) \supset N$ or $Z(S) \cap N = \{e\}$. If $Z(S) \supset N$, then $N = Z(S) \cap U$ and

9.15 (i) shows that U is the direct product of N and of a k-subgroup M, stable under S. Clearly $X \in \mathfrak{m}$, so that we may apply induction. Assume now $Z(S) \cap N = \{e\}$. Then, since N is minimal, 9.8 shows that N is k-isomorphic to G_a. Let c be the weight of S in N. We consider the three cases of the theorem separately.

(i) If c is a multiple of b, we are done. If not, then (9.15) U is k-isomorphic to $U/N \times N$ and we may apply induction.

(ii) By 9.15 (ii), U is commutative. If c is a p-power multiple of b, there is nothing more to prove, so assume it is not. Let $d = p^s \cdot b$ be the greatest p-power multiple of b occurring among the weights of S in U, and let Y be an eigenvector of T in $\mathfrak{u}$ with weight d. By 9.8, Y is tangent to a one-dimensional k-subgroup P, stable under S and k-isomorphic to G_a. If $k \cdot X = k \cdot Y$, we are done; if not, we apply the induction assumption to U/P.

(iii) By (ii) and the assumption, V may be chosen so that S has only the weight b in V. Then (iii) follows from 9.8.

(b) G is reductive. Let $L = (\ker b)^0$. It is a connected k-torus; $Z(L)$ is connected, defined over k (4.6), and reductive ([3], 2.15 (d), p. 70). The element X belongs to the Lie algebra of $Z(L)$. If $L \neq \{e\}$, the result follows, using induction on $Z(L)/L$, and (a). Thus we may again assume S to be one-dimensional, k-split. Choose a maximal k-split torus T of G containing S and defined over k. Fix an ordering on $X^*(T)$ compatible with the ordering of $X^*(S)$ such that $b > 0$ ([3], 3.1, p. 71). Standard facts about parabolic subgroups ([3], 5.12, p. 99) show that X belongs to the Lie algebra of the unipotent radical $R_u(P)$ of the minimal parabolic k-group P containing $Z(T)$ and associated to the given ordering. Since $R_u(P)$ is defined over k ([3], 3.14, p. 80), we are back to case (a).

(c) G is neither solvable nor reductive. If $X \in \mathfrak{u}$, we may apply (a) to $S \cdot U$. If not, we apply (b) to the image of X in the Lie algebra of G/U, under the canonical projection, and get a suitable k-unipotent subgroup V' to which $d\pi(X)$ is tangent. Then we apply (a) to $S \cdot \pi^{-1}(V')$.

References

[1] A. Borel, Groupes linéaires algébriques, Ann. of Math. (2) 64(1956), 20–82.

[2] A. Borel et J.-P. Serre, Théorèmes de finitude en cohomologie galoisienne, Comm. Math. Helv., 39(1964), 111–164.

[3] A. Borel et J. Tits, Groupes réductifs, Publ. Math. I. H. E. S., no. 27(1965), 55–150.

[4] A. Borel and T. A. Springer, Rationality properties of linear algebraic groups, Proc. Symp. Pure Math., vol. IX(1966), 26–32.

[5] H. Cartan and S. Eilenberg, Homological Algebra, Princeton University Press, 1956.

[6] P. Cartier, Questions de rationalité des diviseurs en géométrie algébrique, Bull. Soc. Math., 86(1958), 177–251.

[7] P. Cartier, Isogénies des variétés de groupes, Bull. Soc. Math., 87(1959), 191–220.

[8] C. Chevalley, Théorie des groupes de Lie, Tome II, Act. Sci. Ind., no. 1151, Hermann, Paris, 1951.

[9] C. Chevalley, On algebraic group varieties, J. Math. Soc. Japan, 6(1954), 303–324.

[10] C. Chevalley, Séminaire sur la classification des groupes de Lie algébriques, 2 vol., Paris, 1958.

[11] C. Chevalley, Fondements de la géométrie algébrique, Paris, 1958.

[12] M. Demazure et A. Grothendieck, Schémas en groupes, I. H. E. S., 1964.

[13] J. Dieudonné, Sur les groupes de Lie algébriques sur un corps de caractéristique $p > 0$, Rend. Cir. Mat. Palermo (2), 1(1953), 380–402.

[14] A. Grothendieck et J. Dieudonné, Eléments de géométrie algébrique, III, Publ. Math. I. H. E. S., 11(1961).

[15] J. E. Humphreys, Algebraic groups and modular Lie algebras, Memoirs AMS, 71(1967), 76 p.

[16] S. Lang, Algebraic groups over finite fields, Amer. J. Math., 78(1956), 555–563.

[17] M. Rosenlicht, Some rationality questions on algebraic groups, Annali di Mat. (IV), 43(1957), 25–50.

[18] M. Rosenlicht, Extensions of vector groups by abelian varieties, Amer. J. Math., 80 (1958), 685–714.

[19] M. Rosenlicht, On quotient varieties and the affine embedding of certain homogeneous spaces, Trans. Amer. Math. Soc., 101(1961), 211–223.

[20] M. Rosenlicht, Questions of rationality for solvable algebraic groups over nonperfect fields, Annali di Mat. (IV), 61(1963), 97–120.

[21] J. -P. Serre, Quelques propriétés des variétés abéliennes en caractéristique p, Amer. J. Math., 80(1958), 715–739.

[22] J. -P. Serre, Groupes algébriques et corps de classes, Act. Sci. Ind., no. 1264, Hermann, Paris, 1959.

[23] J. -P. Serre, Cohomologie Galoisienne des groupes algébriques linéaires, Colloque sur la Théorie des groupes algébriques, Bruxelles, 1962, 53–67.

[24] J. -P. Serre, Cohomologie Galoisienne, Lecture Notes in mathematics, no. 5, Springer-Verlag, 1964.

[25] R. Steinberg, Regular elements of semisimple algebraic groups, Publ. Math. I. H. E. S., no. 25(1965), 49–80.

[26] J. Tits, Classification of algebraic semisimple groups, Proc. Symp. Pure Math., vol. IX (1966), 33–62.

[27] A. Weil, Foundations of algebraic geometry, Amer. Math. Soc. Colloq. Publ., vol. XXIX, 2nd. ed., Amer. Math. Soc., Providence, R. I. 1962.

The Institute for Advanced Study
Princeton, New Jersey, U. S. A.

and

University of Utrecht
Netherlands

Commentaires et corrections

Les Notes à chaque article sont numérotées consécutivement. Un symbole tel que XY.z dans la marge de gauche réfère à z, page XY de ce volume.

46. Fixed points of elementary commutative groups

69.1   Les résultats annoncés ici sont démontrés dans [52].

49. Homology theory for locally compact spaces (with J. C. Moore)

102.1   Un complément à 3.8 et une correction à 7.2 sont indiqués le n° 7.10 de [56].

50. Density properties for certain subgroups of semi-simple groups without compact components

126.1   En écrivant cet article, j'avais naturellement en vue des sous-groupes discrets de covolume fini. Cependent j'ai considéré plus généralement des sous-groupes fermés ayant une propriété que A. Selberg avait déduite de la finitude du covolume, et que j'ai appelée propriété (*S*), cela non pour la plus grande généralité, mais parce que cette condition était plus maniable dans les démonstrations. En fait, je dois avouer ne toujours pas savoir si la propriété (*S*) est vraiment plus générale dans un groupe semi-simple. D'après S. P. Wang, elle est équivalente à la finitude du covolume dans un groupe résoluble unimodulaire (Amer. J. M. **92** (1970), 382−397).

Le théorème de densité démontré ici a été étendu aux groupes de points rationnels de groupes algébriques semi-simples sur un corps localement compact non discret par S. P. Wang (Annals of Math. **94** (1971), 325−329).

Pour les sous-groupes de covolume fini de groupes de Lie réels, d'autres démonstrations en ont été données par G. D. Mostow (Annals of Math. **75** (1962), 17−37) et H. Furstenberg (Proc. A.M.S. **55** (1976), 209−212). Cette dernière vaut du reste sur tout corps localement compact non discret de caractéristique zéro.

53. Sous groupes commutatifs et torsion des groupes de Lie compacts connexes

141.1   A la fin de cette introduction, je regrette d'avoir dû établir certaines propositions d'énoncé général par des vérifications cas par cas, utilisant la classification. Certaines ont été depuis remplacées par des démonstrations ne la faisant pas intervenir. Je profite de cette occasion pour faire le point. G désigne un groupe de Lie compact, connexe et simplement connexe, presque simple, T un tore maximal de G

et τ_G l'ensemble des «nombres premiers de torsion de G», défini comme l'ensemble des nombres premiers divisant un coefficient de la coracine associée à la racine dominante, exprimée comme combinaison linéaire de coracines simples. Soient p un nombre premier et $\mathbf{F}_p$ le corps à p éléments. Alors les huit conditions suivantes sont équivalentes:

(1) $H^*(G; \mathbf{Z})$ *n'a pas de p-torsion;*
(2) $H^*(G; \mathbf{F}_p)$ *est une algèbre extérieure à générateurs de degrés impairs;*
(3) $H^*(B_G; \mathbf{F}_p)$ *est une algèbre de polynômes à générateurs de degrés pairs;*
(4) $H^*(B_G; \mathbf{Z})$ *est sans p-torsion;*
(5) $H^*(G/T; \mathbf{F}_p)$ *est engendré par ses éléments de degré deux;*
(6) *Tout p-sous-groupe abélien élémentaire de G est contenu dans un tore;*
(7) *Tout p-sous-groupe abélien élémentaire de G de rang $\leqq 3$ est contenu dans un tore;*
(8) $p \notin \tau_G$.

Les implications qui sont actuellement démontrées sans utiliser de classification sont:

$$(1) \Leftrightarrow (2) \Leftrightarrow (3) \Leftrightarrow (4) \Leftrightarrow (5) \Rightarrow (6) \Leftrightarrow (7) \Leftrightarrow (8) \,.$$

Pour $(1) \Leftrightarrow (2) \Leftrightarrow (3)$, voir la Prop. 5.2 de [37]. En fait, les points essentiels sont $(2) \Rightarrow (3)$ (Théor. 19.1 de [23]) et $(3) \Rightarrow (2)$ (Théor. 6.1 de [29]). L'implication $(3) \Rightarrow (4)$ est claire puisque (3) entraîne que les groupes $H^i(B_G; \mathbf{F}_p)$ sont nuls en dimensions impaires. La démonstration de $(4) \Rightarrow (3)$ est esquissée dans [32], p. $93-94$. La Prop. 5.5 de [37] fournit l'équivalence $(4) \Leftrightarrow (5)$: c'est un argument utilisant la suite spectrale de la fibration $B_T \to B_G$ de fibre type G/T, très facile vu que l'on sait que $H^i(G/T; \mathbf{Z})$ est nul pour i impair. L'implication $(5) \Rightarrow (6)$ est établie dans [52: XII, 5.2] par un argument de points fixes.

Tout cela était connu à l'époque de la rédaction de cet article. Le point nouveau principal acquis depuis est une démonstration *a priori* des équivalences $(6) \Leftrightarrow (7) \Leftrightarrow (8)$ par R. Steinberg (Advances in Math. **15** (1975), $63-92$, Th. 2.28).

Il resterait à établir directement que (8) implique l'une des conditions (1) à (5). Un théorème dans cette direction a été démontré par M. Demazure (Inv. Math. **21** (1973), $287-301$): M. Demazure associe à G un ensemble de nombres premiers, disons σ_G, tel que $p \notin \sigma_G \Rightarrow (5)$ et vérifie (Prop. 8) que $\sigma_G \subset \tau_G$, mais la démonstration utilise la classification. Il s'avère du reste *a posteriori* que $\sigma_G = \tau_G$.

Remarquons que dans cet article et les précédents je considérais, au lieu de τ_G, l'ensemble des nombre premiers divisant un coefficient de la racine dominante. Cet ensemble contient τ_G et lui est presque toujours, mais pas toujours, égal. C'est R. Steinberg qui a remarqué que τ_G fournissait exactement les nombres premiers p divisant les coefficients de torsion de $H^*(G; \mathbf{Z})$.

144.₂ Les algèbres de cohomologie $H^*(G; \mathbf{F}_p)$, et aussi leurs structures d'algèbre de Hopf, sont maintenant toutes connues, *voir* A. Kono, J. M. Kyoto University **17** (1977), $259-298$, qui donne aussi des références à des travaux antérieurs.

La structure d'algèbre de Hopf de $H^*(G; \mathbf{F}_p)$ est aussi liée à la structure d'algèbre de $H^*(B_G; \mathbf{F}_p)$. La détermination de cette dernière pour G exceptionnel et p de torsion ne semble pas encore être terminée. *Voir* M. Mimura and

Y. Sambe, J. Math. Kyoto Univ. **19** (1979), 553−581 et **21** (1981), 203−230 pour des références et des contributions récentes.

146.3 Après avoir vérifié que les coefficients de torsion de $H^*(G; \mathbf{Z})$ sont premiers pour tout G simplement connexe, je demandais s'il existait un H-espace simplement connexe, à cohomologie entière de type fini, ne satisfaisant pas à cette propriété. Il n'en est rien, au moins si l'on suppose de plus que X est un complexe CW de type fini en chaque dimension. En effet, W. Browder (Annals of Math. **74** (1961), 24−51) a montré que si X est connexe, à homologie entière de type fini, alors $\pi_2(X) = 0$. D'autre part, le Théor. 4.6.1 de J. Lin (Annals of Math. **107** (1978), 41−88) entraîne que si X est comme plus haut et 2-connexe, alors les coefficients de p-torsion de $H^*(X; \mathbf{Z})$ sont au plus égaux à p. J. Lin montre également que si X a même cohomologie rationnelle qu'un groupe exceptionnel G, alors les coefficients de torsion de $H^*(X; Z)$ sont contenus dans un petit ensemble de nombres premiers (Théor. 6.1), qui se trouve être τ_G.

147.4 La «Note ultérieure» annoncée ici n'a jamais vu le jour. Mais le résultat énoncé ici a été démontré ultérieurement pour un groupe semi-simple connexe et simplement connexe sur un corps algébriquement clos quelconque par R. Steinberg, Memoirs A.M.S. **80** (1968), Theor. 8.1.

155.5 Pour une généralisation à des groupes algébriques semi-simples en caractéristique quelconque, voir R. Steinberg, Springer Lecture Notes **131** (1970), Part E, Theor. 5.8, p. 208.

156.6 On montre facilement que si p est un nombre premier > 2 le p-rang $l_p(G)$ d'un groupe de Lie compact connexe G est toujours égal au rang $l(G)$ de G, en procédant par récurrence sur la dimension de G et en utilisant le fait que le sous-groupe de congruence principal $\Gamma(p) = \{g \in \mathbf{SL}_n(\mathbf{Z}) \mid g \equiv 1 \bmod p\}$ de $\mathbf{SL}_n(\mathbf{Z})$ est sans torsion (Minkowski). Si G a de la p-torsion, il possède donc au moins deux classes de conjugaison de $[p]$-sous-groupes maximaux. Cela confirme les résultats de 6.4, 6.5 mais je ne sais si ces classes de conjugaison ont été déterminées dans d'autres cas, par exemple pour les groupes exceptionnels autres que $\mathbf{G}_2$, ni s'il existe des groupes simples où l'on rencontre plus de deux classes.

Le 2-rang de **Spin**(n) a été déterminé par D. Quillen (*loc. cit* dans la Note 2 à [29]). Ici aussi, j'ignore si l'on connait les classes de conjugaison des [2]-sous-groupes maximaux pour $n > 10$.

56. La classe fondamentale d'un espace analytique
(avec A. Haefliger)

208.1 Nous nous demandons ici si un cycle dans une variété lisse V est homologiquement équivalent à une combinaison linéaire de sous-variétés lisses, question qui peut être aussi posée pour d'autres types de relations d'équivalence. Des réponses partielles positives ont été données par H. Hironaka (Amer. J. M. **90** (1968), 1−54) et S. Kleiman (Publ. Math. I.H.E.S. **36** (1969), 281−298). En particulier, cela est vrai si Z est de dimension $r \leqq \min(3, (n-1)/2)$, en caractéristique zéro (Hironaka), ou encore si Z est de la forme $(n-r-1)! \, Z'$ et $r < (n+1)/2$, où Z' est un cycle et n la dimension de V (Kleiman). Par contre, R. Hartshorne, E. Rees et E. Thomas ont fourni un exemple de cycle dans une grassmannienne non équivalent à une combinaison linéaire à coefficients entiers de sous-variétés lisses, sans toutefois pouvoir exclure qu'un multiple non nul de ce cycle ne le soit (Bull. A.M.S. **80** (1974), 847−851).

59. Ensembles fondamentaux pour les groupes arithmétiques

287.1   Les résultats du § 1 sur les groupes algébriques linéaires sont englobés dans [66]. Ceux du § 2 sur les ensembles fondamentaux sont démontrés dans [84]. Par contre, je n'ai pas publié de démonstrations des énoncés du § 3 et du Théorème 4.3, mais ce ne sont guère que des variations sur les articles de I. Satake (références [7] et [8] de l'article). On en trouvera dans un article récent de S. Zucker («Satake compactifications»), à paraître, qui met de plus en relations les compactifications de Satake et la compactification par une variété à coins de [98]. Une autre démonstration du Théor. 4.4 figure dans [60].

61. Arithmetic properties of algebraic groups

335.1   Ce programme est réalisé dans [69] et [88]. Des résultats analogues mais *a priori* différents sont établis par I. I. Piatetski-Shapiro (Uspehi Math. **19** (1964), 93−120; Russian Math. Surveys **19** (1964), 83−109). Pour les relations entre les deux, *voir* la remarque à [95].

337.2   Une démonstration se trouve dans un article de T. Ono (Annals of Math. **82** (1965), 88−111).

338.3   Cette conjecture a été démontrée pour les groupes déployés sur un corps de nombres par R. P. Langlands (Proc. Symp. pure math. **9,** 143−148, AMS 1966), dont la méthode a été étendue aux groupes quasi-déployés par K. F. Lai (Comp. Math. **41** (1980), 153−188). Pour les groupes adéliques sur un corps de fonctions à une variable et corps de constantes fini, la théorie de la réduction a été établie par G. Harder (Inv. Math. **7** (1969), 33−54), qui a aussi démontré l'analogue du théorème de Langlands sur le nombre de Tamagawa (Annals of Math. **100** (1974), 249−306).

339.4   Un contre-exemple simple à cette conjecture m'a été indiqué il y a quelques années par G. Harder: Soient p, q deux nombres premiers distincts, $n = p \cdot q$. Il existe une algèbre à division de degré n^2 sur $\mathbf{Q}$ ayant un invariant local p^{-1} en $2p$ places de $\mathbf{Q}$ et q^{-1} en $2q$ places de $\mathbf{Q}$ et trivial en toute autre place. Alors le groupe algébrique G dont les points sur une $\mathbf{Q}$-algèbre commutative A sont les éléments de norme réduite un de $(D \otimes_{\mathbf{Q}} A)$ est une $\mathbf{Q}$-forme anisotrope de $\mathbf{SL}_n$, qui est isotrope en toute place de $\mathbf{Q}$.

339.5   Cela a été confirmé et précisé par T. Ono (*loc. cit* 2).

340.6   Cette conjecture a été démontrée par F. Bruhat et J. Tits dans le cadre d'une théorie générale des sous-groupes compacts de groupes p-adiques réductifs (*Voir* Publ. IHES **41** (1972), 5−251), Proc. Symp. pure math. **33,** Vol. 1, 29−69, AMS 1979 (où l'on trouvera d'autres références), ainsi que l'article mentionné dans la Note 1 à [89].

340.7   Ce qui a été démontré par M. Kneser dans la version finale du travail dont il est question ici (J. f. reine u. angew. Math. **214/15** (1964), 345−349).

341.8   Pour des exemples de sous-groupes arithmétiques de **SO** $(n, 1)$ dont le groupe des commutateurs est d'indice infini, *voir* notamment E. B. Vinberg, Mat. Sbornik **78** (1969), 617−622, J. Millson, Annals of Math. **104** (1976), 235−247. Un théorème de G. A. Margoulis d'une très grande généralité (cité par J. Tits, Sém. Bourbaki, Exp. 482 (1975−76), Théorème 3) montre que cela ne peut se produire essentiellement qu'en rang réel un.

62. Compact Clifford-Klein forms of symmetric spaces

346.1   On renvoie à E. B. Vinberg (Mat. Sbornik **72** (1967), 429−444) pour des exemples de groupes discrets d'isométries d'un espace hyperbolique de dimension $\leqq 6$, arithmétiques ou non, à domaine fondamental relativement compact (pour $n \leqq 5$) ou de volume fini, construits par voie géométrique. Il s'agit de groupes de Coxeter engendrés par des réflexions à des hyperplans géodésiques. *Voir* aussi N. Bourbaki, Groupes et Algèbres de Lie, Chap. 4, 5, 6, Masson Paris 1981, Exercices 11 à 19, Chap. 5, § 4, p. 133−135.

354.2   J'ai en effet écrit un travail sous ce titre, en plusieurs étapes, à la suite d'un séminaire à l'Institute (1953−54) et de cours à MIT (1958) et au Tata Institute de Bombay (1961). Mais il n'a jamais été publié.

65. Cohomologie et rigidité d'espaces compacts localement symétriques

420.1   *Voir* P. A. Griffiths, Math. Annalen **155** (1964), 292−315, **158** (1965), 326−351, en particulier Theorems 5.1 et 7.1 dans le deuxième article.

66. Groupes réductifs (avec J. Tits)

425.1   Des corrections sont publiées dans [94]. Il faut ajouter à cela que G. D. Mostow est l'auteur de [19] dans la bibliographie.

80. Rationality properties of linear algebraic groups, II
(with T. A. Springer)

771.1   T. Nakanumara (Kodai Math. Sem. **23** (1971), 127−130) a montré que 9.15 (ii) n'est pas correct, mais le devient si l'on remplace $(p^i + p^j)a$ $(a \in \psi, i, j \geqq 0)$ par

$$p^i a + p^j b \quad (a, b \in \psi, i, j \geqq 0) .$$

Nous utilisons 9.15 (ii) pour prouver 9.16 (ii). Nakanumara le démontre aussi en utilisant le nouveau 9.15 (ii), et établit de plus 9.16 (iii) sous des hypothèses un peu plus faibles que les nôtres.

Bibliographie

La numérotation de gauche suit l'ordre chronologique de parution (de rédaction pour [30] et [87]). Un chiffre romain I, II ou III à droite d'un titre indique dans quel volume l'article en question est contenu. Les titres suivis du signe 0 sont ceux des travaux non reproduits ici.

1. (avec J. de Siebenthal) Sur les sous-groupes fermés connexes de rang maximum des groupes de Lie clos, C. R. Acad. Sci., Paris **226** (1948) 1662–1664 — I, 1–2

2. Some remarks about Lie groups transitive on spheres and tori, Bull. Amer. Math. Soc. **55** (1949) 580–587 — I, 3–10

3. (avec J. de Siebenthal) Les sous-groupes fermés de rang maximum des groupes de Lie clos, Comment Math. Helv. **23** (1949) 200–221 — I, 11–32

4. Groupes d'homotopie des groupes de Lie, Espaces Fibrés et Homotopie, Sém. H. Cartan, E. N. S. Paris 1949–1950, Exp. 12, 13; Notes polycopiées, 2éme éd. Secrét. Mathématique, Soc. Math. France (1955) — 0

5. Limites projectives de groupes de Lie, C. R. Acad. Sci., Paris **230** (1950) 1197–1199 — I, 33–35

6. Sections locales de certains espaces fibrés, C. R. Acad. Sci., Paris **230** (1950) 1246–1248 — I, 36–38

7. Le plan projectif des octaves et les sphères comme espaces homogènes, C. R. Acad. Sci., Paris **230** (1950) 1378–1380 — I, 39–41

8. Groupes localement compacts, Séminaire Bourbaki, Exp. 29 (1949/50) — I, 42–54

9. (avec J-P. Serre) Impossibilité de fibrer un espace euclidien par des fibres compactes, C. R. Acad. Sci., Paris **230** (1950) 2258–2259 — I, 55–56

10. Remarques sur l'homologie filtrée, J. Math. Pures Appl., (9) **29** (1950) 313–322 — I, 57–66

11. Impossibilité de fibrer une sphère par un produit de sphères, C. R. Acad. Sci., Paris **231** (1950) 943–945 — I, 67–69

12. Sous-groupes compacts maximaux des groupes de Lie, Séminaire Bourbaki, Exp. 33 (1950/51) — I, 70–76

13. Sur la cohomologie des variétés de Stiefel et de certains groupes de Lie, C. R. Acad. Sci., Paris **232** (1951) 1628–1630 — I, 77–79

14. La transgression dans les espaces fibrés principaux, C. R. Acad. Sci., Paris **232** (1951) 2392–2394 — I, 80–82

15. Sur la cohomologie des espaces homogènes de groupes de Lie
compacts, C. R. Acad. Sci., Paris **233** (1951) 569–571   I, 83–85

16. (avec J-P. Serre) Détermination des p-puissances réduites de
Steenrod dans la cohomologie des groupes classiques. Applications, C. R. Acad. Sci., Paris **233** (1951) 680–682   I, 86–88

17. Cohomologie des espaces homogènes, Séminaire Bourbaki,
Exp. 45 (1950/51)   0

18. Cohomologie des espaces localement compacts d'après
J. Leray, Notes. E. P. F. Zürich, 1951; 3ème édition: Lect.
Notes Math. **2** (1964)   0

19. (avec A. Lichnérowicz) Groupes d'holonomie des variétés riemanniennes, C. R. Acad. Sci., Paris **234** (1952) 1835–1837   I, 89–91

20. (avec A. Lichnérowicz) Espaces riemanniens et hermitiens symétriques, C. R. Acad. Sci., Paris **234** (1952) 2332–2334   I, 92–94

21. Les espaces hermitiens symétriques, Séminaire Bourbaki, Exp.
62 (1951/52)   I, 95–103

22. Les fonctions automorphes de plusieurs variables complexes,
Bull. Soc. Math. France **80** (1952) 167–182   I, 104–119

23. Sur la cohomologie des espaces fibrés principaux et des espaces
homogènes de groupes de Lie compacts, (Thèse, Paris, 1952)
Ann. Math., (2) **57** (1953) 115–207   I, 121–216

24. (avec J-P. Serre) Sur certains sous-groupes des groupes de Lie
compacts, Comment. Math. Helv. **27** (1953) 128–139   I, 217–228

25. La cohomologie mod 2 de certains espaces homogènes, Comment. Math. Helv. **27** (1953) 165–197   I, 229–261

26. (avec J-P. Serre) Groupes de Lie et puissances réduites de
Steenrod, Amer. J. Math. **75** (1953) 409–448   I, 262–301

27. Les bouts des espaces homogènes de groupes de Lie, Ann.
Math., (2) **58** (1953) 443–457   I, 302–316

28. Homology and cohomology of compact connected Lie groups,
Proc. Nat. Acad. Sci. USA **39** (1953) 1142–1146   I, 317–321

29. Sur l'homologie et la cohomologie des groupes de Lie compacts connexes, Amer. J. Math. **76** (1954) 273–342   I, 322–391

30. Représentations linéaires et espaces homogènes kähleriens des
groupes simples compacts (inédit, Mars 1954)   I, 392–396

31. Kählerian coset spaces of semi-simple Lie groups, Proc. Nat.
Acad. Sci, USA **40** (1954) 1147–1151   I, 397–401

32. Topics in the homology theory of fibre bundles, Univ. of Chicago 1954 (Notes by E. Halpern), Lect. Notes Math. **36** (1967)   0

33. Topology of Lie groups and characteristic classes, Bull. Amer.
Math. Soc. **61** (1955) 397–432   I, 402–437

34. Nouvelle démonstration d'un théorème de P. A. Smith, Comment. Math. Helv. **29** (1955) 27–39   I, 438–450

35. (with C. Chevalley) The Betti numbers of the exceptional groups, Mem. Amer. Math. Soc. **14** (1955) 1–9 — I, 451–459

36. (with G. D. Mostow) On semi-simple automorphisms of Lie algebras, Ann. Math., (2) **61** (1955) 389–405 — I, 460–476

37. Sur la torsion des groupes de Lie, J. Math. Pures Appl., (9) **35** (1955) 127–139 — I, 477–489

38. Groupes algébriques, Séminaire Bourbaki, Exp. 121 (1955/56) — 0

39. Groupes linéaires algébriques, Ann. Math., (2) **64** (1956) 20–82 — I, 490–552

40. Transformation groups with two classes of orbits, Proc. Nat. Acad. Sci. USA **43** (1957) 983–985 — I, 553–555

41. Travaux de Mostow sur les espaces homogènes, Séminaire Bourbaki, Exp. 142 (1956/57) — I, 556–564

42. The Poincaré duality in generalized manifolds, Mich. Math. J. **4** (1957) 227–239 — I, 565–577

43. (with F. Hirzebruch) Characteristic classes and homogeneous spaces I, Amer. J. Math. **80** (1958) 458–538 — I, 578–658

44. (avec J-P. Serre) Le théorème de Riemann-Roch, d'après Grothendieck, Bull. Soc. Math. France **86** (1958) 97–136 — I, 659–698

45. (with F. Hirzebruch) Characteristic classes and homogeneous spaces II, Amer. J. Math. **81** (1959) 315–382 — II, 1–68

46. Fixed points of elementary commutative groups, Bull. Amer. Math. Soc. **65** (1959) 322–326 — II, 69–73

47. (with F. Hirzebruch) Characteristic classes and homogeneous spaces III, Amer. J. Math. **82** (1960) 491–504 — II, 74–87

48. On the curvature tensor of the hermitian symmetric manifolds, Ann. Math., (2) **71** (1960) 508–521 — II, 88–101

49. (with J. C. Moore) Homology theory for locally compact spaces, Mich. Math. J. **7** (1960) 137–159 — II, 102–124

50. Density properties for certain subgroups of semi-simple groups without compact components, Ann. Math., (2) **72** (1960) 179–188 — II, 125–134

51. Commutative subgroups and torsion in compact Lie groups, Bull. Amer. Math. Soc. **66** (1960) 285–288 — II, 135–138

52. Seminar on transformation groups, Ann. Math. Stud. **46** (1960), (with contributions by G. Bredon, E. Floyd, D. Montgomery, R. Palais) — 0

53. Sous groupes commutatifs et torsion des groupes de Lie compacts connexes, Tôhoku Math. J., (2) **13** (1961) 216–240 — II, 139–163

54. (with Harish-Chandra) Arithmetic subgroups of algebraic groups, Bull. Amer. Math. Soc. **67** (1961) 579–583 — II, 164–168

55. Some properties of adele groups attached to algebraic groups, Bull. Amer. Math. Soc. **67** (1961) 583–585 — II, 169–171

56. (avec A. Haefliger) La classe d'homologie fondamentale d'un espace analytique, Bull. Soc. Math. France **89** (1961) 461–513   II, 172–224

57. (mit R. Remmert) Über kompakte homogene Kählersche Mannigfaltigkeiten, Math. Ann. **145** (1962) 429–439   II, 225–235

58. (with Harish-Chandra) Arithmetic subgroups of algebraic groups, Ann. Math., (2) **75** (1962) 485–535   II, 236–286

59. Ensembles fondamentaux pour les groupes arithmétiques, Colloque sur la Théorie des Groupes Algébriques, Bruxelles 1962, 23–40   II, 287–304

60. Some finiteness properties of adele groups over number fields, Publ. Math., Inst. Hautes Etud. Sci. **16** (1963) 5–30   II, 305–330

61. Arithmetic properties of linear algebraic groups, Proc. Int. Congr. Mathematicians, Stockholm 1962, Uppsala 1963, 10–22   II, 331–343

62. Compact Clifford-Klein forms of symmetric spaces, Topology **2** (1963) 111–122   II, 344–355

63. (with W. L. Baily Jr.) On the compactification of arithmetically defined quotients of bounded symmetric domains, Bull. Amer. Math. Soc. **70** (1964) 588–593   II, 356–361

64. (avec J-P. Serre) Théorèmes de finitude en cohomologie galoisienne, Comment. Math. Helv. **39** (1964) 111–164   II, 362–415

65. Cohomologie et rigidité d'espaces compacts localement symétriques, Séminaire Bourbaki, Exp. 265 (1963/64)   II, 416–423

66. (avec J. Tits) Groupes réductifs, Publ. Math., Inst. Hautes Etud. Sci. **27** (1965) 55–150   II, 424–520

67. Statement of the index theorem. Outline of proof, Chap. I in: Seminar on the Atiyah-Singer index theorem by R. Palais et al., Ann. Math. Stud. **57** (1965) 1–11   II, 521–531

68. A spectral sequence for complex analytic bundles, Appendix Two in: F. Hirzebruch, Topological methods in algebraic geometry, 3rd edition, 202–217, Springer 1966   II, 532–547

69. (with W. L. Baily Jr.) Compactification of arithmetic quotients of bounded symmetric domains, Ann. Math., (2) **84** (1966) 442–528   II, 548–634

70. Density and maximality of arithmetic subgroups, J. Reine Angew. Math. **224** (1966) 78–89   II, 635–646

71. Opérateurs de Hecke et fonctions zêta, Séminaire Bourbaki, Exp. 307 (1965/66)   II, 647–661

72. Class invariants, Chap. III, IV in: Seminar on complex multiplication, with S. Chowla, C. S. Herz, K. Iwasawa, J-P. Serre, Lect. Notes Math. **21** (1966)   0

73. Linear algebraic groups, Proc. Symp. Pure Math. **9,** Amer. Math. Soc. (1966) 3–19   II, 662–678

74. Reduction theory for arithmetic groups, Proc. Symp. Pure Math. **9,** Amer. Math. Soc. (1966) 20–25   II, 679–684

75. Introduction to automorphic forms, Proc. Symp. Pure Math. **9**, Amer. Math. Soc. (1966) 199–210 — II, 685–696

76. (with T. A. Springer) Rationality properties of linear algebraic groups, Proc. Symp. Pure Math. **9**, Amer. Math. Soc. (1966) 26–32 — II, 697–703

77. (with R. Narasimhan) Uniqueness conditions for certain holomorphic mappings, Invent. Math. **2** (1967) 247–255 — II, 704–712

78. Sur une généralisation de la formule de Gauss-Bonnet, An. Acad. Bras. Cienc. **39** (1967) 31–37 — II, 713–719

79. Ensembles fondamentaux pour les groupes arithmétiques et formes automorphes, Notes d'un cours à l'Inst. H. Poincaré 1964, rédigées par H. Jacquet, J.-J. Sansuq et B. Schiffmann, Ecole Normale Supérieure, Paris 1967 — 0

80. (with T. A. Springer) Rationality properties of linear algebraic groups II, Tôhoku Math. J., (2) **20** (1968) 443–497 — II, 720–774

81. On the automorphisms of certain subgroups of semi-simple Lie groups, Proc. Inter. Colloquium on Algebraic Geometry 1968, Tata Institute, Bombay (1969) 43–73 — III, 1–31

82. (avec J. Tits) On 'abstract' homomorphisms of simple algebraic groups, Proc. Colloquium on Algebraic Geometry 1968, Tata Institute, Bombay (1969) 75–82 — III, 32–39

83. Injective endomorphisms of algebraic varieties, Arch. Math. **20** (1969) 531–537 — III, 40–46

84. Introduction aux groupes arithmétiques, Actualités Sci. Ind. no 1341, Hermann, Paris (1969) — 0

85. Linear algebraic groups (Notes by H. Bass), Math. Lecture Notes Series, Benjamin, Inc. New York (1969); Traduction russe Moscou MIR (1972) — 0

86. Sous-groupes discrets de groupes semi-simples (d'après D. A. Kajdan et G. A. Margoulis), Séminaire Bourbaki, Exp. 358, (1968/69), Lect. Notes Math. **179** (1971) 199–216 — III, 47–56

87. On periodic maps of certain $K(\pi, 1)$, (unpublished, 1969) — III, 57–60

88. Pseudo-concavité et groupes arithmétiques, Essays on Topology and Related Topics, Mémoires dédiés à G. de Rham, Springer (1970) 70–84 — III, 61–75

89. Properties and linear representations of Chevalley groups, Seminar on algebraic groups and related finite groups, Lect. Notes Math. **131** (1970) 1–55 — III, 76–108

90. (avec J-P. Serre) Adjonction de coins aux espaces symétriques; Applications à la cohomologie des groupes arithmétiques, C. R. Acad. Sci., Paris **271** (1970) 1156–1158 — III, 109–111

91. (avec J-P. Serre) Cohomologie à supports compacts des immeubles de Bruhat-Tits; Applications à la cohomologie des groupes S-arithmétiques, C. R. Acad. Sci., Paris **272** (1971) 110–113 — III, 112–115

92. (avec J. Tits) Eléments unipotents et sous-groupes paraboliques de groupes réductifs I, Invent. Math. **12** (1971) 95–104   III, 116–125

93. Cohomologie réelle stable de groupes S-arithmétiques, C. R. Acad. Sci., Paris **274** (1972) 1700–1702   III, 126–128

94. (avec J. Tits) Compléments à l'article: 'Groupes réductifs', Publ. Math., Inst. Hautes Etud. Sci. **41** (1972) 253–276   III, 129–152

95. Some metric properties of arithmetic quotients of symmetric spaces and an extension theorem, J. Differ. Geom. **6** (1972) 543–560   III, 153–170

96. Représentations de groupes localement compacts, Lect. Notes Math. **276** (1972)   0

97. (avec J. Tits) Homomorphismes 'abstraits' de groupes algébriques simples, Ann. Math., (2) **97** (1973) 499–571   III, 171–243

98. (with J-P. Serre) Corners and arithmetic groups. With an appendix by A. Douady and L. Hérault: Arrondissement des Variétés à coins. Comment. Math. Helv. **48** (1973) 436–491   III, 244–299

99. Cohomologie de certains groupes discrets et Laplacien p-adique (d'après H. Garland), Séminaire Bourbaki, Exp. 437 (1973/74), Lect. Notes Math. **431** (1975) 12–35   III, 300–314

100. Stable real cohomology of arithmetic groups, Ann. Sci. Ec. Norm. Super., (4) **7** (1974) 235–272   III, 315–352

101. Cohomology of arithmetic groups, Proc. Int. Congr. of Mathematicians, Vancouver, 1974, Vol. 1 (1975) 435–442   III, 353–360

102. Linear representations of semi-simple algebraic groups, Proc. Symp. Pure Math. **29**, Amer. Math. Soc. (1975) 421–439   III, 361–373

103. Formes automorphes et séries de Dirichlet (d'après R. P. Langlands), Séminaire Bourbaki, Exp. 466 (1974/75), Lect. Notes Math. **514** (1976) 183–222   III, 374–398

104. Cohomologie de sous-groupes discrets et représentations de groupes semi-simples, Astérisque **32–33** (1976) 73–112   III, 399–438

105. (avec J-P. Serre) Cohomologie d'immeubles et de groupes S-arithmétiques, Topology **15** (1976) 211–232   III, 439–460

106. Admissible representations of a semi-simple group over a local field with vectors fixed under an Iwahori subgroup, Invent. Math. **35** (1976) 233–259   III, 461–487

107. (with B. M. Schreiber) p-adic linear groups with ergodic automorphisms, Isr. J. Math. **24** (1976) 199–205   III, 488–494

108. Cohomologie de SL_n et valeurs de fonctions zeta aux points entiers, Ann. Sc. Norm. Super. Pisa, Cl. Sci., (4) **4** (1977) 613–636; Correction, ibid. **7** (1980) 373   III, 495–519

109. (with G. Harder) Existence of discrete cocompact subgroups of reductive groups over local fields, J. Reine Angew. Math. **298** (1978) 53–64   III, 520–531

110. (avec J. Tits) Théorèmes de structure et de conjugaison pour les
     groupes algébriques linéaires, C. R. Acad. Sci., Paris **287** (1978)
     55–57   III, 532–534

111. On the development of Lie group theory, Proc. of the Bicenten-
     nial Congr. of the Dutch Math. Soc., Math. Centre Tract
     100/101 (1979) 25–37 and Nieuw Archief voor Wiskunde (3)
     27 (1979) 13–25; Math. Intell. **2.2** (1980) 67–72   III, 535–547

112. (with H. Jacquet) Automorphic forms and automorphic repre-
     sentations, Proc. Symp. Pure Math. **33,** Part 1, Amer. Math.
     Soc. (1979) 189–202   III, 548–561

113. Automorphic L-functions, Proc. Symp. Pure Math. **33,** Part 2,
     Amer. Math. Soc. (1979) 27–61   III, 562–596

114. Symmetric compact complex spaces, Arch. Math. **33** (1979)
     49–56   III, 597–604

115. (with N. Wallach) Continuous cohomology, discrete subgroups
     and representations of reductive groups, Ann. Math. Stud. **94**
     (1980), Introduction   III, 605–613

116. Stable and L^2-cohomology of arithmetic groups, Bull. Amer.
     Math. Soc., (N.S.) **3** (1980) 1025–1027   III, 614–616

117. Commensurability classes and volumes of hyperbolic 3-mani-
     folds, Ann. Sc. Norm. Super. Pisa, Cl. Sci., (4) **8** (1981) 1–33   III, 617–649

118. Stable real cohomology of arithmetic groups II, Prog. Math.,
     Boston **14** (1981) 21–55   III, 650–684

119. Mathematik: Kunst und Wissenschaft, Themen-Reihe der Carl
     Friedrich von Siemens Stiftung XXXIII München 1982   III, 685–701

A paraître:

Cohomology and spectrum of an arithmetic group, Proc. of a Conference on Operator Algebras and Group Representations, Neptun, Rumania (1980), Pitman (1983)

On free subgroups of semi-simple groups, Enseign. Math. (2)

(with H. Garland) Laplacian and the discrete spectrum of an arithmetic group, to appear in the Amer. J. Math.

Regularization theorems in Lie algebra cohomology. Applications

(with W. Casselman) L^2-cohomology of locally symmetric manifolds of finite volume

L^2-cohomology and intersection cohomology of certain arithmetic varieties

Acknowledgements

Springer-Verlag would like to thank the original publishers of Armand Borel's papers for granting permission to reprint them here.

The numbers following each source correspond to the numbering of the articles in the bibliography at the end of each volume.

Reprinted from Amer. J. Math., © by Johns Hopkins University Press: 26, 29, 43, 45, 47

Reprinted from An. Acad. Bras. Cienc., © by Academia Brasiliera de Ciencias: 78

Reprinted from Ann. Math. Stud., © by Princeton University Press: 67, 115

Reprinted from Ann. Math., (2), © by Princeton University Press: 23, 27, 36, 39, 48, 50, 58, 69, 97

Reprinted from Ann. Sc. Norm. Super. Pisa, Cl. Sci., (4), © by Scuola Normale Superiore, Italy: 108, 117

Reprinted from Ann. Sci. Ec. Norm. Super., (4), © by Editions Bordas-Dunod-Gauthier-Villars: 100

Reprinted from Arch. Math.,© by Birkhäuser Verlag, Basel: 83, 114

Reprinted from Astérisque, © by Société Mathématique de France: 104

Reprinted from Bull. Am. Math. Soc., © by The American Mathematical Society: 2, 33, 46, 51, 54, 55, 63, 116

Reprinted from Bull. Soc. Math. France, © by Editions Bordas-Dunod-Gauthier-Villars: 22, 44, 56

Reprinted from C. R. Acad. Sci., Paris, © by Editions Bordas-Dunod-Gauthier-Villars: 1, 5, 6, 7, 9, 11, 13, 14, 15, 16, 19, 20, 90, 91, 93, 110

Reprinted from Comment. Math. Helv., © by The University of Zürich: 3, 24, 25, 34, 64, 98

Reprinted from Essays on Topology and Related Topics, © by Springer-Verlag Berlin Heidelberg New York: 88

Reprinted from Invent. Math., © by Springer-Verlag Berlin Heidelberg New York: 77, 92, 106

Reprinted from Isr. J. Math., © by Weizmann Science Press: 107

Reprinted from J. Differ. Geom., © by The Lehigh University: 95

Reprinted from J. Math. Pures Appl., (9), © by Editions Bordas-Dunod-Gauthier-Villars: 10, 37

Reprinted from J. Reine Angew. Math., © by Walter de Gruyter & Co.: 70, 109

Reprinted from Math. Ann., © by Springer-Verlag Berlin Heidelberg New York: 57

Reprinted from Mem. Amer. Math. Soc., © by The American Mathematical Society: 35

Reprinted from Mich. Math. J., © by The University of Michigan: 42, 49

Reprinted from Proc. Int. Congr. Mathematicians, Stockholm, 1962, © by The American Mathematical Society: 61

MIX
Papier aus verantwortungsvollen Quellen
Paper from responsible sources
FSC® C105338

If you have any concerns about our products,
you can contact us on
ProductSafety@springernature.com

In case Publisher is established outside the EU,
the EU authorized representative is:
Springer Nature Customer Service Center GmbH
Europaplatz 3, 69115 Heidelberg, Germany

Printed by Libri Plureos GmbH
in Hamburg, Germany